Methods in Enzymology

Volume XLIV

IMMOBILIZED ENZYMES

METHODS IN ENZYMOLOGY

EDITORS-IN-CHIEF

Sidney P. Colowick   Nathan O. Kaplan

Methods in Enzymology

Volume XLIV

Immobilized Enzymes

EDITED BY

Klaus Mosbach

BIOCHEMICAL DIVISION
CHEMICAL CENTER
UNIVERSITY OF LUND
LUND, SWEDEN

1976

ACADEMIC PRESS   New York   San Francisco   London
A Subsidiary of Harcourt Brace Jovanovich, Publishers

COPYRIGHT © 1976, BY ACADEMIC PRESS, INC.
ALL RIGHTS RESERVED.
NO PART OF THIS PUBLICATION MAY BE REPRODUCED OR
TRANSMITTED IN ANY FORM OR BY ANY MEANS, ELECTRONIC
OR MECHANICAL, INCLUDING PHOTOCOPY, RECORDING, OR ANY
INFORMATION STORAGE AND RETRIEVAL SYSTEM, WITHOUT
PERMISSION IN WRITING FROM THE PUBLISHER.

ACADEMIC PRESS, INC.
111 Fifth Avenue, New York, New York 10003

United Kingdom Edition published by
ACADEMIC PRESS, INC. (LONDON) LTD.
24/28 Oval Road, London NW1

Library of Congress Cataloging in Publication Data

Main entry under title:

Immobilized enzymes.

  (Methods in enzymology ; v. 44)
  Includes bibliographical references and index.
  1. Immobilized enzymes. I. Mosbach, Klaus.
II. Series, [DNLM: 1. Enzymes. W1 ME9615k v. 44 etc.
/ QU135 I32]
QP601.M49 vol. 44 574.1'925'08s 76-44223
ISBN 0–12–181944–2 [574.1'925]

PRINTED IN THE UNITED STATES OF AMERICA

Table of Contents

Contributors to Volume XLIV xi
Preface xv
Volumes in Series xvii

Section I. General Comments

1. Introduction — Klaus Mosbach — 3

Section II. Immobilization Techniques

A. Covalent Coupling

2. Functional Groups on Enzymes Suitable for Binding to Matrices — Paul A. Srere and Kosaku Uyeda — 11
3. Immobilization of Enzymes to Agar, Agarose, and Sephadex Supports — Jerker Porath and Rolf Axén — 19
4. Enzymes Immobilized to Cellulose — M. D. Lilly — 46
5. Immobilization of Enzymes to Various Acrylic Copolymers — Rolf Mosbach, Ann-Christin Koch-Schmidt, and Klaus Mosbach — 53
6. Immobilization of Enzymes on Hydroxylalkyl Methacrylate Gels — Jaroslava Turková — 66
7. Enzymes Covalently Bound to Polyacrylic and Polymethacrylic Copolymers — R. Epton, Barbara L. Hibbert, and T. H. Thomas — 84
8. Immobilization of Enzymes on Neutral and Ionic Carriers — G. Manecke and J. Schlünsen — 107
9. Immobilization of Enzymes on Nylon — William E. Hornby and Leon Goldstein — 118
10. Covalent Coupling Methods for Inorganic Support Materials — Howard H. Weetall — 134

B. Adsorption

11. Adsorption and Inorganic Bridge Formations — Ralph A. Messing — 148

C. Entrapment and Related Techniques

12. Techniques of Enzyme Entrapment in Gels — K. F. O'Driscoll — 169
13. Immobilization of Steroid-Transforming Microorganisms in Polyacrylamide — Per-Olof Larsson and Klaus Mosbach — 183
14. A Method for Continuous Production of Polyacrylamide-Entrapped Enzyme Beads — A. Köstner and M. Mandel — 191

15. Preparation and Properties of Enzymes Immobilized by Copolymerization	D. Jaworek, H. Botsch, and J. Maier	195
16. Microencapsulation of Enzymes and Biologicals	T. M. S. Chang	201
17. Enzyme Entrapment in Liposomes	Gregory Gregoriadis	218
18. Fiber-Entrapped Enzymes	Dino Dinelli, Walter Marconi, and Franco Morisi	227
19. Collagen-Immobilized Enzyme Systems	W. R. Vieth and K. Venkatasubramanian	243

D. Aggregation

20. Chemically Aggregated Enzymes	Georges B. Broun	263

E. Miscellaneous

21. Photochemical Immobilization of Enzymes and Other Biochemicals	Patrick Guire	280
22. Nonaqueous Synthesis of Polystyryl Lysozyme	H. D. Brown and S. K. Chattopadhyay	288
23. Physical Immobilization of Enzymes by Hollow-Fiber Membranes	R. P. Chambers, W. Cohen, and W. H. Baricos	291

F. Other Techniques and General Classification

24. Immobilized Enzymes—Miscellaneous Methods and General Classification	Oskar R. Zaborsky	317

Section III. Assay Procedures

25. Assay Procedures for Immobilized Enzymes	Bo Mattiasson and Klaus Mosbach	335

Section IV. Characterization by Physical Methods

26. Investigation of the Physical Properties of Immobilized Enzymes	E. Kendall Pye and Britton Chance	357
27. Microfluorometry of Product Formation in Single Enzyme-Sepharose Beads Using Fluorogenic Substrates	M. Sernetz and H. Puchinger	373

Section V. Characterization by Chemical Methods

28. Characterization of Immobilized Enzymes by Chemical Methods	Detlef Gabel and Rolf Axén	383

Section VI. Kinetics of Immobilized Enzymes

29. Kinetic Behavior of Immobilized Enzyme Systems	L. Goldstein	397

30. Activity Correlations between Similarly Modified Soluble and Immobilized Enzymes	DAVID F. OLLIS AND RATHIN DATTA	444

Section VII. Multistep Enzyme Systems

31. Multistep Enzyme Systems	K. MOSBACH AND B. MATTIASSON	453
32. The Glucose Oxidase–Catalase System	JAMES C. BOUIN, MOKHTAR T. ATALLAH, AND HERBERT O. HULTIN	478

Section VIII. The Application of Immobilized Enzymes to Fundamental Studies in Biochemistry

33. Immobilized Subunits	WILLIAM W.-C. CHAN	491
34. Covalently Bound Glutamate Dehydrogenase for Studies of Subunit Association and Allosteric Regulation	HAROLD E. SWAISGOOD, H. ROBERT HORTON, AND KLAUS MOSBACH	504
35. Immobilization as a Means of Investigating the Acquisition of Tertiary Structure in Chymotrypsinogen	H. ROBERT HORTON AND HAROLD E. SWAISGOOD	516
36. Conformational Transitions in Immobilized Proteases	DETLEF GABEL AND VOLKER KASCHE	526
37. Immobilized Hemoproteins	ERALDO ANTONINI AND MARIA ROSARIA ROSSI FANELLI	538
38. Immobilized Proteins in Single Crystals	F. A. QUIOCHO	546
39. Mechanochemistry of Immobilized Enzymes: A New Approach to Studies in Fundamental Enzymology I. Regulation by Mechanical Means of the Catalytic Properties of Enzymes Attached to Polymer Fibers	I. V. BEREZIN, A. M. KLIBANOV, G. P. SAMOKHIN, AND KAREL MARTINEK	558
40. Mechanochemistry of Immobilized Enzymes: A New Approach to Studies in Fundamental Enzymology II. Regulation of the Rate of Enzymic Protein-Protein Interaction in Polyacrylamide Gel	I. V. BEREZIN, A. M. KLIBANOV, V. S. GOLDMACHER, AND KAREL MARTINEK	571

Section IX. Application of Immobilized Enzymes

A. Analytical Area

41. Enzyme Electrodes and Solid Surface Fluorescence Methods	GEORGE G. GUILBAULT	579
42. The Applications of Immobilized Enzymes in Automated Analysis	WILLIAM E. HORNBY AND GEORGE A. NOY	633
43. Monitoring of Air and Water for Enzyme Inhibitors	LOUIS H. GOODSON AND WILLIAM B. JACOBS	647

44. Immobilized Enzymes in Microcalorimetry	Åke Johansson, Bo Mattiasson, and Klaus Mosbach	659
45. Enzyme Thermistor Devices	Bengt Danielsson and Klaus Mosbach	667

B. Medical Area

46. Methods for the Therapeutic Applications of Immobilized Enzymes	Thomas Ming Swi Chang	676
47. Medical Applications of Liposome-Entrapped Enzymes	Gregory Gregoriadis	698
48. Immunoenzymic Techniques for Biomedical Analysis	Stratis Avrameas	709

C. Enzyme Engineering (Enzyme Technology)

49. Immobilized-Enzyme Reactors	M. D. Lilly and P. Dunnill	717
50. Production of L-Aspartic Acid by Microbial Cells Entrapped in Polyacrylamide Gels	Ichiro Chibata, Tetsuya Tosa, and Tadashi Sato	739
51. Production of L-Amino Acids by Aminoacylase Adsorbed on DEAE-Sephadex	Ichiro Chibata, Tetsuya Tosa, Tadashi Sato, and Takao Mori	746
52. Production of 6-Aminopenicillanic Acid with Immobilized *Escherichia coli* Acylase	E. Lagerlöf, L. Nathorst-Westfelt, B. Ekström, and B. Sjöberg	759
53. Process Engineering of Glucose Isomerization by Collagen-Immobilized Whole Microbial Cells	W. R. Vieth and K. Venkatasubramanian	768
54. Scale-Up Studies on Immobilized, Purified Glucoamylase, Covalently Coupled to Porous Ceramic Support	H. H. Weetall, W. P. Vann Wayne H. Pitcher, Jr., D. D. Lee, Y. Y. Lee, and G. T. Tsao	776
55. The Preparation, Characterization, and Scale-Up of a Lactase System Immobilized to Inorganic Supports for the Hydrolysis of Acid Whey	Wayne H. Pitcher, Jr., James R. Ford, and Howard H. Weetall	792
56. Continuous Production of High-Fructose Syrup by Cross-Linked Cell Homogenates Containing Glucose Isomerase	Poul Børge Poulsen and Lena Zittan	809
57. Lactose Reduction of Milk by Fiber-Entrapped β-Galactosidase. Pilot-Plant Experiments	Mauro Pastore and Franco Morisi	822

D. Application of Immobilized Enzymes in Organic Chemistry

58. On the Potential of Soluble and Immobilized Enzymes in Synthetic Organic Chemistry	J. Bryan Jones	831
59. The Synthesis of Porphobilinogen by Immobilized δ-Aminolevulinic Acid Dehydratase	Daniela Gurne and David Shemin	844

60. Properties and Applications of an Immobilized Mixed-Function Hepatic Drug Oxidase	L. L. POULSEN, S. S. SOFER, AND D. M. ZIEGLER	849

Section X. Immobilized Coenzymes

61. Immobilized Coenzymes	KLAUS MOSBACH, PER-OLOF LARSSON, AND CHRISTOPHER LOWE	859
62. Covalent Immobilization of Adenylate Kinase and Acetate Kinase in a Polyacrylamide Gel: Enzymes for ATP Regeneration	GEORGE M. WHITESIDES, ANDRE LAMOTTE, ORN ADALSTEINSSON, AND CLARK K. COLTON	887

Section XI. Miscellaneous

63. Artificial Enzyme Membranes	D. THOMAS AND G. BROUN	901
64. Immobilization of Lactic Dehydrogenase	FRANCIS E. STOLZENBACH AND NATHAN O. KAPLAN	929

AUTHOR INDEX 937

SUBJECT INDEX 959

Contributors to Volume XLIV

Article numbers are in parentheses following the names of contributors.
Affiliations listed are current.

ORN ADALSTEINSSON (62), *Department of Chemistry, Massachusetts Institute of Technology, Cambridge, Massachusetts*

ERALDO ANTONINI (37), *Institute of Chemistry, Faculty of Medicine, University of Rome, Rome, Italy*

MOKHTAR T. ATALLAH (32), *Department of Food Science and Nutrition, University of Massachusetts, Amherst, Massachusetts*

STRATIS AVRAMEAS (48), *Unité d'Immunocytochimie, Departement de Biologie Moleculaire, Institut Pasteur, Paris, France*

ROLF AXÉN (3, 28), *Institute of Biochemistry, University of Uppsala, Uppsala, Sweden*

W. H. BARICOS (23), *Department of Biochemistry, Tulane Medical School, New Orleans, Louisiana*

I. V. BEREZIN (39, 40), *Laboratory of Bioorganic Chemistry, Lomonosov State University, Moscow, U.S.S.R.*

H. BOTSCH (15), *Boehringer Mannheim GmbH., Biochemical Work, Tutzing, West Germany*

JAMES C. BOUIN (32), *Department of Food Science and Nutrition, University of Massachusetts, Amherst, Massachusetts*

GEORGES B. BROUN (20, 63), *Bioengineering Department, Université de Technologie de Compiègne, Compiègne, France*

H. D. BROWN (22), *Cook College, New Jersey Agricultural Experiment Station, Rutgers University, State University of New Jersey, New Brunswick, New Jersey*

R. P. CHAMBERS (23), *Department of Chemical Engineering, Auburn University, Auburn, Alabama*

WILLIAM W.-C. CHAN (33), *Department of Biochemistry, McMaster University Medical Center, Hamilton, Ontario, Canada*

BRITTON CHANCE (26), *Department of Biochemistry and Biophysics and the Johnson Research Foundation, University of Pennsylvania School of Medicine, Philadelphia, Pennsylvania*

THOMAS MING SWI CHANG (16, 46), *Department of Physiology, Faculty of Medicine, McGill University, Montreal, Quebec, Canada*

S. K. CHATTOPADHYAY (22), *Cook College, New Jersey Agricultural Experiment Station, Rutgers University, State University of New Jersey, New Brunswick, New Jersey*

ICHIRO CHIBATA (50, 51), *Research Laboratory of Applied Biochemistry, Tanabe Seiyaku Company, Limited, Osaka, Japan*

W. COHEN (23), *Department of Biochemistry, Tulane Medical School, New Orleans, Louisiana*

CLARK K. COLTON (62), *Department of Chemistry, Massachusetts Institute of Technology, Cambridge, Massachusetts*

BENGT DANIELSSON (45), *Biochemical Division, Chemical Center, University of Lund, Lund, Sweden*

RATHIN DATTA (30), *Department of Chemical Engineering, Princeton University, Princeton, New Jersey*

DINO DINELLI (18), *Laboratori Processi Microbiologici, Snamprogetti, Rome, Italy*

P. DUNNILL (49), *Department of Chemical Engineering, University College London, London, England*

B. EKSTRÖM (52), *AB Astra, Södertälje, Sweden*

R. EPTON (7), *Department of Physical Sciences, The Polytechnic, Wolverhampton, England*

JAMES R. FORD (55), *Corning Glass Works, Corning, New York*

DETLEF GABEL (28, 36), *Institute of Biochemistry, University of Uppsala, Uppsala, Sweden*

V. S. Goldmacher (40), *Laboratory of Bioorganic Chemistry, Lomonosov State University, Moscow, U.S.S.R.*

Leon Goldstein (9, 29), *Department of Biochemistry, George S. Wise Center for Life Sciences, Tel Aviv University, Tel Aviv, Israel*

Louis H. Goodson (43), *Midwest Research Institute, Kansas City, Missouri*

Gregory Gregoriadis (17, 47), *Clinical Research Centre, Harrow, Middlesex, England*

George G. Guilbault (41), *Department of Chemistry, University of New Orleans, New Orleans, Louisiana*

Patrick Guire (21), *Midwest Research Institute, Kansas City, Missouri*

Daniela Gurne (59), *Department of Biochemistry and Molecular Biology, Northwestern University, Evanston, Illinois*

Barbara L. Hibbert (7), *Manchester Regional Blood Transfusion Service, Manchester, England*

William E. Hornby (9, 42), *Miles Laboratories Limited, Stoke Court, Stoke Poges, Slough, Bucks, England*

H. Robert Horton (34, 35), *Department of Biochemistry, North Carolina State University, Raleigh, North Carolina*

Herbert O. Hultin (32), *Department of Food Science and Nutrition, University of Massachusetts, Amherst, Massachusetts*

William B. Jacobs (43), *Midwest Research Institute, Kansas City, Missouri*

D. Jaworek (15), *Boehringer Mannheim GmbH., Biochemical Work, Tutzing, West Germany*

Åke Johansson (44), *LKB-Produkter AB, Fack, Bromma, Sweden*

J. Bryan Jones (58), *Department of Chemistry, University of Toronto, Lash Miller Chemical Laboratories, Toronto, Ontario, Canada*

Nathan O. Kaplan (64), *Department of Chemistry, University of California, San Diego, La Jolla, California*

Volker Kasche (36), *Fachsektion Biologie, Universität Bremen, Bremen, West Germany*

A. M. Klibanov (39, 40), *Laboratory of Bioorganic Chemistry, Lomonosov State University, Moscow, U.S.S.R.*

Ann-Christin Koch-Schmidt (5), *Biochemical Division, Chemical Center, University of Lund, Lund, Sweden*

A. Köstner (14), *Tallinn Polytechnic Institute, Tallinn, Estonia, U.S.S.R.*

E. Lagerlöf (52), *AB Astra, Södertälje, Sweden*

Andre Lamotte (62), *Department of Chemistry, Massachusetts Institute of Technology, Cambridge, Massachusetts*

Per-Olof Larsson (13, 61), *Biochemical Division, Chemical Center, University of Lund, Lund, Sweden*

D. D. Lee (54), *Iowa State University, Ames, Iowa*

Y. Y. Lee (54), *Auburn University, Auburn, Alabama*

M. D. Lilly (4, 49), *Department of Chemical Engineering, University College London, London, England*

Christopher Lowe (61), *Department of Physiology and Biochemistry, Medical and Biological Sciences Building, Southampton, England*

J. Maier (15), *Boehringer Mannheim GmbH., Biochemical Work, Tutzing, West Germany*

M. Mandel (14), *Tallinn Polytechnic Institute, Tallinn, Estonia, U.S.S.R.*

G. Manecke (8), *Institute for Organic Chemistry, Free University of Berlin, Berlin, West Germany*

Walter Marconi (18), *Laboratori Processi Microbiologici, Snamprogetti, Monterotondo, Rome, Italy*

Karel Martinek (39, 40), *Laboratory of Bioorganic Chemistry, Lomonosov State University, Moscow, U.S.S.R.*

Bo Mattiasson (25, 31, 44), *Biochemical Division, Chemical Center, University of Lund, Lund, Sweden*

Ralph A. Messing (11), *Research and Development Laboratories, Corning Glass Works, Corning, New York*

TAKAO MORI (51), *Research Laboratory of Applied Biochemistry, Tanabe Seiyaku Company, Limited, Osaka, Japan*

FRANCO MORISI (18, 57), *Laboratori Processi Microbiologici, Snamprogetti, Monterotondo, Rome, Italy*

KLAUS MOSBACH (1, 5, 13, 25, 31, 34, 44, 45, 61), *Biochemical Division, Chemical Center, University of Lund, Lund, Sweden*

ROLF MOSBACH (5), *Landesimpfanstalt Nordrhein-Westfalen, Düsseldorf, West Germany*

L. NATHORST-WESTFELT (52), *AB Astra, Södertälje, Sweden*

GEORGE A. NOY (42), *Department of Biochemistry, University of St. Andrews, St. Andrews, Fife, Scotland*

K. F. O'DRISCOLL (12), *Department of Chemical Engineering, University of Waterloo, Waterloo, Ontario, Canada*

DAVID F. OLLIS (30), *Department of Chemical Engineering, Princeton University, Princeton, New Jersey*

MAURO PASTORE (57), *Centrale del Latte, Milan, Italy*

WAYNE H. PITCHER, JR. (54, 55) *Corning Glass Works, Corning, New York*

JERKER PORATH (3), *Institute of Biochemistry, University of Uppsala, Uppsala, Sweden*

L. L. POULSEN (60), *Clayton Foundation Biochemical Institute, and the Department of Chemistry, University of Texas, Austin, Texas*

POUL BØRGE POULSEN (56), *Analytical and Chemical Research and Development, Enzymes Division, Novo Industri A/S, Novo Alle, Bagsvaerd, Denmark*

H. PUCHINGER (27), *Battelle-Institut e.V., Frankfurt am Main, Am Römerhof, West Germany*

E. KENDALL PYE (26), *Department of Biochemistry and Biophysics, University of Pennsylvania School of Medicine, Philadelphia, Pennsylvania*

F. A. QUIOCHO (38), *Department of Biochemistry, Rice University, Houston, Texas*

MARIA ROSARIA ROSSI FANELLI (37), *C.N.R. Center for Molecular Biology, Rome, Italy*

G. P. SAMOKHIN (39), *Laboratory of Bioorganic Chemistry, Lomonosov State University, Moscow, U.S.S.R.*

TADASHI SATO (50, 51), *Research Laboratory of Applied Biochemistry, Tanabe Seiyaku Company, Limited, Osaka, Japan*

J. SCHLÜNSEN (8), *Institute for Organic Chemistry, Free University of Berlin, Berlin, West Germany*

M. SERNETZ (27), *Institut für Biochemie und Endokrinologie, Universität Giessen, Giessen, West Germany*

DAVID SHEMIN (59), *Department of Biochemistry and Molecular Biology, Northwestern University, Evanston, Illinois*

B. SJÖBERG (52), *AB Astra, Södertälje, Sweden*

S. S. SOFER (60), *School of Chemical Engineering, University of Oklahoma, Norman, Oklahoma*

PAUL A. SRERE (2), *Pre-Clinical Science Unit, Veterans Administration Hospital, and Department of Biochemistry, University of Texas Southwestern Medical School, Dallas, Texas*

FRANCIS E. STOLZENBACH (64), *Department of Chemistry, University of California, San Diego, La Jolla, California*

HAROLD E. SWAISGOOD (34, 35), *Department of Food Science, North Carolina State University, Raleigh, North Carolina*

D. THOMAS (63), *Bioengineering Department, Laboratory of Enzyme Technology, Université de Technologie de Compiègne, Compiègne, France*

T. H. THOMAS (7), *Department of Chemical Pathology, School of Medicine, University of Leeds, Leeds, England*

TETSUYA TOSA (50, 51), *Research Laboratory of Applied Biochemistry, Tanabe Seiyaku Company, Limited, Osaka, Japan*

G. T. TSAO (54), *Purdue University, Lafayette, Indiana*

JAROSLAVA TURKOVÁ (6), *Institute of Organic Chemistry and Biochemistry, Czechoslovak Academy of Sciences, Prague, Czechoslovakia*

KOSAKU UYEDA (2), *Pre-Clinical Science Unit, Veterans Administration Hospital, and Department of Biochemistry, University of Texas Southwestern Medical School, Dallas, Texas*

W. P. VANN (54), *Corning Glass Works, Corning, New York*

K. VENKATASUBRAMANIAN (19, 53), *Department of Chemical and Biochemical Engineering, Rutgers University, State University of New Jersey, New Brunswick, New Jersey*

W. R. VIETH (19, 53), *Department of Chemical and Biochemical Engineering, Rutgers University, State University of New Jersey, New Brunswick, New Jersey*

HOWARD H. WEETALL (10, 54, 55), *Corning Glass Works, Corning, New York*

GEORGE M. WHITESIDES (62), *Department of Chemistry, Massachusetts Institute of Technology, Cambridge, Massachusetts*

OSKAR R. ZABORSKY (24), *Advanced Productivity Research and Technology Division, National Science Foundation, Washington, D.C.*

D. M. ZIEGLER (60), *Clayton Foundation Biochemical Institute, and the Department of Chemistry, University of Texas, Austin, Texas*

LENA ZITTAN (56), *Analytical and Chemical Research and Development, Enzymes Division, Novo Industri A/S, Novo Alle, Bagsvaerd, Denmark*

Preface

Editing this volume has been both a challenge and a much more difficult task than originally anticipated. The reason for this has been, in part, that because of the broad scope of "immobilized enzymes" contributors were invited from many different disciplines, including chemical engineering and medicine. As a result, this volume does differ somewhat from the usual format. Many sections begin with a general methodological review, and are followed by specific examples. In the introductory paper, as a further guide, the background of each section is given. I hope the volume will prove useful from a methodological point of view and will not only acquaint its readers with the various activities in the field but will also act as a stimulus for further research.

Research on immobilized enzymes is still in its infancy. The reader will probably, therefore, search in vain for the best immobilization technique and the best support for his or her enzyme. Few comparative studies have been made to evaluate the different procedures; much of the research to date has involved quick exploration of new methods and supports. I very much doubt whether any optimal technique or support exists. Requirements for different enzymes may vary, and specific conditions may be needed for a particular application of an immobilized enzyme such as a highly rigid support.

The term "immobilized enzymes" has been used following a recommendation made at the 1973 Enzyme Engineering Conference in Henniker, New Hampshire. However, other terms, such as "matrix-bound" and "insolubilized," to refer to the same type of preparation and "support," "matrix," and "carrier" are used indiscriminately.

I would like to thank all the authors for their contributions and cooperation. I am also indebted to my co-workers for their valuable comments and to Mrs. Scott for her secretarial assistance. The friendly cooperation of the staff of Academic Press is acknowledged.

KLAUS MOSBACH

METHODS IN ENZYMOLOGY

EDITED BY

Sidney P. Colowick and Nathan O. Kaplan

VANDERBILT UNIVERSITY
SCHOOL OF MEDICINE
NASHVILLE, TENNESSEE

DEPARTMENT OF CHEMISTRY
UNIVERSITY OF CALIFORNIA
AT SAN DIEGO
LA JOLLA, CALIFORNIA

I. Preparation and Assay of Enzymes
II. Preparation and Assay of Enzymes
III. Preparation and Assay of Substrates
IV. Special Techniques for the Enzymologist
V. Preparation and Assay of Enzymes
VI. Preparation and Assay of Enzymes (*Continued*)
  Preparation and Assay of Substrates
  Special Techniques
VII. Cumulative Subject Index

METHODS IN ENZYMOLOGY

EDITORS-IN-CHIEF

Sidney P. Colowick Nathan O. Kaplan

VOLUME VIII. Complex Carbohydrates
Edited by ELIZABETH F. NEUFELD AND VICTOR GINSBURG

VOLUME IX. Carbohydrate Metabolism
Edited by WILLIS A. WOOD

VOLUME X. Oxidation and Phosphorylation
Edited by RONALD W. ESTABROOK AND MAYNARD E. PULLMAN

VOLUME XI. Enzyme Structure
Edited by C. H. W. HIRS

VOLUME XII. Nucleic Acids (Parts A and B)
Edited by LAWRENCE GROSSMAN AND KIVIE MOLDAVE

VOLUME XIII. Citric Acid Cycle
Edited by JOHN M. LOWENSTEIN

VOLUME XIV. Lipids
Edited by JOHN M. LOWENSTEIN

VOLUME XV. Steroids and Terpenoids
Edited by RAYMOND B. CLAYTON

VOLUME XVI. Fast Reactions
Edited by KENNETH KUSTIN

VOLUME XVII. Metabolism of Amino Acids and Amines (Parts A and B)
Edited by HERBERT TABOR AND CELIA WHITE TABOR

VOLUME XVIII. Vitamins and Coenzymes (Parts A, B, and C)
Edited by DONALD B. MCCORMICK AND LEMUEL D. WRIGHT

VOLUME XIX. Proteolytic Enzymes
Edited by GERTRUDE E. PERLMANN AND LASZLO LORAND

VOLUME XX. Nucleic Acids and Protein Synthesis (Part C)
Edited by KIVIE MOLDAVE AND LAWRENCE GROSSMAN

VOLUME XXI. Nucleic Acids (Part D)
Edited by LAWRENCE GROSSMAN AND KIVIE MOLDAVE

VOLUME XXII. Enzyme Purification and Related Techniques
Edited by WILLIAM B. JAKOBY

VOLUME XXIII. Photosynthesis (Part A)
Edited by ANTHONY SAN PIETRO

VOLUME XXIV. Photosynthesis and Nitrogen Fixation (Part B)
Edited by ANTHONY SAN PIETRO

VOLUME XXV. Enzyme Structure (Part B)
Edited by C. H. W. HIRS AND SERGE N. TIMASHEFF

VOLUME XXVI. Enzyme Structure (Part C)
Edited by C. H. W. HIRS AND SERGE N. TIMASHEFF

VOLUME XXVII. Enzyme Structure (Part D)
Edited by C. H. W. HIRS AND SERGE N. TIMASHEFF

VOLUME XXVIII. Complex Carbohydrates (Part B)
Edited by VICTOR GINSBURG

VOLUME XXIX. Nucleic Acids and Protein Synthesis (Part E)
Edited by LAWRENCE GROSSMAN AND KIVIE MOLDAVE

VOLUME XXX. Nucleic Acids and Protein Synthesis (Part F)
Edited by KIVIE MOLDAVE AND LAWRENCE GROSSMAN

VOLUME XXXI. Biomembranes (Part A)
Edited by SIDNEY FLEISCHER AND LESTER PACKER

VOLUME XXXII. Biomembranes (Part B)
Edited by SIDNEY FLEISCHER AND LESTER PACKER

VOLUME XXXIII. Cumulative Subject Index Volumes I–XXX
Edited by MARTHA G. DENNIS AND EDWARD A. DENNIS

VOLUME XXXIV. Affinity Techniques (Enzyme Purification: Part B)
Edited by WILLIAM B. JAKOBY AND MEIR WILCHEK

Volume XXXV. Lipids (Part B)
Edited by John M. Lowenstein

Volume XXXVI. Hormone Action (Part A: Steroid Hormones)
Edited by Bert W. O'Malley and Joel G. Hardman

Volume XXXVII. Hormone Action (Part B: Peptide Hormones)
Edited by Bert W. O'Malley and Joel G. Hardman

Volume XXXVIII. Hormone Action (Part C: Cyclic Nucleotides)
Edited by Joel G. Hardman and Bert W. O'Malley

Volume XXXIX. Hormone Action (Part D: Isolated Cells, Tissues, and Organ Systems)
Edited by Joel G. Hardman and Bert W. O'Malley

Volume XL. Hormone Action (Part E: Nuclear Structure and Function)
Edited by Bert W. O'Malley and Joel G. Hardman

Volume XLI. Carbohydrate Metabolism (Part B)
Edited by W. A. Wood

Volume XLII. Carbohydrate Metabolism (Part C)
Edited by W. A. Wood

Volume XLIII. Antibiotics
Edited by John H. Hash

Volume XLIV. Immobilized Enzymes
Edited by Klaus Mosbach

Volume XLV. Proteolytic Enzymes (Part B)
Edited by Laszlo Lorand

Volume XLVI. Affinity Labeling (in preparation)
Edited by William B. Jakoby and Meir Wilchek

Section I
General Comments

[1] Introduction

By KLAUS MOSBACH

General Comments

The most common questions asked about immobilized enzymes (=insolubilized, matrix-bound, carrier-bound, support-bound, entrapped, microencapsulated) relate to their activity and stability. With regard to the latter aspect, apart from the obvious prevention of autodigestion of proteolytic enzymes, few reports have yet appeared which prove beyond any doubt an increased stability (for example, thermostability) for immobilized enzymes. In a number of cases, so-called stability is caused by overloading the support with enzyme, and any prevailing enzyme denaturation will thus not be observed. On the other hand, this author finds it likely that such stabilization can be accomplished by either covalent multiple-point attachment of enzymes or by embedding the enzyme in a stabilizing and protecting microenvironment.

With regard to activity, immobilized enzymes generally show less specific activity than the native enzyme. But with more sophisticated procedures being developed, losses in activity are minimized. In characterizing immobilized enzymes the activity should ideally be expressed as the initial reaction rate in terms of for instances microkatals per milligram of dry immobilized enzyme preparation or any multiple thereof. [A microkatal (μkat) refers to the conversion of 1 μmole of substrate per second.] These suggestions were made by an ad hoc committee on nomenclature (chairman P. V. Sundaram) and are *published* in "Enzyme Engineering," Vol. 2. Plenum, New York, 1974.

The measurement of the kinetic constants of immobilized enzymes may not yield true values equivalent to those obtained in homogeneous reactions, because of the influence of physical factors, such as diffusion. Hence, rate and binding constants should be referred to as apparent V_{max} and apparent K_m and written as $V_{max}(app)$ and $K_m(app)$. An operational definition of $K_m(app)$ would be "that substrate concentration which gives a reaction velocity corresponding to half the $V_{max}(app)$."

However, as has been suggested by J. M. Engasser and C. Horvath [*Biotechnol. Bioeng.* **16**, 919 (1974)], in the presence of diffusional resistance, the experimentally derived "apparent K_m" [either by determination as $S_{0.5}$ (the substrate concentration that yields half of the saturation activity) or by determination of K from the large portion of straight line of slope V_{max}/K on the Burk–Lineweaver plot] can be misleading and should not be employed when $S_{0.5}$ and K have different values. Instead, $S_{0.5}$ and K should be used to characterize the observed kinetic behavior of immobilized enzymes depending on the graphical procedure employed to analyze experimental data. For practical applications, however, determinations of, for instance, V_{max} and K_m are of less importance, and it may often suffice to give the "operational activity," which expresses the enzymic activity at a substrate concentration which is chosen to suit the specific process in question.

The remaining chapters in this volume have been grouped into ten sections (Sections II–XI), each of which is briefly described below.

Section II

Immobilization Techniques

The major part of the volume covers various immobilization techniques since such information is of the most obvious interest to the reader. In contrast to what is the case, for instance, in the isolation and purification of an enzyme, for which often just one optimal method applies, probably no *one* optimal immobilization technique or support can be found for one particular enzyme. This is because the usefulness of an immobilized enzyme preparation will depend on its particular area of application, whether in an industrial process, in medicine or as a biological model system. Thus there is little use in providing experimental data from all of more than a thousand reports found in the literature on the immobilization of more than 200 enzymes. Rather, it is more useful to describe general procedures illustrated with a few typical examples. With this knowledge at hand, the reader should be in a better situation to select a procedure and support "tailor-made" for his purpose. [If he wishes he can, in addition, look up specific references to the enzyme he is working with; these can be found in a compendium of references of immobilized enzymes, up to 1974, prepared for Corning Glass Works, Corning, New York 14830 (editor H. H. Weetall) by the New England Research Application Center, University of Connecticut, Storrs, Connecticut 06268. It should be added, though, that a substantial part of the immobilizations listed there date back to older and less developed procedures than those described in this volume.] The major immobilization techniques known, including covalent binding, adsorption, entrapment, and aggregation, are treated separately in that order, as are the various major supports used for each immobilization technique as given in the titles.

The major immobilization techniques are treated in large sections rather than as a collection of recipes in the hope that the reader will thus be in a position to evaluate the various techniques.

Section III

Assay Procedures for Immobilized Enzymes

Because of the particulate nature of immobilized enzyme preparations, normal assay procedures, in particular spectrophotometric ones, have to be modified. In this section an attempt is made to collect and describe the various procedures used, which are scattered throughout the literature.

Section IV

Characterization by Physical Methods

This section is in the nature of a review and guide, and little methodology is found apart from the chapter on microfluorometry. The reason is simply that very little work has yet been experimentally carried out, but it is hoped that the chapter by Pye and Chance will serve as a guide for future studies on the physicochemical characterization of immobilized enzymes. This is an area that has been utterly neglected, and their account, which has been taken largely from studies with related biological material (i.e., membrane-bound enzymes), should provide potentially useful procedures for the physical characterization of immobilized enzymes and be a stimulus for future studies.

Section V

Characterization of Immobilized Enzymes by Chemical Methods

More work has been carried out on the chemical characterization of immobilized enzymes. A number of procedures are given, first on the chemical characterization of the support itself, and then on the immobilized protein. Whereas good techniques are available to determine the actual amount of protein immobilized, thus obtaining a value of the "apparent" specific activity of the immobilized enzyme in units per milligram of immobilized protein, more sophisticated studies, such as the determination of the number and kind of amino acids involved in covalent binding, are still in their infancy.

Section VI

Kinetics of Immobilized Enzymes

One large chapter is devoted to various aspects of the kinetics of immobilized enzyme systems. In this area there is a great need both to define the various kinetic parameters as applied to immobilized enzymes and to introduce another way of thought necessary when dealing with enzymes bound to matrices, as implied, for instance, in the term "microenvironment."

Section VII

Multistep Enzyme Systems

Multistep enzyme systems are the subject of this section, in which examples are given of the preparation and use of immobilized enzymes both as biological model systems of sequential enzyme reactions as they occur in general metabolism and also in their practical applications.

Section VIII

The Application of Immobilized Enzymes to Fundamental Studies in Biochemistry

This section collects a number of contributions, all of which have in common the application of the immobilization technique itself to solve problems in fundamental biochemistry. It is hoped that the examples given, including their methodology, will stimulate "pure" biochemists to apply these techniques as an aid in their studies.

Section IX

Application of Immobilized Enzymes

The large section on the application of immobilized enzymes covers four major areas. These are analysis, medicine, industrial processes, and organic chemistry. The latter has been included, even though only a few examples are yet known, to serve as a stimulus. It appears, however, that many of the immobilization techniques given can be directly applied to the kind of studies in organic chemistry that are described.

In general, as much methodology as possible has been given. Obviously the methodology of the medical and industrial sections in particular will, by their very nature, differ greatly from what "conventional" biochemists are used to. Here, factors such as immunological tolerance or, in the case of industrial applications, operational stability of support and enzyme have to be considered, and they thus influence various aspects of the methodology of immobilization to be employed. This section thus departs somewhat from the scope of "Methods in Enzymology" and is extended to become in part a "Methods in Enzyme Technology."

It is felt that such a collection as given in Section IX is of value in providing, besides methodology, information as to what has already been achieved and, as such, will act as a catalyst for further studies in these areas.

Section X

Immobilized Coenzymes

Since many enzyme systems of interest require coenzymes, usually at high cost, their possible retention and regeneration are of importance. Therefore, in this section both syntheses of various coenzyme analogs suitable for immobilization and examples of their application in their carrier-bound state are given.

Section XI

Miscellaneous

The miscellaneous section deals particularly with the important area of artificial enzyme membranes.

Section II
Immobilization Techniques

A. Covalent Coupling
   Articles 2 through 10

B. Adsorption
   Article 11

C. Entrapment and Related Techniques
   Articles 12 through 19

D. Aggregation
   Article 20

E. Miscellaneous
   Articles 21 through 23

F. Other Techniques and General Classification
   Article 24

[2] Functional Groups on Enzymes Suitable for Binding to Matrices

By PAUL A. SRERE and KOSAKU UYEDA

Proteins can be immobilized (active but not soluble) in a number of different ways. They can be trapped within gels or microcapsules; they can be adsorbed tightly to insoluble materials (clays, starches, ion-exchange resins); they can be copolymerized with some other repeating monomer; they can be cross-linked with a bifunctional reagent *around* (pellicular) an insoluble matrix; they can be derivatized by the addition of long polyamino chains until an insoluble material is produced; they can be covalently attached to an insoluble carrier. This last method is the one that is primarily discussed in this article. However, the principles discussed here, concerning the reaction of the amino acid residues on protein molecules with an active group on some support material, are also applicable to the other methods involving covalent coupling of the protein.

When a protein is covalently coupled to an insoluble support, two techniques may be employed. One technique is to activate the support material for reaction with groups on the protein. Second, one can use a coupling reagent to link protein to matrix. A third possibility would be to activate the protein molecule for coupling to the support material (see this volume [24]). Throughout this volume and in a preceding one there are specific discussions and details of such procedures. This chapter discusses the groups on the protein that are available for covalent coupling to various matrices and some of the most commonly used procedures.

Conditions for Coupling

It is important that the protein retain its active conformation after the covalent coupling procedure. Protein molecules exist in their active form (enzymes, carriers, receptors, etc.) only in a small number of active conformations. The usual native conformation(s) is determined by its primary amino acid sequence. The bonds involved in maintaining its secondary, tertiary, and quaternary structures (except for the disulfide bond) are noncovalent ones. These noncovalent interactions are salt and hydrogen bonds and hydrophobic interactions; each of these as an individual bond is relatively labile, and it is the sum of a large number of these individually weak bonds that gives stability to the final structure of the protein. However, the fragile nature of each bond limits the solution condition in which a protein can maintain an active conformation. Thus, we usually cannot exceed a temperature of 35°, nor can we use solutions whose acidity is greater than pH 3 or 4 or whose basicity is greater than pH 9 or 10.

There are exceptions to these strictures. Certain proteins can be reversibly denatured. One can carry out the coupling procedure using the unfolded polypeptide chains, then after removal of denaturing conditions the protein assumes its normal active conformation. Most of the commonly employed denaturing conditions are relatively mild in chemical terms, i.e., pH 3.0; 6 M urea; 6 M guanidinium chloride. More drastic conditions may well cleave the peptide bonds or cause SH oxidation, and these often result in irreversible changes in the protein structure.

Protein Groups Available for Covalent Attachment

First let us consider the chemical nature of the amino acid residue on proteins that are amenable to covalent reaction. Table I lists eleven amino acids whose residues could be considered reactive enough to par-

TABLE I
REACTIVE RESIDUES OF PROTEINS

—NH$_2$	ϵ-Amino of lysine (Lys) and N-terminus amino group
—SH	Sulfhydryl of cysteine (Cys)
—COOH	Carboxyl of aspartate (Asp) and glutamate (Glu) and C-terminus carboxyl group
—C$_6$H$_4$—OH	Phenolic of tyrosine (Tyr)
—N(H)—C(NH)(NH$_2$)	Guanidino of arginine (Arg)
(imidazole)	Imidazole of histidine (His)
—S—S—	Disulfide of cystine
(indole)	Indole of tryptophan (Trp)
CH$_3$—S—	Thioether of methionine (Met)
—CH$_2$OH	Hydroxyl of serine (Ser) and threonine (Thr)

TABLE II
AVERAGE PERCENT COMPOSITION OF PROTEINS
(REACTIVE RESIDUES ONLY)[a]

Residue	Percent
Ser	7.8
Lys	7
Thr	6.5
Asp	4.8
Glu	4.8
Arg	3.8
Tyr	3.4
Cys	3.4
His	2.2
Met	1.6
Trp	1.2

[a] From M. Dayhoff and L. T. Hunt, "Atlas of Protein Sequence and Structure." National Biomedical Res. Fdn., Washington, D.C., 1972.

ticipate in covalent coupling to a carrier. Absent from this list are the amino acids with amide bonds (glutamine and asparagine) and those with hydrocarbon residues (glycine, alanine, leucine, isoleucine, valine, phenylalanine, and proline). Of the amino acids listed as having reactive residues, neither the thioester of methionine nor the hydroxyl of threonine has been implicated in the covalent linkage to an insoluble support.

Two other factors should be considered as possible determinants (besides reaction type) of which groups might be more involved in covalent attachment to a solid support. One factor is the relative concentration of the amino acids in the protein molecule. Table II shows the average amino acid composition of a number of proteins for those amino acids with reactive residues.[1] While it is true that each individual protein will have its characteristic composition, the composition shown here can be used for arriving at some general conclusions regarding groups involved in immobilization techniques.

There also exists a difference in the hydrophobic nature of the various amino acid residues. Increased hydrophobicity would tend to increase the chances of a residue being buried inside the protein. Table III lists the change in free energy (ΔG transfer) for the transfer of a residue from water to 100% ethanol (25°), which is the energetic basis for the internalization of the residues.[2] The reactive residues that are not listed have

[1] M. Dayhoff, "Atlas of Protein Sequence and Structure." National Biomedical Res. Fdn., Washington, D.C. 1972.
[2] C. Tanford, J. Am. Chem. Soc. 84, 4240 (1962).

TABLE III
ΔG TRANSFER[a]

Residue	ΔG
Tyr	−2.87
Met	−1.30
Thr	−0.44
Ser	−0.04

[a] C. Tanford, *J. Am. Chem. Soc.* **84**, 4240 (1962).

positive values for ΔG transfer. The more negative this number, the more hydrophobic is the residue, and the more likely it is that it will be found inside the protein. It is more likely, therefore, that tyrosine, methionine, threonine, and serine residues will be internalized than the other amino acid residues, and the likelihood for reaction with buried residues is reduced.

Reactivity of Amino Acid Residues

It is not possible to compare accurately the amino acid residue in terms of reactivity without specifying a particular reaction and reaction conditions. We can, however, compare the number of reactions available for each residue. In a table listing side-chain reactivities, Means and Feeney[3] list the following number of reactions available for modification of these residues: Cys-31, Lys-27, Tyr-16, His-13, Met-7, Trp-7, Arg-6, (Glu, Asp)-4, and (Ser, Thr)-0.

Most of the reactions listed below can be classified as carbonyl-type reactions[4] with the nucleophilic groups on the protein, —NH_2, —SH, and —OH. Many factors are important in determining the rate of nucleophilic reactions. We can consider only one or two points here. Since the reactive species would be —NH_2, —S^-, and —O^-, then at pH values of about 9 or so the concentration of —NH_2 residues would be the highest of the three and therefore the most likely to react. In terms of nucleophilic reactivity the S anion is one or two orders of magnitude larger than that of normal N and O compounds of comparable basicity. However, the stability of the esters formed by these groups in nucleophilic reactions varies widely. Thioesters are much less stable than the oxygen esters; these in turn are less stable than the substituted amines that are formed.

[3] G. Means and R. E. Feeney, "Chemical Modification of Proteins." Holden-Day, San Francisco, California, 1971.
[4] W. P. Jencks, "Catalysis in Chemistry and Enzymology." McGraw-Hill, New York, 1969.

If one considers the factors mentioned—relative concentration, hydrophobicity, and reactivity—it is apparent that Lys would be predicted to be the most likely coupling residue, followed by Cys, Tyr, His, (Asp, Glu), Arg, Trp, (Ser, Thr), and Met.

Active Site Problems

In addition to the factors mentioned above, the reactivity of a certain group in the protein depends on the location and environment in which the amino acid residues are placed. Since the proteins are large molecules, many of the site changes are partially "buried" and may be relatively inert, while some residues that are exposed to the solvent may be highly reactive.

It is well known that amino acid residues at active sites have special properties and reactivities. In fact the overall conformation of the protein (and therefore the primary sequence) has been designed to bring the residues comprising the active site into a precise three-dimensional arrangement of charges and microenvironment so that there are special reactivities of the residues. Thus only the serine residue of the active site of certain esterases will react with diisopropylfluorophosphate; the other serine residues on the molecule are inactive with this compound. It is also known that the pK values of ionizable groups at the active site are often quite different from the value of the same group elsewhere in the molecule. Thus, given the increased reactivity of these residues, one might wonder why these groups are not the first to react with the derivatized support and therefore give rise to immobilized proteins that are inactive. This probably does happen to some degree. If the active site is on the surface of the protein, the chance for a reaction would depend on the ratio of the area of the active site compared to the area of the whole protein surface. Since the coupling reaction is usually a "strong" one, i.e., any reactive residue group on the protein that interacts with the activated matrix group lends to coupling, the active site is not especially favored for reaction. Also, the active sites on enzymes are usually in a recessed part of the protein structure. Entrance to the active site and the groups therein depends upon the size, stereochemistry, and charge characteristics of the reagent. In the case of an active grouping immobilized on a solid support, the geometry of the situation may be an important factor in allowing the active site to survive the coupling procedure. It is possible that in those cases where active site modification causes loss of activity during the immobilization process, the addition of a substrate or a competitive inhibitor to the coupling mixture will protect the active site on the enzyme against loss of activity.

Determination of Group Involved in Coupling

For many purposes it is not necessary to know which groups on the protein are involved in the covalent linkage. Indeed, often it is not important to obtain quantitative recovery of total activity, or to have assurance that the active site is intact. However, there are often instances when it is important to know how many and which linkages on a protein molecule are involved during its insolubilization and whether or not the active and/or regulatory sites on the protein have been modified. Such studies may involve experiments in which questions are asked concerning the behavior of enzymes in particular microenvironments or the behavior of enzyme subunits.

Relatively few measurements have been reported concerning this problem, and often the amino acid residues involved are "guessed at" on the basis of the coupling reaction used. Several methods have been employed for determination of the protein groups involved. In one procedure a complete HCl hydrolysis and amino acid analysis is performed upon the original protein and upon the immobilized protein. The loss of amino acid residues between the two determinations is taken to mean that those residues were involved in the covalent bond. This method has the disadvantage that acid hydrolysis may cleave the covalent linkage in a way that would restore the original amino acid residue so that linkage would not be detected.

Another method of determining which amino acid residue is involved in the covalent bond depends upon the use of analytical methods for specific amino acid residues. For example, cysteinyl residues can be measured easily with dithionitrobenzoate so that the SH groups on the protein can be determined before and after the coupling procedure. The difference in these values would indicate the number of cysteinyl groups involved in the covalent immobilization. Other specific methods exist for many of the other reactive amino acid residues, and each could be applied in a similar manner.

Specific Coupling Reactions[5-7]

Cyanogen Bromide Activation of Sepharose

One of the most useful procedures for the covalent coupling of proteins to a supporting material has been the one devised by Axén et al.[8] The

---

[5] H. D. Brown and F. X. Hasselberger, in "Chemistry of the Cell Interface" Part B (H. D. Brown, ed.), pp. 185–258. Academic Press, New York, 1972.

[6] O. R. Zaborsky, "Immobilized Enzymes." Chem. Rubber Publ. Co., Cleveland, Ohio, 1973.

activation reaction of a polysaccharide (Sepharose) with cyanogen bromide, leads to a reactive imidocarbonate on the Sepharose:

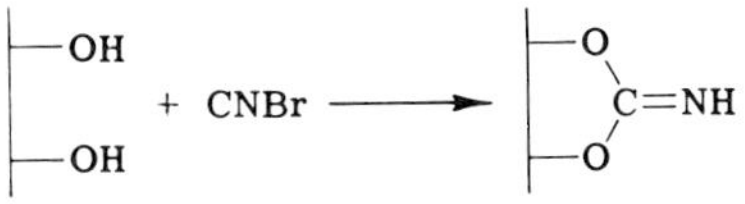

$$\begin{bmatrix} -OH \\ -OH \end{bmatrix} + CNBr \longrightarrow \begin{bmatrix} -O \\ -O \end{bmatrix}C=NH$$

The activated Sepharose will then couple to an amino group on the protein to yield the immobilized product:

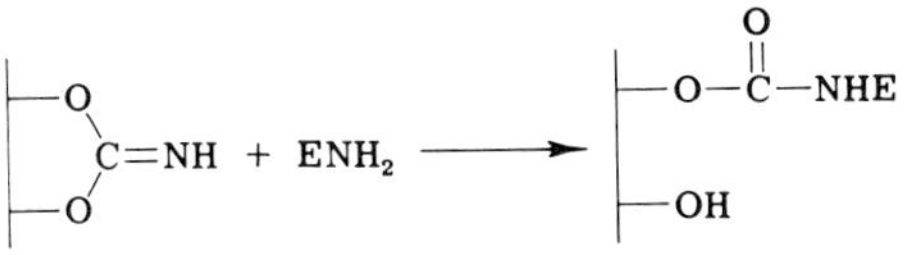

$$\begin{bmatrix} -O \\ -O \end{bmatrix}C=NH + ENH_2 \longrightarrow \begin{bmatrix} -O-\overset{O}{\underset{\|}{C}}-NHE \\ -OH \end{bmatrix}$$

Coupling of the protein is thought to be mainly the ε-amino groups of lysine and the α-amino group of the N-terminal amino acid (see also this volume [3]).

Carbodiimide

The use of carbodiimide for the chemical synthesis of many biological compounds has been available for many years. It was used extensively in the biochemical field by Khorana for the chemical synthesis of nucleotides, coenzymes, and nucleic acids. The reactions usually occur easily at room temperature at neutral pH, and they have a broad specificity. The coupling of an enzyme to a carboxylic polymer would occur as follows:

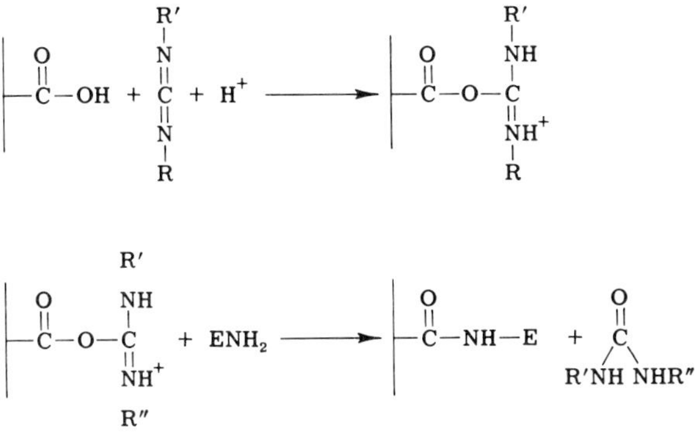

$$\begin{bmatrix} -\overset{O}{\underset{\|}{C}}-OH \end{bmatrix} + \begin{matrix} R' \\ | \\ N \\ \| \\ C \\ \| \\ N \\ | \\ R \end{matrix} + H^+ \longrightarrow \begin{bmatrix} -\overset{O}{\underset{\|}{C}}-O-\overset{NH-R'}{\underset{NH^+}{\underset{|}{C}}}\\ \quad\quad\quad\quad | \\ \quad\quad\quad\quad R \end{bmatrix}$$

$$\begin{bmatrix} -\overset{O}{\underset{\|}{C}}-O-\overset{NH-R'}{\underset{NH^+}{\underset{|}{C}}}\\ \quad\quad\quad\quad | \\ \quad\quad\quad\quad R'' \end{bmatrix} + ENH_2 \longrightarrow \begin{bmatrix} -\overset{O}{\underset{\|}{C}}-NH-E \end{bmatrix} + \overset{O}{\underset{R'NH\quad NHR''}{\underset{\diagdown\quad\diagup}{C}}}$$

---

[7] R. Goldman, L. Goldstein, and E. Katchalski, *in* "Biochemical Aspects of Reactions on Solid Supports" (G. R. Stark, ed.), pp. 1–78. Academic Press, New York, 1971.

[8] R. Axén, J. Porath, and S. Ernback, *Nature (London)* **214**, 1302 (1967).

If the support contains amines or substituted amino groups, then coupling in the presence of carbodiimides would be through the carboxyl group on the protein. Water-soluble carbodiimides would be expected to interact with sulfhydryl and tyrosyl residues as well.

Linkage through Diazonium Salt

Diazonium compounds react extremely well with lysyl, tyrosyl, and histidinyl residues. They react also with cysteinyl, arginyl, and tryptophanyl residues. In the use of this technique for the insolubilization of enzymes a substituted aromatic amine is coupled to a polymer and the amino moiety diazotized using acid and $NaNO_2$. The enzyme is added and the coupling proceeds.

$$\vdash\!\!\bigcirc\!\!-NH_2 \xrightarrow{HNO_2} \vdash\!\!\bigcirc\!\!-N_2^+$$

$$\vdash\!\!\bigcirc\!\!-N_2^+ + E\!\!-\!\!\bigcirc\!\!-OH \longrightarrow \vdash\!\!\bigcirc\!\!-N\!\!=\!\!N\!\!-\!\!\bigcirc\!\!\overset{HO}{-}E$$

In the immobilization of several proteolytic enzymes by this method Goldman et al.[7] found that tyrosyl, lysyl, and arginyl residues had been involved in the linkage.

Use of Supports Containing Anhydride Groups

Anhydrides react extremely rapidly with lysyl, cysteinyl, tyrosyl, and histidyl residues. The reaction with cysteinyl, tyrosyl, and histidyl residues is either a spontaneously or an easily reversible reaction. N-Carboxyl anhydrides, on the other hand, are specific for lysyl residues and have been used to insolubilize enzymes in a copolymerization process:

$$\begin{array}{c}-C-C-\\ |\ \ |\ \\ C\ \ \ C\\ \diagdown\!\!\diagup\!\!\diagdown\!\!\diagup\\ O\ \ O\ \ O\end{array} + ENH_2 \longrightarrow \begin{array}{c}-C-C-\\ |\ \ \ |\\ CO\ \ COO^-\\ |\\ ENH\end{array}$$

Linkage to Acyl Azide-Activated Material

The activation of carboxymethyl cellulose is performed first by esterification to yield the methyl ester; this is followed by hydrazinolysis to form the hydrazide. The hydrazide is allowed to react with nitrous acid to form the acyl azide. The acyl azide can then react with the nucleophilic groups, sulfhydryl, amino or hydroxyl, to yield the thioester, amide, or ester linkage:

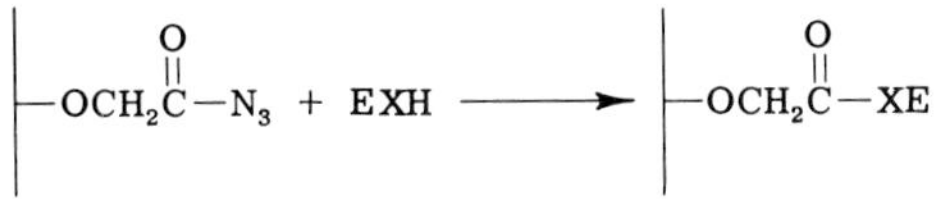

$$-OCH_2\overset{O}{\underset{\|}{C}}-N_3 + EXH \longrightarrow -OCH_2\overset{O}{\underset{\|}{C}}-XE$$

The method is one of the oldest and most frequently used methods of covalent coupling of proteins to a solid support.

Other Methods

Celluloses have been activated by use of cyanuric chloride and some of its dichloro derivatives. Reaction then can occur with amino groups in proteins. In addition, supports containing isothiocyanates have been used for the covalent coupling of proteins.[6]

Acknowledgment

This work was supported in part by Grant No. GB41851 from the National Science Foundation.

[3] Immobilization of Enzymes to Agar, Agarose, and Sephadex Supports

By JERKER PORATH and ROLF AXÉN

Hydroxylic Supports, Pro and Con

When selecting supports suitable for immobilization of enzymes, one has to consider the following factors:

1. Mechanical properties, such as rigidity and durability, while the gel product is subjected to mechanical agitation, compression, etc.
2. Physical form (granules, sheets, inner tube walls, etc.)
3. Resistance to chemical and microbial attacks
4. Hydrophilicity as manifested by the tendency to incorporate water into its structure
5. Permeability
6. Ability to be derivatized with satisfactory retention of factors 1–5
7. Price and availability

From experience it appears that hydrophilicity is a very important factor for the preservation of enzymes in a highly active state after their immobilization to the solid support. This fact places hydrophilic gels in a very favorable position among immobilized enzyme matrices. Among synthetic hydrophilic supports, cross-linked polyacrylic acid amide as yet occupies the most important position, while the polysaccharides and some of their derivatives dominate as hydroxylic supports for enzymes. In the latter category only granular supports of agar and dextran gels will be treated; however, the immobilization techniques described in this survey are not restricted to this type, but can be used independently of the physical form displayed by such supports. They may also be used for other kinds of polymeric supports, e.g., those containing primarily amino or thiol groups.

The most important polysaccharide supports are (1) cellulose, (2) dextran, (3) starch, (4) agar. From a purely economic point of view, cellulose and starch are perhaps the most attractive starting materials for immobilization. Their chemistry is well understood. In spite of these facts they have certain disadvantages that make them less suitable for granular immobilized enzymes, the most serious of which are an improper macroporous structure and a susceptibility to microbial disintegration. Dextran in cross-linked form,[1,2] Sephadex, under certain conditions is superior to cellulose and starch. Enzymes fixed to Sephadex exhibit higher relative activity than the corresponding cellulose-bound enzymes.[3] Presumably the microenvironment within Sephadex gels is less disruptive for the exertion of catalytic action than in fibrous cellulose with its microcrystalline regions. Starch is more easily attacked by microorganisms than is Sephadex. Agarose has been found to be very well suited as a matrix for the production of biospecific adsorbents, although less extensively purified agar may be a better choice for economic reasons in technical applications.

Agar and Agarose Gels

Agar and agarose gels in bead form were introduced as support media in zone electrophoresis and as molecular sieves by Polson[4] and Hjertén.[5] Agarose was somewhat later used as a support for immobilized enzymes.[6] These gels approach in many aspects the ideal solid matrix for

[1] J. Porath and P. Flodin, *Nature (London)* **183,** 1657 (1959).
[2] P. Flodin, Doctoral Dissertation, University of Uppsala, 1962.
[3] R. Axén, J. Porath and S. Ernback, *Nature (London)* **214,** 1302 (1967).
[4] A. Polson, *Biochim. Biophys. Acta* **50,** 565 (1961).
[5] S. Hjertén, *Biochim. Biophys. Acta* **79,** 393 (1964).
[6] J. Porath, R. Axén, and S. Ernback, *Nature (London)* **215,** 1491 (1967).

binding proteins. In a series of studies, the authors and their collaborators have attempted to improve agar and agarose for protein immobilization and chromatography.[7-10] Such improvements include, for instance, cross-linking by appropriate bifunctional reagents, which leads to increased chemical resistance and sometimes to a higher gel rigidity.

Mechanical Stability. Agar and fractionated agar of low charge density, agarose, form rigid gels when solutions of these polygalactans are cooled to temperatures below about 45°. The rigidity is not significantly influenced by a variation in the cooling program or by intermixing with diffusible substances, which when present increase the rigidity. When such substances (e.g., 60–70% sucrose) are removed from the gel, the rigidity returns to normal.

Agar from different suppliers can differ considerably with respect to gel strength. The latter is measured as resistance of the gel to deformation and yield when subjected to different loads under prescribed conditions. At the same concentration of solid substance in the interval of 2–6%, the strongest gels might be as much as three times stiffer than the weakest. Alkali extraction may improve agar considerably to form gels of higher rigidity.

The solubility of agar depends upon its quality. It is often difficult to obtain more than 6–10% agar in solution by autoclaving, and this sets the upper limit for the matrix concentration in the gels prepared in the usual way; however, there are ways to prepare granular agar of higher concentration if an increased gel strength is desired. According to one method, granulated agar is swelled in water and lyophilized or dried *in vacuo* after the water has been displaced by methanol or ethanol. Thereafter it is treated with a cross-linking agent in an alkaline alcohol–water mixture. The alcohol concentration of the reaction mixture determines the extent of swelling and the matrix concentration in the cross-linked water-swelled gel. Extremely dense and hard gels are obtained, but at the expense of decreased permeability. Enzymes can be fixed only to the surface and the outermost layer of the beads of these extreme types of gels. Such highly concentrated agar gels can possibly be used as nuclei in composite gels, where the particles are converted by a thin layer of gel of less dense nature. Together with I. Olsson and C. Eklund, the authors have made such types of rigid gels. Unfortunately it is not easy to cover the beads evenly. The advantages one can gain are (1) decrease in bed elasticity and (2) more rapid substrate permeation of the enzyme-containing gel; both achievements will allow extremely high flow rates.

[7] J. Porath, J.-C. Janson, and T. Låås, *J. Chromatogr.* **60**, 167 (1971).
[8] T. Låås, *J. Chromatogr.* **66**, 347 (1972).
[9] J. Porath, T. Låås, and J.-C. Janson, *J. Chromatogr.* **103**, 49 (1975).
[10] T. Låås. *J. Chromatogr.* **111**, 373 (1975).

Composite gels of another type can also be produced by incorporating other kinds of particles in the agar beads. These "hybrid gels" may have a number of desirable properties, such as high rigidity.

There are still other ways to increase gel rigidity. Cross-linking with bifunctional reagents of suitable molecular chain length can lead to a higher gel strength. Maximum increase is obtained with cross-links consisting of 5 atoms in a row between the two matrix-connecting ether oxygens. Divinyl sulfone and glutaryl dichloride are therefore useful cross-linkers for obtaining agar gels of extremely high rigidity. The maximum flow rate increase under comparable bed geometry and pressure is most pronounced for gels of low matrix density, where it may become 10–100 times as high as that of non-cross-linked gels. The solid content may increase perhaps 2-fold or even 3-fold without much change in permeability. This is perhaps somewhat surprising; in the authors' opinion, it reflects the unique three-dimensional structure of the spontaneously formed agar gels with large cavities that cannot be spanned by cross-links. A cross-linker only locks and stabilizes the supporting denser regions.[9] It should finally be mentioned that the introduction of aromatic groups to a high degree of substitution is also accompanied by an increase in gel strength.

Space-Fitting Properties. Many kinds of hydrophilic matrices shrink or swell substantially upon change of solvent. This is particularly striking when charged groups are introduced. Ion exchangers often expand at low ionic strength and contract at high concentration of salt. Hydrophilic gels swell to a slight extent or not at all in nonpolar organic solvents.

Agar and agarose gels differ from gels based on cross-linked dextran, polyvinyl alcohol, or polyacrylamide in their capacity to maintain an almost constant volume when less polar solvents are substituted for water. For example, it is possible to retain practically the same bed volume by sequential displacement of water with solvents of gradually decreasing polarity. This is probably due to the stabilization of the expanded gel structure by hydrogen bonds and hydrophobic interaction brought about by the swelling in water, whereby matrix segments become associated with nondisplaceable water (the galactose residues containing 3,6-methylene ether bridges are likely to be slightly hydrophobic).

The agar gels are characterized by a favorable permeability–matrix-density relationship. In spite of this they have not yet been used as extensively for enzyme immobilization as the occasion would warrant. Gels in the matrix range 6–10% can probably be used for enzymes of moderate molecular size (less than about 100,000 daltons and with substrates of small molecular size). This means that all reasonable demands for high flow rates can be satisfied before or after appropriate cross-linking. On the other hand, a low matrix density is favorable for the formation of enzyme–substrate complexes of large molecular size (2–6% cross-linked agar can be used for RNA or starch degradation).

Chemical Resistance. Agar is extremely stable in aqueous solutions whose pH is above 3. In the presence of reducing agents or in the absence of oxygen, epichlorohydrin cross-linked agar gels can be treated with 10% or even 30% sodium hydroxide solution at temperatures up to 120° without destruction of the matrix. Agar is less stable in strong acid but can be treated with 1 M HCl or 50% acetic acid for brief periods of time. The acid stability is higher for cross-linked than for non-cross-linked agar or agarose.

EPICHLOROHYDRIN CROSS-LINKING OF AGAR OR AGAROSE.[7] Agar or agarose gel in bead form is washed with distilled water. One liter of swollen gel is suspended in 1 liter of 1 M NaOH solution containing 100 ml of epichlorohydrin and 5 g of $NaBH_4$. The suspension is heated to 60° for 2 hr with stirring. The product is washed with hot water to neutrality and suspended in 500 ml of 2 M NaOH solution containing 2.5 g of $NaBH_4$. The mixture is heated for 1 hr in an autoclave at 120°. The gel is washed with 1.5 liters of hot 1 M NaOH containing 0.5% $NaBH_4$ and then with 1.5 liters of cold solution. The gel is suspended in cold water together with crushed ice, and acetic acid is added to pH 4. The product is finally washed with hot distilled water and stored as a suspension in 0.02% NaN_3 solution.

Biological Resistance. It is often pointed out that synthetic polymers are resistant to bacterial attack. The same is true for agar polysaccharides. The high stability of the latter is in all likelihood due to the unique molecular structure, which is absent from any polysaccharides in terrestrial organisms. Agar-degrading enzymes (agarases) have been found only in certain microbes living on seaweed. Agarases are not very efficient in attacking agar. Cross-linked agars are not significantly attacked at all.[11] If bacterial growth is found in agar gel column, the carbon substrate is most likely a compound of the imbibing solution, not the matrix itself.

Ability to Form Primary Derivatives. Each agarobiose unit has 4 hydroxyl groups. In the polysaccharides the average number of hydroxyls is somewhat lower owing to naturally occurring substituents (sulfate and pyruvic acid) (or in the cross-linked gels, where part of hydroxyls have been consumed in the cross-linking reaction). It has been found that derivatization in aqueous solution never approaches theoretical limits, at least not when the reactions take place in media of alkalinity lower than 2 M sodium or potassium hydroxide. Matrices with 1–1.5 mmoles of very reactive substituents per gram of dry substance, i.e., about one substituent

[11] B. von Hofsten and M. Malmqvist, *J. Gen. Microbiol.* **87,** 150 (1975).

for every second galactose unit, can occasionally be obtained (imidocarbonate, oxirane, vinylsulfonyl).

The reactions used for derivatization of cellulose, starch, dextran, etc., can also be used for preparing derivatives of agar and vice versa.

In those cases where a higher concentration of hydroxyl groups in the matrix substance is desirable, sugar alcohols and polyphenols may be introduced.[12]

Ability to Retain Enzymic Activity. The catalytic activity of enzymes may decrease as a consequence of immobilization resulting from the following mechanisms: (1) damage of the active center by substitution on groups necessary for the catalytic action; (2) blocking of the active center (caused by the spacer-connector unit, or the matrix itself, or both), which prevents the formation of the substrate-enzyme complex; (3) conformation changes in the enzyme induced by interaction with the adsorption centers in the matrix.

Matrix blocking was demonstrated for chymotrypsin-Sephadex by Axén et al.[13,14] (see below). We are lacking direct rigorous proof for the blocking effects in agar, since no agarase has been found that is capable of liquefying enzymes coupled to agar. Experience collected over the years indicates a profound influence of the matrix on the retention of activity. Agar gels seem to interfere less with the normal catalytic function of immobilized enzymes than do most, if not all, other types of supports. Introduction of hydrophobic groups in the matrix tends to increase the risk of denaturation, but under carefully controlled conditions immobilization can be accomplished satisfactorily by "hydrophobic adsorption" (see below).

At very high matrix density, enzymes are immobilized exclusively in a thin surface layer of the polysaccharide gel beads. Such products will perhaps not be technologically attractive in view of the low enzyme concentration attainable. On the other hand, since a low matrix density is associated with low mechanical strength, it will be necessary to compromise. The increase in matrix density may have a beneficial effect on the thermal stability and resistance to denaturing media,[15] as has been found for chymotrypsin-Sephadex G-200.

Economic Considerations. Crude agar is inexpensive. For most analytical purposes it is extensively purified to yield agarose of high quality.

[12] J. Porath and L. Sundberg, *Nature (London) New Biol.* **238**, 261 (1972).
[13] R. Axén, P.-Å. Myrin, and J.-C. Janson, *Biopolymers* **9**, 401 (1970).
[14] R. Axén and P. Vretblad, "Protides of the Biological Fluids" (H. Peeters, ed.), p. 383. Pergamon, Oxford, 1971.
[15] D. Gabel, P. Vretblad, R. Axén, and J. Porath, *Biochim. Biophys. Acta* **214**, 561 (1970).

For industrial applications of immobilized enzymes, such far-reaching purification is probably not necessary. Simple alkali extraction may be sufficient. The conversion of agar into beaded form is now a comparatively expensive process involving suspension in organic solvents and the use of stabilizers that subsequently have to be removed. A new process of preparing agarose beads invented by Pertoft and Hallén greatly simplifies the formation of spherical beads.[16] The gel is formed in an extremely interesting way. By mixing an agarose solution with a colloidal suspension of silica particles and polyethylene glycol, phase separation takes place under specified conditions with the formation of spherical droplets containing silica. The droplets are transformed to gel beads upon cooling. We have now also found that crude agar forms silica-containing spherical beads under similar conditions, and it is possible to regulate droplet formation to yield a narrow size distribution. Such inexpensive agar may be very attractive support material for large quantities of immobilized enzymes to be used in industry.

Further improvement of mechanical properties may be necessary. An increase in gel matrix density is probably most easily accomplished by drying, controlling swelling in aqueous-ethanol (or methanol), and cross-linking. The durability and excellent properties of such agar products may well outweigh the extra costs involved.

Conclusion. It may be stated that agar and agarose seem to afford almost ideal microenvironments for immobilized enzymes. The only drawback from a technological standpoint is perhaps the resulting limited resistance to deformation and breakage of the agar gels most frequently used until now. New types of agar gels seem to better meet the demand for higher rigidity and lower costs.

Methods for Immobilization of Enzymes to Sephadex, Agar, and Agarose Derivatives

It may be convenient to distinguish the following ways of immobilizing enzyme to Sephadex and agar–agarose matrices: (1) covalent binding; (2) adsorption. Inclusion is perhaps possible also under certain circumstances, but no examples have as yet been found in the literature.

COVALENT BINDING

Reactive groups may be incorporated into either the matrix or the protein.

[16] H. Pertoft and A. Hallén, *Z. Chromatogr.*, in press.

Certain gel-forming polymers contain other groups (e.g., carboxylic) in addition to the hydroxyls, which could serve as anchoring centers for immobilization of the enzymes.

We may distinguish between the following two principles for enzyme attachment to the support: (1) the introduction of reactive groups into the gel via the hydroxyls of the matrix; (2) the partial degradation or conversion of the matrix into a more active form. Case 1 will be treated in more detail, and we will consider direct coupling in two steps: activation and coupling, and indirect coupling whereby an intermediate group of higher reactivity than hydroxyl is introduced, e.g., NH_2, COOH, or SH.

Direct Coupling

An electrophilic group is introduced into the matrix in order to make the gel more reactive toward the nucleophiles present in the enzyme. The following groups are of particular interest:

Imido carbonate	$\begin{matrix}-O\\-O\end{matrix}\!\!>\!\!\overset{(+)}{C}\!=\!NH$
Carbonate	$\begin{matrix}-O\\-O\end{matrix}\!\!>\!\!\overset{(+)}{C}\!=\!O$
Oxirane	$-CH\overset{(+)}{-}CH_2$ with O bridge
Aziridine	$-CH\overset{(+)}{-}CH_2$ with NH bridge
Activated double bond, e.g.,	$\overset{(+)}{C}H_2=CH-SO_2-$ $-\overset{(+)}{C}H=CH-CO-$
Activated halogen, e.g.,	$Br-\overset{(+)}{C}H_2-CO-$ $Br-\overset{(+)}{C}O-CH_2-$ $Cl-\overset{(+)}{C}=N$

The electrophilic groups primarily react with the amino and thiol groups of the enzyme, but may also be attached by weaker nucleophiles, such as phenolic hydroxyls of tyrosine residues.

Some reactions, viz., those involving cyanogen bromide, divinyl sulfone, and bisoxirane have been described earlier in this series.[17,18] Although it cannot be avoided, the subsequent cross-linking reaction can be suppressed. The problem is to find the optimum condition for the intro-

[17] J. Porath, this series, Vol. 34, p. 13 (1974).
[18] I. Parikh, S. March, and P. Cuatrecasas, this series, Vol. 34, page 77 (1974).

duction of a maximum number of electrophilic groups for a given concentration of reagent such that the product is suitable for the following enzyme coupling step. Thus each attachment method must be optimized with respect to activation as well as to subsequent coupling. In addition, complications arise by the fact that each enzyme may require specific conditions for coupling depending on the molecular size of the enzyme itself, the molecular bulk of the substrate, and the nature and density of the matrix.

Imido Carbonate

The coupling method most extensively used at present is based on the reaction between cyanogen halides with the hydroxyls of the polysaccharide matrix. The reaction is carried out in alkaline solution in water or mixed water–organic solvents. Hydrolysis of the reagent, usually cyanogen bromide, is an undesirable side reaction that cannot be avoided but can be suppressed by optimizing the conditions for imido carbonate formation. At and below room temperature the pH optimum for the desired reaction is in the interval 11–12.5. The most important types of reactions in the activation step are the following:

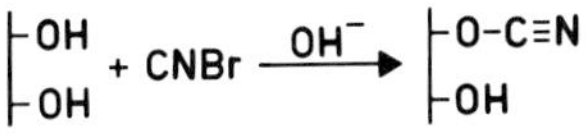

$$\begin{bmatrix} -OH \\ -OH \end{bmatrix} + CNBr \xrightarrow{OH^-} \begin{bmatrix} -O-C\equiv N \\ -OH \end{bmatrix}$$

Cyanates are very labile, and the reaction will proceed to give reactive imidocarbonate (I) and unreactive carbamate (II):

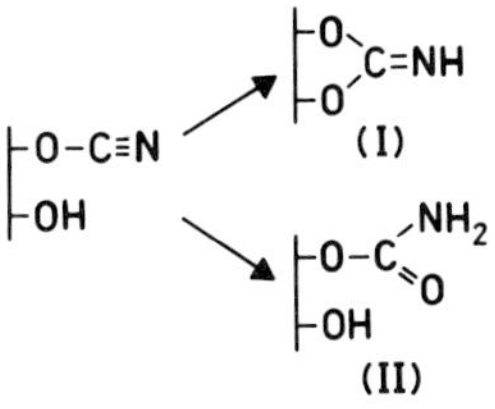

$$\begin{bmatrix} -O-C\equiv N \\ -OH \end{bmatrix} \begin{matrix} \nearrow \begin{bmatrix} -O \\ -O \end{bmatrix} C=NH \quad (I) \\ \searrow \begin{bmatrix} -O-C\begin{smallmatrix} NH_2 \\ O \end{smallmatrix} \\ -OH \end{bmatrix} \quad (II) \end{matrix}$$

A considerable portion of the imidocarbonate may be in the 5-membered ring form if the polymer contains vicinal hydroxyls. In the agar polysaccharides, vicinal groups are believed to be absent. The prevailing form may therefore be the following:

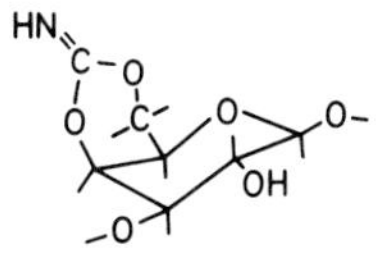

A part of the imidocarbonate is undoubtedly in the form of cross-links, since the gel becomes insoluble in boiling water:

$$\vdash\!O\text{-}C\!\equiv\!N + HO\!\dashv \longrightarrow \vdash\!\!\begin{array}{c}O\text{-}C\text{-}O\\\|\\NH\end{array}\!\!\dashv$$
(III)

The imidocarbonate may be hydrolyzed to carbonate (IV)

$$\vdash\!\!{}^{O}_{O}\!\!>\!\!C\!=\!NH \xrightarrow{\;H^+\;}_{H_2O} \vdash\!\!{}^{O}_{O}\!\!>\!\!C\!=\!O + NH_4^+$$
(IV)

and further back to the original matrix.

$$\vdash\!\!{}^{O}_{O}\!\!>\!\!C\!=\!NH$$
$$\vdash\!O\text{-}C\!\!<\!\!{}^{NH_2}_{O} \xrightarrow{\;OH^-\;}_{H_2O} \vdash\!\!{}^{OH}_{OH} + CO_3^{2-} + NH_4^+$$

A further complication is the undesirable hydrolysis of the reagent itself with the production of bromide, cyanide, and cyanate.

The presence of species I-IV were indicated by infrared (IR) analysis, as pointed out in the early papers in which the method was introduced.[3,6] More comprehensive studies that followed[19-22] corroborated and extended the knowledge of the complicated reaction pattern of the activation step. Hydroxide ions are consumed in most reactions with concomitant decrease in pH. According to the first published versions of the method, the gel suspension is not buffered, and consequently the reaction is followed in a pH stat under continuous addition of alkali. It is more convenient to perform the reaction in the presence of strong phosphate[17,23] or carbonate[18] buffer. The amount of reagent and the buffer concentration to be used are dependent on the density of the agar, agarose, or Sephadex. For 4% agarose gel (Sepharose 4B), the recommended activation procedure follows.

Preparation of Activated Gel with a High Degree of Substitution.[23] Sepharose 4B is washed with 2 M potassium phosphate buffer (pH 12.1) and sucked dry under slightly reduced pressure on a filter. Then 10 ml of cold 5 M potassium phosphate buffer (pH 12.1) are added to each 10 g of gel, followed by 20 ml of distilled water and finally by small portions of a solution of 100 mg of CNBr per milliliter. Sometimes the commercial preparations of CNBr are not easily soluble. Acetonitrile may then be used (2 g of CNBr per milliliter of CH_3CN).[18] The addition

[19] R. Axén and S. Ernback, *Eur. J. Biochem.* **18**, 351 (1971).
[20] R. Axén and P. Vretblad, *Acta Chem. Scand.* **25**, 2711 (1971).
[21] L. Kågedahl and S. Åkerström, *Acta Chem. Scand.* **25**, 1855 (1971).
[22] M. Joustra and R. Axén, *Protides of the Biological Fluids, 23rd Colloq. Bruges, 1975*, in press.
[23] J. Porath, K. Aspberg, H. Drevin, and R. Axén, *J. Chromatogr.* **86**, 53 (1973).

of the CNBr should take about 2 min at a temperature kept between 5° and 10°. The reaction should be allowed to continue for about 10 min under gentle stirring. The gel is then transferred to a glass filter and washed with cold water. The gel may be immediately transferred to a medium for protein coupling and preserved in cold storage for periods of up to a day or longer, but it is advisable to continue with the coupling immediately.

The procedure is the same as above when a low degree of substitution is desired, except that a smaller quantity of cyanogen bromide, e.g. 20 mg in total, is used.

Activated agarose may be preserved in a dry stabilized state by cold storage for long periods of time (such gel material is available from Pharmacia Fine Chemicals).

The coupling of enzyme occurs entirely or at least predominantly through the free amino groups of the ligand protein. As revealed by IR spectra and studies on low-molecular-weight model substances,[19] the following reactions have been found to occur:

$$\begin{array}{c} \vdash O \\ \diagdown C=NH \\ \vdash O \end{array} \quad\longrightarrow\quad \begin{array}{c} \vdash O \\ \diagdown C=N\text{-}R \\ \vdash O \end{array}$$

$$+ H_2N\text{-}R \longrightarrow \vdash O\text{-}\underset{\underset{N\text{-}R}{\|}}{C}\text{-}O\dashv \longrightarrow \begin{array}{c} \vdash O\text{-}C\diagup^{NH\text{-}R}\diagdown O \\ \vdash OH \end{array}$$

$$\begin{array}{c} \vdash O\text{-}\underset{\underset{NH}{\|}}{C}\text{-}O\dashv \end{array} \quad\longrightarrow\quad \begin{array}{c} \vdash O\text{-}C\diagup^{NH\text{-}R}\diagdown_{NH} \\ \vdash OH \end{array}$$

The amine should be unprotonized, which means that the reaction with proteins takes place at an optimum rate in the pH interval 9–10. The rates of the reactions involved are dependent upon pH, and presumably also depend on the type of imidocarbonate polymer (and contaminating carbonate). Evidently the properties of the final product depend on the conditions prevailing both in the coupling step and the subsequent rearrangements occurring during storage. Some of the derivatives are apparently extremely stable[24] while others are not.[25,26] Multipoint attachment between enzyme and matrix increases the stability, since the probability for breakage of all the bonds is very much less than for a single bond. Such an argument favors the use of highly substituted imido carbonate gel. However, a gel of this kind might produce immobilized enzymes with low activity. It seems necessary to work out the most satisfactory condi-

---

[24] T. Kristiansen, L. Sundberg, and J. Porath, *Biochim. Biophys. Acta* **184**, 93 (1969).
[25] G. I. Tesser, H.-U. Fisch, and R. Schwyzer, *FEBS Lett.* **25**, 56 (1972).
[26] G. I. Tesser, H.-U. Fisch, and R. Schwyzer, *Helv. Chim. Acta* **57**, 1718 (1974).

tions in each separate case with respect to the reactivity and alkali stability of the enzyme to be coupled. To illustrate the importance of the parameters involved, the preparation of chymotrypsin-Sephadex by Axén and collaborators is described.[13]

Preparation of Chymotrypsin-Sephadex Conjugate.[13] Chymotrypsin (100 mg) dissolved in 4 ml of 0.3 M potassium bicarbonate and added to CNBr activated Sephadex G-200 having been activated at pH 10.3 and 9.8, respectively. The reaction is carried out at 5° for a period of 24 hr under gently stirring. To deactivate the remaining imido carbonate, 1.5 g of glycine may be added to the suspension, which then is left for another period of 24 hr. The coupling products are thoroughly washed in the sequence: water, 0.1 M sodium bicarbonate, 0.001 M hydrochloric acid, 0.5 M sodium chloride, and finally distilled water.

Comments. Chymotrypsin immobilized as above was subjected to the action of dextranase from a *Cytophaga* species capable of attacking highly swellable Sephadex.[13] Dextranase is available from Sigma Chemical Co., St. Louis, Missouri, and Worthington Biochemical Corporation, Freehold, New Jersey. The results from a comparison between free enzyme and chymotrypsin immobilized to gels activated at two different pH are compiled in Table I. The Sephadex gel was permeable enough to allow the dextranase to attack the glucosidic bonds and liquefy the gel. The solubilization was accompanied by an increase in activity. The derivative prepared at pH 9.8 was more easily solubilized, and the catalytic parameters approached those of the native enzyme. It is very likely that the enzyme molecules still contain covalently bound fragments from the matrix. In this case the immobilization had evidently left the catalytic site intact. The difference in properties between the enzymes coupled to gels activated at different pH is striking. At pH 10.3 the active site may have been damaged, although the most important deactivation factor in this case is also governed by steric hindrance.

The cyanogen halide method was incidentally discovered when attempts were being made to use cyanamide derivatives of Sephadex and agarose to immobilize enzymes. Since the residual amino groups from the reaction with the primary derivative of the gel will necessarily act as nonspecific adsorption centers, the study of the cyanamide gels were abandoned. For immobilizing enzymes, nonspecific adsorption is of little or no importance, and therefore, in view of their higher reactivity at low pH, it may be worthwhile to optimize the cyanamide coupling again:

$$\vdash O\text{-}\!\!\sim\!\!\text{-}NH\text{-}C\equiv N + H_2N\text{-}R \longrightarrow \vdash O\text{-}\!\!\sim\!\!\text{-}NH\text{-}C\begin{subarray}{l}\nearrow NH \\ \searrow NH\text{-}R\end{subarray}$$

One interesting feature of the reaction is the preservation of charge.

TABLE I
PROPERTIES OF CHYMOTRYPSIN-DEXTRAN DERIVATIVES

Enzyme derivative	pH of activation, CNBr method	Time of dextranolytic digest	pH of optimal activity	N-Acetyl-L-tyrosine ethyl ester			Casein (% activity relative to free enzyme)
				Activity (μmoles/min per mg enzyme)	K_μ (mM)	V_{max} (μmoles/min per mg enzyme)	
Free chymotrypsin	—	—	7.8	230	3.3	280	100
Immobilized chymotrypsin–Sephadex G-200 I	10.3	—	9.8	60	25–30	160–180	15
Solubilized chymotrypsin–Sephadex G-200 I	10.3	60	8.3	140	5	190	60
Immobilized chymotrypsin–Sephadex G-200 II	9.8	—	9.8	80	30–40	250–290	35
Solubilized chymotrypsin–Sephadex G-200 II	9.8	6	7.8	230	3	280	100

Oxirane

Highly strained three-membered electrophilic ring systems may be used for the immobilization of enzymes:

$$\text{(M)}\sim\text{CH-CH}_2 + \text{HS-R} \longrightarrow \text{(M)}\sim\text{CH(XH)-CH}_2\text{-S-R}$$
$$\phantom{\text{(M)}\sim\text{CH}}\diagdown\text{X}\diagup$$

$$\text{(M)}\sim\text{CH-CH}_2 + \text{H}_2\text{N-R} \longrightarrow \text{(M)}\sim\text{CH(XH)-CH}_2\text{-NH-R}$$
$$\phantom{\text{(M)}\sim\text{CH}}\diagdown\text{X}\diagup$$

$$\text{(M)}\sim\text{CH-CH}_2 + \text{HO-R} \longrightarrow \text{(M)}\sim\text{CH(XH)-CH}_2\text{-O-R}$$
$$\phantom{\text{(M)}\sim\text{CH}}\diagdown\text{X}\diagup$$

when X may be NH, O, or S. Attempts to use thio analogs have not been successful in this laboratory; however, oxiranes have proved to be extremely useful for the introduction of small ligands in agarose to produce specific adsorbents for biospecific affinity chromatography.[17,27] Oxirane-containing gels can also couple proteins.

Oxirane groups can be introduced in many hydroxyl polymers, particularly those containing vicinal hydroxyls, by means of tosylation followed by nucleophilic displacement of the tosyl groups under formation of fused oxirane ring systems. A more convenient and general procedure, not requiring 1,2-diols, is based on a specific reaction with bisoxiranes, 1-halo-2-hydroxy, or 1-hydroxy-2-halo compounds (halohydrins).

$$\vdash\text{OH} + \text{CH}_2\text{-CH}\sim\text{CH-CH}_2 \longrightarrow \vdash\text{O-CH}_2\text{-CH(OH)}\sim\text{CH-CH}_2$$

As a side reaction cross-linking will occur:

$$\vdash\text{O-CH}_2\text{-CH-CH}_2 + \text{HO}\dashv \longrightarrow \vdash\text{O-CH}_2\text{-CH(OH)-CH}_2\text{-O}\dashv$$

This cross-linking reaction makes the matrix insoluble in boiling water and somewhat more acid stable.

The reactivity toward nucleophilic groups in the proteins follows the usual order SH > NH > OH. Nucleophilic attack on aliphatic hydroxyl groups takes place at a strongly alkaline pH ($\sim$11), while the amino groups react at a lower pH, and SH still lower (around neutrality or slightly above). Aromatic hydroxyls, such as those in tyrosyl residues, occupy a position intermediate between aliphatic hydroxyls and amino groups. Guanidino groups and imidazoles can also react.

Activation Step. Optimization of the oxirane incorporation into Sepharose was made in a comprehensive study.[27] Accordingly, the reaction mixture containing the bisoxirane should be 0.2–0.25 M with respect

[27] L. Sundberg and J. Porath, *J. Chromatogr.* **90**, 87 (1974).

to alkali hydroxide, the temperature around 25°, and the reaction time 10–20 hr. The oxirane content in the product increases continuously with the diglycidyl ether concentration (1,4-n-butanediol diglycidyl ether) up to at least 50%. More than 1000 μmoles of oxirane groups per gram of dry gel may be introduced. Since the introduction of excessive amounts of oxirane groups does not result in a corresponding increase in ability to couple proteins, it is advisable to activate with 2–10% bisoxirane (cf. the imido carbonate coupling reaction). The oxirane group is hydrolyzed at pH extremes but is moderately stable in neutral solution. About 90% of the groups were found intact after 15 days of storage at 4° in 1 M NaCl, pH 6.8. No decrease in oxirane content was observed after storage of the activated gel in dimethyl sulfoxide or acetone for 1 month. Activated agarose gels can be lyophilized with retention of reactivity, but there is risk that the swelling will not return the gel to its original volume. Permeability may therefore decrease, as well as the amount of oxirane available for protein coupling. Activation of Sephadex with bisoxirane has not been extensively studied.

Coupling Step. Optimum conditions for coupling have been found only for low-molecular-weight, amino group-containing substances.[27] The yield increases continuously in the pH interval 8–11. A rise in temperature from 20° to 40° has a marked influence on the extent of coupling. The rate rapidly declines below room temperature; however, a slow rate of reaction can sometimes be compensated by extending the reaction time to several days or even weeks.

Since sulfhydryl groups couple more readily than amino groups at lower pH, proteins containing accessible SH groups are likely to undergo this reaction under rather mild conditions.

Conclusion. In cases where enzymes are sufficiently reactive and stable, the oxirane method offers some very important advantages in comparison to other methods, including imido carbonate coupling. The O—C, N—C, and probably the S—C bonds formed are extremely stable. By selecting a long-chained bisoxirane such as 1,4-n-butanediol diglycidyl ether, one may obtain a long spacer, which separates the enzyme from the matrix, a possibly important consideration for retention of activity. The substitution does not affect the charge of the enzyme around the site of attachment.

Immobilization of Chymotrypsin. Ten grams of washed, suction-dried 2% agarose gel in spherical bead form, is suspended in 5 ml of 2.5 M sodium hydroxide solution. Sodium borohydride (20 mg) is added, followed by the dropwise addition of 1 ml of 1,4-butanediol diglycidyl ether under vigorous stirring. The reaction time is 6 hr at room temperature, after which the gel is washed with water on a glass filter until neutrality is reached, then with 50 ml of acetone, and finally with water. The content of oxirane function as determined by sodium thiosulfate titration,[28] is about 100–120 μmoles per gram of dry product.

The oxirane gel is suspended in 25 ml of 0.5 M sodium bicarbonate solution, and 50 mg of chymotrypsin is added. The reaction occurs at ambient temperature for 5 hr, and the product is washed with the following solutions in order: 0.1 M Na borate buffer, pH 8.5 (1 M in NaCl); 0.1 M Na acetate, pH 4.1 (1 M in NaCl); and finally with 0.01 M Na acetate, pH 4.1. The conjugate is stored in the last buffer at 4°. The content of enzyme, as determined by amino acid analysis, is found to be 50 mg of protein per gram of dry product. The product when assayed against a 0.01 M solution of N-acetyl-L-tyrosine ethyl ester in 0.05 M KCl (5% with respect to ethanol) has an optimum activity at pH 9.3. The activity of bound enzyme at this pH is about 20% that for the soluble enzyme, as determined at pH 7.9.

Activated Double Bonds

Vinylsulfonylethylene Ether

Vinylsulfonyl groups can be introduced into hydroxyl-containing polymers by treatment of the latter with divinyl sulfone in alkali[17]:

$$\vdash OH + CH_2=CH-SO_2-CH=CH_2 \longrightarrow \vdash O-CH_2-CH_2-SO_2-CH=CH_2$$

The unavoidable simultaneous cross-linking is in fact very desirable, since the resultant immobilized enzyme will have bed-flow characteristics far superior to those of the original gel.[9] The rates of nucleophilic attack by the enzymes follow in the same order as for oxirane—SH > NH > OH—but the reactions take place at about 1 pH unit lower.[29] The thiol groups participate in the formation of a rather labile sulfide linkage:

$$\vdash O-CH_2-CH_2-SO_2-CH=CH_2 + HS-R \longrightarrow \vdash O-CH_2-CH_2-SO_2-CH_2-CH_2-S-R$$

The C—N bonds created through coupling with free amino groups are more stable:

$$\vdash O-CH_2-CH_2-SO_2-CH=CH_2 + H_2N-R \longrightarrow \vdash O-CH_2-CH_2-SO_2-CH_2-CH_2-NH-R$$

A slow release of low-molecular-weight ligands occurs at a pH above 8. Since proteins become more firmly attached, the method looks promising as an alternative to oxirane coupling.

[28] R. Axén, H. Drevin, and J. Carlsson, *Acta Chem. Scand.*, **B29**, 471 (1975).
[29] J. Porath and L. Sundberg, unpublished results (1973).

Vinylketo Function

A keto group will enhance the electrophilic character of a double bond adjacent to a carbonyl to a lesser extent than does the sulfone group. Attempts to cross-link agarose with divinyl ketone have failed to yield a heat-insoluble agar product. Quinones, however, are very much more reactive, even more so than divinyl sulfone, and will attack hydroxylic polymers more readily than by any other method discussed so far.[30] The study of model compounds[30] indicate the following sequence of reactions for benzoquinone:

Activation:

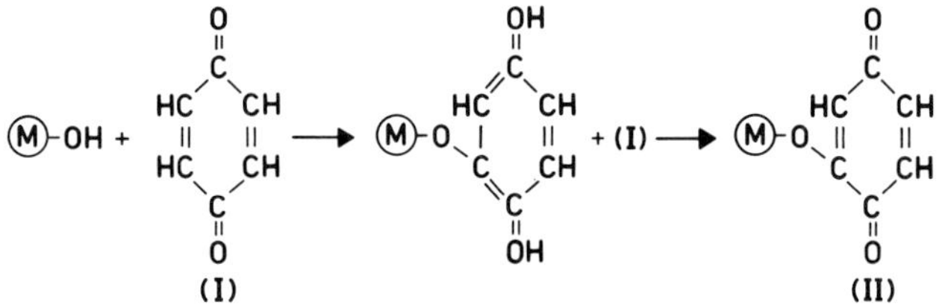

$$\text{\textcircled{M}-OH} + \text{(I)} \longrightarrow \text{\textcircled{M}-O-(I)} + \text{(I)} \longrightarrow \text{\textcircled{M}-O-(II)}$$

Coupling:

$$\text{(II)} + \text{Hx-R} \longrightarrow \text{\textcircled{M}-O-C-X-R} \quad (X= -S-, -O- \text{ or } {\geq}NH)$$

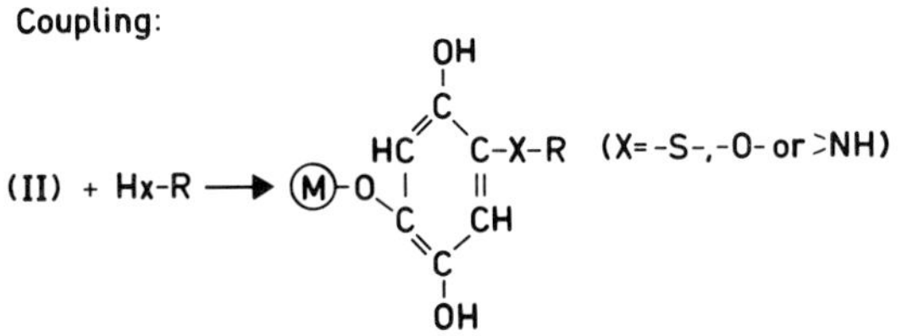

The method is easy to apply, gives high coupling yields, and can be used in a wide pH range as exemplified by serum albumin, which can be coupled between pH 3 and 10. The color of the products indicates undesirable side reactions. The reaction has been used for immobilization of serum albumin to Sepharose 4B and Sephadex G-200.

Activation. Sepharose 6B gel, 100 ml, cross-linked or not cross-linked, is washed on a glass filter with 0.1 M sodium phosphate buffer, pH 8.0, containing 20% ethanol and subsequently suspended in 100 ml of the ethanol phosphate buffer, 100 mM with respect to benzoquinone (10.8 g). The reaction is allowed to proceed under gentle stirring for 1 hr at room temperature. The gel is washed on a glass filter with the following solutions in sequence: 20% ethanol, water, 1 M NaCl, water, and finally with the buffer used for the coupling.

Coupling. Chymotrypsin, 150 mg, is added to 10 ml of activated gel in sodium phosphate buffer at pH 8.0. The coupling is allowed to take

[30] J. Brandt, L.-O. Andersson, and J. Porath, *Biochim. Biophys. Acta* **386**, 196 (1975).

TABLE II
BENZOQUINONE-IMMOBILIZED ENZYMES ON SEPHAROSE 4B[a]

Enzyme	Amount of fixed enzyme (mg/g dry gel)	pH optimum Free enzyme	pH optimum Coupled enzyme	Activity ratio bound to free enzyme (%)
Chymotrypsin[b]	72	7.7	9.8	78
Ribonuclease[c]	78	6.9	7.5	97

[a] Activation and coupling as described in the text.
[b] Chymotrypsin substrate is N-acetyl-L-tyrosine ethyl ester.
[c] Ribonuclease substrate is cytidine-2,3-cyclic phosphate.

place at room temperature for 24 hr under gentle stirring; 1 g of glycine is added for deactivation, and stirring is continued for another 24 hr. The gel is washed on a filter with 0.1 M sodium acetate, pH 4.0, 0.1 M sodium bicarbonate, pH 8.5, 1 M KCl, and water. The gel will contain about 100 mg of enzyme per gram of dry conjugate.

Satisfactory retention of activity is obtained as shown in Table II.

The quinone coupling method is apparently promising but needs further development and more intense studies to assess its merits for immobilizing enzymes.

Activated Halogen

Highly reactive halogens may be introduced into hydroxyl-containing polymers to facilitate enzyme coupling. One of the first applications of these general procedures was made by Jagendorf et al.[31] Cellulose was converted to its bromoacetate derivative:

(Cellulose)-OH - Br-Co-CH$_2$-Br ⟶ (Cellulose)-O-CO-CH$_2$-Br

Halogens can also be activated by other groups, e.g., by hydroxyls in halohydrins. The linkage to the matrix formed by coupling via the bromoacetate with proteins is presumably not very stable. Ethers are preferable and can be formed after activation with dihaloketones, such as dibromoacetone or dibromodiacetyl. Attempts are presently being made to optimize such methods.

Immobilization can also be made with dibasic acid dichlorides, in which a high substitution of acid groups in the final product can be tolerated. Acid dichlorides as bifunctional reagents for enzyme immobilization

[31] A. T. Jagendorf, A. Patchornik, and M. Sela, *Biochim. Biophys. Acta* 78, 516 (1963).

seem to offer the possibility of obtaining inexpensive rigid agar gels. The principle is simply the following:

$$\text{M}-OH + Cl-CO-Y-CO-Cl \rightarrow \text{M}-O-CO-Y-CO-Cl \quad (a)$$

$$\text{M}-O-CO-Y-CO-Cl \underset{H_2O}{\overset{R-NH_2}{\rightleftarrows}} \begin{array}{l} \text{M}-O-CO-Y-CO-NH-R \quad (b) \\ \text{M}-O-CO-Y-COOH \quad (c) \end{array}$$

The reaction (a) is conducted in an aprotic solvent with a tertiary amine catalyst. The transfer of the gel to an aqueous solvent system and the subsequent coupling presents problems, since reaction (c) competes with the desired reaction (b). Since glutaric acid dichloride yields extremely rigid agar gels as a consequence of concomitant cross-linking, it is of interest to attempt the production of agar–enzyme conjugates by means of these reactions.

Halogens in structures such as Cl—C = N—C are also highly reactive. Kay and Crook introduced cyanuric acid chlorides for cellulose-fixed enzymes.[32]

Irradiated-Induced Coupling

Brandt and Andersson have coupled enzymes to agarose and dextran by γ-radiation of protein-containing gel suspensions.[33] Free radicals are produced in the protein as well as in the support, and by radical–radical interaction the protein is fixed to the gel.

$$MH \rightarrow M^{\bullet} + H^{\bullet} \quad (MH = matrix)$$
$$EH \rightarrow E^{\bullet} + H^{\bullet} \quad (EH = enzyme)$$
$$M^{\bullet} + E^{\bullet} \rightarrow M:E$$

The method is obviously nonselective and can take place under widely different conditions. Insensitivity to pH and temperature is an advantage, but low coupling yields and radiation damage are serious drawbacks. The method must be considerably improved before it can be accepted as an alternative to other available techniques.

Indirect Coupling

The nucleophilic character of the hydroxyls in Sephadex, Sepharose, cellulose, and other similar solid enzyme carriers is so weak that drastic reaction conditions often are necessary for derivatization. The introduc-

---
[32] G. Kay and E. M. Crook, *Nature (London)* **216**, 514 (1967).
[33] J. Brandt and L.-O. Andersson, *Acta Chem. Scand.*, in press (1976).

tion of amino, carboxyl, and thiol groups widen the scope of immobilization reactions. In the classical work of Campbell, Lerman, Gurvich, and their associates[34-36] aromatic groups were first introduced via nitro ethers. The immobilization was accomplished by diazonium coupling. Similar coupling procedures were tried with Sephadex as solid support (unpublished). Later the amino group of p-aminophenylglycidyl ether was converted to isothiocyanate, which reacts with free amino groups in proteins, leading to the formation of thiourea derivatives.[37]

Indirect coupling via moderately reactive primary derivatives has been used extensively by Cuatrecasas,[18,38] Wilchek,[39] and others for the introduction of spacers between ligand and matrix for specific-affinity chromatography. There are at least three reasons to reflect upon the use of indirect coupling: (1) a good spacer effect may be obtained; (2) there is wider selection of attachment reactions; (3) the number of different attachment groups in the enzymes is increased.

At present very little is known with certainty about which role the spacer may play for retaining the activity of an enzyme after it has been rendered insoluble. It is conceivable that a long-chain spacer could decrease the steric hindrance and matrix interaction. The effects should be more clearly recognizable for substrates of high molecular weight.

Most of the commonly used methods for direct coupling can also be used for the activation of primary derivatives; for example:

$$\vdash O\text{-}\wedge\wedge\text{-}NH_2 + CNBr \longrightarrow \vdash O\text{-}\wedge\wedge\text{-}NH\text{-}C\equiv N$$

$$\vdash O\text{-}\wedge\wedge\text{-}NH_2 + CH_2\text{-}CH\text{-}\wedge\wedge\text{-}CH\text{-}CH_2 \longrightarrow$$

$$\longrightarrow \vdash O\text{-}\wedge\wedge\text{-}NH\text{-}CH_2\text{-}CH(OH)\text{-}\wedge\wedge\text{-}CH\text{-}CH_2$$

Isocyanide Method

When carboxyl groups are introduced into the matrix, coupling is possible with carbodiimides and isocyanides. The latter method is particularly interesting, since no fewer than four functional groups participate in

---

[34] D. H. Campbell, E. Lenscher, and L. S. Lerman, *Proc. Natl. Acad. Sci. U.S.A.* **37**, 575 (1951).
[35] L. S. Lerman, *Proc. Natl. Acad. Sci. U.S.A.* **39**, 232 (1953).
[36] A. E. Gurvich, *Biochemistry (USSR)* **22**, 977 (1957).
[37] R. Axén and J. Porath, *Nature (London)* **210**, 367 (1966).
[38] P. Cuatrecasas and C. B. Anfinsen, this series, Vol. 22, p. 31 (1972).
[39] M. Wilchek and T. Miron, this series Vol. 34, p. 72 (1974).

the reactions leading to the final stable product: carboxyl, amino, carbonyl, and isocyanide.[40] In spite of the complicated reactions involved, the application is very simple and the method is extremely versatile.[41] An immonium ion (Schiff base) is formed; this ion in a concerted reaction with carboxylate attacks the unsaturated carbon of a suitable isocyanide (e.g., cyclohexyl isocyanide), forming an unstable adduct; the adduct by a sequence of rearrangement reactions forms amide or peptide bonds. Any of the four interacting species may be present in the primary gel derivative or in the ligand to be immobilized. It should be pointed out that the method is complementary to many of the previously described methods, since it extends the pH range available for coupling to acidic solutions. The method has been optimized for proteins by Vretblad and Axén.[42]

Ionic groups do not necessarily have any adverse affect on immobilized enzymes. The isocyanide method may therefore be useful for attaching proteins to carboxy or amino polymers. When acetaldehyde is used as a carbonyl component, the reactions involved for matrices containing amino groups are as follows:

$$(M)\text{-}\!\!\sim\!\!\text{-}NH_2 + CH_3\text{-}CHO \underset{}{\overset{H^+}{\rightleftharpoons}} (M)\text{-}\!\!\sim\!\!\text{-}\overset{+}{\underset{H}{N}}\!\!=\!\!CH\text{-}CH_3 + H_2O$$

$$(I)$$

$$(I) + R\text{-}N\!\!\equiv\!\!C + (Protein)\text{-}COO^- \rightleftharpoons (M)\text{-}\!\!\sim\!\!\text{-}NH\diagdown\!\!\overset{CH\diagup CH_3}{\underset{(Protein)\text{-}CO\text{-}O\diagup}{C=N\text{-}R}}$$

$$(II)$$

$$(II) \xrightarrow{\text{rearrangement}} (M)\text{-}\!\!\sim\!\!\text{-}\underset{(Protein)\text{-}CO}{N}\!\!\diagdown\!\!\overset{CH_3}{\underset{O}{\overset{CH}{\underset{\|}{C}}\text{-}NH\text{-}R}}$$

Preparation of p-Phenylenediamine Agarose.[43] Twenty-five grams of 4% agarose gel in spherical bead form is cross-linked with epichlorohydrin.[7] The cross-linked gel is activated with cyanogen bromide[6] at pH 11. The activated gel is allowed to react with 0.5 g of p-phenylenediamine in 50 ml of 0.1 M sodium bicarbonate solution. Nitrogen is bubbled through the coupling suspension for 15 min, and the reaction is allowed to continue in a closed vessel for 16 hr in the dark at room temperature. The amino polymer is washed carefully with the following solutions in order: 0.1 M Na borate buffer (1 M in NaCl), pH 8.5; 0.1 M Na acetate (1 M in NaCl), pH 4.5; and 0.1 M Na acetate buffer, pH 4.5. The product

---

[40] I. Ugi, *Angew. Chem.* **74**, 9 (1962).
[41] R. Axén, P. Vretblad, and J. Porath, *Acta Chem. Scand.* **25**, 1129 (1971).
[42] P. Vretblad and R. Axén, *Acta Chem. Scand.* **27**, 2769 (1973).
[43] P. Vretblad and R. Axén, *FEBS Lett.* **18**, 254 (1971).

can be stored in the last-mentioned buffer, preferably containing 0.02% NaN_3.

Coupling of Pepsin by Means of Cyclohexyl Isocyanide.[43] Phenylenediamine-agarose (10 g) is suspended in 40 ml of distilled water, and 250 μl of acetaldehyde are added. The reaction suspension is stirred gently. After 15 min, 250 mg of pepsin (2 times crystallized, from Worthington Biochemical Corporation) is added along with 250 μl of cyclohexyl isocyanide. The pH is kept in the range 5.6–6.0 for 6 hr. The product is washed with water and thereafter with the following solutions in order: 0.1 M Na citrate buffer (1 M in NaCl), pH 4.0; 0.1 M Na acetate buffer (1 M in NaCl), pH 5.4; and 0.01 M Na citrate buffer, pH 4.0. The immobilized pepsin is stored in the last buffer containing 0.02% NaN_3 at 4°. The product contains 150–190 mg of enzyme per gram of dried conjugate. The activity of insolubilized pepsin toward denatured hemoglobin at pH 4 is 15% of that for native enzyme at the same pH value.

Periodate Oxidation of Agarose Gel. Agarose gel (2%) in spherical bead form is cross-linked by epichlorohydrin as described on page 23.

The cross-linked agarose gel (100 g) is suspended in 200 ml of water in which 2 g of sodium metaperiodate is dissolved. The temperature is increased to 45° over a 20-minute period and maintained there for 100 min under gentle stirring. The oxo-agarose is washed with water.

Covalent Fixation of β-Amylase to Oxo-Agarose by Means of Cyclohexyl Isocyanide. Oxo-agarose (10 g) is suspended in 20 ml of 0.4 M Na acetate, pH 6.5. β-Amylase (Sigma Type I-B, sweet potato) (100 mg) dissolved in 5 ml of water is added to the suspension, followed by 200 μl of cyclohexyl isocyanide. After reaction under gentle stirring for 6 hr at room temperature, the immobilized enzyme is carefully washed with the following solutions in order: 0.4 M Na acetate, pH 6.5; 0.01 M Na acetate containing 0.5 M NaCl, pH 4.8; 0.01 M Na acetate containing 0.5% starch, pH 4.8; 0.01 M Na acetate. The activity of the immobilized enzyme toward 0.5% starch solution (Zulkowsky starch) at pH 4.8 as compared with the enzyme in solution is about 40%. Native enzyme incubated with 5% starch at 60° loses its activity within 25 hours, but the immobilized enzyme after the same period was 95% active.

Thiol-Disulfide Interchange

A particularly attractive method, also applicable in the acid pH range, was introduced by Brocklehurst *et al.*[44] The coupling occurs

[44] K. Brocklehurst, J. Carlsson, M. P. J. Kierstan, and E. M. Crook, *Biochem. J.* **133**, 573 (1973).

through thiol-disulfide interchange, a method that links protein–cysteine–thiol groups to activated thiolated Sepharose. The principle and its application in covalent chromatography has been described in detail by the inventors.[45] The reactions are the following:

$$\text{(M)-SH} + \text{X-S-S-X} \longrightarrow \text{(M)-S-S-X} + \text{X-SH} \quad (a)$$

$$\text{(M)-S-S-X} + \text{ESH} \longrightarrow \text{(M)-S-S-E} + \text{X-SH} \quad (b)$$

The problem is to select X so that reactions (a) and (b) completely favor the formation of the matrix–enzyme conjugate M-S-S-E. Brocklehurst et al. described the advantages in selecting 2,2'-dipyridine disulfide as an activating agent.

Preparation of Mercapto-Agarose Gel.[28] Agarose gel (6%) in spherical bead form, washed with water and suction-dried, is suspended (30 g) in 24 ml of NaOH solution, followed by dropwise addition of epichlorohydrin at room temperature under stirring. In order to obtain a mercapto gel with a degree of substitution of about 50 μmoles of SH groups per gram of dried product, one should add 0.75 ml of epichlorohydrin (for 700 μmoles of SH groups per gram, add 4.5 ml of reagent). Time for the addition is 15 min. The temperature is then raised to 60°, and the stirring continues for 2 hr. The product is washed with water and with 0.5 M sodium phosphate buffer of pH 6.25. The gel is suction dried and suspended in 30 ml of the above-mentioned buffer, whereupon 30 ml of 2 M sodium thiosulfate solution are added, followed by stirring for 6 hr at room temperature.

The alkyl thiosulfate ester obtained is washed with water and suspended in 0.1 M Na bicarbonate solution (total volume 60 ml). The alkyl thiosulfate ester gel is then reduced by dithiothreitol, which has been dissolved in 1 mM EDTA solution to a concentration of 8 mg/ml. The reduction of agarose gel with a low degree of substitution requires 4 ml of dithiothreitol solution whereas a reduction of agarose gel having a high degree of substitution uses 50 ml of dithiothreitol solution. The mercapto gel obtained is washed with 300 ml of 0.1 M Na bicarbonate solution, which is 1 M in NaCl and 1 mM in EDTA, and finally with 100 ml of 1 mM EDTA solution. The gel is stored in 0.01 M deaerated Na acetate buffer of pH 4 (1 mM in EDTA).

Preparation of 2-Pyridine Disulfide Agarose Gel. Mercapto-agarose gel characterized by a degree of substitution with respect to mercapto group content of 50 μmoles per gram of dried product is washed with water and a solution of 50% acetone–water. The gel is suspended in the acetone–water solution. 2-Pyridine disulfide (100 mg) is dissolved in 50% acetone–water solution and added to the gel suspension (reaction time

30 min). The gel is washed with 50% aqueous acetone and 1 mM EDTA solution.

Comments. Brocklehurst *et al.*[45] described the use of 2-pyridine disulfide agarose gels for the isolation and purification of thiol enzymes. Carlsson and collaborators[46] later utilized the thiol-disulfide interchange reaction for the purpose of immobilizing enzymes with conservation of activity. Urease could thus be attached to mercapto-agarose by means of thiol-disulfide interchange with the conservation of most of its activity. (Such gels are now commercially available from Pharmacia.)

In other cases the enzymes were thiolated by 3-mercaptopropioimidate prior to the immobilization step[47]; however, it is also possible to immobilize disulfide-containing proteins after limited reduction. This is conveniently achieved by passing the enzyme solution through a bed of thiol agarose (reductor bed) and an adsorbator column containing active agarose according to the following scheme.[48]

Immobilization of Amyloglucosidase. Two milliliters of a solution containing 41.4 mg of amyloglucosidase in 6 ml of 0.1 M NaHCO$_3$, 1 mM in EDTA, is passed into a reductor column containing 17 ml of thiol agarose (about 700 µmoles of SH per gram of dry gel), which is operated in tandem with another column (3.2 ml volume) containing 2-pyridine disulfide agarose. The enzyme solution is followed by 20 ml of 0.1 M NaHCO$_3$–1 mM EDTA, 160 ml of 1 M NaAc buffer, pH 4.8, 30 ml of 1 M NaAc containing 1% starch, and finally 30 ml of 0.01 M NaAc, pH 4.8. The flow rate may be kept at about 10 ml per hour. In an experiment conducted as above, Axén and collaborators[48] immobilized 0.32 enzyme activity unit of 1.25 units applied. A control experiment with no reductor column was also made to show that no adsorption occurred unless the enzyme was partially reduced.

[45] K. Brocklehurst, J. Carlsson, M. P. J. Kierstan, and E. M. Crook, this series Vol. 34, p. 531 (1974).
[46] J. Carlsson, R. Axén, K. Brocklehurst, and E. M. Crook, *Eur. J. Biochem.* **44**, 189 (1974).
[47] J. Carlsson, R. Axén, and T. Unge, *Eur. J. Biochem.* **59**, 567 (1975).
[48] R. Axén, J. Carlsson, and H. Drevin, unpublished results (1973).

ADSORPTION METHODS

The most advanced enzyme reactors are based on the immobilization technique of ionic adsorption on Sephadex ion exchangers. They were developed in Japan by Chibata and co-workers, see this volume [51]. Immobilization by adsorption has the advantage of easy recharging with fresh enzyme when the substrate conversion power of the reactor has dropped below the lowest tolerable level. Another attractive feature is the wide latitude of adsorbent for greatly varying enzymes. For detailed treatment of adsorption, see this volume [11]. There are several articles in which the use of hydrophobic adsorbents for enzyme immobilization is described.[49-52] Especially promising are aliphatic and aromatic ether derivatives of agar.

Aromatic adsorption on cross-linked dextran gels was discovered at the same time as the molecular sieve effect.[1] The behavior of model substances on Sephadex was described in early papers by Porath[53] and Gelotte.[54] Since then many contributions have been made. The adsorption is too weak to be of interest for enzyme immobilization; however, by suitable substitution with hydrophobic groups in the hydroxylic gels, nonlinear strong adsorption of enzymes can be accomplished. Since the gel must retain for the most part its capacity to swell and to adsorb from aqueous solutions, it must be amphipathic or amphiphilic in character.[55] Amphiphilic gels for enzyme immobilization and purification are presently under intensive study in many quarters.[49-51,55,56] In those cases where the hydrophobic groups have been introduced by aliphatic amines via imido carbonates, the immobilization is based on mixed ionic-hydrophobic adsorption.[49] When neutral aliphatic ethers are used, only hydrophobic interaction is in operation. Adsorption phenomena of a more complicated nature are probably encountered with aromatic ethers,[55] where, in addition to hydrophobic effects, $\pi-\pi$ interaction is taking part as an essential factor in the adsorption.

It is not possible to assess with confidence the factors contributing to the adsorption of the enzymes and how they affect the activity and

---

[49] B. H. J. Hofstee and N. F. Otillio, *Biochem. Biophys. Res. Commun.* **53**, 1137 (1973).
[50] S. Shaltiel, this series Vol. 34, p. 126 (1974).
[51] K. Dahlgren Caldwell, R. Axén, and J. Porath, *Biotechnol. Bioeng.* **17**, 613 (1975).
[52] S. Hjertén, J. Rosengren, and S. Påhlman, *J. Chromatogr.* **101**, 281 (1974).
[53] J. Porath, *Biochim. Biophys. Acta* **39**, 193 (1960).
[54] B. Gelotte, *J. Chromatogr.* **3**, 330 (1960).
[55] J. Porath, L. Sundberg, N. Fornstedt, and I. Olsson, *Nature (London)* **245**, 465 (1973).
[56] R. I. Yon, *Biochem. J.* **126**, 765 (1972).

long-term stability of the enzyme gels. With the intention of throwing some light upon these questions, Dr. Caldwell and we have embarked upon broad study to develop adsorption processes for enzyme immobilization. In an exploratory study, allyl and hexyl ethers of Sepharose 6B were tested as immobilizing adsorbents for γ-amylase.

Preparation of Hexyl Ether of Agarose.[51] Cross-linked Sepharose 6B (200 ml) is transferred to a 1-liter conical flask provided with a cooler and a magnetic stirrer; 200 ml of 5 M NaOH containing 0.5% $NaBH_4$ is added together with 50 ml of n-hexylbromide. The suspension is heated and kept at 100° for 35 hr under gentle stirring. The contents are cooled in an ice-bath and neutralized with concentrated HCl. The gel is transferred to a glass filter and washed with water, ethanol, water, and finally the buffer to be used for adsorption.

Immobilization of the β-Amylase.[51] Hexyl agarose (0.5 ml) is packed into a shallow bed, 1 cm in diameter. The gel is washed with 0.02 M NaAc buffer (3 M in NaCl) and β-amylase and percolated with a solution of β-amylase (2 mg/ml) in the same buffer until saturation. The bed is then washed first with the adsorption buffer then with 0.02 M NaAc buffer (without NaCl) until no UV-absorbing material (280 nm) can be detected in the eluate.

Starch Conversion in Bed of Adsorption-Immobilized β-Amylase. The following experiment was carried out to determine the efficiency of the enzyme bed prepared as above. The gel, which had adsorbed 41 mg protein per milliliter (based on UV measurements), was percolated for 28 days with 1% Zulkowsky starch at a speed of 3 ml per hour (about 4000 bed volumes). The activity was determined against a 5% starch solution in a column-kinetic experiment where the optimal hourly maltose production was found to be 4.9 g. When the same gel conjugate was assayed as a batch against 100 ml of a 25% starch solution, the maltose production during the first hour amounted to 12 g. The release of enzyme from the bed was very low.

Some Further Aspects of Hydrophobic-Aromatic Adsorption. In some recent studies, Låås[10] found a great variation in enzyme affinity and sensitivity for amphipathic agarose. Dextranase and α-amylase, for example, can be adsorbed and desorbed quantitatively with respect to activity from gels containing as much as 2.5 mmoles of benzyl groups per gram of dry adsorbent. Serious losses of activity were found in similar experiments with lactic dehydrogenase and α-chymotrypsin when chromatographed on benzyl agarose with 900 μmoles per gram of dry gel substance. Låås further found that, whereas LDH was highly active in the adsorbed state, the proteolytic enzymes were not. It may be concluded that some enzymes retain activity both in the adsorbed and desorbed state, whereas others are active when immobilized, but are inactivated when desorbed; still others are inactivated already in the adsorption step. Considering the high benzyl content in some of the gels, it is not surprising that denaturation may occur. The most highly substituted gels contain more than 100 μmoles per milliliter of gel, equivalent to an average concentration of well above 1% benzyl alcohol. Locally the substituent concentration may be several times higher. The mechanism of adsorption must be better understood as well as the subsequent events in the solid phase, which may further change the conformation of adsorbed proteins. The temperature at which adsorption takes place may be very important for the retention of activity. Adsorption at very low temperatures from ethylene glycol–water or glycerol–water mixtures may greatly widen the scope and improve the performance of the amphipathic agar gels.

Conclusions

The immobilization of enzymes on agar(ose), Sephadex, and similar hydroxylic matrices may be accomplished according to two distinctly different techniques: (1) covalent irreversible coupling and (2) reversible enzyme attachment. Either technique has its advantages and disadvantages.

Covalent irreversible coupling in its ideal form precludes leakage of enzyme. On the other hand, when the catalytic activity has dropped to an intolerably low level, the enzyme gel may not be easily regenerated; also the regeneration of the matrix itself may be difficult.

Reversible attachment can be of two different kinds: (a) covalent coupling and (b) adsorptive coupling. An attractive feature of reversible attachment may be pointed out; the starting gel matrix is easily regenerated and possibly the soluble enzymes as well (a fact that may be of economic importance).

The agar(ose) matrices are particularly well suited for the immobilization of enzymes, substrates, and products of large molecular size.

It is difficult *a priori* to evaluate in general terms which of the immobilization methods is to be preferred in a particular situation. The coupling yield or adsorption capacity and retention of activity will depend on the reactivity or adsorption properties, respectively, and on the stability of the particular enzyme to be immobilized.

[4] Enzymes Immobilized to Cellulose

By M. D. LILLY

Cellulose is a vegetable fiber composed of 1,4-linked polymers of β-D-glucose that interact through hydrogen bonds. The aggregated polymers form fibrils that agglomerate into fiber structures. Cellulose contains highly ordered crystalline regions and more accessible amorphous regions of a low degree of order. In general, about 70% of native cotton cellulose is crystalline and 30% amorphous. When native cellulose is regenerated from solution or passed through a highly swollen state following strong alkali treatment, the linear polymers shift to a different, less-ordered configuration and the cellulose is referred to as mercerized.

Cellulose can undergo all the usual reactions associated with polyhydric alcohols so that a wide range of modified celluloses can be prepared. Since the most useful are commercially available (Table I), the methods of preparation will not be described, but they may be found, for instance, in a review by Weliky and Weetall[1] and a book by Ott, Spurlin, and Grafflin.[2] The chemical reactivity and physical properties of cellulose and modified celluloses depend very much on the treatments used. The proportion of hydroxyl groups that are accessible to chemical reagents depends on the size of the reagent molecule and on the degree of crystallinity of the cellulose. The degree of substitution (DS) is defined as the average number of the three available hydroxyl groups in each anhydroglucose unit that have been substituted. Sodium carboxymethyl cellulose is water soluble at DS values above about 0.4. With a less polar substituent, water solubility occurs at higher DS values.

Most commercially available modified celluloses are supplied in dried form and must be swollen so that they become fully accessible to solutes.

TABLE I
COMMERCIALLY AVAILABLE CELLULOSIC DERIVATIVES

p-Aminobenzyl	Diethylaminoethyl- (DEAE-)
Aminoethyl- (AE-)	Epichlorohydrin triethanolamine- (ECTEOLA-)
Benzoyl-DEAE-	Oxy-
Benzoyl-naphthoyl-DEAE-	Phospho-
Benzyl-DEAE-	Sulfoethyl-
Carboxymethyl- (CM-)	Triethylaminoethyl- (TEAE-)

[1] N. Weliky and H. H. Weetall, *Immunochemistry* 2, 293 (1965).
[2] E. Ott, H. M. Spurlin, and M. W. Grafflin, "Cellulose and Cellulose Derivatives," 2nd ed. Wiley (Interscience), New York, 1954.

This treatment breaks those additional hydrogen bonds formed when the cellulose was dried. Except for ester derivatives of cellulose, a precycling with 0.5 N HCl and 0.5 N NaOH with intermediate washing is recommended.[3] For anionic exchangers, treatment with alkali should precede acid treatment.

Modified celluloses were some of the first supports used for immobilized proteins.[1,4] Over 40 different enzymes have now been immobilized successfully to modified celluloses by covalent binding. It should be noted that ion-exchange celluloses have also been used successfully for immobilization by noncovalent-binding. The industrial conversion of glucose to fructose with packed beds of DEAE-cellulose to which glucose isomerase is bound has been reported.[5]

Preparation of Cellulose-Bound Enzymes

The availability of modified celluloses, originally for use as ion exchangers, has permitted a wide range of methods to be developed for covalent binding of enzymes. The more common methods are summarized below. With a few exceptions it is assumed that the enzyme binds mainly through amino groups on its surface, but this has been confirmed in only a few instances.

Attachment via Cellulose Hydroxyl Groups

Using Cyanuric Chloride or s-Triazinyl Derivatives[6,7]

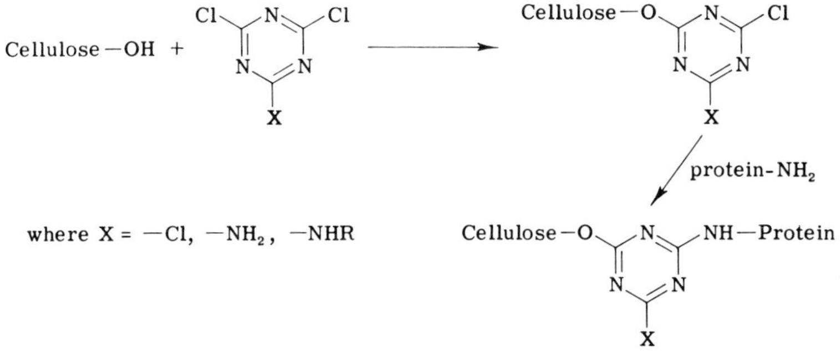

Cellulose—OH + [triazine with Cl, Cl, X] → Cellulose—O—[triazine with Cl, X]

where X = —Cl, —NH$_2$, —NHR

Cellulose—O—[triazine]—NH—Protein (+ protein-NH$_2$)

---

[3] Whatman Biochemicals Ltd., Maidstone, Kent, U.K., Laboratory Manual.
[4] M. A. Mitz and L. J. Summaria, *Nature (London)* **189**, 576 (1961).
[5] B. J. Schnyder, *Die Stärke* **26**, 409 (1974).
[6] G. Kay and E. M. Crook, *Nature (London)* **216**, 514 (1967).
[7] G. Kay and M. D. Lilly, *Biochim. Biophys. Acta* **198**, 276 (1970).

Using Cyanogen Bromide. This method has been used mainly for covalent binding of enzymes to Sephadex and Sepharose (see this volume [3]), but it also works well with celluloses.[8]

Using Transition Metal Salts. This method is based on the activation of celluloses by steeping in a solution of the metal salt (e.g., $TiCl_4$) followed by addition of the enzyme to the washed cellulose–metal salt complex.[9] The exact nature of the binding has not yet been elucidated.

Attachment via Carboxylic Acid Groups on Cellulose Derivative

The Curtius Azide Method. This method was first described by Micheel and Ewers[10] and modified subsequently by Mitz and Summaria.[4] Crook, Brocklehurst, and Wharton[11] have described the experimental procedures in detail.

$$\text{Cellulose—O—CH}_2\text{—COOH} \xrightarrow{CH_3OH, HCl} \text{cellulose—O—CH}_2\text{—COOCH}_3 \downarrow NH_2NH_2$$

$$\text{cellulose—O—CH}_2\text{—CON}_3 \xleftarrow{NaNO_2, HCl} \text{cellulose—O—CH}_2\text{—CO—NHNH}_2$$

$$\text{protein-NH}_2 \downarrow \text{mild alkali}$$

$$\text{cellulose—O—CH}_2\text{—CO—NH—protein}$$

Using Isoxazolium Salts. These are used to synthesize amide bonds between the carboxyl group and enzyme.[12]

Using N, N-Dicyclohexylcarbodiimide. See Weliky et al.[13]

Attachment to Aliphatic Amine Derivative

Glutaraldehyde or other bifunctional agents are used. Habeeb[14] reported the binding of trypsin to aminoethyl cellulose using glutaraldehyde. Many other enzymes have now been immobilized by this method. There is still some doubt about the reaction mechanism.[15]

---

[8] R. Axén and S. Ernback, *Eur. J. Biochem.* **18**, 351 (1971).
[9] A. N. Emery, J. S. Hough, J. M. Novais, and T. P. Lyons, *Chem. Eng. (London)*, p. 71, February (1972).
[10] F. Micheel and J. Ewers, *Makromol. Chem.* **3**, 200 (1949).
[11] E. M. Crook, K. Brocklehurst, and C. W. Wharton, this series, Vol. 19, p. 963 (1970).
[12] T. Wagner, C. J. Hsu, and G. Kelleher, *Biochem. J.* **108**, 892 (1968).
[13] N. Weliky, H. H. Weetall, R. V. Gilden, and D. H. Campbell, *Immunochemistry* **1**, 219 (1964).
[14] A. F. S. A. Habeeb, *Arch. Biochem. Biophys.* **119**, 264 (1967).
[15] P. Monzan, G. Puzo, and H. Mazarguil, *Biochimie* **57**, 1281 (1975).

Attachment to Aromatic Amine Derivative

The method involves diazotization, presumably via a tyrosine residue.[16]

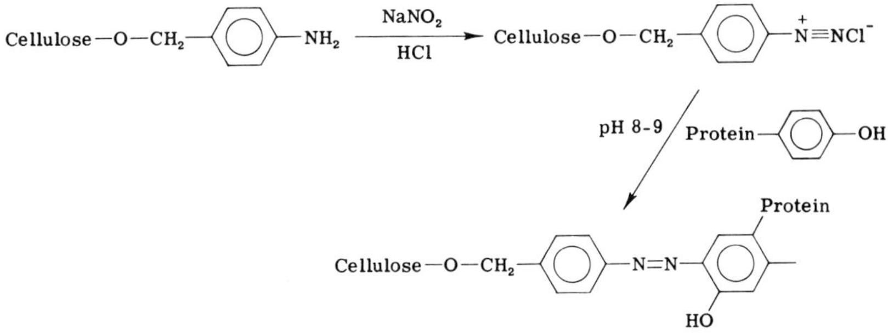

Cellulose—O—CH$_2$—〈◯〉—NH$_2$ $\xrightarrow[\text{HCl}]{\text{NaNO}_2}$ Cellulose—O—CH$_2$—〈◯〉—N≡NCl$^-$

pH 8-9 / Protein—〈◯〉—OH

Protein

Cellulose—O—CH$_2$—〈◯〉—N=N—〈◯〉—
                                       HO

Attachment via Alkyl Halide Derivatives

These derivatives are not available commercially and must be prepared by acylation of cellulose hydroxyl groups by, for example, bromoacetyl bromide.[17] The product will then react with the enzyme by alkylation of an amino group, or presumably a thiol group if this is available. This method suffers from the lability, even at neutral pH, of the ester linkage produced in the first reaction.[11]

Immobilization Procedures: Experimental Procedures

In addition to the azide method, details of which have been published in a previous volume,[11] two other methods have been used for the immobilization of a variety of enzymes to cellulose or modified celluloses. The experimental procedures for these methods, using (1) cyanuric chloride or its derivatives, and (2) glutaraldehyde, are given below. Care must be taken in handling these two reagents.

Preparation of s-Triazinyl-Bound Immobilized Enzymes

It is possible to make dichloro-s-triazinyl derivatives of celluloses using cyanuric chloride. These preparations react rapidly with proteins[6] (e.g., chymotrypsin at pH 7 in a few minutes). The remaining

---

[16] D. H. Campbell, E. Luescher, and L. S. Lerman, *Proc. Natl. Acad. Sci. U.S.A.* **37**, 575 (1951).

[17] A. T. Jagendorf, A. Patchornik, and M. P. Sela, *Biochim. Biophys. Acta* **78**, 516 (1963).

chlorines may be treated with an amine to prevent further reaction, including cross-linking of the cellulose. As the reaction of the second chlorine with protein is so fast, it is preferable to replace this substituent and use the third chlorine for this purpose. Monochloro-s-triazinyl derivatives may be made by allowing the cellulose to react initially with cyanuric chloride and then with an amine, such as N-(3-aminopropyl)-diethanolamine,[18] under controlled conditions to leave the third chlorine unreacted. Usually, however, it is preferable to use 2-amino-4,6-dichloro-s-triazine in the reaction with cellulose or modified cellulose.

Preparation of 2-Amino-4,6-dichloro-s-triazine.[7] A stream of gas, generated by blowing N_2 through a warmed solution of ammonia (sp. gr. 0.88) and dried by passage through NaOH pellets, is passed into a cold slurry (5°) of cyanuric chloride (184 g) in a mixture of 1 liter of dioxane and 0.2 liter of toluene. When the reaction products form a thick suspension, this solid is filtered off, washed with 0.5 liter of dioxane, and discarded. Care is required to ensure that the solids formed do not block the gas stream. The filtrate and washings are mixed and evaporated to dryness under reduced pressure on a rotary film evaporator. The product is recrystallized first by dissolving it in the minimum volume of acetone/water (1:1, v/v) followed by removal of the acetone under reduced pressure and, second, by dissolving the recovered solid in boiling water and cooling to 20° within 10 min to prevent any hydrolysis. The yield is about 90%.

Preparation of Aminochloro-s-triazinyl Derivatives of Cellulose or Modified Celluloses. Cellulose or modified cellulose (10 g dry weight) is added to 50 ml of acetone/water (1:1, v/v) at 50° containing 1 g of 2-amino-4,6-dichloro-s-triazine and stirred for 5 min. Then 20 ml of a 15% (w/v) aqueous solution of Na_2CO_3 to which 0.6 volume of 1 M HCl has been added is poured into the reaction liquor, and the slurry is stirred for a further 5 min at 50°. Concentrated HCl is then added to reduce the pH rapidly to below 7. The product is recovered by filtration, washed 3 times with 100 ml of an equivolume mixture of acetone and water, followed by extensive washing with water and finally 0.05 M phosphate buffer, pH 7.0. It may then be stored at 2°.

Preparation of Immobilized Enzyme. The aminochloro-s-triazinyl cellulose suspension is adjusted to pH 8.0 with alkali. A solution of the enzyme in 0.05 M phosphate, pH 8.0, is added, and the suspension is stirred gently for 18 hr at 20°. The product is recovered by filtration and washed thoroughly with (1) phosphate buffer, pH 8.0, (2) the same buffer containing 1 M NaCl, and (3) the phosphate buffer. The immobi-

[18] A. K. Sharp, G. Kay, and M. D. Lilly, *Biotechnol. Bioeng.* **11**, 363 (1969).

lized enzyme may be stored in buffer at 2°. The concentrations of aminochloro-s-triazinyl cellulose and enzyme in the final reaction mixture should be as high as possible (e.g., 10–50 g dry wt per liter and over 5 g/liter, respectively) to achieve the highest binding of enzyme. For enzymes stable in alkaline pH, it is preferable to do the binding reaction at pH 8.7 in borate buffer.

In the above example, it is assumed that the modified cellulose derivative and enzyme have oposite net charges at the reaction pH. If they have similar net charges, then the ionic strength should be increased to reduce electrostatic repulsion. Alternatively, a different triazinyl derivative and enzyme have opposite net charges at the reaction pH. If they to react with the cellulose to give the cellulose derivative a more negative charge.

Although 18 hr are required to achieve the maximum attachment of enzyme, most of the enzyme is bound within a few hours. For less stable enzymes, the immobilization reaction may be done at 2° and the reaction time reduced.

For immobilization of an enzyme to soluble CM-cellulose using 2-amino-4,6-dichloro-s-triazine, different conditions must be used.[19]

Preparation of Glutaraldehyde-Bound Enzymes

Aminoethyl cellulose is pretreated by washing with 0.5 M NaOH (20 ml per gram dry weight) and then with water until the pH of the washings is close to neutrality. The AE-cellulose (10 g dry weight) is suspended in 0.05 M phosphate buffer, pH 7.0, to give a volume of 240 ml, and 20 ml of an aqueous solution of glutaraldehyde (25%, w/v) are added. After stirring for 2 hr at room temperature, the cellulose derivative is recovered by filtration and washed 5 times with 100 ml of the buffer to remove residual glutaraldehyde. It is then resuspended in the same buffer, the enzyme solution is added, and the volume is made up to 200 ml with more buffer. After gentle stirring at room temperature for 2 hr, the immobilized enzyme is recovered by filtration and washed with buffer containing 1 M NaCl to remove noncovalently bound enzyme.

For maximum binding of enzyme, the concentration of glutaraldehyde in the reaction with AE-cellulose should be 1% (w/v) or greater. Glassmeyer and Ogle[20] used almost 10% to immobilize trypsin, but much more washing of the cellulose derivative is necessary to remove residual glutaraldehyde.

[19] J. R. Wykes, P. Dunnill, and M. D. Lilly, *Biochim. Biophys. Acta* **250**, 522 (1971).
[20] C. K. Glassmeyer and J. D. Ogle, *Biochemistry* **10**, 786 (1971).

TABLE II
EXAMPLES OF CELLULOSE-BOUND ENZYMES

Enzyme	Support	Binding method	Protein bound (% of protein presented)	Protein content of preparation (mg/g)	Activity retention[a] (%)	Reference
Ficin	CM-cellulose	Azide	15	30	13	[b]
Bromelain	CM-cellulose	Azide	73	110	52	[c]
Trypsin	AE-cellulose	Glutaraldehyde	63	70	75	[d]
β-Galactosidase	AE-cellulose	Glutaraldehyde	—	104	27	[e]
Lactate dehydrogenase	DEAE-cellulose	s-Triazine	96	19	42	[f]
Chymotrypsin	DEAE-cellulose	s-Triazine	38	110	42	[g]
	CM-cellulose	s-Triazine	48	52	46	[g]
	Cellulose	s-Triazine	10	4	—	[g]
	Mercerized cellulose	CNBr	84	295	11	[h]
Amyloglucosidase	Cellulose	TiCl$_4$	83	85	46	[i]

[a] Values for proteases are with substrates of low molecular weight.
[b] W. E. Hornby, M. D. Lilly, and E. M. Crook, *Biochem. J.* **98**, 420 (1966).
[c] C. W. Wharton, E. M. Crook, and K. Brocklehurst, *Eur. J. Biochem.* **6**, 565, 572 (1968).
[d] C. K. Glassmeyer and J. D. Ogle, *Biochemistry*, **10**, 786 (1971).
[e] D. L. Regan, P. Dunnill, and M. D. Lilly, *Biotechnol. Bioeng.* **16**, 333 (1974).
[f] J. R. Wykes, P. Dunnill, and M. D. Lilly, *Biotechnol. Bioeng.* **17**, 51 (1975).
[g] G. Kay and M. D. Lilly, *Biochim. Biophys. Acta* **198**, 276 (1970).
[h] R. Axén and S. Ernback, *Eur. J. Biochem.* **18**, 351 (1971).
[i] A. N. Emery, J. S. Hough, J. M. Novais, and T. P. Lyons, *Chem. Eng. (London)*, p. 71, February (1972).

Properties of Cellulose-Bound Enzymes

Some examples of cellulose-bound immobilized enzymes are given in Table II. The usual amounts of enzymes that can be immobilized are about 100 mg per gram of modified cellulose. The retentions of activity given in Table II are only approximate, since many of the preparations are sufficiently active to be subject to internal diffusional limitation.[21] Shifts in the pH activity profiles of enzymes on immobilization have been observed with ion-exchange celluloses when assayed at low ionic strength.[7,19,22] Similarly, changes in K_m due to electrostatic interaction have been observed.[22,23] Two enzymes, ficin[22] and bromelain,[23] had enhanced resistance to oxidation on immobilization. Many of the enzymes have increased thermal stabilities on immobilization.

[21] D. L. Regan, M. D. Lilly, and P. Dunnill, *Biotechnol. Bioeng.* **16**, 1081 (1974).
[22] W. E. Hornby, M. D. Lilly, and E. M. Crook, *Biochem. J.* **98**, 420 (1966).
[23] C. W. Wharton, E. M. Crook, and K. Brocklehurst, *Eur. J. Biochem.* **6**, 565, 572 (1968).

[5] Immobilization of Enzymes to Various Acrylic Copolymers

By ROLF MOSBACH, ANN-CHRISTIN KOCH-SCHMIDT, and KLAUS MOSBACH

Among the large number of supports available for the immobilization of enzymes, the synthetic carriers, in particular the hydrophilic cross-linked polyacrylic polymers, are of great interest. They show good mechanical and chemical stability and are inert to microbial degradation. In one respect they are outstanding: that is, in the ease of preparing various polymers "tailor-made" for a particular enzyme and application, e.g., with regard to stability and specific activity. Parameters that can easily be varied to this end include (1) the degree of porosity and (2) the chemical composition, which can be achieved by either copolymerization of different monomers among the large number available or by chemical modification of preformed polymers.

Acrylic polymers have been used primarily as supports for the covalent attachment of enzymes (see this chapter and Chapters [6]–[8]), but they have also been used for the entrapment of enzymes (see in particular Chapter [12] and also Chapters [13], [14], [50], and [62]). The intermediate situation, i.e., where enzymes are both covalently bound and entrapped in acrylic gels, is treated in Chapter [12] and, in particular, in Chapter [15]. In Table I are listed a number of acrylic polymers that have been applied in the covalent attachment of enzymes.

TABLE I
Various Synthetic Polymers Based on Acrylamide or Acrylate to Which Enzymes Have Been Covalently Bound

Original matrix with functional group	Coupling or activating agents	Intermediate matrix	Major reacting groups of proteins	Re
―COOH Acrylamide/acrylic acid Acrylamide/maleic acid	R'―N=C=N―R'' + H⁺	―COO―C(=NH⁺―R'')―NH―R'	―NH₂ ―SH ―⟨⟩―OH	a, c
―OH ―OH Acrylamide/hydroxyethyl-methacrylate Hydroxyalkylmethacrylate	BrCN	―O―C(=NH)―O―	―NH₂	a b
―CONH₂ Acrylamide Acrylamide/methyl-acrylate	OHC(CH₂)₃CHO	―CON=CH(CH₂)₃CHO	―NH₂ ―⟨⟩―OH	a, c f
―CONH₂ Acrylamide	H₂N·NH₂	―CONHNH₂^i	―NH₂ ―COOH	g, h
―CONH₂ Acrylamide	H₂N·(CH₂)₂NH₂	―CONH(CH₂)₂NH₂^i	―NH₂ ―COOH	g, h
―CONH₂ Acrylamide Acrylamide/methyl-acrylate	H₂N·NH₂, HNO₂	―CON₃	―NH₂ ―SH ―⟨⟩―OH	g, h j

TABLE I (Continued)

Original matrix with functional group	Coupling or activating agents	Intermediate matrix	Major reacting groups of proteins	Refs.
-CO-O-CO- (anhydride) Acrylamide/maleic acid anhydride Methacrylic acid anhydride	—	—	—NH$_2$	c k
-CO-NH-C$_6$H$_4$-NH$_2$ -CONH$_2$ Acrylamide/ anilinoarylamide	HNO$_2$	-CONH-C$_6$H$_4$-N$^+$≡N	—NH$_2$ —SH -C$_6$H$_4$-OH Histidine/ arginine	l
-C$_6$H$_4$-F Methacrylic acid/ 3- or 4-fluorostyrene	HNO$_3$, H$_2$, HNO$_2$	-C$_6$H$_3$(F)-N$^+$≡N	—NH$_2$	m
-CO-NH-C$_6$H$_4$-F Methacrylic acid Methacrylic acid/ m-fluoroanilide	HNO$_3$	-CO-NH-C$_6$H$_3$(F)(NO$_2$)	—NH$_2$	m

This chapter.
This volume, Chapter [6].
H. Lang, N. Hennrich, H. D. Orth, W. Brümmer, and M. Klockow, *Chem.-Ztg.* **96**, 595 (1972).
P. O. Weston and S. Avrameas, *Biochem. Biophys. Res. Commun.* **45**, 1574 (1971).
T. Ternynck and S. Avrameas, *FEBS Lett.* **23**, 24 (1972).
A.-C. Johansson and K. Mosbach, *Biochim. Biophys. Acta* **370**, 348 (1974).
J. K. Inman and H. M. Dintzis, *Biochemistry* **8**, 4074 (1969).
J. K. Inman, this series Vol. 34 [3].
Requiring additional coupling agent.
Y. Ohno and M. A. Stahmann, *Macromolecules* **4**, 350 (1971).
A. Conte and K. Lehmann, *Hoppe-Seyler's Z. Physiol. Chem.* **352**, 533 (1971).
This volume, Chapter [7].
This volume, Chapter [8].

In the first part of this chapter procedures are given for the preparation of various acrylic copolymers in either bead form or as granules. In the second part, a number of examples are given on the binding of enzymes to such preparations; and in the third, the "tailor-made" approach is illustrated with two examples.

Preparation of Various Acrylic Copolymers

Acrylic copolymers are available from commercial sources and also relatively easily prepared in the laboratory. Commercially available crosslinked polymers include Bio-Gel P (polyacrylamide), Bio-Gel CM2 (acrylic lattice containing carboxyl groups), aminoethyl Bio-Gel P and hydrazide Bio-Gel P (all from Bio-Rad Laboratories Ltd., Richmond, California 94804, USA), Enzacryl AA and AH (polyacrylamide containing aromatic amines and hydrazide residues, respectively; Koch-Light Laboratories Ltd., Colnbrook SL3 OBZ, Buckinghamshire, England), Spherons [poly(hydroxyalkylmethacrylate)] (see this volume [6]).

Below are given two general procedures for the preparation of various acrylic copolymers from the great variety of available monomers. By choosing the ratios of the participating monomers, it is possible to prepare a polymer with the desired amount of a specific functional group.

The polymerization is carried out in a liquid medium as bulk, or directly as bead, polymerization. The procedure in liquid medium has the effect that porous polymers of defined structure are obtained in contrast to polymerization in substance. The structure, and thus the size and distribution of porosity within the gel, is dependent besides on polymerization kinetics and other conditions, mainly on the total monomer concentration T (w/v), and the relative concentration, or the proportion by weight, C, (w/w) of cross-linking agent (such as N,N'-methylenebisacrylamide, Bis). Increasing the total monomer concentration, T, while keeping the concentration, C, of the cross-linking agent constant, decreases the resolution of proteins (as tested in gel chromatography[1-3]). T thus determines the pore size of the gel. Conversely, C will also influence pore size. Thus it was found that polyacrylamide with $C = 5\%$ yields the minimal average pore size (using constant total concentration). Above and below approximately 5%, pore size will be larger; such transition points are a characteristic of cross-linked polymers.[4] Preparations with C values above the transition point, again keeping T constant, turn more and more turbid indicating microprecipitation and organization leading to larger pores.[4,5] However, gel

[1] S. Hjertén and R. Mosbach, *Anal. Biochem.* 3, 109 (1962).
[2] S. Hjertén, *Arch. Biochem. Biophys.* Suppl. 1, 147 (1962).
[3] R. Mosbach, U.S. Patent No. 3,298,925 (1967).
[4] J. S. Fawcett and C. J. O. R. Morris, *Sep. Sci.* 1, 9 (1966).
[5] E. G. Richards and C. J. Temple, *Nature (London) Phys. Sci.* 230, 92 (1971).

structure can also still be found in the microparticles of this coherent disperse system. These porous bodies (also called macroreticular gels) are characterized by having interparticular pores which have different physicochemical properties from the pores normally found in gels. There exists a continuous transition, depending on the method of preparation, from gels to hollow-space systems.

Another method of preparing bodies of varying porosity involves polymerization in different solvent systems (see Chapter [6] on Spheron gels).

Preparation of Acrylic Copolymers as Granules ($T:C = 10:5$) through Bulk-Polymerization Followed by Physical Fragmentation

Reagents

Acrylic monomer(s) such as acrylamide, 2-hydroxyethyl methacrylate, methylacrylate, acrylic acid (Eastman Organic Chemicals, New York. Solid monomers are preferably recrystallized and liquid ones distilled before use)

N,N'-Methylenebisacrylamide, Bis (Eastman, once recrystallized in acetone, 45°)

Tris-HCl buffer, 0.1 M, pH 7.0

N,N,N',N'-Tetramethylethylenediamine, TEMED (Eastman)

Ammonium persulfate (peroxido sulfate) (Merck, Darmstadt, West Germany)

Monomer(s) 2.375 g (neutralized to pH 7.0), and Bis, 0.125 g, are dissolved in 0.1 M Tris-HCl buffer, pH 7.0, to a volume of 24 ml in a beaker. After addition of the catalyst system system consisting of 0.5 ml of TEMED and 0.5 ml of ammonium persulfate solution (0.5 g/ml), the solution is placed under slight vacuum (O_2 inhibits the polymerization process) for about 1 hr. The gel block is cut into 1-mm slices and pressed through a 30-mesh sieve. The granules are then thoroughly washed with water in order to remove nonpolymerized monomers (toxic!) and the catalyst system.

Preparation of Spherical Acrylic Copolymers ($T:C = 10:5$) by Bead Polymerization[6,7]

Reagents

Hydrophilic phase:
Acrylic monomer(s) (Eastman)
N,N'-Methylenebisacrylamide, Bis (Eastman)

[6] H. Nilsson, R. Mosbach, and K. Mosbach, *Biochim. Biophys. Acta* **268**, 253 (1972).
[7] A.-C. Johansson and K. Mosbach, *Biochim. Biophys. Acta* **370**, 339 (1974).

0.1 M Tris-HCl buffer, pH 7.0
N,N,N',N'-Tetramethylethylenediamine, TEMED (Eastman)
Ammonium persulfate (Merck)
Hydrophobic phase:
Chloroform (analytical grade)
Toluene (analytical grade)
Sorbitan sesquioleate (Pierce Chemical Company, Rockford, Illinois)

The bead-polymerization process is carried out in a 1-liter round-bottom flask equipped with a stirrer, an inlet and outlet for N_2, and a tap funnel for admission of the monomer solution (see Fig. 1a). The flask is cooled in ice water, and contains 400 ml of the hydrophobic phase, toluene/chloroform (290/110, v/v), in order to achieve the same density as the aqueous phase. This phase also contains 1–3 ml of the suspension stabilizing agent, Sorbitan sesquioleate. Prior to the polymerization, the hydrophobic phase is stirred under N_2 at about 260 rpm; the stirring rate determines the bead size formed. To the cold monomer solution, consisting of 5.7 g of monomer(s) and 0.3 g Bis dissolved in Tris-HCl buffer to a total volume of 58.5 ml, 0.5 ml of TEMED and 1.0 ml ammonium persulfate solution (0.6 g/ml) are added. When the resulting mixture begins to polymerize slightly (as indicated by an increasing viscosity), it is run into the flask. The polymerization of the aqueous phase, when performed at 1–5°, is usually complete within 30 min. The beads are washed on a glass filter, first with toluene to remove the chloroform and then with several liters of distilled water prior to freeze-drying.

It is important that the aqueous and organic phases have the same density. Instead of using toluene it is possible to use other organic solvents, e.g., n-butanol. By changing the stirring rate and/or the amount of stabilizer, different bead diameters (10–500 μm) can be prepared. Hydrophobic monomers, if employed, must be first dissolved in organic solvents (methanol) (to a maximum volume of 50% of the total volume of the hydrophilic phase) and then added to the buffer solution. The purpose of the alcohol is to retain the monomer in the aqueous phase during the polymerization process.

A much simpler alternative arrangement is given in Fig. 1b. This procedure, in which polymerization is carried out in a beaker with magnetic stirring, but otherwise under identical conditions, as outlined above, also yields polymer beads. Whether the bead diameter distribution is as good as when the former technique is used remains to be studied. It should be added that the above (Fig. 1a) bead-polymerization techniques have been used in the entrapment of enzymes[6,7] and

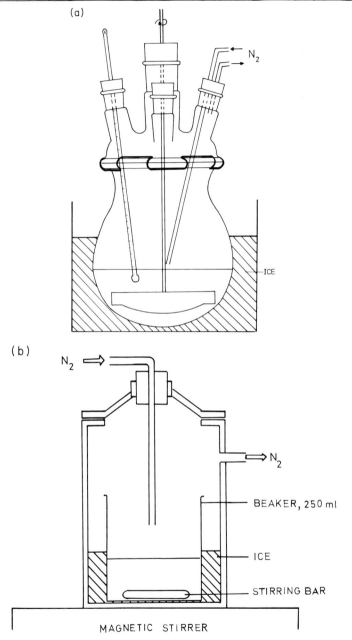

FIG. 1. Apparatus used for the preparation of bead-formed acrylic copolymers.

the latter alternative (Fig. 1b) recently also to whole cells of *Streptomyces albus* (A.-C. Koch-Schmidt and K. Mosbach, to be published).

In some cases, it was found advantageous, in order to avoid leakage, to entrap the enzyme in the presence of a carbodiimide in the acrylic acid–acrylamide monomer solution, leading to a preparation in which the enzyme was both entrapped and covalently bound.[8]

Examples of Covalent Binding of Enzymes to Bead-Formed Acrylic Copolymers

Below are given methods suitable for the covalent binding of enzymes and other ligands to various acrylic copolymers, which have been prepared in bead form (as outlined here) except for one example using a commercially available polymer. The methods given involve (a) glutaraldehyde treatment of polyacrylamide first reported in 1971,[9] (b) CNBr activation of acrylamide/hydroxyethylmethacrylate polymers, first reported in 1970[7,8,10] (see also Chapter [6]), and (c) coupling by the carbodiimide method[7,8,11,12] (see also Chapters [6] and [34]) to acrylamide/acrylic acid polymers. The properties of these preparations are summarized in Table II.

Covalent Binding to Glutaraldehyde-Treated Polyacrylamide Beads ($T:C = 10:5$)

Example: Immobilization of Trypsin and Ribonuclease A to Polyacrylamide Beads. Dry polymer, 100 mg, prepared in the setup given in Fig. 1a, is allowed to swell in water and then activated for 14 hr at 37° using 4 ml of a 6% solution of glutaraldehyde dissolved in 0.2 M phosphate buffer, pH 7.4. The glutaraldehyde-treated beads are thoroughly washed with distilled water, at 4-hr intervals, for 1 day at 4°. Filtered beads are transferred to a tube containing 2 mg of trypsin (Sigma, bovine pancreas, 10,000 BzArg-OEt units/mg) or ribonuclease A (Sigma, bovine pancreas, type IA, 60 Kunitz units/mg) dissolved in 2 ml of 0.1 M phosphate buffer, pH 7.5. The coupling is allowed to proceed for about 12 hr at 4°. The beads are then washed with 0.5 M NaCl, 0.1 M NaHCO$_3$, and distilled water, successively.

In a control experiment run to investigate the binding capacity of

[8] K. Mosbach, *Acta Chem. Scand.* **24**, 2084 (1970).
[9] P. O. Weston and S. Avrameas, *Biochem. Biophys. Res. Commun.* **45**, 1574 (1971).
[10] J. Turková, O. Hubálková, M. Kriváková, and J. Čoupek, *Biochim. Biophys. Acta* **322**, 1 (1973).
[11] K. Mårtensson and K. Mosbach, *Biotechnol. Bioeng.* **14**, 715 (1972).
[12] K. Mårtensson, *Biotechnol. Bioeng.* **16**, 579 (1974).

TABLE II
VARIOUS ENZYMES IMMOBILIZED TO DIFFERENT ACRYLIC COPOLYMERS BY
THREE IMMOBILIZATION TECHNIQUES DESCRIBED

	mg enzyme/g dry polymer	% Bound enzyme of added enzyme	Specific activity of bound enzyme as % of free	Units[a]/g dry polymer
Polyacrylamide (glutaraldehyde)				
Ribonuclease A	20	98	10	60
Trypsin	11	63	14	67
Trypsin[b]	18	90	69	541
Hydroxyethyl methacrylate–acrylamide copolymer (50/50, w/w) (CNBr)				
Ribonuclease A	20	50	36	216
Trypsin	15	85	20	131
Trypsin[b]	17	87	57	422
Acrylic acid–acrylamide copolymer (50/50, w/w) (EDC)				
Ribonuclease A	40	100	9	108
Trypsin	20	57	49	429
Trypsin[b]	26	65	65	736
Bio-Gel CM-100 (CMC)				
pullulanase	19	34	43	1.70

[a] The units given refer to international units.
[b] Recently prepared acrylic gels identical to those given except using a ratio of T:C = 8:30 instead of T:C = 10:5 to increase the porosity of the gels.

the polymer, glycine was bound to the glutaraldehyde-treated beads in the amount of 54 μmoles per gram of dry polymer.

Covalent Binding to CNBr-Treated Copolymers of Acrylamide 2-Hydroxyethylmethacrylate (50/50, w/w; T:C = 10:5)

Example: Immobilization of Trypsin and Ribonuclease A to an Acrylamide-2-hydroxyethylmethacrylate Copolymer. Swollen beads (100 mg dry weight) are activated in 4 ml of distilled water containing 80 mg of CNBr. The activation step proceeds for 6 min at 4°, while the pH is maintained at 10.8 by the addition of 1 M NaOH. Activated beads are thoroughly washed on glass filter with ice cold 0.1 M NaHCO$_3$ and then transferred to the enzyme solution containing 2 mg of trypsin or ribonuclease A dissolved in 2 ml 0.1 M of NaHCO$_3$. The coupling reaction proceeds for about 12 hr at 4°. The beads are then washed in 0.1 M NaHCO$_3$, 0.5 M NaCl, 1 mM HCl, and distilled water.

It appears that the concentration of CNBr used for activation of the copolymer has a critical bearing on the amount of enzymic activity recovered. A concentration of about 80 mg of CNBr per 100 mg of dry polymer appears to be optimal for several enzymes. If the concentration of the activating agent is too high, there is a drastic decrease in the swelling capacity of the polymer.

When coupling glycine to this copolymer and using 500 mg of CNBr per 100 mg of dry polymer in the activation step, 540 μmoles of glycine are bound per gram of dry polymer.

Covalent Binding to Copolymers of Acrylamide and Acrylic Acid (50/50, w/w; T:C = 10:5)

Example: Immobilization of Trypsin and Ribonuclease A to an Acrylamide-Acrylic Acid Copolymer. Swollen beads (100 mg dry weight) are washed in 0.2 M NaHCO$_3$ after which 2 ml of a solution of 1-ethyl-3(3-dimethylaminopropyl) carbodiimide (EDC)·HCl (Sigma Chemical Company, 2 mg of EDC per milliliter of distilled water) and 2 ml of enzyme solution (2 mg of trypsin or ribonuclease A per milliliter of 0.2 M NaHCO$_3$) are added. (Alternatively, the coupling can be performed at pH 3.5 using diluted HCl.) The coupling proceeds for 2–4 hr at 4°. The beads are then washed in 0.1 M NaHCO$_3$, 0.5 M NaCl, distilled water, 1 mM HCl, and buffer.

When coupling glycine and using equimolar amounts of EDC, the coupling yield is 86% resulting in 120 μmoles of bound glycine per gram of dry polymer.

Covalent Binding of Pullulanase to Bio-Gel CM 100 Using Carbodiimide[11]

One hundred milligrams of Dry Bio-Gel CM-100 (high capacity, 100–200 mesh, Bio-Rad Lab.) is treated with 0.5 M HCl and then thoroughly washed in distilled water. The beads are transferred to the enzyme solution, containing 5.5 mg of pullulanase (*Aerobacter aerogenes*) dissolved in 10 ml of distilled water, which also contains 60 mg of the substrate pullulan. After addition of 50 mg of 1-cyclohexyl-3-(2-morpholinoethyl)carbodiimide metho-p-toluenesulfonate (CMC) (Aldrich Chemical Company), the mixture is gently shaken for 18 hr at 4°. The enzyme beads are then washed in 0.1 M NaHCO$_3$, distilled water 1 mM HCl, and 0.5 M NaCl.

Per gram of dry polymer 19 mg (1.70 units) of pullulanase are bound (coupling yield of 34%), and the specific activity of bound enzyme as percent of free is 43.

The same carrier, Bio-Gel CM-100, was also used when coupling the sequentially working enzymes β-amylase and pullulanase (*Aerobacter aerogenes*) on the same matrix[12] (see Table II).

The EDC-mediated coupling reaction when tested with trypsin was found to be more efficient than CMC.

Examples of "Tailor-Made" Acrylic Polymer Preparations

The use of immobilized enzymes as biological model systems has in recent years attracted great interest (see also Chapters [29], [31], and [63]). In order to obtain a relevant model that will mimic the polyelectrolytic nature of biological membranes charged groups can be introduced in the polymer preparation. Furthermore, the high content of lipids in biological membranes indicates the lipid requirement for enzyme action; for example, a hydrophobic medium has been shown to be of importance in the biosynthesis of phospholipids. This would seem to indicate the need to introduce hydrophobic groups in order to obtain relevant models.

Ideally, to be able to correlate the influence of charge or hydrophobicity exerted by a membrane on enzymic activity, it would be necessary to prepare a series of "polymer membranes" in which only one parameter is changing. This possibility is also of interest from a practical point of view. For instance, when forced to carry out enzymic reactions at low substrate concentrations, it should be advantageous to prepare matrices that "attract" substrate creating a higher substrate concentration around the immobilized enzyme than that found in the bulk solution. This then will lead to a lower $K_{m(app)}$, and thereby increase the efficiency of the immobilized enzyme preparation.

Example. When optimizing the hydrolysis of the positively charged substrate α-*N*-benzoylarginine ethyl ester catalyzed by immobilized trypsin, the negatively charged acrylamide/acrylic acid copolymer (50/50, w/w) is a more suitable carrier than the neutral polyacrylamide, since the former immobilized enzyme derivative shows a decrease in $K_{m(app)}$ for this substrate by a factor of 4 over the latter when assayed at pH 8.3 in 0.1 M Tris-HCl buffer, 0.1 M in NaCl. (The $K_{m(app)}$ values found were 0.58 mM and 2.4 mM, respectively.)

Example. The influence of hydrophobicity was studied varying only one parameter, i.e., the amount of the hydrophobic component, methylacrylate, in an acrylic copolymer.[13] The enzyme chosen was liver alcohol dehydrogenase known to have a broad substrate spectrum for alcohols with different degrees of hydrophobicity. The matrices studied, polyacryl-

[13] A.-C. Johansson and K. Mosbach, *Biochim. Biophys. Acta* **370**, 348 (1974).

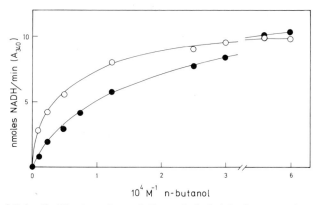

FIG. 2. Michaelis–Menten plots of liver alcohol dehydrogenase immobilized to poly(acrylamide) (●———●) and to copoly(acrylamide/methylacrylate, 75/25) (○———○) using n-butanol as substrate at pH 8.0. The NAD⁺ concentration was 2 mM.

amide and methylacrylate–acrylamide copolymer (25/75, w/w), were prepared by bulk polymerization as described in the methods section and 4 mg of the enzyme was coupled to the glutaraldehyde-treated polymers (300 mg dry weight) as described above.

When the enzyme was bound to the more hydrophobic matrix, methylacrylate–acrylamide copolymer, the $K_{m(app)}$ (40 μM) for the more hydrophobic substrate, n-butanol, decreased by a factor of 4 compared to $K_{m(app)}$ (160 μM) obtained when the enzyme was bound to the hydrophilic polyacrylamide, whereas the $K_{m(app)}$ of ethanol was about the same (0.55 mM) in both cases. Equilibration studies with ¹⁴C-labeled alcohols and the respective polymers showed an enrichment of 6 times of n-butanol to the hydrophobic matrix compared to the hydrophilic one, while such preferential absorption was not found for ethanol.

Figure 2 shows Michaelis–Menten plots for the substrate n-butanol of the enzyme immobilized to both types of copolymers. At a concentration of, for example, 50 μM, the rate of reaction was found to be about 75% higher with the more hydrophobic methylacrylate–acrylamide copolymer compared to that of polyacrylamide. Thus preparations with "substrate attracting" carriers permit enzymic reactions to be carried out more efficiently at a rather low substrate concentration.

Conclusion

The procedures given for the preparation of various acrylic copolymers in bead form are simple to carry out in the laboratory. Usually the simplified procedure (Fig. 1b), where polymerization is carried out in a beaker, should suffice. (In addition, as mentioned, some of the acrylic copolymers are commercially available.) Obviously either bead polymerization process appears to be advantageous compared to the bulk polymerization process, including physical fragmentation, because of the well-defined form of beads per se as well as improved mechanical stability. Further beads are better suited for quantitative measurements. By changing stirring rate or the hydrophobic phase in which the monomers are dispersed, different bead diameters can be prepared (10–500 μm).

In general it can be assumed that the various monomers are present in the formed copolymer roughly in the same ratio as when administered. However, for further characterization infrared spectroscopy is a recommended technique (see also Chapter [28]).

The data for the coupling procedures on the three supports given in Table II for ribonuclease A and trypsin show fairly good binding yields (no optimization has been attempted though) and specific activities of bound versus free, thus providing a general idea of the efficiency of these procedures. In this context, we would like again to draw the attention of the reader who wishes to apply these polymers, for example, in the medical area, to the toxicity of nonpolymerized monomers.

For further reading on the use of CNBr and carbodiimide as coupling aid to acrylic polymers, see in particular Chapter [6].

In addition to the techniques reported here, polyacrylamide beads have been derivatized (see Table I) to yield functional groups, such as amines suitable for subsequent coupling[14,15] (see also Chapter [30]).

It is difficult at the present stage of development to recommend any one particular support or coupling procedure of the three major types described. Not enough enzymes have been tested, and it is likely that even then no such general recommendation can be made. (One recommendation is to use water-soluble carbodiimides when enzyme coupling has to be carried out at low pH.)

However, the general techniques given here should provide the investigator with the means of preparing his own "tailor-made" support material.

[14] J. K. Inman and H. M. Dintzis, *Biochemistry* **8**, 4074 (1969).
[15] J. K. Inman, this series, Vol. 34 [3].

[6] Immobilization of Enzymes on Hydroxyalkyl Methacrylate Gels

By JAROSLAVA TURKOVÁ

Hydroxyalkyl Methacrylate Gels and Their Derivatives

Hydrophilic organic carriers used in chromatography and for the immobilization of enzymes are loosely cross-linked xerogels having a considerable swelling capacity. They are characterized by a high capacity ratio, but the mechanical stability of their particles in the swollen state rapidly decreases with decreasing network density (i.e., with increasing exclusion limit of molecular weight). In addition, gel materials cross-linked by secondary valence forces exhibit a comparatively low stability in concentrated electrolyte solutions and organic solvents.

Macroporous copolymers containing neutral hydroxyl groups and prepared by suspension copolymerization of hydroxyalkyl acrylate and/or hydroxyalkyl methacrylate monomers are a new type of chromatographic material and carriers of biologically active compounds. They are prepared in the form of spherical particles, which in the dry state exhibit measurable porosity and rather high specific surface values. The internal structure of a strongly cross-linked gel has a heterogeneous character; the particle consists of submicroscopic spheroids of roughly the same size, glued together and swelling only insignificantly in contact with the solvent. Pores of the gel are interconnected, and their distribution can be controlled by suitably adjusting the reaction conditions during copolymerization.

The fact that the starting raw materials are commercially available synthetic monomers of comparatively high purity enables a high reproducibility of properties to be achieved. Depending on the route of synthesis, it is possible to obtain spherical particles ranging from 5 μm to 3 mm, films, blocks, and tubes with a varying molecular weight limit of 1000 to about 10 million.

Preparation and Chemical Structure of Gels

Hydroxyalkyl methacrylate gels are formed by radical suspension copolymerization in an aqueous dispersing medium[1] containing a suitable suspension stabilizer. Besides monomeric components, the organic phase also contains mixed solvents, which determine the internal structure of

---

[1] J. Čoupek, M. Křiváková, and S. Pokorný, *J. Polym. Sci., Polym. Symp.* **42**, 185 (1973).

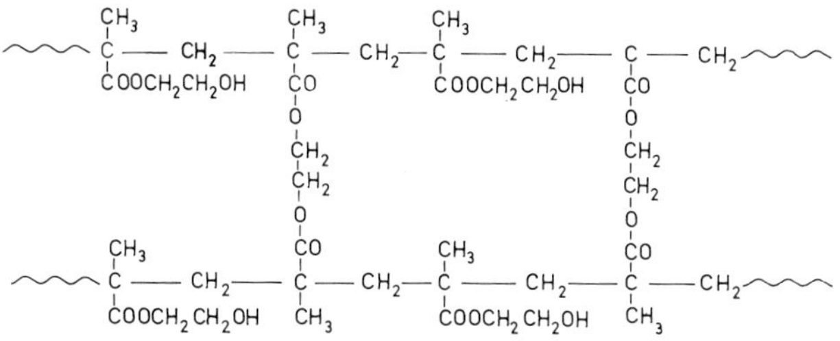

FIG. 1. Schematic formula of hydroxyethyl methacrylate gel.

the gel and hold hydrophilic monomeric fractions within the polymerizing particle during the reaction. The distribution coefficient of the hydrophilic monomer between the aqueous and organic phase approaches zero—the solvents prevent, especially during the initial stages of reaction, its extraction into the aqueous phase.[2]

At the beginning of the reaction, copolymerization proceeds via a homogeneous mechanism; with growing conversion and depending on the thermodynamic quality of the solvent, separation of phases takes place and the reaction proceeds further as a heterogeneous polymerization. Consequently, the resulting porosity of the copolymer depends predominantly on the solvating properties of the mixed solvent with respect to forming copolymer and on the concentration of the cross-linking agent in the polymerization mixture. It also greatly depends on the kinetics of the copolymerization reaction. Particle size distribution is a function of the type and concentration of the stabilizer of the suspension and of the stirring efficiency.

Most widely used are the copolymers of 2-hydroxyethyl methacrylate and ethylene dimethacrylate, produced under the trade name Spheron by Lachema, Brno, Czechoslovakia and available also from Hydron Labs., Inc., New Brunswick, New Jersey and Dr. -Jng. H. Knauer & Co., GmbH, West Berlin. Spheron is a macroporous copolymer of hydroxyalkylmethacrylate and alkylenedimethacrylate; in special cases, a third monomer containing reactive functional groups is present. The pore distribution is controlled by a change in the concentration of cyclohexanol and dodecyl alcohol. The concentration of the cross-linking agent is higher than 20% for the majority of types, and polyvinylpyrrolidone is used as a suspension stabilizer. The chemical structure of the gel is schematically shown in Fig. 1.

The hydrophobic basic skeleton of the carrier is hydrophilized by the presence of hydroxyl groups, the content of which depends on the copolymer composition. Owing to the high degree of cross-linking of submicroscopic particles forming the skeleton of the macroporous gel, it can be

[2] Czech. Patent No. 150, 819.

assumed that some part of the hydroxyl group will be entrapped within the polymer and will not be solvated or made accessible to a chemical reaction with the agents. This assumption was confirmed by a number of reactions. In general, for low-molecular-weight compounds (acetylation, etherification, preparation of alkoxide, etc.) about 80% of all hydroxyl groups present in the gel are available. The amount of hydroxyl groups available for reaction decreases proportionately with increasing molecular weight of reacting compounds. If the reactive functional group is bonded to a spacer, its reactivity is to a certain degree affected by the distance of the group from the gel matrix.[3]

Physical and Chemical Properties

Successful application of the carrier requires a suitable form of the particles, which determines the hydrodynamic conditions of the system. Suspension copolymerization gives rise to perfect spherical particles with a minimum of deformed particles or their fragments. After being packed into columns, they allow work to be conducted at high flow rates with minimum pressure losses.

Since biologically active compounds are immobilized by binding on the inner surface of the pores, the specific surface value will have a decisive influence upon the capacity of the carrier for the given compound besides the pore size distribution of the gel.[4] Desorption of nitrogen has revealed that the size of the specific surface of macroporous hydroxyethyl methacrylates varies from several square meters to several hundred of square meters per gram[5] according to the type of carrier.

An important property of the carrier is the absence of nonspecific sorptions in the chosen system. Since hydroxyalkyl methacrylate gels have a macroporous character, it cannot be expected that there will be no adsorption effects on the large internal surface. However, nonspecific adsorptions are reduced to a minimum by the chemical character of the aliphatic polymeric matrix and the neutral hydrophilic functional group. Where the polymeric matrix could nevertheless adsorb dissolved compounds by hydrophobic interaction, the carrier surface can be further hydrophilized by a suitable chemical reaction, or more-hydrophilic monomers could be used in the copolymerization.[6]

[3] J. Čoupek, J. Turková, O. Valentová, J. Labský, and J. Kálal, Reprints of the 4th Bratislava International Conference on Transformation of Polymers. Vol. 1, P-14 (1975).
[4] J. Štamberg and S. Ševčík, *Collect. Czech. Chem. Commun.* **31**, 1009 (1966).
[5] J. Volková, M. Křiváková, M. Patzelová, and J. Čoupek, *J. Chromatogr.* **76**, 159 (1973).
[6] O. Valentová, J. Turková, R. Lapka, J. Zima, and J. Čoupek, *Biochim. Biophys. Acta* **403**, 192 (1975).

Like other macroporous gels having predominantly the character of an aerogel, hydroxyalkyl methacrylate gels swell only a little in solvents. The degree of swelling depends on the polarity of the carrier and of the solvent used. On changing the ionic strength (within the range of three orders of magnitude) of the aqueous buffer used as the swelling agent, the volume of the gel column of Spheron 300 remains unchanged. The change in pH between pH 3 and 12 also leaves the equilibrium degree of swelling unaffected. Minimum changes in the particle volume in various organic and aqueous solvents (for example: Spheron 300 in H_2O, 4.00 ml swollen gel per gram of dry gel, in benzene 3.80) make possible their exchange without unpleasant effects of dilatation or contraction of the gel column and of the change in the flow properties connected with it, not only for neutral carriers, but also for gels bearing ionogenic functional groups.

One may expect that hydroxyalkyl methacrylate copolymers will exhibit high hydrolytic stability. Their structure resembles esters of pivalic acid, which rank among the most stable esters known. The high hydrolytic stability of polymers of 2-hydroxyethyl methacrylate was proved by Štamberg et al.[4] Highly cross-linked macroporous hydroxyethyl methacrylates were subjected to acid and alkaline hydrolysis. Heating with 20% HCl or with 1 mole Na glycolate did not alter the course of the calibration dependence of the elution volume on the molecular weight of polydextran standards. Even though treatment with concentrated agents at an elevated temperature leads to surface hydrolysis of the hydroxyethyl ester groups,[7] the pore distribution of the carrier remains obviously unaffected by hydrolysis; the flow properties also remain unaltered.

The thermal stability of the gel in the dry state was checked by thermal analysis. At a temperature of about 250° depolymerization takes place, as proved by analysis of the pyrolysis products.

Because they have the character of a synthetic copolymer, hydroxyalkyl methacrylate carriers are not attacked by microorganisms. Their high thermal and hydrolytic stability allows for their sterilization by heat and chemicals.

Chemical Modifications

Like other organic polymeric carriers, chemically modified hydroxyalkyl methacrylate may be obtained in three ways: (1) by polymeranalogous transformation of an already prepared macroporous neutral carrier

[7] Czech. Patent Appl. PV-842-73.

by the reaction of hydroxyl groups with reagents; (2) by copolymerization with monomers containing a reactive or otherwise active functional group in its final form; and (3) by copolymerization with monomers containing a precursor of the functional group followed by a suitable polymer analogous transformation of the precursor.

All three routes yield a wide range of feasible reactions in compliance with the intention of the experimentalist. The individual methods of modification will be discussed briefly.

Polymeranalogous Transformations of Hydroxyalkyl Methacrylate Gels

The hydroxyethyl groups of Spheron gel may be modified by employing virtually all procedures known from the transformations of polysaccharide carriers (Sephadex, Sepharose, cellulose). In contrast to polysaccharides the modification and activation reactions can be conducted under much more drastic reaction conditions—no difficulties are encountered when working in nonaqueous media or using gaseous agents. The gels are activated with cyanogen bromide,[8] diisocyanates, dithioisocyanates, anhydrides and dicarboxylic acid chlorides, phosgene, and bromoacetyl bromide and by a number of other usual procedures. An advantage of using macroporous carriers of the hydroxyalkyl methacrylate type is found in the easy storage of reactive intermediates in the dry state (depending on the character of the active groups, sometimes in an inert atmosphere), so that the activated gels have a real polymeric agent ready for immediate use in attaching a biologically active compound.

The gels are also activated in two steps; in the first reaction the spacer (hexamethylenediamine, ϵ-aminocaproic acid) is attached; the second reaction (for instance with water-soluble carbodiimide) binds the enzyme.[6] Another example involves surface hydrophilization of the gel by transforming it into an alkoxide derivative, followed by a polyaddition reaction with ethylene oxide, yielding surface-bonded polyethyleneglycol chains.

The surface modification of gels is the least demanding procedure from the technological standpoint, because it makes use of a standard basic copolymer and because the reactions employed are generally known from the organic chemistry of low-molecular-weight compounds. In most cases, the basic carrier is so stable that there usually is no need to alter the reaction conditions, which have been described in the literature for low-molecular-weight analogs.

[8] J. Turková, O. Hubálková, M. Křiváková, and J. Čoupek, *Biochim. Biophys. Acta* **322**, 1 (1973).

Copolymerization with Monomers Containing Reactive Groups in the Final Form

The chemical properties of surfaces of macroporous hydroxyalkyl methacrylate gels are modified by their copolymerization with other monomers containing functional groups. It is obvious that the addition of another component to the polymerization mixture leads to a change in conditions under which the heterogeneous phase within the polymerizing particle is separated from the liquid phase. Owing to the different polarity, and thus solvatability, of the third monomer, interactions between the liquid phase and the polymer will also be different; this may lead to changes in the physical structure of the macroporous gel.[9]

Since the copolymerization proceeds practically up to a 100% conversion, the linear dependence of the concentration (capacity) of functional groups on the initial composition of the monomeric phase in the copolymerization of N,N-diethylaminoethyl methacrylate has been confirmed. A number of macroporous ion exchangers suited for separation or ionic binding of proteins were prepared by copolymerization with monomers containing ionogenic functional groups.[10]

Carriers containing reactive ester functional groups were prepared by direct suspension copolymerization[3] of the respective monomers under conditions where they do not hydrolyze to any considerable extent during the synthesis. All monomers used and containing a reactive functional group were crystalline.

Copolymerization with Monomers Containing a Precursor of the Functional Group

The ternary copolymerization of hydrophilic methacrylates, a crosslinking agent, and a monomer containing the precursor of the functional group enabled—as for the synthesis of other carriers of biologically active groups based, e.g., on acrylamide—the modifying procedures to be developed to the largest possible extent. Factors decisive for the internal structure of the gel were determined, then chemical transformations of p-acetaminophenylethoxy methacrylate (APEMA), occurring during its copolymerization and leading to activated carriers, were investigated (Fig. 2). Similarly, copolymers with methacrylanilide yield, after chloromethylation[11] or mercurization,[12,13] reactive gels suitable for binding the spacer or the protein molecule.

[9] M. Křiváková, S. Pokorný, and J. Čoupek, *Angew. Macromol. Chem.*, in press.
[10] Czech. Patent Appl. PV-703-74.
[11] Czech. Patent Appl. PV-7475-72.
[12] Czech. Patent Appl. PV-5742-72.
[13] J. Turková, S. Vavrenová, M. Křiváková, and J. Čoupek, *Biochim. Biophys. Acta* **386**, 503 (1965).

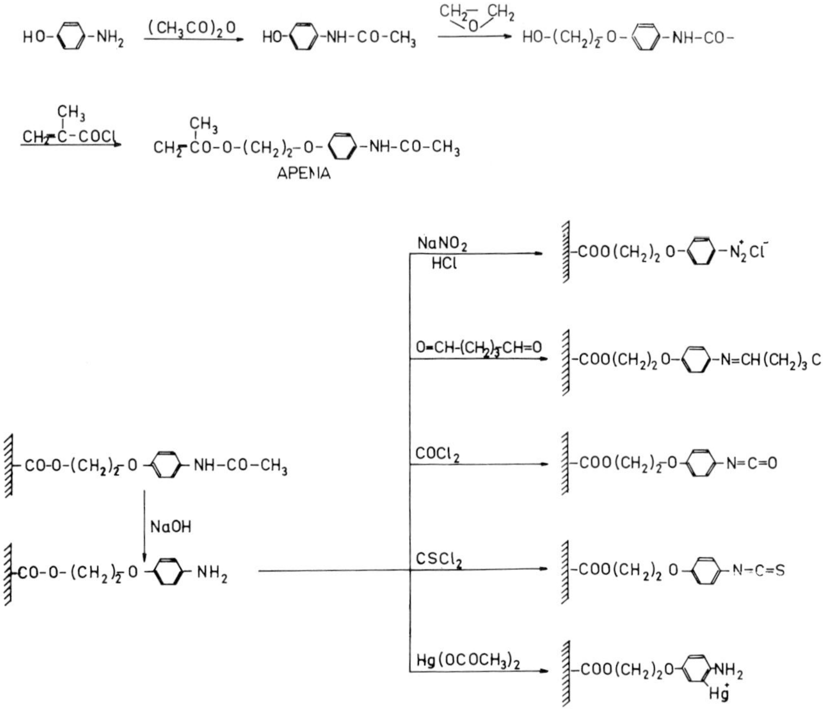

Fig. 2. Schematic representation of *p*-acetaminophenylethoxy methacrylate (APEMA) transformations into reactive intermediates.

Very promising carriers are seen in macroporous copolymers containing glycidyl methacrylate, which after synthesis[13a] can be modified by a number of reactions of the epoxy group with the formation of reactive polymers. (All the polymeric carriers containing reactive functional groups for enzyme immobilization are licensed by National Patent Development Corporation, 375 Park Avenue, New York, N.Y. 10022. Upon request, research quantities can be obtained from Realco/Hydron Laboratories, New Brunswick, N.J.)

Coupling of Chymotrypsin to Hydroxyalkyl Methacrylate Gels after Their Activation by Cyanogen Bromide[8]

Binding of Chymotrypsin (CHT) to Spheron Carriers

The properties of hydroxyalkyl methacrylates allow an activation analogous to that of Sepharose[14] (see also this volume [5]). The activa-

---
[13a] F. Švec, J. Hradil, J. Čoupek, and J. Kálal, *Angew. Macromol. Chem.*, in press.
[14] P. Cuatrecasas, M. Wilchek, and C. B. Anfinsen, *Proc. Natl. Acad. Sci. U.S.A.* **61**, 636 (1968).

tion must be performed in a well-ventilated hood at room temperature. One hundred milliliters of swollen gel are suspended in distilled water (volume ratio 1:1), and 100 ml of freshly prepared 10% cyanogen bromide solution are added gradually with constant stirring. The pH of the suspension is kept at 11 by the addition of 6 N, then 4 N, and finally 2 N NaOH solution. The reaction is completed in 15 min. Activated Spheron is washed quickly with 2 liters of precooled 0.1 M NaHCO$_3$, pH 9.0, on a sintered-glass filter.

The washed activated Spheron (100 ml) is suspended in 100 ml of 0.1 M NaHCO$_3$, pH 9.0, and immediately 10 g of solid CHT are added. The coupling is performed with constant stirring at a lower temperature (4°) for 20 hr. After the coupling is completed, the gel with attached CHT is transferred onto a sintered-glass filter and the solution of unreacted CHT is filtered off. The CHT-gel is suspended in 100 ml of 1 M ethanolamine, pH 9, to block the remaining active groups. After reacting for 2 hr, the excess uncoupled ethanolamine is washed off with a borate buffer solution (0.1 M, pH 8.5) and an acetate buffer solution (0.1 M, pH 4.1), each containing 1 M NaCl. The CHT-gel is then placed in a column and washed alternately with borate and acetate (see above) buffers until the effluent shows no absorbance at 280 nm.

By way of an example, the quantities of CHT, a high-molecular-weight product, and of glycine, a low-molecular-weight compound, bound to seven types of Spheron are shown in Table I. The individual types are listed in the order of decreasing exclusion molecular weights. The quantity of bound CHT is directly proportional to the specific surface of the gels, whereas the quantity of bound glycine indicates only small differences in the number of reactive hydroxyl groups. Table I also gives the proteolytic and esterolytic activities of immobilized chymotrypsins, as well as the pH optima of their esterolytic activities. The amount of attached CHT and glycine is determined by quantitative amino acid analysis according to Spackman, Stein, and Moore[15] after acid hydrolysis of a sample obtained by the procedure described by Axén and Ernback.[16] The proteolytic activity of free and immobilized CHT is determined by a modification of Anson's method using denatured hemoglobin as substrate. The relative proteolytic activity shows the ratio of the activity of bound to free CHT per 1 ml of gel. The esterolytic activity of free and immobilized CHT was determined by the method of Axén and Ernback[16] using acetyl-L-tyrosine ethyl ester as substrate.

[15] D. H. Spackman, W. H. Stein, and S. Moore, *Anal. Chem.* **30**, 1190 (1958).
[16] R. Axén and S. Ernback, *Eur. J. Biochem.* **18**, 351 (1971).

TABLE I
DEPENDENCE OF THE AMOUNT OF CHYMOTRYPSIN (CHT) BOUND TO HYDROXYALKYL METHACRYLATE GELS (SPHERON) ON THEIR SURFACE AREAS[a,b]

Gel Spheron[b]	MW exclusion limit	Specific surface (m^2/ml)	Amount of bound glycine (mg/ml)	Amount of bound CHT (mg/ml)	Proteolytic activity (A_{280}/min·ml)	Relative proteolytic activity (%)	pH optimum of esterase activity	Esterase activity (µmoles/ml per min)	Relative esterase activity (%)
10^5	10^8	0.96	0.5	0.73	—	—	—	—	—
10^3	10^6	5.9	3.1	7.8	1.23	44	9.4	305	29
700	700 000	3.6	2.8	6.7	1.17	49	9.1	392	43
500	500 000	23	2.6	17.1	2.28	37	9.2	810	35
300	300 000	19.5	3.15	17.7	2.8	44	9.1	1320	55
200	200 000	0.6	2.3	6.9	1.33	53	9.0	626	67
100	100 000	0.2	2.6	4.3	0.58	38	9.1	354	61

[a] Proteolytic and esterolytic activities are indicated.
[b] Particle size 40–80 µm.

pH Profiles of Proteolytic and Esterolytic Activity and Stability of Immobilized Preparations

The dependence of the proteolytic activity on pH for free and immobilized CHT is demonstrated in Fig. 3. It can be seen that there is no shift in the pH optimum of the proteolytic activity after the binding of CHT to hydroxyalkyl methacrylate gels. The dependence of the esterase activity on pH for free and immobilized CHT is illustrated in Fig. 4. Similarly to the binding of CHT to Sepharose, here too the pH optimum is shifted to alkaline values by more than 1 unit. The pH optima of esterase activities for the individual gels are given in Table I. Analogous shifts in the pH esterase optima of CHT after binding to Sepharose and Sephadex have been discussed by Axén and Ernback,[16] who demonstrated that this shift is apparent only if caused by the difference in pH on the surface of the matrix and in the surrounding medium. In the case of rapid cleavage of the substrate, the product, acetyltyrosine, accumulates on the internal surface of the gel, causing a decrease in pH as demonstrated by means of an internal indicator.

When CHT-Spheron is suspended in 0.01 M sodium acetate, pH 4.1, and stored at 4° for 25 days, its whole activity is unchanged; after stor-

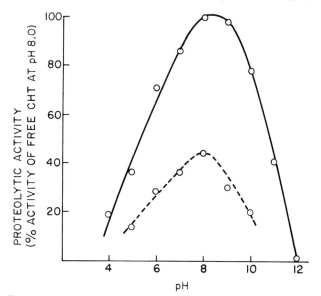

FIG. 3. Dependence of proteolytic activity on pH for chymotrypsin (CHT) and CHT-Spheron 300 with respect to hemoglobin. Amount of fixed enzyme, 70.8 mg per gram of dry gel. Activity of free chymotrypsin, 0.36 unit of A_{280}/min·mg at pH 8. Activity ratio of bound:free enzyme, 44%. ———, Free CHT; ----, CHT-Spheron 300.

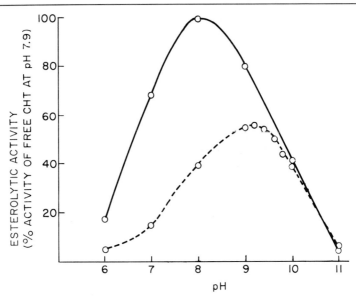

FIG. 4. Dependence of esterolytic activity on pH for chymotrypsin (CHT) and CHT-Spheron 300. Activity was estimated from acetyl-L-tyrosine ethyl ester (0.01 M solution). Amount of fixed enzyme 70.8 mg/g dry gel. Activity of free CHT is 135 μmoles/min·mg at pH 7.9. Activity ratio of bound:free enzyme, 55.3%. ——, free CHT; ----, CHT-Spheron 300.

age at room temperature for the same time its activity drops to 74%. Lyophilization reduces the activity of CHT-Spheron by almost 30%. If the lyophilized sample is stored at 4° or at room temperature, the above value no longer drops within 25 days.

Binding of Pepsin to Amino and Carboxyl Derivatives of Hydroxyalkyl Methacrylate Gels[6]

The coupling of proteins to cyanogen bromide-activated supports occurs in alkaline, or at least in neutral, media. Therefore it cannot be used for the binding of acid proteinases, which are inactivated at pH 5 or higher. Supports with spacers for the attachment of pepsin can, however, be used.

Preparation of NH_2-Spheron and COOH-Spheron

After activation with cyanogen bromide (as described above) 1,6-diaminohexane or ε-aminocaproic acid is bound to Spheron 300 (particle size 120–300 μm) by a modified procedure of Cuatrecasas.[17]

[17] P. Cuatrecasas, J. Biol. Chem. 245, 3059 (1970).

50 ml of activated Spheron is mixed with 50 ml of 10% solution of 1,6-diaminohexane or ε-aminocaproic acid, pH 10. The coupling is performed with constant stirring at room temperature (25°C ± 2°C) for 24 hr, and upon completion the gel with attached 1,6-hexamethylenediamine (NH_2-Spheron) or bound ε-aminocaproic acid (COOH-Spheron) is washed with acetate buffer (0.1 M, pH 4.1) and borate buffer (0.1 M, pH 8.5) solutions, each containing 1 M NaCl, and finally with distilled water. The washing is repeated until the eluate contains no hexamethylenediamine or ε-aminocaproic acid. Seven micromoles of 1,6-hexamethylenediamine bound to 1 ml of Spheron were determined by the Kjeldahl method; 30 μmoles of ε-aminocaproic acid attached to 1 ml of Spheron were determined by titration.

Coupling of Pepsin to NH_2-Spheron and COOH-Spheron Using Soluble Carbodiimide

NH_2-Spheron or COOH-Spheron, 4.8 ml, is suspended in 15 ml of 0.1 M sodium acetate, pH 4.0, containing 300 mg of pepsin. After stirring for 5 min, 100 mg of N-ethyl-N'-(3-dimethylaminopropyl)carbodiimide hydrochloride are added to each suspension and stirring is continued with a magnetic stirrer at 4°. Both samples are then suspended several times in 0.1 M sodium acetate, pH 4.1, containing 1 M sodium chloride and washed in the column alternately with the above buffer and 3 M urea, pH 3.0. The gels are finally washed with 0.01 M sodium acetate, pH 4.1, and with 0.1 M acetic acid and stored in the form of wet cake at 4°. The amount of fixed enzyme is determined from the content of acidic and neutral amino acids in the hydrolyzate after acid hydrolysis of dried gels.[15,16] The proteolytic activity of free and immobilized pepsin is determined by a modification of Anson's method using denatured hemoglobin as substrate. The relative proteolytic activity indicates the ratio of the activity of bound and free pepsin per gram of dry gel. The values obtained are given in Table II.

Dependence of Relative Proteolytic Activity on the Quantity of Bound Pepsin

To study the effect of a relatively hydrophobic medium such as a Spheron gel on the stability of the attached pepsin, we prepared two types of hydroxyalkyl methacrylate gels which contained different amounts of hydroxyl groups. Spheron 1000 BTD contains about 2.5 times more OH groups than Spheron 300. It is therefore more hydrophilic and also has

TABLE II
Properties of Pepsin Immobilized on Hydroxyalkyl Methacrylate Gels

Support	Quantity of pepsin bound (mg/g)	Bed volume (ml/g)	Proteolytic activity of bound pepsin (A_{280}/min·mg bound enzyme)	Proteolytic activity of bound pepsin (A_{280}/min·g conjugate)	Relative proteolytic activity of bound pepsin (%)
NH_2-Spheron P-300	13	3.2	4.02	52.26	92.8
NH_2-Spheron 300	46.8	3.3	2.85	133.5	65.7
COOH-Spheron 300	50.8	3.3	1.97	100	45.3
NH_2-Spheron 1000 BTD	65	4.0	1.64	106.6	37.8

a higher capacity for enzyme immobilization. As expected, a larger quantity of pepsin was bound to this gel. The highly active insoluble derivatives of pepsin thus prepared differed predominantly in the amount of enzyme bound (cf. Table II). Table II also gives data for immobilized pepsin with a lower protein content, obtained under slightly modified coupling conditions. These data demonstrate explicitly how the relative proteolytic activity of immobilized pepsin preparations decreases with increasing quantity of enzyme bound. We observed the same phenomenon previously with immobilized chymotrypsins; it could perhaps be ascribed to sterical hindrances.

pH Profiles for Proteolytic Activity and Stability of Immobilized Pepsins

The dependence of proteolytic activity on pH for free and bound pepsin is shown in Fig. 5. In both cases the pH optimum of bound enzyme is slightly shifted toward the more acidic region.

The stability of immobilized pepsin as a function of pH is shown in Fig. 6. Pepsin is attached to regular amino-Spheron and also to amino-Spheron with an increased content of hydroxyl groups. The higher content of hydroxyl groups does not affect the stability of immobilized pepsin preparations. Both carrier-bound pepsin preparations show practically the same original activity after being kept at 4° for 30 days.

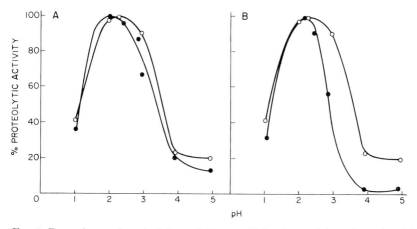

FIG. 5. Dependence of proteolytic activity on pH for free and bound pepsin with respect to hemoglobin. -●-, Pepsin bound to NH_2-Spheron P-300 (A) and NH_2-Spheron 1000 BTD (B); -○-, free pepsin (100% = proteolytic activity at optimum pH).

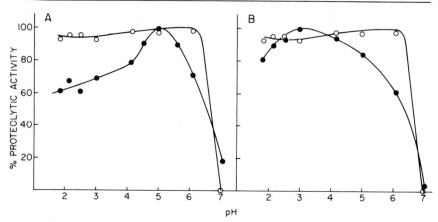

FIG. 6. Dependence of stability of free and bound pepsin on pH. -●-, Pepsin bound to NH₂-Spheron P-300 (A) and NH₂-Spheron 1000 BTD (B); -○-, free pepsin (100% = proteolytic activity at optimum pH).

Binding of Papain to Hydroxyalkyl Methacrylate Gels by Using Active Esters*

Amino acid or protein is bonded to the p-nitrophenol ester derivative of the hydroxyalkyl methacrylate gel (NPAC) according to the reaction

$$\begin{array}{c} CH_3 \\ | \\ -C- \\ | \\ COHN(CH_2)_5CO-O-\!\!\!\!\bigcirc\!\!\!\!-NO_2 \end{array} + NH_2-R$$

$$\downarrow$$

$$\begin{array}{c} CH_3 \\ | \\ -C- \\ | \\ CONH(CH_2)_5CONHR \end{array} + HO-\!\!\!\!\bigcirc\!\!\!\!-NO_2$$

Because of its yellow color, the rate of release of p-nitrophenol can be followed by measuring the absorbancy of the solution at 400 nm. While amino acid is being bound on the NPAC gel, p-nitrophenol ester is hydrolyzed in an aqueous medium by excess OH ions.

Determination of the Capacity of p-Nitrophenol Ester Derivatives of Hydroxyalkyl Methacrylate Gels

The number of active groups is determined by measuring spectrophotometrically the released p-nitrophenol. For each determination, 100 mg of NPAC gel are suspended in 10 ml of dilute ammonia (1:1) and left

*Upon request research quantities can be obtained from Realco/Hydron Laboratories, New Brunswick, N.J.

standing for 1–4 hr with occasional stirring. After the addition of 90 ml of distilled water, the absorbance at 400 nm is measured and the amount of released p-nitrophenol is read from the calibration curve. NPAC gels are prepared having a capacity of 5.8, 18.2, 68.5, and 80 μmoles of releasable p-nitrophenol per gram of dry gel.

pH Dependence of Binding of Amino Acid and Protein on NPAC Gel

For each determination, 250 mg of NPAC gel having a capacity of 68.5 μmoles of releasable p-nitrophenol groups, particle size 100–200 μm, are suspended in a solution of 25 mg of DL-phenylalanine in 40 ml of Britton–Robinson buffers, pH 7, 8, 9, 10, 11, and 12. The rate of release of p-nitrophenol is a function of pH. Practically the same dependence is obtained for simple hydrolyses of the NPAC gel in Britton–Robinson buffers having corresponding pH values. In no case does the presence of phenylalanine markedly accelerate the release of p-nitrophenol, and one cannot therefore refer to a faster course of aminolysis than hydrolysis in this arrangement. The amounts of phenylalanine covalently bound on the gels depend on pH and are determined in the hydrolyzate by amino acid analysis.[15] The maximum amount of phenylalanine is bound at pH 9, 16.1 μmoles/g. Also, the maximum amount of serum albumin coupled to NPAC gel, 0.156 μmole/g, occurs at pH 9.

Amount of Phenylalanine Bonded on NPAC Gel Depending on Concentration

If amino acid or protein is bound in an aqueous medium, the amount of compound bound depends on the ratio of the NH_2 and OH groups.

Binding of Papain to NPAC Gel

To a solution of 50 mg of papain (proteolytic activity 1.01 units A_{280}/min per milligram) in 8 ml of 0.5 M $NaHCO_3$ containing 5 mM mercaptoethanol, pH 9.0, 1 g of dry gel is added having a capacity of 5.8 μmoles of releasable p-nitrophenols, particle size 100–200 μm. The suspension is stirred at 4° for 24 hr. Upon completion, the gel bound with papain is transferred to a column and washed first with the buffer used in binding, then alternately with 0.01 M sodium acetate + 5 mM cysteine, pH 4.1, and 0.1 M phosphate buffer + 0.5 M NaCl + 5 mM cysteine, pH 7.6. The gel is filtered from 0.01 M acetate buffer +5 mM cysteine, pH 4.1, then stored. Owing to the relatively low concentration of papain used in binding, limited by the solubility of the sample used, only 1.5 mg of papain is bound per gram of dry gel; the relative

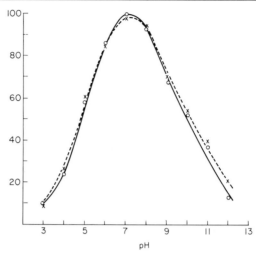

FIG. 7. Dependence of activity of free and bound papain on pH. Carrier p-nitrophenol ester of hydroxymethacrylate gel (NPAC) Spheron-300 NPAC gel. ○――○, free papain; ×---× papain-NPAC gel, 98% relative activity.

proteolytic activity, however, is 99% (0.99 unit A_{280}/min per milligram of bound papain).

The dependence of the proteolytic activity (determined by means of denatured hemoglobin) on pH of immobilized and free papain is shown in Fig. 7. The binding of papain on NPAC gel does not cause any shift in the pH optimum.

Binding of Trypsin and Chymotrypsin on Hydrazide, Diazo, and Anhydride Derivatives of Hydroxyalkyl Methacrylate Gels*

Binding of Trypsin on Hydrazide Derivative of Gel⁷ (HDZ)

One gram of HDZ gel (200–300 μm) is suspended at 0° in 50 ml of 2 N hydrochloric acid. Forty milliliters of an ice-cold solution of 2% sodium nitrite is added with continuous stirring. After stirring at 0° for 15 min, the suspension is transferred on a sintered-glass filter, filtered off, and washed with a cooled 0.1 M phosphate buffer, pH 7.5 (about 1 liter). The activated gel (corresponding to 1 g of the dry gel) is put into a 10-ml solution of 100 mg of trypsin and 0.1 M phosphate buffer, pH 7.5. The capacity of the HDZ gel for binding the low-molecular-weight substances is 38 μmoles per gram of the dry gel (determined on the basis of the bound phenylalanine), and the capacity for binding the

---

* Upon request research quantities can be obtained from Realco/Hydron Laboratories, New Brunswick, N.J.

high molecular substances is 4 μmoles per gram of the dry gel (determined on the basis of the bound trypsin). However, the resulting preparation has a very low relative proteolytic activity.

Binding of Chymotrypsin on Diazotized Gels of the APEMA-Spheron Type[18]

Ten milliliters of diazotized gel (100–200 μm) are washed several times on a sintered-glass filter with 0.2 M phosphate buffer, pH 7.8. After filtration, the gel is transferred into 25 ml of a solution of 2.5 g of chymotrypsin in 0.2 M phosphate buffer, pH 7.8, and the suspension is stirred at 4° for 24 hr. After filtering the chymotrypsin solution, the gel is washed several times with a phosphate buffer, pH 7.8. To remove unreacted diazo groups the gel is transferred into 500 ml of a solution of 0.4 g of β-naphthol in 0.2 M phosphate buffer, pH 7.8, and the suspension is stirred at 4° for another 24 hr. During the binding of β-naphthol the gel gradually becomes red. To remove free diazo groups any low-molecular-weight agent able to form bonds with the diazo group, e.g., bishydroxyethylaniline, can be used. After completion of the reaction the gel is washed on a sintered-glass filter alternately with 2 liters of 0.1 M borate buffer, pH 8.5, and 0.1 M acetate buffer, pH 4.1, both with 1 M NaCl added. The gel containing 10% APEMA monomer has 20.5 mg of chymotrypsin per gram of dry gel with a proteolytic activity of 0.74 unit A_{280}/min per milligram of bonded chymotrypsin. The gel with 20% monomer contained 25 mg of chymotrypsin per gram of dry gel and has a proteolytic activity of 0.46 unit of A_{280}/min per milligram of bonded chymotrypsin.

Chymotrypsin on an Anhydride Derivative of the Gel[7]

One gram of dry gel is suspended in 40 ml of a solution of 200 mg of chymotrypsin in 0.2 M Tris-HCl buffer, pH 7.5. The binding is completed overnight with continuous stirring at 4°. On completion, the gel is washed alternately with a borate buffer (0.1 M, pH 8.5) and acetate buffer (0.1 M, pH 4.1), both with 1 M NaCl added, until the eluate does not exhibit any absorbance at 280 nm as the pH of the buffer is changed. Per gram of gel, 12.2 mg of chymotrypsin is bound, with a proteolytic activity of 0.12 unit of A_{280}/min per milligram of bonded chymotrypsin.

Acknowledgment

The author wishes to thank Dr. Jiří Čoupek and to acknowledge his contributions to the preparation, chemical structure, and modification of hydroxyalkyl methacrylate gels, without which this chapter would not be complete.

[18] Czech. Patent Appl. PV-5741-72.

[7] Enzymes Covalently Bound to Polyacrylic and Polymethacrylic Copolymers

By R. Epton, Barbara L. Hibbert, and T. H. Thomas

Support matrices derived from polyacrylamide, polyacrylic acid, and polymethacrylic acid comprise the largest class of wholly synthetic polymers used for enzyme immobilization.

The first purpose-synthesized supports of this type were macroreticular copolymers of either methacrylic acid, methacrylic acid m-fluoranilide and divinyl benzene,[1,2] or methacrylic acid, 3- or 4-fluorostyrenes, and divinyl benzene.[3] These copolymers, prepared by Manecke et al., were activated for enzyme coupling by nitration. Subsequently, a macroreticular copolymer of methacrylic acid, m-isothiocyanatostyrene, and divinyl benzene was prepared, which was effective in enzyme binding without preactivation.[4,5] See also this volume [8].

The copolymers described above all possess substantial hydrophilicity and react well with enzymes. However, the aromatic residues incorporated in these supports must necessarily constitute sites of localized hydrophobicity. It is recognized that hydrophobic, and sometimes ionic, sites on supports may lead to enzyme instability.[6,7]

Other methacrylate-based supports have been described which possess substantially more hydrophobic character. Examples are the aromatic homopolymer of vanillin methacrylate (Vanacryl) and its copolymer with allylic alcohol (Vanacryl P),[8,9] the aliphatic homopolymers of 2-iodoethyl methacrylate and 4-iodobutyl methacrylate (Poliodal-2 and Poliodal-4)[10–12] and copolymers of these iodinated monomers with methyl methacrylate (Copoliodals).[11,13]

[1] G. Manecke and G. Gunzel, *Makromol. Chem.* **51**, 199 (1962).
[2] G. Manecke, *Pure Appl. Chem.* **4**, 507 (1962).
[3] G. Manecke and H. Forster, *Makromol. Chem.* **91**, 136 (1966).
[4] G. Manecke and G. Gunzel, *Naturwissenschaften* **54**, 531 (1967).
[5] G. Manecke, G. Gunzel, and H. J. Forster, *J. Polym. Sci. Polym. Symp.* **30**, 607 (1970).
[6] I. H. Silman and E. Katchalski, *Annu. Rev. Biochem.* **35**, 873 (1966).
[7] S. A. Barker, P. J. Somers, R. Epton, and J. V. McLaren, *Carbohydr. Res.* **14**, 287 (1970).
[8] E. Brown and A. Raçois, *Tetrahedron Lett.* **50**, 5077 (1972).
[9] E. Brown and R. Joyeau, *Polymer* **15**, 546 (1974).
[10] E. Brown and A. Raçois, *Bull. Soc. Chim. Fr.* **1971**, 4351 (1971).
[11] E. Brown and A. Raçois, *Bull. Soc. Chim. Fr.* **1971**, 4357 (1971).
[12] E. Brown, A. Raçois, R. Joyeau, A. Bonté, and J. Rioual, *C. R. Acad. Sci., Ser. C* **273**, 668 (1971).
[13] E. Brown and A. Raçois, *Tetrahedron Lett.* **1971**, 1047 (1971).

Promising methacrylate esters for enzyme immobilization are the cross-linked poly (hydroxyethyl) methacrylates (Spheron) described by Coupek et al.[14] These supports are intended primarily as matrices for permeation chromatography. Since they are intensely hydrophilic and may be prepared in macroreticular form, they fulfill, coincidentally, the requirements for a good enzyme support. It has been demonstrated that cross-linked copolymers of acrylamide and hydroxyethyl methacrylate ester may be activated for enzyme binding by treatment with cyanogen bromide.[15] This activation and coupling procedure has been used to synthesize enzyme conjugates of Spheron.[16] These derivatives are described in this volume [6].

Polyacrylic acid may be activated for enzyme coupling by the classical reaction sequence of esterification, hydrazinolysis, and nitrous acid treatment.[17] A simpler approach is by activation with N-ethyl-5-phenylisoxazolium 3'-sulfonate (Woodward's Reagent K).[18]

An important type of enzyme conjugate, in which the synthetic polymer may be regarded as a polyacrylic acid comprised of head-to-head and tail-to-tail linked monomer units, is that obtained by reaction of the enzyme with ethylene/maleic anhydride (EMA) copolymers in the presence of 1,6-diaminohexane as cross-linking agent. A detailed account of these derivatives has been given in a previous volume of this series.[19] Related conjugates have been prepared by reaction of enzymes with polymethacrylic acid anhydride and 1,6-diaminohexane.[20]

Insofar as it is extremely hydrophilic and nonionic, cross-linked polyacrylamide is an ideal enzyme carrier. Polyacrylamide supports with reactive functional groups may be prepared either by direct derivatization of the preformed polyacrylamide matrix or by copolymerization of acrylamide with suitable reactive monomers (see also this volume [5]). Adopting the former approach, Inman and Dintzis[21] have subjected the commercial cross-linked polyacrylamide, Bio-Gel P, to hydrazinolysis in order to convert some of the primary amide residues to acid hydrazide groups. After nitrous acid treatment, the polymer was used directly for enzyme coupling.

[14] J. Čoupek, L. Drahoslav, S. Pokorny, and M. Krivakova, Ger. Offen 2,157,627 (1972).
[15] K. Mosbach, Acta Chem. Scand. 24, 2084 (1970).
[16] J. Turková, O. Hubalková, M. Krivaková, and J. Čoupek, Biochim. Biophys. Acta 322, 1 (1973).
[17] B. G. Erlanger, M. F. Isambert, and A. M. Michelson, Biochem. Biophys. Res. Commun. 40, 70 (1970).
[18] B. P. Patel, D. V. Lopiekes, S. P. Brown, and S. Price, Biopolymers 5, 577 (1967).
[19] L. Goldstein, this series, Vol. 19, p. 935 (1970).
[20] A. Conte and K. Lehmann, Hoppe-Seyler's Z. Physiol. Chem. 352, 533 (1971).
[21] J. K. Inman and H. M. Dintzis, Biochemistry 8, 4074 (1969).

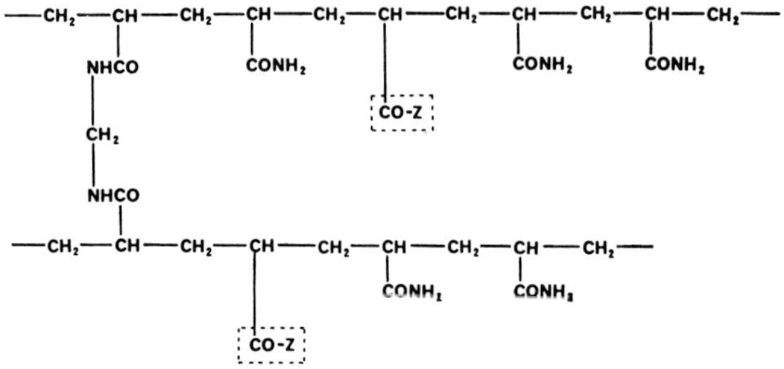

(a) ENZACRYL AA   Z = $-NHC_6H_4NH_2$

(b) ENZACRYL AH   Z = $-NHNH_2$

(c) ENZACRYL Polythiol Z = $-NHCH(COOH)CH_2SH$

(d) ENZACRYL Polythiolactone Z = $-NHCH-CO$
$\qquad\qquad\qquad\qquad\qquad\qquad\quad |\quad\ \ |$
$\qquad\qquad\qquad\qquad\qquad\qquad CH_2-S$

FIG. 1. Structural features of polyacrylamide-based Enzacryls.

It was also demonstrated that the reaction of Bio-Gel P with ethylenediamine effected insertion of aminoethyl substituents. These were condensed with p-nitrobenzoyl chloride. Subsequent reduction of the nitro groups gave aryl amino residues, which were activated for protein coupling by diazotization. Preformed polyacrylamides have also been prepared for enzyme coupling by glutaraldehyde treatment.[22]

Commercial Bio-Gel P is intended primarily as a matrix for aqueous permeation chromatography. Consequently, several bead-polymerized Bio-Gel networks, each differing in molecular weight exclusion limit, are available. These networks form an excellent starting point for the synthesis of enzyme supports. However, the derivatization procedure must be controlled carefully to avoid major changes in porosity.

Copolymerization of acrylamide with suitably reactive monomers provides for the synthesis of polyacrylamides with a wide range of groups functionally active in enzyme binding. Several supports of this type have been commercialized under the trade mark Enzacryl® (Fig. 1).

Enzacryl AA is prepared by copolymerization of acrylamide, p-nitroacrylanilide, and N,N-methylene*di*acrylamide as cross-linker. Reduction of the copolymer with titanous ions results in aryl amino groups.[7] Enzacryl AA has been activated for enzyme coupling both by diazotization and treatment with thiophosgene (Fig. 2). Enzacryl AH is derived

[22] P. O. Weston and S. Avrameas, *Biochem. Biophys. Res. Commun.* **45**, 1574 (1971).

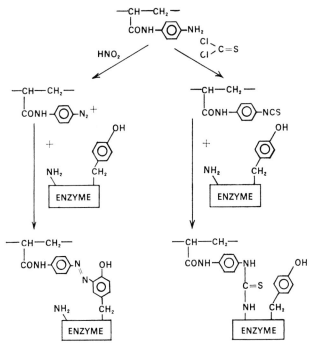

Fig. 2. Enzyme coupling with Enzacryl AA.

from a copolymer of acrylamide and N-acryloyl-N'-t-butyloxycarbonyl hydrazine. The t-butyloxycarbonyl groups are removed by controlled acid hydrolysis to expose acid hydrazide groups.[7,23] A similar support to Enzacryl AH has been prepared by hydrazine treatment of a cross-linked copolymer of acrylamide and methyl acrylate.[24]

Enzacryl polythiol is prepared by acid hydrolysis of a cross-linked copolymer of acrylamide and N-acryloyl-4-carboxymethyl-2,2-dimethylthiazolidine. Carbodiimine treatment of Enzacryl polythiol yields Enzacryl polythiolactone.[25] Enzacryl polythiol may be used for the reversible binding of enzymes. Since most enzymes contain either no free thiol groups or a very small number, prior enrichment by treatment with N-acetylhomocysteine thiolactone is desirable. Subsequently, oxidative coupling with Enzacryl polythiol is brought about with potassium ferricyanide. The enzyme may be detached by washing with a solution of

[23] R. Epton, J. V. McLaren, and T. H. Thomas, Biochim. Biophys. Acta 328, 418 (1973).
[24] Y. Ohno and M. A. Stahmann, Macromolecules 4, 350 (1971).
[25] Koch-Light Technical Bulletin, "Enzacryl Polythiol and Enzacryl Polythiolactone" (1971).

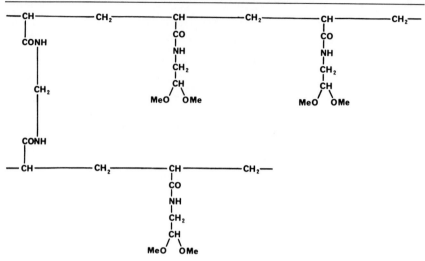

FIG. 3. Structural features of Enzacryl polyacetal.

ethane thiol or cysteine. Enzacryl polythiolactone is effective in coupling enzymes without further activation.

A number of N-substituted polyacrylamides are useful for enzyme immobilization. For example, titanium complexes of poly(4-aminosalicylic acid) and poly (5-aminosalicylic acid) are effective in binding enzymes via chelation.[26]

Another useful N-substituted polyacrylamide is cross-linked poly-(acryloylaminoacetaldehyde dimethylacetal) (Fig. 3).[27,28,29] This carrier, commercialized as Enzacryl polyacetal, is activated for enzyme coupling by acid hydrolysis of some of the dimethylacetal residues. Immobilization takes place within an intensely hydrophilic, gelatinous microenvironment consisting of acetal, hemiacetal, and aldehydrol functions (Fig. 4). Recently two other related gel supports have been described. One is a cross-linked polymer of acryloyl-N,N-bis(2,2-dimethoxyethyl)amine (Fig. 5), and the other is a cross-linked copolymer of this monomer and acryloyl morpholine (Fig. 6).[30] Both supports are activated for enzyme coupling by controlled acid hydrolysis in the presence of tartaric dihydrazide followed by treatment with nitrous acid (Fig. 7).

The typical experimental procedures that follow have all been devised in the author's laboratory. They have been selected to illustrate the syn-

[26] S. A. Barker, J. F. Kennedy, and J. Epton, U.S. Patent No. 3,794,563 (1974).
[27] R. Epton, J. V. McLaren, and T. H. Thomas, *Biochem. J.* **123**, 21P (1971).
[28] R. Epton, J. V. McLaren, and T. H. Thomas, *Carbohydr. Res.* **22**, 301 (1972).
[29] R. Epton, J. V. McLaren, and T. H. Thomas, *Polymer* **15**, 564 (1974).
[30] R. Epton, B. L. Hibbert, and G. Marr, *Polymer* **16**, 314 (1975).

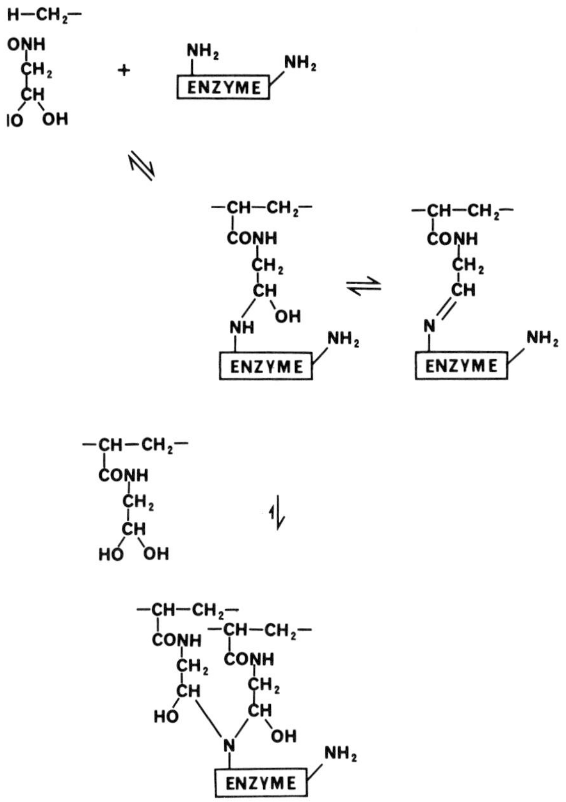

FIG. 4. Possible enzyme immobilization reactions with Enzacryl polyaldehyde. Reproduced from R. Epton, J. V. McLaren, and T. H. Thomas, *Polymer* 15, 564 (1974) by permission of the copyright owners, IPC Business Press Ltd.

thesis of enzyme supports of this class and the subsequent preparation of immobilized enzymes using these and commercial carriers.

Polyacrylamide Immobilized Thermolysin[23]

Thermolysin ex *Bacillus thermoproteolyticus*, 3× crystallized and salt free, is commercially available (Calbiochem Inc.). Thermolysin is assayed spectrophotometrically at 30°, pH 7.25, using either furacryloyl-glycyl-L-leucinamide (FAGLA)[31,32] (10 ml; 2 mM FAGLA in 50 mM Tris-HCl buffer, 10 mM in CaCl$_2$) or casein[33] (10 ml; 1% in 50 mM

[31] J. Feder, *Biochem. Biophys. Res. Commun.* 32, 325 (1968).
[32] J. Feder and J. M. Schuck, *Biotechnol. Bioeng.* 12, 313 (1970).
[33] M. Kunitz, *J. Gen. Physiol.* 30, 291 (1947).

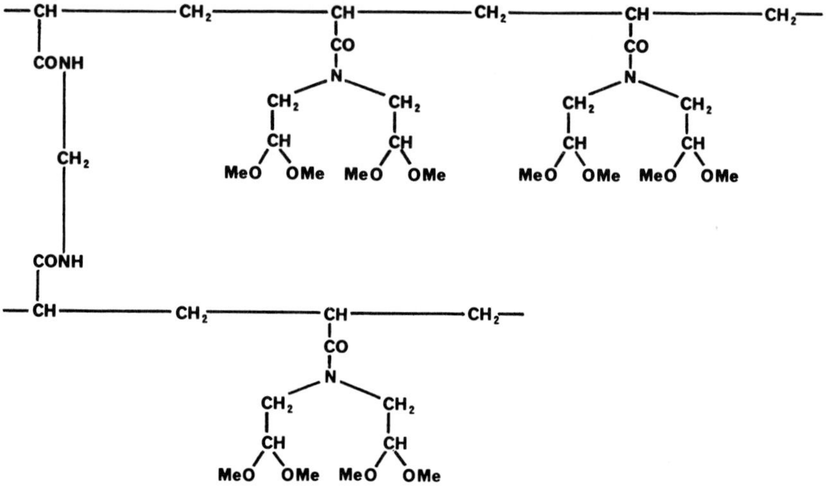

FIG. 5. Cross-linked poly[acryloyl-N,N-bis(2,2-dimethoxyethyl)amine]. Reproduced from R. Epton, B. L. Hibbert, and G. Marr, *Polymer* **16**, 314 (1975) by permission of the copyright owners, IPC Business Press Ltd.

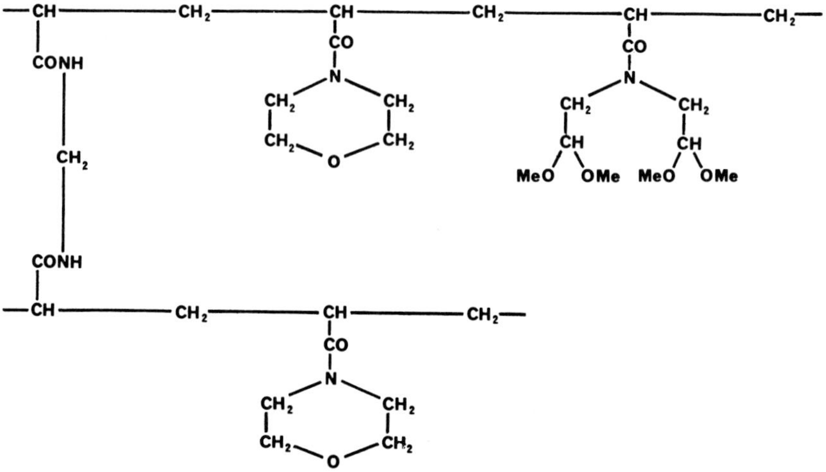

FIG. 6. Cross-linked poly[acryloylmorpholine/acryloyl-N,N-bis(2,2-dimethoxyethyl)amine]. Reproduced from R. Epton, B. L. Hibbert, and G. Marr, *Polymer* **16**, 314 (1975) by permission of the copyright owners, IPC Business Press Ltd.

Tris-HCl buffer, 10 mM in CaCl$_2$) as substrate. In the case of FAGLA, thermolytic activity is recorded directly by following the change in extinction at 345 nm in incubation mixtures. With casein digests, samples (1 ml) are withdrawn at intervals, diluted with 5% (w/v) aqueous trichloroacetic acid (1.5 ml), and centrifuged; the change in extinction at

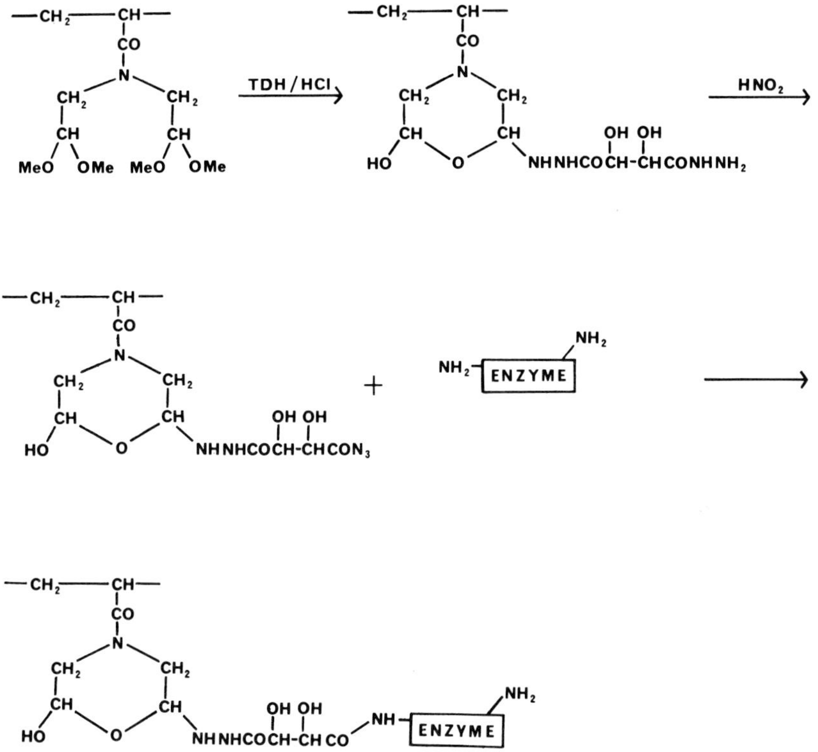

FIG. 7. Derivatization of polyacetals with tartaric dihydrazide (TDH) and enzyme immobilization. Reproduced from R. Epton, B. L. Hibbert, and G. Marr, *Polymer* **16**, 314 (1975) by permission of the copyright owners, IPC Business Press Ltd.

280 nm is recorded. It is important to note that the substrate solution must be prepared by dissolving casein in the Tris-HCl buffer and allowing the preparation to age overnight before adding buffered $CaCl_2$ solution. Prepared in this way, casein solutions are stable for the duration of a working day.

Synthesis of Aminoaryl Derivatized Copolyacrylamide

Reagents. Acrylamide, N,N'-methylene*di*acrylamide, and acidic titanous chloride solution (15% w/v) are all available commercially from several sources. The monomer 1-acryloylamino-2-(4-nitrobenzoylamino)-ethane[23] also is available commercially (Koch-Light Ltd.).

Procedure. A mixture of acrylamide (0.639 g, 9 mmoles), N,N'-methylene*di*acrylamide (0.154 g, 1 mmole), and 1-acryloylamino-2-(4-nitrobenzoylamino)ethane (0.263 g, 1 mmole) is dissolved in ethylene glycol

(5 ml). Nitrogen is continuously bubbled through the solution while it is irradiated from above with an 80-W ultraviolet lamp at a distance of 8 cm over 15 hr. The soft gel produced is broken up by grinding lightly in a mortar, equilibrated with ethanol, collected by filtration, and dried under reduced pressure. On reswelling in aqueous media, this copolymer gives a gel possessing excellent mechanical stability.

Selective reduction of the aryl nitro groups in this copolymer is effected by an adaptation of the procedure of Kolthoff and Robinson.[34] Sodium citrate solution (20% w/v) is added dropwise and with magnetic stirring to an acidic solution of titanous chloride (15% w/v) until the pH of the mixture reaches 5.0. The copolymer (2.5 g) is suspended in this solution (250 ml) with magnetic stirring and left for 3 days at ambient temperature in the absence of light. The resulting aminoaryl-derivatized copolyacrylamide is washed exhaustively with distilled water, equilibrated with ethanol, and dried under reduced pressure.

Characterization of Aminoaryl Copolyacrylamide. The aryl amino and carboxylic acid content of the copolymer is determined by radioestimation and titrimetric procedures.[7] First, the copolymer is washed exhaustively with aqueous ammonia (4 M) followed by distilled water and dried. This procedure ensures that the copolymer is in the base form.

A sample (200 mg) of copolymer is weighed into a stoppered test tube, and an aliquot (15 ml) of 50 mM hydrochloric acid containing radioactive chloride (^{36}Cl, 5×10^{-2} μCi/ml) is added. The suspension is magnetically stirred for 2 hr, then centrifuged, and the decrease in radioactivity of the supernatant is determined. This enables the hydrochloride insolubilized as amine hydrochloride, and hence the aminoaryl groups present in the copolymer, to be estimated. By titrating the supernatant solution with standard sodium hydroxide solution, the total amount of hydrogen chloride neutralized is determined. This is equivalent to the ammonium carboxylate and aminoaryl groups in the copolymer.

The arylamino content of the copolymer is found to be 0.57 mmole/g, and the carboxylate content is 0.11 mmole/g.

Synthesis of Acyl Hydrazide Derivatized Copolyacrylamide

This copolymer is available commercially (Koch-Light Ltd.) as Enzacryl AH.

Reagent. The monomer N-acryloyl-N'-t-butoxycarbonyl hydrazide[7] is available from the same source.

Procedure. A mixture of acrylamide (1.05 g, 15 mmoles), N,N'-methy-

---

[34] I. M. Kolthoff and C. Robinson, *Recl. Trav. Chim. Pays-Bas* **45**, 109 (1926).

lene*di*acrylamide (0.28 g, 1.8 mmoles), and N-acryloyl-N'-t-butoxycarbonyl hydrazide (0.34 g, 2.7 mmoles) is dissolved in ethylene glycol (15 ml) and irradiated from above with an 80-W ultraviolet lamp at a distance of 8 cm over 15 hr. The resulting gel is broken up by grinding lightly in a mortar, washed exhaustively with water, equilibrated with ethanol, collected by filtration, and dried under reduced pressure.

Selective hydrolysis of the t-butoxycarbonyl protecting groups is effected by stirring with an excess of 2 M hydrochloric acid at ambient temperature for 2 days.

Characterization of Acyl Hydrazide Derivatized Copolyacrylamide. Th acid hydrazide and carboxylate content of the copolymer are determined by the combined radioestimation and titrimetric procedure.

The acyl hydrazide content is found to be 0.58 mmole/g, and the carboxylate content 0.42 mmole/g.

NOTE: Inman and Dintzis[21] have described a titrimetric procedure for estimation of acyl hydrazide and carboxylate groups in functionalized polyacrylamides. Equivalent amounts of acetylated and succinylated carrier are each titrated with 1 M NaOH. The titer for the acetylated sample gives the carboxylate content of the carrier, and the difference in titers gives the original degree of acyl hydrazide substitution.

Immobilization of Thermolysin by Diazo Coupling

Carrier. Reduced copolymer of acrylamide, N,N'-methylene*di*acrylamide, and 1-acryloylamino-2-(4-nitrobenzoylamino)ethane (9:1:1) (aminoaryl derivatized copolyacrylamide).

Procedure. Two samples (100 mg) of aminoaryl derivatized copolyacrylamide are suspended in aliquots of 2 M HCl (5 ml) at 0° and ice-cold 10% (w/v) aqueous sodium nitrite solution (2 ml) added to each. Diazotization is allowed to proceed for 30 minutes with magnetic stirring, after which the suspensions are centrifuged and the supernatants are discarded. One of the diazotized samples is equilibrated with 50 mM phosphate buffer (pH 7.5) at 0° and the other with 50 mM carbonate-bicarbonate buffer (pH 10.0) at 0°. These are the buffers appropriate for the enzyme coupling.

Immediately prior to enzyme binding the samples are centrifuged, the supernatants are discarded, and a 0.25-ml aliquot of a solution of thermolysin (20 mg/ml) in the appropriate coupling buffer is added. The reaction is allowed to proceed at 0° for 4 hr with gentle magnetic stirring. Residual diazo groups are destroyed by the addition of 5 ml of a 0.01% (w/v) phenol solution in 10% (w/v) aqueous sodium acetate and stirring for 10 min at 25°.

The polyacrylamide–thermolysin conjugates are then washed alternately with 25 mM phosphate buffer (pH 7.5, 15 ml) and 1% casein solution (15 ml) in the same buffer over 10 cycles. Each washing is of 20 min duration at 0° with vigorous magnetic stirring. After washing, each conjugate is stored in suspension in 50 mM Tris-HCl buffer (pH 7.25, 10 ml) containing 10 mM CaCl$_2$. The conjugate derived by coupling at pH 7.5 is designated copolyacrylamide–thermolysin conjugate A, and that derived by coupling at pH 10.0 is designated copolyacrylamide–thermolysin conjugate B.

Determination of Thermolytic Activity. The copolyacrylamide–thermolysin conjugates A and B are assayed spectrophotometrically using buffered FAGLA and casein solutions of composition similar to those used for the native, soluble enzyme. All assays are carried out with constant-speed magnetic stirring at 30°. In the case of FAGLA assays, it is necessary to withdraw aliquots from the incubation mixture and centrifuge briefly before measuring the changes in extinction.

Determination of Bound Protein. Samples (50 mg) of the copolyacrylamide–thermolysin conjugates A and B are hydrolyzed by heating with aliquots (0.5 ml) of 6 M HCl at 110° for 18 hr (see also this volume [28]). After cooling, each hydrolyzate is centrifuged and the supernatant is decanted from the solid residue. The residue is washed with distilled water and the washings combined with the hydrolyzate. The washings are evaporated to dryness by rotary evaporation, maintained for 18 hr *in vacuo* over a mixture of NaOH and CaCl$_2$, and then reconstituted with distilled water (0.2 ml). Hydrolyzates of the native and arylamino derivatized copolyacrylamide are treated similarly.

The reconstituted hydrolyzates are subjected to ascending paper chromatography with an organic phase derived from a mixture of butanol–acetic acid–water (4:1:5). The detailed chromatographic procedure, including the ninhydrin estimation of the amino acid components, is carried out as described by Kay *et al.*[35] Enzyme hydrolyzates give rise to two bands of ninhydrin-positive material (R_f 0.50 and 0.35), which are clearly separated from bands due to the carrier.[36] These bands are extracted into a mixture of ethanol–water (3:1), and the absorbance (575 nm) of the resulting solution is determined. The enzyme concentration in samples prior to hydrolysis is calculated using a standard graph obtained by chromatographic assay of hydrolyzates of known amounts of the native enzyme.

---

[35] R. E. Kay, D. L. Harris, and C. Entenmann, *Arch. Biochem. Biophys.* **63**, 14 (1956).
[36] R. Epton and T. H. Thomas, "An Introduction to Water-Insoluble Enzymes." Broglia Press, London, 1970.

Immobilization of Thermolysin by Acyl Azide Coupling

Carrier. Partially hydrolyzed copolymer of acrylamide, N,N'-methylenediacrylamide and N-acryloyl-N'-t-butoxycarbonyl hydrazide (9:1:1) (acyl hydrazide derivatized copolyacrylamide or Enzacryl AH).

Procedure. A sample (100 mg) of acyl hydrazide derivatized copolyacrylamide (Enzacryl AH) is suspended in 2 M HCl (5 ml) at 0°, and ice-cold 2% (w/v) aqueous sodium nitrite solution (5 ml) is added. Activation is allowed to proceed for 15 min with magnetic stirring; the suspension is then centrifuged, and the supernatant is discarded. The subsequent thermolysin coupling and the washing procedure are carried out as outlined for the preparation of the copolyacrylamide–thermolysin conjugates A and B except that 25 mM phosphate buffer (pH 8.0) is used for enzyme coupling and the treatment with phenol solution is omitted. The thermolysin derivative is designated copolyacrylamide–thermolysin conjugate C.

Determination of Thermolytic Activity. This is carried out as detailed for the copolyacrylamide–thermolysin conjugates A and B.

Determination of Bound Protein. This is carried out as detailed for the copolyacrylamide–thermolysin conjugates A and B.

Properties of Copolyacrylamide–Thermolysin Conjugates

Kinetic Behavior. The retention of FAGLA hydrolase activity at maximum FAGLA solubility (2 mM) together with K_m and V for the copolyacrylamide–thermolysin conjugates A, B, and C are presented in Table I. K_m and V are both reduced dramatically relative to values determined for the soluble enzyme. The magnitude of the practical activity retention observed at 2 mM FAGLA concentration is primarily a result of this decrease in K_m, since the soluble enzyme is operating well below saturation kinetics at this concentration.

Substantial decrease in K_m on immobilization is usually observed when polyionic carriers of opposite charge to the substrate are involved.[19] This is not the case in the present studies. Decrease in both V and K_m suggest that immobilization leads to changes in protein structure resulting in diminished turnover number and enhanced affinity of the enzyme for its substrate.

Steric and diffusional limitations on the approach of substrate molecules to the active site of the enzyme are accepted to be of importance in governing the activity retained toward a macromolecular substrate such as casein.[6] However, in the case of copolyacrylamide–thermolysin conjugates, it is observed that the percentage retention of caseinolytic activity parallels the percentage retention of V relative to the soluble

TABLE I
Protein Coupled and Activity Parameters for Polyacrylamide Immobilized Thermolysin

Enzyme preparation	Method of preparation	Bound protein (mg/g of carrier)	FAGLA hydrolase activity parameters					Caseinolytic activity	
			Enzyme units/mg of protein[a]	Practical activity retention[b] (%)	K_m (mM)	V (μmoles/ min/mg enzyme)	V as % V for soluble enzyme	Enzyme unit/mg of protein[c]	Practical activity retention[b] (%)
Conjugate A	Diazo coupling, pH 7.5	23.4	7.9	15	0.32	7.8	0.4	2	0.1
Conjugate B	Diazo coupling, pH 10.0	13.4	13.4	31	0.80	24.4	1.6	43	1.8
Conjugate C	Acyl azide coupling	12.1	47.4	110	2.38	139	8.5	186	7.6
Native enzyme	—	—	43.3	100	64	1604	100	2400	—

[a] One unit of furacryloyglycyl-L-leucinamide (FAGLA) hydrolase activity is that which on incubation with 2 mM FAGLA solution promotes hydrolysis of 1 μmole of substrate per minute at 30°.
[b] Use of the term "practical activity retention" is intended to imply that the values are meaningful only at the substrate concentration given in the enzyme unit definition. This arises owing to the large changes in K_m and V. In most cases practical activity retention is measured at substrate concentrations well below that required for saturation kinetics.
[c] One caseinolytic unit is that which, on incubation with 1 ml of 1% casein solution for 1 hr at 30° followed by precipitation with 1.5 ml of trichloroacetic acid solution, leads to an increase in extinction (280 nm) of 1 unit.

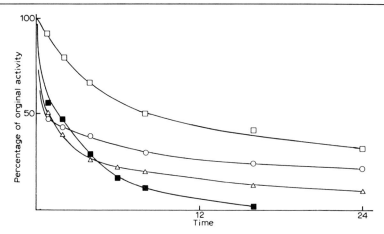

FIG. 8. Heat denaturation of soluble and immobilized forms of thermolysin at 80° in 50 mM Tris-HCl buffer (pH 7.25): conjugate A (○), conjugate B (△), conjugate C (□), soluble enzyme (■). Reproduced from R. Epton, J. V. McLaren, and T. H. Thomas, *Biochim. Biophys. Acta* **328**, 418 (1973) by permission of the copyright owners, Elsevier, Amsterdam.

enzyme (measured with FAGLA). In this case, therefore, it appears that there is little difference in the retention of activity toward low-molecular-weight and macromolecular substrates. Steric and diffusional limitations seem to be relatively unimportant.

Stability. The comparative stabilities of the immobilized thermolysin derivatives to heat denaturation are shown in Fig. 8. Acyl hydrazide derivatized copolyacrylamide, which is the most hydrophilic carrier, is considerably superior in conferring thermal stability on the bound enzyme. The derivative prepared by diazo coupling at pH 10 is the least stable to heat denaturation. However, it has by far the best storage properties, retaining 96% of its original activity after 6 months at 0°–2°. The conjugates prepared by diazo coupling at pH 7.5 and by acyl azide coupling retain 30% and 56%, respectively, of their initial activities when stored under identical conditions.

Katchalski *et al.*[37] have suggested that decrease in the thermal stability of an enzyme on immobilization may be related to a decrease in the probability of returning to an enzymically active conformation after thermal perturbation. If the thermolysin molecules, immobilized by diazo coupling at pH 10, were very tightly bound, then at storage temperature this might aid retention of the active conformation. However, if this conformation were disturbed, as in the heat denaturation process, the enzyme

[37] E. Katchalski, I. Silman, and R. Goldman, *Ad. Enzymol.* **34**, 445 (1971).

molecules would be less likely to regain activity, owing to the constraint imposed upon them by immobilization.

Enzacryl Immobilized α- and β-Amylase

α-Amylase (EC 3.2.1.1), ex *Bacillus subtilus*, 4× crystallized, is commercially available (Sigma Chemical Co. Ltd.). The enzyme is assayed at 20°, pH 6.9, against soluble starch as substrate (10 ml; 1% starch, 0.02 M in phosphate buffer). To follow the course of digestion, the concentration of reducing sugar in samples of the incubation mixture is determined spectrophotometrically at 520 nm, using dinitrosalicylate reagent.[38,39]

β-Amylase (EC 3.2.1.2) ex sweet potato, crystalline, is commercially available (Sigma Chemical Co. Ltd.). The enzyme is assayed at 20°, pH 4.8, against soluble starch as substrate (10 ml; 1% starch, 0.02 M in acetate buffer), using the same spectrophotometric procedure as that outlined for α-amylase.

Amylase Immobilization with Enzacryl AA

Carrier. Enzacryl AA, a reduced copolymer of acrylamide, N,N'-methylene*di*acrylamide, and 4-nitroacrylanilide (8:1:1), aminoaryl content 0.3–0.5 mmole/g, is available commercially (Koch-Light Ltd.).

Diazo Coupling Procedure. Two samples of Enzacryl AA (100 mg) are each suspended in 2 M hydrochloric acid (5 ml) at 0°. Aliquots (4 ml) of an ice-cold aqueous 2% (w/v) solution of sodium nitrite are added, and diazotization is allowed to proceed with magnetic stirring. After 15 min the samples of diazotized Enzacryl AA are each equilibrated with the 75 mM phosphate buffer (pH 7.6) to be used for enzyme coupling.

Prior to enzyme binding the diazotized samples are centrifuged, and the excess buffer is discarded. Solutions of α-amylase (5 mg/ml) and β-amylase (5 mg/ml) in the coupling buffer are prepared, and a 0.5-ml aliquot of each is added to a sample of diazotized Enzacryl AA. Coupling is allowed to proceed for 48 hr at 0° with gentle magnetic stirring.

The diazo Enzacryl/α-amylase conjugate is washed alternately with 20 mM phosphate buffer (pH 6.9, 15 ml) and a 0.5 M NaCl solution (15 ml) in the same buffer over 5 cycles. Each washing is of 20 min duration with vigorous magnetic stirring. The conjugate is stored suspended in phosphate buffer (10 ml). The diazo Enzacryl/β-amylase con-

---

[38] P. Bernfeld, this series, Vol. 1, p. 149 (1955).
[39] S. A. Barker, P. J. Somers, and R. Epton, *Carbohydr. Res.* **8**, 491 (1968).

jugate is washed and stored similarly except that 20 mM acetate buffer is employed.

Isothiocyanato Coupling Procedure. Two samples of Enzacryl AA (100 mg) are stirred vigorously with 3.5 M phosphate buffer (pH 6.9, 1 ml) at ambient temperature. Aliquots (0.2 ml) of a 10% (v/v) solution of thiophosgene in carbon tetrachloride are added. Some 20 min later, further aliquots (0.2 ml) of thiophosgene solution are added, and the stirring is continued for another 20 min. The isothiocyanato derivatives of Enzacryl AA are then washed once with acetone (15 ml) and twice with 0.5 M sodium hydrogen carbonate (15 ml) and equilibrated with the 50 mM borate buffer (pH 8.6) in preparation for enzyme coupling.

With the exception of the coupling reaction, which is performed in borate buffer, the subsequent experimental procedure is similar to that described for the diazo-coupled Enzacryl derivatives. The resulting conjugates are designated isothiocyanato Enzacryl/α-amylase and isothiocyanato Enzacryl/β-amylase.

Determination of Amylolytic Activity. The activities of the conjugates are measured in a similar manner to the corresponding soluble enzymes. Assays are performed with magnetic stirring, the rate of which is kept constant for all determinations.

Determination of Bound Protein. This is carried out as detailed for the copolyacrylamide–thermolysin conjugates.

Amylase Immobilization with Enzacryl AH

Carrier. Enzacryl AH, a partially hydrolyzed copolymer of acrylamide, N,N'-methylenediacrylamide and N-acryloyl-N'-t-butyloxycarbonyl hydrazide (8:1:1), acyl hydrazide content 0.3–0.5 mmole/g, is available commercially (Koch-Light Ltd.)

Procedure. Two samples of Enzacryl AH (100 mg) are treated with nitrous acid as described for the diazotization of Enzacryl AA. The acid azide derivatives of Enzacryl AH are equilibrated rapidly with 50 mM borate buffer (pH 8.6) and centrifuged; the supernatants are discarded. Solutions of α-amylase (5 mg/ml) and β-amylase (5 mg/ml) in the same buffer are prepared, and a 0.5-ml aliquot of each is added to a sample of the acid azide derivative. The subsequent experimental procedure is again similar to that described for the diazo Enzacryl derivatives. The conjugates are designated acyl azide Enzacryl/α-amylase and acyl azide Enzacryl/β-amylase.

Determination of Amylolytic Activity. The enzyme activities of the conjugates are measured in a similar manner to those of the Enzacryl AA–amylase conjugates.

Determination of Bound Protein. This is carried out as detailed for the copolyacrylamide–thermolysin conjugates.

Properties of Enzacryl–Amylase Conjugates

Activity. The amounts of enzyme immobilized, together with the percentages of the free solution activities retained on binding in the various amylase conjugates, are summarized in Table II. Enzacryl AA is a more effective carrier for protein binding than Enzacryl AH. This is possibly because, in the latter, the functional group chemically active in binding is close to the hydrocarbon backbone. In terms of activity retention, the immobilization of α-amylase is more efficient than that of β-amylase. However, much more protein is bound in the case of β-amylase. This enzyme seems much more susceptible to inactivation on binding.

Stability. Immobilization results in a dramatic improvement in heat stability in the case of α-amylase (Fig. 9). The more hydrophilic carrier, Enzacryl AH, confers greatest heat stability. Immobilization of β-amylase results in destabilization. After 3 months of storage in buffer suspension at the optimum pH for enzyme activity, the activity retained by acyl azide Enzacryl/α-amylase (85%) is superior to that of diazo Enzacryl/α-amylase (73%) and isothiocyanato Enzacryl/α-amylase (67%). These results also reflect the greater hydrophilicity of Enzacryl

TABLE II
Protein Coupled and Enzyme Activity of Enzacryl Immobilized
α- and β-Amylase

Enzyme	Enzyme units[a]/mg of free protein	Carrier activated	Functional group active in protein binding	Bound protein (mg/g of carrier)	Enzyme units/mg of bound protein	Enzyme activity (%) retained on coupling
α-Amylase	1118	Enzacryl AA	Diazo	10.6	68.7	6.1
			Isothiocyanato	6.2	106.8	9.5
		Enzacryl AH	Acid azide	2.2	178.3	16.0
β-Amylase	431.2	Enzacryl AA	Diazo	32	6.4	1.5
			Isothiocyanato	26	3.4	0.8
		Enzacryl AH	Acid azide	9	Inactive	Inactive

[a] One amylase unit is that which liberates reducing sugar equivalent to 1 μmole of maltose at 20° at optimum pH. Reproduced from S. A. Barker, P. J. Somers, R. Epton, and J. V. McLaren, *Carbohydr. Res.* **14**, 287 (1970) by permission of the copyright owner, Elsevier, Amsterdam.

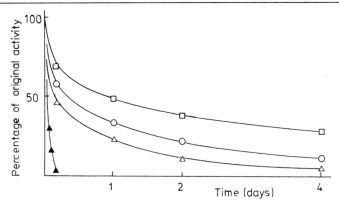

FIG. 9. Heat denaturation of soluble and immobilized forms of α-amylase at 45° in 20 mM phosphate buffer (pH 6.9): diazo Enzacryl/α-amylase (△), isothiocyanato Enzacryl/α-amylase (○), acid azide Enzacryl/α-amylase (□), soluble enzyme (▲). Reproduced from *Carbohydr. Res.* **14**, 287 (1970), by permission of copyright owners, Elsevier, Amsterdam.

AH relative to Enzacryl AA. Activity retention on storage of diazo Enzacryl/β-amylase (73%) and isothiocyanato Enzacryl/β-amylase (51%) are similar to those of α-amylase conjugates of Enzacryl AA.

Poly[acryloyl-*N,N*-bis(2,2-dimethoxyethyl)amine] and Poly[acryloyl morpholine] Immobilized β-Glucosidase[30]

β-D-Glucosidase (EC 3.2.1.21), ex. sweet almonds, salt free, and lyophilized is commercially available (Koch-Light Ltd.). β-Glucosidase is assayed spectrophotometrically at 420 nm at 30°, pH 5.5, using *p*-nitrophenyl β-D-glucopyranoside (10 ml, 25 mM in 50 mM sodium acetate buffer) as substrate.[40] To follow the course of reaction, aliquots (0.25 ml) of the incubation mixture are treated with 200 mM aqueous sodium carbonate (0.7 ml), and the *p*-nitrophenol concentration is determined spectrophotometrically at 420 nm. The substrate solution is freshly prepared prior to each determination.

Synthesis of Cross-Linked Poly[acryloyl-*N,N*-bis(2,2-dimethoxyethyl)-amine] (Molar Ratio 8:1)

Reagents. Acryloyl chloride and *N,N*-bis(2,2-dimethoxyethyl)amine are both available commercially (Koch-Light Ltd.).

[40] K. M. L. Agrawal and O. P. Bahl, *J. Biol. Chem.* **243**, 98 (1968).

Procedure. The monomer, acryloyl-*N,N*-bis(2,2-dimethoxyethyl)-amine[30] is prepared by adding acryloyl chloride (4.52 g, 50 mmoles) in dry, peroxide-free ether (100 ml) to a solution of *N,N*-bis(2,2-dimethoxyethyl)amine (19.3 g, 100 mmoles) in ether (100 ml) at such a rate that the temperature does not rise above 5°. The reaction mixture is stirred for a further 0.5 hr, then the precipitate is removed by filtration and washed with dry peroxide-free ether (50 ml). On evaporation of the combined ether solutions at room temperature, pure acryloyl-*N,N*-bis(2,2-dimethoxyethyl)amine (11.9 g, 97%) is obtained as a mobile oil.

To prepare the copolymer, acryloyl-*N,N*-bis(2,2-dimethoxyethyl)-amine (4.9 g, 20 mmoles) is dissolved with the cross-linker *N,N'*-methylene*di*acrylamide (0.38 g, 2.5 mmoles) in a deoxygenated mixture of ethanol/water (80% v/v, 13 ml). An aliquot (2 ml) of a solution of ammonium persulfate (0.5% w/v) in the same solvent is added. Gelation rapidly occurs, and the reaction is left to proceed to completion overnight. The gel is lightly ground in a mortar, then the particles are pushed twice through a wire-mesh sieve (75 μm) and stored in suspension in ethanol/water (80%, v/v). The weight of cross-linked polymer obtained is estimated by drying out a small amount of the hydrated gel. The yield of cross-linked poly[acryloyl-*N,N*-bis(2,2-dimethoxyethyl)amine] is 2.5 g (40%).

Synthesis of Cross-Linked Poly[acryloyl morpholine/acryloyl-*N,N*-bis(2,2-dimethoxyethyl)amine] (Molar Ratio 4:1)

Reagents. Acryloyl morpholine, sorbitan trioleate (Span 85), and polyoxyethylene(20)sorbitan trioleate (Tween 85) are commercially available (Koch-Light Ltd.).

Procedure. This copolymer is prepared by aqueous suspension polymerization in liquid paraffin. A 500-ml glass polymerization flask, fitted with a mechanically operated semicircular paddle, is batched with liquid paraffin (100 ml) (ρ^{20} 0.85 g/ml, η^{20} 3.5–4.0 m sec/m^2) and a surfactant mixture, consisting of sorbitan trioleate (7 ml) and polyoxyethylene (20) sorbitan trioleate (1 ml). Nitrogen is blown through the paraffin for 1 hr, after which a nitrogen atmosphere is maintained in the flask throughout the polymerization. An aliquot (1 ml) of a saturated aqueous solution of potassium persulfate is added quickly to a magnetically stirring, deoxygenated solution of acryloyl morpholine (2.27 g, 16 mmoles), acryloyl-*N,N*-bis(2,2-dimethoxyethyl)amine (0.99 g, 4 mmoles), and *N,N'*-methylene*di*acrylamide (0.030 g, 0.2 mmole) in distilled water (32 ml). This solution is poured immediately into the stirring paraffin. The rate of stirring is adjusted so that the size of the droplets formed is maintained approximately in the range of 10–40 µm. Polymerization occurs within 1 hr and is assumed to be complete after 5 hr. The gel beads are allowed to settle overnight. Most of the paraffin is then decanted, and the beads are washed thoroughly with petroleum spirit 40°–60° (3 × 500 ml), acetone (3 × 500 ml), and finally distilled water (3 × 500 ml). The beads are stored suspended in water. The estimated total by weight of the copolymer matrix within the gel is 3.1 g (97%).

Characterization. Acid hydrolysis of cross-linked poly[acryloyl-N,N-bis(2,2-dimethoxyethyl)amine] leads to the generation of pyranose rings of the bis hemiacetal type, which condense spontaneously with one another to give an alternative system of acetal linkages. This has been confirmed by a proton magnetic resonance study of the deuterochloric acid-catalyzed deuterolysis of phenylacetyl N,N-bis(2,2-dimethoxyethyl)amine in D_2O/CD_3OD as solvent.[30]

Immobilization of β-Glucosidase by Acyl Azide Coupling

Carriers. Poly[acryloyl-N,N-bis(2,2-dimethoxyethyl)amine] (molar ratio 8:1) and poly[acryloyl morpholine/acryloyl-N,N-bis(2,2-dimethoxyethyl)amine] (molar ratio 4:1).

Procedure. Samples of each hydrated carrier are suspended in separate aliquots of a solution of tartaric acid dihydrazide (5%, w/v) in 1 M HCl for 18 hr at room temperature. Each gel is washed exhaustively with distilled water, until tartaric acid dihydrazide is no longer detectable in the washings when tested with trinitrobenzene sulfonic acid reagent.[21]

Samples of both functionalized gels (each equivalent to 100 mg dry weight of copolymer) are equilibrated with 2 M HCl and centrifuged; the supernatant is discarded. Each gel is resuspended in 2 M HCl (5 ml) with magnetic stirring, and cooled to 0°, and an aliquot (4 ml) of sodium nitrite solution (2% w/v) is added. After 15 min the gels are washed quickly four times with 50 mM sodium acetate buffer (pH 5.5, 10 ml) at 0°. The final washings are discarded, then an aliquot (0.5 ml) of β-glucosidase solution (10 mg/ml) in 50 mM acetate buffer is added to each gel. Reaction is allowed to proceed with magnetic stirring overnight at 0°–4°. The resulting conjugates are centrifuged and subjected to five washing cycles, each consisting of one wash with 50 mM acetate buffer (pH 5.5, 15 ml) and one wash with 1 mM sucrose in 1 M NaCl solution (15 ml). Each washing is of 20 min duration with vigorous magnetic stirring. The immobilized enzyme derivatives, which are designated polyacetal/β-glucosidase and poly(acryloylmorpholine)/β-glucosidase are stored at 0°–2° in suspension in acetate buffer (10 ml).

Determination of β-Glucosidase Activity. The immobilized β-glucosidase derivatives are assayed by adapting the method described for the native, soluble enzyme. Constant-speed magnetic stirring is used throughout. Aliquots of the enzyme digests are removed at regular intervals, diluted with 200 mM aqueous sodium carbonate solution, and centrifuged to remove the immobilized enzyme before the change in extinction is determined at 420 nm.

Determination of Bound Protein. The detailed procedure is similar to that described for the copolyacrylamide–thermolysin conjugates. The protein immobilized in the polyacetal/β-glucosidase conjugate is 12 mg/g; and in the poly(acryloylmorpholine)/β-glucosidase conjugate, 42 mg/g. These values reflect the relative swelling and gelation properties of the copolymer networks from which the conjugates are derived.

Properties of β-Glucosidase Conjugates

Activity. The overall activities of the polyacetal/β-glucosidase and poly(acryloylmorpholine)/β-glucosidase conjugates are 31 and 28 units per gram. This corresponds to respective activity retentions on immobilization of 12.8 and 3.2% of that of the soluble enzyme. The activity of soluble β-glucosidase is 20.5 units/mg. One unit of β-glucosidase activity is taken to be that which releases 1 μmole of p-nitrophenol in 1 min under the standard conditions of the assay.

Stability. Immobilization leads to improved heat stability. Whereas the poly(acryloylmorpholine)/β-glucosidase conjugate is the more stable initially to heat denaturation, it is less stable, over an extended period, than the polyacetal/β-glucosidase conjugate. The latter conjugate exhibits the better activity retention to long-term storage. At 0°–2° in aqueous buffer suspension, the respective activity retentions of poly-(acryloylmorpholine)/β-glucosidase and polyacetal/β-glucosidase are 28 and 79%.

Enzacryl Polyacetal Immobilized Urease[29]

Urease (EC 3.5.1.5), ex jack bean meal and lyophilized (162 units/g) is commercially available (P-L Biochemicals, Inc.). Urease is assayed at 30°, pH 7.5, using urea as substrate (10 ml; 150 mM urea in 100 mM phosphate buffer, containing 5 mM EDTA). To follow the course of hydrolysis, the concentration of ammonia in samples of the incubation mixture is determined spectrophotometrically at 480 nm, using Nessler's reagent.[41] One urease unit is defined as that which produces 1 μmole of ammonia in 1 min at 30°.

[41] K. J. Laidler and J. P. Hoare, *J. Am. Chem. Soc.* **71**, 2599 (1949).

Immobilization of Urease by Aldehydrol Coupling

Carrier. Enzacryl polyacetal, a particulate cross-linked copolymer of acryloylaminoacetaldehydedimethylacetal and N,N'-methylene*di*acrylamide (molar ratio 8:1), is available commercially in aqueous xerogel form (Koch-Light Ltd.).

Procedure. Enzacryl polyacetal is activated for enzyme binding by shaking with aqueous 0.25 M HCl at 30° for 48 hr. The activated xerogel product, Enzacryl polyaldehyde A, has an aldehydrol content of 5.4 mmoles/g (71% theoretical). This is determined by reaction of a sample with buffered hydroxylamine hydrochloride followed by titration.[42] The water content of the Enzacryl polyaldehyde A particles is 10.9 ml/g.

A sample of Enzacryl polyaldehyde A is restructured, without significant diminution in aldehydrol content, by equilibrating the aqueous xerogel particles successively with ethanol and ether, removing the solvent component of the xerogel produced *in vacuo*, and rehydrating. The restructured xerogel so formed, Enzacryl polyaldehyde B, has a water content of 5.0 ml/g.

Samples of hydrated Enzacryl polyaldehydes A and B (each equivalent to 150 mg dry weight of copolymer) are allowed to equilibrate with 25 mM phosphate buffer (15 ml, pH 7.5) for 1 hr. After centrifugation, the excess buffer is discarded, and an aliquot (0.5 ml) of urease solution (20 mg/ml) in 25 mM phosphate buffer (pH 7.5) is added. The suspensions are stirred magnetically at 0°–5° for 18 hr. The resulting urease derivatives are washed six times with 25 mM phosphate buffer (pH 7.5, 20 ml) containing 5 mM EDTA and finally stored in the same buffer at 0°–5°.

Determination of Urease Activity. The Enzacryl polyaldehyde/urease conjugates are assayed in similar manner to the soluble enzyme. After nesslerization of samples withdrawn from incubation mixtures, it is necessary to centrifuge in order to remove traces of conjugate before measuring the extinction at 480 nm.

Determination of Bound Protein. The protein contents of the immobilized urease derivatives are determined as described for the copolyacrylamide–thermolysin derivatives. The urease binding capacities of Enzacryl polyaldehyde A and Enzacryl polyaldehyde B are, respectively, 20 mg/g and 8 mg/g. These figures reflect the relative swelling of the two carriers.

Properties of Polyacetal Immobilized Urease

Kinetic Behavior. The apparent K_m of urease does not change significantly on immobilization of the enzyme (8.5 ± 1.5 mM), but the value

[42] W. M. D. Bryant and D. M. Smith, *J. Am. Chem. Soc.* **57**, 57 (1935).

for V is altered significantly. In the case of the more expanded conjugate, Enzacryl polyaldehyde A/urease, V is 60% of that for the native enzyme, whereas for Enzacryl polyaldehyde B/urease the corresponding figure is 27%.

The most important change in the kinetic behavior of urease following immobilization is in its substrate concentration activity profile. The more expanded conjugate, polyaldehyde A/urease, attains maximum activity at lower substrate concentration than the soluble enzyme and is also less sensitive to substrate inhibition at higher urea concentrations. The reverse is found to be the case for polyaldehyde B/urease, which exhibits a much narrower substrate concentration/activity profile.

In the case of the more expanded conjugate, the larger cavities between the solvated polymer chains at the surface of the gel possibly favor penetration by the enzyme molecules. This may well facilitate multiple linkage between the carrier and each enzyme molecule, thereby providing more general support for the latter and helping to preserve tertiary structure. This multiple linkage might result in a more stable enzyme conjugate. However, the additional constraint would impede allosteric processes. In this connection, it is interesting to note that Laidler and Hoare[41] have suggested that an allosteric mechanism of substrate inhibition is a valid alternative to their more widely accepted hypothesis of urea molecules competing directly for a water site in the active center of the enzyme.

Stability. Once the immobilized enzymes are allowed to become active by contact with urea solution, removal of the latter results in the enzyme activity being permanently impaired. Both conjugates lose approximately 18% activity if they are washed with buffer after being incubated with substrate. Enzacryl polyaldehyde B/urease is damaged merely by diminution of substrate concentration. Enzacryl polyaldehyde A/urease is relatively insensitive to such changes and can be repeatedly transferred between stirred reactors with full activity retention. It is relevant that Enzacryl polyaldehyde A/urease retains full activity at lower substrate concentration.

Enzacryl polyaldehyde A/urease is also superior to Enzacryl polyaldehyde B/urease in its activity retention during continuous use in packed columns. After 20 days of continuous perfusion with substrate solution containing 5 mM EDTA, the respective activity retentions are 20 and 10%. The respective activity retention after 6 months of storage at 0°–2° in buffer suspension are 48 and 17%. Both conjugates are less stable to heat denaturation than the native enzyme in solution. Thus the conjugates are found to retain approximately 15% of their original activity after 24 hr at 60° in buffer suspension whereas, maintained under similar conditions, urease in solution retains 24% activity.

Concluding Remarks

The preparation of the aminoaryl and acyl hydrazide derivatized copolyacrylamides have been described in detail because they represent important representative enzyme carriers of the acrylate type. The methodology of their synthesis and characterization is of general application. They are similar to the commercially available Enzacryl AA and Enzacryl AH also described. Carriers of this type are reliable where total and irreversible covalent immobilization is all important. Often they are effective in conferring improved heat and storage stability on the immobilized enzyme. Poly[acryloyl-N,N-bis(2,2-dimethoxyethyl)amine] and poly[acryloylmorpholine] carriers are described because they provide novel microenvironments for enzyme immobilization. Enzacryl polyacetal and the derived Enzacryl polyaldehydes offer the possibility of multiple enzyme to carrier linkages. Conjugates of this carrier and enzymes whose activity is mediated by allosteric effects are of particular interest. Enzacryl polyaldehyde conjugates suffer from the disadvantage that certain bound enzymes may undergo resolubilization in contact with macromolecular substrates.

Conjugates of thermolysin and urease illustrate the potential of covalent immobilization procedures to bring about real changes in biocatalytic activity.

[8] Immobilization of Enzymes on Neutral and Ionic Carriers

By G. MANECKE and J. SCHLÜNSEN

The immobilization of enzymes by covalent binding on suitable water-insoluble polymeric carriers allows manifold variations. Besides their insolubility, the carriers should have good mechanical and chemical stability and they should bind the substrate and products to only a small extent. Good hydrophilic properties are obtained from neutral, anionic, or cationic groups. We have found the following carriers to be effective in the immobilization of enzymes: carriers with anionic matrices obtained from nitrated copolymers of methacrylic acid with methacrylic acid-3-fluoroanilide[1,2] or fluorostyrenes,[3] copolymers of methacrylic acid and acrylic acid with 3- or 4-isothiocyanatostyrene,[4-6] or copolymers of N-vinylpyrrolidone with maleic acid anhydride[7]; carriers with neutral matrices obtained by chemical modification of poly(vinyl alcohol),[8] poly-(allyl alcohol),[9] or vinylether copolymers.[10]

The effects of composition and of the chemical and physical properties of the carrier on the system can only be mentioned here.[11,12]

The binding ability of the carrier for enzymes is influenced by several parameters. Higher binding abilities and a higher relative activity (activity of the bound enzyme compared with the activity of an equal amount of native enzyme) can be achieved by using smaller grain sizes and macroreticular carriers. These provide better accessibility of the matrix for the enzymes and the substrates. Also the number of reactive groups influences the binding ability of the carrier. An increased number of reactive groups binds larger amounts of proteins provided the parallel decrease in the number the hydrophilic groups in the carrier does not cause extensive deterioration in the swellability. Too high concentrations of reactive groups may cause multiple binding of the enzyme molecules. Increased cross-linking of the polymeric carrier decreases the swellability, and thus the amount of bound enzyme. The swellability of the ionic matrices is pH dependent, so pH also affects the binding ability. The pH-dependent charges of the carriers (in the case of the polyionic carriers) also have an effect on binding. The binding ability for enzymes depends also on the concentration of the enzyme solution during the immobilization procedure. Higher enzyme concentrations lead to greater amounts of bound enzyme. However, using lower concentrations improves the activity yield; i.e., the enzyme activity is better utilized. With the reactive carriers mentioned above, alcohol dehydrogenase, cellulase, chymotrypsin, diastase, glucose oxidase, invertase, papain, alkaline phosphatase, and urease were immobilized with varied success.

[1] G. Manecke and S. Singer, *Makromol. Chem.* **39**, 13 (1960).
[2] G. Manecke and G. Günzel, *Makromol. Chem.* **51**, 199 (1962).
[3] G. Manecke and H.-J. Förster, *Makromol. Chem.* **91**, 136 (1966).
[4] G. Manecke and G. Günzel, *Naturwissenschaften* **54**, 531 (1967).
[5] G. Manecke, G. Günzel, and H.-J. Förster, *J. Polym. Sci., Polym. Symp.* **30**, 607 (1970).
[6] G. Manecke and J. Schlünsen, publication in preparation.
[7] G. Manecke and R. Korenzecher, publication in preparation.
[8] G. Manecke and H.-G. Vogt, *Makromol. Chem.* **177**, (1966) in press.
[9] G. Manecke and R. Pohl, *Makromol. Chem.* (1976) in press.
[10] G. Manecke and D. Polakowski, publication in preparation.
[11] G. Manecke, *Chimia* **28**, 467 (1974).
[12] G. Manecke, *Proc. Int. Symp. Macromolecules,* Rio de Janeiro, July 1974, p. 397. Amsterdam, 1975, (Elsevier).

The binding procedures and investigations with immobilized papain are given as an example. Most of the products show good storage stability in aqueous solution or in the freeze-dried state at 4°.

Reagents

Papain (Merck/Darmstadt and Boehringer/Mannheim)
N-α-Benzoyl-L-arginine amide·HCl·H$_2$O (BAA, Merck)
N-α-Benzoyl-L-arginine ethyl ester·HCl (BAEE, Merck)
Cysteine·HCl (Merck)
Ethylenediaminetetraacetic acid (EDTA, Merck)
1,4-Divinylbenzene (p-DVB: from purification of 50% DVB from Merck-Schuchardt)
Poly(vinyl alcohol) (Merck-Schuchardt, MW 70,500, degree of hydrolysis 97.5–99.5%)
Polyacrolein (Degussa/Hanau, MW 40,000)
Tridistilled water
Dibenzoylperoxide (DBPO) (Merck)
2,2'-Azobisisobutyronitrile (AIBN) (Merck)

All substances for the preparation of the carriers were freshly recrystallized or distilled before being used.

Assay Methods

Determination of Bound Enzyme. Differential determination of the protein content in the enzyme solution before the coupling reaction and the amount of unreacted enzyme after the reaction in the used enzyme and washing solutions (500 ml) is employed using Folin reagent (Lowry et al.,[13] 10-fold amounts). A new standard calibration plot is made for every measurement.

Determination of the Enzymic Activity

WITH BAA AS SUBSTRATE (MODIFIED METHOD OF SMITH et al.[14]) The immobilized papain sample is separated from the supernatant water and suspended in 0.1 M cysteine solution (pH 5.1, 1 ml) and 0.02 M citric acid/phosphate buffer (pH 5.1, 2 ml). After an activation time of 10 min, 0.1 M BAA solution (4 ml) is added. The mixture is incubated for 30 min at 37° with vigorous stirring. (Avoid magnetic stirring and grinding the grains.) Immediately after centrifuging, the solution (5 ml) is added to a neutralized concentrated formalin solution (10 ml). The titration is performed (under nitrogen atmosphere) with 0.02 M NaOH (indi-

---

[13] O. H. Lowry, N. J. Rosebrough, A. L. Farr, and R. J. Randall, *J. Biol. Chem.* **193**, 265 (1951).
[14] E. L. Smith, A. Stockel, and J. R. Kimmel, *J. Biol. Chem.* **207**, 551 (1954).

cator: phenolphthalein). Blank titrations are run to correct the activities. The corresponding activity of the native enzyme is determined under identical conditions.

WITH BAEE AS SUBSTRATE (MODIFIED METHOD OF JACOBSEN et al.[15]) The rate of the enzymic reaction is observed with an automatic titrator (Metrohm, Switzerland consisting of Dosimat E 535-5, pH meter E 500, Impulsomat E 473, recorder E 478, macro electrode measuring chain) using a thermostatic reaction vessel with a stirrer at 30° and pH 6.0 (unless otherwise stated) under purified nitrogen. Suited amounts of immobilized papain are suspended and activated in 0.1 M cysteine solution (pH 6.0, 1 ml in 0.002 M EDTA). After addition of 0.05 M BAEE solution (15 ml) the free amino acid is titrated with 0.05 M NaOH. The corresponding activity of the native enzyme is determined under identical conditions.

The pH/activity profiles of immobilized papains with ionic matrices are measured, the pH being adjusted before the activation. The K_m values, i.e., the apparent K_m values, were determined graphically (Lineweaver-Burk plot).

General Notes

1. Polymerization and Work-Up. Polymerization mixtures are frozen in bomb tubes with liquid nitrogen, evacuated, flushed with nitrogen, and degassed by careful melting. This pretreatment is repeated several times. Then the tube is sealed while frozen and under vacuum. After the indicated heating, the polymer is crushed and extracted exhaustively with an appropriate solvent in a Soxhlet apparatus, dried *in vacuo*, ground, and sieved.

2. Immobilization of the Enzyme, Washing, and Storage. The reactive carrier (grain size 0.1–0.2 mm) is vigorously shaken or stirred in a well closed vessel for a definite time with a solution of papain in aqueous $NaHCO_3$ at room temperature, unless otherwise stated (amount of carrier and duration are stated below for each preparation). A recommended procedure for separating the immobilized enzyme from the supernatant solution involves the use of a peristaltic pump and a tube fitted with a glass frit. Unreacted, adsorbed enzyme is washed out (use vigorous stirring!) using successively small portions of 0.1 M or 1 M $NaHCO_3$ (200 ml), 1 M NaCl (200 ml), and water (80 ml) (typical procedure). The product can be stored under water or in the freeze-dried state of 4°.

[15] C. F. Jacobsen, J. Léonis, K. Linderstrøm-Lang, and M. Ottesen, *Methods Biochem. Anal.* **4**, 171 (1957).

Preparation of Immobilized Papain Derivatives

1. Nitrated Methacrylic Acid/Methacrylic Acid 3-Fluoroanilide Papain

Reactive carriers are obtained by nitration of cross-linked (DVB as cross-linking agent) copolymers of methacrylic acid and methacrylic acid 3-fluoroanilide.[1,2] Carriers having gel or macroreticular structure can be used.

Experimental Procedures

Preparation of Carrier (I) (with Gel Structure)

POLYMERIZATION. Methacrylic acid (0.86 g, 10.0 mmoles), methacrylic acid 3-fluoroanilide (1.79 g, 10.0 mmoles), and DVB (26.5 mg, 0.20 mmole) are mixed with DBPO (2.6 mg, 0.1 weight %) in a bomb tube and polymerized (see General Note 1) for 48 hr at 75° and for additional 48 hr at 90°. The polymer is extracted with methanol (8 hr). The composition of the copolymer (elementary analysis) is F:COOH = 1:2

NITRATION. The copolymer (2.0 g) is added at $-10°$ in small portions with stirring to the nitrating acid (75 ml) consisting of 1 part HNO_3 ($d = 1.52$) and 3 parts H_2SO_4 ($d = 1.84$). During the reaction the temperature rises to $-2°$. The reaction mixture is stirred another 6 hr at this temperature and is poured over ice (500 g). The polymer is filtered off with suction, washed with water to the neutral reaction, and exhaustively extracted with acetone. The dry product is sieved again. The degree of nitration (elementary analysis) is 1.4 NO_2 groups per fluoroaryl group.

Immobilization of the Enzyme. See General Note 2. Use 100 mg of carrier, 20 ml of a 2% papain solution in 0.1 M $NaHCO_3$; stir for 24 hr.

Properties of the Immobilized Papain

Content of protein: 1170 mg per gram of carrier.
Relative activity (BAA as substrate): 25% at pH 5.1.

2. Nitrated Methacrylic Acid/Fluorostyrene Papain

The reactive carriers are obtained by nitration of cross-linked (DVB as cross-linking agent) copolymers of methacrylic acid and fluorostyrenes.[3] Carriers having gel or macroreticular structure can be used.

Experimental Procedures

Preparation of Carrier (II) (with Gel Structure)

POLYMERIZATION. Methacrylic acid (1.72 g, 20.0 mmoles), 3-fluorostyrene (1.22 g, 10.0 mmoles), and DVB (39.0 mg, 0.30 mmole) are mixed with DBPO (5.96 mg, 0.2 weight %) as an initiator in a bomb tube. The polymerization (see General Note 1) is carried out for 48 hr at 70° and for an additional 12 hr at 100°. The polymer is extracted with methanol. The composition of the copolymer (COOH-titration) is F:COOH = 1:1.75.

NITRATION. Nitration is carried out as described for carrier (I). The degree of nitration (elementary analysis) is 2 NO_2 groups per fluoroaryl group.

Immobilization of the Enzyme. See General Note 2. Use 100 mg of carrier, 20 ml of a 2% papain solution in 0.1 M $NaHCO_3$; stir for 24 hr.

Properties of the Immobilized Papain

Content of protein: 1170 mg per gram of carrier.
Relative activity (BAA as substrate): 25% at pH 5.1.

3. Acrylic Acid/Isothiocyanatostyrene Papain

The reactive carriers are cross-linked (p-DVB as cross-linking agent), macroreticular copolymers of acrylic acid and isothiocyanato styrene.[4-6]

Experimental Procedures

Preparation of Carrier (III). Acrylic acid (3.50 g, 50.0 mmoles), 3-isothiocyanatostyrene (805 mg, 5.0 mmoles), and p-DVB (285 mg, 2.2 mmoles) are mixed with petroleum ether (b.p. 120°–180°) (2.30 g, 50 weight %) and AIBN (46 mg, 1 weight %) in a bomb tube and polymerized (see General Note 1). The homogeneous solution is heated for 12 hr at 45°, and the temperature is slowly increased to 80° over 36 hr. The bomb tube is carefully opened. The solvent is pumped away at 80°. The white opaque product is extracted with acetone for 24 hr (4-fold exchange of the extracting agent) and deswelled with petroleum ether (b.p. 40°–80°) (3-fold exchange of the solvent). The composition of the copolymer (elementary analysis) is NCS:COOH = 1:10, i.e., 1.05 mmoles of NCS groups per gram of carrier. Swellability of the carrier (in McIlvaine buffer, ionic strength 0.5 M^{16}): volume expansion from 3.4 ml/g at pH 4 to 7.1 ml/g at pH 9; pK_{app} 6.5.

Immobilization of the Enzyme. The carrier (200 mg) is preswollen in 0.1 M phosphate solution, pH 5, and adjusted to this pH by exchanging the buffer solution, 40 ml of a 0.5% papain solution, pH 5, in 0.1 M phosphate are added. For 6 hr the suspension is well stirred at 20°. Washing procedure: successively use 0.1 M phosphate solution, pH 5 (100 ml), 0.1 M phosphate solution, pH 8 (100 ml), 1.0 M NaCl (100 ml), 0.1 M phosphate solution, pH 6 (100 ml), and water (60 ml) in 20-ml portions.

Analogously, the carrier (100 mg) is allowed to react with a 0.5% papain solution, pH 8 (20 ml) for 4 hr. The coupling reaction and the washing procedure are controlled using a Beckman DB-GT spectrophotometer at 275 nm in a 2-mm flow cuvette.

Properties of the Immobilized Papain

Coupling reaction performed at pH	Content of protein (mg/g carrier)	Relative activity[a] At pH 6 (%)	Relative activity[a] At pH 9 (%)	$K_{m,app}$ at pH 6[b] (moles/liter)
5	493	12	84	1.39×10^{-2}
8	1098	9.5	31.5	—

[a] Determined with 15 mg of freeze-dried immobilized papain referring to the activity of the native enzyme at pH 6.0 (BAEE as substrate).
[b] The K_m value of the native enzyme under the same conditions was 4×10^{-2} mole/liter.

The pH optimum is 9.0–9.4.

4. N-Vinylpyrrolidone/Maleic Acid Anhydride Papain

The reactive carriers are copolymers of N-vinylpyrrolidone and maleic acid anhydride cross-linked with p-DVB.[7]

Experimental Procedures

Preparation of Carrier (IV). N-Vinylpyrrolidone (1.13 g, 10.2 mmoles), maleic acid anhydride (2.00 g, 20.4 mmoles), and p-DVB (0.06 g, 0.46 mmole) are mixed with DBPO (15.1 mg) as the initiator in a bomb

[16] P. J. Elving, J. M. Markowitz, and J. Rosenthal, *Anal. Chem.* 28, 1179 (1956).

tube. The polymerization (see General Note 1) is carried out for 24 hr while increasing the temperature from 95° to 105°. A brown product is formed which is extracted with dimethylformamide (DMF). The content of the reactive groups (elementary analysis) is 1.85 mmoles of anhydride groups per gram of carrier.

Immobilization of the Enzyme. See General Note 2. Use 85 mg of carrier, 20 ml of a 0.5% papain solution in 1 M NaHCO$_3$; stir for 24 hr.

Properties of the Immobilized Papain

Content of protein: 442 mg per gram of carrier.
Relative activity (BAEE as substrate): 6.1% at pH 6.0.

5. Poly(vinyl Alcohol) (PVA)/Diazonium Papain

The reactive carriers based on PVA are obtained by cross-linking soluble PVA with terephthalaldehyde[17] and by allowing this cross-linked polymer to react with different amounts of 2-(m-aminophenyl)-1,3-dioxolane followed by diazotization.[8]

Experimental Procedures

Preparation of the Cross-linked Carrier (V)

CROSS-LINKING. PVA (22.0 g, 0.50 mole of hydroxylic groups) is cross-linked with terephthalaldehyde (3.35 g, 25.0 mmoles). The insoluble polymer is ground and sieved.

INTRODUCTION OF AMINO GROUPS. Cross-linked PVA (1.0 g) is allowed to swell in water (40 ml) and treated with HCl (d = 1.19, 1 ml). Then 2-(m-aminophenyl)-1,3-dioxolane (0.60 ml, 0.72 g, 4.37 mmoles) is added dropwise with stirring.[18] The temperature is raised to 45°–50° in 1 hr; when this temperature is reached stirring is continued for another 15 hr. The yellowish product is filtered off, washed with water until neutral, extracted with acetone (72 hr) in a Soxhlet apparatus, dried *in vacuo*, and sieved again. The content of amino groups (elementary analysis) is 1.40 mmoles of NH$_2$ per gram of carrier.

DIAZOTIZATION. The above amino product (50 mg) is allowed to swell in water (2 ml) and treated at 4° with 0.5 M HCl (10 ml) and 1 M

[17] W. Kuhn and G. Balmer, *J. Polym. Sci.* **57**, 311 (1962).
[18] E. Minami and S. Kojima, *J. Chem. Soc. Jpn, Ind. Chem. Sect.* **57**, 826 (1954); *C.A.* **49**, 10628h (1955).

sodium nitrite solution (1 ml). After stirring for another 1 hr at this temperature, the polymer is filtered off by suction and washed carefully with ice water until neutralized.

Immobilization of the Enzyme. The freshly prepared diazonium product is immediately treated with 20 ml of a 2% papain solution in 1 M NaHCO$_3$ and stirred for 15 hr at 4° (see General Note 2).

Properties of the Immobilized Papain

Content of protein: 685 mg per gram of carrier.
Relative activity (BAEE as substrate): 12.0% at pH 6.0.
pH optimum: 7.5, $K_{m,app}$: 5.6 × 10^{-2} mole/liter at pH 6.0.

6. Poly(allyl Alcohol)/Isothiocyanato Papain

The reactive carriers based on poly(allyl alcohol) are obtained by reduction of polyacrolein with sodium borohydride[19] to poly(allyl alcohol), followed by treatment with different amounts of 1-isocyanato-4-isothiocyanatobenzene sometimes using 1,4-diisocyanatobenzene as a cross-linking agent.[9]

Experimental Procedures

Preparation of a Non-Cross-Linked Carrier (VI). Poly(allyl alcohol) (465 mg, 8.00 mmoles of OH groups) is dissolved in DMF (30 ml, anhydrous) and treated with a solution of 1-isocyanato-4-isothiocyanatobenzene (705 mg, 4.00 mmoles in 30 ml of DMF). After it has been stirred for 2 hr at 80°, the hot reaction mixture is poured into methanol/water (v/v, 1:1; 500 ml). The precipitated white polymer is filtered off with suction. It is repeatedly extracted with methanol, dried *in vacuo*, ground, and sieved. The content of reactive groups (elementary analysis) is 2.92 mmoles of NCS groups per gram of carrier.

Immobilization of the Enzyme. See General Note 2. Use 50 mg of carrier, 20 ml of a 2% papain solution in 1 M NaHCO$_3$; stir for 48 hr.

Properties of the Immobilized Papain

Content of protein: 112 mg per gram of carrier.
pH optimum: 8.1. Relative activity (BAEE as substrate): 3.5% at pH 6.0; 12.1% at pH 8.1.
$K_{m,app}$: 5.3 × 10^{-3} mole/liter at pH 8.1.

[19] R. C. Schulz, J. Kovács, and W. Kern, *Makromol. Chem.* **54**, 146 (1962).

7. Poly(vinyl Ether)/Diazonium Papain

The reactive carriers with poly(vinyl ether) groupings are obtained by cationic polymerization of 2-vinyloxyethyl-4-nitrobenzoate (in some cases by copolymerization with vinylethyl ether) using divinyl ether as cross-linking agent followed by reduction and diazoization.[10]

Experimental Procedures

Preparation of a Cross-Linked Homopolymeric Carrier (VII)

POLYMERIZATION. 2-Vinyloxyethyl-4-benzoate (2.37 g, 10.0 mmoles) and divinyl ether (70 mg, 0.1 mmole) are dissolved in methylene chloride (20 ml, anhydrous) and chilled to $-70°$. Boron trifluoride etherate (0.1 ml) is slowly added. Spontaneous polymerization occurs after 1 hr, and the excess of catalyst is destroyed by adding acetone/25% aqueous ammonia (v/v, 1:1) at $-60°$. Subsequently the reaction mixture is warmed to room temperature. The polymer is filtered off with suction (working up: see General Note 1). Extraction: methylene chloride (2 hr) followed by acetone (2 hr).

REDUCTION. The above product (0.15 g) is heated in diethylene glycol-diethyl ether (10 ml) under nitrogen to 185° and treated with stirring with a solution of phenylhydrazine (1.5 g) in the used glycol ether (3 ml). After another 3 hr of stirring, the suspension is chilled and filtered off with suction. The polymer is extracted with dioxane (5 hr) in a Soxhlet apparatus and dried *in vacuo*. The content of amino groups (perchloric acid titration) is 3.66 mmoles of NH_2 per gram of carrier.

DIAZOTIZATION. The insoluble amino product (60 mg) is suspended in HCl ($d = 1.09$, 6 ml), and a solution of $NaNO_2$ (350 mg in 3 ml of water) is added with vigorous stirring at 0°–4°. After additional stirring for 2 hr at this temperature, the mixture is carefully neutralized (5 M NaOH, cooling), and the product is intensively washed with ice water.

Immobilization of the Enzyme. The freshly prepared diazonium product is shaken with 20 ml of a 0.5% papain solution in 0.1 M $NaHCO_3$ for 48 hr at 4° (working up: see General Note 2).

Properties of the Immobilized Papain

Content of protein: 142 mg per gram of carrier.
Relative activity (BAEE as substrate): 3.4% at pH 6.0.

Acknowledgment

The help of Dipl.-Chem. E. Ehrenthal during the preparation of this manuscript is gratefully acknowledged.

[9] Immobilization of Enzymes on Nylon

By WILLIAM E. HORNBY and LEON GOLDSTEIN

The nylons are a family of linear polymers consisting of repeating assemblies of methylene groups joined together by secondary amide linkages. Several types of nylon are available commercially, differing only in the number of methylene groups in the repeating alkane segments. These different types of nylon are designated according to the number of carbon atoms in their component monomeric units. Thus nylon 6 is so-called because it is prepared by the polymerization of the 6-carbon compound caprolactam. In nylon 6, all the alkane segments contain equal numbers of methylene groups, in this case five. Other types of nylon may contain unequal numbers of methylene groups in the alkane segments. For example, in nylon 610, which is synthesized from the 6-carbon compound 1,6-diaminohexane and the 10-carbon compound sebacic acid, adjacent alkane segments contain eight and six methylene groups, respectively.

On many accounts, it is attractive to consider the use of nylon as a support matrix for the preparation of immobilized enzymes (see also this volume [42]). For example, nylon is readily available in a wide variety of physical forms, such as films, membranes, powders, hollow fibers, and tubes. Nylon structures are mechanically strong and nonbiodegradable, which renders possible their prolonged exposure to biological media without impairment of their structural integrity. Finally, some of the nylons, such as nylon 6 and nylon 66, are relatively hydrophilic; this means that they can support an environment in which to immobilize enzymes that is conducive to the stability of the protein.

Several types of immobilized enzyme derivatives have been prepared using nylon as the support matrix. Thus Chang[1,2] has described the preparation of immobilized enzymes by their microencapsulation within semipermeable membranes of nylon 610. Several derivatives have been prepared in which the protein is adsorbed onto the surface of a nylon structure and secured there by its subsequent cross-linking. For example, Edelman *et al.*[3] immobilized various antigens by their adsorption onto nylon monofilaments followed by cross-linking with a water-soluble car-

---

[1] T. M. S. Chang, *Science* **146**, 524 (1964).
[2] T. M. S. Chang, *Nature (London)* **229**, 117 (1971).
[3] G. M. Edelman, U. Rutishauser, and C. F. Millette, *Proc. Natl. Acad. Sci. U.S.A.* **68**, 2153 (1971).

bodiimide. A similar strategy was adopted by Horvath[4] and Reynolds[5] for the preparation of enzymes immobilized to pellicular nylon and nylon floc, respectively. In this case the enzymes were "fixed" on the surface of the nylon by cross-linking with the bifunctional agent glutaraldehyde. Cross-linking with glutaraldehyde has also been used by Inman and Hornby[6] for the preparation of nylon membrane-supported enzymes. The enzymes glucose oxidase and urease were adsorbed into the pores of the membrane and then cross-linked *in situ* by exposure to the bifunctional agent.

To date, the majority of immobilized enzyme derivatives using nylon as the support material have been made by covalently binding the enzyme protein to the nylon polymer itself. Native nylon of high molecular weight has few free end groups, and so it must be pretreated in order to generate potentially reactive centers that are capable of interacting covalently with enzyme molecules. In principle, this prerequisite can be satisfied by adopting one of three approaches.

The first approach recognizes that the components of the secondary amide linkages of the nylon, namely aliphatic amino and carboxyl groups, are potentially reactive species. Accordingly, by this method the reactive centers are generated by subjecting the nylon to treatments that cleave some of its secondary amide linkages and in the process liberate their component amino and carboxyl groups. By the second approach, the reactive centers are introduced onto the nylon by processes that involve the O-alkylation of the support's secondary amide bonds. Thus this method yields the reactive imidate salt of the nylon without necessitating depolymerization of the support. The third and final approach generates the reactive centers by a process that initially involves partial acid hydrolysis of the nylon. The peptide bonds cleaved in this process subsequently are reconstituted in a four-component condensation that involves reaction of the liberated amino and carboxyl groups with an aldehyde and an isocyanide. Overall, this method yields an intact nylon that contains reactive centers derived from the N-substitution of some of the support's secondary amide linkages.

Covalent Binding of Enzymes to Nylon by Methods Involving Peptide Bond Cleavage

In essence all these methods involve three separate operations. (1) The nylon is partially depolymerized by cleaving some of its secondary

[4] C. Horvath, *Biochim. Biophys. Acta* 358, 164 (1974).
[5] J. H. Reynolds, *Biotechnol. Bioeng.* 16, 135 (1974).
[6] D. J. Inman and W. E. Hornby, *Biochem. J.* 129, 255 (1972).

amide linkages. (2) Either the aliphatic amino groups or the aliphatic carboxyl groups, which are released in the first step, are activated. (3) The enzyme is allowed to react with the activated nylon derived from the second step.

The above procedure has been adopted by several workers for the immobilization of enzymes to the inside surface of nylon tubes. In the majority of cases the initial step, whereby the peptide bonds are cleaved, is effected by subjecting the nylon to partial acid hydrolysis. Thereafter, the resulting free amino and carboxyl groups can be activated by a variety of methods. Hornby and Filippusson[7] activated the free carboxyl groups of acid-hydrolyzed nylon tube by first coupling them to either benzidine or hydrazine using a carbodiimide. The resulting aryl amino and acid hydrazide derivatives were then converted to the corresponding diazonium salt and acid azide, respectively, for reaction with the enzyme. Horvath and Solomon[8] immobilized trypsin and uricase to nylon tube by activating the amino groups. The trypsin was coupled to an isocyanate derivative of nylon, which was prepared by reaction of the amino groups with phosgene, and the uricase was coupled to the amino groups through the trifunctional reagent cyanuric chloride. Coupling through the amino groups of acid-hydrolyzed nylon tube also was used by Sundaram and Hornby[9] for the preparation of nylon tube-immobilized urease. In this case the amino groups were allowed to react with an excess of the bifunctional agent glutaraldehyde, and the enzyme was coupled by its reaction with the free aldehyde groups of the glutaraldehyde-derivatized nylon. A similar process was adopted by Allison et al.[10] and Bunting and Laidler[11] for the immobilization of asparaginase to nylon tube.

The procedures described above, which entail partial acid hydrolysis of the nylon and coupling through the amino groups of the support, all leave a residue of ionized carboxyl groups on the surface of the nylon. By performing the peptide bond cleavage step in nonaqueous conditions it is possible to release the free amino groups and at the same time block the complementary carboxyl groups. For example, Hornby et al.[12] described a method for the immobilization of the enzymes lactate, malate, and alcohol dehydrogenase to nylon tube, which involved the treatment of the nylon in nonaqueous conditions with N,N-dimethyl-1,3-propanediamine. This process breaks some of the peptide bonds on the surface of

[7] W. E. Hornby and H. Filippusson, *Biochim. Biophys. Acta* **220**, 343 (1970).
[8] C. Horvath and B. A. Solomon, *Biotechnol. Bioeng.* **14**, 885 (1972).
[9] P. V. Sundaram and W. E. Hornby, *FEBS Lett.* **10**, 325 (1970).
[10] J. P. Allison, A. Davidson, A. Gutierrez-Hartman, and G. B. Kitto, *Biochem. Biophys. Res. Commun.* **47**, 66 (1972).
[11] P. S. Bunting and K. J. Laidler, *Biotechnol. Bioeng.* **16**, 119 (1974).
[12] W. E. Hornby, D. J. Inman, and A. McDonald, *FEBS Lett.* **23**, 114 (1972).

the nylon, liberating free amino groups and amidating the carboxyl groups, thus leaving cationic dimethylamino groups on the surface of the nylon. The following procedure, which is based on that described by Inman and Hornby,[13] typifies the immobilization of enzymes to nylon tube by methods that involve peptide bond cleavage and coupling of the protein to the liberated primary amino groups of the support.

Procedure

A 3-m length of nylon 6 tube, 1 mm bore (Portex, Hythe, Kent, U.K.) is filled with a mixture of 18.6% (w/w) $CaCl_2$, 18.6% (w/w) water in methanol and incubated at 50° for 20 min. This process etches the inside surface of the nylon tube by dissolving out the regions of amorphous nylon, thereby increasing both the available surface area and its wettability. The amorphous nylon is then purged from the tube by perfusion with water for 30 min at a flow rate of 5 ml min^{-1}.

For nonhydrolytic cleavage of the nylon, the tube is dried first by perfusing it with methanol for 30 min at a flow rate of 2 ml min^{-1}, then filled with N,N-dimethyl-1,3-propanediamine and incubated for 12 hr at 70°. Thereafter, excess amine is removed by washing through the tube with water for 12 hr at a flow rate of about 2 ml min^{-1}. For hydrolytic cleavage of the nylon, the tube is perfused for 40 min with 3.65 M HCl at 45° and a flow rate of 2 ml min^{-1} and then washed through with water for 12 hr at a flow rate of about 2 ml min^{-1}.

The free amino groups, which are released in the above treatments, are activated by perfusion of the tube with 5% (w/v) glutaraldehyde in 0.2 M borate buffer, pH 8.5, for 15 min at 20° and a flow rate of 2 ml min^{-1}. Thereafter, the tubes are washed free of excess bifunctional agent by their perfusion for 10 min with 50 ml of 0.1 M phosphate buffer, pH 7.8, and immediately filled with a solution of the enzyme (0.5–1.0 mg ml^{-1}) in the same buffer. After 3 hr of incubation at 4°, excess protein together with noncovalently bound protein are removed by washing through with 1 liter of 0.5 M NaCl in 0.1 M phosphate buffer, pH 7.0, at a flow rate of 5 ml min^{-1}. Nylon tube-immobilized enzymes prepared in this way are stored filled with their assay buffer at 4°.

The various steps involved in the above procedure are described in Fig. 1. For the sake of simplification, the reaction products of the glutaraldehyde with the amino groups of the nylon and the enzyme are shown as Schiff's bases. However, it must be noted that the exact manner in

[13] D. J. Inman and W. E. Hornby, *Biochem. J.* **137**, 25 (1974).

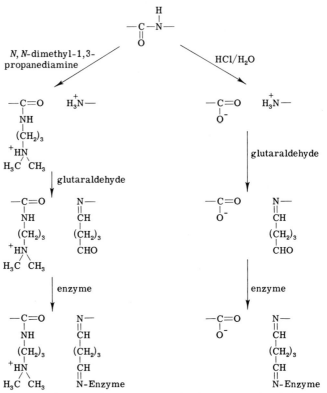

FIG. 1. Covalent binding of enzymes to nylon by methods involving peptide bond cleavage and coupling through the liberated primary amino groups.

which glutaraldehyde reacts is still uncertain, and alternative mechanisms have been proposed.[14-16]

Covalent Binding of Enzymes to Nylon by Methods Involving O-Alkylation of the Nylon

All the methods described in the previous section involved partial depolymerization of the nylon in order to generate the reactive centers necessary for enzyme coupling. Such methods are not entirely satisfactory, since the cleavage of the secondary amide linkages also impairs the mechanical strength of the support. Thus the extent of peptide bond cleavage must represent a compromise between that which generates an

[14] F. Richards and J. Knowles, *J. Mol. Biol.* **37**, 231 (1968).
[15] J. Bowes and C. Carter, *Biochim. Biophys. Acta* **168**, 341 (1969).
[16] P. M. Hardy, A. C. Nicholls, and H. N. Rydon, *Chem. Commun.* **1969**, 565 (1969).

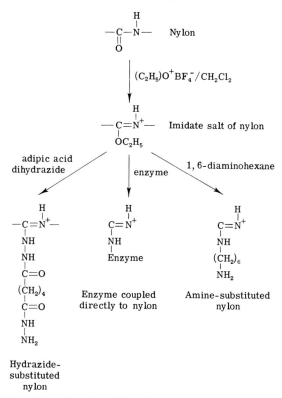

FIG. 2. O-Alkylation of nylon with triethyloxonium salts and reaction of the imidate salt of nylon with acid hydrazides, enzymes, and bisalkyl amines.

adequate supply of free amino groups and that which still maintains the structural integrity of the polymer. In view of these considerations, a more desirable chemistry for immobilizing enzymes to nylon would be one by which the reactive centers are generated without necessitating any depolymerization of the support. In principle, such a chemistry would permit an adequate provision of reactive centers without compromising either the mechanical strength or structural integrity of the nylon.

Secondary amides can be alkylated by powerful alkylating agents to yield the corresponding imidate salts.[17] Thus treatment of nylon with such reagents results in the O-alkylation of some of the support's peptide bonds, thereby producing their imidate salts. Imidate salts and their free bases, imido esters, will react with nucleophiles such as amines and acid hydrazides, yielding the corresponding amidines and amidrazones, respectively (Fig. 2). In this way it is possible to introduce reactive centers

[17] R. E. Benson and T. L. Cairns, *Org. Synth.* 31, 72 (1951).

onto the nylon without breaking any of its secondary amide bonds in the process.

The activation of nylon for enzyme immobilization by O-alkylation of its secondary amide bonds was proposed by Hornby et al.[18] Subsequently, Campbell et al.[19] used this approach for the preparation of nylon tube-immobilized glucose oxidase. In these reports, the O-alkylation was effected by incubating the tube filled with dimethyl sulfate for 4–6 min at 100°. The resulting imidate salt of the nylon then was allowed to react with either lysine or polyethyleneimine, so yielding a derivatized nylon containing free primary amino groups with its secondary amide bonds still intact. The free amino groups on these derivatives finally were activated for enzyme coupling by their reaction with an excess of either of the bifunctional agents, diethyl adipimidate or glutaraldehyde.

Morris et al.[20] described the O-alkylation of nylon using triethyloxonium salts. They found that these reagents brought about a smooth alkylation of the nylon at room temperature and also that the reaction was easier to control than that previously described using dimethyl sulfate. In addition they observed that greater yields of active immobilized enzyme were obtained with this reagent. In view of these observations, together with the noted toxicity of the volatile dimethyl sulfate, it was concluded that O-alkylation with triethyloxonium salts represented the preferred route for modification of the nylon.

Figure 2 depicts some of the modified nylons that can be produced through the initial O-alkylation of the polymer with triethyloxonium tetrafluoroborate. The first product in this scheme, the imidate salt of nylon, represents a versatile intermediate and affords several routes for enzyme immobilization. (1) The enzyme can be allowed to react directly through its free amino groups with the imidate salt of the nylon. However, it has been noted that direct coupling in this manner does not afford the most active immobilized enzyme derivatives.[19] (2) The imidate salt of the support can be allowed to react in nonaqueous conditions with a bis-acid hydrazide, such as adipic acid dihydrazide, to give a hydrazide-substituted nylon. This type of modified nylon is a stable intermediate and can be stored filled with water for at least 10 weeks at room temperature. Hydrazide-substituted nylon can be activated for enzyme coupling in a variety of ways. For example, the free acid hydrazide group

---

[18] W. E. Hornby, J. Campbell, D. J. Inman, and D. L. Morris, "Enzyme Engineering," (E. K. Pye and L. B. Wingard Jr., eds.), Vol. 2, p. 401. Plenum, New York, 1974.

[19] J. Campbell, W. E. Hornby, and D. L. Morris, Biochim. Biophys. Acta **384**, 307 (1975).

[20] D. L. Morris, J. Campbell, and W. E. Hornby, Biochem. J. **147**, 593 (1975).

can be coupled to a bifunctional agent such as glutaraldehyde or a bisimidate and the enzyme coupled in turn to the free functional group of the bifunctional agent. Alternatively, the free acid hydrazide group can be converted into an acid azide through reaction with HNO_2 and then coupled to the enzyme (Fig. 3). (3) The O-alkylated nylon can be allowed to react with a bis-amine, such as 1,6-diaminohexane, to yield amine-

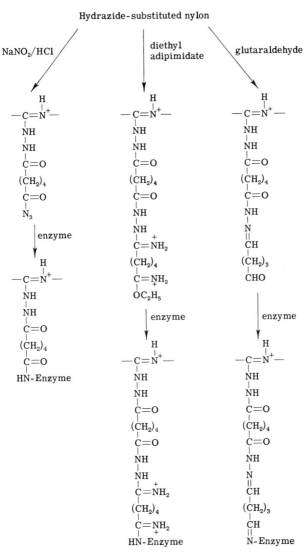

FIG. 3. Coupling of enzymes to hydrazide-substituted nylon by the use of bifunctional agents and of acid azides.

substituted nylon, which can be reactivated for enzyme coupling with bifunctional agents.[20]

All the amino-substituted nylons carry a residue of positive charges on their surface in the form of the protonated amidine groups. In some situations it is possible that the polycationic nature of the amine-substituted derivatives will be troublesome. For example, this property may elicit nonspecific binding of anions present in reaction mixtures using the immobilized enzyme derivative. This problem can be overcome by using the corresponding hydrazide-substituted derivatives. In these preparations the amidine groups are replaced by amidrazone groups, which have lower pK_a values and consequently are not protonated to the same extent. This was shown by Hornby and Morris,[21] using caprolactam as a model for nylon 6. Amine- and hydrazide-substituted caprolactam were prepared by allowing O-methylcaprolactim to react with hexylamine and hexanoic acid hydrazide, respectively. The pK_a values at 25° of the resulting amidine and amidrazone groups were found to be 11.2 and 6.5. This suggests that at most operational pH values of the immobilized enzymes, those derivatives prepared from hydrazide-substituted supports will carry considerably less positive charge than the corresponding derivatives prepared from the amine-substituted supports.

The following procedure, which is based on that described by Morris et al.,[20] typifies the immobilization of enzymes to nylon tube by methods that involve the O-alkylation of the support.

Procedure

Triethyloxonium tetrafluoroborate is prepared according to the method of Meerwein.[22] Fifty milliliters of 10% (v/v) 1-chloro-2,3-epoxypropane in dry ether is added slowly to 200 ml of 5% (v/v) boron trifluoride diethyl etherate in dry ether. The mixture is stirred first under reflux for 1 hr and then at room temperature for a further 3 hr. Thereafter, the precipitated triethyloxonium tetrafluoroborate is washed three times by decantation with 100-ml aliquots of dry ether and finally dissolved in dry dichloromethane to a final concentration of 10% (w/v). This solution can be used for up to 48 hr for the O-alkylation of nylon 6 tube.

A 3-m length of nylon 6 tube is filled with the triethyloxonium tetrafluoroborate solution and incubated at room temperature for 15 min. Excess alkylating agent is removed by suction and the tube is washed through for 2 min with dichloromethane and for 2 min with dioxane. The O-alkylated nylon tube is then filled immediately with a solution of

[21] W. E. Hornby and D. L. Morris, unpublished results.
[22] H. Meerwein, Org. Synth. **46**, 120 (1966).

either the amine (10%, w/v, 1,6-diaminohexane in methanol) or the acid hydrazide (3%, w/v, adipic acid dihydrazide in formamide) and incubated for 3 hr at room temperature. Thereafter excess amine or hydrazide is removed by perfusing the tube with water for 12 hr.

Both the amine- and hydrazide-substituted derivatives can be reactivated with glutaraldehyde and coupled to enzyme as described in the preceding section of this paper. Alternatively, they can be reactivated with a bisimidate as follows. The tube is dried by its perfusion with methanol for 10 min and then filled with a 2% (w/v) solution of diethyl adipimidate dihydrochloride in 20% (v/v) N-ethylmorpholine in methanol and incubated for 45 min at room temperature. Thereafter the tube is washed through with 50 ml of methanol for 5 min, immediately filled with a solution of the enzyme (0.5–1.0 mg ml^{-1}) in 0.1 M phosphate buffer, pH 8.0, and incubated for 3 hr at 4°. Excess protein and noncovalently bound protein are removed by perfusion with 0.5 M NaCl in 0.1 M phosphate buffer, pH 7.0, as previously described.

Covalent Binding of Enzymes to Nylon by a Method Involving N-Alkylation of the Nylon Backbone

This section describes a method for the introduction of chemically reactive side chains on nylon by N-alkylation of some of the secondary amide groups on the polymer. The procedure employed leads to no decrease in molecular weight as estimated from the end group content of the polymer surface. The mechanical strength of nylon is thus presumably not impaired; moreover, the modified polyamide backbone carries no charged groups resulting from the modification procedures.

The method is based on (a) mild acid hydrolysis[6,9] to generate COOH · · · NH$_2$ pairs on the surface of a nylon structure, (b) resealing of the newly formed COOH · · · NH$_2$ pairs by a four-component condensation reaction that involves the neighboring carboxyl and amino groups on the nylon backbone, an aldehyde and an isocyanide. In four-component condensation reactions (4CC)[23–26] between carboxyl, amine, aldehyde, and isocyanide (Fig. 4), the carboxyl and amino components (R^1 and R^2) combine to form an N-substituted amide, the aldehyde and isocyanide components (R^3 and R^4) appearing as the side chain on the amide nitrogen. These reactions allow in principle for considerable versatility since, by proper choice of aldehyde and isocyanide, various func-

[23] I. Ugi, *Angew. Chem. Int. Ed. Engl.* **1**, 8 (1962).
[24] G. Gokel, G. Lüdke, and I. Ugi, in "Isonitrile Chemistry" (I. Ugi, ed.), pp. 145–199. Academic Press, New York, 1971.
[25] R. Axén, P. Vretblad, and J. Porath, *Acta Chem. Scand.* **25**, 1129 (1971).
[26] P. Vretblad and R. Axén, *FEBS Lett.* **18**, 254 (1971).

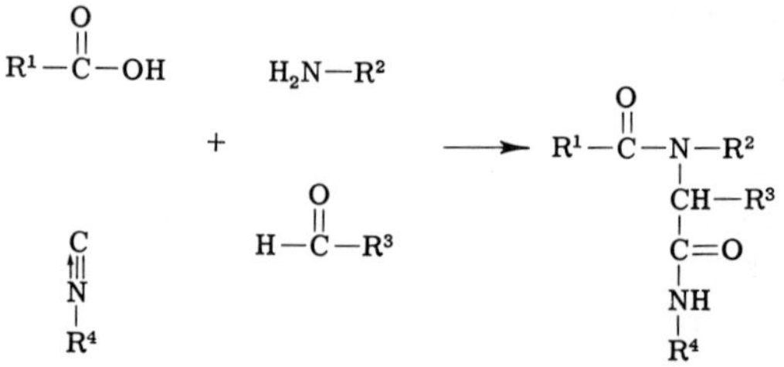

$$R^1-\overset{\underset{\|}{O}}{C}-OH \quad H_2N-R^2$$

$$+ \quad \longrightarrow \quad R^1-\overset{\underset{\|}{O}}{C}-\underset{R^4}{\underset{|}{N}}-\underset{\underset{R^4}{|}}{\underset{|}{\underset{C=O}{\underset{|}{CH-R^3}}}}\text{-}R^2$$

$$\underset{R^4}{\underset{|}{\underset{N}{\overset{\|}{C}}}} \quad H-\overset{\underset{\|}{O}}{C}-R^3$$

FIG. 4. Four-component condensation reaction between amine, carboxyl, aldehyde, and isocyanide.

tional groups can be introduced on the N-alkyl side chains of the reformed amide groups of nylon. The procedure adopted here, utilizing acetaldehyde or isobutyral as aldehyde component and a bifunctional isocyanide, 1,6-diisocyanohexane, leads to nylon derivatives containing isocyanide (isonitrile) functional groups (Fig. 5).[27,28]

The polyisonitrile-nylon obtained in this manner can be coupled to enzymes via the NC group on the polymer, again by a four-component condensation reaction carried out in aqueous buffer at neutral pH in the

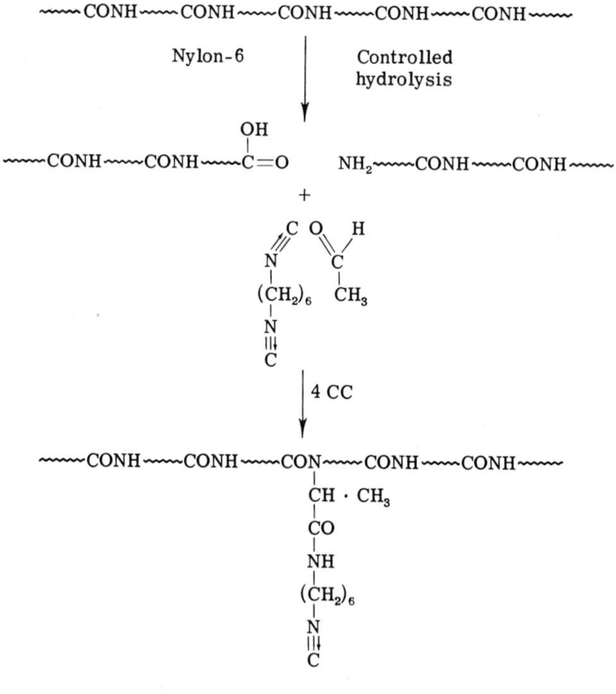

∼∼∼CONH∼∼∼CONH∼∼∼CONH∼∼∼CONH∼∼∼CONH∼∼∼

Nylon-6 ↓ Controlled hydrolysis

∼∼∼CONH∼∼∼CONH∼∼∼C=O   NH₂∼∼∼CONH∼∼∼CONH∼∼∼
                  |
                  OH

+

$$\underset{\underset{\underset{C}{\overset{|||}{N}}}{\underset{|}{(CH_2)_6}}}{\underset{|}{N}}\overset{C}{\underset{}{\overset{\nearrow}{}}}\overset{O}{\underset{}{}}\overset{H}{\underset{CH_3}{\overset{\nwarrow}{C}}}$$

↓ 4 CC

∼∼∼CONH∼∼∼CONH∼∼∼CON∼∼∼CONH∼∼∼CONH∼∼∼
                      |
                      CH · CH₃
                      |
                      CO
                      |
                      NH
                      |
                      (CH₂)₆
                      |
                      N
                      |||
                      C

Polyisonitrile-nylon

FIG. 5. Synthesis of polyisonitrile-nylon.

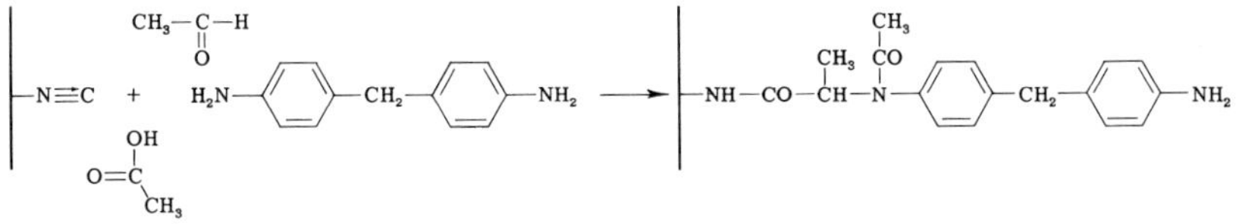

FIG. 6. Synthesis of polyaminoaryl-nylon.

presence of a water-soluble aldehyde (e.g., acetaldehyde); here the protein supplies either the amino or carboxyl component, the isocyanide group on the support being directed mainly toward only one type of functional group on the protein by the addition of the missing fourth component, in excess, to the reaction medium.[27] Thus enzymes can be bound to polyisonitrile-nylon through their amino groups by four-component reactions in the presence of acetaldehyde and acetate; conversely, enzymes can be bound through their carboxyl groups in the presence of acetaldehyde and an amine, such as Tris.

The isocyanide groups on nylon can be easily modified to other functional groups. This is demonstrated here by the conversion of polyisonitrile-nylon to a diazotizable arylamino derivative, polyaminoaryl-nylon, by four-component condensation with a bifunctional aromatic amine, p,p'-diaminodiphenylmethane, in the presence of an aldehyde (acetaldehyde or isobutyral) and a carboxylate compound (Fig. 6).[27] Coupling of an enzyme to this nylon derivative is effected through azo bonds, mainly with aromatic amino acid residues on the protein.

The conversion of the isocyanide group of polyisonitrile-nylon to other functional groups should be useful in cases where coupling of enzymes to nylon via four-component condensation reactions is undesirable owing to sensitivity of the enzyme to aldehydes.

The following procedures, developed by Goldstein et al.[27,28] for the immobilization of enzymes to nylon powders, by methods involving the N-alkylation of the support, can be easily adapted to other nylon structures, such as tubes, fibers, sheets.

Procedure

Synthesis of 1,6-Diisocyanohexane

1,6-Diisocyanohexane is a component required in the preparation of polyisonitrile-nylon given below. The general procedure adopted here is

[27] L. Goldstein, A. Freeman, and M. Sokolovsky, *Biochem J.* **143**, 497 (1974).
[28] L. Goldstein, A. Freeman, and M. Sokolovsky, in "Enzyme Engineering" (E. K. Pye and L. B. Wingard, Jr., eds.), Vol. 2, p. 97. Plenum Press, New York, 1974.

based on the conversion of a primary amine into the appropriate N-alkyl formamide by treatment with ethyl formate, followed by dehydration to the isocyanide.[29,30]

N,N'-Diformyl-1,6-diaminohexane. This compound is prepared by a modification of the procedure of Moffat et al.[31] 1,6-Diaminohexane (60 g, 0.5 mole) is dissolved in ethyl formate (170 ml, 3.3 moles) with stirring over ice. Upon completion of dissolution, a white precipitate is formed; an additional portion of ethyl formate (50 ml) is added, and the mixture is refluxed for 2 hr. In the course of the reaction, solid N,N'-diformyl-(1,6-diaminohexane) separates at the bottom of the flask. The reaction mixture is cooled, then the liquid is decanted and the solid is transferred to a rotatory evaporator to remove residual solvent. The solid material (60 g, 70% yield, m.p. 105°) is used in the next step without further purification.

1,6-Diisocyanohexane. A modification of the procedure of Hertler and Corey[32] is used: p-toluene sulfonyl chloride (150 g, 0.8 mole) is dissolved with stirring in 300 ml of pyridine (KOH dried). To the light yellow solution, N,N'-diformyl-(1,6-diaminohexane) (30 g, 0.2 mole) is added in small portions with stirring. The dehydration reaction occurs as the formamide dissolves; the solution warms up, and the color darkens from yellow to brown and finally to black. After dissolution is complete the reaction mixture is stirred for 1.5 hr at room temperature. Water (250 ml) and crushed ice, to lower the temperature, is then added and the solution is extracted with three 150-ml portions of ether. The combined ether extract is washed with water (three 250-ml portions) and dried over anhydrous Na_2SO_4. The ether is then removed by evaporation and the crude oil is vacuum distilled (pressure < 0.1 mm). The fraction distilling at 85°–92° (9 g, 30% yield) is collected and stored in closed vials at −5°.

Preparation of Polyisonitrile-Nylon

Nylon Powder. Commercial nylon 6 pellets are suspended in a 20% methanolic solution of $CaCl_2$ (30 g/liter) and stirred at room temperature until a homogeneous, though extremely viscous, solution is obtained.

The nylon solution is added dropwise with strong stirring into a large

---

[29] I. Ugi, U. Fetzer, U. Eholzer, H. Knupfer, and K. Offermann, *Angew. Chem. Int. Ed. Engl.* **4**, 472 (1965).
[30] P. Hoffman, G. Gokel, D. Marquarding, and I. Ugi, in "Isonitrile Chemistry" (I. Ugi, ed.), pp. 9–39. Academic Press, New York, 1971.
[31] J. Moffat, M. V. Newton, and G. J. Papenmeier, *J. Org. Chem.* **27**, 4058 (1962).
[32] W. R. Hertler and E. J. Corey, *J. Org. Chem.* **23**, 1221 (1958).

excess of water; the powder thus obtained is separated on a suction filter, washed with water, resuspended in water, and homogenized. The powder is separated, washed again with water and ethanol ether, and air dried. Traces of solvent and moisture are removed in a vacuum desiccator over phosphorus pentoxide; the dry powder is ground in a mortar. The mean carboxyl content of the nylon powders, determined titrimetrically, is about 25 µmoles per gram of dry nylon powder. The nylon powder particles are spherical and range in diameter between 0.2 and 0.7 µm.

Controlled Hydrolysis of Nylon Powder. Nylon 6-powder (10 g) is suspended in 3 N HCl (300 ml) and stirred at room temperature (20°) for 4 hr. The powder is separated on a suction filter, washed exhaustively with water, ethanol, and ether, and air dried. Traces of solvent and moisture are removed in a vacuum desiccator over phosphorus pentoxide. The powders are stored in closed vessels. The carboxyl content of nylon-6 powder samples hydrolyzed for 4 hr is 60–70 µmoles/g.

Polyisonitrile-Nylon. Partially hydrolyzed nylon powder, 2 g (mean carboxyl content 60–70 µmoles/g) is suspended in isopropanol (80 ml). Acetaldehyde (20 ml, 0.3 mole) or isobutyral (32 ml, 0.3 mole) is then added, followed by 1,6-diisocyanohexane (8 ml, 0.06 mole), and the reaction is allowed to proceed in a closed vessel for 24 hr with stirring at room temperature. The polyisonitrile-nylon powder is separated on a suction filter, washed with isopropanol (50 ml) and then with ether (~200 ml), and air dried. Traces of solvent are removed in a vacuum desiccator over phosphorus pentoxide. The polyisonitrile-nylon powder is stored at −5° in a dark, stoppered vial over silica gel. The mean carboxyl content of polyisonitrile-nylon is about 20 µmoles/g. The mean isocyanide content of polyisonitrile-nylon is 40–50 µmoles/g.

Carboxyl Content of Nylon Powders. The carboxyl content of the nylon-powder samples is determined by anhydrous titration with sodium methoxide essentially as described by Patchornik and Ehrlich-Rogozinski.[33] Nylon powder (50 mg) is suspended in dimethylformamide (2 ml) and stirred magnetically in a stoppered test tube for 4 hr (to achieve uniform swelling for all titrated samples). Thymol blue indicator (2 drops of a 1% ethanolic solution) is added, and the stirred sample is titrated with 0.05 N sodium methoxide (in benzene-methanol) delivered from a microburette, until a blue color appears, which persists for 5–10 sec. Appropriate solvent blanks are determined for each titration. Care should be taken to keep the titration vessels covered to prevenet penetration of carbon dioxide.

Isocyanide Content of Nylon Powders. The isocyanide content of

[33] A. Patchornik and S. Ehrlich-Rogozinski, *Anal. Chem.* 33, 803 (1961).

$$\text{R—NC} + 2\,\text{HSCN} \longrightarrow \text{R—N}\underset{\underset{S}{\overset{}{\diagdown}}}{\overset{\overset{N}{\diagup}}{\diagdown}}\!\text{=S}$$

polyisonitrile nylon powders can be determined according to the reaction by an adaptation of the titrimetric method of Arora et al.[34]

METHOD.[35] Polyisonitrile-nylon powder (30 mg; 1.5–2 µmoles of isocyanide groups) is suspended in 1.4 ml of ethyl acetate and cooled over ice. A 5-fold molar excess of isothiocyanic acid in ethyl acetate (0.1 ml of a 0.1 N solution) is added, and the cooled reaction mixture is stirred magnetically for 30 min. Dimethylformamide (1.5 ml) and 1 drop of neutral red/methylene blue indicator are then added. The magnetically stirred sample is titrated at room temperature with 0.05 M triethylamine in ethyl acetate. Blank titration with all reagents except the nylon powder is carried out in parallel. The amount of isocyanide in the polyisonitrile-nylon sample is calculated from the difference between the two titrations, assuming a stoichiometry of 2:1 for isothiocyanic acid to isocyanide.

Reagents[34]

Isothiocyanic acid, 0.1 N. An aqueous solution of NH_4SCN (900 mg in 2.5 ml) is mixed with 100 ml of ethyl acetate; 10% sulfuric acid (5 ml) is added dropwise with stirring and cooling. The organic phase is separated and dried over $MgSO_4$ or Na_2SO_4. To remove traces of Fe(III), powdered $K_4[Fe(CN)_6]$ (100 mg) is added with stirring and then filtered off. The normality is determined on an 0.1-ml aliquot by titration with 0.05 N triethylamine in ethyl acetate as described above.

Indicator: 0.2% solution of neutral red and methylene blue in methanol. Color change from purple to green.

Triethylamine, 0.05 N. Triethylamine, 0.6 ml (distilled over KOH) is dissolved in 100 ml of ethyl acetate. The normality is determined by titrating 50 µl of 0.1 N aqueous HCl in 1.5 ml of ethyl acetate and 1.5 ml of dimethylformamide as above.

Preparation of Polyaminoaryl-Nylon

p,p'-Diaminodiphenylmethane (2 g, 0.01 mole) is dissolved in 200 ml of methanol, and 0.5 ml (0.005 mole) of isobutyral is added. Polyisonitrile nylon powder (2 g) is suspended in the solution, and 1 ml of glacial acetic acid (0.017 moles) is added. The reaction is allowed to proceed in a closed vessel for 24 hr with stirring at room temperature. The poly-

[34] A. S. Arora, E. V. Hinrichs, and I. Ugi, *Fresenius' Z. Anal. Chem.* **269**, 124 (1974).
[35] A. Freeman and L. Goldstein, unpublished results.

aminoaryl-nylon is separated on a suction filter, washed with dimethylformamide, methanol, and then ether, and air dried. The diazotization capacity of polyaminoaryl-nylon thus prepared is about 20 μmoles/g.

Determination of the Diazotization Capacity of Polyaminoaryl-Nylon. Polyaminoaryl-nylon (50 mg) is suspended in cold 0.2 N HCl (5 ml), and aqueous sodium nitrite (25 mg in 1 ml) is added dropwise. The reaction mixture is stirred for 30 min over ice; the diazotized resin is separated on a suction filter, washed with cold water, cold 0.1 M phosphate pH 8, and resuspended in the same buffer (6 ml). An aqueous solution of p-bromophenol (50 mg dissolved in 3 ml of water, by dropwise addition of 2 N NaOH) is then added, and the reaction mixture is left stirring overnight at 4°. The p-bromophenol-nylon conjugate is separated on a filter, washed with 0.05 M carbonate buffer at pH 10.5 or water brought to the same pH, with deionized water, methanol, and ether and dried *in vacuo* over phosphorus pentoxide. The diazotization capacity of polyaminoaryl-nylon is estimated from the bromine content of the p-bromophenol conjugate determined by the Schöniger combustion method.[36]

Coupling of Enzymes to N-Alkylated Nylon Derivatives

Enzymes can be coupled to polyisonitrile-nylon through the amino groups on the protein by four-component condensation reactions in the presence of acetaldehyde and a carboxylate (e.g., acetate); conversely, enzymes can be bound through their carboxyl groups in the presence of acetaldehyde and an amine, such as Tris. In cases where coupling through tyrosyl residues on the enzyme is preferable, polyisonitrile-nylon is converted to the polyaminoaryl derivative.

Coupling of Enzymes to Polyisonitrile-Nylon through Amino Groups on the Protein. Polyisonitrile-nylon (50 mg) is suspended in 2 ml of cold 0.1 M sodium phosphate 0.5 M in sodium acetate, pH 7. A cold aqueous solution of enzyme (5–10 mg in 1 ml) is then added, followed by 0.1 ml of acetaldehyde. The acetaldehyde, b.p. 21°, is pipetted with a precooled pipette to prevent formation of bubbles. The reaction mixture is left stirring overnight at 4°. The water-insoluble enzyme derivative is separated on a filter, washed with water, then with 1 M KCl, 0.1 M in NaHCO$_3$, and again with water, resuspended in water (4 ml), and stored at 4°.

Coupling of Enzymes to Polyisonitrile-Nylon through Carboxyl

[36] A. Steyermark, "Quantitative Organic Microanalysis" (2nd ed.), pp. 177–178. Academic Press, New York, 1961.

Groups on the Protein. Polyisonitrile-nylon (50 mg) is suspended in 1 ml of cold 0.1 M Tris·HCl buffer pH 7.0. A cold solution of enzyme in the same buffer (2–10 mg in 1 ml) is added followed by 0.1 ml of acetaldehyde, pipetted from a precooled pipette. The reaction mixture is left stirring overnight at 4°, washed, and resuspended in water (4 ml) as described above.

Maximum recoveries of immobilized enzymic activity (30–40% of the amount added to the reaction mixture) are obtained by both methods with 4–5 mg of enzyme per 100 mg of support.

Coupling of Enzymes to Polyaminoaryl-Nylon. Polyaminoaryl-nylon (100 mg) is suspended in cold 0.2 M HCl (7 ml) and aqueous sodium nitrite (25 mg in 1 ml) added dropwise. The reaction mixture is stirred for 30 min over ice; the red-brown diazotized polyaminoaryl-nylon is separated on a suction filter, washed with cold water, cold 0.1 M phosphate pH 8, and resuspended in 6 ml of the same buffer. A cold aqueous solution of enzyme (5–15 mg in 2–3 ml) is then added to the magnetically stirred suspension of diazotized support, and the reaction mixture is left stirring overnight at 4°. The insoluble enzyme derivative is separated by filtration, washed with water, 1 M KCl, 0.1 M in NaHCO$_3$, and water, resuspended in water (5 ml), and stored at 4°.

[10] Covalent Coupling Methods for Inorganic Support Materials

By HOWARD H. WEETALL

Choosing the Carrier

Inorganic support materials have been shown to be excellent carriers for immobilized enzymes.[1-8] The characteristics of these inorganic support materials are extremely important and play an definitive role in the choice of the proper carrier for the enzyme to be immobilized.

The carrier must meet several important criteria. Eaton has presented

[1] H. H. Weetall, *Science* **166**, 615 (1969).
[2] H. H. Weetall and L. S. Hersh, *Biochim. Biophys. Acta* **185**, 464 (1969).
[3] H. H. Weetall, *Nature (London)* **223**, 959 (1969).
[4] H. H. Weetall and N. B. Havewala, *Biotechnol. Bioeng. Symp.* 3, 241 (1972).
[5] H. H. Weetall, *Anal. Chem.* **46**, 602A (1974).
[6] R. A. Messing, *Process Biochem.* **1974**, Nov. (1974).
[7] R. A. Messing and H. R. Stinson, *Mol. Cell. Biochem.* **4**, 217 (1974).
[8] C. C. Q. Chin and F. Wold, *Anal. Biochem.* **61**, 379 (1974).

TABLE I
DURABILITY TEST RESULTS FOR SUPPORTS

Material description	Static test (mg/m² per 16 hr)		Dynamic test (mg/m² per day)		
	1% NaOH	5% HCl	pH 4.5	pH 7.0	pH 8.2
TiO_2	0.2	0.8	0.05	0.05	ND[a]
ZrO_2	0.2	1.1	0.004	0.004	0.01
Al_2O_3	0.6–0.8	2.0	0.056–0.086	0.01	0.01
Al_2O_3–SiO_2	1.85	3.65	0.08–0.1	0.02–0.05	0.06
CPG^b–ZrO_{2COAT}	1.3–2.0	0.2	0.03	0.7	0.7–0.9
CPG	3.06	0.08	0.02	0.5	0.3

[a] Not done.
[b] CPG, controlled-pore glass.

what he terms a "Decision Tree."[9] Answering the questions Eaton poses permits one to choose the proper carrier for enzyme immobilization.

Immobilized Enzyme Decision Tree

A. Does the pore morphology permit entry of the enzyme?
B. Can the enzyme be immobilized on the support?
C. Is the immobilized enzyme durable in (1) acid, (2) base, (3) high salt?
D. Can the material be conveniently handled?
E. Does the carrier have compression strength?
F. Is the maximum enzyme loading adequate for the system?
G. What is the maximum tolerable pressure drop?
H. How is the above affected by particle size, flow rate, and particle shape?
I. What is the operational half-life of the system?
J. How is the half-life affected by temperature, pH, and other conditions?
K. Under what conditions can the derivative be stored?

By answering the above questions, one can determine whether one has the best carrier for the enzyme of interest.

Studies in our laboratories have shown that inorganic supports, including porous glass and porous ceramics, have extremely different physical characteristics. Table I gives the results of durability tests on several different support materials.

One can see from these data that under some conditions the ceramic

[9] D. L. Eaton, *in* "Immobilized Biochemicals and Affinity Chromatography" (R. Bruce Dunlap, ed.), pp. 241–258. Plenum, New York, 1974.

TABLE II
COMPARISON OF SURFACE AREA TO PORE VOLUME FOR A
CONTROLLED-PORE INORGANIC SUPPORT

Pore diameter (Å)	Surface area[a] (m^2/g)	Surface area[b] (m^2/g)
75	249	356
125	149	214
175	107	153
240	78	111
370	50	72
700	27	38
1250	15	21
2000	9	13

[a] The pore volume for these calculations was taken at 0.70 ml/g.
[b] The pore volume for these calculations was taken at 1.0 ml/g.

carriers are more durable than the glass. Thus, if one intends to operate an immobilized enzyme system at alkaline pH values, a ceramic, such as TiO_2, might be the carrier of choice rather than controlled-pore glass (CPG).

Another major factor in choosing the carrier is the pore diameter relative to surface area (Table II). There will always be an optimal pore diameter for an enzyme. The relationship between pore diameter and surface area is an inverse relationship. Therefore, one always should choose the lowest possible pore diameter giving the greatest possible surface area and greater enzyme loading. This relationship is shown very nicely in Fig. 1 with amyloglucosidase covalently coupled to porous glass. It is immediately obvious that the optimum pore diameter is 300 Å. Anything less excludes the enzyme.

Porous glass can be prepared at any desired pore diameter over the range of 30 Å to 2000 Å. The pore diameter available on the ceramic carriers developed at Corning[10] are not quite as broad, but they fall within the useful range (Table III).

The choice of particle size also has a major impact on the activity of an immobilized enzyme. The larger the particle, the greater the effect of diffusion control (Fig. 2). Although the data indicate that the smallest particle size is usually the best choice, one must also consider pressure drop vs particle size. The larger the particle, the lower the pressure drop across a bed of immobilized enzymes (Fig. 3).

[10] R. A. Messing, Res. Dev. 25, 32 (1974).

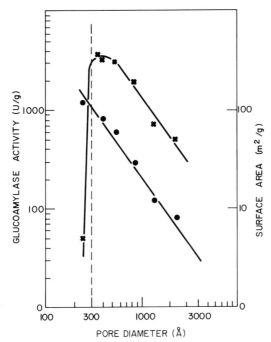

FIG. 1. Relationship of pore size to surface area and enzymic activity. ●, Surface area date; ×, enzymic activity results.

TABLE III
PHYSICAL PARAMETERS OF POROUS CERAMICS

Composition		Pore diameter range (Å)	Pore diameter, average (Å)	Pore volume (ml/g)
SiO$_2$	75%	205–875	465	0.76
TiO$_2$	25%			
SiO$_2$	90%	185–700	435	0.76
ZrO$_2$	10%			
SiO$_2$	100%	185–700	435	0.76
SiO$_2$	84.3%	110–575	235	1.30
ZrO$_2$	15.7%			
SiO$_2$	75%	205–575	435	0.89
Al$_2$O$_3$	25%			
TiO$_2$	98%	205–500	410	0.53
MgO	2%			
Al$_2$O$_3$	100%	150–250	175	0.60
TiO$_2$	100%	220–400	350	0.45
TiO$_2$	100%	300–590	420	0.40
TiO$_2$	100%	725–985	855	0.22

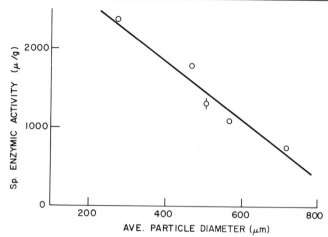

Fig. 2. Typical results expected for a diffusion-restricted material. Enzyme activity increases as the particle size and diffusion path decrease in size or length.

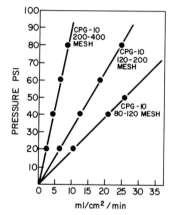

Fig. 3. Experimental results for controlled-pore glass (CPG) showing the effect of particle size distribution on pressure drop and flow rate.

All these parameters must be considered when choosing the inorganic support for enzyme immobilization.

Enzyme Immobilization

Immobilization by adsorption techniques is covered in this volume [11]. We concern ourselves here only with covalent attachment methods.

Immobilization by covalent attachment to inorganic supports involves reactions that are similar to the covalent attachment of enzymes to organic supports. Only methods with which this author has had personal experience are presented (see also this volume [32], [54], [55]).

Preparation of Carrier. Generally an acid wash (5% HNO_3) at 80°–90° for 60 min followed by rinsing with distilled water will hydrate and clean the carrier surface.

Silane Coupling Techniques

Alkylamine Coupling

Aqueous Silanization. This method of silanization appears to couple a monolayer of silane across the carrier surface. The organic solvent techniques give higher amine loadings. However, experience has shown that greater carrier durability with slightly lower enzyme loadings are achieved by aqueous silanization (Fig. 4).

To 1 g of clean inorganic support material is added 18 ml of distilled water plus 2 ml of γ-aminopropyltriethoxysilane (10% v/v). This compound is commercially available from Union Carbide, as are the other silanes. After addition of the silane solution, the pH is adjusted to between pH 3 and 4 with 6 N HCl. The neutralization of the basic silane solution may cause heating. After pH adjustment, place the reactants in a 75° water bath for 2 hr. Remove from bath, filter on a Büchner funnel, and wash with 20 ml of distilled water. Dry in a 115° oven for at least 4 hr. The product may be stored at this point for later use.

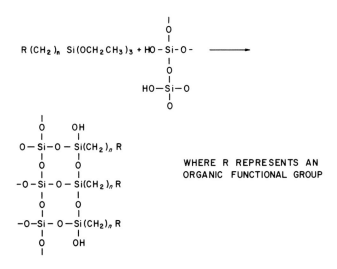

FIG. 4. Silanization of the surface of porous glass.

Organic Silanization. This technique will give much higher loadings of alkylamine than the aqueous method. However, the silane appears to be somewhat patchy, and it is not quite as durable.

To 1 g of clean porous carrier add 50 ml of a 10% solution of γ-aminopropyltriethoxysilane in toluene (v/v). Reflux overnight. Filter on a Büchner funnel, and wash with toluene followed by acetone. Air dry before placing in an oven overnight at 115°. The product may be stored at this point.

Another organic silanization method is an evaporative technique that lays down a more even silane layer. To 1 g of porous inorganic support material, add 25 ml of a 1% solution of γ-aminopropyltriethoxysilane in acetone (v/v). Evaporate to dryness and then heat at 115° overnight. The product may be stored at this point.

Derivatives of Alkylamines and Coupling Techniques

Alkylamine Coupling

Carbonyl Derivative. The technique using glutaraldehyde is simple, gentle, and rapid. It is important to remove all excess glutaraldehyde before adding the enzyme; otherwise cross-linking will occur. The crosslinked enzyme will decrease activity recovery by blocking pores and preventing passage of larger molecules (Fig. 5).

To 1 g of alkylamine carrier, add 25 ml of a 2.5% solution of glutaraldehyde in 0.05 M Na_2HPO_4 buffer adjusted to pH 7.0. Allow the reaction to continue for at least 60 minutes. During this time one should observe some sort of color change either to a magenta or tan. Wash exhaustively with distilled water on a Büchner funnel. To the activated carrier add the enzyme in as small a volume as possible at a minimum of a 1% concentration, if possible using the same buffer as that used in the activation step. The quantity of enzyme added should be between 50 and 100 mg per gram of carrier. Allow 2–4 hr for the reaction. Wash with distilled water.

Soaking in 6 M urea has been used successfully for removing adsorbed enzyme without inactivating the coupled enzyme in most cases. However, before subjecting the entire preparation to a urea soak, it is advisable

$$\text{CARRIER} \left\{ -\overset{|}{\underset{|}{O}} -\overset{|}{\underset{O}{Si}}(CH_2)_3 NH_2 + \overset{CHO}{\underset{CHO}{(CH_2)_3}} \longrightarrow \left\{ -\overset{|}{\underset{|}{O}} -\overset{|}{\underset{O}{Si}}(CH_2)_3 N=CH(CH_2)_3 CHO \right. \right.$$

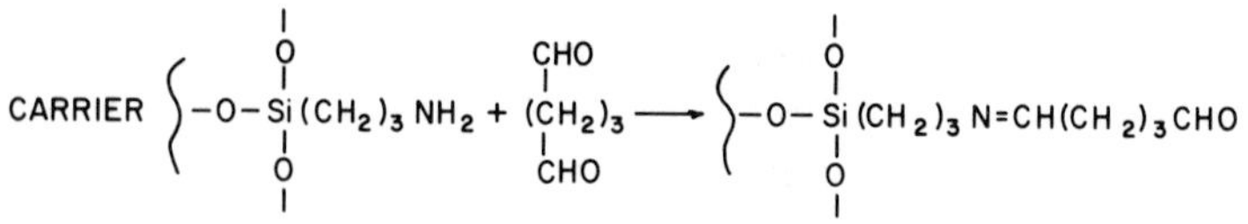

FIG. 5. Preparation of the aldehyde derivative from the alkylamine derivative.

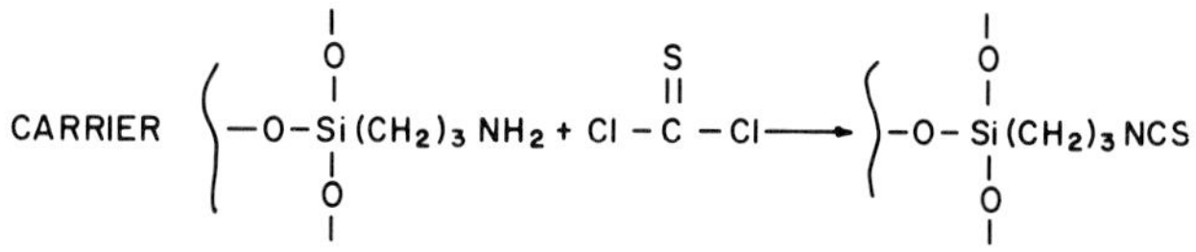

$$\text{CARRIER} \left(-\text{O}-\underset{\underset{\text{O}}{|}}{\overset{\overset{\text{O}}{|}}{\text{Si}}}(\text{CH}_2)_3\,\text{NH}_2 + \text{Cl}-\overset{\overset{\text{S}}{\|}}{\text{C}}-\text{Cl} \longrightarrow \right. \left. -\text{O}-\underset{\underset{\text{O}}{|}}{\overset{\overset{\text{O}}{|}}{\text{Si}}}(\text{CH}_2)_3\,\text{NCS} \right)$$

FIG. 6. Preparation of the isothiocyanate from the alkylamine derivative.

to test a small batch first. Do not dry the sample of immobilized enzyme. Retain as a damp preparation, as recovered from the filter, in a sealed container in the refrigerator.

Isothiocyanate Coupling. To 1 g of alkylamine carrier is added a 10% solution of thiophosgene in chloroform (v/v). All work must be carried out under a hood because of the nauseous and toxic nature of the thiophosgene. The reactants are refluxed for at least 4 hr; generally 12–15 hr is best. Wash with dry chloroform and vacuum-dry in a desiccator. The derivative should be used as soon as possible for coupling to an enzyme (Fig. 6).

The derivative is added slowly to a 1% enzyme solution, containing 50–100 mg of enzyme per gram of carrier, previously adjusted to pH 8.5–9.0. Monitor the pH and maintain until the pH stabilizes. Allow an additional 2 hr before washing and storing the final product.

The reaction can also be carried out with an arylamine glass. The preparation of the arylamine will be presented below.

Carbodiimide Coupling. This technique is useful if the enzyme must be coupled at an acidic pH value, as in the case of pepsin.[11] It is preferable to use a carboxyl derivative, since there will be less chance of cross-linking to the carboxyl-activated carrier. However, coupling via carbodiimide to the alkylamine in many cases is an excellent coupling method (Fig. 7).

To 1 g of alkylamine glass add 50 ml of 0.03 M H_3PO_4 adjusted to pH 4.0. To this add 100–200 mg of a water-soluble carbodiimide. The carbodiimide we prefer to use is 1-cyclohexyl-3-(2-morpholinoethyl) carbodiimide, metho-p-toluene sulfonate. The enzyme, usually 50–100 mg per gram of carrier, is also added directly to the mixture. The reaction is continued overnight at 4°. The product is washed and stored until use.

Triazine Coupling. This reaction is relatively simple and deserves greater attention (Fig. 8).

To 1 g of alkylamine carrier add 10 ml of benzene containing 0.2 ml of triethylamine and 0.3 g of 1,3-dichloro-5-methoxytriazine. Allow the mixture to react for 2–4 hr at 45°–55°. Decant, wash with benzene, and dry

[11] W. F. Line, A. Kwong, and H. H. Weetall, *Biochim. Biophys. Acta* **242**, 194 (1971).

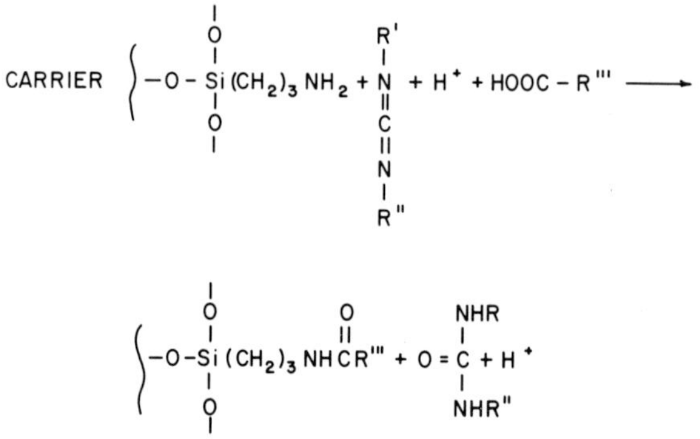

$$\text{CARRIER} \left\{ \begin{matrix} | \\ O \\ | \\ -O-Si(CH_2)_3\,NH_2 \\ | \\ O \\ | \end{matrix} \right. + \begin{matrix} R' \\ | \\ N \\ \| \\ C \\ \| \\ N \\ | \\ R'' \end{matrix} + H^+ + HOOC-R''' \longrightarrow$$

$$\left\{ \begin{matrix} | \\ O \\ | \\ -O-Si(CH_2)_3\,NHCR''' \\ | \\ O \\ | \end{matrix} \right. \;\; + \;\; O=\underset{\underset{NHR''}{|}}{\overset{\overset{NHR}{|}}{C}} + H^+$$

Fig. 7. Preparation of an amide-linked ligand with the alkylamine derivative and a carbodiimide. R' and R'' represent groups on the carbodiimide, e.g., cyclohexyl groups. R''' represents the ligand having a free carboxyl group.

in an evaporator oven at 100°. Coupling is in 0.05 M phosphate buffer pH 8.0 at 4° overnight using 50–100 mg of enzyme per gram of activated carrier.

Preparation of Arylamine Carrier

The arylamine derivative is normally prepared by reaction of p-nitrobenzoylchloride with the alkylamine derivative. The arylamine can be activated by the same processes used for organic arylamines. However, the most common use of arylamines is for coupling via azo linkage (Fig. 9).

To 1 g of alkylamine carrier add 25–50 ml of chloroform containing 5% triethylamine (v/v) and 1 g of p-nitrobenzoylchloride. Reflux for at least 4 hr. Wash with chloroform, dry, and boil in a 5% solution of sodium dithionite in water. Dry and store in a dark place.

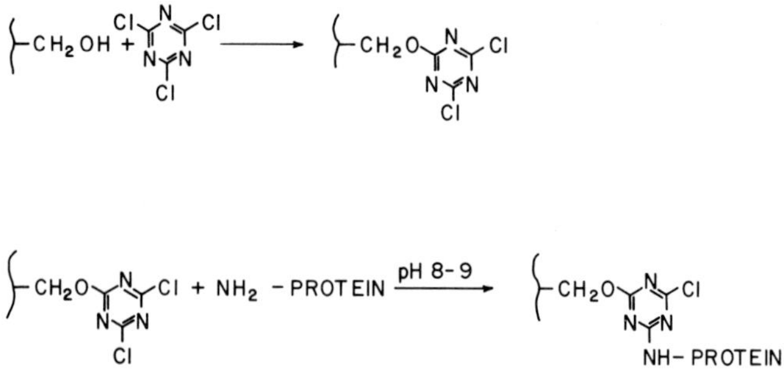

Fig. 8. Coupling through a substituted triazine.

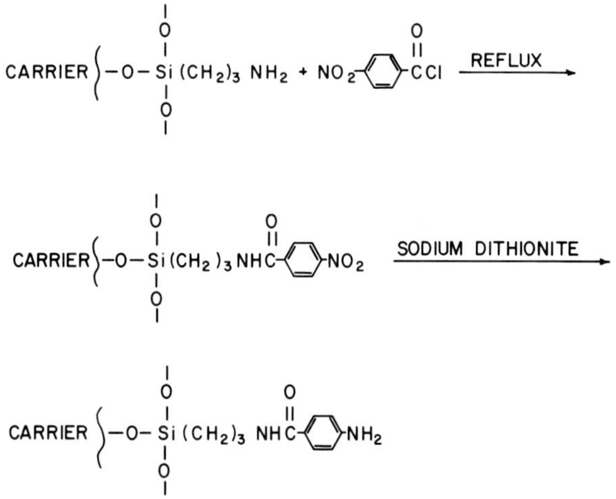

[10] COVALENT COUPLING METHODS 143

CARRIER⎨−O−Si(CH$_2$)$_3$NH$_2$ + NO$_2$−⟨⎯⟩−CCl →REFLUX→

CARRIER⎨−O−Si(CH$_2$)$_3$NHC−⟨⎯⟩−NO$_2$ →SODIUM DITHIONITE→

CARRIER⎨−O−Si(CH$_2$)$_3$NHC−⟨⎯⟩−NH$_2$

FIG. 9. Preparation of the arylamine derivative from the alkylamine derivative.

Azo Coupling. To 1 g of arylamine carrier add 20 ml of 2 N HCl and cool in an ice bath. To the cold preparation add 100 mg of solid NaNO$_2$. Allow diazotization to continue at least 30 min. The carrier should take on a yellowish color. By testing with starch-iodine paper, one can tell whether any additional NaNO$_2$ is necessary during the diazotization procedure. After 30 min remove the derivative, then wash on a Büchner funnel with ice cold water. The derivative is now ready for enzyme coupling (Fig. 10).

The enzyme is dissolved in an appropriate buffer at pH 8.5. Usually 100 mg of enzyme per gram of carrier is a reasonable quantity of enzyme. The higher the enzyme concentration, the more efficient will be the coupling. A range of 0.5–2.0% enzyme solution is an excellent working concentration. For maximum coupling, allow the reaction to continue in

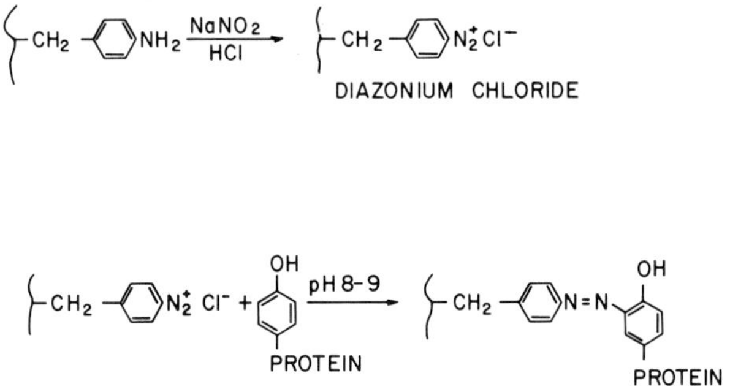

⎨−CH$_2$−⟨⎯⟩−NH$_2$ →NaNO$_2$/HCl→ ⎨−CH$_2$−⟨⎯⟩−N$_2^+$Cl$^-$

DIAZONIUM CHLORIDE

⎨−CH$_2$−⟨⎯⟩−N$_2^+$Cl$^-$ + ⟨⎯⟩−OH (PROTEIN) →pH 8-9→ ⎨−CH$_2$−⟨⎯⟩−N=N−⟨⎯⟩−OH (PROTEIN)

FIG. 10. Coupling through azo linkage.

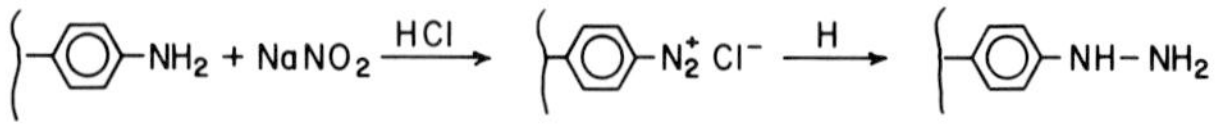

$$\text{\large \{}\!\!-\!\!\bigcirc\!\!-\!\text{NH}_2 + \text{NaNO}_2 \xrightarrow{\text{HCl}} \text{\large \{}\!\!-\!\!\bigcirc\!\!-\!\text{N}_2^+\,\text{Cl}^- \xrightarrow{\text{H}} \text{\large \{}\!\!-\!\!\bigcirc\!\!-\!\text{NH}-\text{NH}_2$$

FIG. 11. Preparation of the phenylhydrazine.

the cold for at least 2 hr for proteases, 18 hr for enzymes not autolyzed. However, maximum coupling and recoverable activity may not be the same. One must determine optimum coupling time.

Phenylhydrazine Coupling. This coupling technique is not very common but can be used for coupling through carbonyl groups (Fig. 11).

One gram of arylamine carrier is diazotized as described above, washed with ice cold water followed by a wash with ice cold 1% sulfamic acid, followed again by ice cold water. The product is reduced by reaction with 10 ml of 1% sodium dithionite solution at pH 8.5 by refluxing. Refluxing is continued with the addition of NaOH for pH maintenance as necessary for 60 min. The product is washed with water and is ready for immediate coupling under slightly alkaline conditions at 0°.

Preparation of the Carboxy Derivative

The carboxy derivative can be prepared by the reaction of the alkylamine carrier with succinic anhydride (Fig. 12).

To 1 g of alkylamine carrier is added 25 ml of 0.05 M phosphate buffer adjusted to pH 6.0. To this is added 1.0 g of succinic anhydride. The preparation is stirred at room temperature with pH adjustment as necessary over the first few hours. Continue the reaction for at least 15 hr at room temperature. The product is washed and can be dried.

Carbodiimide Coupling to Carboxyl Derivative. Carbodiimide coupling to the alkylamine carrier can cross-link; coupling to the carboxyl derivative cannot, if handled properly (Fig. 13).

To 1 g of carboxylated derivative is added 200 mg of 3-(2-morpholinoethyl) carbodiimide previously dissolved in distilled water. The reactants are adjusted to pH 10.0 and allowed to react at room temperature for 2 hr. Successful activation can also be achieved at pH 4.0. The product is washed with distilled water and added to the enzyme solution. Although

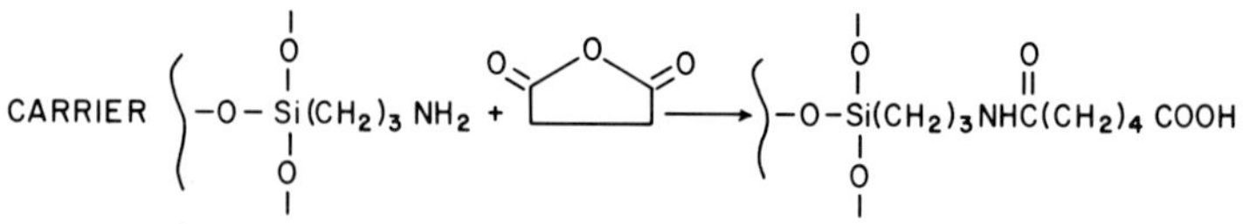

$$\text{CARRIER}\ \Big\{\!-\!\underset{\underset{\text{O}}{|}}{\overset{\overset{\text{O}}{|}}{\text{Si}}}(\text{CH}_2)_3\,\text{NH}_2 + \overset{\text{O}}{\underset{}{\diagdown}}\!\!\!\overset{}{\underset{}{\diagup}}\!\!\!\overset{\text{O}}{\underset{}{\diagdown}}\!\!\!\overset{}{\underset{}{\diagup}}\!\overset{\text{O}}{\underset{}{}} \longrightarrow \Big\}\!-\!\text{O}\!-\!\underset{\underset{\text{O}}{|}}{\overset{\overset{\text{O}}{|}}{\text{Si}}}(\text{CH}_2)_3\text{NHC}(\text{CH}_2)_4\,\text{COOH}$$

FIG. 12. The covalent coupling of an amine functional ligand with the aldehyde derivative followed by reduction with sodium borohydride.

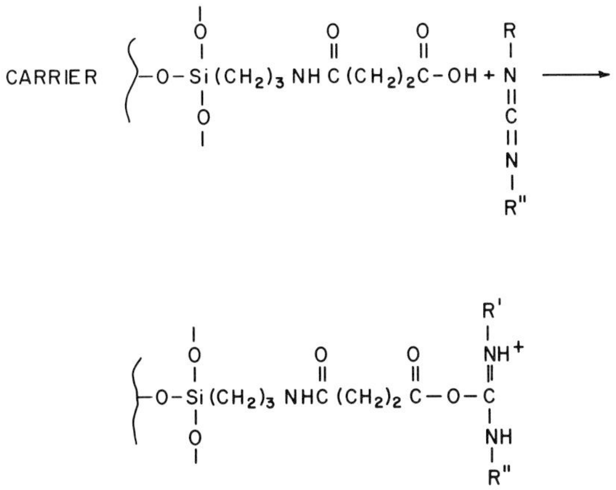

```
                      O          O     O      R
                      |          ||    ||     |
  CARRIER   }-O-Si(CH2)3 NHC(CH2)2C-OH + N
                      |                        ||
                      O                        C
                      |                        ||
                                               N
                                               |
                                               R"
```

```
                                        R'
                                        |
                      O          O     O   NH+
                      |          ||    ||    ||
  }-O-Si(CH2)3 NHC(CH2)2 C-O-C
                      |                        |
                      O                        NH
                      |                        |
                                               R"
```

FIG. 13. The preparation of the pseudourea of a carboxyl derivative by reaction with a carbodiimide.

coupling will occur at pH 4.0–5.0, we have observed coupling at pH 9–10 also. The final product is washed and stored for use.

Acid Chloride Coupling. The acid chloride can be prepared from the carboxyl derivative. However, it is not very stable and should be used soon after preparation (Fig. 14).

To 1 g of carboxylated carrier add 50 ml of a 10% solution of thionylchloride in chloroform. The preparation is refluxed for 4–6 hr, filtered, and dried in a vacuum oven at 60°–80°.

Coupling is in a slightly alkaline solution. The activated carrier is added to the enzyme solution previously adjusted to pH 8.0–8.5 with the appropriate buffer. The pH must be maintained by the addition of NaOH solution. After complete addition of the derivative, permit the reaction to continue for 1–2 hr at room temperature. Wash and store at 4° until use.

N-Hydroxysuccinimide Ester. Active esters are rather easy to prepare. They can be stored for later use and coupled to enzymes by rather gentle procedures (Fig. 15).

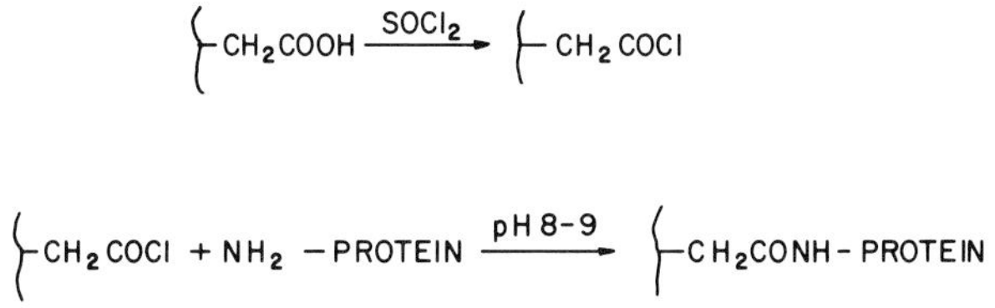

```
  }-CH2COOH  ──SOCl2──▶  }-CH2COCl
```

```
  }-CH2COCl + NH2 -PROTEIN  ──pH 8-9──▶  }-CH2CONH-PROTEIN
```

FIG. 14. Coupling through an acid chloride.

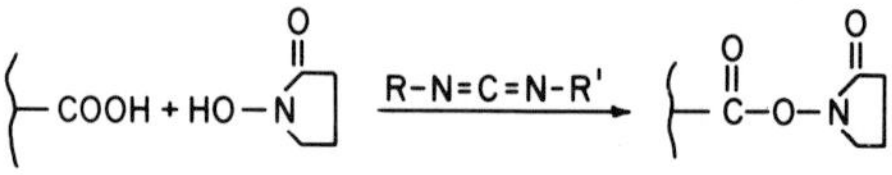

$$\{-COOH + HO-N\overset{O}{\underset{O}{\rceil}} \xrightarrow{R-N=C=N-R'} \{-\overset{O}{\underset{}{C}}-O-N\overset{O}{\underset{O}{\rceil}}$$

FIG. 15. The preparation of the N-hydroxysuccinimide ester.

To 1 g of carboxylated carrier, add 10 ml of dioxane. To this now add 200 mg of N-hydroxysuccinimide and 400 mg of N,N'-dicyclohexylcarbodiimide. The reaction is continued for 4 hr. Wash the derivative with dioxane followed by methanol. The product can be dried in an evacuated oven at 70°–80° for 1 hr. Store refrigerated in an amber bottle. Keep desiccated if possible.

Activation of Inorganic Supports with γ-Mercaptopropyltrimethoxysilane

The preparation of an inorganic support having surface sulfhydryl groups may be useful for coupling through sulfhydryl linkages.

To 1 g of cleaned porous carrier add 5 ml of a 10% aqueous solution (v/v) of γ-mercaptopropyltrimethoxysilane previously adjusted to pH 5.0 with 6 N HCl. The mixture is refluxed for 4 hr, then washed with distilled water. The product is heated in an oven at 120° for 4 hr.

Preparation of Alkylhalide Silane Derivative

To 1 g of clean porous glass is added 25 ml of a 10% solution of γ-chloropropyltriethoxysilane (v/v) in toluene. The mixture is refluxed for 4 hr, washed with toluene, and dried at 120° for an additional 4 hr. The dried product should be stored in a desiccator until use.

Coupling can be accomplished by very slow addition of the alkyl chloride to an enzyme solution at pH 8–9. The pH should be maintained with NaOH. Coupling is carried out at 0° in an ice bath.

Covalent Attachment to Inorganic Supports in the Absence of Silane Coupling Agents

The silanol residues on the surface of glass and the metal oxide groups on ceramic surfaces appear to be capable of reaction with several organic activating groups. The activated carriers will react with enzymes, forming permanent linkages.

Cyanogen Bromide Coupling to Porous Glass

One gram of clean porous glass is suspended in 25 ml of distilled water. The pH is adjusted to 10–11 with NaOH, and 1.0 g of CNBr is very slowly added with maintenance of pH. The reaction temperature

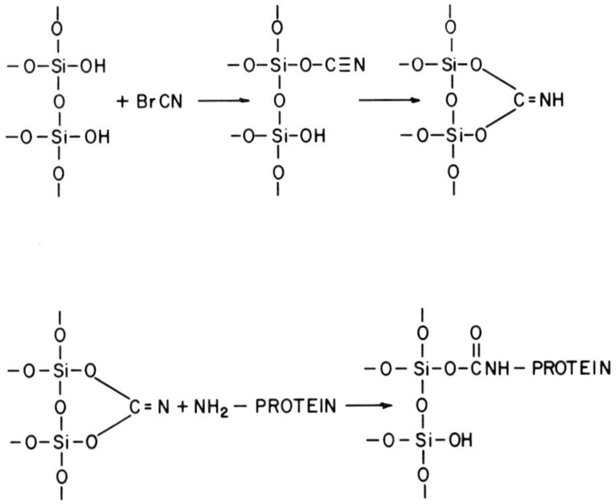

FIG. 16. Possible mechanism of CNBr activities of glass.

should be kept between 15° and 20° by the addition of ice as necessary. The CNBr should be ground into small pieces before addition. Wear gloves and carry out the reaction in a hood (Fig. 16).

After addition of the CNBr is complete, continue the reaction until the pH remains constant. Wash on a filter with cold distilled water and add to an enzyme solution previously adjusted to pH 9.0. Continue reaction in an ice bath for 2–4 hr. Wash and store. Several washes with 6 M urea will remove the adsorbed enzyme. This can be monitored by assay after each 30-min urea soak. When the activity levels off, the adsorbed activity has been completely removed.

Coupling to Porous Glass through a Bifunctional Bis-Diazotized Reagent[12]

A solution of 0.01% 4,4'-bis(2-methoxybenzene diazonium) chloride is prepared by suspension in distilled water. The diazonium salt can be purchased from J. T. Baker Company. The suspended agent is stirred until dissolved at room temperature. To 500 mg of porous glass add 2 ml of 0.01% coupling reagent. Allow the reagent to react with the carrier at room temperature for 20 min. Decant and wash the glass with distilled water. To the activated glass now add 10 ml of a 1% enzyme solution (w/v) dissolved in 0.1 M phosphate buffer at pH 7.8. The reaction is continued at 37° for 3 hr (Fig. 17). The product is then allowed to react overnight at room temperature, washed with distilled water, then with

[12] R. A. Messing and H. R. Stinson, *Mol. Cell. Biochem.* **4** (3), 217 (1974).

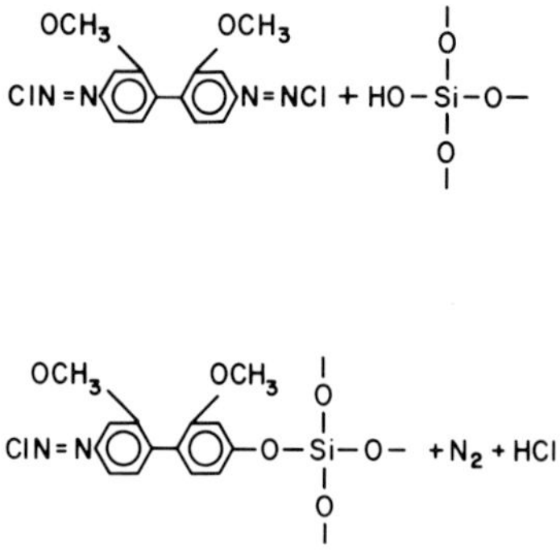

$$\text{ClN=N}\underset{\text{OCH}_3}{\text{\textcircled{ }}}\underset{\text{OCH}_3}{\text{\textcircled{ }}}\text{N=NCl} + \text{HO}-\underset{\underset{|}{\text{O}}}{\overset{\overset{|}{\text{O}}}{\text{Si}}}-\text{O}-$$

$$\text{ClN=N}\underset{\text{OCH}_3}{\text{\textcircled{ }}}\underset{\text{OCH}_3}{\text{\textcircled{ }}}-\text{O}-\underset{\underset{|}{\text{O}}}{\overset{\overset{|}{\text{O}}}{\text{Si}}}-\text{O}- + \text{N}_2 + \text{HCl}$$

FIG. 17. Coupling through a bifunctional reagent.

0.5 M NaCl solution followed again by water. The sample can be stored in water at 4°.

Choice of Coupling Techniques

The choice of coupling technique should be based upon the characteristic of the enzyme to be immobilized. Enzymes with deactivate at pH values greater than pH 8 should be immobilized using techniques which are most effective at lower pH values, e.g., glutaraldehyde or, as in the case of pepsin, carbodiimide. Similarly, enzymes easily denatured at lower pH values should be immobilized with methods most useful at alkaline pH values.

Other factors to be considered should include: groups in the active site that may be capable of binding to the activated carrier, coupling temperature, ionic strength, composition, and any other parameter that could denature the enzyme or interfere with the coupling reaction.

In the case of proteases, one should always couple at low temperatures (0°–4°) for the shortest period of time to decrease the amount of autolysis that occurs.

The simplest rule of thumb is to use a little common sense.

[11] Adsorption and Inorganic Bridge Formations

By RALPH A. MESSING

Adsorption appears in the author's opinion to be the most economical procedure for immobilizing an enzyme on a carrier (See also this volume [3] on hydrophobic adsorption of enzymes and [51] on aminoacylase adsorbed to DEAE-Sephadex.) Although the immobilization procedure may be simple, the reactions involved in adsorption are complex and involve multiple types of bond formations. The stability of an adsorbed enzyme will depend upon the additive strength of those bonds formed under the conditions of immobilization and those bonds maintained under the application conditions that are finally employed for the immobilized enzyme.

Definition

By the generally accepted definition, adsorption is the adhesion or condensation of solute molecules on the surface of solids. This definition is not totally acceptable for the purpose of this discussion. If we were to adopt the definition as it stands, then enzymes immobilized on carriers that have covalent coupling agents attached to the surface would be considered adsorbed, since the enzyme would be accumulated or condensed on the surface of a solid from solution. In order to resolve this dilemma, the following definition will be adopted: adsorption is the adhesion of an enzyme to the surface of a carrier that has not been specifically functionalized for covalent attachment.

Bonding Mechanisms and the Carrier

The bond or bonds formed between the enzyme and the carrier during the adsorption process will be dependent upon the surface properties and chemistry of the carrier. The two types of bonds most frequently reported in literature are (1) the result of charge–charge interaction (ionic, salt bridge); (2) hydrogen bonds. In addition, recent work with titania carriers indicates that a covalent bond may be formed during the adsorption process.[1] In order to examine the three bonding forces that have been identified as participating in the immobilization of enzymes on carriers, three different carriers will be discussed: collagen, controlled-pore glass, controlled-pore titania.

Much of the work recently reported that employs collagen as the carrier for enzymes was performed by Vieth and his colleagues. They have classified their immobilization techniques into two separate procedures[2,3] called complexation and impregnation. Impregnation appears

[1] R. A. Messing, in "Immobilized Enzymes for Industrial Reactors" (R. A. Messing, ed.), p. 95. Academic Press, New York, 1975.
[2] K. Venkatasubramanian, R. Sani, and W. R. Vieth, J. Fermentn. Technol. **52**, 268 (1974).
[3] F. R. Bernath and W. R. Vieth, in "Immobilized Enzymes in Food and Microbial Processes" (A. C. Olson and C. L. Cooney, eds.), p. 157. Plenum, New York, 1974.

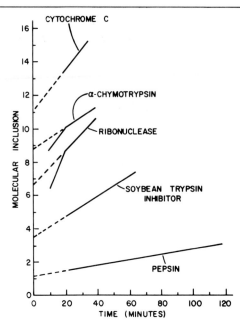

Fig. 1. Molecular inclusion (adsorption) of model proteins in porous glass membranes. ---, Extrapolations to zero time.

to be a simple adsorption process whereby the enzyme is adsorbed on a preformed collagen carrier (see this volume [19] and [53]).

A study of the binding of proteins to the surface of controlled-pore glass was reported by the author under the title "Molecular Inclusion. Adsorption of Macromolecules on Porous Glass Membranes."[4] When porous glass membranes were utilized as a dialysis membrane, it was noticed that protein was rapidly lost from the solution contained by the membrane; however, after several days little or no protein appeared in the buffer surrounding the membrane. Well characterized pure protein models were employed to study spectrophotometrically the kinetics of inclusion of protein in glass pores. The initial inclusion of proteins was too rapid to follow with any degree of accuracy; however, after 20 min, the inclusion rate was considerably diminished. The results were remarkably reproducible, and the inclusion appeared to be linear with respect to time (Fig. 1).

When the linear inclusion rate was plotted against the molecular weight of protein (Fig. 2), it became apparent that the "second inclusion rate" was inversely proportional to the molecular weight of the protein.

[4] R. A. Messing, *J. Am. Chem. Soc.* **91**, 2370 (1969).

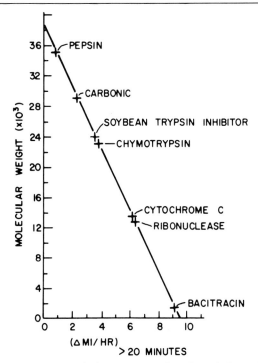

FIG. 2. Second molecular inclusion rate as a function of the molecular weight of the protein.

By extrapolating the linear reaction to zero time, the approximate magnitude of the initial inclusion reaction was determined. These extrapolated values were then plotted against the isoelectric point of the protein molecules (Fig. 3).

These results indicate that the first very rapid inclusion is dependent upon the surface charge of the protein. The higher isoelectric point, or more electropositive, proteins react more vigorously initially with the porous glass surface. The predominant group on these electropositive proteins is NH_3^+; this suggests that this initial inclusion involves the formation of ionic bonds between the amine groups of the protein and the dissociated silanol (SiO^-) groups of the glass. This suggestion is consistent with the Holt and Bowcott[5] studies of the reaction of silicic acid with heptadecylamine and with protein.

The protein glass bond formed during the protein adsorption process is so strong that strong acids, ammonium hydroxide, or a variety of ionic strength buffer solutions are not capable of eluting the protein

[5] P. Holt and J. Bowcott, *Biochem. J.* **57**, 471 (1954).

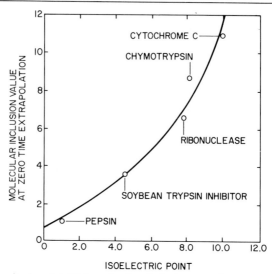

Fig. 3. Magnitude of initial molecular inclusion reaction as a function of isoelectric point.

quantitatively from the glass. This finding indicates that the ionic amine silicate cannot be the sole mechanism involved in the bonding process. The proteins, however, are quantitatively released from the glass by acid solutions of urea, but not by urea solutions alone; thus both hydrogen and amine silicate bonds are involved in the adherence of proteins to glass. Weldes' investigations[6] of the interactions of proteins and alkali metal silicates also indicated the importance of hydrogen bonds in protein silicate complexing. It is apparent, therefore, that both the ionic bond and the hydrogen bond participate in the immobilization of enzymes by adsorption on controlled-pore glass.

Titania, unlike glass (silica), readily undergoes redox reactions. In a study of glucose oxidase and catalase immobilized by adsorption on controlled pore titania,[7] it was reported that the preparation turned blue-violet when stored in water at room temperature and not continuously utilized. When the glucose solution was passed through the immobilized enzyme, however, it turned yellow. Again on standing in water it reverted to a blue-violet color. These observations lead to the following conjectures: The change in color on the surface of the titania is probably due to a change in oxidation state as a result of transfer of electrons. In the case of glucose oxidase, this transfer of electrons is probably mediated by FAD; and in the case of catalase, the iron porphyrin is involved. The

[6] H. Weldes, *Adhes. Age* **10**, 32 (1967).
[7] R. A. Messing, *Biotechnol. Bioeng.* **16**, 1419 (1974).

color change may be attributed to the formation of reduced titania, Ti_2O_3, which is violet. This transfer of electrons would seem to imply that at least a partially covalent bond exists between the carrier and the enzyme. Thus, in addition to simple ion–ion interaction and hydrogen bonding, some covalency may exist with specific carriers.

Factors Affecting Adsorption and the Immobilized Enzyme-Conjugated Proteins

A number of pitfalls should be cited when immobilization by adsorption is chosen. Although conjugated or complex proteins react in the same manner as do simple proteins with respect to adsorption, an additional reaction may occur when the process includes immobilization within pores, namely, the dissociation of the nonprotein moiety from the protein backbone. Stinson[8] has demonstrated that, upon exposure to porous glass, hemoglobin is dissociated and the small porphyrin prosthetic group is adsorbed at a substantially greater rate than the protein. A cleavage of a conjugated enzyme would lead to a loss of activity.

Active Site. An additional hazard that one may confront when adsorption is utilized to immobilize enzymes is reaction at a residue involved in the active site of the enzyme. Such was the case when trypsin was allowed to react with porous glass with the apparent involvement of histidine residue at the active site.[9]

pH, Ionic Strength. Enzymes immobilized by adsorption require close control of pH and ionic strength during application conditions. If the bonding is predominantly ion–ion interaction with very little hydrogen bonding, then a simple shift in pH or ionic strength could exchange the protein ion for another ion, and thus, desorption would occur. This desorption as a result of ionic strength is illustrated by the previously cited work of Bernath and Vieth [3] with catalase adsorbed to collagen. The control of pH and ionic strength is not quite as critical in covalent bonding of enzymes.

Quantity of Enzyme. A major factor influencing the quantity of enzyme adsorbed to a carrier is enzyme concentration exposure to the unit surface or weight of carrier during the immobilization process. Bernath and Vieth [3] developed sorption isotherms for both lactase and lysozyme with collagen membranes. With both enzymes the amount of activity increases with increasing bath enzyme concentrations, approaching a saturation value asymptotically at higher enzyme concentration. The authors claim that the asymptotic approach to a limiting value demon-

---

[8] H. R. Stinson, *J. Biomed. Mater. Res.* 3, 583 (1969).
[9] R. A. Messing, *Enzymologia* 38, 370 (1970).

strates that there are a finite number of binding sites on collagen and that different enzymes are bound to the matrix by a general mechanism involving protein–protein interaction. We have found a similar effect in our laboratories with respect to loading inorganic carriers. However, it has also been noted in many cases that, if the carrier is overloaded with enzyme, the amount of enzyme available for reaction is restricted or limited again by diffusion, since the predominant reaction occurs in the first few layers of enzyme that are seen by the substrate; thus, the efficiency of enzyme utilization would be diminished if all the enzyme is not available for reaction.

Temperature and Time. Both time and temperature are important considerations in immobilization by adsorption. Since diffusion is an important factor when an enzyme is immobilized within a pore, it would appear that the immobilization could be accomplished in a shorter time if the temperature were elevated. However, high temperatures are contraindicated by the denaturing effect upon the protein. In our laboratories, we have found that, if an enzyme is adsorbed at the highest temperature where denaturation is not a major factor, both time requirement and activity losses are minimal (this may be the temperature optimum for an enzyme over an extended period of time).

The contributing factors involved in the immobilization by adsorption may be itemized as follows: (1) concentration and quantity of enzyme vs the unit surface area or weight of carrier; (2) effect of adsorption upon enzyme protein (conjugated protein, enzyme active site); (3) ionic strength; (4) initial pH of adsorption reaction and subsequent pH of application; (5) temperature of adsorption reaction; (6) isoelectric point of protein and adsorption sites of carrier.

Optimizing Carriers for Adsorption

There are three major points to be considered in the optimization of a carrier for adsorption of enzymes: (1) preconditioning the surface with respect to pH; (2) preconditioning the surface with respect to cofactors; (3) selection of the optimum pore diameter for immobilizing the enzyme.

If one chooses to immobilize an enzyme either on a surface that does not contain pores or if immobilization is to take place external to the pores, then the point 3 need not be considered.

pH

The importance of preconditioning surface with respect to pH becomes very acute when an enzyme is particularly pH sensitive. Controlled-pore silica has a rather acid surface, pH 4, thus an enzyme that would normally be denatured at pH 4 could not withstand that environment. Under these conditions, the controlled-pore silica should be treated with a buffer prior to the exposure of that enzyme. The buffer should be chosen such that its pH coincides with the optimum stability of the enzyme. Approximately 1 hr of equilibration at approximately 37° usually suffices to precondition the surface.

Cofactors

Most carriers selectively bind such cofactors as metal ions. The carrier, therefore, is in competition with the enzyme requiring metal ions for activation. Hence, these metal ions will be removed from the functional site of the enzyme and bound to the carrier. The enzyme will then be reduced in activity or become inactive. We have found in a number of instances that it is appropriate to precondition the surface of the carriers with a solution of metal ion that is required for the enzyme prior to the exposure of the enzyme. Pretreatment of the carrier with a concentration of metal ions approximately ten times higher than that required for activation of the enzyme in a volume of at least twice the volume of the carrier for approximately 1 hr at 37° has generally been effective for saturating the carrier requirements for the metal ion. This pretreatment will generally avoid the competition by the carrier for the enzyme metal ions.

Pore Diameters

This topic is of concern whether one has chosen to immobilize within a dimensionally stable pore, such as controlled-pore ceramics, or within a flexible pore, such as exists within a collagen membrane. In some of our early work with controlled-pore glass,[10] it became rather apparent that the amount of enzyme adsorbed was a function not only of the pore diameter, but also of the surface area (Table I). Even though a carrier material may have a large surface area, the availability of some of that surface for attachment is limited by the restrictive openings. An excellent example of a material with a very great surface area, but with little available internal surface for attachment, is bentonite. Another such material is controlled-pore glass with an average pore diameter of 40 Å.

A rather simple procedure was utilized in recent work with controlled-pore ceramics to gain a more substantial fix upon the relationship of pore diameter to molecular size. This procedure may be utilized if the

[10] R. A. Messing, *Enzymologia* **39**, 13 (1970).

TABLE I
Relation of Pore Size and Surface Area to Quantity of Adsorbed Stabilized Enzyme

Glass (Corning designation)	Pore diameter (Å)	Enzyme (MW)	Surface area (m/g²)	Milligram of active bound enzyme per gram glass[a]
7740 (test tube)	Not porous	Papain (21,000)	Very low	0
7740 (F-frit)	50,000	Papain (21,000)	0.2	0.01
Controlled-pore glass	900	Papain (21,000)	20	0.33
7930	92	Papain (21,000)	60	0.08
7930	68	Papain (21,000)	118	0.07
Controlled-pore glass	900	Glucose oxidase (150,000)	20	0.12
7930	68		118	0.0001

[a] Activity determined on 1 g of glass in terms of the original free enzyme as calculated from an activity curve relating milligrams of enzyme to optical density.

unit cell dimensions either of the enzyme or of the substrate are unknown. The objective of this program was to utilize all the internal surface of a porous material for coupling enzyme and yet maintain as great a surface area as possible. This means that the smallest pore diameter with a very narrow pore distribution is required to just allow the entrance of the limiting molecule (the enzyme when it is larger than the substrate). A study of glucose oxidase and catalase[11] with relation to pore diameter indicated that the optimum pore diameter for immobilizing an enzyme was approximately twice the major axis of the unit cell of the enzyme. We have chosen to call this factor of twice the major axis the spin diameter.

Example: Adsorption of Glucose Oxidase–Catalase[12] on Controlled-Pore Ceramics

The approach that we chose for optimizing the adsorption of glucose oxidase and catalase on controlled-pore ceramics was that of exposing the carrier to an excessive quantity of enzyme (far more than would be expected to bind to the surface) at the optimum temperature of the enzyme for a short time and then at room temperature for an additional longer period of time. This approach required the minimum of preparative work. The particle size and size distribution of controlled-pore ceramics of various pore diameters were so chosen that they were in a similar

[11] R. A. Messing, *Biotechnol. Bioeng.* **16**, 897 (1974).
[12] See also this volume [32] on the same enzyme system.

particle size range. The particle mesh sizes were either 25/60 mesh, 30/80 mesh, or 25/80 mesh. The identical weight of controlled-pore ceramics was used for each determination.

For each pore diameter, a 300-mg quantity of controlled-pore ceramic carrier (the ceramic carrier is a product of Corning Glass Works produced according to R. A. Messing, U.S. Patent No. 3,892,580) was transferred to a 10-ml cylinder. A 9-ml volume of 0.5 M sodium bicarbonate was added to precondition (remove fines and neutralize surface) the carrier surface. The cylinder was then placed in a shaking water bath at 35°. After shaking in the bath for approximately 3 hr, the cylinder was removed from the bath, the bicarbonate solution was decanted, and 9 ml of distilled water was added to the cylinder. The cylinder was stoppered and inverted several times, and the water wash was then decanted. At this point the carrier was ready for the adsorption and immobilization of the enzyme.

The glucose oxidase–catalase enzyme used for this study was purchased from the Marshall Division of Miles Laboratories, Inc., under the name DeeO liquid; it contained 750 glucose oxidase units (GOU) per milliliter and 225 EU per milliliter with respect to catalase activity. This preparation is stabilized with glycerol. The GOU is based on that of the Scott[13] procedure and is defined as the quantity of enzyme that will cause the uptake of 10 mm^3 of oxygen per minute at 30° in the presence of excess oxygen with the substrate containing 3.3% of glucose monohydrate and phosphate buffer, pH 5.9. In our laboratories, we have correlated this unit with a circulating column assay performed with a differential conductivity bridge as the sensor.[11,14] The GOU reported for this study are therefore equivalent to the Scott unit.

The enzyme solution was prepared for the immobilization process by dialyzing 20 ml of the DeeO liquid against four 3500-ml charges of distilled water in a 4-liter beaker stirring at room temperature for approximately 24 hr. This dialysis was necessary both to remove the glycerol, which interferes with the immobilization, and to reduce the ionic strength of the fluid, which improves the immobilization.

The immobilization process was performed by exposing 300 mg of the preconditioned carrier either to a volume of 8 ml containing 2400 GOU or to a volume of 21 ml containing 4500 GOU in a shaking water bath at 35° for 2 hr and 20 min. The vessels containing the carrier and the enzyme solutions were removed from the bath and allowed to stand at room temperature for approximately 15 hr. The enzyme solutions were then decanted from the immobilized enzymes; each immobilized enzyme was

[13] D. Scott, *J. Agric. Food Chem.* **1**, 729 (1953).
[14] R. A. Messing, *Biotechnol. Bioeng.* **16**, 525 (1974).

washed first with 9 ml of distilled water, then with 9 ml of 0.5 M NaCl, followed by a 9-ml wash of 0.2 M acetate buffer, pH 6.1, and finally with a 9-ml distilled water rinse. Each of these washings was performed with agitation and inversion. The immobilized enzyme preparations were then transferred to small columns, in which they were assayed by the circulation of glucose solutions; they were stored in water at room temperature between assays.

Comments

The physical parameters for the carriers used in this study are recorded in Table II. A sample of the assay results are depicted in Table III, and the cumulative data for stable loading are plotted in Fig. 4. A particular point should be noted in Table III: the assay values generally become higher after the first and second exposures to substrate. We have noticed this quite frequently with enzymes immobilized by adsorption. This higher activity may be due to the fact that with exposure to the substrate more of the active sites of the enzyme become exposed, since it is probable that the active site has a greater affinity for the substrate than for either the carrier or the adjoining neighbor protein. That is, additional active sites are exposed with exposure to substrate.

Our studies indicated that, although we could load the 175 and the 220 Å controlled-pore ceramics, the activity of these immobilized enzymes was exceedingly unstable; the activity diminished rapidly with storage, so that within 18 days less than half the activity was retained. On the other hand, a very stable glucose oxidase–catalase system was achieved with a titania having an average pore diameter of 350 Å and a maximum

TABLE II
CARRIERS[a,b]

Parameter	Al_2O_3	44% TiO_2 56% Al_2O_3	TiO_2	TiO_2	TiO_2	TiO_2
Average pore diameter (Å)	175	220	350	420	820	855
Minimum pore diameter (Å)	140	140	220	300	760	725
Maximum pore diameter (Å)	220	300	400	590	875	985
Pore volume (cm^3/g)	0.6	0.5	0.45	0.4	0.2	0.22
Surface area (m^2/g)	100	75	48	35	7	9
Particle mesh size	25/60	25/60	25/60	30/80	25/80	25/80

[a] From R. A. Messing, *Biotechnol. Bioeng.* **16**, 897 (1974).
[b] Source of inorganic supports: the controlled-pore inorganic carriers are manufactured according to a process evolved by the author (patent pending) and may be obtained from Corning Glass Works.

TABLE III
COMPARISON OF 420, 820, AND 855 Å TITANIA-IMMOBILIZED ENZYMES[a]

	Apparent activity (GOU/g)		
Assay day	420 Å	820 Å	855 Å
0	56.4	43.9	37.8
3	77.7	43.9	39.7
8	84.5	46.2	40.3
13	84.5	46.2	44.3
42	80.5	40.3	42.0

[a] From R. A. Messing, *Biotechnol. Bioeng.* **16,** 897 (1974).

pore of 400 Å. This can be related to the unit cell dimensions in the following manner. The largest dimension of glucose oxidase is 84 Å; thus, this enzyme would have substantial access to a pore diameter having 168 Å, twice the major axis. Catalase, on the other hand, having a major axis of 183 Å, would not be restricted by a pore having an approximate diameter of 366 Å. The 175 Å and the 220 Å carriers were therefore loaded with glucose oxidase, but catalase was excluded because it would

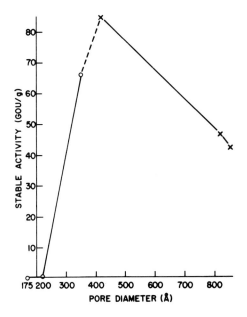

FIG. 4. Cumulative data relating stable glucose oxidase activity (glucose oxidase units, GOU) to pore diameter.

require a pore of approximately 366 Å. The glucose oxidase contained within the small-pore ceramics, 175 and 220 Å, when exposed to glucose solutions accumulated a good deal of peroxide within the pores. In excess quantities of peroxide, glucose oxidase is inactivated. Thus, an unstable immobilized enzyme was evolved. With ceramics having pore dimensions greater than 350 Å, remarkable stability was achieved owing to the presence of catalase, which destroyed the accumulated peroxide.

The results recorded in Fig. 4 indicate that a maximum loading of stable enzyme is achieved between 350 and 420 Å. This brackets the spin diameter of the limiting molecule, catalase, at 366 Å. The maximum utilization of the surface area for adsorption is then achieved by selecting a pore diameter that closely approximates the spin diameter of the largest molecule involved in the enzyme reaction. In cases such as proteases and carbohydrases of large molecular weight, the limiting molecule in all probability will be the substrate; therefore, the maximum dimension of the substrate should be chosen to guide the selection of the pore diameter for the full utilization of the enzyme within the pore.

Methods Utilized for Immobilizing Enzymes by Adsorption

There are four general methods for immobilizing an enzyme by adsorption: (1) the static procedure; (2) electrodeposition; (3) in-column loading; (4) mixing or shaking-bath loading.

The static procedure requires the most time and is the most inefficient of the four general methods of immobilization by adsorption. This procedure simply allows the carrier to be loaded in a solution containing the enzyme without agitation and relies completely upon diffusion of the enzyme to the carrier surface and into the structure of the carrier itself. Generally, the loading is not uniform and is rather low, unless the carrier has been exposed for many days to the enzyme solution. It is not a process of choice.

Electrodeposition is an interesting process whereby the carrier is placed proximal to one of the electrodes in an enzyme bath, the current is turned on, the enzyme migrates to the carrier and deposits on its surface as the result of the electromigration of the enzyme. This process is rather simple, but for achieving an economical procedure for handling large quantities of enzyme and immobilizing in minimal processing areas, it is not a prime choice in the author's opinion (see this volume [19] on electrocodeposition).

For commercial use, the in-column loading procedures are perhaps the prime candidates. Two general procedures fall within this category. One is plug-flow loading, in which the enzyme is circulated through the column starting from the top and leaving from the base of the column on a continuous basis. The process may be performed in water-jacketed columns such that the temperature can be controlled. The flow rate can be controlled by adjusting the pump. The second procedure utilizes the fluidized bed. In this procedure, the enzyme solution is flowed from the base of the column and leaves from the top. The particles are kept in continuous motion by the flow of the fluid. The pump should be adjusted to the flow rate at which the particles are in motion. Again the temperature can be controlled by a water-jacketed column. These processes offer marked advantages since they can be employed not only for loading the column with enzyme but, in addition, for washing the immobilized enzyme within the same column. The amount of space required either in a plant or in a laboratory to perform these procedures are minimal.

The mixing- or shaking-bath process is the technique probably employed most in laboratory preparations. In this process the carrier is placed in the enzyme solution and either mixed with a stirrer or placed in a shaking water bath and continually agitated. This process is effective, and normally the results are rather uniform; however, the in-column processes are greatly preferred for commercial preparation, since the equipment required for those processes is the same as that employed in the final application of the immobilized enzyme.

Example: Method for Immobilizing Glucose Isomerase by a Combination of Mixing and Static Procedure on Controlled Pore Alumina

An actual procedure and the results of that procedure for preparing an immobilized glucose isomerase follows.

Materials. The carrier used for this preparation was controlled-pore alumina (product of Corning Glass Works, U.S. Patent #3,892,580), 25–60 mesh, average pore diameter 175 Å, minimum pore diameter 140 Å, maximum pore diameter 230 Å, surface area 100 m²/g, and porosity 63.5%.

The glucose isomerase used for this preparation was derived from *Streptomyces* sp.

Assay and Monitoring Procedure. The conversion of glucose to fructose was followed with a Bendix recording automatic polarimeter. The glucose isomerase unit (GIU) is defined as the quantity of enzyme required to convert 1 μmole of glucose to fructose per minute at 60°, pH 6.85, in a solution containing 50% w/v glucose.

Preconditioning of Carrier. An 11-g batch of carrier was transferred to a 100-ml glass-stoppered cylinder. A 100-ml volume of solution containing 0.05 M magnesium acetate and 0.01 M cobalt acetate, pH 7.5, was added to the carrier in the cylinder. The cylinder was stoppered and placed in a 60° bath after inversion. After 15 min of reaction, the cylinder was inverted and the fluid was decanted. An additional volume of the magnesium–cobalt acetate solution, described previously, was added to the cylinder; the cylinder was inverted and allowed to stand at room temperature for 2.5 hr. The preconditioning solution was then decanted from the carrier, and the carrier was ready for the immobilization process.

Preparation of Glucose Isomerase for Adsorption. A glucose isomerase solution containing 590 GIU/ml in 0.6 saturated ammonium sulfate solution was treated as follows for adsorption: To 40 ml of glucose isomerase solution, 1.4 g of ammonium sulfate was added and the slurry was stirred at room temperature for 20 min. The sample was then transferred to the centrifuge and centrifuged at 6000 rpm for 30 min at 2°. The supernatant fluid was decanted and discarded. Twelve milliliters of a solution containing 0.05 M magnesium acetate and 0.01 M cobalt acetate, pH 7.5, were added to the precipitate and stirred. An additional 3 ml of 0.5 M sodium bicarbonate was then added to this solution and mixed. The total solution was placed in a 60° water bath for 15 min. After removal from the bath, the solution was centrifuged for 15 min at 16,000 rpm and 2°. The clear supernatant enzyme solution was decanted, and the precipitate was discarded. The total volume of the enzyme solution was 28 ml; the pH was 7.5.

Preparation of Immobilized Enzyme. The 28 ml of enzyme solution was added to the preconditioned 11 g of alumina carrier in the 10-ml cylinder. The cylinder was stoppered and inverted and then placed in a 60° water bath. The enzyme was permitted to react with carrier for 2.5 hr at 60°. During this period, the cylinder was inverted every 15 min to mix the alumina with the enzyme solution. After removal from the 60° bath, the reaction was continued at room temperature with inversion at 30-min intervals for the next 2 hr. The reaction was continued without inversion overnight at room temperature. The enzyme solution was decanted and now had a volume of 28.5 ml with a pH of 7.1. This solution was saved for subsequent assay. The immobilized enzyme was then washed in the cylinder first with 60 ml of distilled water by inverting the cylinder four times, then with 40 ml of 0.5 M sodium chloride, and finally with 40 ml of solution containing 0.05 M magnesium acetate, 0.01 M cobalt acetate, pH 7.5. One gram of immobilized enzyme was removed from the batch for subsequent assay. The remaining 10 g of immobilized enzyme were transferred to a column which was water-jacketed for additional column studies.

General Comments on Inorganic Carriers and the Operating Environment

The carriers that perform well under conditions below pH 7.0 are glass and silica. The carrier that may be employed for immobilization in the intermediate pH range of 3 to 9 is titania. Alumina should be chosen for the more alkaline operating conditions. This carrier is effective in the pH range of 5 to 11. For more information on covalent binding of enzymes to inorganic supports, see this volume [10].

Assay

The 1-g sample that was removed from the batch was assayed in a shaking water bath and found to contain an activity of 381 GIU per gram.

The column containing the 10 g of immobilized enzyme was thermostated to 60°. It was fed with a solution that contained 50 g of glucose per 100 ml and 0.005 M magnesium sulfate buffered with sodium sulfite to a pH between 7.7 and 8.0 (approximately 0.01 M sodium sulfite). The initial flow rate of the column was approximately 190 ml per hour. This flow rate was reduced with time so that between 40 and 43% of the glucose was converted to fructose over the column life studies. The product of the column was sampled daily and analyzed for fructose production over the column life studies. These results are plotted in Fig. 5.

Since the 1-g sample of immobilized enzyme removed from the batch contained 381 GIU/g the 11-g batch contained 4191 GIU; hence, 17.8% of the offered activity was recovered in the immobilized enzyme batch. The reacted enzyme solution, 28.5 ml, was found to contain 161 GIU/ml, which represented 19.4% of the enzyme offered. Thus, the coupling effi-

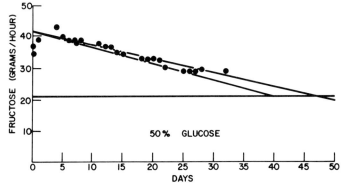

FIG. 5. Column performance at 60° of glucose isomerase (10.0 g, 3810 GIU, pH 7.5) immobilized by adsorption on controlled-pore alumina.

ciency, or the percent of active enzyme immobilized on the carrier as a function of the amount of enzyme consumed in the process, is approximately 22%.

Discussion of Results and the Procedure

During the studies with this enzyme, it was noted that the immobilization of the enzyme was ineffective below pH 6.8. In addition, activity was lost during extended periods of exposure below pH 6.3. Although the optimum activity for this enzyme in solution appears to be in the neighborhood of pH 6.8, the immobilized enzyme was found to have an optimum pH of approximately 7.5. In our early studies we noted that the alumina carrier picked up considerable quantities of cobalt and magnesium. Therefore, it was found that preconditioning in the manner described with cobalt and magnesium was necessary for maintaining an active preparation.

It will be observed in Fig. 5 that the activity appeared to increase in the first 24 hr of operation. This may be the result of opening additional enzyme active sites by the presence of the substrate, which may formerly have been bound to either the alumina or an adjoining protein molecule. The half-life of this column was estimated to be approximately 42 days after 31 days of continuous operation.

Although the column was fed at a pH of between 7.7 and 8.0, the product solution was found to be approximately 7.5.

Regeneration

There are three separate entities to be considered under this topic: (1) reactivation of the immobilized enzyme; (2) regeneration of the immobilized enzyme; (3) regeneration of the carrier.

Reactivation of the Immobilized Enzyme. In this case, the immobilized enzyme is normally reversibly inactivated. This may be due to a plugging of pores in controlled-pore ceramics or a film formation on the surface of the immobilized enzyme. Generally this can be handled by simple washing procedure for the immobilized preparation.

Regeneration of the Immobilized Enzyme. In this case, the enzyme has irreversibly lost activity. Such was the case with aminoacylase immobilized on DEAE-Sephadex reported by Chibata et al.[15] (see also this volume [51]). These authors found that, if they exposed the some-

---

[15] I. Chibata, T. Tosa, T. Sato, T. Mori, and Y. Matuo, *in* "Fermentation Technology Today" (G. Terui, ed.), p. 383. Society of Fermentation Technology, Yamada-Kami, Suita-shi, Osaka, Japan, 1972.

what inactivated carrier to fresh enzyme solution, they could regenerate the full activity of the original immobilized enzyme preparation.

Regeneration of the Carrier. When an immobilized enzyme loses activity and cannot be regenerated simply by exposing the preparation to fresh enzyme and the useful level of activity for this preparation has been passed, then it may be wise to remove the inactive enzyme and prepare the carrier surface for a fresh immobilization. This carrier regeneration can be accomplished either by changing the pH such that the enzyme is desorbed from the carrier, or by chemically stripping the enzyme from the surface by such reagents as hydrogen peroxide or other oxidizing agents. Another approach that we have found to be rather useful with controlled-pore ceramics is pyrolysis. In this case, the immobilized enzyme is simply burned off in a furnace at elevated temperatures (generally above 400° in the presence of air or oxygen). The carrier is then ready for preconditioning and immobilization of fresh enzyme.

Promising Applications of Adsorbed Immobilized Enzyme Systems

Although there is but one commercial success employing an adsorbed immobilized enzyme, L-aminoacylase, a number of other adsorbed systems appear to be evolving. The immobilized aminoacylase commercially employed in Japan is the result of the efforts of Tosa, Mori, Fuse, Chibata, and Matuo,[15-18] which were directed toward the separation of racemic mixtures of amino acids (see this volume [51]).

Another immobilized process based upon adsorption has been recently announced by Clinton Corn Products Co. for converting glucose to fructose utilizing immobilized glucose isomerase.[19] The carrier, DEAE-cellulose, and the immobilization procedure is based upon the techniques developed by Tosa *et al.* for aminoacylase.

A system involving the adsorption of glucose isomerase in controlled-pore ceramics based on the author's work[20,21] is currently being evaluated for commercial exploitation by CPC International, Inc., Englewood Cliffs, New Jersey.

A very promising effort is that of Mandel *et al.*[22] with cellulase.

[16] T. Tosa, T. Mori, N. Fuse, and I. Chibata, *Enzymologia* 32, 163 (1967).
[17] T. Tosa, T. Mori, N. Fuse, and I. Chibata, *Agric. Biol. Chem.* 33, 1047 (1969).
[18] T. Tosa, T. Mori, and I. Chibata, *Agric. Biol. Chem.* 33, 1053 (1969).
[19] *Chem. Eng.* (*N.Y.*), 81, 52 August 19, 1974.
[20] U.S. Patent No. 3,868,304, R. A. Messing, February 25, 1975.
[21] U.S. Patent No. 3,850,751, R. A. Messing, November 26, 1974.
[22] M. Mandel, J. Kostick, and R. Parizek, *J. Polym. Sci. Part C, Cellulose* 36, 445 (1971).

This work is directed toward the production of glucose from cellulose. The carrier utilized for the enzyme, cellulose, is also the substrate. This research is so promising that at this time the production of ethanol and single cell protein utilizing cellulose as a starting material is being evaluated. The cellulase utilized in this process is obtained from the fungus *Trichoderma viride*. A very interesting and novel aspect of this particular technology is that it is the only instance, currently, in which it is the intent that the immobilized enzyme is to utilize its carrier. The enzyme, initially soluble when added to the reactor, forms a tight complex with the insoluble cellulose; thus, it becomes immobilized. As the carrier is consumed by the enzyme, it is converted to a soluble glucose. The enzyme can be reimmobilized by the addition of fresh cellulose.

Immobilization by Inorganic Bridge Formations

This technique of immobilization should not be considered as a simple adsorption process. It is being discussed under the general subject of adsorption, however, since many of the elements involved in the technique are similar to those involved with controlled-pore titania and other controlled-pore ceramics.

Emery et al.[23] and Barker et al.[24] have evolved this technique, which they term metal-linked enzymes. The process employs primarily transition metal salts for linking an enzyme to the carrier. Cellulose, nylon, borosilicate glass, soda glass beads, and filter paper have been used as carriers. $TiCl_4$, $TiCl_3$, $SnCl_4$, $SnCl_2$, $ZrCl_4$, VCl_3, $FeCl_2$, and $FeCl_3$ were the salts employed to form the metal link between the carrier and the enzyme.

The activated carrier was prepared by steeping the solid carrier in a solution of the metal salt for 5–10 min; the complex was then filtered and/or dried at 45°. The dry solid was then washed with buffer to remove excess unreacted salt. The immobilized enzyme is then prepared by stirring the activated carrier in the buffered enzyme solution. The pH of the buffer selected is that of the optimum pH for the free enzyme. The product is then washed to remove the excess unattached enzyme. The investigators were unable to couple enzymes to the surfaces of nylon tubes by this procedure. Another problem that they encountered was due to the low pH, 0 to 1, of the titanic chloride which caused

---

[23] A. N. Emery, J. S. Hough, J. M. Novais, and T. P. Lyons, *Chem. Eng.* (*London*). p. 71 (February 1972).

[24] S. A. Barker, A. N. Emery, and J. M. Novais, *Process Biochem.* 5, 11 (1971).

deterioration of some of the support materials. Most carriers produced functional products.

Example 1: Bridge Formation of Glucoamylase to Cellulose

A 10-g quantity of microcrystalline cellulose was transferred to 50 ml of a solution of 15% titanic chloride, mixed well, and placed in a desiccator containing NaOH. The desiccator was evacuated and placed in an incubator at 45°. After drying, the treated carrier was washed three times with 0.02 M acetate buffer, pH 4.5. The activated carrier was then added to the amyloglucosidase solution, which was composed of the same buffer previously cited and contained 1020 mg of protein having a specific activity of 78 units/mg. (A unit of activity liberates reducing sugars equivalent to 1 μmole of glucose in 1 min at 45° from a 1% solution of soluble starch in 0.02 M acetate buffer, pH 4.5.) The enzyme and carrier were stirred at 2° for 18 hr; the suspension was then centrifuged, and the immobilized enzyme was washed first with 0.02 M acetate buffer, pH 4.5, containing 1 M NaCl, then with the acetate buffer without NaCl. The immobilized enzyme was resuspended in the same buffer and stored at 4°. The product had an activity of 2900 units per gram of solid.

Example 2: Bridge Formation of Urease to Controlled-Pore Ceramics

Recently, the author[7] has employed the stannous bridge for immobilizing urease to controlled-pore titania (product of Corning Glass Works, U.S. Patent No. 3,892,580). The carrier used for this study was a controlled-pore titania of 420 Å diameter. The control (adsorption) carrier was preconditioned by shaking 500 mg of the carrier in 11 ml of 0.5 M sodium bicarbonate at 37° for 1 hr and 40 min, after which the sodium bicarbonate was decanted. The stannous bridge carrier was prepared by exposing 500 mg of titania to 20 ml of 1% stannous chloride dihydrated solution in a shaking water bath at 37° for 45 min. The unreacted stannous chloride solution was then decanted, and the carrier was washed with 3 aliquots containing 20 ml of distilled water.

The immobilized enzymes were prepared by exposing each of the above carriers to 80 Sumner Units (SU) in 20 ml of water for 2 hr in a shaking water bath at 37°. The enzyme solutions were further allowed to react with the carrier at room temperature for 22 hr without shaking. The enzyme solutions were then decanted, and the immobilized preparations were washed, successively, with 0.5 M NaCl and distilled water. The immobilized urease preparations were then transferred to columns and stored in water at room temperature. Each preparation was assayed periodically with 1 M urea solution.

Properties of the Preparations

Besides these two preparations, an additional scaled-up version of the stannous bridge urease was prepared and evaluated. The storage stability of these preparations in water at room temperature is plotted in Fig. 6. In that figure, it can be noted that, with extrapolation back to 0 days, the adsorbed enzyme was loaded to 3.37, the stannous bridge enzyme to 2.79, and the scaled-up version of the stannous bridge urease to 3.47 SU/g. It can therefore be said that the stannous bridge enzyme was loaded to approximately the same level as the adsorbed urease; however, the half-life of the adsorbed enzyme was 10 days whereas that of the stannous bridge urease was 30 days for small preparations and 36 days for scaled-up version. It is probable that this increased stability of the stannous bridge preparation is a result of the protective effect of the stannous ion upon the sulfhydryl groups of the enzyme.

The adsorbed urease control described in the figure was a cream to yellow color identical to that of the stannous bridge urease. However, the stannous bridge enzyme turned a blue-violet with storage in water, whereas the control enzyme did not. The bridged enzyme reverted to the yellow-cream color when urea solution was passed through the preparation. It again turned blue-violet upon storage in water. This color change could be attributed to the formation of a reduced titania, Ti_2O_3, which

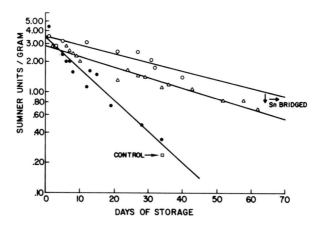

Fig. 6. Storage stability in water at room temperature of adsorbed urease and stannous bridge urease. ●, Control; △, Sn bridge; ○, scaled-up Sn bridge inorganic bridge formation. From R. A. Messing, *Biotechnol. Bioeng.* **16**, 1419 (1974).

is blue-violet. It is probable that the change in oxidation state of the titania is due to a transfer of electrons that occurs as a result of the titania interfacing with the stannous ion, which in turn probably interfaces with the sulfhydryl group of the protein. The sulfhydryl group of urease is probably the only functional group on the surface of the protein that is capable of readily undergoing redox reactions. This appears to indicate that the bridge between transition metals and enzymes may be at least partially covalent in nature.

$$SnCl_2 + H-O-\overset{\overset{\displaystyle |}{O}}{\underset{\underset{\displaystyle |}{O}}{Ti}}-O- \rightarrow Cl-Sn-O-\overset{\overset{\displaystyle |}{O}}{\underset{\underset{\displaystyle |}{O}}{Ti}}-O- + HCl$$

$$Protein-SH + Cl-Sn-O-\overset{\overset{\displaystyle |}{O}}{\underset{\underset{\displaystyle |}{O}}{Ti}}-O- \rightarrow protein-S-Sn-O-\overset{\overset{\displaystyle |}{O}}{\underset{\underset{\displaystyle |}{O}}{Ti}}-O- + HCl$$

Evaluation of Inorganic Bridge Formations as a Coupling Technique

The utilization of inorganic bridges for coupling enzymes to carriers offers additional versatility with respect to surface contributions, and the bridges are very simply applied for the purposes of coupling. This technique, in the author's opinion, is perhaps second only to direct adsorption insofar as economic considerations and ease of handling are concerned.

[12] Techniques of Enzyme Entrapment in Gels

By K. F. O'DRISCOLL

General Considerations

A variety of techniques have been developed to immobilize enzymes by entrapment. We can distinguish between those techniques where a polymerization causes a gel or membrane to be formed around an enzyme and those techniques where an enzyme is introduced to a preformed polymer and subsequent reactions cause a cross-linked network to be formed, thereby decreasing the enzyme's mobility.

Any consideration of the techniques for creating an immobilized enzyme system must be made in the light of an understanding of the following attributes of the system: (1) enzyme loading; (2) kinetic behavior; (3) stability; (4) reactor configuration. Although these attributes are highly coupled, they can be discussed in a general sense as separate factors. What follows is written primarily with entrapped enzymes in mind, but some of it is necessarily applicable to other types of immobilized enzymes.

Enzyme loading refers not only to the actual amount of enzyme that is entrapped, but it also includes the fact that in an entrapment procedure some enzyme may be lost owing to leakage or denaturation. As a consequence of this, the immobilized preparation may have a specific activity per unit weight that is less than might be expected from the amount of enzyme used in the preparation. Furthermore, the amount of enzyme that can be entrapped in some cases is seriously limited by the solubility of the enzyme in the entrapping reaction mixture.

Other factors that may determine the actual specific activity of the immobilized preparation include unequal partitioning of the substrate between the immobilized phase and the supernatant liquid, slow diffusion of the substrate to the active site, microenvironment effects such as localized pH or ionic strength inside the entrapping matrix, and diffusive or partitioning effects on inhibitors of the enzyme activity. All these factors are here lumped under the term kinetic behavior.

The stability of an immobilized enzyme in use or storage may be one of the most important parameters in the cost estimation of an immobilized enzyme process. An immobilized enzyme can denature while immobilized, but it may do so at a rate higher or lower than the native enzyme. In the case of enzymes immobilized by entrapment the worry also exists that the enzyme may continuously leak out of the matrix, thus reducing the specific activity of the preparation and possibly contaminating the product.

Finally, consideration must be given to the configuration in which the immobilized enzyme is to be used. This factor is highly coupled to the others, in that the specific activity, kinetic behavior, and stability are all quantitative factors that must be employed in optimizing a reactor configuration. In addition, the mechanical characteristics of the immobilizing material will determine its suitability for use in a given reactor: a soft, compressible material might cause too large a pressure drop across a tubular reactor and plug it; a hard, friable material might not withstand the shear field in a well stirred batch or continuous tank reactor; a hydrated membrane in which an enzyme is entrapped may need to be supported. From a more positive viewpoint, an entrapped enzyme is not subject to bacterial attack; the mechanical strength of the immobilizing material can be controlled by, for example, preparing a polymer gel having a desired degree of hydration; multiple enzyme systems can easily be coentrapped, or even whole cells may be entrapped.

From the above discussion it is apparent that careful attention must be given to a number of interrelated factors if an entrapped enzyme system is to be used, or considered for use, in a real process. Similar considerations, of course, must be made if a surface-immobilized enzyme is to be used. A choice of a particular immobilization technique, whether for a surface-immobilized enzyme or an entrapped enzyme, should be made on the basis of knowledge of all these factors. No single type of immobilization procedure can be expected to be perfect, and therefore the choice of an immobilization procedure ought to depend on a careful balancing of its known advantages and disadvantages.

In the recent past, almost all entrapped enzymes were immobilized in poly(acrylamide) gels. These gels are quite weak in a mechanical sense and have an open network that often permits the enzyme to leak out. As a consequence, entrapped enzymes have a poor reputation. This reputation is undeserved since enzymes may be entrapped in a variety of ways, some to which can lead to preparations that are most desirable in that their properties are superior for the particular use for which the immobilized enzyme is intended.

Gel Entrapment by Polymerization

To entrap an enzyme one must either form a cross-linked polymeric network around the enzyme molecule, or place an enzyme inside a polymeric material and then cross-link the polymer chains. Bernfeld and Wan[1] appear to have been the first to use the former technique when they dissolved one of several enzymes (e.g., trypsin, ribonuclease, β-amylase) in an aqueous solution of N,N'-methylenebisacrylamide (Bis) and carried out a low-temperature, free-radical polymerization of this crosslinking monomer. The choice of this particular monomer is easy to understand, since the preparation of hydrated gels for electrophoresis using acrylamide and Bis is a common biochemical laboratory procedure. Subsequent workers followed this lead and used these two monomers exclusively.[2-4] In retrospect, it is unfortunate that other monomers were

---

[1] P. Bernfeld and J. Wan, *Science* **142**, 678 (1963).
[2] G. P. Hicks and S. J. Updike, *Anal. Chem.* **38**, 762 (1966); also U.S. Patent No. 3,788,950 (1974).
[3] K. Mosbach and R. Mosbach, *Acta Chem. Scand.* **20**, 2807 (1966).
[4] T. Wieland, H. Determann, and K. Buennig, *Z. Naturforsch. Teil B* **21**, 1003 (1966).

not used, since poly(acrylamide) gels cross-linked with Bis have extremely high (about 80–95%) water content and are therefore structurally weak and easily leak some enzymes. Nevertheless, much has been learned about gel-entrapped enzymes using poly(acrylamide) gels, and this knowledge can be carried over to preparations using other, more desirable monomer/polymer systems. So far, the only other vinyl monomer that has been extensively investigated for entrapping enzymes is 2-hydroxyethyl methacrylate (HEMA) cross-linked with ethyleneglycol dimethacrylate (EGDMA). This is surprising when the large number of water-soluble, easily polymerized vinyl monomers is considered.

Solution Polymerization

The ingredients common to all gel-entrapping polymerizations which have been reported are: monomer, cross-linking agent, initiator, enzyme, and aqueous buffer. If a polyfunctional monomer, such as Bis or EGDMA, is used, the monomer and cross-linker are the same. Some preparations have included a substrate for the enzyme or other material (such as albumin) added to protect the enzyme.

Polymerizations are commonly carried out in the absence of oxygen and at low temperatures, 0° to 25°, in order to guard against thermal denaturation. Vinyl polymerizations are exothermic, and so the polymerization must be carried out slowly or be well thermostated to minimize temperature rise. At the conclusion of the polymerization, which may take from 30 min to several hours, the hydrated gel contains the entrapped enzyme and, possibly, nontrapped enzyme, residual monomer, and unused initiator. Common practice is to cut the gel into small pieces and wash them with a buffer. Depending on the intended use, the gel may then be used as is, or dried, ground into small particles, sieved into known size ranges, and stored prior to rehydration and use.

Any or all of the components of the polymerization mixture can serve to denature the enzyme while the reaction is going on. Therefore their concentrations and relative proportions should be carefully chosen. Some general principles can be discerned from reported work, but each system is different.

One might suspect, *a priori*, that the free radicals which initiate the polymerization might also attack and denature the enzyme. Fortunately, this does not appear to be the case, probably because the concentration of free radicals in a vinyl polymerization seldom goes above 10^{-6} M. The substance of greatest danger to the enzyme activity is probably the monomer itself, which suggests that monomer concentration should be kept low in a polymerization recipe. However, the amount of protein entrapped and the gel strength will decline as the monomer concentration is decreased in the entrapping reaction mixture. The specific activity of the enzyme may pass through a maximum because of these two countering influences of monomer concentration. Typically, the monomer is 30–60% of the polymerization reaction mixture. For similar reasons, the cross-linking agent concentration may also be optimized. The optimum cross-link concentration is usually about 5%. Figure 1 shows two sets of experimental data that confirm this.

The choice of initiator and its optimal concentration does not lend itself to easy generalization. So-called "redox" systems, such as potassium persulfate and tetramethylethylenediamine (TEMED), are useful for low-temperature work and do not appear to injure enzymes in the low concentrations necessary for initiation of polymerization. Photochemical initiation is also practicable at low temperatures.

The amount of enzyme dissolved in the polymerization reaction mixture is usually limited only by its solubility. Since the entrapping mixture typically contains 20–50% monomer, the enzyme solubility is not necessarily as high as it would be in pure buffer. This limitation on the maximum loading of enzyme in the final gel may prove to be the most serious factor limiting the use of gel-entrapped enzymes in some commercial processes. However, in those cases where a high loading is not necessary or where only a small amount of enzyme is available, gel

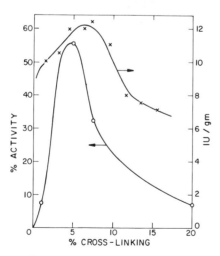

FIG. 1. Effect of cross-linker content on immobilized enzyme activity; ×, glucose oxidase in poly(2-hydroxyethyl methacrylate) with ethylene glycol dimethacrylate (unpublished work, this laboratory); O, cholinesterase in poly(acrylamide) with N,N'-methylenebisacrylamide. After Y. Degani and T. Miron, *Biochim. Biophys. Acta* **212**, 362 (1970).

entrapment provides a facile means for immobilizing enzymes with a minimum of preparative effort.

Example of Procedure for Immobilizing Glucose Oxidase in Poly(acrylamide)

Updike and Hicks[2] have described the following procedure. Stock solutions are prepared by dissolving 40 g of acrylamide monomer in 100 ml of 0.1 M phosphate buffer (pH 7.4) and 2.3 g of N,N-methylenebisacrylamide (Bis) in 100 ml of the same buffer. Solutions are stored and reactions carried out at 0°–4°. One milliliter of the monomer solution is then mixed with 4 ml of the cross-linker solution plus 1 ml of enzyme solution (containing 0.1–20 mg of enzyme). To this is added a photochemical initiator such as riboflavin and potassium persulfate (0.3 mg of each).

The reaction mixture is deoxygenated by bubbling nitrogen through it for about 15 min, and the reaction vessel is then closed by means of a stopcock or seal. Polymerization is then initiated by means of a strong light (the original paper suggests a No. 2 photoflood lamp; we have found a fluorescent desk lamp to be satisfactory). After 15 min, the polymerization is essentially complete, and the gel may be fragmented by any high shear action such as forcing it through a syringe needle or mechanically chopping it.

The gel particles are washed in distilled water, lyophilized, and sieved to obtain particles of a defined size.

Comments

The effect of varying the ratio of cross-linking agent (Bis) to monomer and of varying the water content of the polymerization reaction mixture is shown in Table I. Unfortunately, only relative activities were reported in this early work. It can be seen by comparing gels 1, 2, and 3 that total polymer content can be optimized, and, by comparing gels 2, 4, and 5 that cross-link content can also be optimized.

Many workers have used this procedure to immobilize a variety of enzymes. However, the hydrogels formed by the polymerization of acrylamide are water rich (about 80–95% water) and therefore mechanically weak; enzyme may also leak out through the "pores" of such an open gel network. By contrast, the polymerization of 2-hydroxethyl methacrylate (HEMA) using ethyleneglycol dimethacrylate (EGDMA) as a cross-linker gives gels which, on a wet basis, contain as little as 40% water. Such gels are not "porous," but small molecules may move

TABLE I
GLUCOSE OXIDASE ACTIVITY AS A FUNCTION OF GEL COMPOSITION[a]

Gel	Polymer content (g of monomer + Bis[b] in 100 ml of gel)	Bis as % of polymer	Relative activity
1	8.2	19	1.0
2	5.0	19	~0
3	11.1	10	0.32
4	5.0	10	0.66
5	5.0	5	0.60

[a] Data from references cited in text footnotes 2-4.
[b] Bis = N,N-methylenebisacrylamide.

through them by a process of activated diffusion.[5] The water content of poly (HEMA) gels may be controlled by the addition of comonomers or another hydrophilic polymer, such as poly(vinylpyrrolidone) (PVP). A method for preparing such a gel has been described by O'Driscoll et al.[6] and applied to a variety of enzymes. The following formulation was used for trypsin.

Example of Trypsin Formulation

In an ampule were placed 1 ml of 0.5 M Tris buffer, pH 8.0, 1 ml of HEMA containing 2% (by weight) of EGDMA, 0.2 g of PVP (molecular weight 40,000), 0.2 ml of a solution containing 2 mg of 2× crystallized trypsin in Tris buffer, and 0.2 ml of a polymerization initiator solution. The initiator is di(sec-butyl)peroxydicarbonate 1.25% in methanol (obtained as Lupersol 225 from Lucidol Co., Buffalo, New York; this initiator has a 10-hr half-life at about 45° and a 1-hr half-life at about 65°). Nitrogen was bubbled through this mixture for 5 min, and the ampule was then sealed and placed in a 27° water bath for 14 hr. A hard, translucent gel was formed, which was chopped, washed, and dried *in vacuo* at 5° for 3 days. The yield of polymer was 80%, and the apparent retention of activity in the rehydrated gel was 6%. The latter number must be interpreted cautiously, since the activity of a gel-entrapped enzyme is diffusion controlled and is determined by the

[5] H. Yasuda, C. E. Lamaze, and A. Peterlin, *J. Polym. Sci., Polym. Phys. Ed.* **9**, 1117 (1971).
[6] K. F. O'Driscoll, M. Izu, and R. Korus, *Biotechnol. Bioeng.* **14**, 847 (1972); also U. S. Patent No. 3,859,169 (1975).

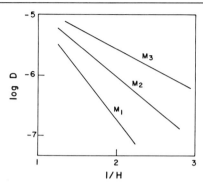

FIG. 2. Variation of substrate diffusivity, D, in a hydrogel as a function of fractional water content, H, and molecular weight of substrate $M_1 > M_2 > M_3$. After H. Yasuda, L. D. Ikenberry, and C. E. Lamaze, *Makromol. Chem.* **125**, 108 (1969).

particle size and substrate concentration as well as the absolute amount of active sites present in the gel.[7]

Gels formed by the above process contain 40–60% water (wet basis) when hydrated. In general, the higher the water content, the weaker the gel strength will be, but the greater will be the substrate diffusivity in the gel as shown in Fig. 2. The retention of enzyme activity by the above process of immobilization has varied, in the author's laboratory, from as low as 5% to as high as 89%. Enzymes that have been immobilized by this procedure include β-galactosidase, glucose oxidase, asparaginase, urease, invertase, uricase, and galactose oxidase.

Suspension Polymerization

Instead of carrying out an aqueous solution polymerization as described above, the hydrogel entrapping the enzyme can be formed in a suspension polymerization. In such a polymerization, an aqueous phase containing monomer, initiator, and enzyme is dispersed in a continuous, hydrophobic liquid. The resulting hydrogel consists of spherical particles, whose size can be controlled, to some extent, by the polymerization reaction conditions. The spherical particles have a potential advantage over the irregular, broken particles obtainable from a solution polymerization, in that the spheres pack more neatly in a tubular reactor.

One difficulty with suspension polymerization is that a detergent is used to stabilize the suspension. The use of such a detergent may denature some enzymes, so care must be taken in adopting existing recipes for suspension polymerization.

[7] K. F. O'Driscoll and R. Korus, *Can. J. Chem. Eng.* **52**, 775 (1974).

Mosbach and co-workers have described[8] a suspension polymerization procedure for the formation of "enzyme beads" consisting of cross-linked polyacrylamide in which an enzyme is entrapped (see also this volume [14] and [5]).

Example: Procedure for β-Galactosidase[9]

A yeast lactase (100 mg) was dissolved in a solution of buffer and acrylamide of 56 ml total volume. The buffer was 0.05 M phosphate, pH 7.3, and the monomer proportion was 15–30%, containing 5% Bis (based on total monomer) as a cross-linker. The initiator used was 4 ml of buffer containing 0.6 ml of N,N,N',N'-tetramethylethylenediamine and 200 mg of ammonium persulfate. The initiator and monomer solutions were mixed, then rapidly added to a reaction vessel containing 400 ml of a toluene–chloroform mixture (290:110) containing 1 ml of Sorbitan sequioleate (Pierce Chemical Co., Rockford, Illinois) as a suspension stabilizer. The reaction vessel consisted of a 1-liter flask, containing a paddle stirrer that rotates at about 240 rpm and just clears the sides of the flask. A nitrogen atmosphere is maintained in the reaction vessel, and the reaction is carried out at 4° for 30 min.

The beads are filtered on sintered glass, washed with 500 ml of buffer, and then, with stirring, washed for 60 min in 0.1 M NH_4HCO_3. The washed beads are frozen and lyophilized.

Comments on Suspension Procedure

The effect of varying the monomer content of the aqueous portion of the reaction mixture is seen in Table II. The fraction of the enzyme bound into the gel varies only from 50 to 75%, but the fraction of activity retained varies by more than 4-fold. No estimate of the effect of monomer content on particle size is given, so it is difficult to be certain whether the results on activity retention reflect denaturation of the enzyme or diffusion control of the activity. Similar results in solution polymerization[10] suggest the former. Particle sizes of 100–250 μm are expected,[8] and for such particles, activity may be strongly a function of size.[7]

Gel Entrapment by Cross-Linking Polymers

If an enzyme is dispersed in a linear polymer or polymer solution and the polymer is then cross-linked to form an insoluble network, one

[8] H. Nilsson, R. Mosbach, and K. Mosbach, *Biochim. Biophys. Acta* **268**, 253 (1972).
[9] A. Dahlqvist, B. Mattiasson, and K. Mosbach, *Biotechnol. Bioeng.* **15**, 395 (1973).
[10] Y. Degani and T. Miron, *Biochim. Biophys. Acta* **212**, 362 (1970).

TABLE II
EFFECT OF VARYING MONOMER CONTENT ON SUSPENSION
POLYMERIZED, GEL ENTRAPPED β-GALACTOSIDASE[a]

Percent of monomer content of aqueous phase	Percent of protein trapped in gel	Percent of free enzyme activity measured in gel
15	50	36
20	75	45
25	60	63
30	60	14

[a] From A. Dahlquist, B. Mattiasson, and K. Mosbach, *Biotechnol. Bioeng.* **15**, 395 (1973).

can expect the enzyme to be entrapped. At least four different types of preparations of this sort have been reported. Poly(dimethylsiloxane) was cross-linked with tin octoate by Brown and co-workers[11-13] and formed silicone membranes to immobilize several different enzymes. The success of the immobilization is difficult to establish from the reported data, since only relative activities are given. Guilbault and Das, however, have reported[14] that the immobilization conditions were too rigorous for cholinesterase since 80% activity was lost in their silicone preparations. Considering the hydrophobic character of this rubber, it is most likely that in these preparations low activity is a result of restricted permeability of the membrane to the water-soluble substrate.

Guilbault and his co-workers have also immobilized a variety of enzymes in starch.[15,16] They prepared the gel by dissolving the enzyme in a warm starch solution and then expressed the water from the cooled starch gel in a polyurethane pad that was used in analytical work.

Maeda and co-workers have used γ-irradiation and electron beam techniques to cross-link an aqueous poly(vinyl alcohol) solution containing glucoamylase or invertase.[17] In the case of γ-irradiation no leakage of protein from the gel was observed, but there was not much enzyme

[11] H. D. Brown, A. B. Patel, and S. K. Chattopadhyay, *J. Biomed. Mater. Res.* **2**, 231 (1968).
[12] S. N. Pennington, H. D. Brown, A. B. Patel, and C. O. Knowles, *Biochim. Biophys. Acta* **167**, 479 (1968).
[13] S. N. Pennington, H. D. Brown, A. B. Patel, and S. K. Chattopadhyay, *J. Biomed. Mater. Res.* **2**, 443 (1968).
[14] G. G. Guilbault and J. Das, *Anal. Biochem.* **33**, 341 (1970).
[15] E. K. Bauman, L. H. Goodson, G. G. Guilbault, and D. N. Kramer, *Anal. Chem.* **37**, 1378 (1965).
[16] G. G. Guilbault and D. N. Kramer, *Anal. Chem.* **37**, 1675 (1965).
[17] H. Maeda, H. Suzuki, and A. Yamauchi, *Biotechnol. Bioeng.* **15**, 607, 827 (1973).

activity either. Electron beam irradiation seemed more promising, since they recovered 50% of the enzyme activity and still had no enzyme leakage. The gels formed in this fashion are cross-linked because the high-energy irradiation creates free radicals on the polymer chain which couple with each other. The amount of radiation required is somewhat greater than that needed for initiation of polymerization, and there is, therefore, more likelihood of damage to the enzyme.

Although a considerable body of knowledge exists on the formation of gels by mixing polyvalent ions with polyelectrolytes,[18-20] there have been only two reports[21,22] of the use of this technique for immobilizing enzymes. The technique, termed coarctation, would appear to be one of the more gentle ways of cross-linking polymers and ought to be given more attention.

Example of Coarctation Technique

In the work of Horvath and Sovak,[21] a trypsin–polycarboxylic acid conjugate was formed by allowing 60 mg of a dry maleic anhydride–vinyl methylether copolymer (GAF Co.) to make contact with 5 ml of a 0.1 M phosphate buffer, pH 7.0, containing 100 mg of the enzyme, overnight in a shaker bath at 4°. Protein contents as high as about 40% of the dry conjugate were found. The conjugate would swell to take up amounts of water that varied with the Ca^{2+} content of the supernatant. The gel character was also affected by calcium ion concentration, changing from a "soft mucilage" containing more than 97% water (wet basis) at 10^{-4} M calcium ion concentration to a hard gel at 10^{-2} M. The apparent diffusivity of substrate in the gel was similarly affected, and this resulted in a marked change in the kinetic behavior of the immobilized enzyme.

In their work, Horvath and Sovak prepared gel particles so as to be pellicular; i.e., the gel is a relatively thin shell around a rigid spherical core. The core in this case consisted of a glass bead coated with a very thin layer of anion exchange resin. Practical advantages of the use of pellicular catalyst particles have been discussed.[23] It is easy to recognize that such particles will have better packing qualities in a column and less diffusive control of reaction when compared to particles comprised solely of a gel.

[18] F. T. Wall and J. W. Drenan, *J. Polym. Sci.* **7**, 83 (1951).
[19] A. Ikegami and N. Imai, *J. Polym. Sci.* **56**, 133 (1962).
[20] R. Y. M. Huang and N. R. Jarvis, *J. Polym. Sci., Polym. Symp.* **41**, 117 (1973).
[21] C. Horvath and M. Sovak, *Biochim. Biophys. Acta* **298**, 850 (1973).
[22] C. Horvath, *Biochim. Biophys. Acta* **358**, 164 (1974).
[23] C. Horvath and J. Engasser, *Ind. Eng. Chem., Fundam.* **12**, 229 (1973).

Interpenetrating Networks

A novel variation on the usual gel entrapment procedure has been introduced by Stegemann and his co-workers.[24] They isolated an esterase (from potatoes) by gel electrophoresis in a lightly cross-linked poly(acrylamide) gel and cut the electrophoretic gel portion containing the enzyme into small (3 mm) pieces. These pieces of gel were then soaked in another polymerization reaction mixture containing about 5–7% acrylamide and varying levels of Bis cross-linker. After the second polymerization, about 40% of the original activity was retained. An optimum level of about 5% Bis was observed. This technique is of particular value for the immobilization of minute quantities of highly purified enzyme. The interpenetration of two separately formed poly(acrylamide) networks as reported in this work provides a tighter matrix for immobilizing the enzyme and prevents leakage.

A variant on the technique, which might be of value, would be to change the nature of the monomer for the formation of the second network, thus giving a measure of control over the physicochemical properties of the matrix. The general subject of interpenetrating networks is of considerable current interest.[25]

Copolymerization

The principle of using more than one monomer (besides the cross-linking agent), and thereby introducing a variety of functional groups, is easy to espouse as a method of optimizing the gel structure. In this fashion, one expects to exercise some control over the microenvironmental effects of the gel on the entrapped enzyme. In practice, few gel entrapments have been described that deliberately do so. An example of the potential benefits is the work of Beck and Rase,[26] who demonstrated a broadening of the pH profile for glucoamylase immobilized in poly(acrylamide); this could be achieved by incorporating 15% of acrylic acid in the monomer portion of the entrapping reaction.

It is important to recognize that although copolymerization offers the opportunity to control the microenvironment, the reaction may proceed in a manner that can yield a heterogeneous polymer. This is so because the copolymer initially formed from a given reaction mixture will usually *not* be identical in composition to that of the monomer

[24] K. N. Dinish, K. N. Shivaram, and H. Stegemann, *Fresenius' Z. Anal. Chem.* **263**, 27 (1973).
[25] V. Huelck, D. A. Thomas, and L. H. Sperling, *Macromolecules* **5**, 340, 348, (1972).
[26] S. R. Beck and H. F. Rase, *Ind. Eng. Chem., Prod. Res. Dev.* **12**, 260 (1973).

feed. This, in turn, will cause a drift in the copolymer composition as the reaction advances and give a final product which, although its overall composition is that of the initial feed, will have a broad distribution of compositions. This compositional heterogeneity may result in a phase separation, thus providing pores in the gel and enabling leakage of the enzyme. The relative reactivities of most vinyl monomers are tabulated.[27]

Multiple Enzyme Systems

If gel-entrapped enzymes have one outstanding attribute, relative to other immobilization techniques, it is that they come closer to mimicking the natural enzyme in the cell. Since multiple enzyme systems are so common in cells, it is probable that complex syntheses will be carried out by gel-entrapped enzymes better than by enzymes immobilized in other ways.

Published work on multienzyme immobilizations is scarce. A recent paper of Mosbach et al.[28] will serve as an example: a three-enzyme system (malate dehydrogenase, citrate synthase, lactate dehydrogenase) was used as a model for oxaloacetate production and utilization in mitochondria. The simultaneous gel entrapment of three enzymes in polyacrylamide was done in essentially the manner of their early work,[3] like that described above by Updike and Hicks.[2] When NADH was reoxidized in this system (by the addition of pyruvate in slurry reactor) the rate of the coupled, three-enzyme system immobilized in poly(acrylamide) was 4-fold faster than that for the free enzymes in solution, and twice as fast as the same system covalently bonded onto Sephadex. Care was taken to ensure equivalent single-enzyme activities and the absence of partitioning effects.

Simply put, rate enhancement is understood to occur in gel entrapped, multienzyme systems because the intermediates, products of one enzyme and substrates for the next, are at an abnormally high local concentration relative to what would be the case in free solution. A quantitative appreciation of this phenomenon can be gained from the theoretical work by Lawrence and Okay.[29] They examined the equations describing the system of glucose oxidase and catalase coimmobilized and homogeneously distributed throughout a porous support. The glucose would be oxidized to produce H_2O_2, and the catalase would regenerate oxygen. It

[27] L. J. Young, in "Polymer Handbook" (J. Brandrup and E. H. Immergut, eds.), pp 11–105. Wiley, New York, 1975.
[28] P. A. Srere, B. Mattiasson, and K. Mosbach, Proc. Natl. Acad. Sci. U.S.A. 70, 2534 (1973).
[29] R. L. Lawrence and V. Okay, Biotechnol. Bioeng. 15, 217 (1973).

was shown that, under conditions where kinetics are not strongly affected by diffusion and at a nonzero ratio of peroxide to oxygen, there would be an expected rate enhancement for glucose oxidation of the order of 2- to 5-fold.

Some very early workers on gel-entrapped enzymes also had an appreciation of the relationship between naturally occurring enzymes and the gel system. As early as 1964, van Duijn et al. studied enzymes immobilized in poly(acrylamide) films as models for cytochemical determination of enzymes.[30,31] They appear to have been the first to appreciate the diffusion control of gel-entrapped enzyme kinetics and attempted to utilize it in their work.

Qualitative Considerations

The variety of work done to date on the gel entrapment of enzymes by polymerization is fairly large with regard to the number of enzymes entrapped, but it is still quite small in respect to the different entrapping monomers that have been used. Nevertheless, it is possible to make some generalizations about this technique. In the author's opinion, two facts stand out: (1) the technique is extremely easy to use; and (2) the enzyme loading that can be achieved is quite low, usually of the order of 10–100 mg of protein per gram of gel.

These facts taken together imply that gel entrapment is a technique of high convenience in the laboratory. The ease of, and the stability imparted by gel entrapment may make the technique of value in analytical applications (see this volume [41]) or where the enzyme is difficult to isolate.

Many quantitative data that have been reported on the activities, K_m values, and stabilities of gel-entrapped enzymes must be interpreted with caution. The effectiveness factor (i.e., the enzyme activity per unit volume of gel relative to the activity of the same concentration of enzyme in free solution) can be a strong function of all the variables in a given preparation and the characterization of it. Figure 3 shows the theoretical variation[7] of the effectiveness factor as a function of the substrate concentration [S] and the Michaelis constant in the gel, K_m', for different values of the Damköhler number β where

$$\beta = (v_{max}'/K_m'D_s)r^2$$

D_s is the substrate diffusivity in the gel of radius r having an enzyme

[30] M. van der Ploeg and P. van Duijn, J. R. Microsc. Soc. 83, 415 (1964).
[31] P. van Duijn, E. Pascoe, and M. van der Ploeg, J. Histochem. Cytochem. 15, 631 (1967).

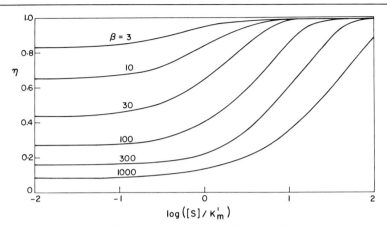

FIG. 3. Variation of effectiveness factor, η, with dimensionless substrate concentration at various Damköhler numbers, β.

concentration in gel that gives rise to a maximum reaction rate v_{max}'. Before comparisons of kinetic parameters can be made between immobilized and free enzymes, it is important to recognize at what effectiveness factor the comparison is being made: i.e., is diffusion confounding a physical chemical interpretation of results? It has been shown that diffusion effects can even cause a pseudo increase in stability.[32,33]

[32] D. F. Ollis, *Biotechnol. Bioeng.* **14**, 871 (1972).
[33] R. A. Korus and K. F. O'Driscoll, *Biotechnol. Bioeng.* **17**, 441 (1975).

[13] Immobilization of Steroid-Transforming Microorganisms in Polyacrylamide

By PER-OLOF LARSSON and KLAUS MOSBACH

The immobilization of intact microbial cells has recently attracted marked attention, in particular because of their potential for industrial applications (see especially this volume, Chapters [50] and [53]). This interest stems mainly from the fact that immobilized whole cells favorably combine the advantages inherent in the use of immobilized enzymes with those of microbial fermentation. For instance, they have suitable mechanical properties, they are reusable, the need for costly enzyme isolation is eliminated, and, finally, the enzyme stability is increased, probably from the localization of the enzyme in its natural surroundings.

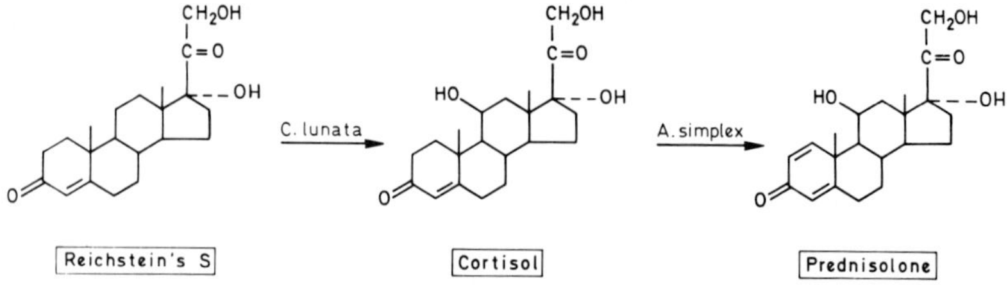

FIG. 1. Steroid transformation by polyacrylamide-entrapped microorganisms.

The major part of the work reported on whole-cell entrapment,[1-13] besides initial studies, has focused on the production of organic acids including amino acids[6-9] and conversions of carbohydrates.[10-12] Spore entrapment has also been tried.[13]

Another interesting area that appears to be suitable for application of immobilized cells lies in steroid transformation,[14] which conventionally is carried out by fermentation processes. Two important reactions in steroid transformation involve 3-ketosteriod Δ^{1-2} dehydrogenation and 11-β-hydroxylation (Fig. 1). The preparation of polyacrylamide-entrapped *Arthrobacter simplex* and *Curvularia lunata* cells carrying out these reactions is described in this chapter, and some of their properties are discussed.

Immobilization

Since acrylamide polymerization is an exothermic reaction, means must be provided to efficiently dissipate the developed heat, which might otherwise denature the steroid-converting enzyme system. A sandwichlike polymerization vessel having good heat-transfer properties was prepared from two thin-layer chromatography glass plates (0.2 × 20 × 20 cm) and a piece of latex tubing (o.d., 5 mm; i.d., 3 mm). The

[1] K. Mosbach and R. Mosbach, *Acta Chem. Scand.* **20**, 2807 (1966).
[2] S. J. Updike, D. R. Harris, and E. Shrago, *Nature (London)* **224**, 1122 (1969).
[3] N. E. Franks, *Biochim. Biophys. Acta* **252**, 246 (1971).
[4] W. Slowinski and S. E. Charm, *Biotechnol. Bioeng.* **15**, 973 (1973).
[5] R. R. Mohan and N. N. Li, *Biotechnol. Bioeng.* **17**, 1137 (1975).
[6] I. Chibata, T. Tosa, and T. Sato, *Appl. Microbiol.* **27**, 878 (1974).
[7] T. Tosa, T. Sato, T. Mori, and I. Chibata, *Appl. Microbiol.* **27**, 886 (1974).
[8] K. Yamamoto, T. Sato, T. Tosa, and I. Chibata, *Biotechnol. Bioeng.* **16**, 1589 (1974).
[9] K. Yamamoto, T. Sato, T. Tosa, and I. Chibata, *Biotechnol. Bioeng.* **16**, 1601 (1974).
[10] W. R. Vieth, S. S. Wang, and R. Saini, *Biotechnol. Bioeng.* **15**, 565 (1973).
[11] R. Saini and W. R. Vieth, *J. Appl. Chem. Biotechnol.* **25**, 115 (1975).
[12] K. Toda and M. Shoda, *Biotechnol. Bioeng.* **17**, 481 (1975).
[13] D. E. Johnson and A. Ciegler, *Arch. Biochem. Biophys.* **130**, 384 (1969).
[14] K. Mosbach and P. O. Larsson, *Biotechnol. Bioeng.* **12**, 19 (1970).

glass plates were kept 1–2 mm apart by the latex tubing and were firmly pressed against the tubing by means of metal clamps. The volume of the vessel could conveniently be adjusted between 10 and 60 ml by merely repositioning the latex tubing. The construction of the vessel also allowed easy transfer to cooling-baths of desired temperatures.

Arthrobacter simplex[15]

Solutions

A. *Arthrobacter simplex* cells[16] (5 g wet weight; centrifuged at 10,000 g) suspended in ice-cold 0.05 M Tris-HCl buffer, pH 7.5 to a final volume of 25 ml

B. Acrylamide, 7.23 g and N,N'-methylene bis(acrylamide), 0.37 g, dissolved in ice-cold 0.05 M Tris-HCl buffer, pH 7.5, to a final volume of 23 ml

C. N,N,N',N'-Tetramethylenediamine, 0.10 g dissolved in 1 ml of water

D. Ammonium persulfate, 0.05 g, dissolved in 1 ml of water

Solutions A and B are rapidly mixed, and solutions C and D are added immediately. The mixture is poured without delay into the polymerization vessel, and nitrogen gas is bubbled through the mixture (the gas is preferably introduced via a steel capillary). Polymerization starts within 2 min, and the vessel is transferred to an ice-bath and kept there for 5 min to allow efficient cooling during the initial phase of the polymerization process. The vessel is then transferred to a water bath at room temperature; after 1 hr, polymerization is considered complete.

Curvularia lunata

Solutions

A. *Curvularia lunata* mycelium[17] (20 g wet weight, 1.0 g dry weight) suspended in 0.7% NaCl, 0.02% Tween 80 to a final volume of 30 ml

---

[15] Synonymous with *Corynebacterium simplex*.

[16] A concentrated cell culture (500 ml) was obtained according to F. Kondo and E. Masuo, *J. Gen. Appl. Microbiol.* **7**, 113 (1961). It was used as inoculum for the final medium consisting of 10 liter of 0.25% yeast extract, pH 7. After 12 hr at 28° in an aerated fermentor, 600 ml of 3.3 mM cortisol in methanol was added, and induction was allowed to proceed for 10 hr, when the cell mass was harvested by centrifugation.

[17] *C. lunata* was grown essentially as described by M. H. J. Zuidweg, W. F. van der Waard, and J. De Flines, *Biochim. Biophys. Acta* **58**, 131 (1962). It is advantageous to use a mycelium that has been gently ground in a mortar, since this will ensure a fairly even distribution of mycelium in the final polyacrylamide gel.

B. Acrylamide, 4.25 g, and N,N'-methylene bis(acrylamide), 0.25 g, dissolved in water to a final volume of 18 ml
C and D as above

The same procedure as that given for *A. simplex* may be followed. The polymerization process is markedly slower, however, and initial cooling in an ice-bath is thus not necessary.

Granulation and Washing

The gel sheet is removed from the polymerization vessel and granulated either by passing it through a 30-mesh nylon net or by treating it in a homogenizer together with buffer (50 g of gel + 200 ml of buffer was treated for 30 sec in a Sorvall Omnimixer at highest speed setting). Both methods produce fines, which may be removed by decantation. The gel granules are poured into a column and washed overnight with buffer to remove any remaining monomers and cell debris.

Assays

3-Ketosteriod-Δ^1-Dehydrogenase Activity

The conversion of cortisol to prednisolone was followed spectrophotometrically at 285 nm; at this wavelength, the molar absorption coefficient of cortisol is 420 M^{-1} cm^{-1} and of prednisolone is 3240 M^{-1} cm^{-1} (aqueous solutions). The transformation is thus characterized by a $\Delta\epsilon$-value of 2820 M^{-1} cm^{-1} and, provided a spectrophotometer with a narrow bandwidth (2 nm) is used, the reaction can accurately be followed at 285 nm up to absorbance readings of approximately 1.5.

Standard Procedure. Arthrobacter simplex gel (1.0 g, wet weight) was mixed with 8.5 ml of air-saturated Tris-HCl buffer, pH 8.0, in a stirred and thermostated (25°) plastic beaker. The supernatant was pumped via a flow-through cell of the spectrophotometer and back to the beaker, the gel granules being prevented from entering the circulating stream by a nylon net fused to the inlet tubing (see also chapter [25]). The blank reaction, if any, was checked, and the Δ^1-dehydrogenation reaction was then initiated by adding 0.5 ml of 20 mM cortisol dissolved in 95% methanol–5% water. The progress of the reaction was followed on a recorder.

The Δ^1-dehydrogenase activity of free *A. simplex* cells was also determined spectrophotometrically, but using an ordinary 3-ml cell.

11-β-Hydroxylase Activity

The conversion of Reichstein's compound S[18] to cortisol was followed by thin-layer chromatography.

Curvularia lunata gel granules (5 g) were suspended in 20 ml of 0.12 M NaCl containing 0.02% Tween 80; 0.5 ml of 30 mM compound S dissolved in 0.5 ml of dimethyl sulfoxide was added, and the incubation proceeded under shaking (60 rpm) in a 100-ml Erlenmeyer flask in the dark at 28°. At suitable time intervals (1–3 hr) aliquots (0.25 ml) were removed and extracted with ethyl acetate; the solvent was evaporated off, and the steroid mixture was dissolved in acetone and applied to silica gel thin-layer plates. The plates were developed twice in the solvent system acetone–dichloromethane (1:4), and the steroid spots were located under UV light. The amount of cortisol formed was estimated by comparison with reference spots containing known amounts of steroids. R_f values: Reichstein's compound S = 0.39, cortisol = 0.18.

General Considerations

The immobilization technique described in the experimental section is simple and yields gel granules with good retention of steroid converting activity.

It should be pointed out that the precise mode of entrapping microbes in polyacrylamide may be critical for a satisfactory retention of activity. In the case of *A. simplex*, the steroid dehydrogenase activity was satisfactorily retained only in gels characterized by a rather high bacterial content. Gels containing 10% (wet weight/wet weight) bacteria showed a rather high activity retention of approximately 40%, whereas gels with less than 5% bacterial content were almost void of dehydrogenase activity. The reason for this protective effect on enzyme activity at high bacterial content in unclear but poses at least no practical problem, since densely loaded gels are anyhow desired and, furthermore, bacteria-rich gels are mechanically superior to less densely substituted gels, which are somewhat brittle.

Another important, and perhaps related, parameter is the buffering of the polymerization medium. The presence of Tris or phosphate buffer (pH 7.5) during polymerization thus preserved the Δ^1-dehydrogenase activity of the bacteria, whereas unbuffered polymerization media and media with sodium chloride (0.15 M) as sole additive yielded preparations with poor activity.

The concentration of the catalyst system (ammonium persulfate +

---

[18] Reichstein's compound S = pregna-4-ene-17α,21-diol-3,20-dione.

tetramethylenediamine) was varied but seemed at least not per se to influence the retention of activity. An important factor, however, was the time span between mixing of monomers with bacteria and the start of polymerization. For good retention of activity it was necessary to minimize this time, and in the experimental procedure the conditions are thus chosen to afford polymerization as soon as conveniently is possible (approximately 1 min after mixing). The polymerization vessel also allows efficient cooling, which is necessary to avoid heat-denaturation of the steroid dehydrogenase. It was found that the bacteria rapidly lose their activity at temperatures higher than 40°.

For most purposes the mechanical properties of the polyacrylamide-entrapped microorganisms were satisfactory, so that easy filtration and resistance to mechanical agitation (generation of fines) was achieved. In column operations, however, care had to be exercised and only moderate pressure drop could be allowed in order to prevent compression of the gel bed with an accompanying drop in the flow rate. At least for column operations, it may be advantageous to use immobilized cells obtained by bead polymerization (see also chapter [5]). For large-scale production of such material the procedure described in chapter [14] may be applicable. Bead polymerization, however, involves the use of organic solvents, which might denature the sensitive cofactor system associated with the Δ^1-dehydrogenase.[19] The electron acceptor menadione (2-methyl-1,4-naphthoquinone) is an efficient cofactor for both isolated and cell-bound Δ^1-dehydrogenase.[19,20] Recent observations[21] indicate that also polyacrylamide-entrapped cells are capable of utilizing menadione as a cofactor. Furthermore, immobilized cells that have been inactivated by organic solvents regain their Δ^1-steroid-dehydrogenase activity upon addition of menadione,[21] and in analogy it may be possible that polyacrylamide-entrapped *A. simplex* obtained by bead polymerization also should have fair Δ^1-dehydrogenase activity.

Investigations of the influence of substrate concentration on activity revealed that 1 mM cortisol was optimal for gel-bound *A. simplex*. At higher concentrations a slight substrate inhibition occurred, which is in agreement with data obtained with isolated Δ^1-dehydrogenase.[22,23] Pre-

---

[19] B. K. Lee, W. E. Brown, D. Y. Ryu, and R. W. Thoma, *Biotechnol. Bioeng.* **13**, 503 (1971).

[20] V. Stefanovic, M. Hayano, and R. I. Dorfman, *Biochim. Biophys. Acta* **17**, 429 (1963).

[21] S. Ohlson, P. O. Larsson, and K. Mosbach, unpublished observations.

[22] L. Penasse and M. Peyre, *Steroids* **12**, 525 (1968).

[23] L. Penasse and E. E. Baulieu, *Abh. Dtsch. Akad. Wiss. Berlin, Kl. Med.* **1968** (2), 201–211 (1969).

liminary studies with gel-entrapped *C. lunata* showed optimal activity with 0.6 mM Reichstein's compound S and slight substrate inhibition at higher concentrations.

In Table I the storage stability of gel-entrapped *A. simplex* is listed. As to be expected, the stability is good at $-20°$ and also in freeze-dried preparations. The lyophilization process, however, is accompanied by a drastic reduction of activity, which may be only in part overcome by freeze-drying in the presence of a stabilizing medium. Gel-entrapped *C. lunata* was best preserved by first equilibrating it with saturated sucrose solution followed by storage at $-20°$. The sucrose apparently prevented damage to the cells judged from the fact that untreated *C. lunata* gel lost its activity at $-20°$.

From a practical viewpoint, steroid transformations such as 3-ketosteroid-Δ^1-dehydrogenation and 11-β-hydroxylation should be as complete as possible, since otherwise tedious separation procedures must follow to ensure pure products. Both gel-entrapped *A. simplex* and *C. lunata* are capable of complete transformations and in Table II the capacities are listed. Table II shows two sets of data for each immobilized microbe. One set was obtained from freshly prepared gels and the other set from gels that had been treated with a reactivating medium, containing nutrients and a steroid inducer. As can be seen, the activation process rendered the gels three to five times as active as before. The process could be repeated several times and could be used whenever the transformation capacity had decreased to an impractically low level. The process is thus an illustration of a feature unique for immobilized whole cells catalysts; namely, the possibility of restoration of catalytic power *in situ*.

TABLE I
STORAGE STABILITY OF Δ^1-DEHYDROGENASE ACTIVITY OF GEL-ENTRAPPED
Arthrobacter simplex CELLS

Storage conditions	Activity (%) after			
	0	12	21	120 days
$+4°$	100[a]	80	60	0
$-20°$	100	100	90	80
Freeze-dried	40	—	—	30
Freeze-dried + stabilizer[b]	50	—	—	40

[a] The activity of freshly prepared gel = 100%.
[b] Before freeze-drying the gel granules were equilibrated with a solution containing 5% Dextran T40 (Pharmacia, Sweden), 5% sucrose, and 5% sodium glutamate.

TABLE II
TRANSFORMATION CAPACITY OF POLYACRYLAMIDE-ENTRAPPED MICROORGANISMS

Polymer-entrapped microorganism	Steroid substrate	Transformation rate (μmoles/hr/g wet gel)	
		Initial	Effective transformation rate for 100% conversion
Curvularia lunata	Reichstein's S, 0.3 mM	0.15	0.05
C. lunata reactivated[a]	Reichstein's S, 0.3 mM	0.50	0.15
Arthrobacter simplex	Cortisol, 1.0 mM	9.0	4.0
A. simplex reactivated[b]	Cortisol, 1.0 mM	41	18

[a] Reactivation was carried out as follows: 100 g of gel (wet weight) was suspended in 400 ml of medium containing 0.5% corn steep liquor, 0.67% sucrose, 0.7% NaCl, 0.02% Tween 80, and 0.010% Reichstein's compound S. The suspension was shaken on a rotary shaker in a 2-liter Erlenmeyer flask for 16 hr at 28°. The granules were then washed with 0.7% NaCl.

[b] Reactivation was carried out as follows: 1g of gel (wet weight) was suspended in 20 ml 1 mM cortisol, dissolved in 0.5% peptone, pH 7.0. The suspension was shaken for 48 hr at 25°. The gel granules were then washed with 0.05 M Tris-HCl buffer, pH 8.0. (Conditions have not yet been optimized.)

It is interesting to compare the catalytic efficiency of immobilized whole cells and immobilized enzymes. With respect to 3-ketosteroid-Δ^1-dehydrogenase activity, the whole cell approach seems decidedly advantageous. Cell disintegration, enzyme isolation, and handling of a labile enzyme system are nonexisting problems, and furthermore, the resulting immobilized steroid dehydrogenating activity is considerably more satisfactory with the immobilized whole cell method. For example, whereas 1 g of *A. simplex* cells when entrapped showed a Δ^1-dehydrogenase activity of 100 μmoles/hour, the isolated Δ^1-dehydrogenase from the same amount of cells and subsequently entrapped yielded an activity of only about 10 μmoles/hour.[14] A similar result would probably be found when entrapped 11-β-hydroxylase/entrapped *C. lunata* comparisons were made, especially since isolated 11-β-hydroxylase is reportedly very unstable.[17]

[14] A Method for Continuous Production of Polyacrylamide-Entrapped Enzyme Beads

By A. KÖSTNER and M. MANDEL

Principles

The immobilization of enzymes by entrapping them into polyacrylamide gel was proposed by Bernfeld and Wan[1] in 1963. Since then this method has proved efficient and convenient for many enzymes. In most cases the enzyme gel is produced by block polymerization with subsequent disruption of the formed gel. The gel particles produced are irregular in form and size. These irregular particles cause high hydraulic resistance in packed beds and the gel in batch reactors wears out quickly. Exact kinetic measurements are also difficult under these conditions. The production of polyacrylamide gel in microspherical form by dispersing the polymerizable solution into an organic solvent immiscible in water is well known (e.g., Hjertén and Mosbach[2]). The same method has been employed for polymerizing enzyme-containing solutions.[3-5] Trypsin entrapped in spherical beads of gel showed higher activity than that in irregular particles of block gel.

In order to increase the productivity of bead polymerization and to regulate gel size, we have employed a special apparatus for carrying out this process.[6,7]

Apparatus

This apparatus (see Fig. 1) consists of a column (8 in Fig. 1B) in which a solution of monomers, initiators, and enzyme is dispersed in oil and polymerized. The initial solutions are prepared and deaerated in the usual way and stored in reservoirs (1). To avoid untimely polymeri-

---

[1] P. Bernfeld and J. Wan, *Science* **142**, 678 (1963).
[2] S. Hjertén and R. Mosbach, *Anal. Biochem.* **3**, 109 (1962).
[3] H. Nilsson, R. Mosbach, and K. Mosbach, *Biochim. Biophys. Acta* **268**, 253 (1972).
[4] K. Mosbach, *Biotechnol. Bioeng. Symp.* **3**, 189 (1972).
[5] S. Gestrelius, B. Mattiasson, and K. Mosbach, *Eur. J. Biochem.* **36**, 89 (1973).
[6] A. Köstner, K. Kivisilla, M. Mandel, and E. Siimer, Method for Production of Matrix Bound Enzyme. U.S.S.R. Patent 414301 (Cl. C 12d 13/10), 29.05.1973, Appl. 30.10.1971. From *Otkrytiya, Izobret. Prom. Obraztsy, Tovarnye Znaki* **5**, 92 (1974).
[7] A. Köstner. Apparatus for Production of Spherical Microbeads. U.S.S.R. Patent 458323 (Cl. B. 01j 2/06), 5.07.1974, Appl. 26.03.1973. From *Otkrytiya, Izobret. Prom. Obraztsy, Tovarnye Znaki* **4**, 7 (1975).

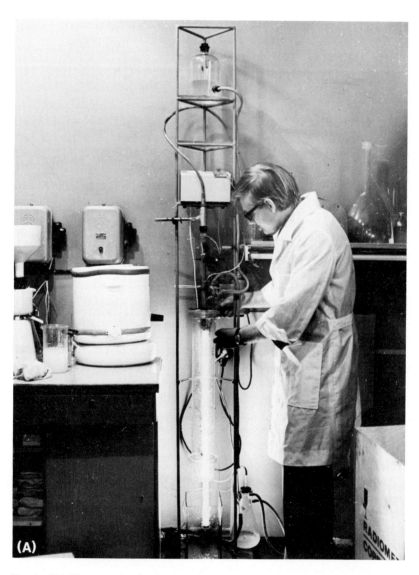

Fig. 1. (A) Photograph of column apparatus. (B) Diagram of polymerization apparatus. 1, Reservoirs for the reaction mixtures; 2, peristaltic pump; 3, flow-through mixer; 4, vibrating capillary tube; 5, eccentric, mounted to the shaft of variable speed motor for generating vibrating movement; 6, stroboscope, a controlled flashlight source for measuring the frequency of vibration; 7, elastic diaphragm; 8, column; 9, reservoir for oil; 10, water jacket; 11, luminescence lamps; 12, collection reservoir.

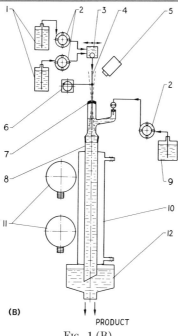

FIG. 1.(B).

zation in the connections, it is necessary to prepare a solution of one of the initiators [preferrably $(NH_4)_2S_2O_8$] separately and to mix it with other components just before dispersion. This point is especially important in the case of redox polymerization without light initiation using the tetramethylethylenediamine (TEMED)–persulfate system. All solutions are measured and injected by means of a peristaltic pump (2) into a small volume flow-through mixer (3), which is connected to the upper part of a vibrating capillary tube (4). The vibrating capillary tube permits a simple and reproducible means of dispersion. The upper end of the capillary tube is driven by an eccentric, connected to a variable-speed motor (5). The vibration frequency is controlled by means of a stroboscope (6). The elastic diaphragm (7) across the top of the column enables the vibrations to be transferred to the lower end of the capillary tube, which is submerged in the oil in the column. Before the process is started the column is filled with deaerated paraffin oil and the lower end is submerged in oil in the collection vessel (12) to avoid rising air bubbles in the column. During the process a permanent flow of deaerated oil is generated through the column. The capillary tube acts as a disperser, and by varying the frequency and amplitude of vibration, droplets having the required diameter can be obtained. As the reaction mixture passes through the column, polymerization occurs and gel beads sus-

OPERATION CONDITIONS AND RESULTS

Parameters and conditions	Photo-polymerization	Chemical polymerization
Operation conditions		
Column length (mm) × diameter (mm)	900 × 31	700 × 310
Capillary tube diameter (mm)	0.6	1.0
Vibration frequency (sec^{-1}) × amplitude (mm)	35 × 15	10 × 20
Total power of lamps (W)	1200	—
Water thermostating	Yes	No
Flow rate of oil × reaction mixture (ml/hr)	300 × 300	200 × 1000
Composition of reaction mixture		
Enzyme used	Invertase	Penicillin amidase[a]
Activity (U/ml)	400	104
Acrylamide (%)	6.3	9.0
N,N-Methylenebisacrylamide (%)	0.7	1.0
$(NH_4)_2S_2O_8$ (%)	0.25	1.0
Riboflavin (%)	0.005	—
β-Dimethylaminopropionitrile (%)	0.25	—
N,N,N',N'-Tetramethylethylenediamine (%)	—	1.0
Properties of gels		
Yield of gel by weight (%)	84	81
0.1–0.2 mm (%)	—	7.6
0.25–0.5 mm (%)	14.8	52.0
0.5–1.0 mm (%)	41.3	36.8
1.0 mm (%)	44.0	4.8
Activity of gel (U/g)	93	65.2
Total yield of activity (%)	19.5	50.6
Productivity (g/hr)	256	810

[a] Chemically modified by cross-linking with glutaraldehyde.

pended in oil leave the column and are collected in the reservoir (12).

Polymerization depends on column dimensions and reaction mixture flow rate. The column (8) must have a hydrophobic inner surface and may be made from silanized glass or polymethyl methacrylate. Hydrophilic untreated glass causes coalescence of the emulsion, which may lead to polymerization on the walls, thus clogging the column.

In the case of photochemically initiated polymerizations, installation of powerful luminescence lamps (11) near the column and a thermostated water-jacket (10) are necessary. Most of the oil can be removed from the gel beads in a basket centrifuge, and the process is completed by washing the beads in a nonionic detergent solution and petroleum ether. It is convenient to fractionate gel beads by wet sieving or by using sedimentation equipment.

Operation Conditions

In our experiments we have used columns with the following dimensions: 11 × 600, 31 × 900, 100 × 700 mm. Maximum productivity of swollen gel has been obtained: 80, 600, 2000 g/hr.

Examples of the operation conditions and results from two typical experiments are described in the Table.

The gel preparations were used in laboratory and pilot-plant reactors for sucrose and benzylpenicillin hydrolyses. The gels had good enzymic stability. The rate of mechanical wear-out was determined to be about 10 times lower than that of the irregular particles from block gel having the same composition.

The method described above can be used for the production of other entrapped enzymes once the optimum conditions for gel formation have been determined.

[15] Preparation and Properties of Enzymes Immobilized by Copolymerization

By D. JAWOREK, H. BOTSCH, and J. MAIER

In the "protein copolymerization" process[1,2] (Fig. 1), proteins are vinylated with acylating and alkylating monomers and copolymerized with comonomers. For vinylation of the enzyme, several heterobifunctional acylating and alkylating reagents containing a copolymerizable double bond have been investigated[3] (Table I). Depending on the concentration of the alkylating monomer, soluble-to-insoluble protein polymers are obtained with acrylamide as comonomer and may be separated on molecular sieves.[4] If copolymerization of the enzyme is also carried out in the presence of cross-linkers, for example N,N'-methylenebisacryl-

[1] D. Jaworek, *in* "Enzyme Engineering," Vol. 2 (E. K. Pye and L. B. Wingard, Jr., eds.), pp. 105–114. Plenum, New York, 1974.

[2] D. Jaworek, *et al.*, Ger. Offen (Boehringer Mannheim GmbH) 22,60,185 and 21,28,743.

[3] D. Jaworek, H. Botsch, J. Maier, and E. Schoen, Enzyme Engineering Conference, Portland, 1975.

[4] D. Jaworek, *in* "Insolubilized Enzymes" (M. Salmona, C. Saronio, and S. Garattini, eds.), pp. 65–76. Raven, New York, 1975.

Step I: Preincubation with alkylation or acylation monomer

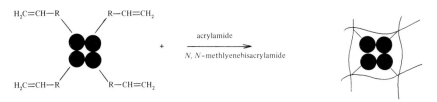

Step II: Copolymerization of the vinylated protein with acrylamide and N,N-methylenebisacrylamide

Fig. 1. Protein copolymerization. Vinylation of the protein with alkylating or acylating monomer and copolymerization with acrylamide and N,N'-methylenebisacrylamide.

amide, "insoluble" immobilized enzymes are obtained from the outset. Mechanical properties, specific activity, and activity yield may be influenced by the polymerization conditions, the monomers, and the cross-linkers used. The conditions observed for mechanical inclusion into cross-linked polyacrylamide may for the most part be adopted for the covalent immobilization in polyacrylamide gels described here; the concentration of the alkylating or acylating monomer, based on acrylamide, amounts to between 0.3 and 3%.

Method of Preparation

Material. Materials include acrylamide, N,N' methylenebisacrylamide, and 3-dimethylaminopropionitrile from Fluka, Buchs, Switzerland; allyloxy-2,3-epoxypropane, ethyleneimine ammonium persulfate and acrylic acid chloride from Merck, Darmstadt. Proteins include trypsin (bovine pancreas), glucose oxidase (*Aspergillus niger*), and amino acid acylase (hog kidney) from Boehringer Mannheim.

The monomer, 1-allyloxy-3-(N-ethyleneimine)-2-propanol (b.p. 60° at 1 mm Hg), is synthesized by mixing 50 ml of allyloxy-2,3-epoxypropane with 35 ml of ethyleneimine dropwise while stirring slowly at room temperature (about 30 min). The preparation is then reflux heated for 3 hr and fractionally distilled *in vacuo*.

TABLE I
ALKYLATING AND ACYLATING MONOMERS USED FOR "PROTEIN COPOLYMERIZATION"

Structure	Name
$CH_2=CH-CH-CH_2$ with epoxide O	3,4-Epoxybutene
$CH_2=CH-C(=O)-O-CH_2-CH-CH_2$ with epoxide O	Acrylic acid-2,3-epoxypropyl ester
$CH_2=CH-C(=O)-O-CH_2-CH-CH_2$ with S (thiirane)	Acrylic acid-2,3-thioglycidyl ester
$CH_2=CH-CH_2-O-CH_2-CH(OH)-CH_2-N(CH_2)(CH_2)$ (ethylenimine)	1-Allyloxy-3-(N-ethylenimine)-2-propanol
$CH_2=CH-C(=O)-O-N$ (succinimide with two C=O and CH$_2$-CH$_2$)	Acrylic acid-O-succinimide ester
$CH_2=CH-C(=O)-Cl$	Acrylic acid chloride
maleic anhydride ring ($HC=CH$, two $C=O$, bridging O)	Maleic acid anhydride
chloromaleic anhydride ring ($HC=C-Cl$, two $C=O$, bridging O)	Chloromaleic acid anhydride
maleic azide ($HC=CH$, two $C=O$, two N_3)	Maleic acid azide
$CH_2=CH-CH_2Br$	3-Bromopropene

Immobilized Trypsin. Trypsin, 600 mg (specific activity 0.4 U/mg), is dissolved in 20 ml of 1 M triethanolamine buffer and 2 mM EDTA, pH 8.0; the resulting solution is transferred to a 100-ml Erlenmeyer flask in which it is cooled to 10° and gassed with nitrogen while being stirred. A solution of 0.15 ml of acrylic acid chloride diluted with 10 ml of cold ether is slowly added dropwise to the enzyme solution, which is stirred for about another 30 min.

The reaction vessel is cooled with ice, then to the incubation mixture is added a solution of 6 g of acrylamide and 0.6 g of N,N'-methylenebisacrylamide in 20 ml of distilled water. After saturation with nitrogen, polymerization is initiated by the addition of 1 ml of 5% 3-dimethylaminopropionitrile and 1 ml of 5% ammonium peroxydisulfate.

After 2–3 min, the polymerization solution hardens into a gel, which subsequently solidifies into a solid block. The polymerization mixture is left standing and cooled for about 5 hr before further working up.

For granulation, the polymer is forced through a 0.4-mm mesh metal sieve, transferred to a 0.2-mm mesh sieve, and washed with water, thus separating it from the fine gel particles.

The immobilized enzyme is introduced into a column with an internal diameter of 20–30 mm and eluted with 5 liters of 0.5 M sodium chloride solution, 0.05 M triethanolamine buffer, pH 7.5, from the enzyme bound in heteropolar form or by adsorption. The enzyme is washed with distilled water until it is free from salt, then is lyophilized. The yield is 5 g (lyophilyzate). The specific activity (enzyme polymer) is 12–15 U/g at 25°; N-benzoyl-L-arginine-p-nitroanilide is the substrate.

Immobilized Glucose Oxidase. To prepare immobilized glucose oxidase, 100 mg of enzyme (specific activity 210 U/g) are preincubated with 0.06 ml of acrylic acid chloride, dissolved in 5 ml of dichloromethane, as described for the preparation of immobilized trypsin. The vinylated enzyme is copolymerized with a solution of 6 g of acrylamide and 0.4 g of N,N'-methylenebisacrylamide and further processed accordingly. The yield is 5 g (lyophilyzate). The specific activity (enzyme polymer) is 300 U/g at 25°; glucose is the substrate.

Immobilized Amino Acid Acylase. Amino acid acylase 600 mg (specific activity 26 U/mg) is dissolved in 23 ml of ice-cold 1-tris(hydroxymethyl)aminomethane (Tris) buffer, pH 6.7, followed by the addition with gentle stirring over a period of 10 min of 2 ml of 1-allyloxy-3-(N-ethylenimine)-2-propanol in the reaction vessel described in reference to the preparation of immobilized trypsin. The solution is further stirred after 20 min at 4°.

A solution of 6 g of acrylamide, 0.72 g of N,N'-methylenebisacrylamide, and 24 mg of $CoCl_2$ in 21 ml of 1 M Tris buffer, pH 6.7, is then added to the preparation. Polymerization is initiated by the addition of 2.4 ml of a 5% ammonium peroxydisulfate solution and 2.4 ml of a 5% 3-dimethylaminopropionitrile solution.

The polymer is granulated as described above and packed into a column; it is washed salt-free with 2 liters of 0.5 M phosphate buffer, pH 7.5, and then with distilled water. The yield is 5 g (lyophilyzate). The specific activity (enzyme polymer) is 160 U/g at 25°; N-acetyl-L-methionine is the substrate.

Properties

The acylation or alkylation of the enzymes with reactive monomers is governed to a large extent by the nucleophilicity of the amino acid side radicals and hence by the pH of the reaction solution and the reactivity of the acylating or alkylating monomers used.

If, for example, the vinylation of acylase from kidneys is carried out with allyloxy-(ethylenimine)-2-propanol at different pH values, and if thereafter copolymerization is carried out under the usual conditions, it is possible to obtain enzyme granulates of different specific activity (Fig. 2).

The optimum is between 6 and 7. This value indicates that the protein

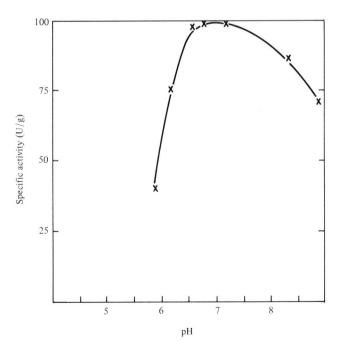

FIG. 2. Dependence of specific activity of immobilized amino acid acylase on pH during vinylation of the protein with 1-allyloxy-3-(ethylenimine)-2-propanol.

TABLE II
SPECIFIC ACTIVITIES (U/G) OF ENZYMES VINYLATED WITH VARIOUS ACYLATING AND ALKYLATING MONOMERS AND IMMOBILIZED BY PROTEIN COPOLYMERIZATION

Vinylating reagent	Glucose oxidase	Glucose isomerase	Chymo-trypsin	Catalase	Glucose-6-phosphate dehydro-genase	Hexokinase	Uricase	Trypsin	Per-oxidase	Acylase
1	350	—	—	—	—	—	—	—	—	—
2	240	—	0.03	5900	8	80	3.8	—	10	70
3	170	—	—	—	—	—	—	—	—	—
4	300	—	3.3	7000	20	48	3.3	16	21	160
5	310	—	—	—	—	100	—	13	—	—
6	300	5300	0.12	8500	20	100	7.7	5.7	50	0
7	200	—	—	—	—	—	—	—	—	—
8	—	850	—	—	—	—	—	7.8	0	—
9	240	—	0.5	—	—	—	3.2	—	—	—
10	270	—	—	—	—	—	3.0	—	—	—
None[a]	90	500	—	5200	8	40		0.5	—	50

[a] Mechanical inclusion.

content of the polymer is responsible for the specific activity and hence for the activity yield.

By comparison with mechanical inclusion, the enzymes listed in Table II were immobilized in covalent form by protein copolymerization under the same conditions. As can be seen from Table II, the specific activities obtained with the active monomers under investigation here differ in some cases. With some monomers, it was even possible to observe complete inactivation of acylase from kidneys and peroxidase from horse radish.

The microkinetic properties of the immobilized enzymes remain substantially unchanged; thus, the pH optimum of carrier-bound glucose oxidase is only slightly different from that of the native enzyme; nor is there any change in enzymic activity that depends on temperature.

The temperature stability of enzymes bound by protein copolymerization is in some cases higher than that of the soluble enzyme; carrier-bound glucose oxidase is more stable than the soluble enzyme. After heating for 30 min to 60°, 70% of the original activity is still intact, whereas the soluble enzyme has a residual activity of only 18%.

In the authors' opinion the advantages of "protein copolymerization" over mechanical inclusion are covalent binding of the enzyme, no bleeding, higher specific activity, and higher activity yield in some cases.

Compared with immobilization through activated matrices, the products obtained by the above-described process have a neutral matrix; do not adsorb substrate or reaction product; retain kinetic properties substantially unchanged; and show statistical distribution of the protein in the polymer.

[16] Microencapsulation of Enzymes and Biologicals

By T. M. S. CHANG

In nature, most enzymes are present in an intracellular environment, either in solution with a high concentration of cytoplasmic protein or in association with intracellular membranes. For a long time, enzyme research has been concentrated on the extraction, isolation, and purification of enzymes from biological sources and the analysis of the kinetics of these enzymes in dilute solution. The next stage in enzyme research is the use of "immobilized enzymes." Immobilized enzyme[1] is a term that

[1] P. V. Sundaram, E. K. Pye, T. M. S. Chang, V. H. Edwards, A. E. Humphrey, N. O. Kaplan, E. Katchalski, Y. Levine, M. D. Lilly, G. Manecke, K. Mosbach, A. Patchornik, J. Porath, H. Weetall, and L. B. Wingard, Jr., *Biotechnol. Bioeng.* **14**, 18 (1972).

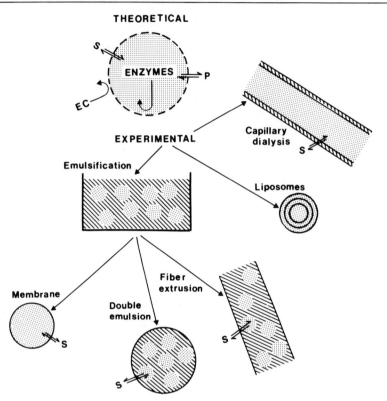

Fig. 1. *Top:* Theoretical representation of microencapsulated enzyme within semipermeable membrane. The small dots represent enzyme molecules. S, substrate; P, product; EC, "extracellular environment." *Lower:* Experimental systems. Dots represent enzyme molecules. Shaded lines represent polymer material. The basic system was prepared by emulsification followed by spherical ultrathin membrane formation. Extensions include double emulsion and fiber extrusion. Others include liposomes and dialysis membranes.

has been used to describe (a) the insolubilization of enzymes by adsorption, covalent linkage, and matrix entrapment as models for enzymes associated with biological membranes in a "microenvironment" and (b) microencapsulation of enzymes as artificial cell models for enzymes in the "intracellular environment." Thus immobilized enzymes have been classified into four major types[1]: (a) adsorption, (b) covalent linkage, (c) matrix entrapped, and (d) microencapsulated.

Microencapsulated enzymes are formed by enveloping enzymes within various forms of semipermeable membranes (Fig. 1). The enveloped enzymes cannot leak out, but external substrates can dialyze freely across the membrane to be acted on by the enveloped enzymes. The first type prepared are "artificial cells" consisting of spherical ultrathin membranes of cellular dimensions enveloping an aqueous interior of enzyme solution or suspension.[2-6] The enveloping semipermeable membrane separates the enzymes from the "extracellular" environment, preventing them from leaking out or coming into contact with external impermeant macromolecules. Microencapsulated enzymes can thus be retained in an "intracellular" environment with a high concentration of cytoplasmic protein. However, permeant substrates can equilibrate rapidly across the ultrathin membrane to be acted on by the enveloped enzymes. It is very easy to enclose any combination of pure or crude enzymes, multiple enzyme systems, cofactors, cell extract, whole cell, protein, adsorbents, magnetic material, multiple compartmental systems of enzymes, or other material.[2-6] The concentration of enzymes in the microcapsule depends only on the solubility of the particular enzyme system. Furthermore, with an ultrathin membrane of 200 Å, the polymer-to-enzyme ratio is negligible.

The basic methods of microencapsulation of enzymes are very simple procedures involving no special technical background in enzyme immobilization. However, the basic methods can be extended to become extremely complicated approaches and combinations of various systems. This chapter describes some typical basic procedures that could easily be carried out without background training in enzyme immobilization. Further details of the theory, modifications, and extensions of the basic procedures are described in detail elsewhere.[6]

First Basic Method

The first basic method will give a very high recovery of enzyme activity and is applicable to all enzymes tested so far. The second basic method, which will be described later, is not as good since enzyme activity recovered is lower and some enzymes are inactivated by the chemical reaction.

Principle

The first method to be described is a simple physical method involving no chemical reactions.[2-6] With this method, all enzymes tested have been

[2] T. M. S. Chang, Hemoglobin corpuscles. Report of research project for B. Sc. Honours Physiology, McGill University, Montreal, Canada, 1957.
[3] T. M. S. Chang, F. C. MacIntosh, and S. G. Mason, *Proc. Can. Fed. Biol. Sci.* **6**, 16 (1963).
[4] T. M. S. Chang, *Science* **146**, 524 (1964).
[5] T. M. S. Chang, F. C. MacIntosh, and S. G. Mason, *Can. J. Physiol. Pharmacol.* **44**, 115 (1966).
[6] T. M. S. Chang, "Artificial Cells." Thomas, Springfield, Illinois, 1972.

found to retain their activity. In this procedure, a high concentration of hemoglobin from red blood cells is used to provide an intracellular environment. Enzymes (single, multiple, pure, and crude extracts) are dissolved or suspended in the hemoglobin solution, then enveloped within microscopic, spherical, ultrathin membranes by the following procedure. The exact steps, timing, and hemoglobin contents are very important.

Material

Hemoglobin solution: 1 g of hemoglobin (hemoglobin substrates, Worthington) is dissolved in 10 ml of distilled water, then filtered through Whatman No. 42 filter paper. In place of hemoglobin substrate, fresh red blood cell hemolysate can also be used as long as the final concentration of hemoglobin is exactly 10 g/100 ml.

Enzyme solution: Enzymes (single, multiple, insolubilized, cell extract, cells, or combinations of enzymes with other material) are dissolved or suspended in the 10 g per 100 ml of hemoglobin solution. If a solution or suspension is added to the hemoglobin solution, suitable adjustment is required to ensure a final concentration of 10 g of hemoglobin per 100 ml. Also, if an acidic enzyme preparation is added to the hemoglobin solution, a minimal pH of 8.5 should be maintained by sufficient Tris buffer.

Organic solution: 100 ml of ether (analytical grade) is saturated with water by shaking with distilled water in a separating funnel, then discarding the water layer. 1 ml of a water/oil emulsifying agent, Span-85 (Atlas Powder Company, Canada Ltd., Montreal, Quebec, Canada), is added to 100 ml of water-saturated ether just before use.

Cellulose nitrate solution: This is prepared by evaporating collodion (USP, 4 g of cellulose nitrate in a 100-ml mixture of 1 part alcohol and three parts ether) to exactly 20% of its original weight, making it up to its original volume by adding ether (analytical grade) and redissolve.

N-Butyl benzoate–Span-85 solution: 100 ml n-butyl benzoate (Eastman) with 1 ml of Span-85 added just before the procedure

Tween-20 solution: 50% Tween-20 solution is prepared by dissolving 50 ml of Tween-20, an oil/water emulsifying agent (Atlas Powder Company Canada Ltd., Montreal, Quebec, Canada) in an equal volume of distilled water; 1% Tween-20 solution is prepared by dissolving 1 ml of Tween-20 in 99 ml of water pH is adjusted to 7.

Magnetic stirrer: It is important to have a magnetic stirrer with sufficient power to give the speed (rpm) stated, especially when stirring the very viscous Tween-20 mixture. The Jumbo Magnetic Stirrer (Fisher Scientific Company, Montreal, Quebec, Canada) is used in the procedures described below. Also needed is a 4-cm magnetic stirring bar and a 150-ml glass beaker with an internal diameter of less than 6 cm.

Procedure

To a 150-ml glass beaker containing 2.5 ml of Tris-buffered hemoglobin solution, add 25 ml of the organic solution. The mixture is immediately stirred with the magnetic stirrer at a speed setting of 7 (2600 rpm). After stirring for 5 sec, 25 ml of the cellulose nitrate solution is added and stirring is continued for another 60 sec. The beaker is then covered and alowed to stand unstirred at 4° for 45 min.

If the microcapsules have been completely sedimented by the end of 45 min, all but about 4 ml of the supernatant can be conveniently removed. Otherwise, the suspension will have to be centrifuged at 350 g for 5 min before the supernatant can be removed. Centrifugation at greater speed or longer duration may adversely affect the formation of the microcapsules.

Immediately after the removal of most of the supernatant, 30 ml of the n-butyl benzoate–Span-85 solution is added to the beaker and stirred with the suspension on the Jumbo Magnetic Stirrer at speed 5 (1200 rpm) for 30 sec. The beaker is then allowed to stand uncovered and unstirred at 4° for 30 min.

The final step is to transfer the microencapsulated enzyme from the organic liquid phase into an aqueous phase. First, butyl benzoate supernatant should be removed completely. In the case where microcapsules greater than 50 μm in diameter are prepared, this can be done readily if the microcapsules have sedimented completely by 30 min. If the microcapsules have not sedimented completely, then centrifugation at 350 g for 5 min is required.

After removal of supernatant, 25 ml of the 50% Tween-20 solution is added. The suspension is dispersed by stirring with the Jumbo Magnetic Stirrer at a speed setting of 10 for 30 sec. It is important to maintain a speed of 2900 rpm in the presence of the 50% Tween-20 solution. Stirring is then slowed down to speed 5 (1200 rpm), and 25 ml of water is added. After a further 30 sec of stirring, the suspension is further diluted with 200 ml of water. The slightly turbid supernatant may now be removed by centrifugation of the suspension at 350 g for 5 min.

The microencapsulated enzymes so obtained are washed repeatedly in

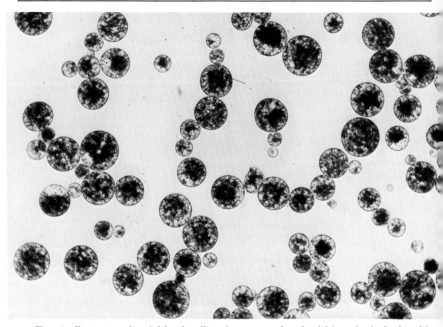

Fig. 2. Contents of red blood cells microencapsulated within spherical ultrathin membranes of cellulose nitrate prepared by the first basic method for "artificial cells." Mean diameter is about 80 μm.

1% Tween-20 solution until no further leakage of hemoglobin takes place and the smell of butyl benzoate is no longer detectable. The removal of n-butyl benzoate with 1% Tween-20 solution is very important because, unless it is completely removed, butyl benzoate may affect the permeability characteristics of the semipermeable microcapsules. The final preparation is suspended in a solution containing 0.9 g sodium chloride per 100 ml of water (Fig. 2).

Variation

Variation of the diameter of the microcapsules depends on the degree of emulsification.[4-6] For the preparation of microcapsules of less than 10 μm mean diameter, more complicated procedures have to be followed. The 10 g of hemoglobin per 100 ml, besides its necessity in the microencapsulation procedure, also stabilizes the microencapsulated enzymes. This is the basic procedure in using cellulose nitrate, however, extensions and modifications of other types of membranes may require extensive modifications of the procedure.[6]

Second Basic Method

Principles

This is an example of the preparation of microencapsulated enzymes by a chemical method.[3-6] This method is an extension of the basic method described above. The semipermeable microcapsule membrane is formed from nylon (Poly 610). (See also Chapter [9] on immobilization to nylon.) Although their appearance is much better than the cellulose nitrate membrane microcapsule, their membranes are not as stable, and furthermore some enzymes like catalase are inactivated by this procedure.

Two examples are given below, one is a drop technique and the other is the emulsification technique.[3-6] The drop technique is extremely simple and can be used to prepare large microcapsules containing enzymes without the necessity of using the 10 g/100 ml hemoglobin solution. However, if a large number of semipermeable microcapsules are required, the drop technique is more time-consuming, and the diameter is usually larger than 1 mm. The second procedure using emulsification would prepare small semipermeable microcapsules of the order of 50 μm mean diameter. In the emulsification procedure, a hemoglobin solution of 10 g/100 ml is required.

Material

Hemoglobin solution: 1 g of hemoglobin (hemoglobin substrate, Worthington) is dissolved in 10 ml of distilled water. It is then filtered through Whatman No. 42 filter paper. In place of this, fresh red blood cell hemolysate containing 10 g per 100 ml can also be used.

Enzyme solution: Enzymes (single, multiple, insolubilized, cell extract, cells, or combinations of enzyme with other material) are either dissolved or suspended in the 10 g/100 ml hemoglobin solution. The final solution has a concentration of 10 g% hemoglobin.

Alkaline diamine solution: 4.4 g of 1,6-hexamethylenediamine (Eastman); 1.6 g of $NaHCO_3$; and 6.6 g of Na_2CO_3 are dissolved as a 100-ml aqueous solution. This stock solution is filtered and then stored and refrigerated at 4°, but should be discarded after 24 hr.

Stock organic Span-58 liquid: This is prepared just before use, made up of 40 ml of chloroform and 160 ml of cyclohexane. In the emulsification procedure, 2 ml of Span-85 (Atlas Powder Company Canada Ltd., Montreal, Quebec, Canada) is added to 200 ml of the stock organic liquid. Span-85 is not required for the drop technique.

Sebacyl chloride liquid (0.018 M): This is prepared immediately (within 2 min) before use by adding 0.4 ml of sebacyl chloride (Eastman) to each 100 ml of the stock organic liquid. Failure to prepare satisfactory microcapsules is frequently due to the deterioration of sebacyl chloride after the seal on the bottle has been opened. A new bottle of sebacyl chloride will usually solve this problem.

50% Tween-20 solution: This is made up by dissolving 50 ml of Tween-20 (Atlas Powder Company Canada Ltd., Montreal, Canada) in 50 ml of distilled water.

Magnetic stirrer, 4 cm-magnetic stirring bar, 150-ml glass beaker as described for the above procedure.

Drop Technique Procedure[4-6]

Sebacyl chloride organic liquid (150 ml; 0.018 M) is prepared as described, but Span-85 is omitted from the organic solution. This sebacyl chloride organic liquid is placed in a flat-bottom glass petri dish 14 cm in diameter. In this case, enzymes or other material to be microencapsulated are added to 3 ml of the alkaline diamine solution if no hemoglobin is required. If hemoglobin is required, then the enzyme could be added to the 1.5 ml aklaline diamine and 1.5 ml of hemoglobin solution mixture. The 3 ml of the aqueous phase containing enzymes is placed in a glass syringe and added dropwise through its attached steel needle placed 1 cm above the surface of the sebacyl chloride solution. It is important to use a glass syringe and a steel needle with no polymer components, so that the equipment will be affected by the organic solution. The diameter of the needle would decide the diameter of the droplets and thus the diameter of the microcapsules. The dropwise addition could be done manually or by using a perfusion pump for continuous drop formation. The drop has to be added throughout the surface of the container with no droplets overlapping. After the addition of the droplets, they settle to the bottom of the container and the reaction is allowed to take place for 5 min with gentle shaking. At the end of 5 min, the semipermeable microcapsules are formed. The organic solution can be decanted and any trace amount allowed to evaporate completely before the saline is added to suspend the droplets. It is important to allow the organic solution to evaporate completely before adding this solution, otherwise the permeability of the membrane may be affected by the organic solution. This is an extremely easy method for forming semipermeable microcapsules, especially as a preliminary step toward the emulsification procedure, which is more complex.

Emulsification Procedure

Of the hemoglobin solution containing the enzyme, 2.5 ml is added to 2.5 ml of the alkaline diamine solution in a 150-ml beaker placed in an

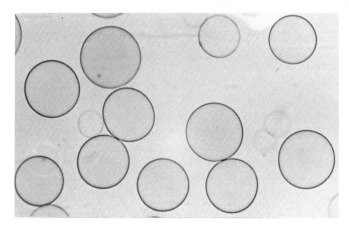

FIG. 3. Microencapsulation of enzymes within spherical ultrathin membranes of nylon prepared by the second basic method for "artificial cells." Mean diameter is about 100 μm.

ice bath. The two solutions are gently mixed. Within 10 sec after mixing the two solutions, 25 ml of the stock organic–Span-85 liquid is added and then stirred at speed 5 (1200 rpm) on the Jumbo Magnetic Stirrer for 60 sec. While stirring is continued at the same speed, 25 ml of the sebacyl chloride organic–Span-85 liquid is added and stirring is continued for 3 min. After exactly 3 min, the reaction is quenched by the addition of 50 ml of stock organic–Span-85 liquid; stirring is discontinued and the suspension is centrifuged for 30 sec at 350 g.

All the supernatant is discarded, and 50 ml of the 50% Tween-20 solution is added. The suspension is stirred at Jumbo Magnetic Stirrer speed setting of 10 (2900 rpm) for 30 sec. The speed is then decreased to 5 (1200 rpm); 50 ml of distilled water is added, and stirring is continued for another 30 sec. The suspension is then added to 200 ml of distilled water, and stirring is continued at the same speed for another 30 sec. This suspension is allowed to sediment or centrifuge for 1 min at 350 g; all the supernatant is discarded. Alternatively, the suspension may be sieved in a mesh screen to remove the supernatant. The microcapsules so prepared are suspended in 0.9 g/100 ml sodium chloride solution (Fig. 3). Initially, the microcapsules may be "crenated" after being in contact with the hypertonic Tween-20 solution, but they regain sphericity if contact with the strong Tween-20 medium has been kept at a minimum.

Variation

Extensive variation is possible in diameter, membrane thickness, and membrane material, including protein or charged membranes.[6]

Extensions and Variations

The original microencapsulation of enzymes was the result of attempts to prepare artificial cells, each consisting of an ultrathin spherical membrane enveloping enzyme solutions, suspensions, or cell extract.[2-6] Besides their use as artificial cells, their other advantages for *in vivo* application include an ultrathin membrane and a large surface-to-volume ratio, allowing for rapid exchange of substrate and product, negligible amounts of polymer introduced into the body, and an injectable and implantable cell-like system. However, for other applications, these factors may not be important. In these cases, the principle of microencapsulation of enzymes can be further extended.

An extension of the first basic method depends on the principle of microencapsulation by secondary emulsion.[7,8] Aqueous solution containing the enzyme is emulsified in an organic polymer solution as described in the first basic method. The organic polymer solution containing the microdroplets of aqueous enzyme solution is then dispersed in an aqueous phase. The organic phase containing the polymer solution is then solidified to form solid spheres each containing a number of aqueous microdroplets. Unlike the earlier two basic methods, this procedure does not produce spherical ultrathin membranes enclosing an aqueous interior; instead, it produces solid polymer microspheres, each of which contains several aqueous microdroplets (Fig. 1).

A typical example is the use of silicone rubber (silastic) to microencapsulate the content of red blood cells.[8] Very briefly, hemolysate is emulsified in 2 volumes of Silastic (Dow Corning, S-5392). While stirring is continued, 0.025 volume of stannous octoate catalyst (Dow Corning) is added, followed within 2 sec by 20 volumes of 50% Tween-20 solution containing 50 mg sodium hydrosulfite per 100 ml. Stirring is then continued at a slower speed for 20 min or more to attain the "tack-free time" of the Silastic polymer.

This principle of secondary emulsion has also been used to microencapsulate enzymes and proteins within cellulose polymer material[7] and other polymers.[9,10] In this case, an aqueous solution of enzymes is emulsified in an organic polymer solution as described under the first basic method. A second emulsion is then formed in an aqueous solution. The polymer is allowed to solidify by removal of the organic solvent.

An extension of this approach is to form the emulsion of enzyme in the organic polymer solution; in this case, instead of microspheres, fine

[7] T. M. S. Chang, Semipermeable aqueous microcapsules. Ph.D. Thesis, Physiology, McGill University, Montreal, Canada, 1965.

[8] T. M. S. Chang, *Trans. Am. Soc. Artif. Intern. Organs* **12**, 13 (1966).

[9] M. Kitajima, S. Miyano, and A. Kondo, *J. Chem. Soc. Jpn.* (*Kogyo Kagaku Zasshi*) **72**, 493 (1969).

[10] K. Miyamoto, N. Takamatsu, M. Okazaki, and Y. Miura, *J. Fermentn. Technol.* **52**, 899 (1974).

threads are formed in the form of fiber[11,12] (Fig. 1). The details are discussed in this volume [18].

Another extension of the double-emulsion approach makes use of a special hydrocarbon liquid to form a "liquid membrane." Li's group microencapsulated enzymes in liquid membrane emulsions.[13,14] A brief outline of a typical procedure is as follows: 82 ml of an aqueous enzyme solution (e.g., 0.064% urease) is emulsified by dropwise addition with vigorous agitation (350 rpm) to 100 g of an organic mixture consisting of 2% Span-80 (Atlas Chemical Industries), 3% ENJ-3029 (a high-molecular-weight amine, ENJAY Additives Laboratories), and 95% S100N (a high-molecular-weight isoparaffin having an average molecular weight of 386.5, a cloud point of 93°F, a pour point of 90°F, and a gravity of 0.836 g/cm^3 between 18° and 140°F). This water-in-oil emulsion is dispersed in an aqueous solution to form a double emulsion. Each organic droplet of 100 μm to several millimeters in diameter contains many smaller enzyme aqueous droplets, 1–10 μm in diameter. The hydrocarbon separates the enzyme from the external aqueous phase. The permeability characteristics of the hydrocarbon membrane to different substrates would determine the ability of the microencapsulated enzyme to act on the substrate. The product so formed can diffuse out, or, if not permeable, can remain inside.

Microencapsulated enzymes are in fact a microdialysis system in which the enclosed enzymes act on external substrates dialyzing across the enclosing semipermeable membrane (Fig. 1). Thus, it has been suggested that many of the aims that might be achieved by the microencapsulation of enzymes within artificial cells can be fulfilled by placing the enzymes in the external dialysis fluid of an artificial kidney[8] (Fig. 1). The main advantages of enclosing them within microcapsules in the form of artificial cells is the enormous surface-to-volume ratio and the ultrathin membrane that would be achieved this way. This large surface area and small volume relationship should be useful where the biological material is expensive or where a portable or wearable device is wanted. This suggestion to extend the principle of microencapsulated enzymes using the standard dialysis membrane system[8] has been demonstrated to be feasible for actual experimentation and theoretical analysis.[15-18]

[11] D. Dinelli, *Process Biochem.* **7**, 9 (1972).
[12] W. Marconi, S. Gulinelli, and F. Morisi, in "Insolubilized Enzymes" (M. Salmona, C. Saronio, and S. Garattini, eds.), pp. 51–64. Raven, New York, 1974.
[13] S. W. May and N. N. Li, *Biochem. Biophys. Res. Commun.* **47**, 1179 (1972).
[14] S. W. May and N. N. Li, in "Enzyme Engineering" (E. K. Pye and L. B. Wingard, Jr., eds.), Vol. 2, p. 77. Plenum, New York, 1974.
[15] S. Rogers, *Nature (London)* **220**, 1321 (1968).
[16] M. Apple, *Proc. West. Pharmacol. Soc.* **14**, 125 (1971).
[17] J. R. Rony, *J. Am. Chem. Soc.* **94** (23), 8247 (1972).
[18] J. C. Davis, *Biotechnol. Bioeng.* **16**, 1113 (1974).

In addition to those mentioned, other extensions include the formation of microcapsule membranes using lipids,[19] proteins, [3-6] lipid–protein complex,[6,20,21] heparin-complexed membrane,[22,23] and membrane with different surface charges.[2-6]

Enzymes have been directly microencapsulated into red blood cells by a process of reverse hemolysis,[24] and can also be microencapsulated into liposomes[25,26] (Fig. 1), as described in this volume [17].

Stabilization and Cross-Linkage of Microencapsulated Enzymes

Microencapsulated enzymes are stabilized by enclosure with the 10 g/100 ml hemoglobin solution. In this way, unlike native enzymes in a dilute solution the microencapsulated enzyme is at all times retained in an intracellular environment with a high concentration of intracellular protein.

Insolubilized enzymes in the form of covalent binding, adsorption, or gel-matrix entrapment could also be microencapsulated in suspension[27] (Fig. 4). An easier way to stabilize the enzyme is to directly cross-link the microencapsulated enzyme solution by the following method[28] (Fig. 4). For example, after the 10 g/100 ml hemoglobin solution containing catalase, asparaginase, or lactase is microencapsulated within cellulose nitrate membranes, the following procedure is carried out: 100 ml of a solution (0.1 M sodium metaborate and 0.001 M benzamidine HCl) is added to a 20-ml suspension of the microencapsulated enzyme. To the suspension 200 μl of 50% glutaraldehyde is added; the suspension is kept slightly agitated at 20° for 1 hr, then 100 ml of 0.05 M sodium borohydride is added, and the suspension is left at 4° for 20 min. The supernatant is then removed, and the microcapsules are washed twice with 200 ml of a 0.9 g/10 ml sodium chloride solution.

When asparaginase or catalase are stored at 4° as a dilute solution, only 50% of the activity is left after 15–20 days. However, when these

[19] P. Mueller and D. O. Rudin, *J. Theor. Biol.* **18**, 222 (1968).
[20] T. M. S. Chang, *Fed. Proc., Fed. Am. Soc. Exp. Biol.* **28**, 461 (1969).
[21] A. Rosenthal and T. M. S. Chang, *Proc. Can. Fed. Biol. Sci.* **14**, 44 (1971).
[22] T. M. S. Chang, L. J. Johnson, and O. Ransome, *Proc. Can. Fed. Biol. Sci.* **10**, 30 (1967).
[23] T. M. S. Chang, L. J. Johnson, and O. Ransome, *Can. J. Physiol. Pharmacol.* **45**, 70 (1967).
[24] G. M. Ihler, R. H. Glew, and F. W. Schnure, *Proc. Natl. Acad. Sci. U.S.A.* **70**, 2663 (1973).
[25] G. Sessa and G. Weissmann, *J. Biol. Chem.* **245**, 3295 (1970).
[26] G. Gregoriadis, P. D. Leathwood, and B. E. Ryman, *FEBS Lett.* **14**, 95 (1971).
[27] T. M. S. Chang, *Sci. Tools* **16**, 33 (1969).
[28] T. M. S. Chang, *Biochem. Biophys. Res. Commun.* **44**, 1531 (1971).

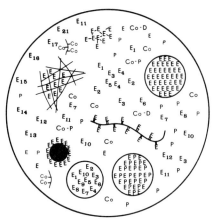

FIG. 4. Schematic representation of the microencapsulation of any variation of other enzyme (E) systems. For example, multiple enzyme systems, adsorbed enzymes, covalently linked enzymes, gel-entrapped enzymes and other microencapsulated enzymes, cross-linked enzymes, and soluble and insoluble cofactors.

enzymes are microencapsulated with a high concentration of protein with or without further cross-linking and stored at 4°, more than 90% of the original activity is retained after more than 100 days (Table I). At a body temperature of 37°, the microencapsulated enzymes were significantly more stable if cross-linked. Further details are given in this volume [46].

Enzymes, Cell Extracts, Cells, Vaccines, Antigens, and Antiserums

Table II shows the different types of enzymes microencapsulated. This list does not cover microencapsulation by fibers or liposomes since they are dealt with in this volume [17] and [18]. Enzyme systems can be added in any combination and concentration by dissolving or suspending in the hemoglobin solution before microencapsulation. Furthermore, the

TABLE I
PERCENTAGE OF INITIAL ACTIVITY (4° STORAGE)

Enzyme	Free solution	Microencapsulated with 10 g/100 ml hemoglobin	Microencapsulated with 10 g/100 ml hemoglobin, then cross-linked
Catalase	50% at 15 days 10% at 35 days	>90% at 100 days	>90% at 100 days
Asparaginase	50% at 20 days 10% at 60 days	>90% at 100 days	>90% at 100 days

TABLE II
ENZYMES AND PROTEINS IMMOBILIZED BY MICROENCAPSULATION

Enzymes	References
Enzymes and hemoglobin of red blood cell hemolysate	a–i
Urease	a–h, j–s
Carbonic anhydrase	a–e, t
Uricase	b–e
Trypsin	b–e
Catalase	e, u–w
Asparaginase	w–z, aa–cc
Albumin	c–f, dd
Lipase	h
α-Glucosidase	ee
Zymase complex (yeast)	ff
Muscle extract	ff
Phenolase	gg
Nitrate reductase	gg
β-Fructofuranosidase	hh
Lactase	ii, jj
Lysozyme	kk
Alkaline phosphotase	ll, mm
Glucose oxidase	i
Multistep enzyme system of hexokinase–pyruvate kinase	nn

[a] T. M. S. Chang, Hemoglobin corpuscles. Report of research project for B.Sc. Honours Physiology, McGill University, Montreal, Canada, 1957.
[b] T. M. S. Chang, F. C. MacIntosh, and S. G. Mason, *Proc. Can. Fed. Biol. Sci.* **6**, 16 (1963).
[c] T. M. S. Chang, *Science* **146**, 524 (1964).
[d] T. M. S. Chang. F. C. MacIntosh, and S. G. Mason, *Can. J. Physiol. Pharmacol.* **44**, 115 (1966).
[e] T. M. S. Chang, "Artificial Cells." Thomas, Springfield, Illinois, 1972.
[f] T. M. S. Chang, Semipermeable aqueous microcapsules. Ph.D. Thesis, Physiology, McGill University, Montreal, Canada, 1965.
[g] T. M. S. Chang, *Trans. Am. Soc. Artif. Intern. Organs* **12**, 13 (1966).
[h] M. Kitajima, S. Miyano, and A. Kondo, *J. Chem. Soc. Jpn.* (*Kogyo Kagaku Zasshi*) **72**, 493 (1969).
[i] K. Miyamoto, N. Takamatsu, M. Okazaki, and Y. Miura, *J. Fermentn. Technol.* **52**, 899 (1974).
[j] S. W. May and N. N. Li, *Biochem. Biophys. Res. Commun.* **47**, 1179 (1972).
[k] S. W. May and N. N. Li, *in* "Enzyme Engineering" (E. K. Pye and L. B. Wingard, Jr. eds.), Vol. 2, p. 77. Plenum, New York, 1974.
[l] T. M. S. Chang, L. J. Johnson, and O. Ransome, *Can. J. Physiol. Pharmacol.* **45**, 70 (1967).
[m] T. M. S. Chang and S. K. Loa, *Physiologist* **13**, 70 (1970).
[n] S. N. Levine and W. C. LaCourse, *J. Biomed. Mater. Res.* **1**, 275 (1967).

entire cell content of erythrocytes, with its complex enzyme systems, has been microencapsulated.[2-6] In the same way, muscle extract too has been microencapsulated.[29] Whole cells—for instance, erythrocytes, leukocytes, and liver cells—have also been microencapsulated.[4-6] In all cases, the microencapsulated enzymes did not leak out but acted efficiently on permeant substrates, equilibrating rapidly across the membrane.

We have also microencapsulated vaccines for vaccination, antiserums for immunosorbents, and hormones for slow release.

Suspensions or Particulates

Enzymes could be microencapsulated with any type of particulate matter. Thus, various types of adsorbents, such as ion-exchange resin,

[29] M. Kitajima and A. Kondo, *Bull. Chem. Soc. Jpn.* **44**, 3201 (1971).

[o] R. E. Sparks, R. M. Salemme, P. M. Meier, M. H. Litt, and O. Lindan, *Trans. Am. Soc. Artif. Intern. Organs* **15**, 353 (1969).
[p] R. D. Falb, P. G. Anapakos, H. Nack, and B. C. Kim, Annual Report, Artificial Kidney Program, National Institutes of Health, 1968.
[q] P. V. Sundaram, *Biochim. Biophys. Acta* **321**, 319 (1973).
[r] D. L. Gardner, *Trans. Am. Soc. Artif. Intern. Organs* **17**, 239 (1971).
[s] W. J. Asher, K. C. Bovée, J. W. Frankenfeld, R. W. Hamilton, L. W. Henderson, P. G. Holtzapple, and N. N. Li, *Kidney Int.* **7**, S409 (1975).
[t] R. C. Boguslaski and A. N. Janik, *Biochim. Biophys. Acta* **250**, 260 (1971).
[u] T. M. S. Chang, *Sci. J.* **3**, 63 (1967).
[v] T. M. S. Chang and M. Poznansky *Nature* (*London*) **218**, 243 (1968).
[w] T. M. S. Chang, A. Pont, L. J. Johnson, and N. Malave, *Trans. Am. Soc. Artif. Intern. Organs* **15**, 163 (1968).
[x] T. M. S. Chang, *Sci. Tools* **16**, 33 (1969).
[y] T. M. S. Chang, *Proc. Can. Fed. Biol. Sci.* **12**, 62 (1969).
[z] T. M. S. Chang, *Nature* (*London*) **229**, 117 (1971).
[aa] E. Siu Chong and T. M. S. Chang, *Enzyme* **18**, 218 (1974).
[bb] T. Mori, T. Sato, Y. Matuo, T. Tosa, and I. Chibata, *Biotechnol. Bioeng.* **14**, 663 (1972).
[cc] T. Mori, T. Tosa, and I. Chibata, *Biochim. Biophys. Acta* **321**, 653 (1973).
[dd] M. Shiba, S. Tomioka, M. Koishi, and T. Kondo, *Chem. Pharm. Bull.* **18**, 803 (1970).
[ee] B. E. Ryman, personal communications (1968).
[ff] M. Kitajima and A. Kondo, *Bull. Chem. Soc. Jpn.* **44**, 3201 (1971).
[gg] O. R. Zaborsky, "Immobilized Enzymes." Chem. Rubber Publ. Co. Cleveland, Ohio, 1973.
[hh] G. Gregoriadis, P. D. Leathwood, and B. E. Ryman, *FEBS. Lett.* **14**, 95 (1971).
[ii] J. C. W. Østergaard and S. C. Martiny, *Biotechnol. Bioeng.* **15**, 561 (1973).
[jj] T. M. S. Chang, *in* "Enzyme Engineering" (E. K. Pye and L. B. Wingard, Jr., eds), vol. 2, p. 419. Plenum, New York, 1974.
[kk] G. Sessa and G. Weissmann, *J. Biol. Chem.* **245**, 3295 (1970).
[ll] M. Apple, *Proc. West. Pharmacol Soc.* **14**, 125 (1971).
[mm] P. R. Rony, *J. Am. Chem. Soc.* **94** (23), 8247 (1972).
[nn] J. Campbell and T. M. S. Chang, *Bioch. Biophy. Acta* **397**, 101 (1975).

zirconium phosphate, and activated charcoal, can be added to the aqueous phase and microencapsulated together with the enzyme solution.[3-6,8]

Magnetic Microencapsulated Enzymes

Iron filings or magnetically active material can also be microencapsulated together with enzymes, resulting in magnetically active semipermeable microcapsules which, when used in extracorporeal shunt-type reactor systems, could control the location and stirring of the microencapsulated enzymes by the application of external magnetic fields.[8] A mechanical stirrer, rather than a magnetic one, must be used in the preparation of magnetic microencapsulated enzymes.[8]

Multiple Compartmental Systems

Smaller semipermeable microcapsules containing different enzyme systems could be microencapsulated into large microcapsules, resulting in semipermeable microcapsules with multicompartmental systems, each containing a special type of single enzyme or multiple enzyme system[3-6] (Fig. 5).

Multistep Enzyme System for Recycling of Coenzymes

The first basic method has been used to microencapsulate a multistep enzyme system consisting of hexokinase and pyruvate kinase, which was demonstrated to act sequentially in recycling ADP/ATP.[30] Microencapsulated multistep enzyme systems have also been used to recycle NAD^+/NADH.[31] This includes the use of coenzymes immobilized to soluble or insoluble macromolecules to allow their retention within the microcapsules.[31]

Microencapsulation of Radioactive Labeled Enzymes and Proteins

Enzymes or proteins that had been previously isotope-labeled could be microencapsulated as described in the basic procedure. Enzymes and proteins could also be microencapsulated first, then labeled with isotope (e.g., hemoglobin with ^{51}Cr).[6-7]

Biodegradable Microcapsules

The ultrathin polymer membranes in microcapsules prepared by the 2 basic methods are negligible in amount. The ultrathin nylon and cross-

[30] J. Campbell and T. M. S. Chang, *Bioch. Biophys. Acta* **397**, 101 (1975).
[31] J. Campbell and T. M. S. Chang, *Bioch. Biophys. Res. Commun.* (In press).

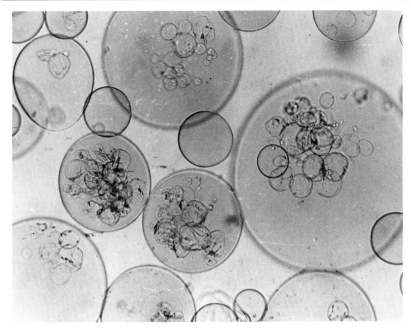

FIG. 5. Multiple compartmental system. Spherical ultrathin membrane microcapsules each containing a large number of smaller microcapsules. The smaller microcapsules can each contain a different enzyme or multiple enzyme system. Artificial cell model for complex intracellular organelles.

linked protein membranes are biodegradable. Liposomes and erythrocyte-entrapment are biodegradable, but the amount of entrapped enzymes carried by these systems is small. Polylactic acid, a polymer which is metabolized in the body into carbon dioxide and water, has been used successfully in our laboratory for microencapsulation of biologically actiave materials including enzymes, proteins, cell extracts, vaccines, antigens, hormones, and others. Other similar biodegradable polymers can also be used for microencapsulation.

Conclusion

As discussed earlier,[6] the microencapsulation of enzymes and cell contents within spherical ultrathin membrane is an initial physical system to demonstrate the principle of artificial cells. The work reviewed in this chapter demonstrates that, on an *in vitro* biochemical basis, much progress has been made on a simple biochemical "artificial cell" as a result of work being carried out in an increasing number of laboratories to extend this basic principle. Its therapeutic applications are discussed in this

TABLE III
CONDITIONS FULFILLED BY MICROENCAPSULATED ENZYMES

1. Artificial cells with intracellular environment
2. Action on dialyzable substrates up to the size of polypeptides, but not proteins
3. Rapid equilibration of substrates due to the ultrathin membrane with large surface-to-volume ratio
4. Carrier negligible in comparison to enzymes (1: 10,000)
5. Immobilized complex system: cell extract, organelles, cells, multistep enzymes, cofactors, vaccines, antigens, antiserums, and hormones
6. Stabilization of enzyme
7. Avoidance of immunological and hypersensitivity reactions
8. Multicompartmental enzyme systems
9. Enzymes coimmobilized with adsorbents, magnetic materials, and others
10. Biodegradable membranes possible

volume [46]. Conditions fulfilled by the microencapsulation of enzymes are summarized in Table III.

[17] Enzyme Entrapment in Liposomes

By GREGORY GREGORIADIS

Description of Liposomes

Liposomes have been described as assemblages of phospholipids and other lipids sustaining a bimolecular configuration and not requiring mechanical support for their stability.[1] They form when water-insoluble polar lipids, such as phosphatidylethanolamine, -choline, and -serine, cardiolipins, phosphatidic acids, are confronted with water and undergo a sequence of conglomerations that reflect the thermodynamic perturbation of increasing water–water, water–oil, and oil–oil interactions. The highly ordered structures that finally emerge (liposomes, also known as smectic mesophases) persist in the presence of excess water, which being associated with unfavorable entropy leads to a further arrangement of lipid molecules. The final result is a system of concentric closed membranes each one of which represents an unbroken bimolecular sheet of molecules (Fig. 1).

There are three principal types of liposomes, namely multilamellar, monolamellar, and macrovesicular.[2] Multilamellar liposomes consist of

[1] A. D. Bangham, M. W. Hill, and N. G. A. Miller, in "Methods in Membrane Biology" (E. D. Korn, ed.), Vol. 1, p. 1. Plenum, New York, 1974.
[2] G. Gregoriadis, in "Enzyme Therapy in Lysosomal Storage Diseases" (J. M. Tager, G. J. M. Hooghwinkel, and W. T. Daems, eds.), p. 131. North-Holland Publ., Amsterdam, 1974.

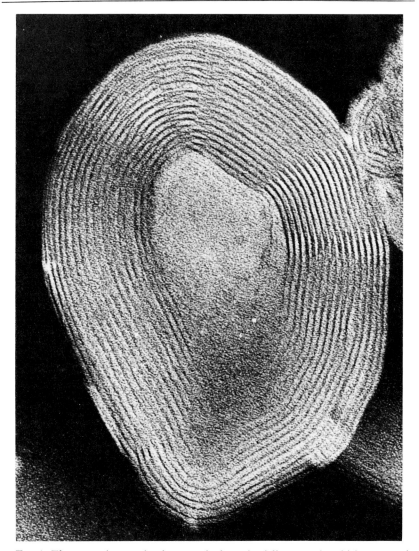

FIG. 1. Electron micrograph of a negatively stained liposome in which concentric lipid bilayers separated by electron-opaque layers can be seen. Negatively stained with silicotungstate. 272,000×. From A. W. Segal, E. J. Wills, J. E. Richmond, G. Slavin, C. D. V. Black, and G. Gregoriadis, *Br. J. Exp. Pathol.* **55**, 320 (1974).

several phospholipid bilayers in the form of concentric spheres alternating with aqueous compartments. Upon sonication these can break up to form smaller monolamellar structures. Macrovesicles, on the other hand, are a different variation of the smectic mesophase model. Their walls consist of only a few bimolecular sheets, yet their diameter is believed to be about 1 μm.

Criteria for Enzyme Entrapment

The usefulness of liposomes as vehicles for the administration of enzymes and other therapeutic agents[2] (see this volume [47]) derives from the fact that, as lipid molecules undergo rearrangements prior to the formation of closed structures, there is unrestricted entry of solutes (e.g., enzymes) between the planes of polar head groups. It is therefore possible to entrap in the aqueous phase of liposomes a wide range of substances, provided that such substances are soluble in water and that they do not interact with lipids in a way that upsets liposome formation. For unknown reasons, only nonelectrolytes can be entrapped in the macrovesicle version of liposomes, since in the presence of electrolytes or soluble proteins these vesicles do not form. On the other hand, macrovesicles, because of their size, will accommodate in their innermost aqueous phase substances of very high molecular weight. This is not the case with either multilamellar liposomes of which the central aqueous core is much smaller (about 0.15 μm in Fig. 1) and the distance between the planes of hydrophilic head groups of lipid layers (i.e., the width of the aqueous phase) is only about 75 Å,[3] or with monolamellar liposomes of which the diameter of the aqueous space is about 85 Å.[1]

In discussing enzyme entrapment, two aspects that are relevant to medical applications will be considered. These are (a) efficient entrapment procedures giving a low liposomal lipid-entrapped enzyme ratio, and (b) criteria for real entrapment as opposed to external binding onto the liposomal surface. Entrapment of an enzyme or of any other solute in liposomes is a passive process: as liposomes form, they "embrace" and trap volumes of water containing the solutes. In is therefore apparent that the extent to which solutes can be entrapped is proportional to the volume of the aqueous phase (V_{H_2O}) within liposomes, which in turn is related to the charge of the inner or outer surface of the lipid bilayer and to the ionic strength of the medium used for the formation of liposomes[4]:

$$V_{H_2O} = k(\text{charge/ionic strength}) \tag{1}$$

It follows that in order to increase the volume of entrapped water one must use media of low ionic strength or incorporate a proportion of charged lipid into the lipid structure or both.

[3] A. D. Bangham, J. De Gier, and G. D. Greville, *Chem. Phys. Lipids* **1**, 225 (1967).
[4] D. Papahadjopoulos and N. Miller, *Biochim. Biophys. Acta* **135**, 624 (1967).

Enzyme entrapment in liposomes by no means implies that the enzyme exists entirely within the aqueous phase. It is possible that, during the formation of the lipid bilayers, hydrophobic regions of the enzyme penetrate the lipid phase with the hydrophilic portions of the protein extending extraliposomally or intraliposomally into the aqueous phase.[5] Adsorption of some of the enzyme onto the liposomal surface as well as electrostatic enzyme–lipid interaction, which can be detrimental to liposome formation, is also possible. The latter can often be avoided by adjusting the pH so that the net charges of both entities (lipid and protein) are homologous.[6] Depending on the purpose for which a particular enzyme is entrapped, there could or could not be a need for the demonstration of "real" entrapment. For instance, if one is solely interested in directing an enzyme to a particular tissue *in vivo*, then, providing that the enzyme follows the carrier to its destination, the exact position of the enzyme molecules in the liposomal carrier could be irrelevant. On the other hand, it could be that expression in the circulation of the immunological or enzymic identity of the liposome-entrapped enzyme is biologically unacceptable. Thus, contact of antienzyme antibodies in blood with the extraliposomal antigenic portions of the enzyme could promote dangerous allergic reactions,[7] and enzymically active portions of the enzyme on the liposomal surface could lead to untoward interactions with circulating substrates.[8]

Prior to their particular use in biological experiments, liposome-entrapped enzymes should be tested (a) enzymically and/or (b) immunologically. Assay of the liposomal enzyme activity, described below, should always be carried out with substrates that do not penetrate the lipid bilayers, and the extent to which liposome-entrapped enzymes are immunologically active can be tested with methods described by the author elsewhere in this volume [47]. If necessary, enzymic or immunological activity on the liposomal surface can be eliminated by the treatment of such liposomes with a highly purified nonspecific protease (e.g., *Streptomyces griseus* protease).

Methods for Enzyme Entrapment

The preparation of enzyme-containing liposomes is exceedingly simple (Fig. 2). The appropriate lipids are dissolved in an organic solvent, which is subsequently eliminated under reduced pressure, leaving a thin lipid

---

[5] H. K. Kimelberg and D. Papahadjopoulos, *Biochim. Biophys. Acta* 233, 805 (1971).
[6] G. Gregoriadis, *FEBS Lett.* 36, 292 (1973).
[7] G. Gregoriadis and A. C. Allison, *FEBS Lett.* 45, 71 (1974).
[8] G. Gregoriadis, D. Putman, L. Louis, and E. D. Neerunjun, *Biochem. J.* 140, 323 (1974).

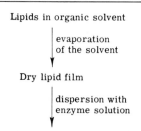

Fig. 2. General scheme for enzyme entrapment.

layer on the walls of the flask. This is dispersed with the enzyme solution, and liposomes containing the enzyme form spontaneously. Prior to their separation from nonentrapped enzyme, liposomes can be reduced in size by sonication down to monolamellar structures. In an alternative method[1] for the preparation of monolamellar liposomes, a solution of phospholipids in an organic solvent is layered onto the surface of the solute-containing aqueous phase and the solvent is removed with a stream of inert gas. However, contact of the solvent with the aqueous phase could lead to enzyme denaturation.

Entrapment Procedure

Solutions of lipids in chloroform or other appropriate organic solvents are kept in sealed ampoules at $-30°$ under argon. Glassware must be cleaned with detergents and thoroughly washed with tap and distilled water. To a dry spherical round-bottom 50-ml flask (Quickfit) 40 μmoles of phospholipid (e.g., egg lecithin, grade I, Lipid Products, Crab Hill Lane, South Nutfield, Nr. Redhill, Surrey, England), 11.4 μmoles of a sterol (e.g., cholesterol, British Drug Houses Chemicals Ltd., Poole, England), and 5.7 μmoles of a charged amphiphile (e.g., dicetyl phosphate, Sigma, London, Chemical Company) (usually the contents of individual ampoules) are added in a total volume of about 2–3 ml; to avoid oxidation of unsaturated lipids, the air in the flask is replaced by oxygen-free argon. The organic solvent is then removed in a rotary evaporator under reduced pressure at about $37°$ until a thin uniform film is visible on the walls of the flask. This takes 1–2 min, but, to ensure complete removal of the solvent, evaporation is continued for 10 min more. Traces of solvent (e.g., chloroform) can, at the later stage of sonication, cause changes in the pH leading to the chemical degradation of lecithins.[9]

[9] L. Erdei, F. Joó, I. Csorba, and C. Fajszi, *Acta Biochim. Biophys. Sci. Hung.* 9, 121 (1974).

The flask is removed from the evaporator by flushing through argon instead of air, and then immediately the enzyme solution in 1–2 ml of the appropriate buffer is added (e.g., 2 ml of 3 mM sodium phosphate buffer, pH 7.2, containing 15.4–44.0 units or 14–40 mg of *Clostridium perfringens* neuraminidase, grade VI, Sigma, London, Chemical Company). The flask is covered with a Quickfit stopper and hand-shaken gently until there is no lipid film on the walls of the flask. Disruption of lipid film can be facilitated by the addition of a few clean glass beads. The milky liposomal suspension is then kept at room temperature (or at 4° if necessary) for about 1 hr so that multilamellar liposomes can form to completion.

Separation of such liposomes, containing some of the enzyme from the nonentrapped enzyme, is carried out by centrifugation of the suspension, diluted to 10 ml with buffer, at 100,000 g for 30 min. The pellet is covered with 10 ml of buffer and disrupted with the tip of a 5-ml pipette, which is used for resuspension of the liposome debris by pipetting in and out about 5–10 times. The preparation is centrifuged again, and the pellet is finally suspended in an appropriate volume of buffer containing 1% NaCl (buffered saline, BS).

These liposomes because of their size (up to several micrometers in diameter) are not appropriate for intravenous injection. Smaller multilamellar liposomes are prepared as follows: At the end of the time interval (1 hr) required for the formation of liposomes, a sonication probe (0.75-inch diameter) attached on a MSE sonicator tuned for maximum power output is inserted into the wide neck of the flask containing the liposomal suspension and immersed about 1 mm beneath the surface of the fluid. Thin tubing carrying argon at low pressure is inserted into the air compartment of the flask, which is half-way submerged into crushed ice. Sonication is carried out at 30-sec time intervals, interrupted by 15-sec pauses, for a specified period of time depending on the desired size of liposomes. It is advisable to observe the suspension during sonication because frothing of the enzyme solution, which could be detrimental to the enzyme, can occur if the probe is not adjusted at the right depth within the fluid. In addition, certain enzymes will, upon sonication, precipitate in the presence of low ionic strength buffer, and it will then be necessary to add a few drops of concentrated buffer to bring the proteins into solution. A 5–10 sec sonication will produce liposomes tolerated by animals after intravenous injection. Experience in this laboratory has shown that, under the conditions described, after about 10 min of sonication time there is no further significant decrease in the size of liposomes as monitored by the measurement of turbidity at 410 nm.

After sonication, titanium fragments are removed by brief centrifugation at 3000 rpm, and liposomes are again kept for about 1 hr at room temperature (or 4°) and then separation of the liposome-entrapped from the nonentrapped enzyme is carried out by molecular sieve chromatography using a Sepharose 6B column (1 × 22 cm) equilibrated in BS. With relatively stable enzymes such as horseradish peroxidase, *Aspergillus niger* amyloglucosidase, and yeast invertase, chromatography can be carried out at room temperature. Liposomes appearing as a milky suspension are eluted at the end of the void volume (Fig. 3), and with highly sonicated liposomes a second peak of small liposomes emerges several fractions later. Because of the possibility that the elution profile of small liposomes might overlap with that of nonentrapped enzyme, it is advisable to measure in addition to the absorbance at 410 nm, or lipid-bound phosphorus, the enzyme activity or the radioactivity in the case of radiolabeled enzymes. Ideally, eluted liposomes should be rechromatographed on a similar column, but this will depend on how important is the absence of any trace of free enzyme in the liposome suspension. Double chromatography will increase the initial volume of the suspension 3–4 times. Alternatively, separation of the liposome-entrapped from the nonentrapped enzyme can be achieved by centrifugation at 100,000 g for 60 min, although, with a highly sonicated suspension, small liposomes will be present in the supernatant.

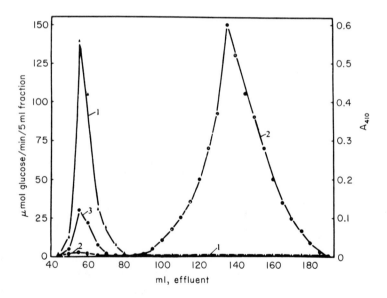

FIG. 3. Separation of liposomes containing amyloglucosidase from nonentrapped enzyme on Sepharose 6B. Curve 1: lipid (A410); curve 2: nonentrapped enzyme; curve 3: latent enzyme. From G. Gregoriadis, P. D. Leathwood, and B. E. Ryman, *FEBS Lett.* **14**, 95 (1971).

Having obtained a "pure" suspension of enzyme-containing liposomes, total enzyme activity will be measured to ascertain how much of the material originally used was entrapped and also to what extent the entrapped enzyme exists behind the outer lipid bilayer, and it is not therefore measurable without disrupting liposomes (latent activity). Assay of the latter enzyme activity requires the use of substrates which, because of their size or physical properties, cannot penetrate the lipid bilayers. Such substances will be attacked only by enzyme molecules adsorbed on the outer liposomal surface or by molecules which, although entrapped, have enzymically active hydrophilic regions extending extraliposomally.

Recent work[10] has shown that it is possible to increase the *in vitro* or *in vivo* stability of a liposomal enzyme. In the case of entrapped invertase, this was achieved by treating liposomes with a bifunctional reagent, such as glutaraldehyde, which can penetrate the lipid bilayers and reach the enzyme: a suspension of liposomes composed of phosphatidylcholine, cholesterol, and phosphatidic acid (18–20 mg total lipid) and containing 1.5–5.0 mg of invertase (bakers' yeast β-fructofuranosidase, grade VI, Sigma, London, Chemical Company) was adjusted to pH 4.5 with freshly prepared 0.2 M sodium acetate buffer pH 4.5 (50 mM final molarity) containing glutaraldehyde (British Drug Houses Chemicals Ltd., Poole, England) at a final concentration of 0.8–2.5%. The preparation was left overnight at room temperature and then dialyzed exhaustively at 4° against 1% NaCl. It should be borne in mind, however, that with other enzymes the conditions described might need adjustment.

Assay of Liposomal Enzymic Activity

In measuring enzyme entrapment in liposomes, unless the enzyme is radioactively labeled, it is difficult to predict the extent of entrapment and decide on the volume of the preparation to be used in the assay. Experience in this laboratory has shown that entrapment of any given enzyme can vary from 1.0% to up to 15.0% of the material originally used. Even with the same enzyme, entrapment values in different preparations can vary significantly. It is therefore advisable to use a range of volumes for entrapped enzymes assayed for the first time.

Having decided on the volume of the liposomal preparation to be used, samples are mixed with equal volumes of a 0.1% aqueous solution of Triton X-100 in two test tubes marked "control" and "total." A third sample is added to a tube marked "apparent" and containing an equal

[10] G. Gregoriadis and E. D. Neerunjun, *Eur. J. Biochem.* **47**, 179 (1974).

volume of BS. Depending on the specific activity of the enzyme, these samples can be diluted and then sampled or used as such. It is important to keep the amount of Triton X-100 to a minimum since this detergent can interfere with the assay. All tubes are incubated at 37° for about 10 min or until the liposomal samples in the relevant tubes turn from a milky suspension to a clear solution in the presence of Triton X-100. At this point, the enzyme in the tube marked "control" is inactivated by bringing the contents of the tube to the boiling point, or by other means. To all tubes the appropriate substrate in a buffer compatible with the preservation of the liposomal structure is added, and samples are incubated according to the conditions of the particular enzyme assay. At the end of the incubation period the reaction in tubes "total" and "apparent" is stopped by enzyme inactivation, and the tube "apparent" is supplemented with 0.1% Triton X-100. The product of the reaction in tubes "total" and "apparent" is then measured against the "control." Latent liposome-entrapped enzyme activity (i.e., activity in intact liposomes not available to the substrate) can be computed as the difference between total and apparent liposomal enzyme activity, the latter being derived from enzyme adsorbed on the liposomal surface or from enzymatically active extraliposomal portions of the entrapped molecules.

Comments on Enzyme Entrapment

The main lipid component of liposomes is phospholipids and of these, the most commonly used is natural (egg) lecithin. Synthetic lecithins are also used, and in work with human patients[11] these are preferable to natural lecithins, which can be contaminated with immunogenic substances. Synthetic lecithins with one long-chain saturated fatty acid are preferable to those with both chains saturated because they can, as a dry film, be dispersed with the enzyme solution at room temperature instead of at 50° or more, which would be unsuitable for many enzymes. In connection with heat-labile enzymes, it should also be remembered that formation of liposomes by ultrasonic irradiation can be carried out only above the liquid crystalline transition temperature of the lipids employed. Beef brain sphingomyelin and some synthetic phosphonyl and phosphinyl analogs of lecithins can also be used instead of lecithin, but not bacterial phosphatidylethanolamine as the sole phospholipid.[12]

Although lecithins alone are sufficient for the formation of liposomes, these can be improved by the addition of other lipid compounds. For

---

[11] G. Gregoriadis, P. C. Swain, E. J. Wills, and A. S. Tavill, *Lancet* **1**, 1313 (1974).
[12] S. C. Kinsky, this series, Vol. 32, p. 501.

instance, the presence of a sterol, such as cholesterol or ergosterol, can promote the stability of lipid bilayers, especially those composed of synthetic lecithins. Further, the inclusion of a charged amphiphile can greatly increase the volume of entrapped enzyme solution and hence the absolute amount of entrapped enzyme (Eq. 1). This effect is produced by the electrostatic repulsion between any two charged bilayers on either side of an aqueous channel. The addition of a charged amphiphile confers a positive or negative charge on the surface of liposomes, and this modifies or alters some of their biological properties. For instance, liposomes rendered positive by the inclusion of stearylamine or L-hexadecylamine acquire a longer half-life in the circulation of injected animals.[10] The opposite effect is produced by negatively charged lipids such as dicetyl phosphate, phosphatidic acid, or gangliosides.[10] Another biological property of liposomes, namely their adjuvant effect in immunization with liposome entrapped antigens, can be drastically improved by the addition of negatively charged lipids.[13]

In deciding on the choice of the charged amphiphile in the preparation of enzyme-containing liposomes, consideration should also be given to the nature of the enzyme. To avoid electrostatic attraction and the formation of precipitable enzyme–lipid complexes, it is preferable to use amphiphiles and enzymes of similar charge. If this is not possible, then the pH should be adjusted so that the net charge of the protein can be temporarily altered. For instance, entrapment of albumin in positively charged liposomes can occur without precipitate formation at pH 4.2, which is below the isoelectric point of the protein. After entrapment and separation from nonentrapped albumin, the pH of liposomes can be adjusted.

The relative concentrations of the lipids used for the preparation of liposomes can vary widely. Having in mind that the major constituent, and a necessary one, of the liposomal structure is phospholipids, then—according to the particular requirements of the preparation in regard to stability, solute permeability, and surface charge—appropriate quantities of sterols and charged amphiphiles can be utilized.

[13] A. C. Allison and G. Gregoriadis, *Nature (London)* **244**, 170 (1974).

[18] Fiber-Entrapped Enzymes

By DINO DINELLI, WALTER MARCONI, and FRANCO MORISI

The method of immobilizing enzymes developed by Snamprogetti consists of the physical entrapment of enzymes inside the microcavities of porous fibers.[1-3] The entrapment is achieved by (1) dissolving a fiber-forming polymer in an organic solvent not miscible in water; (2) emulsifying this solution with the aqueous solution of the enzyme; (3) extruding the emulsion through fine holes into a liquid coagulant which precipitates the polymer in filamentous form. At the end of the process, a bundle containing several parallel continuous individual filaments of unlimited length is obtained; each individual filament consists of a macroscopically homogeneous dispersion of small droplets of the aqueous enzyme solution in the porous polymer gel. The porosity is such as to allow the free diffusion of the low-molecular-weight compounds, but does not permit the escape of the enzyme because of the relatively large size of its molecules.

The fiber entrapment of enzymes is very similar to the wet-spinning procedure for the manufacture of manmade fibers, and the apparatus are very similar to those used in the textile industry.[4] A simple apparatus,[5] particularly suitable for laboratory experiments and useful when only small amounts of expensive enzymes are available, is shown in Fig. 1. The apparatus consists of a nitrogen bottle; a jacketed round-bottom flask provided with a stirrer for dissolving the polymer and preparing the emulsion; a spinneret, which is a plate made of a platinum alloy, containing an orifice 80–125 μm in diameter through which the emulsion is extruded under pressure; a coagulating bath containing the polymer precipitant; and a driven roll to wind up the filament.

Entrapment of Enzymes in Fibers

Many enzyme preparations having various levels of purity have been successfully entrapped in fibers such as cellulose triacetate, and the technique is essentially the same. More complex systems, involving a number of enzymes, can also be entrapped in these fibers. As the technique applies to any high-molecular-weight compound, problems involving those enzymes whose activity depends on the presence of small-molecular-weight coenzymes can be solved by entrapping, together with the parent enzyme, coenzymes chemically bonded to large inert molecules. Proteins other than enzymes, particulate material, microorganisms, cells, and spores can also be entrapped under the same conditions. Depending on the nature of the enzyme, a buffer solution containing stabilizing agents can be used

[1] D. Dinelli, *Process Biochem.* **7**, 9 (1973).
[2] W. Marconi, S. Gulinelli, and F. Morisi, in "Insolubilized Enzymes" (M. Salmona, C. Saronio, and S. Garattini, eds.), p. 51. Raven, New York, 1974.
[3] D. Dinelli and F. Morisi, in "Enzyme Engineering" (E. K. Pye and L. B. Wingard, Jr., eds), p. 295. Plenum, New York, 1974.
[4] R. W. Moncrieff, "Man-Made Fibres." Heywood, London, 1970.
[5] F. Bartoli, unpublished data, 1974.

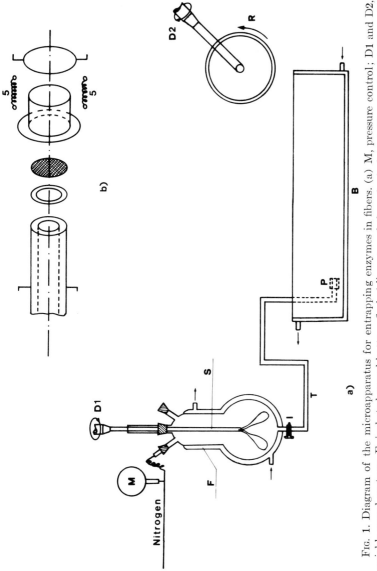

FIG. 1. Diagram of the microapparatus for entrapping enzymes in fibers. (a) M, pressure control; D1 and D2, variable-speed motors; F, jacketed round-bottom flask (diameter of the spherical part, 28 mm; height, 50 mm); S, stirrer with chain paddle shaft; I, glass tubing (diameter, 2 mm); P, spinneret with a single central hole (diameter, 80 μm) (Italfidel, Milan, Italy); B, jacketed coagulating bath; R, roller (diameter, 65 mm). (b) spinneret. 1, glass tubing; 2, Teflon O ring (thickness, 1 mm; external diameter, 9 mm; internal diameter, 3 mm); 3, wire disk (100 mesh; diameter, 9 mm); 4, spinneret; 5, assembly spring; 6, assembly ring.

instead of an aqueous solution in order to improve the enzyme stability toward solvent denaturation. In the following examples, entrapment of invertase is reported, but it must be stressed that any other enzyme can be entrapped using the same procedure.

Procedures

Cellulose Triacetate. Place 2.1 ml of methylene chloride, and 200 mg of cellulose triacetate (Fluka A. G., Buchs, Switzerland) in a round-bottom flask. The mixture is gently stirred at room temperature until complete solution is obtained. After the solution has cooled to 0°, add dropwise 0.4 g of concentrated invertase solution (British Drug Houses Ltd, Poole, England). The rate of stirring is gradually increased to 900 rpm. After 5 min of stirring at this speed, an emulsion consisting of small droplets of aqueous enzyme solution dispersed in the polymer phase is obtained. As soon as the emulsion is ready (under microscopic examination, a uniform distribution of aqueous droplets not greater than 2 μm in diameter must appear), the stirrer is removed and the flask is stoppered. The emulsion is allowed to stand for 20 min at 0° to permit air bubbles trapped during the emulsification to rise. The stopcock is then opened, and using a nitrogen bottle a pressure of 0.7 atm is applied to the flask. Under nitrogen pressure, the emulsion passes through the tubing connecting the flask to the spinneret and flows from the spinneret, which is immersed in the coagulating bath containing toluene. As soon as the emulsion flows from the spinneret, the coagulated polymer is pulled away and wound on the driven roll rotating at 60 rpm. The filament is then continuously pulled away and wound until all the emulsion is extruded. The fibers are collected from the roll in skein form and allowed to stand for 15 min in the open air to remove organic solvents. After spinning, the spinneret must be immediately washed in methylene chloride. The other parts of the apparatus must also be washed in methylene chloride or acetone.

Nitrocellulose. The apparatus and the procedure are the same as those described for cellulose triacetate. The spinneret used in this case has an opening 125 μm in diameter. Nitrocellulose (Snia Viscosa, Colleferro, Rome, Italy) having the following characteristics is employed: nitrogen content, 11.5%; complete solubility in alcohol–ether 99%; viscosity in 3% acetone solution (Hoppler, 25°), 42.26 cps. The organic phase consists of 200 mg of nitrocellulose dissolved in a mixture of 1.56 ml of *n*-butyl acetate and 1.06 ml of toluene. The coagulating solvent is petroleum ether (b.p. 40°–60°).

Cellulose. Cellulose fibers are easily obtained by converting cellulose nitrate fibers back into cellulose without destroying the fibrous structure.

In a 100-ml round-bottom flask, place approximately 50 ml of 2% aqueous ammonia solution. A slow stream of hydrogen sulfide is passed through the solution (which is cooled to about 4°) in a hood until a pH of 8.5 is reached. Add 500 mg of invertase nitrocellulose fibers and allow the reaction to stand at room temperature for 6 hr with occasional stirring. After 6 hr the denitration reaction is practically complete. The cellulose fibers are well rinsed with 0.1 M sodium phosphate solution, pH 4.5.

Ethyl Cellulose. The apparatus and procedure are the same as those described for cellulose triacetate. The polymer used is ethyl cellulose N 200 (Hercules Inc., Wilmington, Delaware) having the following characteristics: ethoxyl content, 48.3%; viscosity (5% in 80:20 toluene–ethanol by weight, at 25°), 166 cps. The polymer solution is obtained by dissolving 200 mg of ethyl cellulose in 2.6 ml of toluene; the coagulating bath consists of petroleum ether.

Poly-γ-methyl-L-glutamate. The apparatus and the procedure are the same as those that described for cellulose triacetate. Use a spinneret with an opening 125 µm in diameter. The polymer solution consists of 10% (w/v) poly-γ-methyl-L-glutamate PLG-30 (Kyowa Hakko Kogyo Co., Ltd., Tokyo, Japan) in 1,2-dichloroethane. The coagulating solvent is petroleum ether.

For particular applications, it might be useful to prepare fibers in which some of the methyl ester groups are hydrolyzed, resulting in a polymer having COOH groups. To prepare such fibers, use the following procedure: Dilute 10-fold a commercial solution of poly-γ-methyl-L-glutamate with dichloroethane. To the diluted solution, add dropwise an amount of alcoholic potassium hydroxide (3.5 N, in methyl alcohol) equivalent to 15–20% of the total ester groups. Stir the reaction mixture at room temperature for 20 hr. Remove the insoluble material by centrifugation and precipitate the polymer by adding methyl alcohol (2 volumes per 1 volume of polymer solution) to the supernatant while stirring. The partially hydrolyzed polymer collected by filtration is redissolved in dichloroethane at a concentration of about 7% (w/v), and this solution is used for spinning under the same conditions as those described previously.

In most of our studies we have used cellulose triacetate for the following reasons: (1) low cost; (2) good biological resistance; (3) it is unaffected by dilute solutions of weak acids; and (4) the cellulose triacetate fibers are resistant to many common solvents.

Analytical Determinations of Fiber-Entrapped Enzymes

Enzyme Activity Assay. The activity of a fiber-entrapped enzyme is determined by incubating the enzyme fibers and the substrate solution while stirring. For the invertase fibers prepared as described above, the following procedure is suggested: Place about 200 mg of invertase fibers into a 100-ml Erlenmeyer flask; add 50 ml of 0.1 M sodium phosphate solution, pH 4.5, and shake the flask in a reciprocating water bath shaker at 25° for 3 hr. Remove and discard the liquid by suction with a pipette, add another 50-ml portion of the same phosphate solution, and shake again for 3 hr. Discard the liquid and repeat the washing. To the flask containing the washed invertase fibers, add 50 ml of substrate solution (20%, w/v, sucrose solution in 0.1 M phosphate solution, pH 4.5) previously thermostated at 25°. Shake the flask in the reciprocating shaker at 150 strokes per minute. After 10 min, withdraw a 1.0 ml sample and determine its glucose content by the glucose oxidase procedure. Determine the activity from the dry weight of the fibers as described below.

Dry Weight of the Fibers. Since the wet weight of the enzyme fibers varies according to the storage conditions, it is necessary to refer all analytical data on the fibers back to their dry weight. The dry weight includes the polymer support plus all the macromolecular substances contained in the aqueous enzyme solution. Dry weight determination is carried out on the same sample of enzyme fibers used for the activity assay. The sample is rinsed many times with distilled water, dried in an oven at 105° until a constant weight is obtained, and weighed.

Nitrogen Content of the Fibers. The nitrogen content of the fiber-entrapped enzymes is determined according to the Kjeldahl procedure after exhaustive washing of the fibers with distilled water in order to remove the loosely absorbed enzyme and the low-molecular-weight compounds originally entrapped. A parallel determination of nitrogen on control fibers, i.e., fibers not containing the enzyme, is suggested.

Using Enzyme Fibers

Fiber-entrapped enzymes can be used in every type of catalytic reactor. Because of their peculiar form, enzyme fibers can be easily removed from the reaction mixture without using complicated physical methods of solid–liquid separation. For laboratory procedures, the long filaments of the enzyme fibers are put together with the reaction mixture in a flask and the flask is shaken in either a rotating or a reciprocating water-bath shaker. For industrial applications, slurry reactors are more convenient: the fibers are cut into small pieces 1–3 cm in length and maintained in the reaction mixture by mechanical stirring. At the end of the reaction, the enzyme fibers are recovered by filtration or centrifugation and resuspended in fresh reaction mixture to form a new batch.

A recycling type of reactor is particularly useful for enzyme fibers, especially when the substrate solution is not perfectly clear and contains suspended particles. This reactor consists of a column made of glass, plastic, or stainless steel depending on the operating conditions; the fibers, in the form of skeins, are arranged parallel to the long axis of the column and fixed at the two extremes; the packing density is 0.05–0.15 g (dry weight basis) of fibers per milliliter of column volume. A pump continuously recirculates a fixed amount of the reaction mixture through the column until the desired conversion is reached. Because of the particular arrangement of the fibers and the relatively low packing density, a suspension can also be recirculated through the fibers without observing a filtering action of the catalyst bed. For these reasons this type of reactor is useful when natural complex mixtures, such as milk, fruit juices, or liquefied starch slurries, have to be treated.

Enzyme fibers can also be used in fixed-bed continuous reactors. In this case, the fibers either as long filaments or small pieces are packed at high density in the reactor. To demonstrate the flexibility in using enzyme fibers, it is worthwhile to mention the possibility of making fabrics. Enzyme fibers are sufficiently mechanically resistant so as to be woven, or in certain cases they can be reinforced with nylon filaments. The fabrics obtained are easily used for particular purposes, e.g., in the frames of a filter press.

Finally a useful centrifugal reactor was developed in our laboratory; it is shown in Fig. 2. The reactor consists of a basket containing a layer of enzyme fibers. The basket is connected to a variable-speed motor through a collet-type chuck; when the basket is rotating, the flow pattern of the liquid is that schematized in Fig. 2. To fill the reactor with enzyme fibers, the basket is immersed in a beaker containing 1 liter of water; the motor is turned on, and, at low stirring speed, small pieces (1–2 cm in length) of enzyme fibers are added to the beaker. The fibers are held on the basket by a stainless steel net. When about 1 g (dry weight) of fibers is packed on the reactor, the water is removed from the beaker and the stirring speed is increased to complete the removal of water from the catalyst bed. Then the reactor is immersed in the substrate solution and stirred to carry out the catalytic reaction. This reactor is very efficient in overcoming external diffusional limitations. One other advantage is that it removes almost all the reaction mixture from the catalyst bed by centrifugal force at the end of the reaction.

Physical Properties of Enzyme Fibers

The physical properties of enzyme fibers depend on the chemical nature of the polymer and are generally different from those of fibers nor-

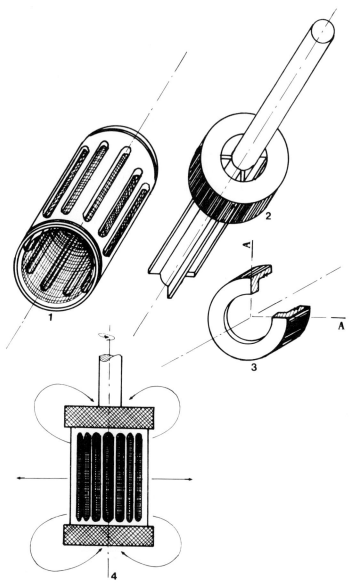

Fig. 2. Laboratory reactor for enzyme fibers: 1, basket with wire gauze; 2, rotor with ring nut; 3, ring nut; 4, view of the whole reactor (the arrows indicate the flow directions).

mally obtained in the textile industry. These differences arise chiefly from the fact that, while in the normal wet-spinning procedure a homogeneous solution of the polymer is spun, in the technology of enzyme entrapment an emulsion is extruded. Moreover, for enzyme entrapping, the filaments are not stretched, and the polymer undergoes little or no orientation. These facts result in a more porous fiber structure, lowered mechanical properties, and discontinuous structure owing to the presence of numerous microcavities contaning the aqueous solution of the enzyme. For example, the linear density, expressed in denier (the weight in grams of 9000 meters of the filament), of the enzyme fibers is lower than for textile fibers. Figure 3 shows a micrograph of a monofilament cross section. The white areas are the microcavities containing the enzyme; the black areas consist of the porous polymer material. The size distribution of the microcavities depends on how finely dispersed the original emulsion was. On the average, the long axis of cross section for the microcavities is not greater than 1 μm. Notwithstanding their particular structure, the enzyme fibers retain a good mechanical strength so that they can be reused many times in different types of reactors and under conditions where the mechanical stresses are particularly severe. For example, enzyme fibers can tolerate the normal manufacturing processing into yarns and fabrics.

One of the most important properties of the fibers as a catalyst support is undoubtedly their enormous surface area.

Other physical properties of the fibers are of great importance for the action of the enzyme entrapped in them. According to their chemical structure, hydrophilic or hydrophobic fibers can be prepared. For example, in cellulosic fibers their ability to absorb water is gradually reduced as the degree of acetylation of the hydroxyl groups on each anhydroglucose unit is increased.

The ability of enzyme fibers to absorb chemical compounds other than water from an external solution is also important. A typical example is given by tryptophan synthetase entrapped in cellulose triacetate fibers. Indole, a substrate of this enzyme, is also strongly absorbed by cellulose triacetate, and this absorption completely changes the properties of the enzyme. The importance of these sorption phenomena must be considered, especially when the immobilized enzymes must operate in complex solutions or natural mixtures.

Enzyme fibers are resistant to many chemical agents, including weak acids or alkalies, high ionic strength solutions, and some organic solvents. Depending on the polymer used, the fibers show good resistance to microbiological attack.

Catalytic Properties of Fiber-Entrapped Enzymes

The catalytic properties of fiber-entrapped enzymes are generally different from those of the corresponding free enzymes. The principal differ-

Fig. 3. Micrograph of a monofilament cross section.

ences are ascribed to the restricted diffusion of substrates and products inside the porous enzyme support.

Activity. The measurement of activity displayed by a fiber-entrapped enzyme is the first step in its characterization. The activity per gram of fiber (dry weight) is obtained, and since the amount of enzyme originally added per gram of polymeric support is known, it is possible to calculate the ratio of this activity to the activity of the entrapped enzyme, defined as efficiency, η. For example, for the invertase cellulose triacetate fibers prepared as described above, we emulsified 2 g of invertase solution, corresponding to about 6140 U (1 µmole/min, at 25°) per gram of polymer. The activity displayed by 1 g of fibers (dry weight) is about 1840 U; the percentage ratio of normal to entrapped activity, i.e., the efficiency, η, is (1840/6140) × 100. Provided the enzyme is stable under the spinning conditions, the efficiency of a fiber-entrapped enzyme depends on the external (i.e., rate of stirring) and internal (i.e., porosity of the support) diffusional limitations because the entrapment is practically complete.

Effect of Enzyme Density on the Support. If the amount of enzyme per gram of polymeric support is increased, the efficiency of the resulting fibers decreases because by increasing the activity density, much more substrate must diffuse through the porous support and the rate of this diffusion process becomes the rate-determining step.

For this reason, in order to achieve the best utilization of the enzyme activity, the optimal enzyme density must be chosen in relation to the particular application of the fibers and the type enzyme. The table shows

ENTRAPPED AND DISPLAYED ACTIVITY OF SOME FIBER-ENTRAPPED ENZYMES

Enzyme	Entrapped activity (units/g dry fibers)	Measured activity (units/g dry fibers)
Penicillin acylase[a]	25000	8750
Invertase[b]	6140	1840
Glucose isomerase[c]	1100	440
Tryptophan synthetase[d]	4100	3030

[a] Penicillin Acylase Unit = 1 µmole of penicillin G per hour (37°, pH 8.5).
[b] Invertase Unit = 1 µmole of sucrose per minute (25°, pH 4.5).
[c] Glucose Isomerase Unit = 1 µmole of glucose per minute (60°, pH 7.0).
[d] Tryptophan Synthetase Unit = 0.1 µmole of indole per hour (25°, pH 7.8).

the normal amount of enzyme used for fiber entrapping of some enzymes.

Effect of pH. The efficiency of a fiber-entrapped enzyme is at a minimum at the optimum pH for activity, because when the activity is higher, the diffusion of substrate is more limiting. The effect of pH on the efficiency of fiber-entrapped glucose isomerase is shown in Fig. 4.

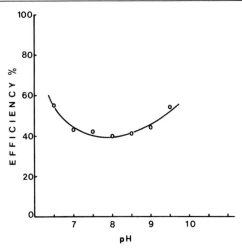

Fig. 4. Effect of pH on the efficiency of glucose isomerase fibers obtained by entrapping 380 GIU (μmoles of glucose isomerized per min at 60°) per gram of cellulose triacetate.

When H^+ or OH^- ions are produced in the reaction catalyzed by the enzyme, the concentration of these ions in the immediate proximity of the entrapped enzyme is greater than that of the external solution, and apparent shifts in the optimum pH are observed.

Effect of Substrate Concentration. An interesting example using invertase fibers illustrates the effect of substrate concentration on the efficiency of the entrapped enzyme. The activity of the free enzyme follows the Michaelis–Menten law up to a sucrose concentration of about 5%, then substrate inhibition occurs and the activity decreases. The efficiency of invertase fibers as a function of substrate concentration is shown in Fig. 5. The efficiency increases as the sucrose concentration in the external solution increases; the effect of substrate inhibition is attenuated because the sucrose concentration is always lower than that of the bulk solution surrounding the entrapped enzyme.

Effect of Temperature. The efficiency of the fiber-entrapped enzymes decreases with increasing temperature, as illustrated in Fig. 6 for glucose isomerase fibers. This is due to the diffusion limitation which increases because of the increased enzyme activity owing to the rise in temperature. The best utilization of the activity of enzyme fibers is at low temperatures.

Stability of Fiber-Entrapped Enzymes

Storage Stability. Most of the fiber-entrapped enzymes can be stored for a long time at 4°. This is of interest if one considers that an aqueous

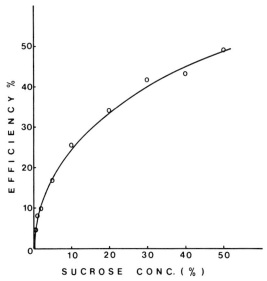

Fig. 5. Effect of substrate concentration on the efficiency of invertase fibers obtained by entrapping 6140 U per gram of cellulose triacetate.

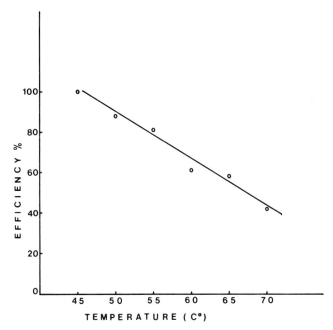

Fig. 6. Effect of temperature on the efficiency of glucose isomerase fibers obtained by entrapping 380 GIU (μmoles of glucose isomerized per min at 60°) per gram of cellulose triacetate.

solution of an enzyme is normally unstable, even at low temperature, and that enzymes are normally stored as a dry powder or suspensions in highly concentrated ammonium sulfate solution. It is therefore surprising that fiber-entrapped enzymes show excellent storage stability even if the active protein is present in the fibers as an aqueous solution. When the fibers are spun out, they contain small amounts of organic solvents and the aqueous phase. Under these conditions, the enzyme fibers are generally stable; the presence of organic solvents helps to prevent the growth of microorganisms on the external surface of the fibers. In order to preserve their morphological structure, the fibers must not be dried; the presence of water inside the microcavities is absolutely necessary for the retention of enzyme activity.

Operational Stability. The stability of an enzyme displaying catalytic activity depends upon the nature of the enzyme and the conditions under which the catalytic activity takes place.

The principal causes of enzyme deactivation are those related to its thermodynamic conformation, such as high temperatures and extreme pH values. Entrapping in fibers does not improve the stability of an enzyme toward adverse pH and temperature conditions. The improved stability of many fiber-entrapped enzymes is most probably due to the fact that the enzymes generally remain in the microcavities of the filamentous support as highly concentrated aqueous solutions and do not undergo dilution on repetitive use. Another important cause of enzyme deactivation arises from the hydrolytic attack of proteases or microbial contamination of the enzyme solution. If the enzymic solution to be entrapped is free from proteases, after entrapment the enzymes are protected against external proteolytic attack or microbial digestion because neither proteases nor microorganisms can penetrate the porous support. In fact, many fiber-entrapped enzymes do not show any loss of activity when the external substrate solution is heavily contaminated. When microbial contamination occurs, the reaction mixture is usually lost, but the entrapped enzyme, after suitable decontaminating treatments, can be recovered unaffected.

Additional factors, causing enzyme deactivation, such as poisoning and the loss of small molecules loosely attached to the enzyme and necessary for its activity, have to be taken into account. The latter factor has a significant effect on entrapped enzymes, which, in operation, undergo a continuous dialysis and might lose important substances, such as coenzymes and metal ions, on which the activity depends. Generally speaking, enzyme fibers have proved to be quite stable under operating conditions. It follows that a fixed amount of an active protein, immobilized in fibers and repeatedly used, transforms a large amount of substrate into products. This fact is obviously of the utmost importance for the technical feasibility and the economic success of industrial applications.

The stability of fiber-entrapped enzymes is illustrated most spectacularly by a sample of invertase fibers, which for more than 7 years has been continuously used to hydrolyze sucrose and shows only a 10% loss in activity. Perhaps many other enzymes, especially of the hydrolytic type, would also give similar results.

Applications of Fiber-Entrapped Enzymes

Because of their excellent activity and stability, fiber-entrapped enzymes have potential technological application in industry, medicine, and analytical chemistry. The easy and inexpensive preparation procedures for enzyme fibers show a clear economic advantage in a large number of applications, which include both replacement of currently used free enzymes and the use of sophisticated enzymes considered too expensive in the soluble form.

Some fiber-entrapped enzymes, such as penicillin acylase or lactase, are no longer laboratory curiosities and are used industrially. Other industrial applications include: (1) the enzymatic conversion of preliquefied starch syrups to glucose by entrapped glucoamylase,[6] (2) the isomerization of glucose to fructose by glucose isomerase fibers,[7] (3) the debittering of fruit juices by hydrolyzing naringin,[8] (4) the reduction of milk lactose by entrapped β-galactosidase[9-11] (see this volume [57]), and (5) the hydrolysis of sucrose by invertase fibers.[12]

Some processes of great interest for pharmaceutical industries have been developed using enzyme fibers. They include (1) the hydrolysis of benzyl penicillin, by entrapped penicillin acylase, to give 6-aminopenicillanic acid as a starting material for the production of semisynthetic penicillins[13]; (2) the resolution of racemic mixtures of amino acids by fiber-entrapped aminoacylase; and (3) the enzymic synthesis of L-tryptophan from indole and serine by tryptophan synthetase fibers.[14,15]

[6] C. Corno, G. Galli, F. Morisi, M. Bettonte, and A. Stopponi, *Staerke* **24**, 420 (1972).
[7] S. Giovenco, F. Morisi, and P. Pansolli, *FEBS Lett.* **36**, 57 (1973).
[8] Snamprogetti, S. p. A., Ital. Appl. 20,645 (1971).
[9] F. Morisi, M. Pastore, and A. Viglia, *J. Dairy Sci.* **56**, 1123 (1973).
[10] M. Pastore, F. Morisi, and A. Viglia, *J. Dairy Sci.* **57**, 269 (1974).
[11] M. Pastore, F. Morisi, and D. Zaccardelli, in "Insolubilized Enzymes" (M. Salmona, C. Saronio, and S. Garattini, eds.), p. 211. Raven, New York, 1974.
[12] W. Marconi, S. Gulinelli, and F. Morisi, *Biotechnol. Bioeng.* **16**, 501 (1974).
[13] W. Marconi, F. Cecere, F. Morisi, G. Della Penna, and B. Rappuoli, *J. Antibiot.* **26**, 228 (1973).
[14] P. Zaffaroni, V. Vitobello, F. Cecere, E. Giacomozzi, and F. Morisi, *Agric. Biol. Chem.* **38**, 1335 (1974).
[15] W. Marconi, F. Bartoli, F. Cecere, and F. Morisi, *Agric. Biol. Chem.* **38**, 1343 (1974).

In biomedical applications, fiber-entrapped urease has been used for hydrolyzing blood urea.[16] Asparaginase fibers inserted into the peritoneal cavity of mice have been found efficient in reducing the L-asparagine level.[16] Fiber-entrapped enzymes have proved to be useful tools in chemical analysis, both for manual and for automated procedures.[17,18]

Finally enzyme fibers have been used as aids in particular organic syntheses, where the purity of the products is of the utmost importance.[19,20]

General Remarks on the Characteristics and the Use of Fiber-Entrapped Enzymes

Compared with other types of immobilized enzymes, fiber-entrapped enzymes have peculiar characteristics that must be considered when they are used.

Their use is limited to low-molecular-weight substrates because high-molecular-weight compounds cannot diffuse through the porous gel of the fibers and reach the enzyme inside the microcavities. On the other hand, enzymes in fibers are very efficiently protected against the action of microorganisms and proteases; consequently thermodynamically stable enzymes retain their activity for a very long time. In biomedical applications, immunological reactions can be avoided.

The high ratio of surface to volume maintains the effect of diffusion phenomena at a relatively low level. Since it is possible to entrap high percentages of protein (200 mg of protein and more per gram of dry fiber can be entrapped) by using enzymes having a high level of purity, fibers displaying very high enzymic activity can be obtained.

Where high enzymic activity is not needed, enzyme solutions at every level of purity can be entrapped.

Perhaps the greatest advantage of this technique is that many different enzymes can be entrapped under similar conditions. It is possible to entrap aqueous solutions containing more than one enzyme, which will catalyze successive reactions. In such cases the small dimension of the microcavities and the high concentration of enzymes minimize the effects of the diffusion. Therefore the overall kinetics are high.

[16] M. Salmona, C. Saronio, I. Bartosek, A. Vecchi, and E. Mussini, in "Insolubilized Enzymes" (M. Salmona, C. Saronio, and S. Garattini, eds.), p. 189. Raven, New York, 1974.
[17] W. Marconi, F. Bartoli, S. Gulinelli, and F. Morisi, *Process Biochem.* **9**, 22 (1974).
[18] C. Saronio, M. Salmona, E. Mussini, and S. Garattini, in "Insolubilized Enzymes" (M. Salmona, C. Saronio, and S. Garattini, eds.), p. 143. Raven, New York, 1974.
[19] C. Fuganti, D. Ghizinghelli, D. Giangrasso, P. Grasselli, and A. Santopietro Amisano, *La Chimica e l'Industria* **56**, 424 (1974).
[20] C. Fuganti, D. Ghizinghelli, D. Giangrasso, and P. Grasselli, *JCS Chem. Comm.* **18**, 726 (1974).

For special purposes high-molecular-weight compounds other than proteins can be entrapped in fibers too. Macromolecular polyelectrolytes entrapped together with enzymes extend the range of pH at which an enzyme can be used. Recently our laboratory succeeded in transforming NAD into a water-soluble macromolecular compound which maintains coenzymic activity[21] (see also this volume [61]). This compound can be entrapped in fibers together with NAD-dependent enzymes so that no coenzyme is needed in the solution to be treated.

[21] P. Zappelli, A. Rossodivita, and L. Re, *Eur. J. Biochem.* **54**, 465 (1975).

[19] Collagen-Immobilized Enzyme Systems

By W. R. VIETH and K. VENKATASUBRAMANIAN

Choice of Collagen as Support Matrix

The biomaterial collagen offers a number of advantages as a carrier material for enzyme and whole microbial cell immobilization. It is the most abundant protein constituent of higher vertebrates; its main *in vivo* function is to act as a structural component to support and hold individual cells together in tissues. It can readily be isolated from a number of biological sources and reconstituted into various forms, such as membranes, without losing its native structure. Since reconstituted collagen is now widely used as an edible sausage casing, a well-developed technology exists already for its reconstitution and shaping. This, in conjunction with its ready availability from a large number of biological species—from fish to cattle—makes it an inexpensive carrier.

Collagen has an open internal structure and contains a large number of potential binding sites for enzyme attachment. Its proteinaceous nature contributes polar and nonpolar amino acid residues for strong, cooperative noncovalent interactions with enzyme molecules. The accessibility of the enzyme to the binding sites on collagen is facilitated by its hydrophilicity. At neutral pH, it can sorb water at a level of 100% of its dry weight and the water sorption capacity can increase up to 500% at very acidic and basic pH values. Such high degrees of swelling provide an aqueous environment for the bound enzyme and minimize intracarrier transport resistances to the diffusion of substrate and products of enzymic catalysis. The hydrophilic, proteinaceous nature of the support matrix also tends to stabilize the conjugated enzyme.

Another important forte of collagen is its fibrous nature, which is well suited in strength and form to enzyme immobilization. A microfibrillar collagen dispersion can be used to prepare collagen–enzyme complexes in membranous form. Many enzymes *in vivo* are known to be localized on or within the various membranous structures of the cell. Hence enzymes attached to a membranous material of biological origin, such as collagen, can constitute a more realistic model system for studying *in vivo* enzyme reactions. Furthermore, the membranes can be used to construct different reactor configurations with high surface-to-volume ratios and contact efficiencies.[1] In particular, high throughputs in the reactor can be achieved, with concomitant decrease in channeling and dispersion of the substrate, which are otherwise responsible for decreased yields and process efficiencies. Even in the swollen state, collagen has a shear modulus of about 10^8 dynes/cm (which is in the range of plasticized PVC). This implies that there should not be any pressure-drop problems encountered with collagen in particulate form due to excessive compaction.

The procedure for the complexation of enzyme to collagen is simple and it is applicable to a wide variety of enzymes. It utilizes what the two biological macromolecules naturally offer in their unmodified forms: the cooperative interactive potentials of the amino acid residues. In addition, immobilization takes place under mild conditions—usually at ambient temperature in an aqueous environment—thus reducing the possibility of enzyme inactivation.

It is possible to impart a wide variety of desired characteristics to collagen via controlled chemical modification. For instance, a cross-linking agent can be used to obtain a stronger and/or more constricted microstructure; it can be treated with specific proteolytic enzymes to eliminate antigenic activity; its surface charge can be altered by blocking its carboxyl or amino groups. Similarly, its hydrophile:hydrophobe ratio can be changed so as to facilitate immobilization of lipophilic systems, such as steroid modifying enzymes. Thus, we have coupled a large number of enzymes and whole cells to collagen (see also this volume [53]).

Another point deserves mention here. The use of a proteinaceous material, such as collagen, as a carrier matrix for enzyme binding raises the question: Should its applicability be necessarily limited to nonproteolytic enzymes? The answer is no—since many commonly used, commercially important proteases, such as papain, rennin, and ficin, do not attack collagen specifically, they can readily be complexed with collagen.[2]

[1] W. R. Vieth, S. G. Gilbert, S. S. Wang, and K. Venkatasubramanian, U.S. Patent No. 2,809,613 (1974).
[2] K. Venkatasubramanian, R. Saini, and W. R. Vieth, *J. Food Sci.* **40**, 109 (1975).

Preparation Methods for Collagen–Enzyme Complexes

Three different procedures can be used for preparing collagen–enzyme membranes, as illustrated schematically in Fig. 1: (a) membrane impregnation, (b) macromolecular complexation, and (c) electrocodeposition. These methods are applicable to the immobilization of single as well as multiple enzymes. When cow tendon is used as the source of collagen, a 1% (w/w) dispersion is prepared in a methanol–water solvent (1:1 by

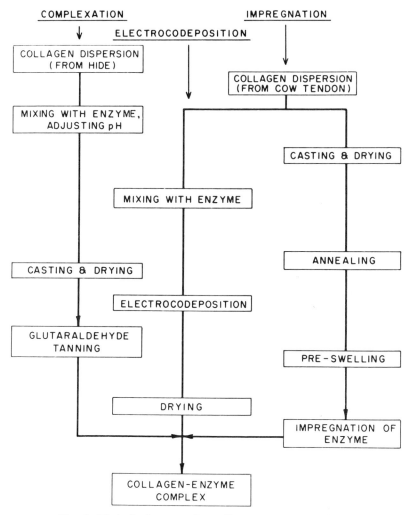

FIG. 1. Methods for preparing collagen–enzyme complexes.

volume) according to the method of Hochstadt et al.[3] Ground cattle hide pulp (water to hide ratio 1.3:2.0) can be obtained from the USDA Eastern Regional Research Laboratories, Philadelphia, Pennsylvania.

Membrane Impregnation Method

In this method, the enzyme is impregnated into a preformed solid membrane. The driving forces that bring about the penetration of the enzyme into the membrane are the concentration and pH gradients set up between the enzyme bath and the membrane during impregnation. Both of these gradients facilitate the molecular diffusion of the enzyme into the membrane. The ensuing drying process removes the water clustered around the charged groups on collagen as well as the enzyme, thus reducing the intermolecular distance between them. This promotes the completion of the complex of interactions between them, which are incipient in the impregnation step.

Reagents

Collagen
Enzyme
Buffer solutions
Lactic acid (95%)
Sodium hydroxide, 1 N
Glutaraldehyde solution

Procedure. Sixty grams of hide collagen (~30% solids) are dispersed in 4 liters of distilled water. Lactic acid is added dropwise to lower the pH to 3.0. The pulp is homogenized in a Waring blender at low speed during about fifteen 1-min treatments. The temperature of the dispersion should be maintained below 25°. This can usually be achieved by cooling the blender and its contents in an ice bath before starting each homogenization treatment. At the end of the homogenization steps, the dispersion should be free from any lumps and should flow smoothly. The dispersion is then degassed under a vacuum of at least 25 inches of mercury until all the trapped air bubbles are removed. Next, the dispersion is placed on a Mylar (E.I. duPont de Nemours Company, Wilmington, Delaware) sheet and is contained within an aluminum frame, 8 feet $\times$ 1 foot. A Gardner knife, 1 foot in width, is drawn by a motor at a uniform speed over the dispersion guided by the metallic frame. The membrane thickness can be adjusted by two knurled screws on the knife. A wet thickness of about 6 mm can normally be used (this would yield a membrane of about 0.1 mm dry thickness). After the casting operation, the dispersion is left to dry at room temperature. The drying process may be accelerated by blow-

[3] H. R. Hochstadt, F. Park, and E. R. Lieberman, U.S. Patent No. 2,920,000 (1961).

ing a stream of air over the membrane. When the membrane is dry, it is peeled off the Mylar sheet; it is now ready for enzyme immobilization.

The membrane is swollen in a suitable buffer solution for 2 hr at a pH which is optimal for the enzyme to be immobilized. It is then soaked in an enzyme impregnation bath, containing the enzyme solution at its optimum pH. Enzyme concentration in the bath should be in the range of 10–20 mg/ml. Sufficient enzyme solution should be used to immerse the membrane completely. Impregnation can be carried out at room temperature for most enzymes; if the enzyme is very temperature sensitive, lower temperatures could be employed. Duration of impregnation would depend on the particular enzyme and the temperature of impregnation; 24–36 hr are sufficient for most enzymes. The impregnated membrane is layered on a Mylar sheet and allowed to dry completely at room temperature.

NOTES. (1) Collagen membranes can also be prepared on the basic side of the pH scale. In this case, 1 N sodium hydroxide is used to bring up the pH of the dispersion to 10.5–11.0. (2) The enzyme–collagen membrane can be tanned with glutaraldehyde, if desired, as described later in this section. (3) A simpler procedure to cast the membrane is to pour the dispersion on a glass plate and spread it uniformly with a glass rod.

Macromolecular Complexation

In this process, the enzyme is added directly to an aqueous collagen dispersion and comixed before casting the membrane. In the dispersion, collagen macromolecules exist in their native microfibrillar state and therefore can participate in interactions with the enzyme, further facilitated by the subsequent drying step. This technique reduces the number of steps involved in preparing the collagen–enzyme complex. However, good molecular dispersions of collagen can be obtained only in the pH ranges of 2.5 to 4.5 and 10.0 to 11.5. If the enzyme to be coupled is not stable at these pH values, the impregnation process should be resorted to.

Procedure. To a collagen dispersion, prepared as described above, a desired amount of enzyme is added and homogenized. The dry weight ratio of the enzyme to collagen can be between 0.1 and 0.3. The enzyme is first dissolved in distilled water and then added to the dispersion. After sufficient blending and deaeration, the dispersion is cast to form a membrane, as outlined above.

Tanning

The dried membrane can be tanned with a suitable bifunctional crosslinking agent. The tanning treatment imparts excellent mechanical strength to the membrane. In addition, it may increase the amount of enzyme attached to the membrane by covalently cross-linking some of the fraction of weakly bound enzyme molecules residing in the structure. While different cross-linking agents could be used to accomplish this, we have been using primarily glutaraldehyde. Some of the other possible reagents we have tried are glyoxal, chromium sulfate, and ultraviolet radiation. Formaldehyde and chrome are detrimental to most enzymes. The effectiveness by glutaraldehyde treatment depends on its concentration, reaction pH and temperature, and contact time. It must be emphasized here that glutaraldehyde tanning is not a mandatory step in preparing stable complexes.

Procedure. Dilute glutaraldehyde solution in the concentration range of 0.1 to 5% (w/v) is used. pH of the tanning solution is adjusted to the required pH (i.e., the optimum pH for enzyme stability) by adding solid sodium bicarbonate. The enzyme membrane is exposed to the tanning solution at room temperature for 0.5 to 2 min, followed immediately by washing in running cold tap water for 1 hr.

NOTE. Recently, we have developed a one-step procedure for making enzyme membranes at any pH. By adding the required amount of dilute glutaraldehyde solution directly to the dispersion along with the enzyme, membranes can be prepared by the macromolecular complexation procedure at any pH value.

Electrocodeposition Method

This method is based on the well-known principle of electrophoresis—the phenomenon of the migration of amphoteric polyelectrolytes (e.g., proteins) under the influence of an external electrical field gradient at pH values different from their isoelectric pH values. When an enzyme is added to collagen dispersion, the two proteins interact under appropriate pH conditions to form macromolecular complexes. These complexes can be made to migrate to—and deposit on—an electrode surface by imposing an electrical potential. The rate of electrophoretic mobility and the amount of collagen–enzyme complex deposited depend on the applied voltage gradient, the surface area of the electrodes, and the pH and viscosity of the dispersion. By suitably adjusting these parameters, membranes of different thickness can be formed.[4] Suzuki and co-workers have also reported a process for the electrolytic preparation of collagen–enzyme films.[5]

Procedure. Collagen dispersion is prepared at a concentration of 0.3–

[4] W. R. Vieth, S. G. Gilbert, S. S. Wang, and R. Saini, U.S. Patent No. 3,758,396 (1973).

[5] S. Suzuki, I. Karube, and Y. Watanake, *in* "Fermentation Technology Today" (G. Terui, ed.), p. 375. Society of Fermentation Technology, Japan, 1972.

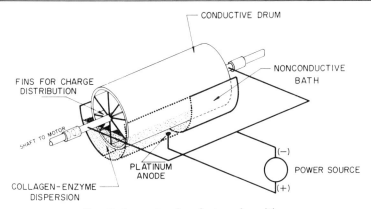

FIG. 2. Apparatus for electrocodeposition.

0.4% (w/v), preferably using deionized water. The enzyme to be coupled is also added to the dispersion at a concentration of 2–4 mg/ml and the pH of the mixture is adjusted to 3.0 by dropwise addition of lactic acid. The mixture is then thoroughly stirred for 1 hr and introduced into a nonconductive Lucite electrodeposition vessel, provided with an anode and a cathode. A stainless steel plate of appropriate dimensions is used as the cathode and a platinum plate or disc is used as the anode. The two electrodes are separated by a distance of 2 cm by means of a nonconducting rod. A dc power source is used to supply a potential gradient of 25–50 V/cm. A cam arrangement is employed to alternately immerse the electrodes in and raise them out of the mixture at 10 cycles per minute to provide layer-by-layer deposition and drying of the membrane on the cathode surface. It also helps to prevent overheating of the electrolytic solution. This procedure is continued for 45–90 min, and the cathode is left to dry completely at room temperature. The dried membrane can be readily peeled off the electrode and used without further processing. Electrocodeposition can also be carried out to deposit collagen–enzyme membranes on preformed configurations, such as a helix. In this case, a stainless steel vessel is used directly as the anode, and a stainless steel helix serves as the cathode. Further details can be found in other publications.[6,7] In some cases, substantial purification of the enzyme results during electrocodeposition, chiefly owing to preferential protein–protein interactions between collagen and the enzyme.

For large-scale, continuous electrocodeposition of enzyme–collagen membranes, the apparatus shown in Fig. 2 is employed. A Lucite half-

[6] A. Constantinides, W. R. Vieth, and P. M. Fernandes, *Mol. Cell. Biochem.* 1, 127 (1973).
[7] K. Venkatasubramanian, W. R. Vieth, and S. S. Wang, *J. Ferment. Technol.* 50, 600 (1972).

cylinder is used as the deposition vessel. A platinum sheet is fitted to the inside of the half-cylinder, and wired as the anode. The cathode is a revolving, stainless steel motor-driven drum. The revolving drum is mounted on a conductive axle shaft, insulated at its end from an electric motor which powers the rotation in such a way that a constant spacing between the electrodes is achieved. A plurality of conductive fins radiating from the axle to the drum outer periphery ensures uniform charge distribution over the drum surface. For large-scale operation, the deposited membrane may be continuously stripped off the drum and layered onto a supporting base film, such as Mylar, for drying and further treatment.

Immobilization of Whole Microbial Cells

The basic technique for immobilizing whole cells is essentially the same as described above under the macromolecular complexation method. The enzyme is replaced by whole cells in the dispersion. The ratio of collagen to cells (dry weight) is usually between 1.0 and 5.0. A specific example is described below for the attachment of microbial cells of *Streptomyces venezuelae* (glucose isomerase activity) to collagen.

Procedure I. *Streptomyces venezuelae* cells are grown under conditions that induce good isomerase activity. The cells are separated from the fermentation broth by centrifugation and washed twice with distilled water. The washed cells are then resuspended in distilled water, and the enzyme is fixed interiorly by heating the aqueous suspension of cells at 80° for 90 min with shaking. This is followed by the addition of cooled cells to a homogenized collagen dispersion. After initial flocculation, the pH of the aggregated mixture is changed slowly to 11.0, using 1 N sodium hydroxide. The well-dispersed mixture is then cast on a Mylar sheet and allowed to dry at room temperature. The dried membrane, 2–10 mils thick, is tanned with 10% (w/v) glutaraldehyde solution at pH 7.5 for 0.5–5 min and washed with running cold water for 1 hr to remove any unreacted glutaraldehyde.

Procedure II. A one-step procedure, mentioned earlier, can also be used to prepare a cell–collagen membrane at pH 8.5. At this pH, the loss in activity due to exposure to unfavorable pH is significantly reduced. Fourteen grams of wet collagen are mixed with 4 g of dry cells in 90 ml of distilled water. While the mixture is being blended, 1 ml of 50% glutaraldehyde (Eastman Kodak, Rochester, New York) and enough 1 N sodium hydroxide are added to the mixture dropwise to change its pH to 8.5. The dispersion is cast to form a membrane and allowed to dry at room temperature. A total of 8 g of dry product is obtained in which the cells constitute 50% of the dry weight.

Assay and Use of Collagen–Enzyme Membranes

Batch Reactor

The catalytic activity of collagen–enzyme membranes can be assessed in one of several reactor types. The features of different reactor types have been reviewed elsewhere (see this volume [49]).[8] Perhaps the simplest way to assay the membrane is to chip it up and use the chips in a stirred batch reactor to contact the substrate. Untanned collagen chips, when wet, have a tendency to overlap and adhere together, resulting in poor contact efficiency. By employing sufficiently high agitation rates, the chips can be retained in suspension individually. Such high shear rates also reduce the bulk phase diffusional resistance. However, it must be borne in mind that too high shear rates might denature the enzyme. Small-scale batch contacts can also be accomplished by mounting the membrane on an aluminum frame and contacting the frame with substrate solution.[7]

Continuous Biocatalytic Reactor Module

Spirally wound multipore biocatalytic modules have been shown to provide an excellent reactor configuration for the use of collagen–enzyme membrane systems.[7] The spiral-wound reactor offers a number of advantages. It guides the fluid flow through a number of organized flow channels which offer low and essentially equal hydraulic resistances. Therefore, all the fluid elements spend the same residence time in contact with the membrane unit thus ensuring good contact efficiency. The small fluid channels provide a large membrane surface area per unit reactor volume and high throughput. The mesh spacer acts as a local turbulence promoter and improves the radial mixing characteristics in the flow compartments.

Procedure. The constructional details of the module are shown in Fig. 3. The spiral-reactor geometry is formed by coiling alternate layers of the enzyme-membrane and a backing material around a central spacer element. An inert polymeric netting, Vexar, is used as the backing material. It segregates the successive layers of the membrane, thus preventing their overlapping. The coiled cartridge is fitted into an outer shell, which is affixed to two end plates provided with an inlet and outlet for substrate flow over the membrane surface. A distributor plate on the upstream side routes the flow properly to achieve a uniform axial flow distribution. This design can readily be modified to permit forced permeation of the

---

[8] W. R. Vieth and K. Venkatasubramanian, *Chemtech.* 4, 434 (1974).

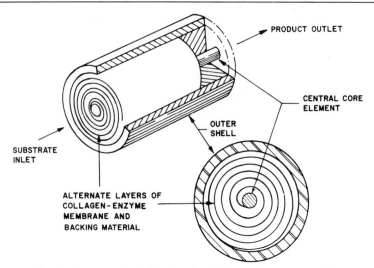

FIG. 3. Constructional details of spiral-wound reactor module.

TABLE I
CONSTRUCTIONAL DETAILS OF COLLAGEN-INVERTASE REACTOR

Material of construction (reactor shell)	Plexiglas
Inner diameter	$1\frac{3}{4}$ inches
Outer diameter	$1\frac{7}{8}$ inches
Overall length	2 feet, 6 inches
Total volume	1.18 liters
Void volume	630 ml
Total membrane surface	4700 square inches
Amount of catalyst	100 g
Membrane thickness	4 mils
Backing material	Vexar (E. I. duPont Company, Rochester, New York)

substrate under a hydraulic pressure gradient.[9] This offers as well a means of reducing intramembrane diffusional impedance. Typical reactor construction data are shown in Table I.

Chips of collagen–enzyme membrane can also be used to construct a packed bed reactor. An example of this reactor type is employed in the process for isomerizing glucose using immobilized whole cells.[10]

[9] K. Venkatasubramanian and W. R. Vieth, *Biotechnol. Bioeng.* **15**, 583 (1973).
[10] R. Saini and W. R. Vieth, *J. Appl. Chem. Biotechnol.* **25**, 115 (1975).

Preshaped forms, such as a helix, prepared by the electrocodeposition process, can be nested into an outer shell to obtain a flow-through reactor. Several helices can be arranged concentrically to improve the packing density of the reactor.[6]

Recycle Reactor

In laboratory reactors, necessary substrate flow rates over the catalyst surface for measurable conversion in single pass operation are often too low, resulting in a poor mass transfer rate from the fluid bulk to the catalyst surface. Therefore, the measured kinetic parameters may be obscured by the diffusional resistance. However, by operating the reactor at high flow rates in a batch recycle mode, this difficulty can be eliminated. The advantages of this type of reactor operation have been discussed in detail elsewhere.[11]

Procedure. The substrate is continuously recirculated from a substrate reservoir through the reactor by means of a suitable pump. The reservoir vessel should be provided with temperature and agitation control. The reactor is maintained at the desired reaction temperature by either immersing it in a constant temperature bath or by jacketing the reactor shell. The concentration change of the substrate and/or product in the reservoir is followed with time by means of suitable assay procedures. In a recycle reactor, the substrate spends only a fraction of the total time of the experiment in the reactor. Therefore the observed reaction rates, based on the concentration history in the reservoir, must be converted to true reaction rates by means of a correction factor, which accounts for the residence time distribution of the substrate between the reactor and the reservoir. The value of this factor is directly dependent on the degree of mixing prevailing in the recycle loop, which, in turn, is a function of the ratio of the volumes of reactor and reservoir, the recirculation rate, the amount of enzyme–membrane complex present in the reactor, and the kinetic parameters of the enzyme–catalyzed reaction. Detailed mathematical analysis has revealed that most collagen–enzyme recycle reactor systems behave like perfectly mixed system.[12] [This condition can be achieved by ensuring that the following conditions are met: (1) a reservoir:reactor volume ratio of 15 or more, (2) a high circulation rate, and (3) a small amount of enzyme in the reactor, such that the extent of substrate conversion in a single pass through the reactor is less than

---

[11] J. R. Ford, A. H. Lambert, W. Cohen, and R. P. Chambers, *Biotechnol. Bioeng. Symp.* **3**, 267 (1972).

[12] K. Venkatasubramanian and R. Shyam. Unpublished results, 1974.

0.1%.] Under these conditions, the true reaction rate is given by[6]

$$r_t = r_{obs}(\tau_{Res} + \tau_R)/\tau_R \qquad (1)$$

$$\tau_{Res} = V_{Res}/Q; \qquad \tau_R = V_R/Q \qquad (2)$$

where r_t = true reaction rate; r_{obs} = observed reaction rate; τ_{Res} = residence time in the reservoir; τ_R = residence time in the reactor; V_{Res} = volume of reservoir; V_R = fluid volume of reactor; Q = recirculation rate.

Properties of Collagen–Enzyme Membranes

Activity and Membrane Loading Capacity

When in contact with its substrate, a collagen–enzyme complex membrane may lose some of the activity initially, before reaching a stable limit of activity. This loss is often attributable to the desorption of the weakly bound fraction of enzyme residing in the structure. However, once the stable limit is reached, the activity level remains constant thereafter for a very long period of time. In untanned membranes, stable limits of activity representing 20–35% of the complex's initial (overloaded) activity have been observed.[13,14] These correspond to the saturation limits of the sorption isotherms for the given enzyme–collagen systems. The use of a cross-linking agent such as glutaraldehyde increases the stable loading capacity of the membrane; 60–70% of the initial activity is retained in these cases.[15] By loading capacity is meant herein the expressed enzymic activity of the bound enzyme per unit weight of the carrier–enzyme complex (e.g., in International Units per gram of collagen–enzyme membrane). It may be noted that this is the most useful parameter for design purposes. Table II summarizes the loading factors of several collagen–enzyme membranes.

Amount of Enzyme Bound to Collagen

In addition to the loading capacity, it is necessary to find the amount of enzyme actually conjugated to the carrier membrane, in order to ascertain the extent of enzyme inactivation during the process of immobilization. In addition, the apparent specific activity of the bound

---

[13] F. R. Bernath and W. R. Vieth, in "Immobilized Enzymes in Food and Microbial Processes" (A. C. Olson, and C. L. Cooney, eds.), p. 157. Plenum, New York, 1974.
[14] S. S. Wang and W. R. Vieth, *Biotechnol. Bioeng.* 15, 93 (1973).
[15] K. Venkatasubramanian, W. R. Vieth, and F. R. Bernath, in "Enzyme Engineering" (E. K. Pye, and L. B. Wingard, Jr., eds.), p. 439. Plenum, New York, 1974.

TABLE II
"Loading Capacities" of Several Collagen–Enzyme Membranes

Enzyme	Specific activity of free enzyme (U/mg protein)	Expressed activity or "loading capacity" of collagen–enzyme complex (U/g complex)
Catalase	3000	7140
β-Galactosidase	50	3500
Lysozyme	8500[a]	2500[a]
Papain[b]	12.7	600
Glucose oxidase	118	475
Invertase	100	450
Asparginase	300	200
Tannase[c]	4	70
Urease	10	250

[a] Activity expressed in lysozyme units; 1 unit of lysozyme activity is defined as the amount of enzyme that causes 0.001 unit decrease in absorbance at 25°, pH 7.0.
[b] Assayed with benzoyl-l-arginine ethyl ester as substrate.
[c] Activity expressed in tannase units; 1 unit is the amount of enzyme that can hydrolyze 1 μmole of ester bonds per minute.

enzyme (e.g., in IU/mg) can also be evaluated from this. The amount of enzyme conjugated to collagen cannot be determined by common methods of protein analysis owing to the interference of the proteinaceous carrier material. Therefore, a method that would distinguish between the enzyme protein and the carrier protein must be resorted to. This is accomplished by taking advantage of the fact that most enzymes contain tryptophan and cysteine, while collagen does not. We have employed successfully two different methods for tryptophan analysis and a method for cysteine determination. Of these, a method based on ninhydrin reagent for assessing tryptophan content has been published in some detail elsewhere.[16] The other methods are described below.

Determination of Bound Enzyme by Tryptophan Analysis

The method of Blackburn[17] has been modified.

Reagents

Sodium hydroxide, 1 N
Sulfuric acid, 21.4 N

[16] A. Eskamani, T. Chase, J. Fruendenberger, and S. G. Gilbert, *Anal. Biochem.* **57**, 421 (1974).
[17] H. Blackburn, "Amino Acid Determination." Dekker, New York, 1968.

p-Dimethylaminobenzaldehyde
Sodium nitrite, 0.04% (freshly prepared)
Collagen–enzyme membrane, 20–30 mg (dry weight), finely chipped

Procedure. The membrane sample is dissolved in 1 ml of sodium hydroxide. To this, 9 ml of sulfuric acid and 30 mg of p-dimethylaminobenzaldehyde is slowly added and mixed, while cooling in an ice bath. The mixture is stored in the dark for 1 hr. Sodium nitrite, 0.1 ml, is then added to develop a blue color. After 30 min, the absorbance of the mixture is read at 600 nm using a Beckman-DBG spectrophotometer. The absorbance readings are obtained against a blank prepared in the same manner as above, but omitting the p-dimethylaminobenzaldehyde reagent. A calibration curve is prepared by using different concentrations of (soluble) enzyme. Blank collagen membranes containing no enzyme do not give any detectable background absorbance.

NOTE. In the case of tanned membranes, the membrane is incubated in 9 ml of sulfuric acid in the dark until the membrane is dissolved completely. This step must be done prior to adding NaOH and aldehyde reagent. With relatively high degrees of cross-linking, blank collagen membranes (without the enzymes) might give a small absorbance value, which should be subtracted from the observed absorbance readings.

Determination of Bound Enzyme by Cysteine Analysis

Cysteine analysis is based on a modification of the method of Roston.[18] In this method, cysteine reacts with noradrenochrome in an equimolar basis to produce a stable, yellow compound. It is very specific for cysteine; other sulfhydryl compounds are insensitive.

Reagents

Norepinephrine
Ascorbic acid, 7.6 mM
Potassium ferricyanide, 7.6 mM
Sodium bisulfite, 1.9 mM
Sodium phosphate buffer, 0.1 M, pH 7.5

Procedure. Noradrenochrome is formed by the addition of 1 ml of potassium ferricyanide to 2 ml (400 mg) of norepinephrine in a solution buffered by 5 ml of sodium phosphate buffer and 15 ml of water. The formation of noradrenochrome is complete within 2 min as observed by the appearance of a yellowish-pink color. The yellow color of excess potassium ferricyanide is eliminated by the addition of 1 ml of ascorbic

[18] S. Roston, *Anal. Biochem.* **6**, 486 (1963).

acid. The pink color of noradrenochrome is not affected. Five milliliters of the sulfhydryl compound ($\sim 10^{-8}$ M) are added, and the stopwatch is restarted. Formation of a yellow color (due to reaction of cysteine with noradrenochrome) increases to a maximum intensity within 7 min and remains stable for an additional 8 min. Sodium bisulfite is added during the period of color stability 10 min after the addition of the sulfhydryl compound. The absorbance of the mixture is measured at 415 nm in a colorimeter between 0.5 min and 1 min after the addition of sodium bisulfate. The absorbance increases linearly quite rapidly with increasing cysteine concentration and then levels off around 3×10^{-8} M.

NOTES. (1) The stability of the norepinephrine reagent is a critical factor in the test. Experience has shown that the best method is to weigh out 2 mg of the reagent into each of several stoppered test tubes and then to store them in a refrigerator. Just prior to the test, the sample should be dissolved in 10 ml of water, and 2 ml of it be used for the analysis. A fresh sample is necessary for each test.

(2) Two different yellow colors are present at the beginning and end of the test. After the initial 2 min, the yellow color is due to excess potassium ferricyanide. Care should be taken to ensure that this color is completely removed by the ascorbic acid before the test sample is introduced. It is probably better to conduct the entire test in a 50-ml Erlenmeyer flask equipped with a magnetic stirring bar to aid in the mixing of the reagents.

(3) The test is very sensitive to changes in pH of the solution.

(4) When testing for cysteine in immobilized enzyme-collagen membrane, the enzyme and support must be hydrolyzed to their constituent amino acids before the assay can be done. This can be achieved by the addition of 10 M sodium hydroxide to a sample of fine chips of membrane. After incubation for about 96 hr at 4°, a clear solution is obtained. The pH of the sample is then brought down to 7.5 by the dropwise addition of 10 M HCl. Five milliliters of the resulting solution is then used for the analysis.

Effect of Tanning on Enzymic Activity

The bifunctional reagent glutaraldehyde is, relatively, very reactive. As the intensity of cross-linking is increased, structurally superior membranes are obtained with a concomitant decrease in catalytic activity resulting from denaturation of a portion of the enzyme. Therefore, the tanning treatment (i.e., aldehyde concentration, pH, reaction temperature, and contact time) must be carefully adjusted for each system to obtain a membrane of good mechanical strength and high enzymic activity. Other publications from our laboratory have addressed themselves to this question in considerable detail.[2,19,20]

Stability and Reusability

Storage Stability. Collagen–enzyme and collagen–whole cell complexes retain their activity for more than a year when stored in the cold under dry conditions. Even at room temperature, the membranes are stable for a number of months.[21]

Operational Stability. This is perhaps the most important engineering parameter in the design of an immobilized enzyme reactor. Operational stability refers to the life of the conjugated enzyme under conditions of its actual use. It is often expressed in terms of catalyst half-life, the time required for 50% of the enzymic activity to be lost. Collagen–enzyme systems possess very high half-life values (50–300 days, depending on the particular enzyme). The operational stabilities of several collagen–enzyme membranes have been listed elsewhere.[21]

pH and Temperature Stabilities. In many instances, thermal and pH stabilities of collagen–enzyme membranes are found to be at least as good as the free enzyme.[13] A slightly decreased thermal stability has been observed in one case (bound hesperinidase).[22] However, at temperatures below 80°, the same enzyme has far greater storage and operational stabilities compared to its soluble counterpart. The same preparation also exhibits a remarkable pH stability. It is worth mentioning here that, even in the dispersed state, collagen has a significant protective effect on the stability of enzymes.[5,22] Collagen dispersion containing the enzyme has far greater storage stability compared to the stabilities of enzyme solutions in water, buffer, or substrate.

Microbial Attack. Contrary to what is sometimes advocated as a drawback of collagen as a carrier, viz., susceptibility to microbial attack, we have very seldom observed the degradation of the matrix under actual conditions of use or in storage. The membrane structure is retained very well, particularly when it is toughened by a mild tanning treatment. It must be pointed out that the extent of microbial growth in a reaction system is often a function of the nature of substrate itself and thus must be confronted, irrespective of whether a soluble or an immobilized enzyme is employed.[21] An intensive investigation at our laboratory

[19] D. Strumeyer, A. Constantinides, and J. Fruendenberger, *J. Food Sci.* **39**, 492 (1974).
[20] M. F. P. Wang, S. S. Wang, and F. R. Bernath, *J. Fermentn. Technol.* (in press).
[21] W. R. Vieth and K. Venkatasubramanian, *Chemtech.* **4**, 309 (1974).
[22] S. Balaji, M.S. Thesis, Rutgers University, New Brunswick, New Jersey, 1975.

has shown that for practical applications involving microbially contaminated substrates, collagen–enzyme reactors could be sanitized very easily with low dosages of reagents such as hydrogen peroxide, and quaternary ammonium salts, which are commonly used in industrial practice.[23]

pH and Temperature-Activity Profiles

For almost all the collagen–enzyme systems we have studied, the pH optimum values for enzymic activity are the same in both free and fixed forms.[13] In some instances, maximal catalytic activity of the membrane-bound enzyme is realized over a broader range of pH values; i.e., the pH-activity profile is flattened around the optimal pH value. These results indicate that the microenvironmental pH of the collagen matrix is not significantly different from the bulk pH. However, the chemical composition of collagen microstructure lends itself to incorporation of a desired environmental effect. For example, by blocking amino or carboxyl groups or by introducing additional charged groups to the carrier, the surface charge could be altered; this, in turn, could modify the local pH. A highly negatively charged membrane would attract hydrogen ions, maintain a lower environmental pH than the solution bulk, and thus stabilize the enzyme in a highly alkaline substrate stream and vice versa.

Enzymes attached to collagen and their soluble counterparts exhibit the same temperature optima in most instances. However, the conjugated enzymes have lower temperature coefficients of deactivation at higher temperatures than the soluble enzymes. For example, in the temperature range of 40°–45°, the absolute values of the temperature coefficients for soluble and immobilized glucose oxidase are 9.9 and 3.3 kcal/g·mole, respectively (at lower temperatures, the values are almost the same).[6] This clearly indicates that the enzyme is more stable in the conjugated form. Activation energy values, estimated from Arrhenius-type plots of temperature-activity data, are about the same for soluble and immobilized forms of the enzyme.[7]

Brief Consideration of Mass Transfer and Kinetic Behavior

Kinetic data obtained from several collagen–enzyme reactors, tested in batch recycle systems, have been found to obey pseudo first-order kinetics, and apparent kinetic parameters; i.e., Michaelis-Menten constants, K_m', and maximal reaction velocities, V_m', have been evaluated.[21] The kinetic constants for collagen-immobilized enzymes are highly de-

[23] R. L. Barndt, J. G. Leeder, J. R. Giacin, and D. H. Kleyn, *J. Food Sci.* **40**, 291 (1975).

TABLE III
APPARENT MICHAELIS CONSTANTS FOR COLLAGEN-IMMOBILIZED ENZYMES

Enzyme	Immobilization technique	Apparent Michaelis constant (K'_m)	
		Soluble form	Immobilized form
β-Galactosidase	Impregnation	77 mM	100 mM
Glucose oxidase	Electrocodeposition	22 mM	72 mM
l-Asparaginase	Direct complexation	0.01 mM	0.833 mM
Urease	Impregnation	4 mM	12.1 mM
Invertase	Direct complexation	160 mM	136 mM
Lysozyme	Impregnation	120 mg/l	6000 mg/l

pendent on the extent of external and internal mass transfer resistances. As shown in Table III, apparent Michaelis constant values for collagen-immobilized enzymes are higher than those for soluble enzymes. These parameters actually have little value other than providing some gross indication of the existence of transport impedances.

For the efficient design of industrial-scale enzyme reactors, it is necessary to consider the combined diffusion and reaction steps. Such an analysis should describe the effects of reactor configuration, reactor size, carrier packing density, size of catalyst particles, etc., on the extent of substrate conversion and the reactor throughput. We have developed several mathematical models for quantitating the mass transfer-kinetic behavior of collagen–enzyme systems. Models for (a) simplified first-order approximations,[24] (b) Michaelis–Menten kinetics,[25] (c) steady and unsteady state analyses with detailed mechanistic reaction steps,[26] and (d) multienzyme systems[27] have been reported in the literature. All the aspects of the design and analysis of enzyme reactors have been reviewed in depth recently.[28] Based on these models, mass transfer correlations have been developed for quantitating the effects of external diffusion in collagen–enzyme reactors.[8,24] Intramembrane diffusional aspects are described in terms of effectiveness factors.

[24] W. R. Vieth, A. K. Mendiratta, A. O. Mogensen, R. Saini, and K. Venkatasubramanian, *Chem. Eng. Sci.* **28**, 1013 (1973).
[25] B. Davidson, W. R. Vieth, S. Zwiebel, and R. B. Gilmore, in "Water—1974" (G. F. Bennett, ed.), p. 182. American Institute of Chemical Engineers, New York, 1974.
[26] R. Shyam, B. Davidson, and W. R. Vieth, *Chem. Eng. Sci.* **30**, 669 (1975).
[27] P. M. Fernandes, A. Constantinides, W. R. Vieth, and K. Venkatasubramanian, *Chemtech.* **5**, 438 (1975).
[28] W. R. Vieth, K. Venkatasubramanian, A. Constantinides, and B. Davidson, in "Advances in Applied Biochemistry and Bioengineering" (L. B. Wingard, L. Goldstein, and E. Katchalski, eds.). Academic Press, New York, in press.

Procedure for the Estimation of True Kinetic Parameters. Parameter estimation techniques employing statistical analysis that can be used to obtain the true kinetic parameters (i.e., without diffusional disguises) of collagen–enzyme membranes have been outlined in detail elsewhere.[25,28] The true parameters can also be estimated much more easily by proper experimental design. Since internal diffusional resistance is dependent on the thickness of the membrane, its limitation can be relieved by reducing the membrane thickness sufficiently. Intrinsic maximum reaction velocity (V_m) of the bound enzyme can be determined by plotting the overall reaction rate as a function of decreasing membrane thickness and extrapolating the profile to the limiting case of zero membrane thickness. The true K_m value is best approached by employing very thin membranes in a flow system, utilizing a flow rate high enough to eliminate bulk mass transfer resistance. Since this is sometimes difficult, a batch system may be used instead. In this case, a sufficiently high agitation rate is employed which reduces the bulk transport resistance to an insignificant level.

Effect of Metal Ions

The effects of metal ions on the catalytic activity of the bound enzyme have been explored briefly.[29] Collagen-immobilized lactase shows inhibited activity when acid whey is used as the substrate, but not in aqueous lactose solutions. This inhibition is caused by metal ions, such as magnesium and the protein constituent of whey. Magnesium ions were found to inhibit the activity of an untanned membrane, but not a tanned one.

Binding Mechanism

Immobilization of enzymes on collagen is accomplished through the formation of a network of noncovalent bonds between collagen and the enzyme. Such bonds are salt linkages (ionic interactions), hydrogen bonds, and van der Waals' interactions. The complexation (i.e., network forming) process involves the formation of multiple, cooperative noncovalent bonds acting in concert to establish a stable network. Evidences to corroborate this postulated binding mechanism have been obtained.[30]

The amount of enzyme bound to the matrix is highly dependent on the pH of the impregnation process; maximal enzyme linkage is achieved at a pH value bracketed by the isoelectric points of the carrier-protein and

---

[29] J. Jakubowski, J. R. Giacin, D. H. Kleyn, S. G. Gilbert, and J. G. Leeder, *J. Food Sci.* **40**, 467 (1975).
[30] K. Venkatasubramanian, R. Saini, and W. R. Vieth, *J. Ferment. Technol.* **52**, 268 (1974).

the enzyme-protein. This implies that the complexation process involves ionic interactions between the charged group of the two protein moieties, the ionization states of which are pH dependent. The number of salt linkages formed can also be expected to be dependent on the ionic strength of the impregnation bath. As the bath ionic strength is increased, the amount of enzyme immobilized decreases, since the smaller microions preferentially interact with the available binding sites on collagen. In the presence of compounds such as sucrose and sorbitol, which have a great propensity to form hydrogen bonds, the amount of enzyme coupled decreases, alluding to the role of hydrogen bonds in the complexation process.

The mechanism of enzyme complexation is most probably the same for membranes prepared by the three different techniques (see section on preparation methods for collagen–enzyme complexes). They all appear to exhibit the same catalytic activity, kinetic behavior, and sorption isotherms. Sorption data for different enzymes attached to collagen obey a common, Langmuirian-type isotherm. This demonstrates that there are a finite number of binding sites on collagen and that different enzymes are bound to the matrix by a general mechanism involving protein–protein interactions.

That the complexation is different from other methods of enzyme immobilization, such as entrapment, adsorption, and covalent binding, has also been established. Unlike adsorption, which involves just surface interactions, complexation involves bulk-phase interactions. This would also explain the equivalency of the three different means of enzyme immobilization. When subjected to washing by solutions of high electrolyte concentrations, enzymes immobilized by simple physical adsorption leach out from the matrix. However, complexed enzymes reach a limit beyond which there is no further desorption of the attached enzyme, meaning that a stable, complexed, network state of matter has been formed. During this process, it is conceivable that the network is further stabilized by quasi-reversible covalent bonds of the type encountered in the relatively slow, self-cross-linking of collagen structures.[31] Furthermore, the complexation process leads to a reduction in the average molecular weight between elastically active linkages by about 35% (relative to a control membrane without any enzyme), showing that a supranetwork is formed. However, when the same enzyme is linked to collagen by a procedure that leads to simple adsorption, no significant change in the average molecular weight between the elastically active linkages is observed.

If the enzyme were physically entrapped within the matrix, then reswelling of a membrane, after the immobilization process, should result in

[31] M. L. Tanzer, *Science* **180**, 561 (1973).

the desorption of the trapped enzyme. However, no reduction in the amount of bound enzyme is seen even at very high degrees of swelling, thus ruling out a physical entrapment mechanism. Data to differentiate between covalent binding and complexation have also been published.[2]

[20] Chemically Aggregated Enzymes

By GEORGES B. BROUN

Enzyme immobilization can be obtained by chemical aggregation with bifunctional or multifunctional agents. Symmetrical and nonsymmetrical agents were assayed either for the aggregation of active proteic molecules or for the binding of these molecules onto specific functions of an adequate matrix.

When acting on a homogeneous solution of proteins, symmetrical bifunctional agents can either react with identical functions of the same molecule, resulting in an intramolecular cross-link, or with those in different molecules, resulting in a cross-linked aggregate of molecules.

Survey of Some Published Methods of Enzyme Aggregation

Many polyfunctional agents have been examined for the binding of an enzyme to a matrix. Very few were examined for enzyme aggregation.

In the literature on cross-linking of proteins,[1] especially in the early literature,[2] several homo- and heterobifunctional agents have been described. Some of the most frequently described agents are p,p'-difluoro-m,m'-dinitrophenylsulfone, bisdiazobenzidine and its derivatives, hexamethylene diisocyanate, N,N'-(1,2-phenylene)bismaleimide, phenol-2,4-disulfonyl chloride, p-nitrophenyl chloroacetate, and hexamethylene bisiodoacetamide. Multifunctional agents, such as chlorotriazines, were also used. Heterobi- and multifunctional agents were used more for the binding of enzyme to an insoluble carrier than for chemical aggregation. As for the homobifunctional agents, diazo compounds react with lysyl, histidyl, tyrosyl, arginyl, or cysteinyl residues of proteins; isocyanates with primary amino groups; alkyl iodides with nucleophilic groups; iodoacetamides with cysteinyl residues; and carbonyl-containing compounds with lysyl residues.

Glutaraldehyde was first applied to the cross-linking of enzyme crys-

[1] O. R. Zaborsky, "Immobilized Enzymes." Chem. Rubber Publ. Co., Cleveland, Ohio, 1973.
[2] I. H. Silman and E. Katchalski, *Annu. Rev. Biochem.* **35**, 872 (1966).

tals by Quiocho and Richards[3] and later by Habeeb and Hiramoto to soluble proteins.[4] This agent was used extensively by Avrameas for the cross-linking of enzymes and other proteins.[5] See also [38], [48], and [56] in this volume.

Among the bifunctional agents tested by the author's group,[6] this dialdehyde appeared to give the most satisfactory and reproducible results. It is known to react under mild conditions with free amino groups of proteins. For this reason, all examples described in this chapter will concern the use of glutaraldehyde, although several other bifunctional agents can be used for specific applications.

According to Richards and Knowles,[7] glutaraldehyde reacts with amino groups of lysine through double bonds of its oligomer. This hypothesis explains (1) the stability of the bond, which cannot be due to a simple Schiff-base formation, and (2) the lower reactivity on proteins of solutions of freshly distilled glutaraldehyde. Nevertheless, this hypothesis gives no information on the nature of the links either between the glutaraldehyde molecules or between this dialdehyde and the proteic amino groups. This ignorance of its mechanism of reaction is a limitation in the use of glutaraldehyde. But the technical ease and versatility of its application explains its use by many authors, especially those lacking facilities and expertise in organic chemistry.

When applied to solutions of low proteic concentration, glutaraldehyde gives rise initially to water-soluble oligomers. When applied to more concentrated solutions, aggregation rapidly gives rise to high molecular weight, water-insoluble polymers. The nature of the protein also has an influence on its insolubilization by glutaraldehyde. Lysine-rich proteins are readily insolubilized whereas the aggregation of proteins near their isoelectric point favors insolubilization. The pH of the medium is a significant factor when lysine-poor proteins at relatively low concentrations are concerned. In order to insolubilize low concentrations of an active protein, such as an enzyme, it is often necessary to add inert lysine-rich protein to the enzyme solution. When glutaraldehyde is added, cross-linking between the enzyme and the inert protein is observed. If the concentration of the auxiliary protein in high enough, it can be considered as the insoluble supporting matrix of the enzyme. Often bovine serum albumin was chosen as the auxiliary soluble protein to the enzyme solution

[3] F. A. Quiocho and F. M. Richards, *Biochemistry* **1966**, 4062 (1966).
[4] A. F. S. A. Habeeb and R. Hiramoto, *Biochem. Biophys. Acta* **126**, 16 (1968).
[5] S. Avrameas and T. Ternynck, *Immunochemistry* **6**, 53 (1969).
[6] G. Broun, D. Thomas, G. Gellf, D. Domurado, A. M. Berjonneau, and C. Guillon, *Biotechnol. Bioeng.* **15**, 359 (1973).
[7] F. M. Richards and C. O. Knowles, *J. Mol. Biol.* **37**, 231 (1968).

because of its lysyl residue content. When employed at concentrations higher than 50 mg/ml, it is easily insolubilized at any pH between 5 and 7 (see example 2).

Glutaraldehyde can also be used to bind a protein to matrices possessing free amino groups. Since this application is described in other chapters, it is only mentioned here. Aggregation has also been used in order to retain proteins on the surface of a porous material. In this case, the aggregation by glutaraldehyde results in "anchoring" the proteic polymer into its matrix. The insoluble mass of cross-linked aggregate remains stuck to the support, as if covalently bound to it. With a support having few binding sites, or when dense binding of enzymes on a matrix is required, an aggregation of enzymes by glutaraldehyde on proteins already bound to the matrix is sometimes desirable. Enzymes, or supporting proteins, are bound to the matrix as a first step, and enzyme and glutaraldehyde are added in a second step.

Chemical aggregation can also be used together with inclusion in order to increase the mean size of enzyme molecules entrapped in a microcell or in a gel. Resulting aggregates permit larger pore sizes of the gel, thus reducing diffusion constraints. Conversely, microcells or matrices can be cross-linked in order to reduce pore size and avoid leaching of the entrapped protein into the solution.

Procedures

Immobilization can be obtained for many enzymes in a range of conditions. For each problem, the conditions should be chosen according to the specific result required. The following aspects should be taken into account: (1) nature of the enzyme; (2) physicochemical properties of the support; (3) density and distribution of active sites; (4) initial activity versus life-time; (5) storage and operating conditions.

In order to help the reader in the choice of an adequate technique for any specific application, *one example* is given for each set of conditions, i.e., chemical aggregation of an isolated enzyme; chemical aggregation of an enzyme with an inert proteic feeder (co-cross-linking) either to obtain a membrane or a particle; chemical aggregation in a porous matrix (Cellophane and acrylamide gel).

Example 1. Immobilization of Enzyme by
   Glutaraldehyde (Chymotrypsin)[8]

Chymotrypsin powder, 500 mg, is dissolved in 5 ml of 0.2 M acetate buffer adjusted to pH 6 (alternatively, 0.1 M phosphate buffer at pH 6.8

[8] S. Avrameas and T. Ternynck, *J. Biol. Chem.* 242, 651 (1967).

can be used). Two milliliters of a 2.5% aqueous solution of glutaraldehyde are added dropwise while gently stirring the protein solution. The best activity yields are generally obtained with concentrations ranging from 0.3 to 0.6% glutaraldehyde. The reaction mixture is left at bench temperature without stirring for 1–3 hr. A gel appears after 10–30 min, depending on the enzyme. The gel can be cut with a blender, homogenized in a Potter homogenizer, and rinsed with distilled water long enough to remove free molecules of glutaraldehyde and protein (both are checked in the rinse water in a spectrophotometer at 280 nm). Alternatively, the gel can be dispersed in 50 ml of water, or buffer, through the needle of a syringe.

The aggregated enzyme is then centrifuged (5000 g, 15 min) and resuspended several times until the supernatant no longer shows absorption at 280 nm. It is suspended and left overnight in 20 ml of 0.1 M lysine or ethanolamine in buffer, then rinsed again as described above. The gel should be stored suspended in water in a refrigerator. In some cases, the enzyme can be frozen and lyophilized, but often freezing after cross-linking results in a denaturation of the enzyme. Before use, it has to be equilibrated for a few hours with the buffer chosen for the assay. The assay of the aggregated chymotrypsin is performed in a pH stat using N acetyl-L-tyrosine ethyl ester (ATEE) as a substrate.

Comments on Example 1

1. The reaction works also at lower temperatures ($+4°$), which are preferred for more labile molecules, but immobilization requires more time (6–18 hr).

2. The cross-linking reaction can be stopped by an inhibitor, such as tris(hydroxymethyl)aminomethane added to a final concentration of 0.7 M, or by a small molecule that reacts with glutaraldehyde (for instance, lysine or ethanolamine at a final concentration of 0.1 M). The inhibitor can be released, for instance by dialysis, and the cross-linking potential of excess glutaraldehyde be restored. On the other hand, the aminated molecule—once bound to glutaraldehyde—cannot be easily released. It can be routinely used to block the remaining cross-linking sites of glutaraldehyde. By this method, soluble oligomers can be obtained at an early stage of the reaction, but later the mechanical properties and porosity of the gel can be controlled by stopping the cross-linking at the chosen moment. Mechanical resistance increases while porosity decreases for longer periods of aggregation.

3. The speed of immobilization is controlled by (a) *protein concentration:* usually, final concentrations of 50–200 mg/ml are optimal for homogeneous immobilization of proteins; (b) *glutaraldehyde concentration:* with increasing glutaraldehyde concentration in the mixture, the yield of immobilized activity increases and then decreases, giving rise to a maxi-

mum. The optimal ratio of glutaraldehyde to protein concentration is usually in the range of 10% (w/w). Commercially available glutaraldehyde is a 25% solution in water. It has to be diluted to 2.5% before use to obtain a homogeneous insolubilization.

4. The assay of activity can be performed using the same technique as for the soluble enzyme. Each aliquot of suspension is to be centrifuged before assay. Preferentially, adapted continuous methods should be used. When possible, pH stat methods are easily adaptable to the assay of insolubilized enzymes. When photometric methods are preferred, the best device consists of continuous pumping in a loop between a spectrophotometric flow-through cuvette and the vessel containing the enzyme suspension in the reactive mixture.

Activity yields are calculated as the ratio of the determined activity of an aliquot of insolubilized enzyme to the same aliquot of initial active solution. It takes into account both the chemical yield of the immobilization and the specific activity of the immobilized enzyme. It is expressed as:

$$\rho \text{ (as percent)} = \frac{\text{overall activity of the insolubilized enzyme} \times 100}{\text{overall activity of the initial enzyme solution}}$$

Table I shows activity yields of bovine trypsin and chymotrypsin immobilized either at pH 5 or 7 (it is observed that pH spontaneously decreases in the gel during cross-linking by glutaraldehyde if the medium is not efficiently buffered).

5. The lifetime of the enzyme should be determined both under storage conditions and under operating conditions. Since the best stability performances were obtained with insolubilized enzymes when inert proteic feeders were used, the procedure for lifetime determination will be dealt with in the comments on example 2.

TABLE I
Specific Activities of Soluble and Insoluble Bovine Trypsin and Chymotrypsin Cross-Linked by Glutaraldehyde at pH 5 and 7[a]

Enzyme	pH of immobilization	Activity before immobilization (IU/mg)	Activity after immobilization (IU/mg)	Percentage of remaining activity
Trypsin	5	42	8	19
Trypsin	7	42	7.4	17.6
Chymotrypsin	5	243	138	56.8
Chymotrypsin	7	243	106	43.6

[a] According to S. Avrameas, personal communication.

Example 2. Insolubilization of Enzymes by Aggregation with an Inert Protein Added as a Feeder (Co-cross-linking)[6]

For several reasons, the direct aggregation by glutaraldehyde often gives poor activity yields: (a) excess of cross-linking agent which acts on amino groups located at (or close to) the active site; (b) the production of compact pseudocrystalline structures that multiply steric hindrances; (c) excess cross-links, which hinder the conformational adaptation of the enzyme to the substrate and result in an inactivation of the enzyme molecules.

The presence of a proteic feeder permits sufficiently high protein concentrations for immobilization, while reducing the above-mentioned constraints. Better activity yields can thus be obtained with the same enzyme, especially when aggregating the enzyme molecules on the surface of the previously cross-linked proteic feeder.

As previously mentioned, bovine serum albumin (BSA) is convenient for this use. This soluble protein is commercially available in various grades of purification and is relatively cheap. It is suitable for the immobilization of a variety of enzyme preparations, but not for proteolytic enzymes. The latter are often leached from the aggregate after a while, as a result of proteolysis of the supporting protein. Other soluble proteins can be used as feeders: for instance, hemoglobin, ovalbumin, and fribrinogen.

Two procedures are given as examples.

Procedure 1. Proteinic Gel.[8] A mixture of 100 mg of *Aspergillus niger* glucose oxidase and 400 mg of purified BSA are dissolved in 100 ml of 0.2 M acetate buffer at pH 5. The steps of the reaction are the same as in example 1.

The assay of the immobilized glucose oxidase activity in the presence of catalase is performed in a pH stat, using 0.01 M solution of sodium hydroxide in the syringe.

Procedure 2. Proteinaceus Membrane.[9] Catalase (400 IU) is added to 5 ml of a 60 mg/ml solution of BSA in 0.02 M phosphate buffer at pH 6.8. Glutaraldehyde solution, 2.5%, is added up to a final concentration of 0.7%. When the viscosity of the mixture begins to increase, the solution is spread on a perfectly flat glass plate, between the limits of a square drawn with a glass pencil. The presence of air bubbles should be avoided. After a couple of hours at 20° (or overnight in a refrigerator), a yellowish flat membrane is obtained; this is easily removed from the glass plate. It is rinsed by soaking the plate in water for 1 or 2 hr, then in a 0.1 M lysine solution, and then again in distilled water with gentle stirring. The

[9] D. Thomas, G. Broun, and E. Sélegny, *Biochimie* **54**, 229 (1972).

membrane is checked for enzyme leaching (by an enzymic test in the rinse water) and assayed for activity by the measurement of absorption of hydrogen peroxide at 240 nm. The membrane can be stored for months in a Petri dish in a refrigerator. Desiccation should be avoided. Microbial contamination is prevented by addition of sodium azide (1%) to the last rinsing solution.

Comments on Example 2

1. The addition of a feeder protein has a favorable effect on the insolubilized activity yields for most enzymes tested. It allows the variation of activities of the gel, or of the membrane, by modifying the ratios of the enzyme to the proteic feeder.

2. The distribution of the enzyme can be homogeneous in the aggregate, or alternatively an enzyme coating of previously aggregated inactive proteins can be obtained. In the latter case, the enzyme is dissolved in a few drops of buffer and added with a pipette or a syringe after a preliminary cross-linking of albumin solution, when it turns yellow and viscosity starts increasing. The solution is immediately stirred in order to disperse the enzyme molecules superficially on the preformed proteic oligomers.

3. The membrane shape is especially designed for the determination of diffusion constraints in the proteic matrix. These determinations and their interpretation are given in this volume [63].

Example 3. Porous Proteinic Particles Produced by Aggregating Proteins in the Frozen State[6]

Aggregation of enzymes with a proteic feeder in the frozen state, by glutaraldehyde, gives rise to spongelike porous structures that can be broken into porous elastic particles permitting favorable diffusivities of substrate (S) and product (P) when processed in a column (or in a packed-bed reactor).

The insolubilization of β-galactosidase (from *Escherichia coli*) as porous particles is given as an example. In 5 ml of 0.02 M phosphate buffer, pH 6.8, 5,000 IU of β-galactosidase are mixed with BSA (final concentration 60 mg/ml) and glutaraldehyde (2 mg/ml). The solution is frozen after stirring at $-30°$, then slowly warmed in a refrigerator ($+4°$). After 4 hr, a spongelike proteic copolymer is easily removed from the beaker. After rinsing with water, then with 0.1 M lysine solution, and with water again, the insoluble aggregate is lyophilized and ground. A light hydrophilic powder is obtained, which can be suspended in buffer and used either in a batch or in a column reactor.

The assay of enzymic activity is performed using o-nitrophenyl-β-

galactoside (ONPG) as a substrate. The absorption of the product is observed at 405 nm.

Comments to Example 3

1. For still undetermined reasons, the freeze-drying of such porous structures gives rise to less denaturation of the enzyme than when starting from a gel. The different conditions of crystallization of water in each case could explain this discrepancy.

2. Porous particles having various kinds of enzyme activities can be stored for months in a refrigerator without any noticeable loss of activity. Furthermore, they are far less subject to microbial contamination than are gels or membranes.

3. For practical purposes, the assays of the activity of the aggregated enzymes should be repeated under the conditions of expected utilization, in storage, and in operation. For instance, the activity should be checked daily the first week, weekly the first month, then monthly.

4. When applied to processing of industrial substrates, the suspended particles are often contaminated during operation. It is then necessary periodically to impregnate the suspension with antibiotics. The same is true when the system is stored between two periods of operation.

Table II gives some activity yields produced by the methods described in example 2 (membranes) and example 3 (porous particles).

Example 4. Cross-Linking of Enzyme Molecules inside a Porous Matrix[10]

This method was extensively used by Goldman *et al.* to immobilize papain inside a collodion sheet.[11] The enzyme molecule, which is small enough to diffuse through pores of a hydrophilic matrix, such as collodion, Cellophane, cellulose, or of a hydrophobic matrix, such as porous Teflon, silicone, or glass, is no longer able to do so when insolubilized in the pores of the matrix. Thus, the protein is "anchored" in the porous matrix without any chemical reaction between the support and the cross-linking agent.

The immobilization of a system of three sequential enzymes (β-glucosidase, glucose oxidase, and catalase) in a Cellophane sheet is given as an example: Cellophane sheets, devoid of glycerol, 30 μm thick, can be obtained from the manufacturer (Société la Cellophane, Bd Haussmann, Paris, France). Alternatively, Cellophane sheets are continuously rinsed for 7 days with water in a stirred batch to remove glycerol. Squares, 100 cm^2, are impregnated with a mixture of β-glucosidase, glucose oxidase, and catalase to give an activity ratio (in IU) of 1:1:5. The final proteic

[10] G. Broun, E. Sélegny, S. Avrameas, and D. Thomas, *Biochim. Biophys. Acta* **185**, 260 (1969).

[11] R. Goldman, I. H. Silman, S. R. Caplan, O. Kedem, and E. Katchalski, *Science* **150**, 758 (1965).

TABLE II
ACTIVITY RATIOS OF ENZYMES AGGREGATED WITH GLUTARALDEHYDE

Enzymes	Ratios (%)		
	In a porous matrix	With a proteic feeder[a] (co-cross-linking)	
1. *Oxidoreductases*			
Glucose oxidase	10 (Cellophane)	80	Alb
Urate oxidase	5 (Cellophane)	30	Alb
L-Amino acid oxidase	—	50	Alb
Xanthine oxidase	—	60	Alb
Catalase	5 (activated carbon)	80	Alb
Peroxidase	5 (Whatman 3)	60	Alb
Alcohol dehydrogenase	—	2	56* Alb
Lactate dehydrogenase	—	5	43* Alb
Glucose-6 phosphate dehydrogenase	—	5	Alb
2. *Transferases*			
Hexokinase	3 (aminated paper)	30	Alb
Ribonuclease	—	30*	Alb
Pyruvate kinase	—	15	Alb
Phosphofructokinase	—	10	Alb
3. *Isomerases*			
Glucose-6 phosphate isomerase	5 (silicon sheet)	50	Alb
Triose phosphate isomerase	—	50	Hb
4. *Lyases*			
Carbonic anhydrase	5 (silicon sheet)	—	—
Tyrosine decarboxylase	—	50	Alb
Phenylalanine decarboxylase	—	60	Alb
Arginine decarboxylase	—	50	Alb
Lysine decarboxylase	—	40	Alb
5. *Hydrolases*			
α-Amylase	2 (silk)	80*	Alb
β-Galactosidase	—	50	80* Alb
Trypsin	30 (Cellophane)	—	
Chymotrypsin	30 (Cellophane)	—	
Urease	—	60	Alb
Asparaginase	—	30*	Alb
β-Glucosidase	—	20	Alb

[a] Alb, albumin; Hb, hemoglobin. Figures with an asterisk (*) are given when aggregation in a porous spongelike structure gives higher ratios than sheets or coatings.

concentration is 6 mg/ml in water. Three milliliters of this solution are necessary to impregnate such an area of Cellophane. The sheets are stretched on a glass plate and left, until desiccated, in a refrigerator. With several impregnations, more active preparations can be obtained.

The dry sheet is then impregnated with 2.5% glutaraldehyde in 0.02 M phosphate buffer, pH 6.8. After desiccation at 4°, the sheet is rinsed with water, then 0.1 M glycine solution in 0.02 M phosphate buffer, pH 6.8, then with water again for 3–4 hr. Elution of glutaraldehyde and proteins can be monitored at 280 nm in the supernatant. Activity of catalase is assayed with hydrogen peroxide at 240 nm. Then, glucose oxidase activity is assayed in a pH stat to detect the production of gluconic acid, and finally β-glucosidase is assayed with salicin, the production of saligenin being monitored at 280 nm.

Comments to Example 4

1. The choice of the matrix for aggregate enzymes in pores depends on the expected application. The following factors should be considered. (a) The shape of the matrix is important—it may be particulate for packed bed and tank reactors, membranous for sheet reactors and diffusion-reaction modeling, or fibrous for flow-through reactors. (b) Its mechanical properties should be taken into account. (c) The porosity of the support affects the necessary time of impregnation during its preparation, and the resulting concentration gradient of the enzyme active sites inside the matrix. (d) Hydrophilic supports are penetrated better by enzyme molecules than a hydrophobic one of the same average pore size. Enzyme molecules appear to be more readily denatured by the contact with hydrophobic supports than with hydrophilic ones.

2. Because of the slow diffusion of the enzyme in the support, it is impossible, in practice, to obtain homogeneous distribution of sites in the matrix.

3. After anchoring proteins in the porous support by cross-linking, it is possible subsequently to obtain a coating with another layer of proteins. In this case, proteins are first aggregated in the matrix as described in example 4, without rinsing with glycine solution. After desiccation, the matrix is again impregnated with 2.5% glutaraldehyde and an enzyme solution, and desiccated again. Rinsing is performed as previously described.

4. It must be recognized that it can be advantageous to implement enzymes with glutaraldehyde after binding them chemically onto a matrix. This two-step procedure can result in a better stabilization, or in an increase in the number of enzymically active sites if the reactive functions present on the matrix are a limiting factor. The first layer of enzyme molecules provides binding sites for further molecules.

Example 5. Inclusion in a Gel of Soluble Aggregated Enzyme Molecules[12]

Aggregation can be performed without insolubilization, either to obtain a better operational stability of the enzyme, or to prevent any leaching

[12] B. Paillot, M. H. Remy, D. Thomas, and G. Broun, *Pathol.-Biol.* **22**, 491 (1974).

of the enzyme after inclusion. The aggregation of urease before inclusion in a polyacrylamide-agarose gel is given as an example.

As a first step, glutaraldehyde is used to aggregate enzyme molecules in solution. Aggregation is then stopped (by lysine, glycine, or ethanolamine) early enough to obtain soluble oligomers, which are included in an agarose–polyacrylamide gel according to the method described by Uriel.[13]

This method of enzyme immobilization is a compromise between inclusion and chemical insolubilization. Like all inclusion methods, it provides good homogeneity of the sites in the system, but the prior enzyme aggregation allows the use of poorly reticulated gel matrices, which are more permeable to substrates and products, and thus give rise to higher enzyme activity ratios. Nonaggregated enzyme molecules would leak out from such loosely cross-linked gels.

Step 1. Aggregation in a solution. A 1 mg/ml (60 IU/ml) urease solution in 0.02 M phosphate buffer, pH 6.8, is mixed in a test tube with an equal volume of 2.5% glutaraldehyde solution in the same buffer. Aggregation proceeds at 4° for 10 hr. The cross-linking process is then interrupted by the addition of glycine up to a concentration of 0.1 M.

Step 2. Inclusion. Polyacrylamide–agarose gel is prepared according to Uriel. In order to obtain 100 cm² of a 1.5 mm-thick gel, 1 ml of a solution of 60% acrylamide and 1.6% bisacrylamide are mixed with 5 ml of enzyme oligomer solution. The mixture is heated at 37° (solution A). A gel containing 96 mg of agarose in 6 ml of water is boiled for 5 min, then cooled to 40° with constant stirring (solution B). To solution B is added 5 ml of N,N,N',N'-tetramethylethylenediamine solution, 3.6 mg of sodium persulfate, followed immediately by the bulk of the solution A. The entire mass is spread between two glass plates. Polymerization occurs for 30 min at 4°. The gel is removed and then rinsed with distilled water until the rinse water no longer absorbs at 280 nm.

The preparation described above remains active for several weeks if kept in a refrigerator at 4°C. Its high porosity permits a sufficient diffusion of urea and of the products of the reaction. If the enzyme has not been aggregated before it is added to the acrylamide gel, it quickly loses its activity by leakage.

Comments to Example 5

1. Such preparations can be of use for implantable materials (medical, pharmaceutical, and physiological applications).

2. This example illustrates the fact that different immobilization methods described in this volume can be usefully combined in order to solve specific problems.

[13] J. Uriel, *Bull. Soc. Chim. Biol.* **48**, 969 (1966).

General Suggestions for the Optimization of Enzyme Cross-linking

The techniques mentioned above can be successfully used for the enzymes mentioned. For the methods used for insolubilization by chemical aggregation, one must take into account some common principles of protein chemistry and enzyme kinetics in order to adapt the techniques to one's specific needs.

1. *In the choice of the cross-linking agents,* chemical functions giving rise to mild reactions are to be preferred. Reactions that are to be performed at extremes of pH, or in apolar solvents, are to be excluded. Those cross-linking agents which allow a good molecular distance between the aggregated molecules are to be preferred to agents giving tight cross-links. The mildness of glutaraldehyde may be due to the formation of an oligomer that adapts itself to the feasible intermolecular distances.

2. *In the choice of the cross-linking conditions,* all conditions during the procedure that may result in a denaturation of the proteic molecule are to be avoided. Low-temperature techniques, buffering near neutral pH, hydrophilic supports and solvents, and gentle stirring are to be preferred. Shear forces, such as exposure to liquid-gas and water-lipid interfaces, or slow freezing and thawing, are to be avoided. The aggregation reactions have to be fully stopped in order to prevent unstable preparations and side reactions.

3. *Several modifications of the aggregation technique* aim to protect the enzyme molecule against denaturation by the aggregation procedure. Addition of inert proteins and cross-linking in the frozen state have been described above. The addition of substrate (lactose for β-galactosidase), or of only one of them (urate but not oxygen for urate-oxidase), or of the coenzyme (NAD for alcohol dehydrogenase) often result in an increase of the activity ratio after immobilization. In some cases, SH groups have to be protected (Hg complex of papain).

4. *After cross-linking, enzyme molecules are generally stabilized against denaturation,* thus sometimes permitting more drastic procedures, such as contact with apolar solvents or a certain degree of heating. Paradoxically, cross-linked enzymes often become more fragile to freezing-thawing cycles.

5. *Insolubilization by aggregation is favored under conditions that develop physical aggregation of proteins,* hence high concentrations of proteins, low concentrations of ionized salts, and vicinity to the isoelectric point facilitate insolubilization. With respect to glutaraldehyde, activity yields are maximal for the chosen concentrations. With low concentrations of glutaraldehyde (0.25%), no insolubilization occurs even with concentrated protein solutions. With increasing glutaraldehyde concentrations,

enzyme is partly insolubilized. The ratio of immobilized activity to total activity increases with increasing glutaraldehyde concentrations, up to a maximum. In the approximate concentration range of 0.3 to 1%, activity of the insoluble phase remains nearly constant. With higher glutaraldehyde concentrations, more rapid insolubilization and tighter material are obtained. When the concentration of the cross-linking agent increases above this optimum range, the activity of the immobilized enzyme decreases. Two factors seem to cooperate in a negative way: the aggravation of diffusion constraints and the denaturation of enzyme molecules due to the multiplication of cross-links. The optimum range varies with different enzymes and operating conditions. The above-mentioned values are those most frequently employed.

6. *Protein concentration affects both the aggregation kinetics and the texture of the aggregate.* The higher the protein concentration, the shorter the reaction delay and the more compact the membrane or the particle. Higher protein concentrations increase the diffusion constraints and cause a rapid approach to the limiting operational enzyme activity of the aggregate. Such dense protein aggregates can be used for superficial coatings by enzyme molecules of previously aggregated protein cores.

7. *Methods for the determination of enzyme activity* have to be adapted to the specificity of insolubilized enzymes (see this volume [25]). The use of current procedures in enzymology is feasible, if some specific errors are prevented. When taking an aliquot for measurement, a thorough stirring is necessary in order to avoid settling; the volume of the suspended enzyme in the substrate must sometimes be taken into account in the calculations: the activity yield should be compared with the overall activity introduced into the reaction, not with the weight of insolubilized proteins. For the determination of aggregated enzyme activity, continuous methods are to be preferred, since their use prevents several misinterpretations that arise in discontinuous measurements.

Titrimetric and polarographic methods are easily adaptable to insoluble enzyme kinetics. When photometric methods are employed, the best device uses continuous pumping in a loop between the vessel containing the enzyme in the reactive mixture and a spectrophotometric flow-through cuvette. The reactive mixture must be thermostated and continuously stirred. A filter prevents the escape of enzyme particles into the loop. If the volume of the loop and the cuvette is more than $1/20$ of the total volume, or if the temperature of operation exceeds bench temperature by more than $10°$, the whole device should be thermostated (Fig. 1).

When working in V_{max} conditions, especially when the turnover number of the enzyme is very high, diffusion into the aggregate is a limiting factor, leading to an underestimation of V_{max} of the immobilized enzyme.

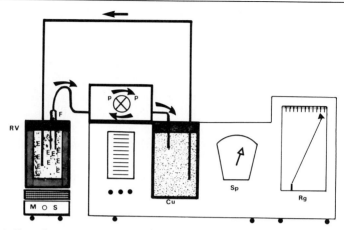

FIG. 1. In order to determine the activity of chemically aggregated enzymes, continuous procedures are preferred. A device consisting of a loop of tubing between the reaction vessel and the spectrophotometer is shown. The solution is pumped continuously through the cuvette while the immobilized enzyme is retained by a sintered-glass filter. Sp, spectrophotometer; Cu, cuvette; Rg, recorder; RV, thermostated reaction vessel; pp, peristatic pump; F, sintered-glass filter; E, immobilized enzyme; MS, magnetic stirrer.

Dispersion by mincing or grinding reduces the error due to diffusion constraint. It is advisable to test the resulting powders in order to determine the granulosity with which the error becomes negligible. The underestimation of initial activity can lead to an overestimation of the stability of the preparation. It is important to check accurately the lifetime of the enzyme preparation both on storage and in operation. Ideally, the lifetime of the immobilized enzyme itself and of the particle should be assayed separately.

8. *A partial solubilization, or an enzyme leakage, are to be checked for* in the supernatant or in the first and subsequent rinsing effluents of the aggregate. This can be done either by a rapid colorimetric test (for instance, the precipitation of the tetrazolium derivatives with phenazine methosulfate for NAD–NADP linked dehydrogenases), by the absorption test at 280 nm for proteins, or by a colorimetric test (phenol test) after precipitation or adsorption of protein.

Applications

The applications of chemically aggregated enzymes are not basically different from those of enzymes immobilized by other methods. Three examples are given to illustrate applications in (1) an affinity chromatographic separation, (2) the measurement of a specific metabolite by an enzyme electrode, and (3) the transformation of a biological molecule in a fluidized bed reactor.

Illustration 1. Use of an Insoluble Enzyme Derivative for the Isolation of Inhibitors[14]

Insoluble trypsin can be used for the isolation, by affinity, of trypsin inhibitors. The insoluble trypsin derivative is dispersed in 50 ml of 0.2 M HCl-glycine buffer, pH 2.8, stirred for 15 min at 4°, and centrifuged. The precipitate is dispersed in 50 ml of 0.1 M phosphate buffer, pH 6.8, and centrifuged once more. The gel is mixed in the centrifuge tube with an appropriate volume (usually 20 ml containing 10–20 mg of proteins per milliliter) of the inhibitor—containing extract in phosphate buffer, then stirred gently for 2 hr at 4°. Next the suspension is centrifuged for 15 min at 5,000 g at the same temperature. The precipitate is suspended in the same buffer and packed by centrifugation until the absorption of the washings at 280 nm is less than 0.020. The absorbed proteins are eluted with 0.1 N HCl-glycine buffer, pH 2.8–3.0. The eluates are neutralized with 1 M K_2HPO_4, dialyzed overnight at 4° against 0.01 M phosphate buffer, pH 6.8, and tested for inhibitory activity. Depending on the source of inhibitor used by this procedure, purifications of the initial product ranging from 2 to 450 times can be achieved.

Illustration 2. Use of Cross-Linked Lysine Decarboxylase in the Measurement of Lysine with an Enzyme Electrode [15,16]

The application of a membrane containing a specific decarboxylase onto an electrode sensing pCO_2 allows the measurement of the concentration of one amino acid in a complex mixture. The direct determination of lysine in a biological solution is possible using a lysine decarboxylase (from *E. coli*) (for general treatment of enzyme electrodes, see this volume [41]).

The commercially available enzyme (Sigma) is first purified by affinity chromatography through a 20 × 1 cm column containing lysine bound to the surface of an albumin-glass packed bed. The packed bed is prepared in two steps.

Step 1. Bovine serum albumin is dissolved in 20 ml of 0.02 M phosphate buffer, pH 6.8, to give a final concentration of 1.3%. 3 g of glass wool are suspended in this solution. Glutaraldehyde is added up to 0.3%.

[14] S. Avrameas and B. Guilbert, *Biochimie* **53**, 603 (1971).
[15] A. M. Berjonneau, D. Thomas, and G. Broun, *Pathol.-Biol.* **22**, 497 (1974).
[16] C. Tran-Minh and G. Broun, *Anal. Chem.* **7**, 1359 (1975).

The mixture is frozen at −30° overnight. The porous spongelike product is packed in the column and rinsed in water, then impregnated with an excess of glutaraldehyde (250 ml of a 2.5% solution). The column is rinsed again with water then with a saturated (about 1.4%) solution of tyrosine buffer, pH 10. Excess tyrosine is rinsed away with water until no more absorption at 280 nm can be observed.

Step 2. A 10 mg/ml lysine decarboxylase solution in 0.02 M buffer phosphate, pH 5.8, containing 3 M KCl is passed through the column (thermostated at +4°) at a rate of 70 ml/hr. Fractions are collected and their activity is assayed.

The eluted enzyme is assayed at 25° in a thermostated (Metrohm EA 876-1) cell. CO_2 liberated by the enzyme in the presence of 0.01 M lysine in 0.2 M phosphate buffer, pH 5.8, is determined by a pCO_2 electrode Radiometer E 5036 with an adapted pH meter (Radiometer 27). From the slope of the pCO_2 curve, the decarboxylase activity is obtained.

A membrane can be produced as described in example 2 (procedure 2) by introducing 0.2 IU of lysine decarboxylase in a 0.02 M buffer phosphate, pH 6.8, containing 1.5% BSA and 0.3% glutaraldehyde. The membrane is layered on top of the pCO_2 electrode special silicone membrane. Both are applied to the electrode, with the enzyme membrane on the solution side. Determinations are performed as indicated above. The electrode is first checked with several known lysine solutions. A steady state is obtained after 7–10 minutes. An average of one lysine determination every 10 min can be obtained.

The range of concentrations over which the enzyme electrode can be used is limited by the sensitivity of the pCO_2 electrode itself, and the linearity of the response of the enzyme ($\ll K_m$). For lysine, a reproducible response could be obtained from 0.2 to 1.5 mM. The same membrane can be used for many determinations over a period of 1–2 weeks. But, for some unknown reason, the lysine electrode cannot be used for many sequential determinations in a long series; "fatigue" sets in. But after a few hours in water, the same electrode is effective again.

Illustration 3. Chemical Aggregation of Enzymes on a Magnetic Carrier for Use in a Fluidized Bed Reactor[17]

Magnetic iron–manganese oxide particles, 100–200 μm in diameter, (purchased from "La Radiotechnique," 27000 Evreux, France) are added as an inert charge to a solution containing an enzyme and BSA, as in example 3. Glucose oxidase, papain, and invertase are efficiently immobilized by this method. The procedure for the production of papain

[17] G. Gellf and J. Boudrant, *Biochim. Biophys. Acta* **334**, 467 (1974).

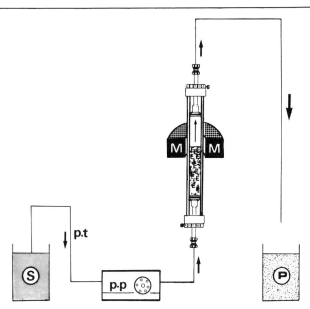

FIG. 2. A fluidized bed can be generated using enzyme chemically aggregated on magnetic particles. Substrate solution is forced through the bed. Particles are retained in the reactor by use of an external electromagnet. S, substrate reservoir; P, product; p.p, peristatic pump; E, immobilized enzyme; M, magnet; p.t, polyethylene tubing.

(EC 3.4.22.2) active particles is given as an example: 30 g of magnetic iron oxide are mixed with 25 ml of 0.02 M buffered phosphate, pH 6.8, containing 8% plasma albumin, 0.4% glutaraldehyde, and 100 mg of crystalline papain (Miles-Yeda Ltd). A thick suspension is obtained, which is frozen at −30° for at least 3 hr, then slowly warmed overnight in a refrigerator at 4°. The resulting protein copolymer is thoroughly rinsed, then freeze-dried, ground, and sieved in order to obtain particle sizes ranging from 170 to 250 μm in diameter. These active particles can be used in a fluidized-bed enzyme reactor. A flow of from 4 to 40 ml of substrate solution per minute can be pumped into a glass column, 30 × 1 cm. A turbulent flux results in the column. No enzyme leakage is observed when a circular magnet is mounted around the column in order to retain the enzyme-coated iron oxide particles (Fig. 2).

General Remarks

Chemical aggregation of enzymes by cross-linking agents can be used in quite unrelated circumstances. Their ease of operation and versatility is a good argument for the use of such techniques in laboratory tests, model-building, and some medical applications. When scaling-up, however, new limitations to these procedures appear: among these are poor mechanical properties of the support and, often, high cost.

One advantage of chemical aggregation over binding of enzymes onto a carrier is that the carrier adds to the reactor a volume that can be reduced significantly by aggregation. Over encapsulation, one advantage is the possibility of greatly reducing diffusional constraints by particle division and superficial coating. The efficiency of these methods depends on the cross-linking conditions, on the availability of active functions (e.g., NH_2 of lysine, or SH of cysteine) at the surface of the enzyme molecule, on the interference of the functions that play a part at its active site, and on its adaptability to the substrate.

Some enzymes give very poor activity yields by chemical aggregation. This seems to be the case with enzymes of higher molecular weight and with enzymes made up of subunits. This is not a general rule, and it may be due to the increased probability of molecular hindrances or to unfavorable regulation of these sophisticated molecules. Other techniques of immobilization should be preferred in such cases. Many difficulties in the use of aggregation procedures should be partly or completely overcome in the future by a more systematic choice of suitable procedures and by a deeper insight into the potential application of other bi- or multifunctional chemical agents.

At present, a good knowledge of the technical mistakes to be avoided and of parameters to be taken into account, and also a certain degree of expertise in the use of one set of cross-linking conditions, help to solve many of the problems in the immobilization of a large variety of enzymes.

[21] Photochemical Immobilization of Enzymes and Other Biochemicals

By Patrick Guire

Electromagnetic radiation offers some distinct advantages over thermal energy for the activation of chemical reactions involving biological and other chemical compounds. Compared to most thermochemical reactions (activated by kinetic energy uptake), photochemical activation is a relatively rapid (essentially instantaneous) process, which is easily and quickly initiated and terminated and can produce a highly energetic and relatively homogeneous reactive species capable of the rapid formation of covalent bonds with target molecules with relatively little dependence upon the temperature and pH of the reaction mixture or even the chemical character of the target group. The quantum character of the interaction of electromagnetic energy with molecules or groups of atoms provides an additional advantage in the design and use of bifunctional or cross-linking agents for the covalent coupling of one molecule or functional group with another. Since cross-linking requires reaction with at least two groups, probably differing appreciably in chemical reactivity and/or spatial location, much better control over the reaction process is provided to the operator through the use of a stepwise cross-linking process. Such a time-controlled stepwise reaction process is available through the use of a reagent containing both a thermally activatible group and a light activatible (photochemical) group or a reagent with two or more photochemical groups with little overlap in their photoactivation (action) spectra.

For the coupling of enzymes and most other biochemicals to each other or to relatively inert water-insoluble carrier materials, one would seek a photochemical group that is quite stable in the dark under most conditions that are compatible with the enzyme or other biochemical and stable under reaction conditions required for activating the thermochemical group. The photochemical group should be subject to activation by light of those wavelengths (e.g., visible light) that are harmless to most enzymes and other biochemicals. The photoactivated species thus produced should exhibit minimum dependence of its reactivity upon temperature, pH, target group character, and other reaction parameters of importance to the maintenance of the biological activity of the target molecules. In 1969 Knowles et al.[1-3] demonstrated the 2-nitro-4-azidophenyl (ANP) group to exhibit these desired characteristics to a useful degree.

Fluoro-2-nitro-4-azidobenzene (FNAB) is an example of a combined thermochemical-photochemical bifunctional reagent useful for the stepwise coupling of even quite fragile biochemical and other molecules or functional groups to one another or to water-insoluble carrier materials. The aryl azide group is quite stable in the dark under physiological conditions and even under many harsher reaction conditions useful for the thermochemical replacement of the fluorine by nucleophilic groups on carrier derivatives or on soluble compounds containing other functional groups differing from the nitrophenyl fluoride in conditions of thermal activation and target specificity. The basic structure of the stepwise cross-linking reagents emphasized herein is presented in Fig. 1, where the R

[1] G. W. J. Fleet, R. R. Porter, and J. R. K. Knowles, *Nature* (*London*) **224**, 511 (1969).
[2] G. W. J. Fleet, J. R. Knowles, and R. R. Porter, *Biochem. J.* **128**, 499 (1972).
[3] Jeremy R. Knowles, *Acc. Chem. Res.* **5**, 155 (1972).

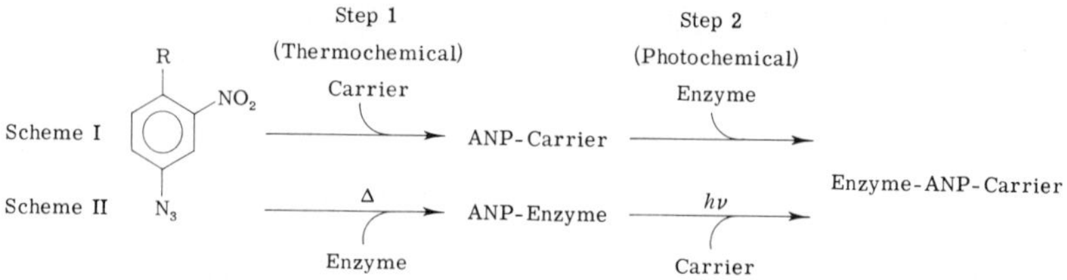

```
                    Step 1              Step 2
         R      (Thermochemical)    (Photochemical)
              NO₂   Carrier             Enzyme
Scheme I  ⬡                ⟶  ANP-Carrier  ⟶
                                                      Enzyme-ANP-Carrier
Scheme II  N₃       Δ      ⟶  ANP-Enzyme   hν  ⟶
                  Enzyme              Carrier
```

Fig. 1. Alternative schemes for the stepwise thermophotochemical coupling of enzymes with carrier materials. ANP, azidonitrophenyl group.

group represents in our use fluorine or other thermochemically activated groups with better activation conditions and/or target specificities.

The dark-stable aryl azide (ANP) group is activated by visible light to generate N_2 and the aryl nitrene diradical, which is a short-lived species capable of insertion into even carbon-hydrogen bonds[3] with little dependence upon the temperature or pH of the reaction mixture over the ranges useful for enzyme couplings.[4]

If the thermochemical group is on a side chain substituted for the fluorine of FNAB, the thermochemical and the photochemical activations may be performed with essentially complete independence of each other. While this allows the operator to choose either time sequence for the two reaction steps, the thermochemical stability and low level of target specificity of the photochemical group recommends it for the second or final step of the cross-linking process for most immobilization systems. Therefore, we outline in Fig. 1 alternative schemes for the stepwise immobilization of enzymes with thermophotochemical bifunctional reagents, in both of which the dark reaction is executed first.

Whereas the potential value of Scheme II is great and is currently being demonstrated in several laboratories[1-17] for the coupling of hor-

[4] P. Guire, M. Yaqub, C. Chirpich, R. Blake, and E. Podrebarac. Manuscript in preparation.
[5] F. Richards, J. Lifter, C. L. Hew, M. Yoshioka, and W. H. Konigsberg, *Biochemistry* **13**, 3572 (1974).
[6] C. A. Converse and F. F. Richards, *Biochemistry* **8**, 4431 (1969).
[7] I. Schwartz and J. Ofengand, *Fed. Proc., Fed. Am. Soc. Exp. Biol.* **34**, Abst. No. 2042 (1975).
[8] W. G. Hanstein, *Fed. Proc., Fed. Am. Soc. Exp. Biol.* **34**, Abst. No. 2126 (1975).
[9] G. Rudnick, R. Weil, and H. R. Kaback, *Fed. Proc., Fed. Am. Soc. Exp. Biol.* **34**, Abst. No. 1525 (1975).
[10] D. Levy and T. Trosper, *Fed. Proc., Fed. Am. Soc. Exp. Biol.* **34**, Abst. No. 2404 (1975).
[11] D. Levy, *Biochim. Biophys. Acta* **332**, 329 (1973).
[12] B. E. Haley, *Fed. Proc., Fed. Am. Soc. Exp. Biol.* **34**, Abst. No. 2250 (1975).
[13] L. W. Yielding, W. E. White, and K. L. Yielding, *Fed. Proc., Fed. Am. Soc. Exp. Biol.* **34**, Abst. No. 2084 (1975).

mones, haptens, and other biologically active ligands to their receptor sites on macromolecules, subcellular organelles, and intact cells and tissues, we shall describe only our experience with Scheme I in this treatise on immobilization techniques. We have demonstrated the usefulness of FNAB and various related stepwise cross-linking reagents containing the nitrophenylazide photoreactive group for the thermochemical preparation of dark-stable photoreactive derivatives of various carrier materials. These photoreactive carrier derivatives may be prepared, washed, and stored wet or dry at ambient temperatures in the dark. The user need merely choose the preferable carrier derivative, mix it in water with his subject enzyme at the temperature and pH chosen for optimizing the enzyme stability and/or interaction with the carrier, and execute the covalent coupling by mere illumination with visible light. The photoimmobilized enzyme product may then be processed and used by the usual procedures.

Materials

Essentially all the hydrophilic carrier materials currently being used for enzyme immobilization are expected to be compatible with the photoimmobilization process. Suitable azidonitrophenyl (ANP) derivatives may be coupled to these carriers, in most cases by the same thermochemical reactions used to couple these carriers to enzymes. Approximately 15 ANP derivatives differing in the length, chemical character, and mode of coupling of the extender arm between the polymer backbone and the ANP group have been prepared and demonstrated to be useful for the photoimmobilization of enzymes. These derivatives are based on five basic materials, which are commercially available: controlled-pore glass (aminoalkyl, with and without zirconium), cellulose (aminoethyl and carboxymethyl), Sephadex (carboxymethyl), agarose (aminopropyl, aminohexyl, and aminoethylaminopropyl), and polyacrylamide (aminoethyl). Most other natural and synthetic polymers useful as enzyme carriers are expected to be compatible with this Scheme I process. Scheme II is expected to be preferable for the coupling of enzymes to carriers that are effective competitors to the enzymes as targets for the photochemical aryl nitrene coupling (e.g., insoluble proteins, such as collagen, and carriers with a high aromatic content, such as polystyrene).

Fluoro-2-nitro-4-azidobenzene (FNAB). This thermophotochemical bifunctional reagent is now commercially available (e.g., FNPA from

---

[14] H. Kiefer, J. Lindstrom, E. S. Lennox, and S. J. Singer, *Proc. Natl. Acad. Sci. U.S.A.* **67**, 1688 (1970).
[15] B. A. Winter and A. Goldstein, *Mol. Pharmacol.* **6**, 601 (1972).
[16] U. Das Gupta and J. S. Rieske, *Biochem. Biophys. Res. Commun.* **54**, 1247 (1973).
[17] J. A. Katzenellenbogen, *Annu. Rep. Med. Chem.* **9**, 222 (1974).

Pierce Cat. No. 23700, or ICN-K&K Cat. No. 28064). It may be prepared from 4-fluoro-3-nitroaniline by the diazotization–azide substitution methodology described by Fleet, Knowles, and Porter.[2]

4-Azido-2-nitrophenyl-aminoalkylcarboxyl derivatives may be prepared by allowing FNAB to react with the chosen aminocarboxylic acid compound (e.g., glycine, β-alanine) in dimethyl sulfoxide according to Levy,[11] in aqueous ethanol with Na_2CO_3 according to Fleet,[2] or in aqueous ethanol with $Ca(OH)_2$ according to the procedure used in our laboratory.[4]

ANP-aminoalkylamine derivatives may be obtained by allowing FNAB to react with equimolar or excess diamines (e.g., ethylenediamine, 1,3-diaminopropanol).[4]

Other thermochemically functional groups more useful for coupling to the particular carrier material of choice may be expected to be substituted onto the photoreactive ANP group by these or other known thermochemical reaction procedures.

Procedures

In the limited space available for this introduction to photoimmobilization methodology, the use of a commercially available thermophotochemical bifunctional reagent (FNAB) with commercially available aminoalkyl carrier materials will be described. Only Scheme I, the preparation and use of dark-stable, photoreactive carrier derivatives for the gentle and facile photoimmobilization of enzymes, will be described in detail.

Step 1. Dark Reaction. Thermally stable photoreactive derivatives of aminoalkyl matrix or carrier materials (ANP–AAM) may be prepared by allowing FNAB to react in the dark or in dim light with a suspension of an aminoalkyl matrix (AAM) or carrier material in alkaline aqueous ethanol. FNAB exhibits thermochemical reactivity properties for fluoride displacement similar to those of Sanger's reagent (fluorodinitrobenzene), the latter being significantly more reactive. The azide group of FNAB appears to be thermochemically stable in many solvents at temperatures up to at least 50° and pH up to at least 10.5. The useful conditions (except for low light intensity) for the FNAB + AAM reaction, and subsequent processing and storage of the ANP-AAM products, are expected to be subject to wide variation. A specific set of process conditions that are expected to be useful for a large number of aminoalkyl carrier products is described below.

The aminoalkyl carrier material (e.g., aminoethyl cellulose, aminoalkyl agarose, aminoalkyl glass) is suspended in and equilibrated with an aqueous alkaline buffer (e.g., 0.5 M borate, pH 9.5–10). The washed equilibrated material is recovered as a packed bed of gel or a slurry of particles. FNAB of sufficient quantity to give the desired molar ratio over carrier amino groups (a range of 1–3 is usually satisfactory) is dissolved in ethanol of two times the volume of the aqueous carrier suspension or gel and the two are mixed together. This and subsequent exposure of the photochemical materials prior to photolysis (step 2) is executed in dim daylight or darker. The reaction flask is stoppered and the contents mixed at 37°–40° for 16–64 hr.

The reaction mixture solid is then separated by filtration or centrifugation, washed sequentially with 95% ethanol until the washings are colorless, with 1 M NaCl in H_2O or a buffer of the pH desired for the planned photoimmobilization use, then with this buffer or with deionized water. If the material is to be dried for storage before use, the fibrous or particulate materials should be washed finally with H_2O, the gels with ethanol. These washed derivatives may then be dried under vacuum at room temperature and stored in the dark at ambient temperature until used.

Step 2. Photoimmobilization Procedure. The reaction conditions useful with these photoreactive carrier materials are subject to large variation without harm to the photochemical reaction. Such variables as pH, temperature, and ionic strength may be chosen for the optimum activity stability of the enzyme to be immobilized.

The reaction rate depends mostly upon the intensity of illumination with visible light. Satisfactory illumination, mixing, temperature control, and measurement of the N_2 evolved may be obtained in an illuminated Warburg apparatus. Although the photochemical immobilization reaction can be activated to completion in several hours by a regular 200 or 300 W tungsten filament light bulb, or a focused microscope illuminator, still better light intensity at the target and improved light-to-heat ratio can be obtained with parallel-beam tungsten–halide projector lamp bulbs [e.g., Eumig 711R (EU 1172) 12 V, 100 W, reflector lamp bulb] mounted on the manometer-flask carrier for movement with the target flask during the motion for sample stirring (Fig. 2). $t_{1/2}$ of N_2 evolution in photoimmobilizations with this illuminated Warburg setup, operated at 10 V on the lamp, usually fell in the range of 20–40 min.

The photoreactive carrier material is equilibrated to the pH chosen for the enzyme and washed with water to remove organic buffer ions if any have been used. The carrier material is suspended to form a slurry thin enough for efficient mixing, with water or aqueous inorganic buffer containing enzyme up to 10% the dry weight of the carrier material. The system is then flushed with N_2 before illumination as a precaution against possible O_2 interference with the desired nitrene reactions.

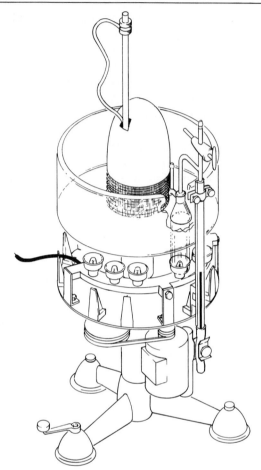

Fig. 2. Modified Warburg apparatus for photochemical coupling reactions. The parallel-beam tungsten–halide projector lamps are mounted on the manometer platform of a Bronwill Warburg Apparatus Model UV so that the light beams may be directed continuously through the clear Plexiglas temperature control bath container into the Warburg flasks during reciprocal motion in the water bath.

An example of a useful experimental protocol follows: 20–25 mg dry of ANP carrier material previously equilibrated to pH 4.5–5.0 is placed in an approximately 5-ml Warburg flask without a center well. H_2O (usually 0.5–2 ml, depending upon the type of carrier material being used) containing 2–3 mg of invertase (β-fructofuranosidase from bakers' yeast) is added, and the flask is attached to the Warburg manometer. The sample is then placed on the Warburg apparatus with the flask immersed in ice-water. The manometer-flask system is flushed with a stream of N_2 for a few minutes while temperature equilibration is occurring. Work up to this point is done in dim daylight. The flask-manometer system is then closed, a zero-time manometer reading is taken, and illumination with mixing motion is initiated and continued until N_2 evolution ceases, as indicated by the periodic manometer readings. When the parallel-beam 12 V, 100 W projector bulb is operated at 10 V, the reaction is usually continued from 2 to 5 hr illumination time. Further processing of the immobilized enzyme need be no different from that of enzyme immobilized by other methods.

Satisfactory photoimmobilization without measurement of the rate or extent of N_2 evolution may be obtained by placing such an enzyme + ANP-AAM reaction mixture in an approximately 5-ml round-bottom flask blown from glass tubing. The mouth is covered with a gum rubber dropper bulb, and the flask is flushed with N_2 or other inert gas. The flask is then placed in a bath for temperature control and stirred magnetically for about 16 hr illumination with a focused 8 V, 5 amp microscope illuminator light passing through about 0.5 cm of 1 M aqueous sodium nitrite in a petri dish.

Discussion

Many advantages are obvious in the use of stepwise cross-linking reagents for coupling carrier materials to enzymes and other biochemicals. The high level of reaction independence of the two reactive groups of certain thermophotochemical bifunctional reagents allows the chemist to prepare dark-stable photoreactive derivatives of his chosen carrier material and to further "tailor" it (e.g., modify its ionic character, add ligands, and/or otherwise modify the "microenvironment" of the carrier surface) for the particular enzyme system(s) to be immobilized. These dark-stable photoreactive carrier materials can be removed from storage and used in the facile covalent coupling of enzymes by merely mixing them together in an aqueous slurry at the temperature and pH chosen for the enzyme and illuminating the mixture with visible light. Since most enzymes are transparent to light of wavelengths required to activate the nitrophenyl azide, and the aryl nitrene thus produced exhibits a relatively low level of specificity for the target amino acid, the enzyme molecule is expected to be covalently immobilized with exceptionally little stress by this process. The photoimmobilized enzyme product is thus expected to exhibit relatively high levels of catalytic capacity and specific activity. Otherwise its properties are found to be quite similar to those of the enzyme-carrier couple formed by established thermochemical processes.[4,18,19]

Although the low level of target specificity exhibited in the nitrene coupling reaction provides significant advantages for the immobilization of enzymes and other biological macromolecules, this may present a limitation for the photochemical immobilization of smaller biologically active molecules in which target group specificity for the coupling reaction may be necessary for the maintenance of biological activity. In addition the susceptibility of the carrier material itself to reaction with the nitrene may limit the Scheme I (Fig. 1) process to those carriers containing little aromatic or other hydrophobic character. One must be cautious also in the use of organic solvents or organic buffer ions in the photolysis mixture, since these may provide competitive targets for the nitrene reactions.

On the basis of experience with the photoimmobilization of approximately 10 enzymes with approximately 15 different ANP-carrier derivatives,[4] one would expect an ANP-carrier containing 50–100 μmoles of ANP per dry gram to photoimmobilize 20–70 mg of enzyme with a free enzyme activity equivalent of 5–60 mg of enzyme per dry gram. Appreciably higher values may be expected from a carrier with larger exposed surface area and improved carrier-enzyme compatibility for the immobilization step (e.g., ionic attraction for concentrating the enzyme at the carrier surface) and subsequent catalytic function.

[18] M. Yaqub and P. Guire, *J. Biomed. Mater. Res.* **8**, 291 (1974).
[19] P. Guire, *Fed. Proc., Fed. Am. Soc. Exp. Biol.* **34**, Abst. No. 2674 (1975).

[22] Nonaqueous Synthesis of Polystyryl Lysozyme

By H. D. BROWN and S. K. CHATTOPADHYAY

Reaction utilizng organic solvation in enzyme polymer complex syntheses have a number of potential advantages over methods limited to aqueous conditions: increase of available reactive moieties of the protein; increase in the available polymeric matrices (by virtue of appropriate solubility as well as available alternative reactive groups); use of the solvent system to control the configurational properties of the enzyme. A representative technique is lysozyme binding to polystyrene under anhydrous conditions using N,N'-carbonyldiimidazole as an activating agent. Enzyme complexes synthesized by this technique have enhanced thermal and chemical stability and modified pH and substrate optima; the protein moiety may have an altered configuration.[1]

Synthesis

Copoly(styrene-4-vinylbenzoic acid) lysozyme can be prepared under anhydrous conditions, in N,N-dimethylformamide, using N,N-carbonyldiimidazole as an activating agent.

Reagents

Benzoyl peroxide
N,N'-Carbonyldiimidazole
4-Chlorostyrene
N,N-Dimethylformamide
Lysozyme (EC 3.2.1.17)
Maleate buffer, 0.1 M, pH 6.2
Styrene
4-Vinylbenzoic acid[2]

Preparation of the Copolymer. A polymerization tube is charged with 12.4 g (0.119 mole) of styrene and 5.90 g (0.039 mole) of 4-vinylbenzoic acid. The mixture is treated with 250 mg of benzoyl peroxide and stirred until solution of the catalyst is complete. The tube is sealed and immersed in an oil bath at 80° for 24 hr. A clear, colorless polymer, removed from the polymerization vessel, is ground to a fine powder in a mortar. The product (18.1 g) is characterized by IR (KBr); 1680 cm^{-1} (C=O) 1420, 1280, 1175, 1017, and 825 cm^{-1}.

Preparation of Lysozyme Polymer Complex. A 50-ml flask equipped for magnetic stirring and fitted with a drying tube is charged with 200 mg of copolymer and 5 ml of N,N-dimethylformamide. The mixture is stirred to dissolve the polymer, and 70 mg (0.432 mmole) of N,N'-carbonyldiimidazole are added. Carbon dioxide is evolved immediately, and the mixture is stirred for 30 min to complete the conversion to the imidazolide. Twenty milligrams of crystalline lysozyme are added in one portion, and the two-phase reaction mixture is stirred for 18 hr at 25°. At the end of the stirring period unreacted lysozyme is removed by centrifugation, and the clear, colorless organic supernatant is decanted. The dimethylformamide solution of the polymer–enzyme complex is slowly poured into 50 ml of 0.1 M, pH 6.2, maleate buffer. A fine, white precipitate separates immediately and is removed by centrifugation. This product is washed with fresh buffer and suspended in 10 ml of buffer for assay.

[1] H. D. Brown, G. J. Bartling, and S. K. Chattopadhyay, in "Enzyme Engineering" (E. K. Pye and L. B. Wingard, Jr., eds.), Vol. 2, p. 83. Plenum, New York, 1974.
[2] J. R. Leebrick and H. E. Ramsden, *J. Org. Chem.* **23**, 935 (1958).

Assay

The enzymic hydrolysis of N-acetylhexosaminidic linkages of chitin-derived oligomers is the basis of the assay described here. Activity is determined by colorimetric measurement of the formation of reducing groups (Park and Johnson,[3] modified). Hydrolysis of the β $(1 \rightarrow 4)$ linkage in linear poly-N-acetylglucosamine of *Micrococcus* cell walls, measured densitometrically,[4] is also useful though less appropriate to the immobilized enzyme.

Reagents

Maleate buffer, 0.1 M, pH 6.2.
Hexa-(N-acetylglucosamine)[5]
Ferricyanide solution, 50 mg potassium ferricyanide in 100 ml of water. Store in brown bottle
Carbonate-cyanide solution, 530 mg of sodium carbonate and 65 mg of potassium cyanide (KCN) in 100 ml of water
Ferric iron solution, 150 mg of ferric ammonium sulfate and 100 mg of sodium lauryl sulfate in 100 ml of 0.05 N sulfuric acid

Procedure. The reaction mixtures consist of 0.12 mg of bound protein and 1.0 mg of hexa-(N-acetylglucosamine) in a total of 2.0 ml of 0.1 M maleate buffer, pH 6.2. Appropriate controls are also used. Samples are incubated at 37° with constant stirring for 30 min. The reaction is stopped by immersion into boiling water. After cooling to 25° the polymer–enzyme complex is removed by centrifugation and a 0.2-ml aliquot is removed from each tube. Each aliquot is diluted with 0.3 ml of water; 0.5 ml of carbonate-cyanide solution and 0.5 ml of ferricyanide solution are added. Assay tubes are placed in a boiling water bath for 5 min and then cooled in an ice bath. One milliliter of ferricyanide solution and 5.0 ml of ferric iron solution are added. After 15 min the absorbency of the solution is measured against a reagent blank at 690 nm. Reducing groups equivalent is calculated from a standard curve using glucose[6] as reference. Bound protein is determined by the method of Lowry *et al.*[7] using lysozyme as standard.

[3] J. T. Park and M. J. Johnson, *J. Biol. Chem.* **181**, 149 (1949).
[4] D. Shugar, *Biochim. Biophys. Acta* **8**, 302 (1952).
[5] From chitin; J. A. Rupley, L. Butler, M. Gerring, F. J. Hardegen, and R. Pecoraro, *Proc. Natl. Acad. Sci. U.S.A.* **57**, 1088 (1967).
[6] A unit of activity is defined as the production of 1 µg of glucose per minute per milligram of protein at 37°.
[7] O. H. Lowry, N. J. Rosebrough, A. L. Farr, and R. J. Randall, *J. Biol. Chem.* **193**, 265 (1951).

[23] Physical Immobilization of Enzymes by Hollow-Fiber Membranes

By R. P. CHAMBERS, W. COHEN, and W. H. BARICOS

Hollow-fiber semipermeable membranes provide a ready means for the physical immobilization of enzymes. Immobilization is achieved without the necessity of chemical techniques by physically confining the enzyme by the semipermeable membrane of the hollow fiber. This membrane is impermeable to the enzyme, but permeable to substrates and products. Physical immobilization, the containment of enzyme by noncovalent means, can be effected by a variety of techniques discussed in this volume including entrapment within cross-linked gels and other polymers, adsorption by various supports, and confinement within liposomes and microcapsules.

Hollow-fiber immobilization offers several distinct advantages relative to other *physical* immobilization methods. Immobilization can be accomplished quickly and easily. Chemical alteration of the enzyme is not required although it may be desirable (see the section on modified enzymes). Hollow-fiber immobilization is a potentially valuable method for the study of soluble enzymes since, in contrast to many other immobilization methods, the enzyme may be immobilized without any alteration of its kinetic properties. Confinement by the semipermeable membrane prevents microbial and antibody access to the enzyme. Other advantages that accrue from this method are simplicity, versatility, capability for coimmobilizing several enzymes, selectivity control of substrates and products through membrane selectivity, an inherently large ratio of surface area to volume, the absence of enzyme leakage (with properly chosen fibers), and continuous operation at low pressures. Among the disadvantages, one may list the possible limitation in reaction velocity as a result of the permeability resistance of the hollow-fiber membrane, the inherent and somewhat inflexible geometry, and the difficulty of working with very low substrate concentrations due to substrate adsorption or adsorption by the membrane.

In this chapter we will discuss the preparation, analysis, and performance of hollow-fiber immobilized enzymes. While the discussion will be limited primarily to commercially available hollow fibers, it should be pointed out that many membrane materials can be spun into hollow fibers, and the discussion presented below is theoretically applicable to such custom-made hollow fibers as well as to those commercially available.

Hollow-Fiber Membranes and Devices

Commercially Available Hollow Fibers and Hollow-Fiber Devices

Laboratory-scale hollow-fiber devices are currently available from a number of manufacturers. Table I presents manufacturers' data on some representative commercially available laboratory-scale hollow-fiber devices. As opposed to devices, hollow fibers themselves may be obtained from Abcor, Amicon, BioRad, Dow, DuPont, Gulf South Research Institute, Monsanto, North Star Research Institute, and Romicon.

The geometry of the tubular hollow-fiber device is depicted in Fig. 1. This shell and tube geometry provides for two compartments within the device: a lumen compartment within the fibers and a shell compartment in the annular space between the fibers and the exterior wall. Beaker-type devices consist of a U-shaped bundle of fibers sealed in a beaker. Again the lumen compartment refers to the volume within the fibers and the shell compartment refers to the extra-fiber space within the beaker.

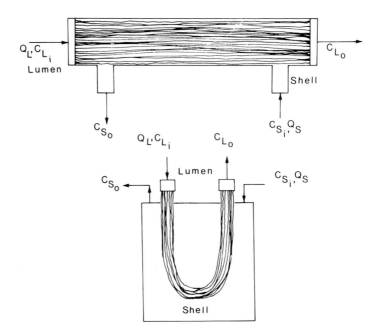

FIG. 1. Schematic illustration and nomenclature for hollow-fiber devices: tubular device in upper figure, beaker device in lower. Q_L and Q_S refer to volumetric flow rates in the lumen and shell, respectively. C_L and C_S refer to solute concentration in the lumen and shell, respectively.

TABLE I
Hollow-Fiber Devices

Manu-facturer	Designation	Nominal cutoff (daltons)	Surface area (cm^2)	Membrane	Type	Inside diameter (μm)	Geometry
Amicon[a]	P5	5,000	560[e]	Polysulfone	Anisotropic	500	Tubular
	P8	8,000	560[e]	Polysulfone	Anisotropic	500	Tubular
	P10	10,000	930[e]	Polysulfone	Anisotropic	200	Tubular
	X50	50,000	930[e]	Acrylic copolymer	Anisotropic	200	Tubular
	P100	100,000	740[e]	—	Anisotropic	200	Tubular
BioRad[b]	B-20	200	50[f]	Cellulose acetate	Isotropic	200	Tubular
	B-50	5,000	50[g]	Cellulose	Isotropic	200	Tubular
	B-80	30,000	50[f]	Cellulose acetate	Isotropic	200	Tubular
	B-20	200	1,000	Cellulose	Isotropic	200	Beaker
	B-50	5,000	900[h]	Cellulose	Isotropic	200	Beaker
	B-80	30,000	1,000	Cellulose acetate	Isotropic	200	Beaker
Cordis[c]	CDAK	5,000	10,000	Cellulose	Isotropic	225	Tubular
Romicon[d]	20-PM10	10,000	1,700[i]	Aromatic polymer	Anisotropic	500	Tubular
	20-XM50	50,000	2,300[i]	Polyolefin	Anisotropic	500	Tubular
	45-XM50	50,000	1,000[i]	Polyolefin	Anisotropic	1100	Tubular
	20-GM80	80,000	2,300[i]	—	Anisotropic	500	Tubular

[a] Amicon Corporation, Lexington, Massachusetts 02173.
[b] BioRad Laboratories, Richmond, California 94804. These fiber devices are manufactured by Dow and distributed by BioRad under the name Biofiber.
[c] Cordis Laboratories, Miami, Florida 33137.
[d] Romicon, Inc., Woburn, Massachusetts 01801.
[e] The designation of the fiber device listed is prefixed by H1; a 9300 cm^2 device, 500 μm in diameter, is also available, which is prefixed by H10.
[f] The fiber device listed is referred to as a Minitube; a 10,000 cm^2 device called a Miniplant is also available.
[g] The fiber device listed is referred to as a Minitube, a 15,000 cm^2 device called a Miniplant is also available.
[h] A smaller beaker (Minibeaker) containing 80 cm^2 is also available.
[i] Larger devices with approximately 12 times the surface area are also available.

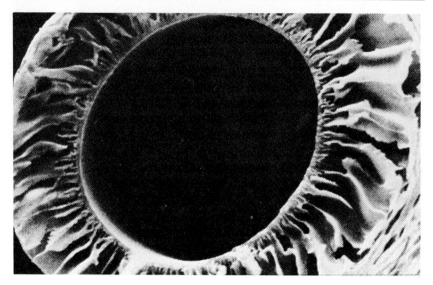

FIG. 2. Scanning electron micrograph of the cross section of anisotropic Amicon and Romicon hollow fibers.

Two major types of hollow fibers are manufactured: the more nearly isotropic cellulosic and nylon fibers by Dow and DuPont, respectively, and the anisotropic synthetic polymer fibers of Amicon and Romicon. In the Dow and DuPont fibers the properties near the surface of these fibers are reasonably representative of the bulk of the fiber. However, the Amicon and Romicon fibers have a pronounced dense, thin skin, less than 1.5 μm thick, next to the interior surface, which contains the controlled-pore region. The remainder of the fiber, referred to as the "spongy layer," serves as a supporting structure and consists of a much thicker layer of the same polymer with considerably larger openings (Fig. 2).

Preparation of a Single-Fiber Device

One drawback to the commercially available devices is the lack of availability of one device in which a variety of different hollow-fiber membranes may be evaluated. Rony and co-workers[1,2] developed a simple, economical, low-surface area, reusable, tubular single-fiber device,

[1] W. K. Lo, S. Putcha, B. U. Kim, L. Griffith, S. Bissell, and P. R. Rony, in "Enzyme Engineering" (E. K. Pye and H. H. Weetall, eds.), Vol. 3. Plenum, New York, 1976.

[2] W. K. Lo, M. S. thesis with P. R. Rony, Chemical Engineering Department, Virginia Polytechnic Institute, December, 1975.

which is amenable for use with a wide variety of commercially available hollow fibers. The following method was used to prepare 1.0 cm^2 nominal surface area devices containing X50 fibers. A glass tube, of 1.0 mm internal diameter, approximately 20 cm long, and tapered at each end to accommodate $\frac{1}{8}$-inch Teflon tubing, is fitted with side tubes approximately 2 cm long, one near each end of the main glass tube. The tapered ends are individually heated to red heat, and 2-cm lengths of Teflon tubing are quickly pressed on. The Teflon tubing, so applied, softens and expands, giving when cool a snug watertight seal. A single fiber is threaded through the Teflon tubing and attached by applying a small amount of Epon (Shell Chemical) epoxy cement at the junction of the fiber and Teflon tubing. The cementing fluid flows into the annular region between the fiber and the Teflon tubing, forming a watertight seal. A proper viscosity epoxy, such as the above, should be used so that the cementing fluid does not creep by capillary action into the lumen of the fiber. After 15–24 hr curing time, the module is ready for use. These hollow-fiber devices may be used singly or more often a number may be connected in series to increase the available surface area.

Selection of a Hollow Fiber

The selection of the appropriate hollow-fiber membrane should take into account a number of factors. These factors include the nominal molecular weight cutoff, permeability of the substrates and products, solubility of the substrates and products within the hollow-fiber membrane, hydraulic permeability, membrane anisotropy, and membrane compatibility with enzyme and substrate solutions.

Molecular Weight Cutoff. The molecular weight cutoff is an empirically determined parameter obtained from ultrafiltration experiments. A solution containing a known concentration of the solute in question is circulated through the lumen of the hollow fiber under pressure. The fluid permeating through the hollow-fiber membrane from the lumen to the shell is collected on the shell side (outside of the fibers). The solute concentration in the shell is denoted the outlet concentration. The concentration ratio, (inlet-outlet)/(inlet), is reported as F, the fraction retained. Since the fraction retained depends on the many factors (the permeation rate, operating conditions, pressure drop across the fiber, flow rate, and concentration), it is only a relative measure of the molecular separating pore structure of the membrane. Keeping in mind these limitations, retention data for several hollow fibers may be found in Table II. Note that the retention capability of a given membrane depends not only on the molecular weight but also on the molecular configuration.

TABLE II
Hollow-Fiber Retention Data[a]

Solute	Fraction retained, F				
	B-20	P10	B-50	X50	B-80
Urea (60)[b]	50	0	0	0	0
Creatinine (113)	79	0	0	0	0
Sucrose (342)	95	0	0	0	0
Raffinose (594)	—	0	0	0	0
Vitamin B_{12} (1355)	—	—	2	0	0
Bacitracin (1400)	—	55	—	0	0
Inulin (5200)	—	<10	82	—	—
Cytochrome c (12,400)	—	90	95	80	10
Ribonuclease (13,600)	—	90	—	30	—
Myoglobin (17,800)	—	90	—	40	30
α-Chymotrypsinogen (24,500)	—	90	—	75	—
Pepsin (35,000)	—	>95	—	—	85
Ovalbumin (45,000)	—	>98	—	90	—
Bovine serum albumin (67,000)	—	>95	—	>90	98
Dextran 110 (110,000)	—	30	—	10	—
Aldolase (142,000)	—	>98	—	>95	—
IgG (160,000)	—	>98	—	>98	—

[a] Retention data evaluated from ultrafiltration experiments reported in the manufacturer's literature at the following applied-pressure differentials: P10, 1.75 atm; X50, 1.0 atm; B-20, B-50, B-80, unstated.
[b] In parentheses, molecular weight.

As can be seen from the tabulated data, linear or branched polysaccharides behave quite differently from globular proteins. Apparently, molecular configuration, relative ease of deformability, and fluid shear stresses in the vicinity of membrane pores act to cause the behavior of flexible linear or branched molecules to be more representative of their smaller cross-sectional area rather than their molecular weight (in a similar fashion to gel filtration). In addition, the ionic strength and pH of the solution being ultrafiltered also have a strong influence on the substrate retention. Increased retention has been reported for polyelectrolytes at lower ionic strengths.[3] It is important to emphasize the influence of molecular configuration and the lack of 100% retention of many macromolecules. It is necessary to determine the retention of the enzyme of interest under the experimental conditions to be employed.

[3] A. S. Michaels, L. Nelsen, and M. C. Porter, in "Membrane Processes in Industry and Biomedicine" (M. Bier, ed.), p. 197. Plenum, New York, 1971.

Dialytic Permeability. The permeability of commercially available laboratory-sized hollow fibers for a number of substrates of interest may be found in Table III. In contrast to the retention data, permeability refers only to diffusive dialytic permeation due to the molecular diffusion of solute molecules across the membrane. For solutes whose molecular weights are much less than the molecular weight cutoff, note that the dialytic permeability is quite similar for all membranes regardless of material and anisotropy. Permeabilities in all cases are based on the geometrical surface area available for diffusion using the dry membrane surface area. Note that in some cases the net overall permeability is tabulated. The net overall permeability includes the effect of the dialytic permeability of the membrane plus the effect of the Nernst diffusion layer in the flowing fluid in both the lumen and shell of the hollow fiber. The contribution of the fluid Nernst layer varies with the flow conditions. It was reported that for the Amicon fibers the effect of the Nernst layers

TABLE III
DIALYTIC PERMEABILITIES[a]

Solute	Permeability (cm/min × 1000)				
	B-20[b]	B-50[b]	P30[c]	X50[c]	B-80[b]
Ethanol	1.5[d]	—	—	—	—
Acetaldehyde	0.42[d]	—	—	—	—
Acetic acid	0.030[d]	—	—	—	—
Sodium chloride	0.042	29	49	46	—
Urea	1.8	25	31	27	32
Creatinine	—	16	34	22	17
Uric acid	—	14	18	17	—
Raffinose	—	4.2	—	—	8.0
NAD	0[d]	1.8[e]	—	—	—
Vitamin B_{12}	—	1.5	—	—	6.0
Bacitracin	—	—	4.1	11.4	—
Inulin	—	0.4	1.8	10.4	—
Bovine serum albumin	—	0	—	—	0.014

[a] The values in the table represent true dialytic permeabilities except where noted.
[b] Personal communication from J. Landau, Dow Chemical, Walnut Creek (1973) for Biofibers 20, 50, and 80 except for data from references cited in footnotes d and e.
[c] R. A. Cross, W. H. Tyson, and D. S. Cleveland, *Trans. Am. Soc. Artif. Intern. Organs* **17**, 279 (1971); overall permeability of Amicon fibers P30 and X50.
[d] J. Ford, Ph.D. dissertation with R. Chambers, Tulane University, *Diss. Abstr. Int. B* **33**, 3624 (1972).
[e] D. J. Fink and V. W. Rodwell, *Biotechnol. Bioeng.* **17**, 1029 (1975); overall permeability.

TABLE IV
HYDRAULIC PERMEABILITIES

Hollow fiber	Deionized water fluxes (ml/cm² min atm × 10,000), H
Biofiber 20	1.0[a]
Biofiber 50	9.5[b]
Biofiber 80	86[b]
P5	300[b]
20-PM10	800[b]
P10	1500[c]
20-GM80	2500[b]
X50	3000[c]
P100	6000[b]

[a] Personal communication from J. Landau, Dow Chemical, Walnut Creek (1973).
[b] Current manufacturers literature.
[c] R. A. Cross, *AIChE Symp. Ser.* **120,** 15 (1972).

was to reduce the ×50 fiber dialytic membrane permeability of 0.029 to an overall permeability of 0.025 cm/min (at relatively high flow rate).[4] This relatively small difference between overall and dialytic permeabilities increases slowly as flow rate is decreased and can be quite significant for stagnant fluid. The percentage difference also varies with membrane permeability, markedly decreasing as the membrane permeability decreases.

Ultrafiltration. Ultrafiltration provides the potential for much more rapid solute transport through the membrane than does dialytic permeation. Hydraulic permeability, the flux of solvent through the membrane in response to an applied pressure differential, is an important property of candidate hollow-fiber membranes. Table IV lists deionized-water hydraulic permeabilities for a number of the available laboratory hollow fibers. Note in particular that the anistropic, thin-"skinned" fibers have much higher relative hydraulic permeabilities than do the isotropic cellululosics.

Ultrafiltration can be produced by any factor that will create an effective pressure driving force: by a pressure difference between the lumen and the shell, by the frictional pressure drop caused by fluid flowing in the lumen of the fiber, and by differences in osmotic pressure between the lumen and the shell. The frictional pressure differential can provide significant ultrafiltration for those fibers which have high hydraulic permeabilities. Lambert, Chambers, and co-workers[5] have found that even

[4] R. A. Cross, *AIChE Symp. Ser.* **120,** 15 (1972).
[5] A. H. Lambert, G. A. Swan, W. H. Baricos, W. Cohen, and R. P. Chambers, *Meeting Am. Inst. Chem. Eng.,* Philadelphia, Pa., Paper 64c (1973).

when net ultrafiltration is eliminated by completely filling the shell side with fluid, ultrafiltration can aid the transport of solute across the membrane. As the fluid flows down fibers, the pressure in the fiber lumen continuously decreases. Near the entrance of the hollow fiber this pressure in the fibers is higher than that in the beaker, causing ultrafiltration of solvent (and solute) into the beaker. On the other hand, near the exit, the lumen pressure is less than in the beaker, and ultrafiltration and solute transport occur in the opposite direction. Thus, the friction-produced pressure differential causes an ultrafiltration phenomenon to occur that can materially aid the transport of substrate into the fibers in the upstream portion and the transport of product out of the fibers in the downstream part.

Concentration Polarization. As a result of solvent ultrafiltration a phenomenon known as concentration polarization[6,7] of retained solute occurs. As solvent is forced by the action of the pressure differential through the fiber membrane, solutes of sufficient molecular weight are either totally or partially retained by the membrane. This retention causes their local concentration to increase. This localized solute in turn can dramatically affect membrane performance, often greatly reducing the ultrafiltration capacity of the membrane. The concentration level attained by the retained solute depends both on the shear forces near the membrane surface and on the solute solubility.

For very dilute concentrations of retained solute in the inlet solution, solubility effects are usually not encountered, but at larger retained solute concentration, a macromolecular "gel" or "cake" builds up on the hollow-fiber membrane surface. This gel causes two dramatic effects on membrane performance: First, the ultrafiltration rate is markedly decreased by the gel layer becoming independent of pressure differential; and second, the membrane selectivity is frequently altered by the gel (now a secondary membrane), resulting in a lower and more diffuse molecular weight cutoff. The gel-layer thickness may be reduced, thereby increasing ultrafiltration rate, by increasing the shear rate at the membrane surface (by means of stirring rate or velocity of flow). The thickness of the gel may also be decreased by lowering the concentration of the retained solute or by increasing the total surface area. For example, as a result of the finer pore structure of the secondary gel membrane, the effective cutoff for membranes ultrafiltering protein mixtures, such as human plasma, is considerably lower than that of the membrane itself.

[6] M. C. Porter, *in* "Enzyme Engineering" (L. B. Wingard, ed.), Vol. 1, p. 115. Wiley (Interscience), New York, 1972.
[7] M. C. Porter and L. Nelsen, *in* "Recent Developments in Separation Science" (N. N. Li, ed.), Vol. II, p. 227. Chem. Rubber Publ. Co., Cleveland, Ohio, 1972.

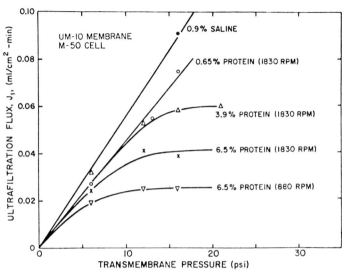

Fig. 3. Ultrafiltration rates per square centimeter of membrane surface area versus pressure differential across the membrane for bovine serum albumin solution in an Amicon stirred cell containing a flat UM-10 semipermeable membrane. From W. F. Blatt, A. Dravid, A. S. Michaels, and L. Nelsen, in "Membrane Science and Technology" (J. E. Flinn, ed.), p. 47. Plenum, New York, 1970.

Figure 3 illustrates some of the principles discussed above from a study in which an Amicon stirred ultrafiltration cell was employed. As the concentration of protein (bovine serum albumin) increases, the hydraulic ultrafiltration rate is decreased. At 3.9% protein, the protein forms a gel on the surface of the planar semipermeable membrane in the stirred cell. Gel formation may be identified by the lack of dependence of the ultrafiltration rate on the applied pressure differential. The two lower curves with 6.5% protein illustrate the effect of stirring rate on the gel layer (fluid shear at the membrane surface) showing the increased ultrafiltration rates attainable at higher shear rates. Blatt et al.[8] made a comprehensive study of this gel polarization phenomena.

Membrane Compatibility. A major limitation to the use of hollow fibers is compatibility of the membrane. Both the cellulosic and synthetic polymer membranes are highly labile in many solution environments. Before a membrane is selected, considerable care should be given to examining the durability of the membrane toward the solutions to be employed during preparation and use of the immobilized enzyme and during sterilization.

[8] W. F. Blatt, A. Dravid, A. S. Michaels, and L. Nelsen, in "Membrane Science and Technology" (J. E. Flinn, ed.), p. 47. Plenum, New York, 1970.

Measurement of Hollow-Fiber Membrane Permeability

The rate of dialytic permeation of the substrate across the hollow-fiber membrane can be expressed in terms of the overall permeability as follows:

$$\text{Moles substrate permeated/time} = PA\Delta C \tag{1}$$

where P is the overall membrane permeability (cm/min), A is the hollow-fiber membrane surface area (cm^2), and ΔC is the difference in concentration between the lumen and shell side of the hollow fiber (mmoles/ml).

The moles of substrate permeated per unit time may also be expressed in terms of flow of substrate in the lumen and in the shell using the tubular hollow fiber device flow configuration and nomenclature found in Fig. 1.

$$\text{Moles substrate permeated/time} = Q_L(C_{L_i} - C_{L_o}) = Q_S(C_{S_o} - C_{S_i}) \tag{2}$$

where Q represents flow rate (ml/min) and C the substrate concentration (mmoles/ml).

For measuring the overall permeability, P, a buffered solution containing the substrate is continuously pumped through the lumen at flow rate Q_L, typically in the middle of the manufacturer's recommended flow rate range. The same buffer without substrate ($C_{S_i} = 0$) is pumped through the shell countercurrent to the flow in the lumen at flow rate Q_S (see Fig. 1). Q_S should nominally be 2–10 times Q_L. Next, the flow rate in the lumen, Q_L, is adjusted to eliminate ultrafiltration. Ultrafiltration may be checked for by temporarily recirculating the lumen side exit fluid back to the substrate buffer solution reservoir and Q_L adjusted until the substrate solution volume remains constant. The substrate concentration in the shell-side exit, C_{S_o}, is monitored until steady state is achieved. At that time, Q_S, C_{S_o}, C_{L_i}, and C_{L_o} are measured. The overall permeability may be calculated using the relation

$$\Delta C = [(C_{L_i} + C_{L_o})/2] - (C_{S_o}/2) \tag{3}$$

together with Eqs. (1) and (2) to give

$$P = Q_S C_{S_o}/A\Delta C \tag{4}$$

Care should be exercised to avoid two experimental artifacts. First, substrates often have a very appreciable solubility within the membrane matrix. Experiments in which the permeability is derived from the rate of change of substrate concentration in a reservoir solution in contact with the fiber lumen suffer from solubility and membrane conditioning (to be discussed later) artifacts. The decrease in substrate concentration in the solution in contact with the lumen is greater, and the rate of increase in substrate concentration on the shell side is smaller, as a result of substrate absorbed by the membrane matrix. This solubility-caused artifact can be overcome by carrying out permeability experiments at the elucidated steady-state, continuous-flow conditions.

At very low flow rates the outlet lumen concentration may be considerably less than the inlet. In that case the second artifact arises because Eq. (3) is no longer valid. Equation (3) is valid as long as $C_{L_o}/C_{L_i} > 0.9$ and $C_{S_o}/C_{L_i} < 0.1$. These concentrations should be checked, and, if C_{L_o} is too low or C_{S_o} too high, the flow rates Q_L and Q_S should be increased sufficiently to overcome this second artifact.

The permeability of the beaker-type hollow fibers may be determined in the same manner as for the tubular hollow-fiber device with several exceptions. First, the beaker stirring rate should be adjusted sufficiently high that a further increase has no effect. Second, Eq. (3) should be modified, since the concentration on the shell side of the beaker device is the exit concentration, C_{S_o}. Thus P can be calculated from beaker data using Eq. (4) when ΔC is represented by

$$\Delta C = (C_{L_i} + C_{L_o})/2 - C_{S_o} \tag{5}$$

Preparation of Semipermeable Liquid-Membrane Hollow Fibers

Semipermeable hollow fibers permit discrimination on the basis of solute molecular weight and configuration. Liquid membranes, on the other hand, have been used to provide permeation selectivity.[9] The marriage of these two techniques has been accomplished by Rony et al.,[1,2] who has developed selective semipermeable hollow fibers by impregnating semipermeable X50 fibers (Amicon) with a FC43 fluorocarbon liquid (3M Company). The fluorocarbon acts as a liquid membrane, preventing the permeation of charged or highly polar molecules. For example, the permeability of the FC43-impregnated fiber for methyl acetate is twice that for acetaldehyde and 70 times that for ethanol, whereas the permeability of each is approximately equal in the native X50 fibers. The permeability of NAD is sufficiently reduced by the fluorocarbon so that NAD permeation is undetectable. Rony et al. have successfully used this FC43/X50 fiber to investigate the hollow fiber-immobilized ethanol oxidation multienzyme system of Ford[10] and Chambers et al.[11]

[9] W. J. Ward, in "Recent Developments in Separation Science" (N. N. Li, ed.), Vol. I, p. 153. Chem. Rubber Publ. Co., Cleveland, Ohio, 1972.
[10] J. R. Ford, Ph.D. dissertation with R. Chambers, Tulane University, 1972; Diss. Abstr. Int. B 33, 3624.
[11] R. P. Chambers, J. R. Ford, J. H. Allender, W. H. Baricos, and W. Cohen, in "Enzyme Engineering" (E. K. Pye and L. B. Wingard, eds.), Vol. 2, p. 195. Plenum, New York, 1974.

The method that Rony and co-workers employed for impregnating hollow fibers with liquid membranes follows. One end of a single-fiber device, described earlier, is immersed in a small beaker containing the water-immiscible impregnation liquid, for example FC43. As a result of capillary action, the impregnation liquid slowly rises in the vertical hollow fiber. Entrapped air is eliminated in the nominal 4 hr required to completely fill the fiber. FC43 is added so that both the lumen and shell side of the fiber device are filled with FC43; the FC43-loaded fiber is then incubated in a horizontal position for 2 days at room temperature. The impregnated fiber device is then thoroughly washed with distilled water followed by the buffer of interest.

The liquid-membrane impregnated hollow fiber should not be subjected to pressure differentials, as the impregnating fluid can be displaced. There is also a slow continuous loss of impregnating liquid from the fiber when it is subjected to fluid flowing in the shell or lumen. It is sometimes useful to presaturate the inflowing solution with impregnating liquid.

Hollow-Fiber Immobilized Enzymes

Theory

The performance of an enzyme immobilized by hollow fibers is determined by the interaction of kinetics with dialytic and hydraulic permeation. For enzymes immobilized in soluble form, the interaction of enzyme kinetics and dialytic permeation in the absence of ultrafiltration can be easily described for Michaelis–Menten kinetics. The overall reaction rate v (mmoles/min) can be written as

$$v = (\eta V V_{max} S_1)/(S_1 + K_m) \qquad (6)$$

where V is the total volume of the enzyme-containing fluid (ml), including the enzyme solution in any recirculation tubing; S_1 is the substrate concentration in the enzyme-free substrate compartment (mmoles/ml); V_{max} and K_m have their usual meaning (mmoles/min/ml and millimoles/ml); and η is the effectiveness factor. The effectiveness factor is the ratio of the reaction rate, including the effect of dialytic permeation, to the rate in its absence. Thus η takes into account the reduction in substrate metabolism caused by the presence of the membrane. The effectiveness factor can be readily calculated by equating the reaction rate in the enzyme compartment to the rate of substrate permeation through the hollow-fiber membrane

$$v = (V V_{max} S_2)/(S_2 + K_m) = PA(S_1 - S_2) \qquad (7)$$

where S_2 is the substrate concentration in the enzyme compartment and $(S_1 - S_2)$ is the difference in substrate concentration causing substrate to permeate across the membrane. The effectiveness factor can be calculated by setting the v's equal to one another in Eqs. (6) and (7).

$$\eta = \frac{VV_{\max}[S_2/(S_2 + K_m)]}{VV_{\max}[S_1/(S_1 + K_m)]} \tag{8}$$

By algebraic manipulation of Eqs. (6), (7), and (8), S_2 can be eliminated and a relation for η in terms of Φ and $K_m{}^*$ obtained.

$$\Phi\eta^2 - (K_m{}^* + 1)(\Phi + 1)\eta + (K_m{}^* + 1) = 0 \tag{9}$$

where $K_m{}^*$ is a dimensionless Michaelis constant $(K_m{}^* = K_m/S_1)$ and Φ, the permeation modulus, is equal to

$$\Phi = \frac{VV_{\max}[S_1/(S_1 + K_m)]}{PAS_1} \tag{10}$$

The permeation modulus is the ratio of the reaction rate in the absence of the membrane to the permeation rate when S_2 is equal to zero. The effectiveness factor in Eq. (9) is plotted as a function of the dimensionless variables Φ and $K_m{}^*$ in Fig. 4.

Figure 4 can be used in the selection of a proper hollow-fiber device or in analyzing experimental results. In the former situation, estimated permeabilities and kinetic constants can be used to estimate the permeation modulus Φ. The magnitude of Φ determines the controlling factor in the performance of the hollow-fiber immobilized enzyme. For η approaching 1.0, the rate in the absence and in the presence of the hollow-fiber membrane are equal, thus kinetics controls the performance. Note that for Φ less than about 0.3, η approaches 1.0; thus kinetics controls for $\Phi < 0.3$. Similarly, for $\Phi > 3.0$, membrane permeation controls the performance of the hollow-fiber immobilized enzyme.

In the analysis of experimental data, Fig. 4 plus Eqs. (6) and (8) are useful in the interpretation of experimentally observed effectiveness factors, permeabilities, and substrate concentrations in the enzyme compartment. Equation (6) indicates that η can be determined experimentally by dividing the reaction rate of the immobilized-enzyme hollow-fiber device by the reaction rate under the same conditions in the absence of the hollow-fiber membrane (the cuvette reaction rate). A sure method for determining permeation effects is to measure the substrate concentration in both the enzyme and substrate compartments, S_2 and S_1, and then compare the theoretical effectiveness factor calculated from Eq. (8) with the experimental value described in the previous sentence. If experimental

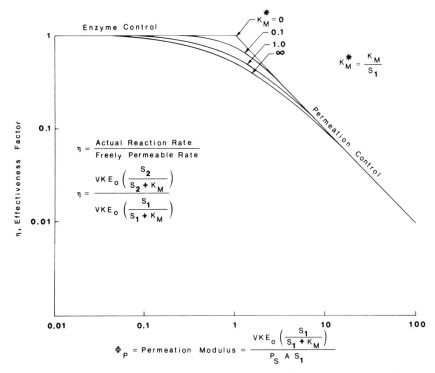

FIG. 4. Effectiveness factor for hollow-fiber immobilized enzymes with dialytic substrate permeation as a function of the dimensionless variables Φ_p and $K_m{}^*$. KE_o in the figure is V_{max}. From A. H. Lambert, G. A. Swan, W. H. Baricos, W. Cohen, and R. P. Chambers, *Meeting Am. Inst. Chem. Eng.* Philadelphia, Pa., Paper 64c (1973).

results show that S_2 is small relative to S_1, it is conclusive that membrane permeation controls the observed phenomena.

Some amount of ultrafiltration (hydraulic permeation) normally accompanies dialytic permeation. The effects of ultrafiltration on the total substrate permeation can be neglected if the rate of hydraulic permeation of substrate is significantly less than the rate of dialytic substrate permeation. This will be true if the following inequality holds:

$$US_1 \ll PA(S_1 - S_2) \tag{11}$$

where U is the ultrafiltration rate (ml/min) and S_1 and S_2 are the substrate concentrations (mmoles/ml) in the substrate and enzyme compartments, respectively. U can be readily measured or determined from the tabulated hydraulic permeability by the relation:

$$U = HA\Delta P \tag{12}$$

where H is the hydraulic permeability (ml/min cm² atm), A is the hollow-fiber surface area (cm²), and ΔP is the pressure differential (atm) causing the ultrafiltration. Note that H in practice can be significantly less than the deionized water hydraulic permeability (Table IV) as a result of concentration polarization, membrane fouling, etc. The pressure differential, ΔP, can be obtained experimentally, by connecting pressure gauges (e.g., gas cylinder gauges) to the tubing leading to the lumen and shell-side compartments.

If enzyme kinetics controls, the performance of immobilized hollow-fiber enzymes can be improved by increasing enzyme loading, but it will be independent of the hollow-fiber properties. However, if dialytic permeation is found to control, the performance can be improved by increasing the surface area or employing a more permeable hollow fiber. If the performance is still inadequate, ultrafiltration may be superimposed on dialytic permeation.

Soluble Enzymes

Enzymes physically immobilized by hollow fibers in their native soluble configuration have been studied by Rony,[12] Ford et al.,[10,13] Chambers et al.,[11] Fink and Rodwell,[14] Davis,[15] and others. The results of several of these studies indicate that the enzyme kinetic characteristics remain unaltered within the hollow fibers. Yet, as previously discussed, the permeation properties of the membrane often cause the overall performance of enzyme-containing hollow fibers to be quite different from that of the soluble enzyme.

Several different techniques have been employed for immobilization of soluble enzymes: the enzyme-containing solution may be placed as a stationary fluid in the lumen[12] or in the shell compartment[15] of hollow fibers. However, a preferred technique is to slowly recirculate the enzyme, whether it is placed in the lumen[10] or shell compartments, in such a way that the enzyme solution also passes through a small reservoir. When the enzyme is immobilized in the latter configuration, enzyme activity can be readily measured by withdrawing aliquots from the enzyme solution. This configuration also allows the measurement of the concentration of substrates and products in the enzyme compartment as well as in the substrate compartment.

[12] P. R. Rony, J. Am. Chem. Soc. **94**, 8247 (1972).
[13] J. R. Ford, W. Cohen, and R. P. Chambers, Meeting Am. Inst. Chem. Eng., Detroit, Mich., Paper 24a (1973).
[14] D. J. Fink and V. W. Rodwell, Biotechnol. Bioeng. **17**, 1029 (1975).
[15] J. C. Davis, Biotechnol. Bioeng. **16**, 1113 (1974).

Several points should be kept in mind when using soluble enzymes physically immobilized by hollow fibers. First, in order to protect the enzyme from surface-caused inactivation, surface treatment of the fibers is often required. In this laboratory, we have found that addition of 1.0 mg of bovine serum albumin per milliliter to the circulating enzyme solution is usually beneficial.[10] Second, since the Amicon and Romicon fibers are anisotropic with large pores on the outside surface, enzyme contact should be avoided with the outside fiber surface if significant capillary sorption of the enzyme is to be avoided. Third, the experimental configuration and operating conditions should be chosen to minimize shear rate-produced enzyme inactivation. Fink and Rodwell[14] found that the half-life ($t_{1/2}$) for native, soluble horse liver alcohol dehydrogenase was unaffected by flow rates up to 0.4 ml/min in the lumen of Biofiber 80 Minitubes. However, these workers found that the half-life was decreased by 85% when the recirculation rate was increased to 1.5 ml/min. Fourth, the hollow fiber should be incubated with a protein-containing solution and then checked for enzyme leakage. The membrane properties undergo change during initial use as a result of polymer swelling, protein and substrate adsorption, sealing of microdefects, etc. Lo and Rony[2] have extensively studied the change in permeability of Amicon X50 fibers with time and have found that small changes in permeability occur for up to 5–7 days while the fibers are incubated in water before the dialytic permeability becomes completely constant. In our laboratory, we have noted that the hollow-fiber immobilized enzyme may yield nonreproducible data during the initial use of the hollow fiber. However, if that fiber is cleaned and reused, satisfactory performance is obtained for a number of repetitive reuses with membrane-compatible enzyme solutions. Fifth, enzyme leakage can occur before the fiber has been allowed to swell or be incubated with a protein solution. Even after proper incubation, enzyme leakage may occur in very high molecular weight-cutoff hollow fibers owing to the lack of 100% retention of higher molecular weight molecules (see Table II). Sixth, many hollow fibers have a very delicate microstructure that is irreversibly damaged if the fibers are allowed to dry out after use or come in contact with nominal concentrations of a number of common organic solvents (contact the manufacturer for proper precautions). Seventh, rigid particulates should be avoided in the hollow fiber lumen, as they may cause plugging or abrasion of the fibers. All particulates should be avoided on the spongy side of the Amicon or Romicon fibers since this spongy structure acts as a depth filter for most particulates. Last, the fiber should be sterilized after use. Davis[15] noted that the half-life of alkaline phosphatase was 3080 min in a reused sterilized Biofiber 80 Minitube versus only 30 min in the same reused fiber device without sterilization. The fiber was sterilized overnight with 1.5% formaldehyde followed by vigorous flushing with deionized water prior to use.

Immobilization Procedure and Rate Measurement. After initial preparation of the hollow fibers as described by the manufacturer, the hollow-fiber bundle is then checked for leaks. The lumen of the fibers is filled with inert gas at about 5 psi, and the shell with distilled water. A continuous stream of bubbles will be seen in the shell compartment if leaks or macrodefects are present. The inert gas is then allowed to purge through the fibers to loosen any particulates that may be present. Distilled water is then purged through the fibers at high flow rates to eliminate any residual gas bubbles, particulates, etc.

EXAMPLE: The enzyme solution is added to one compartment of the hollow-fiber device, and substrate solution is continuously pumped through the other. In the study by Davis,[15] 31.6 units of alkaline phosphatase (Sigma type III) in Tris buffer, pH 10.4, were placed in the 1.0-ml shell-side compartment of a tubular Biofiber 50 Minitube. This hollow-fiber device contained 96 cellulose fibers of 200 μm internal diameter with an active length of 6.5 cm. Substrate solution containing 0.56 mM p-nitrophenyl phosphate buffered with Tris to pH 10.4 was continuously pumped through the fibers at a flow rate of 0.23 ml/min using a syringe pump. The p-nitrophenolate product was continuously monitored at 410 nm with a spectrophotometer. The results of several experiments are shown in Fig. 5. Note that 20–40 min are required before steady-state rates are achieved in this tubular hollow-fiber device.

The use of continuous-flow, steady-state experiments gives results unencumbered by artifacts associated with substrate solubility in the hollow-fiber membrane (see discussion of hollow-fiber permeability measurements). For substrates where membrane solubility is not a problem, a convenient technique for reaction rate measurement is to connect the substrate compartment to a reservoir and continuously recirculate the solution through both the substrate compartment of the hollow fiber device and the reservoir. The substrate (or product) concentration in the reservoir may be monitored and from the rate of change of substrate (or product) concentration, the reaction rate of the hollow-fiber immobilized enzyme may be determined. It is also useful to measure substrate (or product) concentrations and enzyme activities in the enzyme compartment, whether continuous flow-steady state or recirculation-reservoir experiments are employed. Enzyme compartment measurements significantly assist in the interpretation of experimental results, the interpretation of which in their absence may be difficult[15] or tedious.

Immediately at the end of the experiment, both lumen and shell sides

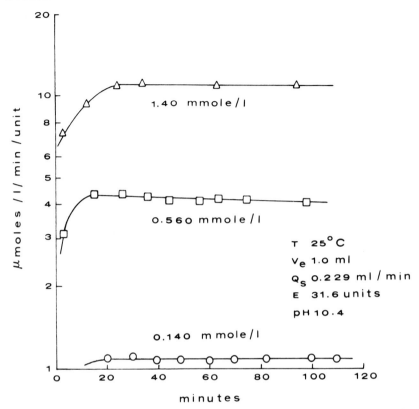

FIG. 5. Reaction velocity as a function of time for the conversion of p-nitrophenylphosphate to p-nitrophenolate by alkaline phosphate (31.6 units) in the shell compartment (1.0 ml) of a Biofiber 50 Minitube. Substrate at 0.14, 0.56, and 1.40 mmoles/liter is continuously pumped through the lumen at 0.229 ml/min and 25°. From J. C. Davis, *Biotechnol. Bioeng.* **16**, 113 (1974).

are purged thoroughly with buffer. Buffer or sodium chloride is ultrafiltered into the enzyme-containing compartment with about 5 psi pressure differential for several hours. This ultrafiltration procedure, commonly called back-flushing, serves to dislodge adsorbed protein, substrate, salt, and bound particulates. After vigorous flushing of each compartment with deionized water, the device is stored in the refrigerator in a solution containing a membrane-compatible bacteriostat. If more extensive cleaning is required, incubating overnight in a solution containing trypsin, papain, or a commercial enzyme detergent (Gain) will often act to loosen any bound debris sufficiently so that it can then be removed by vigorous flushing. Several fibers are resistant for short time periods to dilute sodium hydroxide (Amicon and Romicon fibers) or 100-mg/liter sodium hypochlorite solutions (Romicon GM80). These solutions when back-flushed through the fibers have been found to be extremely effective in cleaning clogged fibers.[16,17]

Modified Soluble Enzymes

In order to improve the half-life of soluble enzymes, a number of techniques have been used. Prominent among these are the covalent coupling of macromolecules to the enzyme. O'Neill et al.[18] increased the stability of chymotrypsin by attaching it to a soluble dextran (mw 2×10^6) and also to insoluble DEAE-cellulose. The activity decay curve for each preparation is given in Fig. 6. These authors also found that chymotrypsin attached to the soluble polymer exhibited considerably higher activity than the enzyme bound to the insoluble carrier. Although the modified soluble enzyme did show a slightly higher rate of inactivation than that bound to the insoluble support, its stability is far superior to that of the native enzyme, probably owing to the prevention of autodigestion as well as some conformational stabilization.

A small Amicon ultrafiltration reactor in which the dextran-linked chymotrypsin was retained by a flat 100,000 mw cutoff membrane (Amicon XM-100) was operated continuously on a casein substrate solution. In 2 weeks of operation, there was no leakage of the polymer-linked enzyme through the nominal 100,000 mw membrane, whereas the native enzyme would readily pass through this membrane. Thus, polymer-linked enzymes provide increased stability over native enzymes and make available a wider selection of hollow-fiber membranes for complete enzyme retention.

Immobilized-Enzyme Slurry

An alternative technique for stabilization of enzymes for use with hollow fibers is immobilization of the enzyme on particles small enough to be circulated through or placed within the hollow-fiber lumen. The use of a slurry of enzymes immobilized on small particles circulated through a hollow-fiber lumen should virtually eliminate the Nernst diffusion layer on the internal surface of the hollow fiber.

[16] B. R. Breslau, E. A. Agranat, A. J. Testa, S. Messinger, and R. A. Cross, *Meeting Am. Inst. Chem. Eng.*, Houston, Tex. (1975).
[17] C. van Alteena, *Proc. Biochem.* **10**(2), 26 (March, 1975).
[18] S. P. O'Neill, J. R. Wykes, P. Dunnill, and M. D. Lilly, *Biotechnol. Bioeng.* **13**, 319 (1971).

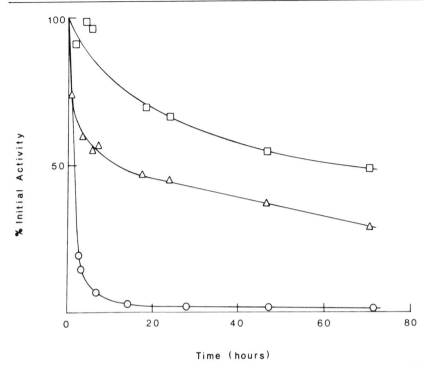

FIG. 6. Enzyme activity—time profiles for (○) native soluble chymotrypsin, (△) soluble dextran-coupled chymotrypsin, and (□) chymotrypsin bound to DEAE-cellulose particles. From S. P. O'Neill, J. R. Wykes, P. Dunnill, and M. D. Lilly, *Biotechnol. Bioeng.* 13, 319 (1971).

Cohen et al.[19] immobilized trypsin on Affi-Gel 703, 1–3 μm polyacrylamide beads containing hydrazide functional groups. Twenty milligrams of the Affi-gel were washed with 0.3 N HCl then resuspended in the same solvent and converted to the azide derivative by incubation with 0.6 ml of 1 M sodium nitrite for 20 min. The azide was rapidly washed with 5 mM sodium chloride and incubated for 4 hr with 25 mg of trypsin in 5 mM barbital buffer, pH 8.0. The covalently immobilized enzyme was washed and stored in 10 mM Tris-HCl buffer, pH 7.5. Each operation was performed in 10 ml of cold solvent (0°–4°). Recovery of the gel after each washing and incubation required centrifugation for 5 min at 55,000 g. The gel was dispersed as necessary during each step of the immobilization procedure via sonication in a Cole Palmer Ultrasonic Cleaner model 8845-6. The aggregation of the gel was monitored by microscopic examination.

[19] W. Cohen, W. H. Baricos, and R. P. Chambers, unpublished data, 1975.

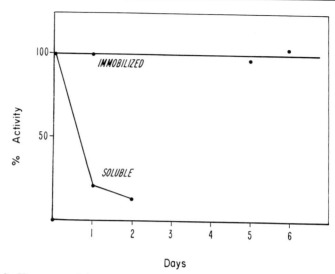

Fig. 7. Enzyme activity-time profiles for a 0.4% slurry of immobilized trypsin (300 units per gram of solid) which was circulated through the lumen of a Biofiber 50 beaker, and for native enzyme similarly treated. The trypsin was immobilized on 1–3 μm particles of Affi-gel 703 polyacrylamide beads by covalent coupling with the hydrazide functional groups on the beads. W. Cohen, W. H. Baricos, and R. P. Chambers, upublished data, 1975.

Ten milliliters of a 0.4% slurry of immobilized trypsin (300 units per gram of solid) were circulated through the lumen of a Biofiber 50 hollow-fiber beaker containing 100 ml of 1 mM DL-benzoylarginine p-nitroanilide in 10 mM Tris buffer, pH 8.0. Each day the partially hydrolyzed substrate was removed from the beaker and replaced by fresh substrate solution. A parallel control experiment with native trypsin was also run. The results, plotted in Fig. 7, show that the immobilized enzyme slurry retained 100% activity for 6 days whereas the soluble enzyme rapidly deteriorated. The stability enhancement resulted since the Affi-gel-immobilized enzyme was protected against autodigestion. It is also important to note that the Affi-gel-immobilized enzyme retained excellent activity even thought it was circulated through a peristaltic pump and the lumen of the fibers. Therefore, it appears feasible to combine in this way the stability advantages of particulate immobilized enzyme with the semipermeability properties of hollow fibers.

Enzyme Absorbed within Hollow-Fiber Membrane

An enzyme-containing solution in contact with the spongy layer of anisotropic membranes will fill the large pores of this layer, and the semipermeable membrane on the thin, inner skin will prevent this enzyme solution from penetrating to the hollow-fiber lumen. Utilizing this behavior, enzymes have been immobilized by Waterland, Michaels, and Robertson.[20,21] The enzyme is physically absorbed within the macroporous region of the hollow fiber, a 75-μm thick spongy layer with 80–90% void space and 5–10 μm pores (see Fig. 2). Waterland and co-workers immobilized the enzyme by simply soaking the shell side of the device in an enzyme solution and then draining the liquid from the shell compartment. Specifically, the shell compartment of a clean X50 hollow-fiber device was filled with 25 ml of a β-galactosidase (Worthington) enzyme solution, 2.5 units/ml containing 1 mM MgCl$_2$, 0.5 mM disodium EDTA, and 0.1 M phosphate buffer at pH 7.2. After overnight incubation, the device was drained and refilled with enzyme solution of the same concentration. The device was then gently agitated at room temperature for 3 hr. The fluid was drained from the shell compartment, then the device containing membrane-immobilized enzyme was brought in contact with a O-nitrophenyl-β-D-galactopyranoside substrate solution pumped through the lumen. The hollow-fiber device was found to retain 42 units of enzyme by this procedure, most of which was apparently in free solution within the pore structure of the membrane.

Dramatically higher enzyme loading has been achieved by a significant advancement to this technique. Breslau and Kilcullen[22] reported the achievement of 5.96 mg lactase/cm^2 Romicon XM50 membrane surface (117 mg/ml membrane matrix volume) using their new ultrafiltration enzyme-loading technique. At this high loading, the enzyme that is absorbed within the membrane matrix has been found to have gellike properties. The enzyme is immobilized by pumping enzyme solution into the shell side of a tubular hollow-fiber device. The shell-side exit valve is closed, and the enzyme solution is forced through the membrane. Permeate is returned to the original enzyme solution. Since the enzyme solution makes contact with the spongy side of the anisotropic fibers first and cannot permeate the skin on the inside of the fibers, the enzyme protein builds up in the spongy layer. Enzyme absorption in the spongy layer is followed by monitoring enzyme activity in the reservoir. High enzyme loading is achieved by employing high concentrations in the enzyme solution and successive repetitions of the loading process.

The membrane-absorbed enzyme is brought in contact with substrate

[20] L. R. Waterland, A. S. Michaels, and C. R. Robertson, *AIChE J.* **20**, 50 (1974).
[21] L. R. Waterland, C. R. Robertson, and A. S. Michaels, *Chem. Eng. Commun.* **2**, 37 (1975).
[22] B. R. Breslau and B. M. Kilcullen, in "Enzyme Engineering" (E. K. Pye and H. H. Weetall, eds.), Vol. 3. Plenum, New York, 1976.

TABLE V
LACTOSE HYDROLYSIS BY MEMBRANE-ABSORBED ENZYME[a]

Mode of operation		Conversion (%)
Ultrafiltration[b]	Number of times device contacted	
	1	21.1
	2	30.8
	3	36.0
	4	40.0
Ultrafiltration superimposed on dialytic permeation[c]	Time (min)	
	15	3.8
	60	10.6
	120	17.7
	180	22.0

[a] Wallerstein lactase (200 LU/mg) absorbed in 45-XM50 by ultrafiltration loading technique (5.96 mg/cm^2 membrane area), 10.0% lactose substrate.

[b] Substrate ultrafiltered from shell to lumen across enzyme-absorbed fibers. Solution from first pass was measured giving a steady-state conversion of 21.1%. This solution was again contacted with the enzyme-absorbed fiber, giving a second steady-state conversion of 30.8%, etc.

[c] A recirculation system was employed with fluid entering the lumen at 25 psi and leaving the lumen at 23.5 psi; shell-side fluid at 24.25 psi, no flow.

solution by either of two methods: ultrafiltration in which substrate enters the shell side and leaves the lumen side of the hollow-fiber device in the same direction as when the enzyme was loaded; and ultrafiltration superimposed on dialytic permeation, in which the substrate enters and leaves on the lumen side with the shell-side valves turned off. It should be noted that in this second method, the enzyme can be washed out of the spongy layer by ultrafiltration. In order to prevent this enzyme loss, Breslau and Kilcullen[22] fixed the membrane-absorbed enzyme in place by ultrafiltering a glutaraldehyde-containing solution through the membrane containing the absorbed enzyme.

Table V shows experimental data of Breslau and Kilcullen[22] using β-galactosidase (Wallerstein, 200 LU/mg), which was ultrafiltration-loaded into a Romicon 45-XM50 hollow fiber and used for the hydrolysis of lactose.

Complex Multiple Enzyme and Cofactor Immobilization by Hollow Fibers

The investigation of the simultaneous interaction of many enzymes is facilitated by the use of hollow-fiber immobilization techniques. Ford[10] and Chambers et al.[11] studied the behavior of the cofactor-containing four-enzyme system—alcohol dehydrogenase (ADH) and aldehyde dehydrogenase (ALDDH), diaphorase (DIA), and catalase—immobilized as native soluble enzymes in a recirculating solution in the lumen compartment of a Biofiber 20 beaker device. The low molecular weight cutoff (200 MW nominal) of the device permitted the complete retention of all enzymes and cofactor (NAD), while allowing dialytic permeation of the substrates ethanol, acetaldehyde, and acetate (see Fig. 8 and Table VI).

Continuous-flow, steady-state ethanol oxidation revealed that experimentally measured and predicted oxidation rates and acetate selectivities were in very close agreement. Predicted values were based on native enzyme kinetic constants and separately measured hollow-fiber permeabilities. Table VI presents data for ethanol oxidation rates and acetate selectivity (the ratio of the rate of acetate production to the rate of ethanol oxidation). The favorable agreement is an excellent confirmation that physical immobilization of soluble enzymes does not alter their kinetic behavior. In parallel studies the investigators showed how hollow-fiber enzyme immobilization could be used to cause the selective production of acetate or acetaldehyde and how the optimum proportion of each enzyme can be determined in order to maximize the ethanol oxidation rate.

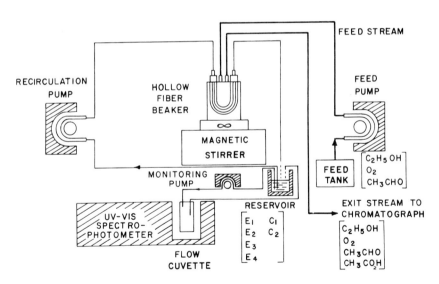

Fig. 8. Schematic illustration of experimental system for the study of hollow-fiber immobilized enzymes [J. R. Ford, Ph.D. dissertation with R. Chambers, Tulane University, 1972, Diss. Abstr. Int. B 33, 3624; R. P. Chambers, J. R. Ford, J. H. Allender, W. H. Baricos, and W. Cohen, in "Enzyme Engineering" (E. K. Pye and L. B. Wingard, eds.), Vol. 2, p. 195. Plenum, New York, 1974)].

TABLE VI
HOLLOW-FIBER IMMOBILIZED MULTIPLE-ENZYME SYSTEM[a,b]

Enzymes present (units)[c]			Ethanol oxidation rate (μmoles/min)		Acetate selectivity	
ADH	ALDDH	DIA	Experimental	Predicted	Experimental	Predicted
120	10	13	1.40	1.49	1.00	0.99
120	10	10	1.22	1.30	1.00	0.99
120	2	10	0.73	0.66	0.86	0.92
30	7	30	0.67	0.59	0.68	0.68
2.5	0.5	2.5	0.10	0.11	0.26	0.23

[a] R. P. Chambers, J. R. Ford, J. H. Allender, W. H. Baricos, and W. Cohen, in "Enzyme Engineering" (E. K. Pye and L. B. Wingard, eds.), Vol. 2, p. 195. Plenum Press, New York, 1974. J. R. Ford, Ph.D. dissertation with R. Chambers, Tulane University, 1972, Diss. Abstr. Int. B **33**, 3624.
[b] Feed solution contained 0.2 M ethanol, 1 mM dithiothreitol, 0.2 M potassium phosphate buffer, pH 8.0.
[c] Hollow-fiber lumen solution contained yeast alcohol dehydrogenase (ADH; Sigma), yeast aldehyde dehydrogenase (ALDDH; Sigma), pig heart diaphorase (DIA ; Sigma) in the units specified plus 10,000 units of catalase (Sigma), 1 mg/ml bovine serum albumin, 1.4–2.4 mM NAD, 0.2 M ethanol, 1 mM dithiothreitol, 0.2 M potassium phosphate buffer, pH 8.0, 25°.

The use of hollow-fiber immobilized cofactor-requiring enzymes presents several unique problems, which have been recently discussed by Baricos et al.[23,24] Hollow fibers, which are permeable to cofactors yet retain enzymes, have been utilized by Fink and Rodwell,[14] who investigated multiple cycling of NAD in a hollow-fiber system.

The recent development of various soluble, macromolecular cofactor derivatives is of particular interest.[11,25–30] These derivatives provide potential improvement over the earlier approach of Ford and Chambers

[23] W. H. Baricos, R. P. Chambers, and W. Cohen, Anal. Lett. **9**, No. 3 (1976).
[24] W. H. Baricos, R. P. Chambers, and W. Cohen, Enzyme Technol. Dig. **4**, 39 (1975).
[25] P. Larsson and K. Mosbach, FEBS Lett. **46**, 119 (1974).
[26] C. R. Lowe and K. Mobach, Eur. J. Biochem. **49**, 511 (1974).
[27] H. F. Voss, C. Y. Lee, and N. A. Kaplan, Meeting Am. Inst. Chem. Eng., Los Angeles, Calif., Paper 37d, Fiche 67 (1975).
[28] M. K. Weibel, C. W. Fuller, J. M. Stadel, A. F. E. P. Buckmann, T. Doyle, and H. J. Bright, in "Enzyme Engineering" (E. K. Pye and L. B. Wingard, eds.), Vol. 2, p. 217. Plenum, New York, 1974.
[29] J. R. Wykes, P. Dunnill, and M. D. Lilly, Biochim. Biophys. Acta **286**, 260 (1972).
[30] J. R. Wykes, P. Dunnill, and M. D. Lilly, Biotechnol. Bioeng. **17**, 51 (1975).

et al., since macromolecular cofactors permit the use of higher molecular weight-cutoff hollow fibers which do not restrict substrate permeation. However, one must recognize that the enzyme kinetic characteristics of these macromolecular cofactors may be quite inferior to that of the native cofactor.[27]

A recent novel use of hollow-fiber immobilized enzymes is the immobilization of the liver microsomal enzyme system by Cohen, Baricos, Chambers, and co-workers.[31]

[31] W. Cohen, W. H. Baricos, P. R. Kastl, and R. P. Chambers, *in* "Biomedical Applications of Immobilized Enzymes and Proteins" (T. M. S. Chang, ed.). Plenum, New York, 1976.

[24] Immobilized Enzymes—Miscellaneous Methods and General Classification

By OSKAR R. ZABORSKY

Miscellaneous Methods for Immobilizing Enzymes

This section deals with various methods for immobilizing enzymes that either are based on this author's work or are not extensively covered by others in this volume. Some of these methods may not be the method of choice for every immobilization or for every enzyme, but they do have their distinctive advantages, which may turn out to fulfill the need. They are described under the following headings:

Immobilization of Glycoenzymes via Activation of Carbohydrate Residues
Immobilization with Imidoester-Containing Polymers
Immobilization on Magnetic Supports
Immobilization via Copolymerization with N-Carboxyamino Acid Anhydrides
Immobilization with Nonpermanent Microcapsules—Liquid Surfactant Membranes

Immobilization of Glycoenzymes via Activation of Carbohydrate Residues

Prior to our report,[1] all chemical techniques of immobilization involved only the modification of the amino acid residues of an enzyme, even

[1] O. R. Zaborsky and J. Ogletree, *Biochem. Biophys. Res. Commun.* **61**, 210 (1974).

though the molecule may have contained other functional groups that could have been employed. For example, glycoenzymes such as glucose oxidase or glucoamylase had been immobilized by chemical modification of only their amino acid residues, not by their carbohydrate moieties. Because our work and that of others on the function of the carbohydrate residues of glucose oxidase revealed that the sugar moieties are not implicated in catalysis, we reasoned that it would be more desirable to bond glycoenzymes covalently to water-insoluble polymers by these catalytically nonessential carbohydrate groups rather than by amino acid groups, some of which are responsible for substrate binding and catalysis. The principle of the method is outlined in Eq. (1) and consists of two steps:

$$\text{Native} \xrightarrow{H_5IO_6} \text{Oxidized} \xrightarrow[\text{amino polymer}]{P-NH_2 \; / \; \text{water-insoluble}} \text{Water-insoluble} \quad (1)$$

(1) activating a glycoenzyme via periodate oxidation and (2) contacting the modified "aldehydic" enzyme with an amino-containing water-insoluble polymer. In this chemical method, as in others, it is desirable to bond the enzyme to the polymer through only nonessential groups; in this instance it is dependent on being able to oxidize carbohydrate residues of the glycoenzyme without losing activity. Although the exact role of carbohydrate residues in glycoenzymes remains to be established, the evidence gathered by us and others argues against their involvement in catalysis.

Periodate oxidation is a well-established procedure for modifying carbohydrate residues of polysaccharides and glycoproteins, although with the latter there may be some oxidation of Cys-, Cys, Tyr, Trp, and Met. Oxidation of glucose oxidase with periodic acid, as described below, resulted in the oxidation of approximately 60% of the carbohydrate residues of the enzyme, a molecule which is composed of 16% carbohydrate having

TABLE I
SPECIFIC ACTIVITIES OF GLUCOSE OXIDASES

Glucose oxidase	Specific activity (units/mg protein)
Native	91.9[a,b]
Oxidized	92.5[a,b]
p-Aminostyrene bound	93.1[c]

[a] Based on the amount of protein as determined by anaerobic spectral titration with glucose. This method measures only catalytically active protein.
[b] No difference in specific activity between the native and oxidized enzyme was noted when the protein concentration was determined by using the extinction coefficient at 450 nm. This method measures total protein.
[c] Based on protein as determined by amino acid analysis. Protein loading was 5.87 mg of enzyme per gram of enzyme–polymer conjugate.

128 mannose, 19 glucosamine, and 3 galactose residues per 150,000 daltons. Amino acid analysis revealed that no periodate-sensitive amino acids were oxidized. Likewise, no changes were observed in the spectral properties (ultraviolet, visible, or circular dichroism) of the oxidized enzyme compared with the native enzyme.[2] Most important, the specific activity of the oxidized enzyme was identical to that of the native enzyme (Table I). There was neither a loss of activity nor precipitation of the enzyme as had been reported by Pazur et al.[3] Because no activity was lost upon oxidation, it is tempting to suggest that perhaps carbohydrate residues are not involved in catalysis. However, this may not be the case at all, for 40% of the residues were *not* oxidized (presumably the more sterically hindered or periodate-resistant carbohydrate residues). Further, although the activity of glucose oxidase was not affected, the activity of other glycoenzymes is diminished by periodate oxidation, e.g., horseradish peroxidase[4] or α-amylase.[5]

Coupling of the oxidized glucose oxidase to amino-containing supports such as p-aminostyrene, aminoethylcellulose, aminoalkyl derivatives of glass and agarose, and hydrazides of carboxymethyl cellulose and polyacrylamide can be accomplished either in slightly acidic or alkaline solutions (pH 5–9) and results in active enzyme–polymer conjugates, presumably through imine or hydrazone linkages. The protein loading (milligrams of enzyme per gram of enzyme–polymer conjugate) with p-aminostyrene was 5–8 mg, but the activity of the immobilized enzyme was high. In fact, within experimental error, it was equivalent to the native and soluble

[2] O. R. Zaborsky, unpublished results, 1973.
[3] J. H. Pazur, K. Kleppe, and A. Cepure, *Arch. Biochem. Biophys.* **111**, 351 (1965).
[4] I. Weinryb, *Biochem. Biophys. Res. Commun.* **31**, 110 (1968).
[5] K. Maekawa and K. Kitawa, *Proc. Jpn. Acad.* **29**, 353 (1953).

oxidized enzymes. With other supports, the loading was much higher. The thermal stability of the oxidized glucose-p-aminostyrene conjugate was also superior to that of the native enzyme or the soluble oxidized form. This method has been extended to other glycoenzymes such as peroxidase (where even some partial inactivation occurs). Periodate oxidation followed by coupling to amino-containing supports could be used also for immobilizing carbohydrate-containing hormones, proteins, or immunoglobulins. The approach could be adapted to other heteroproteins.

Besides being a relatively simple method for immobilizing natural and perhaps synthetically produced glycoproteins, the most outstanding advantage of this technique is that the immobilization occurs via catalytically nonessential residues of the enzyme.

Procedure

As an example of this procedure, the oxidation and coupling of glucose oxidase is described.

Oxidation of Glucose Oxidase with Periodic Acid. To a stirred solution of glucose oxidase (40.2 mg, 0.268 μmole, in 20 ml of 50 mM acetate buffer, pH 5.60) in a thermostated vessel at 25° protected from light is added 0.4 ml of a periodic acid solution (9.12 mg, 40.0 μmole). The yellow solution is stirred for 4 hr, after which 0.025 ml of ethylene glycol, 0.448 mmole, is added and stirred for an additional 0.5 hr. The solution is transferred to an Amicon Model 202 ultrafiltration cell equipped with an XM-50 filter at 50 psi N_2 pressure and dialyzed with 50 mM acetate buffer, pH 5.59, until no further dialyzable material comes forth. The conversion of periodic acid after 4 hr of reaction with glucose oxidase, based on the absorbance change at 223 nm, is 61.6%. The oxidized enzyme is stored at 5°.

Coupling of Oxidized Glucose Oxidase to p-Aminostyrene. To a 5-ml solution of oxidized glucose oxidase (10 mg) is added 250 mg of finely powdered p-aminostyrene. The suspension is adjusted to pH 9 with 100 mM NaOH, stirred at about 25° for 1 hr, and then filtered with a Millipore 0.45-μm filter. The solid is washed with 1 liter of H_2O and 1 liter of 1 M NaCl in 50 mM phosphate buffer, pH 6.4. After several milliliters of the H_2O wash, no further activity is detected in the wash. The original filtrate exhibits high activity; the NaCl wash exhibits no activity.

Immobilization with Imidoester-Containing Polymers

The imidoester functionality has numerous proved and potential desirable characteristics with regard to protein modification and immobilization.

Imidoesters react selectively with only the terminal α-amino group and the ε-amino groups of lysyl residues in proteins to form amidines[6] [Eq. (2)]. With enzymes, high catalytic activity is usually retained upon modification because only negligible conformational changes are induced into the protein—a consequence of the fact that the charge on lysyl

$$\begin{array}{c} R-\overset{+}{C}=NH_2 \\ | \\ OR' \end{array} + E-NH_2 \underset{-H^+}{\overset{}{\rightleftharpoons}} \begin{array}{c} R-\overset{+}{C}=NH_2 \\ | \\ NH-E \end{array} + R'OH \quad (2)$$

$$E-\overset{+}{N}H_3$$

residues is not eliminated. Furthermore, the original lysyl-containing protein can be regenerated by reversing the amidination reaction; removal of the amidino group can be accomplished by treating the modified protein with an ammonia–ammonium acetate solution or with hydrazine (Eq. 3).

$$\begin{array}{c} R-\overset{+}{C}=NH_2 \\ | \\ NH-E \end{array} + NH_3 \rightleftharpoons \begin{array}{c} R-\overset{+}{C}=NH_2 \\ | \\ NH_2 \end{array} + E-NH_2 \quad (3)$$

This reversibility, known for low-molecular-weight amines and ammonia, would permit either the replacement of inactivated, bound enzyme with fresh, active enzyme or the simultaneous attachment of different enzymes, E_1, E_2, etc. (Eq. 4).

$$\begin{array}{c} |-\overset{+}{C}=NH_2 \\ | \\ NH-E \end{array} \xrightarrow{E'-NH_2} \begin{array}{c} |-C=NH_2 \\ | \\ NH-E' \end{array} + E-NH_2 \quad (4)$$

Low-molecular-weight imidoesters exhibit varying degrees of stability in aqueous solution. Depending on the pH, hydrolysis produces esters, amides, or nitriles (Eq. 5). Yet even this apparent disadvantage (i.e., loss of coupling power) can be viewed as beneficial with regard to immobilization. The hydrolysis of any unreacted imidoester group of an enzyme-

$$\begin{array}{c} \overset{H^+/H_2O}{\nearrow} \begin{array}{c} R-C=O \\ | \\ OR' \end{array} + NH_4^+ \\ R-\overset{+}{C}=NH_2 \\ | \\ OR' \\ \overset{H_2O}{\searrow}_{OH^-} \\ \searrow R-C=O \\ | \\ NH_2 \end{array} + R'OH \text{ or } R-C\equiv N + R'OH \quad (5)$$

[6] M. J. Hunter and M. L. Ludwig, this series, Vol. 25B, p. 585 (1972).

polymer conjugate is desirable because the positively charged imido group is transformed into a uncharged group (ester, amid, or nitrile), and nonspecific adsorption of proteins or other charged molecules would be reduced. Charge elimination would occur only at the nonenzymically modified positions of the imidoester polymer. In contrast to low-molecular-weight imidoesters, imidoester-containing polymers derived from polyacrylnitrile have considerable stability in aqueous media.

Imidoesters can also be readily synthesized from a variety of inexpensive starting materials. Most often, low-molecular-weight imidoesters are synthesized from nitriles and alcohols via the Pinner synthesis (Eq. 6) or from amides with dimethyl sulfate or triethyloxonium tetrafluoroborate (O-alkylations), and numerous different esters have been so prepared

$$\text{R}-\text{C}\equiv\text{N} + \text{R}'\text{OH} + \text{HCl} \rightarrow \text{R}-\underset{\text{OR}'}{\text{C}}=\overset{+}{\text{N}}\text{H}_2\text{Cl}^- \qquad (6)$$

and reported. Imidoester-containing polymers can also be prepared through similar procedures, but their synthesis has been reported by only several groups. We favored the synthesis of imidoester-containing polymers from the appropriate nitrile-containing precursors because they are available and inexpensive; they have good mechanical, chemical, and microbial stability; they come in a variety of forms (powder, fiber, film, etc.); and they offer a wide range of microenvironments for the to-be-immobilized enzyme. A rather large selection of nitrile-containing supports (homopolymers, copolymers, graft polymers, and cyanoethylated materials) makes different microenvironments possible.

Although the preparation of imidoester-containing polymers via the Pinner synthesis was reported in the 1960s by Tazuke et al.,[7] immobilization of enzymes to these supports was demonstrated first by Zaborsky in the early 1970s.[8-10] In turn, the immobilization of enzymes to imidoester-containing polymers prepared from polyamides and O-alkylating reagents was demonstrated by Hornby et al.[11] (see Hornby and Goldstein, this volume [9]).

[7] S. Tazuke, K. Hayashi, and S. Okamura, *Kobunshi Kagaku* **22**, 259 (1965).
[8] O. R. Zaborsky, U.S. Patent No. 3,830,699 (1974).
[9] O. R. Zaborsky, in "Immobilized Enzymes in Food and Microbial Processes" (A. C. Olson and C. L. Cooney, eds.), pp. 187–203. Plenum, New York, 1974.
[10] O. R. Zaborsky, in "Enzyme Engineering" (E. K. Pye and L. B. Wingard, Jr., eds.), Vol. 2, p. 115–122. Plenum, New York, 1974.
[11] W. E. Hornby, J. Campbell, D. J. Inman, and D. L. Morris, in "Enzyme Engineering" (E. K. Pye and L. B. Wingard, Jr., eds.), Vol. 3, pp. 401–407. Plenum, New York, 1974.

A brief summary of pertinent facts about the preparation and properties of methyl imidoester-containing polyacrylonitrile (PANIE), the most prevalent ester used by us, and about the properties of the trypsin and chymotrypsin conjugates follows.

Various methyl imidoester-containing polymers were prepared from the corresponding nitrile polymers, methanol and HCl gas being used, and the extent of modification was increased at higher temperatures and longer reaction times. Under the conditions described below, the extent of modification for the polyacrylonitrile powder was usually 2–4%. Although the value is rather low, it represents only surface modification for polyacrylonitrile does not swell in methanol. Likewise, the lower modification obtained with fibers (usually <1%) is the result of an even lower surface area. Cosolvents can be used to swell nitrile-containing polymers and to increase the extent of modification.

The coupling of enzymes such as trypsin and α-chymotrypsin to the imidoesters-containing polymers was performed at alkaline pH (from 8.5 to 10) usually for 1–2 hr at room temperature. As expected, the amount of loading increased with increasing pH of the coupling solution and with the time of coupling. Typical protein loading for these enzymes ranged from 1.5 to 8 mg of enzyme per gram of enzyme–polymer conjugate. Specifically, the protein loading for trypsin on the PANIE powder coupled at pH 9.5 for 0.5 hr and coupled at pH 10.0 for 5 hr was 2.3 and 6.8 mg of trypsin per gram of enzyme PANIE conjugate, respectively. The low enzyme loading capacity is, of course, a direct reflection of the low modification of the parent polymer.

The activity of several enzymes immobilized on methyl imidoester-containing polyacrylonitrile powder was quite good, with trypsin and α-chymotrypsin exhibiting no loss in activity. A slight activation was even observed toward low-molecular-weight substrates. The thermal stability of immobilized trypsin and α-chymotrypsin was also superior to the soluble enzymes, and good storage stability was also noted. With the trypsin-PANIE conjugate, no loss of activity was observed even after one year of storage at 5° in 100 mM phosphate buffer, pH 6.7. In addition, the trypsin conjugate lost no activity upon lyophilization, nor was there any enzyme "leakage" which could have been caused conceivably by degradation of the support or by physical desorption. Ammonolysis of trypsin-PANIE conjugates revealed a reduction in protein, but the loss was rather negligible.[9] Reversibility of the amidination reactions was more easily demonstrated by using a higher protein-containing *bis*-imidoester-treated α-chymotrypsin conjugate.

Although imidoester-containing polymers are good supports for immobilizing enzymes and, in principle, have certain characteristics which make them potentially superior to presently available ones, the disadvantages of this approach are the slowness of the synthesis reaction, the low modification, and the need for protein coupling under alkaline pH (the latter a characteristic of the imido functionality).

Procedure

A typical procedure for the preparation of a methyl imidoester-containing polymer and its subsequent use for immobilization is described.

Preparation of Methyl Imidoester-Containing Polyacrylonitrile (PANIE). Dry hydrogen chloride gas is slowly introduced into a cooled ($-10°$ to $5°$), magnetically stirred suspension of 10 g (0.19 mole of nitrile) of polyacrylonitrile powder and 100 ml of methanol (previously dried over 4 Å molecular sieves and distilled) in a 250-ml 3-necked round-bottom flask. The flask is equipped with a low-temperature thermometer and a drying tube. Addition of the HCl gas is carefully monitored because heat is evolved at the initial stage of the reaction. After addition of the hydrogen chloride gas (usually at least a 10 M excess, 68.5 g), the mixture is allowed to come to room temperature and stirred for 2 to 3 days. The modified polyacrylonitrile powder is filtered with a sintered-glass funnel, washed thoroughly with dry methanol and anhydrous ether, and vacuum dried over KOH flakes. The imidoester content of the resulting material is determined by titration of the hydrochloride salt with base, potentiometric titration of the soluble chloride with silver nitrate, or infrared analysis.

Coupling of Trypsin to PANIE. A solution of trypsin (100 mg dissolved in 10 ml of 100 mM CaCl$_2$) is placed into a titration vessel equipped with a magnetic stirring bar. The pH is adjusted to 9.0, and 1.0 g of PANIE powder is added portionwise at room temperature keeping the pH constant by continued addition of base. After the final portion of the activated polymer is added (at about 25 min), the mixture is stirred for 2–3 hr at room temperature. The suspension is then filtered with a Millipore filtration unit (0.45 µm filter) and the enzyme conjugate is thoroughly washed with 100 mM CaCl$_2$, water, and various buffer and salt solutions until no enzymic activity is detected. The enzyme conjugate is stored in buffers or in water (for trypsin, usually in a 50 mM Tris-HCl buffer, pH 7, in 50 mM CaCl$_2$).

Immobilization on Magnetic Supports

A recent innovation for immobilizing enzymes is the use of magnetic materials. First referred to by Hedén,[12] and more recently expanded by

[12] C.-G. Hedén, *Biotechnol. Bioeng. Symp.* **3**, 173 (1972).

Robinson et al.,[13] Van Leemputten and Horisberger,[14] and Gellf and Boudrant,[15] the method involves simply the use of magnetic materials instead of conventional water-insoluble supports. Variations exist in the methodology: aminoalkylation of magnetite by 3-aminopropyltriethoxysilane followed by activation and bonding of the enzyme[13]; adsorption of the enzyme onto magnetite followed by cross-linking[14]; composite formation of cellulose and magnetite followed by activation of the carbohydrate with cyanogen bromide and subsequent coupling of the enzyme to the activated cellulose–magnetite complex,[13] simultaneous microencapsulation of the enzyme with magnetite in permanent membrane microcapsules, incorporation of magnetic supports in a copolymeric gel. Although any magnetic material could be used, only magnetite, magnetic iron oxide (Fe_3O_4), has been employed to date. The diameter of the particles normally used has been in the range of 100–700 μm.

Advantages of the magnetic supports are ease of separation and process control. An enzyme immobilized on a magnetic support can be separated magnetically from viscous solution or suspension (even if the substrate is itself insoluble) or separated from a multienzyme component mixture if other enzymes are immobilized on nonmagnetic supports. This latter operation could conceivably be used to advantage for separating and recharging an immobilized multienzyme reactor if the enzymes inactivate at different rates. In addition, enzymes immobilized on magnetic particles seem to be ideally suited for fluidized bed operations because of their density and ease of retainment. Even at high flow rates, carryover can be prevented by surrounding the outlet with a magnetic field.[15]

Although no extensive characterization of the properties of these derivatized enzymes has been done, the evidence presented by Robinson et al.[13] indicates that the magnetic support does not change the activity of the immobilized enzyme; the activity of an immobilized α-chymotrypsin was comparable to that of other reported preparations.

Procedure

The procedure for the activation and coupling of enzymes to magnetite by aminoalkylation is identical to that described by Weetall (this volume [10]), and the procedure for adsorption followed by cross-linking with glutaraldehyde (enzyme only or with a coprotein) is similar to that

[13] P. J. Robinson, P. Dunnill, and M. D. Lilly, *Biotechnol. Bioeng.* **15**, 603 (1973).
[14] E. Van Leemputten and M. Horisberger, *Biotechnol. Bioeng.* **16**, 385 (1974).
[15] G. Gellf and J. Boudrant, *Biochim. Biophys. Acta* **334**, 467 (1974).

described by Broun (this volume [20]). The preparation of the cellulose–iron oxide composite reported by Robinson et al.[13] is detailed below.

Ten grams of α-cellulose is added into 100 ml of freshly prepared, dark-blue cuprammonium hydroxide solution, and the mixture is stirred until the polysaccharide dissolves. Precipitated magnetic iron oxide power (10 g) is stirred into the cellulose solution, and the suspension is extruded under pressure through a 2-mm nozzle into 200 mM HCl and allowed to equilibrate for 1 hr. The thread is washed, dried, crushed, and sieved to obtain a range of particle sizes. Residual Cu^{2+} is removed by treating the powder with 200 mM EDTA in alkaline solution for 10 hr. The composite is then washed with water.

α-Cellulose is prepared by treating 50 g of cellulose (Whatman grade CF2) with 15% NaOH at 50° for 2 hr, followed by filtration and washing. The cuprammonium hydroxide solution is prepared by adding freshly precipitated cupric hydroxide in excess to 100 ml of cold ammonia solution (SG 0.88). The resulting dark-blue solution is filtered with a glass fiber filter paper, cooled on ice, and subsequently used.

The cellulose–iron oxide composite is activated with cyanogen bromide in the manner described by Porath and Axén (this volume [3]).

Immobilization via Copolymerization with N-Carboxyamino Acid Anhydrides

Stahmann and Becker[16] showed that the polymerization of reactive N-carboxyamino acid anhydrides can be initiated by the amino groups of proteins to yield the corresponding polypeptidyl derivatives. Peptide polymerization involves the α- and ε-amino groups of enzymes and consists of a chain elongation on these residues [Eq. (7)]. However, the polypeptidyl derivatives originally described by Stahmann and Becker and subsequently by Bar-Eli and Katchalski[17] and Glazer et al.[18] were still

$$H_2N\overset{NH_2}{\underset{NH_2}{E}} + HN\overset{CH(R)}{\underset{C(O)}{\diagdown C\diagup}}O \longrightarrow A-HN\overset{NH(A)}{\underset{NH-A}{E}}$$

where $A = -\left(\underset{O}{\overset{R}{C}}-\overset{|}{CH}-NH\right)_n \underset{O}{\overset{R}{C}}-\overset{|}{CH}-NH_2$

(7)

[16] M. A. Stahmann and R. R. Becker, *J. Am. Chem. Soc.* **74**, 2695 (1952).
[17] A. Bar-Eli and E. Katchalski, *Nature (London)* **188**, 856 (1960).
[18] A. N. Glazer, A. Bar-Eli, and E. Katchalski, *J. Biol. Chem.* **237**, 1832 (1962).

soluble. Nevertheless under appropriate conditions and with suitable
N-carboxyamino acid anhydrides, it is possible to obtain water-insoluble
enzyme derivatives. For example, Glazer et al.[18] reported the preparation
of a water-insoluble trypsin derivative prepared by the copolymerization
of the enzyme with N-carboxy-L-tryosine anhydride. The polytyrosyl
derivative dissolved readily in aqueous solution of pH values lower than
5 or higher than 9.5, but had a rather limited solubility between these
limits. The partially insoluble polytyrosyl trypsin derivative, insolubilized
by precipitation from solution with acetate, phosphate, or borate anions,
retained enzymic activity upon redissolution.

A more thorough study of this method of enzyme immobilization was
reported by Kirimura and Yoshida[19] in 1966. In their patent, they disclosed the preparation of water-insoluble aminoacylases through the
copolymerization with a host of N-carboxyamino acid anhydrides. Typical
amino acids whose N-carboxy anhydride derivatives could be employed
for the preparation of *water-insoluble* derivatives included glycine, alanine, α-amino-n-butyric acid, valine, leucine, isoleucine, phenylalanine,
and several β- and γ-esters of aspartic and glutamic acid. N-Carboxy
anhydrides of proline, serine, and threonine could not be used because the
resulting derivatives were still soluble. With regard to the degree of
insolubility of these modified polypeptidyl acylases, it should be noted
that the term "practically insoluble in water" was used several times by
these investigators. The immobilized acylases were active and exhibited
no loss in activity upon storage in buffer solutions at room temperature
for 2 weeks.

Although this immobilization technique is experimentally simple, its
major drawback is the production of enzyme derivatives whose insolubility varies with the conditions employed for preparation and use.

Procedure

The following experimental for the preparation of polytyrosyl trypsin
was reported by Glazer et al.[18]

To 0.5 g of trypsin dissolved in 72 ml of ice-cold buffer, pH 7.2 (in
the case of Glazer et al. consisting of 0.5 g of $MgSO_4$, 36 ml of 2.5 mM
HCl, and 36 ml of 100 mM phosphate buffer, pH 7.6) is added dropwise
and with stirring 0.8 g of N-carboxy-L-tryosine anhydride dissolved in 16
ml of anhydrous dioxane. The cloudy mixture is gently stirred at 4° for
16 hr and dialyzed for 7 days at 4° against a daily change of 6 liters of
2.5 mM HCl. The resulting clear solution containing the polytyrosyl
trypsin is lyophilized and stored at 4°.

[19] J. Kirimura and T. Yoshida, U.S. Patent No. 3,243,356 (1966).

Immobilization with Nonpermanent Microcapsules— Liquid Surfactant Membranes

The immobilization of enzymes with nonpermanent microcapsules has been developed and reported by a number of investigators, notably Li, May, and Mohan, from the Exxon Research and Engineering Company.[20-23] The method is similar to that developed by Chang (this volume [16]), but it has some potentially interesting and useful characteristics of its own.

The immobilization of an enzyme by this method involves encapsulating the protein solution within a semipermeable liquid-surfactant membrane. The term "liquid-surfactant membrane" refers to a water-immiscible phase composed of surfactants, additives, and a hydrocarbon solvent that contains emulsion-size aqueous droplets of various reagents or the enzyme. The size range of the emulsion-size droplets is approximately 1–100 μm in diameter; membrane thickness is usually 1–10 μm.

The general procedure of immobilization with liquid-surfactant membranes is similar to that used for microencapsulation with permanent membranes. The aqueous enzyme-containing solution is emulsified with a surfactant to form the liquid membrane encapsulated enzyme. The microcapsules are transferred to an aqueous solution to which is added the substrate. Substrates that can diffuse through the semipermeable liquid membrane are acted on by the entrapped and nondiffusible enzyme, and the product, once formed, can then diffuse through the membrane to the external aqueous phase. If desired, the enzyme can be immobilized in an aqueous membrane layer between two hydrocarbon-based interior and exterior phases.

The composition of the hydrocarbon membrane-forming solution can be varied extensively to give the appropriate membrane needed for a particular application. Usually the membrane is composed of a surfactant, additives, and a high-molecular-weight paraffin. The most often used surfactant is Span 80 (sorbitan monooleate). Additives are added in order to give the semipermeable liquid membrane both altered selectivity and physical stability. Polyethylene glycol, polyvinyl alcohol, and cellulose derivatives may be used as membrane-strengthening agents. The high-molecular-weight paraffin most often employed is an isoparaffin, Enjay S100N, having an average molecular weight of about 390.

[20] N. N. Li, R. R. Mohan, and D. R. Brusca, U.S. Patent No. 3,740,315 (1973).
[21] S. W. May and N. N. Li, *Biochem. Biophys. Res. Commun.* **47**, 1179 (1972).
[22] R. R. Mohan and N. N. Li, *Biotechnol. Bioeng.* **16**, 513 (1974).
[23] S. W. May and N. N. Li, in "Enzyme Engineering" (E. K. Pye and L. B. Wingard, Jr., eds.), Vol. 2, pp. 77–82. Plenum, New York, 1974.

Figure 1 shows a representative example of hydrocarbon-based liquid-surfactant membranes. The microcapsules are spherical, and the individual emulsion-size aqueous droplets within the assembly are usually 1–100 μm in diameter. The size of the microdroplets (the entire assembly) also varies and is typically in the range of 500–2000 μm in diameter. As with any emulsion, the size of these liquid membrane microcapsules is dependent on such variables as the speed of mechanical emulsification, the concentration of the emulsifying agent, the viscosity of the liquids, and the chemical nature of the reagents used. Membrane thickness is determined by the size of the reagent droplets and the ratio of reagent to organic phase.

The permeability and stability of liquid membrane microcapsules are dependent on the chemical nature of the membrane phase, the type and concentration of additives, and the temperature of the system. The permeability of substances across the liquid membrane microcapsules is

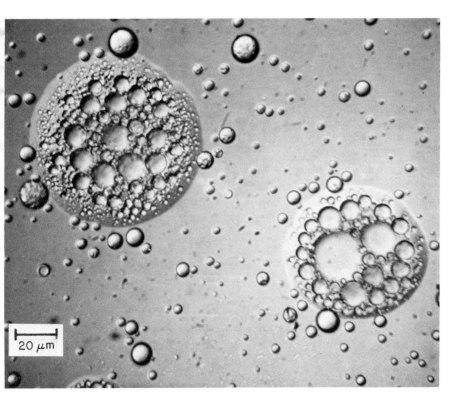

FIG. 1. Photomicrograph of liquid-surfactant membrane microcapsules containing an enzyme.

solubility dependent, and no diffusion of substrate or product will take place unless both have some solubility in the hydrocarbon-based liquid-surfactant membrane. This solubility dependency is in contrast to permanent microcapsules, where the diffusion of substrate or product is pore size dependent. However, additives can be introduced to enhance the solubility of the permeate in a liquid membrane and facilitate transfer. The stability of liquid membrane microcapsules can be varied considerably depending on the nature of the membrane-forming solution, temperature of the system, etc. By appropriate formulation, liquid-membrane microcapsules can be made to be stable for years, but yet can be broken easily by heat, centrifugation, or other emulsion-breaking techniques.

The liquid-membrane microencapsulation method for immobilizing enzymes has certain advantages additional to those of the permanent microcapsule technique. Besides providing for an extremely large surface area to volume ratio and for the simultaneous immobilization of several enzymes in one simple step, the technique also provides for a completely nonchemical and reversible methodology. The immobilized enzyme can be recovered by breaking the emulsified microdroplets through either chemical or physical means. Disadvantages of the method are possible leakage of the enzyme and the fact that the transfer of substrates and products through the membrane is solubility dependent.

Procedure

The following experimental for the immobilization of urease was reported by May and Li.[21]

To a rapidly stirred mixture of 100 g of 2% Span 80, 3% ENJ-3029 and 95% S100N are added dropwise to 82 ml of an aqueous urease solution (0.64 mg/ml in water). The immobilized enzyme emulsion so prepared is then dispersed in an aqueous solution of substrate. Span 80 is sorbitan monoleate; ENJ 3029 is a high-molecular-weight amine with an average molecular weight of 2000 manufactured by ENJAY Laboratories; and S100N is isoparaffin having an average molecular weight of 387.

Classification of the Methods

There are several ways of classifying the diverse array of methods for immobilizing enzymes. The classification can be based on the kind of interaction responsible for the immobilization, the nature of the support material, or the nature of the resulting conjugate. This author's preference for classifying the various methods is based on the nature of the inter-

TABLE II
CHEMICAL AND PHYSICAL METHODS FOR IMMOBILIZING ENZYMES

Chemical methods (covalent bond formation dependent)	Physical methods (noncovalent bond formation dependent)
Attachment to water-insoluble matrix Incorporation into growing polymer Intermolecular cross-linking with multifunctional, low-molecular-weight reagent	Adsorption onto water-insoluble matrix Entrapment within water-insoluble gel matrix (lattice entrapment) Entrapment within permanent and nonpermanent semipermeable microcapsules Containment within semipermeable membrane-dependent devices

action responsible for the immobilization, i.e., whether it is due to chemical bond formation or physical, noncovalent forces.[24]

Chemical methods of immobilization include procedures that involve the formation of at least one covalent bond (or partially covalent bond) between groups of an enzyme and a functionalized, water-insoluble support or between two or more enzyme molecules. In reality, more than one covalent bond between the reacting components is usually formed. Thus the immobilization of an enzyme by coupling it to a preformed polymer, by treating it with a bifunctional reagent, or by incorporating it into a growing polymer chain are all chemical methods. Until recently, chemical methods have resulted in irreversible coupling, and the original enzyme, once attached to a polymer or another enzyme molecule, could not be regenerated. However, this irreversibility is not an inherent feature of chemical methods and is only due to the limited nature of the chemical reactions employed for immobilizations. Recent examples of reversible covalent immobilization procedures are the coupling of enzymes to aldehyde-containing polymers (e.g., Enzacryl Polyacetal, described in this volume by Epton et al. [7]) and to sulfhydryl-containing water-insoluble polymers (e.g., Enzacryl Polythiol, described in this volume by Epton et al. [7] or by Porath and Axén [3]).

Physical methods, by elimination, include procedures that involve localizing the enzyme in any manner that is not dependent on covalent bond formation. In this group, the immobilization is dependent on the operation of physical forces (e.g., electrostatic interactions), the entrapment of enzymes within microcompartments, or the containment of the catalysts in special prefabricated membrane-containing devices. In principle, physical methods are reversible in that the enzyme can be completely

[24] O. R. Zaborsky, "Immobilized Enzymes." CRC Press, Inc., Cleveland, Ohio, 1973.

recovered in an active state. However, many examples of the individual methods do not follow idealized behavior, and the enzyme is usually not recovered completely or is recovered somewhat inactivated. The subdivision of the physical methods in Table II is quite arbitrary but has been retained from a historical basis. It is also to be recognized that many current methods of immobilization involve both chemical and physical treatments, and the exact nature or the magnitude of the forces responsible for the immobilization are indeed multivariant or not easily categorizable.

Section III
Assay Procedures

[25] Assay Procedures for Immobilized Enzymes

By Bo Mattiasson and Klaus Mosbach

Immobilized enzyme systems (matrix-bound or support-bound) are usually particulate in nature. For this reason conventional, soluble enzyme assay procedures can rarely be applied without modification.

It is the aim of this chapter to summarize and exemplify various procedures that are suited for such preparations. It may very well be that the methods that have been developed for the assay of immobilized enzymes may also find applications in the future in the study of naturally occurring particulate cell components, such as membranes and organelles.

The mere fact of immobilizing enzymes on particulate systems renders them easier to handle; such preparations can, for instance, be packed into columns. There are at least two fundamentally different methods that may be used in the assay of particulate enzymes, one alternative involving columns, the other the use of homogeneous suspensions of the enzyme-support in the incubation medium.

By having a homogeneous suspension, the substrate concentration is kept uniform around all the beads, each of which is thus in a milieu equivalent to that of any other in the suspension. In the column alternative, the enzyme beads at the top of the column are in contact with pure substrate, whereas farther down the column the substrate concentration decreases and the product concentration increases. This causes a concentration gradient in the column resulting in a range of different microenvironmental conditions between the enzymes at the top and those at the bottom. Columns operating either at a high percentage conversion or in systems with product inhibition or in multistep enzyme-catalyzed processes may suffer from the problem of heterogeneity, whereas at a low percentage conversion of the added substrate, such heterogeneity may be negligible, at least for single enzyme-catalyzed processes with no product inhibition or stimulation.

Once again there are two principal approaches to the problem of assaying immobilized enzymes. In cases where a discontinuous measurement is permissible, it suffices to remove aliquots at intervals then, after filtration, assay according to conventional techniques. For continuous measurements, however, complications occur, such as inhomogeneity and light scattering caused by the support necessitating the modification of conventional procedures before they can be applied.

It is essential to carry out the assay of an immobilized enzyme under well-defined conditions if the kinetic data obtained are to provide useful information in the overall characterization of a bound enzyme. For example, in a stirred-batch assay, care must be taken to obtain optimal stirring rates to avoid diffusional hindrance of the substrate or product.

In other chapters of this volume, dealing with enzyme membranes [63] and in particular enzyme technology [49]–[57] and analysis [41]–[45], the reader will find several of the assay procedures given here which provide useful additional information and therefore are not covered here. The principal difference is that whereas in the present chapter the immobilized enzyme *per se* is in focus, in the others the interest lies mainly in the conversion of substrate or analysis.

Spectrophotometric Methods

The most common and often the easiest way of following an enzyme reaction is to use spectrophotometry, thereby studying changes in absorbance caused by consumption of substrate or generation of product.

The systems to which spectrophotometric measurements can be applied may be conveniently divided into two main groups: that in which the enzyme-matrix is in the light path and that in which it is not.

Enzyme-Matrix in the Light Path

The activity of enzymes immobilized on matrices that are not too optically dense can be measured in a cuvette provided the particles are kept in suspension by means of stirring during the assay procedure.

Stirring in the Cuvette during the Assay

A homogeneous suspension of immobilized enzyme may be obtained in the cuvette if the particles are adequately stirred. It is then possible to measure the activity almost as easily as that of the free enzyme. The method was introduced by Weliky *et al.* for determining the activity of peroxidase immobilized on carboxymethyl cellulose[1] and has since been developed further by Mort *et al.*[2]

Equipment. Besides a photometer and recorder facilities, the method requires a thermostated stirring device. Such units can be obtained as accessories to some of the more elaborate commercially available instruments.*

Example. In the paper presented by Mort *et al.*[2] aldolase bound to Sepharose was studied using a Beckman Acta V double-beam spectro-

[1] N. Weliky, F. S. Brown, and E. C. Dale, *Arch. Biochem. Biophys.* **131**, 1 (1969).
[2] J. S. Mort, D. K. K. Chong, and W. W.-C. Chan, *Anal. Biochem.* **52**, 162 (1973).

* In this context we would like to stress that whenever in the following a specific commercially available model of equipment is given, this should be considered as an example taken usually out of several possible.

photometer equipped with stirring facilities and a Beckman spherical stirrer. For differentiation between soluble and immobilized enzymes, fines were removed from Sepharose 4B by several decantations. A typical incubation mixture (final volume 2.0 ml), containing 50 mM triethanolamine, pH 7.4, 10 mM EDTA, 0.1 mM fructose 1,6-diphosphate, 0.15 mM NADH, triosephosphate isomerase (10 µg/ml), and α-glycerophosphate dehydrogenase (10 µg/ml), was stirred with a Teflon-coated bead. The reaction was started by addition of an aliquot of gel. The aldolase–Sepharose sample was suspended in buffer in a beaker, and, while stirring was maintained, the aliquot for assay was removed rapidly using a Biopet (Schwarz-Mann). In general the gel was diluted 10-fold with 10 mM sodium phosphate at pH 7, containing 1 mM EDTA. Aliquots of 0.1 ml were assayed. When greater amounts of gel suspension were used, assay mixtures were made more concentrated and diluted to give a constant final concentration.

The enzymic activity was registered on a recorder. As shown in Fig. 1, a continuous, linear reaction rate was observed as long as stirring continued. However, when stirring ceased, the reaction rate dropped to zero, indicating that all the enzyme activity was associated with the matrix, which had sedimented, leaving no soluble enzyme in the supernatant. Resumption of stirring led to a linear decrease in absorbance at the same rate as that observed during the previous period of stirring.

Discussion. That the light scattering of the enzyme beads in the light beam causes small perturbations is apparent from the noise on the recorder diagram. These disturbances in the registered signal are more pronounced the bigger the particles used.

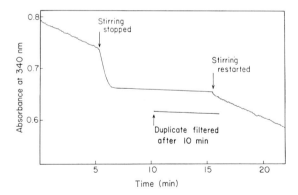

Fig. 1. Spectrophotometric recording of the reaction catalyzed by matrix-bound aldolase (1.7 mU, 0.1 ml of 1-in-10 gel suspension), with and without stirring. Reproduced with permission from J. S. Mort, D. K. K. Chong, and W. W.-C. Chan, *Anal. Biochem.* **52**, 162 (1973).

The use of this assay technique is restricted to those matrices that are, at most, only slightly fragmented by the stirring action. A Sepharose-bound enzyme, as is shown by Mort et al.,[2] is suitable for measurement, whereas more fragile matrices, such as porous glass beads, give a "background activity" due to increased absorbance and light scattering from the ground particles.

Cessation of stirring during the incubation gives a fairly strong indication of whether any free enzyme activity or soluble matrix-fragments containing bound enzyme are present.

The method is simple and gives accurate measurements of the activity of immobilized enzymes, provided the support is not too fragile.

Enzyme-Matrix Outside the Light Path

Keeping the enzyme-matrix outside the light path circumvents the problems of light scattering. This is achieved by running the reaction outside the photometer and pumping only the clear product-containing solution through a flow-cuvette for spectrophotometric analysis. Two main approaches may be distinguished: analysis of an eluate from a packed bed; recirculation of the supernatant from a stirred-batch procedure.

Method 1. Continuous Spectrophotometric Analysis of the Eluate from a Packed Bed of Immobilized Enzyme

This method has been one of the most frequently used of the continuous methods for assaying the activity of immobilized enzymes. A kinetic evaluation of this method was presented by Lilly et al.[3] (see also this volume [49]).

Equipment. The equipment needed comprises a spectrophotometer, a flow-cuvette, a column, and a pump. The enzyme-matrix is packed in the column, through which fresh substrate solution is pumped for subsequent analysis in the photometer.

Example. Lilly et al.[3] studied ficin immobilized to CM-cellulose. The final preparation, containing 4.2% of protein, was packed into columns 1.0 cm in diameter and of various lengths, which, like the perfusion solutions, were thermostated at 25°. The columns were equilibrated by pumping phosphate-saline, $I = 0.4$. The esterolytic activity was determined by following the hydrolysis of N-α-benzoylarginine ethyl ester (BAEE) hydrochloride (see the Table), measuring the absorbance of the eluate at 251 nm. The effect of variation in flow rate on the degree of hydrolysis as well as on the $K_{m,app}$ was studied (see Fig. 2).

[3] M. D. Lilly, W. E. Hornby, and E. M. Crook, *Biochem J.* **100**, 718 (1966).

VALUES OF THE MICHAELIS CONSTANT FOR N-α-BENZOYLARGININE ETHYL
ESTER (BAEE) HYDROLYZED BY FICIN ATTACHED TO CM-CELLULOSE[a]

CM-cellulose-70-ficin ($I = 0.5$)	
Flow	$K_{m,app}$
30 ml/hr	10.0
140 ml/hr	5.4
Very fast	~5.2

[a] Reproduced with permission from M. D. Lilly, W. E. Hornby, and E. M. Crook, *Biochem. J.* **100**, (1966).

Discussion. The method is straightforward in that it is easy to handle, but as Lilly *et al.* stated, all kinetic constants obtained are severely influenced by the flow rate. A high flow rate results in a thin Nernst diffusion layer and hence improved diffusional conditions for the entrance of substrate and the exit of product as compared to a situation with a slower flow rate. Thus when kinetic data such as K_m and V_{max} are discussed, care should be taken to specify the exact assay conditions used.

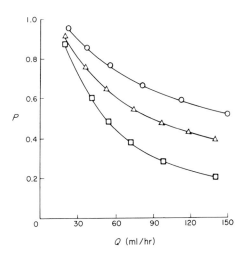

FIG. 2. Relationship between degree of hydrolysis, P, of N-α-benzoylarginine ethyl ester (BAEE) and flow rate, Q, for a column (4.0 cm $\times$ 1.0 cm diameter) packed with 670 mg of CM-cellulose-90-ficin containing 4.2% of protein (K_m 2.5 mM and K_I 0.18 mole of BAEE hydrolyzed per second per mole of bound ficin) with initial substrate concentrations 0.5 mM (○), 5 mM (△), and 15 mM (□) in phosphate-NaCl, $I = 0.4$. Reproduced with permission from M. D. Lilly, W. E. Hornby, and E. M. Crook, *Biochem. J.* **100**, 718 (1966).

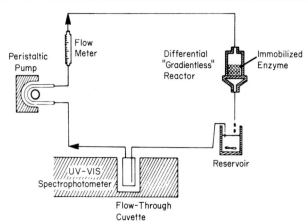

Fig. 3. Recirculation packed-bed reactor. Reproduced with permission from J. R. Ford, A. H. Lambert, W. Cohen, and R. P. Chambers, *Biotechnol. Bioeng. Symp.* 3, 267 (1972).

Method 2. Continuous Enzymic Assay Using a "Gradientless" Recirculation Packed-Bed Reactor

Circulating a substrate solution through a column of immobilized enzyme, under conditions of very low conversion of added substrate on each cycle, gives rise to a continuous enzymic process. With shortening of the column bed and increase of the flow rate, the system gets more and more identical to the stirred-batch procedure.

This reactor type was to the best of our knowledge first described by Ford et al.[4] and has later been used, among others, by Gould and Goheer.[5]

Equipment. A small column, spectrophotometer, peristaltic pump, magnetic stirrer, tubing, and glassware.

Example. The equipment was arranged according to Fig. 3 for assay of glucose-6-phosphate dehydrogenase immobilized on Sepharose.[5]

The incubation solution contains glucose 6-phosphate (2 mM), NADP$^+$ (0.08 mM), MgCl$_2$ (10 mM), and glycylglycine buffer, pH 7.4 (42 mM), final pH 7.2, total volume 20 ml, flow rate 1–40 ml/hr. The amount of immobilized enzyme selected ensures that the conversion rate never exceeds 2% per pass. The effect of recirculation flow rate on activity of immobilized glucose-6-phosphate dehydrogenase is shown in Fig. 4.

Discussion. The method is easy to handle. The enzyme preparation may be used repeatedly. Solutions can be changed easily in the recirculation reactor system.

[4] J. R. Ford, A. H. Lambert, W. Cohen, and R. P. Chambers, *Biotechnol. Bioeng. Symp.* 3, 267 (1972).
[5] B. J. Gould and M. A. Goheer, *Biochem. Soc. Trans.* 1, 1284 (1973).

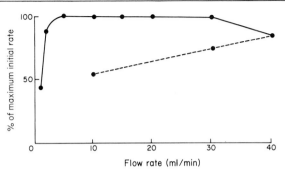

FIG. 4. Effect of increase in recirculation flow rate on activity of immobilized glucose-6-phosphate dehydrogenase (●——●). Reproduced with permission from B. J. Gould and M. A. Goheer, *Biochem. Soc. Trans.* 1, 1284 (1973).

Method 3. The Fluidized-Bed Reactor

The fluidized bed, a modification of the packed-bed reactor, was recently introduced into the field of immobilized enzymes; it improves the homogeneity throughout the column. Instead of pumping the substrate solution down through the packed bed, the substrate flow is applied from below. This results in a "bed" in which the enzyme particles are suspended in the medium by the pressure of the flow, but are retained in the column by the force of gravity. In many cases improved efficiency is obtained with this reactor type because of both better mixing and the avoidance of channels in the reactor bed. Additionally, the risk of precipitation of protein or lipids in a packed bed, which can lead to clogging, is reduced.[6,7] In order to run such fluidized beds with good flow rate, high-density particles have to be used; recent developments on the binding of enzymes to magnetic supports (stainless steel and iron oxide) have in part been stimulated by this need[8,9] (see also this volume [24]).

To date no reports directed toward the kinetic analysis of such immobilized enzymes have appeared. The use of low enzyme activities in reactors permitting good mixing, however, creates conditions quite similar to those in the stirred batch reactor (to be discussed below), thus permitting kinetic studies.

The eluate from such a fluidized bed may then be analyzed using a flow cuvette placed in a spectrophotometer.

[6] S. A. Barker and R. Epton, *Process Biochem.* Aug., p. 14 (1970).
[7] J. M. Novais, Ph.D. thesis, University of Birmingham, Birmingham, U.K., 1971.
[8] G. Gellf and J. Boudrant, *Biochim. Biophys. Acta* 334, 467 (1974).
[9] F. X. Hasselberger, B. Allen, E. K. Rarachuri, M. Charles, and R. W. Coughlin, *Biochem. Biophys. Res. Commun.* 57, 1054 (1974).

Method 4. Continuous Spectrophotometric Analysis of the Supernatant from a Stirred Batch Reactor

Principle. The reaction is carried out in a stirred vessel containing the enzyme beads suspended in the incubation solution. With a peristaltic pump, this solution is continuously pumped out of the reactor, passed through a flow cuvette in a spectrophotometer, and recycled through the reaction vessel.

The method was first developed by Mosbach and Mattiasson[10] to determine the enzyme activities in an immobilized two-step enzyme system (hexokinase–glucose-6-phosphate dehydrogenase); several similar reports have since been published[11,12] (see also this volume [31]).

Equipment. Magnetic stirrer, magnetic bar (ideally the length of the Teflon-coated magnetic bar should correspond to the inner diameter of the reaction vessel to ensure homogeneity), peristaltic pump, flow cuvette, photometer, water bath, Teflon and silicon tubing with an inlet covered with nylon net (200 mesh), E-flask 25 ml, or beaker.

Example. Reagents for the assay of immobilized hexokinase: 11.6 ml of 0.05 M Tris-HCl buffer with 7 mM $MgCl_2$, pH 7.6, containing enzyme polymer (bound to Sepharose, polyacrylamide, or acrylic acid–acrylamide copolymer) (equivalent to 20 mg of dry polymer containing about 10^{-6} enzyme units), 26.64 μmoles of glucose, 5.23 μmoles of $NADP^+$, and 3.3 U of soluble glucose-6-phosphate dehydrogenase. The enzymic test was started by addition of 7.26 μmoles of ATP dissolved in 400 μl of the Tris-HCl–$MgCl_2$ buffer.

The experiment was arranged according to Fig. 5. Prior to starting the assay by the addition of ATP, all reagents were well mixed and circulating through the pumping system.

Discussion. The "dead" volume of the recirculating system, i.e., tubing and flow cuvette has to be minimized to be only a small fraction of the total. In the experiments cited, the total volume was 12 ml whereas the recirculating volume was 2 ml, but this has since been halved. In cases of large excess of substrate and relatively low enzymic activity, no effects were observed that were due to isolation of a fraction of the incubation solution from the site of catalysis—the enzyme matrix—during the time of recirculation. By maintaining a high flow rate, any such effects should be even further reduced.

As in the case of the column situation, the enzyme-catalyzed reaction is highly diffusion sensitive unless very little enzyme activity is present,

[10] K. Mosbach and B. Mattiasson, *Acta Chem. Scand.* **24**, 2093 (1970).
[11] D. L. Marshall, J. L. Walter, and R. D. Falb, *Biotechnol. Bioeng. Symp.* **3**, 195 (1972).
[12] I. C. Cho and H. E. Swaisgood, *Biochim. Biophys. Acta* **258**, 675 (1972).

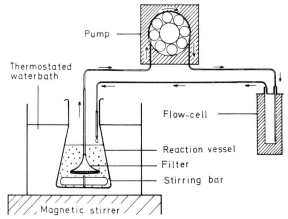

FIG. 5. The stirred-batch reactor with auxiliary pump and flow system.

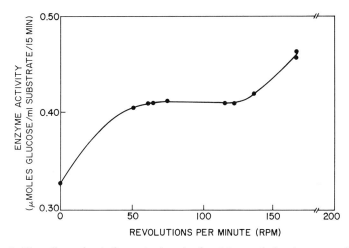

FIG. 6. The effect of rotation rate (rpm) of a 1.3-cm stirring bar on activity of β-galactosidase at 37°. Little or no change was observed at stirring rates between 50 and 125 rpm. Reproduced with permission from G. O. Hustad, T. Richardson, and N. F. Olson, *J. Dairy Sci.* **56**, 1111 (1973).

and thus it is very dependent upon the flow rates used. On increasing the stirring speed, the thickness of the unstirred layer is gradually decreased. This leads to a higher mass transfer in the diffusion-restricted system and hence to more rapid catalysis. The effect due to increased stirring gradually plateaus, resulting in a constant rate of catalysis above a certain stirring speed (see Fig. 6).[13]

[13] G. O. Hustad, T. Richardson, and N. F. Olson, *J. Dairy Sci.* **56**, 1111 (1973).

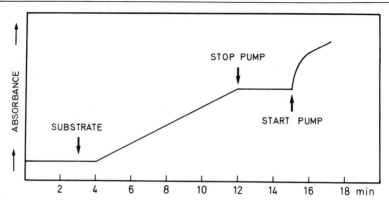

FIG. 7. A generalized recorder diagram obtained from a stirred-batch procedure. Arrows indicate the addition of initiating substrate, stopping of the pump, and restarting the pump.

This approach presents an alternative method with great accuracy, high reproducibility, and good possibilities of obtaining homogeneity in the system. A further advantage lies in the possibility of using the same gel-enzyme preparation for many assays with washings in between the measurements without removing the matrix from the reaction vessel. Control assays for the presence of soluble enzyme are readily carried out by stopping the pump and monitoring any free enzyme in the flow cell[14,15] (Fig. 7). Drawbacks include the grinding of fragile supports, e.g., glass, and also in some cases a short time delay (see Fig. 7) before the results of addition of substrate or other reactants to the incubation solution can be observed.

Recently Horton et al.[16] developed a modified assay system for analysis of enzymes suited for fragile matrices, such as glass beads. Instead of being stirred, the incubation solution was kept as a homogeneous suspension by vibrating on a Vortex mixer according to Widmer et al.[14] (see also this volume [34]).

Intermittent Assay of Immobilized Enzymes in a Stirred Batch (Fig. 8)

Example. The reaction is carried out in a thermostated vessel. The incubation solution is kept homogeneous by means of a Vortex mixer.[14]

[14] F. Widmer, J. E. Dixon, and N. O. Kaplan, *Anal. Biochem.* **55**, 282 (1973).
[15] P. A. Srere, B. Mattiasson, and K. Mosbach, *Proc. Natl. Acad. Sci. U.S.A.* **70**, 2534 (1973).
[16] R. Horton, H. E. Swaisgood, and K. Mosbach, *Biophys. Biochem. Res. Commun.* **61**, 1118 (1974).

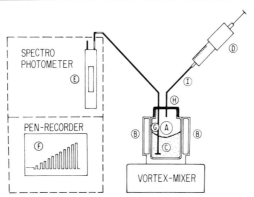

FIG. 8. Arrangement of the equipment for intermittent assay of immobilized enzymes. A, Reaction vessel; B, water-jacketed holder; C, reaction medium; D, air-driven syringe; E, flow cell; F, recording chart; G, needle with nylon strainer; H, rubber septum cap; I, plastic tubing. Reproduced with permission from F. Widmer, J. E. Dixon, and N. O. Kaplan, *Anal. Biochem.* **55**, 282 (1973).

To the reaction vessel inlet is connected an empty syringe; the outlet leads through a nylon mesh to a flow cuvette placed in a spectrophotometer. Intermittently the syringe is actuated, thereby displacing a small volume of the reaction medium into the flow cell. After the absorbance has been read, the syringe piston is released, withdrawing the aliquot into the reaction vessel.

The reaction is performed in 0.1 M potassium phosphate pH 7.5 containing 0.1 mM NADH and varying concentrations of pyruvate.

Discussion. The syringe-driven assay method is simple to handle, permitting the repeated use of a gel preparation (after washings) without removing it from the reaction vessel. This minimizes the experimental errors, since the weighing out of the enzyme gel is usually a difficult step to reproduce. The mixing technique almost eliminates grinding and fracturing of the support.

There are, however, some drawbacks. The intermittent mode of operation necessitates either manual operation or the development of some auxiliary automatic unit to govern the pushing and withdrawal of the reaction medium to and from the flow cuvette. The problem of having part of the incubation solution out of contact with the enzyme has been discussed in relation to the continuous stirred-batch procedure (see above). This source of error seems also to be negligible in this case.

Adaptation of these spectrophotometric assay methods to a fluorimetric procedure using, for example, the flow cell described in this volume [36], seems an obvious extension to make.

Titrimetric Methods

Many enzyme-catalyzed reactions produce or consume protons. Such reactions dominated research in the field of immobilized enzymes until the recent development of spectrophotometric methods. This may have been due to various factors, for example, the availability of esterolytic enzymes, but also the simplicity of assaying these enzymes when immobilized.

In one of the first studies, Levin et al.[17] studied trypsin immobilized on copolymers based on maleic anhydride and ethylene.

Equipment. An automatic titrator (TTT1C, Radiometer, Copenhagen), a recorder, and a thermostated reaction vessel.

Example. The enzymic activity of the immobilized trypsin was determined at pH 9.5 whereas the free enzyme was assayed at pH 7.5. This because the immobilized trypsin showed maximum esterase activity in the pH range 9.5–10. (See further the section General Considerations at the end of this article and, in particular, chapter [29].)

The reaction mixture (5 ml) was 0.01 M in phosphate and 5.8 mM in substrate, N-α-benzoyl-L-arginine ethyl ester (BAEE) hydrochloride; 0.1 M NaOH was used as titrant. Trypsin at a concentration of 2.0–20 μg per milliliter of reaction mixture gave a specific activity of 35 μmoles per minute per milligrams of enzyme per milliliter. The rate registered was obtained from the rate of addition of titrant to the reaction mixture.

Disucssion. The assay is easy to handle, accurate, and reproducible. Since, however, it is sensitive to exchange of carbon dioxide with the surroundings, the reaction should preferentially be run under nitrogen.

In the presence of strong buffering substances, for example, charged matrices or buffers, titration is impossible. The sensitivity is lower than, for example, that obtained from using spectrophotometric methods when a suitable chromophore is present. Titrimetry is a realistic alternative in cases with excess of substrate and good enzyme activities, provided the reaction consumes or produces protons.

Electrode-Monitored Reactions

Photometric method of assaying immobilized enzymes have to be modified versions of the standard procedures. All methods, however, that are based on the use of specific electrodes to follow depletion of substrate or generation of products, can be used almost unmodified.

[17] Y. Levin, M. Pecht, L. Goldstein, and E. Katchalski, *Biochemistry* 3, 1905 (1964).

Oxygen Electrodes

Any reaction that consumes or produces oxygen may be followed using oxygen electrodes. From *in vitro* experiments on the metabolism of mitochondria, it is known that oxygen-sensitive polarographic electrodes may be used with great success.[18] In the particulate suspensions, any change in oxygen concentration is registered. This method turned out to be convenient in the study of immobilized enzymes. Weibel and Bright[19] studied the kinetics of glucose oxidase immobilized on porous glass.

Equipment. Clark electrode, electrode assembly and power supply, an electrode chamber designed to allow rapid stirring (necessary to uniformly suspend the glass beads) (Yellow Springs Instrument Co., Model 5301 bath assembly), and a recorder.

Example. Typically 0.4–0.6 ml of the immobilized enzyme preparation was used in each assay. Buffer, 0.1 M potassium acetate, pH 5.5, containing 0.5 mM EDTA and 0.1 mM KCN, was added up to 5 ml; the reaction was started by the addition of an aqueous solution of glucose (25–200 μmoles). Oxygen was recorded as a function of time.

Discussion. In this case, in which a glass-bound enzyme was used, grinding of the support took place, necessitating the replacement of the enzyme preparation for each assay. However, fracturing of the beads was not accompanied by destruction of active enzyme. The method may be used also for optically dense suspensions.

Ion-Selective Electrodes

The combination of ion-specific electrodes and enzymes has been used in the development of enzyme electrodes, i.e., instruments used in routine analysis of substrate levels (see this volume [41]).

The ion-selective electrodes have also been applied to the determination of the activity of soluble enzymes.[20] The methods have been found to be less sensitive than the photofluorimetric methods where these are available, but it presents an alternative method for especially opaque solutions.

Example. An NH_4^+-sensitive monovalent cation electrode has been used for assaying the activity of urease immobilized on glass.[21] Equipment required includes a pH meter, NH_4^+-sensitive monovalent cation

---

[18] R. W. Estabrook, this series Vol. 10, p. 41 (1967).
[19] M. K. Weibel and H. J. Bright, *Biochem. J.* **124**, 801 (1971).
[20] G. G. Guilbault, R. K. Smith, and J. G. Montalvo, Jr., *Anal. Chem.* **41**, 600 (1969).
[21] H. H. Weetall and L. S. Hersh, *Biochim. Biophys. Acta* **185**, 464 (1969).

electrode (Corning Glass Works, Cat. No. 476220), pump, column, and tubing.

The preparation of immobilized urease was packed in a column 1 × 10 cm. Urea, dissolved in 0.5 M Tris-HCl, at various concentrations and pH, was passed through at a constant flow rate of 2.5 ml/min. The eluate was assayed for NH_4^+ ions.

Discussion. The electrode (calibrated against increasing concentrations of NH_4Cl and NH_4HCO_3) gave a nearly Nernstian response over the range 10^{-4} to 10^{-1} M.

Maximum velocity of the enzyme-catalyzed reaction was registered for 0.17 M urea, whereas at concentrations higher than 0.34 M substrate inhibition was observed.

Warburg Apparatus

The laborious Warburg apparatus may also be used to measure reactions consuming or producing gases. It was originally demonstrated using immobilized orsellinic acid decarboxylase.[22]

Examples. Orsellinic acid, 36 μmoles dissolved in 3 ml of 0.02 M phosphate buffer at pH 6.2, was placed in the main chamber of the Warburg vessel. The side arm contained 0.4 g of the polyacrylamide-entrapped enzyme together with 0.1 ml of the same buffer. After equilibration and mixing, readings were made for a total of 10 min and enzymic activity was calculated as micromoles of CO_2 evolved per minute.

Discussion. The apparatus is laborious to use, but accurate. In most cases, however, pCO_2 or pO_2 electrodes will fulfill the same purpose.

Differential Conductivity

The change in conductivity of a reaction mixture during enzyme catalysis provides a good method for monitoring the process. In two recent reports,[23,24] Messing demonstrated the method on the system glucose oxidase–catalase. A conductivity flow cell, Model 219-020, having a cell constant $K = 80$, was obtained from Wescan Instruments Inc.

Equipment. Differential conductivity meter (Wescan Instruments Inc., Model 211), recorder, magnetic stirrer, pump, column, flasks, and tubing. The equipment was arranged as shown in Fig. 9.

---

[22] K. Mosbach and R. Mosbach, *Acta Chem. Scand.* **20**, 2807 (1966).
[23] R. A. Messing, *Biotechnol. Bioeng.* **16**, 525 (1974).
[24] R. A. Messing, *Biotechnol. Bioeng.* **16**, 897 (1974).

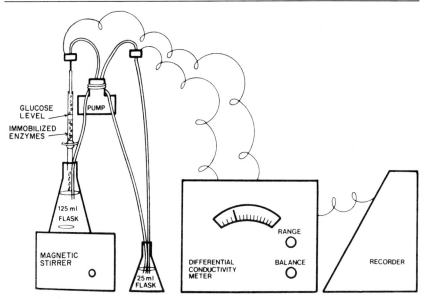

Fig. 9. Differential conductivity equipment with immobilized enzyme column. Reproduced with permission from R. A. Messing, *Biotechnol. Bioeng.* **16**, 897 (1974).

Example. Glucose solution, 6%, containing 0.0045% hydrogen peroxide, pH 5.7–6.4, unbuffered. Volume of reference and reaction mixture was 25 and 100 ml, respectively. The flow rate was 145 or 390 ml/hr.

The column assay was performed by recirculating the reaction mixture as well as the blank solution and thereby passing conductivity flow cells. The difference in conductivity between the two sensors was registered on the recorder. The results were then corrected for cell constants and dilution effects.

Discussion. The differential conductivity assay method offers an alternative that in some cases may add to the total knowledge of the system studied details that are not measurable by any of the conventional techniques. In the cited case, glucose oxidase was studied in the presence of catalase. Hydrogen peroxide added in the medium produced oxygen, which is necessary for the glucose oxidase-catalyzed reaction. Most other methods for glucose oxidase assay are based either on O_2 consumption or on the conversion of hydrogen peroxide using peroxidase.

An advantage of using the conductivity method is that the reaction *per se* may be followed without any additions of auxiliary enzymes, etc. This fact may be even more stressed in connection with practical applications.

Polarimetry

In a recent study[25] of α-galactosidase, polarimetry was applied to continuous monitoring of the reaction.

Example. To follow the development of products by assaying the specific rotation of the eluate from a packed-bed reactor requires a polarimeter with high sensitivity, e.g., a Perkin Elmer 141, flow cell, recorder, column, pump, and tubings.

A solution of 1.85% raffinose in 0.1 M phosphate buffer at pH 7.0 was pumped through the column filled with α-galactosidase immobilized on nylon microfibrils. The enzyme catalyzes the conversion of raffinose to galactose and sucrose.

The initial rotation (α_0) of the raffinose solution was 2.279°. At 100% conversion to galactose (37%) and sucrose (63%), the rotation was 1.351°.

The approximate percent conversions was calculated from the ratio

$$(2.279 - \alpha)/(2.279 - 1.351) \times 100 = \% \text{ hydrolysis}$$

Discussion. The polarimetric assay is not very accurate because of mutarotation. The experimental errors inherent in this fact may be markedly reduced either by using long retention times to complete mutarotation or, preferentially, retention times short enough to allow the assumption of zero mutarotation.

Simultaneous Registration of More Than One Enzymic Activity

The above-discussed assay procedures for immobilized enzymes may be used in combination with each other for the simultaneous registration of more than one enzymic activity. This is achieved by recirculating the incubation solution through two or more different monitoring systems.

For example, the esterolytic activity of trypsin has been assayed titrimetrically in a stirred-batch procedure with the simultaneous spectrophotometric assay of glucose oxidase activity[34]. In an extended system, two different enzymic activities (glucose oxidase and hexokinase) were registered simultaneously spectrophotometrically with the titrimetric assay of trypsin using a double-beam spectrophotometer equipped with a repetitive scanning device.

General Considerations

The kinetic behavior of immobilized enzymes is, in many respects, different from that of free enzymes in solution.

[25] J. H. Reynolds, *Biotechnol. Bioeng.* **16**, 135 (1974).

Diffusional Restrictions

Diffusional problems are pronounced in all particulate enzyme preparations—naturally membrane bound as well as artificially immobilized. This matter is discussed in more detail in this volume by Goldstein [29], but the importance of creating homogeneous conditions for the assay of immobilized enzymes cannot be overstressed. There are several categories of diffusional problems of which the diffusion from the external solution into the particle has already been mentioned. The thickness of the diffusional layer (Nernst layer) governs the mass transfer of substrates and products and thereby also the observed catalytic rate. The thickness of the diffusion layer is influenced by changes in the stirring and flow rate. An increase in the latter reduces the thickness, and thus the effect of this diffusional hindrance.

Internal diffusional restrictions may be observed on immobilized enzyme preparations with high catalytic activities per volume and also when the gel beads are large enough to generate pronounced diffusional gradients.

In kinetic analysis of immobilized enzymes, therefore, preparations containing rather low enzyme activities per matrix volume should be used. The matrix beads in the preparation should be homogeneous—or at least within a narrow size distribution—and not too big.

Diffusional hindrance—external as well as internal—will influence all kinetic parameters determined for the immobilized enzyme. $K_{m,app}$ and $K_{i,app}$ values will increase as compared to the corresponding values for the soluble enzymes.[26] On decreasing the diffusional hindrance by increasing the flow around the particles (increased stirring or flow rate) the discrepancy between the apparent values and those obtained in free solution will decrease.

On studying extremely heavily loaded enzyme-matrices it must be realized that the characteristics measured relate only to the operating fraction of all enzyme molecules. The reaction rate measured is in such cases strictly dependent upon diffusion of substrate. A system with such an excess of latent catalytic capacity will behave as though it were almost insensitive to changes in the external medium. On changing pH so that the activity of these "effective" enzyme molecules decreases, substrate will diffuse into the matrix, thus coming into the proximity of the "latent" enzyme molecules. This "buffering" effect of the overloaded gels may be observed as an enzyme activity almost independent of pH. What is being studied is pH dependence of diffusion.

[26] J. Carlsson, D. Gabel, and R. Axén, *Hoppe Seyler's Z. Physiol. Chem.* **353**, 1850 (1973).

A similar effect is observed on using enzyme preparations overloaded with respect to enzyme stability, as has previously been pointed out.[27]

Leakage from the Matrix

The immobilization procedure is usually followed by a careful washing routine to eliminate all enzyme not tightly bound to the matrix. This is important, since otherwise some enzyme may leach out into the assay solution giving rise to a mixed system containing both bound and free enzyme and thus to misleading values on characterizing the preparation. To ensure that the enzyme is firmly bound to the matrix even during and after the assay, controls have to be run. In the case of recirculation systems using a flow cell, the flow can be stopped and any free enzyme activity present in the cell at that instant can be measured[14,15] (Fig. 7). In all other situations the incubation must be filtered through double filter paper and then tested for free enzyme activity.

If the assay for free enzyme is positive, various possibilities must be checked: (a) insufficient washing of the enzyme gel after coupling; (b) fragmentation of the gel during the assay, thereby liberating either free enzyme or enzyme bound to small soluble fragments of the matrix; (c) instability of the coupling bonds under the assay conditions chosen.

Mechanical Stability and Grinding Effects

The mechanical stability of the particles has been mentioned several times in connection with various methods described. In cases of high catalytic activity per gel volume, the particle size is extremely important to the expressed catalytic activity. Regan *et al.*[28] have shown that in reactor experiments using β-galactosidase attached to aminoethyl cellulose, the smaller particles had a higher specific activity. Also, on stirring for a prolonged time, attrition of the support material will decrease the particle size and simultaneously increase the specific enzymic activity. This means that if severe grinding takes place during the assay, the results obtained may be misleading. Therefore it is important to choose a method of suspending the enzyme beads in the medium to suit the fragility of the support.

Buffer Capacity and Ionic Strength of the Assay Medium

Enzymes immobilized on charged matrices show pronounced pH shifts at low ionic strength.[17] This effect is caused by the creation of a

[27] D. F. Ollis and R. Carter, Jr., *in* "Enzyme Engineering" (E. K. Pye and L. B. Wingard Jr., eds.), Vol. 2, pp. 271–278. Plenum, New York.
[28] D. L. Regan, P. Dunnill, and M. D. Lilly, *Biotechnol. Bioeng.* **16**, 333 (1974).

microenvironment around the immobilized enzyme that differs from the conditions in the bulk solution. The effect may be considerably diminished by raising the ionic strength of the surrounding medium.[17,29]

Proton production or consumption catalyzed by the immobilized enzyme itself may create local proton concentrations different from that of the bulk solution. The effect of these changes is controlled by diffusion of product between the site of catalysis and the surrounding medium and also depends on the pK value of the buffer used.[30,31] Such microenvironmental pH effects may result in altered pH activity profiles.[32-34] The effect of locally generated proton concentrations may be reduced and even eliminated either by increasing the buffer capacity of the medium[33,34] or by reducing the size of the enzyme particles.[30,31] Independently of which method is used, some or all of the following points should be observed: (a) stirring and flow rate, (b) activity loading on the gel, (c) particle size and particle size distribution, (d) control measurements of free enzymes in the assay mixture, (e) appropriate suspension method, (f) buffer capacity, (g) ionic strength.

In summarizing, the following points should be stressed:

1. The choice of assay procedure will for obvious reasons be governed by the nature of the product formed, e.g., carbon dioxide gas, protons, etc.

2. Likewise the properties, such as fragility, of the matrix, should be taken into consideration.

3. On strict kinetic characterization of immobilized enzymes, all the above considerations should be taken into account. However, for practical applications the assay is often carried out under conditions identical to those of the process studied. Consequently the demand for saturating substrate concentration, avoidance of overloading the matrix with enzyme, etc., do not always have to be met. However, the assay conditions used should always be given completely, to permit proper comparison and evaluation.

All the methods presented here represent possible ways of assaying the activity of immobilized enzymes. Which method is likely to be appropriate in a particular case is difficult to predict, since not only the enzymic reaction, but also the properties of the support, have to be considered. It may be that also other procedures not discussed here, such as the versatile thermoanalysis, may be useful in the assay of the immobilized enzyme *per se* (see also this volume [44] and [45]).

[29] W. E. Hornby, M. D. Lilly, and E. M. Crook, *Biochem. J.* **98**, 420 (1966).
[30] I. H. Silman and A. Karlin, *Proc. Natl. Acad. Sci. U.S.A.* **58**, 1664 (1967).
[31] R. Axén, P. A. Myrin, and J. C. Jansson, *Biopolymers* **9**, 401 (1970).
[32] R. Axén and S. Ernback, *Eur. J. Biochem.* **18**, 351 (1971).
[33] R. Goldman, O. Kedem, I. H. Silman, S. R. Caplan, and E. Katchalski, *Biochemistry* **7**, 486 (1968).
[34] S. Gestrelius, B. Mattiasson, and K. Mosbach, *Eur. J. Biochem.* **36**, 89 (1973).

Section IV
Characterization by Physical Methods

[26] Investigation of the Physical Properties of Immobilized Enymes

By E. KENDALL PYE and BRITTON CHANCE

The physical properties of immobilized enzymes will differ from those displayed by the native enzymes primarily because of (a) the influence of their matrix-dependent microenvironment, (b) the restraining influence of the covalent or weaker bonds by which they are bound to the matrix, and (c) the steric constraints caused by the proximity of the polymer matrix and other bound proteins. The degree to which each of these factors influences the enzyme will depend on the method of immobilization employed. Since an immobilized enzyme can be defined as an enzyme whose free diffusional mobility is restricted, the term covers any enzyme that is constrained through microencapsulation, gel entrapment, adsorption, and covalent bonding, as well as any combination and modification of these.[1] Certain types of immobilization, especially simple microencapsulation, can be expected to have minimal effects on the physical properties normally seen for the enzyme in free solution. In contrast, other types of immobilization, especially covalent bonding, can be expected to significantly alter the physical properties of the free enzyme. Other methods of immobilization, such as gel entrapment and adsorption, will probably have effects intermediate between these two extremes.

Two of the physical properties that are of perhaps the greatest interest with respect to immobilized enzymes are (a) the changes in conformation and (b) the restrictions on conformational change that result from immobilization. Restrictions on conformational flexibility caused by the multiple cross-linking of an enzyme to a stable and inert support have been postulated to be the reason for the dramatic increase in thermal stability and instability to denaturing agents, such as urea, observed with a number of covalently immobilized enzyme preparations. Additionally, the proximity of charged groups on the matrix and adjacent proteins, the high local concentration of enzyme, the change in ion concentration caused by Donnan-type effects within the matrix microenvironment as compared to the bulk solution, and the possible change in water structure caused by hydrophobic groups on the matrix should all be expected to make the conformation of the immobilized enzyme different from that seen in the native state, whether it be normally a freely soluble or a membrane-associated enzyme. It is of interest to note that if the enzyme is normally

[1] P. V. Sundaram and E. K. Pye, in "Enzyme Engineering" (L. B. Wingard, Jr., ed.), Vol. 1, p. 15. Wiley, New York, 1972.

an intracellular enzyme, it is possible that the above phenomena may make the properties of the immobilized enzyme more similar to those in the *in vivo* state, where the enzyme is in a highly confined and restricted environment that is greatly modified by the density and proximity of other macromolecules.

The Problem of Heterogeneity within the Immobilized Enzyme Preparation

The primary interest in the effects of immobilization on conformation and conformational flexibility arises because of the impact these factors might have on the reaction kinetics, regulatory properties, and stability of the immobilized enzyme. However, there are other physical properties of immobilized enzyme systems, such as packing density and relative distribution of the enzyme on the matrix surface, that should be examined in order to characterize and, it is hoped, to optimize them with respect to their catalytic properties. This raises the issue of the severe difficulty concerning the unknown degree of heterogeneity that exists in many immobilized enzyme preparations. It is well recognized that covalent bonding of a homogeneous enzyme to a supporting matrix can give rise to a highly heterogeneous preparation.[2] This heterogeneity arises in many ways and can be a major problem in any investigation of the physical properties of immobilized enzymes. No covalent immobilization procedure employed so far allows for a specific orientation of the enzyme on the surface of even a flat matrix. Consequently, the enzyme can be bound with its active site facing outward toward the bulk solution, partly occluded by other proteins or the matrix, or bound in such a way that the active site is totally obscured by cross-linkages and matrix (Fig. 1). This should give rise to a preparation having complex reaction kinetics and only partial activity as compared to the free-enzyme activity.

Heterogeneity can also arise from the number of cross-linkages that bind each enzyme molecule to the matrix (Fig. 2a). As the orientation of the enzyme molecule varies with respect to the matrix surface, the number of reactive surface groups on the enzyme capable of taking part in the cross-linking chemistry will change. Hence, some enzyme molecules may have multiple cross-linkages whereas others may have only one, or a few (Fig. 2b). Consequently, it might be expected that such preparations will show highly heterogeneous physical properties. In addition, the nature of the matrix can also influence the degree of heterogeneity observed. A very flat surface with regular placement of binding sites will obviously

[2] R. Kallen, M. A. Diegelman, T. L. Newirth, and D. D. Perlmutter, *AIChE 66th Annu. Meeting, 1973*, Abstract 64d.

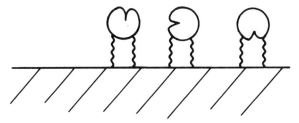

Fig. 1. Random covalent bonding of an enzyme to the surface of a solid matrix could result in multiple orientations of the enzyme on that surface. Three possible orientations of the enzyme are shown here, where the active site (cleft) is oriented toward the bulk solution, partly occluded by the other bound proteins, and totally obscured by the cross-linkages. If all extremes of orientation are possible in a particular preparation, the result will be a heterogeneous preparation of the immobilized enzyme arising from a homogeneous sample of soluble enzyme. The potentials for this type of heterogeneity are even greater in fibrous matrices.

provide a lesser degree of heterogeneity than a highly irregular surface or even a porous or fibrous structure with random binding sites (Fig. 3). It is complexities such as these that cause us to advise caution in the interpretation of data from physical property measurements of immobilized

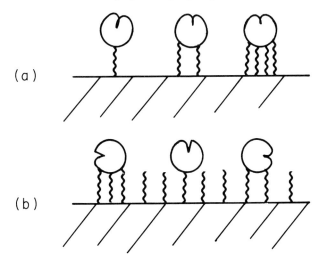

Fig. 2. (a) Uneven distribution of cross-linking "arms" on the matrix surface could lead to heterogeneity in the preparation because some enzyme molecules will be attached to the matrix surface by only one arm, others by two, and others by three, etc. (b) Uneven distribution of specific reactive groups on the enzyme surface could also lead to heterogeneity in the preparation even when the cross-linking "arms" are evenly distributed on the matrix surface. Hence the orientation of the enzyme to the surface might well determine the number of cross-linkages by which it is attached.

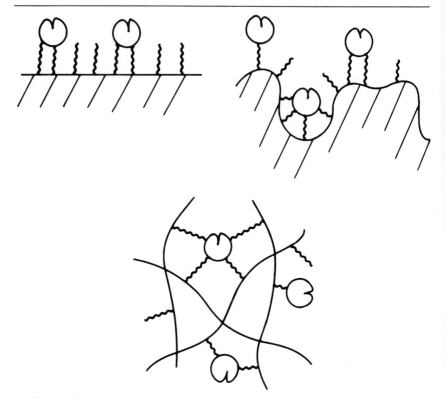

Fig. 3. The nature of the matrix surface will also determine the degree of heterogeneity in the immobilized enzyme preparation. A flat, solid surface should provide a greater degree of binding uniformity than a highly irregular, solid surface. A porous, fibrous matrix has the greatest potential for binding heterogeneity.

enzymes, especially those immobilized by covalent bonding to highly irregular matrices. It might be expected that less heterogeneity would be encountered in preparations of enzymes immobilized by gel entrapment or by ionic adsorption, where a considerable degree of reorganization to a preferred orientation could take place, but it should also be recognized that these immobilization procedures are of less intrinsic interest than immobilization by covalent bonding. It is perhaps because of these considerations that the literature on the physical characterization of immobilized enzymes is so sparse.

The Problem of Matrix Variation

In addition to the major complexity of bonding heterogeneity, it must also be recognized that enzymes can be covalently bound, or adsorbed, to many different types and forms of matrices, ranging from reasonably inert, relatively low-molecular-weight, water-soluble polymers, such as dextran T-40, through water-insoluble, small beads of natural polymer or glass, to the large surfaces of nylon tubing and glass plates. Obviously, techniques for physical characterization that can be applied to some of these immobilized enzyme preparations will most certainly be impossible to apply to others. In this context it should be recognized that many of the techniques currently used to investigate enzymes bound to biological membranes can be employed only because these preparations can be dispersed—a characteristic clearly not shared by all enzyme preparations immobilized to some of the large inorganic supports. Furthermore, certain supports (e.g., magnetic particles, collagen) can interfere with some of the physical techniques that might otherwise be usefully employed. Consequently, those studies which have been performed so far have been highly selective with respect to the immobilized enzyme preparation employed, and it appears that very few, if any, general methods can be described for investigating all immobilized enzyme preparations.

Physical Properties to be Investigated

Before moving on to a discussion of the methods for physical characterization of immobilized enzymes, it should be stated that the aim of this chapter is not to be encyclopedic, but rather to highlight certain examples of methods that have appeared in the literature or might reasonably be employed in the future. Furthermore, it should be realized that a complete characterization will entail an investigation not only of the physical properties of the immobilized enzyme itself, but also of the physical properties of the immobilized enzyme *preparation*. Consequently, studies related to conformational change, conformational flexibility, thermal stability, stability to denaturing agents and proteolysis, and physical dimensions of an immobilized enzyme should be supplemented by investigations of the density and packing distribution of the enzyme on the matrix surface. It is not within the scope of this article to discuss the physical properties of the support itself, although it is realized that this is of extreme interest in many commercial and medical applications of immobilized enzymes. These aspects are considered elsewhere in this volume.

Studies on the Conformation of Immobilized Enzymes

The great impetus to the study of the conformation of immobilized enzymes and the comparison with the native enzyme in aqueous solution has been the remarkable degree of thermal stability and protection from denaturing agents and proteolytic enzymes reported for many immobilized enzymes.[3-5] There is a general feeling that the explanation for this lies within the concept that the tertiary structure of the enzyme is stabilized and made less flexible by strong ionic or covalent interaction with a relatively rigid matrix. The fact that this increase in stability is not a general phenomenon—instances of unchanged or even decreased stability on immobilization being commonly reported[6]—could be explained by the notion that on occasions the protein–matrix interactions could also stress the tertiary structure of the protein into a less stable conformation. Certainly it can be envisioned that multiple bonding and consequent stressing of the surface of a protein could also seriously influence the conformation of the interior of the protein, which might, or might not, affect its overall stability as well as its catalytic properties. Although many instances of apparently improved thermal stability can be entirely explained by prevention of autodigestion in the case of immobilized proteases, or by prevention of digestion by contaminating proteases in the case of other enzymes, the above concepts appear now to have strong validity.

Initially, however, many investigators sought to explain the observed differences in kinetic properties between the immobilized and native enzymes on the basis of the matrix-induced modification of the microenvironment of the enzyme. For a long time there seemed to be a tacit assumption that the conformation of an immobilized enzyme, and consequently its intrinsic catalytic properties, were identical to the native enzyme in solution. This assumption has now been invalidated by a series of studies over the past 5 years or so, in which direct and indirect evidence for significant conformational change of an enzyme on immobilization has been obtained. These studies have made use primarily of fluorescence, spectrophotometry, circular dichroism, and electron spin resonance techniques. We would suggest that these techniques could be supplemented by application of techniques employing fluorescent probes and hydrogen exchange in order to confirm or extend these earlier studies.

Fluorescence Methods

In 1971, in a pioneering paper, Gabel et al.[7] pointed out that of all the physicochemical methods so far developed to study protein confirma-

[3] D. Gabel, P. Vretblad, R. Axén, and J. Porath, *Biochim. Biophys. Acta* **214**, 561 (1970).
[4] W. F. Line, A. Kwong, and H. H. Weetall, *Biochim. Biophys. Acta* **242**, 194 (1971).
[5] H. H. Weetall, *Science* **166**, 615 (1969).
[6] E. Katchalski, I. Silman, and R. Goldman, *Adv. Enzymol.* **34**, 455 (1971).
[7] D. Gabel, I. Z. Steinberg, and E. Katchalski, *Biochemistry* **10**, 4661 (1971).

tion in solution, fluorescence techniques seem to be most readily adaptable to conformational studies of immobilized proteins. (See this volume [27], [35], [36].) The basis for fluorescent methods lies in the fact that the degree of polarity of the environment of the fluorescent groups affects both the quantum yield and the emission maximum of the emitted fluorescent light. In the case of small-molecular-weight fluorescent probes, such as 1-anilino-8-naphthalene sulfonate (ANS), it is also known that changes in excimer concentration can cause spectral changes, which can thus be a useful measure of local probe concentration. Furthermore, measurements of fluorescence lifetime and fluorescence polarization can provide information on environmental constraint, mobility of the fluorescent molecule, its orientation and microviscosity of its environment, but it remains to be seen whether these other aspects of fluorescence measurement can be employed successfully and to any useful degree in the investigation of immobilized-enzyme systems.

In one of the earliest reported attempts to investigate conformational changes of an immobilized enzyme by fluorescence, Gabel et al.[7] employed chymotrypsin and trypsin covalently bound to cyanogen bromide-activated Sephadex G-200 or Sepharose 2B. The purpose of their study was to investigate the conformational changes in these two immobilized enzymes caused by urea, heat, and specific ligands, such as 2-p-toluidinylnaphthalene-6-sulfonic acid (TNS). In order to accomplish these goals, they built a special flow-through-type microcuvette that could be packed with the immobilized enzyme and by which the fluorescent light could be monitored through the same face used for irradiating the sample with incident light. This was felt to be important because the scattering of the exciting light by the polymeric matrix—a major problem in studies with immobilized enzymes—might cause difficulties in the fluorescence measurement. Furthermore, by having a high concentration of enzyme bound to the particulate matrix, in this case 10 mg/ml of immobilized enzyme preparation, the absorption of light at the relevant wavelengths was high and most of the exciting light was absorbed within a very thin layer at the face of the packed bed of immobilized enzyme.

Several other important precautions and controls were followed by these authors. To minimize the possibility of undesirable photoreactions, the samples were exposed to the exciting light only during the time needed to record the fluorescence spectrum. Unpolarized exciting light was used throughout, and fluorescence was measured at 90° to the excitation beam so that changes in fluorescence intensity in this mode approximately reflected changes in total emitted light. No fluorescence was detectable when Sephadex or cyanogen bromide-activated Sephadex was irradiated at 280 or 360 nm, thus indicating the absence of any detectable fluorescent impurities in the matrix material. This is of prime importance, since fluorescence associated with the matrix, or the cross-linking "arm," could severely complicate the interpretation of the fluorescence data.

The fluorescence of proteins is primarily the result of the fluorescence of the aromatic amino acid tryptophan and, less so, that of tyrosine, the contribution from phenylalanine being almost negligible. In the case of tyrosine residues at neutral pH, the absorption band at 275 nm is used almost exclusively for excitation, the fluorescence maximum at room temperature occurring at about 303 nm. Tryptophan and tryptophan-containing compounds in neutral solution absorb in a broad range about 280 nm, with a fluorescence maximum at room temperature at about 350 nm. Since the fluorescence emission band of tyrosine overlaps the absorption band of trytophan, energy transfer from tyrosine to tryptophan is possible and tryptophan fluorescence with a maximum in the range of 350 nm becomes the major emission. Hence these wavelength ranges become the important ones in protein fluorescence studies. Fortunately, the excitation wavelength range is sufficiently separated from the fluorescence emission range to avoid interference of the fluorescence signal from scattered excitation light. But, serious interference with the fluorescence measurement could arise from any chromophore present in the carrier matrix of the immobilized-enzyme preparation that absorbs or emits light in the wavelength range where the enzyme fluoresces. For this reason, cyanogen bromide-activated Sephadex or Sepharose appear to be excellent matrices for these types of studies.

The results from the studies of Gabel et al.[7] were very interesting. It was shown that, whereas trypsin covalently bound to agarose completely loses its enzymic activity in 8 M urea and the fluorescence peak shifts to the red by about 20 nm, trypsin covalently bound to Sephadex retains most of its enzymic activity in 8 M urea and neither its fluorescence peak nor its intensity showed any significant change under these conditions. On the other hand, chymotrypsin covalently bound to Sephadex or agarose loses all its enzymic activity in 8 M urea and also shows the 20 nm shift to the red in its fluorescence peak. The interpretation of this is that whenever the effect of urea is sufficient to impair the enzymic activity of these immobilized enzymes, changes occur in the conformation of the proteins such that the environment of the tryptophan residues is significantly altered. In the case of the trypsin-Sephadex there must be sufficient stabilization of the conformation of the enzyme by the matrix so that 8 M urea is unable to alter either the active site of the enzyme or the environment of the tryptophan residues. This appears to be evidence for the notion that under certain conditions covalent immobilization of an enzyme to a matrix can impart considerable stability to the conformation of the enzyme is the presence of denaturing agents.

To determine whether conformational flexibility of an enzyme was impaired by covalent immobilization, these same workers monitored the fluorescence ($\lambda_F = 455$ nm) of the dye TNS, which binds to chymotrypsin, but not at its active site. It is known that binding of substrate analogs, such as 3-phenylpropionic acid, to the TNS–chymotrypsin complex in free solution causes a complete quenching of the TNS fluorescence. However, in the case of the TNS–chymotrypsin–Sephadex analog, binding causes only 35% quenching. With other explanations for this behavior eliminated, the authors were led to the conclusion that while binding of the enzyme to the Sephadex matrix does not appreciably modify the binding sites for TNS or substrate, it does interfere with the interaction between these sites, conceivably by limiting the conformational flexibility of the bound enzyme.

This same study appears to provide evidence that immobilization of an enzyme to a matrix might sufficiently stress the enzyme so that a new conformation is preferred. On raising the temperature of a sample of chymotrypsin covalently bound to Sephadex, it was observed that the emission peak shifted from 335 nm at room temperature to 352 nm at temperatures above 50°, which correlated with an inflection point in the enzymic activity curve which occurred at about 47°. Unlike chymotrypsin in free solution, which is almost completely denatured at 50° for 5 min, chymotrypsin–Sephadex is relatively stable and retains considerable activity even when maintained at 60° for 15 min. On cooling the still partially active chymotrypsin–Sephadex conjugate the transmission peak did not revert to its initial position at 335 nm, but instead remained at 352 nm. This finding suggests that despite the fact that significant activity is retained in the heat-treated chymotrypsin–Sephadex, there is a considerable difference in conformation between the heat-treated immobilized enzyme and the native enzyme, or the non-heat-treated immobilized enzyme, as evidenced by the fact that the red-shifted fluorescence indicates that the tryptophan residues in the heat-treated immobilized enzyme are in a more hydrophilic environment. Thus the heat-treated immobilized enzyme appears to have taken up a preferred conformation different from that of the native enzyme.

The results from the work of Gabel et al.[7] have been described in detail because they serve to illustrate the many aspects of immobilized enzyme conformation that can be studied by fluorescence techniques. Furthermore, their work firmly established the concept that covalent immobilization of an enzyme to a matrix via multiple cross-linkages can significantly alter its conformation and conformational flexibility. In a more recent paper, Lasch et al.[8] have also used tryptophan fluorescence emission of α-chymotrypsin and of chymotrypsin covalently bound to two soluble, transparent carriers to study the influence of carrier matrix on thermal unfolding and catalytic activity. They showed that chymotrypsin, when bound to an anionic copolymer of maleic acid anhydride and acrylic acid, showed considerably decreased structural stability as compared to chymotrypsin bound to dextran or to native chymotrypsin. The conclusion was that ion-pair formation between the enzyme and the polyanionic matrix was responsible for this labilization. This provides additional evidence that covalent immobilization to a matrix can, under certain circumstances, be a destabilizing force on the conformation of the protein, whereas under other circumstances it can be a stabilizing force.

ORD, CD, NMR, and ESR

Fluorescence measurements primarily provide information on the nature of the environment surrounding a particular chromophore, whether this chromophore be tryptophan residues in a protein or a particular fluorescent dye. Consequently fluorescence changes simply provide indirect evidence of conformational changes in a protein. More direct evidence concerning the nature of these conformational changes can be obtained by the use of techniques such as optical rotatory dispersion (ORD), circular dichroism (CD), and nuclear magnetic resonance (NMR) spectroscopy. The application of some of these techniques to any type of covalently immobilized enzyme system is extremely difficult because of interference by the carrier matrix material. This is especially true for NMR techniques, although there is some possibility that refinements and modifications of NMR spectroscopy might be profitably employed with certain specially designed immobilized enzyme preparations, such as small proteins bound to small matrices. The combination of proton magnetic resonance (PMR) and ^{13}C NMR spectroscopy of the same protein might prove valuable in some special cases, as might observations on the changes in both relaxation times and chemical shifts caused by the introduction into the protein of atoms that perturb the normal spectrum. The use of ^{19}F-substituted cofactors, or substrate or inhibitor analogs, might also provide some direct information on conformation changes in certain systems. The fact that PMR techniques

[8] J. Lasch, L. Bessmertnaya, L. V. Kozlov, and V. K. Antonov, *Eur. J. Biochem.* **63**, 591 (1976).

are now being applied to the study of lipids and natural biomembrane systems[9] suggests that they may also be applied, albeit with considerable difficulty, to the study of certain immobilized enzyme systems.

Circular dichroism, on the other hand, has already been employed by a number of investigators to determine the degree of α or β structure in covalently immobilized proteins, as well as to monitor minor conformational changes of certain side-chain residues. Unfortunately, the method requires that the sample be transparent, and consequently it has only been used for enzymes covalently immobilized to soluble polymers, such as dextrans.

Lasch et al.[8] monitored the CD spectra over the range 195–300 nm of native α-chymotrypsin and chymotrypsin covalently bound to two soluble transparent polymers, dextran and maleic acid anhydride–acrylic acid copolymer. In spite of the inherent difficulties with the method, which they discuss in detail, they were able to obtain valuable information on structure-building and structure-breaking influences as well as on structure flexibility. They noted, but were unable to explain, a reproducible finding of a temperature-dependent increase in apparent α-structure at the expense of apparent β-structure in all three materials over the range from 25° to 45°. Above that temperature there was an apparent reversal of this process.

Zaborsky[10] has also used CD to examine both native RNase and RNase covalently coupled to cyanogen bromide-activated dextran. No differences were observed in the CD spectra of the two compounds over the wavelength range 235–325 nm, thus indicating that with this enzyme and this coupling procedure no change in conformation was likely to have occurred on immobilization.

Another physical technique used for the examination of changes in the conformation of native enzymes is ORD. In this method the optical rotation of an enzyme is scanned over a range of incident light wavelengths, usually 220–280 nm. Most proteins show a sharp change in the magnitude of the optical rotation, the so-called Cotton effect, in the range 225–235 nm. It is possible to recognize even minor conformational changes in a protein by observing the optical rotation at or near the Cotton effect wavelengths. As with NMR techniques it would appear that ORD methods may be applicable to immobilized enzymes only in a few highly specialized instances. However, modifications of this technique, particularly MORD, promise much more general applicability.

[9] See H. W. E. Rattle, Prog. Biophys. Mol. Biol. 28, 1 (1974).
[10] O. R. Zaborsky, in "Enzyme Engineering," (E. K. Pye and L. B. Wingard, Jr., eds.), Vol. 2, p. 161. Plenum, New York, 1975.

Reiner and Siebeneick,[11] recognizing the problems of opacity and turbidity in the application of other physical techniques to immobilized enzymes, have used electron spin resonance (ESR) techniques to examine the conformation of two different ESR-spin-labeled RNase preparations. ESR techniques allow the examination of heterogeneous mixtures and suspensions of solid particles in water. In their studies RNase I was allowed to react once on the active site at pH 5.5 with the ESR-spin label N-4-(2.2,6.6-tetramethyl-1-oxyl-piperidinyl) bromacetamide and with RNase II, once on the enzyme surface, at pH 8.5. The enzymes were either absorbed separately to Sepharose and Bio-Gel, or covalently bound to cyanogen bromide-activated Sepharose. The ESR spectrum of the label in the active site of RNase I was very similar to that for covalently immobilized enzyme but different from that of nonabsorbed RNase I in the presence or in the absence of gel. Only minimal differences were observed with RNase II, indicating that absorption itself and the gels had virtually no effect on the signal. The results were interpreted[11] as suggesting that, owing to the gel microenvironment, the active site cleft is somewhat narrowed and thus hinders the mobility of the label in RNase I.

Spectrophotometry

It is perhaps worth mentioning that other techniques can provide information on protein conformation and conformational change, and, although they have not all been applied to immobilized enzymes as yet, they may provide additional or confirmatory data. Among the spectrophotometric techniques that might be applied is infrared spectroscopy. This provides information about the vibrational modes of the protein molecule, which are themselves very sensitive to changes in chemical structure, conformation, and environment. Additionally, low-temperature luminescence measurements, like fluorescence measurements, are capable of providing information on the static structural features of proteins as well as their structural transitions. Spectrophotometric studies in the visible wavelength range can also be used with certain suitable immobilized proteins. Ferri- and ferrocytochrome c, for example, has been immobilized onto an agarose gel, and the 695-nm absorption band of the ferri form has been compared to that of the native protein.[12] (See this volume [37].) Despite the fact that the immobilized form was protected

[11] R. Reiner and H.-U. Siebeneick, *in* "Enzyme Engineering," (E. K. Pye and L. B. Wingard, Jr., eds.), Vol. 2, p. 179. Plenum, New York, 1975.
[12] C. Greenwood and T. A. Moore, *Biochem. J.* **153**, 159 (1976).

to a varying degree from denaturation processes that affect the native form, the presence of the 695-nm absorption band, as well as the influence of pH on this band, suggested that immobilization by this technique did not significantly change the conformation of the molecule.

Nonspectrophotometric Techniques

Several nonspectrophotometric techniques also have considerable promise in providing information on conformational changes of immobilized enzymes. Hydrogen exchange is one such technique.[13] It relies on the fact that when denatured proteins are dissolved in deuterium oxide some of the hydrogen atoms, especially those bound to nitrogen, oxygen, and sulfur, will exchange rapidly with deuterium. In native proteins, structural factors can influence this rate of exchange and slow it down, sometimes to a rate where a potentially exchangeable hydrogen can remain unchanged for a number of days at room temperature. The exchangeable hydrogens on the amino acid side chains of the protein exchange much faster than the hydrogen of the peptide bond. Exchange rates of a particular hydrogen will clearly be related to the conformation of the protein and the proximity of the hydrogen to solvent water. Hence determination of hydrogen-deuterium exchange curves at various pH values provides information about the capacity of the protein conformation to unfold and expose peptide bonds to exchange with the solvent. The technique appears to be immediately applicable to certain immobilized enzyme preparations, expecially a modification of this method in which hydrogen-tritium exchange is followed in tritium-enriched water.

Certain destructive techniques might also be of great value in recognizing conformational changes in immobilized enzymes. Susceptibility to attack by proteolytic enzymes is a method used to monitor conformational changes in soluble proteins, and despite the dramatic decrease in proteolytic action on immobilized enzymes, major conformational changes might be detected in this way. It has already been shown that a number of covalently immobilized zymogens are capable of being proteolytically cleaved to the active enzyme. As an example, chymotrypsinogen, bound to a surface matrix, can be readily activated, as can immobilized plasminogen. (See this volume [35].) Equally well, the susceptibility of disulfide bonds to reduction and reoxidation can provide information on protein conformation. The use of protein denaturation techniques has also been valuable in looking at immobilized enzymes having quaternary structure. Cho and Swaisgood[14] have co-

[13] S. W. Englander, N. W. Downer, and H. Teitelbaum, *Annu. Rev. Biochem.* **41**, 903 (1972).

[14] I. C. Cho and H. Swaisgood, *Biochim. Biophys. Acta* **334**, 243 (1974).

valently immobilized rabbit muscle lactic dehydrogenase to porous glass beads by a method that resulted in only one of the four subunits of each tetramer being bound. (See this volume [33].) By washing the attached subunits with strong denaturants, such as 7 M guanidinium chloride, or 1% sodium dodecyl sulfate (SDS) and then returning them to nearly physiological conditions, the attached subunits were refolded and shown to be capable of enzymic activity and recombination with native subunits in solution.

Physical Methods for Estimation of Protein Density on Matrix

We are unaware of any physical studies having been performed that attempt to quantitate the concentration and distribution of a protein immobilized onto a surface, and yet it appears that in many circumstances these are highly important questions. Ellipsometry and electron microscopy are both methods that may be valuable in this regard, although it is doubtful that either could provide good numerical data. Graves[15] has developed a laser microspectrophotometer that can determine reaction rates on the surface or in the subsurface of a gel containing a suitable immobilized enzyme. Presumably scanning sections of the gel with such an instrument would, with suitable analysis, reveal any inhomogeneity in the concentration of bound enzyme.

Comparison of Immobilized Enzymes and Biomembrane-Associated Enzyme

It is clear from the foregoing discussion that many of the physical techniques that may be applied to immobilized enzymes have already been applied to monitor activities and conformational changes in biomembrane-associated enzymes. However, not all these techniques can be directly applied to immobilized enzymes, since interference problems caused by the matrix can be much greater because of their greater rigidity than is the case with the lipids of biomembranes. Furthermore, the enzymes in biomembranes have a much greater degree of freedom than the multi-cross-linked covalently immobilized enzymes, being capable of rotation within the bilayer. Nevertheless, some of the difficulties encountered in applying physical techniques to biomembrane-associated enzymes are similar, but possibly of lesser magnitude, than those that will be encountered with immobilized enzymes.

[15] D. J. Graves, in "Enzyme Engineering," (E. K. Pye and L. B. Wingard, Jr., eds.), Vol. 2, p. 253. Plenum, New York, 1975.

Perpectives on the Physical Properties of Immobilized Enzymes

Immobilization of enzymes in membranes is nature's way of obtaining stability, orientation, and efficient transfer reactions. Nature seems to have solved the problem of heterogeneity by an appropriate process of assembly of membrane proteins, so that, in a particular case, all the cytochrome c is on one side of the membrane and all the cytochrome a_3 is on the other side.[16]

The properties of the membrane-bound system are remarkable; electron transfer rates of 500 sec^{-1} at 23°[17] are characteristic of electron flow through a single assembly or set of cytochromes, while electron transfer between sets of cytochromes is much slower (by a factor of over 10^2). Thus, the assemblies of enzymes operate effectively as a single chain and inefficiently in interchain electron transfer.[18]

The exact dependence of the kinetic properties upon the immobilization is not clear; similar kinetic constants—within a factor of two[19]— can be obtained with the detergent-solubilized enzyme, but in this case the enzyme may lie in detergent micelles. Even the lipid-depleted enzyme exhibits the partial reactions of the membrane-bound enzyme. Thus, in the enzyme demonstrated to be oriented in the membrane-bound state, we fail to find a large effect of immobilization upon the kinetics of its reactions. Instead, the membrane orientation is required for the "sidedness" of the reaction of cytochrome c and oxygen.

Methods of measuring the activity of membrane-bound enzyme have much in common with those immobilized in other matrices. The basic questions are measurement of the rate of reaction and the amount of catalyst involved. The precise measurement of the rate of reaction may require the avoidance of diffusion limitations in the matrix or membrane. This is partially satisfied by flash photolysis methods, which allow the reaction ligand to diffuse to the reaction site prior to flash activation of the reaction. This can be done with ferric carbonyl compounds as in the reaction of oxygen with cytochrome oxidase–CO_5.[19]

The determination of the amount of reactive intermediates is more difficult and requires direct measurement of the intermediate compound with ligand involved in the above-mentioned case, the compound of

---

[16] V. P. Skulachev, in "Energy Transducing Mechanisms" (E. Racher, ed.), Vol. 3, p. 31. Buttersworth, London, 1975.
[17] B. Chance, in "The Molecular Basis for Electron Transport" (J. Schultz and B. F. Cameron, eds.), p. 65. Academic Press, New York, 1972.
[18] B. Chance, B. Schoener, and D. DeVault, in "Oxidases and Related Redox Systems" (T. E. King, H. S. Mason, and M. Morrison, eds.), p. 907. Wiley, New York, 1965.
[19] B. Chance in "Oxygen," p. 163. Little, Brown, Boston, Massachusetts, 1965.

cytochrome oxidase and oxygen, or the turnover oxidation of cytochrome c.[20,21] In such cases, both the rate of reaction and the amount of reactive intermediate may be determined.

The problem of heterogeneity may also be faced by a separate set of experiments which employs a nondiffusable substrate and measures the fraction of the total intermediate reactive in that fashion as compared with that oxidized with a diffusible substrate that can react with all forms of the bound immobilized enzyme. An example is afforded by studies of cytochrome c oxidase immobilized in biological membranes in which the side of the molecule binding cytochrome c is detected by its reaction with the membrane impermeant protein yeast cytochrome c peroxidase, which can oxidize membrane-bound cytochrome c.[21] All the cytochrome oxidase present is oxidized by oxygen, which readily permeates both sides of the membrane and gives the total amount present. The results obtained with this method show a homogeneity of orientation of the cytochrome c oxidase in the natural membrane and a practically random orientation of the cytochrome c oxidase in the disrupted and reconstituted membrane.[22]

In summary, a variety of methods have been developed for determining the activity and the heterogeneity of enzyme systems immobilized in natural and artificial membranes. Presumably, these approaches will be of key importance in the generalized study of immobilized enzymes.

[20] B. Chance and M. Erecinska, *Arch. Biochem. Biophys.* **143**, 675 (1971).
[21] B. Chance, M. Erecinska, and E. M. Chance, in "Oxidases and Related Redox Systems" (T. E. King, H. S. Mason, and M. Morrison, eds.), 2nd ed., p. 851. University Park Press, Baltimore, Maryland, 1973.
[22] B. Chance, M. Erecinska, and C. P. Lee, *Proc. Natl. Acad. Sci. U.S.A.* **66**, 928 (1970).

[27] Microfluorometry of Product Formation in Single Enzyme-Sepharose Beads Using Fluorogenic Substrates

By M. SERNETZ and H. PUCHINGER

Investigations into the properties and kinetic parameters of immobilized enzymes have so far been based in most cases on experiments using stirred-batch reactors, reactor columns, or immobilized enzyme-membrane systems. The overall kinetic behavior of the reactor or of the enzyme in heterogeneous catalysis, however, has been deduced and calculated from model assumptions on the combined action of diffusion and catalysis within an immobilized enzyme membrane or the single catalytically active particle. Detailed results of analytical investigations into the reaction kinetics for the special, geometrically simple, and most relevant case of spherical catalyst particles as the basic units of the reactor have been reported.[1-10] These investigations not only provide information about the reactor kinetics, but also postulate and describe the substrate and product profiles within the gel matrix of the particles. But so far there has been no direct experimental approach to relate these postulates directly to the kinetic behavior of the single microspheres.

A method for the measurement of turnover characteristics in single beads of immobilized enzyme gels is here described.[11,12] It is based on

[1] C. N. Satterfield, "Mass Transfer in Heterogeneous Catalysis." MIT Press, Cambridge, Massachusetts, 1970.
[2] L. B. Wingard, Enzyme Engineering. In "Advances in Biochemical Engineering" (T. K. Ghose, A. Fichter, and N. Blakethrough, eds.). Springer-Verlag, Berlin, Heidelberg, and New York, 1972.
[3] J. Lasch, FEBS, Proc. 8th Meeting, North-Holland Publ. Amsterdam, 1972.
[4] J. Lasch and R. Koelsch, Mol. Cell. Biochem. 2, 72 (1973).
[5] J. Lasch, Mol. Cell. Biochem. 2, 79 (1973).
[6] D. J. Fink, Tsu Yen Na, and J. S. Schultz, Biotechnol. Bioeng. 15, 879 (1973).
[7] D. J. Fink, Diss. Abstr. B. 34, 1489 (1973); Thesis Michigan, Order No. 73-24, 562 (1973).
[8] C. Horvath and J. M. Engasser, Ind. Eng. Chem., Fundam. 12, 229 (1973).
[9] V. Kasche, H. Lundquist, R. Bergman, and R. Axén, Biochem. Biophys. Res. Commun. 45, 615 (1971).
[10] J. P. Kernevez and D. Thomas, IRIA rapport de recherche No. 28, Sept. 1973. Institut de Recherche d'Informatique et d'Automatique, Rocquencourt, Le Chesnay.
[11] M. Sernetz and H. Puchinger, Proc. Conf. Enzyme Technol. (Stockholm, Nov. 28, 1973), IVA-rapport, 70, ISBN 917082 0554 (1974).
[12] M. Sernetz, H. Puchinger, C. Couwenbergs, and M. Ostwald, Anal. Biochem. 72, 24 (1976).

applying microfluorometric techniques, originally developed for measuring intracellular enzymic reactions in single cells,[13-15] using fluorogenic substrates.

It should be noted that microscope fluorometry not only produces an integral fluorescence signal (in analogy to conventional fluorometry) of a tiny specimen, but also yields its image or projection owing to the image-forming function of the optical path of the microscope. This method thus permits a fluorometric signal to be correlated with the geometry of the specimen. Therefore, it is also possible to obtain information on concentration profiles of substrates and products within the catalytic particle. In addition to the experiments described below, some cell-physiological techniques,[16] especially microspectrophotometry,[17] open up new opportunities for research on immobilized enzymes.

Methods and Experimental Examples

Principle

The method described here consists in focusing single immobilized enzyme gel-matrix beads (in our example esterase-Sepharose) into the excitation, and measuring and imaging beam path of a microscope fluorometer (Fig. 1). This is achieved by trapping the beads separately, according to their diameter, from a dilute suspension into a wedge-shaped cuvette. This cuvette then serves as a flow-through cuvette for the introduction of fluorogenic substrates and variation of the experimental conditions. The fluorogenic reactions within the beads are shown in Fig. 1, with fluorescein diacetate as substrate (S) and fluorescein (F) as the reaction product and measuring parameter.

Method

Pig liver esterase (Boehringer carboxylic ester hydrolase, EC 3.1.1.1, specific activity 100 U/mg) is suspended in 3.2 M ammonium sulfate, pH 6, 10 mg/ml. Dialyzed esterase, 2 ml (10 mg/ml) in borate buffer, pH 8.6, $I = 0.5$, is allowed to react with 1 g of CNBr-Sepharose 4B

[13] B. Rotman and M. Papermaster, *Proc. Natl. Acad. Sci. U.S.A.* **55**, 134 (1966).
[14] M. Sernetz, *in* "Fluorescence Techniques in Cell Biology" (A. A. Thaer and M. Sernetz, eds.), p. 243. Springer-Verlag, Berlin, Heidelberg, and New York, 1973.
[15] M. Sernetz and G. v. Sengbusch, *Acta Histochemica*, Suppl. Vol. **14**, 107 (1975).
[16] K. Mosbach, *Sci. Am.* **224**, 26 (1971).
[17] D. J. Graves, *Proc. Eng. Found. Conf. Enzyme Eng.* (Henniker, New Hampshire, Aug. 5–10, 1973), *in* "Enzyme Engineering" (E. K. Pye and L .B. Wingard, eds.), Vol. 2. Plenum, New York, 1974.

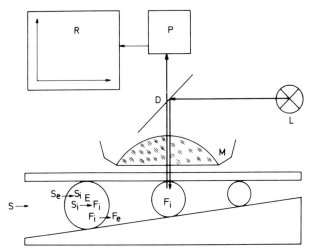

FIG. 1. Diagram apparatus setup for method of focusing single immobilized enzyme gel-matrix beads in the measuring and imaging beam path of a microscope fluorometer. See the text for details.

(Pharmacia), overnight at 4°. After inactivation with 10 ml of 0.1 M ethanolamine, pH 8.0, at 4° for 2 hr, the product was washed several times with 0.1 M acetate buffer, pH 4.0, and 0.1 M borate buffer, pH 8.6, and stored in 0.15 M phosphate buffer, pH 7.2, at 4°. Fluorescein diacetate (FDA; Sigma) served as fluorogenic substrate (stock solution 10^{-2} M in acetone, diluted in 0.15 M phosphate buffer, pH 7.2, to 0.1 mM–1 μM.

The wedge-shaped cuvette was built by cementing a cover glass onto a glass plate, giving a clearance of 200 to 10 μm, and a useful area of 5 mm × 20 mm. Two oblique drillings in the plate permit a homogeneous laminar substrate flow. The cuvette can easily be emptied and cleaned under reversed-flow conditions. All tubings are made of polyethylene to avoid adsorption of substrate. Pumping is performed using an LKB peristaltic pump (0.3 ml/min).

The microscope fluorometer (see Fig. 1) is based on a Leitz Ortholux microscope (M), equipped for epiillumination with Knott photometer MFLK BN 601, Multiplier EMI 6525 (P), xenon lamp XBO 100 W (L), excitation filters Calflex BG 12 (3 mm), BG 38 (3 mm), dichroic mirror K 480 (D), barrier filter K 530, objectives Leitz apo 25/0.65, "Köhler" conditions for excitation and measuring light beams. Excitation was performed intermittently by means of an automatic mechanical shutter (Variotimer Wild), exposure time 0.8–1.0 sec, intervals 10–30 sec. Epiillumination was preferred for reasons of constancy of common excitation and emission focal planes and apertures. In addition, this technique makes no restrictions with respect to the mechanical arrangement and the dimensions of the cuvette under the objective.

Experimental Examples

Figures 2–5 illustrate typical examples of applications of this method, together with results that have been reported and discussed in detail elsewhere.[12]

Discussion

The microscope fluorometer used in the experiments provides the following information: (1) in planes correlated to the object plane (= bead in the cuvette), the image of the particle with its fluorescence intensity pattern for fluorescence microphotography (Fig. 2), microdensitometry, fluorescence intensity scanning, or TV projection including line selection on an oscilloscope (Fig. 1); (2) in planes correlated to the aperture planes as measuring plane of the photomultiplier (Fig. 1, P), the integral fluorescence signal of the particle within the field area defined by the measuring stop.

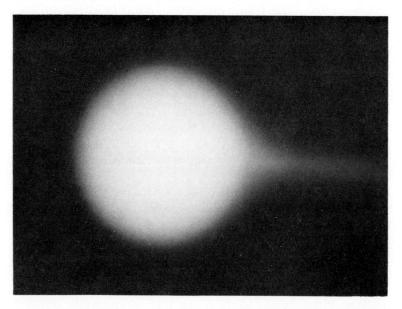

FIG. 2. Fluorescence photomicrograph of a single esterase Sepharose 4B bead incubated in 10^{-5} M fluorescein diacetate. A fine laminar thread of released fluorescein can be seen at the right side.

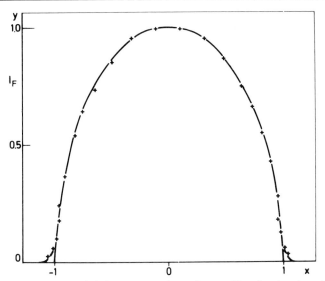

FIG. 3. Normalized radial fluorescence intensity profile of a bead under steady-state conditions, measured by microdensitometry of a photonegative. The fact that the intensity profile obeys the function $y = 2\sqrt{1-r^2}$ indicates a constant product concentration within the bead.

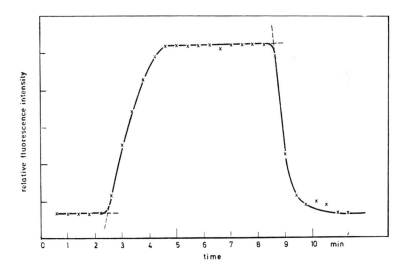

FIG. 4. Relative fluorescence intensity of a single bead as a function of time, starting with 10^{-5} M fluorescein diacetate, reaching a steady state of product concentration in the bead (plateau), and vanishing upon washing the bead with substrate-free buffer.

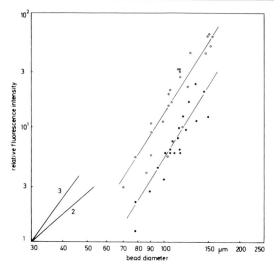

FIG. 5. Intensity of single beads as a function of their diameter, suggesting surface proportional turnover in the steady state[5] (compare Fig. 3). In the case of merely glass surface-bound enzymes [C. Horvath and J. M. Engasser, *Ind. Eng. Chem., Fundam.* **12,** 229 (1973)], surface proportional intensity would be expected, as indicated in the figure (bottom left) by the index 2.

In microscope fluorometry the quantitative correlation of the fluorescence signal with the fluorochrome in the specimen requires strict observation of special experimental conditions, which differ fundamentally from the conventional situation of fluorometry on a normal "macro" scale.[18]

This applies in particular to the observation of carefully controlled "Köhler" conditions independently for the exciting and the fluorescence measuring beam. This requires projection of the image of the light source into the rear aperture plane of the objective to achieve constant homogeneous luminance across the object plane.

The second difference consists in the fact that the dimensions of the objects (in our case the beads), equivalent to the fluorescence cuvette in conventional fluorometry, are smaller than the diameter of the excitation- and emission-measuring light beams. Thus, in microscope fluorometry the fluorescence intensity is proportional to the mass of the excited fluorochrome, rather than to its concentration, which then has to be calculated by volume determination of the object.

The third difference is the entirely different measuring range. According to the general relation

$$I_F = I_o \cdot k \cdot \phi \cdot (1 - e^{-\epsilon \cdot c \cdot d})$$

[18] M. Sernetz and A. Thaer, *Anal. Biochem.* **50,** 98 (1972).

the useful linear measuring range is shifted to much higher concentrations (c) (10^{-6} to 10^{-3} M) owing to the equivalent reduction of the cuvette thickness d (= bead diameter) to the range of 10–100 µm.[18, 19]

The microfluorometric threshold of detectability is 10^{-18} mole of fluorescein. An absolute calibration of the fluorescence intensities in terms of moles of excited fluorochrome can be achieved by use of fluorescent glass fibers or capillary cuvettes.[18,19] By this technique, it is also possible to control homogeneity of the luminance in the object plane.

An alternative to microfluorometry would be microscope absorption photometry.[17] This method, however, is less sensitive than fluorometry and requires chromophores with extremely high extinction coefficients or in high concentrations to ensure sufficient density at the low object thickness involved.

It is obvious that the described microfluorometric technique opens up a wide range of new experimental approaches and offers new measuring parameters for immobilized enzyme research.

The technique is not limited to Sephadex or Sepharose, as described here, but is applicable to other matrix materials in bead or membrane form, provided that these exhibit sufficient optical indifference, including homogeneous refractivity,[11] to avoid light scattering, high transmittance, possibly even in the ultraviolet, and noninterference by their own fluorescence. By proper selection of optical conditions and beam splitting, substrates and products may be measured simultaneously.

The availability of a wide range of fluorogenic substrates, for example, esters and glycosides of fluorescein, methylumbelliferone, or naphthol derivatives (as summarized, e.g., by Udenfriend[20]), and natural fluorogenic substrate systems, such as NAD$^+$/NADH, allows this technique to be applied to many enzymes. It should be noted that chromogenic substrates developed for histochemical enzyme localization techniques, with the special property of low solubility of the product, are not considered here because measurements under steady-state conditions are not possible, since the product precipitates.

The experimental system of microfluorometry on single beads could also be applied to rapid flow-through techniques and pulse microfluorometry, as already employed in determining intracellular esterase turnover, using fluorogenic substrates.[15]

[19] M. Sernetz and A. Thaer, *in* "Fluorescence Techniques in Cell Biology" (A. A. Thaer and M. Sernetz, eds.), p. 41. Springer-Verlag, Berlin, Heidelberg, and New York, 1973.

[20] S. Udenfriend, "Fluorescence Assay in Biology and Medicine." Academic Press, New York, 1969.

Section V
Characterization by Chemical Methods

[28] Characterization of Immobilized Enzymes by Chemical Methods

By DETLEF GABEL and ROLF AXÉN

For immobilization of proteins by covalent fixation to carriers, chemically reactive groups in the carrier and in the protein are required. In general, those functional groups normally being present in proteins, especially amino groups (see this volume [2]), are being utilized for the attaching reactions and only few procedures are described to introduce new reactive groups into the protein prior to immobilization.[1] It has therefore been accepted that the success in immobilizing a given protein would depend on the functionalized carrier and the coupling reaction chosen. The assumption has been made that available functional groups of the protein would behave as to be expected from classical organic chemistry.

In this chapter, methods for determining the protein contents of resulting conjugates are discussed, and ways of assessing the proportion of active enzyme are described. Specific aspects of the choice of support are covered in Section II of this volume. However, characterization of the carrier as an organic chemical reagent is considered here.

The Carrier as Chemical Reagent

The properties of the final enzyme–carrier conjugate will depend, among other factors, on the number of chemically reactive structures on the carrier. Whereas the total amount of protein bound increases with increasing concentration of reactive groups, the specific activity of the enzyme after immobilization tends to go down with increasing protein contents in the conjugates.[2] Thus it may be of no great use to maximize the amount of enzyme bound to a gel, as the total activity may increase only fractionally. In other instances, for example, when trying to immobilize subunits of polymeric enzymes,[3] the concentration of reactive groups in the gel should be low, in order not to bind the polymeric protein by more than one covalent link. Also, the number of links by which a protein is bound can influence its physical-chemical or enzymic properties, for example, its resistance against denaturation.[4]

Methods have been described to determine the concentration of,

---

[1] A. Bar-Eli and E. Katchalski, *Nature (London)* **188**, 856 (1960).
[2] R. Axén and S. Ernback, *Eur. J. Biochem.* **18**, 351 (1971).
[3] W. W.-C. Chan and H. M. Mawer, *Arch. Biochem. Biophys.* **149**, 136 (1972).
[4] D. Gabel, *Eur. J. Biochem.* **33**, 348 (1973).

among others, the following groups in polymers: carboxyl groups,[5] amino groups,[6,7] diazo groups,[5] oxirane groups,[8] halogenoacetyl groups,[9] carbonyl groups,[10] sulfhydryl groups, [11,12] dye-triazine groups,[13] anhydride groups.[14]

Determination of Carboxyl Groups[5,14]

Carboxyl groups can be determined by direct titration of the polymer either in 0.5 M sodium chloride with sodium hydroxide[14] or in anhydrous dimethylformamide with sodium methoxide in methanol-benzene.[15]

Alternatively, the carboxyl groups can be allowed to react with glycine ethyl ester, and the amount of glycine incorporated into the polymer determined.[5] To this end, 325 mg (2.5 mmoles) of glycine ethyl ester·HCl is dissolved in 2.5 ml of dimethyl sulfoxide; 350 µl of triethylamine (2.5 mmoles) are added, and the mixture is stirred for 1 hr on ice. After filtration, the solution is added to 50 mg of resin. Dicyclohexyl carbodiimide (325 mg in 3 ml of dimethyl sulfoxide) is added, and the mixture is stirred at room temperature overnight. The gel is recovered by centrifugation or filtration, washed extensively with dimethyl sulfoxide and acetone, and dried. An aliquot is subjected to amino acid analysis after acid hydrolysis.

Determination of Amino Groups[7]

Amino groups can be determined by the use of radioactive anions. The carrier is suspended in 2 M NaOH, filtered off, washed extensively with distilled water and dried after washing with acetone; 250 mg of the carrier thus prepared are suspended in 15 ml of 0.1 M HCl containing a known amount of ^{36}Cl$^-$. After stirring for 1 hr and centrifugation or filtration, the radioactivity of aliquots of the supernatant is determined. A

[5] L. Goldstein, *Biochim. Biophys. Acta* **315**, 1 (1973).
[6] P. J. Robinson, P. Dunnill, and M. D. Lilly, *Biochim. Biophys. Acta* **242**, 659 (1971).
[7] S. A. Barker, P. J. Somers, R. Epton, and J. V. McLaren, *Carbohydr. Res.* **14**, 287 (1970).
[8] L. Sundberg and J. Porath, *J. Chromatogr.* **90**, 87 (1974).
[9] T. Sato, T. Mori, T. Tosa, and I. Chibata, *Arch. Biochem. Biophys.* **147**, 788 (1971).
[10] R. Epton, J. V. McLaren, and T. T. Thomas, *Carbohydr. Res.* **22**, 301 (1972).
[11] B. H. Anderton, F. W. Hulla, H. Fasold, and H. A. White, *FEBS Lett.* **37**, 338 (1973).
[12] K. Brocklehurst, J. Carlsson, M. P. J. Kierstan, and E. M. Crook, *Biochem. J.* **133**, 573 (1973).
[13] R. J. H. Wilson, G. Kay, and M. D. Lilly, *Biochem. J.* **108**, 845 (1968).
[14] Y. Levin, M. Pecht, L. Goldstein, and E. Katchalski, *Biochemistry* **3**, 1905 (1964).
[15] A. Patchornik and S. Ehrlich-Rodozinski, *Anal. Chem.* **33**, 803 (1961).

blank correction for nonspecific absorption of Cl⁻ by similar gels not containing amino groups has to be done. Alternatively, amino groups can be allowed to react with suitable carboxylic acids as described for determining carboxyl groups.

Determination of Sulfhydryl Groups[12,16]

An amount of the gel to be analyzed containing about 0.015–0.5 µmoles of SH groups is suspended in 1.5 ml of 0.1 M Tris-HCl buffer, pH 7.6, and the suspension is mixed with 1.5 ml of 0.002 M 2,2′-dithiodipyridine (obtained by iodine oxidation of commercially available 2-thiopyridone). After 30 min at room temperature, the solution is filtered, and its absorbance at 343 nm is determined. A molar extinction coefficient of 8.08 × 10³ is used to convert absorbance readings into concentration.[16a]

Determination of Carbonyl Groups[10]

A 0.5 M hydroxylamine hydrochloride solution is prepared by dissolving 35 g of hydroxylamine hydrochloride in 160 ml of water and adding 95% ethanol to 1 liter. Twenty milliliters of pyridine and 0.25 ml of a 4% alcoholic bromophenol blue solution are mixed with 95% ethanol to 1 liter. A 0.5 M sodium hydroxide solution in 90% methanol is prepared. In a pressure flask, 30 ml of the hydroxylamine solution and 100 ml of the pyridine solution are mixed. About 1 g of the carrier is added; the flask is capped and heated to 100° for 2 hr. The flask is allowed to cool spontaneously and then titrated with sodium hydroxide to the color of a blank solution. For each equivalent of carbonyl groups, one equivalent of hydrochloric acid is liberated.

An operational concentration of reactive groups can be derived from data of the binding of low-molecular-weight substances similar in chemical reactivity to those of the side chains of the proteins participating in the coupling reaction, as described by Axén and Ernback.[2]

Determination of Diazo Groups[5]

The amount of diazo groups present in a carrier can be determined by assessing the amount of low-molecular-weight phenols that can be introduced. About 50 mg of the carrier are suspended in 10 ml of cold 0.2 M

[16] D. R. Grassetti and J. F. Murray, *Arch. Biochem. Biophys.* **119**, 41 (1967).
[16a] T. Stuchbury, M. Shipton, R. Norris, Y. P. G. Malthouse, K. Brocklehurst, Y. A. L. Herbert and H. Suschitzky, *Biochem. J.* **151**, 417 (1975).

phosphate buffer, pH 7.8. *p*-Bromophenol, 75 mg in 2 ml of water (dissolved by adding dropwise 2 M NaOH), is added, and the mixture is stirred in the cold overnight. The resin is filtered off and washed successively with 0.1 M carbonate buffer, pH 10.5, water, and methanol, and dried. The amount of phenol introduced can be determined by elemental analysis for bromine.

Comments

In general, it will be found that the concentration of reactive groups by far surpasses the coupling capacity of those gels for proteins. Often, this will be due to a mere steric effect, in that the average distance between reactive groups is smaller than the dimensions of proteins (see, e.g., Gabel[4]). Also, such groups may be prevented from reaction with proteins because of steric restrictions of the matrix. Third, the reactivity of protein side chains may be lower than those of model compounds, increasing side reactions, e.g., hydrolysis.[15]

The chemical reaction between the carrier and the protein leading to immobilization is often inferred from studies on model systems. Only in a few cases has there been a positive demonstration of the pathway of the reaction with proteins, and only then have the side chains involved in the coupling reaction been identified. One of these cases is the four-component reaction[16b] where the amino acid synthesized during the coupling reaction can be separated from other material and identified. In general, those amino acids participating in the covalent link often show reduced recoveries after acid hydrolysis. Thus, when dipeptides are coupled to cyanogen bromide-activated Sephadex, the amino acid bearing the free amino group (which participates in the coupling reaction) is recovered to only 80–85%. In another instance, that of binding proteins to thiol gels,[17] treatment of the resulting conjugate with reductive agents leads to elution of the protein, which must be regarded as strong evidence for the participation of thiol groups of the protein in the coupling reaction. Few attempts have been reported to assess directly the number of links between the carrier and the protein.[18]

Determination of Coupling Yield and Bound Protein

The amount of protein bound to a carrier will generally be correlated to the final enzymic properties of the conjugate. It will also have influence

[16b] P. Vretblad and R. Axén, *Acta Chem. Scand.* **27**, 2769 (1973).
[17] R. Axén and P. Vretblad, *Acta Chem. Scand.* **25**, 2711 (1971).
[18] G. P. Royer and R. Uy, *J. Biol. Chem.* **248**, 2627 (1973).

on the most economical way of immobilizing a given enzyme for preparative use of its catalytic properties. Also, it must be known how much enzyme is present in a conjugate when kinetic properties are to be investigated (see Section VI of this volume). The determination of protein immobilized is therefore a necessary part of the characterization of an immobilized enzyme. By coupling yield, we mean the amount of protein bound in relation to the amount of protein added to the immobilization mixture.

A variety of methods have been developed to determine the amount of protein present in protein–carrier conjugates.

Difference between Protein Added and Protein Recovered

The simplest way of assessing quantitatively the coupling of a protein to a carrier is to measure the disappearance of protein material from the reaction mixture, either by measuring ultraviolet absorption (or absorption at any other wavelength where the protein in question absorbs) or reagents that can measure protein quantitatively, e.g., the Folin reagent.

Usually, the measurement is done in the combined washings of the conjugate. To this end, the conjugate is filtered off and washed briefly with suitable buffer solutions to elute most of the nonfirmly bound protein. In order not to dilute the washing solutions too much, only small quantities of eluents are taken. The washing must therefore be continued in a more rigorous way in order to deplete the conjugate completely from the last traces of adsorbed protein.[2]

Colorimetric Determination of Proteins. SOLUTIONS. A: 2% (w/v) Na_2CO_3 in 0.1 M NaOH; B: 0.5% (w/v) $Cu_2SO_4 \cdot 5H_2O$ in 1% (w/v) sodium citrate; C: a mixture of 50 ml of A and 1 ml of B; D: commercial Folin–Ciocalteau reagent, diluted with 1 part H_2O.

DETERMINATION. To 0.1–0.2 ml of a sample containing 20–200 μg of protein, 2.0 ml of solution C are added and mixed. After 10 min, 0.2 ml of solution D are added and mixed immediately. The color is read at 750 nm after 0.5–2 hr. The amount of protein present is determined from a calibration curve.

As this method is both rapid and inexpensive, it has been widely used for determining coupling yield and protein bound to the conjugate.

Elemental Analysis

Nitrogen Analysis. As proteins contain a rather constant proportion of nitrogen, standard analysis for nitrogen will give an indication of the amount of protein bound (see, e.g., Levin et al. [14]). Corrections must be applied for the nitrogen contents of the activated carrier taken through the coupling and washing procedures without addition of protein. This method gives reliable values only when most of the nitrogen originates from protein material. Thus, it is not readily adaptable for gels like polyacrylamide, cyanogen bromide-activated carbohydrates, diazotized polymers, or nitrogen-containing ion exchangers. In such polymers, the amount of nitrogen present in the reacting polymer can vastly exceed the amount of nitrogen introduced by proteins.[17]

Sulfur Analysis. Analysis for sulfur has not been carried out routinely in immobilized proteins. It has, however, been applied to the determination of low-molecular weight substances present in conjugates.[17]

Phosphate Analysis. The contents of nucleotides[19] in such conjugates has been determined by means of analysis for phosphate. Reportedly it has not been applied to phosphate-containing immobilized proteins.

Metal Analysis. Similarly to sulfur and phosphate analysis, metal analysis has not been applied to immobilized enzymes. It may, however, develop into a powerful tool to determine the contents of a conjugate in one specific protein in the presence of other proteins, as may be present in multistep enzyme systems.

Amino Acid Analysis

Analysis of the amino acids released from a protein–carrier conjugate by acid hydrolysis is the most widely used method to determine protein contents and coupling yield of immobilized enzymes.

A small amount of the conjugate (0.1–10 mg, depending on the sensitivity of the amino acid analyzer and the expected contents of protein) are hydrolyzed with 2 ml of 6 M HCl at 110° during 24 hr. Under these conditions, most organic carriers will dissolve. After hydrolysis, the contents of the hydrolysis tube are evaporated to dryness and dissolved in the buffer used for application to the amino acid analyzer. The solution is clarified by centrifugation, where any particulate material (usually degradation products of the carrier) is removed, and an aliquot is applied to the analyzer. No additional treatment of the sample is necessary prior to analysis.

Recovery of most amino acids is good, except for those participating in the coupling reaction. Usually, the enzyme content of the conjugate is calculated from only a few amino acids. Side reactions of the polymer with amino acids can decrease the recovery of certain residues. When poly-

[19] K. Mosbach and S. Gestrelius, *FEBS Lett.* **42**, 200 (1974).

acrylamide (or other carriers releasing great amounts of ammonia) is hydrolyzed, the amino acids eluting after ammonia are usually difficult to determine quantitatively. Also, extreme care must be taken not to plug the reaction coil of the analyzer or to fog the windows of the photometer.

Other Methods

When proteins are bound reversibly to carriers with noncovalent bonds (as is the case with ion-exchange resins, resins substituted with aliphatic carbon chains, or minerals) or covalent bonds (in the case of thiol gels), the amount of protein present in the gel can be determined by elution of the protein material from the conjugate. Such a method has been used to immobilize and purify urease.[20]

Proteins containing removable prosthetic groups can be measured quantitatively by transferring the conjugate to conditions where the groups are released from the protein. With such a method, glucose oxidase (containing flavine adenine dinucleotide) could be determined in the presence of large quantities of trypsin.[21] This is the only instance reported so far where one enzyme has been quantitized in the presence of extraneous protein material present.

The contents of protein in soluble protein–carrier conjugates can be determined by absorption measurements at suitable wavelengths as reported for nucleotides.[19]

Quantitative determination of immobilized chromophores can be achieved by direct absorption measurements of a suspension of the conjugate in viscous media, like a 1% solution of polyethylene-glycol,[22] using unsubstituted carriers to correct for light scattering. With thin suspensions, care must be taken of absorption flattening.[23]

Color reactions specific for certain amino acids have been used to determine the amount of protein in conjugates. Thus, a reaction specific for arginine has been used after hydrolysis in 6 M HCl.[24] Also, a reaction specific for tryptophan has been used for the same purpose.[25] The latter method does not seem to be applicable universally, as several support materials give disturbingly high background values. Attempts have been made to release protein material from insoluble conjugates by treatment

[20] J. Carlsson, R. Axén, K. Brocklehurst, and E. M. Crook, *Eur. J. Biochem.* **44**, 189 (1974).
[21] S. Gestrelius, B. Mattiasson, and K. Mosbach, *Eur. J. Biochem.* **36**, 89 (1973).
[22] M. Lindberg, P.-O. Larsson, and K. Mosbach, *Eur. J. Biochem.* **40**, 187 (1973).
[23] L. N. M. Duysens, *Biochim. Biophys. Acta* **19**, 1 (1956).
[24] W. F. Line, A. Kwong, and H. H. Weetall, *Biochim. Biophys. Acta* **242**, 194 (1971).
[25] A. Eskamani, T. Chase, J. Freudenberger, and S. G. Gilbert, *Anal. Biochem.* **57**, 412 (1974).

with pronase[26] and then to determine the amount of ninhydrin colorable material rendered soluble. Such a method will have to be calibrated separately for every protein immobilized, as proteins may differ widely in susceptibility to pronase attack. More important, the exclusion of large-molecular-weight substances from tighter conjugates may render a varying amount of the bound protein inaccessible to contact with the pronase enzymes.[27,28]

Conclusions

Of the methods described above for determining the amount of protein in immobilized protein derivatives, two seem to be particularly useful, one by virtue of its simplicity and the other because of its versatility and accuracy. The determination of protein added to the reaction mixture and that washed out from the conjugate after coupling is done very rapidly and easily and therefore gives a direct answer to whether an immobilization has been carried out successfully. Amino acid analysis after acid hydrolysis is nowadays a routine procedure in most laboratories. Also when applied to immobilized protein, it is reproducible and requires only small amounts of material.

Any of the figures calculated from the results of the above-described methods have a definite meaning only when pure proteins are taken to immobilization. Thus, one cannot *a priori* assume that every protein in a mixture will be bound in the relative amount present in solution, but rather that some proteins couple preferentially as compared to others. For the determination of bound proteins of deliberate mixtures, specific methods will have to be applied, and no one will be generally applicable.

Active-Site Titration of Immobilized Enzymes

For soluble enzymes, methods have been described to determine the concentration of active sites in enzyme solutions by photometry[29,30] or fluorometry.[31] Equivalent and alternative methods for assessing the concentration of active enzyme in insoluble conjugates will be described below. Such methods are of importance as not all protein material in immobilized enzyme conjugates necessarily exhibit enzymic activity.

[26] P. D. Chantler and W. B. Gratzer, *FEBS Lett.* **34**, 10 (1973).
[27] P. Cresswell and A. R. Sandersson, *Biochem. J.* **119**, 447 (1970).
[28] R. J. Knights and A. Light, *Arch. Biochem. Biophys.* **160**, 377 (1974).
[29] F. J. Kédzy and E. T. Kaiser, this series Vol. 19, p. 3 (1970).
[30] T. Chase and E. Shaw, this series Vol. 19, p. 20 (1970).
[31] G. W. Jameson, D. V. Roberts, R. W. Adams, W. S. A. Kyle, and D. T. Elmore, *Biochem. J.* **131**, 107 (1973).

Photometric Titration

Photometric methods have been applied to immobilized enzymes.[28,32] Because of the light scattering of the conjugate, which would interfere with the signal to be observed, special precautions have to be taken to prevent enzyme particles from entering the optical system. Ford et al.[32] have therefore used a specially designed column reactor packed with immobilized trypsin, through which a solution with the active-site titrand p-nithophenyl p'-guanidine benzoate (specific for trypsin) is pumped. The solution then passes a spectrophotometer, and the burst of p-nitrophenol is calculated from extrapolation of the obtained curve to zero time.

Photometric Titration of Immobilized Trypsin.[28] An amount of immobilized trypsin corresponding to about 25 μmoles of enzyme is suspended in 3.0 ml of 0.1 M Veronal buffer, pH 8.3. Thirty microliters of a 0.01 M solution of p-nitrophenyl-p'-guanidinobenzoate in dimethylformamide/acetonitrile (1:4 v/v) are added, and the suspension is stirred for 0.5–3 min. Then the solution is cleared by filtering off the conjugate, and its absorbance at 410 nm is measured against a blank prepared by adding the titrand to buffer only. A molar extinction coefficient $\epsilon_{410} = 16,600$ for p-nitrophenol is used to convert absorbance readings into concentration.

Fluorometric Titration

Fluorometric titrations of active sites are more easily applied to immobilized enzymes, and require less material. Immobilized chymotrypsin could be determined with 4-methylumbelliferyl p-(N,N,N-trimethylammonium)cinnamate,[33] which is not fluorescent, but from which fluorescent 4-methylumbelliferone is released.

Fluorometric Titration of Immobilized Chymotrypsin.[33] To a fluorescence cuvette containing a solution of 25 μM 4-methylumbelliferyl p-(N,N,N-trimethylammonium)cinnamate in 0.1 M sodium phosphate buffer, pH 7.5, a proportion of the conjugate containing about 1 nmole of enzyme is added. The contents of the cuvette are stirred for 30 sec, and the increase in fluorescence at 450 nm, after excitation at 360 nm, is measured. The concentration of free 4-methylumbelliferone is calculated from the increase of fluorescence extrapolated to zero time and a calibration curve of 4-methylumbelliferone in the same buffer.

As the observed signal is separated from the scattered exciting signal, the conjugate can be present in the cuvette without impeding the results. Also, because concentration determinations by fluorometry are much more

[32] J. R. Ford, R. P. Chambers, and W. Cohen, *Biochim. Biophys. Acta* **309**, 175 (1973).
[33] D. Gabel, *FEBS Lett.* **49**, 280 (1974).

sensitive than photometric methods, as little as 0.1 nmole of enzyme can easily be determined.

Spin-Labeling Method

Spin-labeling of immobilized trypsin has been used to determine the concentration of active sites in the conjugate.[34] Here, the spin label attached to the enzyme was hydrolyzed off in 0.5 M NaOH and the resulting free nitroxide radical was quantitatively determined from the electron resonance spectrum.

Other Methods

Determination of Sulfhydryl Groups in Immobilized Proteins.[35] The number of SH groups in immobilized papain has been determined by allowing the conjugate to react with iodoacetic acid and determining carboxymethyl cysteine after hydrolysis.[35]

In the reaction vessel of a pH stat, an amount of immobilized enzyme corresponding to about 5 mg of protein is suspended in 10 ml of water. Iodacetic acid, 200 mg dissolved in 0.7 ml of 1 M NaOH, is added; the reaction vessel is wrapped with aluminum foil and the pH is kept constant at 6.85 for 15 min. Nitrogen gas is used to eliminate air. Then the gel is washed on a filter with 1 M acetic acid and water, dried, and analyzed for carboxymethyl cysteine after acid hydrolysis.

Binding of High-Molecular-Weight Inhibitors. This method can be used to determine the amount of enzyme accessible to large substrates.[28] Here, generally a proportion of the total amount of active enzyme can retain its activity against small substrates. Reaction of immobilized enzymes with radioactive compounds yielding either covalent or noncovalent complexes would be an alternative way of determining the concentration of active enzyme in conjugates.[36]

Binding of Radioactive Compounds to Immobilized Enzymes.[36] The procedure described here has been elaborated for the binding of NAD to immobilized lactate dehydrogenase. A known amount of immobilized enzyme is suspended in 3 ml of a NAD solution in 0.1 M phosphate buffer, pH 8.0, containing a known amount of ^{14}C-labeled NAD. After 1 hr, 2 ml of the supernatant are removed and counted in a liquid scintillation counter. A control experiment is done with the plain carrier. The data are

[34] L. J. Berliner, S. T. Miller, R. Uy, and G. P. Royer, *Biochim. Biophys. Acta* **315**, 195 (1973).
[35] J. Carlsson and R. Axén, unpublished results, 1971.
[36] I. C. Cho and H. Swaisgood, *Biochim. Biophys. Acta* **334**, 243 (1974).

analyzed in a Scatchard plot, where the proportion of bound NAD is plotted against the proportion bound divided by the bound NAD concentration (see also this volume [34]).

Conclusion

The protein contents of immobilized enzyme conjugates, determined with methods described above, gives a maximum figure of the amount of active enzyme in the conjugate. Inactive protein can arise from three processes: inactivation prior to coupling (by, e.g., autolysis in the case of proteases); inactivation because of the immobilization reaction; and coupling in such a way that the catalytic site of the enzyme is not accessible to the substrate (thus being intact, but not active). Thorough investigation of the enzyme kinetic behavior of a conjugate will give results that are influenced by changes of the enzyme upon immobilization as well as by changes in the microenvironment of the enzyme. On condition that an active site-directed reagent for a soluble enzyme will react with the insoluble derivative, its application to probe the static condition of the conjugate will give information that complements that obtained from the dynamic kinetic investigation.

Section VI
Kinetics of Immobilized Enzymes

[29] Kinetic Behavior of Immobilized Enzyme Systems

By L. GOLDSTEIN

The study of enzymes artificially bound onto or within solid supports has attracted considerable attention owing to the potential of these water-insoluble preparations as highly specific reusable and removable reagents, as industrial catalysts, and as components of various analytical devices, electrodes, and monitoring systems. Moreover, immobilized enzymes can serve as simple, relatively well-characterized model systems by means of which some of the principles underlying the kinetic behavior of membrane-bound enzymes in nature can be deduced.

The effects of immobilization on the kinetic behavior of an enzyme can be classified as follows:

1. Conformational and steric effects: the enzyme may be conformationally different when fixed on a support; alternatively it may be attached to the solid carrier in a way that would render certain parts of the enzyme molecule less accessible to substrate or effector. These effects are the least well understood.

2. Partitioning effects: the equilibrium substrate, or effector concentrations within the support, may be different from those in the bulk solution. Such effects, related to the chemical nature of the support material, may arise from electrostatic or hydrophobic interactions between the matrix and low-molecular-weight species present in the medium, leading to a modified microenvironment, i.e., to different concentrations of substrate, product or effector, hydrogen and hydroxyl ions, etc., in the domain of the immobilized enzyme particle.

3. Microenvironmental effects on the intrinsic catalytic parameters of the enzyme: such effects due to the perturbation of the catalytic pathway of the enzymic reaction would reflect events arising from the fact that enzyme-substrate interactions occur in a different microenvironment when an enzyme is immobilized on a solid support.

4. Diffusional or mass-transfer effects: such effects would arise from diffusional resistances to the translocation of substrate, product, or effector to or from the site of the enzymic reaction and would be particularly pronounced in the case of fast enzymic reactions and configurations, where the particle size or membrane thickness are relatively large. An immobilized enzyme functioning under conditions of diffusional restrictions would hence be exposed, even in the steady state, to local concentrations of substrate product or effector different from those in the bulk

solution. This would be reflected in the values of the kinetic parameters usually employed to characterize enzymic reactions.

In view of the above effects, in the case of immobilized enzymes, where the apparent kinetic behavior can be controlled by both microenvironmental and mass-transfer effects, it is useful to distinguish between[1] (1) *intrinsic* rate parameters of the enzymic reaction, i.e., the kinetic parameters characteristic of the native enzyme in solution; (2) *inherent* rate parameters, those pertaining to the immobilized enzyme in the absence of any diffusional limitations; and (3) *effective* rate and kinetic parameters, observed in an immobilized enzyme when diffusional limitations are significant.

Several recent books and review articles summarize various aspects of the kinetic behavior of heterogeneous enzyme systems.[1-15]

Notation

D_s Substrate diffusivity in solution
D_s' Substrate diffusivity in porous medium
h_s Transport coefficient for substrate
h_p Transport coefficient for product
I Ionic strength
J_s Rate of flow of substrate
J_p Rate of flow of product
K_m Michaelis constant

[1] J. M. Engasser and C. Horvath, in "Applied Biochemistry and Bioengineering" (L. B. Wingard, E. Katchalski, and L. Goldstein, eds.), Vol. 1, in press. Academic Press, New York, 1976.
[2] R. Goldman, L. Goldstein, and E. Katchalski, in "Biochemical Aspects of Reactions on Solid Supports" (G. R. Stark, ed.), p. 1. Academic Press, New York, 1971.
[3] L. Goldstein and E. Katchalski, Z. Anal. Chem. **243**, 375 (1968).
[4] E. Katchalski, I. Silman, and R. Goldman, Adv. Enzymol. **34**, 445 (1971).
[5] O. R. Zaborsky, "Immobilized Enzymes." Chem. Rubber Publ. Co., Cleveland, Ohio, 1973.
[6] K. J. Laidler and P. V. Sundaram, in "Chemistry of the Cell Interface" (H. D. Brown, ed.), Part A, p. 255. Academic Press, New York, 1971.
[7] I. B. Berezin, A. M. Klibanov, and K. Martinek, Usp. Khim. **44**, 17 (1975).
[8] D. Thomas and S. R. Caplan, in "Membrane Separation Processes" (P. Meares, ed.), in press. Elsevier, Amsterdam, 1976.
[9] E. Selegny, in "Polyelectrolytes" (E. Selegny, ed.), p. 419. Reidel, Dordrecht, Holland, 1974.
[10] R. Goldman, Biochimie **55**, 953 (1973).
[11] A. Goldbeter and S. R. Caplan, Adv. Biophys. Bioeng. **5**, in press (1976).
[12] D. Thomas and G. Broun, Biochimie **55**, 975 (1973).
[13] A. D. McLaren and L. Packer, Adv. Enzymol. **33**, 245 (1970).
[14] P. A. Srere and K. Mosbach, Annu. Rev. Microbiol. **28**, 61 (1975).
[15] L. B. Wingard, in "Advances in Biochemical Engineering" (T. K. Ghose, A. Fiechter, and N. Blakebrough, eds.), Vol. 2, p. 1. Springer-Verlag, Berlin and New York, 1972.

K_m' Apparent Michaelis constant, equal to S_o when $V' = V_{max}/2$
k Boltzmann constant
k' Effective first-order rate constant
L Thickness of membrane
l Dimensionless distance in membrane (χ/L)
P Local concentration of product
P_o Bulk concentration of product
$\mathcal{P}$ Partition coefficient
$\mathcal{P}_{H^+}$ Partition coefficient for hydrogen ions
$\mathcal{P}_s$ Partition coefficient for substrate
R Radius of sphere
S_i Equilibrium local concentration of substrate
S Effective local concentration of substrate
S_o Bulk concentration of substrate
T Absolute temperature (°K)
χ Longitudinal distance from membrane surface
V' Effective rate of reaction
V_{max} Saturation rate of an enzymic reaction
V'''_{max} Saturation rate of enzymic reaction per unit volume
z Valence of ion
γ Activity coefficient
δ Thickness of diffusion boundary layer
ϵ Effectiveness factor for first-order kinetics
e Electronic charge
η Effectiveness factor
η_I Effectiveness factor in the presence of inhibition
μ Substrate modulus for external diffusion $(V_{max}/h_s K_m)$
ϕ Substrate modulus for internal diffusion [modified dimensionless Thiele modulus, $L(V_{max}/K_m D_s')^{1/2}$]
Φ_L Dimensionless modulus defined by Eqs. (55) and (57)
ρ Dimensionless Michaelis constant (K_m/S)
σ_o Dimensionless bulk substrate concentration (S_o/K_m)
σ Dimensionless effective substrate concentration (S/K_m)
ψ Electrostatic potential

Microenvironmental Effects

When an enzyme is embedded in a solid support, the protein is in effect removed from its native aqueous milieu, being now exposed to a different local *microenvironment*, whose characteristics are determined by the chemical nature of the support material. This may have rather far-reaching kinetic consequences, since microenvironmental effects may lead to unequal distribution, viz. partitioning, of substrate, product, and low-molecular-weight effectors between the two phases—the immobilized enzyme particles and the bulk solution.[2-4,13,16-18] Such a phenomenon—

[16] L. Goldstein, Y. Levin, and E. Katchalski, *Biochemistry* **3**, 1913 (1964).
[17] A. D. McLaren and E. F. Esterman, *Arch. Biochem. Biophys.* **68**, 157 (1957).
[18] A. C. Johansson and K. Mosbach, *Biochim. Biophys. Acta* **370**, 339 (1974).

related, for example, to electrostatic or hydrophobic interactions between the matrix and low-molecular-weight species in solution—would be reflected in the concentration-dependence of the rate parameters. Furthermore, the intrinsic kinetic parameters of an enzyme, as well as its apparent specificity, might be changed as a result of perturbations of the catalytic pathway of the enzymic reaction induced by a modified microenvironment.[19,20]

The microenvironmental effects arising from the electrostatic field generated in highly charged supports have been most thoroughly investigated[2-4,13,16,17,21-24] and will be discussed here in greater detail.

Partitioning Effects

The observed kinetic behavior of an enzyme embedded in a charged support may differ from that of the same enzyme in solution even in the absence of diffusional effects. This modified behavior can be attributed to the fact that the concentration of charged species, substrates, products and effectors, hydrogen and hydroxyl ions, etc., in the domain of the immobilized enzyme is different from that in the outer solution, owing to electrostatic interactions with the fixed charges on the support.[2-4,13,16,17,21-24] Such differences in the equilibrium concentrations of charged solutes may be conveniently described by the partition coefficient, $\mathcal{P}$, given by

$$\mathcal{P} = c_i/c_o \quad (1)$$

where c_i and c_o are the local and bulk concentrations, respectively.

Assuming the Boltzmann distribution, the partitioning of hydrogen ions between a charged immobilized-enzyme particle and the outer solution can be described by[16]

$$a_i^{H^+} = a_o^{H^+} \exp(-e\psi/kT) \quad (2)$$

where $a_i^{H^+}$ and $a_o^{H^+}$ are the hydrogen ion activities in the charged solid phase and the outer solution, respectively; ψ is the electrostatic potential; e, the electronic charge; k, the Boltzmann constant; and T, the absolute temperature.

[19] L. Goldstein, *Biochemistry* 11, 4072 (1972).
[20] L. Goldstein, *Isr. J. Chem.* 11, 379 (1973).
[21] J. M. Engasser and C. Horvath, *Biochem. J.* 145, 431 (1975).
[22] M. L. Shuler, R. Aris, and H. M. Tsuchiya, *J. Theor. Biol.* 35, 67 (1972).
[23] W. E. Hornby, M. D. Lilly, and E. M. Crook, *Biochem. J.* 98, 420 (1966).
[24] C. M. Wharton, E. M. Crook, and K. Brocklehurst, *Eur. J. Biochem.* 6, 572 (1968).

Introducing the partition coefficient for hydrogen ions, $\mathcal{P}_{H^+}$, Eq. (2) can be rewritten as

$$\mathcal{P}_H = a_i^{H^+}/a_o^{H^+} = \exp(-e\psi/kT) \tag{2a}$$

Equations (2) and (2a) show that the local hydrogen ion concentration in the domain of a polyanionic enzyme derivative would be higher than that measured in the bulk solution. The local pH would hence be lower. The reverse would be true for a polycationic enzyme derivative. Consequently the pH-activity profile of an enzyme immobilized within a charged carrier would be displaced toward more alkaline or toward more acidic pH values for a negatively or positively charged carrier, respectively.[16] Quantitatively this may be expressed in the form

$$\Delta pH = pH_i - pH_o = 0.43e\psi/kT \tag{3}$$

where pH_i and pH_o are the local and outer (bulk) pH values.

The displaced pH-activity profiles of a polyelectrolyte enzyme derivative can be alternatively represented in terms of changes in the values of the apparent acidic dissociation constant of an active-site ionizing group BH^+, given by $K = [B][H^+]/[BH^+]$. In analogy to Eq. (3) we can therefore write[2,16]

$$\Delta pK = pK - pK' = 0.43e\psi/kT \tag{3a}$$

where pK' is the dissociation constant of the ionizing group on the bound enzyme. It can be easily seen that for a positively charged support ($\psi > 0$), $pK' < pK$; conversely for a negatively charged support ($\psi < 0$), $pK' > pK$. Active-site ionizing group pK's can be estimated from the midpoints of the appropriate pH-rate curves and by a number of other procedures.[25,26]

By a similar argument, the partitioning of charged substrate between a charged enzyme particle and the outer solution can be represented in the form[2,16,27]

$$S_i = S_o \exp(-ze\psi/kT) \tag{4}$$

where S_i and S_o are the substrate concentrations in the domain of the polyelectrolyte enzyme particle and the outer solution and ze is the sub-

[25] M. Dixon and E. C. Webb, "Enzymes," 2nd ed. Longmans, Green, New York, 1964.
[26] K. J. Laidler and P. S. Bunting, "The Chemical Kinetics of Enzyme Action," 2nd ed. Oxford Univ. Press (Clarendon), London and New York, 1973.
[27] It should be pointed out that the notation used by Goldstein et al.[16] in their original article has led to some concern regarding the self-consistency of signs in the equations describing the partitioning of hydrogen ions and substrate [analogous to Eqs. (2) and (4) in this article]. For the sake of clarity Eqs. (2) and (4) are written here in a more conventional form.

strate charge. In terms of the partition coefficient for substrate, $\mathcal{P}_s$, Eq. (4) becomes

$$\mathcal{P}_s = S_i/S_o = \exp(-ze\psi/kT) \quad (4a)$$

Examination of Eqs. (4) and (4a) shows that $S_i > S_o$ when the immobilized enzyme and substrate are of opposite charge (ze and ψ opposite in sign); conversely $S_i < S_o$ when enzyme and substrate are of the same charge (ze and ψ of the same sign).

Assuming that the Michaelis–Menten scheme is obeyed by the charged enzyme derivative and that diffusional limitations can be neglected, the observed rate of the reaction, V', will depend on the local equilibrium concentration of substrate, S_i, hence

$$V' = (V_{max}S_i)/(K_m + S_i) \quad (5)$$

Insertion of Eq. (4) into Eq. (5) leads to

$$V' = [V_{max}S_o \exp(-ze\psi/kT)]/[K_m + S_o \exp(-ze\psi/kT)] \quad (6)$$

where the observed rate, V', is expressed in terms of the bulk concentration of substrate. From Eq. (6) it follows that $V' = \frac{1}{2}V_{max}$ when $S_o = K_m \exp(ze\psi/kT)$. Thus the bulk concentration of substrate, S_o, at which the half-maximal rate is attained, leads to an apparent Michaelis constant, K_m', related to the intrinsic Michaelis constant of the native enzyme, K_m, by the expression

$$K_m' = K_m \exp(ze\psi/kT) \quad (7)$$

Equation (7) shows explicitly that in the case of similar charge on substrate and supported enzyme (ze and ψ of the same sign), $K_m' > K_m$; conversely for dissimilar charges (ze and ψ of opposite signs) $K_m' < K_m$.

In analogy to Eqs. (3) and (3a), Eq. (7) can be written in the form[16]

$$\Delta pK_m = pK_m - pK_m' = \log(K_m'/K_m) = 0.43ze\psi/kT \quad (7a)$$

Equations (3), (3a), (7), and (7a) have been verified in numerous experiments: Goldstein and co-workers[16] have shown that the pH-activity profiles of polyanionic derivatives of several enzymes acting on their specific low-molecular-weight substrates at low ionic strength were displaced toward more alkaline pH values by 1–2.5 pH units, as compared to the native enzymes. Polycationic derivatives of the same enzymes exhibited the reverse effect, i.e., displacement of the pH-activity profiles toward more acidic pH values. As expected, the anomalies were abolished at high ionic strength. This is illustrated in Figs. 1 and 2 for polyanionic and polycationic derivatives of trypsin and chymotrypsin. Goldstein and co-workers[16] have also shown that the apparent Michaelis constant of a polyanionic derivative of trypsin acting on the positively charged substrate benzoyl-L-arginine amide (BAA) at low ionic strength was lower by more than one order of magnitude as compared to the native enzyme ($K_m = 6.9$ mM; $K_m' = 0.2$ mM; see Figs. 3 and 4). The perturbation of the apparent Michaelis constant was again abolished at high ionic strength as expected from theoretical considerations. Similar effects have been reported for a number of charged enzyme derivatives.[2,13,16,17,23,24]

Using the Donnan relation to describe the distribution of charged substrate between the outer solution and the polyelectrolyte enzyme phase, Wharton and co-workers[24] deduced an expression in which the effects of ionic strength, I, on the apparent Michaelis constant, K_m', are given explicitly:

$$(K_m')^2 = \gamma^2 K_m[K_m - K_m' Z m_c/I] \tag{8}$$

In Eq. (8), Z is the modulus of the number of charges on the matrix, and m_c the concentration of the matrix in its own hydrated volume. $Z m_c$ thus denotes the effective concentration of fixed charged groups in the polyelectrolyte phase. The activity coefficient, γ, is given by $\gamma = \gamma_\pm^i/\gamma_\pm^o$,

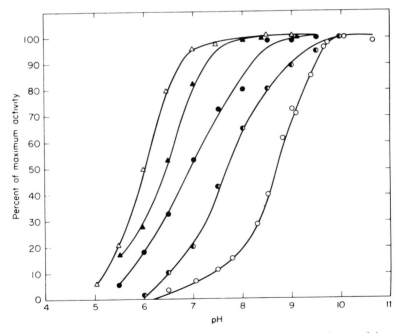

FIG. 1. pH-activity curves for trypsin and a polyanionic, ethylene-maleic acid (EMA) copolymer derivative of trypsin (EMA-trypsin) at different ionic strengths: Trypsin—△, 3.5×10^{-2}; ▲, 1.0; EMA-trypsin—○, 6.0×10^{-3}; ◐, 3.5×10^{-2}; ●, 1.0. The substrate was benzoyl-L-arginine ethyl ester. Redrawn from data of L. Goldstein, Y. Levin, and E. Katchalski, *Biochemistry* **3**, 1913 (1964).

where $\gamma_{\pm}^{i}$ and $\gamma_{\pm}^{o}$ are the mean ion activity coefficients in matrix and outer phase, respectively.

Rearrangement of Eq. (8) gives

$$K_m' = \gamma K_m[1 - K_m'Zm_c/K_mI]^{1/2} \qquad (9)$$

Binomial expansion and truncation after the first term yields

$$K_m' = \gamma K_m[1 - K_m'Zm_c/2K_mI] \qquad (10)$$

Equation (10) is valid only if $0 < (K_m'Zm_c/2K_mI) < 1$.

Since $K_m/K_m' = S_i/S_o$ [see Eqs. (4) and (7)] a comparison of Eqs. (10) and (7) shows that both equations will yield identical expressions

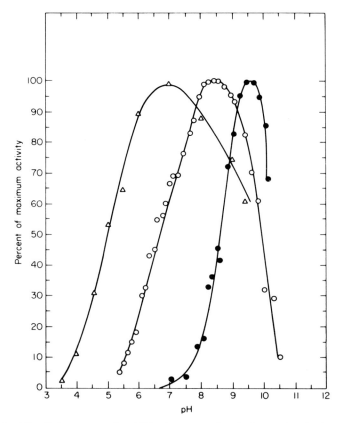

Fig. 2. pH-activity curves for (○) chymotrypsin, (●) a polyanionic ethylene-maleic acid (EMA) copolymer derivative of chymotrypsin (EMA-chymotrypsin), and (△) a polycationic derivative, polyornithyl chymotrypsin. The substrate was acetyl-L-tyrosine ethyl ester. Ionic strength, 0.01. Redrawn from data of L. Goldstein [*Biochemistry* **11**, 4072 (1972)] and L. Goldstein and E. Katchalski [*Z. Anal. Chem.* **243**, 375 (1968)].

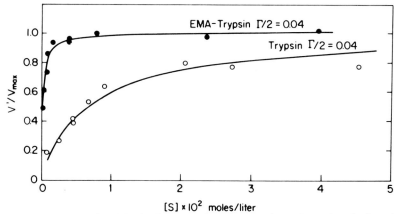

FIG. 3. Normalized Michaelis–Menten plots for trypsin and a polyanionic, ethylene-maleic acid (EMA) copolymer derivative of trypsin (EMA-trypsin) acting on benzoyl-L-arginine amide. Ionic strength 0.04. Redrawn from data of L. Goldstein, Y. Levin, and E. Katchalski, *Biochemistry* **3**, 1913 (1964).

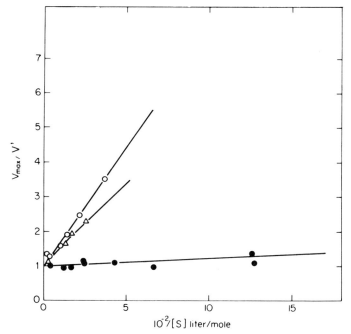

FIG. 4. Normalized Lineweaver–Burk plots for trypsin (○; $I = 0.04$; $K_m = 6.85 \pm 1.0$ mM) and a polyanionic ethylene-maleic acid copolymer derivative of trypsin (EMA-trypsin) (△, $I = 0.5$, $K_m' = 5.2 \pm 0.50$ mM; ●, $I = 0.04$, $K_m' = 0.2 \pm 0.05$ mM) acting on benzoyl-L-arginine amide. Data of Fig. 3. From L. Goldstein, Y. Levin, and E. Katchalski, *Biochemistry* **3**, 1913 (1964).

for K_m/K_m' when the exponential term in Eq. (7) equals $1/\gamma + Zm_c/2I$. Equation (7) is the more general of the two; it does not, however, permit an assessment of the effects of charge on the apparent Michaelis constant, K_m', in terms of readily measurable quantities. Equation (10), on the other hand, relates K_m' to the ionic strength of the medium, I, and the electrostatic parameter, Zm_c, characteristic of the matrix material. Zm_c, the effective concentration of fixed charged groups in the hydrated volume of the polyelectrolyte matrix, can be estimated by suspending a given amount of matrix material, of known net charge, in water and allowing it to settle in a measuring cylinder. The hydrated volume of the solid can then be read off on the cylinder.

Equation (10) can be rearranged into Eqs. (11) and (12)

$$1/K_m' = I/\gamma K_m + Zm_c/2K_m \tag{11}$$

$$K_m' = \gamma K_m I/[(\gamma Zm_c/2) + I] \tag{12}$$

Equation (11) predicts that a plot of $1/K_m$ against I will be linear assuming that γ is constant, with intercept $Zm_c/2K_m$ and slope $1/\gamma K_m$ (see Fig. 5). Equation (12) is the hyperbolic form of Eq. (10) (see Fig. 6).

The validity of Eqs. (10), (11), and (12) was illustrated in a study of the kinetics of hydrolysis of benzoyl-L-arginine ethyl ester (BAEE) by carboxymethyl cellulose-bromelain.[24] The value of K_m obtained for the hydrolysis of BAEE by native bromelain was invariant with ionic strength. The value of K_m' at low ionic strength ($K_m' = 0.007\ M$ at $I = 0.023$) was found to be lower by about one order of magnitude than the value of K_m for native bromelain under similar conditions ($K_m = 0.11\ M$). K_m' increased with increasing ionic strength, approaching a value which is somewhat lower than that of the native enzyme. The data plotted according to Eqs. (11) and (12) are shown in Figs. 5 and 6.

Equation (12) shows that $K_m' = \gamma K_m/2$ when $I = \gamma Zm_c/2$. Thus, if $\gamma = 1$, $K_m' = K_m/2$ when $I = Zm_c/2$, i.e., the apparent K_m is equal to half its limiting value when the ionic strength of the outer solution equals half the concentration of CM-cellulose carboxylate groups in their own hydrated volume.

The finding that the limiting value of K_m' (at very high ionic strength) is lower than the value of K_m for the native enzyme suggested that interactions of matrix and substrate other than those of the charge–charge type might be of some significance in determining the magnitude of the apparent Michaelis constant.[24]

The effects of a hydrophobic microenvironment on the partitioning of substrate were recently investigated by Johansson and Mosbach.[28] These authors coupled horse liver alcohol dehydrogenase to cross-linked co-

[28] A. C. Johansson and K. Mosbach, *Biochim. Biophys. Acta* **370**, 348 (1974).

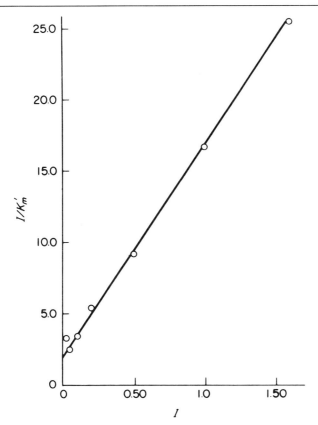

FIG. 5. Dependence of the apparent Michaelis constant, K_m', upon the ionic strength (I) for the carboxymethyl cellulose-bromelain catalyzed hydrolysis of benzoyl-L-arginine ethyl ester. The solid line was calculated according to Eq. (11) using the values $\gamma K_m = 0.066$ and $\gamma Z m_e = 0.26$. From C. M. Wharton, E. M. Crook, and K. Brocklehurst, *Eur. J. Biochem.* **6**, 572 (1968).

polymers of acrylamide and methylacrylate of varying composition, using the glutaraldehyde method[18]; in these polymers acrylamide served as the hydrophilic component and methylacrylate as the hydrophobic component; increasing the hydrophobicity of the matrix by changing the ratio of methyl acrylate to acrylamide in the copolymerization mixture caused a decrease by a factor of 4 in the apparent Michaelis constant of the bound enzyme for the more hydrophobic substrate n-butanol, whereas K_m' for ethanol was essentially unaffected. The assumption that the more hydrophobic matrix will adsorb preferentially the more hydrophobic substrate was substantiated by equilibrium studies with n-[^{14}C]-butanol. The more hydrophobic acrylamide-methylacrylate copolymer showed about 6 times higher binding of radioactivity.[28]

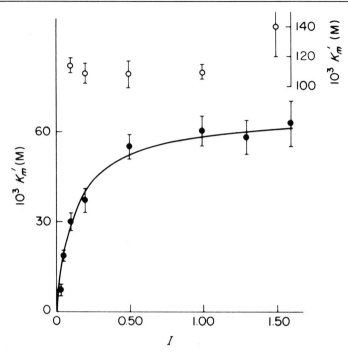

Fig. 6. Dependence of the apparent Michaelis constant, K_m', upon the ionic strength (I) for the hydrolysis of benzoyl-L-arginine ethyl ester by bromelain (○) and by carboxymethyl cellulose-bromelain (●). The solid line was calculated by Eq. (12), with $\gamma K_m = 0.066$ and $\gamma Z m_c = 0.26$. From C. M. Wharton, E. M. Crook, and K. Brocklehurst, *Eur. J. Biochem.* **6**, 572 (1968).

Similar findings were reported by Filippusson and Hornby,[29] who showed that the inhibition constant of aniline for β-fructofuranosidase was about three times higher when the enzyme was bound to a hydrophobic support, poly(*p*-aminostyrene).

Microenvironmental Effects on the Intrinsic Kinetic Parameters of Immobilized Enzymes

In the simple electrostatic models employed heretofore to explain the perturbed kinetics of charged immobilized enzyme derivatives, it was tacitly assumed that the intrinsic catalytic properties of the bound protein were independent of the charge characteristics of the microenvironment and identical with those of the native enzyme.[2,13,16,17,21-24] The generality of such assumptions is, however, questionable. In the following, a brief account will be given of some experimental evidence,

[29] H. Filippusson and W. E. Hornby, *Biochem. J.* **120**, 215 (1970).

suggesting that at least in some cases the intrinsic catalytic properties of an enzyme might be affected by the chemical characteristics of the microenvironment.[19,20,28]

The kinetic behavior of charged enzyme derivatives was recently reinvestigated, using a series of water-soluble polyanionic and polycationic derivatives of chymotrypsin; these derivatives could be prepared by growing polyornithyl or polyglutamyl side chains on the enzyme and by coupling chymotrypsin to a copolymer of ethylene and maleic acid (EMA).[19] The amount of active enzyme in these preparations could be determined by the conventional spectrophotometric "active-site" titration methods; thus the intrinsic kinetic parameters of the polyelectrolyte derivatives of chymotrypsin and those of the native enzyme could be compared on an absolute basis. The kinetic data indicated that the overall rate constants (k_{cat}) of the polyanionic derivatives of chymotrypsin were higher than the k_{cat} value of the native enzyme; conversely the k_{cat} values of the polycationic derivatives were lower (Fig. 7). The perturbation of k_{cat} was greater with amide than

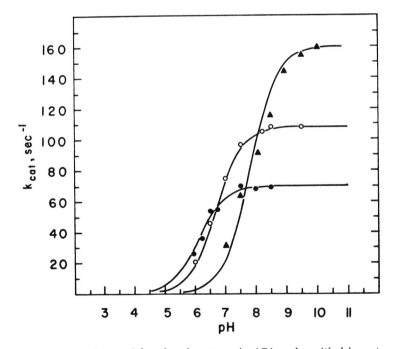

FIG. 7. pH dependence of k_{cat} for chymotrypsin (○), polyornithylchymotrypsin (●), and an ethylene-maleic acid copolymer derivative of chymotrypsin (EMA-chymotrysin) (▲) acting on acetyl-L-phenylalanine methyl ester. Ionic strength is 0.05. Redrawn from data of L. Goldstein, *Biochemistry* 11, 4072 (1972).

with ester substrates, suggesting that the acylation step[30] was more strongly affected by electrostatic perturbations. Moreover, an increase in the ionic strength of the medium caused an increase in the values of k_{cat} of both native chymotrypsin and the positively charged polyornithyl chymotrypsin derivatives; the k_{cat} values of the negatively charged EMA-chymotrypsin and polyglutamyl chymotrypsin, on the other hand, were not affected by ionic strength (see the table for representative data).[19]

KINETIC PARAMETERS OF CHYMOTRYPSIN, POLYORNITHYLCHYMOTYPSIN, AND ETHYLENE-MALEIC ACID (EMA) DERIVATIVE OF CHYMOTRYPSIN ACTING ON ACETYL-L-TYROSINE ETHYL ESTER AND ACETYL-L-TYROSINAMIDE[a]

		Ac-L-TyrOEt		Ac-L-TyrNH$_2$	
Enzyme	Ionic strength	k_{cat} (sec^{-1})	$K_{m(app)}$ (mM)	$k_{cat} \times 10^2$ (sec^{-1})	$K_{m(app)}$ (mM)
Chymotrypsin	0.05	185	0.74	8	34
	1.00	230	0.55	14	21
EMA-chymotrypsin	0.05	300	2.50	33	30
	1.00	280	1.90	29	35
Polyornithylchymotrypsin	0.05	120	7.10	3.9	38
	1.00	165	5.80	5.9	25

[a] From L. Goldstein, *Biochemistry* **11**, 4072 (1972).

The simplest kinetic model that would accommodate these findings could be based on the assumption that a kinetically significant step in chymotypsin catalysis involves the interaction of two positively charged residues.[19,31] By this model, the high k_{cat} values of the polyanionic derivatives of chymotrypsin could be related to screening of the above interactions by the negative charges on the support, this being analogous to the effect of high salt concentration on the native enzyme. In the case of a polycationic derivative, however, the charge–charge interaction between the two positive residues would be enhanced and k_{cat} would be lower.[19]

The model thus assumed, in effect, that catalytically significant steps in an enzymic reaction could in principle be modified by perturbations

[30] G. P. Hess, in "The Enzymes" (P. D. Boyer, ed.), 3rd ed., Vol. 3, p. 185. Academic Press, New York, 1971.
[31] These two residues have been tentatively identified as the active-site histidine (His$_{57}$) and Arg$_{145}$, the penultimate C-terminal residue on the B-chain of chymotrypsin.[19]

arising from the chemical nature of the matrix. This view found indirect support in experiments originally intended to establish the magnitude of product inhibition effects in systems where both product and support are charged.[20] In these studies the hydrolysis of acetyl-L-tyrosine ethyl ester by α-chymotypsin, by the polyanionic polyglutamyl- and EMA-chymotrypsin derivatives, and by a polycationic polyornithylchymotypsin derivative was investigated in the presence of increasing amounts of acetyl-L-tyrosine anion (AcTyr), the product of the enzymic reaction. As expected, the kinetic parameters of native chymotrypsin and of the polyanionic chymotrypsin derivatives were not significantly affected even at fairly high concentrations of acetyltyrosine. In the case of polyornithylchymotrypsin, however, both K_m' and k_{cat} increased hyperbolically with increasing AcTyr concentration (Fig. 8), suggesting a saturation phenomenon. Several aliphatic and aromatic anions, e.g., acetate, propionate, benzoate, and 1-anilino-8-naphthalene sulfonate (ANS), caused a similar increase in the k_{cat} values of polyornithylchymotrypsin. The common feature of all anionic modifiers tested was that they all contained an apolar hydrocarbon core attached to the ionizing carboxyl or sulfonate group. The data could be explained by assuming that the strong binding of AcTyr (and the other anions) was the result of a combination of electrostatic and hydrophobic interactions between the relatively hydro-

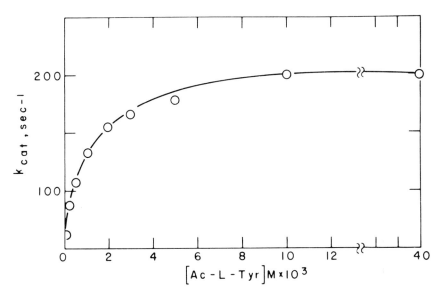

FIG. 8. Variation of k_{cat} of polyornithylchymotrypsin with the concentration of acetyl-L-tyrosine. The substrate was acetyl-L-tyrosine ethyl ester. Ionic strength is 0.01. From L. Goldstein, *Isr. J. Chem.* **11**, 379 (1973).

phobic acetyltyrosine anion and the positively charged polyornithyl side chains on the enzyme.[20] This could also explain the hyperbolic increase of k_{cat} with increasing AcTyr concentration; strong and essentially stoichiometric binding of AcTyr anions onto the cationic ornithyl residues of the side chains would result in the, at least partial, cancellation of the positive charge on the matrix, hence abolishing its deleterious effects on the catalytic activity of the bound enzyme.

The presentation of the data in this section may be considered as an attempt to delineate some of the effects of a microenvironment imposed by the chemical nature of the support on the intrinsic kinetic parameters of an immobilized enzyme and the possible abrogation or modification of such effects through interactions between a modifier present in the bulk solution and a chemical component of the support. The fragmentary nature of the data at our disposal does not allow us, however, to draw general conclusions at this stage.

Mass-Transfer Effects

When an enzyme is immobilized on or within a solid support, the substrate has to diffuse from the bulk solution to the site of the enzymic reaction. Hence when the rate of diffusion of substrate is slower than the rate of its transformation by the enzyme, the observed rate of the reaction would be lower than that expected for a given amount of enzyme in solution, since not all enzyme molecules would be in contact with substrate at a concentration level identical with that of the bulk solution.[1,2,4,6,8-12,15,32-44] Following chemical engineering practice, this phenomenon can be expressed quantitatively by the effectiveness factor,

---

[32] J. M. Engasser and C. Horvath, *J. Theor. Biol.* **42**, 137 (1973).
[33] C. Horvath and J. M. Engasser, *Biotechnol. Bioeng.* **16**, 909 (1974).
[34] R. Goldman, O. Kedem, I. H. Silman, S. R. Caplan, and E. Katchalski, *Biochemistry* **7**, 486 (1968).
[35] R. Goldman, O. Kedem, and E. Katchalski, *Biochemistry* **7**, 4518 (1968).
[36] R. Goldman, O. Kedem, and E. Katchalski, *Biochemistry* **10**, 165 (1971).
[37] P. V. Sundaram, H. Tweedale, and K. J. Laidler, *Can. J. Chem.* **48**, 1498 (1970).
[38] T. Kobayashi and K. J. Laidler, *Biochim. Biophys. Acta* **302**, 1 (1973).
[39] V. Kasche, H. Lundquist, R. Bergman, and R. Axén, *Biochem. Biophys. Res. Commun.* **41**, 615 (1971).
[40] V. Kasche and M. Bergwall, in "Insolubilized Enzymes" (M. Salmona, C. Saronio, and S. Garattini, eds.), p. 77. Raven, New York, 1974.
[41] E. Selegny, G. Broun, J. Geffroy, and D. Thomas, *J. Chim. Phys.* **66**, 391 (1969).
[42] D. Thomas, G. Broun, and E. Selegny, *Biochimie* **54**, 229 (1972).
[43] J. A. DeSimone and S. R. Caplan, *J. Theor. Biol.* **39**, 523 (1973).
[44] B. K. Hamilton, C. R. Gardner, and C. K. Colton, *AIChE J.* **20**, 503 (1974).

η,[45-51] defined as the ratio of the actual reaction rate, V' to the rate, V_{kin}, which would obtain if no mass-transfer limitations were present and hence all enzyme molecules were exposed to the same substrate concentration as that in the bulk solution. This relationship is expressed by

$$V' = \eta V_{\text{kin}} \tag{13}$$

For an enzymic reaction obeying Michaelis–Menten kinetics,[25,26] V_{kin} is given by

$$V_{\text{kin}} = V_{\text{max}} S/(K_{\text{m}} + S) \tag{14}$$

where V_{max}, K_{m}, and S have their usual meaning.

Since, in the presence of diffusional limitations, the observed rate, V', is not the true rate of the reaction, the conventional methods of treating enzyme kinetic data, whereby the characteristic rate parameters V_{max} and K_{m} are deduced from the slopes and intercepts of the appropriate linear plots, would be misleading. This can be easily seen by introducing the effectiveness factor, η, from Eq. (13) into the Michaelis–Menten expression:

$$V' = \eta[V_{\text{max}} S/(K_{\text{m}} + S)] \tag{15}$$

and rewriting the latter in the common linear forms:

$$1/V' = 1/\eta V_{\text{max}} + K_{\text{m}}/\eta V_{\text{max}} S \tag{16}$$
$$S/V' = K_{\text{m}}/\eta V_{\text{max}} + S/\eta V_{\text{max}} \tag{17}$$
$$V' = \eta V_{\text{max}} - V' K_{\text{m}}/\eta S \tag{18}$$

Unlike the corresponding linearized forms of the Michaelis–Menten equation for enzymes in solution, Eqs. (16)–(18) need not represent straight lines because η is a function of the substrate concentration, as will be shown later.[32,33,37,38,44,48]

The kinetics of an enzyme embedded in a porous medium are governed by two types of diffusional resistances to transport of substrate: (1) *external* diffusional resistances arising from the fact that substrate must be transported from the bulk solution to the catalyst's surface

---

[45] K. B. Bischoff, *AIChE J.* **11**, 351 (1965).
[46] C. W. Knudsen, G. W. Roberts, and C. N. Satterfield, *Ind. Eng. Chem., Fundam.* **5**, 325 (1966).
[47] G. W. Roberts and C. N. Satterfield. *Ind. Eng. Chem., Fundam.* **4**, 288 (1965).
[48] C. N. Satterfield, "Mass Transfer in Heterogeneous Catalysis." MIT Press, Cambridge, Massachusetts, 1970.
[49] J. M. Smith, "Chemical Engineering Kinetics," 2nd ed. McGraw-Hill, New York, 1970.
[50] E. W. Thiele, *Ind. Eng. Chem.* **31**, 916 (1939).
[51] R. Aris, *Chem. Eng. Sci.* **6**, 265 (1957).

across a boundary layer of liquid[1,15,22,33,36,52-56]; (2) *internal* diffusional resistances, i.e., diffusional limitations pertaining to the transport of substrate inside the porous catalytic medium. [15,32-44,47-51,57-70]

In the case of external diffusional limitations, the chemical reaction occurs *after* the substrate has reached the catalyst's surface. The depletion of substrate across the boundary layer can in many cases be approximated by a linear gradient. Moreover, partial cancellation of external diffusional limitations can be effected by increasing the rate of stirring of a suspension of immobilized enzyme particles or increasing the rate of flow of substrate solution through an enzyme column. With internal transport limitations the diffusion process occurs *simultaneously* with the chemical reaction, so that the two events are coupled in the mathematical sense and would normally give rise to nonlinear substrate concentration gradients within the porous catalyst (q.v.). Hence the theoretical approaches to the analysis of the interplay of enzyme catalyzed reaction with external or internal transport of substrate are different.[1,7,8]

In the following, the effects of external and internal diffusion on a solid-phase enzyme reaction will be discussed using slab geometries (i.e., flat surfaces or membranes), where mass transport can be treated as a unidimensional problem. The system will be described in terms of dimensionless quantities, independent of the substrate concentration, the most important parameter being a dimensionless substrate modulus analogous

[52] W. E. Hornby, M. D. Lilly, and E. M. Crook, *Biochem. J.* **107**, 669 (1968).
[53] S. P. O'Neill, *Biotechnol. Bioeng.* **14**, 675 (1972).
[54] T. Kobayashi and K. J. Laidler, *Biotechnol. Bioeng.* **16**, 77 (1974).
[55] A. O. Mogensen and W. R. Vieth, *Biotechnol. Bioeng.* **15**, 467 (1973).
[56] P. V. Sundaram, *Biochim. Biophys. Acta* **321**, 319 (1973).
[57] P. Van Duijn, E. Pascoe, and M. Vander Ploeg, *J. Histochem. Cytochem.* **15**, 631 (1967).
[58] J. J. Blum, *Biochim. Biophys. Acta* **21**, 155 (1956).
[59] J. J. Blum and D. J. Jender, *Arch. Biochem. Biophys.* **66**, 316 (1957).
[60] D. R. Marsh, Y. Y. Lee, and G. T. Tsao, *Biotechnol. Bioeng.* **15**, 483 (1973).
[61] P. R. Rony, *Biotechnol. Bioeng.* **13**, 431 (1971).
[62] T. Kobayashi, G. Van Dedem, and M. Moo Yang, *Biotechnol. Bioeng.* **15**, 27 (1973).
[63] W. J. Blaedel, T. R. Kissel, and R. C. Boguslaski, *Anal. Chem.* **44**, 2030 (1972).
[64] J. Lasch, *Mol. Cell. Biochem.* **2**, 79 (1973).
[65] M. Moo Yang and T. Kobayashi, *Can. J. Chem. Eng.* **50**, 162 (1972).
[66] W. R. Vieth, A. K. Mendirata, A. O. Mogensen, R. Saini, and K. Venkatasubramanian, *Chem. Eng. Sci.* **28**, 1013 (1973).
[67] D. J. Fink, T. Na, and J. S. Schultz, *Biotechnol. Bioeng.* **15**, 879 (1973).
[68] E. Selegny, G. Broun, and D. Thomas, *Physiol. Veg.* **9**, 25 (1971).
[69] D. Thomas and G. Broun, *Biochimie* **55**, 975 (1973).
[70] D. Thomas, D. Bourdillon, G. Broun, and J. P. Kernevez, *Biochemistry* **13**, 2995 (1974).

to the modified Thiele modulus[48-50] employed by the chemical engineers to describe diffusional limitations in heterogeneous catalysis. By this approach, numerical solutions of the appropriate differential equations can serve to illustrate the effects of diffusion on the plots commonly used in enzyme kinetics.[1,6,32,33,37,42]

External Diffusion

When an enzyme is attached to a fluid-impervious solid surface,[33,38,52-56] which can be considered equiaccessible[71] to substrate, the rate of flow of substrate, J_s, from the bulk solution to the catalytically active surface can be described by the product of a transport coefficient and the corresponding driving force which is the concentration difference between surface and bulk:

$$J_s = h_s(S_o - S) \tag{19}$$

S_o and S are the bulk and surface concentrations of substrate, respectively, and h_s is the transport coefficient for substrate. The transport coefficient, h_s, can in principle be replaced by an effective boundary layer of thickness δ, the so-called unstirred, or Nernst, layer,[72] which satisfies the relation[73]

$$h_s = D_s/\delta \tag{20}$$

Where D_s is the diffusivity of substrate in the bulk solution. The flow of product, J_p, away from the surface can be similarly expressed by

$$J_p = h_p(P - P_o) \tag{21}$$

In a surface reaction, the flow of substrate to the catalytic surface, and its transformation by an enzyme reaction, which obeys Michaelis–Menten kinetics, take place consecutively. At the steady state, the two processes proceed at the same rate, hence from Eqs. (14) and (19)

$$h_s(S_o - S) = V_{max}S/(K_m + S) \tag{22}$$

In the particular case where the overall reaction obeys first-order kinetics, i.e., $K_m \gg S$, Eq. (22) can be rewritten as

$$h_s(S_o - S) = (V_{max}/K_m)S \tag{23}$$

[71] D. A. Frank-Kamenetskii, "Diffusion and Heat Transfer in Chemical Kinetics." Plenum, New York, 1969.
[72] W. Nernst, Z. Phys. Chem. 47, 52 (1904).
[73] V. G. Levich, "Physicochemical Hydrodynamics." Prentice-Hall, Englewood Cliffs, New Jersey, 1972.

Hence the effective substrate concentration at the catalytic surface, S, would be

$$S = h_s S_o / (h_s + V_{max}/K_m) \tag{24}$$

and the experimentally observed rate, V', is

$$V' = (V_{max}/K_m) S = k' S_o \tag{25}$$

The effective first-order rate constant, k', in the above equation is given by

$$k' = (V_{max}/K_m) h_s / [h_s + (V_{max}/K_m)] = 1/[1/h_s + 1/(V_{max}/K_m)] \tag{26}$$

It can be seen from Eq. (26) that when $h_s \gg V_{max}/K_m$, i.e., mass transport is much faster than the chemical reaction $k' = (V_{max}/K_m)$, and the reaction is kinetically controlled [$V' = V_{kin}$; see Eq. (14)]. Conversely, if $h_s \ll V_{max}/K_m$, i.e., the chemical reaction proceeds much faster than mass transport, $k' = h_s$, and the reaction is diffusion controlled [$V' = V_{diff}$; see Eq. (19)].

At intermediate effective concentrations of substrate, i.e., $K_m \sim S$, either V_{diff} or V_{kin} will play predominant roles, depending on their relative magnitudes. In the limit of sufficiently high substrate concentrations, when $S \gg K_m$ and the reaction is zero order in substrate, V' will always equal V_{max}, the inherent saturation value of the enzymic reaction. The values of the apparent Michaelis constant, defined as the substrate concentration which gives a reaction velocity corresponding to one-half V_{max}, will, however, be considerably higher when diffusional limitations are present.[33] These arguments are illustrated schematically in Fig. 9.

The graphical representation of Eq. (22) can be facilitated by introducing two dimensionless quantities: the dimensionless substrate concentration, σ, and the dimensionless substrate modulus, μ:[33]

$$\sigma = S/K_m \tag{27}$$

$$\mu = \frac{V_{max}}{h_s K_m} \tag{28}$$

Equation (22) can then be rewritten as

$$\sigma_o - \sigma = \mu[\sigma/(1+\sigma)] \tag{29}$$

and the effective rate of the reaction, V', as a function of σ can be expressed after normalizing to V_{max} by

$$V'/V_{max} = \sigma/(1+\sigma) \tag{30}$$

The dependence of V'/V_{max} on σ_o for different values of μ, shown in Fig. 10, demonstrates the relative importance of external diffusional limitations on the observed rates of the reaction.

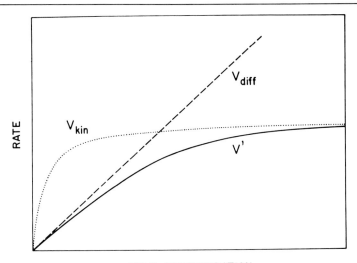

FIG. 9. Schematic plot of the overall rate of reaction, V, catalyzed by a surface-bound enzyme against the bulk substrate concentration. From C. Horvath and J. M. Engasser, *Biotechnol. Bioeng.* **16**, 909 (1974).

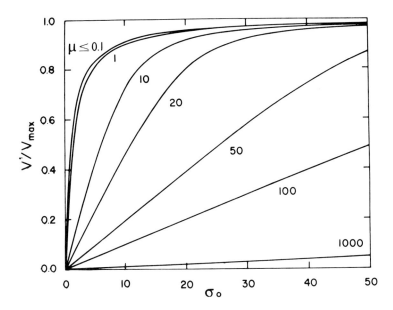

FIG. 10. V'/V_{max} against the dimensionless bulk concentration, σ_0, for different values of the substrate modulus, μ, for external diffusion. From C. Horvath and J. M. Engasser, *Biotechnol. Bioeng.* **16**, 909 (1974).

The effects of diffusional limitations on the catalytic activity of an immobilized enzyme can be quantitatively expressed by the effectiveness factor, η, defined by Eq. (13).

Figure 11 shows the dependence of the effectiveness factor on σ and the substrate modulus, μ. The value of η is unity when the reaction is kinetically controlled, but decreases with increasing diffusional limitations. The straight lines obtained for high values of μ, represent the diffusion-controlled reaction domain. At low concentrations, $\sigma < 0.1$, where the reaction is first order in substrate, η approaches a limiting value ϵ which can be shown to obey the following dependence on μ:

$$\epsilon = 1/(1 + \mu) \quad (31)$$

Under these conditions, the apparent first-order rate constant, k', is given by

$$k' = V_{\max}/\kappa \quad (32)$$

where

$$\kappa = K_m/\epsilon = K_m(1 + \mu) \quad (33)$$

The parameter κ, as defined by Eq. (33), is the effective Michaelis constant, i.e., the limiting value of the apparent Michaelis constant in the presence of diffusional resistances. Figure 12 shows the normalized Eadie-

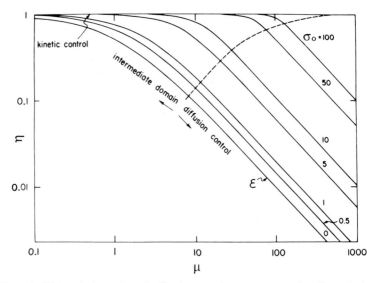

FIG. 11. Plots of the external effectiveness factor, η, as a function of the substrate modulus, μ, for different values of the bulk substrate concentration, σ_0; ϵ is the limiting first-order effectiveness factor, attained at sufficiently low concentrations. From C. Horvath and J. M. Engasser, *Biotechnol. Bioeng.* **16**, 909 (1974).

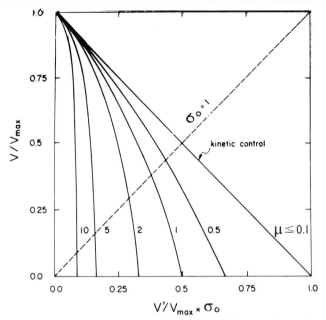

Fig. 12. Normalized Eadie–Hofstee type plots for different values of μ, the substrate modulus for external diffusion. From C. Horvath and J. M. Engasser, *Biotechnol. Bioeng.* **16**, 909 (1974).

Hofstee plots, V'/V_{max} vs $V'/(V_{max}\sigma_o)$, for different values of the substrate modulus, μ.[33] This plot yields straight lines when V obeys the Michaelis–Menten laws as seen for $\mu \to 0$. Significant departures from linearity are, however, observed for the V'/V_{max} vs $V'/(V_{max}\sigma_o)$ plots with increasing diffusional limitations, particularly when a wide range of concentrations is examined.

The effects of external diffusional limitations on the observed kinetic parameters of immobilized enzymes have been demonstrated in several studies.[36,55,56,74–80] The reaction rate was found to depend on the rate of stirring in batch reactions or on the rate of flow of substrate through an enzyme column.[74–77,79,80] Moreover the substrate concentration required for the attainment of half the saturation rate of the reaction, $\frac{1}{2}V_{max}$,

[74] C. Horvath and B. A. Solomon, *Biotechnol. Bioeng.* **14**, 885 (1972).
[75] M. D. Lilly, W. E. Hornby, and E. M. Crook, *Biochem. J.* **100**, 718 (1968).
[76] J. H. Wilson, G. Kay, and M. D. Lilly, *Biochem. J.* **108**, 845 (1968).
[77] A. K. Sharp, G. Kay, and M. D. Lilly, *Biotechnol. Bioeng.* **11**, 363 (1968).
[78] J. B. Taylor and H. E. Swaisgood, *Biochim. Biophys. Acta* **284**, 268 (1972).
[79] B. J. Rovito and J. R. Kittrell, *Biotechnol. Bioeng.* **15**, 143 (1973).
[80] T. Kobayashi and M. Moo Yang, *Biotechnol. Bioeng.* **15**, 47 (1973).

was found to be larger than the intrinsic Michaelis constant, K_m, of the enzyme[36,55,56,75,80]; alternatively the limiting first-order rate constant for the bound enzyme was smaller than the value of the ratio V_{max}/K_m for the native enzyme.[36,75-77,80]

Internal Diffusion

When an enzyme is embedded within a porous medium, the substrate has to diffuse inside the pores to reach the sites of the enzyme catalyzed reaction. Hence, unlike external transport, internal diffusion proceeds *in parallel* with the chemical reaction. The rate of the chemical reaction per unit volume element of immobilized enzyme will therefore decrease with increasing distance (in depth) from the surface of the immobilized enzyme structure owing to the progressive depletion in substrate as the latter is consumed in the course of its translocation through the catalytically active medium.[32-44,57-70] By the same reasoning, the formation of product within the porous structure will lead to its local accumulation, owing to diffusional resistances, and to the generation of product concentration gradients.

The model usually employed to study the effects of internal diffusion, both experimentally and theoretically, consists of a porous membrane containing uniformly distributed enzyme. The membrane, placed in a diffusion cell of the classical type, separates two compartments. Substrate solution of a given concentration can then be placed either in one compartment only or in both compartments; the coupled reaction-diffusion process can hence be studied under symmetric or asymmetric boundary conditions.[1-12]

Since the functioning of an enzyme membrane involves two simultaneous phenomena, enzyme reaction and metabolite (substrate and product) translocation, the rate of change of substrate concentration with time in a volume element of membrane, $\partial S/\partial t$, depends both on substrate diffusion in the membrane matrix and on enzymic activity:[1-12,15,35]

$$\partial S/\partial t = (\partial S/\partial t)_{\text{diffusion}} + (\partial S/\partial t)_{\text{reaction}} \qquad (34)$$

Considering variations in concentration as taking place only in the x-direction, perpendicular to the plane of the membrane, and introducing Fick's second law and the Michaelis–Menten relation, Eq. (34) becomes

$$\partial S/\partial t = D_s'(\partial^2 S/\partial x^2) - [V_{max}'''S/(K_m + S)] \qquad (35)$$

where the local diffusion coefficient of substrate, D_s', is assumed to be concentration independent, and the maximum local reaction rate, V_{max}, depends only on the (uniform) local concentration of enzyme. Similarly we can write for the product

$$\partial P/\partial t = D_p'(\partial^2 S/\partial x^2) + [V_{max}'''S/(K_m + S)] \quad (36)$$

The system is thus ruled by partial differential equations of the second order. For stationary state conditions, $\partial S/\partial t = 0$, Eq. (35) can be written as

$$D_s'(\partial^2 S/\partial x^2) = V_{max}'''S/(K_m + S) \quad (37)$$

Equation (37) described the decrease in stationary state substrate concentration from the membrane surface inward. Closed analytical solutions of Eq. (37) have been obtained in the first- and zero-order kinetic domains only.[37,61,63,64,66] Several analytical solutions in the form of series expansions have been proposed for the full range of Michaelis–Menten type kinetics.[42,58,69,68] Most often the appropriate differential equations have been solved by numerical calculation, the results being presented in terms of dimensionless quantities.[32,35-40,44,47,54,60,67-70] This procedure is adopted here.[1,32,33]

Equation (37) can be written in dimensionless form as

$$d^2\sigma/dl^2 = \phi^2[\sigma/(1 + \sigma)] \quad (38)$$

In Eq. (38) l is the position in the membrane given by

$$l = x/L \quad (39)$$

where L is the thickness of the membrane and x the distance from the surface; l is thus a dimensionless pore distance; ϕ, the substrate modulus for internal diffusion, known as the Thiele modulus,[46-50] is given by

$$\phi = L(V_{max}'''/K_m D_s')^{1/2} \quad (40)$$

where V_{max}''' is the saturation rate per unit volume of membrane.

For spherical particles of radius R, ϕ can be written as

$$\phi = R/3(V_{max}'''/K_m D_s')^{1/2} \quad (40a)$$

The Thiele modulus as defined by Eq. (40) contains the three factors that determine the substrate concentration profile within an enzyme membrane: the membrane thickness, the facility of the substrate to diffuse through the support, and the intrinsic activity of the catalyst. The modulus ϕ in the form given by Eq. (40) is analogous to the parameter αl employed by Goldman,[2,4,10,34-36] and by Laidler and Sundaram[6,26,37] to describe coupled reaction-diffusion in enzyme membranes. Numerical integration of Eq. (38) with asymmetric boundary conditions, i.e.,

$$\sigma = \sigma_o \quad \text{at} \quad l = 0 \tag{41}$$

and

$$d\sigma/dl = 0 \quad \text{at} \quad l = 1 \tag{42}$$

yields the substrate concentration profiles shown in Fig. 13. In Eqs. (41) and (42), it is implicitly assumed that external diffusion is negligible, i.e., that the substrate concentration at the membrane surface is essentially identical to the bulk concentration. The concentration profiles plotted in Fig. 13 reveal that at any given value of l, the effective substrate concentration in the membrane, σ, decreases with increasing values of ϕ. Moreover, the steepness of the substrate concentration gradient increases with increasing ϕ; thus, for large values of ϕ, most of the membrane will be completely devoid of substrate. In other words, as ϕ increases, the thickness of the enzyme layer participating in the catalytic reaction decreases, owing to substrate depletion.

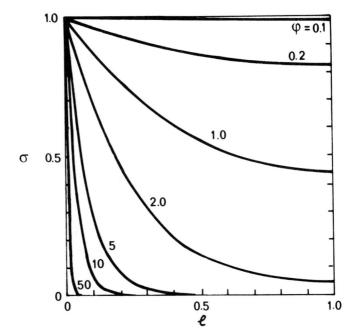

Fig. 13. Substrate concentration profiles in an enzyme membrane exposed to asymmetric boundary conditions. The dimensionless substrate concentration in the membrane, σ, is plotted against the dimensionless distance from the membrane surface, l, for different values of the substrate modulus for internal diffusion, ϕ. The concentration of substrate at the membrane surface, σ_0, is unity. Redrawn from J. M. Engasser and C. Horvath, *J. Theor. Biol.* **42**, 137 (1973).

The effective rate of the reaction in the membrane, V', normalized to the saturation rate, V_{max}, is plotted in Fig. 14 as a function of the external concentration, σ_0, for different values of the modulus, ϕ. For $\phi \leq 1$ the reaction is essentially kinetically controlled; at higher values of ϕ, the rate of the reaction is lower due to substrate depletion. Comparison of Figs. 10 and 14 shows that the effects of internal diffusional limitations on the observed rates of the enzymic reaction are much more pronounced than those arising from external diffusion for comparable values of the substrate moduli.[1,10,33,36] It should be noted that the V/V_{max} vs σ_0 curves in Fig. 14 resemble in overall shape the rectangular hyperbolas expected for a system obeying the Michaelis–Menten scheme even at large values of the modulus, ϕ, and do not display the sigmoidal form predicted by Selegny[68] and Thomas.[42]

The magnitude of diffusional limitations can be characterized by the effectiveness factor, η, defined in Eq. (13). As pointed out earlier the effectiveness factor is a function of both the substrate modulus, and the concentration σ_0;[45-49] at very low substrate concentrations ($\sigma_0 < 0.1$), the effectiveness factor becomes independent of σ and approaches a

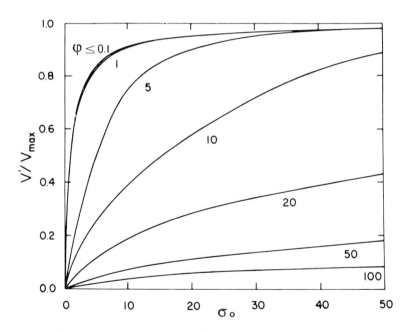

FIG. 14. Normalized overall rate as a function of the dimensionless bulk concentration of substrate, σ_0, for different values of ϕ, the substrate modulus for internal diffusion in an enzyme membrane. From C. Horvath and J. M. Engasser, *Biotechnol. Bioeng.* **16**, 909 (1974).

limiting value ϵ, given by [33]

$$\epsilon = (\tanh \phi)/\phi \qquad (43)$$

This limiting value of the first-order effectiveness factor delineates the magnitude of diffusional perturbations on the kinetic parameters of an immobilized enzyme [compare Eqs. (31) and (43)]. At low substrate concentrations, $\sigma_o < 0.1$, where the reaction obeys first-order kinetics, the effective first-order rate constant, k', is given by

$$k' = V_{\max}/\kappa \qquad (44)$$

where κ, the effective Michaelis constant, is defined by

$$\kappa = K_m/\epsilon = (K_m\phi)/(\tanh \phi) \qquad (45)$$

Equations (44) and (45) imply that in the case of a good substrate, i.e., relatively fast enzymic reaction [high values of ϕ by Eq. (40)], the full activity of an enzyme membrane might not be realizable in the domain of first-order reaction kinetics owing to complete depletion in substrate before the whole thickness of the enzyme membrane has been traversed (compare Fig. 14). The value of ϕ could be decreased in practice by choosing a poor substrate or by employing a thin membrane, thereby increasing the effectiveness factor, η, of the membrane. This practical aspect of diffusion-limited enzyme kinetics has been demonstrated by Goldman et al.,[34-36] and by others.[32,33,37-44] Needless to say that, in the limit of sufficiently high substrate concentration where zero-order kinetics are obeyed, the effects of internal diffusion would vanish and the overall rate, V', will equal $V_{\max}$. The value of the apparent Michaelis constant will, however, always be higher in the presence of diffusional limitations. In other words, although the observed rate will show a limiting behavior similar to that expected for the Michaelis–Menten scheme, the functional dependence of V', on the concentration, σ_o, would be different. This is illustrated in the Eadie–Hofstee plots of Fig. 15. Such plots yield a straight line only when diffusion limitations are negligible, i.e., for $\phi < 0.1$. Otherwise sigmoidal curves are obtained with increasing values of ϕ.[32,33]

Comparison of Figs. 12 and 15 suggests that Eadie–Hofstee plots could in principle be used to distinguish between external and internal diffusional limitations since only in the latter case would such plots yield sigmoidal curves.[1,33]

The validity of the conclusions arrived at from theoretical considerations has been demonstrated experimentally in numerous studies. The effects of internal diffusion on the effective kinetic parameters of immo-

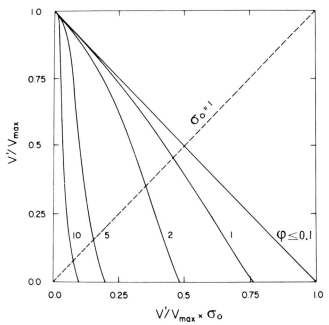

Fig. 15. Normalized Eadie–Hofstee plots for different values of ϕ, the substrate modulus for internal diffusion. From C. Horvath and J. M. Engasser, *Biotechnol. Bioeng.* **16**, 909 (1974).

bilized enzymes have been elucidated using a variety of support configurations, e.g., polyacrylamide films[57,81] or collodion membranes[34-36] of varying thickness and porous particles glass,[60] Sephadex,[82] agarose,[39,40] and ion-exchange resins[80] of different diameters. To give a few examples: Goldman and co-workers[34,35] studied the kinetic behavior of papain-collodion membranes of thickness 470, 156, and 49 μm, using a good substrate, benzoyl-L-arginine amide (BAA), and a poor substrate, acetyl L-glutamic acid diamide (AGDA). All three papain membranes showed linear increase in activity with increase in enzyme membrane thickness when the poor substrate, AGDA, was used; the activities of the different membranes with BAA, as substrate, on the other hand, were similar. Axén and co-workers[82] showed that in the case of particulate chymotrypsin–Sepharose conjugates of highly perturbed values of the effective Michaelis constant, the values of the latter dropped to essentially those of the K_m of the native enzyme following solubilization with dextranase. In the case of a very fast enzyme, glucose oxidase, immobilized on porous

[81] P. S. Bunting and K. J. Laidler, *Biochemistry* **11**, 4477 (1972).
[82] R. Axén, P. A. Myrin, and J. C. Jansson, *Biopolymers* **9**, 401 (1970).

glass beads, only about 6% of the total available activity could be detected by a rate assay.[83]

All these studies thus showed that, as predicted by theory, an increase in the characteristic length (essentially the particle diameter or membrane thickness)[84] was associated with a decrease in the effectiveness factor; alternatively, the substrate concentration required to obtain half the saturation activity was considerably larger than the K_m of the corresponding native enzyme. Furthermore the slopes of Eadie–Hofstee plots for alkaline phosphatase immobilized in polyacrylamide films of varying thickness were found to be similar to the curves shown in Fig. 15.[57] Nonlinear Lineweaver–Burk plots [compare Eq. (16) and references (32) and (44)] have been reported for β-galactosidase entrapped in relatively thick polyacrylamide films.[81]

When external and internal diffusional resistances affect simultaneously the rate of the enzymic reaction the relative contributions of each factor have to be estimated separately (compare Figs. 10 and 14 for the relative weight of these effects).[1,2,7,36,44] It can be shown, however, that the combined effects of internal and external diffusion must give the same limiting kinetic behavior for an enzyme membrane at sufficiently low and sufficiently high substrate concentration where the reaction becomes first and zero order, respectively. Thus the enzyme activity will reach the saturation rate, V_{max}, at sufficiently high substrate concentrations; at sufficiently low substrate concentrations it will show first-order kinetics with a rate constant V_{max}/κ.[1,85-87] The parameter κ is related to K_m and to the first-order overall effectiveness factor ϵ by

$$\kappa = K_m/\epsilon$$

and ϵ is given[1] by

$$\epsilon = (\tanh \phi)/\{\phi[1 + (D_s'\phi \tanh \phi)/h_s L]\} \tag{46}$$

A more detailed treatment of the combined effect of external and internal diffusion on Michaelis–Menten kinetics is outside the scope of this review article. The reader is referred to several reviews and original papers dealing with this problem both theoretically and experimentally.[1,2,7,36,44,48,49,85-87]

---

[83] M. K. Weibel and H. J. Bright, *Biochem. J.* **124**, 801 (1971).
[84] The characteristic length has been defined by Aris[51] as the ratio of the volume of a sphere to its external surface area.
[85] C. Horvath and J. M. Engasser, *Ind. Eng. Chem., Fundam.* **12**, 229 (1973).
[86] C. Horvath, L. H. Shendalman, and R. T. Light, *Chem. Eng. Sci.* **28**, 375 (1973).
[87] L. R. Waterlands, A. S. Michaels, and C. R. Robertson, *AIChE J.* **20**, 50 (1974).

Inhibition in Diffusionally Constrained Systems

In the presence of diffusional limitations, enzyme inhibition manifests itself differently than in free solution, since in a heterogeneous, enzymic reaction inhibition would lower not only the inherent activity of the enzyme, but also the extent of substrate depletion.[1,12,34-36,42,88-91]

The interplay of enzyme inhibition and diffusional resistance for both external and internal diffusion can be described by procedures similar to those outlined in the preceding sections, provided that the appropriate dimensionless moduli are used.[1,12,65,70,88-90] The combined effects of inhibition and diffusion can in principle be expressed by the effectiveness factor for inhibition, η_I, which is defined as the ratio of the effective enzyme activity, V', to the activity of the enzyme in the absence of both diffusional and inhibition effects:

$$V' = \eta_I[V_{max}S/(K_m + S)] \qquad (47)$$

In the absence of diffusional limitations the magnitude of enzyme inhibition as reflected in the observed rate, V', is expressed by the factor $(1 + i/K_I)$. When inhibition and diffusional resistance occur concurrently, the effectiveness factor would depend on both the dimensionless substrate modulus, μ or ϕ [compare Eqs. (40) and (43)], and on the factor i/K_I.[12,65,70,88-90]

In this section the effects of inhibition in the presence of diffusional limitations will be described qualitatively for only two cases: (1) inhibitors that are neither substrates nor products, and (2) inhibition by the product of the enzymic reaction. The reader is referred to the original literature for quantitative treatments of coupled inhibition–diffusion phenomena.[1,12,34-36,42,65,70,88-90]

When diffusional inhibition and chemical inhibition by a substance that is neither substrate nor product act simultaneously their combined effects in decreasing the rate of the reaction are smaller than the sum of the effects that would be obtained if each of the two inhibition phenomena were acting independently.[88] This can be qualitatively interpreted as follows: substrate depletion in an immobilized enzyme structure occurs when the diffusion rate is slow relative to the inherent enzymic activity; an enzyme inhibitor that decreases the inherent enzymic activity would hence reduce the difference between the rate of substrate consumption and the rate of its translocation to the site of the enzyme

---

[88] J. M. Engasser and C. Horvath, *Biochemistry* **13**, 3845 (1974).
[89] J. M. Engasser and C. Horvath, *Biochemistry* **13**, 3849 (1974).
[90] J. M. Engasser and C. Horvath, *Biochemistry* **13**, 3855 (1974).
[91] J. Carlsson, D. Gabel, and R. Axén, *Hoppe-Seyler's Z. Physiol. Chem.* **353**, 1850 (1972).

reaction. Chemical inhibition in heterogeneous enzymic reactions is thus characterized by two antagonistic phenomena affecting the rate simultaneously: the decrease of the inherent enzymic activity and the reduction of the extent of diffusional inhibition.

The antagonistic nature of the interrelationship between chemical inhibition and diffusional limitations has been demonstrated experimentally in several publications. Carlson and co-workers[91] studied the effect of benzamidine, a competitive inhibitor of trypsin. Both soluble and Sephadex-bound trypsin showed the same inhibition profiles when N-benzoyl-L-arginine p-nitroanilide (BAPA), a substrate of a low turnover number, was used. With benzoyl-L-arginine ethyl ester (BAEE), a good substrate, on the other hand, immobilized trypsin was inhibited to a markedly lesser extent than trypsin in free solution. This finding could be explained by the strong diffusional limitations for BAEE, which is hydrolyzed rapidly by trypsin. Similar observations have been reported by Thomas et al.,[70] who compared the inhibition by tartrate, for lactate dehydrogenase in free solution and when immobilized in collodion membranes.

In this context, it is of interest to point out that denaturation of an enzyme can be considered as a special case of irreversible inhibition with time.[92] Applying the concepts about the antagonistic interrelationship between diffusion and inhibition outlined above would lead to the conclusion that diffusional resistances would enhance the *apparent* stability of an enzyme, since inhibition of the activity through denaturation would decrease the extent of diffusional limitations in an immobilized enzyme system. In the limiting case when the system is diffusion controlled, the observed activity of a bound enzyme may remain constant, despite the fact that the protein may have undergone considerable denaturation.[92] Caution should hence be exercised when activity vs time or activity vs temperature data are used to deduce inherent stability parameters of immobilized enzymes.

When the product of an enzymic reaction is a competitive inhibitor, both product accumulation and substrate depletion in the domain of the immobilized enzyme have to be taken into account.[12,42,70,89] The concurrent effects of substrate depletion and product accumulation would be reflected in the magnitude of the effectiveness factor, η_I. The relatively low values obtained for the observed reaction rate under such conditions can be attributed to three separate effects: (1) the intrinsic effect of inhibition by product, (2) the effect of diffusional limitations leading to substrate depletion, and (3) the effect of product accumulation that

---

[92] D. F. Ollis, *Biotechnol. Bioeng.* **14**, 871 (1972).

would manifest itself in enhanced inhibition.[89] By a similar argument, it can be shown that the decrease in enzymic activity due to substrate depletion and product accumulation would be greatest at low concentrations of product, i.e., when the activity of the enzyme is highest for a given system. Since the enzymic activity and hence the magnitude of diffusional limitations are reduced with increasing product concentration, the contributions of both substrate depletion and product accumulation would vanish at sufficiently high concentrations of product. The antagonistic interrelationship between diffusional and chemical inhibition would thus result in the relative insensitivity of the enzymic activity to changes in the bulk concentration of product in the presence of diffusional limitations. The results would be much the same if the product were a noncompetitive inhibitor. When the substrate concentration varies, however, the combined effects of diffusion and product inhibition would be different for competitive and noncompetitive inhibitors, especially at high substrate concentrations; in the case of competitive inhibition only the "apparent Michaelis constant" of the enzyme would increase, the saturation rate, V_{max}, being always reached at sufficiently high substrate concentrations where both diffusional and product inhibition effects vanish.

The effects of inhibition due to product accumulation are of particular interest when the product is acid or base; in such cases, the enzyme reaction would generate local changes of pH. The extent of the deviations of the pH-rate dependence in immobilized enzyme systems would thus depend on the rate of substrate transformation in the enzyme membrane or particle.[34,35,42,93-95]

Goldman and co-workers[34-36,95] have found that the pH dependence of the rates of hydrolysis of various substrates by papain and alkaline phosphatase embedded in collodion membranes deviated from that observed with the corresponding native enzymes; thus the rate of hydrolysis of benzoyl-L-arginine ethyl ester (BAEE) by papain membrane showed a monotonous increase with pH up to pH 9.6 when assayed in the absence of buffer in contradistinction to the bell-shaped pH-activity profile of the native enzyme acting on the same substrate (Fig. 16). Grinding the membrane to a fine powder, or forcing a buffered substrate solution through the membrane, effected partial cancellation of the anomalies in the pH-rate dependence (Fig. 17). The data could be explained by assuming that, owing to diffusional limitations on the translocation of product, the local steady-state concentration of hydrogen

[93] J. M. Engasser and C. Horvath, *Biochim. Biophys. Acta* **358**, 178 (1974).
[94] J. Konecny and J. Slanicka, *Biochim. Biophys. Acta* **403**, 573 (1975).
[95] R. Goldman, I. H. Silman, S. R. Caplan, O. Kedem, and E. Katchalski, *Science* **150**, 758 (1965).

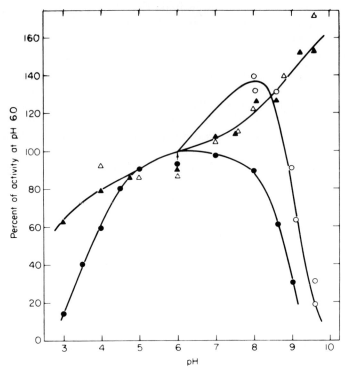

FIG. 16. pH-activity profile for papain and for a papain–collodion membrane using benzoyl-L-arginine as substrate. △, ▲, Papain collodion membranes; ○, papain membranes ground to powder; ●, native papain. Data from R. Goldman, O. Kedem, I. H. Silman, S. R. Caplan, and E. Katchalski, *Biochemistry* **7**, 486 (1968); R. Goldman, H. I. Silman, S. R. Caplan, O. Kedem, and E. Katchalski, *Science* **150**, 758 (1965).

ions generated by hydrolysis of the ester substrate (Eq. 48) in the domain of the papain membrane was higher, and hence the local pH lower than that measured in the bulk solution

$$R^1COOR^2 + H_2O \to R^1COO^- + H^+ + R^2OH \qquad (48)$$

The optimal effective pH for papain (~pH 7) inside the membrane would then be attained at considerably higher bulk pH values. When benzoylglycine ethyl ester (BGEE), which is hydrolyzed by papain at a much lower rate, was used as substrate the anomalies in the observed pH-rate dependence were significantly smaller. In the case of a urease membrane studied by Thomas et al.[42] where a base is liberated in the course of the enzymic reaction, the apparent pH optimum of the membrane-bound enzyme was shifted to lower pH values; here the local

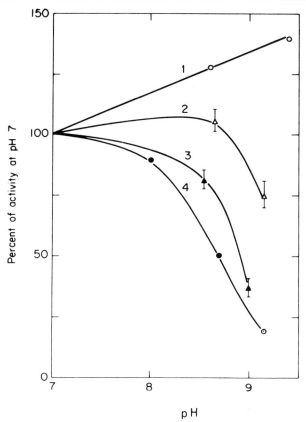

FIG. 17. pH-activity curves for a papain–collodion membrane acting on benzoyl-L-arginine ethyl ester under different conditions: curve 1, papain membrane in absence of buffer; 2, papain membrane in presence of buffer; 3, papain membrane through which substrate dissolved in buffer was forced under a pressure difference of five atmospheres; 4, native papain. From R. Goldman, O. Kedem, I. H. Silman, S. R. Caplan, and E. Katchalski, *Biochemistry* **7**, 486 (1968).

steady-state pH was more alkaline than that of the external solution. Similar effects have been reported by other authors.[93,94]

The existence of local pH gradients generated by the enzyme catalyzed reaction was directly demonstrated by Goldman and co-workers[34]: When an indicator, neutral red, was added to a solution of BAEE containing an inactive papain membrane, both membrane and solution were yellow at pH values above 7.0. On activating the bound enzyme by the addition of a thiol reagent, the membrane became immediately red at all external pH values up to 10.0 although the indicator in the bulk solution remained yellow, suggesting that the difference in pH between the membrane and the external solution was at least 3 pH units.

Immobilized Multienzyme Systems

Immobilized multienzyme systems in which several enzymes carrying out consecutive reactions are bound together on solid particles or membranes have recently attracted considerable attention.[8,9,68,96–102] (See also this volume [31] and [32].) Such systems are of both practical and theoretical interest. In most industrial and analytical processes that could in principle be carried out enzymically, sequences of enzymic reactions would be required.[100] Moreover coimmobilized multienzyme systems could serve as much more realistic models of metabolic processes than enzymes immobilized singly, since diffusional resistances in a multienzyme system would lead to local accumulation of intermediates, thus allowing the study of some of the physicochemical aspects of regulation and control.[8,9,11] The results of the so far rather limited number of studies in this field are briefly summarized below.

Mosbach and Mattiasson[96] coimmobilized hexokinase and glucose-6-phosphate dehydrogenase. These enzymes catalyze the reaction

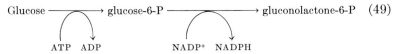

$$\text{Glucose} \xrightarrow{\text{ATP} \;\; \text{ADP}} \text{glucose-6-P} \xrightarrow{\text{NADP}^+ \;\; \text{NADPH}} \text{gluconolactone-6-P} \quad (49)$$

The steady-state activity of the coimmobilized enzymes was essentially the same as that observed for the two enzymes in solution. The initial rate of appearance of the last product was considerably enhanced, however, in the case of coimmobilized hexokinase–glucose-6-phosphate dehydrogenase; moreover, the lag in the appearance of the last product observed with the soluble enzymes and attributed to the buildup of glucose-6-P was absent in the immobilized two-enzyme systems. The data suggested that due to the spatial proximity of the two enzymes on the supporting matrix and the diffusional resistances arising most probably from unstirred layers, high local concentration of the intermediate product could be attained in the immobilized two-enzyme system.[100,101] A theoretical analysis based on these assumptions gave predictions that

[96] K. Mosbach and B. Mattiasson, *Acta Chem. Scand.* **24**, 2093 (1970).
[97] B. Mattiasson and K. Mosbach, *Biochim. Biophys. Acta* **235**, 253 (1971).
[98] P. A. Srere, B. Mattiasson, and K. Mosbach, *Proc. Natl. Acad. Sci. U.S.A.* **70**, 2534 (1973).
[99] R. Goldman and E. Katchalski, *J. Theor. Biol.* **32**, 243 (1971).
[100] K. Mosbach, B. Mattiasson, S. Gestrelius, and P. A. Srere, *in* "Enzyme Engineering" (E. K. Pye and L. B. Wingard, eds.), Vol. 2, p. 143. Plenum, New York, 1974.
[101] P. A. Srere and K. Mosbach, *Annu. Rev. Microbiol.* **28**, 1 (1974).
[102] D. Lecoq, J. F. Hervagault, G. Broun, G. Joly, J. P. Kernevez, and D. Thomas, *J. Biol. Chem.* **250**, 5496 (1975).

were in good agreement with the experimental observations.[99] With an immobilized three-enzyme system, β-galactosidase–hexokinase–glucose-6-phosphate dehydrogenase, which catalyzes the reaction

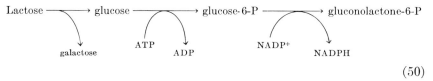

(50)

the reduction of the transient, presteady-state period was even more pronounced.[97]

Using coimmobilized β-glucosidase-glucose oxidase, Lecoq et al. were able to demonstrate feedback effects.[8,102] With salicin as substrate for β-glucosidase, the first enzyme in the sequence, the end product for the overall reaction is gluconolactone [Eq. (51)].

Gluconolactone is, however, an inhibitor of β-glucosidase. As expected, feedback inhibition was more pronounced in the coimmobilized system. The kinetic behavior of the system was examined by computer simulation; good agreement was obtained between experimental and simulated curves.[102]

Diffusional resistances leading to the accumulation of an intermediate, in a sequence of enzymic reactions occurring in a solid phase, could also explain the displaced pH optima of immobilized multienzyme systems.[103,104] When amyloglucosidase (pH optimum 4.8) and glucose oxidase (pH optimum 6.4) were "bound" separately on Sepharose particles, the pH-activity profiles of the immobilized enzymes remained essentially unchanged. In the reaction catalyzed by the two coimmobilized enzymes [Eq. (52)],

$$\text{Maltose} \xrightarrow{\text{amyloglucosidase}} \text{glucose} \xrightarrow{\substack{\text{glucose}\\\text{oxidase}}} \text{gluconolactone} \quad (52)$$

the pH-optimum was shifted to a higher value than that obtained with the enzymes in free solution.[103] The magnitude of the shift in pH opti-

---

[103] S. Gestrelius, B. Mattiasson, and K. Mosbach, *Biochim. Biophys. Acta* **276**, 339 (1972).
[104] S. Gestrelius, B. Mattiasson, and K. Mosbach, *Eur. J. Biochem.* **36**, 89 (1973).

mum was found to be strongly dependent on the ratio of the activities of the two enzymes. With a large excess of glucose oxidase, the first reaction catalyzed by amyloglucosidase became rate limiting and the pH dependence of the combined reaction followed the pH dependence of the first enzyme in the sequence (optimum pH 4.6) with both soluble and immobilized enzymes. Decreasing the relative amount of glucose oxidase led to an increase in the pH optimum of both soluble and bound enzymes, as the second reaction became rate limiting. The pH optimum of the bound enzyme was higher, however, than that of the enzymes in solution, since, owing to diffusional limitations, in the case of the immobilized two-enzyme system the local concentration of glucose would be higher; this would induce an increase in glucose oxidase activity and would pull the pH dependence of the overall reaction toward that of the second enzyme.[103]

In the case when the first of two consecutive enzymic reactions is reversible, coimmobilization of the two enzymes was shown to result in an increase in the overall rate of the reaction.[98] In the system

$$\text{Malate} \xrightarrow{\text{malate dehydrogenase}} \text{oxaloacetate} \xrightarrow{\text{citrate synthase}} \text{citrate} \quad (53)$$

the first reaction is thermodynamically unfavorable in the direction of oxaloacetate formation; by coimmobilization of the two enzymes, the equilibrium was shifted to the right, resulting in a 2-fold increase in the steady-state rate of citrate formation as compared to the rate of the reaction in solution.

When the individual enzymes of a reaction sequence are confined in an asymmetric structure, the overall kinetic behavior of the system would depend on the spatial arrangement of the enzymes as well as on their catalytic properties. Broun et al.[105] and Thomas et al.[70] described a structured multilayer bienzyme membrane by means of which transport of glucose against a concentration gradient, fulfilling the requirements of "active transport," could be demonstrated. The membrane was composed of two protein layers consisting of hexokinase and phosphatase. The enzyme layers were covered on their external surfaces with selective films permeable to glucose, but impermeable to glucose 6-phosphate. Glucose entering the hexokinase layer is phosphorylated in the presence of ATP to glucose 6-phosphate; the latter diffuses into the phosphatase layer where it is promptly dephosphorylated:

$$\text{Glucose} \xrightarrow[\substack{\nearrow \searrow \\ \text{ATP} \quad \text{ADP}}]{\text{hexokinase}} \text{glucose-6-P} \xrightarrow{\text{phosphatase}} \text{glucose} \quad (54)$$

[105] G. Broun, D. Thomas, and E. Selegny, *J. Membr. Biol.* **8**, 313 (1972).

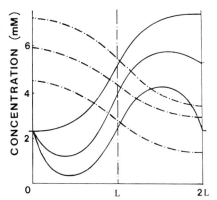

FIG. 18. "Active transport" of glucose (hexokinase/phosphatase system). The membrane separates two compartments. The glucose concentration (S_1) within the first (donor) compartment can be considered constant, the volume being much larger than that of the second (receptor) compartment in which the concentration, S_2, changes with time; ———, glucose concentration profiles; –·–·–, glucose 6-phosphate concentration profiles; lowest profile represents initial conditions $S_1 = S_2$, evolving over a period of time to the uppermost profile, which represents stationary state conditions. From D. Thomas, D. Bourdillon, G. Broun, and J. P. Kernevez, *Biochemistry* **13**, 2995 (1974); G. Broun, D. Thomas, and E. Selegny, *J. Membr. Biol.* **8**, 313 (1972).

Since glucose is consumed in the first enzyme layer and then regenerated in the second, the interplay of reaction and diffusion yields a sinusoidal profile of glucose concentration across the membrane (Fig. 18). The concentration gradient thus established at the two surfaces causes glucose to enter the membrane on the hexokinase side and to leave it on the phosphatase side. The overall effect of the two enzymic reactions is thus the "active" transport of glucose across a spatially and metabolically sequential enzyme array, driven by the ATP supplied to the system.

Conformational and Steric Effects

The decrease in specific activity often encountered when enzymes are covalently bound to solid supports is usually attributed to changes in the tertiary structure of the protein and to steric effects resulting from limitations on the accessibility of substrate.[1-6,13,106-110] When the activity

[106] H. H. Weetall, ed., "Immobilized Enzymes, Antigens, Antibodies and Peptides. Preparation and Characterization." Dekker, New York, 1975.
[107] Y. Levin, M. Pecht, L. Goldstein, and E. Katchalski, *Biochemistry* **3**, 1905 (1964).
[108] D. Gabel, I. Z. Steinberg, and E. Katchalski, *Biochemistry* **10**, 4661 (1971).
[109] L. Goldstein, this series Vol. 19, p. 935 (1970).
[110] L. J. Berliner, S. T. Miller, R. Uy, and G. P. Royer, *Biochim. Biophys. Acta* **315**, 195 (1973).

is measured with low-molecular-weight substrate, these effects are difficult to separate except in the cases when the number of catalytically active bound molecules can be determined by an independent method, e.g., active-site titration (see, for example, Goldstein,[19,20] Gabel,[111] and Ford et al.[112]). Conformational changes arising from strains on the tertiary structure of the immobilized protein have been invoked by several authors to explain the low specific activities observed with bound enzymes.[108,109,113–115] Such approaches, however, did not always take into account denaturation effects, induced by contact with the support, which in the case of hydrophobic carriers may be of considerable importance.[116] The data available on the subject are still rather fragmentary.

The reduced enzymic activity observed with high-molecular-weight substrates can in most cases be related to steric limitations on the penetration of substrate.[4,13,23,95,117–136] Thus a water-insoluble derivative of

[111] D. Gabel, *FEBS Lett.* **49**, 280 (1974).
[112] J. R. Ford, R. P. Chambers, and W. Cohen, *Biochim. Biophys. Acta* **309**, 575 (1973).
[113] D. Gabel and B. V. Hofsten, *Environ. J. Biochem.* **15**, 410 (1970).
[114] T. C. Cho and H. Swaisgood, *Biochim. Biophys. Acta* **334**, 243 (1974).
[115] H. Swaisgood and H. R. Horton, in "Enzyme Engineering" (E. K. Pye and L. B. Wingard, eds.), Vol. 2, p. 169. Plenum, New York, 1974.
[116] L. Goldstein and G. Manecke, in "Applied Biochemistry and Bioengineering" (L. B. Wingard, E. Katchalski, and L. Goldstein, eds.), Vol. 1, in press. Academic Press, New York, 1976.
[117] J. Porath, R. Axén, and S. Ernback, *Nature (London)* **215**, 1491 (1967).
[118] R. Axén and S. Ernback, *Eur. J. Biochem.* **18**, 351 (1971).
[119] J. Degani and T. Miron, *Biochim. Biophys. Acta* **212**, 362 (1970).
[120] R. Haynes and K. A. Walsh, *Biochem. Biophys. Res. Commun.* **36**, 235 (1969).
[121] C. K. Glassmeyer and J. D. Ogle, *Biochemistry* **10**, 786 (1971).
[122] A. Bar-Eli and E. Katchalski, *J. Biol. Chem.* **238**, 1690 (1963).
[123] I. H. Silman, M. Albu-Weissenberg, and E. Katchalski, *Biopolymers* **4**, 441 (1966).
[124] C. W. Wharton, E. M. Crook, and K. Brocklehurst, *Eur. J. Biochem.* **6**, 565 (1968).
[125] L. Goldstein, *Biochim. Biophys. Acta* **315**, 1 (1973).
[126] L. Goldstein, M. Pecht, S. Blumberg, D. Atlas, and Y. Levin, *Biochemistry* **9**, 2322 (1970).
[127] P. Cresswell and A. R. Sanderson, *Biochem. J.* **119**, 447 (1970).
[128] S. Lowey, L. Goldstein, C. Cohen, and S. M. Luck, *J. Mol. Biol.* **23**, 287 (1967).
[129] S. Lowey, H. S. Slayter, A. G. Weeds, and H. Baker, *J. Mol. Biol.* **42**, 1 (1969).
[130] T. W. Wolodko and C. M. Kay, *Can. J. Biochem.* **53**, 175 (1975).
[131] E. B. Ong, Y. Tsang, and G. E. Perlmann, *J. Biol. Chem.* **241**, 5661 (1966).
[132] C. K. Colton, K. A. Smith, E. W. Merrill, and P. C. Farrell, *J. Biomed. Mater. Res.* **5**, 459 (1971).
[133] O. Kedem and A. Katchalsky, *J. Gen. Physiol.* **45**, 143 (1961).
[134] G. Thau, R. Bloch, and O. Kedem, *Desalination* **1**, 129 (1966).
[135] P. Smulders and E. M. Wright, *J. Membr. Biol.* **5**, 297 (1971).
[136] E. E. Petersen, "Chemical Reaction Analysis." Prentice-Hall, Englewood Cliffs, New Jersey, 1965.

trypsin hydrolyzed casein at 15% of the rate that would have been expected on the basis of the amount of the enzymically active bound protein, determined by rate assay using low-molecular-weight substrate.[122] Similar results have been reported for immobilized derivatives of ficin,[23,75] bromelain,[124] papain,[109,123,126] and subtilopeptidase A.[109,125,126] Additional support for the view that lower activities may result from steric interferences of the support could be found in the work of Cresswell and Sanderson,[127] who reported that the rate of hydrolysis of proteins of different molecular weights by immobilized Pronase was inversely proportional to the square root of the molecular weight of the substrate. The data could be explained by assuming that substrate molecules were excluded from the pores of the supported enzyme by a mechanism analogous to that effective in molecular sieving. Similar effects could also explain the observation that the degree of inhibition of insoluble trypsin derivatives was inversely related to the molecular weight of the inhibitor.[120,121]

Additional effects might become prominent in the case of enzymes immobilized on charged supports, e.g., electrostatic interactions between charged carrier and charged high-molecular-weight substrate[2-4,107]; in the digestion of casein by polyanionic derivatives of trypsin, the rate of hydrolysis was found to depend on the carrier-to-enzyme ratio. Preparations of high enzyme content (60–70% protein by weight) showed caseinolytic activities close to those of native trypsin, while preparations of low enzyme content (5–10% protein by weight) had only about 20% of the expected proteolytic activity. Raising the pH, and thus increasing the charge density on the support, led to a marked decrease in the caseinolytic activity of both polyanionic-trypsin preparations.[107] The data suggested that the number of peptide bonds split could be monitored by controlling the magnitude of charge-charge interactions between charged substrate and charged immobilized enzymes by varying the charge density on the support. Indications that the sites of attack as well as the rates of cleavage might be affected by the chemical nature of the support could be found in the literature[1-7]: For example, Lowey and co-workers[128] reported that the first-order rate constants estimated for the digestion of myosin by a polyanionic derivative of trypsin were about 50-fold lower than those of native trypsin, and only about half as many peptide bonds were split by the polyanionic derivative. Moreover, different protein fragments were obtained on digesting myosin and heavy meromyosin with the negatively charged derivatives of trypsin.[128-130] It was also shown that while native trypsin hydrolyzed 15 lysyl peptide bonds in pepsinogen, the maximal number of bonds split by polyanionic trypsin did not exceed 10. The same difference was observed with reduced and carboxymethylated pepsinogen.[131] Again the various factors determining the modified, and in a sense more restricted, specificity of charged enzyme derivatives toward high-molecular-weight substrates cannot be separated. A trial-and-error approach, however, allows considerable flexibility in designing experiments aimed at the controlled transformation of biological macromolecules.

Determination of Kinetic Parameters from Experimental Data

The measured rate and kinetic parameters of an immobilized enzyme (V' and K_m') do not in most cases reflect the intrinsic or inherent kinetic parameters. These can be extracted from the experimental data only when both microenvironmental and diffusional effects can be evaluated.

Microenvironmental partitioning effects are relatively easy to demonstrate when they stem from electrostatic interactions, which are canceled at high ionic strength.[2,13,16,17,19,20,23,24,28] In the case of hydrophobic interactions the use of labeled materials appears to be the simplest method of detecting partitioning effects.[28,29]

The techniques commonly employed for the characterization of diffusional resistances and the evaluation of the intrinsic or inherent kinetic parameters of immobilized enzyme systems, which will be the main concern of this section, can be classified in three main groups, discussed below.

1. Direct Determination of Kinetic and Transport Parameters. The most straightforward way to characterize diffusional effects in immobilized enzyme systems is undoubtedly through the direct evaluation of the external and internal substrate moduli, μ and ϕ. For this, however, the kinetic and transport parameters of the system have to be known [see also Eqs. (55) and (57)]. In the absence of partitioning and specific microenvironmental effects, or if such effects can be eliminated or estimated separately, one can assume that the intrinsic kinetics of an immobilized enzyme will follow the form of the irreversible Michaelis–Menten rate law with the same values of V_m and K_m. The substrate diffusivity, D_s', and the transport coefficient for substrate, h_s, must then be measured independently by the established procedures[34-36, 132-135] under conditions where the immobilized enzyme is inactivated or an essential cofactor is absent. To separate the contributions of external and internal diffusion, such experiments must be carried out under conditions where external mass transfer resistance is negligible or can be estimated by standard techniques[1,33]; this would permit the calculation of the effective surface substrate concentration from the known bulk concentration [compare Eqs. (19) and (20)]. This approach, originally described by Goldman,[34-36] has been used by several authors to characterize enzyme membranes.[1,4-7]

2. Variation of Substrate Concentration. The evaluation of the intrinsic kinetic and rate parameters of an enzyme by means of the conventional linearized forms of the Michaelis–Menten rate law may be misleading when applied to immobilized enzyme systems.[1,32,33,37,38,44,48] This would be particularly true when a relatively narrow range of substrate concentrations is involved. As shown in the preceding sections, the overall kinetics of such systems do not follow the Michaelis–Menten scheme when diffusional limitations occur. Yet inspection of Eqs. (16)–(18) and Fig. 19 reveals that the three classical plots may yield, even at large values of the substrate moduli, μ or ϕ, curves that can be approximated by straight lines in certain intervals (see also Figs. 10 and 14). The results of kinetic measurements with immobilized enzymes may therefore give straight lines on such plots when a relatively narrow concentration range is investigated. Although an apparent K_m can be obtained under these circumstances, it has no theoretical meaning.

When overall rates are measured with immobilized enzymes at a sufficiently wide range of substrate concentrations, the curves obtained by the classical graphic methods can, however, be characterized by the parameters V_{max} and V_{max}/κ in the limit of infinitely high and infinitely low substrate concentrations, respectively.[1,32,33] Different approaches to obtain the values of the effective Michaelis constant, κ, and of V_{max}, from plots based on Eqs. (16), (17), and (18), are illustrated schematically on Fig. 19 for a system in which only internal diffusional resistances occur[32]: The saturation rate, V'''_{max}, can in principle be determined from data obtained at effectiveness factors close to unity, usually reached at high enough concentrations. The value of V'''_{max}/κ can be extracted from plots based on Eqs. (16)–(18) (Fig. 19) by one of the following methods[1,32,33]: (1) On plots of type (a), at low concentrations, $\sigma_s < 1$, the curve can be approximated by a straight line whose slope is equal to κ/V_{max} (see Fig. 19). (2) On plots of types (b) and (c), κ/V_{max} or V_{max}/κ can be determined by extrapolation of curves obtained from measurements at $0.1 < \sigma_s < 1$. Here again the interpretation of the data is greatly facilitated if the experiments are carried out under conditions where external mass transfer resistances are negligible or can be evaluated separately.

The curves on Fig. 19 were calculated for internal diffusional limitations only, employing a spherical geometry. A similar overall pattern can be expected, however, for other geometries, since the qualitative behavior

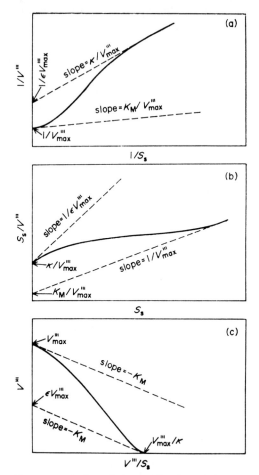

FIG. 19. Schematic representation of the application of the conventional graphical methods for the evaluation of the kinetic and rate parameters of immobilized enzymes, in the presence of internal diffusional limitations. (a) According to Eq. (16); (b) according to Eq. (17); (c) according to Eq. (18). From J. M. Engasser and C. Horvath, *J. Theor. Biol.* **42**, 137 (1973).

of the effectiveness factor, η, is not very sensitive to geometry.[136,137] In fact when ϕ is sufficiently large so that the reaction zone is confined to a relatively thin shell near the surface, the term accounting for curvature in spherical and cylindrical geometries can be dropped and the appropriate differential equations reduce to Eq. (37). It should be pointed

[137] G. R. Gavalas, "Non-Linear Differential Equations of Chemically Reacting Systems." Springer-Verlag, Berlin and New York, 1968.

out that when V_{max} and K_m are evaluated from plots of the general form given by Eqs. (16), (17), and (18), their reliability is limited by the accuracy of the extrapolation procedure used.[38,44,47] It is thus useful for the experimentalist to have a criterion of validity which contains only quantities that can be measured directly or predicted. Such a criterion is provided by the dimensionless modulus, Φ_L, given by [138,139]

$$\Phi_L = L^2 V'''/D_s' S_s' \tag{55}$$

where V''' is the effective rate per unit volume of catalyst, and S_s is the effective surface substrate concentration. For a sphere, L is replaced by $R/3$. The substrate concentration at the surface can be calculated from the bulk concentration, S_o, the measured activity, V''', and the external transport coefficient, h_s, by

$$S_s = S_o - (V'''/h_s) \tag{56}$$

For Michaelis–Menten kinetics, Φ_L is also given by[38,39]

$$\Phi_L = \eta[\phi^2/(1 + S_s/K_m)] \tag{57}$$

hence if the inherent K_m of the bound enzyme is available, the modulus, Φ_L, can be calculated. Internal diffusional limitations are generally negligible when $\Phi_L < 1$. For larger values of Φ_L the effectiveness factor, η, can be determined graphically from charts obtained by Roberts and Satterfield.[47]

3. Variation of Characteristic Support Dimensions or V_{max}. The most direct way to demonstrate the presence of diffusional limitations involves the change of transport conditions for substrate and product. As demonstrated previously, external diffusion plays a role if the activity of the bound enzyme depends on the efficiency of mixing. Internal diffusional limitations can be detected by comparing the effective reaction rates obtained with membranes of different thickness or particles of different diameters.[1,32-36,42-44,47-49,79,105,140] Two methods will be described to illustrate this approach, for internal diffusional limitations.

METHOD I. To obtain K_m and D_s', kinetic measurements are made with particles or membranes having the same K_m and D_s' but different moduli ϕ_1 and ϕ_2.[32] The preparation of two such kinds of immobilized enzymes should be relatively simple when using the same support material and the same binding technique. In the case of spherical particles, such

[138] C. Wagner, *Z. Phys. Chem. Abt. A* **193**, 1 (1943).
[139] P. B. Weisz and C. D. Prater, *Adv. Catal. Relat. Subj.* **6**, 143 (1954).
[140] J. Meyer, F. Sauer, and D. Woermann, *Ber. Bunsenges. Phys. Chem.* **74**, 245 (1970).

preparations can have different radii R_1 and R_2 as well as different enzyme concentrations corresponding to V_{max_1} and V_{max_2}. From the definition of ϕ and κ by Eqs. (40), (40a), and (45), the following simple relations can be derived[32]:

$$\epsilon_1/\epsilon_2 = \kappa_1/\kappa_2 \tag{58}$$

and

$$\phi_2/\phi_1 = R_2/R_1(V_{max_2}/V_{max_1})^{1/2} \tag{59}$$

The dependence of ϵ_2/ϵ_1 as a function of the ratio ϕ_2/ϕ_1 for different values of ϕ, can be obtained from calculated curves.[32] Since the corresponding κ and V_{max} values can be obtained from kinetic data, and since the particle radii are easily measurable, the ratios ϵ_2/ϵ_1 and ϕ_2/ϕ_1 can be obtained by Eqs. (58) and (59). With these ratios at hand, ϕ can be extracted from calculated graphs.[32]

METHOD II. This method, based on the "triangle method" of Weisz and Prater,[139] can be applied to experimental rate data for three or more support geometries at constant substrate concentration.[38] The principles of the method can be illustrated as follows: for spherical particles of radii R_1, R_2, and R_3, consider first one pair of particles of radii R_1 and R_2; the ratio of the rates (per unit volume of support) on these particles is the ratio of the affectiveness factors ϵ_1/ϵ_2, and the ratio R_1/R_2 is ϕ_1/ϕ_2.[38] The simultaneous equations expressing ϵ_1 as a function of ϕ_1 and ϵ_2 as a function of ϕ_2 can then be solved to obtain ϵ_1 and ϵ_2 for an assumed constant value of $\rho = K_m/S$. The triangle method is based on the fact that, on a logarithmic plot of ϵ vs ϕ for different values of ρ, the ratio ϵ_1/ϵ_2 forms a line of fixed length on the ordinate and the ratio ϕ_2/ϕ_1 a fixed length on the abscissa (see, for example, Fig. 11). These two lengths form a triangle, which can be fitted on the calculated ϵ vs ϕ plots for an assumed value of ρ, resulting in the determination of ϵ_1, ϵ_2, ϕ_1, and ϕ_2.[38] The same procedure is then carried out for another pair of particles of radii R_2 and R_3, and again individual values of ϵ_2, ϵ_3, ϕ_2, and ϕ_3 are obtained in the same way. If the assumed value of ρ is correct, the values of ϕ_2 and ϵ_2 should coincide with each other. From the values of ρ and ϕ_2, the values of K_m and V_{max} can be then calculated.

The presence of diffusional effects can also be inferred from the temperature dependence of the rate constants and the shape of the appropriate Arrhenius plots, since the temperature dependence of diffusion is relatively small as compared to that of most enzymic reactions.[49] Thus, at sufficiently low temperatures, in the domain of kinetic control of the reaction the true activation energy is observed. At higher temperatures, when diffusional limitations become significant, the apparent activation energy is lower; this is reflected as a break in the log rate vs

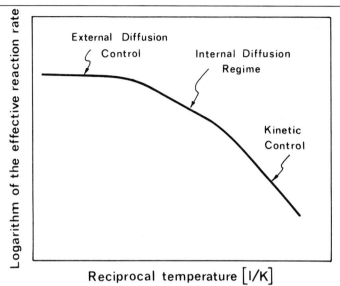

Fig. 20. Schematic Arrhenius plot for an immobilized-enzyme-catalyzed reaction in the presence of diffusional limitations.

$1/T$ curves.[38,141-144] At sufficiently high temperatures, when the reaction becomes bulk-diffusion controlled, the observed rate is essentially independent of temperature; the apparent activation energy under these conditions is close to zero. These arguments are illustrated in Fig. 20.[47,49] It should be kept in mind, however, that nonlinear Arrhenius plots might also arise from temperature-dependent conformational changes (i.e., inactivation).

In concluding this section it should be mentioned that the merits of the various techniques have still to be demonstrated. Further work would undoubtedly establish the relative accuracy of the different methods, thus allowing the determination of the limits within which the intrinsic kinetic parameters of an enzyme can be altered following immobilization.

[141] T. T. Ngo and K. J. Laidler, *Biochim. Biophys. Acta* 377, 303 (1975).
[142] P. Bernfeld and R. E. Bieber, *Arch. Biochem. Biophys.* 131, 587 (1969).
[143] T. Sato, T. Mori, T. Tosa, and I. Chibata, *Arch. Biochem. Biophys.* 147, 788 (1971).
[144] K. Buchholz and W. Rüth, *Biotechnol. Bioeng.* 18, 95 (1976).

[30] Activity Correlations between Similarly Modified Soluble and Immobilized Enzymes

By DAVID F. OLLIS and RATHIN DATTA

Successful recipes for enzyme immobilization by chemical coupling on solid supports[1,2] are clearly derived from previous experience in amino acid chemistry and chemical modification of soluble enzymes.[3] It is surprising, then, that no effort has been made to obtain correlations at the next level of sophistication, that of maximizing the specific catalytic activity (units per milligram of immobilized enzyme) of an active bound enzyme. We report here such similarities in behavior between specific activity of three enzymes as a function of their degree of chemical derivatization by soluble and analogous insoluble protein coupling groups.

Materials and Methods

Materials used included the polyacrylamide support Bio-Gel P-2 (Bio-Rad Laboratories[4]), lysozyme (hen egg white), α-chymotrypsin (bovine pancreas), lipase (hog pancreas), and dried cells of *Micrococcus lysodeikticus* (Worthington Biochemical Corp.). All other chemicals were reagent grade.

Protein, cell, and benzoyl-L-tyrosine ethyl ester (BTEE) concentrations were determined by a Cary-14 spectrophotometer. Specific activities of both soluble and immobilized lipase were measured in a Sargent recording pH stat. The circular dichroism (CD) spectra of protein solutions were obtained with a Cary 60 spectropolarimeter and a CD attachment.

Experimental Procedure[5]

Preparation and Activity of Lysozyme–Polyacrylamide. The procedure of Datta, Armiger, and Ollis[6] was used. The beads were aminoethylated

[1] R. Goldman, L. Goldstein, and E. Katchalski, in "Biochemical Aspects of Reactions on Solid Supports" (G. R. Stark, ed.), pp. 1–72. Academic Press, New York, 1971.
[2] H. H. Weetall and R. A. Messing, in "Chemistry of Biosurfaces" (M. L. Hair, ed.), Vol. 2, p. 563. Dekker, New York, 1972.
[3] G. F. Means and R. E. Feeney, "Chemical Modifications of Proteins." Holden-Day, San Francisco, California, 1971.
[4] Bio-Rad Laboratories, Berkeley, California.
[5] R. Datta and D. F. Ollis, "Immobilized Biochemicals and Affinity Chromatography" (R. B. Dunlap, ed.), p. 293. Plenum, New York, 1974.
[6] R. Datta, W. Armiger, and D. F. Ollis, *Biotechnol. Bioeng.* 993 (1973).

for 10 min, 30 min, and 2 hr. An additional support aminoethylated for 6–8 hr was purchased from Bio-Rad laboratories.[4] The different aminoethylation times result in different surface concentrations of the aminoethyl groups (which are determined by direct titration[7]), and consequently also in the final surface diazo coupling groups. This useful result occurs because the aminoethylation reaction is slow, and not yet complete after 8 hr.[7]

The polyacrylamide diazonium salt intermediate was quickly suspended in a 2 mg/ml solution of lysozyme and 1 mg of N-acetylglucosamine (NAG) per milliliter in pH 9 buffer. The NAG was used to protect the active site of the enzyme during the covalent coupling process.[8] The reaction was carried out for 6 hr at 0°–5° and the pH was maintained at 9.0 by addition of 1 M NaOH solution. The beads and the protein were washed first with borate buffer and then with 2 M NaCl to remove noncovalently attached enzyme. From the wash (OD at 2890 Å) and the wash volume, the total enzyme removed by washing was calculated. A total enzyme balance between original solution, wash solution, and original solution after enzyme attachment gave the immobilized enzyme loading by difference, expressed as milligrams of enzyme attached per milligram dry weight of polyacrylamide support.

The activity of the enzyme beads was measured by a simple batch reaction. One sample of *Micrococcus lysodeikticus* found to be fairly stable toward autolysis was used for all the experiments. Five Nalgene beakers, each containing 10 ml of 0.5 g of the dried cells per liter suspended in pH 6.8 phosphate buffer, were stirred at identical speeds. To beakers 1 through 4, 0.2 g of the wet gel–enzyme derivative was added. To beaker 5, which served as a control to measure the cell autolysis in the presence of beads, 0.2 g of wet Bio-Gel P-2 beads was added. Samples from beaker 1 were removed after 30 min, from beaker 2 after 60 min, from beaker 3 after 75 min, and from beaker 4 after 120 min. The OD at 440 nm of these samples was measured against that of the samples removed from the control beaker 5 at the same times. The difference in OD was converted to concentration units by a calibration curve. In this manner, a reaction rate was measured, which was presumed to be free from effects of cell autolysis in the absence of lysozyme.

Diazotization of Soluble Lysozyme. Lysozyme was diazotized by reaction with diazobenzene sulfonic acid[9]: 0.01 mole of sulfanilic acid was dissolved in 250 ml of cold 0.4 N HCl solution, and 0.05 mole of sodium nitrite was added. The reaction was allowed to proceed for 5 min under constant stirring at 0°. At the end of 5 min, 1 ml, 0.5 ml, 0.1 ml, 0.05 ml,

[7] J. K. Inman and H. M. Dintzis, *Biochemistry* **8** (10) 4074 (1969).
[8] A. Neuberger and B. M. Wilson, *Nature (London)* **215**, 524 (1967).

and 0.01 ml of the reaction solution was added to different beakers, each containing 10 ml of 1 mg/ml lysozyme solution in pH 9 borate buffer. A few drops of 2 N NaOH were added to the beakers containing 1 ml and 0.5 ml of the diazobenzene sulfonic acid in order to neutralize the excess acid. As a control, 1.0 ml of cold 0.4 N HCl containing 0.2 mN sodium nitrite was added to 10 mg of lysozyme in pH 9 buffer. All beakers were stored overnight at 5°.

Activity of Soluble Lysozyme. The activity of the diazotized soluble lysozyme and the control was determined from the rate of lysis of one batch of *Micrococcus lysodeikticus* in a pH 7 phosphate buffer. The rate was assumed to be proportional to the decrease in absorbance at 450 nm of a 0.3 mg/ml suspension of the cells in the buffer. To 2.9 ml of the cell suspension, 0.1 ml of the enzyme solution was added. One unit is taken as the decrease in absorbance of 0.001 per minute.

Preparation and Activity Measurement of α-Chymotrypsin-Polyacrylamide. The acyl azide intermediate method of Inman and Dintzis[7] was used to immobilize α-chymotrypsin to Bio-Gel P-2. The polyacrylamide beads were allowed to react with hydrazine hydrate (6 M) for 2 hr, 4 hr, 6 hr, 15 hr, 18 hr, and 24 hr to give different surface concentrations of hydrazide groups on the bead surfaces as determined by titration.[7] After formation of the acyl derivative by reaction with 0.1 M sodium nitrite in 0.25 N HCl solution at 0° for 2 min, the acyl azide derivative was washed with cold 0.001 N HCl–0.1 M CaCl$_2$ solution and suspended in a 1 mg/ml solution of α-chymotrypsin in 0.001 N HCl–0.1 ml CaCl$_2$. Then 1 M NaOH was added to bring the pH to ~9.0; this value being maintained during reaction by continuous addition of 1 M NaOH. The coupling reaction was continued for 1 hr.

The beads were washed by 0.001 N HCl–0.1 M CaCl$_2$ solution and then by 2 N NaCl to remove the noncovalently attached protein. The amount of α-chymotrypsin in the wash was measured by its absorbance at 280 nm in the spectrophotometer. From the initial amount of enzyme and the final amount in the wash, the attachment of enzyme was calculated.

The activity of immobilized α-chymotrypsin was measured by its action on benzoyl-L-tyrosine ethyl ester (BTEE).[9] Twenty-five milliliters of 1.07 mM BTEE in 50% (w/w) methanol were mixed with 25 ml of pH 7.9 Tris buffer; 1.6 ml of 0.001 N HCl–1 M CaCl$_2$ was also added. After stirring, 1.6 ml of the solution was pipetted into the spectrophotometer reference cell. The solution in the reaction beaker was then connected through a fritted-glass filter to a continuous-flow cell in the sample compartment through a peristaltic pump. The solution was pumped

[9] B. C. W. Hummel, *Can. J. Biochem. Physiol.* **37**, 1393 (1959).

through the cell at a constant flow rate, and the instrument was balanced to zero reading at 256 nm. Then 0.5 g of wet α-chymotrypsin-polyacrylamide beads were added to the reaction beaker and stirred vigorously. Care was taken that the fritted-glass filter immersed in the beaker was not clogged by the beads. Absorbance readings were taken every 30 sec for 10–15 min. After reaction, the enzyme beads in the beaker were filtered and dried to give the exact weight of beads added. The specific activity was calculated by the formula:

$$\frac{\text{units}}{\text{mg of attached enzyme}} = \frac{\Delta A_{256}/\text{min} \times 100 \times 50}{964 \times \text{mg of attached enzyme}}$$

since 964 equals the molar extinction coefficient for N-Benzoyl-L- tyrosine.

Acylation and Activity of Soluble α-Chymotrypsin. α-Chymotrypsin was acylated by the method of Oppenheimer et al.[10] Five beakers of 100 mg of α-chymotrypsin in pH 8.5 borate buffer containing 0.05 M $CaCl_2$ were cooled to 4°. Then 0.25, 0.1, 0.05, and 0.01 ml, respectively, of acetic anhydride was added to four different beakers; pH was maintained between 6.7 and 7.0 by addition of 2 N NaOH. After reaction for 1 hr, excess acetic anhydride (if any) was neutralized by the alkali and the reacted enzyme solutions were dialyzed overnight against 0.01 M sodium borate solution at 4°. The fifth beaker served as control. The solutions were later diluted to 1 liter in pH 9 borate buffer, and the activities of 0.1 mg/ml protein solutions were measured by the reaction on BTEE according to the method of Hummel.[9]

Preparation and Activity Measurement of Immobilized Lipase. Lipase was immobilized on polyacrylamide by Lieberman[11] using the diazonium intermediate method of Inman and Dintzis[7] described earlier. The immobilized enzyme was assayed at pH 8 using sonically dispersed tributyrin substrate with NaOH solutions as the pH-stat titrant.

Diazotization and Activity Measurement of Soluble Lipase. The diazotization of soluble lipase followed the method used in the diazotization of soluble lysozyme. The activity of this modified lipase was assayed by its reaction of tributyrin as follows: A mixture of 1 volume of tributyrin was diluted by 50 volumes of deionized water and emulsified by sonication for 1 hr. A 5-ml volume of the emulsion was then diluted with 5 ml of deionized water, and to it were added 2 ml of 75 mM $CaCl_2$ solution, 1 ml of 15 mg/ml sodium taurocholate solution, and 2 ml of 1 M NaCl solution. The 15 ml of reactant solution were adjusted to pH 8, and 1 ml of the solution of the soluble diazotized enzyme of appropriate concentration was added. Rate measurements were again taken with a recording pH stat.

[10] H. L. Oppenheimer, B. Labouesse, and G. P. Hess, *J. Biol. Chem.* **241**, 2720 (1966).
[11] R. Liberman and D. F. Ollis, *Biotechnol. Bioeng.* **XVII**, 1401 (1975).

Results and Discussion

The total concentration of protein coupling groups, expressed as coupling groups per gram dry weight of gel support, and the corresponding protein loading achieved, is summarized in Table I. The coupling groups were diazonium species for lysozyme and lipase and acyl azide species for chymotrypsin.

The relative specific activities of *immobilized* lysozyme, pancreatic lipase, and α-chymotrypsin are plotted in Fig. 1 versus the molar ratio surface coupling groups:enzyme immobilized. These data indicate that the specific activity of lysozyme and lipase are strongly dependent upon the degree of surface derivatization; note that the activity declines logarithmically with higher derivatization densities for both enzymes. The third enzyme, α-chymotrypsin, bound via an acyl azide intermediate, is less sensitive to such variation of surface group concentrations.

A plot of the relative specific activity of the *modified soluble* enzymes is given in Fig. 2. The ordinate is now the molar ratio soluble coupling reagent:soluble enzyme.

Comparison of Figs. 1 and 2 shows a strong similarity in the sequence by which the relative specific activity of these three enzymes is modified with increasing coupling group concentration. In particular, the susceptibility of the enzymes to activity modification following covalent derivatization is lysozyme > lipase > chymotrypsin for both figures.

TABLE I
CHARACTERIZATION OF ENZYME–POLYACRYLAMIDE CATALYSTS

Enzyme	Coupling group concentration (mmoles/g dry wt gel)	Protein loading (mg/g dry wt gel)
Lysozyme	0.0284	2.85
	0.135	4.9
	0.808	5.8
	2.0	10.8
Lipase	0.04	3.38
	0.058	2.05
	0.16	2.54
Chymotrypsin	0.112	2.0
	0.845	2.18
	1.31	1.64
	2.49	2.38
	4.46	2.84

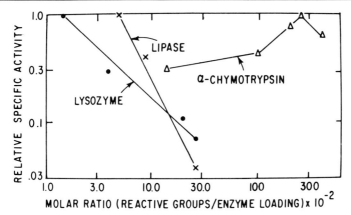

FIG. 1. Relative specific activity of immobilized enzyme versus the molar ratio (immobilized surface coupling groups:enzyme immobilized). ●, Diazotized lysozyme; ×, diazotized lipase; △, acylated α-chymotrypsin. From R. Datta and D. F. Ollis, in "Immobilized Biochemicals and Affinity Chromatography" (R. B. Dunlap, ed.), p. 293. Plenum, New York, 1974.

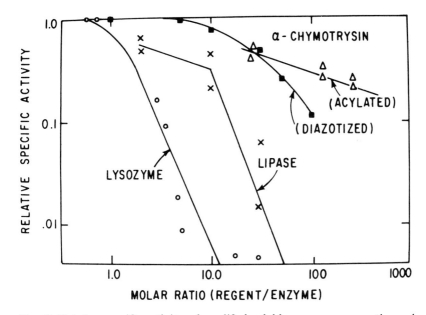

FIG. 2. Relative specific activity of modified soluble enzymes versus the molar ratio (soluble coupling reagent:enzyme). ○, Diazobenzenesulfonic acid–lysozyme; ×, diazobenzenesulfonic acid–lipase; ■, diazobenzenesulfonic acid–chymotrypsin; △, acetic anhydride–chymotrypsin. From R. Datta and D. F. Ollis, in "Immobilized Biochemicals and Affinity Chromatography" (R. B. Dunlap, ed.), p. 293. Plenum, New York, 1974.

TABLE II
INTERCEPTS OF ENZYME DATA

Enzyme	Soluble	Immobilized
Lysozyme	1.0	1.4×10^2
Lipase	5.0	5.2×10^2

The horizontal axis of the two plots are not, we emphasize, identical. However, we expect that the *total* number of surface coupling groups in the porous polyacrylamide gel (which is the quantity determined by the acid titration of aminoethylated precursor groups) ought to be proportional to the number of external or superficial bead coupling groups. Only these later reagent groups are presumed to be available for protein immobilization, since the claimed molecular weight exclusion for the polyacrylamide beads (2000) is far below the molecular weight of the enzymes under investigation. Thus, the similar proportions of these intercepts (Table II) and the previously noted sequence of suceptibility to activity modification lead us to argue that the same processes are effective in multiple binding immobilization as in multiple derivatization by covalent modification of these two enzymes. In particular, the intercept data for the soluble enzymes modified provide a clear indication of the number of reagent–enzyme coupling reactions that may occur before serious deactivation results.

In summary, there appears to be a major potential for increasing our understanding of the behavior of immobilized enzymes if a similar parallel or prior examination is made of soluble enzymes modified by soluble analogs of the immobilization reagents (as typified by *p*-diazobenzenesulfonic acid and acetic anhydride in the present examples).

Section VII
Multistep Enzyme Systems

[31] Multistep Enzyme Systems

By K. MOSBACH and B. MATTIASSON

There has been great interest over the last few years in the immobilization of enzymes acting in sequential sets of reactions. This interest stems from the fact that such systems constitute valuable models for the understanding of metabolism in general. They also are of potential practical interest (for the latter aspect, see in particular chapter [32] by Bouin, Atallah, and Hultin on an immobilized glucose oxidase–catalase system).

It is a general feature of metabolic processes that the product of one enzymic reaction is the substrate for the next enzyme in sequence, and so on. Increasing evidence is accumulating that in nature the various enzymes are either associated in more or less tight aggregates, or are kept membrane bound, or are active in more or less gellike surroundings.

It has been demonstrated that after centrifugation of extracts of *Euglena* cells, practically all the enzymes were found in the various strata within the cell, from which it was concluded that there were practically no free or unbound enzymes in the cell.[1]

Enzyme systems may be either somewhat tightly associated or form true multienzyme complexes. There appears to be no single type or mechanism characteristic of the latter. Some are simply comprised of two partial reactions (e.g., acetyl-CoA carboxylase); others, such as ATCase, consist of enzymic and regulatory subunits. Fatty acid synthetase consists of seven enzymes which, in the case of the complex from yeast, are inactive when separated; conversely, the compounds of α-keto acid dehydrogenases remain active on dissociation. In fatty acid synthetase and the related 6-methyl salicylic acid synthetase the intermediates are covalently bound, whereas this may not be the case with tryptophan synthetase.

We have deliberately chosen to designate the examples described in this chapter as "multistep enzyme systems," thereby reserving the term "multienzyme complex" for natural systems in order to avoid confusion. The term "multienzyme complex" as normally used refers to tightly agglomerated complexes occurring in the cell. Immobilization techniques, however, enable sequentially acting enzymes not naturally aggregated to be coimmobilized artificially, thus forming tightly coupled enzyme systems.

[1] E. S. Kempner and J. H. Miller, *Exp. Cell. Res.* **51**, 141 (1968).

There are basically two ways of creating an enzyme sequence: (a) by immobilizing the different enzymes on the same particle, thereby creating a multistep enzyme system; (b) by immobilizing the different enzymes on separate particles, subsequently to be either mixed in a bed or arranged in sequential reactors.

Since the latter approach of immobilization is treated elsewhere in this volume, only its practical application will be discussed and illustrated with a few examples, whereas methodological emphasis will be placed on the coimmobilization of sequential enzymes on the same particles.

Examples

1. *Preparation of an Immobilized Two-Step Enzyme System, Hexokinase–Glucose-6-Phosphate Dehydrogenase*

The procedures given below follow essentially previously published work.[2]

Sepharose-Bound. One milliliter of Sepharose 4B suspension (20 mg/ml) was activated using 2 ml of CNBr solution (50 mg per milliliter of water). Activation proceeded with slight stirring for 8 min, the pH of the suspension being maintained at 11 by continuous addition of 4 M NaOH. The gel was then washed on a glass filter with cold water and 0.1 M NaHCO$_3$. The activated Sepharose was then transferred to a small beaker containing 1 ml of cold 0.1 M NaHCO$_3$. After addition of 15.9 U (100–130 U/mg) of hexokinase (HK) (the enzyme had been dialyzed prior to coupling against 0.1 M sodium phosphate buffer (pH 7.0, 4°, for 20 hr, 497 U/ml) together with 2.5 U (300 U/mg) of glucose-6-phosphate dehydrogenase (G-6-PDH) (dialyzed as above, 227 U/ml), coupling proceeded with gentle stirring for 12 hr at 4°. The enzyme gel was subsequently washed for 30 min each in cold 0.1 M NaHCO$_3$, 0.001 M HCl, and 0.5 M NaCl. After a final wash in 0.05 M Tris-HCl, 0.007 M MgCl$_2$ buffer, pH 7.6, the preparation was ready for use.

Polyacrylamide–Polyacrylic Acid-Bound (20% w/v, Acrylamide/Acrylic acid, 1:1). Twenty milligrams of dry gel granules of the cross-linked copolymer of acrylamide/acrylic acid, T:C = 20:5, (see also chapter [5]) in a small beaker were permitted to swell for 30 min in 1 ml of 0.1 M sodium phosphate buffer (pH 6.5); 12.4 U of HK (dialyzed as above, 497 U/ml) together with 3.4 U of G-6-PDH (dialyzed as above, 227 U/ml) were added to the suspension. After slight stirring for 5 min, 50 mg of 1-cyclohexyl-3(2-morpholinoethyl)-carbodiimidemetho-p-tolu-

---

[2] K. Mosbach and B. Mattiasson, *Acta Chem. Scand.* **24**, 2093 (1970).

enesulfonate were added to the suspension, and stirring was continued for 12 hr at 4°. The enzyme–gel granules were washed as above.

Polyacrylamide-Coentrapped (10% w/v). Two stock solutions were prepared. Solution I: 0.5 ml of 0.1 M sodium phosphate buffer (pH 6.5) containing 95 mg of acrylamide, 2 mg of ammonium persulfate, and 15 µl of N,N,N',N'-tetramethylethylenediamine (TEMED). Solution II: 0.5 ml of the above buffer containing 5 mg of N,N'-methylenebis(acrylamide) (Bis), 100 U of HK (333 U/ml), and 33.3 U of G-6-PDH (333 U/ml). Both solutions were kept at 0° for a few minutes prior to mixing in a small test tube. The solution was then gassed with nitrogen and kept at 0°. Polymerization proceeded within 1 hr. The preparation of granules from the gel obtained was carried out by cutting the gel into 1 mm-thick slices, which were pressed through a nylon net (30 mesh). The enzyme granules were subsequently washed as above and were ready for use.

Determination of Enzymic Activities. The following general procedure was applied (for a more detailed treatment of the topic see chapter [25]).

A 25-ml Erlenmeyer flask, containing the incubation mixture, was kept in a water bath at 25° placed above a magnetic stirrer. A speed setting was chosen to keep the suspension stirred at a rate found optimal for the enzymic reaction to be studied (120 rpm, dimensions of Teflon bar used: 0.4×2.7 cm). With a peristaltic pump (LKB Perpex, type 12001-2) the incubation solution was continuously pumped out of the system (2 ml/min) by way of a plastic tubing, the inlet of which had been covered with a piece of nylon net to keep off any gel particles. The incubation solution was passed through a flow-cuvette (Zeiss MT 4 D), placed in a spectrophotometer (Zeiss P M Q II), and returned to the reaction vessel.

The incubation solution present in tubing and cuvette—and thus not in contact with the enzyme gel granules—amounted in the present studies to 2 ml out of a total of 12 ml of incubation mixture.

The determination of the enzymic activities of gel particles carrying both HK and G-6-PDH proceeded in the following way. First the formation of NADPH was determined in the coupled reaction in an incubation mixture containing 26.64 µmoles of glucose, 5.23 µmoles of NADP⁺, and 7.26 µmoles of ATP in 12 ml of 0.05 M Tris-HCl, 0.007 M MgCl₂ buffer, pH 7.6 and using an amount of wet gel corresponding to 20 mg dry weight. Subsequently, the incubation solution was sucked off and the enzyme gel particles were washed thoroughly with Tris-MgCl₂ buffer. Immediately thereafter the separate G-6-PDH activity was determined by incubation with glucose 6-phosphate and NADP⁺. Finally, after thorough washing of the gel particles as above, the separate HK activity was determined through incubation of the same gel particles with an excess (3.3 U) of free G-6-PDH, glucose, ATP, and NADP⁺ as described above. No loss of activity of either bound HK or G-6-PDH was observed through the different assays following each other. To check whether the different polymer matrices affect the enzymic reactions studied through adsorption of substrate or cofactor, incubations were carried out with soluble HK and G-6-PDH in the presence of varying amounts of "blank gel." No such interference, however, was observed.

In order to permit a comparison between the enzymic activity of free HK and free G-6-PDH in the coupled reaction and that of the corresponding immobilized two-step enzyme system, the following assay was run. The same total number of enzyme units of free HK and G-6-PDH per milliliter, corresponding to the obtained separate enzymic activities (determined as nanomoles of product formed per minute per milliliter by HK and G-6-PDH in the matrix-bound two-step enzyme system), has been added to an incubation solution containing 6.66 μmoles of glucose, 1.83 μmoles of ATP, and 1.31 μmoles of NADP$^+$ in 3 ml of Tris-MgCl$_2$ buffer. The formation of NADPH was determined at 340 nm.

2. Preparation of an Immobilized Three-Step Enzyme System, β-Galactosidase–Hexokinase–Glucose-6-Phosphate Dehydrogenase

The procedure given below follows essentially previously published work.[3]

After the activation of 100 mg of Sephadex G-50 coarse with 100 mg of CNBr. 1.0 ml of a 0.1 NaHCO$_3$ solution of the three enzymes were added in the following proportions: 0.75 U of β-galactosidase (*Escherichia coli*), 2.0 U of hexokinase (baker's yeast), and 20 U of glucose-6-phosphate dehydrogenase (*Torula* yeast).

Coupling proceeded in a rotating test tube for 12 hr at 4°. The obtained Sephadex matrix, with the three enzymes bound probably at random, was subsequently washed for 30 min each at 4° in the following solutions to assure removal of enzymes not covalently bound: 0.1 M NaHCO$_3$, 0.1 M sodium acetate buffer (pH 5.0) (twice), 0.5 M NaCl, and 0.05 M triethanolamine-HCl with 0.07 M MgCl$_2$ (pH 7.6).

Determination of Enzymic Activities. The enzymic activity of the bound three-step enzyme system in coupled reaction was then determined by monitoring the absorbance at 340 nm as described for the above two-enzyme system. The assay proceeded in a 25-ml Erlenmeyer flask with the immobilized enzyme suspended in 10.0 ml of 0.05 M triethanolamine-HCl, 0.07 M MgCl$_2$ (pH 7.6) containing 5.23 μmoles of NADP$^+$ and

[3] B. Mattiasson and K. Mosbach, *Biochim. Biophys. Acta* **235**, 253 (1971).

7.26 μmoles of ATP. The reaction was carried out with stirring at 25° and was initiated by addition of a solution of 466 μmoles of lactose in 2.0 ml of the above buffer.

The rate of reaction of this system was then compared with that of the three participating enzymes in free solution. To this end a solution was prepared containing the equivalent number of enzyme units per incubation volume as measured from the immobilized preparation. The individual enzymic activity of each bound enzyme was determined from the rate of production of NADPH.

The incubation mixture, 12.0 ml, was modified from that of the coupled system described above in the following manner: glucose-6-phosphate dehydrogenase, omission of lactose and ATP, addition of glucose 6-phosphate (11.18 μmoles); hexokinase, omission of lactose, addition of glucose (26.64 μmoles) as well as excess of glucose-6-phosphate dehydrogenase (0.5 U); β-galactosidase, addition of excess of soluble hexokinase (0.5 U) and glucose-6-phosphate dehydrogenase (0.5 U).

3. Preparation of an Immobilized Three-Step Enzyme System, Malate Dehydrogenase–Citrate Synthase–Lactate Dehydrogenase

The procedures given below follow essentially previously published work.[4]

Sephadex-Bound. Sephadex G-50 (100 mg) was activated at pH 11 with 4.0 ml of CNBr (25 mg/ml of water). The pH was kept constant by the addition of 4 M NaOH. After 8 min, the activated gel was washed on a glass filter with 200 ml of cold 0.1 M NaHCO$_3$. The enzymes were added to the activated gel, and coupling proceeded for 14 hr at 4° in a rotating test tube. The reaction mixture consisted of 2.0 ml of 0.1 M NaHCO$_3$, activated gel, and various quantities of the ammonium sulfate suspensions of the enzymes. For instance: 25 μl of malate dehydrogenase (pig heart, 10 mg/ml, 1200 U/mg), 25 μl of lactate dehydrogenase (beef heart, 5 mg/ml, 250 U/mg) and 1μl of citrate synthase (pig heart, 10 mg/ml, 110 U/mg). Similar enzyme concentrations were also used in the below given examples. The enzyme–gel preparation was then washed 30 min with each of the following cold solutions: 0.1 M NaHCO$_3$ that was 50 mM in mercaptoethanol, 0.5 M NaCl that was 50 mM in mercaptoethanol, and finally 0.05 M potassium phosphate buffer (pH 7.5). Later preparations were washed without addition of mercaptoethanol.

Sepharose-Bound. One gram of well packed and thoroughly washed Sepharose 4B was suspended in 8.0 ml of CNBr solution (25 mg/ml) and

[4] P. A. Srere, B. Mattiasson, and K. Mosbach, *Proc. Natl. Acad. Sci. U.S.A.* **70**, 2534 (1973).

activated at pH 11.0. Activation proceeded at pH 11 for 10 min; the coupling of enzymes to the gel and the washing was then done as described above for the activated Sephadex G-50.

1,6-Diaminohexamethylene-Sepharose-Bound. One gram of well washed and packed derivatized gel (CNBr-activated Sepharose 4B reacted with excess of 1,6-diaminohexane at pH 8.5) was treated for 4 hr at room temperature (25°) with 20 ml of 0.1 M potassium phosphate buffer (pH 6.9) containing 4% (v/v) glutaraldehyde. One aldehyde group of glutaraldehyde reacts with the amino group of the derivatized Sepharose, the other terminal aldehyde group remains free. This new gel material was carefully washed with water and could then be used for enzyme coupling. The aldehyde–gel (0.5 g) was suspended in 2.0 ml of 0.1 M potassium phosphate buffer (pH 7.5) together with the enzymes. Coupling of the enzyme to the gel and washing of the final product was achieved under the same conditions as described above for the Sephadex system.

Polyacrylamide-Coentrapped (10% w/v). The enzymes in potassium phosphate buffer (pH 7.5) and 5 mg of cross-linking monomer (Bis) dissolved in the same buffer were mixed to a total volume of 0.50 ml. Three hundred microliters of the acrylamide solution (340 mg/ml of buffer) and the catalyst system, 100 μl of TEMED and 150 μl of ammonium persulfate solution (10 mg/ml of water), were mixed separately in a small test tube. About 1 min after the catalyst system was added to the acrylamide solution, the enzyme–Bis mixture was also added, and the solution was carefully mixed. All solutions were kept cold during the whole procedure. A gel was formed within 1–3 min, but the polymerization was allowed to proceed for 4 hr at 4°. The gel was then cut into 1-mm slices, each of which was pressed through a nylon net (30 mesh). The granules obtained were washed for 30 min in each of the following cold solutions: 0.25 M potassium phosphate buffer (pH 7.5), 0.1 M potassium phosphate buffer (pH 7.5), 0.25 M potassium phosphate buffer (pH 7.5), and 0.1 M potassium phosphate buffer (pH 7.5).

Determinations of the Enzymic Activities. Citrate synthase activity was assayed in 0.1 M Tris-HCl (pH 8.1) containing 4.8 μmoles of dithiobis(2-nitrobenzoic acid) and 1.44 μmoles of acetyl coenzyme A (acetyl-CoA). The reaction was initiated by addition of 60 μmoles of oxaloacetate. Activity was recorded at 412 nm, the absorption maximum of the thiophenol chromophore produced. Malate dehydrogenase was determined by the rate of oxidation of NADH to NAD$^+$ at 340 nm in the presence of oxaloacetate. The measurements were done in 0.1 M potassium phosphate buffer (pH 7.5) containing 2.12 μmoles of NADH and 2.40 μmoles of oxaloacetate. Lactate dehydrogenase activity was measured in a similar manner at 340 nm by measurement of the oxidation of NADH to NAD^+ in the presence of pyruvate. The incubation solution used was 0.1 M potassium phosphate (pH 7.5) containing 2.12 µmoles of NADH and 12.0 µmoles of pyruvate.

The gel was suspended in 12 ml of 0.1 M Tris-HCl (pH 8.1) containing 4.80 µmoles of dithiobis(2-nitrobenzoic acid), 3.6 µmoles of NAD^+, 1.44 µmoles of acetyl-CoA. After the reaction mixture had reached a constant temperature, the reaction was initiated by the addition of 6.0 µmoles of malate. The amount of citrate produced (equivalent to CoA produced) was continuously recorded at 412 nm as the chromophore was formed. The influence on the rate of citrate production of lowering the NADH concentration (formed in the malate dehydrogenase reaction) was investigated by the addition of 12.0 µmoles of pyruvate; the lactate dehydrogenase also present in the gel caused a reoxidation of NADH.

After measurements on the two- and three-enzyme systems, the incubation solution was removed, the enzyme-gel particles were thoroughly washed with 0.1 M Tris-HCl buffer (pH 8.1), and the separate enzyme activities were determined. Citrate synthase activity was measured as described above. After this determination the gel was washed again, but in 0.1 M potassium phosphate buffer (pH 7.5), and malate dehydrogenase activity was determined. After a final wash of the gel, the activity of lactate dehydrogenase was assayed. No loss of enzyme activities in these assays was observed.

Activities of the free enzymes were determined with the same concentrations of reagents as described for the bound enzymes, but in a total volume of 1.0 ml. The two dehydrogenases were measured at 340 nm as described above, but they were also assayed under the same conditions used for the coupled enzyme assay, so that activity ratios could represent activities under the same conditions rather than V_{max} ratios. Thus malate dehydrogenase was also measured when working in the forward reaction (i.e., converting malate to oxaloacetate) in the presence of excess of both acetyl-CoA and citrate synthase and either in the absence of dithiobis(2-nitrobenzoic acid), at 340 nm as NADH formed, or in its presence at 412 nm, as the amount of chromophore formed. Lactate dehydrogenase was measured in analogous fashion, except as already described, with excess malate dehydrogenase and citrate synthase as auxiliary enzymes. The incubation solution was 0.1 M Tris buffer (pH 8.1) containing 0.4 µmoles of dithiobis(2-nitrobenzoic acid), 0.5 µmoles of malate, 1.0 µmoles of pyruvate, 0.12 µmoles of acetyl-CoA, 0.17 µmoles of NADH, 3.6 U of citrate synthase, and 16 U of malate dehydrogenase.

So that valid comparisons could be made between the matrix-bound and the free systems, the activities on the bound system were determined and then the identical ratios of free enzymes were carefully reconstructed. Routinely, the same total number of units per volume of citrate synthase, malate dehydrogenase, and lactate dehydrogenase as measured on the matrix-bound systems were mixed in a 1-ml cuvette, and the overall reaction was measured. After the system had reached constant rate, lactate dehydrogenase activity was initiated by addition of 1.0 µmole of pyruvate.

4. Preparations of the Immobilized Two-Step Enzyme System: Amyloglucosidase–Gluose Oxidase

The procedures given below follow essentially previously published work.[5]

Two milligrams of amylo-α-1,4-α-1,6-glucosidase (*Aspergillus niger*, 14 U/mg) and 0.2 mg of glucose oxidase (*Aspergillus niger*, 200 U/mg) were bound to Sepharose 4B (corresponding to 400 mg dry weight) by the CNBr method. Coupling proceeded in a rotating test tube for 12 hr at 4°. The immobilized enzyme preparation, with the different enzymes bound probably at random, was washed for 30 min each with the following solutions to ensure removal of enzymes not covalently bound: 0.5 M NaHCO$_3$, 0.1 M β-D(+)-glucose, 1 M NaCl, and 0.05 M sodium maleate–acetate buffer (pH 5.3), 0.25 M in NaCl. For assay the described stirred batch assay method was applied. The incubation proceeded in a 25-ml Erlenmeyer flask with enzyme matrix (30 mg) suspended in 0.05 M sodium maleate–acetate buffer (11.4 ml) 0.25 M in NaCl, and peroxidase (100 U). To this suspension were added ethanolic Triton X-100 (100 µl, 80:20, v/v), o-dianisidine (4 µmoles) in ethanol (100 µl), and finally a solution of maltose (220 µmoles) in the above buffer (400 µl). The commercially available maltose had to be purified prior to use by acetylation to β-maltose octacetate, which was recrystallized to constant rotation α$_D$ = +62.5° (c 1.0; chloroform), m.p. 159°–160°C, and deacetylated.[6]

The enzymic reaction of the two-enzyme system was carried out with stirring (120 rpm; dimensions of Teflon bar used: 0.4 cm × 1.5 cm) at 25°C and assayed at 450 nm as o-dianisidine oxidized by the peroxidase reaction on the hydrogen peroxide formed.[7]

[5] S. Gestrelius, B. Mattiasson, and K. Mosbach, *Biochim. Biophys. Acta* **276**, 339 (1972).
[6] M. L. Wolfrom and A. Thompson, *in* "Methods in Carbohydrate Chemistry" (R. L. Whistler and M. L. Wolfrom, eds.), Vol. 1, p. 334. Academic Press, New York, 1962.
[7] A. Dahlqvist, *in* "Methoden der Enzymatischen Analyse" (H. Bergmeyer, ed.), Vol. 1, p. 877. Verlag Chemie, Weinheim, 1970.

The total activities of the two different enzymes of the immobilized system are defined as those determined separately at their respective optimal pH. The above buffer was used over the whole pH range investigated. Amyloglucosidase activity was determined by addition of excess soluble glucose oxidase (110 U) to the above incubation mixture at pH 4.8, and glucose oxidase activity was determined at pH 6.4 by substituting maltose with excess D-glucose (312 μmoles). The activities of the soluble enzymes, separate or in the two-enzyme reaction, were determined as above except for assaying in a normal cuvette in a total volume of 3 ml.

General Considerations on Preparation and Assay of Coimmobilized Multistep Enzyme Systems

Choice of Support and Immobilization Technique. Virtually any support or technique could be used for the preparation of such systems. On using porous supports, pore diffusional "problems" will occur; similarly, preparations in which the enzymes are immobilized by entrapment will show such diffusional limitations. Although this may not be a decisive factor for practical application since in return relatively large amounts of enzyme can be immobilized, kinetic analysis becomes complicated. Therefore, in two of the models for biological systems given (examples 2,3), Sephadex G-50 was used, leading to enzymes bound only to the surface of the matrix.

Enzyme Activity Ratio. With a simultaneous coupling of all the participating enzyme species, it is difficult to predict the activity ratio on the final product. There seems to be competition between the enzymes to bind to the gel, however, a decrease in the amount added of one enzyme will not automatically lead to a comparable increase in coupling yield of the others. To obtain preparations with given predicted activity ratios may involve much trial and error, though perhaps not at the two-enzyme level, but when three enzymes are involved the situation gets complicated. Very stable enzymes may be bound one after the other with successive coupling procedures.[8] It should be much easier under such circumstances to get preparations with predicted activity ratios. A drawback is that a preactivated gel cannot be used; this means that the coupling agent has to be added to the enzyme solution. By these successive procedures protein–protein couplings cannot be avoided.

Obviously the ratio between the activities of the participating enzymes

[8] K. Mårtensson, *Biotechnol. Bioeng.* **16**, 567 (1974).

is of importance for the kinetic behavior of the system.[9] Be aware that nonspecific adsorption of the intermediate to the matrix is a possibility. Also, the higher the activity of the immobilized enzymes, the sooner the steady-state level is obtained.

Assay. When assaying the separate activities on the gel, always begin with the last enzyme in the sequence, thus avoiding disturbing influences from the added free auxiliary enzymes. Thus with the coimmobilized system β-galactosidase–hexokinase–glucose-6-phosphate dehydrogenase, start by assaying glucose-6-phosphate dehydrogenase adding an excess of glucose 6-phosphate and NADP$^+$, then assay hexokinase in excess of glucose, ATP, and free glucose-6-phosphate dehydrogenase, etc. Make sure always to use excess of auxiliary enzyme and substrate.

When working with systems containing a thermodynamically unfavorable step, care must be taken to assure that the intermediate concentration is not saturating the enzyme next in line, because then no differences between the two-enzyme systems (the immobilized and the free) will be observed. This is the case with the system malate dehydrogenase–citrate synthase. When saturating concentrations of malate were used, the rate of the free system was the same as that of the immobilized system owing to the fact that citrate synthase in the free system now has the same oxaloacetate concentration as that found in the microenvironment of immobilized citrate synthase.

The effect of the volume of the assay solution in determining the rates of consecutive enzymic reactions cannot be readily determined in explicit form. A marked increase in volume will have little effect on the product formation since the local enzyme concentration will not be effected in such heterogeneous systems, whereas obviously in a soluble system an increase in volume is accompanied by a corresponding decrease in the rate of accumulation of product.

Reconstitution of the Corresponding Soluble System. After determination of the different enzymic activities found on the support, the corresponding soluble system is made up. To permit true comparisons, obviously the same number of units per incubation volume must be used.

Protein Determinations in Multistep Enzyme Systems. In the characterization of a preparation of immobilized enzymes, determination of the amount of protein coupled is a normal and important step. On working with multifunctional preparations such determination is of little value, since no conclusions concerning, for example, specific activities of the enzymes, can be drawn. The determination method applied must have the potential of discriminating between the various participating enzymic entities.

[9] R. Goldman and E. Katchalski, *J. Theor. Biol.* **32**, 243 (1971).

One possible method, although difficult and to our knowledge not yet tested, is based on a comparison of the ratio of individual amino acids found from total hydrolysis of a multistep enzyme preparation with that of each individual enzyme. Alternatively, the total protein content is determined first, followed by separate determination of one, or preferably more, of the enzymes in the system through assaying radioactivity of enzymes prelabeled with radioactive groups (e.g., ^{125}I), or by determination of a prosthetic group, such as FAD as in example 6. In the latter case, FAD was dissociated from glucose oxidase by treatment with trichloroacetic acid, after which the amount of FAD was measured fluorometrically as FMN at 520 nm upon irradiation at 278 nm. From the values obtained, the amount of each protein present can be calculated.

Theoretical Aspects

Kinetic Behavior of the Immobilized Multistep Enzyme Systems

The immobilization of sequentially working enzymes on the same matrix has been carried out to provide simplified model systems for *in vivo* situations, where enzymes are arranged in sequences on membranes or within gellike structures. Such systems should give rise to higher local concentrations of the intermediate substrates within the microenvironment of the enzymes as a result of tight coupling of the enzymes acting in sequence on the support and the Nernst diffusion layer,[10] present around enzyme polymer particles in stirred solutions. This then will establish a more favorable operating condition for the second enzyme in the sequence as obtained with the two-step enzyme system described:

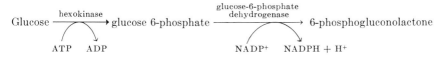

As described, the total activity of the coupled enzyme sequence was measured first, then followed determination of the activities of the separate enzymes. The observed rate of the overall reaction catalyzed by the above system was compared to blank runs consisting of the same amount of units of free enzyme per volume of incubation solution as found for each of the enzymes of the immobilized system. Another reference system consisted of the individual enzymes immobilized on separate polymer particles.[11] Figure 1 shows the results from typical experiments, where the

---

[10] F. Helfferich, *in* "Ion-Exchange," p. 267. McGraw-Hill, New York, 1962.
[11] B. Mattiasson and K. Mosbach, unpublished results.

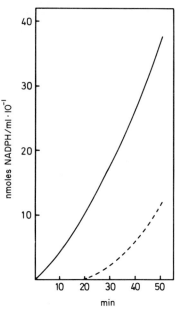

Fig. 1. Time dependence of NADPH formation in the coupled reaction of the hexokinase-glucose-6-phosphate dehydrogenase system. Two-enzyme system attached to acrylamide/acrylic acid copolymer, using water-soluble carbodiimide (———) (activity ratio 1:4). Corresponding free enzymes of the same total number of enzyme units/ml (for both enzymes) as in the immobilized system or when bound to separate particles (---). Reproduced with permission from K. Mosbach and B. Mattiasson, *Acta Chem. Scand.* **24**, 2093 (1970).

formation of NADPH is plotted against time. Both the system in which the enzymes are bound to the same matrix and the equivalent free system show a lag phase prior to reaching steady state. However, the immobilized system reaches steady state much faster than the corresponding soluble system, although both systems eventually reach the same steady-state rate. Enzymes immobilized to separate particles behaved like the soluble system. The length of the lag phase is dependent upon the activity ratio of both enzymes. The higher the total enzyme activity of the first enzyme in the sequence, the more will the soluble system behave like the immobilized one (see also chapter [32]). We interpret these results in the following way. In the immobilized system, the product from the first enzyme-catalyzed reaction is available in a higher concentration around the second enzyme than is the case in the corresponding free system. This is due in part to the close proximity of the enzymes in the immobilized state and in part to the impeded out-diffusion of the intermediate caused by the Nernst diffusion layer. The next enzyme in the sequence will thus work more efficiently and increase the rate of the overall reaction. Theoretical calculations on the behavior of an immobilized two-step enzyme system[9] are in good agreement with the above data obtained. (Preparation of the two-enzyme system xanthine oxidase–uricase entrapped within a membrane is described in chapter [63].)

Extension to the three-step enzyme system

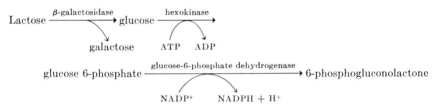

allows a study on the relationship of the length of the lag period and the efficiency of the overall reaction as influenced by the number of participating enzymes within the sequence.

Such an immobilized three-step enzyme system is more efficient than the corresponding soluble system during the initial phase before reaching steady state. The total reaction rate for all three immobilized enzymes as well as for the last two enzymes in the sequence is first determined and then compared with the corresponding soluble three- and two-step enzyme systems. These studies demonstrate that on extending the number of enzymes participating in a consecutive set of reactions, provided that the intermediate concentrations are rate limiting, binding of several enzymes to the same matrix has a cumulative effect on the efficiency of the overall reaction of the system during its initial phase.

It should be pointed out that, following the procedure given, individual bound enzymic activities were measured at a ratio of 1.0 (β-galactosidase):2.1 (hexokinase):3.5 (glucose-6-phosphate dehydrogenase). With such ratios, the observed lag phases are long (≥ 70 min) prior to reaching steady state. In Fig. 2, a generalized picture of the behavior of immobilized versus soluble multistep enzyme systems is given.

In accordance with these studies it was later calculated that for a hypothetical unaggregated system of the aromatic complex of *Neurospora crassa*, lags, (transient times) would be obtained that would be 10–15 times longer than for the complex. From this it was concluded that the aggregated multienzyme system compartmentalizes intermediate substrates during the course of the overall reaction.[12] (On general aspects of metabolic compartmentalization see Srere and Mosbach.[13]) Therefore in addition to "channeling" intermediates of competing pathways, reduc-

[12] G. R. Welch and F. H. Gaertner, *Proc. Natl. Acad. Sci. U.S.A.* **72**, 4218 (1975).
[13] P. A. Srere and K. Mosbach, *Annu. Rev. Microbiol.* **28**, 61 (1974).

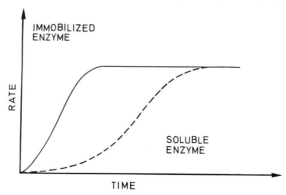

Fig. 2. Generalized picture of the system depicted in Fig. 1 and similar systems comparing the overall transient rates of consecutive enzymic reactions catalyzed by the coimmobilized enzymes and the enzymes in free solution. Reproduced with permission from J. M. Engasser and C. Horvath, "Applied Biochemistry and Bioengineering." Academic Press, New York, 1976.

tion of the transient time may be an equally important consequence of the confinement of intermediates within a physically associated enzyme sequence. In addition the aromatic complex exhibits a second catalytic property unique to an aggregated system, i.e., "coordinate activation," indicating that the physical association of these enzymes may have more than one physiological function.

Enzyme systems containing thermodynamically unfavorable reactions behave differently. In the other three-step enzyme system described

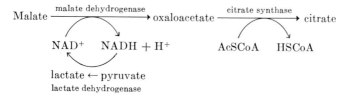

the malate dehydrogenase-catalyzed step is thermodynamically unfavorable in the direction of oxaloacetate, and hence concentration of the intermediate is not to be expected. Catalysis by malate dehydrogenase creates a constant concentration of oxaloacetate in the microenvironment of the enzyme and a decreasing concentration gradient of oxaloacetate away from its site of production. Differences between the immobilized system for the coupled reaction with malate dehydrogenase and citrate synthase and the corresponding free system are observed (at low concentrations of malate). Figure 3 shows that practically no lag phases are observed. However, the steady-state rates differ quite markedly between

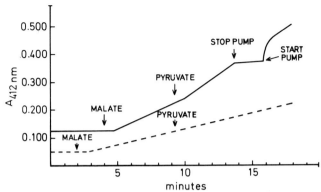

FIG. 3. Coupled reactions of a three-enzyme system: malate dehydrogenase–citrate synthase–lactate dehydrogenase. The enzymes immobilized in polyacrylamide (——) and the equivalent free enzyme system (---). Arrows indicate additions. The continuous-flow system was interrupted by stopping the peristaltic pump. Reproduced with permission from P. A. Srere, B. Mattiasson, and K. Mosbach, *Proc. Natl. Acad. Sci. U.S.A.* **70**, 2534 (1973).

the two systems, the immobilized system being up to 100% more efficient. The statistical mean distance between molecules of malate dehydrogenase and citrate synthase is shorter when the enzymes are matrix bound than when they are free in solution and may in the immobilized system lead to a steeper oxaloacetate concentration gradient and a higher mass transfer, making citrate synthase condense oxaloacetate with acetyl-SCoA more effectively. This removal of oxaloacetate from the unfavorable equilibrium will in turn result in a higher rate of catalysis by malate dehydrogenase and hence increase the rate of the overall reaction. This effect is not a result of the Nernst diffusion layer, but a consequence of the close proximity of the participating enzymes. When pyruvate is added to the system at a point indicated by an arrow in the figure, lactate dehydrogenase, also bound to the matrix, starts catalyzing the oxidation of NADH to NAD^+. This increases the efficiency of the immobilized system compared to the free, by as much as 400% in some preparations. In addition to the effect on oxaloacetate concentration, creation of a favorable NADH gradient as well as the enrichment of NAD^+ in the microenvironment of the enzyme (as compared to the situation in free solution) takes place also, resulting in an even more pronounced accelerated conversion of malate to oxaloacetate.

The favorable arrangement of having the reduction and the oxidation processes of NAD^+/NADH in close proximity to each other facilitates a recycling of the coenzyme between the two dehydrogenases *per se* which also contributes to the increase in the production of citrate.

The above system was studied to create a model and check for proposed compartmentalization of Krebs' cycle enzymes.[14] These model studies were subsequently verified by studies with mitochondria in which it was shown that in fact compartmentalization of the enzymes of the Kregs' cycle in the mitochondrial matrix is present.[15] This was accomplished by treating rat liver mitochondria with increasing concentrations of digitonin and following the loss of latency of enzymes as the inner membranes became permeable to substrates or acceptors.

pH-Optima of Coupled Systems. The two-enzyme system described as

$$\beta\text{-D-Maltose} \xrightarrow[\text{H}_2\text{O}]{\text{amylo-}\alpha\text{-1,4-}\alpha\text{-1,6-glucosidase}} 2\ \beta\text{-D-glucose} \xrightarrow[2\ \text{O}_2]{\text{glucose oxidase}} 2\ \text{D-glucono-}\delta\text{-lactone} + 2\text{H}_2\text{O}_2$$

with well-separated pH optima of the different enzyme activities when bound on the same matrix particle, results in an apparent displacement of the pH optimum of the overall reaction, although the individual pH optima may remain unaffected. The relative position of the pH optimum of such a system is dependent upon the activity quotient of the enzymes, i.e., for a two-enzyme system, the ratio of the activity of the first enzyme to the activity of the second.

Amylo-α-1,4-α-1,6-glucosidase (pH optimum, 4.8) and glucose oxidase (pH optimum, 6.4) coimmobilized to Sepharose show a difference between the pH optima of the coupled reaction of the immobilized and the soluble system of 0.3 pH unit (Fig. 4a,b). On variation of the activity quotient of the enzymes, differences of up to 0.75 pH units were obtained.

The shift in pH optima for the bound systems is due to the enrichment of the intermediate glucose in the microenvironment leading to a higher activity of the second enzyme as compared to that in free solution. This change is expressed as a changed ratio between the enzyme activities, which leads to a new optimization at a new different pH value as compared to the soluble system.

Practical Aspects

Some Considerations Relating to Application

Technical and analytical processes involving multistep enzyme catalysis may be performed in columns, hollow fibers, or so-called enzyme

[14] P. A. Srere, *in* "Energy Metabolism and the Regulation of Metabolic Processes in Mitochondria" (M. Mehlman and R. W. Hanson, eds.), pp. 79–91. Academic Press, New York, 1972.

[15] M. A. Matlib and P. J. O'Brien, *Arch. Biochem. Biophys.* **167**, 193 (1975).

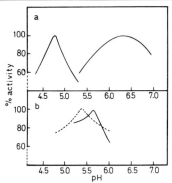

FIG. 4. Graphical illustration of pH-activity profiles of the two enzymes amyloglucosidase and glucose oxidase individually (a) and together (b). The individually determined pH optima of each enzyme (amyloglucosidase, pH 4.8; glucose oxidase, pH 6.4) as shown in (a) are the same in the free and Sepharose-bound state. The pH optima obtained in the subsequent reaction, however, are different in free solution (---, pH 5.4) and in the coimmobilized form (——, pH 5.7). The activity at the pH optima was arbitrarily set at 100%. One hundred percent activity corresponds to 0.30 ΔA/min for amyloglucosidase and 0.49 ΔA/min for glucose oxidase. Reproduced with permission from S. Gestrelius, B. Mattiasson, and K. Mosbach, Biochim. Biophys. Acta **276**, 339 (1972).

reactors. The different enzymes participating may either (a) be attached together to the same supporting particle or (b) be bound to separate particles. The method of choice in each individual case depends on specific parameters of the processes to be catalyzed and on the characteristics of the participating enzymes.

a. Enzymes Coimmobilized to the Same Support Particle

i. Generally it can be said that systems using alternative (a) with the enzymes bound to the same particles are more efficient in the initial phase of the reaction as is the case in the examples given here. This can be utilized, for instance, when they are used in reactors permitting higher flow rates.

ii. In cases of multistep enzyme reactions involving an enzyme-catalyzed step which is thermodynamically unfavorable, such preparations are also to be preferred. If the next enzymic step is thermodynamically favorable, then this will, owing to existing proximity effects, pull the equilibrium in the right direction and give a higher rate of the overall process during the course of the reaction when compared to the equivalent system with the enzymes bound to separate beads. The immobilized malate dehydrogenase–citrate synthase–lactate dehydrogenase system is a good example. Another such example is the two-enzyme system glyceraldehyde-3-phosphate dehydrogenase–phosphoglycerate kinase.[16]

iii. In processes where total conversion of the substrate is required, arrangement of enzymes on the same supporting particle may be advantageous because of higher efficiency when operating at rate-limiting substrate concentrations.

iv. Alternative (a) may also be advantageous in cases where the pH of the surrounding medium cannot be changed, e.g., potentially in replacement therapy, where immobilized enzymes may be injected into the bloodstream. Optimal pH-activity conditions may be obtained by adjusting the ratio of activities of the participating enzymes.

v. In coenzyme-dependent enzyme systems, in which the coenzyme may be regenerated by the immobilized enzyme system itself, arrangement according to (a) is certain to be superior to alternative (b), since the coenzyme will be kept within the vicinity of the particle flipping back and forth in a great number of cycles, thus overcoming the otherwise limiting diffusion rate of the coenzyme,[17] ultimately leading to improved economics.[18]

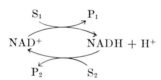

vi. Another case where alternative (a) is advantageous involves sequentially working enzymes acting intermittently on large-substrate molecules; i.e., when the catalytic action of the first enzyme in sequence makes the substrate molecule sterically available for the second enzyme, which then works until the first enzyme can interact again, and so on. Such is the case in the enzymic hydrolysis of starch to maltose. Here a two-step enzyme system, β-amylase–pullulanase, is covalently bound to a cross-linked copolymer of acrylamide and acrylic acid using water-soluble carbodiimide.[8] Hydrolysis of the α-(1-4) linkages by β-amylase renders the α-(1-6) linkages of the amylopectin available for hydrolysis by pullulanase.

b. Enzymes Immobilized on Separate Support Particles

Binding of enzymes to separate particles may be the procedure of choice for some processes. Several such examples can be found in the

[16] D. L. Marshall, in "Immobilized Biochemicals and Affinity Chromatography" (R. B. Dunlap, ed.), pp. 345–368. Plenum, New York, 1974.
[17] M. Dixon and E. C. Webb, in "Enzymes," p. 546. Longmans, Green, New York, 1964.
[18] J. R. Wykes, P. Dunnill, and M. D. Lilly, Biotechnol. Bioeng. **17**, 51 (1975).

literature; for example, the glycolytic enzyme sequence hexokinase–phosphoglucoisomerase–phosphofructokinase–aldolase has been assembled in this fashion.[19]

i. In general, it can be said that such preparations are easier to accomplish, since better control of the binding of individual immobilized enzyme activities can be maintained.

ii. Separate immobilization can be advantageous in reaction sequences in which one of the enzymes participating is more easily denatured than the others. The system may then be supplemented by addition of more of this particular polymer-bound enzyme preparation.

iii. Many enzyme sequences involve proton production or consumption, leading to extreme, and for some enzymes inhibiting, proton concentrations in the micromilieu of the immobilized enzyme sequence. In these cases it may be more viable to operate with the enzymes bound to separate beads (see example 6).

iv. In the cases hitherto considered, the immobilized enzymes can be applied mixed in stirred tank reactors or in hollow fibers. In certain cases, however, arrangement of the enzymes in separate volumes is necessary. This can readily be accomplished using column processes. This will be so when the substrate of one enzyme in the sequence inhibits another enzyme, when the separate enzyme-catalyzed reactions require very different conditions for optimal activity, and when strict control of each enzyme step is required in the course of the overall conversion. A process that might require such an arrangement for optimal conditions is the two-step steroid transformation process (see chapter [13]):

$$\text{Reichstein's compound S} \xrightarrow{11\beta\text{-hydroxylase}} \text{cortisol} \xrightarrow{\Delta^{1-2}\text{-dehydrogenase}} \text{prednisolone}$$

In conclusion, the various factors discussed above should be considered in the application of immobilized multistep enzyme systems. It is to be expected that studies of the behavior of such systems will provide useful information leading to future enzyme applications, as well as contributing to the understanding of the dynamics of integrated metabolic sequences.

Examples of Multistep Enzyme Systems (Coimmobilized on the Same Support Particles) Applied in Analysis

In a recent study a coupled system tryptophanase–lactate dehydrogenase was used for the spectrophotometric assay of L-tryptophan according to L-tryptophan $\rightleftharpoons$ indole $+ NH_4^+ +$ pyruvate, pyruvate $+$ NADH $+ H^+ \rightleftharpoons$ lactate $+ NAD^+$.[20] Both enzymes were coimmobilized

[19] H. D. Brown, A. B. Patel, and S. K. Chattopadhyay, *J. Chromatogr.* **35**, 103 (1968).
[20] S. Ikeda and S. Fukui, *FEBS Lett.* **41**, 216 (1974).

to CNBr-activated Sepharose. The advantage of such a system lies in the proximity between the participating enzymes, which is observed as a more rapid response in the initial phase and a higher sensitivity toward low concentrations of substrate for the first enzyme.

A somewhat similar system is the use of coimmobilized aspartate aminotransferase–malate dehydrogenase for the microassay of L-aspartate following the reaction scheme: L-aspartate → oxaloacetate + NH_4^+, oxaloacetate + NADH + H^+ → malate + NAD^+.[21] Owing to the proximity effects discussed above, the sensitivity of the immobilized coupled system was superior by approximately 10-fold to a system with aspartate aminotransferase immobilized alone and malate dehydrogenase free in solution.

Sequentially working enzyme systems may be used in thermal analysis, for example, by using enzyme thermistors[22] (see also chapter [45]) to amplify the heat response of the primary reaction, thereby increasing the sensitivity. The amplification of the response is achieved by the heat produced by the additional step(s) *per se*, but simultaneously a faster response and a higher sensitivity at low substrate concentration is obtained. Thus, the heat signals obtained from reaction with glucose in an enzyme thermistor filled with coimmobilized glucose oxidase–catalase preparations were significantly higher than those obtained with immobilized glucose oxidase alone.

It should be added that in other sequential systems studied, coimmobilization was either not attempted or shown not to be advantageous. In one study belonging to the former category, a multiple immobilized enzyme reactor using lactate dehydrogenase and pyruvate kinase was used to determine spectrophotometrically pyruvate and phosphoenolpyruvate.[23] The enzymes were bound separately to glass beads by diazotization and arranged separately in two packed bed reactors. In the other example that will be given here, coimmobilization appeared to be of no advantage. For the analysis of sucrose, invertase and glucose oxidase were covalently bound to the inner surface of nylon tubes.[24] Preparations with the enzymes coimmobilized were compared with combinations of the tubes each containing one of the enzymes. The analytical data obtained indicated no difference between the systems, the authors explaining the results in terms of the fact that, on hydrolysis of sucrose, α-D-glucose is produced, which has to mutarotate to β-D-

[21] S. Ikeda, Y. Sumi, and S. Fukui, *FEBS Lett.* **47**, 295 (1974).
[22] B. Mattiasson, B. Danielsson, and K. Mosbach, *Anal. Lett.* 9 (3), 217 1976.
[23] T. L. Newirth, M. A. Diegelman, E. K. Pye, and R. G. Kallen, *Biotechnol. Bioeng.* **15**, 1089 (1973).
[24] J. K. Inman and W. E. Hornby, *Biochem. J.* **137**, 25 (1974).

glucose before oxidation by glucose oxidase takes place. As mutarotation is the rate-limiting step, the coimmobilized system offered no advantage.

Other Systems; Examples 5 and 6

5. *Preparation of the Soluble Bifunctional Enzyme Aggregates of β-Glucosidase–Glucose Oxidase and Malate Dehydrogenase–Citrate Synthase*

The procedure given follows previously published work.[25]

β-Glucosidase–Glucose Oxidase. β-Glucosidase, 20 U (almonds, 5.1 U/mg), and glucose oxidase, 1820 U (*Aspergillus niger*, 200 U/mg), were dissolved in 0.2 M K_2HPO_4 to give a total volume of 1800 μl. The pH was adjusted with 4 M NaOH to 7.5. Glutaraldehyde (200 μl of 4% v/v) was distilled before use and was added slowly to the vigorously stirred enzyme solution. The coupling proceeded with stirring at 0°–4° for 8 hr and was stopped by the addition of 200 μl of saturated sodium bisulfite solution. Gel chromatography was performed in a column filled with Sephadex G-200 (15 × 250 mm) at room temperature using 0.1 M potassium phosphate buffer, 0.25 M in NaCl, pH 6.8, as eluent.

Malate Dehydrogenase–Citrate Synthase. Malate dehydrogenase and citrate synthase were coupled for 3 hr as described in the previous paragraph. The reaction products were separated by gel chromatography at 4° using 0.1 M potassium phosphate, pH 7.6, as eluting buffer.

Preparation of Monofunctional Enzyme Aggregates. Monofunctional aggregates of β-glucosidase and of glucose oxidase were prepared as references for the isoelectric focusing process. The coupling procedure and gel chromatography were performed as described above. Fractions of glucose oxidase eluting with an apparent molecular weight of the native enzyme were shown to have a molecular weight of 180,000 by dodecyl sulfate gel electrophoresis, indicating intramolecular cross-links in the native dimer enzyme.

Characterization. ISOELECTRIC FOCUSING. Native and glutaraldehyde-modified β-glucosidase and glucose oxidase as well as the β-glucosidase–glucose oxidase aggregate were focused in the pH ranges 3–5 and 4–6. Ampholine (2.5 ml of 40%) was dissolved in a 200-ml sucrose density gradient (0 to 20%).[26] The diameter of the column used was 24 mm. The focusing proceeded for 3 days at 4° and 2–2.5 W.

[25] B. Mattiasson, A. C. Johansson, and K. Mosbach, *Eur. J. Biochem.* **46**, 431 (1974).
[26] O. Vesterberg, this series, Vol. 22, p. 389 (1971).

DODECYL SULFATE-GEL ELECTROPHORESIS. Dodecyl sulfate gel electrophoresis[27] of the β-glucosidase–glucose oxidase aggregate was performed using polyacrylamide gels (T:C = 5:1.4) with β-glucosidase, glucose oxidase, bovine serum albumin, and β-galactosidase as references.

Determination of Enzymic Activities. Glucose-oxidase activity was measured at 460 nm in 0.1 M potassium phosphate buffer pH 6.8, 0.25 M in NaCl (total volume 1 ml) containing 100 μmoles of glucose, 0.5 μmoles of o-dianisidine dissolved in 10 μmoles of methanol, 10 μl of ethanolic Triton X-100 (80/20, v/v), 10 U of peroxidase, and varying amounts of polymer. β-Glucosidase activity was assayed at 400 nm in 1 ml of the above buffer containing 3.6 μmoles of p-nitrophenyl-β-D-glucoside and varying amounts of polymer.

The coupled reaction was assayed at 475 nm, where changes in the absorbance caused by the oxidation of o-dianisidine were not influenced by the p-nitrophenol produced.

$$p\text{-Nitrophenyl-}\beta\text{-D-glucoside} \xrightarrow{\beta\text{-glucosidase}} \text{glucose} \xrightarrow{\text{glucose oxidase}} \text{gluconolactone} + H_2O_2$$
$$\searrow p\text{-nitrophenol} \qquad + O_2$$

The reaction was run in 1 ml of incubation solution containing 3.6 μmoles of p-nitrophenyl-β-D-glucoside, 10 U of peroxidase, 10 μl of ethanolic Triton X-100 (80/20, v/v), 0.5 μmole of o-dianisidine dissolved in 10 μl of methanol.

Citrate synthase activity was assayed in 0.1 M Tris-HCl buffer, pH 8.1 (1 ml), containing 0.4 μmole of 5,5'-dithiobis-(2-nitrobenzoic acid), 0.12 μmole of acetyl-CoA, and 5 μmoles of oxaloacetate. Activity was recorded at the absorption maximum of the thiophenol chromophore produced, 412 nm.

Malate dehydrogenase activity was determined by the rate of oxidation of NADH to NAD⁺ at 340 nm in the presence of oxaloacetate. Measurements were made in 1 ml of M potassium phosphate buffer, pH 7.5, containing 0.176 μmole of NADH and 0.2 μmole of oxaloacetate.

The activity of the coupled reaction was determined in 0.1 M Tris-HCl buffer, pH 8.1 (1 ml), containing 0.40 μmole of 5,5'-dithiobis-(2-nitrobenzoic acid), 0.3 μmole of NAD⁺, 0.12 μmole of acetyl-CoA, and, as initiator of the reaction, 0.5 μmole of malate. The amount of citrate produced (equivalent to CoA produced) was assayed continuously at 412 nm as the chromophore was formed.

Soluble Bifunctional Enzyme Aggregates. Preparations of soluble bifunctional enzyme aggregates, such as β-glucosidase–glucose oxidase and malate dehydrogenase–citrate synthase, may be of interest for

[27] K. Weber and M. Osborn, *J. Biol. Chem.* **244**, 4406 (1969).

the following reasons. First, in the interpretation of kinetic data of such soluble systems, interference effects caused by factors such as unstirred layers should be minimized, thus permitting evaluation of the "pure" proximity effect. In addition, water-soluble or insoluble multifunctional enzyme aggregates may, from a practical point of view, be useful alternatives to the more normal matrix-bound multistep enzyme systems. In the case of the better-studied β-glucosidase–glucose oxidase system, bifunctional aggregates with molecular weights in the range 200,000–300,000 have been prepared. In contrast to the matrix-bound systems, no pronounced differences in the length of the lag phase between aggregates and free enzymes were found. This may be due either to the fact that the latter enzyme in the sequence was in vast excess in terms of activity or the fact that the two enzymes are bound at random such that most of their active sites will be unfavorably oriented relative to each other, thus lowering the probability that the intermediate will react with the second enzyme present in the aggregate. However, following essentially the techniques described, it should be possible to achieve sequentially operating enzyme aggregates with the desired increase in efficiency observed with matrix-bound systems. It should also be of interest to subsequently bind such aggregates to supports leading to preparations combining the advantages inherent in using solid supports with that of structured enzyme systems (compared to the normal random mosaic situation found in the preparations described here).

6. Polyacrylamide Coentrapped (25% w/v) Trypsin–Glucose Oxidase–Hexokinase

The procedures given below follow essentially previously published work.[28]

Preparation. The immobilization technique used was entrapment of the enzymes within polymer beads of 25% (w/v) polyacrylamide (see also chapter [12]).

In order to protect hexokinase from trypsin digestion, the enzymes were dissolved in separate monomer solutions and were thereby exposed to each other only for a short while prior to polymerization. Moreover, the hexokinase was protected by addition of glucose 6-phosphate and ATP to the monomer solution and also from trypsin digestion by the addition of the trypsin inhibitor benzamide. In a typical preparation of enzyme beads containing glucose oxidase, hexokinase, and trypsin, the following solutions were prepared. Solution I contained 7.13 g of acryl-

[28] S. Gestrelius, B. Mattiasson, and K. Mosbach, *Eur. J. Biochem.* **36**, 89 (1973).

amide in 0.05 M K-phosphate buffer, pH 7.5 (final volume 12 ml), 300 mg trypsin, and 36 mg of benzamide. Solution II contained 0.38 g of methylene-bis(acrylamide) in 15.5 ml of buffer, 75 U of glucose oxidase, 1500 U of hexokinase (baker's yeast, 420 U/mg) in 1 ml of buffer, 60 mg of ATP, 100 mg of glucose 6-phosphate, 50 mg of cysteine; the pH of this solution was adjusted to 7.5. Tetramethylethylenediamine (0.75 ml) and 0.75 ml of ammonium persulfate solution (25 mg/ml) were added to solution I, which was mixed with solution II after 30 sec; the mixture was poured into the reaction vessel containing 400 ml of the hydrophobic phase (toluene:chloroform, 290:110, v/v) and stirred together with 0.35 ml of the stabilizer Sorbitan sesquioleate (Pierce Chemical Company, Rockford, Illinois). The polymerization started in the stirred dispersion within a few minutes under a nitrogen atmosphere at 4°. After 15 min the gel beads were washed on a glass filter with 0.1 M $NaHCO_3$ and then stirred for 30 min each in 0.1 M $NaHCO_3$, 1 mM HCl, and H_2O. The beads were quickly frozen and lyophilized. The preparations were then stored at $-20°C$.

Determination of Enzyme Activities. Assays were carried out using a double-beam spectrophotometer (Beckman Acta III) permitting continuous repetitive scanning together with titration equipment (Radiometer titrator TTT lc and burette ABU 112). The electrodes and the burette delivery tubing were placed in the reaction vessel, a 50-ml glass beaker (inner diameter 36 mm). The incubation solution was cycled through a flow cuvette and back, using a tubing, the inlet of which was covered with a nylon net to keep gel particles from the flow cell. The reaction was carried out at 25° and the pH was kept constant by titrating with 0.10 M NaOH. The incubation solution was 15 ml and contained enzymematrix (100 mg dry weight) in 14.2 ml of buffer (usually 5 mM McIlvaine buffer, sodium citrate phosphate buffer with constant ionic strength, 0.15 M in $NaNO_3$ in place of the usual halide ions), 100 µl of ethanolic Triton X-100 (80:20, v/v) to solubilize colored products, 4 µmoles of o-dianisidine dissolved in 100 µl of ethanol, and 100 µl of peroxidase solution (1000 U/ml H_2O); 9 µmoles of ATP, 11 mmoles of $MgCl_2$, 7 µmoles of $NADP^+$, and 6.25 U of glucose-6-phosphate dehydrogenase (baker's yeast, 346 U/mg) were added. After equilibration for 5 min, the glucose oxidase and hexokinase activities were initiated by addition of 0.89 mmole of glucose in 0.5 ml of buffer. Trypsin activity was initiated by addition of 100 µl of 0.10 M Bz-Arg-OEt solution. The reaction proceeded while stirring (120 rpm; dimensions of bar used: 0.4 cm $\times$ 2.7 cm) at constant pH of the bulk solution (pH stat used), and the increase in absorbance at 450 nm (glucose oxidase) and 366 nm was continuously recorded. The hexokinase activity was determined by measuring NADPH formed at 366 nm in the coupled assay with glucose-6-phosphate dehydrogenase. The true NADPH absorbance was obtained after a correction for the o-dianisidine was made. Peroxidase present in the assay mixture did not interfere with NADPH. The time between two consecutive recordings at the same wavelength was 90 sec when the fastest scanning speed (2 nm/sec) was used.

Kinetic Behavior of a Coentrapped Nonconsecutive Three-Enzyme Preparation. The coentrapped nonconsecutive three-enzyme preparation, trypsin–hexokinase–glucose oxidase, was prepared to obtain a model system simulating the flow of substrate through different metabolic pathways whereby hexokinase and glucose oxidase were taken as models of two enzymes both competing for the same substrate, glucose. Their relative activities are followed under rate-limiting concentration of glucose as they are influenced by protons formed through the activity of the hydrolytic enzyme. As depicted in Fig. 5, at a chosen outer pH of 8.5, the portion of glucose phosphorylated by hexokinase (pH optimum, 8.5) decreased as a consequence of trypsin activity from 13% of the total amount of metabolized glucose found prior to addition of trypsin substrate to almost zero. This decrease in hexokinase activity was accompanied by an increase in glucose oxidase activity (pH optimum, 6.6) resulting in almost total conversion of glucose by glucose oxidase during

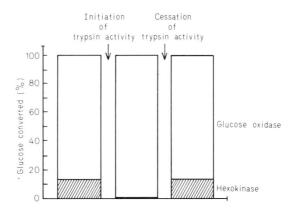

FIG. 5. Regulatory effect of trypsin on hexokinase and glucose oxidase activities when the three enzymes are immobilized together in polyacrylamide. The pH of the incubation solution was kept constant at 8.5, the pH optimum of hexokinase. The pH optimum of glucose oxidase under these conditions is 6.6. The glucose-converting enzyme activities at 0.65 mM glucose are expressed as a percentage of the total amount of glucose converted per minute (mean value 21.5 nmole/min). The columns represent the steady states before, during, and after trypsin activity. Reproduced with permission from S. Gestrelius, B. Mattiasson, and K. Mosbach, *Eur. J. Biochem.* 36, 89 (1973).

trypsin activity. The activities returned to their original values when all trypsin substrate was exhausted.

A proton-generating system was chosen, since in many reactions including kinases, phosphatases, and dehydrogenases in such metabolic pathways as glycolysis and Krebs' cycle, protons are produced or consumed and can, through their accumulation in the microenvironment, affect enzymes in their neighborhood, e.g., other components within a multienzyme complex. In these and other cases, local proton concentrations may regulate the activity of some of the participating enzymes.

[32] The Glucose Oxidase–Catalase System

By JAMES C. BOUIN, MOKHTAR T. ATALLAH, and HERBERT O. HULTIN

Glucose oxidase (GO_x) catalyzes the oxidation of beta-D-glucopyranose to δ-gluconolactone utilizing O_2 as the hydrogen acceptor with the production of H_2O_2. The lactone hydrolyzes spontaneously to gluconic acid. Catalase (Cat) brings about the mutual oxidation and reduction of two molecules of H_2O_2 to produce one molecule of O_2 and two molecules of water. These reactions of GO_x and Cat are shown in Eqs. (1) and (2), respectively.

$$O_2 + \beta\text{-D-glucose} \xrightarrow{\text{glucose oxidase}} H_2O_2 + \delta\text{-gluconolactone} \downarrow \atop \text{gluconic acid} \quad (1)$$

$$2H_2O_2 \xrightarrow{\text{catalase}} O_2 + 2H_2O \quad (2)$$

Although the principal interest has centered on the reaction sequence initiated by GO_x, as a cyclic pathway either enzyme may be used to initiate the overall reaction sequence.

Glucose Oxidase

Glucose oxidase is a glycoprotein having a molecular weight of approximately 150,000–180,000[1]; it contains 2 moles of FAD per mole of enzyme and 11–16% carbohydrates.[2,3] It has a pH optimum of approximately 5.5 to 5.7.[1] The rate of oxidation of β-D-glucose is about 157 times

[1] R. Bentley, in "The Enzymes" (P. D. Boyer, H. Lardy, and K. Myrbäck, eds.), 2nd ed., Vol. 7, pp. 567–576. Academic Press, New York, 1963.
[2] J. H. Pazur, K. Kleppe, and A. Cepure, Arch. Biochem. Biophys. 111, 351 (1965).
[3] J. H. Pazur, this series Vol. 9, pp. 82–87 (1966).

faster than that of α-D-glucose.[1] Important considerations in any assay technique, therefore, are the anomeric form of glucose present, the rate of mutarotation in the system compared to the rate of the reaction, and the extent of the reaction being measured. Glucose oxidase has been assayed manometrically,[1] polarographically,[4] by differential conductivity,[5] and by a coupled colorimetric assay involving peroxidase and o-dianisidine.[6,7] A sensitive spectrofluorometric analysis of H_2O_2 has been developed using diacetyl dichlorofluorescein,[8] which has possibilities of being adapted for the assay of GO_x. The polarographic and conductivity methods avoid the difficulties of direct spectrophotometric assays when using particular matter in the assay mixture.

Catalase

Catalase has a molecular weight range of 225,000 to 250,000[9] and contains four hematin groups. The kinetic behavior of Cat depends on the formation of several complexes between substrate and enzyme.[10,11] Jones and Suggett[12] have discussed proposed mechanisms for Cat reaction. On the basis of a reexamination of the kinetics of formation of the intermediate complex between Cat and H_2O_2,[13] it was concluded that the peroxidatic mechanism is the correct model for Cat activity.

The disappearance of H_2O_2 caused by Cat can be measured titrametrically with permanganate[14,15] or with acid iodide.[16] Catalase has also been assayed using sodium perborate as substrate followed by titration of excess perborate with permanganate.[17] The decrease of absorption in the region from 230 to 250 nm has been used as a direct measurement of the

[4] M. K. Weibel and H. J. Bright, *J. Biol. Chem.* **246**, 2734 (1971).
[5] R. A. Messing, *Biotechnol. Bioeng.* **16**, 525 (1974).
[6] A. St. G. Huggett and D. A. Nixon, *Lancet* **2**, 368 (1957).
[7] Worthington Enzyme Manual, p. 19. Worthington Biochemical Corporation, Freehold, New Jersey, 1972.
[8] M. J. Black and R. B. Brandt, *Anal. Biochem.* **58**, 246 (1974).
[9] A. C. Maehly and B. Chance, in "Methods of Biochemical Analysis" (D. Glick, ed.), Vol. 1, pp. 357–376. Wiley (Interscience), New York, 1969.
[10] P. Nicholls, and G. R. Schonbaum, in "The Enzymes" (P. D. Boyer, H. Lardy, and K. Myrbäck, eds.), 2nd ed., Vol. 8, pp. 147–158. Academic Press, New York, 1963.
[11] A. S. Brill and R. J. P. Williams, *Biochem. J.* **78**, 253 (1961).
[12] P. Jones and A. Suggett, *Biochem. J.* **110**, 621 (1968).
[13] E. Zidoni and M. L. Kremer, *Arch. Biochem. Biophys.* **161**, 658 (1974).
[14] G. Cohen, D. Dembiec, and J. Marcus, *Anal. Biochem.* **34**, 30 (1970).
[15] R. K. Bonnichsen, B. Chance, and H. Theorell, *Acta Chem. Scand.* **1**, 685 (1947).
[16] D. Herbert, this series Vol. 2, pp. 784–788.
[17] R. M. Feinstein, *J. Biol. Chem.* **180**, 1197 (1949).

TABLE I
Activities of Immobilized Glucose Oxidase and Catalase on Dual Catalyst Prepared at Different Enzyme Concentrations

Sample	Enzyme concentration in immobilizing solution (M)		Activities of the dual catalyst (pmoles O_2/sec-mg)	
	Catalase	Glucose oxidase	Catalase	Glucose oxidase
1	3.44×10^{-9}	5.33×10^{-6}	18.6	84.4
2	3.44×10^{-8}	5.33×10^{-6}	72.0	83.0
3	3.44×10^{-7}	5.33×10^{-6}	182.5	81.0
4	3.44×10^{-7}	5.33×10^{-7}	230.0	50.7
5	3.44×10^{-7}	5.33×10^{-8}	264.0	9.1

decrease of H_2O_2 caused by Cat action,[18] and production of O_2 has been measured with the oxygen electrode.[19,20]

Immobilization Procedure

Porous silica alumina pellets impregnated with 5% nickel (Harshaw Chemicals, No. Ni 0901) are ground, and particles in the range of 125–250 μm in diameter are used. The support is refluxed in toluene containing 10% α-aminopropyltriethoxysilane for 24 hr, washed with chloroform, then acetone, and dried.[21] The functional amino support is treated with a 2.5% solution of glutaraldehyde (Fisher) at room temperature for 30 min under vacuum followed by 60 min at atmospheric pressure. After washing the particles, the enzyme(s) are added in citrate-phosphate buffer, pH 7.0, in the ratio of 1 ml of buffer per gram of support.[22] Immobilization is allowed to proceed for 30 min under vacuum, then for 60 min at atmospheric pressure. By varying the concentration of enzyme in the immobilizing solution (Table I) and the order of addition of the enzymes and glutaraldehyde (Table II), the relative activities of the enzymic activities on the support particles can be controlled.[23] The procedure described below for assay of this system should be adaptable to

---

[18] B. Chance and A. C. Maehly, this series Vol. 2, pp. 764–788 (1955).
[19] D. B. Goldstein, *Anal. Biochem.* **24**, 431 (1968).
[20] M. Rorth and P. K. Jensen, *Biochim. Biophys. Acta* **139**, 171 (1967).
[21] H. H. Weetall and L. S. Hersh, *Biochim. Biophys. Acta* **206**, 54 (1970).
[22] Corning Manual: Biomaterial Supports. Corning Biological Products Group, Medfield, Massachusetts, 1973.
[23] J. C. Bouin, M. T. Atallah, and H. O. Hultin, *Biotechnol. Bioeng.* **18**, in press (1976).

TABLE II
EFFECT OF ORDER OF ADDITION OF ENZYMES AND
GLUTARALDEHYDE ON IMMOBILIZED ACTIVITIES

Additions[a]	pH	Activities (pmoles O_2/sec-mg)
GHO → GO_x	4.5	10.3
GHO + GO_x	4.5	31.0
GHO + GO_x	7.5	51.1
GHO → CAT	7.5	109.0
CAT → GHO	7.5	74.8
GHO + CAT	7.5	55.0

[a] The → indicates sequential additions, and the plus sign (+) indicates simultaneous addition of enzyme and glutaraldehyde (GHO). GO_x, glucose oxidase.

any immobilized system of GO_x and Cat. Glucose oxidase from *Aspergillus niger* was obtained from Worthington Biochemical Corporation. We have used Cat from both *A. niger* (Calbiochem) and beef liver (Worthington). Significant stability problems were encountered with the liver Cat. Thus, we recommend Cat from *A. niger* as the preferred material.

Assay Procedures

Glucose Oxidase. Glucose oxidase is assayed by following the decrease in O_2 in solution at 25°, using the oxygen electrode (Gilson Medical Electronics Oxygraph). The reaction medium consists of citrate–phosphate buffer[24] (pH 5.5), 13.9 mM D-glucose, and the appropriate amount of catalyst particles. The reaction is started by the addition of D-glucose. The final volume in this particular apparatus is 1.8 ml. Air should be bubbled through the incubation mixture for a minute or two before beginning the assay to completely saturate the solution. The glucose solution should be allowed to stand for about 2 hr at room temperature to ensure that an equilibrium mixture of the α- and β-anomers occurs. The K_m for glucose is 15.4 mM; thus, the glucose concentration used is not saturating. However, it is in large excess of the concentration of O_2, and so the reaction can be considered to be zero order with respect to glucose under the above conditions. The concentration of O_2 in solution at 25° and standard pressure is approximately 0.25 mM. The ionic strength of the assay medium has a negligible effect on O_2 solubility.

The size of the sample assayed should be such that the total activity

[24] G. Gomori, this series Vol. 1, pp. 138–146 (1955).

is approximately 0.25 unit, where a unit is defined as the amount of enzyme capable of utilizing 1.0 μmole of O_2 per minute under the described conditions. The sample should be stirred at a speed such that sufficient mixing is ensured to prevent localized gradients of substrate or product from forming. However, excess stirring may result in some mechanical attrition of the catalyst particles.

The activity is calculated as the first-order rate constant. Most preparations of GO_x contain some contaminating Cat (even the most highly purified commercially available samples). These impurities may be removed by a column technique,[4] or the assay may be carried out in the presence of excess soluble Cat. This converts each mole of H_2O_2 produced to 0.5 mole of O_2. Thus the true GO_x content can be calculated by taking the first-order rate constant in the presence of excess Cat and multiplying by 2. We have found that 7.0 units of Cat are sufficient to utilize all the H_2O_2 produced by GO_x of the activity described above. Care should be taken to rinse out all traces of Cat from the reaction compartment after using large amounts of soluble Cat. Repeated rinsing with distilled water or with the buffer is not sufficient. Leaving a diluted bleach solution (1:4 v/v) in the reaction compartment for 1–3 min, followed by several rinsings with distilled water and buffer, eliminates the adsorbed Cat.

Catalase. Catalase is assayed by following the production of O_2 at 25° utilizing the oxygen electrode. The assay medium contains citrate–phosphate buffer (pH 5.5), 0.5 mM H_2O_2, and a suitable amount of catalyst particles. The reaction is started by the addition of H_2O_2. The substrate and enzyme concentrations used are limited by the fact that the rate of generation of O_2 cannot be greater than what can be dissolved in the assay medium. With the substrate concentrations used, this reaction demonstrated pseudo first-order kinetics, and the activity was calculated as the first-order rate constant. To measure the production of O_2, the initial concentration in the reaction medium has to be reduced. This is accomplished by bubbling N_2 through the buffer and substrate solutions.

Two-Enzyme Systems. The methodology for the two-step sequential reaction is the same as for the single reactions except that conditions are used such that the intermediate product is supplied only by the first enzyme. In the case of the GO_x-initiated overall reaction, this is done by addition of glucose without adding any excess soluble Cat. In the case of the Cat-initiated two-step reaction, glucose (13.9 mM) is included in the reaction mixture, and H_2O_2 (0.5 mM) is used to initiate the reaction. Care must be taken in the latter case to remove all the O_2 from solution. As discussed later, this is critical in the correct evaluation of the results. Oxygen remaining in solution after bubbling with N_2 can be removed by allowing the two-enzyme system to proceed in the presence of glucose, until all O_2 is utilized, before addition of H_2O_2. Generally, pH 5.5 was used for the assay since fungal Cat exhibits a broad pH spectrum over the range from 2.5 to 7.0 while GO_x has a pH optimum around 5.5–5.7. Both the Cat-initiated and the GO_x-initiated two-step reactions were observed to be pseudo first order.

If the dual catalyst (each particle contains both enzymes) is to be compared to a mixed catalyst (each particle contains only one of the two enzymes), or to a soluble system, it is convenient to run the assays on the dual system first, since it is easier to adjust the activities of the mixed catalyst or the soluble enzymes to match those of the dual catalyst. To assay for the activities of the individual enzymes and both the GO_x-initiated and the Cat-initiated two-step overall reactions on the same sample of the dual catalyst, the following sequence of assays is used: First the sample is assayed for the GO_x-initiated two-step overall reaction followed by assaying for Cat activity alone (in absence of glucose). Then the Cat-initiated two-step overall reaction is assayed in the presence of glucose and H_2O_2. Finally, GO_x activity in the presence of excess soluble Cat is determined. Since it is impractical to remove all the added soluble Cat from the particles and reaction cell of this last assay, the sample is usually discarded.

To rinse the particles between assays, the sample is suspended in H_2O or buffer by stirring, then allowed to settle before removing the excess liquid with suction. This procedure should be repeated until the removal of all soluble reactants from the pores of the sample is assured.

Unsatisfactory results have been obtained by us in the Cat-initiated two-step reaction where beef liver Cat incubated with slowly generating H_2O_2, i.e., as the result of GO_x activity, was observed to have a higher initial rate of O_2 production than nonincubated enzyme. This is believed to be due to conformational changes reported by other investigators.[12]

Data Evaluation

Typical data obtained for the assay of GO_x and the GO_x-initiated 2-step reaction are shown in Fig. 1. Corresponding data obtained for assays of Cat and Cat-initiated overall reactions are shown in Fig. 2. The data are evaluated in two ways.

Substrate Converted. Since a cyclic reaction is involved, the final product is identical to the initial substrate. The product (O_2) of the GO_x-initiated reaction is measured as indicated by the arrow in Fig. 1. It is the difference between the O_2 utilized in the single GO_x-catalyzed reaction (obtained by determining the reaction rate in the presence of excess soluble Cat and multiplying by two) and that observed with the two-

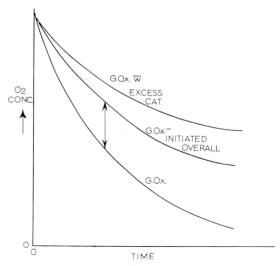

Fig. 1. Oxygen consumption in the glucose oxidase (GO_x) and glucose oxidase-initiated overall reactions. CAT, catalase.

enzyme sequential system. This measured difference is an approximation that is only valid until the recycled O_2 significantly changes the O_2 concentration that would exist in a noncyclic two-step reaction with the same rate constants.

Likewise, the H_2O_2 produced in the reaction sequence initiated by Cat is measured as the difference in O_2 produced by Cat alone and that produced by the Cat-GO_x two-enzyme system (indicated by the arrow in Fig. 2). The difference in O_2 contents in the solutions between these two is equivalent to what has been converted to H_2O_2 by GO_x. Again, this is accurate only as long as the H_2O_2 produced does not contribute in a significant way to the concentration of substrate that would occur in a similar two-step sequential reaction.

Some examples of the errors that occur in the procedures above as a function of the fraction of initial substrate recycled (fraction of product of reaction sequence to initial substrate) are shown in Table III. The minimum error occurs when the second enzyme is present in excess ($k_2 = \infty$). The higher the ratio of $k_2:k_1$, the lower is the error of estimation, since it is the concentration of intermediate that determines the error for any given fraction of conversion. Although the percentage of error in the approximation is theoretically independent of the absolute values of k_1 and k_2 if the ratio is constant, practical considerations in the use of a particular instrument will govern the range of values of k_1 and k_2 that can be used.

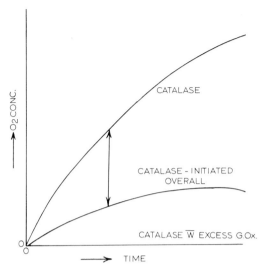

FIG. 2. Oxygen production in the catalase and catalase-initiated overall reactions. GO_x, glucose oxidase.

Typical results of the measurement of fraction of substrate converted for the GO_x-initiated overall reaction are shown in Fig. 3 for three different sets of activities of GO_x and Cat on the support. Similar results are obtained for the Cat-initiated overall reaction. Using glucose oxidase and catalase bonded by glutaraldehyde to supports similar to those described here, 99% conversion of a 16% solution of glucose to gluconic acid has been attained with residence times of 8–10 hr on a trickle-bed reactor which allows for concurrent flow of liquid and gaseous phases.

Efficiency. The other method of evaluation of the data is based on the efficiency of the reaction. We define efficiency as how well the second enzyme in the overall reaction sequence utilizes the product of the first reaction compared to its utilization by an excess amount of second enzyme. Catalase efficiency is calculated as the difference in rate constants

TABLE III
ERROR IN MEASUREMENT OF PRODUCT FORMED OWING TO RECYCLING

Fraction converted	Ratio of rate constants, $k_2:k_1$				
	0.1	1	3	10	∞
0.02	29.6	10.0	6.0	3.6	1.4
0.05	46.5	16.7	10.5	6.9	4.4

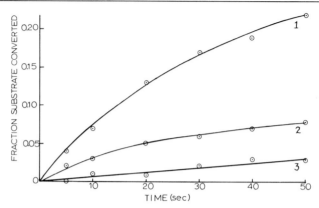

Fig. 3. Initial substrate converted by the dual immobilized system with three different activities of glucose oxidase and catalase. (1) $k_1 = 1.9 \times 10^{-2}$ sec^{-1}, $k_2 = 3.3 \times 10^{-2}$ sec^{-1}; (2) $k_1 = 2.4 \times 10^{-2}$ sec^{-1}, $k_2 = 7.8 \times 10^{-3}$ sec^{-1}; (3) $k_1 = 1.5 \times 10^{-3}$ sec^{-1}, $k_2 = 4.0 \times 10^{-2}$ sec^{-1}.

between the GO_x reaction and the GO_x-initiated overall reaction, divided by the rate constant of the GO_x reaction in the presence of excess catalase (top curve in Fig. 1), times 100 [Eq. (3)]:

$$\text{Cat Eff} = \left(\frac{k_{GOx} - k_{overall}}{k_{GOx}(x\text{'s Cat})}\right) 100 \qquad (3)$$

In a similar manner, GO_x efficiency (Fig. 2) is calculated as 1, minus the rate constant of the Cat-initiated overall reaction, divided by the rate constant of the Cat reaction, times 100 as shown in Eq. (4).

$$GO_x \text{ Eff} = [1 - (k_{overall}/k_{catalase})]100 \qquad (4)$$

An excess amount of soluble second enzyme was sufficient to completely utilize the product produced by the first enzyme with both the Cat- and GO_x-initiated overall sequences. In the determination of efficiency, it is particularly important that the intermediate be equal to 0 at the beginning of the reaction since the rate of the second reaction is dependent on the concentration of the intermediate. This is a very practical problem in the case of the Cat-initiated reaction, where all the O_2 must be removed.

Typical results for Cat efficiencies in the GO_x-initiated two-step sequential reactions are shown in Table IV. At constant activity of GO_x, there is an increase of efficiency with increasing activities of the second enzyme (Cat) in the sequence. At a constant ratio of the activities of the two enzymes, increasing absolute activities give increasing efficiencies, as had been predicted by Goldman and Katchalski.[25] Similar results were

[25] R. Goldman and E. Katchalski, *J. Theor. Biol.* **32**, 243 (1971).

TABLE IV
Efficiencies of Dual Immobilized System[a]

Activity[b] ratio cat:GO_x	Enzymic activity[c]		Cat eff
	Cat	GO_x	
1.7	3.2×10^{-2}	1.9×10^{-2}	89
7.0	3.8×10^{-2}	5.4×10^{-3}	99
27	4.0×10^{-2}	1.5×10^{-3}	98
1.8	1.1×10^{-2}	6.0×10^{-3}	47
1.6	1.5×10^{-3}	9.4×10^{-4}	5
0.4	7.8×10^{-3}	2.1×10^{-2}	24
0.05	1.0×10^{-3}	2.2×10^{-2}	14

[a] J. C. Bouin, M. T. Atallah, and H. O. Hultin, *Biochim. Biophys. Acta* in press (1976).
[b] Activity = first-order rate constant (sec^{-1}).
[c] Cat, catalase; GO_x, glucose oxidase.

observed for GO_x efficiencies in the Cat-initiated two-step reaction. The efficiencies obtained will vary with many factors including the size of the catalyst support, pore diameter, and enzymic activities.

Lag Times

Major differences in reaction rates of immobilized versus soluble enzymic sequences occur in the early stages. Mosbach and Mattiasson[25,26,27] have shown experimentally in both a two- and a three-enzyme system that the rate of production of final product is higher in a matrix-bound system than in a soluble system before reaching steady state. They ascribed this to the buildup of intermediates in the microenvironment of the matrix-bound enzymes. A theoretical treatment of membrane-bound enzymes by Goldman and Katchalski[27] reached a similar conclusion. The time required before maximal rate of production of end product is reached is termed the lag time,[28] which is defined arbitrarily as the time required for the intermediate to reach its maximal concentration. An iterative computer program can be used to determine lag times for a two-enzyme cyclic reaction with various rate constants. Some typical lag times are shown in Table V. The importance of the use of lag time is to know the general nature of the reaction over the period of time when the reaction

[26] K. Mosbach and B. Mattiasson, *Acta Chem. Scand.* **24,** 2093 (1970).
[27] B. Mattiasson and K. Mosbach, *Biochim. Biophys. Acta* **235,** 253 (1971).
[28] A. A. Frost and R. G. Pearson, "Kinetics and Mechanism: A Study of Homogeneous Chemical Reactions." Wiley, New York, 1961.

TABLE V
RELATION OF LAG TIME TO RATE CONSTANTS IN A CYCLIC REACTION[a]

k_1	k_2	$k_1:k_2$	Lag time (sec)
10^{-3}	10^{-3}	1	1246
10^{-2}	10^{-2}	1	125
10^{-1}	10^{-1}	1	13
10^{-3}	10^{-2}	0.1	307
10^{-2}	10^{-3}	10	307

[a] First-order rate constants (sec^{-1}).

will be studied. If the lag time is long relative to the time of assay, differences between a two-enzyme immobilized system and a soluble system should be higher than if the lag time is short compared with the assay period. Switching rate constants in a two-step cyclic (or sequential) reaction does not change the lag time. Therefore, with an immobilized two-enzyme cyclic system, initiating the reaction with either enzyme does not affect the lag time of the reaction.

Section VIII
The Application of Immobilized Enzymes to Fundamental Studies in Biochemistry

[33] Immobilized Subunits

By WILLIAM W.-C. CHAN

As the term "immobilized enzyme" indicates, the freedom of movement of enzyme molecules in these derivatives is restricted. This property opens up a number of obvious possibilities in the study of molecular interactions. Since 1970, immobilized derivatives have been used to prevent the spontaneous association between subunits of an oligomeric protein.[1] With this approach, it is possible to determine whether the subunit form of an enzyme is catalytically active. If the immobilized subunit is active, then comparison of its enzymic properties with those of the corresponding immobilized oligomer can yield valuable information regarding the effects of subunit interactions on enzyme function. Such information is not available for many enzymes without the use of special techniques because their native oligomeric structures are sufficiently stable that severe conditions are necessary to cause dissociation. In these cases, subsequent return to nondissociating conditions leads to spontaneous reassociation into the native structure. Therefore the oligomeric and monomeric forms cannot be compared under normal assay conditions. It is for such enzymes that the present method is most applicable. A prerequisite for the method is that renaturation with good recovery of activity can be achieved after dissociation.

Choice of Experimental Conditions

Sepharose 4B has been found to be a suitable carrier for the study of immobilized subunits. The nonionic and hydrophilic nature of its polysaccharide matrix may be expected to give derivatives whose properties are not affected very significantly by the presence of the matrix. Thus results obtained with these derivatives are likely to be applicable to conditions in ordinary buffered solutions. The highly porous nature of the Sepharose matrix allows the easy diffusion of macromolecules in and out of the gel during the preparation of the immobilized derivatives. If other carriers are used which do not permit the entry of macromolecules into the interior of the gel particles, then immobilization can occur only at the outer surface. In such cases, it is much more difficult to obtain sufficient amount of bound subunits per unit volume of gel without having the subunits so close together that interaction between adjacent subunits

---

[1] W. W.-C. Chan, *Biochem. Biophys. Res. Commun.* **41**, 1198 (1970).

may occur. The matrix structure of Sepharose, however, is not completely rigid and therefore extended storage of immobilized subunits or incubation at moderately elevated temperatures should be avoided.[2] Sepharose that has been treated with cross-linking reagents such as divinyl sulfone[3] has been reported to give more stable immobilized subunits.[2]

One of the basic features of this approach is that the *oligomeric* form of the enzyme is coupled to the carrier via only one of its subunits, and the monomeric form is then generated by exhaustive washing under conditions that lead to dissociation. This method offers a number of important advantages over the alternative way of coupling the monomeric form directly under dissociating conditions. The main reason is that, depending on the severity of the dissociating conditions, some change in the tertiary structure of the dissociated subunit is generally unavoidable. (In many cases, high concentrations of protein denaturants are necessary to obtain dissociation, and the individual polypeptides then exist essentially as random coils.) Although such changes in tertiary structures are often reversible (and indeed this is a prerequisite for the preparation of immobilized subunits), coupling of the monomer under dissociation conditions will lead to polypeptides bound in such a way that regain of the correct structure is no longer possible. This failure of bound monomers to renature can be caused by (1) coupling via a side chain of the polypeptide which is essential for maintaining the correct tertiary structure or (2) coupling to a point on the matrix that does not provide sufficient space for the renatured conformation of the monomer. Monomers coupled directly under dissociating conditions are also unlikely to be able to reassociate with added soluble subunits and return to the oligomeric form because again the coupling may have occurred with a side chain at or near the normal subunit contacts or the particular location of the bound monomer on the matrix has insufficient space. As will be discussed in a later section, the comparison between the subunit derivative and the corresponding immobilized oligomer, with regard to protein content and stability characteristics, can indicate whether subunits have in fact been obtained by the procedure. The specific ability of the immobilized monomer (produced by coupling the oligomer and subsequent dissociation) to interact with added subunits in solution constitutes strong evidence for the existence of monomers in the derivative. These valuable experimental possibilities are not available when subunits are directly coupled to the matrix under dissociating conditions.

The immobilization of the oligomer is best achieved by activating the carrier and then coupling the protein to it. This procedure avoids the

[2] N. M. Green and E. J. Toms, *Biochem. J.* **133**, 687 (1973).
[3] J. Porath and L. Sundberg, *Nature (London) New Biol.* **238**, 261 (1972).

risk of cross-linking among protein molecules, or among the subunits. The cyanogen bromide method[4] of activating Sepharose is quite suitable, although the linkage formed between the protein and the carrier may undergo slow cleavage.[5] In order to prevent the coupling of oligomers via more than one subunit to Sepharose or the interaction between the immobilized subunits after preparation, it is necessary to limit the density of activated points on the matrix. In practice, a ratio of 1 to 5 mg of CNBr per milliliter of packed Sepharose has given preparations with acceptably low contamination by immobilized dimers. The low density of coupled molecules has a further advantage in minimizing some characteristic matrix effects, such as the artificial increase in K_m caused by the limited rate of substrate diffusion.

Examples

Preparation of Immobilized Subunits of Rabbit Muscle Aldolase

A convenient amount of Sepharose 4B (e.g., 5 ml) is measured out by centrifugation at moderately low speeds (1,500 g) in a swinging-bucket centrifuge. A slurry of Sepharose and deionized water (1:1 v/v) is then activated according to Axén et al.[4] using 1 mg of CNBr per milliliter of packed gel. More recent, improved procedures for activation using buffered alkaline solutions[6,7] should also be suitable, although they have not yet been tested for the preparation of immobilized subunits. The activated gel is washed with cold 0.1 M sodium bicarbonate, pH 9.0, and excess liquid is removed by filtration. The moist gel is then transferred to 25 mg of native (tetrameric) aldolase in 5 ml of the above bicarbonate buffer and kept at 4° for 24 hr with occasional gentle swirling. High protein concentration during coupling has been found to increase substantially the amount of bound enzyme. To increase the coupling yield further, the pH chosen (9.0) is the highest possible without inactivation or dissociation of the protein. Continuous mechanical stirring during coupling is avoided as some denaturation of the protein would occur, giving a product of lower specific activity. After washing briefly to remove the bulk of the excess aldolase in solution, the Sepharose derivative is treated overnight with ethanolamine hydrochloride (0.1 M, pH 7.5) to block any re-

[4] R. Axén, J. Porath, and S. Ernback *Nature* (*London*) **214**, 1302 (1967).
[5] H. J. Kolb, R. Renner, K. D. Hepp, L. Weiss, and O. H. Wieland, *Proc. Natl. Acad. Sci. U.S.A.* **72**, 248 (1975).
[6] J. Porath, K. Aspberg, H. Drevin, and R. Axén, *J. Chromatogr.* **86**, 53 (1973).
[7] S. C. March, I. Parikh, and P. Cuatrecasas, *Anal. Biochem.* **60**, 149 (1974).

maining activated groups on the Sepharose. (More recent experience suggests that aromatic amines, such as aniline, might be more suitable than ethanolamine on account of their lower pK_a values.) The product, "immobilized aldolase," is washed exhaustively by a batchwise procedure in which the gel is suspended with gentle overhead stirring in 10 times its volume of Tris-HCl buffer (0.1 M, pH 7.5) containing 1 M NaCl and 1 mM EDTA and then drained. For moderate to large quantities of gel (>2 ml) this washing is most conveniently performed in a sintered-glass funnel whose outlet is controlled with a stopcock. After seven repeated washing and draining cycles, all the activity in the product is found to be tightly bound to the Sepharose.

Dissociation to obtain immobilized subunits is achieved by washing seven times at room temperature with ten times its volume of 6 M guanidine hydrochloride in Tris-HCl buffer (0.1 M, pH 7.5, containing 1 mM EDTA and 14 mM mercaptoethanol). The above described procedure of suspension by overhead stirring in a sintered-glass funnel followed by draining is also suitable for this step. However, since the immobilized subunits are not stable to extended storage, it might often be advantageous to prepare small amounts of these subunits as needed. For example, 1 ml (packed volume) of immobilized aldolase may be dissociated by suspending in 5 ml of the guanidine HCl solution contained in a 15-ml centrifuge tube. A vortex mixer can be used to obtain good mixing, and the sides of the tube can then be washed down with a further 5 ml of the solution. After a few minutes the slurry is centrifuged and the supernatant is carefully removed with a Pasteur pipette. This procedure is performed seven times to ensure the complete removal of noncovalently bound subunits. No loss of gel particles during the operation is detectable. Renaturation of the immobilized subunit is then accomplished by washing several times with the above Tris-HCl buffer containing EDTA and mercaptoethanol but no guanidine hydrochloride, using the same washing procedure. All immobilized aldolase derivatives are stored conveniently as 1:5 (v/v) suspensions in sodium phosphate buffer (10 mM, pH 7.0) containing 1 mM EDTA.

Preparation of Immobilized Subunits of Yeast Transaldolase

The procedure is similar to the above method for preparing immobilized subunits of aldolase except that 5 mg of CNBr per milliliter of Sepharose is used for activation, and 0.5 M bicarbonate solution, pH 8.5, is used for coupling of the protein. The buffers for washing and for storage of the immobilized transaldolase derivatives contain triethanolamine–EDTA, pH 7.6, instead of Tris-HCl, pH 7.5.

Assay of Immobilized Derivatives of Aldolase

The coupled spectrophotometric assay for aldolase[8] is adapted for immobilized derivatives by stirring the assay mixture in the cuvette.[9] No special equipment is necessary. The mixture contains triethanolamine–hydrochloride buffer (40 mM, pH 7.4), EDTA (1 mM), fructose 1,6-biphosphate (1 mM), NADH (0.1 mM), triosephosphate isomerase (1 μg/ml), and α-glycerophosphate dehydrogenase (10 μg/ml). A convenient aliquot (e.g., 50 μl) is withdrawn from the standard suspension of the Sepharose derivative containing immobilized aldolase using an automatic pipette (either of the fixed-volume type, such as those made by Eppendorf, or of the adjustable type, such as those made by Gilson). To ensure uniformity, the standard suspension is kept well stirred while the aliquot is removed. The plastic tip used on the automatic pipette also needs to be cut to enlarge the opening so that the entry of gel particles is not hindered. If spectrophotometers with built-in magnetic stirring motors are available, a continuous assay is easily carried out using 2 ml of assay mixture in an ordinary 1-cm² cuvette. The spherical stirring magnets (about 0.7 cm in diameter) from Beckmann Instrument Co. are particularly suitable. At the wavelength of 340 nm used for the assay, the effect of light scattering by the suspended gel particles is negligible. Recording spectrophotometers without stirring facilities are, however, perfectly adequate, since the cuvette containing the reaction mixture and the magnet may be transferred at intervals (e.g., of 1 min) by hand from its position on an ordinary magnetic stirring motor to the sample chamber of the spectrophotometer. The gel particles remain well suspended during the few seconds necessary to obtain a recording trace. If the recorder is allowed to run continuously, several such traces will allow the linear rate to be determined with ease. Using a thermostat-jacketed cuvette holder mounted on a magnetic stirring motor, it has been possible to assay eight cuvettes at the same time by putting each cuvette into the spectrophotometer in turn.

Assay of Immobilized Derivatives of Transaldolase

The method is based on the coupled spectrophotometric assay described by Tchola and Horecker.[10] The assay mixture (final volume 2.0 ml) contains triethanolamine–HCl (pH 7.6, 40 mM), EDTA (10 mM), erythrose 4-phosphate (0.2 mM), fructose 6-phosphate (2.8 mM), NADH

[8] E. Racker, *J. Biol. Chem.* **167**, 843 (1947).
[9] J. S. Mort, D. K. K. Chong, and W. W.-C. Chan, *Anal. Biochem.* **52**, 162 (1973).
[10] O. Tchola and B. L. Horecker, this series Vol. 9, p. 499 (1966).

(0.1 mM), triosephosphate isomerase (1 µg/ml), α-glycerophosphate dehydrogenase (10 µg/ml). The above modifications for immobilized derivatives also apply to the transaldolase assay.

Criteria for the Presence of Immobilized Subunits

From the method of preparation, one may assume that immobilized subunits are present in the final product. However, additional evidence would be highly desirable and there are several ways for obtaining such evidence. These criteria for the presence of immobilized subunits are summarized in Table I.

TABLE I
CRITERIA FOR THE PRESENCE OF IMMOBILIZED SUBUNITS

Test criterion	Type of immobilized subunits to which test is applicable	Expected result if subunits are present
Protein content	Active or Inactive	$= 1/n \times$ protein content of immobilized oligomer (n = number of subunits)
Stability to denaturants	Active	Inactivation occurring at a lower concentration of denaturant
Stability to heat or proteolysis	Active	Inactivation may occur at a different rate
Extent of chemical modification[a]	Active or Inactive	Possible modification of some previously "buried" residues
Rate of chemical modification	Active	Inactivation may occur at a different rate
Interaction with soluble subunits	Active or Inactive	Increase in activity up to plateau level close to original activity of immobilized oligomer; properties of product revert to those of the oligomeric form
Interaction with soluble subunits inactivated at the active site	Active	No change in activity if subunits act independently, properties of product revert to those of the oligomeric form
	Inactive without subunit interaction	Appearance of activity
	Inactive owing to modification of essential residue during immobilization	No appearance of activity

[a] Can be determined only when immobilization is reversible and the modified product can be subsequently removed from the carrier.

Protein Content

In the first place, the protein content of the immobilized oligomer can be compared with that of the immobilized subunit prepared from it. For this purpose the Sepharose may be hydrolyzed in 6 N HCl and then analyzed for amino acids. From the values of the more stable amino acids and the amino acid composition of the protein, the protein content can be calculated. Table II shows values for some typical preparations. It can be seen that increasing the pH or the CNBr:Sepharose ratio both increase the amount of protein bound. However, the protein content of the immobilized monomer approaches the theoretical value (25% of that of the immobilized tetrameric aldolase) only when the CNBr:Sepharose ratio used for coupling the oligomer is low. The coupling yield tends to vary substantially depending on the nature of the protein, and in the case of yeast transaldolase only 25.5 µg/ml was bound even when 5 mg of CNBr per milliliter of Sepharose was used and coupling was done at pH 8.5. Since Sepharose itself contains a small amount of amino acids (e.g., up to 3 nmoles of alanine per milliliter of packed gel), protein contents of less than 10 µg/ml cannot be determined with sufficient accuracy. Thus, for transaldolase it is not possible to reduce the CNBr:Sepharose ratio further, and therefore it has been necessary to work with a preparation of immobilized monomers whose protein content (64.6%) departs significantly from the theoretical (50% of the value for immobilized dimeric transaldolase).

TABLE II
EFFECT OF COUPLING CONDITIONS ON PROTEIN CONTENT OF IMMOBILIZED DERIVATIVES[a]

Enzyme	Amount of CNBr for activation (mg/ml Sepharose)	pH of coupling	Protein content			
			Immobilized oligomer		Immobilized monomer	
			µg/ml	%	µg/ml	%
Rabbit muscle aldolase (tetramer)	1.0	8.0	96.8	100	24.0	24.8
	2.5	8.0	166	100	45.7	27.5
	5.0	8.0	206	100	63	30.6
	1.0	9.0	810	100	218	26.9
Yeast transaldolase (dimer)	5.0	8.5	25.5	100	16.5	64.6

[a] Data reproduced from references 14 through 16 with permission of the publishers.

Differences in Stability

In cases where the immobilized subunits are active, a highly desirable criterion is to show that the bound subunits have properties distinct from those of the bound oligomer. Since the concentration of bound enzyme in these derivatives is limited by the need to employ low density of coupling points, the only available experimental tool is activity measurements. In spite of this limitation, a number of valuable tests are possible. In the author's opinion, the threshold concentration for inactivation by a general protein denaturant such as urea is likely to be widely applicable in distinguishing between the oligomeric and monomeric forms of the same enzyme.

Urea appears to be particularly useful in this respect, since it is hydrophilic and nonionic and therefore often has little effect on activity until a critical concentration is reached, at which unfolding of the polypeptide begins. The unfolding process requires the simultaneous binding of a number of urea molecules and is consequently highly cooperative. Thus inactivation may often occur rather sharply at a critical concentration of urea. When an enzyme normally exists in an oligomeric form, the interactions at the subunit surface must contribute to the stability of the whole structure. Therefore a higher concentration of urea is likely to be required for the inactivation of the oligomer than for that of the monomer. The presence in the monomer of additional exposed surface on which more urea molecules will bind, should probably also contribute to its greater instability.

For the tetrameric rabbit muscle aldolase, the urea concentration for half-inactivation of the immobilized monomer is 1 to 1.5 M lower than the corresponding value for the immobilized tetramer. Thus it is possible to choose an intermediate urea concentration (e.g., 2.3 M) at which the oligomer is essentially unaffected while the immobilized monomer is largely inactivated (Table III). This preferential inactivation of the aldolase monomer was also found for aldolase monomers in solution formed as intermediates during renaturation[11] and for monomers stabilized by chemical treatment during renaturation.[12] The correlation of properties of immobilized monomers with those of monomers in solution provides good indication that these properties are not unduly affected by coupling to the Sepharose matrix. However, minor differences between immobilized and soluble monomers have been observed. For example, the immobilized monomer of aldolase retains in 2.3 M urea 22% of its normal activity

[11] W. W.-C. Chan, J. S. Mort, D. K. K. Chong, and P. D. M. Macdonald, *J. Biol. Chem.* **248**, 2778 (1973).

[12] W. W.-C. Chan, C. Kaiser, J. M. Salvo, and G. R. Lawford, *J. Mol. Biol.* **87**, 847 (1974).

TABLE III
DIFFERENCES IN PROPERTIES BETWEEN IMMOBILIZED OLIGOMER AND MONOMER[a]

Enzyme	Condition for distinction between oligomer/monomer	Ratio, activity under special conditions /activity in normal assay			
		Immobilized oligomer	Immobilized monomer	Soluble oligomer	Soluble monomer
Rabbit muscle aldolase	2.3 M Urea	0.90	0.22	0.85	0
	pH 10	1.03	0.20	0.90	0
	0.45 Guanidine HCl	0.62	0.20	0.50	NT[b]
Yeast transaldolase	1.5 M urea	0.80	0.15	0.75	—
	pH 9	0.68	0.10	0.52	—

[a] Data reproduced from references 14 through 16 with permission of the publishers.
[b] NT, not tested.

whereas soluble monomers[11,12] are known to be completely inactivated at this urea concentration. This residual activity in 2.3 M urea is likely to reflect significant heterogeneity in the immobilized monomer preparation. Thus, depending on the CNBr:Sepharose ratio, some dimers might arise from molecules originally attached to the matrix via two adjacent subunits. It is interesting that immediately after preparation, immobilized subunits of aldolase are completely inactivated by 2.3 M urea, and the partial loss of urea sensitivity develops only gradually.[13] Green and Toms[2] have reported changes in the stability characteristics of avidin subunits upon incubation at 37° and have concluded that the matrix may be sufficiently flexible to permit some reassociation.

Although some interaction to form dimers appears entirely plausible, it is inconceivable to the present author that the precise alignment of subunits necessary for reassociation into a tetramer can take place from four subunits originally attached to widely separated parts of the matrix. The gradual partial conversion into a more stable form need not necessarily be attributed to reassociation but may be explained by noncovalent interaction between the backbone of Sepharose and the amino acid side chains on the surface of the subunits. Thus some subunits may be visualized as gradually settling into a pocket formed by the movement of the Sepharose matrix. The additional noncovalent linkages should confer the same kind of stabilization as the establishment of the subunit contacts but should not require stereospecific and highly precise alignment.

Immobilized monomers of yeast transaldolase are also more sensitive

[13] W. W.-C. Chan, unpublished results.

to urea[14] with an inactivation threshold occurring at a urea concentration some 2 M lower than that for the immobilized oligomer (Table III). Another means of distinction between the oligomer and monomer forms of these two enzymes is the threshold for inactivation by increasing pH. Inactivation of the immobilized monomer occurs at pH 9 for yeast transaldolase[14] and at pH 10 for muscle aldolase,[15] respectively (Table III).

Specific Interaction with Soluble Subunits Generated in Situ

Another valuable criterion for the existence of immobilized monomers is the specific ability of the latter to pick up subunits in solution. For this purpose, a sample of the native enzyme is first denatured and dissociated in 6 M guanidinium chloride. A very small aliquot of this denatured enzyme is then diluted into a much larger volume of buffer containing the immobilized subunit. It can then be expected that, as the guanidinium chloride is diluted out, the denatured enzyme refolds in the first instance to give monomers in solution. These soluble monomers can reassociate either with the immobilized monomers or among themselves. In order to increase the probability of the former process, the denatured enzyme can be added in small amounts with a sufficient time interval (5–10 min) between each addition. It should be noted that the immobilized subunit remains under nondenaturing conditions during this process. It is therefore possible to show by control experiments that if the immobilized oligomer is used instead of immobilized monomers, no increase in activity occurs. It is also possible to demonstrate that the immobilized monomer is unable to pick up native oligomeric enzyme added to it.

The highly specific nature of the interaction between the Sepharose derivative being tested and soluble monomers generated *in situ* strongly suggests that immobilized monomers are indeed present in the derivative. Such a demonstration has been realized in the case of both rabbit muscle aldolase and yeast transaldolase.[14,16] Additional evidence for the specific nature of interaction between immobilized monomers and soluble monomers comes from the fact that the amount of bound activity increases in the process, reaching a plateau level that is not substantially lower than the original activity of the immobilized oligomeric form. This result indicates that each of the immobilized monomers has now regained the required number of subunits to regenerate the original quaternary structure. The properties (e.g., urea sensitivity) of this immobilized renatured en-

[14] W. W.-C. Chan, H. Schutt, and K. Brand, *Eur. J. Biochem.* **40**, 533 (1973).
[15] W. W.-C. Chan and H. M. Mawer, *Arch. Biochem. Biophys.* **149**, 136 (1972).
[16] W. W.-C. Chan, *Can. J. Biochem.* **51**, 1240 (1973).

zyme should now revert back to those of the original oligomer. This was in fact observed for muscle aldolase and for yeast transaldolase.

Interaction with Soluble Subunits Inactivated at the Active Site

If the immobilized subunits are intrinsically inactive, then their presence obviously cannot be demonstrated with methods based on differential inactivation (e.g., by increasing concentrations of urea). In this case, the above test for the ability of the immobilized subunits to pick up soluble subunits is of crucial importance. Although the regain of activity resulting from such an interaction indicates the presence of immobilized subunits, it does not demonstrate unequivocally that subunit interactions are essential to catalytic activity. It is possible that the immobilized subunits are inactive because, in their coupling to the Sepharose matrix, an amino acid side chain essential for activity has been modified. This explanation is unlikely if CNBr-activated Sepharose is used for coupling, since binding could occur relatively nonspecifically with one of the many lysine side chains on the surface of the proteins. However, this possibility can be ruled out by interaction experiments using soluble subunits which had been previously inactivated by modification at the active site. If the addition of inactive modified subunits (produced by renaturation of the denatured modified enzyme) to the inactive immobilized subunits leads to significant bound activity, then one must conclude that the immobilized subunits were previously inactive because subunit interactions were absent rather than because an essential group was modified by coupling to Sepharose. An important test of this type was carried out in the case of the inactive immobilized monomers of glycogen phosphorylase.[17]

Scope and Limitations of the Approach

The techniques described above are generally capable of deciding whether the monomeric form of an enzyme is catalytically active. This question is of some interest in view of the widespread occurrence of oligomeric enzymes with identical subunits. Many such enzymes display strictly Michaelis–Menten kinetics indicating no cooperativity among the active sites. The advantage conferred by the quaternary structure is not immediately obvious in these cases.

If the immobilized subunit turns out to be active, then considerable information concerning the effects of subunit interactions can be obtained.

[17] K. Feldman, H. Zeisel, and E. Helmreich, *Proc. Natl. Acad. Sci. U.S.A.* **69**, 2278 (1972).

The stabilizing effect (e.g., toward urea) has already been discussed and the change in pK of an active site residue has been inferred from the results on transaldolase.[14] Possible differences between the oligomeric and the subunit forms in their rate of inactivation by chemical modification might be particularly interesting for enzymes (e.g., rabbit muscle glyceraldehyde-3-phosphate dehydrogenase) for which half-of-the-sites reactivity has been found[18] and for which an active subunit form (in this case, the dimer) can be prepared.[19] If the oligomeric enzyme is known to display cooperativity in substrate binding, the study of immobilized monomers (if these are active) could demonstrate the direct effects of subunit interactions in cooperative behavior. In cases where an activator or inhibitor is known to cause dissociation or aggregation, the use of immobilized derivatives should make it possible to distinguish whether the change in activity is due to the change in quaternary structure or whether they are simply two concurrent but independent effects of the ligand.

So far we have considered only enzymes with identical subunits (ignoring for the moment the change from Asn to Asp in the C-terminal region of aldolase, which gives rise to the α and β subunits). For enzymes containing nonidentical subunits, the different subunits may possibly be separated under dissociating conditions, and each type of subunit can be tested for activity. However, if the assay conditions are different from those used for dissociation, subunits of each kind may reassociate among themselves. For example, the β subunit of tryptophan synthetase when separated from the α subunit tends to exist as β_2, and thus the question of whether β itself is active can be answered only with special techniques, such as immobilized subunits. Derivatives containing only part of the native quaternary structure (e.g., an immobilized $\alpha\beta$ dimer of the normally tetrameric $\alpha_2\beta_2$ enzyme) might also be studied using immobilization techniques here. A complex of the nonidentical subunits of aspartate transcarbamylase has been studied with immobilized derivatives.[20]

One of the chief limitations of the approach is that few physicochemical studies of the immobilized monomers can be made. The particulate state of the bound enzyme precludes any examination of its hydrodynamic properties. Spectroscopic techniques are handicapped by the light-scattering effects of the gel particles and also by the very low protein content in these derivatives. (Spectroscopic methods of assays, however, are not affected significantly by light scattering, owing to the very low amounts of the immobilized derivative needed.) Another experimental problem in the method is the somewhat heterogeneous nature of the immobilized mol-

[18] A. Levitzki, *J. Mol. Biol.* **90**, 451 (1974).
[19] N. K. Nagradova, T. O. Golovina, and A. T. Mevkh, *FEBS Lett.* **49**, 242 (1974).
[20] W. W.-C. Chan, *FEBS Lett.* **44**, 178 (1974).

ecules. The coupling points on the matrix do not have identical microenvironments, and the attachment may occur with one of many possible side chains on the protein surface. This heterogeneity is compounded by small amounts of contamination with dimers (depending on the CNBr:Sepharose ratio) and by the lack of complete rigidity of the Sepharose matrix. However, the correlation of results from studies in solution with those from immobilized derivatives in the case of aldolase indicates that reliable conclusions regarding subunit interactions can be made with immobilized derivatives.

Desirable Future Developments

Modification of the techniques used here to overcome some of the above limitations would seem to be highly desirable. If the covalent linkage to the matrix (e.g., via an —S—S— bridge) can be subsequently cleaved under mild conditions, then the enzyme can be detached at an appropriate stage for characterization in solution. If the immobilized subunit is first allowed to pick up soluble subunits with different net charge (from a different isozyme or by modification, e.g., with succinic anhydride), the hybride can then be detached and identified electrophoretically or by ion-exchange chromatography. This would provide a direct demmonstration of the presence of immobilized monomers. Residues on the subunit contact surface may be modified to produce subunits, which are therefore prevented from reassociation when detached from the matrix. Modification of these residues, which are normally buried at the intersubunit surface, may allow their subsequent identification and location in the amino acid sequence, thus providing structural information. These possible developments have been discussed in greater detail elsewhere.[21] The study of immobilized subunits would also greatly benefit from the development of a more rigid matrix material, where coupling may occur at sites of identical microenvironment.

---

[21] W. W.-C. Chan, P. Davies, and K. Mosbach, in "Ion-Selective Electrodes and Enzyme Electrodes in Biology and Medicine" (M. Kessler, ed.). Urban & Schwarzenberg, Berlin, in press.

[34] Covalently Bound Glutamate Dehydrogenase for Studies of Subunit Association and Allosteric Regulation[1]

By Harold E. Swaisgood, H. Robert Horton, and Klaus Mosbach

The possession of quaternary structure has been a characteristic feature of nearly all enzymes identified to date which exhibit regulation by allosteric mechanisms. Several models for allosteric regulation (control of enzymic activity by metabolites which are neither substrates nor products of the particular reaction being catalyzed) have been proposed: viz., the concerted model of Monod et al.,[2] the ligand-induced conformational model of Koshland and co-workers,[3-5] and a model by Weber[6] based on free-energy conservation. These models attribute cooperative binding to some form of interaction between subunits, such that the presence of ligand in an adjacent protomer is "recognized" by the neighboring protomer. It has also been noted that subunit interactions may not be an absolute requirement for cooperativity.[7] In addition, regulation of enzymic activity through coupling of ligand-binding to association-dissociation reactions of oligomeric enzymes is also possible.[8,9]

The technique of enzyme immobilization by covalent attachment to a surface poses a means for investigating various perturbations of the quaternary structure of oligomeric enzymes. Covalent attachment of a single protomer per molecule of oligomeric enzyme could allow experimental separation of subunits, and thus could provide a means for directly measuring the regain of secondary and tertiary structures within protomers in the absence of complicating reassociation reactions. In principle, one could examine the reassociation process, alone, by incubating matrix-bound protomers with equilibrium mixtures of native oligomeric enzyme. (Solution studies have relied on kinetic analysis of changes in optical rotation, light scattering, and enzymic activities to distinguish these pro-

---

[1] Research carried out at North Carolina State University and Avdelningen för Biokemi, Kemicentrum, Lunds Universitet, Sweden and supported in part by National Science Foundation Grant GI-39208.
[2] J. Monod, J. Wyman, and J.-P. Changeux, *J. Mol. Biol.* **12**, 88 (1965).
[3] D. E. Koshland, Jr., G. Némethy, and D. Filmer, *Biochemistry* **5**, 365 (1966).
[4] D. E. Koshland, Jr., *in* "The Enzymes" (P. D. Boyer, ed.), 3rd ed., Vol. 1, p. 341. Academic Press, New York, 1970.
[5] A. Cornish-Bowden and D. E. Koshland, Jr., *J. Biol. Chem.* **245**, 6241 (1970).
[6] G. Weber, *Biochemistry* **11**, 864 (1972).
[7] G. Weber and S. R. Anderson, *Biochemistry* **4**, 1942 (1965).
[8] C. Frieden and R. F. Colman, *J. Biol. Chem.* **242**, 1705 (1967).
[9] A. Levitzki and D. E. Koshland, Jr., *Biochemistry* **11**, 247 (1972).

cesses.[10]) Moreover, by using immobilization techniques to prevent the formation of kinetically trapped, low free-energy aggregates which can result from interchain associations (see this volume [35]), it might be possible to examine such reactions under equilibrium conditions, as well as under essentially irreversible conditions, as used for kinetic analysis. Such studies have been possible for certain single-chained proteins[11] in solution, but not for oligomeric enzymes.[10]

If one considers the inclusion of a labile bond in the chains that link enzyme molecules to a surface, then various procedures can be envisioned for investigating the acquisition of quaternary structure, such as reassociation of immobilized protomers with chemically or radioactively labeled soluble protomers, followed by selective cleavage of the labile bond. Such experiments could provide a useful alternative to hybridization studies that employ native and chemically modified enzyme species,[12] since, in the latter approach, enzymes must be fairly extensively modified and yet must possess a quaternary structure similar to that of the native protein.

Immobilization of an allosteric enzyme molecule by covalent attachment to an insoluble matrix through a single protomer, followed by dissociation of subunits so as to yield the immobilized protomer, could also provide a means for distinguishing between properties which result from an altered affinity for an effector molecule or ligand *per se* and properties that result from subunit interactions that occur during the binding of an effector molecule or ligand. Such a technique could also offer a direct means of testing for allosteric regulation and cooperativity within a single protomer.

Many oligomeric enzymes are associated with cellular superstructures *in vivo*; hence, studies of their kinetic and regulatory properties in homogeneous solution may, in fact, yield results that are significantly different from those manifested *in situ*.[13] Enzyme immobilization provides a means for investigating such properties with enzymes attached to well defined surfaces or included in well defined synthetic membranes (cf. Mattiasson and Mosbach[14] and Thomas *et al.*[15]). In instances where ligand binding is accompanied by association or dissociation of an oligomeric enzyme, immobilization offers a means for "uncoupling" these interactions so that those changes associated with ligand binding can be assessed independently from those resulting from protein subunit association-dissociation.

[10] J. W. Teipel, *Biochemistry* **11**, 4100 (1972).
[11] C. Tanford, *Adv. Protein Chem.* **23**, 121 (1968).
[12] E. A. Meighen and H. K. Schachman, *Biochemistry* **9**, 1163 (1970).
[13] P. A. Srere and K. Mosbach, *Annu. Rev. Microbiol.* **28**, 61 (1974).
[14] B. Mattiasson and K. Mosbach, *Biochim. Biophys. Acta* **235**, 253 (1971).
[15] D. Thomas, C. Bourdillon, G. Broun, and J. P. Kernevez, *Biochemistry* **13**, 2995 (1974).

Owing to the infancy of these procedures, many of the studies suggested above are as yet incomplete; however, some results have been obtained with two oligomeric dehydrogenases—glutamate dehydrogenase and lactate dehydrogenase.

Studies with Glutamate Dehydrogenase

Chemical Attachment. Bovine liver glutamate dehydrogenase [EC 1.4.1.3, L-glutamate:NAD$^+$(P) oxidoreductase (deaminating)] was coupled covalently to porous succinamidopropyl-glass beads. Chemical linkage was obtained by succinylating amino groups of γ-aminopropyl-silanized glass beads, then activating the resulting matrix-bound carboxyl groups with a water-soluble carbodiimide, washing the beads free of excess reagent, and finally, exposing the activated beads to a solution of the enzyme. Two lots of porous glass beads (80–120 mesh, 654 Å mean pore diameter; and 200–400 mesh, 705 Å mean pore diameter) obtained from Electro-nucleonics (Fairfield, New Jersey) were derivatized for 3 hr at 70° (pH 4) with a 10% aqueous solution of 3-aminopropyltriethoxysilane. The solution was removed, and the beads were heated overnight at 125°. They were then washed with acetone and water to remove excess silane. The aminopropylsilanized beads were then succinylated with succinic anhydride at pH 6, yielding roughly 260 μmoles of carboxyl groups per gram in the case of the 200–400 mesh beads. Assuming an even distribution of sites and a surface area of 36 m^2/g, each site would be enclosed by a square 5 Å by 5 Å. The number of activated sites would depend on the method of activation.

The immobilized carboxyl groups were activated by treatment with 1-ethyl-3-dimethylaminopropyl carbodiimide (EDC)[16] to form reactive O-acylisourea derivatives. The amount of enzyme attached, and possibly the number of linkages per enzyme molecule, can be controlled by adjusting the carbodiimide concentration relative to the enzyme concentration. Both attrition of the glass beads and surface denaturation of enzyme can be minimized by placing the derivatized beads in a column and recycling the reaction mixtures through the column with a peristaltic pump. A typical immobilization procedure is as follows:

1. Degassed succinamidopropyl-glass beads (0.5 g) are equilibrated with 5 ml of 0.2 M NaH$_2$PO$_4$ adjusted to pH 4.75 (all solutions, including the buffers, should be filtered through 0.2-μm pore diameter filters before their addition to the beads). The beads are placed in a jacketed, 6-mm

[16] Abbreviations used are EDC, 1-ethyl-3-dimethylaminopropyl carbodiimide; EDTA, ethylenediamine tetraacetate; GDH, glutamate dehydrogenase; LDH, lactate dehydrogenase.

diameter column, the excess buffer is placed in a reservoir-mixing flask, and the column and flask are fitted with tubing for recycling with a peristaltic pump.

2. After the thermostatic column is equilibrated at 25°, sufficient EDC is added to the mixing flask to provide a concentration of 0.1 M. This solution is then recycled through the bead column at a flow rate of 10–20 ml/min for 30 min.

3. The reaction mixture in the reservoir is then replaced with 200 ml of ice-cold 0.1 M sodium phosphate (pH 7.0), ice-cold water is passed through the column jacket, and the beads are rapidly washed free of excess carbodiimide by pumping through the phosphate buffer.

4. Excess buffer is removed from the column, and 5 ml of a solution of enzyme (1 mg/ml) in pH 7 sodium phosphate are placed in the reservoir and recycled through the column for 24 hr at 4°.

Glutamate dehydrogenase is a hexameric enzyme composed of identical protomers whose primary structure has been established.[17,18] The 6-chain structure is commonly referred to as the "monomer,"[17] and this convention will be followed herein. Protomer interactions are sufficiently strong that oligomers smaller than the monomer (336,000 daltons) have not been observed except in the presence of strong protein solvents.[19] On the other hand, the monomer readily associates to form linear aggregates,[20] and this equilibrium is perturbed in specific directions by the various allosteric effectors of the enzyme.[17,20] For example, in the presence of coenzyme, GTP perturbs the equilibrium in favor of monomer, whereas ADP shifts the equilibrium toward polymeric species. Hence, immobilization of this enzyme could allow examination of the effect of allosteric ligands independent of association-dissociation interactions.

Various studies of immobilized GDH have been reported.[21,22] In contrast to the method of immobilization described herein, that reported by Julliard et al.[21] (coupling to formaldehyde-"tanned" collagen) did not result in an active immobilized enzyme preparation except when "protectors" such as GTP or ADP were included in the reaction mixture. Furthermore, the resulting immobilized preparation did not exhibit Michaelis–Menten kinetics or a pH-activity profile like that of native enzyme. By

[17] B. R. Goldin and C. Frieden, *Curr. Top. Cell. Regul.* **4**, 77 (1971).
[18] T. J. Langley and E. L. Smith, *J. Biol. Chem.* **246**, 3789 (1971).
[19] M. Cassman and H. K. Schachman, *Biochemistry* **10**, 1015 (1971).
[20] R. Josephs, H. Eisenberg, and E. Reisler, *in* "23d Colloquium der Gesellschaft für Biologische Chemie, 13–15 April 1972, Mosbach/Baden" (R. Jaenicke and E. Helmreich, eds). Springer-Verlag, Berlin, Heidelberg, and New York, 1972.
[21] J. H. Julliard, C. Godinot, and D. C. Gautheron, *FEBS Lett.* **14**, 185 (1971).
[22] H. R. Horton, H. E. Swaisgood, and K. Mosbach, *Biochem. Biophys. Res. Commun.* **61**, 1118 (1974).

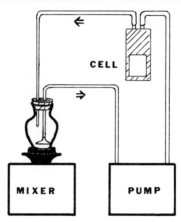

Fig. 1. Diagrammatic illustration of the system designed for assaying enzymic activity. The "scoop" inserted into the solutions incorporates a nylon mesh which excludes the glass beads [K. Mosbach and B. Mattiasson, *Acta Chem. Scand.* **24**, 2093 (1970)]. A 25-ml pear-shaped flask was used as the reaction vessel.

contrast, the pH-activity profile of glass bead-immobilized enzyme was indistinguishable from that of soluble enzyme.[22]

Binding of Coenzymes to Immobilized Glutamate Dehydrogenase. All kinetic assays of immobilized GDH were obtained from initial rate measurements performed with a system that incorporated the methods of Mosbach and Mattiasson[23] and Widmer et al.[24] (see also this volume [25] on assay procedures). This system, illustrated in Fig. 1, provided measurable rates identical to those obtained using the conventional cuvette assay in tests with soluble GDH and, owing to the limited abrasion of the glass beads, allowed numerous assays to be performed on the same beads. Measured activities were independent of pumping rates or vortex-mixing speeds (within the limits selected). Standard assays were conducted using 10 ml of total solution volume consisting of sodium phosphate (ionic strength 0.1, pH 7.8), 0.1 mM EDTA, 100 μM NADH, 50 mM NH$_4$Cl, and 5 mM α-ketoglutarate (the latter substrate added last, to initiate the enzyme-catalyzed reaction). Volume-specific activities, as used herein, are defined as units of activity per unit of settled bead volume of immobilized enzyme.

Whereas NADPH binds only to the active site of GDH, NADH binds to both the active site and, with lower affinity, to a second, regulatory site.[17,25] Binding to the latter site causes enzyme inhibition; however, such inhibition is not evident in solutions of higher enzyme concentration

[23] K. Mosbach and B. Mattiasson, *Acta Chem. Scand.* **24**, 2093 (1970).
[24] F. Widmer, J. E. Dixon, and N. O. Kaplan, *Anal. Biochem.* **55**, 282 (1973).
[25] C. Frieden, *J. Biol. Chem.* **238**, 3286 (1963).

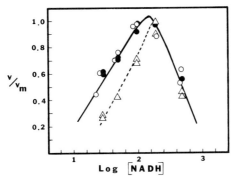

FIG. 2. Relative activities of immobilized and soluble GDH using NADH as the coenzyme. ○, Data of C. Frieden [*J. Biol. Chem.* **234**, 809 (1959)] for soluble GDH; ●, the authors' results for the same lot number of soluble GDH as that used to obtain the immobilized preparation; △, values for immobilized GDH.

(>1 mg/ml) where the enzyme is aggregated.[17] Thus, it was of interest to determine whether surface-bound GDH monomers exhibited a similar NADH-inhibition pattern, even though the local enzyme concentration was greater than 1 mg/ml. (Calculated local enzyme concentrations ranged from 5 to 15 mg/ml, depending upon whether pore volume or total void volume was used for the calculations.) Figure 2 shows the results, in comparison with those obtained using soluble enzyme. Binding of NADH to the regulatory site, resulting in inhibition, was found to occur in the case of the immobilized enzyme, in spite of the apparently high local enzyme concentrations. The small differences between the data from soluble and immobilized enzyme preparations, seen in the left-hand portion of the curve in Fig. 2, may be the result of diffusional effects at these low NADH concentrations or may reflect an actual change in the Michaelis constant resulting from immobilization.

Simple Michaelis–Menten kinetics are observed when NADPH is the coenzyme for soluble GDH preparations.[26] Likewise, a hyperbolic substrate-saturation curve over a 50-fold NADPH concentration range was observed for GDH immobilized on beads of both the large and small diameters. These data are plotted according to the Hofstee–Eadie method in Fig. 3, since such plots are more sensitive to diffusional effects than are Lineweaver–Burk plots.[27] The apparent Michaelis constant obtained for

[26] C. Frieden, *J. Biol. Chem.* **234**, 809 (1959).
[27] J.-M. Engasser and C. Horvath, *J. Theor. Biol.* **42**, 137 (1973). Equation (18) of this reference is apparently in error. Comparison of their theoretical development of these equations with that of Kobayashi and Laidler [see their Eq. (8)] indicates that the correct relationship should be

$$\epsilon(\phi) = 1/\phi[(1/\tanh 3\phi) - (1/3\phi)]$$

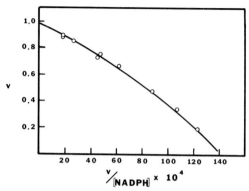

FIG. 3. Hofstee–Eadie plot of initial velocities of NADPH oxidation catalyzed by GDH immobilized on 200–400 mesh porous glass beads. All measurements were performed at 25° in pH 7.8, 0.1 ionic strength sodium phosphate containing 0.1 mM EDTA, 50 mM NH$_4$Cl, and 5 mM α-ketoglutarate as the other substrates. The slight curvature was most likely caused by diffusional effects.

NADPH was corrected for diffusional effects both by the methods of Engasser and Horvath[27] and of Kobayashi and Laidler.[28] These small corrections gave K_m (NADPH) values of 30 μM and 40–50 μM by the former and latter methods, respectively, which are in agreement with that obtained for the same lot of soluble GDH (30 μM) and with those given by Frieden[25,26] (15–35 μM). Thus, the binding of NADPH to glutamate dehydrogenase does not seem to be greatly affected by its immobilization to glass beads.

Denaturation in the Presence of Urea. The similarity of the structural characteristics of glass bead-immobilized GDH to those of soluble enzyme is further implied by comparison of the inactivation of each in solutions of urea. Values for the initial rates of NADH oxidation, catalyzed by soluble GDH (reactions initiated by addition of enzyme), are shown in Fig. 4 as a function of urea concentration. Results obtained with immobilized enzyme are shown in the same figure; however, in this case, the *same* enzyme molecules (i.e., beads) were used for each subsequent measurement, and reactions were initiated by addition of α-ketoglutarate. These data show that the GDH activities of both immobilized and soluble enzyme preparations are lost at urea concentrations between 1 and 2 M. Furthermore, in 1.5 M urea, the initial exposure of immobilized enzyme caused a 50% loss of activity, and a second exposure of the same beads (after washing with buffer and assaying in the absence of urea) caused an additional 50% loss of activity. Since these assays involved a uniform

---

[28] T. Kobayashi and K. J. Laidler, *Biochim. Biophys. Acta* **302**, 1 (1973).

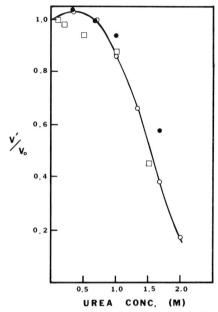

FIG. 4. Effect of varying urea concentrations on the relative velocities of NADH oxidation by soluble and immobilized GDH. All assays were performed using the standard assay method. ●, Soluble GDH in the presence of 0.26 mg/ml bovine serum albumin; ○, soluble GDH in the absence of bovine serum albumin; □, immobilized GDH.

exposure time per assay, the loss in activity appeared to be a function of the time of exposure of the immobilized enzyme to urea in the presence of substrates.

That this is also the case for soluble enzyme was illustrated by experiments at urea concentrations of 1.67 and 1.33 M in which (a) reaction was initiated by addition of enzyme, and rates were determined initially and after 1 and 2 min of reaction; and (b) reaction was initiated with α-ketoglutarate after various periods (1–6 min) of preincubation of soluble enzyme with urea, NH_4^+, and NADH, and rates were determined initially and after 1 and 2 min of reaction. In each case, the spectrophotometric traces were markedly curvilinear, indicating that activity losses had occurred during catalysis, such losses being more rapid in the presence of all substrates than in the presence of only two.

Incubation of either immobilized or soluble enzyme in urea, alone, at these concentrations did not result in substantial losses of activity. For example, incubation of soluble GDH in 2 M urea for 30 min at 25° led to a loss of only 12% of the initial activity. Such findings clearly point to increased sensitivity to urea-induced denaturation of both forms of the enzyme in the presence of substrates, apparently reflecting aspects of conformational flexibility which accompany enzymic catalysis.

Regulation of Immobilized Glutamate Dehydrogenase Activity by Allosteric Effectors. Binding of allosteric effectors to soluble GDH in the presence of coenzyme perturbs the equilibrium between monomer and polymeric forms.[17,20,25,26] Originally, it was believed that monomer was less active than polymer, since GTP, which inhibits glutamate oxidation, causes dissociation, whereas ADP, which activates the reaction, promotes aggregation. However, it is now recognized that the degree of association-dissociation *per se* does not affect the activity, although it has been suggested that ADP preferentially binds to the polymer whereas GTP preferentially binds to the monomer.[8] Further, it was suggested that the monomeric and polymeric forms could be identified as the two conformational forms proposed by Monod *et al.*,[2] which could also account for the cooperativity noted for GTP binding.[8] Based on later stopped-flow measurements, a model was developed predicting a coenzyme-induced isomerization prior to depolymerization induced by GTP.[29]

Immobilization of the enzyme in monomer form could permit measurements of coenzyme- and allosteric ligand-binding in the absence of perturbations caused by association and dissociation of subunits. This technique also provides a means for the direct determination of the reversibility of allosteric ligand-induced changes, since the same enzyme molecules can be alternately exposed to various solutions without altering the local enzyme concentration.

In the presence of coenzyme, both GTP and ADP bind to immobilized GDH, resulting in inhibition and activation, respectively, and further confirming the view that the degree of association of monomers does not directly affect the activity.[22] The reversibility of these effects on enzymic activity was demonstrated by the results of a sequence of assays using the same enzyme beads (see Fig. 5). After an initial treatment of freshly prepared immobilized glutamate dehydrogenase with coenzyme and GTP, the volume-specific activity as measured in standard assays was always 5–15% lower than that measured before exposure of the beads to GTP. Thereafter, subsequent treatment did not further lower the volume-specific activity, suggesting that the initial treatment had completely depolymerized the enzyme, leaving only monomer covalently attached to the beads. Hence, immobilized GDH preparations were routinely treated in this manner before continuing with other studies.

When GDH was immobilized in the presence of 50 μM NADH and

[29] C. Y. Huang and C. Frieden, *J. Biol. Chem.* **247**, 3638 (1972).

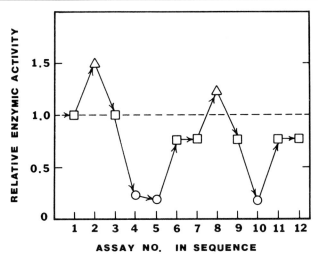

FIG. 5. Effects of ADP and GTP on activity of immobilized glutamate dehydrogenase at pH 7.75. Relative enzymic activity of 1.0 represents 1.93 μmoles of NADH oxidized per minute per gram of beads. After each consecutive assay, the beads were washed with three 10-ml volumes of sodium phosphate ($\Gamma/2 = 0.1$, pH 8.0) containing 0.1 mM EDTA; the beads were then immediately resuspended in the appropriate medium for the next assay. Assays 1–3: beads which had never been exposed to GTP were assayed in the absence and in the presence of ADP as shown; they were subsequently assayed in the presence of GTP (4 and 5), and then in the absence and in the presence of ADP and GTP, respectively. □, Standard assays; △, 75 μM ADP; ○, 20 μM GTP.

18 μM GTP (conditions for depolymerization), the resulting preparation exhibited an allosterically controlled pattern similar to that shown in Fig. 5, except that standard assay activities remained constant before and after exposure of the glass-bound enzyme to GTP.

Stimulation of activity by ADP was examined using 100 μM NADPH as coenzyme. With NADPH, the activity of soluble GDH increases approximately 4-fold at saturating concentrations of ADP[17]; however, in the case of immobilized enzyme, the increase observed was slightly less than 2-fold. The dissociation constants of the immobilized enzyme for ADP, evaluated kinetically,[25] were 10 μM in the presence of NADPH and 5–6 μM in its absence. These values are similar to those reported for soluble enzyme (10–20 μM in the presence of and 2–5 μM in the absence of coenzyme),[17,25] which suggests that binding of this effector may not be perturbed by association. However, such interpretation raises a question regarding the earlier conclusion that ADP binds more tightly to polymeric forms than to monomeric forms of glutamate dehydrogenase.

Studies with Lactate Dehydrogenase[30]

Reconstitution of Oligomeric Lactate Dehydrogenase. In many cases, sufficient dilution of oligomeric enzymes can result in their dissociation to provide an equilibrium mixture of the oligomeric form(s) and smaller subunits or protomers. If dilution *per se* is not sufficient to cause some dissociation, addition of agents that increase the solvent power at concentrations just sufficient to slightly "loosen" the quaternary structure may be necessary. Incubation of immobilized subunits with soluble enzyme in such "equilibrium solutions" can then result in reconstituted oligomeric enzyme on the matrix surface (see this volume [33]). It may be necessary to slowly alter conditions so as to favor subunit association and the attainment of quaternary structure during such incubation; however, in many cases, the local concentration of covalently immobilized enzyme will be greater than that in solution, thereby favoring formation of quaternary structures on the surface.

Rabbit muscle lactate dehydrogenase (LDH) [EC 1.1.1.27, L-lactate:NAD^+ oxidoreductase] has been shown to dissociate at concentrations below 1 mg/ml. Hence, immobilized subunits were incubated with a solution containing 0.1 mg of LDH per milliliter and 0.05 M dithiothreitol.[30] These immobilized subunits were prepared by treatment of immobilized tetrameric enzyme with 7 M guanidinium chloride containing 0.05 M dithiothreitol. Amino acid analyses indicated that three-fourths of the protein was removed by this treatment. After the incubation and washing to remove nonspecifically adsorbed enzyme, these reconstituted preparations exhibited essentially complete return of enzymic activity. (In experiments in which the incubation mixture contained LDH at concentrations greater than 1 mg/ml, the return of enzymic activity was considerably lower.) Thus, the ability to regenerate functional quaternary structure from surface-immobilized protomers appears to be quite good in the case of lactate dehydrogenase.

Measurement of Ligand Binding to Immobilized Enzyme. Use of an equilibrium method for measuring the dissociation constants of enzyme-specific ligands, either cosubstrates or effectors, offers a major advantage over kinetic methods of evaluation, in the case of immobilized enzymes, in that diffusional effects within the matrix do not alter the results obtained in the former case as they do in the latter. The fact that the enzyme is immobilized facilitates application of a subtractive method[31] for determination of the amount of ligand bound.

[30] I. C. Cho and H. Swaisgood, *Biochim. Biophys. Acta* **334**, 243 (1974).
[31] J. Steinhardt and J. A. Reynolds, "Multiple Equilibria in Proteins," p. 34. Academic Press, New York, 1969.

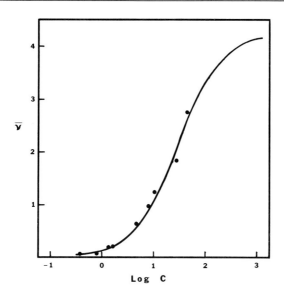

FIG. 6. Binding of [^{14}C]NAD$^+$ to immobilized LDH. The line was computed by a program designed to give the best least-squares' fit of the data to a hyperbolic equation for one class of binding sites.

Equilibrium-binding of NAD$^+$ to immobilized LDH was measured using [^{14}C]NAD$^+$ and liquid scintillation counting to provide sufficient sensitivity. Immobilized enzyme beads (6–10 nmoles of LDH) were equilibrated with [^{14}C]NAD$^+$ solutions of various concentrations for 2 hr at 23°. Controls, consisting of similar beads without covalently bound LDH, were incubated in an identical manner. After equilibration, supernatants were removed and the NAD$^+$ concentration of each was determined by scintillation counting. From the total liquid volume (determined from total weight, dry bead weight, and solution density), the total amount of NAD$^+$ in solution was calculated; the amount of NAD$^+$ bound to the enzyme beads was then determined by subtracting this value from the total NAD$^+$ added.

The data obtained are shown in Fig. 6, plotted as suggested by Cornish-Bowden and Koshland.[5] The computer-fitted curve provides a K_{diss} of 29 μM and a value of 4.27 binding sites per tetramer. These results suggest that each protomer is functional, i.e., capable of binding NAD$^+$, with an affinity that is 30-fold greater than that observed for native rabbit muscle LDH in solution.[32]

[32] V. Zewe and H. J. Fromm, *J. Biol. Chem.* **237**, 1668 (1962).

[35] Immobilization as a Means of Investigating the Acquisition of Tertiary Structure in Chymotrypsinogen[1]

By H. ROBERT HORTON and HAROLD E. SWAISGOOD

Experiments on the reversibility of reductive denaturation of bovine pancreatic ribonuclease A provided the first dramatic evidence that led to general acceptance of the "thermodynamic hypothesis" of protein structure—that acquisition of biologically functional tertiary structure in proteins is determined by the inherent thermodynamics of interactions among the linear array of amino acid residues comprising their polypeptide chains, within the environment of their normal physiological milieu.[2-6] Complete denaturation of the ribonuclease molecule through reductive cleavage of its four intrachain disulfide bonds, in the presence of 8 M urea, resulted in complete loss of both enzymic activity and residual specific three-dimensional structure.[7-9] After the subsequent removal of reducing agent and urea, the reduced, inactive protein was found to undergo spontaneous oxidation when dilute solutions of neutral to mildly alkaline pH were exposed to air. Concomitant with reoxidation of the protein was the appearance of essentially full enzymic activity[3,6] and "native" crystallographic structure.[10] The significance of this finding can be appreciated when it is recognized that the native conformation represents only 1 of 105 possible intramolecular pairings of the eight sulfhydryl groups of reduced ribonuclease A; i.e., totally random pairing would be expected to yield less than 1% of native enzymic activity. (Methodology applicable to the reduction and reoxidation of disulfide bonds in ribonuclease A has been published in this series.[11])

Similar experiments have been performed with a variety of other

[1] Research carried out at North Carolina State University, and supported in part by National Science Foundation Grant GI-39208.
[2] M. Sela, F. H. White, Jr., and C. B. Anfinsen, *Science* **125**, 691 (1957).
[3] F. H. White, Jr., *J. Biol. Chem.* **236**, 1353 (1961).
[4] C. B. Anfinsen and E. Haber, *J. Biol. Chem.* **236**, 1361 (1961).
[5] C. B. Anfinsen, E. Haber, M. Sela, and F. H. White, Jr., *Proc. Natl. Acad. Sci. U.S.A.* **47**, 1309 (1961).
[6] C. J. Epstein, R. F. Goldberger, and C. B. Anfinsen, *Cold Spring Harbor Symp. Quant. Biol.* **28**, 439 (1963).
[7] W. F. Harrington and M. Sela, *Biochim. Biophys. Acta* **31**, 427 (1959).
[8] F. H. White, Jr., *J. Biol. Chem.* **235**, 383 (1960).
[9] D. M. Young and J. T. Potts, Jr., *J. Biol. Chem.* **238**, 1995 (1963).
[10] J. Bello, D. Harker, and E. De Jarnette, *J. Biol. Chem.* **236**, 1358 (1961).
[11] F. H. White, Jr., this series Vol. 11, p. 481 (1967); Vol. 25, p. 387 (1972).

disulfide-containing proteins, with variable degrees of success. Egg white lysozyme, which, like ribonuclease A, contains four intrachain disulfide bonds, could be reactivated to an extent of 50–80% after its reductive denaturation.[12-14] "Successful" reoxidations have also been reported for reduced Taka-amylase A[15] and reduced alkaline phosphatase from *Escherichia coli*.[16]

However, various difficulties have been encountered in attempts to reoxidize reductively denatured proteases, such as trypsin, chymotrypsin, and pepsin.[17-20] It has been concluded that loss of requisite "folding information" accompanies proteolytic activation of zymogens, so that the resulting proteases would not necessarily be expected to reform functional tertiary structures upon reoxidation of their reductively denatured chains.[21,22] Both the pancreatic enzyme, chymotrypsin, and the pancreatic hormone, insulin, exist as multichain proteins in their biologically functional states, as a result of proteolytic cleavage of single-chained precursors. Reduced insulin chains do not combine appreciably to yield functional molecules under conditions that are optimal for reoxidation of sulfhydryl groups to disulfide bonds, whereas reductively denatured proinsulin can be "properly" reoxidized in high yield.[23] Denatured α-chymotrypsin, upon dilution into a medium containing β-mercaptoethanol and a thiol, disulfide-exchange enzyme,[24,25] rapidly lost its capacity for reactivation.[21] Based on the findings with insulin and proinsulin, it seemed likely that the reductively denatured zymogen, chymotrypsinogen A, should retain the "folding information" inherent in its primary sequence and thus be subject to spontaneous oxidation to its biologically functional (i.e., trypsin-activatable) structure.[26]

[12] C. J. Epstein and R. F. Goldberger, *J. Biol. Chem.* **238**, 1380 (1963).
[13] K. Imai, T. Takagi, and T. Isemura, *J. Biochem. (Tokyo)* **53**, 1 (1963).
[14] R. F. Goldberger and C. J. Epstein, *J. Biol. Chem.* **238**, 2988 (1963).
[15] T. Isemura, T. Takagi, Y. Maeda, and K. Yutani, *J. Biochem. (Tokyo)* **53**, 155 (1963).
[16] C. Levinthal, E. R. Signer, and K. Fetherolf, *Proc. Natl. Acad. Sci. U.S.A.* **48**, 1230 (1962).
[17] C. J. Epstein and C. B. Anfinsen, *J. Biol. Chem.* **237**, 2175 (1962).
[18] C. J. Epstein and C. B. Anfinsen, *J. Biol. Chem.* **237**, 3464 (1962).
[19] Y. Nakagawa and G. E. Perlmann, *Arch. Biochem. Biophys.* **140**, 464 (1970).
[20] Y. Nakagawa and G. E. Perlmann, *Arch. Biochem. Biophys.* **144**, 59 (1971).
[21] D. Givol, F. De Lorenzo, R. F. Goldberger, and C. B. Anfinsen, *Proc. Natl. Acad. Sci. U.S.A.* **53**, 676 (1965).
[22] D. B. Wetlaufer and S. Ristow, *Annu. Rev. Biochem.* **42**, 135 (1973).
[23] D. F. Steiner and J. L. Clark, *Proc. Natl. Acad. Sci. U.S.A.* **60**, 622 (1968).
[24] R. F. Goldberger, C. J. Epstein, and C. B. Anfinsen, *J. Biol. Chem.* **238**, 628 (1963).
[25] P. Venetianer and F. B. Straub, *Biochim. Biophys. Acta* **67**, 166 (1963).
[26] J. C. Brown and H. R. Horton, *Proc. Soc. Exp. Biol. Med.* **140**, 1451 (1972).

As a test of this hypothesis, chymotrypsinogen A was reductively denatured with β-mercaptoethanol in 8 M urea to yield reduced protein containing 10 sulfhydryl groups per polypeptide chain. Dilute solutions of the reduced chymotrypsinogen (10–100 µg/ml) were exposed to air after the removal of reducing agent and denaturant by gel filtration or dialysis, or after direct dilution. After various periods of reoxidation under a wide variety of conditions, aliquots of the reductively denatured zymogen were treated with trypsin and then assayed for chymotryptic activity. Results varied with conditions employed for reoxidation, but "optimal" conditions resulted in a maximum recovery of 1.4% of trypsin-activatable chymotrypsin.[26] By contrast, nonreductive denaturation of chymotrypsinogen, under conditions otherwise identical to those employed for complete reduction, was found to be completely reversible, so that 100% recovery of chymotryptic activity was attained by activation with trypsin. Although 1.4% recovery of biologically functional conformation is more than an order of magnitude greater than that expected from entirely random intramolecular oxidative pairing of 10 sulfhydryl groups, it is not sufficiently high to provide convincing evidence that the biologically functional conformation of the polypeptide chain comprising chymotrypsinogen A is that of greatest thermodynamic stability in solution. Further investigation of the oxidized material revealed that low recoveries of activity appeared to stem from aggregation of the reductively denatured polypeptide chains, as has also been found in studies with trypsin[18] and with pepsin and pepsinogen.[19]

Thus, immobilization of chymotrypsinogen through covalent attachment to an insoluble matrix provided the key for further investigation of the regeneration of biologically functional tertiary structure from completely reduced, denatured zymogen chains under conditions that would minimize interactions leading to nonspecific aggregation and formation of intermolecular disulfide bridges.[27]

Studies of Reduction and Reoxidation of Immobilized Trypsin and Ribonuclease[17]

Epstein and Anfinsen utilized the azide coupling procedure of Micheel and Ewers[28] as modified by Mitz and Summaria[29] to prepare CM-cellulose-bound derivatives of trypsin and ribonuclease A, in an attempt to circumvent problems of nonspecific aggregation of reductively denatured

[27] J. C. Brown, H. E. Swaisgood, and H. R. Horton, *Biochem. Biophys. Res. Commun.* **48**, 1068 (1972).
[28] F. Micheel and J. Ewers, *Makromol. Chem.* **3**, 200 (1949).
[29] M. A. Mitz and L. J. Summaria, *Nature (London)* **189**, 576 (1961).

trypsin chains and trypsin autolysis. The immobilized CM-cellulose derivatives of each enzyme were packed into respective columns, reduced with β-mercaptoethanol in 8 M urea, and then slowly allowed to reoxidize by perfusion with air-equilibrated Tris buffer, pH 8.3. Such reduction and reoxidation procedures resulted in a loss of 60% of the trypsin content of CM-cellulose-trypsin columns, as determined by amino acid analysis of acid hydrolyzates.

After complete reduction of the immobilized trypsin, reoxidation at pH 8.3 resulted in recoveries of approximately 3–4% of the specific activity toward benzoyl-L-arginine ethyl ester exhibited by the urea-washed immobilized enzyme prior to its reduction. The immobilized enzyme appeared to be less active in catalyzing hydrolysis of peptide bonds in a macromolecular substrate (oxidized ribonuclease) than it was in catalyzing hydrolysis of the low-molecular-weight substrate. (Similar findings have been reported for a number of immobilized enzymes; see the review by Goldman et al.[30]) Subsequent reexposure of the reoxidized CM-cellulose–trypsin columns to reduction in urea followed by reoxidation led to further losses in enzymic activity and additional losses of enzyme protein from the columns.

By comparison to the results with immobilized trypsin derivatives, recoveries of 34–42% of native enzymic activity toward uridine 2′,3′-cyclic monophosphate were obtained when CM-cellulose–ribonuclease A preparations were reductively denatured and then allowed to reoxidize.

The authors concluded that, since the recovery of 3–4% of native tryptic activity toward benzoyl-L-arginine ethyl ester was much greater than that expected for an entirely random reoxidation of 12 sulfhydryl groups (0.01%), and greater than was actually observed when reoxidation was allowed to occur in solutions of 8 M urea, the three-dimensional conformation of at least the catalytically active portion of the trypsin molecule "is determined by the amino acid sequence alone."[16]

Studies of Activation of Immobilized Pepsinogen[31]

In order to evaluate the degree to which pepsinogen can be activated to pepsin *intramolecularly* upon exposure to acid (pH 2), as compared to *autocatalytically* (through bimolecular reaction with pepsin at pH 2), Bustin and Conway-Jacobs[31] have investigated the activation of Sepharose-bound pepsinogen. Using a slight modification[32] of the cyanogen

[30] R. Goldman, L. Goldstein, and E. Katchalski, *in* "Biochemical Aspects of Reactions on Solid Supports" (G. R. Stark, ed.), p. 1. Academic Press, New York, 1971.
[31] M. Bustin and A. Conway-Jacobs, *J. Biol. Chem.* **246**, 615 (1971).
[32] D. Givol, Y. Weinstein, M. Gorecki, and M. Wilchek, *Biochem. Biophys. Res. Commun.* **38**, 825 (1970).

bromide-activation procedure of Porath et al.,[33] they coupled pepsinogen to Sepharose at pH 8.8. The immobilized pepsinogen derivatives were packed into columns. Under the conditions selected, no detectable pepsin activity was eluted from the columns upon exposure to HCl solutions at pH 2; i.e., all the pepsinogen appeared to be covalently bound to the matrix.

Exposure of the columns to pH 2 resulted in activation of the immobilized pepsinogen to form immobilized pepsin, as evidenced by the appearance of proteolytic activity, assessed by measurements of the appearance of trichloroacetic acid-soluble peptides from globin at pH 2. The proteolytic activity of Sepharose-bound pepsinogen upon activation was found to be about 30% of that obtained for pepsinogen in solution. The specific proteolytic activity of Sepharose-bound pepsinogen columns was found to be independent of the ratio of pepsinogen to Sepharose, leading the authors to conclude that, at pH 2, intramolecular activation of pepsinogen can occur in the absence of autocatalytic activation.

It is interesting to note that the amino acid composition of such Sepharose-pepsinogen preparations was not altered by exposure to pH 2. The authors concluded that, under the conditions selected for immobilization, multiple covalent attachment of the pepsinogen molecules to the insoluble matrix had occurred, so that all the peptide fragments resulting from exposure of Sepharose-pepsinogen to acid remained covalently bound to the Sepharose. Nevertheless, after exposure to acid at pH 2, the Sepharose-bound protein could be irreversibly inactivated by exposure to pH 8.5,[31] a property characteristic of pepsin, but not pepsinogen, in solution.[34]

Immobilization of Chymotrypsinogen A[27]

By analogy to the presumed folding of nascent polypeptide chains attached to ribosomes in the *in vivo* synthesis of chymotrypsinogen, we postulated that refolding of reductively denatured chymotrypsinogen chains that were covalently attached to an insoluble matrix should meet with greater success *in vitro* than had previously been attained with reductively denatured zymogens in solution,[26] provided the microenvironment of the surface did not adversely affect the structural interactions of the polypeptide chain.[35]

Factors involved in the choice of an appropriate method for immobilizing chymotrypsinogen chains include: (a) selection of a relatively

---

[33] J. Porath, R. Axén, and S. Ernback, *Nature* (London) **215**, 1491 (1967).
[34] Z. Bohak, *J. Biol. Chem.* **244**, 4638 (1969).
[35] H. E. Swaisgood and H. R. Horton, in "Enzyme Engineering" (E. K. Pye and L. B. Wingard, Jr., eds.), Vol. 2, p. 169. Plenum, New York, 1974.

rigid, stable, insoluble matrix that is not likely to undergo significant chemical or structural changes through swelling, compaction, or other deformation during reductive denaturation and reoxidation procedures; (b) selection of a means of attachment that would be stable to conditions employed in reductive denaturation and reoxidation of the protein, and yet would provide little or no restriction to complete unfolding of polypeptide chains and reduction of disulfide bridges; (c) selection of a matrix environment that could be reasonably expected to allow the *bimolecular* activation of immobilized chymotrypsinogen by trypsin (i.e., exposure of the Arg_{15}–Ile_{16} peptide bond to tryptic attack), to provide a sensitive means of assessing the acquisition of biologically functional conformation in the unfolded polypeptide chains. Thus, it appeared desirable to select a procedure different from that used to immobilize trypsin and ribonuclease[17] or that used to immobilize pepsinogen and pepsin[31] as the former resulted in losses of protein during reductive denaturation and the latter resulted in multiple covalent bonds that could hinder complete unfolding and reduction.

Accordingly, we selected a procedure whereby chymotrypsinogen A could be covalently bound to carbodiimide-activated succinamidopropyl-glass beads, without direct exposure of the enzyme to the carbodiimide. A column containing 1.0 g of γ-aminopropyl-silanized porous glass beads (40–60 mesh, 522 Å mean pore diameter, obtained from Corning Glass Works) was washed with 1 M NaCl, and then distilled water (12–24 hr), and then degassed in a vacuum desiccator. The column was equilibrated with 6 ml of a degassed solution brought to 0.5 M in succinic acid, pH 4.5. Solid 1-ethyl-3-dimethylaminopropyl carbodiimide was added to a concentration of 0.1 M, and the solution was recycled through the column by continuous pumping for 12 hr at room temperature. Additional solid reagent was then added to double the nominal carbodiimide concentration (to 0.2 M), and recycling was continued for an additional 5 hr. The succinylated beads were then washed with 1000 ml of distilled water to remove residual succinic acid and carbodiimide.

The carboxyl groups of 0.5 g of the succinamidopropyl–glass beads were then activated by similar treatment of a column of degassed beads with two additions of solid 1-ethyl-3-dimethylaminopropyl carbodiimide (each corresponding to a concentration of 0.1 M), with 30 min of recycling at room temperature after each addition. The column was then rapidly washed with 200 ml of 0.1 M sodium phosphate, pH 7.0, at 0° (ice bath), to remove excess carbodiimide. These washing conditions had been found to result in minimal hydrolysis of the activated O-acylisourea derivatives.[36] Immediately after the ice-cold washing, 10 ml of a solution

[36] H. Swaisgood and M. Natake, *J. Biochem. (Tokyo)* **74**, 77 (1973).

of chymotrypsinogen A (1.35 mg/ml) in 0.1 M sodium phosphate, pH 7.0, was pumped through the column and recycled for 24 hr at 4° (cold room). The beads were then washed with 1000 ml of 1 M NaCl–0.08 M Tris-Cl, pH 7.8, to remove any noncovalently bound protein. The efficacy of this washing procedure was ascertained by amino acid analysis of acid hydrolyzates of control beads, which consisted of succinamidopropyl–glass beads which had been similarly treated with chymotrypsinogen, but without prior activation of their carboxyl groups with carbodiimide.

Reductive Denaturation and Reoxidation of Immobilized Chymotrypsinogen

The immobilized zymogen (succinamido–glass–chymotrypsinogen) was reductively denatured using either β-mercaptoethanol[17] or dithiothreitol.[27] In the first procedure, a column containing 0.6 ml of the immobilized zymogen beads was washed with a solution of 0.05 M Tris-Cl, pH 8.6, and then 8 ml of a solution of 8 M urea, 0.05 M in Tris-Cl, pH 8.6, and 10% (v/v) in β-mercaptoethanol was passed slowly through the column. Approximately 1 ml of the reducing solution was allowed to remain on top of the column, the space above the meniscus was flushed with N_2, and the column was sealed and allowed to stand at room temperature for 18 hr. In the second procedure, a similar quantity of beads was equilibrated with 0.1 M sodium phosphate, pH 7.0, and then treated in a similar manner with a solution 8 M in urea, 0.1 M in sodium phosphate, pH 7.0, and 3–4 mM in dithiothreitol. (The concentration of dithiothreitol used represented a 50-fold molar excess of the reagent over the potentially available sulfhydryl groups of the protein.)

Completeness of reduction was assessed by titration of the sulfhydryl groups of the immobilized protein as follows: A column containing a portion of the beads was washed with 0.1 M acetic acid to remove denaturant and reductant, and then treated with a solution of 5,5'-dithiobis(2-nitrobenzoic acid) in 0.5 M sodium acetate–0.01 M sodium phosphate, pH 7.5. The beads were then thoroughly washed with 0.5 M sodium acetate–0.01 M sodium phosphate (pH 7.5) to remove excess reagent. The thiolnitrobenzoate anion was then displaced from the immobilized protein by eluting the column with 10% (v/v) β-mercaptoethanol in like buffer. The absorbance of the eluate (5.0 ml) at 412 nm was measured to determine the concentration of the thiolnitrobenzoate anion[37] ($\epsilon = 1.36 \times 10^4$ M^{-1}cm^{-1}). Protein content of the beads was determined by amino acid analysis of acid hydrolyzates. Such titration provided a value of 10.2

[37] G. L. Ellman, *Arch. Biochem. Biophys.* **82**, 70 (1959).

moles SH groups/mole reduced succinamido–glass–chymotrypsinogen.[27,38]

Air-reoxidation of reductively denatured succinamido–glass–chymotrypsinogen was effected by recycling air-equilibrated 0.05 M Tris–Cl, pH 8.6, through the beads for 24 hr at room temperature.

To activate the immobilized zymogen to immobilized chymotrypsin, a solution of trypsin in 0.1 M $CaCl_2$–0.08 M Tris–Cl, pH 7.8, was recycled through the succinamido–glass–chymotrypsinogen bead preparations (with or without prior reductive denaturation and air-reoxidation) for 2 hr at room temperature. Results obtained with the immobilized zymogen, which had not been exposed to reductive denaturation, revealed that most of the matrix-bound chymotrypsinogen was accessible to trypsin under these conditions, and it was suggested that π-chymotrypsin was probably the major activation product (since the autocatalytic action of chymotrypsin molecules, as generated in solution, should be minimal in the case of the immobilized protein).[27]

Chymotryptic activity of the trypsin-treated, glass-immobilized preparations was then measured using a variety of substrates. Kinetic data obtained from column assays were analyzed by least squares' fitting to an integrated form of the Michaelis–Menten equation.[38–40]

Comparison of the bimolecular rate constants (k_o/K_m) of the air-reoxidized preparation to those of "native" preparation (matrix-bound zymogen which had not been exposed to reductive denaturation prior to activation with trypsin) revealed 76–80% recovery of esterolytic activity toward Cbz-L-tyrosine p-nitrophenyl ester, and 53% recovery of activity toward benzoyl-L-tyrosine ethyl ester.[38,39] The air-reoxidized preparation exhibited a specific activity toward casein (as measured by the appearance of trichloroacetic acid-soluble peptides) that was 30% of that of nonreduced matrix-bound enzyme.

In contrast to these findings concerning the efficacy of refolding of denatured, matrix-bound chymotrypsinogen chains, similar treatment of reductively denatured, matrix-bound chymotrypsin failed to produce any measurable activity, thus implying that the entire zymogen molecule is needed for generation of biologically functional three-dimensional structure.

Two methods have been developed to permit further characterization of the refolding process in chymotrypsinogen: incorporation of a thioester linkage to bind the protein to the succinamidopropyl-glass beads, which permits subsequent, selective chemical cleavage of the zymogen from the

---

[38] J. C. Brown, Ph.D. dissertation, North Carolina State University, Raleigh, 1973.
[39] H. R. Horton and H. E. Swaisgood, in "Immobilized Biochemicals and Affinity Chromatography" (R. B. Dunlap, ed.), p. 329. Plenum, New York, 1974.
[40] M. D. Lilly, W. E. Hornby, and E. M. Crook, Biochem. J. **100**, 718 (1966).

insoluble matrix[41]; and construction of a solid-phase fluorimetry cell to permit examination of the emission spectra of glass-bound proteins.[38,39] In the first procedure, aminopropylsilanized glass beads were succinylated as previously described, and then activated with 1-ethyl-3-dimethylaminopropyl carbodiimide. To 1.0–1.5 g of activated succinamidopropyl–glass beads was then added 15 ml of 1.0 M thioglycolic acid, pH 4.75, and the mixture was stirred continuously under N_2 for 5 hr at room temperature. The beads were then washed with distilled water, prior to coupling the protein through the same carbodiimide-activation procedure as previously described. In this case, the chymotrypsinogen molecules were attached through amide bonds to the carboxyl groups of immobilized thioglycolate moieties, rather than to the carboxyl groups of succinate moieties, per se:

$$\text{(glass)}-CH_2CH_2CH_2-\overset{H}{\underset{|}{N}}-\overset{O}{\underset{\|}{C}}-CH_2CH_2\overset{O}{\underset{\|}{C}}-S-CH_2-\overset{O}{\underset{\|}{C}}-\overset{H}{\underset{|}{(N}}-\text{chymotrypsinogen)}$$

The chymotrypsinogen, thus bound, could then be reductively denatured with dithiothreitol and urea, and subseqently allowed to reoxidize in the presence of air-equilibrated 0.1 M sodium phosphate, pH 7.0. The chymotrypsinogen was solubilized by treating the beads with a neutral solution of 1.0 N hydroxylamine, which effected cleavage of the thioester bonds linking the protein to the matrix. The thioglycolyl-protein, thus released, was found to contain 1.0–1.2 moles of SH groups per mole; this implies that the conditions employed in immobilizing the protein had resulted in the formation of only one covalent bond per chymotrypsinogen molecule.[39] The released, thioglycolated protein was treated with trypsin and then titrated with ^{14}C-labeled diisopropylphosphofluoridate. Measurements of specific radioactivity revealed that 0.70 mole of diisopropylphosphate had become incorporated per mole of protein, implying that the specific conformation of the Asp(102)-His(57)-Ser(195) "charge relay system" had been regenerated in 70% of the molecules which had undergone reductive denaturation and air-reoxidation.[38]

Evidence for acquisition of overall conformation in reductively denatured succinamidopropyl–glass-bound chymotrypsinogen was obtained by measurements of fluorescence emission spectra of the immobilized preparation, in a manner analogous to that developed by Gabel, Steinberg, and Katchalski[42] (see also this volume [36]). The cell was modified from that described by Gabel et al. in that it consisted of a 5.0-mm o.d. cylindrical quartz tube, mounted in an aluminum block 12.5 mm square and 45 mm in height. Immobilized protein beads were packed into the tube,

[41] J. C. Brown and H. R. Horton, Fed. Proc., Fed. Am. Soc. Exp. Biol. 32, 496 (1973).
[42] D. Gabel, I. Z. Steinberg, and E. Katchalski, Biochemistry 10, 4661 (1971).

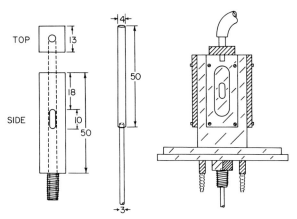

FIG. 1. Diagram of flow-cell assembly for measuring changes in fluorescence emission spectra of immobilized, reductively denatured proteins. *Left:* Aluminum cell holder (all dimensions in millimeters), which houses quartz cylindrical cell. *Center:* Quartz cell, fitted with rigid polyethylene tubing. *Right:* Thermostated cell holder assembly of the Aminco-Bowman spectrophotofluorimeter modified to accommodate flow-through quartz cell in cell holder.

and excited at 280 nm in an Aminco-Bowman spectrofluorimeter fitted with a Texas Instruments Function-Riter X-Y recorder. The reoxidized, immobilized chymotrypsinogen preparations exhibited an emission spectrum that appeared identical to that of immobilized chymotrypsinogen that had not been subjected to reductive denaturation.[39]

The dynamics of refolding can be conveniently followed through a further modification which permits solutions to flow through the bead column contained in the quartz fluorescence cell, illustrated in Fig. 1. Typical spectra are given in Fig. 2. It can be seen that refolding of matrix-bound, reductively denatured chains under conditions of spontaneous air-reoxidation (Tris-Cl, pH 8 6, 25°) requires about 20 hr as judged by return of native fluorescence emission characteristics, with a half-time of 65 min. Titration of sulfhydryl groups under identical conditions showed that their oxidation proceeded at a rate identical to that of refolding as judged by fluorescence (half of the sulfhydryl groups oxidized in 67 min). That the rate of reoxidation of the sulfhydryl groups was limiting in the regeneration of three-dimensional structure of glass-bound chymotrypsinogen chains was established by adding purified sulfhydryl oxidase to the Tris-Cl solution; sulfhydryl oxidase is an enzyme that catalyzes the formation of disulfide bonds using O_2 as oxidant.[43] Again, return of "native" fluorescence characteristics paralleled the loss of sulf-

[43] V. G. Janolino and H. E. Swaisgood, *J. Biol. Chem.* **250**, 2532 (1975).

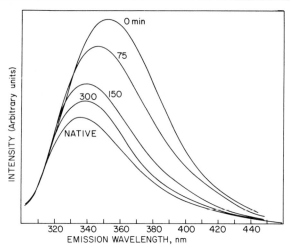

FIG. 2. Fluorescence emission spectra of immobilized chymotrypsinogen prior to reductive denaturation (native), and of reductively denatured immobilized chymotrypsinogen prior to air-reoxidation (0 min) and after various periods of reoxidation at 25°, pH 8.6 (75, 150, and 300 min). After 20 hr reoxidation, emission spectrum appeared indistinguishable from native. Excitation wavelength, 280 nm.

hydryl groups, but in this case, refolding was complete within 60 min rather than 20 hr. (Similar effects on the refolding of reductively denatured ribonuclease A chains in solution have been reported for sulfhydryl oxidase.[43])

In conclusion, immobilization of chymotrypsinogen A has provided a means for the study of structural regeneration in which problems of intermolecular interactions leading to aggregation could be circumvented. Similar procedures may be applicable to a variety of other proteins. On the basis of the results obtained with chymotrypsinogen, it appears that the "thermodynamic hypothesis" of Anfinsen and his colleagues may be extended to the formation of biologically functional structure in proteolytic zymogens.

[36] Conformational Transitions in Immobilized Proteases

By DETLEF GABEL and VOLKER KASCHE

Reversible conformational transitions induced by physical agents (light, heat) are of importance in biological systems, as the activity of these systems can be controlled by such transitions. One protein in which this has been observed is the conjugated plant protein phytochrome.[1,2] The different conformations of this photoreceptor control different physiological activities in an all-or-none fashion. The transitions were found to be similar in *in vivo* (immobilized protein)[1] and *in vitro* (solution)[2] systems. These light-induced transitions are completely reversible in solution.

Reversible transitions with considerable unfolding of the polypeptide chain are, however, difficult to follow in solution because of irreversible and reversible intermolecular interactions (such as aggregation and autolysis). Immobilization can greatly extend the range in environmental conditions (temperature, concentration of denaturing solutes and solvents) under which reversible conformation transitions in proteins can be studied.

Reversible conformational transitions in immobilized proteins have been studied with the following aims:

1. To study their *functional importance*, i.e., the mechanism of unfolding; the relation between the tertiary structure and biological activity; the importance of these transitions in the regulation of biological processes. In these studies possible perturbations of protein and enzyme parameters due to the immobilization reaction must be considered.

2. To study their *applicational importance*, i.e., whether certain active or inactive conformations of the enzyme are stabilized that increase or decrease the range of application of immobilized enzymes compared to the free enzyme.

This contribution summarizes some studies[3-9] on reversible conformation transitions, using immobilized proteases as model proteins. The aim of these studies was to investigate the functional as well as the applicational importance of such transitions.

[1] R. E. Kendrick and C. J. P. Spruit, *Plant Physiol.* **52**, 327 (1972).
[2] D. R. Gross, H. Linschitz, V. Kasche, and J. Tenenbaum, *Proc. Natl. Sci. U.S.A.* **61**, 1095 (1968).
[3] D. Gabel, P. Vretblad, R. Axén, and J. Porath, *Biochim. Biophys. Acta* **214**, 561 (1970).
[4] D. Gabel, I. Z. Steinberg, and E. Katchalski, *Biochemistry* **10**, 4661 (1971).
[5] D. Gabel and R. Svantesson, unpublished results, 1971.
[6] D. Gabel and V. Kasche, *Biochem. Biophys. Res. Commun.* **48**, 1011 (1972).
[7] D. Gabel and V. Kasche, *Acta Chem. Scand.* **27**, 1971 (1973).
[8] D. Gabel, *Eur. J. Biochem.* **33**, 348 (1973).
[9] A. W. Burgess, L. I. Weinstein, D. Gabel, and H. A. Scheraga, *Biochemistry* **14**, 197 (1975).

Experimental Methods Used to Study Conformational Transitions in Immobilized Enzymes

Static Methods

A number of methods have been developed to follow conformational transitions in immobilized proteins. These include equilibrium studies on binary and ternary complexes of the protein with small and large molecules, fluorescence spectroscopy, and electron spin resonance (see Section IV of this volume). These methods will give information about the static conditions of the bound protein, but will not directly give information about the enzymic properties of the conjugate.

Kinetic Methods

The rate of a reaction catalyzed by immobilized enzymes is perturbed compared to the situation using the free enzymes, owing, among other factors, to the diffusion of the substrate to the immobilized enzyme.[10] This effect is of importance for substrate concentrations less than or around K_m.[11] The value of K_m determined for immobilized enzymes will depend, among other factors, on the enzyme content, diffusion constants for substrate and product, catalytic rate constant of the enzyme reaction, physical dimension of the conjugate, and steric hindrance. The apparent K_m value determined for an immobilized enzyme is also a function of the thickness of the unstirred diffusion layer around the particles with the immobilized enzyme.[12] The temperature dependence of K_m includes the effect of temperature on the diffusion in this layer and therefore does not represent an intrinsic molecular property.[13] Thus, the temperature dependence of the apparent K_m value cannot be used as a measure for possible conformational transitions.

The catalytic constant, k_{cat}, does, however, represent a molecular property also for immobilized enzymes.[10,11,14] For proteolytic enzymes, k_{cat} can be described by $k_2 \cdot k_3/(k_2 + k_3)$, where the acylation rate constant k_2 and the deacylation rate constant k_3 are both temperature dependent. The apparent activation energy of k_{cat} should decrease with temperature when the acylation and deacylation rates are rate determin-

---

[10] L. Goldstein, this volume [29].
[11] V. Kasche and M. Bergwall, in "Insolubilized Enzymes" (M. Salmona, C. Saronio, and S. Garattini, eds.), p. 77. Raven Press, New York, 1974.
[12] V. Kasche and A. Kapune, unpublished results, 1974.
[13] K. Buchholz and W. Rüth, unpublished results, 1974.
[14] R. J. Knights and A. Light, *Arch. Biochem. Biophys.* **160**, 377 (1974).

ing in different temperature regions.[15] On the other hand, increases of the activation energy with temperature are caused by conformational transitions. Thus kinetic evidence for conformation changes can be obtained from the temperature dependence of k_{cat}, i.e., the rate of enzymic reaction at substrate concentrations much larger than K_m. Then the diffusion effect is negligible. The data reviewed here were obtained under such conditions.

It is also necessary to work in buffers of sufficient capacity to avoid pH gradients around the immobilized enzyme, which generate an unknown and uncontrollable microenvironment,[10,11] or, when measuring in a pH stat, at pH values where k_{cat} is independent of pH.

Description of Experimental Procedures

Coupling of Proteins[16]

Sephadex G-200 is swollen in water for 24 hr and suspended to a final concentration of 20 mg/ml in a beaker equipped with a stirrer, a pH electrode connected to a pH stat, and a pH stat burette filled with 2 M NaOH. An amount of CNBr equal to the weight of the carrier, dissolved in water to a concentration of 25 mg/ml, is added, and the pH is adjusted to 10.0 and kept there for 6 min. (Reaction at higher pH values with manual addition of NaOH or activation with solid CNBr results in extensive cross-linking of the carrier, which thereby is rendered impenetrable to high-molecular-weight substances.) The gel is filtered off and washed with 200 ml of cold water and 100 ml of cold coupling buffer [0.1 M NaBO$_3$–HCl, 0.02 M CaCl$_2$, pH 8.5, for α- and β-trypsin and chymotrypsin, 0.01 M NaBO$_3$–NaOH, 0.02 M CaCl$_2$, 0.1 M NaCl, 0.01 M benzamidine, pH 10.0, for acetylated trypsin derivatives (their preparation is described below), and 0.1 M NaHCO$_3$, 1 mM ZnCl$_2$ for carboxypeptidase A]. The enzyme (50–100 mg per gram dry gel) is added to the gel suspended in the coupling buffer (50 ml per gram of gel), and the reaction mixture is rotated end-over-end in a capped vessel overnight. The gel is then washed extensively in a column alternately with buffers of high and low pH, and a buffer, pH 8.0, containing glycine in order to block remaining reactive groups of the polymer.

Sepharose 4B is activated with CNBr at pH 11.0 as described for Sephadex, and soybean trypsin inhibitor (STI) is coupled to it in 0.1 M NaHCO$_3$ by adding 100 mg STI to a suspension of 1 g of activated gel.

[15] M. H. Han, J. Theor. Biol. 35, 543 (1972).
[16] R. Axén and S. Ernback, Eur. J. Biochem. 18, 351 (1971).

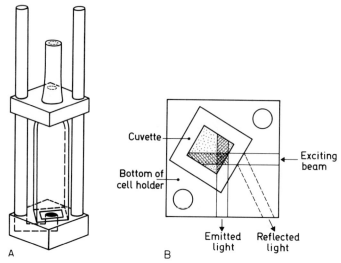

FIG. 1. Cell for measuring fluorescence of immobilized proteins. (A) View of the cell. Reprinted from D. Gabel, *in* "Enzyme Engineering" (E. K. Pye and L. B. Wingard, Jr., eds.), Vol. 2, p. 165. Plenum, New York, 1974. (B) Sketch of the optical pathway, viewed from above.

Activity Measurements in a pH Stat[17]

Trypsin is assayed in 2 ml of 0.1 M NaCl with tosyl-L-arginine methyl ester as substrate at concentrations between 0.01 and 0.05 M. Chymotrypsin is assayed in 2 ml of 0.3 M KCl, 5% ethanol with 0.01 M acetyl-L-tyrosine ethyl ester as substrate. The apparent pH optimum of the reaction is determined under the conditions (ionic strength, temperature, urea concentration) used, and the assay is conducted at optimum pH. An aliquot of a suspension of the immobilized enzyme is added to the salt solution, and after equilibration at the desired temperature for 15 min, substrate is added and the uptake of alkali is followed. A weak stream of nitrogen is used to exclude carbon dioxide from the air. Nonenzymic uptake of alkali is corrected for with a blank not containing the enzyme.

Fluorescence Measurements

A special cell[4,18] for measuring the fluorescence emitted from the surface of packed beds of immobilized enzymes is used (see Fig. 1).

[17] B. Mattiasson and K. Mosbach, this volume [25].
[18] D. Gabel, *in* "Enzyme Engineering" (E. K. Pye and L. B. Wingard, Jr., eds.), Vol. 2, p. 165. Plenum, New York, 1974.

The cell consists of a square 1 × 1 cm piece of Teflon, 3mm thick, which can accommodate in a groove the bottom of a fluorescence microflow cell, from which the lower outlet has been cut off. The cell is positioned at an angle of 60° with respect to the edge of the Teflon square in order to divert the reflected light from the emission optics. The center of the Teflon square is enclosed by the microcell. A boring leads from the space enclosed by the microcell to one of the two stainless steel tubes positioned at opposite corners of the Teflon plate. The bottom of the microcell is glued to the plate with a water-resistant glue in order to avoid leakage. Another plate, with two borings for the steel tube, one for a thermistor, and a fourth one for the top of the microcell, is used to keep the cell in a fixed position. The cell is filled with a suspension of the gel with a Pasteur pipette so that a height of the packed bed is about 1 cm. Liquid can be withdrawn from the bottom of the cell and replaced through a tubing attached to the top. The whole assembly is inserted into an ordinary 1 × 1 cm fluorescence cell.

Fluorescence spectra are recorded in the conventional way in a spectrofluorometer. The light scattered by the carrier is usually well separated from the fluorescent light. Binding of dyes to the immobilized protein is followed by equilibrating the gel with an appropriate solution of the dye in a suitable buffer (0.01 M NaBO$_3$–HCl buffer, pH 8.1, in the case of chymotrypsin), measuring the fluorescence, and reequilibrating the gel with a different dye concentration by passing a solution of the dye through the bed. The binding experiment can be repeated many times with the same gel. The data are analyzed according to the equation[19]

$$I = I_{max} - K_{diss}(I/c)$$

where I is the measured intensity at the concentration, c, of the dye; I_{max} is the intensity at saturation; and K_{diss} is the dissociation constant of the dye–protein complex. A plot of I vs I/c then gives a straight line with the slope $-K_{diss}$ and the intercept I_{max}.

Autolysis of β-Trypsin[6,7]

A β-trypsin solution in 0.1 M NaBO$_3$–HCl buffer, pH 8.1, or 0.01 M NaBO$_3$–HCl, 0.02 M CaCl$_2$, 0.04 M NaCl, pH 8.1 is incubated at the desired temperature. At time intervals, aliquots are withdrawn and assayed against tosyl-L-arginine methyl ester or introduced into a STI-Sepharose column equilibrated with 0.05 M sodium acetate buffer

[19] W. O. McClure and G. M. Edelman, *Biochemistry* **6**, 559 (1967).

pH 5.0, containing 0.5 M NaCl. After washing, a linear gradient of the initial buffer and 0.05 M glycine–HCl buffer, pH 2.7, containing 0.5 M NaCl, is applied. The relative amounts of inactive material (not adsorbed to the column) and α- and β-trypsin (eluting at different positions in the gradient) are measured. A second-order rate constant of inactivation is determined by plotting $1/c$ (c being the remaining concentration of active enzyme) vs time. Identical results are obtained with both methods.

Acetylation of Trypsin[8]

Acetylation with N-Acetyl Imidazole. β-Trypsin, 100 mg, is dissolved in 5 ml of 0.02 M CaCl$_2$, 0.01 M benzamidine at 0°; 20–100 mg of N-acetylimidazole[20] is added, and a pH of 7.0 is kept for 1 hr by addition of 1 M NaOH from a pH stat. Then 350 mg of hydroxylamine hydrochloride are added to deacetylate tyrosines, and the pH is readjusted to 7.0 and maintained for 3 hr. The reaction mixture is applied to a Sephadex G-100 column in 0.1 M NaBO$_3$–HCl buffer, pH 8.0, and the protein-containing fractions are pooled and applied to a STI-Sepharose column in the same buffer. Inactive material (up to half of the starting amount of trypsin) is eluted with the same buffer, and active protein is desorbed with 0.01 M HCl, 0.2 M NaCl.[21]

Acetylation of Trypsin·STI Complex. β-Trypsin, 100 mg, dissolved in 2 ml of 0.2 M NaCl, is added to 100 mg of STI in the same volume of NaCl solution. Acetylation is carried out at room temperature by adding 10 portions of 10 μl of acetic anhydride at 5-min intervals and keeping the pH at 7.0 in a pH stat. Hydroxylamine hydrochloride, 350 mg, is added as described above. After passage through a Sephadex G-100 column, the protein-containing fractions are transferred to 0.02 M CaCl$_2$, 2.5 mM HCl by gel filtration on Sephadex G-25. The enzyme is separated from the inhibitor on SE-Sephadex in the same buffer, using a linear salt gradient from 0 to 1 M. The trypsin-containing fractions are made 0.1 M in NaCl, adjusted to pH 8.0 by addition of 0.1 M NaBO$_3$–NaOH buffer, pH 10.0, and freed from inactive material on STI-Sepharose as described above.

[20] H. Staab, *Chem. Ber.* **89**, 1927 (1956).
[21] The inactive material arises probably from autolysis products and has a very high ratio of amino groups to ultraviolet absorption. It is important to point out that separation according to charge, for example, is difficult to carry out with these products, the modified enzyme bearing a total charge different from the native one.[8]

Determination of Amino Groups[22]

Commercial trinitrobenzene sulfonic acid (TNBS) is dissolved in 1 part water by heating, and concentrated HCl is added to a final concentration of about 2 M. The solid crystallized after cooling is collected on a filter, washed with cold 1 M HCl, and dried. A fresh stock solution (1.8 M) of TNBS is prepared by dissolving 100 mg of reagent in 100 μl water. A sample of acetylated trypsin as prepared above is added to 0.5 ml of 0.1 M Na$_2$B$_4$O$_7$, 0.1 M NaOH, and the volume is adjusted to 1 ml. TNBS solution, 20 μl, is added, and the solution is rapidly mixed. After 5 min, 2 ml of a freshly prepared Na$_2$SO$_3$ solution (1.5 mM) in 0.1 M NaH$_2$PO$_4$ (prepared by adding 1.5 ml of a 0.1 M Na$_2$SO$_3$ solution to 100 ml 0.1 M NaH$_2$PO$_4$) is added. The solution is read at 420 nm against a reagent blank. A molar extinction coefficient of 19,200 M^{-1} cm^{-1} and 22,000 M^{-1} cm^{-1} is used for ϵ- and α-amino groups, respectively.

Digestion of Ribonuclease A with Carboxypeptidase A-Sephadex[9]

Ribonuclease, 30 mg in 3 ml 0.16 M KCl, 1 mM ZnCl$_2$, pH 6.83, is brought to the desired temperature, and 100 μl of a suspension of carboxypeptidase A-Sephadex (prepared as described above) containing 10 mg of conjugate per 1 ml is added. After 10 min the reaction is stopped by addition of 1 ml of a 50% (w/v) solution of trichloroacetic acid. After clarification on a bench centrifuge, the supernatant is analyzed directly for free valine (the C-terminal amino acid of ribonuclease A) on an amino acid analyzer.

Transitions of Functional Importance

Temperature-Induced Transitions in Trypsin

The temperature dependence of the following intrinsic molecular properties of α- and β-trypsin and those enzymes immobilized on Sephadex was studied in the temperature interval 20°–75°: (i) kinetic properties—k_{cat} for the hydrolysis of tosyl-L-arginine methyl ester; rate of autolysis of the free enzyme in the presence and in the absence of calcium ions; (ii) structural properties—fluorescence spectra of the tryptophan residues in trypsin.

The experimental data[6,7] are summarized in Fig. 2. "Breaks" in the Arrhenius plots of the kinetic data and changes in the fluorescence with

[22] R. Fields, *Biochem. J.* **124**, 581 (1971).

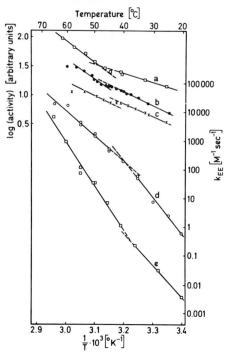

FIG. 2. Curves a–c (left ordinate): Arrhenius plots for activity against tosyl-arginine–methyl ester at pH 9.5, measured in a pH stat in 0.1 M NaCl of β-trypsin-Sephadex (a, substrate concentration 0.05 M) and of β-trypsin (b) and α-trypsin (c, substrate concentration for b and c, 0.01 M). Curves d and e (right ordinate): Arrhenius plots for the second-order rate constant of inactivation of β-trypsin in 0.01 M NaBO$_3$–HCl buffer, pH 8.1, 0.1 M in NaCl (curve d) or 0.02 M in CaCl$_2$ and 0.04 M in NaCl (curve e). From D. Gabel and V. Kasche, *Biochem. Biophys. Res. Commun.* **48**, 1011 (1972); *Acta Chem. Scand.* **27**, 1971 (1973).

temperature are found. The increase in activation energy of k_{cat}, observed when going from temperatures below 40° to those higher than 50°C, exclude the conclusion that a shift from one rate-determining reaction step to another is occurring. In the absence of Ca^{2+}, these "breaks" appear at about the same temperature for the free and the immobilized enzyme. At this temperature, a sudden shift in the wavelength for maximum fluorescence emission is also observed.[7]

These observations allow the conclusion that the "breaks" in the Arrhenius plots represent reversible conformational transitions. The similarity in the temperature behavior of the free and the immobilized enzyme shows that with trypsin the perturbation of the enzyme kinetic parameters due to the immobilization is of minor importance. This conclusion is also supported by the negligible perturbation in k_{cat} for a mixture of immobilized α- and β-trypsin observed by Knights and Light.[14] These studies show that the unfolding of trypsin is a multistate unfolding process,[23] in which the biological function is gradually modified and disappears only at rather high temperatures ($\sim 80°$). It is almost trivial to state that these conclusions could only be obtained using the immobilized enzyme, as free trypsin is autolyzed very rapidly at elevated temperatures.

Not all immobilized enzymes, however, have unperturbed enzyme and protein parameters. Both immediate and latent modifications of properties have been found upon immobilization.

Modification of Chymotrypsin by Immobilization

Chymotrypsin bound to Sephadex exhibits the same catalytic properties as the free enzyme.[11,24] It also binds the dye toluidinyl naphthalene sulfonic acid (TNS) with the same binding constant.[4] The dye is rendered fluorescent by the complex formation. The dye binding site differs from the catalytic site. When the binary complex of the enzyme with TNS in solution is allowed to form a ternary complex by additions of competitive inhibitors, nearly complete quenching of the TNS fluorescence is observed. When, however, the same ternary complex is formed with the immobilized enzyme, only a fraction of the dye fluorescence is extinguished. Thus, it has been suggested[4] that, although the catalytic and the dye-binding sites in chymotrypsin are not altered by the fixation of the enzyme to the matrix, the interaction of the two sites has been impaired, possibly by introduction of slight rigidity into the protein by the coupling to the carrier.

Modification of Immobilized Chymotrypsin by Heat Treatment

Another modification of the properties of chymotrypsin-Sephadex does not appear until the conjugate is heated to temperatures above 50°, where the enzyme is inactive. Upon cooling, considerable activity is recovered[4] (see Fig. 3), as is observed for other insoluble derivatives of chymotrypsin.[5,25] The properties of the enzyme thus obtained are considerably modified. Thus, the temperature optimum of the hydrolysis of

[23] D. C. Poland and H. A. Scheraga, *Biopolymers* 3, 401 (1965).
[24] R. Axén, P.-Å. Myrin, and J.-C. Janson, *Biopolymers* 9, 401 (1970).
[25] V. I. Surovtzev, L. V. Koslov and V. K. Antonov, *Dokl. Akad. Nauk SSSR* 195, 1463 (1970).

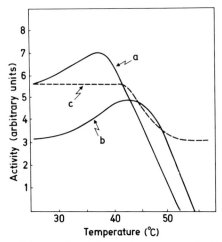

FIG. 3. Enzymic activity and stability of chymotrypsin–Sephadex at different temperatures. Acetyl tyrosine ethyl ester (0.01 M) in 0.3 M KCl was used as substrate at pH 9.5, measured in a pH stat. Curve a: enzymic activity of chymotrypsin–Sephadex measured at the temperatures specified. Curve b: enzymic activity of chymotrypsin–Sephadex after heat treatment for 15 min at 60°, measured at the temperatures specified. Curve c: enzymic activity of chymotrypsin–Sephadex at 25°, measured after exposure to the indicated temperature for 15 min. Reprinted with permission from D. Gabel, I. Z. Steinberg, and E. Katchalski, *Biochemistry* **10**, 4661 (1971) (copyright by the American Chemical Society) and D. Gabel and R. Svantesson, unpublished results (1971).

acetyl tyrosine ethyl ester is shifted to higher temperatures, and even at temperatures where the original conjugate exhibits no activity, the heat-treated one is considerably active (see Fig. 3). The dye TNS is bound with a different binding constant and a different maximum fluorescence intensity, and its fluorescence is not influenced by competitive inhibitors.[4] The catalytic properties of the heat-treated enzyme are altered, and amides are rendered very poor substrates, despite the high activity against esters. Thus, although the conjugate obtained is a mixture of active and inactive enzyme forms (the concentration of active sites[26] has decreased), its properties indicate that at least part, if not all, of the active enzyme has acquired new properties after the heat treatment.

Transitions of Applicational Importance

It has been observed in a number of cases that immobilized enzymes show stability patterns different from those of the corresponding soluble

[26] D. Gabel, *FEBS Lett.* **49**, 280 (1974).

enzymes. Both stabilization[3,8] and destabilization[27] have been observed. In the case of stabilization referred to here, the effect cannot be explained in terms of hindrance of intermolecular reactions, but must be due to other reasons.

Activity of Trypsin-Sephadex in Urea

Trypsin bound to cyanogen bromide activated Sephadex, prepared as described above, exhibits activity in 8 M urea.[3] By acetylation prior to immobilization, the number of amino groups participating in the coupling reaction can be decreased. When such partially acetylated trypsin derivatives are coupled to Sephadex, the insoluble derivatives become susceptible to urea denaturation. When a single amino group is present the insoluble derivative behaves very much like the soluble enzyme.[8] An increasing amount of amino groups available for coupling results in stabilization, and activity is lost only at higher concentrations of urea. There is thus evidence that in this case (and in some other cases; see above) cross-linking[8] between the carrier and the enzyme reduces the internal mobility of the peptide chain. A decreased number of cross-links results in derivatives behaving more like the soluble enzyme.

Destabilization of an active enzyme conformation has also been observed, e.g., for alkaline phosphatase bound to collodion membranes.[27] In such cases, it may be assumed that the environment of the enzyme provided by the matrix will favor denatured conformations under much less drastic conditions than in water. *Vice versa*, stabilization can be due to the same reason. When solutes are taken as denaturants, salting-in and salting-out effects of the carrier can result in apparent destabilization or stabilization, respectively, of the conjugate.

Although the prevention of intermolecular reactions in immobilized conjugates is trivial, it can be of great importance for the application of the enzyme under conditions lethal to the soluble enzyme. Carboxypeptidase A bound to cyanogen bromide activated Sephadex has been shown to retain a considerable amount of its activity at temperatures up to 70°, where the soluble enzyme rapidly loses activity,[9] partly because of aggregation. The immobilized enzyme could therefore be used as a proteolytic probe to study the thermally-induced unfolding of the C-terminus of ribonuclease A,[9] which does not occur until about 65°, and which therefore cannot be studied with the soluble carboxypeptidase A.[28] Such effect has also been taken advantage of in the study of the

[27] R. Goldman, O. Kedem, and E. Katchalski, *Biochemistry* **10**, 165 (1971).
[28] W. A. Klee, *Biochemistry* **6**, 3736 (1967).

catalytic action of chymotrypsin in concentrated dioxane solutions, where the soluble enzyme aggregates and eventually precipitates.[29]

Conclusion

Immobilized enzymes are not subject to intermolecular interactions as are enzymes in solution. Therefore, conformational changes can, as has been reviewed here for proteases, be studied in immobilized enzymes, and information about the corresponding transitions in soluble enzymes can be gained.

By immobilization, it is possible to induce new properties into the bound enzyme. Such properties can be of subtle influence on the enzymic activity or other parameters related to protein conformation and its changes, and will then be of little importance for the use of the conjugate as catalyst. When, however, the influence of the microenvironment on the action of the enzyme is to be studied, or when the conjugate is intended to be a model system, for example, for membranes, these changes in property will have to be taken into account. It is evident that major changes of the molecular properties of the enzyme upon immobilization are of importance not only in the type of studies reviewed here. They may also lead to new operational aspects in the application of such conjugates.

[29] K. Tanizawa and M. L. Bender, *J. Biol. Chem.* **249**, 2130 (1974).

[37] Immobilized Hemoproteins

By ERALDO ANTONINI and MARIA ROSARIA ROSSI FANELLI

Although immobilized enzymes and proteins have been studied mainly for technological applications, the behavior of proteins attached to solid matrices may help us to understand the general properties of these systems and to give us a better insight into their reactivity in solution. A comparison of the behavior of these systems in solution with those in immobilized forms may also be useful in interpreting structure–function relationships.

Hemoproteins, especially myoglobin, hemoglobin, and cytochromes, are well suited for studying proteins in the semisolid state because their visible and near-ultraviolet absorption spectra can be easily observed and the reactions of these immobilized proteins can be followed spectrophotometrically.

Recent investigations of hemoproteins bound to solid matrices will be briefly described. Information, not easily gained otherwise, can be obtained for (a) conformational transitions, (b) ligand binding equilibria, (c) subunit interaction, and (d) electron transfer processes of immobilized systems.

General Properties of Matrix-Bound Hemoproteins

Hemoproteins may be bound to cross-linked insoluble polysaccharides, such as Sephadex or Sepharose, using standard procedures such as activation of the resin with CNBr. The amount of protein bound to the resin may be several milligrams per milliliter of packed gel. If the linking process is carried out under mild conditions, with respect to pH and temperature, the matrix-bound protein will maintain the characteristic spectra and basic reactivity of the protein in solution. Once bound to the matrix, hemoproteins generally show increased resistance to denaturation and autoxidation.

Equilibrium and kinetic studies can be performed readily by direct measurement of the optical absorption properties of the gel in a suitable light path after proper "blank" corrections.

One of the major difficulties in interpreting the behavior of matrix-bound vs soluble-state proteins is the distinction between effects due to immobilization and those due to the chemical modification involved in the linkage of the protein to the matrix. These difficulties can sometimes be overcome by appropriate control experiments on soluble proteins.

Conformational Transitions in Matrix-Bound Myoglobin[1]

Myoglobin from *Aplysia* (a mollusk) shows a fully reversible transition, induced by temperature or solvent, from a native to a denatured form.[2] The two forms differ in spectral as well as in other properties, and the transition can be followed spectrophotometrically. By altering the conditions under which coupling occurs, the protein may be bound to the solid matrix in mainly the native or denatured state. At pH 9.1 and 35° *Aplysia* myoglobin occurs in the native form in the absence of butanol and in the denatured form in 5% butanol. The reaction of Sephadex with CNBr may be carried out in the presence or in the absence of butanol, and when the two reaction products are compared

---
[1] G. M. Giacometti, A. Colosimo, S. Stefanini, M. Brunori, and E. Antonini, *Biochim. Biophys. Acta* **285**, 320 (1973).

[2] M. Brunori, E. Antonini, P. Fasella, J. Wyman, and A. Rossi Fanelli, *J. Mol. Biol.* **34**, 497 (1968).

under identical conditions both the native and the denatured forms are stabilized, though to a different extent, by the linkage to the matrix. Both products undergo the reversible transitions that occur in the soluble protein, but the equilibria are shifted, indicating that the native and denatured conformations tend to be fixed by immobilization on the matrix.

The example given above shows the potentialities of linking a protein in a given conformation in solution to a solid matrix in order to stabilize that conformation and the reactivity associated with it. Apart from obvious practical applications, this may allow a detailed study of a relevant state of the protein, which, in solution, may be present only under special conditions.

Binding of Aplysia Myoglobin in the Native Form. Swollen suspensions of Sephadex G-100 or A-50 (from Pharmacia, Uppsala) are mixed with a freshly prepared solution of CNBr (50 mg/per milliliter of water). The ratio of CNBr to dry resin by weight is 1.25:1. The activation reaction is carried out by stirring the suspension for 10 min at room temperature and pH 10.3; the pH is kept constant by adding 4 M NaOH.

Excess CNBr is removed by washing the suspension on a Büchner funnel with 2% borate buffer (pH 9.1). A protein solution containing myoglobin (10 mg/ml) is added to the washed resin, resuspended in borate buffer. Stir the suspension until a ratio of 5:1 (by weight) of dry resin to protein is reached. The reaction is allowed to proceed for 1–4 hr; then the gel is washed with buffer several times and centrifuged to remove the unreacted myoglobin. The final gel suspension contains 0.5–1 mg myoglobin per milliliter of packed gel and is suitable for spectrophotometric determinations using a cuvette with a 2-mm light path.

Binding of Aplysia Myoglobin in the Denatured Form. The denatured form of *Aplysia* myoglobin is bound to the resin using the same procedure described above for the native form except that the reaction between the activated gel and the protein is carried out at 35° in the presence of 5% (v/v) butanol.

The amount of denatured protein bound to the resin is always considerably larger (by a factor of two) than that of the native protein.

Ligand Equilibria in Matrix-Bound Hemoglobin[3]

Hemoglobin appears to be a most suitable system for studying the effects of immobilization on the allosteric properties of polymeric pro-

[3] A. Colosimo, S. Stefanini, M. Brunori, and E. Antonini, *Biochim. Biophys. Acta* **328**, 74 (1973).

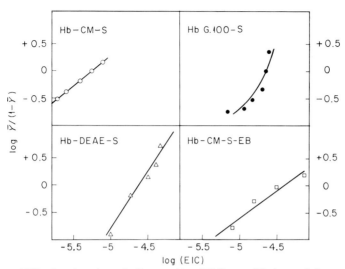

FIG. 1. Hill plot for the ethylisocyanide (EIC) equilibrium of human hemoglobin (Hb) bound to solid matrices: ○, hemoglobin coupled to Sephadex CM-C50 (CM-S), pH 7.4; △, hemoglobin coupled to Sephadex DEAE-A50 (DEAE-S), pH 7.9; ●, hemoglobin coupled to Sephadex G-100 (G100-S), pH 7; □, hemoglobin electrostatically bound to Sephadex CM-C50 (CM-S-E.B), pH 6.

teins.[4] The equilibrium of human hemoglobin covalently and noncovalently bound to different polysaccharide gels has been studied under different conditions.

In experiments the ferrous form of ethyl isocyanide is used as the ligand because of the advantages it offers over gaseous ligands. The basic reactivity of hemoglobin does not change when it is covalently bound to the various gels. The spectra of the deoxy, oxy, and ethylisocyanide derivatives are similar to those of the untreated protein, and the ligand reactions are completely reversible, yielding the deoxygenated derivative. The equilibrium of the matrix-bound hemoglobin with ethylisocyanide is, however, very different from that of the protein in solution. Both homotropic (cooperativity) and heterotropic (Bohr effect) interactions are also reduced (Fig. 1).

These changes may be due to (a) chemical modifications of the protein involved in the linking process to the polysaccharide; (b) dissociation of the protein into subunits, which may be the prevailing species attached to the resin; and (c) restriction of conformational mobility as-

---

[4] E. Antonini and M. Brunori, in "Hemoglobin and Myoglobin in Their Reactions with Ligands," Vol. 21. North-Holland Publ., Amsterdam, 1971.

sociated with immobilization. Apart from the detailed mechanism involved in the alterations of the ethylisocyanide equilibrium, the data obtained show that both the homotropic and the Bohr effects (although not at the same time) can remain unchanged on attachment of hemoglobin to the dextran gels. Depending on pH and the type of matrix employed, the ligand affinity may be higher or lower than that of the protein in solution.

Chemical Activation of Matrix Groups. Activation of carboxyl groups in Sephadex CM-C50 and binding to the protein are carried out by simply adding CCDI (0.5 g per gram of dry resin) to the protein-matrix mixture in phosphate buffer (pH 6.5).

The hydroxyl groups of Sephadex G-100 and Sephadex DEAE-A50 are activated as follows: CNBr (50 mg/ml) in water is added dropwise to the resin. The suspension is slowly stirred overnight at 25°; the pH is maintained at 10.5 by adding 4 M NaOH. The final amount of CNBr present is 1.25 g per gram of dry resin. After 8–10 min the resin is freed from the excess CNBr by washing on a Büchner funnel with bicarbonate buffer (0.1 M). This operation should be carried out in the shortest possible time because of the instability of the activated groups. The activated resin is then resuspended in a minimum volume of buffer, mixed with the protein solution, and allowed to react at 4° after adjusting the pH to 9.5 with NaOH.

Preparation of Matrix-Bound Hemoglobin. Chemical coupling between the active groups of the matrices and hemoglobin is achieved by the methods commonly used for other proteins with only minor modifications. The resins are allowed to swell overnight in the proper buffer, and, once activated, they are mixed with a 7% hemoglobin solution to give a ratio of 5 g, dry weight, of resin per gram of hemoglobin. Coupling is always carried out at 4° for (a) 12–18 hr for Sephadex G-100 and Sephadex DEAE-A50, and (b) 6–10 hr for Sephadex CM-C50. Most of the protein is found to be coupled to the activated matrix; the protein which has not reacted is washed off by suspending and centrifuging the complex several times with the desired buffer until the supernatant liquid is colorless.

The amount of hemoglobin bound to the resin is determined by spectrophotometric analysis of the supernatant after the coupling reaction.

Preparation of a Noncovalent (Electrostatic) Complex between Hemoglobin and Sephadex CM-C50. At sufficiently low ionic strength, hemoglobin binds strongly to Sephadex CM-C50. Two grams of resin, swollen overnight in phosphate buffer (0.01 M, pH 6), are mixed with 0.5 g of hemoglobin in the same buffer. The suspension is gently stirred for a few minutes at room temperature and then washed several times with the buffer.

Ethylisocyanide-Binding Curves for Matrix-Bound Hemoglobin. The formation of the protein–ligand complex after each addition of ethylisocyanide is followed spectrophotometrically.

Special care must be taken to minimize the errors in evaluating the free ligand concentration.

The most satisfactory procedure is as follows: A series of different samples (approximately 1.5 ml) of matrix-bound hemoglobin are suspended in buffer containing $Na_2S_2O_4$ and the volume is adjusted to 20 ml. Ethylisocyanide solution is added with an AGLA microsyringe to obtain the desired ligand concentration (1–100 μM). After equilibration (about 10 min), the supernatant is eliminated and each sample is transferred to a quartz cuvette with the minimum volume of liquid. The absorbance of the samples is measured in the region between 600 and 370 nm after allowing the suspension to settle for a well-defined time. By using appropriate opaque glass in the reference beam, it is possible to obtain satisfactory spectra with reasonable band widths (1–2 nm). However, quantitative data are difficult to obtain by spectral methods because of the variability in the packing of the suspension. Variations in the packing modify both the base line of the spectrum (owing to changes in scattering) and the protein absorbance (owing to changes in the apparent concentration). These errors can be minimized (a) by recording the spectra at a constant time after the beginning of the packing, and (b) by normalizing the observed absorbance changes using absorbance values taken at the isosbestic points.

Subunit Exchange Chromatography[5]

A multisubunit protein may be linked to a solid matrix under conditions in which the subunits rather than the complete protein are bound to the resin. If an equilibrium exists between the polymer and subunits in solution, the soluble subunits will also interact with those covalently attached to the resin to form a matrix-bound polymer. This is the case for a number of polymeric proteins, and it forms the basis for "subunit exchange chromatography." A system in which the matrix-bound subunits are in equilibrium with a soluble associating–dissociating system containing the same subunits may be quantitatively analyzed in simple terms using chromatographic techniques. Consider the simplest case of a reversible dissociation of a dimer into monomers. Two equilibria can

[5] E. Antonini, M. R. Rossi Fanelli, and E. Chiancone, *in* "Protein-Ligand Interactions," p. 45. de Gruyter, Berlin, 1975.

be established in the presence of monomers attached to a solid matrix.

$$X + X \underset{L_2}{\overset{L_1}{\rightleftharpoons}} X_2, \qquad (X)^2 L_1 = (X_2) \tag{1}$$

$$R\text{-}X + X \rightleftharpoons R\text{-}X_2, \qquad (R\text{-}X)(X)L_2 = (R\text{-}X_2) \tag{2}$$

where X and X_2 are the monomer and dimer in solution, R-X is the matrix-bound monomer, $R\text{-}X_2$ is the dimer formed by the binding of a monomer in solution to R-X, and L_1 and L_2 are the association constants in deciliter per gram.

If (y) is defined as the initial concentration of the protein in solution, which is bound at equilibrium to the matrix, and $(R\text{-}X_t)$ is the total concentration of monomer initially bound to the matrix,

$$(y) = (R\text{-}X_2)/2 \tag{3}$$

$$(R\text{-}X_t) = (R\text{-}X) + (R\text{-}X_2)/2 = (R\text{-}X) + (y) \tag{4}$$

From Eqs. 1 and 2 the following relationship can be obtained:

$$(y) = \frac{[(R\text{-}X_t)/2]}{\left[\dfrac{L_1}{L_2}\dfrac{(X)}{(X_2)} + \tfrac{1}{2}\right]} \tag{5}$$

The ratio $(X)/(X_2)$ corresponds to the ratio $\alpha/(1-\alpha)$ for the protein in solution, α being the degree of dissociation, which is a function of the total concentration in solution at equilibrium (c):

$$(1-\alpha)/\alpha^2 = L_1 c \tag{6}$$

Thus, Eqs. 5 and 6 relate the quantity (y) to the variables in the system. Similar treatments may be developed for polymerization equilibria beyond the dimer stage, i.e., for values of n, the degree of polymerization, higher than 2.

Experimentally, the quantity (y) can easily be determined by chromatography (Fig. 2). Difficulties may arise from the heterogeneous properties of the covalently bound subunits, which reflect various sites of attachment. In any case, the procedure may be useful as an analytical tool for measuring association–dissociation processes in polymeric proteins under conditions that are not easily determined by other methods. This method may also be useful for evaluating the relative tendency of monomers of homologous proteins to associate, i.e., to form hybrids.

Hemoglobin also provides a typical case for studying the behavior of protein systems in general. For chromatographic experiments similar to those reported in Fig. 1, Sepharose 4B is activated with cyanogen bromide, and oxyhemoglobin is coupled to it at pH 9. After the reaction, the resin is washed with 0.3 M $KHCO_3$ and 2.0 M NaCl in 0.01 M phos-

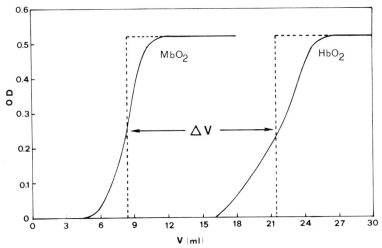

Fig. 2. Elution profiles of MbO$_2$ and HbO$_2$ from a column containing Sepharose-bound oxyhemoglobin. The protein concentration in the solid phase is 4.7 mg/ml, that in the eluting solution 1 mg/ml. The column (8.5 ml bed volume) was equilibrated with 0.1 M phosphate buffer at pH 7.0 + 0.1 mM EDTA. Dashed lines represent the position of the hypersharp front of the boundaries. ΔV is related to (y) of Eq. 5 by $(y) = (\Delta V \times c)/v$, where c is the concentration of the protein in the plateau region and v is the volume of the column.

phate buffer at pH 7. After the washings, the amount of oxyhemoglobin bound to the resin is about 4–5 mg per milliliter of packed resin.

Electron-Transfer Processes in Immobilized Proteins[6]

Cytochrome c bound to a solid matrix such as Sepharose (by CNBr procedures) is still capable of undergoing electron-transfer reactions with proteins of low and high molecular weight, such as soluble cytochrome c. It maintains unaltered redox potentials, but fails to undergo rapid autoxidation in the presence of cytochrome oxidase. This is probably due to the chemical change associated with the coupling procedure rather than to immobilization *per se*.

The kinetics of electron transfer between soluble and matrix-bound cytochrome c may be followed in a system consisting of Sepharose-bound cytochrome c, catalytic amounts of cytochrome oxidase, and soluble cytochrome c. The oxidation of the reduced immobilized cytochrome c occurs by electron transfer to the soluble cytochrome c maintained in a steady oxidized form by the presence of oxygen and cytochrome oxidase.

[6] A. Colosimo, E. Antonini, and M. Brunori, *Biochem. J.* **153**, 657 (1976).

The kinetics is complex, with much slower rates than those expected for a similar system in solution. These results may serve as a model for a better understanding of the behavior of the electron transfer processes in the respiratory chain.

[38] Immobilized Proteins in Single Crystals

By F. A. Quiocho

The intermolecular cross-linking of carboxypeptidase A molecules within single crystals in 1962 represents one of the earliest successful attempts to immobilize (insolubilize) enzymes. This success was most instrumental in the extensive and detailed study which followed of the physicochemical and enzymic properties of such crystals over an otherwise inaccessible range of conditions. Such a study was primarily undertaken in order to assess, under similar or identical conditions, the relation of the structure and function of enzymes in solution to the same molecule in the solid form. This assessment was deemed necessary following the spectacular success of X-ray diffraction technique in solving the three-dimensional structure of myoglobin[1] and lysozyme.[2]

Immobilized single protein crystals were first achieved by cross-linking carboxypeptidase A with glutaraldehyde effectively producing completely insoluble infinite three-dimensional net.[3] The dialdehyde was previously introduced and is, in fact, extensively used as a fixative agent in the preparation of specimens for electron microscopy. The effectiveness of glutaraldehyde in the preservation of the fine structure of fixed tissues must stem from *in situ* insolubilization of many proteins.[4]

The purpose of this chapter is to describe briefly the technique of immobilizing single protein crystals, present data on the properties of such crystals, and discuss the uses of insolubilized protein crystals in X-ray diffraction analysis and in studies of the physicochemical properties of enzymes in the solid state.

Protein Crystals

Protein crystals that diffract X-ray radiation can be obtained by a variety of crystallization conditions (for example, see this series Vol. 22,

[1] J. C. Kendrew, *Science* **139**, 1259 (1963).
[2] C. C. F. Blake, D. F. Koening, G. A. Mair, A. C. T. North, D. C. Phillips, and V. R. Sarma, *Nature* (*London*) **106**, 757 (1965).
[3] F. A. Quiocho and F. M. Richards, *Proc. Natl. Acad. Sci. U.S.A.* **52**, 833 (1964).
[4] D. D. Sabatini, K. Benesch, and R. J. Barrnett, *J. Cell Biol.* **17**, 19 (1963).

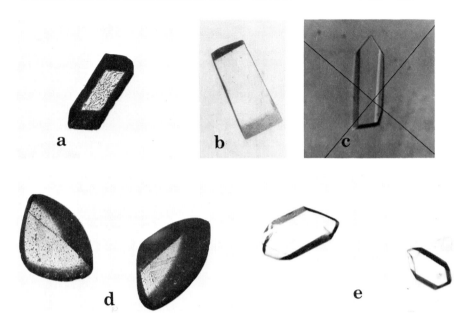

FIG. 1. Protein crystals: (a) carboxypeptidase Aα (elongated along a axis); (b) carboxypeptidase Aα (elongated along b axis); (c) carboxypeptidase Aγ (elongated along b axis); (d) concanavalin A; (e) L-arabinose-binding protein. Crystals shown in (a), (c), and (d) were treated with glutaraldehyde.

p. 24). However, unlike crystals of inorganic or organic small molecules, protein crystals like those shown in Fig. 1 have large liquid-filled channels surrounding individual protein molecules. As a consequence of the existence of these channels, protein crystals commonly contain about 40% by volume of solvent or liquid of crystallization; values vary from about 30% to 75%.[5] Since these channels are very permeable to low-molecular-weight solutes, it is possible to study the chemical behavior of proteins in much the same way as one does in solution.

Other properties of protein crystals are that they are very fragile and are easily cracked or chipped during even gentle mechanical manipulation. Protein crystals also either dissolve or become disordered if the composition of the immersion medium is changed much from that in which they were grown. Often, enzyme crystals soaked in solution containing heavy-atom derivatives or substrates or inhibitors would become disordered. The inherent problems associated with the fragile nature and

[5] B. W. Matthews, *J. Mol. Biol.* **33**, 491 (1968).

instability of protein crystals precluded a detailed study of the properties of enzymes in the crystalline state over a wide variety of conditions or identical conditions used in solution chemistry. A method of insolubilizing crystals or immobilizing protein molecules within single crystals was primarily developed to circumvent these problems. In retrospect, it would seem that insolubilization of protein crystals could be easily accomplished since the molecules within a crystal are ordered in three dimensions and any reactive amino acid side chain between molecules, which would be few, would be effectively juxtaposed for cross-linking with bifunctional reagents. X-ray structure analysis of a number of proteins indicate that only a few polar side chains (i.e., lysine) at the surface of protein molecule would have any appreciable flexibility in the crystalline state. Crystallographic analysis of glutaraldehyde-treated carboxypeptidase Aα-crystals indicates that intermolecular cross-links are most likely found between lysine residues of different molecule and are few in number (Quiocho and Lipscomb, unpublished data).

Insolubilization of Protein Crystals

The successful use of the bifunctional reagent glutaraldehyde as a fixative in the preparation of specimens for electron microscopy[5] presages its use to insolubilize single protein crystals or to immobilize proteins onto solid matrix or support (glutaraldehyde was first used in cross-linking of proteins in the tanning of leather[6]). A number of enzymes have been found to maintain sufficient activity after such fixative treatment of tissues to permit their histochemical localization by specific substrate stains. These successful applications of glutaraldehyde as a cross-linking agent provided a reasonable basis for its utilization in insolubilizing single protein crystals.

Besides carboxypeptidase A crystals,[3] other protein crystals, among which are lysozyme,[7] rennin,[8] β-lactoglobulin,[9] and concanavalin A (Quiocho, unpublished data), have also been immobilized by glutaraldehyde. The cross-linking with the dialdehyde confers several advantages: the crystals, which are normally very fragile, become much more sturdy and robust, so that there is much less chance of damage in handling; they become completely insoluble under a variety of solution conditions but

[6] M. L. Fein, E. H. Harris, J. Naghaski, and E. M. Filacrione, *J. Am. Leather Chem. Assoc.* **54**, 488 (1959).
[7] D. J. Haas, *Biophys. J.* **8**, 549 (1968).
[8] C. W. Bunn, N. Camerman, L. T. T'sai, P. C. Moews, and M. E. Baumber, *Philos. Trans. R. Soc. London Ser. B* **257**, 153 (1970).
[9] W. H. Bishop and F. M. Richards, *J. Mol. Biol.* **33**, 415 (1968).

remain permeable to solute; pH and temperature can be varied over a wide range without dissolution or deterioration of the cross-linked crystals.

Insolubilized amorphous proteins can also be obtained by treatment of proteins initially in solution with glutaraldehyde.[10,11]

Reaction of Glutaraldehyde with Protein Crystals

The reaction of glutaraldehyde with crystals of carboxypeptidase A has been extensively studied,[3,10,12] and consequently some results of these studies will be presented.

Crystals of carboxypeptidase A, with differing morphology (Fig. 1), can be obtained at pH 7.5 adapting basically the condition used by Anson.[13] Suspension of these crystals in solutions of high ionic strength due to the presence of salt or ionic substrate would disorder or dissolve the crystals to varying degrees. Following treatment with glutaraldehyde, carboxypeptidase A crystals became insoluble and are no longer susceptible to disorder under the conditions stated above or in a wide range of conditions, such as high or low pH, or high ionic strength.[3] The cross-linked crystals can no longer be chipped or cracked even under vigorous agitation with a glass rod. Normally untreated native carboxypeptidase crystals would dissolve in solution of high ionic strength ($MgCl_2$ or NaCl) and, like all other protein crystals, are also very fragile and easily cracked or pulverized during mechanical manipulation. The stability conferred on crystals following immobilization was notable. One particular column of cross-linked crystals was intermittently used in a variety of experiments and 67 different assays over a period of 3 months at room temperature without loss of activity.[10] Moreover, prolonged and repeated suspension of insolubilized carboxypeptidase A crystals in 8 M urea did not inactivate the enzyme. However, 8 M urea treatment resulted in the complete loss of crystalline diffraction of the insolubilized crystals without any evidence of solubilization of the protein.

Glutaraldehyde-insolubilized carboxypeptidase Aγ crystals showed substantial peptidase and esterase activity when assayed in the crystalline state. Between 30 and 70% of the initial peptidase activity obtained prior to cross-linking was observed for particular crystal batches.[10]

[10] F. A. Quiocho, and F. M. Richards, *Biochemistry* **5**, 4062 (1966).
[11] E. F. Jansen, Y. Tomimatsu, and A. C. Olson, *Arch. Biochem. Biophys.* **144**, 394 (1971).
[12] F. A. Quiocho, W. H. Bishop, and F. M. Richards, *Proc. Natl. Acad. Sci. U.S.A.* **57**, 525 (1967).
[13] M. L. Anson, *J. Gen. Physiol.* **20**, 663 (1937).

Procedure for Cross-Linking of Carboxypeptidase A Molecules in Crystals

The treatment of crystals of carboxypeptidase Aγ, prepared by the method of Anson[13] and elongated along the b axis[3] with 6% glutaraldehyde at pH 7.5 for 3 hr, did not alter the diffraction pattern of the crystal.[3] However, with carboxypeptidase Aα crystals, which are elongated along the a axis and were exclusively used for the crystallographic structure analysis,[14] this treatment disordered the diffraction pattern of the crystals. However, by using lower concentration of the aldehyde, pretreating the commercial aldehyde with activated charcoal and cross-linking in the presence of a salt, insolubilized carboxypeptidase Aα crystals can be obtained that are suitable for X-ray diffraction studies even up to 2 Å resolution. The procedure is described in detail below.

UNIT CELL DIMENSIONS OF CARBOXYPEPTIDASE Aα UNDER VARIOUS CONDITIONS

Crystal	Cell dimension (Å)			
	a	b	c	β^a
1. Native enzyme in mother liquor	50.9	57.9	45.0	94°40'
2. From same batch as No. 1, treated with 6% glutaraldehyde for 3 hr	50.8	58.1	45.2	94°40'
3. Native enzyme in water in flow tube	51.1	57.8	45.1	95°10'
4. Same crystal as No. 3 after treatment with 1% glutaraldehyde for 3 hr	50.6	58.1	45.1	95°10'
5. Cross-linked crystal in flow tube in water	50.7	58.0	44.8	94°40'
6. No. 5 after 2 hr in 0.01 M piperidine, pH 11	50.8	58.4	45.7	94°40'
7. No. 6 after 2 hr in 0.01 M acetate buffer, pH 5	50.6	58.1	45.1	94°40'
8. No. 7 after 2 hr in 2 M MgSO$_4$	50.6	57.9	44.9	94°40'
9. No. 8 after 18 hr in water	50.8	58.0	45.4	94°40'

[a] Only in the first case was the angle directly measured in an (hol) projection. The others are assumed to be identical or slightly different on the basis of settings required for proper orientation of (hko) and (okl) planes on a precession camera. Data taken from F. A. Quiocho and F. M. Richards, *Proc. Natl. Acad. Sci. U.S.A.* **52**, 833 (1964).

Carboxypeptidase Aα crystals are placed in a small test tube (4 × 50 mm) and equilibrated in 100 mM LiCl–20 mM sodium Veronal buffer pH 7.5. (All operations are done at 4°.) The crystals are subsequently suspended in the same salt–buffer solution, but containing 0.1% dialdehyde. (The commercial 25% aqueous solution of glutaraldehyde is treated with fine-mesh activated charcoal just prior to use.) The test tube containing

[14] M. L. Ludwig, I. C. Paul, G. S. Pawley, and W. N. Lipscomb, *Proc. Natl. Acad. Sci. U.S.A.* **50**, 282 (1963).

the crystals is attached to the shaft of a motor, positioned horizontally, and gently rotated at 1 rpm. (Surface tension in the narrow test tube is sufficient to prevent spilling.) The glutaraldehyde solution is changed at least twice during a period of 6 hr. Excess aldehyde is removed after the reaction by repeated washing with the salt–buffer solution.

Crystals of carboxypeptidase Aα used in the X-ray studies are originally obtained in Tris-HCl buffer pH 7.5.[14] Since Tris buffer has primary amine groups that could react with glutaraldehyde, it is necessary in the cross-linking reaction to exchange this buffer with sodium Veronal buffer, which does not have amine groups. The exchange is not deleterious to the crystal. Intermolecular cross-linking by glutaraldehyde is ideally suited with single protein crystals that are obtained by "isoelectric point crystallization" (i.e., carboxypeptidase A, concanavalin A, aspartate transcarbamylase), primarily because these crystals are obtained in plain buffer at or near their isoelectric points, but not with the use of high concentration of ammonium sulfate, the reagent most commonly used to crystallize proteins. For crystals that are grown in ammonium sulfate, it might be possible, as in the case of aldolase (Quiocho, unpublished result), to exchange the ammonium sulfate, which would interfere in the cross-linking reaction, with high concentration of potassium phosphate buffer prior to treatment. Furthermore, for other protein crystals, optimal cross-linking conditions that will give cross-linked crystals suitable for X-ray analysis and will not inactivate protein function may have to be determined, taking into consideration variables such as aldehyde purity and concentration, ionic strength and salt specie, temperature, pH, and possible inclusion of substrates or other ligands.

Nature of Glutaraldehyde Reaction

The chemical nature of the reaction of glutaraldehyde with proteins is not clearly understood. Glutaraldehyde predominantly reacts with amino groups of proteins. This reaction has generally been assumed to form a Schiff base. This assumption, however, is not clearly substantiated by several facts obtained from studies concerning the reaction of glutaraldehyde with proteins.[3,10,12] These facts are as follows: (1) The reaction is rapid in aqueous solution at room temperature. The reaction occurs at a fairly wide range of pH, from 5 to 9.[15] However, the pH value for the

[15] Protein crystals usually take on a yellow color after reaction with glutaraldehyde at pH 7.5; the intensity of this color is dependent on the time of reaction and/or concentration of the aldehyde.[3] Treatment with NaBH$_4$ causes the disappearance of the yellow color. Subsequently, it was found that treatment of commercial glutaraldehyde with fine mesh activated charcoal prior to use gives less yellow color upon reaction.

most rapid reaction of glutaraldehyde with various proteins in solution to produce insoluble cross-linked products is at their isoelectric points.[11] (2) The modification is apparently irreversible and survives treatment with urea, semicarbazide, and wide ranges of pH, ionic strength, and temperature. (3) Amino acid analyses of hydrolyzates of modified protein indicate partial loss of lysine. The loss increases with time of exposure or an increase in aldehyde concentration. The standard amino acid chromatogram of the glutaraldehyde-treated protein hydrolyzate did not reveal new ninhydrin-positive material. (4) Comparison of treated and untreated protein reveals an acid shift of the hydrogen ion titration curve in the alkaline region.[9] A decrease in apparent pK_a of 0.5 to 1 unit is seen near pH 9 with γ-lactoglobulin.

By analogy with the formation of Schiff bases in the reaction of aniline with aryl aldehydes, the reaction of glutaraldehyde with lysine residues has been generally assumed to involve the same formation of Schiff bases. However, this assumption is not corroborated by data presented above. The failure of semicarbazide to reverse the cross-linking reaction mitigates the suggestion of the existence of a simple, α,ω-bis-Schiff bases in the protein. Schiff bases formed from aniline and aryl aldehydes have been shown to be rapidly attacked by semicarbazide to yield the semicarbazone and the corresponding free amine.[16] Moreover, Schiff base formation is readily reversible; the equilibrium constant is usually unfavorable for nonconjugated amine formation in aqueous solution.[17] Such Schiff-base compounds would not survive a wide range of pH and temperature, much less the acid hydrolysis required for amino acid analysis.

Richards and Knowles,[18] in an attempt to identify the reactive cross-linking specie or species and possibly postulate a reaction mechanism, have subjected aqueous solution of glutaraldehyde to nuclear magnetic resonance (NMR) spectroscopic studies. These investigators initially observed from NMR spectra peaks at τ 4.8, 6.6, 7.4, and 8.3, which were ascribed to the presence of a significant amount of unsaturated polymeric compound (I), in commercial aqueous solution of glutaraldehyde, and which they further suggested to be the effective cross-linking agent.

$$\text{OCH—CH}_2\text{—CH}_2\text{—CH}_2\text{—CH}=\overset{\overset{\text{CHO}}{|}}{\text{C}}\text{—CH}_2\text{—}\overset{\overset{\text{CHO}}{|}}{\text{C}}=\text{CH—CH}_2\text{—CH}_2\text{—CH}_2\text{—CHO}$$
(I)

Furthermore, this study indicates that commercial aqueous glutaraldehyde

[16] E. H. Cordes and W. P. Jencks, *J. Am. Chem. Soc.* **84**, 827 (1962).
[17] W. P. Jencks, *in* "Progress in Physical Organic Chemistry," p. 63. Wiley (Interscience), New York, 1964.
[18] F. M. Richards and J. R. Knowles, *J. Mol. Biol.* **37**, 231 (1968).

consists of a mixture of oligomers and higher polymeric unsaturated materials.

The findings of Richards and Knowles[18] seem to be at variance with the NMR studies conducted by Hardy et al.[19] The major findings of these investigators are as follows: (1) Although the NMR spectra at 60 MHz, 33.5°, of two commercial aqueous solutions of glutaraldehyde showed peaks in approximately the same position as those initially observed by Richards and Knowles[18]; however, the relative areas, which differed in the two commercial samples, are different from those found by Richards and Knowles.[18] (2) Unsaturated polymeric compounds, such as (I), which have a relatively weak Uv absorption at about 235 nm, constituted only a very minor component of the commercial glutaraldehyde solution. (3) Purification of the glutaraldehyde yielded a product that was exclusively the monomeric form of the dialdehyde and devoid of minor unsaturated polymeric aldehyde components. (4) The NMR spectra of aqueous solution of purified glutaraldehyde were those expected of glutaraldehyde itself and were different from those observed with the commercial solution. These spectra were interpreted on the basis of equilibria between the monomeric glutaraldehyde and three hydrated forms. The NMR spectra of the aqueous solution of the redistilled or purified glutaraldehyde and the molecular weight did not change at all even after several weeks of incubation at room temperature. On the basis of the above results, Hardy et al.[19] concluded that cross-linking must involve the glutaraldehyde itself (monomeric specie) or one of its hydrates, not an unsaturated polymer. However, they further observed that at weakly alkaline solution (pH 8), the purified glutaraldehyde rapidly polymerized to an insoluble solid compound. The spectroscopic properties of this compound could be attributed to unsaturated structures, such as compound (I). Therefore, the possibility still exists that cross-linking of proteins in alkaline solution may involve unsaturated polymers of glutaraldehyde. However, the most rapid insolubilization of various proteins by glutaraldehyde was found to be at their isoelectric points, which vary from pH 4.8 for bovine serum albumin to pH 10.5 for lysozyme.[11] The pH for the maximum uptake and cross-linking of collagen with glutaraldehyde occurred at pH 8,[20] and for the preservation of ultrastructure in tissues with the aldehyde between pH 6.8 and 7.6.[4]

Richards and Knowles[18] proposed a possible cross-linking mechanism, which is predicated on the existence of a significant amount of α,β-unsaturated polymer (I) in commercial aqueous solution of glutaraldehyde (see Scheme 1).

[19] P. M. Hardy, A. C. Nicholls, and H. N. Rydon, *Chem. Commun.* **1969**, 565 (1969).
[20] J. H. Bowes, C. W. Cater, and M. J. Ellis, *J. Am. Leather Chem. Assoc.* **60**, 275 (1965).

(a) Aldol condensations:

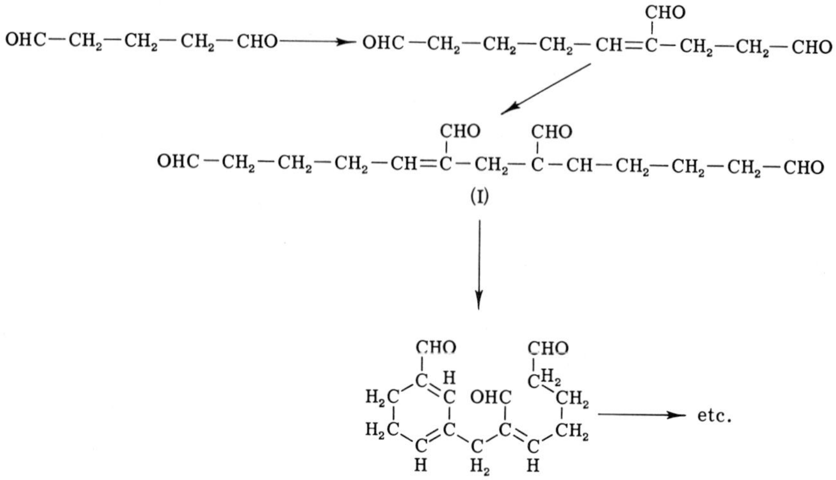

$$\text{OHC}-\text{CH}_2-\text{CH}_2-\text{CH}_2-\text{CHO} \longrightarrow \text{OHC}-\text{CH}_2-\text{CH}_2-\text{CH}_2-\overset{\overset{\text{CHO}}{|}}{\text{CH}}{=}\overset{}{\text{C}}-\text{CH}_2-\text{CH}_2-\text{CHO}$$

$$\text{OHC}-\text{CH}_2-\text{CH}_2-\text{CH}_2-\text{CH}{=}\overset{\overset{\text{CHO}}{|}}{\text{C}}-\text{CH}_2-\overset{\overset{\text{CHO}}{|}}{\text{C}}-\text{CH}-\text{CH}_2-\text{CH}_2-\text{CH}_2-\text{CHO}$$
$$(\text{I})$$

[cyclized aldol product] $\longrightarrow$ etc.

(b) Cross-linking reactions:

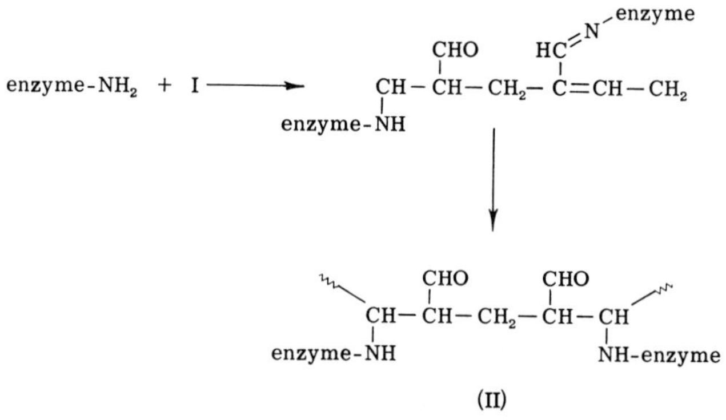

$$\text{enzyme-NH}_2 + \text{I} \longrightarrow \underset{\underset{\text{enzyme-NH}}{|}}{\text{CH}}-\overset{\overset{\text{CHO}}{|}}{\text{CH}}-\text{CH}_2-\overset{}{\text{C}}{=}\text{CH}-\text{CH}_2 \quad (\text{HC}{=}\text{N-enzyme})$$

$$\underset{\underset{\text{enzyme-NH}}{|}}{\text{CH}}-\overset{\overset{\text{CHO}}{|}}{\text{CH}}-\text{CH}_2-\overset{\overset{\text{CHO}}{|}}{\text{CH}}-\underset{\underset{\text{NH-enzyme}}{|}}{\text{CH}}$$
$$(\text{II})$$

Scheme 1

As indicated in Scheme 1, the reactive derivatives of glutaraldehyde (I) will give a Michael-type adduct with amino groups of a protein. The proposed mechanism accounts to some extent for most of the data on the nature of the reaction of glutaraldehyde with proteins as enumerated above. For instance, the adduct (II) will be expected to be stable to acid hydrolysis and should not be attacked by semicarbazide, in conformity with the above data. Further, the product (II) will have an appropriate pK_a similar to the pK_a determined for cross-linked β-lactoglobulin[9] but different from the one expected for a Schiff base.

Since there are discernible differences in the contents of commercial aqueous glutaraldehyde, it may very well be that most of these samples may contain a sufficient amount of compound (I) to account for its cross-linking properties. Besides NMR measurements, chromatographic studies have also indicated that commercial glutaraldehyde contains polymeric components. Glutaraldehyde tends to polymerize when left at room temperature and turns yellow, which is indicative of polymerization. Extensive loss of aldehyde moiety accompanies polymerization at alkaline pH.[21] Several methods have been developed with varying degree of success for the purification of glutaraldehyde.[19,22,23]

X-Ray Analysis of Insolubilized Protein Crystal

Glutaraldehyde-treated crystal of carboxypeptidase A,[3,24] lysozyme,[7] and rennin[8] gave a very similar diffraction pattern to those obtained with corresponding noncross-linked native crystal. Furthermore, comparisons of the diffraction pattern of cross-linked crystals and native carboxypeptidase Aγ crystals showed small intensity changes, which may be attributed to the additional atoms from the glutaraldehyde.[24] Some unit cell dimension data of the native and the cross-linked carboxypeptidase Aγ crystal, alone or suspended in a variety of solutions, are shown in the table.[3] The close similarity of the diffraction patterns of the cross-linked and the native protein crystal provides the strongest evidence that the reaction of glutaraldehyde does not alter the native conformation of the protein in the crystalline state. Moreover, in the case of carboxypeptidase Aα, 3 to 4 cross-links have been located using difference Fourier technique (Quiocho and Lipscomb, unpublished data). All the cross-links were at the surface of the molecule, evidently between lysine residues of adjacent molecules, not at or anywhere near the active site region of the enzymes. The introduction of these cross-links does not change the conformation of the enzyme.

Other Cross-Linking Agents

While glutaraldehyde has been extensively used to cross-link proteins in single crystals and immobilized proteins onto solid supports, other bifunctional reagents have also been tried with very limited success.[10]

[21] L. F. Fieser and M. Fieser, "Advanced Organic Chemistry." Van Nostrand-Reinhold, Princeton, New Jersey, 1961.
[22] H. D. Fahimi and P. Rochmans, *J. Microsc. (Paris)* **4**, 725 (1965).
[23] D. Hopwood, *Histochemie* **11**, 289 (1968).
[24] T. A. Steitz, M. L. Ludwig, F. A. Quiocho, and W. N. Lipscomb, *J. Biol. Chem.* **242**, 4662 (1967).

Recently a new class of cross-linking agent, cleavable bifunctional alkyl imidates, has recently been introduced with tremendous success in studies of protein–protein interactions.[25-27] (Insolubilization of carboxypeptidase A crystals can be effectively produced by treatment with methyl-4,4'-dithiobisbutyrimidate.) The cleavable dithiobisimidates offer several advantages as cross-linking agents: (1) the reaction of the imidate function with lysine specifically is very well understood and characterized as opposed to the reaction of glutaraldehyde[28]; (2) the reaction leads to a retention of positive charge; (3) the cross-links can be cleaved very easily and quantitatively by mild reduction; and (4) the reaction of imidates with proteins is mild and often can be carried at or near physiological pH. The cleavable bisimidates and other similar cross-linking agents may very well be better than with glutaraldehyde as the agent, especially in terms of obtaining insolubilized protein crystals that are suitable for X-ray diffraction studies. However, the fact that imidoesters are unstable and slowly hydrolyze in aqueous solution to the corresponding amide and alcohol[28] poses a serious drawback in that monosubstitutions may occur rather than cross-links.

Uses and Applications of Insolubilized Protein Crystal

Study of the Enzymic Behavior of Enzyme in the Solid State

Insolubilized carboxypeptidase Aγ crystals were extensively used in the study of the enzymic and chemical properties of the solid enzyme in a wide variety of conditions and, more important, in identical conditions used in solution studies.[10,12,29] It would have been impossible to use native and noncross-linked crystals under most of these conditions, primarily because the crystals would dissolve. These studies indicated that many of the properties of carboxypeptidase Aγ that are seen in solution are also observed in the crystalline state, but many changes in the detailed behavior of the enzyme occur in passing from solution to the solid state.

X-Ray Analysis of Protein Structure

Many of the preliminary X-ray diffraction studies of rennin have been carried out on crystals cross-linked or insolubilized with glutaraldehyde.[8]

[25] T. T. Sun, A. Bollen, L. Kahan, and R. R. Traut, *Biochemistry* **13**, 2334 (1974).
[26] K. Wang and F. M. Richards, *Isr. J. Chem.* **12**, 375 (1974).
[27] K. Wang and F. M. Richards, *J. Biol. Chem.* **249**, 8005 (1974).
[28] M. J. Hunter and M. L. Ludwig, *J. Am. Chem. Soc.* **84**, 3491 (1962).
[29] W. H. Bishop, F. A. Quiocho, and F. M. Richards, *Biochemistry* **5**, 4077 (1966).

The use of insolubilized rennin crystals very much extended the scope of experiments aimed at obtaining heavy-atom derivatives. The ionic strength and pH could be changed over a wide range without affecting the quality of the crystal but facilitating the uptake of heavy atom.

Crystallographic Studies of the Binding of Substrates and Inhibitors to Insolubilized Enzyme Crystals

Lipscomb and co-workers[24,30,31] have extensively used carboxypeptidase Aα crystals intermolecularly cross-linked with glutaraldehyde in the crystallographic studies of the binding of substrates and inhibitors. Cross-linking had to be performed in order to keep the crystals suspension free of dissolved enzyme, which would hydrolyze substrates, or to prevent solubilization of crystals at moderately high ionic strengths. The glutaraldehyde-insolubilized carboxypeptidase Aα crystals were of excellent quality and were further used to study the binding of glycyl-L-tyrosine, a substrate with an extremely low turnover number, at atomic resolution (2 Å resolution).[30,31] The binding of glycyl-L-tyrosine to immobilized carboxypeptidase Aα crystal caused numerous structural changes, particularly residues Arg-145, Glu-270, and Tyr-248, and a short segment of the polypeptide chain to which Tyr-248 is attached.[30] It is thus particularly significant that any conformational changes concomitant to ligand binding can still be accommodated in these cross-linked crystals.

Cross-linked crystals were used in the crystallographic study of aspartate transcarbamylase with bound 5'-iodocytidine triphosphate, an allosteric inhibitor.[32]

Diffusion of Solutes in Cross-Linked Protein Crystals and Model Membrane

Bishop and Richards[33] have investigated the time course of the diffusion of a series of bromine-containing solutes into single, cross-linked or insolubilized crystals of β-lactoglobulin using X-ray fluorescence technique to follow the diffusion. This investigation was primarily an attempt to define the nature of the solvent-containing part of a protein, using

[30] W. N. Lipscomb, G. N. Reeke, J. A. Hartsuck, F. A. Quiocho, and P. H. Bethge, *Philos. Trans. R. Soc. London Ser. B* **257**, 177 (1970).
[31] F. A. Quiocho and W. N. Lipscomb, *Adv. Protein Chem.* **25**, 1 (1971).
[32] B. F. P. Edwards, D. R. Evans, S. G. Warren, H. L. Monaco, S. M. Landfear, G. Eisele, V. L. Crawford, D. C. Wiley, and W. N. Lipscomb, *Proc. Natl. Acad. Sci. U.S.A.* **71**, 4437 (1974).
[33] W. H. Bishop and F. M. Richards, *J. Mol. Biol.* **38**, 315 (1968).

diffusion behavior of test solutes of graded size as a parameter. The nature of the solvent near dissolved macromolecules is of great interest since the conformations assumed by the latter are intimately related to the properties of the former.[34,35] The result of this investigation indicated that the major portion of the liquid of crystallization can be considered identical to water in bulk. Diffusion coefficients and pore radius in the crystal were also determined. The pore radius in the crystal of β-lactoglobulin was estimated to be 8–13 Å.

Protein crystal may be considered as an artificial membrane with completely regular and accurately definable pores. Not only the size of the pores, but, for a protein of known X-ray structure, the amino acid residues or functional groups lining these pores would be known. A study of such preparations might be of interest in defining the properties of membrane models. For immobilized large crystals (such as those shown in Fig. 1) with appropriate space groups and with sectioning procedures now applicable, the orientation of the molecules with respect to the plane of the membrane should be adjustable at will.

Acknowledgment

Much of the research on insolubilized single protein crystals was undertaken in the laboratory of Professor F. M. Richards, Yale University, and the X-ray analysis was carried out in the laboratory of Professor W. N. Lipscomb, Harvard University. This author is grateful to both professors for their guidance and help. Current studies are supported by the Robert A. Welch Foundation C-581.

[34] W. J. Kauzman, *Adv. Protein Chem.* 14, 1 (1959).
[35] O. Sinanoglu and S. Badulnur, *Fed. Proc., Fed. Am. Soc. Exp. Biol.* 24 (2), Suppl. No. 15, p. 12.

[39] Mechanochemistry of Immobilized Enzymes: A New Approach to Studies in Fundamental Enzymology

I. Regulation by Mechanical Means of the Catalytic Properties of Enzymes Attached to Polymer Fibers

By I. V. BEREZIN, A. M. KLIBANOV, G. P. SAMOKHIN, and KAREL MARTINEK

There are at least two reasons why the study of immobilized enzymes has attracted so much attention in the last decade. First, immobilized enzymes, which are more stable and technologically more expedient than natural biocatalysts, are of great value economically. Second, the investigation of immobilized enzymes is becoming a promising tool for fundamental studies in enzymology and biochemistry. The first aspect of the problem has been extensively dealt with by the authors cited in footnotes[1-4]; the second, much less explored, aspect is the topic of this chapter.

Immobilized enzymes may contribute to the development of theoretical enzymology in the following ways.

1. As a result of immobilization the molecules of the enzyme acquire a fixed position in space with respect to one another; this allows the behavior of the enzyme to be studied without interference from autolysis or aggregation.[5-7]

2. Definite mutual localization of the subunits of the enzyme molecule allows the fine mechanisms of their interactions to be studied.[8,9]

3. Rigidification of the enzyme molecule due to its being covalently bound to a support permits the study of protein denaturation.[10]

4. The effect on the enzyme of high external fields (electrostatic,[11,12] magnetic,[13] etc.) arising from the binding between the enzyme and support can be studied.

5. Immobilization of multienzyme systems permits a realistic simulation of *in vivo* enzyme systems in membranes, which may help our understanding of the problems of regulation, compartmentation, and enzyme-enzyme interrelationships.[14-17]

Immobilized enzymes possess another interesting feature that makes them useful in studying fundamental problems of enzymology. On immobilization, a *microobject*, i.e., a molecule of the enzyme, becomes attached

[1] G. Kay, *Process Biochem.* 3(8), 36 (1968).
[2] M. D. Lilly and P. Dunnill, *Process Biochem.* 6(8), 29 (1971).
[3] K. Mosbach, *Sci. Am.* 224, 26 (1971).
[4] S. P. O'Neill, *Rev. Pure Appl. Chem.* 22, 133 (1972).
[5] D. Gabel and V. Kasche, *Biochem. Biophys. Res. Commun.* 48, 1011 (1972).
[6] K. Tanizawa and M. L. Bender, *J. Biol. Chem.* 249, 2130 (1974).
[7] K. Martinek, V. S. Goldmacher, A. M. Klibanov, and I. V. Berezin, *FEBS Lett.* 51, 152 (1975).
[8] W. W. Chan, *Biochem. Biophys. Res. Commun.* 41, 1198 (1970).
[9] S. Ikeda and S. Fukui, *Eur. J. Biochem.* 46, 553 (1974).
[10] I. V. Berezin, V. K. Antonov, and K. Martinek, eds., "Immobilized Enzymes," Chapter 5. Moscow State University Press, Moscow, 1976.
[11] L. Goldstein, *Biochemistry* 11, 4072 (1972).
[12] J. B. Taylor and H. E. Swaisgood, *Biochim. Biophys. Acta* 284, 268 (1972).
[13] P. J. Robinson, P. Dunnill, and M. D. Lilly, *Biotechnol. Bioeng.* 15, 603 (1973).
[14] S. Gestrelius, B. Mattiasson, and K. Mosbach, *Eur. J. Biochem.* 36, 89 (1973).
[15] P. A. Srere, B. Mattiasson, and K. Mosbach, *Proc. Natl. Acad. Sci. U.S.A.* 70, 2534 (1973).
[16] D. Thomas and G. Broun, *Biochimie* 55, 975 (1973).
[17] A. David, M. Metayer, D. Thomas, and G. Broun, *J. Membr. Biol.* 18, 113 (1974).

to *a macroobject*, i.e., the support. The support, unlike the enzyme, can be subjected to direct mechanical deformation (stretching, compression, etc), which in principle can induce two kinds of effects: alteration of the conformation of the enzyme molecule and change in the properties of the microenvironment in which the molecules of the enzyme are localized. In other words, external mechanical action applied to the support can change the major parameters of the enzyme process. In our laboratory we are engaged in developing this new mechanochemical approach to the mechanistic problems of (a) enzyme catalysis and (b) functioning of natural mechanochemical enzymic systems (see below).

Study of the Mechanisms of Enzyme Action

Regulation of the Catalytic Properties of Enzymes Chemically Attached to Nylon Fiber, Induced by Mechanical Stretching of the Support

Principle. The structure–function relationship of biocatalysts is a major problem in enzymology. The simplest approach to this problem is to deform the molecule of the enzyme in a certain way (other conditions being constant) to see how this will affect the enzyme. Unfortunately, the small size of the molecule hardly allows an external deformation force to be applied directly to the enzyme. But, on being rigidly attached to a water-insoluble elastic support by several bonds, the enzyme molecule may become altered if the support is deformed. This is the underlying idea of our approach to studying the relationship between the structure of the enzyme and its catalytic properties.[18]

Materials. Use was made of α-chymotrypsin and trypsin, both well studied enzymes. Nylon fiber (140 0.03-mm filaments plaited together) was chosen as a support because of its elasticity, which increases further when the fiber is subjected to a preimmobilization treatment. Another elastic support are blond human hairs and viscose (cellulose) fiber.

Enzyme-Support Binding. The covalent binding of the enzymes to nylon fibers was performed according to Sundaram and Hornby,[19] who described the binding of urease to a nylon tube (see also this volume [9]). A nylon fiber 6 m long is folded into a ring and treated with 3 N HCl at 45° for 1.5 hr with stirring. Then the partially hydrolyzed fiber is thoroughly washed with water, 0.1 M NaHCO$_3$, and again with water, after which, at pH 9.4 (0.2 M NaHCO$_3$) and 4°, it is modified for 15 min with

[18] A. M. Klibanov, G. P. Samokhin, K. Martinek, and I. V. Berezin, *Biochim. Biophys. Acta,* in press.
[19] P. V. Sundaram and W. E. Hornby, *FEBS Lett.* **10**, 325 (1970).

2.5% glutaraldehyde, which forms Schiff's bases with —NH_2 groups of the support. The fiber obtained as a result of all these manipulations is rather elastic—it can be reversibly stretched by ~30%. The enzymes are bound both to unstretched and stretched (by 25%) fiber. To this end, the fiber, after glutaraldehyde has been washed off with water, is treated for 1.5 hr at 4° with 0.1% enzyme solution at pH 8.0 (0.03 M KH_2PO_4), which via its —NH_2 groups binds to the carbonyl groups of the support-bound glutaraldehyde. Noncovalently bound enzyme is washed off with water, 1 mM HCl, 0.1 M $NaHCO_3$, 1 M NaCl, and again with water. The resulting enzyme-carrying fibers are stored in water at 4°.

Covalent binding of the enzymes to viscose (cellulose) fiber is carried out according to a modification of a procedure described previously (see references cited by Klibanov et al.[18]). Viscose fiber, 2.5 m, is wound around a glass rod and incubated, with stirring, for 30 min at room temperature in a 0.4 M aqueous solution of $NaIO_4$. This procedure results in cleavage of part of the saccharide rings, with aldehyde groups being formed. Then the fibers, washed with water, 3% solution of $Na_2S_2O_3$, and water again, are incubated at 4° for 18 hr in 0.1% solution of the enzyme at pH 8.5 (0.2 M H_3BO_3). The noncovalently bound enzyme is washed off as described above.

The covalent binding of the enzymes to human hair is performed as follows. A 40 cm-long strand of hair (~20 hairs) is treated with glutaraldehyde and then, after washing, is incubated with a solution of the enzyme (see above).

Determination of the Activity of Support-Bound Enzymes. The enzymic activity of the support-bound α-chymotrypsin and trypsin is followed by the steady-state rate of hydrolysis of the specific substrates in a pH stat (Radiometer TTT 1c). Enzyme-carrying fibers (40–60 cm) are wound around a special stretching device (see Fig. 1). This device consists of a rod with right-hand threading on the upper half and left-hand threading on the lower. Both halves have matching rollers with the corresponding threading. When the rod rotates, the rollers move along it in different directions, now toward and now away from each other. Fiber wound between the rollers would be stretched, the degree of stretching being determined by the distance between the rollers.

The stretching device with an enzyme-carrying fiber wound around it is placed in a thermostated cuvette of a pH stat, containing 30 ml of 6 mM solution of the substrate (N-acetyl-L-tyrosine ethyl ester for α-chymotrypsin; N-tosyl-L-arginine methyl ester or N-benzoyl-L-arginine ethyl ester for trypsin) in 0.1 M KCl. The acid liberated as a result of the enzymic reaction is titrated with 20 mM KOH. The reaction is performed at pH 8.0 and 25°.

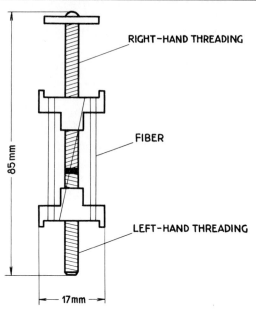

Fig. 1. Stretching device for enzyme-carrying fibers.

Study of the Kinetics of Interaction of Nylon Fiber-Bound Enzymes with Acylating Agents. We studied the effect of stretching nylon fiber on the kinetics of acylation of bound α-chymotrypsin by phenylmethyl sulfonylfluoride and p-nitrophenyl trimethylacetate. The α-chymotrypsin-carrying fiber is incubated in a 50 μM aqueous solution of the acylating agent ($+1\%$ isopropanol) at pH 7.0 (0.02 M KH_2PO_4) and 25°. The fiber is taken out of the solution and thoroughly washed with water; the relative enzymic activity of the unstretched fiber is determined in a pH-stat.

Major result. If the fiber is stretched, the immobilized enzymes undergo a 3- to 6-fold decrease in activity (Fig. 2). After the stretching, the fiber relaxes (this occurs almost instantaneously) and the catalytic activity of the enzymes returns to the initial level. This "stretch-relax" cycle may be repeated many times, the "decrease–increase" of the enzyme activity being recurrent.

We found, by varying the concentration of the substrates (N-acetyl-L-tyrosine ethyl ester for α-chymotrypsin and N-tosyl-L-arginine methyl ester for trypsin), that the Michaelis constants of the enzymic reactions do not change on stretching, but the maximum reaction rate decreases.

The result obtained (Fig. 2) cannot be attributed to diffusion phenomena, as we found that the presence and the magnitude of the mechano-

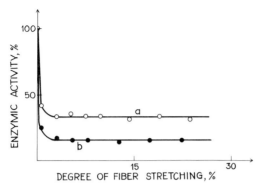

FIG. 2. Dependence of enzymic activity of (a) α-chymotrypsin and (b) trypsin attached to nylon fiber on the degree of stretching of the fiber. (a) Substrate, N-acetyl-L-tyrosin ethyl ester (6 mM). (b) Substrate, N-tosyl-L-arginine methyl ester (3 mM). Other conditions: pH 8, 0.1 M KCl, 25°.

chemical effect do not depend on the rate of stirring during the catalytic activity measurements, the temperature (15°–35°), the concentration of the substrate (as V_{max}, but not $K_{m,app}$, changes on stretching), or the initial activity of the immobilized enzyme (which changes 6-fold).[18] All these factors should have affected the diffusion-controlled process.[20]

The effect observed could in principle be explained by the alteration in the microenvironment of the immobilized enzyme on stretching the fiber. But this can also be ruled out for the following reasons. The presence and magnitude of the mechanochemical effect (Fig. 2) do not change as does ionic strength (from 0.01 to 3 M KCl), pH (7 to 8.5), or concentration of added isobutanol (from 0 to 10 vol %), nor does it depend on the character of the surface to which the enzyme is attached. To change the surface of the nylon fiber, we irreversibly inactivate the immobilized α-chymotrypsin by incubating it in water at 100°. Then the fiber is treated with a glutaraldehyde solution (which interacts with NH_2 groups of denatured immobilized enzyme) and after this the native enzyme is attached to it again. The procedure is repeated three times, so that in the end the enzyme is attached not to the surface of the fiber, but to its protein "coat." It is apparent that in this case the environment of the immobilized enzyme is different. Nevertheless, here too, the dependence of the activity on the degree of stretching is the same as in Fig. 2a. In another experiment of this kind, trypsin is covalently attached to a human hair with the help of glutaraldehyde (via NH_2 groups of the keratin molecules). When the hair is stretched, the catalytic activity of

[20] I. V. Berezin, A. M. Klibanov, and K. Martinek, *Usp. Khim.* **44**, 17 (1975).

trypsin immobilized on it decreases 2-fold, and when the hair is relaxed to initial length, the enzymic activity also returns to the former level. A similar mechanochemical effect is at work with α-chymotrypsin covalently bound to viscose (cellulose) fibers.[18]

This means that the observed decrease in the enzyme activity (Fig. 2) can only be attributed to the fact that the stretching of the fiber induces the deformation of the protein molecule (Fig. 3).

Molecular Model. The following model for the enzyme-support attachment is suggested. First, the molecule of the enzyme can be linked in its regular, catalytically active conformation (Fig. 4a). Second, we assume that in the solution, under the action of thermal motion, there is an equilibrium between the catalytically active (folded) and inactive (unfolded) conformations (it cannot be excluded that such an equilibrium is to a great extent induced by the support). This means that in principle the molecule may also attach itself in a catalytically inactive conformation (Fig. 4b). The probability of this very immobilization is rather high, as the area of the support occupied by the unfolded protein molecule (and, correspondingly, the number of potentially available points of attachment) is larger than if the molecule was folded.

In terms of such a model (Fig. 4) one can understand why deformation of the support (i.e., changing the distance between the points of attachment) may modify the number of enzyme molecules attached to the support in one or the other state, and hence regulate the catalytic activity.

This molecular model was substantiated in experiments where α-

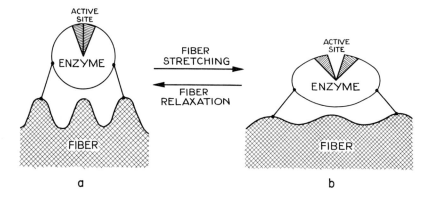

Fig. 3. Schematic representation of reversible deformation of the molecule of the enzyme attached to the fiber induced by the stretching of the fiber. The striated region of the protein globule shows the active site. (a) Normal fiber; (b) stretched fiber.

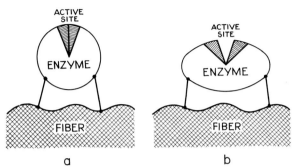

FIG. 4. Schematic representation of rigid attachment of the enzyme molecule to the support (a) in its regular, catalytically active (folded) conformation; (b) in an irregular, catalytically inactive (unfolded) conformation.

chymotrypsin was attached to *prestretched* fiber in the presence of a specific inhibitor, N-acetyl-D tryptophan. The three following effects were to be expected: (1) The relative enzymic activity of α-chymotrypsin attached to the fiber should be higher than in the absence of the substratelike inhibitor. This follows from the assumption that N-acetyl-D-tryptophan binds the enzyme only when the latter is in the catalytically active conformation and should therefore shift the equilibrium toward the folded form of the enzyme. This means that, with a high concentration of inhibitor, the enzyme molecules should be linked predominantly in a catalytically active conformation. (2) As the proportion of enzyme molecules attached to the support in the unfolded (inactive) conformation is rather small, the activity increase on relaxation of the fiber should be insignificant. (3) Moreover, on relaxation of the fiber, the enzyme molecule attached to it in the active conformation may undergo deformation and its relative activity may decrease, as does the degree of deformation of the support. All three effects were shown in the experiment (see Fig. 5).

Dependence of the Catalytic Activity of Enzymes Attached to the Nylon Fiber on the Degree of Fiber Stretching. The data on Figs. 2 and 5a show that both with α-chymotrypsin and trypsin attached to nylon fiber, alteration of enzymic activity is observable at a low (less than 1%) degree of deformation of the support; as the degree of deformation increases, the apparent activity of the immobilized enzyme remains constant.

Assuming that, on stretching the fiber, the molecules of the enzyme deform to the same degree as the support, one may conclude that, if the size of a molecule of α-chymotrypsin or trypsin[21] is ~ 50 Å, the effect of

[21] V. V. Mosolov, "Proteolytic Enzymes." Nauka Press, Moscow, 1971.

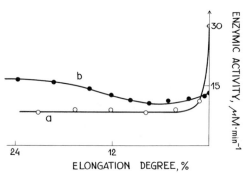

Fig. 5. Alteration of enzymic activity of α-chymotrypsin attached to nylon fiber on its relaxation. The enzyme was attached to stretched fiber, which was washed and placed in a pH-stat cuvette, where the degree of stretching of the fiber was decreased. For conditions, see the legend to Fig. 2. Curve a: The attachment was performed in a buffer solution, pH 8; curve b: the attachment was performed in the same buffer in the presence of 0.2 M N-acetyl-D-tryptophan.

alteration of the catalytic activity is fully realized if their molecules are deformed by less than 50 Å $\times$ 0.01 = 0.5 Å.

One may object that because of the properties of the surface of the fiber on its being stretched, the enzyme molecules are stretched to a higher degree than the polymer chains of the support. To examine this possibility, we drastically changed the surface of the fiber, covering it with several layers of inactivated protein (see above). The enzyme was then attached to this protein "coat." On stretching this modified fiber, the activity of immobilized enzymes also decreased in a fashion resembling that of Fig. 2 (i.e., the effect is fully realized with the fiber deformed by less than 1%); for details see Klibanov et al.[18] Furthermore, on stretching a human hair, to which trypsin is attached, or a viscose (cellulose) fiber, to which α-chymotrypsin is attached, the effect of reversible inactivation of immobilized enzyme is also fully realized when the elongation of both supports is less than 1%.

The result obtained shows that even very small alterations in the native conformation of the enzyme (<0.5 Å) strongly affects its catalytic effectiveness. It is noteworthy that resolution of X-ray analysis (the most important tool for studying conformational changes of protein associated with its catalytic function) does not usually exceed 2 Å.

Alteration in the Substrate Specificity of Immobilized α-Chymotrypsin on Stretching the Nylon Fiber. It has been demonstrated above (Fig. 2) that, as the fiber stretches, the catalytic activity of the immobilized α-chymotrypsin (measured with respect to N-acetyl-L-tyrosine ethyl ester, a *specific* substrate) decreases. The results are different if one makes use

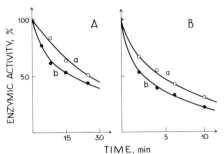

FIG. 6. Kinetics of acylation of α-chymotrypsin attached to nylon fiber by 50 μM p-nitrophenyl trimethylacetate (A) and 50 μM phenylmethyl sulfonylfluoride (B). Acylation (pH 7) was performed with the enzyme attached (a) to an unstretched or (b) to a stretched fiber; then the fiber was placed in a pH-stat cuvette, and its activity was determined in an unstretched state. For conditions of activity determination, see the legend to Fig. 2.

of p-nitrophenyl trimethylacetate or phenylmethyl sulfonylfluoride, well-known[22,23] quasi substrates acylating this enzyme. It is seen in Fig. 6 that, as a result of stretching the fiber, the rate of interaction of α-chymotrypsin with these acylating agents even increases. These results (cf. Figs. 2 and 6) should be interpreted as indicating that, owing to deformation of the enzyme, its substrate specificity strongly changes.

The above evidence exemplifies how the mechanochemical approach allows the study of intimate features of enzyme action mechanisms.

Simulation of Natural Enzyme Systems

In nature, mechanochemical processes play a very important role. This is evident even from simple enumeration of such processes as muscle contraction, functioning of elementary motile structures (flagella of bacteria and spermatozoids, cilia of unicellular organisms, the spindle of mitotic apparatus, bacteriophage sheaths), and movement of protoplasm. Processes of perception occurring in mechanoreceptors of animals, i.e., hearing, touch, gravitational perception, are also mechanochemical.

The functioning of most mechanochemical systems *in vivo* is continuously associated with proteins possessing enzymic (usually ATPase) activity.[24] According to certain theories of mechanochemical processes,

[22] M. L. Bender, M. L. Begue-Canton, R. L. Bleakeley, L. J. Brubacher, J. Feder, C. R. Gunter, F. J. Kezdy, J. V. Killheffer, T. H. Marshall, C. G. Miller, R. W. Roeske, and J. K. Stoops, *J. Am. Chem. Soc.* **88**, 5890 (1966).
[23] D. E. Fahrney and A. M. Gold, *J. Am. Chem. Soc.* **85**, 997 (1963).
[24] B. F. Poglazov, "Structure and Function of Contractile Proteins." Nauka Press, Moscow, 1965.

e.g., those of muscle contraction,[25] audial perception,[26,27] or oxidative phosphorylation,[28,29] one of the indispensable steps of such processes is alteration in the catalytic activity of some enzymic proteins under the action of mechanical force. It is believed that the mechanical action applied to an enzyme molecule should induce in it a conformational change. It is natural to ask then if the sensitivity to mechanical action is a unique specific property of only certain enzymes or if this ability is not inherent in the enzymes but is determined by nonenzymic templates (structural proteins, membranes), in (or on) which the biocatalyst is functioning. This was actually the subject of our study.

Experimental. The materials, the method of binding α-chymotrypsin to nylon fiber, the method of stretching the support, and the method of determining a relative catalytic activity of the immobilized enzyme are described above. In mechanochemical experiments, protein inhibitors (∼20 nM) were introduced directly into the cuvette of the pH stat containing the stretching device in the substrate solution.

Inhibition of α-Chymotrypsin, Chemically Bound to Nylon Fiber, by Protein Inhibitors. Regulation of Inhibition by Mechanical Stretching of the Fiber. Free α-chymotrypsin may be inhibited almost totally by equimolar ratios of protein inhibitors.[21] With the enzyme immobilized on nylon fiber, its catalytic activity (with respect to N-acetyl-L-tyrosine ethyl ester) decreases by as little as 17 or 12% on addition of even a 20-fold molar excess of pancreatic or soybean inhibitors, respectively.[30] This cannot be accounted for in terms of an immobilization-induced decrease in the association constant between the enzyme and the protein inhibitor, as even another 20-fold increase in the inhibitor concentration does not markedly enhance the degree of inhibition.

The phenomenon of incomplete inhibition of the catalytic activity of immobilized enzymes by protein inhibitors has been described in the literature.[31-35] The explanation is that the support creates a steric hindrance

[25] M. V. Volkenstein, *Dokl. Akad. Nauk SSSR* **146**, 1426 (1962).
[26] Y. A. Vinnikov, "Cytological and Molecular Basis of Reception," Chapter 8. Nauka Press, Leningrad, 1971.
[27] D. E. Goldman, *Cold Spring Harbor Symp. Quant. Biol.* **30**, 59 (1965).
[28] P. D. Boyer, *in* "Oxidases and Related Redox Systems" (T. E. King, H. S. Mason, and M. Morrison, eds.), Vol. 2, pp. 994–1017. Wiley, New York, 1965.
[29] D. E. Green and H. Baum, "Energy and the Mitochondrion." Academic Press, New York, 1969.
[30] A. M. Klibanov, G. P. Samokhin, K. Martinek, and I. V. Berezin, *Dokl. Akad. Nauk SSSR* **218**, 715 (1974).
[31] A. Bar-Eli and E. Katchalski, *J. Biol. Chem.* **238**, 1690 (1963).
[32] Y. Levin, M. Pecht, L. Goldstein, and E. Katchalski, *Biochemistry* **3**, 1905 (1964).
[33] R. Haynes and K. A. Walsh, *Biochem. Biophys. Res. Commun.* **36**, 235 (1969).
[34] C. K. Glassmeyer and J. D. Ogle, *Biochemistry* **10**, 786 (1971).
[35] R. J. Knights and A. Light, *Arch. Biochem. Biophys.* **160**, 377 (1974).

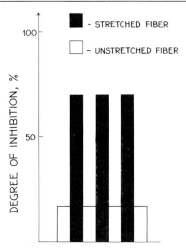

FIG. 7. The degree of inhibition of α-chymotrypsin attached to nylon fiber in the reaction with N-acetyl-L-tyrosine ethyl ester by pancreatic trypsin inhibitor. Concentration of the protein inhibitor is 20 nM; for other conditions see the legend to Fig. 2.

preventing access of a high-molecular-weight inhibitor to the active site of the enzyme, although it does not prevent the enzyme from interacting with the low-molecular-weight substrate.

We have shown[30] that the incomplete inhibition of esterase activity of α-chymotrypsin attached to nylon fiber by protein inhibitors is not associated (a) with electrostatic factors, as the degree of inhibition does not depend on ionic strength (from 0.1 to 3 M KCl); (b) with diffusion factors, as the degree of inhibition does not depend on the rate of stirring during incubation; or (c) with protein–protein contacts between the molecules of the immobilized enzyme, as the degree of inhibition does not depend on the degree to which the surface of the support is occupied by the enzyme (its concentration on the fiber surface was changed 200-fold). Thus, with α-chymotrypsin immobilized on nylon fiber, incomplete inhibition of the enzyme by pancreatic or soybean inhibitors appears to be due to steric hindrances arising from the presence of the support.

With the fiber reversibly stretched[36] by 30% the degree of inhibition of α-chymotrypsin by pancreatic inhibitor increases from 17 to 70%. The effect is fully reversible; i.e., if the fiber is relaxed, the degree of inhibition goes back to the initial low level. Such a "stretch–relax" procedure inducing an immediate "increase–decrease" response may be repeated several times (Fig. 7).

[36] All the inhibition experiments were carried out in a 2–30% fiber stretching range, where the enzymic activity does not depend on support deformation (cf. Fig. 2).

This phenomenon is readily explained in terms of a model presented in Fig. 8. In the initial state of the fiber (before stretching), most of the active sites of immobilized α-chymotrypsin are inaccessible to the pancreatic inhibitor. On stretching the fiber, the system passes from state a to state b; i.e., the surface of the fiber becomes smoother. This may involve elimination of some of the support-induced steric hindrances; thus more of the active sites of the enzyme becomes accessible to the inhibitor.

Concerning the effect of soybean inhibitor on the fiber being stretched, the degree of inhibition of the esterase activity of immobilized α-chymotrypsin increases from 12 to 60%. But this mechanochemical increase in the degree of inhibition is irreversible; i.e., after relaxation of the fiber the degree of inhibition is still 60%. Apparently, after the relaxation, the pancreatic inhibitor may, owing to its sufficiently high mobility, go into solution, and the degree of inhibition of the immobilized enzyme is in equilibrium under these conditions. Soybean inhibitor (whose molecular weight is ~3.3 times higher than that of pancreatic inhibitor[21]) is less mobile, and no degree of inhibition which is in equilibrium with a given extent of fiber stretching may be achieved rapidly.

Comparison with Natural Mechanochemical Enzymic Systems. As seen in the studies of the above systems, the sensitivity of the catalytic activity to external mechanical action may be induced artificially, even with simple enzymes, by attaching them to a suitable support. As we have demonstrated, subsequent mechanical deformation of the support may result either in an alteration of the conformation of the enzyme

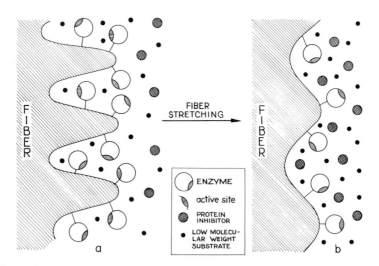

FIG. 8. Schematic representation of interaction of nylon fiber attached α-chymotrypsin with a high-molecular-weight inhibitor and a low-molecular-weight substrate. (a) Unstretched fiber; (b) stretched fiber.

globule, or in an alteration of the microenvironment, e.g., at the expense of steric effects arising during interaction of the immobilized enzyme with protein inhibitors. In both cases the change in the enzymatic activity observed when the fiber is stretched is analogous to, or even exceeds, the range of all reversible changes in ATPase activity, which may be induced by the action of an external mechanical field on natural mechanochemical enzymic systems such as intact muscle fiber,[37-39] muscle fiber in which protein–protein cross-links are reinforced by glutaraldehyde,[40] or the specific muscle enzymes, myosin and actomyosin.[41,42]

[37] Ts. Ohnishi and To. Ohnishi, *Nature (London)* **197**, 184 (1963).
[38] V. I. Vorob'ev and L. S. Ganelina, *Cytologia* **5**, 672 (1963).
[39] R. A. Chaplain, *Pfluegers Arch.* **307**, 120 (1969).
[40] N. M. Morozov, A. E. Bukatina, and S. E. Shnol, *Stud. Biophys.* **42**, 99 (1974).
[41] S. E. Shnol, *in* "Primary Processes in Receptor Elements of Sensory Organs," pp. 185–186. Nauka Press, Moscow-Leningrad, 1966.
[42] V. I. Vorob'ev and L. V. Kukhareva, *Dokl. Akad. Nauk SSSR* **165**, 435 (1965).

[40] Mechanochemistry of Immobilized Enzymes: A New Approach to Studies in Fundamental Enzymology

II. Regulation of the Rate of Enzymic Protein-Protein Interaction in Polyacrylamide Gel

By I. V. BEREZIN, A. M. KLIBANOV, V. S. GOLDMACHER, and KAREL MARTINEK

In nature most enzymic processes (including mechanochemical ones) occur in membrane gellike media.[1,2] With this in mind we investigated the possibility of regulating by mechanical means the catalytic activity of enzymes physically entrapped in a gel support.

We have undertaken a kinetic study of two enzymic protein–protein reactions: the tryptic activation of chymotrypsinogen[3] and the autolysis (bimolecular autodegradation) of trypsin[4] occurring in polyacrylamide gel (which is known to be chemically inert[5]).

Preparation of Gel. The gel is prepared by the riboflavin (10 mg/liter)

[1] T. W. Goodwin and O. Lindberg, eds. "Biological Structure and Function." Academic Press, New York, 1961.
[2] A. L. Lehninger, "Biochemistry," p. 295. Worth Publ., New York, 1970.
[3] I. V. Berezin, A. M. Klibanov, and K. Martinek, *Biochim. Biophys. Acta* **364**, 193 (1974).
[4] I. V. Berezin, A. M. Klibanov, V. S. Goldmacher, and K. Martinek, *Dokl. Akad. Nauk SSSR* **218**, 367 (1974).
[5] H. R. Maurer, "Disk-Electrophorese," Chapter 1. de Gruyter, Berlin, 1968.

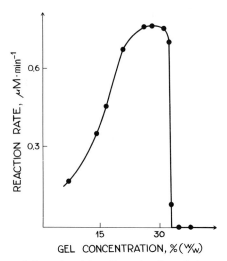

Fig. 1. Dependence of the rate of tryptic activation of chymotrypsinogen in polyacrylamide gel on the gel concentration. Conditions: 0.2 μM trypsin, 0.1 mM chymotrypsinogen, 20°, pH 7.9 (10 mM Tris-HCl buffer) [I. V. Berezin, A. M. Klibanov, and K. Martinek, *Biochim. Biophys. Acta.* **364**, 193 (1974)].

initiated photochemical polymerization of a mixture of acrylamide and N,N'-methylene-bisacrylamide (95:5, w/w) in the presence of all the components of the enzymic process (see the rate determination for the process studied and also this volume [12] and [14]). Polymerization is carried out while cooling with ice-cold water. The time of illumination (20 min) should be amply sufficient to ensure that total polymerization occurs.

Determination of the Rate of Chymotrypsinogen Activation by Trypsin. The quantity of chymotrypsin formed is determined by titration with p-nitrophenyl trimethylacetate (0.15 mM; 2% acetonitrile, v/v).[6] The gel is prepared directly in the optical cuvette of a Hitachi-356 double-wave spectrophotometer (l = 1 cm). The rate of formation of p-nitrophenol is determined in a gel block at 400 nm (by subtracting the absorbance at 550 nm from the absorbance at 400 nm). For the reaction conditions see Fig. 1.

Determination of the Rate of Trypsin Autolysis. The kinetics of trypsin autolysis ($[E]_o$ = 20 μM, pH 7.8, 0.01 M Tris-HCl) is followed[4] by incubating the gel blocks wrapped in a thin rubber film in a thermostat

[6] M. L. Bender, M. L. Begue-Canton, R. L. Bleakeley, L. J. Brubacher, J. Feder, C. R. Gunter, F. J. Kezdy, J. V. Killheffer, T. H. Marshall, C. G. Miller, R. W. Roeske, and J. K. Stoops, *J. Am. Chem. Soc.* **88**, 5890 (1966).

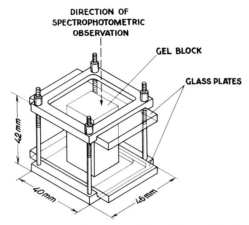

FIG. 2. Device for compression of the gel block.

at 45°. At various intervals pieces of gel are taken from the blocks and triturated. The relative enzymic activity (with respect to N-tosyl-L-arginine methyl ester, 10 mM) is determined in a pH stat (pH 8.0, 25°, 0.1 M KCl).

Experimental Procedure for Compression. A 1 cm thick parallelepiped cut from a piece of gel is placed in a special compressing device (see Fig. 2) between two parallel framed glass plates having a variable distance between them. This system is then placed either in the spectrophotometer cell (in the case of tryptic activation of chymotrypsinogen) or in a thermostat (in the case of trypsin autolysis). In the latter case the gel block is coated with rubber film.

Dependence of the Rate of Protein–Protein Interaction on Polyacrylamide Gel Concentration. Figure 1 shows the dependence of the rate of interaction between trypsin and chymotrypsinogen in gel on the polyacrylamide concentration.[3] As the concentration of the gel is raised to 25%, the rate of the process increases. This is due to the fact that there is less unpolymerized monomer and movable polymer chains, which usually produce an inhibiting effect.[7]

A further increase in the gel concentration [over 32% (w/w)] rapidly decreases the rate of the process almost to zero. The effect of the rate alteration has a threshold character; i.e., if the concentration of the gel is further raised by only 2%, the activation of the zymogen is almost arrested.

We obtained similar results[4] with trypsin autolysis (45°, pH 7.8). At first, when the gel concentration is raised, autolysis is somewhat enhanced,

[7] Y. Degani and T. Miron, *Biochim. Biophys. Acta* **212**, 362 (1970).

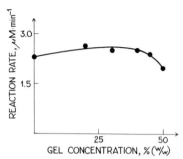

FIG. 3. Dependence of the initial rate of the reaction of trypsin with p-nitrophenyl acetate in polyacrylamide gel on gel concentration. Conditions: 20 μM trypsin, 30 μM p-nitrophenyl acetate, 20°, pH 7.0 (10 mM Tris-HCl buffer).

but then [at a gel concentration of 39% (w/w)] the rate constant goes down sharply (approximately 50-fold, if the gel concentration rises by as little as 0.5%).

Such a dramatic effect of gel concentration on the enzymic processes cannot be attributed to denaturation.[7] The fact that the reaction rate of trypsin with p-nitrophenyl acetate, a low-molecular-weight substrate, depends but slightly on the polyacrylamide gel concentration over a 20–50% (w/w) concentration range (Fig. 3) supports this statement. An explanation of this phenomenon should evidently be sought in terms of the general theory dealing with the effect of diffusion on chemical reaction kinetics.[8] At a polyacrylamide gel concentration exceeding 32% in tryptic activation of chymotrypsinogen (or 39% in the case of trypsin autolysis; the difference in the threshold concentrations seems to depend on the temperature regimes of the processes), the protein–protein reaction becomes a diffusion-controlled process; i.e., it occurs at very low rates. This agrees with experimental evidence[4] that the swelling of concentrated gel (where the reaction is almost nil) in water leads to an increase in the autolysis rate constant to a level typical for this enzymic process in aqueous solutions. In all probability, the swelling of the gel eliminates diffusional hindrances for protein–protein interaction and autolysis again becomes a kinetic-controlled process.

Effect of Mechanical Deformation of the Gel Support on the Rate of Protein–Protein Interaction Occurring in It. We have found[3,4] that the rate of protein–protein reaction may be greatly increased by the mechanical compression of a sufficiently dense gel support, where the enzymic processes are diffusion-controlled. For example, compression of a 33.5%

[8] R. M. Noyes, in "Progress in Reaction Kinetics" (G. Porter, ed.), Vol. 1, pp. 129–160. Pergamon Press, London, 1961.

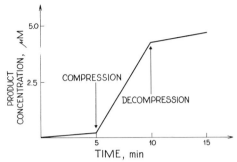

FIG. 4. The product concentration vs time dependence for the reaction of trypsin with chymotrypsinogen in 33.5% (w/w) polyacrylamide gel. A piece of gel is placed in the spectrophotometer cuvette; 5 min after the beginning of the reaction the gel is compressed by 35%, and in another 5 min the load is removed [I. V. Berezin, A. M. Klibanov, and K. Martinek, *Biochim. Biophys. Acta* **364,** 193 (1974)].

gel by 40% increases the rate of the reaction about 20-fold. After decompression, the rate of the process goes back to the initial low level (see Fig. 4). The rate of trypsin autolysis in a concentrated (over 39%) gel may be regulated in the same way.[4]

The "compression–decompression" procedure may be repeated many times, the "increase–decrease" response in the rate of protein–protein interaction always being present. The value of the mechanochemical effect increases with the degree of gel compression.[3]

In trying to explain the observed mechanochemical effect, one should exclude the possibility of a direct mechanical effect on immobilized proteins, as neither compression of the gel having a concentration below 32% in tryptic activation of chymotrypsinogen, nor compression of a gel having a concentration below 39% in the case of trypsin autolysis, nor compression of a gel with a 20–50% concentration in the case of trypsin *p*-nitrophenyl acetate interaction, induces a noticeable change in the reaction rates.

The mechanochemical effect may be explained in terms of the following model for diffusion of proteins in polyacrylamide gel. Let us assume, after Tombs,[9] that with a sufficiently high concentration of gel (when the size of its pores approaches that of a protein), the diffusion coefficient depends not on the mean size of a pore (as in the case of low-concentration gels), but on the number of pores whose size exceeds the critical size (in the first approximation, the effective diameter of a protein globule). Let us assume that in the initial state (before compression) a protein molecule

[9] M. P. Tombs, *Anal. Biochem.* **13,** 121 (1965).

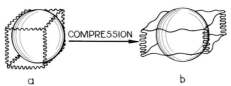

Fig. 5. Schematic representation of an elementary unit of polyacrylamide gel with a molecule of the enzyme entrapped in it (a) before and (b) after compression.

is entrapped in a certain elementary unit of a three-dimensional lattice of polymer gel (schematically represented in Fig. 5a). This unit does not permit the protein globule to come out of it, i.e., there is almost no diffusion. Now, if pressure is applied perpendicular to the upper facet one will have a parallelepiped as shown in Fig. 5b. The volumes of these two figures are equal, because gel, like water, is practically incompressible. This means that the compression-induced decrease in the side facets of the elementary unit should result in the upper and lower facets being larger. This should drastically change the rate of diffusion (and hence the diffusion-controlled reaction rate) because this mechanical compression of the gel results in a greater number of pores larger than the critical size. The conclusion drawn from the model in Fig. 5 is that mechanical compression of the gel may, in principle, facilitate diffusion, thereby increasing the rate of the protein–protein interaction, which in fact we observe experimentally (see Fig. 4).

We have also demonstrated that the regulation of the catalytic activity of trypsin entrapped in polyacrylamide gel can be effected not only by compression, but also by other kinds of mechanical deformation of the polymer support. For example, if a small, cylindrical piece of gel containing trypsin is bent, the autolysis of the enzyme localized along the axis of the cylinder (where there is no deformation of the elementary units of the gel lattice) proceeds at the same rate as in the "pristine" gel, whereas the autolysis of trypsin on the external arc of the bent cylinder increases 5-fold.[4]

The above diffusion mechanism for the mechanochemical regulation of the rate of protein–protein interaction may function *in vivo*, e.g., in some enzymic processes in membranes of ordered gellike structures (the mechanism of hearing?). It would then be expected that if the membrane is subjected to different kinds of mechanical deformation, its structure may undergo alterations ranging from a free diffusion state to one having steric hindrances (and vice versa). As we have demonstrated, this may produce a pronounced effect on the catalytic properties of enzymes immobilized in the membrane.

Section IX
Application of Immobilized Enzymes

A. ANALYTICAL AREA
   Articles 41 through 45
B. MEDICAL AREA
   Articles 46 through 48
C. ENZYME ENGINEERING (ENZYME TECHNOLOGY)
   Articles 49 through 57
D. APPLICATION OF IMMOBILIZED ENZYMES IN ORGANIC CHEMISTRY
   Articles 58 through 60

[41] Enzyme Electrodes and Solid Surface Fluorescence Methods

By GEORGE G. GUILBAULT

Introduction

General

The specificity of enzymes, and their ability to catalyze reactions of substrates at low concentrations, make them attractive as analytical reagents. Enzyme-catalyzed reactions have been used for analytical purposes for some time for the determination of substrates, activators, and inhibitors. However, the disadvantages associated with the use of enzymes have seriously limited their usefulness for analytical purposes. Such frequently cited objections have been their lack of availability, instability, poor precision, and the time and effort required to perform an analysis. While objections were valid earlier, numerous enzymes are now available in purified form, with high specific activity, at reasonable prices. The instability of enzymes is, of course, always a potential problem; yet, if this instability is recognized and reasonable precautions are taken, this problem may be minimized. Again, the poor precision, slowness, and labor that have made enzyme-catalyzed reactions unappealing as a means of analysis have been more of a consequence of the methods and techniques than the fault of the enzymes themselves. With the advent of new techniques, fluorometric and electrochemical, many of the previous difficulties have been resolved. In addition, the automation of enzymic reactions has increased the speed, ease, and reproducibility of assays utilizing enzymes.

In this chapter two modern aspects of the analytical uses of immobilized enzymes will be discussed, the enzyme electrode and fluorometric analysis by a reagentless solid surface method.

The Immobilized Enzyme

One of the primary objections to the use of enzymes as analytical reagents is the high cost of these materials. A continuous or semicontinuous routine analysis using enzymes would require large amounts of these materials, quantities greater than can be reasonably obtained, and quantities that would represent a prohibitive expenditure in many cases. If, however, the enzyme could be prepared in an immobilized (insolubilized) form without loss of activity, so that one sample could be used continuously for many analyses, a considerable advantage would be realized. The immobilized enzyme can be used analytically in much the same way that the soluble enzyme is used, that is, to determine the concentration of a substrate that is acted upon by the enzyme, an inhibitor that inactivates the enzyme, or an activator that provides an acceleration in enzyme activity.

Two major techniques used to immobilize an enzyme are (1) the chemical modification of the molecule by the introduction of insolubilizing groups; this technique, resulting in a chemical "tying down" of the enzyme, is in practice sometimes difficult to achieve because the insolubilizing groups can attack across the active site, destroying the activity of the enzyme; and (2) the physical entrapment of the enzyme in an inert matrix, such as starch or polyacrylamide gels. Physical entrapment techniques offer advantages of speed and ease of preparation over many chemical methods, except some such as the glutaraldehyde technique. The major difference between the entrapped and the attached enzymes is that the former are isolated from large molecules, which cannot diffuse into the matrix. The attached enzyme may be exposed to molecules of all sizes. Hence, the two types of immobilized enzymes will differ in the form of the kinetics observed and in the kinds of interferences observed. Thus, for the assay of large substrates as proteins with proteolytic enzymes, an attached enzyme must be used, not an entrapped enzyme. Either enzyme could be used for the assay of small substrates, such as urea. The chemical binding method produces a product that can be used for several thousand assays; the physically entrapped enzyme can be used for only a few hundred such analyses.

Enzyme Electrodes

Introduction

General

One of the most exciting new fields of potentiometry is the enzyme electrode. During the past decade we have witnessed the advent of almost fifty new "ion-selective electrodes" onto the market, electrodes that work well for all types of cations and anions, but only for inorganic species with any selectivity at all. The ion-selective electrodes for organics have been totally nonselective, irregular in response, and not useful. For this reason none are available commercially, with one exception, the acetylcholine electrode sold by Corning.

However, by the simple union of an ion-selective electrode with an enzyme, that biological catalyst which acts specifically and sensitively with almost all organic and inorganic compounds in nature, one can obtain an electrode, called an enzyme electrode, which is now useful for the assay of organic compounds. The principle here is quite simple: An enzyme is chosen, which acts, either specifically or highly selectively, with the compound to be assayed. This enzyme is then mounted, in an immobilized or insolubilized form so as to be useful over and over for long periods of time, onto a conventional ion-selective electrode which measures either a product of the reaction (NH_3 from urea or acid from penicillin) or a reactant (O_2 used in the oxidation of glucose or uric acid). The substance assayed diffuses into the enzyme layer, produces products or consumes a reactant, which is then measured by the ion-selective electrode. A potential is produced which is a log function of the concentration of the substance assayed. Thus, no longer is it necessary to prepare reagents, make a standard curve, effect a reaction, and measure. A single self-contained electrode is placed into a solution, and the concentration is read directly, without any difficulty and with only a buffer required as the reagent. Furthermore, the same electrode can be used for several hundred to a thousand assays, decreasing the cost per test to fractions of a cent.

In this chapter the reader will be introduced to the concept of the enzyme electrode, if a novice, and will be made an expert, if he has had some experience in this area. He will be taught how to make and use an enzyme electrode, what he can expect from such an electrode, and what types of electrodes have been described.

Since the enzyme electrode is a union of an ion-selective electrode and an enzyme, we shall now discuss each of these elements in some detail. From a good knowledge of each part, we should be able to recognize what can be expected from the total—the enzyme electrode.

Ion-Selective Electrodes

One of the most commonly made measurements in the chemical laboratory is that of pH. The glass electrode, selective for hydrogen ion, a reference electrode, and a pH meter combine to form an extremely useful analytical tool. The advantages of this measuring system, often taken for granted, are speed, sensitivity, cost, reliability, and the fact that sample is not destroyed or consumed in the process. The same advantages apply to other ion-selective electrodes that have become available for the past few years. At this time, ion-selective electrodes that are sensitive to particular cations and anions can be purchased or constructed at moderate cost. The analytically useful range of these sensors is generally from 10^{-1} M to 10^{-5} M although there are several sensors that are useful at much lower concentrations. Since the response of ion-selective electrodes is logarithmic, the precision of measurements is constant over their dynamic range.

TABLE I
LIST OF COMMERCIALLY AVAILABLE ELECTRODES[a]

Cations		Anions		Gases
Ammonium	G, S, L[b]	Bromide	S	CO_2
Hydrogen	G	Chloride	S, L	NH_3
Cadmium	S	Cyanide	S	NO_x
Calcium	L	Fluoride	S	SO_2
Copper	S	Fluoroborate	L	H_2S
Divalent ions	L	Hydroxide	G	HCN
Lead	S	Iodide	S	HF
Monovalent ions	G	Nitrate	L	O_2
Potassium	G, S, L	Perchlorate	L	
Silver	G, S	Sulfide	S	
Sodium	G, L	Thiocyanate	S	
Rubidium	L			

[a] Gas electrodes are all available from Orion (Cambridge, Massachusetts) and CO_2 from Radiometer (Copenhagen, Denmark) or Instrumentation Labs (Massachusetts). Other electrodes are available from many companies such as Orion, Corning (Corning, New York), Beckman (Fullerton, California), or Radiometer.
[b] G = glass; S = solid; L = liquid membranes.

Potentiometry with ion-selective electrodes is finding wide application in chemical studies. Particularly useful applications are in the fields of clinical and environmental chemistry, where a large number of samples and the need for a rapid method of analysis rule out many slower, more involved methods.

Presently ion-selective electrodes for many ions are available. These are listed in Table I. The electrodes are available from several manufacturers and, as newer methods of preparation of ISE have been developed, several kits for the preparation of different electrodes using a common body or housing have been introduced. With the exception of the single crystal F⁻ electrode and the glass electrodes, all electrodes can be easily prepared by oneself with a minimum of time and effort.

The field of ion-selective electrodes was reviewed by Pungor[1] and by Rechnitz.[2] More recent developments and applications of ion-selective electrodes may be found in the report of the Symposium held at the National Bureau of Standards, Gaithersburg, Maryland[3] or the article by Thomas.[4]

[1] E. Pungor, *Anal. Chem.* **39**(13), 28A (1967).
[2] G. A. Rechnitz, *Chem. Eng. News*, June 12, 1967, p. 146.
[3] R. A. Durst, ed., "Ion Selective Electrodes," Special Publication 314, National Bureau of Standards, Washington, D.C., 1969.
[4] G. J. Moody and J. D. R. Thomas, *Talanta* **19**, 623 (1972).

An ion-selective electrode may be defined as a device that develops an electrical potential proportional to the logarithm of the activity of an ion in solution. The term "specific" is sometimes used to describe an electrode. This term indicates that the electrode responds to only one particular ion. Since no electrode is truly specific for one ion, the term ion-selective is recommended as more appropriate.

In summary, many commercially available (or home-made) electrodes have been used and can be used in the construction of enzyme electrodes: (1) glass electrodes—H^+ and monovalent cation; (2) gas—NH_3, CO_2, and O_2; (3) solid and "solid" liquid-membrane electrodes—NH_4^+, S^{2-}, CN^-, and I^-; and (4) Pt electrodes.

The Enzymes as Reagent

The second half of the enzyme electrode team is the enzyme, that biological catalyst, which is ever present in nature and enables the many complex chemical reactions, upon which depends the very existence of life as we know it, to take place at ordinary temperatures.

Because enzymes work in complex living systems, one of their outstanding properties is specificity. An enzyme is capable of catalyzing a particular reaction of a particular substrate, even though other isomers of that substrate or similar substates may be present.

An example of the specificity of enzymes with respect to a particular substrate is found in luciferase, which catalyzes the oxidation of luciferin to oxyluciferin.[5] A rather complete study of many compounds similar in structure to luciferin showed that the catalytic oxidation resulting in the production of the green luminescence occurs only with luciferin. Substitution of an amino group for a hydroxyl group or addition of another hydroxyl group to the luciferin molecule alters the enzymic action, and the green luminescence is not produced. Another example of the specificity of enzymes is glucose oxidase, which catalyzes the oxidation of β-D-glucose to gluconic acid. A rather complete study of about 60 oxidizable sugars and their derivatives showed that only 2-deoxy-D-glucose is catalyzed at a rate comparable to that of β-D-glucose. The anomer α-D-glucose is oxidized catalytically less than 1% as rapidly as the β-anomer.[6] Urease, which catalyzes the hydrolysis of urea, is even more specific.

$$\text{Luciferin} \xrightarrow[\text{Mg}^{2+} \text{ATP}]{\text{luciferase}} \underset{\text{(green luminescence)}}{\text{oxyluciferin}} \quad (1)$$

Enzymes exhibit specificity with respect to a particular reaction. If one attempted to determine glucose by oxidation in an uncatalyzed way,

[5] E. W. White, F. McCapra, and G. F. Field, *J. Am. Chem. Soc.* **85**, 337 (1963).
[6] This series Vol. 3, p. 107 (1957).

for example, by heating a solution of glucose and an oxidizing agent like ceric perchlorate, other side reactions would occur uncontrollably to yield products in addition to gluconic acid. With glucose oxidase, on the other hand, catalysis is so effective at room temperature and a neutral pH that the rates of the other thermodynamically possible reactions are negligible.

There are over two-hundred purified enzymes available today from companies such as Boehringer-Mannheim (Mannheim, Germany and New York City), Calbiochem (Los Angeles), Nutritional Biochemicals (Cleveland), Pierce (Rockford, Illinois), Sigma (St. Louis, Missouri) and Worthington (Freehold, New Jersey), plus others. A listing of enzymes available through 1968 can be found in a book by Guilbault on enzymic methods of analysis.[7] In addition to these, there are many other hundreds of enzymes known which can be isolated and purified for use in constructing enzyme electrodes.

To be useful in an enzyme electrode, the enzyme must be immobilized. There are several ways that this immobilization of the enzyme can be effected, providing a product useful in an enzyme electrode. These are discussed elsewhere in this book (Section II). Furthermore, there are a number of immobilized enzymes sold commercially by companies such as Boehringer (catalase, chymotrypsin, glucose oxidase, papain, ribonuclease, trypsin, urease); Koch Light (Buckinghamshire, England) or Aldrich (Milwaukee, Wisconsin)—catalase, dextranase, glucoxe oxidase, β-glucosidase, and uricase; Miles Laboratories (Elkhart, Indiana) and Corning (Corning, New York, who offers to provide any enzyme insolubilized on glass beads).

In constructing an enzyme electrode one need only (a) pick an enzyme that reacts with the substance to be determined, (b) obtain that enzyme from commercial sources or isolate it oneself, (c) immobilize the enzyme by standard techniques or buy it already immobilized, if possible, and (d) place the immobilized enzyme around the appropriate electrode to monitor the reaction that occurs (note: this will probably be the limiting factor in the construction of an enzyme electrode since steps a–c are always possible). These steps will be discussed more thoroughly below in the experimental section.

Let us now review the current literature to see what electrodes have been designed and used.

Enzyme Electrodes—A Survey

Enzyme electrodes represent the most recent advance in analytical chemistry. These devices combine the selectivity and sensitivity of en-

[7] G. Guilbault, "Enzymatic Methods of Analysis." Pergamon, Oxford, 1971.

TABLE II
SOME ENZYME ELECTRODES AND THEIR CHARACTERISTICS

Type	Sensor	Stability	Response time (min)	Range (M)	Key to references[a]
1. Urea	Cation	> 4 Months	1–2	10^{-2} to 10^{-4}	21
	pH	3 Weeks	5–10	5×10^{-3} to 5×10^{-5}	16
	Gas (NH$_3$)	> 4 Months	1–4	5×10^{-2} to 5×10^{-5}	23
2. Glucose	Pt (O$_2$)	> 4 Months	1	10^{-1} to 10^{-5}	34
	Pt (H$_2$O$_2$)	> 14 Months	1	2×10^{-2} to 10^{-4}	13
	Gas (O$_2$)	3 Weeks	2–5	10^{-2} to 10^{-4}	9
	pH	1 Week	5–10	10^{-1} to 10^{-3}	16
3. L-Amino acids (general)	Pt	4–6 Months	0.2	10^{-3} to 10^{-5}	30
	NH$_4^+$	> 1 Month	1–3	10^{-2} to 10^{-4}	27
L-Tyrosine	Gas (CO$_2$)	3 Weeks	1–2	10^{-1} to 10^{-4}	25
L-Asparagine	Cation	1 Month	1	10^{-2} to 5×10^{-5}	28
4. D-Amino acids	Cation	1 Month	1	10^{-2} to 5×10^{-5}	28
5. Lactic acid	Pt	< 1 Week	3–10	2×10^{-3} to 10^{-4}	11
6. Alcohols	Pt (O$_2$)	> 4 Months	0.5	0.5 to 100 mg/100 ml	34
7. Penicillin	pH	1–2 Weeks	1–2	10^{-2} to 10^{-4}	16, 38, 39
8. Uric acid	Pt (O$_2$)	> 4 Months	0.5	10^{-2} to 10^{-4}	33
9. Amygdalin	CN$^-$	1 Week	1–3	10^{-1} to 10^{-5}	35–37

[a] The numbers refer to text footnotes that give the full reference.

zymic methods of analysis with the speed and simplicity of ion-selective electrode measurements. The result is a device that can be used to determine the concentration of a given compound in solution quickly and a method that requires a minimum of sample preparation. A summary of available enzyme electrodes appears in Table II.

Glucose Electrodes. The first report of an enzyme electrode was given by Clark and Lyons,[8] who proposed that glucose could be determined amperometrically using soluble glucose oxidase held between Cuprophane membranes. The oxygen uptake was measured with an O$_2$ electrode:

$$\text{Glucose} + O_2 + H_2O \xrightarrow{\text{glucose oxidase}} H_2O_2 + \text{gluconic acid} \quad (2)$$

The term "enzyme electrode" was introduced by Updike and Hicks,[9] who coated an oxygen electrode with a layer of physically entrapped glucose oxidase in a polyacrylamide gel. The decrease in oxygen pressure was equivalent to the glucose content in blood and plasma. A response

[8] L. C. Clark and C. Lyons, *Ann. N.Y. Acad. Sci.* **102**, 29 (1962).
[9] S. J. Updike and G. P. Hicks, *Nature (London)* **214**, 986 (1967).

time of less than a minute was observed. Clark[10] proposed measuring the hydrogen peroxide produced in the enzymic reaction with a Pt electrode. An instrument based on this concept, using glucose oxidase held on a filter trap, is now sold by Yellow Springs Instrument Co. (Yellow Springs, Ohio), and by Radiometer (Copenhagen, Denmark). Two platinum electrodes are used, one to compensate for any electrooxidizable compounds in the sample, such as ascorbic acid, the second to monitor the enzyme reaction.

Williams, Doig, and Korosi[11] used quinone as the hydrogen acceptor in place of oxygen and described enzyme electrodes for blood glucose:

$$\text{Glucose} + \text{quinone} + \text{H}_2\text{O} \xrightarrow{\text{glucose oxidase}} \text{gluconic acid} + \text{hydroquinone} \quad (3)$$

$$\text{Hydroquinone} \xrightarrow{\text{Pt}} \text{quinone} + 2\text{H}^+ + 2\text{e}^- \quad (4)$$
$$(E = 0.4 \text{ V vs standard calomel electrode})$$

Using glucose oxidase trapped in a porous or jellied layer and covered with a dialysis membrane over a Pt electrode, glucose could be determined by monitoring the electrooxidation of quinone. About 3–10 min were required to obtain a steady-state current.

Guilbault and Lubrano[12,13] described a simple, stable, rapid reading electrode for glucose. The electrode consists of a metallic sensing layer (Pt or Pt-glass[14]), covered by a thin film of immobilized glucose oxidase held in place by means of cellophane. When poised at the correct potential, the current produced is proportional to the glucose concentration. The time of measurement required with this amperometric approach is less than 12 sec using a kinetic method. The electrode is stable for over a year when stored at room temperature with only a 0.1% change from maximum response per day. The enzyme electrode determination of blood sugar compares favorably with commonly used methods with respect to accuracy, precision, and stability, and the only reagent needed for assay is a buffer solution.

Nagy, von Storp, and Guilbault[15] described a self-contained electrode for glucose based on an iodide membrane sensor:

$$\text{Glucose} + \text{O}_2 \xrightarrow{\text{glucose oxidase}} \text{gluconic acid} + \text{H}_2\text{O}_2 \quad (5)$$

$$\text{H}_2\text{O}_2 + 2\text{I}^- + 2\text{H}^+ \xrightarrow{\text{peroxidase}} 2\text{H}_2\text{O} + \text{I}_2 \quad (6)$$

[10] L. C. Clark, U.S. Patent No. 3,539,455 (1970).
[11] D. L. Williams, A. R. Doig, and A. Korosi, *Anal. Chem.* **42**, 118 (1970).
[12] G. G. Guilbault and G. J. Lubrano, *Anal. Chim. Acta* **60**, 254 (1972).
[13] G. G. Guilbault and G. J. Lubrano, *Anal. Chim. Acta* **64**, 439 (1973).
[14] G. G. Guilbault, G. J. Lubrano, and D. Gray, *Anal. Chem.* **45**, 2255 (1973).
[15] G. Nagy, L. H. von Storp, and G. G. Guilbault, *Anal. Chim. Acta* **66**, 443 (1973).

The highly selective iodide sensor monitors the local decrease in the iodide activity at the electrode surface. The assay of glucose was performed in a flow stream and at stationary electrode. Pretreatment of the blood sample was required to remove interfering reducing agents, such as ascorbic acid, tyrosine, and uric acid.

Nilsson, Åkerlund, and Mosbach[16] described the use of conventional hydrogen ion glass electrodes for the preparation of enzyme-pH electrodes by either entrapping the enzymes within polyacrylamide gels around the glass electrode or as a liquid layer trapped within a cellophane membrane. In an assay of glucose, based on a measurement of the gluconic acid produced, the pH response was almost linear from 10^{-4} to 10^{-3} M with a ΔpH of about 0.85 per decade. Electrodes of this type were also constructed for urea and penicillin (see below). The ionic strength and pH were controlled using a weak (10^{-3} M) phosphate buffer, pH 6.9, and 0.1 M Na_2SO_4.

Urea Electrodes. Guilbault and Montalvo[17,18] prepared several electrodes for urea by physically entrapping urease in a polyacrylamide gel held over the surface of a monovalent cation electrode by cellophane film:

$$\text{Urea} + 2H_2O \xrightarrow{\text{urease}} 2NH_4^+ + 2HCO_3^- \quad (7)$$

The urea diffuses into the urease layer, where it is converted to ammonium ions, which are sensed by the cation electrode. The electrode could be used for up to 3 weeks with no loss of activity and responded to urea in the concentration range 5×10^{-5} to 1.6×10^{-1} M with a response time of about 35 sec.

Because sodium and potassium ions interfered in the measurement, Guilbault and Hrabankova[19] used an uncoated NH_4^+ ion electrode as reference electrode to the urease-coated NH_4^+ electrode, and added ion-exchange resin in attempts to develop an urea electrode useful for assay of blood and urine. Good precision and accuracy were obtained.

In attempts to improve the selectivity of the urea determination, Guilbault and Nagy[20] used a silicone rubber-based nonactin ammonium-ion-selective electrode as the sensor for the NH_4^+ ions liberated in the urease reaction. The selectivity coefficients of this electrode were 6.5 for NH_4^+/K^+, 7.50×10^2 for NH_4/Na^+, and much higher for other cations. The reaction layer of the electrode was made of urease enzyme chemically

[16] H. Nilsson, A. Åkerlund, and K. Mosbach, *Biochim. Biophys. Acta* **320**, 529 (1973).
[17] G. G. Guilbault and J. G. Montalvo, *J. Am. Chem. Soc.* **91**, 2164 (1969).
[18] G. G. Guilbault and J. G. Montalvo, *J. Am. Chem. Soc.* **92**, 2533 (1970).
[19] G. G. Guilbault and E. Hrabankova, *Anal. Chim. Acta* **52**, 287 (1970).
[20] G. G. Guilbault and G. Nagy, *Anal. Chem.* **45**, 417 (1973).

immobilized on polyacrylic gel. A still further improvement was described by Guilbault, Nagy, and Kuan[21] using a three-electrode system, which allowed dilution to a constant interference level. Analysis of blood serum showed good agreement with spectrophotometric methods, and the enzyme electrode was stable for 4 months at 4°.

Still further improvement in the selectivity of this type of electrode was obtained by Anfalt, Granelli, and Jagner,[22] who polymerized urease directly onto the surface of an Orion ammonia gas electrode probe by means of glutaraldehyde. Sufficient NH_3 was produced in the enzyme reaction layer even at pH's as low as 7–8 to allow direct assay of urea in the presence of large amounts of Na^+ and K^+. A response time of 2–4 min was observed.

Guilbault and Tarp[23] described a still better, total interference-free, direct-reading electrode for urea, using the air-gap electrode of Ruzicka and Hansen.[24] A thin layer of urease chemically bound to polyacrylic acid was used, at a solution pH of 8.5, where good enzyme activity was still obtained, yet where sufficient NH_3 is liberated to yield a sensitive measurement with the air-gap NH_3 electrode. The urea diffuses into the gel; the NH_3 produced diffuses out of solution to the surface of the air-gap electrode, where it is measured. A linear range of 3×10^{-2} to 5×10^{-5} M was obtained with a slope of 0.75 pH unit per decade. The electrode could be used continuously for blood serum analysis for up to 1 month (at least 500 samples) with a precision and accuracy of less than 2%. The response time is 2–4 min at pH 8.5, and the electrode is washed under a water tap for 5–10 sec after each measurement. Absolutely no interference from any levels of substances commonly presented in blood was observed (Na^+, K^+, NH_4^+, ascorbic acid, etc.).

A urea electrode using physically entrapped urease and a glass electrode to measure the pH change in solution were described by Mosbach et al.[16] The response time of the electrode to urea was about 7–10 min and had a linear range from 5×10^{-5} to 10^{-2} M with a change of about 0.8 pH unit per decade. The electrode could be kept at room temperature for about 2–3 weeks. The ionic strength and pH were controlled using a weak (10^{-3} M) Tris buffer and 0.1 M NaCl.

Still another possibility for a urea electrode is the use of a CO_2 sensor to measure the second product of the urea–urease reaction, HCO_3^-. Guilbault and Shu[25] evaluated the use of the CO_2 sensor and found that a

---

[21] G. G. Guilbault, G. Nagy, and S. S. Kuan, *Anal. Chim. Acta* **67**, 195 (1973).
[22] T. Anfalt, A. Granelli, and D. Jagner, *Anal. Lett.* **6**, 969 (1973).
[23] G. G. Guilbault and M. Tarp, *Anal. Chim. Acta* **73**, 355 (1974).
[24] J. Ruzicka and E. H. Hansen, *Anal. Chim. Acta* **69**, 129 (1974).
[25] G. G. Guilbault and F. Shu, *Anal. Chem.* **44**, 2161 (1972).

urea electrode, prepared by coupling a layer of urease covered with a dialysis net with a CO_2 electrode, had a linear range of 10^{-4} to 10^{-1} M, a response time of about 1–3 min and a slight response to only acetic acid. Na^+ and K^+ ions had no interference.

Owens-Illinois (Toledo, Ohio) now markets a urea enzyme electrode system that permits assay of 60 samples per hour. Details will be procided below.

Amino Acid Electrodes. The CO_2 sensor was evaluated by Guilbault and Shu[25] for response to tyrosine when coupled with tyrosine decarboxylase held in an immobilized form by a dialysis membrane. A linear range of 2.5×10^{-4} to 10^{-2} M was observed with a slightly faster response time than that observed with the urea electrode mentioned above. A slope of 55 mV per decade was obtained, compared to 59 mV per decade for the urea electrode.

Enzyme electrodes for the determination of l-amino acids were developed by Guilbault and Hrabankova,[26] who placed an immobilized layer of L-amino acid oxidase over a monovalent cation electrode to detect the ammonium ion formed in the enzyme-catalyzed oxidation of the amino acid. These electrodes are stable for about 2 weeks, and have a response time of 1–2 min.

Two different kinds of enzyme electrodes were prepared by Guilbault and Nagy for the determination of L-phenylalanine.[27] One of the electrodes used a dual enzyme reaction layer—L-amino acid oxidase with horseradish peroxidase—in a polyacrylamide gel over an iodide-selective electrode. The electrode responds to a decrease in the activity of iodide at the electrode surface due to the enzymic reaction and subsequent oxidation of iodide.

$$\text{L-Phenylalanine} \xrightarrow{\text{L-amino acid oxidase}} H_2O_2 \qquad (8)$$

$$H_2O_2 + 2H^+ + I^- \xrightarrow[\text{peroxidase}]{\text{horseradish}} I_2 + H_2O \qquad (9)$$

The other electrode was prepared using a silicone rubber-based, nonactin-type ammonium ion-selective electrode covered with L-amino acid oxidase in a polyacrylic gel. The same principle of diffusion of substrate into the gel layer, enzymic reaction, and detection of the released ammonium ion applied to this system. Linear calibration plots were also obtained for L-leucine and L-methionine in the range 10^{-4} to 10^{-3} M.

Electrodes specific for D-amino acids, which are oxidatively catalyzed by D-amino acid oxidase, were reported by Guilbault and Hrabankova.[28]

[26] G. G. Guilbault and E. Hrabankova, *Anal. Lett.* **3**, 53 (1970).
[27] G. G. Guilbault and G. Nagy, *Anal. Lett.* **6**, 301 (1973).
[28] G. G. Guilbault and E. Hrabankova, *Anal. Chim. Acta* **56**, 285 (1971).

The NH_4^+ ion produced is monitored with a cation electrode:

$$\text{D-Amino acid} + O_2 \xrightarrow{\text{oxidase}} NH_4^+ + HCO_3^- \qquad (10)$$

The stability of these electrodes could be maintained for 21 days if they are stored in a buffered flavine adenine dinucleotide (FAD) solution, since the FAD is weakly bound to the active site of the enzyme and is needed for its activity. Electrode probes suitable for the assay of D-phenylalanine, D-alanine, D-valine, D-methionine, D-leucine, D-norleucine, and D-isoleucine were developed. An electrode for asparagine was also developed using asparaginase as the catalyst.[28] No cofactor was necessary.

Guilbault and Shu[29] described an enzyme electrode for glutamine, prepared by entrapping glutaminase on a nylon net between a layer of cellophane and a cation electrode. The electrode responds to glutamine over the concentration range 10^{-1} to 10^{-4} M with a response time of only 1–2 min.

Guilbault and Lubrano[30] prepared an electrode for L-amino acids by coupling chemically bound L-amino acid oxidase to a Pt electrode which senses the peroxide produced in the enzyme reaction:

$$\text{L-Amino acid} + O_2 + H_2O \rightarrow R\text{—COCOOH} + NH_3 + H_2O_2 \qquad (11)$$

The time of measurement is less than 12 sec, using a kinetic measurement of the rate of increase in current per unit time, and the only reagent required is a phosphate buffer. The L-amino acids cysteine, leucine, tyrosine, phenylalanine, tryptophan, and methionine were assayed.

Alcohol Electrodes. Alcohol oxidase catalyzes the oxidation of lower primary aliphatic alcohols.

$$RCH_2OH + O_2 \xrightarrow[\text{oxidase}]{\text{alcohol}} RCHO + H_2O_2 \qquad (12)$$

The hydrogen peroxide produced in these reactions may be determined amperometrically with a platinum electrode as in the determination of glucose above. Guilbault and Lubrano[31] used the alcohol oxidase obtained from a basidiomycete to determine the ethanol concentration of 1-ml samples over the range 0–10 mg/100 ml, with an average relative error of 3.2% in the 0.5–7.5 mg/100 ml range. This procedure should be adequate for clinical determinations of blood ethanol, since normal blood from individuals who have not ingested ethanol ranges from 40 to 50 mg/100 ml. Methanol is a serious interference in the procedure, since the alcohol oxidase is more active for methanol than ethanol. However, the

[29] G. G. Guilbault and F. Shu, *Anal. Chim. Acta* **56**, 333 (1971).
[30] G. G. Guilbault and G. J. Lubrano, *Anal. Chim. Acta* **69**, 183 (1974).
[31] *Ibid.*, p. 189.

concentration of methanol in blood is negligible compared to that of ethanol.

A vastly improved alcohol electrode, selective for ethanol, was described by Guilbault and Nanjo.[32] The decrease in current as O_2 was depleted from solution in the enzymic reaction [Eq. (12)] was measured at an applied potential of -0.6 V vs SCE.

Uric Acid Electrode. A self-contained rapid reading electrode for uric acid was described by Nanjo and Guilbault.[33] The electrode was prepared by placing a layer of glutaraldehyde-bound uricase over the tip of a Beckman Pt electrode, the enzyme was then covered for support with a thin layer of dialysis membrane. The decrease in the level of dissolved oxygen in solution due to the enzymic reaction was maintained at an applied potential of -0.6 V vs SCE.

$$\text{Uric acid} + O_2 \xrightarrow{\text{uricase}} \text{allantoin} \cdot H_2O_2 + H_2O \qquad (13)$$

The current observed is proportional to the level of uric acid at concentrations of 10^{-5} to 10^{-1} M. By measuring the initial rate of change in current an assay can be performed in less than 30 sec.

Further studies indicated the electrode could be used for the assay of glucose and amino acids.[34]

It was found that the peroxide produced in the reaction could not be monitored at $+0.6$ V vs SCE as described in the method of Guilbault and Lubrano for glucose,[12,13] amino acids,[30] and alcohols,[31] since (a) the polarographic curves for uric acid and peroxide are too close to be separated by any pH useful for the enzyme reaction, and (b) an allantoin-peroxide complex is the product of the oxidation of uric acid, not free peroxide. Additionally, the oxygen uptake method was found to be more sensitive allowing the assay of lower concentrations of substrates.

The use of a Pt electrode rather than the Clark type oxygen electrodes eliminates all the problems associated with gas membrane electrodes, namely slow response and blockage of the membrane by substances present in blood.

Lactic Acid Electrode. Williams, Doig, and Korosi[11] used ferricyanide as a hydrogen acceptor for lactic acid, and described an enzyme electrode for lactate based on the following reaction:

$$\text{Lactate} + 2\text{Fe(CN)}_6^{3-} \xrightarrow{\text{LDH}} \text{pyruvate} + 2\text{Fe(CN)}_6^{4-} \qquad (14)$$

$$2\text{Fe(CN)}_6^{4-} \xrightarrow{\text{Pt}} 2\text{Fe(CN)}_6^{3-} + 2e^- + H^+ \qquad (15)$$

[32] G. G. Guilbault and M. Nanjo, *Anal. Chim. Acta* **75**,(2) (1975).
[33] G. G. Guilbault and M. Nanjo, *Anal. Chem.* **46**, 1769 (1974).
[34] G. G. Guilbault and M. Nanjo, *Anal. Chim. Acta* **73**, 367 (1974).

By monitoring the electrooxidation of the ferrocyanide produced at +0.4 V vs SCE at a Pt electrode covered with a porous or jelled layer of lactate dehydrogenase and a dialysis membrane, a current was produced proportional to the concentration of lactic acid. About 3–10 min were required for measurement. Because of the low K_m value of this enzyme ($K_m = 1.2 \times 10^{-3}$ M), it was necessary to dilute the sample with buffered ferricyanide. A linear plot was obtained over the range 10^{-4} to 10^{-3} M.

Amygdalin Electrode. An electrode specific for amygdalin based on a solid-state cyanide electrode was reported by Rechnitz and Llenado.[35,36] The enzyme β-glucosidase, immobilized in acrylamide gel, was used:

$$\text{Amygdalin} \xrightarrow{\beta\text{-glucosidase}} \text{HCN} + 2C_6H_{12}O_6 + \text{benzaldehyde} \quad (16)$$

A linear range of 5×10^{-3} to 10^{-5} M was reported, with a slope of about 40 mV per decade and a response time of about 10 min at concentrations of 10^{-2} and 10^{-3} M amygdalin and 30 min at 10^{-4} to 10^{-5} M. The electrode rapidly lost activity, an indication that an incomplete physical entrapment had been effected.

One reason for this long response time and poor stability was the high pH used (10.4), a pH at which the enzyme has low activity and is denatured. This was recognized by Mascini and Liberti,[37] who improved the response time and other electrode characteristics by working at a pH of 7. The electrode was prepared by spreading the enzyme directly onto the membrane surface and covering it with a thin dialysis membrane. Since the enzyme was not immobilized a stability of less than a week is obtained, but the response time was only about 1–2 min at 10^{-1} to 10^{-3} M and 6 min at 10^{-4} M. Furthermore, a linear calibration was obtained from 10^{-1} to 10^{-4} M with a slope of 53 mV per decade (compared to about 40 mV per decade at pH 10).

Penicillin Electrode. The first attempt at design of a penicillin electrode was made by Papariello, Mukherji, and Shearer.[38] The electrode was prepared by immobilizing penicillin β-lactamase (penicillinase) in a thin membrane of polyacrylamide gel molded around, and in intimate contact with, a glass (H$^+$) electrode. The increase in hydrogen ion from the penicilloic acid liberated from penicillin is measured:

$$\text{Penicillin} \xrightarrow{\text{penicillinase}} \text{penicilloic acid} \quad (17)$$

The response time of the electrode was very fast (<30 sec) and had a slope of 52 mV/decade over the range 5×10^{-2} to 10^{-4} M for sodium am-

[35] G. A. Rechnitz and R. Llenado, *Anal. Chem.* **43**, 283 (1971).
[36] *Ibid.*, p. 1457.
[37] M. Mascini and A. Liberti, *Anal. Chim. Acta* **68**, 177 (1974).
[38] G. J. Papariello, A. K. Mukherji, and C. M. Shearer, *Anal. Chem.* **45**, 790 (1973).

picillin. The reproducibility of the electrode was very poor, probably because no attempt was made to control the ionic strength and pH.

Mosbach et al.[16] prepared a penicillin electrode by entrapping penicillinase as a liquid layer trapped within a cellophane membrane around a glass (H^+) electrode, yet controlled the ionic strength and pH by using a weak 0.005 M phosphate buffer, pH 6.8, and 0.1 M NaCl. Good results were obtained, in comparison to the results of Papariello et al.; the calibration plot was linear from 10^{-2} to 10^{-3} M with a ΔpH of 1.4 and as little as 5×10^{-4} M sodium penicillin could be determined. The electrode could be stored for 3 weeks and the average deviation was $\pm 2\%$ with a response time of about 2–4 min.

Papariello and co-workers[39] reported a revised model of their original penicillin electrode. The authors claimed it was critical that a membrane be placed between the enzyme layer and the glass electrode to achieve satisfactory results.

Substrate Electrodes for the Assay of Enzymes

Urease Electrode. Attempts have been made to determine the activity of enzymes using "immobilized" substrates. Such electrodes have two limitations that the enzyme electrodes above do not have: (1) the substrate, unlike the enzyme, is used up in an assay; hence, a limiting factor will be the amount of substrate available; (2) the enzyme, since it is not consumed, will continually act on the substrate, to produce a product continually; hence, the analysis must involve a kinetic rather than an equilibrium method.

Montalvo[40,41] designed an electrode for urease by continually passing a layer of soluble urea between the tip of a NH_4^+ cation electrode and a dialysis membrane. Urea diffuses through the membrane and is hydrolyzed by urease in the dilute aqueous solutions. The ammonium ion diffuses back through the membrane to the cation electrode, where it is sensed. Although an interesting approach, it is one that lacks practicality.

Cholinesterase Electrode. In another study an enzyme sensing electrode system for serum cholinesterase was prepared by coupling a pH-sensing electrode to a thin polymer membrane with a low molecular weight cutoff.[42] The electrode system utilized two thin-layer solutions to form a microelectrochemical cell. One layer contained the serum to be

---

[39] L. F. Cullin, J. F. Rusling, A. Schleifer, and G. J. Papariello, *Anal. Chem.* **46**, 1955 (1974).
[40] J. G. Montalvo, *Anal. Biochem.* **38**, 359 (1970).
[41] J. G. Montalvo, *Anal. Chem.* **42**, 2093 (1969).
[42] K. L. Crochet and J. G. Montalvo, *Anal. Chim. Acta* **66**, 259 (1973).

assayed; the second, the acetylcholine substrate which had been stabilized to balance the nonenzymic decay of substrate by using a high-molecular-weight buffer. A pseudo-linear curve was obtained from 10–70 units/ml of cholinesterase, and assays could be performed in 1.5–4.5 min.

A more practical approach to a cholinesterase substrate electrode was proposed by Gibson and Guilbault,[43] who prepared the insoluble Reineckate salt of acetylcholine, and placed this on the tip of a pH electrode covered with a nylon net permeable to enzyme for support. The enzyme diffuses into the substrate layer producing acid, which is then sensed by the pH electrode:

$$\text{Acetylcholine Reineckate} \xrightarrow{\text{ChE}} \text{acetic acid} + \text{choline-Reineckate} \quad (18)$$

It is believed by the author that such substrate electrodes do have a future, although a limited one, and will become generally accepted only if preparable in simple, self-contained systems, such as those developed for the enzyme electrodes.

Experimental

In the introductory section under Enzyme Electrodes, we discussed the basic principles of the enzyme electrode and surveyed the literature as to which electrodes have been made and used up to this time. From this discussion, it appears that enzyme electrodes can be made that are highly specific using appropriate enzymes, and a high degree of accuracy and sensitivity is obtained. Applications of such electrodes to the clinical, the pharmaceutical, industrial and pollution laboratories are quite feasible and sound. The electrodes have the further advantages of simplicity, requiring only one solution—a buffer—and economy, being usable for as many as 500–1000 determinations per electrode.

In this section, we shall discuss the experimental parameters involved in the construction of enzyme electrodes and the observed characteristics and limitations of these electrodes.

Immobilization Methods

Cost is a factor in enzymic analysis, because continuous or semicontinuous analysis require large amounts of unrecoverable enzymes. Immobilized enzymes can be used continuously and expensive enzymes can be recovered, thus resolving the problem of cost. Also, most solubilized enzymes are very unstable, often must be stored at low temperatures, and

[43] K. Gibson and G. G. Guilbault, "Insolubilized Substrates for Assay of Cholinesterase." Unpublished results, Technical University of Denmark, 1974.

then retain activity for only a short period of time. In cytochemical systems most, if not all, of the enzymes are attached to cell surfaces or entrapped within cell membranes. Immobilization of enzyme puts them in a more natural environment, with the result that they usually are most stable and efficient.

Within the last 6 years, a new technology based on enzyme immobilization has rapidly emerged. Five methods have been used for the preparation of water-insoluble derivatives of enzymes: (1) microencapsulation within thin-wall spheres; (2) adsorption on inert carriers; (3) covalent cross-linking by bifunctional reagents into macroscopic particles; (4) physical entrapment in gel lattices; and (5) covalent binding to water-insoluble matrices. Let us now consider some of the different methods of enzyme immobilization and the procedures followed to immobilize enzymes used in electrodes. For more technical details one is referred to the various articles in this book.

Construction of Enzyme Electrodes

In the subsection on the enzyme as reagent, above, it was mentioned that there are 4 steps to follow in the construction of an enzyme electrode. Let us now discuss each of these factors in more detail:

Step 1. Pick an Enzyme. The enzyme selected must react with the substance to be determined. From standard reference books on enzymology, such as "Biochemists Handbook" (C. Long, ed., Van Nostrand-Reinhold, Princeton, New Jersey), find an enzyme system that is suitable for your determination. In one case, this will involve the use of the primary function of the enzyme, i.e., the main substrate–enzyme reaction. For example, for a penicillin electrode, penicillinase would be used; for a urea electrode, urease; for a glucose electrode, glucose oxidase. In other cases, this might necessitate using an enzyme that acts on the compound of interest as a secondary substrate, i.e., urease for N-methylurea or malic dehydrogenase for acetic acid.[44] Of course, this latter case will introduce more interferences and less selectivity into the assay. If you cannot find a suitable system yourself, consult with a colleague who is a biochemist for assistance.

NOTE: In some cases, there are several enzymes that act on the substrate of interest via different reactions. For example, L-tyrosine could be determined using L-tyrosine decarboxylase and measuring the CO_2 liberated,[25] or using L-amino acid oxidase using a Pt electrode,[30] or an NH_4^+ electrode.[26,28] The latter enzyme, although less selective, can be

[44] G. G. Guilbault, R. McQueen, and S. Sadar, *Anal. Chim. Acta* **45**, 1 (1969).

obtained commercially in high purity, the former is available in low purity and would have to be purified before use. Hence, the scientific capabilities of one's laboratory might dictate the choice of enzyme.

Step 2. Obtain the Enzyme. Once you have found the enzyme to be used for your application, check the catalogs of commercial suppliers (Sigma, Boehringer, Calbiochem, Worthington, etc.) to see if it can be purchased, and the purity. The latter may or may not present a problem. Many enzymes are stable in an impure state, i.e., jack bean urease, glucose oxidase from the food industry (General Mills), and can be used satisfactorily in a pseudo "immobilized form," i.e., as a liquid covered with a dialysis membrane, for up to a week. In other cases, the impure enzyme has too low an activity to be useful in the low-purity state without further purification, i.e., some of the decarboxylases available from Sigma. In the latter case, the enzyme must be purified, which although not difficult, would involve further work and even assistance from others.

In still other cases, one might find that the enzyme that one wants to use is not available commercially. In this case there are two possibilities: (1) inquire of a large biochemical supply house whether it will isolate and purify the enzyme you want; many will, if the price is right; (2) look up the enzyme in the literature or standard biochemistry-enzymology reference books, obtain the isolation and purification methods used, and do it yourself. We have done this ourselves in many cases with excellent results, and in most cases the techniques are simple enough to be carried out by a person with reasonable scientific training.

Step 3. Immobilize the Enzyme. A simple rule of thumb to follow is that the better the enzyme is immobilized, the more stable it will be, and hence, the longer it will be useful and the more assays will be possible from one batch. Let us now consider the various possibilities and the characteristics of the product.

COMMERCIALLY AVAILABLE IMMOBILIZED ENZYME. This is the ideal case and is the first choice, if possible. There are a number of enzymes available in the immobilized form—most of these are fine products, and as good or better than most scientists produce themselves. The author has personally used the products of Boehringer and Aldrich with good success. Of the enzymes available that are likely to be of most use to the reader, one could mention urease (Boehringer), glucose oxidase (Boehringer or Aldrich), ribonuclease (Boehringer), and uricase (Aldrich). Furthermore, Corning has offered to sell almost any enzyme bound to glass, under certain conditions, and the reader is invited to write for details.

SOLUBLE "IMMOBILIZED" ENZYME. The second choice, which is the easiest for the novice if the commercial enzyme is not possible, is to simply use the soluble enzyme in construction of the electrode. A thick paste of the enzyme powder is made with a little water (1–2 μl); this paste is spread over the surface of the electrode, and the layer is covered with a dialysis membrane of 20–25 μm thick cellophane (Will Scientific, Inc., or Arthur H. Thomas, U.S.A.). Such soluble enzyme electrodes are stable for up to about a week, if kept in a 5°–10° refrigerator, between use. Electrodes with the more crude enzymes, urease or glucose oxidase mentioned above, might be stable for longer periods of time. One example taken from Nilsson et al.[16] is quoted below.

Enzyme—pH-electrodes for the substrates glucose, urea and penicillin were made with the enzyme trapped in a liquid layer around the electrode. 100 μl of enzyme solution was poured in a small groove of a 5 cm × 5 cm piece of dialysis membrane (Union Carbide Corp., U.S.A.). The membrane containing the enzyme solution was placed tightly over the sensitive glass bulb of the electrode and held in place with an "O" ring (about half of the enzyme solution remaining entrapped within the membrane). The electrodes used were either Beckman pH-electrodes (No. 39301) or Radiometer pH-electrodes (G 202 C). The following enzyme solutions were used: Urease (EC 3.5.1.5, jackbean, Sigma, type III, 3500 units/g) 100 mg/ml in 0.1 M Tris buffer, pH 7.0. Glucose oxidase (β-D-glucose: O_2 oxidoreductase, EC 1.1.3.4, Sigma, type 2, 13000–18000 units/g) 200 mg/ml in 0.1 M sodium phosphate buffer, pH 7.0. Penicillinase (β-lactamase, EC 3.5.2.6, 40000 units/mg from Calbiochem) 50 mg/ml in 0.01 M sodium phosphate buffer, pH 6.8. The enzyme electrodes were soaked and stirred for about an hour prior to use in their respective buffer solutions. pH measurements were carried out with either a Beckman Century SS pH-meter or a Radiometer pHM 26, each connected with a recorder. Standard fiber junction saturated calomel electrodes were used as reference electrodes. Measurements were carried out in stirred buffer solutions, and unless otherwise stated at 25°C. Magnetic stirring of approximately 150 rev/min in the outer solution was used in these studies to reach the necessary conditions of equilibrium. On the average, solutions of 100 ml were analysed; however using a Radiometer pH-combination electrode (GK 2301 C), samples down to a few ml could be analysed. Prior to each determination the enzyme-electrode was equilibrated in buffer (approx. 10 min) to retain its original pH. After reaching steady state the pH was read; alternatively pH measurements were displayed on a recorder.

PHYSICALLY ENTRAPPED ENZYMES. For ease of preparation, this is the next choice. Many enzymes have been physically bound in polyacrylamide gels, which are cross-linked polymers that have the enzyme held inside. A typical preparation is mentioned below, and similar preparations can be effected by anyone with a minimum of effort. The stability of the final product depends on the degree of care taken and the control of experimental conditions, as was carefully pointed out by Guilbault and Montalvo,[18] but can be as long as 3–4 weeks or about 50–100 determinations.

CHEMICALLY BOUND ENZYMES. These are the most difficult to prepare, although the preparation can be effected by anyone who has had a year's course in organic chemistry. The products are most stable, and can be used for 200–1000 assays, and stored at room temperature for over a year between assays.[13] The best actual method involved will depend on the individual enzyme, as will be discussed in other parts of this book. In the author's experience, the polyacrylacid diazo coupling[13,30] and glutaraldehyde methods[33,34] have yielded extremely satisfactory products. The covalent binding to polyacrylamide cross-linked polymer, used by Boehringer in its commercial preparations, is also satisfactory.

Reactive intermediates for direct coupling of enzymes via only 1–2 steps are available from Corning (Corning, New York), Aldrich (Milwaukee, Wisconsin), Koch Light (England) and Pharmacia Fine Chemicals (Sweden). These are recommended to anyone interested in making chemically bound enzymes. The glutaraldehyde method is also quite simple to effect (glutaraldehyde is available from Sigma, St. Louis, Missouri).

Step 4. Place the Enzyme around the Appropriate Electrode. In order to develop an electrode for the substrate of interest, one must have as the base sensor an electrode that responds to either one of the reactants, A or B, in Eq. (19)

$$A + B \xrightarrow{\text{enzyme}} C + D \tag{19}$$

or to one of the products, C or D. The sensor can be a gas electrode (to measure all O_2-consuming reactions, NH_3- or CO_2-liberating enzymes), a glass electrode (to follow H^+ changes in reactions that liberate acid, or NH_4^+-producing enzymes), a Pt electrode (to follow all enzyme reactions involving electroactive species or O_2), or some other ion-selective electrode (i.e., the CN^- electrode for amygdalin, a NH_4^+ antibiotic electrode for deaminase enzymes, an I^- electrode for oxidative enzymes coupled with the $I^- \to I_2$ indicator reaction, the S^{2-} electrode for cholinesterase substrates, etc.). In most cases, the limiting factor in design of an enzyme electrode will be the availability of a sensor to monitor the reaction. There are other possibilities for monitoring enzyme reactions: for example, a thermistor could be covered with enzyme and the temperature change resulting from the enzyme reaction monitored[45] (see also chapter [45]). Considerable research is being performed in this area, but as yet no satisfactory systems have been developed.

Assuming that a sensor is available, and the enzyme has been obtained and immobilized, let us now describe the preparation of typical enzyme electrodes (see Fig. 1).

[45] K. Mosbach and B. Danielsson, *Biochim. Biophys. Acta,* Report 364, 140 (1974).

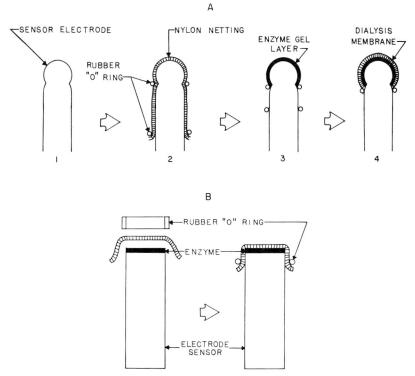

Fig. 1. Preparation of enzyme electrodes. (A) With physically entrapped enzymes. (B) With chemically bound or soluble enzymes.

Type A. Dialysis membrane electrode. Take 10–15 units of the soluble enzyme, physically entrapped enzyme, or chemically bound enzyme (after immobilization of the enzyme, the preparation should be freeze-dried to form a powder) onto a piece of cellophane dialysis membrane (20–25 μm thick, obtained from either Will Scientific or Arthur H. Thomas, U.S.A.) which has been cut into a circular piece with a diameter about twice the size of the electrode sensor. Wrap the cellophane around the electrode, taking care that the powder is evenly spread over the surface of the electrode in a thin layer (this might be as conveniently done by placing a thick paste of the enzyme in water onto the tip of the flat electrode while it is held upside down (Fig. 1B) and coating the enzyme onto the surface with a spatula). Place a rubber O ring, with a diameter that fits the electrode body snugly, around the cellophane (Fig. 1B), and gently push it onto the electrode body so that the cellophane-enzyme lies on the bottom of the electrode and is held tight and flat

(Fig. 4B). Place the electrode in buffer solution for a few hours or overnight to allow penetration of buffer into the enzyme layer and loss of entrapped air. Store the electrode in buffer between use (for references, see footnotes 13, 16, 21, 23, 30, 31, 37).

TYPE B ELECTRODE. Physical entrapment onto surface. Place the electrode sensor upside down (Fig. 1A) and cover it with a thin nylon net (about 90 μm thick—a sheer nylon stocking is satisfactory) which is secured with a rubber O ring in the same manner as above. This serves as a support for the enzyme gel solution. Prepare the enzyme gel solution by mixing 0.1 g of enzyme (purity about 10–15 units/mg) with 1.0 ml of gel solution [1.15 g of N,N'-methylenebisacrylamide (Eastman Organics, Rochester), 6.06 g of acrylamide monomer (Eastman), 5.5 g of potassium persulfate, and 5.5 mg of riboflavin in 50 ml of water]. Gently pour the enzyme gel solution onto the nylon net in a thin film, making sure that all the pores of the net are saturated; 1 ml of this solution should be enough for several electrodes. Place the electrode in a water-jacketed cell at 0°–5° and remove oxygen, which inhibits the polymerization, by purging with N_2 before and during polymerization. Complete the polymerization by irradiating with a 150-W Westinghouse projector spot light for 1 hr. At the end of this time the enzyme layer should be dry and hard. Place a piece of dialysis membrane over the outside of the nylon net for further protection, and secure with a second rubber O ring. Soak the electrode in buffer solution overnight and store in buffer between use (for references, see footnotes 9, 13, 16–20, 26–29, 35, 36, 38).

TYPE C ELECTRODE. Direct polymerization onto the membrane. This can be effected by a direct attachment of the enzyme to the surface of the electrode, if glass, by the "Corning technique" (see chapter [10]) or by direct chemical attachment on the electrode surface, as was done by Anfalt, Granelli, and Jagner[22] in the case of the Orion NH_3 electrode. In the latter study membranes were prepared by dropping 0.1 ml of soluble urease solution (0.5 μ) onto the surface of the gas diffusion membrane. The membrane was set aside for 12 hr at 4° to allow evaporation of the solvent, and glutaraldehyde solution was then added dropwise (2.5% in phosphate buffer, pH 6.2). The membrane was set aside for a further 1.5 hr at 4° and then was rinsed carefully with water in order to remove free enzyme and buffer. Note: Some activity was lost over a 20-day period, an indication that insufficient enzyme was used.[22] At least 5–10 units of urease would have been better in this case, again in 0.1 ml of solution.

Of the three types of electrodes described above, I prefer type A for ease of preparation (once the bound enzyme is obtained) and for long-term stability (if the enzyme is chemically bound). The type B electrodes are not difficult to prepare, but are time consuming, requiring 1 hr of polymerization time per electrode, and have a maximum stability of about 3–4 weeks or 50–100 assays. I do not like the type C electrode with direct attachment to the electrode because it essentially commits the electrode sensor to that one enzyme electrode. In the type A and B electrodes, one can easily replace the enzyme layer, when it is no longer useful, with a new layer—the sensor can be reused until its lifetime is exhausted. The type C electrode prepared by Anfalt et al.[22] is really a type A electrode.

Response Characteristics of Electrodes

Now that the electrode has been made, let us consider some of the factors that affect the response and stability of the electrode. Table II gives a listing of all enzyme electrodes prepared to date, in which the enzyme is held "immobilized" in the vicinity of the electrode, together with some response characteristics: response time, linear range, stability, and type of sensor used.

Stability. The stability of an enzyme electrode is difficult to define, since an enzyme can lose some activity, resulting in a downward shift of the calibration curve. Yet, if the slope remains constant, as is frequently the case, the electrode is still useful, needing only calibration daily. This is seldom a problem since all who use electrodes of any type, i.e., glass electrode, reset the pH or potential of their electrode at least once a day. This should be done with the enzyme electrode also, using serum (i.e., Monitrol, Dade, Miami). Another problem in the definition of stability is that many workers measure the potential of their electrode occasionally over a long period of time, and report the data as the stability. This may mean that the electrode was used 1 time a day or week, 10 times a day, or 100 times a day. Naturally, the more frequent the use, the shorter will be the overall lifetime. As a general rule, a "soluble" electrode is useful for about 1 week or 20–50 assays, provided the electrode is kept refrigerated between uses. The physically entrapped "polyacrylamide" electrodes are good for about 3 weeks or 50–200 assays, depending crucially on the degree of care exercised in the preparation of the polymer. The chemical enzyme can be kept indefinitely if not used very much, even at room temperature (see Table II—as long as 14 months for glucose oxidase, more than 4–6 months for L-amino acid oxidase or uricase). One can expect to get about 200–1000 assays per each electrode, again depending on how good a synthesis is effected. Although the electrode can be stored at room temperature, it is recommended that all electrodes be kept in a refrigerator and covered with a dialysis membrane to prevent the action of bacteria, which tend to feed on the enzyme, destroying its activity. The dialysis membrane (molecular weight exclusion about 1500) prevents the enzyme from getting out and bacteria from getting in.

The stability of the physically entrapped enzyme varies greatly with experimental conditions, and a thorough study of these factors was made by Guilbault and Montalvo.[18] The maximum stability that could be achieved with such enzyme electrodes was obtained with the following immobilization parameters: photopolymerizing for 1 hr at 28° with a No. 1 150-W photoflood lamp; gel-layer thickness of 350 μm; and an enzyme concentration in the gel of about 10 units.

Another factor that affects the apparent stability of all electrodes, especially the "soluble" and physically entrapped electrodes, is the content of enzyme in the reaction layer. As will be shown later, a certain amount of enzyme is required to yield a Nernstian calibration curve. Many times it is advantageous to add more enzyme, say twice as much; in this case more enzymic activity can be lost, yet a linear Nernstian plot will still be obtained.

Still another factor that will affect the stability of an electrode is the choice of operating conditions. An example of this is the comparison of the results obtained by Rechnitz and Llenado[35,36] and those of Mascini and Liberti[37] for the amygdalin electrode. Amygdalin is cleaved by β-glucosidase to give CN^- ions, which are sensed by a CN^- ion-selective electrode. Since this electrode responds best to free CN^- ions, obtainable only at pH's >10, Rechnitz and Llenado used this pH for the operation of their electrode. Even though the enzyme was physically bound, it lost activity continually and had a lifetime of only a few days. It is known that almost all enzymes will lose activity at pH's <3 and >9, and undoubtedly this was one contributant to the poor stability. Mascini and Liberti used only a soluble enzyme at a pH of 7 and found not only better stability (1 week, which is all that can be expected from a soluble enzyme), but also faster response times. Another reason for the poor stability of Rechnitz and Llenado[36] is the "sausage" polymerization these authors tried, in which large pieces of physically entrapped enzyme are made and slices are cut for each assay. From our own experience and that of others with such a technique, the sausage obtained is like a roast beef placed in an oven for 30 minutes—it is well done on the outside and raw on the inside. The reader is advised not to attempt such a large-scale entrapment, but to prepare individual small batches of polyacrylamide enzyme gels.

A thorough comparison of the stability of the three types of "immobilized" electrodes, soluble (type 1), physically entrapped (type 2), and chemically bound (type 3) was shown by Guilbault and Lubrano[13] for glucose oxidase. The long-term stabilities of the types 1, 2, and 3 electrodes were studied by testing the response of each type of electrode to 5 mM glucose in phosphate buffer, pH 6.6, at least once a week for several months. When not in use, the electrodes were stored in phosphate buffer at 25°. The results show that the long-term stability decreases in the order: chemically bound > physically bound > solubilized. Not only did the type 1 electrode response decrease drastically with time, it also decreases with each determination of glucose. This is a serious problem and, as a result, the type 1 electrode is of little use analytically, except with frequent calibration and use of a large excess of enzyme. This problem is not encountered with the types 2 and 3 electrodes consisting of immobilized enzyme. The activity of these two electrodes actually increases for the first 20–40 days before beginning to decrease—this is probably due to the establishment of diffusion channels in the matrix over a period of time with concomitant increase in apparent activity until the channel formation ceases and only denaturation is observed. Or it could be due to changes in the conformation of the fraction of enzyme immobilized in a nonactive conformation to the more stable and preferred conformation. Immobilization in an unfavorable conformation can be due to pH, temperature, or stirring effects during the immobilization process. The decrease in response is due to a decrease in activity of the enzyme layer because of slow denaturation, and possibly also slow irreversible inhibition. The types 2 and 3 electrodes eventually reach a stability change of −0.25 and −0.08% of maximum response per day, respectively. The physically bound enzyme lost half its activity in 7 months, but the chemically bound had lost only 30% of its activity in 400 days (13 months). Of course, this stability would have been much less if the electrodes had been subjected to considerable use each day—about 200 assays for the type 2 electrode, and almost 1000 for the type 3 enzyme electrode are possible.

Still another factor affecting the stability of some enzyme electrodes is the leaching out of a loosely bound cofactor from the active site, a cofactor that is needed for the enzymic activity. Such was found by Guilbault and Hrabankova[28] in the case of D-amino acid oxidase in a polyacrylamide membrane. The bond between protein and coenzyme (flavine adenine dinucleotide, FAD) is very weak in D-amino acid oxidase, and FAD is easily removed by dialysis against buffer without FAD. Without FAD in the solution used to store the electrode, all activity is lost in 1 day; using a 0.4 mM solution of FAD in tris buffer, pH 8.0, to store the electrode between use, resulted in a 3-week stability, with little loss in activity.

Finally, the stability of the enzyme electrode will depend on the stability of the base sensor. This, in most cases, is not the limiting factor in the stability, the sensor having a longer stability than the immobilized enzyme. This factor should be considered, however, when using some of the shorter lifetime electrodes, such as the liquid-membrane electrodes.

Response Time. (A faster response is defined as a decrease in response time.) There are many factors that affect the speed of response of an enzyme electrode. For a response to be obtained (1) the substrate must diffuse through solution to the membrane surface, (2) diffuse through the membrane and react with enzyme at the active site, and (3) the products formed must then diffuse to the electrode surface, where they are measured. Let us consider these factors in detail, and see how the response time can be optimized.

RATE OF DIFFUSION OF THE SUBSTRATE. A mathematical model describing this effect can be derived, as was done by Blaedel et al.,[46] but in simplest practical terms the rate of substrate diffusion will depend on the stirring rate of the solution, as was shown experimentally by Mascini and Liberti[37] for the amygdalin electrode. In an unstirred solution the substrate gets to the membrane surface, albeit slowly, so that long response times are observed. At high stirring rates, the substrate quickly diffuses to the membrane surface, where it can react (Fig. 2). The difference can be as much as a decrease of response time from 10 min to 1–2 min, or less. With rapid stirring for the urea electrode[17,18] a response time of less than 30 sec was achieved. Of importance also is the relationship of stirring rate to the equilibrium potential observed. As shown in Fig. 2, the potential shifts as a function of stirring rate owing to the changes in the amount of substrate brought to the electrode surface and the degree of its reactivity. Hence, for fast response time and steady, reproducible values, it is recommended that a fast stirring of the solution be effected, yet a constant stirring rate (i.e., set the speed on your stirrer and use this same setting for all readings).

REACTION WITH ENZYME IN MEMBRANE. The rate of reaction will depend, according to the Michaelis–Menten equation,

$$V = \frac{k_3[E][S]}{K_m + [S]} \quad (20)$$

on the activity of enzyme and factors that affect it, i.e., pH, temperature, inhibitors, and the concentration of substrate. The equilibrium potential obtained, however, should be dependent only on the substrate concentration and the temperature (since this term appears in the Nernst equa-

---

[46] W. J. Blaedel, T. R. Kissel, and R. C. Bogulaski, *Anal. Chem.* **44**, 2030 (1972).

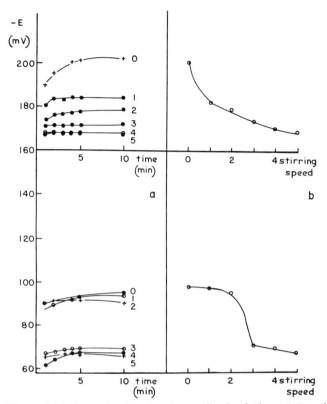

FIG. 2. Effect of stirring rate: 0 refers to unstirred solutions, 1–5 to increasing stirring speeds. [Amygdalin] = (a) 10^{-2} M, (b) 10^{-4} M. Enzyme = 1 mg. From M. Mascini and A. Liberti, *Anal. Chim. Acta* **68**, 177 (1974), by permission of the authors.

tion). The response rate will also depend on the thickness of the membrane layer in which reaction occurs and on the size of the dialysis membrane used to cover the enzyme layer, if one is used. Let us consider each of these factors separately.

1. Effect of substrate. A typical example of the effect of substrate concentration on response rate is shown in Fig. 3, the response of a β-glucosidase membrane electrode to amygdalin.[37] The rate of reaction increases (as indicated by the increased inflection of the E-time curve) as the substrate concentration increases and a faster response time is observed, i.e., 1 min for 10^{-1} M amygdalin, 5 min for 10^{-4} M amygdalin. As an alternative to waiting until an equilibrium potential or current is reached, the rate of change in the current or potential can be measured and equated to the concentration of substrate. This was done by Guil-

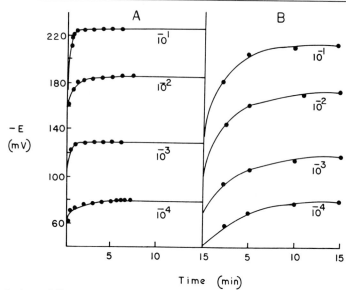

FIG. 3. Amygdalin response-time curves for an electrode containing 1 mg of β-glucosidase immobilized by a dialysis paper. (A) pH 7; (B) pH 10. From M. Mascini and A. Liberti, *Anal. Chim. Acta* **68**, 177 (1974), by permission of the authors.

bault and Lubrano[13] in the case of the glucose electrode, and a result for glucose is obtained in 12 sec.

2. *Effect of enzyme concentration.* The activity of enzyme in the gel will have two effects on an enzyme electrode: (1) it will ensure that a Nernstian calibration plot is obtained, as will be discussed below, and (2) it will affect the speed of response of the electrode. However, this effect is a tricky one, inasmuch as an increase in the amount of enzyme also affects the thickness of the membrane. This is demonstrated in Fig. 4, taken from the results of Mascini and Liberti,[37] for the amygdalin electrode. As the amount of enzyme is increased from 0.1 to 2.5 mg of β-glucosidase a shorter time is observed, yet when 5 mg of enzyme was used the response time became considerably longer. This latter effect is due to a further thickening of the membrane layer by the use of more weight of enzyme, resulting in an increase in the time required for the substrate to diffuse through the membrane. If one weight of enzyme had been chosen and the activity of enzyme increased at constant mass, a steady increase in response rate would be observed, and then a gradual leveling off in response time. Hence, for best results, it is recommended that as active an enzyme be used as possible, to ensure rapid kinetics, in as thin a membrane as obtainable.

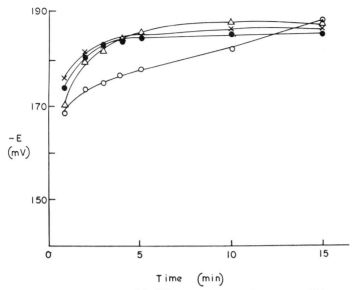

FIG. 4. Response-time curves with different amounts of enzyme; pH 7, amygdalin, 1 mM. Amounts of enzyme: ●, 0.1 mg; ×, 1 mg; △, 2.5 mg; ○, 5.0 mg. From M. Mascini and A. Liberti, *Anal. Chim. Acta* **68**, 177 (1974), by permission of the authors.

3. Effect of pH. Every enzyme will have a maximum pH at which it is most active, and a certain range of pH in which it demonstrates any reactivity. The immobilized enzyme will have a different pH range from the range of the soluble enzyme because of its environment, as was discussed above. The pH range for immobilized glucose oxidase is about 5.8 to 8.0[13] (solution enzyme pH 5 to 7), β-glucosidase about pH 5 to 8.[37] Hence, for fastest responses one should work at the pH optimum. This is not always possible, however, since the sensor electrode might not respond optimally at the pH of the enzyme reaction. Thus, a compromise is generally reached between these two factors. However, one should be careful not to be trapped into forcing the enzyme system to conform with the requirements of the sensor, as was done by Llenado and Rechnitz in the case of the amygdalin electrode.[35,36] These authors tried a pH of 10, which has been shown to be optimum for the electrode sensor, the CN^- electrode. Longer response times were obtained at pH 10 (Fig. 3) compared to pH 7, since the enzyme has very little activity at this pH. Furthermore, the enzyme rapidly loses activity at high pH's (>9–10) and Rechnitz and Llenado found their immobilized enzyme electrode very unstable, changing in response curve downward every day. Yet by working at pH 7, Mascini and Liberti[37] found their soluble enzyme electrode

still useful for a week. Similar effects are noted in other studies; Guilbault and Shu,[29] for example, found that the response time of a glutamine electrode decreased from pH 6 to 5, the optimum for the enzyme reaction.

Further examples of making the electrode conform to the enzyme system, instead of vice versa, are the results of Anfalt et al.[22] and Guilbault and Tarp[23] on the NH_3 sensor for the product of the urea–urease reaction. Although at the optimum pH for this enzyme reaction, 7–8.5, there is very little free NH_3 present to be sensed by the gas-type electrode, and one would predict poor results for urea assay, both groups found that the sensitivity of the sensor was more than sufficient for each measurement at these low pH's. This is partly ascribed to the fact that there is a buildup of larger amounts of product in the reaction layer than in solution, and hence, an increase in sensitivity is obtained for the sensor, which sits close to the enzyme layer.

4. Effect of temperature. One would predict a dual effect of temperature, an increase in the rate of reaction resulting in a faster response time, and also a shift in the equilibrium potential by virtue of the temperature coefficient in the Nernst and van't Hoff equations.

This was demonstrated by Guilbault and Lubrano[13] for the glucose electrode, in which the effect of temperature on the electrode response was studied from 30° to 50°. Linear plots of the log rate and log total current vs $1/T$ were observed as predicted by the van't Hoff ($\ln K = \ln C - (\Delta H/RT)$, K being the equilibrium constant), and the Arrhenius ($\ln k = \ln A - (Ea/RT)$, k being the rate constant) equations. Practically, this means that the temperature of the enzyme electrode should be carefully controlled for best results, although the effect of temperature is most pronounced on reaction rate measurement. Similar effects were noted by Guilbault and Lubrano for amino acid oxidase.[30]

Guilbault and Hrabankova[28] found that the response of the D-amino acid oxidase electrode to D-methionine showed only very small effects at increasing temperatures (25°–40°) although the theoretical Nernstian slope is 61.74 mV per decade at 37°. Similarly, Papariello et al.[38] found that although the response time of his penicillin electrode was somewhat more rapid at 37° than at 25°, no great improvement was observed.

Hence, the user of enzyme electrodes is advised to control the temperature if he is making kinetic measurements, but not to bother when making equilibrium measurements; simply use room temperature or about 25° for convenience.

5. Thickness of the membrane. The time required to reach a steady-state potential or current reading is strongly dependent on the gel-layer thickness. This is due to an effect on the rate of diffusion of the substrate

through the membrane to the active sites of the enzyme, and on the rate of diffusion of the products through the membrane to the electrode sensor, where they are measured. Guilbault and Montalvo[18] observed that the time interval for 98% of the steady-state response was about 26 sec with a net of urease 60 μm thick and about 59 sec with a 350-μm net for 8.33 × 10^{-2} M urea and an enzyme concentration of 176 mg/ml of gel. Similarly, Anfalt et al.[22] in the case of a urea electrode with glutaraldehyde-bound enzyme, and Mascini and Liberti,[37] using a β-glucosidase amygdalin electrode, observed an increase in response time in going from thin to thick membranes.

Thus, it is recommended that as thin a membrane as possible be used for best results. This can be achieved using highly active enzyme.

6. Effect of dialysis membrane. In most cases, it is advantageous to use a dialysis membrane to cover the electrode, as was previously pointed out. This membrane serves to protect the enzyme and prolong the stability of the electrode. Guilbault and Montalvo[18] noted that the cellophane coatings had little effect on the response time of the urea electrode.

A thorough study of the effect of the thickness of the dialysis membrane on the response rate was made by Mascini and Liberti[37] for the amygdalin electrode. Their results indicate that the response time and the equilibrium values are altered by varying the thickness of the membrane. A thin 20-μm membrane (Arthur H. Thomas) or a 25-μm membrane (Will Scientific) will have essentially no effect on the response time and are recommended for use.

MEASUREMENT OF PRODUCTS AT THE ELECTRODE SURFACE. The final factor that will affect the speed of response is the electrode sensor itself, and how fast it will give a potential or current proportional to the amount of product or reactant it sensed.

In most cases of enzyme electrodes, in which rapid stirring is used to minimize the effect of rate of diffusion of the substrate and a thin membrane of highly active enzyme is used under optimum conditions to minimize the effect of reaction with enzyme in the membrane, the determining factor will be the response time of the sensor.

Guilbault and Montalvo[18] observed a response time of 26 sec with a 60-μm thick enzyme layer of urease in a urea electrode, compared to a response time of 23 sec for an uncoated cation electrode. Anfalt, Granelli, and Jagner[22] observed response times of their urea probe of 30 sec to 1 min, almost the same as those for the uncoated NH_3 gas electrode. Mascini and Liberti[37] noted a 1-min response time for their CN^--coated amygdalin electrode at high substrate concentration (10^{-1} M), quite similar to the uncoated CN^- electrode.

At low substrate concentrations (10^{-2} to 10^{-5}), the reaction with enzyme in the membrane becomes the rate-limiting factor on the response time.

Effect of Concentration of Enzyme in Membrane. The concentration of enzyme in the membrane will have two effects on the electrode response. The first, on the response time, was discussed above. The second is on the shape of the response curve. This is best indicated in Fig. 5, which is taken from the work of Guilbault and Montalvo[18] on the urea electrode.

To study the effect of enzyme concentration on the enzyme gel layer activity, gels were prepared with enzyme concentrations ranging from 3 to 110 units of urease per milliliter of gel. The steady-state response of each enzyme-coated electrode when dipped in urea solutions from 5×10^{-5} to 1.6×10^{-1} M was measured. The results are shown in Fig. 5. The slope of each curve increases with the amount of enzyme in the gel layer on the electrode until, with larger enzyme concentrations, only a small increase in activity of the gel membrane is obtained. A plot of the steady-state response for 8.33×10^{-2} M urea against the amount of urease in the gel membrane[18] showed a rapid increase in response or activity up to 7.5 units of urease per milliliter of gel. Above 7.5 units of urease per milliliter of gel, a large increase in enzyme concentration gives only a small increase in activity of the enzyme gel membrane on the cation electrode. Optimum enzyme concentration in the gel, considering only the

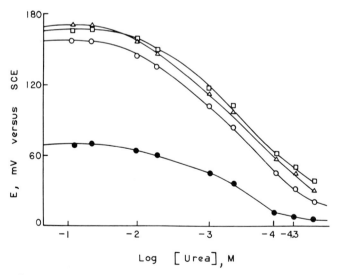

Fig. 5. Effect of enzyme concentration on electrode response. Units of urease per ml of gel: ●, 3.0; ○, 52.5; △, 65.6; □, 87.4. From G. G. Guilbault and J. G. Montalvo, *J. Am. Chem. Soc.* **92**, 2533 (1970), by permission of the author.

economy of enzyme, is obtained at about 7.5 units of urease per milliliter of gel.

Similar results are obtained in the case of every enzyme electrode. The amount of enzyme to be used depends on the individual electrode, however, as a rule of thumb 10–20 units per membrane will generally be sufficient to give an excellent response curve. In the case of the more unstable "soluble" and physically entrapped enzyme electrodes, an excess of enzyme should be used so that a loss of enzyme will not affect the potentials observed.

Likewise, purified enzymes should be used (at least 1 unit/mg) so as to keep the thickness of the membrane to a minimum for fast response rates.

Wash Time of the Electrode. Because there is a buildup of product in the enzyme membrane, every type of enzyme electrode described requires a washing after use in order to re-reach the base-line potential. This wash time varies from only 20 sec in the case of urease and the air-gap electrode[23] to as long as 10 min for urease with a pH electrode sensor.[16] The wash time will increase with the thickness of the enzyme membrane, as expected, and also with the nature of enzyme and the base sensor. This latter effect is due to the charge on the enzyme, which in the case of urease, for example, is negative, with a resulting attraction for the positive NH_4^+ ions formed. The glass, likewise, has an attraction for these +ions, resulting in longer wash times required.

The electrode can be washed in an automatic electrode washer, as described by Montalvo and Guilbault,[47] by simply placing the electrode in water or rinsing it under a distilled-water tap. The type of washing will depend on the type of electrode. Soluble and physically entrapped enzymes must be washed gently since they can be easily washed out of their matrix. The more sturdy chemically bound enzymes can be washed under a water tap.[23] The latter washing yields a much quicker return to base line. Oxygen-based electrodes[33,34] will show a quick return to base line if simply placed in a fresh buffer solution. Other electrodes, such as the Pt-based amperometric enzyme electrodes,[12,13,30,31] have to be pretreated before first use by applying a potential (+0.6 V vs SCE) until the anodic current decays to a low value; after each run the electrode is washed in a stirring phosphate buffer solution until the current decays to a low value (0.5–1 min) indicating the removal of unreacted substrate and reaction products.

Range of Substrate Determinable. As indicated in Table II, all enzyme electrodes sense substrate in the range of 10^{-2} to 10^{-4} M, some electrodes

[47] J. G. Montalvo and G. G. Guilbault, *Anal. Chem.* **41**, 1897 (1969).

being useful to as high as 10^{-1} M or as low as 10^{-5} M. In all cases, curves similar to those in Fig. 5 are obtained, approximately Nernstian in the linear range with a slope of close to 59.1 mV per decade. All curves level off at high substrate concentration, as predicted by the Michaelis–Menten equation, which states that the reaction becomes independent of substrate at high substrate concentration. A leveling off of the curve at low substrate concentration is also observed; this is due to the limit of detection of the electrode sensor used.

Effect of Interferences

Any enzyme electrode will only be as good as the selectivity it possesses. Hence, a consideration of the possible interferences is now in order. Interferences fall into two categories: (1) interferences in the electrode sensor, and (2) interferences with the enzyme itself.

Interferences in the Electrode Sensor. The electrode used to sense the products of, or reactants in, the enzyme reaction should be one that will not react with other substances present in the sample to be assayed, if possible.

The urea electrode probes originally described by Guilbault and Montalvo[17,18] could not be used for assay of urea in blood or urine because the cation glass sensor used to measure the NH_4^+ also responds to Na^+ and K^+ present in blood or urine.

A cell using a glass electrode (Beckman Electrode 39137 or 39047) as the reference electrode was tried by Guilbault and Hrabankova,[19] in an attempt to eliminate the effect described above. Calibration curves for urea were found to be the same when the cell with the SCE reference electrode was used. Also, the interferences of monovalent cations in solution are smaller in this case because both electrodes are sensitive to these ions. However, concentrations of Na^+ and K^+ higher than 10^{-4} M considerably decreased the electrode response. This effect could be explained as a decrease of activity coefficients in the presence of other ions or as decrease of enzymic reaction rate caused by a higher ionic strength.

Combining the cell with the uncoated glass reference electrode with a cation exchanger, the determination of urea in blood and urine is possible with a deviation of less than 3%.

A still further improvement was described by Guilbault and Nagy,[20] in which a solid antibiotic nonactin electrode was used as sensor. This electrode has a selectivity of NH_4^+/K^+ of 6.5/1 and NH_4^+/Na^+ of 7.5 × $10^2/1$, thus partially eliminating any response to these ions by the sensor. By using a three-electrode system (a chemically bound urease over a nonactin solid electrode vs an uncoated nonactin electrode and a calomel electrode as reference) and dilution to a constant interference level of K^+ (Na^+ does not interfere because of the high selectivity coefficient of the sensor), urea in blood was assayed with an accuracy of 2–3% or better. In this procedure a standard calibration plot of E vs urea was prepared in the presence of KCl, at its highest level in blood. Before each run the blood sample to be assayed was brought to the same potential as that observed in the KCl-buffer solution by addition of more KCl to the sample using the uncoated electrode vs SCE. The potential of the urease sensor vs SCE was then read and the urea determined.

Anfalt, Granelli, and Jagner[22] used an Orion NH_3 electrode, which is a glass electrode with a layer of NH_4Cl and a gas-permeable membrane over the outside, to sense the NH_3 produced from the urea–urease reaction. In theory, the use of the gas NH_3 electrode, which senses only NH_3, but not K^+, Na^+ or other ions or substrates, should provide the desired selectivity of a sensor. However, at the pH's optimal for the urea-urease reaction (7–8.5), the free NH_3 present is quite small compared to NH_4^+ ions. Yet, since the enzyme layer is placed directly onto the gas electrode there was sufficient NH_3 produced at the electrode surface to give nice linear calibration plots of pH's 7, 7.4, 8, and 9 with 69.5 mV per decade (actually the slope at pH 7 was 69.5 mV per decade, higher than Nernstian).

In a similar type of study, Guilbault and Tarp[23] showed that the air-gap electrode for NH_3 developed by Ruzicka and Hansen[24] could be used to provide a totally specific urea assay directly at pH 8.5. The air-gap electrode is designed on a principle similar to the gas electrodes, except minus the gas membrane, which usually causes many practical problems such as long response times and clogging of the membrane pores by proteins and other substances in blood, thus tremendously decreasing the useful lifetime. In the air gap, NH_3, generated in solution by the enzyme reaction at pH 8.5, diffuses out of solution to the surface of the glass electrode, which is coated with a layer of 10^{-3} M NH_4Cl and surfactant. A change in pH is observed due to the reaction,

$$NH_3 + H^+ \rightleftharpoons NH_4^+ \tag{21}$$

which is monitored, and is proportional to the log urea concentration. By calibration, the response curve vs Monitrol II (freeze-dried blood serum from Dade, Miami, Florida) the urea content of 30 blood samples from Riggs Hospital in Copenhagen was assayed with an accuracy of about 2%. The method requires only 2–4 min per assay, and, using a chemically bound urease, over 500 assays per electrode are possible.

In the Pt electrode devices described for measuring oxidative enzyme systems, two approaches have been taken: to measure the peroxide produced by monitoring the total current change,[10,12,13] or the rate of change in the current[12,13] at $+0.6V$ vs SCE, or to measure the uptake of oxygen by the enzymic reaction[33,34] by measuring the reduction of O_2 at -0.6 V vs SCE. In either system, compounds present that are either oxidized or reduced at a Pt electrode at ± 0.6 V vs SCE would interfere. Clark[10] eliminates this problem by using a second uncoated Pt electrode, held at the same potential as the enzyme electrode, to measure any compounds present in blood. His system, as modified, forms the basis for instruments to measure glucose sold commercially by Yellow Springs Instrument Co. (Yellow Springs, Ohio), and Radiometer (Copenhagen). However, the normal values of such interferences (ascorbic acid or uric acid are examples of oxidizable compounds) are generally $<5\%$ of the glucose signal. Another way to eliminate the effect is to measure very quickly (~ 21 sec) the rate of change in the current as was done by Guilbault and Lubrano,[12,13] who found that assays of over 200 blood samples could be performed with an accuracy and precision of better than 2%. Still better selectivity is obtained using the measurement of O_2 at -0.6 V since all other compounds present in blood that consume O_2 can be subtracted out before measurement with the enzyme sensor.[33,34]

In other electrodes, such as the CN^- solid-precipitate electrode used by Rechnitz and Llenado[35,36] and Mascini and Liberti[37] for amygdalin, any ions capable of forming insoluble silver salts will interfere because of formation of a precipitate on the electrode surface. Also, substances capable of reducing silver ion will interfere, as will heavy metal ions and transition metal ions capable of forming cyanide complexes. In the use of an iodide sensor to measure glucose, Nagy, von Storp, and Guilbault[15] found several types of interferences making this type of measurement approach very limited: interferences at the iodide electrode (thiocyanate, sulfide, CN^-, and Ag^+) and interferences from oxidizable compounds present in blood, such as uric acid, tyrosine, ascorbic acid, and Fe(II), which compete in the oxidation of iodide to iodine in the peroxide–peroxidase system. These compounds had to be removed by sample pretreatment.

Finally, in the glass electrode systems for pH measurements as described by Papariello et al.[38,39] and Mosbach and co-workers,[16] any acidic or basic components present would interfere in the measurement. However, by adjustment to a definite pH and providing buffering capacity before initiation of the enzyme reaction, and assuming that only the enzyme reaction will give rise to a pH change, these effects can be minimized or eliminated.

Interferences with the Enzyme Reaction. Such interferences fall into two classes: (1) substrates that can catalyze the reaction in addition to the compound to be measured; (2) substances that either activate or inhibit the enzyme.

In the case of some enzymes, such as urease, the only substrate that reacts at a reasonable rate is urea—hence, the urease-coated electrode is specific for urea.[17,18] Uricase, likewise, acts almost specifically on uric.[33] Others, like penicillinase,[38,39] react with a number of substrates: ampicillin, naficillin, penicillin G, penicillin V, cyclibillin, and dicloxacillin. All these can be determined with a penicillinase electrode.

Similarly, D-amino acid oxidase[28] and L-amino acid oxidase[26,30] are less selective in their responses: the former in an electrode yields a good response to D-phenylalanine, D-alanine, D-valine, D-methionine, D-leucine, D-norleucine, and D-isoleucine; the latter for L-leucine, L-tyrosine, L-phenylalanine, L-tryptophan, and L-methionine. Alcohol oxidase[31,32] responds to methanol, ethanol, and allyl alcohol. Hence, in use of electrodes of these enzymes, either a separation must be used if two or more substrates are present, or the total be determined. In the case of L-amino acid assay, an attractive alternative exists, the use of decarboxylative enzymes[25] which act specifically on different amino acids. Such are known for L-tyrosine, L-phenylalanine, L-tryptophan, and others.

Glucose oxidase acts on a number of sugars:[15] glucose and 2-deoxyglucose are the main substrates, but cellubiose and maltose also react, probably due to the presence of other hydrolytic enzymes in the glucose oxidase preparation.

The activity of the enzyme can be adversely affected by the presence of certain compounds, called inhibitors. These are generally heavy metal ions, such as Ag^+, Hg^{2+}, and Cu^{2+}, and sulfhydryl-reacting organic compounds, such as p-chloromercuribenzoate and phenyl mercury(II) acetate (due to their reaction with the free SH groups present at the active site of many enzymes, especially the oxidases).[13] One important point to consider, however, is that the immobilized enzyme is much less susceptible to inhibitors, especially weak or reversible inhibitors, owing to the protection of the immobilization matrix. Thus, by using the enzyme in an immobilized form, most of one's worries about inhibitors are eliminated. From personal experiences, in the design and use of almost 20 different enzyme electrodes, I have not experienced a case of enzyme inhibition interfering with an assay. However, one should always be aware of this problem, especially in assaying solutions containing heavy metal ions, and especially pesticides.

Comparison of Electrodes

Table II presents a listing of the various enzyme electrodes that have been published in the literature from the beginning to the present. Under each type of electrode is listed the enzyme(s) used to make the electrode, the stability of the electrode, its response time, the range of assay, and the reference. The reader is invited to use this list as a general guide in developing his or her electrode.

Let us now discuss these various electrodes, compare them, and see which designs are best. First, it can be said that none of these electrodes are available as self-contained units that can be purchased separately. There are already three instruments out for glucose (Yellow Springs Instrument Co., Radiometer, and Leeds & Northrup), and one for urea (Owens-Illinois, Kimble, Toledo, Ohio) that use an enzyme electrode as an integral part of the instrument. The latter uses a chemically bound enzyme, urease; the ammonium ion liberated is converted to ammonia with sodium hydroxide, and the free ammonia is detected with a gas electrode. Sixty assays per hour are possible with results of 2–3% accuracy and precision. Only 100 μl of blood sample are required. Each enzyme cartridge can be used for at least 500 assays. There are no plans at present for the sale of individual electrodes. However, these are easy to make and use, as we have seen.

Many electrodes have been described for urea[16–18,20–23,25] based on pH, cation (NH_4^+), CO_2, and NH_3 electrode probes. From a practical point of view, the best is the NH_3 electrode-based probe,[22,23] since it eliminates any interference problems from Na^+ and K^+ in solution.[17,18,20,21] With it and a chemically bound enzyme, over 500 assays per electrode are possible in 1–4 min/assay.[23] The CO_2 based sensor[25] is also quite selective, although slower in response, and has the disadvantage of compensation for the CO_2 present in blood before assay. The use of a glass electrode as probe for urea[16] is a sound idea, except that long times are required to get a stable pH change (5–10 min) and the electrode has a more limited range of concentrations in which it is useful. I believe all urea electrodes of the future will be based on some type of NH_3 sensor, the principle used in the O-I Kimble urea system.

Undoubtedly, the best electrodes for glucose are those based on the use of a Pt sensor. Those based on the iodide electrode are definitely not useful, being subject to many interferences, slow in response and with a narrow linear range.[15] The glass electrode-based sensor,[16] though interesting, does have a long response time and long wash time and is not as generally useful as the Pt-based sensor. However, this paper by Mosbach and co-workers[16] does represent a new vista of electrodes, in that these authors have shown it is possible to measure any acid-producing or proton-producing enzyme reaction with good accuracy. The concept should be useful in many systems, e.g., penicillin.

Much credit in the field of enzyme electrodes is due Leland Clark, one of the pioneers in this field, who described the first "enzyme" electrodes, the glucose electrode, using a Pt sensor and the soluble enzyme.[8,10] His electrode is available in glucose apparatus sold by Yellow Springs Instruments and Radiometer. He has also used the polarographic H_2O_2 monitoring system for other electrodes, for example, alcohols[40] and amino acids,[42] and has provided stimulation to others around him.

The Williams, Doig, and Korosi[11] electrode, based on measurement of the oxidation of hydroquinone formed when quinone replaces O_2 as the hydrogen acceptor, is interesting, but only theoretically so.

The Pt based sensors, based on peroxide measurement[8,10,12,13] and on measurement of O_2 consumption,[34] are the most sound and generally useful and may form the basis for future instruments, although the use of an O_2 gas electrode with immobilized glucose oxidase[9] will attract some attention by instrument builders.

The Pt electrode,[30] the gaseous CO_2 electrode,[25] the cation (NH_4^+) electrode,[26-29] the I^- electrode,[27] the antibiotic NH_4^+ electrode,[27] and the Pt-based O_2 electrode[34] have been tried as the base sensor for measurement of amino acids. Of these, I believe the electrodes of the future will be based on the CO_2 electrode, either regular gas type or air gap, with the use of specific decarboxylases to build electrodes for the various amino acids. An alternative will be the ammonia sensors, using specific lyases, such as l-phenylalanine lyase, a useful electrode is obtained.

The Pt electrode-based sensor for lactic acid electrode[11] is of much too limited a usefulness and will not attract any attention.

For alcohols, the Pt-based sensors will be used, the best results probably being attained with the O_2 measurement at -0.6 V vs SCE. The electrode is fast in response, and more sensitive than the H_2O_2 measurement system at $+0.6$ V vs SCE.

Penicillin electrodes based on the pH sensor will continue to attract attention, and it is hoped that the excellent preliminary work of Mosbach and Papariello et al[16,38,39] will be followed up by others.

The uric acid electrode based on O_2 measurement using a Pt electrode without a membrane is a fine electrode and should be widely studied.[33] Alternatively, a CO_2 sensor could be used to measure the other product of this reaction.

For amygdalin the electrode of Mascini and Liberti[37] is the best; further studies with chemically bound enzymes should give a good final product, although measurements of amygdalin are of very limited analytical usefulness.

Other electrodes in the near future will probably be a creatinine electrode using creatinase with measurement of the NH_3 liberated with an air gap electrode or Orion gas electrode; a L-phenylalanine electrode using a lyase from sweet potato and a nitrate electrode using nitrate reductase,[48,49] both using a gaseous NH_3 detection sensor for the product of the reaction; and a cholesterol electrode using cholesterol oxidase and the Pt sensor. The number of new sensors is limited only by the imagination.

See also this volume chapter [63] for further enzyme electrode work and chapter [61] for application of immobilized coenzymes in enzyme electrodes.

Solid Surface Fluorescence Methods

Introduction

In the past, manometric methods, pH procedures, and spectrophotometry have been used for determining enzymic activity. Spectrophotometry has been generally preferred because of its simplicity, its rapidity, and the capability of measuring lower enzyme and substrate concentrations. Spectrophotometry embraces the use of colorimetric methods where chromogenic products are produced as a result of enzyme activity, and fluorescent methods where fluorescent compounds are produced as a result of enzymic activity. Fluorescent procedures are several orders of magnitude more sensitive than colorimetric methods and thus have replaced the colorimetric ones in numerous instances. Further advantages are greater selectivity (since two wavelengths are used) and an accuracy independent of region of measurement.

Previous fluorometric methods, although they have been improvements over other prior art methods of determining enzyme activity, have not eliminated all the problems associated with enzymic analyses. Fluorometric analysis depends on the production of a fluorescent compound as a result of enzyme activity between a substrate and enzyme. The rate of production of the fluorescent compound is related to both the enzyme concentration and substrate concentration. This rate can be quantitatively measured by exciting the fluorescent compound as it is produced and by recording the quantity of fluorescence emitted per unit of time with a fluorometer.

Fluorometrically measuring enzyme activities or enzymic reactions is usually done by wet chemical methods that rely on the reaction of a substrate solution with an enzyme solution. Unfortunately, wet chemical methods involve the preparation of costly substrates, cofactors, coenzymes, and enzyme solutions.

[48] W. R. Hussein and G. G. Guilbault, *Anal. Chim. Acta* **72**, 381 (1974).
[49] *Ibid.*, **72**(2) (1975).

For example, when determining the presence and concentration of an enzyme, a substrate must be accurately measured and dissolved in a large amount of buffer solution, usually about 100 ml, to prepare a stock solution. The enzyme reaction is then usually carried out by measuring a certain volume of stock substrate solution into an optical cuvette, adding a measured amount of enzyme solution to the substrate solution, and recording the change in absorbance emitted from the resultant solutions.

When determining the concentration of a specific substrate kinetically in an enzymic catalyzed reaction, the procedure is more cumbersome and costly. A relatively large amount of expensive enzyme, usually 0.1 ml of stock solution, is needed to make the reaction proceed at a conveniently measurable rate. These enzyme solutions must be prepared fresh daily. This standard wet chemical method requires considerable technician time and relatively large quantities of expensive substrates or enzymes.

Guilbault and co-workers have developed solid-surface fluorometric methods, using a "reagentless" system, for the assay of enzymes, substrates, activators, and inhibitors. An attachment to an Aminco filter fluorometer has been adapted to accept, instead of a glass cuvette, a metal slide (a cell), painted black to reduce the background. A silicone rubber pad is placed on the slide. All the reagents for a quantitative assay are placed in a form of solid reactant film on the surface of the pad. The sample of the fluid containing the substance to be assayed is then added on the pad. The change in fluorescence with time is measured and equated to the concentration of the substance determined.

These reagent pads are simple to prepare, and hundreds could be conveniently manufactured at one time. They are stable for months or longer when stored under specified conditions. There is no need for the cumbersome, time-consuming preparation of reagents when performing an analysis, since essentially all the reagents for a quantitative assay are already present on the pad. If samples are hard to obtain, the pad method could be another great advantage, since only 3–25 μl of sample are required.

Experimental

Preparation of Silicone Rubber Pads

Silicone rubber (Dow-Corning Glass and Ceramic Adhesive, Dow-Corning, Midland, Michigan) pads were prepared by pressing uncured silicone rubber between a glass plate and a stainless steel mold (Fig. 6), both of which were lined by a piece of glassine paper (Eli Lilly and Co.). The surfaces contacted with the silicone rubber were prelubricated with

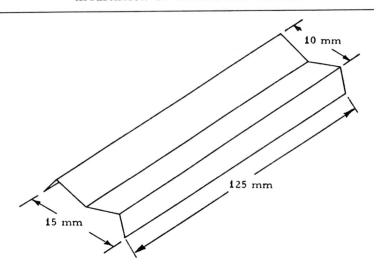

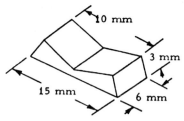

FIG. 6. Silicone rubber pads of 6 mm width (lower left) are cut from a 125 mm strip (upper right).

a thin layer of Dow-Corning silicone stopcock grease (Dow-Corning Co., Midland, Michigan). The silicone rubber was kept in the mold at room temperature for 2 days to cure. The cured strips were then removed, wiped, and washed briefly with concentrated KOH solution to remove the grease. Next the pads were washed with H_2O and dried at 80° for 1 hr. The strips were cut to individual pads 6 mm in width. About 20 individual pads can be obtained from one strip.

The reactant film on the pad may be formed by dissolving the reagents and buffer in a suitable solvent, depositing the reactant solution on the silicone pad (so that the solution spreads evenly over the pad) and evaporating off the solution by vacuum or lyophilization. The reagent may also be applied to the pad in a polymeric film such as polyacrylamide or some stabilizer can be added if an enzyme is present in

TABLE III
COMPARISON OF PAD COLORS[a]

Color	Background	Rate (F/min)
Translucent	1.29	4.6
Gray	0.30	4.3
Black	0.31	1.3

[a] Pad composition was 20 μl of both β-NAD and lithium lactate solution; also 20 μl of glycerine buffer and 30 μl of the 3850 U/ml of commercially available lactate dehydrogenase enzyme solution were used in the analysis.

the reagent. Either substrate or enzyme and/or coenzyme can be deposited on the pad in film form, depending upon whether the substance to be assayed is an enzyme or a substrate. The pads can be stored in a dark, cold place, or in a refrigerator or desiccator before use.

Two methods were used to apply the reactant film to the pad. The first method—a batch method—involved putting a solution of the reagent onto the strip of silicone rubber, allowing the solution to evaporate, and then cutting the strip into the desired pad dimensions. The prepared pads were then stored. The second method involved cutting the individual pad from the strip, and then applying the reagent to the pad. This latter method was found to be preferable.

The color of the silicone rubber affects both the background and rate of change of fluorescence. With any of the filter systems, the background and rate of change of fluorescence increase in the order: black <gray <clear <white. Each possible combination of pad color and filter was examined, and it was found that the most accurate results could be obtained if a gray silicone-rubber pad were used (Table III). The silicone materials can retain a reactant film on their surfaces for an indefinite time and permit the direct measurement of fluorescence from their surface when an appropriate second reagent solution is applied onto the first reactant film. Background interference due to light scattering and nonspecific fluorescence is minimal compared to other materials.

Preparation of Cell and Cell Holders

An Aminco filter fluorometer, set on its end (Fig. 7) and equipped with the cell and cell holder described below (Fig. 8), was used for all the fluorometric measurements. The fluorometer was supported by two wooden blocks placed parallel to the primary filter holder to prevent electric

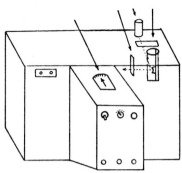

FIG. 7. An Aminco filter fluorometer equipped with a cell and cell holder.

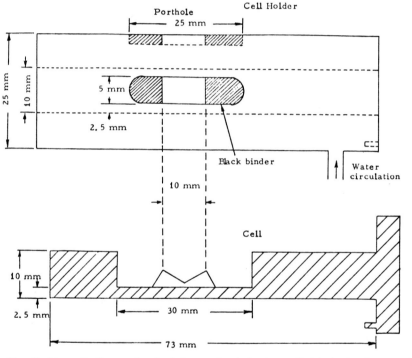

FIG. 8. The cell (lower drawing) is constructed of a cylindrical aluminum rod with a slot, approximately twice the length of the pad, located toward the end of the rod. The cell holder (top drawing) consists of an Aminco cuvette adapter with water circulating around it to maintain constant temperature.

noise and make it convenient to change the primary filter when needed. A 5-inch linear recorder was used to display the result obtained.

Two kinds of cells and cell holders were specially constructed to hold the pad in the optical path of the fluorometer. The first kind, which was

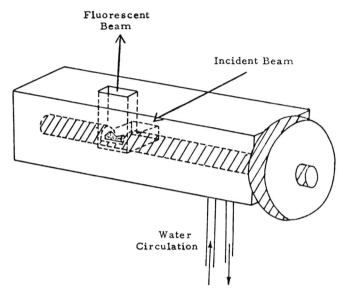

FIG. 9. This diagram shows the pad inside the cell holder and the relationship of the pad to the incident beam.

easier to construct, was designed by Guilbault and Vaughan[50] to study reactions primarily at room temperature. The second kind, which was a modified version of the first kind but had temperature control, was designed to study reactions above 30°. Only the second model will be discussed here, since the same device can be used for room temperature or high-temperature studies.

The cell holder consisted of an Aminco cuvette adapter (Cat. No. J4-7330) with water circulating around it to maintain constant temperature. The temperature of the circulating water was regulated by a constant temperature water bath capable of maintaining ±0.1°. Black binders were placed on both sides of the two entrance and exist slits so that smaller slits, about two-thirds the length of the pad used, would restrict the radiation that entered and left the cell cavity. The cell was constructed of a cylindrical aluminum rod with a slot, approximately twice the length of the pad, located toward the end of the rod. The depth of the slot was such that the pad, with its contents, received the full beam of incident radiation. The cell was painted a dull black to avoid scattered light. A drawing of the cell and cell holder is depicted in Fig. 8. A drawing of the pad inside the cell holder and its relationship with the incident beam is given in Fig. 9.

[50] G. G. Guilbault and A. Vaughan, *Anal. Chim. Acta* **55**, 107 (1971).

Method

The concentration of substrate participating in an enzyme reaction can be calculated in one of two general ways. The first method measures, by chemical, physical, or enzymic analysis, either the total change that occurs in the end product or the unreacted starting material. In this method, large amounts of enzyme and small amounts of substrate are used to ensure a complete reaction. In the second method, which is a kinetic method, the initial rate of reaction is measured, in one of many conventional ways, by following the production of product or the disappearance of the substrate. In this method, the rate of reaction is a function of the concentration of substrate, enzyme, inhibitor and activator.

On the other hand, because enzymes are catalysts, and as such affect the rate, but not the equilibrium of reactions, their concentration and activity must be measured by the rate or kinetic method. Similarly, activators and inhibitors that affect the enzyme's catalytic effect can be measured only by the rate method. Although, as pointed out above, the substrate can be measured either by a total change or a rate method, the latter method is faster because the initial reaction can be measured without waiting for equilibrium to be established. The accuracy and precision of both methods are comparable. The following is a general description of the rate method using fluorometric silicone rubber pad procedures.

The refrigerator stored pads are allowed to reach room temperature inside a desiccator. Then 25–90 μl of water or buffer solution maintained at a certain temperature is added to each pad placed in the cell holder, followed by 3–25 μl of the sample solutions to be assayed, also at the same temperature. After mixing with the tip of the syringe used to deliver the sample, the cell is placed into the cell holder and into the light path of the fluorometer.

The recorder is turned on immediately, and the change in fluorescence (Δf) is recorded. A calibration plot of Δf/min versus concentration is prepared and used for subsequent analyses of enzymes or substrates. The silicone rubber pads can be reused, after they are cleaned with detergent and water.

The fluorometric systems that can be monitored by this type of pad method are as numerous as the fluorometric systems known. Some of the more important systems that have been studied, and some for which methods have been developed, are presented here.

Advantages of the Silicone Rubber Pad Method

The pads with solid reactant film on them are not yet commercially available. In order to use this pad method at the present time, one has to prepare one's own pads and reactant films. This is bothersome for a smaller clinical laboratory. However, the silicone rubber pad method has many advantages and should find wide acceptance for many applicacations.

Time. Since essentially all the reagents that are necessary for a quantitative assay are present on the pads, there is no need for further reagent preparation. If a sample for LDH assay is presented at 2 AM, the technician would only have to open a bottle marked "LDH test," pick up a pad, add 10 µl of buffer solution and a 20-µl sample of serum, and read out the results directly in units per milliliter. Total time is about 4 min using an initial rate method. With the pad method, this would require about 10 bottles of pads, one for each of 10 common tests.

Sample Size and Cost. The pad method is a micro procedure and, as such, micro amounts of reagents and sample are required. On the CPK pad method, for example, the cost of the reagents is only 1/27 that of a regular spectrophotometric method, and as little as 3 µl of serum sample are needed.

Sensitivity and Linearity. Fluorescent procedures are several orders of magnitude more sensitive than colorimetric methods. The accuracy of colorimetric methods is limited to an error of about 2–3% over a narrow range (0.15–0.85 absorbance units). Because fluorescence methods are based on the production of a signal over zero signal, the accuracy of these methods is independent of scale reading and remains the same over a 3- to 4-fold linear range of concentration. The linear range of concentration is further improved in the pad method. Self-quenching interferences in fluorometric assays are minimized, because of the small distance the fluorescence beam travels in the sample drop of the pad—less than 2 mm.

Temperature. In conventional methods, temperature must be critically controlled in all enzyme assays based on a kinetic approach. This can be inconvenient and bothersome. The fluorescent pad method proposed here is temperature independent for those reactions run at 25°, since the silicone rubber pad used to support the sample is non-heat conductive. Thus, provided the sample of serum, blood, or urine is at the same temperature as used to prepare the calibration curve (generally room temperature, 25°), the temperature of the environment does little to affect the results.

Stability. Many substrates and enzyme solutions used in the present clinical procedures are unstable, and new solutions must be prepared fresh daily. *o*-Toluidine and peroxidase in glucose assay, NAD in LDH assay, hexokinase and G-6-PDH in CPK assay are examples of this. Yet, when the reagents are placed in a solid form on the silicone rubber matrix, they can be kept for weeks or months without deterioration.

Assay of Important Enzymes

Cholinesterase in Blood[51]

Low levels of cholinesterase (ChE) are found in individuals with anemia, malnutrition, and pesticide poisoning. High levels indicate a nephrotic syndrome.

Pads for the assay of cholinesterase were prepared by placing 10 μl of a 10^{-2} M solution of N-methylindoxyl acetate (Isolabs, Akron, Ohio) in acetone onto a silicone-rubber pad and allowing evaporation to dryness.

$$\underset{\text{(Nonfluorescent)}}{\underset{\underset{CH_3}{|}}{\text{indole-OCOCH}_3}} \xrightarrow{ChE} \underset{\text{(Fluorescent)}}{\underset{\underset{CH_3}{|}}{\text{indole-OH}}} \quad (22)$$

To start an assay, 20 μl of a sample enzyme solution were applied to the pad, being careful to get the enzyme solution to cover as much of the substrate film as possible. The recorder was started immediately after the pad was placed into the light beam of the fluorometer, and the rate was recorded. From a calibration plot of the change in fluorescence units per minute versus enzyme concentration, activity of cholinesterase in sample solution was obtained.

Linear calibration plots were obtained for cholinesterase concentrations from 10^{-6} to 10^{-2} units/ml with a precision of about 2% and an accuracy of 2.2%.

Preliminary studies have indicated that pads made were stable for at least 30 days if kept in a cold, dark place. Further research is under way.

This stability compares very favorably with that of the solution methods. N-Methylindoxyl acetate solutions are very unstable and must be prepared fresh daily. In contrast, the substrate on the pad, since it is present in a dry state, was quite stable for more than 30 days.

Alkaline Phosphatase in Blood[50]

High levels of alkaline phosphatase are observed in rickets, Paget's disease, obstructive jaundice, and metastatic carcinoma.

Onto a series of gray silicone-rubber pads (adhesive sealant 3845 RTV) was placed 10^{-2} M naphthol AS-BI phosphate (0.01 ml). The ethanol was allowed to evaporate off so that the naphthol AS-BI phosphate was deposited on the surface of the pad. These pads can be used

[51] G. G. Guilbault and R. Zimmerman, *Anal. Lett.* **3**, 133 (1970).

immediately or stored below 0° for at least 2 months without any deterioration.

The pad was placed on the blackened metal strip, which was then slid into the cell holder in such a way that the pad was centrally positioned below the filled aperture of the cell holder. Onto the pad were placed from syringes, pH 9.8, 1 M 2-amino-2-methylpropanol-HCl buffer (0.02 ml) and then serum (0.01 ml). The drop formed was mixed with the needle of a syringe so that the drop covered half the area of the pad.

$$\text{Naphthol AS-BI phosphate (nonfluorescent)} \xrightarrow{\text{alkaline phosphatase}} \text{Naphthol AS-BI (fluorescent)} + H_2O \quad (23)$$

The cell holder was placed in the fluorometer, and the rate of reaction was recorded. It was sometimes necessary to wait 15 sec before the reaction started; this was particularly noticeable with serum samples having low alkaline phosphatase values. The serum alkaline phosphatase value was obtained from the initial rate of the reaction.

A plot of serum alkaline phosphatase values obtained from an alternative procedure versus initial rate of reaction gave a straight line. In the study carried out the rate of reaction was plotted against units of alkaline phosphatase obtained from a standard laboratory procedure using phenolphthalein phosphate. The results indicate good agreement between the two methods over a wide range of serum values.

Lactate Dehydrogenase in Serum[52]

Levels of this enzyme are elevated in individuals with acute and chronic leukemia in relapse, myocardial infarctions, and carcinomatosis.

[52] R. L. Zimmerman and G. G. Guilbault, *Anal. Chim. Acta* **58**, 75 (1972).

The determination of serum LDH is based on the following enzyme-catalyzed reaction.

$$\text{Lithium lactate} + \text{NAD} \underset{}{\overset{\text{LDH}}{\rightleftarrows}} \text{NADH} + \text{pyruvic acid} \qquad (24)$$
$$\text{(nonfluorescent)} \qquad\qquad \text{(fluorescent)}$$

The rate of production of NADH fluorescence is equated to LDH activity.

A 50-μl aliquot of the NAD solution was applied to the pads, and evaporated over silica gel under reduced pressure, to produce a thin film of solid NAD on the surface of the pad. Then a 20-μl aliquot of the lactate solution was added to the pad and evaporated similarly, to produce a thin film of solid lactate. Thus a thin film of solid NAD and lithium lactate on the silicone-rubber surface was the final pad composition. The pad could be stored over silica gel at atmospheric pressure. There was no noticeable decomposition of the reactant film over a period of a month.

The pad was placed in the proper position on the cell. A 10-μl aliquot of pH 9 glycine buffer with semicarbazine hydrochloride was added to the pad and spread over it so that the entire substrate film was dissolved in the drop of solution. Then the cell was placed into the fluorometer, and a background rate was recorded.

After recording the background for about 1 min, the cell was removed from the fluorometer, and a 20-μl aliquot of either the commercially available enzyme solution or the serum solution, whichever was to be analyzed, was put onto the pad. The cell was immediately placed back into the instrument, and the fluorescence rate was recorded. A calibration plot of the change in fluorescence units per minute versus the enzyme concentration could then be constructed and used for future assay of unknown solutions.

From 160 to 1000 units of lactate dehydrogenase in serum were assayed with an average relative error of 2.3%. Normal range of LDH content is 150–630 units per milliliter of serum.

The complete analysis from the measurement of a volume of buffer solution to the end of recording the rate took only 3–5 minutes.

Creatinephosphokinase in Serum[53]

Serum creatine phosphokinase (CPK) activity is of great value in the diagnosis of myocardial and skeletal muscular dystrophy. In a preliminary study by Sigma Co., it was noted that the CPK level is elevated an average of almost 1000% following an infarction.

[53] G. G. Guilbault and H. Lau, *Clin. Chem.* **19**, 1045 (1973).

The basic reaction used is the "reverse" reaction of CPK coupled with two indicator reactions as follows:

$$\text{Creatine phosphate} + \text{ADP} \xrightleftharpoons{\text{CPK}} \text{creatine} + \text{ATP} \quad (25)$$

$$\text{ATP} + \text{Glucose} \xrightleftharpoons{\text{hexokinase}} \text{ADP} + \text{glucose 6-phosphate} \quad (26)$$

$$\text{Glucose 6-phosphate} + \text{NAD} \xrightleftharpoons{\text{G-6-PDH}} \text{6-phosphogluconate} + \text{NADH} + \text{H}^+ \quad (27)$$

The rate of production of NADH fluorescence is measured at a $\lambda_{ex} = 340$ nm, and a $\lambda_{em} = 460$ nm.

In the procedure, for convenience, the Worthington Statzyme CPK kit, which contains all the enzymes, cofactors, and substrates for CPK assay was used. To prepare the pads, the contents of a Worthington CPK vial was dissolved in 0.29 ml of ice-cold distilled water. At least 27 pads were obtained from each CPK vial and 100 pads were easily placed into a freeze drier where the excess liquid was removed. After about 1 hr a white crystalline film of CPK reagent resulted on each pad. The pads were stored in a refrigerator before use, either in the same desiccator or in individual vials under dessication.

A calibration curve for analysis was prepared by diluting a standard serum sample of known elevated CPK value (approximately 560 U/liter) with water. The pads were removed from the refrigerator and allowed to reach room temperature (about 30 min). Then 87 µl of water at 30° was added to each pad placed in the cell holder, followed by 3 µl of the serum sample. The holder was maintained at $30 \pm 0.1°$. The recorder was turned on, and the fluorescence at 460 nm ($\lambda_{ex} = 365$ nm) was measured. About 3–6 min elapsed before the reaction rate became linear, then the change in fluorescence per 5 min, $\Delta f/5$ min, was measured and plotted vs serum CPK activity, to obtain a calibration curve which was linear from 0 to 504 IU/liter.

For assay of CPK activity in unknown serum, the same procedure as above was followed. The activity of the sample was read from the calibration plot. If the sample contained greater than 504 U, less sample (1–2 µl) should be diluted with water to make a total of 90 µl. A comparison of the standard spectrophotometric method using CPK Statzyme vials and the pad method for serum CPK activities indicated excellent agreement.

The stability of the lyophilized CPK pads was tested for 32 days. Results obtained from assays on day 32 agreed with those on day 1, showing that the pads were stable for at least this period of time, with less than $\pm 5\%$ activity change, and possibly for as long as 3 months.

The cost of each analysis using the pad method was less than 5 cents, only 1/27 the cost of each current spectrophotometric method for CPK.

Assay of Important Substrates with Immobilized Enzymes

Urea Nitrogen in Serum[54]

The determination of serum urea nitrogen is presently the most popular screening test for the evaluation of kidney function. Elevated urea nitrogen values are found in patients with acute glomeralonephritis, chronic nephritis, polycystic kidney, nephrosclerosis, and tubular necrosis.

A fluorometric method for the determination of urea in serum was developed using a coupled enzyme system:

$$\text{Urea} \xrightarrow{\text{urease}} 2\text{NH}_4^+ \qquad (28)$$

$$\text{NADH} + 2\text{NH}_4^+ + \alpha\text{-ketoglutarate} \xrightarrow{\text{glutamate dehydrogenase}} \text{NAD} + \text{glutamic acid} \qquad (29)$$

The rate of disappearance of NADH fluorescence ($\lambda_{ex} = 365$ nm; $\lambda_{em} = 460$ nm) was proportional to the content of urea in serum. Reaction conditions and reagent concentrations were similar to those produced by the spectrophotometric reagent kit by Calbiochem, but less NADH was used. The rate of decrease in the fluorescence of NADH, triggered by the addition of urea, was measured. The system worked fine in solution for serum urea. The calibration plot is linear up to 25 mg of urea per deciliter. The method affords a rapid, simple, and inexpensive means for urea assay, the results of which correlate well with the automatic diacetylmonoxime method (correlation coefficient 0.998).

Creatine in Urine[55]

Creatine tolerance tests, which measure the ability of an individual to retain a test dose of creatine, are of great diagnostic value in indicating extensive muscle destruction and diseases of the kidney.

The following reaction scheme was investigated to develop a sensitive fluorometric method and later a fluorometric pad procedure for creatine in biological samples:

$$\text{Creatine} + \text{ATP} \xrightarrow{\text{CPK}} \text{creatine phosphate} + \text{ADP} \qquad (30)$$

$$\text{Phosphoenolpyruvate (PEP)} + \text{ADP} \xrightarrow{\text{PK}} \text{pyruvate} + \text{ATP} \qquad (31)$$

$$\text{Pyruvate} + \text{NADH} \xrightarrow{\text{LDH}} \text{lactate} + \text{NAD} \qquad (32)$$

[54] J. W. Kuan, H. K. Y. Lau, and G. G. Guilbault, *Clin. Chem.* **21**, 67 (1975).
[55] G. G. Guilbault and H. Lau, *Clin. Chim. Acta* **53**, 209 (1974).

The rate of fluorescence change of NADH Δf/min was proportional to the amount of creatine in the sample.

This scheme was successfully applied to urine creatine. The content of creatine in urine is about ten times that in serum. Fluorogenic substances in urine interfere but can be easily removed by activated carbon. There was little or no loss of creatine as determined by a spectrophotometric method, when 10 ml of urine were treated with 0.5 g of activated carbon followed by filtration. The overall reaction occurred at pH 9 with Tris buffer. A linear calibration curve was obtained for 0–20 and 0–100 mg per 100 ml of creatine. The 0–20 mg per 100 ml curve covered the normal range of urine creatine. The amount of NADH, ATP, PEP, Mg, buffer mixture, and LDH-PK mixture were optimized. Two-fifths of the amount of NADH, as in the spectrophotometric method, gave a reasonable slope change, Δf/min, and about 3 min of linearity of the slope was obtained.

The complexity of this solution mixture made its preparation difficult; for many laboratories it would be impractical. A silicone pad method in which all the reagents were lyophilized or immobilized would be an ideal answer to this problem. However, the crystalline lyophilized reagent is stable only for about 48 hr under refrigeration. Attempts are now being made to add some stabilizer to the mixture, such as mannitol, gum, ammonium sulfate, and/or to lyophilize the reagents separately on a pad.

Uric Acid in Serum[56]

Determination of serum uric acid is most helpful in the diagnosis of gout. Elevated levels of uric acid are found in patients with familial idiopathic hyperuricemia and in decreased renal function.

A direct reaction rate method was investigated for the sensitive determination of serum uric acid, based on a fluorometric Schoenemann reaction. The rate of formation of indigo white, Δf/min ($\lambda_{ex} = 395$ nm, $\lambda_{em} = 470$ nm), is a measure of the uric acid content in the sample.

Optimum conditions such as pH, buffer system and organic solvent of the hydrogen peroide-phthalic anhydride were studied. The optimum pH of the uricase reaction (pH 9.4) coincided with the optimum pH using phosphate buffer for the hydrogen peroxide-phthalic anhydride reaction (pH 9.1). Therefore, an overall pH of 9.4 was used. Phosphate buffer was found to be superior to Tris and borate buffers. Acetone or

[56] G. G. Guilbault and H. Lau, presented at NIAMDD Contractor's Conference, Bethesda, Md., Feb., 1973.

$$\text{Uric acid} \xrightarrow{\text{uricase}} H_2O_2 \quad (33)$$

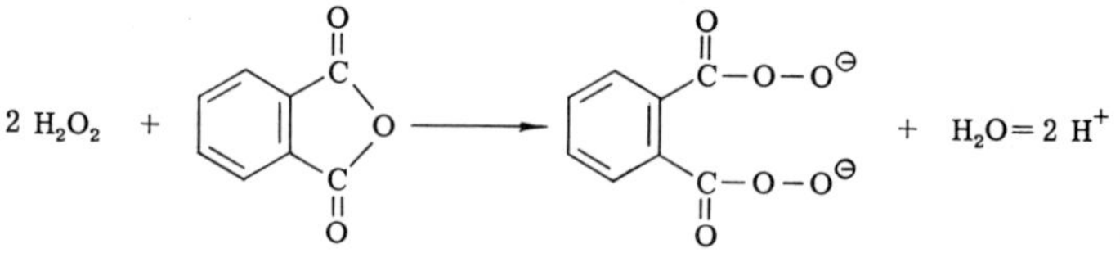

$$2\ H_2O_2 + \underset{\substack{\|\\O}}{\overset{\substack{O\\\|}}{\underset{C}{\bigcirc}}}\!\!\!\!\!\!\!\!\!\!\overset{C}{\underset{\|}{\|}}\!\!\!>\!\!O \longrightarrow \underset{\substack{C-O-O^{\ominus}\\\|\\O}}{\overset{\substack{O\\\|\\C-O-O^{\ominus}}}{\bigcirc}} + H_2O = 2\ H^+ \quad (34)$$

(35)

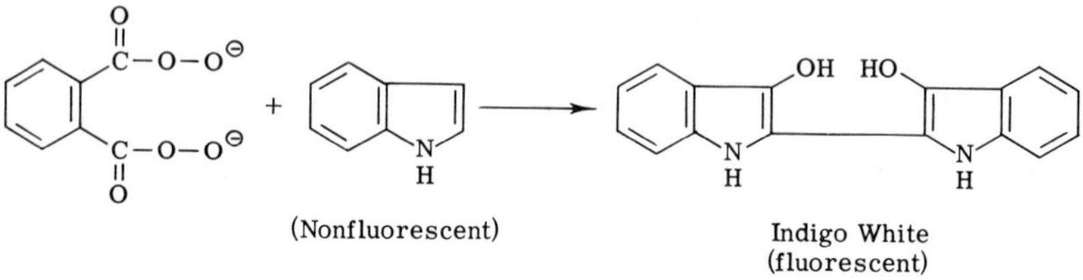

(Nonfluorescent) → Indigo White (fluorescent)

methyl Cellosolve both were good solvents for phthalic anhydride. The rate of formation of hydrogen peroxide in the uricase–uric acid reaction was a function of uricase enzyme. Enough uricase enzyme was used so that in about 5 min the formation of hydrogen peroxide was completed, before the addition of the coupling fluorogenic reagents. The system works fine and gives a linear response of 0–5 mg of uric acid per milliliter. Similar results were observed with the silicone pad method on which the enzyme and the buffer were placed.

Difficulties were encountered in serum samples, however. Protein had to be removed either by trichloroacetic acid, acetone, or tungstate method; still there were inhibitions. Amino acids and ammonium hydroxide were found to inhibit the reaction so that lower values of uric acid were found. Further studies will be made on this system to circumvent present difficulties. Another fluorometric reaction system that holds great promise for a specific analysis of serum uric acid on silicone pads is also being considered.[57]

The method consists of oxidizing uric acid with uricase to hydrogen peroxide, which in turn oxidizes homovanillic acid to a high fluorescent material. The rate of production of fluorescence is proportional to the amount of uric acid present.

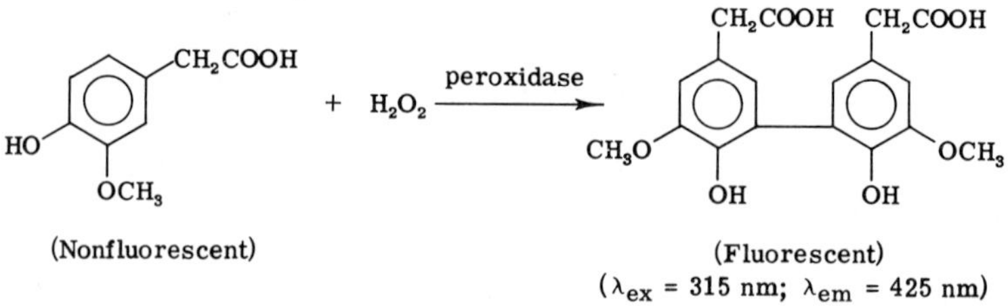

(36)

(Nonfluorescent) (Fluorescent)
(λ_{ex} = 315 nm; λ_{em} = 425 nm)

[57] J. W. Kuan, S. S. Kuan, and G. G. Guilbault, *Clin. Chem.*, in press.

A calibration curve was constructed from a series of standards and was linear from 0.1 to 14 mg/dl. The method is simple and has good precision and accuracy. Comparison of the results obtained by this method with the standard enzymic spectrophotometric method (disappearance of uric acid at 293 nm) gave a coefficient of variation of 0.983.

Cholesterol Assay[58]

A method has been developed based on the following sequential enzymic reactions:

$$\text{Cholesterol esters} + H_2O \xrightarrow{\text{cholesterol ester hydrolase}} \text{fatty acid} + \text{cholesterol} \quad (37)$$

$$\text{Cholesterol} + O_2 \xrightarrow{\text{cholesterol oxidase}} \text{cholest-4-en-3-one} + H_2O_2 \quad (38)$$

$$H_2O_2 + \text{homovanillic acid} \xrightarrow{\text{peroxidase}} \text{fluorescent dimer} \quad (39)$$

The initial rate of fluorescent dimer formation is monitored and is proportional to the concentration of total cholesterol. A linear relationship was found in the range from 0 to 400 mg/dl. Results had good precision and accuracy.

Acknowledgment

The author gratefully acknowledges the financial support of the National Institutes of Health, Grants GM-17268, GM-18646, and NIAMDD-72-2216, for the research described herein.

[58] N. F. Huang, J. W. Kuan, and G. G. Guilbault, *Clin. Chem.*, in press.

[42] The Applications of Immobilized Enzymes in Automated Analysis

By WILLIAM E. HORNBY and GEORGE A. NOY

Enzyme-based analysis embraces those techniques in which enzymes are used as analytical reagents for the specific determination of their respective substrates. In this type of analysis the substrate (analyte) is chemically modified selectively by the enzyme and is determined by measuring the emergence of a specific reaction product. Over recent years enzymes have been accepted widely as important analytical tools, and the techniques of enzyme-based analysis have been applied successfully to a variety of problems. Enzyme-based analysis has made its greatest impact in the area of clinical chemistry, where large numbers of important metabolites and end products are routinely assayed as aids to diagnosis using enzyme-analytical reagents. Because there is an increasing demand for enzyme-based analysis, particularly in the area of clinical chemistry, significant developments have been made over recent years toward the automation of these techniques. In this paper the use of immobilized enzymes as analytical reagents in automated analysis is discussed.

Although the usefulness of enzymes as analytical tools is widely accepted, there still remain some problems associated with their routine application as analytical reagents, and it is because of these problems that immobilized enzymes have begun to be considered as substitutes for enzymes in analysis. There are three features of enzyme-based analysis wherein improvements in existing methodologies are desirable. The first of these features relates to the economics of the analysis. Generally enzymes of analytical grade purity tend to be very expensive because such materials have to be extensively purified in order to remove contaminating enzyme activities that might interfere in and impair the specificity of the analysis. Thus enzyme-analytical reagents are more costly than conventional chemical-analytical reagents, and correspondingly the analyses in which they are involved are more expensive. The high cost of enzyme-based analyses is compounded when it is recalled that current trends in analysis are moving toward automated techniques whereby large numbers of samples can be processed routinely. Thus, in the performance of such automated methods, large quantities of expensive enzyme reagents are consumed concomitantly.

The second feature of enzyme-based analysis wherein enhanced performance is possible is associated with the stability of enzyme-analytical reagents. An important characteristic of any analytical method is its reliability and, consequently, the reproducibility of the results that it generates. In order to achieve reliability it is important to use stable analytical reagents. Unfortunately, some purified enzymes tend to have a limited stability in solution, and thus their reliability as analytical reagents under operational conditions will parallel this lability.

The final feature to be considered interrelates with both features discussed previously and is a reflection of the need to simplify enzyme-based analyses. Many enzyme-based analytical techniques require the provision of an assortment of auxiliary reagents, which may function either as second substrates and coenzymes in the enzyme reaction itself or may serve to stabilize the enzyme under the operational conditions of the assay. Furthermore, because of the intrinsic lability of some enzyme-analytical reagents it might not be possible always to construct robust analytical systems. Clearly, the complexity of some of these analyses together with the lability of the reagents can involve extra preparation time for the analysis. This would entail necessarily increased operator time and further increase the cost of the analysis. Thus in summary it can be argued that the more economical use of more stable enzymes in simplified analytical systems is desirable. It is toward this goal that the application of immobilized enzymes in analysis is directed.

There are several characteristics of immobilized enzymes in general that both warrant the consideration of their application as enzyme substitutes in analysis and at the same time suggest that the above objectives are realizable. For example, the physical form of immobilized enzymes makes possible an improvement in the economics of enzyme utilization in analysis. Conventionally, soluble or free enzymes are used in analysis on a "one-off" basis; in general, a fixed amount of enzyme is dispensed for each assay and is not recovered after the assay for further use. In this way the full catalytic potential of the enzyme is not being realized. On the other hand, an immobilized enzyme can be deployed easily in such a way that a fixed amount of the derivative can be used for the performance of many assays, thereby effectively reducing the amount of enzyme required for each assay. In practice this can be achieved in one of two ways: The immobilized enzyme can be used in a batch mode and recovered after each assay by processes such as filtration and centrifugation for application in further assays. Alternatively, the derivative can be used in a continuous-flow mode, such as a packed bed or an open tube, and sequential samples of the analyte can be perfused continuously through it. In this way a fixed amount of immobilized enzyme can be used for repetitive execution of a large number of analyses.

A second characteristic of immobilized enzymes that has influenced their consideration as analytical reagents relates to the stability of many of these artifacts. The immobilization of an enzyme, particularly by those methods in which the enzyme is either cross-linked or covalently bound to a polymeric matrix, often confers upon it an enhanced stability. Thus in many cases immobilization offers a convenient means whereby enzymes can be stabilized. Since the intrinsic lability of some enzymes represents a drawback to their routine application as analytical reagents, then their use in the same context, but in an immobilized form, is very attractive.

The combination of the above two characteristics of immobilized enzymes constitute a third feature of these materials that promotes their use in analysis. It has already been pointed out that immobilized enzymes can be used in analysis more economically than their water-soluble counterparts. This emanates from the ability to use these materials either repetitively or continuously, owing to their physical form and enhanced stability. In practical terms this represents what might be called a "convenience feature"; i.e., because the same derivative can be used over prolonged periods it obviates the need in the analytical laboratory to prepare enzyme reagents. At the same time this also abates some of the concern ensuing from the lability of some soluble enzyme analytical reagents.

Some of the applications of immobilized enzymes in analysis have been reviewed already[1,2] (see also chapters [41] and [43]–[45] of this volume). It transpires that two different strategies have been followed when using these artifacts in automated enzyme-based analytical procedures. By the first of these strategies the auxiliary analytical hardware is constructed *de novo* around the immobilized enzyme derivative, thereby generating some novel analytical systems. All the analytical systems that fall into this category utilize immobilized enzyme derivative in a continuous-flow mode, i.e., the analyte is perfused through an immobilized structure and determined by measuring either the emergence of a reaction product or the disappearance of a substrate in the effluent stream. The first use of immobilized enzymes in automated analysis by this approach was reported by Hicks and Updike.[3] They described systems for the automated determination of glucose and lactate that utilized small packed beds of polyacrylamide-entrapped glucose oxidase and lactate dehydrogenase, respectively. In each case the formation of specific reaction products in the effluent from the packed beds was measured colorimetrically. Subsequently, the same authors described a "reagentless" assay for glucose in which the utilization of oxygen was used to measure the glucose in the sample[4]; this was achieved by continuously monitoring the concentration of dissolved oxygen in the column effluent with a polarographic oxygen electrode. This concept of a reagentless enzyme-based assay has been pursued by other workers; for example, Wiebel et al.[5] described a prototype apparatus for the analysis of glucose in biological fluids that used a polarographic oxygen sensor on the effluent side of a packed bed of glucose oxidase covalently bound to

[1] H. H. Weetall, *Anal. Chem.* **46**, 602A (1974).
[2] G. G. Guilbault, *in* "Enzyme Engineering" (L. B. Wingard, Jr., ed.), Vol. 1, p. 361. Wiley (Interscience), New York, 1972.
[3] G. P. Hicks and S. J. Updike, *Anal. Chem.* **38**, 726 (1966).
[4] S. J. Updike and G. P. Hicks, *Science* **158**, 270 (1967).
[5] M. K. Weibel, W. Dritschilo, H. J. Bright, and A. E. Humphrey, *Anal. Biochem.* **52**, 402 (1973).

porous glass particles, and Bergmeyer and Hagen[6] have described for the determination of glucose a recirculating assay system that used glucose oxidase coupled to an acrylic polymer. A novel apparatus for the detection of an assortment of insecticides has been reported by Goodson and Jacobs.[7] Their system is different from those described above in that the analyte is not a substrate of the enzyme, but an inhibitor, and is determined by measuring the inhibition of the enzyme cholinesterase. In practice, this was achieved by entrapping the enzyme in starch gel on the surface of open-pore polyurethane foam and using it in an electrochemical cell that monitored the hydrolysis of the substrate butyrylthiocholine iodide.

The application of immobilized enzymes in analysis by the second strategy involves their consideration on the same basis as soluble enzymes, i.e., as analytical reagents. According to this approach, the immobilized enzyme derivative is tailor-made so as to be easily accommodated into an existing analytical system. Thus the objective of this strategy is to make a straightforward substitution of the free enzyme with an appropriate immobilized preparation derived from it. The use of immobilized enzymes in automated analysis by this approach is considered in the remainder of this paper.

Enzyme-based analyses can and have been automated in a variety of ways, and consequently several different types of analytical hardware, operating according to different principles, are commercially available.[8-10] In view of the diverse nature of these automated systems, it is not possible to formulate a general protocol for their operation with immobilized enzymes. Thus it seems likely that each system will work optimally only with a particular type of immobilized enzyme derivative, which will have to be constructed so that it is compatible with the operational features of that machine. The ensuing discussion reviews some of the factors that have to be considered when immobilized enzymes are used in continuous-flow analyzers of the type pioneered by Technicon. It can be argued that comparable considerations will have to be undertaken when the use of immobilized enzymes in alternative systems is reviewed, and hence in this respect the following may serve as a model.

[6] H. U. Bergmeyer and A. Hagen, Z. Anal. Chem. **261**, 333 (1972).
[7] L. H. Goodson and W. B. Jacobs, in "Enzyme Engineering" (K. Pye and L. Wingard, Jr., eds.), Vol. 2, p. 393. Plenum, New York, 1974.
[8] H. U. Bergmeyer and S. Klose, in "Methods of Enzymatic Analysis" (H. U. Bergmeyer, ed.), 2nd ed., Vol. I, p. 221. Academic Press, New York, 1974.
[9] N. G. Anderson, in "Methods of Enzymatic Analysis" (H. U. Bergmeyer, ed.), 2nd ed., Vol. I, p. 213. Academic Press, New York, 1974.
[10] H. G. Netheler, in "Methods of Enzymatic Analysis" (H. U. Bergmeyer, ed.), 2nd ed., Vol. I, p. 205. Academic Press, New York, 1974.

Continuous-flow analyzers of the Technicon type operate by continuously collecting discrete samples of the analyte and pumping them into an air-segmented liquid stream that contains the various reagents required for the analysis. The continuously flowing stream then is pumped through a series of coils, thereby inducing a delay in the system that allows the component reactions of the analysis to go to completion. At this point in the system the liquid stream is pumped through a monitor, which usually is a recording spectrophotometer, where the reaction products are measured. When enzyme based analyses are carried out in these automated systems, the enzyme is included in the continuously flowing reagent stream, and thus it is continuously consumed whenever the instrument is operating. The air segmentation of the liquid stream represents a critical feature of these instruments. It serves to facilitate mixing of the analyte and the various reagents as well as providing a "scouring" effect on the walls of the transmission tube, thus minimizing adjacent sample interaction. Clearly the inclusion of an immobilized enzyme into this type of system must not impair the air-segmentation; i.e., it must be designed to assume a configuration that permits perfusion with an air-segmented liquid stream.

With the above considerations in mind, several types of immobilized enzyme reactors have been used in continuous-flow analyzers. For example, Inman and Hornby[11] used derivatives of glucose oxidase and urease, immobilized by cross-linking the enzymes through glutaraldehyde in the pores of nylon membranes, for the automated determination of glucose and urea, respectively; small packed beds of glucose oxidase, urease, and uricase likewise have been used for the determination of glucose, urea, and urate, respectively[11,12]; and several fiber-entrapped enzyme derivatives have been used by Marconi et al.[13] in similar systems. The use of the above types of immobilized enzyme structures in continuous-flow analyzers has not proved entirely satisfactory from the point of view of their overall operational performance. Although these derivatives were capable of sustaining air segmentation of the liquid stream, by virtue of their physical form they demonstrated acceptable precision only at lower rates of analyte sampling.

Arguably, from an overall operational standpoint open-tubular enzyme reactors have proved to be very useful enzyme reagents in continuous-flow analysis. This type of immobilized enzyme structure is prepared by covalently binding the enzyme to the inside surface of open

---

[11] D. J. Inman and W. E. Hornby, *Biochem. J.* **129**, 255 (1972).
[12] H. Filippusson, W. E. Hornby, and A. McDonald, *FESB Lett.* **20**, 291 (1972).
[13] W. Marconi, F. Bartoli, S. Gulinelli, and F. Morisi, *Process Biochem.* **9**(4), 22 (1974).

tubes made of such materials as polystyrene[14,15] and nylon.[16,17] These structures can be inserted easily in series with the flow system, and thus the enzyme itself is effectively fixed *in situ* in the system. The system using the immobilized enzyme differs from the corresponding system using the free enzyme only in that the enzyme is no longer included in the continuously flowing reagent stream. In this system the enzyme reaction takes place and the products are generated when the analyte comes into contact with the enzyme on the wall of the tube in the course of its passage through the latter. In effect the enzyme is recovered after each analysis, and so it can be used on a continuous basis for the repetitive analysis of many samples of the analyte.

Experimental

The use of a nylon tube-immobilized enzyme in a Technicon AA1 system is illustrated in Fig. 1. This describes the flow system used for the automated determination of blood glucose with nylon tube-glucose dehydrogenase. In principle, repeated samples containing glucose are continuously pumped in turn into an air-segmented stream containing the assay buffer. After passing through a small mixing coil, the stream enters a dialyzer module in which the glucose is dialyzed away from blood proteins and collected in an air-segmented recipient stream containing the same buffer. The coenzyme, NAD^+, is pumped continuously into the recipient stream as it emerges from the dialyzer, and after passing through a small mixing coil it is pumped into the nylon tube-immobilized glucose dehydrogenase. Finally the NADH formed in the enzyme tube is measured automatically by passing the liquid stream, after removal of the air, through a flow cell in either a recording spectrophotometer or spectrofluorometer.

The nylon tube-immobilized glucose dehydrogenase was prepared using the methods described by Hornby and Goldstein (this volume [9]). The enzyme was isolated from *Bacillus cereus* by the method of Sadoff et al.,[18] and used in the form of a 2 mg ml^{-1} of solution in 0.1 M phos-

[14] W. E. Hornby, H. Filippusson, and D. J. Inman, in "Automation in Analytical Chemistry," p. 56. Techicon Instruments Co. Ltd., 1974.
[15] W. E. Hornby, H. Fillipusson, and A. McDonald, *FEBS Lett.* 9, 8 (1970).
[16] W. E. Hornby, J. Campbell, D. J. Inman, and D. L. Morris, in "Enzyme Engineering" (K. Pye and L. Wingard, Jr., eds.), Vol. 2, p. 401. Plenum, New York, 1974.
[17] W. E. Hornby and D. L. Morris, in "Immobilized Enzymes, Antigens, Antibodies and Peptides" (H. H. Weetall, ed.), p. 141. Dekker, New York, 1974.
[18] H. L. Sadoff, J. A. Bach, and J. W. Kools, in "Spores" (L. L. Campbell and H. O. Halvorson, eds.), Vol. III, p. 97. Am. Soc. Microbiology, Ann Arbor, Michigan, 1965.

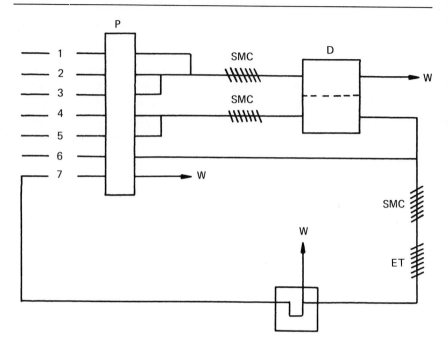

Spectrophotometer (340 nm)
or
Spectrofluorometer
(λ_{ex} 360 nm; λ_{em} 465 nm)

FIG. 1. Flow system used for the automated determination of glucose with nylon tube-immobilized glucose dehydrogenase. The pump tubing lines, 1, 2, 3, 4, 5, 6, and 7 gave flow rates of 0.16, 2.00, 0.60, 2.00, 0.60, 0.23, and 2.00 ml/min, respectively. Sample; 0.1 M-phosphate, 0.5 M NaCl, pH 7.5; air; 0.1 M phosphate, 0.5 M NaCl, pH 7.5; air and 1.5 mM NAD$^+$ were pumped through the tubing lines 1, 2, 3, 4, 5, and 6, respectively. Samples were assayed at the rate of 60 per hour using a 2:1 (v/v) sample/wash ratio. The pump (P) and dialyzer (D) were standard Technicon AA1 modules. The small mixing coils (SMC), the dialyzer module, and the enzyme tube (ET) were maintained at 30°. The spectrophotometer was a Beckman DBGT spectrophotometer fitted with a 1-cm light path flow cuvette, and the spectrofluorometer was a Perkin-Elmer Model 1000 fluorescence spectrophotometer fitted with a 1.6 mm-diameter flow cuvette. W, waste.

phate, 0.5 M NaCl, pH 8.0. The nylon tube (2 m × 0.1 cm) was alkylated on its inside surface using triethyloxonium tetrafluoroborate and then derivatized by substituting it with adipic acid hydrazide. The hydrazide-substituted nylon tube was prepared for coupling by activating it with dimethyl suberimidate, and the enzyme was coupled to it by reaction for 3 hours at pH 8.0 and 4°. The final product was washed free of noncovalently bound protein and stored filled with 0.1 M phosphate, 0.5 M NaCl, pH 7.0, at 4°. In practice, this derivative was deployed in the form of a helix, which was made by coiling it around a 2 cm in diameter plastic former. The coiled derivative was inserted in the flow system shown in Fig. 1 by connecting it through standard transmission tubing (0.16 cm bore). In this way the immobilized enzyme can be easily removed from the circuit between runs for storage at 4°.

In general there are several practical considerations to be borne in mind when using nylon tube-immobilized enzymes in automated analysis. The most important of these are summarized below.

1. Since the methods used for covalently coupling the enzyme to the activated nylon tube involve reaction of nucleophilic groups on the protein, it is important that the enzyme preparation be free of contaminating nucleophiles. Thus, dialyzed or salt-free lyophilized enzymes are preferable. When the starting enzyme is available only as a crystalline suspension in ammonium sulfate, it is necessary to remove the latter before using it in the coupling reaction. The necessary operations for this can be illustrated in the case of the preparation of lactate dehydrogenase for binding to glutaraldehyde-activated nylon tube.

Lactate dehydrogenase from rabbit muscle was obtained from the Boehringer Corporation (London). This material has a specific activity of 550 IU/mg and was in the form of a crystalline suspension (5 mg of protein per milliliter of suspension) in 3.2 M $(NH_4)_2SO_4$. Of the suspension, 1 ml containing 5 mg of protein was centrifuged in a bench-top centrifuge for 5 min at 4°; the supernatant was then carefully decanted off, and the enzyme was dissolved in 2.0 ml of 0.1 M phosphate, pH 7.5. The lactate dehydrogenase solution so obtained was dialyzed for 60 min at 4° against 2 liters of 0.1 M phosphate, pH 7.5, and then used immediately in the coupling reaction.

2. The bore of the nylon tube used for the preparation of the derivatives must be of a size that supports the air-segmented liquid stream. For applications in Technicon AA1 and AA2 systems, a bore of 0.10–0.15 cm is desirable. The length of derivatives used depends on a variety of factors, principal among which are the activity of the immobilized enzyme; the total flow rate through the tube, when inserted in the flow system; and the concentration of analyte in the sample. These factors are discussed in greater detail in a later section of this chapter. In general, however, tube lengths in the range of 1.0–2.0 m are satisfactory. Preferably, the nylon tube should be extruded from Type 6 nylon and free of plasticizers. Nylon tube satisfying these requirements is obtainable from Portex Limited, Hythe, Kent, U.K.

3. When not in use, the nylon tube-immobilized enzymes should be

stored at 4° filled with an appropriate buffer. In most cases storage in the assay buffer is acceptable. When derivatives are stored over long periods, it is important to ensure that they do not dry out. If this occurs, then the immobilized enzyme can lose part of its activity. In order to obviate this risk, extended storage is best achieved by sealing the tubes in plastic bags containing a small piece of cotton wool soaked in water.

4. Although it can be demonstrated that nylon tube-immobilized enzymes display an enhanced stability in relation to the corresponding free enzymes, nevertheless it is still necessary to exercise some care in their usage. Thus, if the run-down sequence on the analyzer requires washing through the system with a strong detergent, the derivatives should be removed before such a step. Furthermore, if reactions in the method, subsequent to the enzyme-catalyzed step, require the involvement of extremes of pH or the participation of compounds that can inactivate the enzyme, these conditions must be set up in the liquid stream after it emerges from the enzyme tube. In practice this is achieved easily by pumping such components into the liquid path at some point "downstream" from the enzyme tube.

Operational Examination of Immobilized Enzymes in Automated Analysis

Several criteria should be evaluated when the overall operational performance of an immobilized enzyme in automated analysis is assessed. The principal criteria in this respect are discussed below.

Activity. The enzymic activity of the derivative in part controls the overall sensitivity of the analysis. Clearly, this must be sufficient to permit detection of the analyte in its expected concentration range in the sample.

The time course of an enzyme-catalyzed reaction is described by the integrated form of the Michaelis–Menten equation, i.e.,

$$FS - K \ln(1 - F) = k_s E_c t; \qquad (1)$$

where F represents the fractional conversion of substrate to product; S is the initial substrate concentration (mM); K is the substrate concentration required to produce half the maximum reaction velocity (mM); k_s is the specific activity of the enzyme (μmoles min^{-1} mg^{-1}); E_c is the concentration of enzyme (mg cm^{-3}); and t is the elapsed reaction time (minutes).

For an open tube enzyme reactor the corresponding equation is

$$FS - K \ln(1 - F) = k_t L Q^{-1} \qquad (2)$$

where F, S, and K have the same significance as above; k_t is the tube specific activity (μmoles min^{-1} cm^{-1}); L is the length of the tube (cm); and Q is the rate of flow of substrate through the tube (cm^{-3} min^{-1}). The term k_t, the tube specific activity, can be determined by using Eq. (2). In practice this is done by perfusing a fixed length of derivative at a flow rate the same as that to which it will be exposed in the analysis with substrate solutions of different concentration. At each substrate concentration the fractional conversion of substrate to product is measured, and a graph is plotted of FS as a function of $\ln(1 - F)$, when k_t is determined from the value of the intercept on the ordinate.

For a given length of derivative and flow rate, k_t determines the amount of reaction that takes place. Thus a minimum value of k_t will be demanded if a given fractional conversion of substrate to product is required in the analysis. This can be explained with reference to the automated determination of urate with nylon tube-immobilized uricase in a Technicon AA1 system. In this system the immobilized enzyme operates essentially under zero-order conditions with respect to the analyte because the concentration of urate after dilution in the reagent stream is substantially greater than the value of K in Eq. (2). In this particular case, Eq. (2) approximates to

$$FS = k_t L Q^{-1} \quad (3)$$

Thus, if a fractional conversion of urate in excess of 0.9 is required, the following relationship must hold:

$$k_t L > 0.9 SQ$$

This inequality can be satisfied either by increasing the length of the tube or by increasing its specific activity. However, on other accounts it might not be practically expedient to increase the length of the tube beyond a certain length, in which case a minimum value of k_t will be demanded.

Stability. One of the most potent arguments in favor of using immobilized enzymes in analysis issues from their enhanced stability compared to the corresponding free enzymes. Among other things, it was shown that this enhanced stability makes it possible to use immobilized enzymes over prolonged periods; this in turn makes it possible to improve the economics of enzyme utilization in the assay. Therefore the operational stability of these derivatives must be carefully evaluated. The stability of immobilized enzymes can be studied in a variety of ways. One of the most common of these methods involves subjecting the enzyme to exaggerated denaturing conditions wherein the loss of enzyme activity is accelerated. It is doubtful whether such accelerated-stability studies are of any real value in assessing the practical operational stability of an immobilized enzyme. In the particular case of the use of immobilized enzymes in analysis, the preferred way to test their operational stability is to subject them to the operational conditions that they will encounter in the analytical laboratory and to measure intermittently their residual activity.

The procedure used for assessing the operational stability of nylon tube-immobilized enzymes in automated analysis can be illustrated with reference to the determination of glucose with nylon tube-glucose dehydrogenase in a Technicon AA1 system. The derivative is inserted into the system as shown in Fig. 1, and pooled bovine plasma is sampled repeatedly at the rate of 60 samples per hour. At intervals of between 5 and 10 hr, in which time 300–600 samples are processed, the tube specific activity, k_t, is measured as described above by sampling aqueous glucose standard solutions. In this way the durability of the derivative is assessed under realistic operational conditions that truly represent those it will encounter in the routine analytical laboratory. Acceptable operational stabilities of nylon tube enzymes should permit the performance of at least 10,000 separate analyses with a single derivative.

It is most unlikely that several thousand determinations of a single analyte would be made in a single uninterrupted operation. What is more likely is that a few hundred analyses at the most will be processed on a daily basis. In this case, in order to capitalize on the full operational stability, the immobilized enzyme will have to be usable over a period of several weeks. Hence, operational stability alone is inadequate in defining the stability of the derivative, and therefore, in addition to assessing the operational stability, the shelf life or storage stability should be tested also. This is done conveniently by storing derivatives filled with assay buffer at 4° and measuring their k_t values as described above at intervals over a period of several weeks.

Sample Rate. When a series of samples are processed in a continuous-flow system, there is an interaction between them that arises from the trailing edge of one sample interfering with the leading edge of the following sample. This phenomenon sometimes is called carry-over, and it governs the maximum sample rate. Clearly, if the carry-over is bad, then the amount of analyte in one sample will influence the value obtained for the amount of analyte in the following sample. One way of minimizing carry-over is to lower the sample rate. However, this might not be possible always—for instance, when a minimum sample rate is needed to get through a given work load in a set time. On Technicon AA1 systems, this rate corresponds to 60 samples per hour.

Another way of reducing carry-over in the case of open-tube enzyme reactors is to decrease their length. Again this might not be possible always, since the tube length governs the reaction capacity of the derivative, which in turn governs the overall sensitivity of the assay. Thus tube length can be decreased insofar as it does not compromise the overall sensitivity. With nylon tube-immobilized enzymes, carry-over can ensue if either the substrate or product of the enzyme binds to the wall of the tube. One of the commonest causes of such binding arises from ionic interactions in the case of charged substrates and products interacting with residual charges on the surface of the nylon. Effects of this type can be minimized by increasing the ionic strength of the liquid stream or by changing the chemistry used for attaching the enzyme to the nylon so that the residual charge density on the surface of the tube is reduced. In this respect the use of bis-acid hydrazide spacers has been most useful when enzymes are coupled to Q-alkylated nylon tube (this volume [9]).

It has been seen that carry-over is affected by a variety of factors, such as sample rate, tube length, ionic strength, and the method used for immobilizing the enzyme. Thus, all these features can be changed within limits in order to generate a system with acceptable carry-over. Clearly, in attempting to optimize the system with respect to carry-over, it is very useful to be able to make a quantitative assessment of the extent of this effect. The method described by Broughton et al.[19] affords a simple means of doing this. Three aqueous standards containing a high concentration of analyte are sampled, then a further three samples containing a low concentration of the analyte. The value of the analyte in each of the samples is recorded, and a carry-over coefficient is expressed according to the expression,

$$\text{Carry-over coefficient} = \frac{\text{value of 1st low sample} - \text{value of 3rd low sample}}{\text{value of 3rd high sample} - \text{value of 3rd low sample}}$$

Improved carry-over is manifested in lower values of this carry-over coefficient, and thus those systems and effects that yield minimum values for this constant can be selected for further evaluation.

Accuracy. Accuracy is a measure of how the values of samples estimated by the method compare with their true value. Assessment of accuracy is achieved by assaying at least 50 samples of the analyte, both by the method being evaluated and by an established method, and the results thus obtained are used to determine the correlation coefficient.

---

[19] P. M. G. Broughton, M. A. Buttolph, A. H. Gowenlock, D. W. Neill, and R. G. Skentelbergy, *J. Clin. Pathol.* **22**, 278 (1969).

For example, in testing the accuracy of the system shown in Fig. 1 for the automated determination of glucose, the following procedure was adopted: Over seventy samples of blood were collected in fluoride-oxalate bottles and centrifuged; the plasma was collected. The concentration of glucose in each of the samples was then determined both using the flow-system shown in Fig. 1 and using standard glucose oxidase–peroxidase method.

Precision. The precision of an analytical method is a measure of the ability of the method to reproduce the same result repeatedly for one sample. The precision of an assay method can be tested in the following way: Three batches of pooled plasma are each assayed at least 25 times in a random order. From the results so obtained, the standard deviation and coefficient of variation are calculated from the replicate values obtained for each of the plasma pools.

The preceding discussion has reviewed and described the most important criteria that have to be complied with in establishing the validity of an analytical method. Only when such trials have been carefully undertaken is it possible to project the usefulness of the system as a viable analytical method. The table summarizes the results obtained

OPERATIONAL CHARACTERISTICS OF NYLON TUBE-IMMOBILIZED GLUCOSE DEHYDROGENASE[a]

Characteristic	Value
Tube specific activity	4.5 μmoles of glucose oxidized per minute per meter
Analyte range assayable	1–25 mM glucose
Operational stability	Over 5000 assays with a 1 m length of derivative
Storage stability	Not more than 10% loss in activity after 4 months at 4°
Accuracy	Correlation coefficient = 0.978; compared with AA2 glucose oxidase–peroxidase method
Precision	Coefficient of variation = 5.1%; 25 glucose samples; 10.5 mM

[a] The nylon tube-immobilized glucose dehydrogenase was prepared by coupling the enzyme through glutaraldehyde to adipic acid hydrazide-substituted nylon tube (this volume [9]). The derivative was used in the flow system shown in Fig. 1 for the automated determination of glucose.

when the Technicon AA1 system using nylon tube-immobilized glucose dehydrogenase shown in Fig. 1 was tested as an automated method for the determination of glucose. The table represents the "assay credentials" of this particular immobilized enzyme and establishes the usefulness of this structure as an analytical tool.

[43] Monitoring of Air and Water for Enzyme Inhibitors

By LOUIS H. GOODSON and WILLIAM B. JACOBS

Immobilized cholinesterase is useful for the collection and detection of enzyme inhibitors from both air and water.[1-5] The immobilized enzyme product functions like a dosimeter which, after exposure to the inhibitors in air or water, is then analyzed for its residual enzyme activity. A loss in enzyme activity, if any, can then be related to the quantity of inhibitor to which the enzyme was exposed. A detection system based on this principle has several potential advantages. First, there is great sensitivity due to the "biological amplification" by the enzyme itself. Enzymes that have large turnover numbers are quite sensitive detectors of enzyme inhibitors, since inhibition of one active site can produce a large reduction in the amount of substrate reaction products formed. For example, in the case of butyryl cholinesterase (acylcholine acyl-hydrolase, EC 3.1.1.8) in which the turnover number is ~84,000, inhibition of one active site by one inhibitor molecule could reduce the formation of substrate reaction products by 84,000 molecules in tests lasting 1 min. Second, the immobilized enzyme sensor possesses specificity for its enzyme inhibitors; this makes possible the construction of detection systems with relatively few interference problems. Third, immobilized enzyme monitoring systems can provide detection capability on a real-time basis. Fourth, the enzyme system is readily automated so that operation of the system may proceed while the system is unattended. Fifth, in many cases the responsiveness of the system to pollutants is related to their animal toxicities. Sixth, the system can be adapted to monitoring air through the use of a concentrator which absorbs the inhibitors from a large volume of air into a small stream of water. With such a concentrator it has been possible to detect subnanogram quantities of Sarin per liter of air in 9 min—a sensitivity level not readily attained on a real-time basis with the usual air-monitoring equipment.

[1] E. K. Bauman, L. H. Goodson, G. G. Guilbault, and D. N. Kramer, *Anal. Chem.* **37**, 1378 (1965).
[2] L. H. Goodson and W. B. Jacobs, *Proc. Natl. Conf. Control of Hazardous Material Spills*, Houston, Texas, March 21-23, 1972, pp. 129- 136.
[3] L. H. Goodson and W. B. Jacobs, *Proc. Natl. Conf. Control of Hazardous Material Spills*, San Francisco, California, August 25-28, 1974, pp. 292-299.
[4] L. H. Goodson and W. B. Jacobs, "Enzyme Engineering" (E. K. Pye and L. B. Wingard, Jr., eds.), Vol. 2, pp. 393-400. Plenum, New York, 1974.
[5] L. H. Goodson, W. B. Jacobs, and A. W. Davis, *Anal. Biochem.* **51**, 362 (1973).

For the air- and water-monitoring systems constructed in our laboratories, we chose to use immobilized butyryl cholinesterase as the sensor and an electrochemical (potentiometric) cell for monitoring the activity of the enzyme product automatically. Information explaining how such an immobilized enzyme product can be made and used in an electrochemical cell for monitoring air and water for organophosphates and carbamate pesticides is described.

Detection Principle

As mentioned above, immobilized cholinesterase has been used for the concentration and detection of inhibitors for air and water. In order to determine the relative activity of the enzyme product before and after exposure to the potential enzyme inhibitors, a substrate for the enzyme was chosen that would be easily detectable by electrochemical means. In this case butyrylthiocholine iodide (BuSChI) was chosen because the unhydrolyzed substrate was stable in the presence of platinum electrodes to which a constant current was applied, whereas this same substrate after hydrolysis by cholinesterase was readily oxidized electrochemically:

$$C_3H_7COSCH_2CH_2-\overset{+}{\underset{CH_3}{N}}(CH_3)_2\ I^- \xrightarrow{ChE} C_3H_7COOH + HS-CH_2CH_2-\overset{+}{\underset{CH_3}{N}}(CH_3)_2\ I^-$$

In this reaction a thioester is cleaved by cholinesterase (ChE) to give thiocholine iodide. Under the influence of an electric current a part of it is converted to the disulfide:

$$2\ RSH - 2\ e \xrightarrow{\text{anode reaction}} RS{-}SR + 2[H]$$

Through the use of gel-entrapped butyryl cholinesterase on an open-pore polyurethane foam pad, it is possible to obtain good contact of the enzyme with the inhibitors or the substrate solution. The activity of the enzyme pad can be determined electrochemically by passing air and substrate through the enzyme pad while about 2 μA of current is applied to the electrodes in contact with the upper and lower surfaces of the enzyme pad. If the enzyme pad is active, the substrate will be hydrolyzed as it passes down through the pad with the resulting presence of thiol at the lower platinum electrode (the anode) where it is sensed by the generation of a potential of $\sim$250 mV. On the other hand, if the enzyme is inhibited by an organophosphate or carbamate, the thiol is not formed and a potential of $\sim$500 mV is generated at the electrodes. Thus, by observing the changes in the cell potentials before and after exposure to enzyme inhibitors, it is possible to observe whether none, part, or all of the enzyme has been inhibited by the materials brought into contact with it. A simple electronic circuit can be used to signal an alarm when the voltage rise exceeds a preset threshold voltage.

Two explanations have been proposed to explain the voltage change that occurs on the inhibition of the enzyme. One theory states that the electrodes in the cell are merely monitoring the redox potentials of the solutions in contact with the electrodes. The other theory says that the platinum anode becomes partially polarized owing to the formation of a coating such as platinum sulfide (composition unknown) by operating the electrodes in the presence of thiocholine iodide. Inhibition of the immobilized enzyme results in a reduction or elimination of thiol formation whereupon further polarization of the anode occurs to give the observed voltage increase. In this system the thiol serves the dual function of forming a coating on the anode and also causing a partial depolarization to lower the cell voltage when it is present. Evidence for the presence of the polarized coating on the anode is good.

In order to convert this electrochemical cell into an air or water monitor, it is necessary to establish a detection cycle and to add some components; these are shown in Fig. 1. The objective of the cycle is to introduce a time interval during which the enzyme pad is exposed to potential inhibitors (2 min) and an enzyme activity test period (1 min). Comparison of cell voltages from successive cycles is necessary to determine the presence or the absence of enzyme inhibitors. That is, increases in voltages from cycle-to-cycle are evidence for the presence of inhibitors; if the cycle-to-cycle increases are large, then the amount of inhibitor is also large. Before the activity of the enzyme pad is completely gone it is necessary to add a fresh enzyme pad.

Figure 1 shows the fluid pathways in the monitoring system and the sequence of events making up the detection cycle. For 2 min of each 3-min detection cycle, water (with or without inhibitors) is pumped through the enzyme pad. During the last 1 min of the detection cycle, air is blown through the pad at the same time that the substrate solution is introduced; the constant current is applied for the last 40 sec of the 3-min cycle. At this time a voltage is generated that is characteristic of the amount of enzyme activity present on the enzyme pad. By comparison of cell voltages from successive cycles it is possible to determine whether enzyme activity has been lost due to inhibitors in the sample.

Figure 2 shows a strip chart recording obtained from an operating monitoring system. In this experiment the chart paper moved at 1 inch/hr and each individual voltage excursion corresponds to a 3-min detection

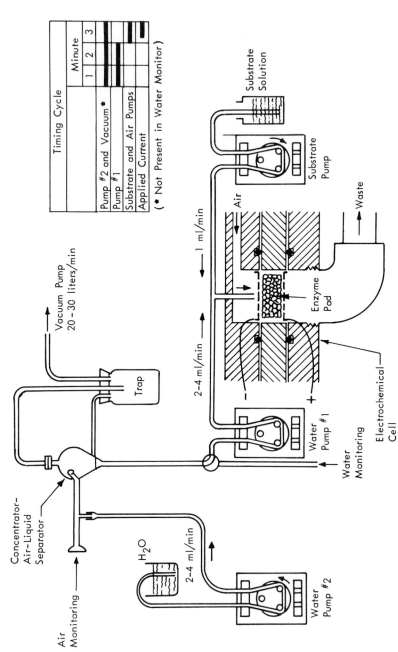

FIG. 1. Fluid flow and timing cycles in the air- and water-monitoring systems. (The concentrator assembly in the upper left is not used for water monitoring.)

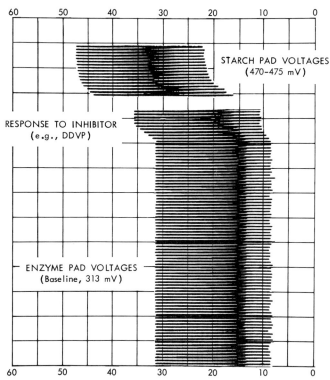

FIG. 2. Strip chart recording from immobilized enzyme monitoring system.

cycle. The base-line voltage was stabilized at approximately 313 mV, and the peak voltages changed very little during the 3¾ hr of operation when inhibitors were absent. Exposure of the enzyme pad to a very dilute solution of a cholinesterase inhibitor, e.g., 1.2 ng of Sarin or 200 ng of 2,2-dichlorovinyl dimethyl phosphate (DDVP) per milliliter, for five cycles produces a voltage increase of about 9 to 10 mV/cycle. After removal of the inhibitor solution a new base line was established at around 360 mV. The gap in the tracing indicates the replacement of the enzyme pad with a starch pad after which the cell voltage reached 470–475 mV. In this particular case the alarm potential (i.e., the maximum voltage increase which can be expected when the enzyme pad is completely inhibited) is the difference between 470 mV and 313 mV or 157 mV.

When the system is used as a water monitor, water suspected of possessing inhibitors is often pumped through the pad at speeds of 500–1500 ml/min. On the other hand, when the system is used as an air monitor, the inhibitor is extracted from air into a small volume of water which may be pumped through the enzyme pad at 2–8 ml/min. Little sensitivity is gained by increasing the speed of pumping the inhibitor solution through the enzyme pad since diffusion of the inhibitor to the enzyme and its subsequent inhibition of the enzyme in the pad become time dependent at the higher pumping speeds.

Components of the Monitoring System

In order to construct a working detection system based upon gel entrapped cholinesterase it will be necessary for each investigator to optimize the many variables for his own system. However, in the following paragraphs, suggestions for the construction of a working system using readily available laboratory supplies are given. An effort is being made to point out a number of the critical elements in the system so that certain of the pitfalls may be avoided.

Enzyme Pad Preparation Procedure[5]

Horse serum cholinesterase (ChE) is immobilized for use in electrochemical systems for the detection of low levels of inhibitors in air or water as described below:

The following materials are used: Open-pore polyurethane foam sheets, 44–45 ppi × ¼ inch, Scott Industrial Foam, Scott Paper Company, Chester, Pennsylvania; partially hydrolyzed potato starch recommended for gel electrophoresis, Connaught Medical Research Laboratory, Toronto, Canada; Chlorhydrol (aluminum chlorhydroxide complex, 50% w/w solution), Reheis Chemical Company, Chicago, Illinois; horse serum cholinesterase, Sigma Chemical Company, Type IV, ~15 μM units/mg; and Tris buffer, Fisher Scientific THAM.

Step 1. A solution of Tris buffer, 0.08 M, is prepared by dissolving 9.7 g of tris (hydroxymethyl) aminomethane (Tris) in 900 ml of distilled or deionized water, adjustment of the pH to 7.4 with concentrated HCl, and then adjusting the volume to 1 liter.

Step 2. Horse serum cholinesterase (21 mg) is dissolved in 6 ml of Tris buffer, and to this is added with mechanical stirring 0.6 ml of a dilute solution of aluminum chlorohydroxide complex (12 g of Chlorhydrol in 100 ml of water). At this point aluminum hydroxide gel precipitates and absorbs the enzyme from the solution (check pH and adjust to 7.4 to make sure that the precipitation is complete). This suspension is set aside at ambient temperature until needed.

Step 3. Potato starch (4 g) is suspended in 10 ml of cool Tris buffer and added to 34 ml of boiling Tris buffer and heated until the suspension clears. Care is taken to avoid the formation of scum or lumps (start step 3 over if lumps are obtained) and the starch slurry is stirred with a magnetic stirrer while it cools spontaneously to 45°. At this point aluminum hydroxide gel entrapped cholinesterase (step 2) is added all at once and quickly mixed. Step 4 and its four replications are done quickly before the starch has a chance to gel.

Step 4. A 10-ml portion of the warm starch-gel slurry from step 3 is deposited on a precut sheet of open-pore urethane foam (4-inch × 4-inch × ¼-inch sheet) lying on a warmed glass or plastic surface (usually over a pan of warm water). The starch-enzyme material is now distributed throughout the sheet as uniformly as possible with a plastic rolling pin filled with warm water; the sheet is rolled in all directions and turned over several times. In this same manner, four other 10-ml portions of the starch-enzyme product are distributed over four additional urethane foam sheets.

Step 5. The coated sheets are placed on edge in a wooden rack, dried at least 1 hr at room temperature, and finally dried overnight at 40°.

Step 6. The resulting sheets are examined carefully to be sure that all areas of all pads are evenly coated. Poorly coated areas are trimmed away. The sheets are handled carefully to avoid breakage of the starch film on the dried sheets. The enzyme sheets are now cut into ⅜-inch pads with a stainless steel cutter (a hole in side of cutter permits cut pads to move up and out) mounted in an electric drill press. The procedure yields approximately 350 enzyme pads which are packaged in screw-capped jars and stored in the refrigerator until needed. Pads retain their activity for more than a year at 5°.

Step 7. It is recommended that three or four individual enzyme pads be analyzed in an automatic titrator or with a pH meter with external electrodes by the procedure described in the Worthington Manual.[6] In this procedure a single enzyme pad is cut into small pieces with a pair of scissors and placed in a small titration vessel or beaker with a magnetic stirrer to which has been added 7 ml of H_2O, 3 ml of $MgCl_2$ (0.2 M), 3 ml Tris (0.02 M), and finally 1.0 ml of acetylcholine chloride (2.2 M). The pH is maintained at 7.4 through addition of 0.008–0.017 M NaOH. The volume of NaOH added versus time for 5 min after constant rate is attained is recorded. The activity is expressed as micromoles of substrate hydrolyzed per minute per pad. Pads made in the procedure described above should have an activity of ~0.1 unit/pad.

[6] Worthington Enzyme Manual, Worthington Biochemical Corp., Freehold, New Jersey, 1972.

The Starch Pad

Starch pads are like enzyme pads except they possess no enzyme and can readily be identified by their red color. These pads are made in the same manner as the enzyme pads described above with the following changes: (1) In step 2 the enzyme solution is replaced with 21 mg of gelatin dissolved in 6 ml of hot Tris buffer; (2) In step 3 sufficient water-soluble red dye is added to make the solution red; the section of the dye is not critical, and any one of the following would be suitable: safranin O, Congo red, FD & C Red No. 2 food coloring. These starch pads are used in electrochemical cells to produce voltages which simulate voltages obtainable from an enzyme pad which has been completely inactivated by inhibitors. The alarm potential from a working system is readily calculated by subtracting the enzyme pad voltage from the starch pad voltage, and it represents the voltage change to be expected when all the enzyme is inhibited.

The Electrochemical Cell

Flow-through electrochemical cells suitable for use with enzyme pads are not readily available, so they must be made from glass, Lucite, Delrin, or other plastic materials. The electrochemical cell shown in Fig. 1 is more complex than needed for preliminary experiments; if desired, one O ring may be omitted and the plastic anode holder can be used to position both the enzyme pad and the platinum anode. Clamps are satisfactory for holding the anode and cathode holders together, although a means for automatic alignment of the two electrodes is quite helpful. The diameter of the cell should be chosen so that the pads fit snugly inside and the electrodes should be arranged so that they are in contact with the upper and lower surfaces of the enzyme pad without mashing it flat ($3/16$ inch between electrodes is a satisfactory distance for the $1/4$-inch thick pads).

Platinum Electrodes

Bright platinum electrodes $3/8$ inch in diameter (either flat disk or cup shaped) with 15 0.046-inch holes are spot welded to a short lengths of 26 or 28 B&S gauge platinum wire to serve as the electrical connectors for the electrodes inside of the cell body. The two electrodes, after cleaning in boiling HNO_3 and a flame, are immersed in a beaker containing 3% H_2PtCl_6 and 0.03% lead acetate; a coating of platinum black is applied by applying 2.5 V and reversing the polarity of the electrodes once each 15 sec for a total of 5 min. The resulting platinized platinum electrodes are washed in water and then heated to redness in a bunsen burner to make platinum gray electrodes. These electrodes are alike when first made; however, after use the anode becomes coated with a material (probably platinum sulfide) that makes it different from the cathode. Care must be taken to observe the polarity of the electrodes if they are to remain stable with time.

Buffer Selection and Concentration

Of the buffers tested Tris buffer (Fisher Scientific THAM) gives the largest and most reproducible responses. Although other buffer concentrations are satisfactory, 0.08 M solutions are recommended. The pH is adjusted to 7.4 with HCl.

Substrate Solution

Butyrylthiocholine iodide (either from Sigma or Pierce) is dissolved in Tris buffer, pH 7.4, immediately prior to use, and its useful life at room temperature exceeds 24 hr. Concentrations ranging from 1.25 to 2.5 $\times$ 10^{-4} M can be used in the water-monitoring system, but greater base-line stability is obtained with the higher concentration. The usual substrate pumping rate is 0.7–1.4 ml/min.

The Constant Current Supply

The greatest differences in voltages observed when hydrolyzed or unhydrolyzed substrate reach the electrodes is observed when the least current is applied to the electrodes. However, when currents much below 2 μA are used the system requires much time for equilibrium voltages to be reached. On the other hand when currents much in excess of 2 μA are used the magnitude of the voltage change is reduced. The constant current is conveniently supplied with a 9 V battery in series with a 4.7 megohm resistor. The lower platinum electrode must be positive.

Voltage Measurements

It is important that a high-impedance voltage measuring equipment be used or it will load the circuit and change the voltage. The Kiethley Model 602 electrometer with recorder output is suitable. Some vacuum tube voltmeters are satisfactory for this use. Also, a zero gain operational amplifier can be used directly to match the impedance of the cell to a recorder. A strip chart recorder (e.g., Hewlett-Packard Model 680 or the Heath/Schlumberger Recorders) is a big help in observing base-line stability and/or response when challenging the system with enzyme inhibitors.

Mechanical Components

Some information about flow rates and the peristaltic pumps for moving water and substrate solutions is given in Fig. 1. Variable-speed pumps (e.g., Scientific Industries) are used, and they offer the flexibility necessary for optimizing the detection system. Reliability and reproducibility of flow rates of liquid and air through the cell are important for obtaining stable base-line voltages. The air pump for blowing the excess water out of the pad while the substrate is flowing is an inexpensive but reliable vibrator pump normally used for aquaria. It should have a capacity of 1–2 liters/min. The vacuum pump in the concentrator is a rotary carbon-vane pump (Gast) with a free air capacity of 1.1 cfm and is satisfactory for this purpose. A mechanical timer driven by a 1/3 rpm motor with four cam-operated switches (Hayden, North American Phillips Controls) is satisfactory for achieving the detection cycle shown in Fig. 1.

System Operation

Upon completion of the fabrication and assembly of the various parts of the fluid flow system shown in Fig. 1, the timer is adjusted to provide the desired timing cycle. An electrometer (preferably with a strip chart recorder) is connected to the electrochemical cell so that cell voltages can be monitored and recorded at the end of each detection cycle. The flow rates of the substrate and water pumps are adjusted and freshly prepared substrate solution is added to the system. A starch pad is placed in the electrochemical cell and the timing cycle is started. Within a cycle or two after the substrate solution reaches the starch pad, the maximum voltages (i.e., starch pad voltages) begin to level out at ~500 mV (the exact voltage varies from system to system and also with the age of the electrodes; new electrodes give lower voltages for the first several days of use). The starch pad is now replaced with an enzyme pad, and, after a few cycles, the maximum voltages again level out ~200 mV lower than the starch pad voltage. The system is now operational but should be checked for its response to the presence of an enzyme inhibitor in the water sampled. A convenient method of doing this is to place 2 ppm of 2,2-dichlorovinyl dimethyl phosphate (DDVP) in water at 25° and pump this through the enzyme pad for several cycles. Voltage maxima should increase by more than 10 mV/cycle until the DDVP solution is replaced with water or until the enzyme pad is inactivated. Aqueous solutions of other organophosphates and carbamates may be used for checking the system, but the concentration of the inhibitor may need to be increased or decreased since the sensitivity of the system is a function of the affinity of the inhibitor for the enzyme.

Discussion

An experimental water-monitoring instrument embodying the principles described above has been built and designated as CAM-1 (continuous aqueous monitor) (Fig. 3). This unit is designed for unattended operation and uses computer logic to determine whether the cycle-to-cycle voltage increase exceeds the alarm threshold level (sensitivity is usually preset at 10 or 20 mV). Also, CAM-1 possesses a voltage sensor which determines when most of the enzyme on a pad has been inhibited; it activates an automatic enzyme pad changer which inserts a fresh enzyme pad when needed. One enzyme pad is often suitable for signaling 10 alarms before it is automatically replaced. The CAM-1 instrument was designed to

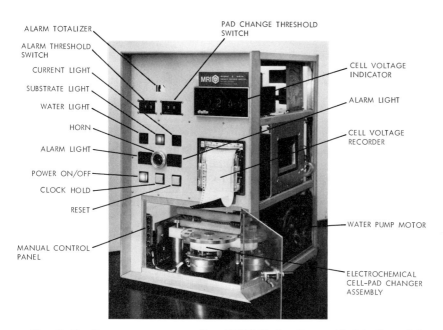

FIG. 3. Continuous aqueous monitor (CAM-1) for the rapid detection of low levels of organophosphates and carbamates in water.

minimize the consumption of enzyme while monitoring water supplies for the presence of pesticide spills. In the absence of inhibitors, one enzyme pad will operate the instrument for up to 3 days. Daily servicing of the instrument with new pads and substrate is recommended.

The sensitivity of CAM-1 to a number of pesticides in water has been reported,[3] and some of these are tabulated here.

Compound	Type	Detectable level (ppm)	Temperature (°C)
Paraoxon	Organic phosphate	0.1	7
Diazinon	↓	1.2	23
Systox	↓	1.4	20
Dursban	↓	4.5	21
Sevin	Carbamate	20	25
Temik	↓	0.5	25

These compounds at the concentrations shown produced voltage responses of 10 mV or more for each cycle they were pumped over the enzyme pad. Compounds which inhibit cholinesterase but which are readily dissociated from it produce voltage responses in CAM-1 during the first or second detection cycle but do not produce a cumulative voltage response, since they are then in equilibrium with the enzyme. Sutan is an herbicide that produces this type of response.

CAM-1 can be used for monitoring air for the presence of these same types of enzyme inhibitors through the use of a concentrator, which collects the inhibitor from air into a small volume of water, and then passing this sample to the electrochemical enzyme cell. Through the use of a concentrator with an air flow rate of 25 liters/min and water flow of 2 ml/min, it has been possible to detect 0.0001 μg of Sarin per liter of air in 9 min (the slow flow of liquid in the concentrator causes the delayed response).

In the present system, horse serum cholinesterase has been used with butyrylthiocholine iodide as the substrate. It should be noted that this system also works with immobilized electric eel cholinesterase and immobilized bovine erythrocyte cholinesterase with acetylthiocholine iodide as the substrate.

Both covalently bonded and gel entrapped cholinesterase products have been used in the electrochemical cell; however, the gel-entrapped cholinesterase products were more easily inhibited than the covalently bound products; thus the gel-entrapped products gave the best sensitivity in the monitoring system.

[44] Immobilized Enzymes in Microcalorimetry

By ÅKE JOHANSSON, BO MATTIASSON, and KLAUS MOSBACH

The calorimetric technique is one of the earliest applied methods for quantitative studies of chemical reactions.[1] Its potential for monitoring biochemical and biological processes was recognized very early, but a wider use of calorimetry for these purposes was not possible until the development of the microcalorimetric technique,[2-4] allowing very sensitive heat measurements on small sample volumes. This has led to the development of several instruments during the past decade.

Calorimetry is by principle a nonspecific technique, which makes it suitable as a general monitor for complex processes, such as cell metabolism. On the other hand, in combination with highly specific enzymic reactions, calorimetry offers a very general principle of detection, totally independent of the optical properties of the system investigated and with the additional advantage of eliminating the need for auxiliary enzymes and their cofactors. The combination of microcalorimetric detection and the well-recognized advantages of immobilized enzymes for analytical purposes offers a very powerful analytical technique.[5-9] In addition, the microcalorimetric method is well suited for the assay of immobilized enzyme preparations and also for thermodynamic investigations of such systems. The scope of this article is limited to the analytical aspects of flow microcalorimetry applied to immobilized enzymes.

Theoretical Considerations

The heat associated with the reaction of an immobilized enzyme can be measured by two different principles based on the flow technique as illustrated in Fig. 1a and b. In the first case, the reaction cell is in good

---

[1] A. L. Lavoisier and P. S. de Laplace, "Sur la Chaleur." Paris, 1780.
[2] E. Calvet and H. Prat, "Recent Progress in Microcalorimetry." Pergamon, Oxford, 1963.
[3] T. H. Benzinger and C. Kitzinger, *Methods Biochem. Anal.* **8**, 309 (1960).
[4] I. Wadsö, *Acta Chem. Scand.* **22**, 927 (1968).
[5] Å. Johansson, *Protides Biol. Fluids, Proc. Colloq.* **20**, 567 (1973).
[6] Å. Johansson, J. Lundberg, B. Mattiasson, and K. Mosbach, *Biochim. Biophys. Acta* **304**, 217 (1973).
[7] L. J. Forrester, D. M. Yourtree, and H. D. Brown, *Anal. Lett.* **7**, 599 (1974).
[8] N. Rehak, J. Everse, R. L. Berger, and N. O. Kaplan, *Fed. Proc. Fed. Am. Soc. Exp. Biol.* **33**, 556 (abst.) (1974).
[9] S. N. Pennington, *Enzyme Technol. Digest* **3** (2), 105 (1974).

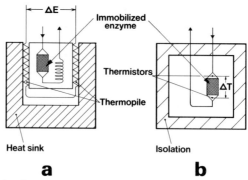

Fig. 1. Principles for measurement of heat of reaction of immobilized enzymes. (a) Heat flow measurement. The heat flows from the reaction cell via thermopiles to the heat sink, generating a voltage, ΔE, over the thermopiles, which is directly proportional to the heat flow. (b) Temperature measurement. The reaction cell is thermally insulated from its surroundings, and the temperature increase, ΔT, of the fluid passing the cell is measured.

thermal contact with a heat sink, and the heat flux from the cell to the heat sink via a series of thermocouples is measured under isothermal conditions. In the second case (Fig. 1b) the reaction cell is isolated from its surroundings, and the temperature change of the liquid passing the cell is measured (see this volume [45]).

The substrate can either be continuously fed into the microcalorimeter (later referred to as steady-state experiments) or introduced as small-volume injections into the buffer flow (later referred to as heat-pulse experiments).

The isothermal principle appears to be advantageous for determination of the total heat of a reaction under both pulse and steady-state conditions. Then the heat flux is simply equal to the number of moles of substrate converted per unit time multiplied by the enthalpy change per mole. Under comparable conditions in Fig. 1b, the temperature level will depend also on the heat capacity of the fluid and on the heat loss to the surroundings.

From an analytical point of view it is advantageous to use a small volume of substrate-solution leading to a short heat pulse from the cell. In order to obtain maximum sensitivity, the time duration of the heat pulse must be as short as possible. Owing to the shorter heat diffusion paths in the case of temperature measurements (Fig. 1b) a minimal duration of the heat pulse can be obtained, leading to maximum sensitivity and minimum time for analysis. However, when maximum speed and sensitivity are not of primary importance, the heat flow principle according to Fig. 1a offers certain advantages in relative independence of flow rate on the base-line level.

The account of experimental results that follows will consider only measurements according to the heat-flow principle, as previously described by Johansson,[5] Johansson et al.[6] and Schmidt and Krisam.[10] Experimental results from using the temperature measurement principle are given in this volume [45].

Apparatus

The experiments referred to in this chapter have been carried out on a LKB 2107 Sorption Microcalorimeter, which is a twin heat-flow calorimeter based on the original development by Wadsö.[4] The sensitivity of the heat measurements is <1 μW (continuous heat effect) and <1 mJ (heat pulse). The principle design of the calorimeter cell is identical to that shown in Fig. 1a. The inner dimensions of the microcolumn are $\emptyset$ 6×17 mm corresponding to a volume of ca 0.5 ml. The short heat-exchanging coil following the microcolumn produces a complete heat transfer from the liquid. The flow was generated by means of an LKB 10200 Perpex pump, and the flow rate used varied between 6 and 10 ml/hr.

The microcolumn is filled with the immobilized enzyme in the form of a preswollen liquid suspension, which is allowed to settle; the microcolumn is inserted into the instrument, which is left to equilibrate with a constant buffer flow until a stable base line is recorded (normally after 2–3 hr). In the pulse experiments a standard sample injection valve for liquid chromatography is used. The heat generated is recorded and, by means of electrical calibration, the heat output can be given in μW (for continuous flow experiments) or mJ (after integration of the heat pulse in injection experiments). The maximum sample rate with the equipment used is about 4 samples per hour.

Preparation of Immobilized Enzymes

Besides the normal requirements for immobilization of enzymes, i.e., high specific activity and long-term stability, the flow resistance has to be kept low in order not to introduce frictional heat disturbances. Two different methods of preparation have been used for the two different enzymes investigated.

In the case of glucose oxidase the enzyme was entrapped in polyacrylamide gel, which was then forced through a mesh, leading to the formation of irregularly shaped particles. In the case of trypsin, a bead

[10] H.-L. Schmidt and G. Krisam, *Biochim. Biophys. Acta* (in press).

polymerization technique[11] was utilized, producing spherical particles having a diameter of about 0.1 mm. The latter technique, which is definitely to be preferred, is given a more detailed description.

The general principle of the bead polymerization technique is to dissolve the enzyme in buffer with the monomers, add the catalyst system to initiate the polymerization, and then pour the "water phase" into a precooled, vigorously stirred N_2-fluxed organic phase containing a stabilizer. The mixing of the two mainly immiscible phases leads to the creation of droplets of one phase in the other. In the system used, the volume of the organic phase greatly exceeded the volume of the water phase, thereby favoring the generation of water droplets in the organic phase. As the polymerization proceeds the droplets of the water phase containing the enzyme and the monomers become spherical gels beads (see also this volume [5] and [12]–[15]). The immobilization of the enzyme in 25% (w/v) polyacrylamide gel was carried out as follows.

Trypsin, 600 mg [Sigma type III, twice crystallized, 10,000 benzoyl-L-arginine ethyl ester (BAEE) units/mg] was dissolved in 0.05 M potassium phosphate buffer (pH 7.3) together with the monomers, 14.25 g of acrylamide and 0.75 g of the cross-linking agent N,N'-methylenebisacrylamide.[6] Final volume was 56 ml. The solution was cooled with ice. The catalyst system (600 μl of $N,N,N'N'$-tetramethylethylenediamine and 200 mg of the ammonium persulfate dissolved in 4 ml of the buffer) was rapidly added to the cold solution. The total mixture was then poured into a reaction vessel containing a precooled, N_2-fluxed, well-stirred organic phase consisting of 290 ml of toluene, 110 ml of chloroform, and 3 ml of the stabilizer Sorbitan sesquioleate (Pierce Chemical Co., Rockford, Illinois). In the water phase of the formed dispersion, polymerization started within a few minutes and was completed after 15 min. In order to protect the enzyme from thermal denaturation, the temperature was controlled (kept within the region 4°–7°) by cooling with ice.

The reaction mixture was poured onto a glass filter and washed carefully with cold 0.1 M $NaHCO_3$, followed by additional washings in cold solutions (while being stirred for 30 min) with 0.1 M $NaHCO_3$, 0.001 M HCl, 0.5 M NaCl, and finally distilled water. The gel beads were then quickly frozen and lyophilized.

Besides the preparation of 25% acrylic gel beads, tighter gels were also prepared. For studies on the permeation of the high-molecular-weight inhibitor, soybean trypsin inhibitor, 30% and 40% gels were prepared. The procedure was the same as described except for the use of more monomers. The ratio of acrylamide to the cross-linker N,N'-methylenebisacrylamide was always kept constant (19:1). Prior to use, the

[11] H. Nilsson, R. Mosbach, and K. Mosbach, *Biochim. Biophys. Acta* **268**, 253 (1972).

lyophilized enzyme beads were left to swell overnight at $+4°$ in an appropriate buffer.

Both the glucose oxidase and the trypsin preparations showed an excellent stability, the decline in response after continuous use at 25° during 1 month not exceeding about 15%. This apparent stability does not, however, exclude partial denaturation of the enzyme taking place. Such denaturation is not registered owing to overcapacity of the preparation. The calorimetric base-line stability was considerably better for the enzyme immobilized according to the bead polymerization technique, owing to its better flow properties.

In similar experiments by Schmidt and Krisam[10] commercially available matrix-bound urease and glucose oxidase have been used. For this application the polyacrylamide-bound enzymes were preferred to such cellulose-triazine-bound enzymes primarily owing to their better mechanical and/or flow properties. Other enzyme supports discussed in this volume should also be applicable in flow microcalorimetry.

Results

Heat-Pulse Experiments

In the case of glucose oxidase,[5] 13- and 60-μl samples of glucose solutions of varying concentration were introduced. The integrated heat responses are plotted in Fig. 2. The experiments were carried out in a 0.2 M phosphate buffer, pH 6.0, also containing 0.5 M NaCl, 0.02 M

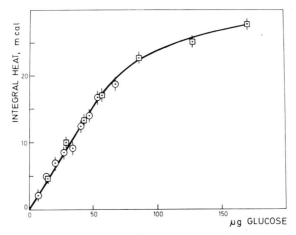

FIG. 2. Integral heat from immobilized glucose oxidase as a function of the amount of glucose introduced. ⊙, 13-μl sample; ⊡, 60-μl samples. From Å. Johansson, *Protides Biol. Fluids, Proc. Colloq.* **20**, 569 (1973).

$(NH_4)Mo_7O_{24} \cdot (H_2O)_4$, 0.017 M KI, and 0.001 M NaN_3. The integral heat increases linearly with increasing amount of substrate up to about 50 µg of glucose, indicating a complete substrate turnover. At higher substrate amounts the curve levels off, indicating incomplete substrate turnover. The relatively big scattering of the points is primarily due to base-line instability arising from the high flow resistance of the irregular shaped gel particles. Similar experiments have since been carried out on immobilized lactate dehydrogenase by Schmidt and Krisam[10] showing similar results.

The corresponding results for trypsin[6] are shown in Fig. 3a together with a typical curve, Fig. 3b. Benzoylarginine ethyl ester (BAEE) was used as substrate and the injection volume was 135 µl. The two curves correspond to two different systems: 0.1 M phosphate (pH 7.9) and 0.1 M Tris-HCl (pH 7.9). The different slopes observed illustrate a very important feature of the calorimetric technique. It is well known that the reaction heat of ester hydrolysis is normally close to zero. The heat observed is therefore mainly due to protonization of the buffer. By choosing a buffer with large heat of protonization, such as Tris buffer, the analytical sensitivity can be increased approximately 10-fold compared to the phosphate buffer (see Fig. 3a) allowing accurate determination of substrate amounts below 1 µmole. (The heats of protonization of the phosphate and Tris buffers are 4.74 kJ/mole[12] and 47.48 kJ/mole[13], respectively.)

An alternative technique for thermal amplification has been demonstrated for glucose analysis using glucose oxidase and peroxidase immobilized on the same gel.[10]

Steady-State Experiments

When applying a continuous flow of substrate through the calorimeter cell, a steady-state heat production will be reached[6] as depicted in Fig. 4a. The steady-state heat production is plotted against the substrate concentration in Fig. 4b. A linear relationship is found up to approximately 10 mM BAEE, and at higher concentrations the heat production passes a maximum, the decrease at high concentrations probably being due to substrate/product inhibition.

Choice of Heat Pulse or Steady-State Conditions

In the pulse experiment, the magnitude of the heat pulse at constant flow rate is determined by the injection volume and the sample concen-

[12] G. D. Watt and J. M. Sturtevant, *Biochemistry* **8**, 4567 (1969).
[13] G. Öljelund and I. Wadsö, *Acta Chem. Scand.* **22**, 2691 (1968).

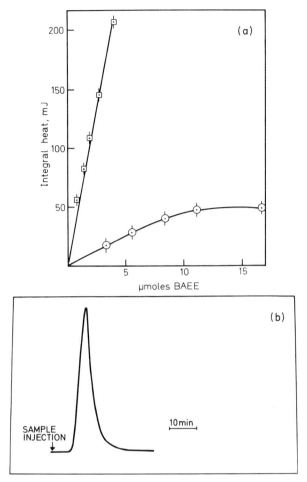

FIG. 3. Heat pulse experiments on immobilized trypsin. (a) Integral heat as a function of the amount of substrate injected. ⊙, Phosphate buffer; ▫, Tris buffer. (b) Experimental curve obtained after injection of 135 μl of a 0.041 M benzoyl-L-arginine ethyl ester (BAEE) solution. Integral heat corresponds to 28.2 mJ. From Å. Johansson, J. Lundberg, B. Mattiasson, and K. Mosbach, *Biochim. Biophys. Acta* **304**, 219 (1973).

tration. If the sample concentration is too low to allow substrate determination in the normal injection volume, the determination may be done in a larger volume. With an increasing sample volume, the maximum response from the thermal detector will increase up to the ultimate level corresponding to the steady state. Therefore, the steady-state conditions should be used when maximum sensitivity in terms of sample *concentra-*

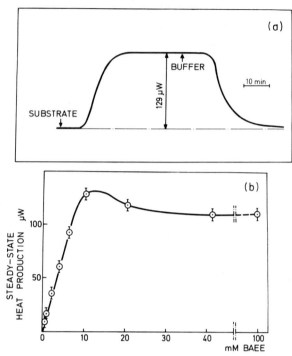

FIG. 4. Steady-state experiments on immobilized trypsin. (a) Recorded calorimetric response at a continuous flow of 10 mM Benzoyl-L-arginine ethyl ester (BAEE) in phosphate buffer. (b) Steady-state heat production as a function of BAEE concentration in phosphate buffer. From Å. Johansson, J. Lundberg, B. Mattiasson, and K. Mosbach, *Biochim. Biophys. Acta* **304**, 219 (1973).

tion is needed. At higher sample concentrations the heat pulse experiment is advantageous, since both sample volume and time can be saved.

Steric Exclusion of Macromolecular Inhibitors

In the general use of immobilized enzymes for analytical purposes, the gel can be used for protection of the enzyme from high-molecular-weight inhibitors that may be present in samples to be analyzed. This principle has been demonstrated by measuring the possible reduction of steady-state heat production from trypsin, immobilized in two different gels (40 and 25% dry weight) when subjected to a large excess of soybean trypsin inhibitor.[6] In the 40% gel, an insignificant reduction in response was observed, indicating an efficient exclusion of the inhibitor molecules from the tight network of the gel. In the 25% gel, no such efficient exclusion was obtained. These conclusions are supported by later investigations by Berezin et al.,[14] who studied trypsin activation of immobilized chymotrypsinogen as dependent upon the gel concentration of the entrapping matrix. It was shown that gels with less than 32% (w/v) acrylamide permit diffusion of the enzyme into the gel. The immobilized zymogen was converted, under those circumstances, into the active enzyme. Gels tighter than 32% protected the immobilized chymotrypsinogen from proteolytic attack.

[14] I. V. Berezin, A. M. Klibanov, and K. Martinek, Biochim. Biophys. Acta 364, 193 (1974).

[45] Enzyme Thermistor Devices

By BENGT DANIELSSON and KLAUS MOSBACH

The use of enzymes, free or immobilized, in various kinds of thermal analysis enhances the specificity of a relatively unspecific technique.[1,2]

The application of immobilized enzymes to microcalorimetric studies of biochemical reactions is described in this volume chapter [44]. The time required for a measurement and the relative complexity and expense of the equipment, however, makes microcalorimetry at its present state of development unattractive for multiple determinations in which accuracy is not of overriding importance.

Recently several simple, cheap devices have been developed that employ thermistors as temperature sensors. The cost is further reduced by using immobilized enzymes.[3-8] In this laboratory we have developed the so-called "enzyme thermistor."[3] In the prototype a small cup was formed by coiling a thin-walled tubing containing the immobilized enzyme and sealing it with epoxy resin, with the thermistor placed in a low-

[1] I. Wadsö, Sci. Tools 21, 18 (1974).
[2] S. N. Pennington, Enzyme Technol. Digest 3, 105 (1974).
[3] K. Mosbach and B. Danielsson, Biochim. Biophys. Acta 364, 140 (1974).
[4] K. Mosbach, B. Danielsson, A. Borgerud, and M. Scott, Biochim. Biophys. Acta 403, 256 (1975).
[5] B. Mattiasson, B. Danielsson, and K. Mosbach, Anal. Lett. 9, 217 (1976).
[5a] B. Mattiasson, B. Danielsson, and K. Mosbach, Anal. Lett., in press.
[6] C. L. Cooney, J. C. Weaver, S. R. Tannenbaum, D. V. Faller, A. Shields, and M. Jahnke, in "Enzyme Engineering" (E. K. Pye and L. B. Wingard, Jr., eds.), Vol. 2, p. 411. Plenum, New York, 1974.
[7] K. Mosbach, B. Mattiasson, S. Gestrelius, P. A. Srere, and B. Danielsson, in "Enzyme Engineering" (E. K. Pye and L. B. Wingard, Jr., eds.), Vol. 2, p. 151. Plenum, New York, 1974.
[8] L. M. Canning, Jr., and P. W. Carr, Anal. Lett. 8, 359 (1975).

heat-capacity fluid in the cup. We here describe two alternative developments having some advantages over the original. In these devices the thermistor is positioned in closer proximity to the site of reaction and in the path of the heat flow, with consequent improvement of the efficiency.

Apparatus

Figure 1a illustrates one version of the enzyme thermistor designed for use with a single probe.[3-4] The holder can accept columns of various sizes, but usually those that hold 0.8–1.0 ml of an immobilized enzyme preparation are employed. The leakproof, adjustable holder permits the thermistor to be placed in the column at an optimal level, which depends on the amount of enzyme used, the actual range of substrate concentrations, and the flow-rate of solution through the column. Samples are introduced into the continuous buffer flow via a three-way valve. With this device the flow rate can be fairly high, usually 60 ml per hour, with sample pulses of 1 ml. Satisfying results have been achieved with peristaltic pumps (Varioperpex, LKB-Produkter, Sweden), although pumps

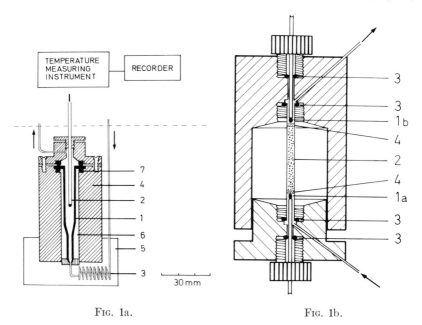

FIG. 1a.   FIG. 1b.

FIG. 1a. Single thermistor unit. 1, Microcolumn containing the immobilized enzyme preparation; 2, steel tube with thermistor; 3, heat exchanger; 4, Perspex cylinder; 5, water-jacket; 6, air space; 7, O-rings.

FIG. 1b. A differential enzyme thermistor unit. 1a, Reference thermistor; 1b, sensing thermistor; 2, Teflon column containing the immobilized enzyme; 3, O-rings; 4, porous polyethylene disks.

of the piston type might reduce fluctuations in the base line. The heat-exchange coil (thin-walled stainless steel tube, 0.8 mm inner diameter, at least 250 mm long) makes it unnecessary to keep solutions thermostated. The influence of short-term temperature fluctuations in the water bath is reduced by shielding the heat exchanger with a plastic container.

Variations in the base line due to fluctuations in the ambient temperature can also be reduced by using a reference thermistor, which can be positioned in the buffer stream so that the solution passes it just before entering the column, as is illustrated in Fig. 1b.[5] Another useful procedure uses a split-flow arrangement.[5a] A heat exchanger of the same type as that used with the single thermistor apparatus can be recommended. The immobilized enzyme preparation (0.2–0.3 ml) is held between porous Teflon plugs in a thin-walled Teflon tubing (3 mm i.d.), the thermistors being positioned in each end of the column close to the plugs.

In both cases the assemblies are completely immersed in a water bath (e.g., Hetotherm Model 05 PG 623 UO, Heto, Birkeroed, Denmark), which allows control of temperature with accuracy better than ±0.01°. The thermistor resistance variation is registered as the unbalance signal from a Wheatstone bridge equipped with a low-noise, high-sensitivity amplifier (Knauer Temperature Measuring Instrument, Wissenschftlicher Geraetebau, W. Berlin, Germany) connected to a potentiometric recorder (Kontron, Switzerland). At the most sensitive measuring range, a change of the output from the bridge of 100 mV corresponds to a temperature change of 0.02° (0.08% change in thermistor resistance). On some occasions a more sensitive recorder range can be used with subsequent improvement of the temperature resolution.

Thermistor types commonly used are the YSI type 44 106 (Yellow Springs Instruments, Yellow Springs, Ohio) with a resistance of 10 kΩ at 25° and the Veco type 35A36, 5 kΩ or 41A28, 10 kΩ (Victory Engineering Corporation, Springfield, New Jersey). The YSI precision thermistors have epoxy-coated beads and are delivered mounted at the tip of a 5 cm long Teflon tubing. They can be used without further modification. The small, glass-encapsulated Veco thermistors (1.5 mm in diameter, 6 mm long) are fastened in the end of 2 mm (o.d.) stainless steel tubes with epoxy adhesive after Teflon insulation of the leads. The single thermistor device has also been used, with excellent results, with inexpensive, standard thermistors viz. ITT type F 23 D (4 mm in diameter, 75 mm long; 2 kΩ at 20°).

When the apparatus has been placed in the water bath, it takes 1–2 hr before a stable base line is obtained. During this time, buffer is constantly pumped through the system. In all experiments described here, the water bath temperature was held at 27°. Sample introduction without stopping the pump is accomplished by means of a three-way valve at the inlet side of the pump. The sample pulses last 1–2 min depending on the flow rate. To avoid solvation effects, it is essential that the substrate is dissolved in the same buffer as that circulated through the system. The height, slope, and area of the resulting peak can all be used as measures of the substrate concentration in the sample.[3] Since it is easiest to determine, we prefer to use the peak height Δt.

If desired, electrical calibration can readily be carried out by placing a small coil of manganin wire near the inlet in the column (e.g., 10 cm of wire with a resistance of about 50 Ω per meter).[4] The low-resistivity leads to the coil can be drawn through the column outlet to a constant voltage source. The heating effect of the manganin wire is varied by changing the current. The current and voltage are determined, and the power is calculated. The buffer and flow rate should be the same as in the enzymic experiments.

Enzyme Immobilization

Alkylamino glass was prepared by treating controlled-pore glass beads (40–80 mesh, pore diameter 55 nm; Corning Glass Works, Corning, N.Y.) with γ-aminopropyltriethoxysilane (Union Carbide, Illinois) following the procedure developed by Weetall (see this volume [10]). To 1 g of glass beads were added 18 ml of distilled water and 2 ml of 10% (v/v) of γ-aminopropyltriethoxysilane. The pH was adjusted to between 3 and 4 with 6 M HCl. The mixture was heated for 2 hr on a water bath at 75°. The glass beads were filtered and washed with 20 ml of distilled water, then dried in an oven at 115° for at least 4 hr.

The immobilized enzymes used in the experiments described here were prepared by coupling of the enzymes to alkylamino glass after activation with 2.5% glutaraldehyde. During the coupling the preparation was cooled in ice. The reaction time was 18 hr, the first hour under reduced pressure. See this volume [10].

Penicillinase

Penicillinase (B grade, 1360 Units/vial, 25 mg) was purchased from Calbiochem (San Diego, California). To 1 ml activated sedimented glass was added one vial (25 mg) of penicillinase dissolved in 3 ml of 0.1 M sodium phosphate buffer, pH 7.0.

Urease

Urease (jack bean, type III, 28 units/mg) was obtained from Sigma (St. Louis, Missouri). To 1 ml activated sedimented alkylamino glass were added 100 mg of urease dissolved in 3 ml of 0.1 M sodium phosphate buffer, pH 6.0.

Glucose Oxidase

Glucose oxidase (*Aspergillus niger*, type V, 200 units/g) was obtained from Société Rapidase (Seclin, France). To 1 ml of activated, sedimented glass beads were added 800 units of glucose oxidase dissolved in 1.5 ml of 0.1 M potassium phosphate buffer, pH 7.0.

Catalase

Catalase (type C-100 from beef liver, 30,000–40,000 U/mg) was purchased from Sigma. Activated, sedimented glass beads (0.5 ml) were mixed with 1.0 ml of 0.1 M potassium phosphate buffer pH 6.5 and 500 μl (230,000 U) of catalase suspension.

Results

Single Thermistor Measurements

Analysis of Penicillin G. The determination of penicillin G has been described previously.[4] Figure 2 shows the recorded temperature peaks (Δt) as a function of the amount of penicillin G in 1.5-ml sample pulses applied to the microcolumn. The linearity was excellent over a broad range of substrate concentrations. The shape of the curve at penicillin G amounts below 1.5 μmoles is probably due to interference by nonspecific heating effects, the influence of which would be relatively more pronounced at lower concentrations. The reproducibility over successive runs or between runs at 1 day's interval was better than $\pm 2\%$ of the recorded Δt.

In these experiments the lower limit for a reliable determination of penicillin G was below 0.25 mM.

Determination of Urea in Standard Solutions and in Serum. Figure 3 depicts the measured peak heights plotted against the urea concentration in 1.5-ml sample pulses. Again the response was linear over a broad range of substrate concentrations. The curvature found in the response curve above 20 mM is due to insufficient amount of urease being present in the microcolumn. This was verified by assaying the effluent from the enzyme thermistor for unreacted urea.

The sensitivity was high enough to permit determination of urea in

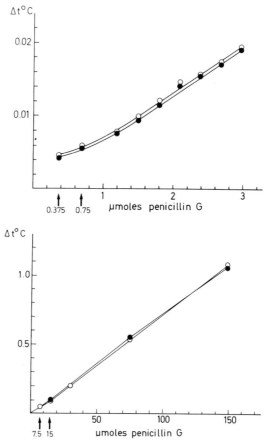

Fig. 2. Measured peak heights (Δt) as a function of the amount of penicillin G present in 1.5 ml pumped through a single thermistor unit containing glass-bound penicillinase at a flow-rate of 60 ml/hr. The values were obtained from two successive runs at one day's interval: ●, first day; ○, second day.

the normal physiological concentration range in human blood serum, even with samples diluted 4-fold with buffer. Figure 3 also shows the response (Δt) obtained after pumping different dilutions of a lyophilized standard serum (Dade, Division of American Hospital Supply Corporation, Miami, Florida) through the microcolumn of the enzyme thermistor. The response is linear, with good correlation between the value of urea concentration obtained from the standard curve and the stated value. Figure 3 shows, thirdly, the results obtained upon addition of known amounts of urea to standard serum diluted with 3 volumes of buffer. It

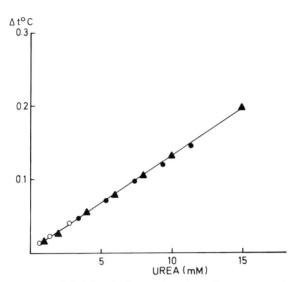

FIG. 3. Measured peak heights (Δt) as a function of urea concentration. Sample pulses of 1.5 ml were pumped through a urease-containing single thermistor unit with a flow rate of 60 ml/hr. -▲-▲-, Standard curve for urea dissolved in 0.1 M sodium phosphate pH 7.0. In the diagram are plotted the results obtained with standard serum diluted with buffer 8-, 4-, and 2-fold (○). Also plotted are the Δt values obtained for standard serum diluted 4-fold (the urea concentration in the standard serum was 5.6 mM) and with addition of urea to increase the concentration by 2, 4, 6, 8, and 10 mM (●).

is essential to use such calibration curves when assaying components in mixtures as complex as blood serum, in which nonspecific reactions are likely to occur.

This type of enzyme thermistor has also been tested with a larger series of human sera of known urea concentration. The correlation between these values and those obtained with the enzyme thermistor was good, and the reproducibility of measurements with the enzyme thermistor seems to be superior to that obtainable with clinical procedures normally employed.

Differential Thermistor Measurements

Glucose Determinations. One of the metabolites successfully studied with the differential enzyme thermistor shown in Fig. 1b is glucose.[5] Glucose can be determined in several ways by following the heat production from an enzymic reaction for which glucose directly or indirectly serves as a substrate. In one method, utilizing immobilized glucose oxidase in the microcolumn, the heat produced when glucose is oxidized to gluconic acid is measured. The temperature response obtained when pulses of glucose solutions of different strengths were pumped through such a column is shown in Fig. 4. The response curve falls off at glucose concentrations exceeding 0.5 mM, oxygen dissolved in the buffer limiting the reaction, but below 0.5 mM glucose the response is linear. Glucose concentrations can be reliably determined down to about 0.03 mM. This excellent performance extends to serum samples, in which glucose can be accurately determined in the physiological range, even at 100-fold dilution.

A precolumn of the glucose oxidase preparation can also be coupled in sequence with the enzyme thermistor, the column of which contains immobilized catalase. This arrangement provides a more general detector system for all reactions in which hydrogen peroxide is evolved. In addition in our experiments with this system, we have obtained a considerable extension of the useful range to higher concentrations of glucose.

Catalase enzyme thermistors have also been successfully utilized in combination with precolumns containing other oxidases, such as cholesterol oxidase and urate oxidase, permitting analysis of cholesterol and uric acid, respectively.[5] Lactose has been analyzed with lactase in a

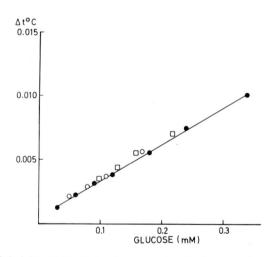

Fig. 4. Peak heights (Δt) obtained on pumping 1-ml pulses through a differential enzyme themistor with glucose oxidase at a flow rate of 60 ml/hr. -●-●-, Standard curve for glucose dissolved in 0.1 M potassium phosphate, pH 6.5. Standard serum (stated glucose concentration 4.9 mM) diluted with buffer 100 times (○) and 50 times (□) with addition of glucose to increase the concentration by 0.03, 0.06, and 0.12 mM.

precolumn linked to either of the systems for glucose analysis, in both standard solutions and skim milk.[5]

A third approach to glucose analysis is to mix preparations of glucose oxidase and catalase or to coimmobilize the two enzymes on the same matrix. This approach should reduce the problem of a limited supply of oxygen which would be regenerated in the reaction:

$$\text{Glucose} + O_2 \xrightarrow{\text{glucose oxidase}} \text{gluconic acid} + H_2O_2$$

$$H_2O_2 \xrightarrow{\text{catalase}} \tfrac{1}{2}O_2 + H_2O$$

(see also this volume chapter [32]). This is valid also for other oxidases producing hydrogen peroxide. A more thorough discussion of the various possibilities for arranging two or more enzymes to be used together in enzyme thermistor analysis has previously been published.[5] The use of amplifying multiple-step reactions is discussed in this volume [31].

Conclusion

The enzyme thermistor units described are fairly simple to build. They can be used in conjunction with other Wheatstone bridges or devices similar to the one used here, provided about the same resolution is obtained. Less sophisticated water baths can be employed. Thus, we have obtained base lines sufficiently good for substrate determinations in the concentrations range of 1 mM using a water bath with a temperature stability no better than ±0.1°. Current studies include a system employing a reference column without enzyme, placed parallel to the enzyme column, to eliminate the influence of unspecific heat due to e.g., solvation effects, pH-variations or interactions with the matrix.

Although the differential enzyme thermistor will give better baseline stability than the single thermistor unit, and therefore permit lower substrate concentrations to be determined, single thermistor devices of the type described here may be preferable because of their simplicity in cases where the demand for sensitivity is less.

In the examples given, the same flow rate, 60 ml/hr, was employed, giving a sample size of 1 ml for a 60-sec pulse. It is advantageous to use an excess of enzymic activity to ensure complete reaction, in order to obtain a linear relationship between the amount of substrate and the temperature response. In addition it will allow prolonged use of the enzyme-thermistor units since possible denaturation of the enzyme is compensated for during a longer period of time. Generally the enzyme column can be used for several weeks without any noticeable reduction in response. The sample sizes applied here are large, leading to near equilibrium (steady-state) conditions, with nearly the same temperature response as for a continuous sample flow through the column. Obviously smaller sample volumes (under non-steady-state conditions) can be employed, although the temperature response will be reduced. It should be pointed out that care must always be taken to use, within a series of determinations, the same flow rate, pH, sample size, and composition of the solution (cf. this volume [44]).

In the experiments described the average duration of an analysis was about 5 min, after which the original base line was obtained; however, a rate of analysis of one sample per minute is a realistic proposition. The average precision obtained of $\pm 2\%$ is sufficient for the demands in clinical analysis. Finally, it is worth mentioning that continuous processes may also be monitored with this type of unit.

[46] Methods for the Therapeutic Applications of Immobilized Enzymes

By THOMAS MING SWI CHANG

The potential of immobilized enzymes in the general biomedical area includes possible applications in therapy, clinical analysis, preventive and environmental medicine, and biochemical and biophysical research. These have been reviewed in detail elsewhere.[1] This chapter will deal more specifically with the methods of experimentation in animals for the therapeutic applications of immobilized enzymes carried out in this laboratory since 1957, and also the exciting work of other laboratories which has been published since 1970. Emphasis will be on the experimental methods and procedures for testing the *in vivo* therapeutic feasibility of immobilized enzymes, including some typical examples of the dersurface of the diaphram. Those that are small, for example, flexible red

At present, enzymes are available in large amounts, mostly from heterogeneous sources. *In vivo* introduction may therefore result in possible hypersensitivity reactions, immunological reactions, and rapid inactivation and removal. Furthermore, enzyme in free solution is often not stable, especially at a body temperature of 37°. Studies were first started in this laboratory to investigate the therapeutic applications of immobilized enzymes mostly in the form of microencapsulated enzymes. Immobilized enzymes have been classified into four major types: ad-

[1] T. M. S. Chang, "Biomedical Applications of Immobilized Enzymes and Proteins." Plenum, New York, in press.

sorbed, covalently linked, matrix-entrapped, and microencapsulated[2] (see also this volume [16]). In more recent years, extensive studies have been carried out to investigate in animal studies all these different types of immobilized enzymes.

Routes of *in Vivo* Administration (Table I)

It is obviously impossible to include all the detailed methods of experimentation. Thus, a general introduction will be given on the various possible routes of *in vivo* administration of immobilized enzymes. This will be followed by typical detailed examples of experimental therapy.

Implantation of Immobilized Enzymes by Local Injections

Immobilized enzymes in the form of suspensions could be implanted by injection intramuscularily, subcutaneously, intraperitoneally, or elsewhere. The decision for the sites of injection would depend on the total amount of immobilized enzymes injected and the availability of the substrate for the immobilized enzymes at the site of implantation. The equilibration of substrates at the subcutaneous site with other body compartments is rather slow. Much faster exchange of substrates is possible at the intramuscular sites. Many substrates can exchange quite rapidly between the circulating blood and the peritoneal cavity. Particulates, when introduced into the peritoneal cavity, drift slowly to the undersurface of the diaphram. Those that are small, for example, flexible red blood cells of 7–8 μm, pass through the right lymphatic duct and into the systemic circulation. Large particulates (>20 μm), especially if they are rigid, do not pass through the lymphatic system but stay in the peritoneal cavity. With time, they would be surrounded by phagocytes and, depending on the chemical properties of the particles, they might persist in this state for many months. Compared to the other sites of local injection, a much larger amount of immobilized enzymes could be introduced into the peritoneal cavity. However, one drawback is that intraperitoneal introduction of foreign material carries a greater risk than subcutaneous and intramuscular injections. For animal experimentation, however, this is a convenient route of administration.[3-5]

[2] P. V. Sundarum, E. K. Pye, T. M. S. Chang, V. H. Edwards, A. E. Humphrey, N. O. Kaplan, E. Katchalski, Y. Levin, M. D. Lilly, G. Manecke, K. Mosbach, A. Patchornik, J. Porath, H. Weetall, and L. B. Wingard, Jr., *Biotechnol. Bioeng.* **14**, 15 (1972).
[3] T. M. S. Chang, *Science* **146**, 524 (1964).
[4] T. M. S. Chang, F. C. MacIntosh, and S. G. Mason, *Can. J. Physiol. Pharmacol.* **44**, 115 (1966).
[5] T. M. S. Chang, "Artificial Cells." Thomas, Springfield, Illinois, 1972.

TABLE I
IMMOBILIZED ENZYMES: ROUTES OF ADMINISTRATION[a]

Implantation by Injection
Chang (1964, 1966, 1969, 1971, 1972, 1973)[3,5,9-11,18]
Chang and MacIntosh (1964)[6]
Chang and Poznansky (1968)[7]
Updike, Prieve, and Magnuson (1973)[13]
Poznansky and Chang (1974)[8]
Siu Chong and Chang (1975)[12]
O'Driscoll, Korus, Ohnuma, and Waxczack (1975)[14]
Intravenous Injection
Chang (1964, 1966, 1972)[3,5,18]
Sekiguchi and Kondo (1966)[53]
Gregoriadis and Ryman (1972, 1974)[16,17]
Updike, Prieve, and Magnuson (1973)[13]
Extracorporeal Perfusion
Chang (1966, 1972, 1973, 1975)[5,11,18,46]
Chang, Pont, Johnson, and Malave (1968)[19]
Chang and Poznansky (1968)[7]
Hydén (1971)[20]
Apple (1971)[27]
Sampson, Hersh, Cooney, and Murphy (1972)[22]
Allison, Davidson, Gutierrez-Hartman, and Kitto (1972)[24]
Horvath, Sardi, and Woods (1973)[25]
Hersh (1974)[23]
Salmona, Saronio, Bartosek, Vecchi, and Mussini (1974)[28]
Venkatasubramanian, Vieth, and Bernath (1975)[26]
Venter, Venter, Dixon, and Kaplan (1975)[21]
Gastrointestinal Administration
Chang and Poznansky (1968)[7]
Chang and Loa (1970)[30]
Gardner, Falb, Kim, and Emmerling (1971)[31]
Chang (1972)[5]
Local Applications
Chang (1972)[32]

[a] Superscript numbers following year refer to text footnotes that give full reference.

Biocompatibility of materials varies over a wide range. It is thus impossible to make any generalization for all types of immobilized enzymes. Those that are less biocompatible may illicit acute adverse reactions and make any valid *in vivo* testing impossible. However, it has been demonstrated that when nylon or collodion microencapsulated enzymes were injected intraperitoneally, they did not produce any acute adverse effects in the animals after intraperitoneal injection. Furthermore, a large volume of more than 40 ml of immobilized enzymes per kilogram body weight can be injected intraperitoneally. This approach of intraperitoneal injection is a convenient route in animal experimentation for estimating and testing the *in vivo* action of immobilized enzymes to act on substrate that is present freely in the body fluid and can equilibrate across the peritoneal cavity to be acted on by the immobilized enzymes. As will be discussed later, the *in vivo* action of immobilized enzymes was tested by the intraperitoneal administration of immobilized urease,[3,5,6] catalase,[7,8] and asparaginase.[9-14]

Methods for Studying the Effects of Intravenous or Intra-arterial Injections

Another way of introducing immobilized enzymes into the body is by intravenous injection or intra-arterial injection. This is a more complicated approach than the intraperitoneal route of injection, since small particles of narrow size distribution are required. The effect, site, and fate of intravenously injected immobilized enzymes were first tested in the form of microencapsulated red blood cell hemolysate.[3,5,15] Recently, another type of microencapsulated enzyme system in the form of liposomes containing enzymes has been injected by the intravenous route.[16,17]

Extracorporeal Perfusion by Blood or Body Fluid

In the implantation of immobilized enzymes by injection or the introduction of immobilized enzymes into the bloodstream to be localized at target organs, one is introducing a foreign material into the body. The introduction of immobilized enzymes into the body can be avoided by the use of an extracorporeal perfusion system if the main function of the enzymes is to act on a substrate that is present in the body fluid.[18] A large number of approaches are possible, but fundamentally, immobilized enzymes are placed in an enclosed system through which blood or body fluid is allowed to circulate continuously to come in contact with immobilized enzymes before returning to the body. Thus, the recirculating blood or body fluid would carry the substrates to the immobilized

[6] T. M. S. Chang and F. C. MacIntosh, *Pharmacologist* **6**, 198 (1964).
[7] T. M. S. Chang and M. J. Poznansky, *Nature (London)* **218**, 243 (1968).
[8] M. J. Poznansky and T. M. S. Chang, *Biochim. Biophys. Acta* **334**, 103 (1974).
[9] T. M. S. Chang, *Proc. Can. Fed. Biol. Sci.* **12**, 62 (1969).
[10] T. M. S. Chang, *Nature (London)* **229**, 117 (1971).
[11] T. M. S. Chang, *Enzyme J.* **14**, 95 (1973).
[12] E. Siu Chong and T. M. S. Chang, *Enzyme J.* **18**, 218 (1974).
[13] S. Updike, C. Prieve, and J. Magnuson, *Birth Defects, Orig. Art. Ser.* **9**, 77, (1973).
[14] K. F. O'Driscoll, R. A. Korus, T. Ohnuma, and I. M. Waxczack, *J. Pharmacol. Exp. Ther.* **195**, 382 (1975).
[15] T. M. S. Chang, Semipermeable Aqueous Microcapsules. Ph.D. Thesis, McGill University, Montreal, Canada, 1965.
[16] G. Gregoriadis and B. E. Ryman, *Eur. J. Biochem.* **24**, 485 (1972).
[17] G. Gregoriadis and B. E. Ryman, *Biochem. J.* **129**, 123 (1974).
[18] T. M. S. Chang, *Trans. Am. Soc. Artif. Intern. Organs* **12**, 13 (1966).

enzymes to be acted on by the enzymes. The advantage of this approach is that the immobilized enzymes are not introduced into the body and could be replaced at any time.[18] The first extracorporeal immobilized enzyme system tested *in vivo* was in the form of a microencapsulated urease system to remove blood urea.[18] Microencapsulated enzymes with heparin attached to the microcapsule membranes, shunt chambers, and tubings have also been used to prevent clotting of blood and avoid the necessity of using systemic heparinization.[19] Peritoneal dialyzing fluid was also used to recirculate the extracorporeal shunt containing microencapsulated catalase.[7] This way, substrate diffusing across the peritoneal cavity can be acted on by the microencapsulated enzyme. With the demonstration of the feasibility of using extracorporeal immobilized enzymes in the form of microencapsulated enzymes and with the demonstration that this approach would prevent introduction and accumulation of foreign material in the body,[18] a large number of other types of extracorporeal immobilized enzyme systems have been designed and tested for *in vivo* applications: for instance, enzymes immobilized in glass[20,21]; polymethacrylic plates[22,23]; nylon tubings[24,25]; multipore collagen spiral systems[26]; dialysis systems[27]; fibers[28]; and combined artificial cell–synthetic capillary systems.[29] For animal experimentation, the extracorporeal approach is more complicated than the intraperitoneal route of administration.

Immobilized Enzymes to Act in the Gastrointestinal Tract

The gastrointestinal tract is another area in the body where there is good exchanges of many metabolites and substrates from the body.

[19] T. M. S. Chang, A. Pont, L. J. Johnson, and N. Malave, *Trans. Am. Soc. Artif. Intern. Organs* **14**, 163 (1968).
[20] H. Hydén, *Arzeneim.-Forsch.* **21**, 1671 (1971).
[21] J. C. Venter, B. R. Venter, J. E. Dixon, and N. O. Kaplan, *Biochem. Med.* **12**, 79 (1975).
[22] D. Sampson, L. S. Hersh, D. Cooney, and G. P. Murphy, *Trans. Am. Soc. Artif. Intern. Organs* **18**, 54 (1972).
[23] L. S. Hersh, *J. Polym Sci.* **547**, 55 (1974).
[24] J. P. Allison, L. Davidson, A. Gutierrez-Hartman, and G. B. Kitto, *Biochem. Biophys. Res. Commun.* **47**, 66 (1972).
[25] C. Horvath, A. Sardi, and J. S. Woods, *J. Appl. Physiol.* **34**, 181 (1973).
[26] K. Venkatasubramanian, W. R. Vieth, and F. R. Bernath, *in* "Enzyme Engineering" (K. Pye and L. Wingard, Jr., eds.), Vol. 2, p. 439. Plenum, New York, 1975.
[27] M. Apple, *Proc. West. Pharmacol. Soc.* **14**, 125 (1971).
[28] M. Salmona, C. Saronio, I. Bartosek, A. Vecchi, and E. Mussini, *in* "Insolubilized Enzymes" (M. Salmona, C. Saronio, and S. Garattini, eds.), p. 189. Raven, New York, 1974.
[29] T. M. S. Chang, *Biomed. Eng. J.* **8**, 334 (1973).

The first immobilized enzyme tested by gastrointestinal administration was in the form of microencapsulated urease either by direct introduction into the intestine or by oral administration into animals.[5,7,30,31] However, this is limited to a few cases where the substrates can equilibrate rapidly across the intestinal tract and are present in the GI tract.

Local Application

In some cases, it might be undesirable to apply enzyme repeatedly in free solution to local lesions or tissues, since the application of free enzyme may result in the adsorption of the enzyme into the body, resulting in immunological and hypersensitivity reactions. Here, immobilized enzyme may be applied directly to the lesion. This has been tested using microencapsulated catalase.[32] The detailed method of a typical example will be described in a later section.

Immobilized Urease as a Model for *in Vivo* Therapeutic Action

In this study, microencapsulated urease was used as a model system since its substrate, urea, is a major endogenous constitute of the body fluid.[5,6,15] It was thought that this could be used as a model system for a feasibility study for the therapeutic applications of immobilized enzymes. Urease was microencapsulated by the method described in the chapter on microencapsulation (this volume [16]). *In vitro* studies show that microencapsulated urease did not leak out, but acted effectively on urea diffusing into the microcapsules, converting urea to ammonia. Because of the large concentration of urea in the biological fluid and the easy diffusibility of urea, the study of injection of immobilized urease *in vivo* to see its effect on the urea was carried out (see Table II).

Methods of Intraperitoneal Introduction

Acute experiments were carried out on male dogs weighing about 10 kg and anesthetized with intravenous sodium phenobarbital (25 mg/kg). Carotid blood pressure, respiratory rate, and electrocardiograph were recorded continuously on a Gilson polygraph. The femoral vein was cannulated for obtaining blood samples. Three control blood samples were taken at hourly intervals. This was then followed by an intraperitoneal

---

[30] T. M. S. Chang and S. K. Loa, *Physiologist* **13**, 70 (1970).
[31] D. L. Gardner, R. D. Falb, B. C. Kim, and D. C. Emmerling, *Trans. Am. Soc. Artif. Intern. Organs* **17**, 239 (1971).
[32] T. M. S. Chang, *J. Dent. Res.* **2**, 319 (1972).

TABLE II
Immobilized Enzymes: Experimental Therapy[a]

Red Blood Cell Substitutes
  Chang (1964, 1966, 1972)[3,5,18]
  Sekiguchi and Kondo (1966)[53]
Model System for Enzyme Therapy
  Chang and MacIntosh (1964)[6]
  Chang (1964, 1966, 1972)[3,5,18]
Congenital Enzyme Deficiency
  Chang and Poznansky (1968)[7]
  Chang (1972)[5,32]
  Poznansky and Chang (1974)[8]
  Gregoriadis and Ryman (1972)[16]
Substrate-Dependent Tumour
  Chang (1969, 1971, 1972, 1973)[5,9-11]
  Chang, Pont, Johnson and Malave (1968)[19]
  Hydén (1971)[20]
  Sampson, Hersh, Cooney, and Murphy (1972)[22]
  Allison, Davidson, Gutierrez-Hartman, and Kitto (1972)[24]
  Horvath, Sardi, and Woods (1973)[25]
  Mori, Sato, Matuo, Tosa, and Chibata (1972, 1973)[40,41]
  Siu Chong and Chang (1974)[12]
  Hersh (1974)[23]
  Salmona, Saronio, Bartosek, Vecchi, and Mussini (1974)[28]
  Venkatasubramanian, Vieth, and Bernath (1975)[26]
  Cooney, Weetall, and Long (1975)[42]
  O'Driscoll, Korus, Ohnuma and Waxczack (1975)[14]
Organ Failure and Metabolic Disorders
  Chang (1966, 1972, 1975)[5,18,46]
  Levine and LaCourse (1967)[43]
  Gordon, Greenbaum, Marantz, McArthur, and Maxwell (1969)[44]
  Chang and Loa (1970)[30]
  Broun, Selegny, Minh, and Thomas (1970)[48]
  Gardner, Falb, Kim, and Emmerling (1971)[31]
  Kusserow, Larrow, and Nichols (1971, 1973)[51,52]
  May and Li (1972)[45]
  Updike, Shults, and Magnuson (1973)[49]
  Bessman and Schultz (1973)[50]
  Venter, Venter, Dixon, and Kaplan (1975)[21]

[a] Superscript numbers following year refer to text footnotes that give full reference.

injection of 2.5 ml/kg of a 10% suspension of 29 μm mean diameter nylon microcapsules containing only hemolysate prepared as described. Three hourly blood samples were then taken. After this, each dog was given an intraperitoneal injection of 2.5 ml/kg of a 10% suspension of 29 μm mean diameter nylon microcapsules containing hemolysate and urease prepared as described.[3-5] The enzyme injected is equivalent to an assayed activity of 100 Sumner units (SU) per kilogram. After this, three hourly blood samples were taken. For long-term studies, blood samples for ammonia estimation were taken from a leg vein before and after the intraperitoneal injection of 5 ml/kg of a 10% suspension of nylon microcapsules containing hemolysate only, and further blood was taken twice a day for 2 days. The test was repeated a week later, but on this occasion the administered microcapsules contained NBC soluble urease (100 SU per milliliter of microcapsules) in addition to hemoglobin; the dose was again 5 ml/kg of a 10% suspension corresponding to an enzyme dosage of 50 SU/kg. The blood ammonia level was followed for 6 days after the injection and the behavior of the animal was closely observed.

As a test for leakage of protein from the injected microcapsules, nylon microcapsules containing ^{51}Cr-tagged hemoglobin were injected intraperitoneally at a dosage of 2.5 ml/kg of a 10% suspension. Four venous samples were taken at hourly intervals and tested for radioactivity; at the end of the experiment, the liver, spleen and kidney were removed for counting. At the same time, the abdominal cavity was irrigated with saline, and the recovered irrigation fluid was also tested for radioactivity. In these studies, intraperitoneal injection of controlled microcapsules containing no urease produced no significant changes in the blood level of ammonia. However, intraperitoneal injection of microencapsulated urease resulted in a rise of the blood ammonia level as urea in the body is converted enzymically to ammonia. Since ^{51}Cr-labeled hemoglobin did not leak out of the microcapsules after intraperitoneal injection, it is likely that the much larger urease molecules would not leak out from the microcapsules after intraperitoneal injection. The microencapsulated urease thus acts on urea diffusing into the peritoneal cavity and then into the microcapsules. In the long-term studies, the ammonia level increased to a height and then fell, with a half-time of about 2.5 days. The half-time of the microencapsulated urease stored *in vitro* at 37° is 7 days.

The results of these studies show conclusively that the model system of microencapsulated urease, when introduced intraperitoneally, can act effectively on the model metabolite, urea, converting it into ammonia. These results lead to the suggestion that immobilized enzymes could be used for enzyme replacement therapy, since in this form the enzyme can act on external permeant substrates, but does leak out to become involved in immunological reactions.[5,6,15]

Methods of Extracorporeal Perfusion

Ten milliliters of microencapsulated urease were retained in an extracorporeal shunt chamber. When the blood of dogs recirculated through the extracorporeal shunt to come in direct contact with the microencapsulated urease, blood urea was effectively converted into ammonia, and the systemic blood urea level fell with a $T_{1/2}$ of 60 min.[18] To prevent the need for systemic heparinization, the microencapsulated enzymes, shunt chamber, and connecting tubing were complexed with heparin so that blood could recirculate without clotting.[19] In this heparinized form, microencapsulated urease still acts effectively in converting urea in the blood to ammonia.[19]

Summary

Thus, it was demonstrated that the model enzyme urease immobilized by microencapsulation can act effectively in converting the metabolite urea in the body into its product, ammonia. At the same time, the enzyme remains in the immobilized state without leaking out. This basic study using the model enzyme urease has led to further studies in this and other laboratories in actual experimental enzyme therapy for more specific disorders.

Immobilized Enzyme Systems for Hereditary Enzyme Deficiencies

One line of research in this laboratory involves the use of immobilized enzymes for experimental replacement in congenital enzyme deficiencies.[7,8]

Example of a Method of Enzyme Replacement by Implantation

The first experiment was carried out in this laboratory with the Feinstein strain of mice with hereditary deficiency in catalase (acatalasemia). Acatalasemia is a condition that is present also in men. These acatalasemic mice (C_s^b) have a blood catalase activity approximately 2% normal and a total body catalase activity approximately 20% normal. The detailed methods for the *in vivo* enzyme replacement therapy experiments were carried out using catalase immobilized by microencapsulation within collodion membranes as follows.

Crystalline lyophilized beef catalase with an activity of 3000 units/mg was purchased from the Nutritional Biochemical Company. Semipermeable microcapsules were prepared by the updated method[5] of the original method for the preparation of collodion microcapsules.[3,4] 300 mg of catalase were dissolved in 6 ml of Tris-buffered hemoglobin solution containing per 100 ml 180 mg of 2-amino-2-(hydroxymethyl)-1,3-propanedial and 10 g of hemoglobin (Worthington). The solution was filtered with Whatman No. 42 to remove any undissolved particles. This final solution was microencapsulated following the standard procedure of interfacial coacervation to make microcapsules with a mean diameter of 80 μm.[3-5] The prepared microcapsules were washed in cold 0.9% saline until a completely clear supernatant was attained with no trace of catalase activity.

The method of Feinstein[33] using $NaBO_3$ as substrate was used for determining catalase activity (expressed as perborate units per milliliter of blood or per gram body weight). Eight milliliters of 1.5% $NaBO_3 \cdot 4H_2O$ previously adjusted to pH 6.8 with concentrated HCl are added to a flask containing 1.5 ml of 0.067 M phosphate buffer, pH 6.8. After 20 min of incubation at 37°, 0.5 ml of each enzyme source blank or control is then added and mixed by shaking. After exactly 5 min, 10 ml of 1 M H_2SO_4 are added and the flasks are titrated with 0.05 M $KMnO_4$. The results are expressed as perborate units per milliliter or per gram of enzyme source.

Acatalasemic $C_s{}^b$ mice and normal $C_s{}^a$ mice were obtained from Dr. R. Feinstein at the Argonne National Laboratory, Chicago, Illinois, and bred in this laboratory. For the determination of blood catalase levels, blood drawn from the orbital sinus was diluted in cold saline and assayed as above. Total body catalase was measured by blending the carcass of each mouse in four parts of cold saline in a Waring Blender and then filtering it through gauze. The filtrate was diluted with saline and assayed as above for catalase activity. Total body $NaBO_3$ was measured by blending the carcass of each mouse in four parts of 0.05 M H_2SO_4. After filtering through gauze, the filtrate was precipitated with 10% trichloroacetic acid, and the supernatant was assayed for $NaBO_3$.

Instead of using $KMnO_4$ titration for the determination of the total body $NaBO_3$ as described in the earlier study,[7] a more sensitive colorimetric method[34] was modified slightly and used in the present experiments. The procedure was as follows: each 4 ml of standard $NaBO_3$ solution was mixed with 3 ml of saturated $Ti(SO_4)_2$ and read on a Bausch and Lomb colorimeter at 420 nm in order to get a standard curve. Four milliliters of the supernatant for the assay of total body $NaBO_3$ was added to 3 ml of saturated $Ti(SO_4)_2$ and read on the colorimeter at 420 nm to determine the $NaBO_3$ concentration.

Acatalasemic mice $C_s{}^b$, normal mice $C_s{}^a$, and rabbits were immunized according to the approach of Feinstein et al.[35] over a period of 90 days with the modification that, in the present study, crystalline lyophilized

[33] R. Feinstein, *J. Biol. Chem.* **180**, 1197 (1949).
[34] P. Bonet-Maury, *C. R. Soc. Biol.* **135**, 941 (1941).
[35] R. Feinstein, H. Suter, and B. N. Jaroslow, *Science* **159**, 638 (1968).

beef catalase was used. Injections, each consisting of Freund's adjuvant and catalase solution (0.1 ml containing 4.8 perborate units of catalase), were given. Another group of acatalasemic mice received repeated injections of microencapsulated catalase with the same dosage, schedule, and duration as above. The Ouchterlony agar gel double-diffusion technique[36] for antigen–antibody precipitation reactions was used to detect the presence and concentration of antibodies.

In *in vivo* studies of nonimmunized mice, each animal received one of the following intraperitoneal injections: (1) saline as control; (2) catalase in free solution (12 perborate units per gram body weight); or (3) catalase-loaded microcapsules (assayed activity of 12 perborate units per gram body weight). This is followed by a subcutaneous injection of 0.014 mM NaBO$_3$ per gram body weight. At predetermined intervals, the animals were sacrificed and assayed for total body NaBO$_3$ as described above. For the *in vivo* immunological studies, in the first series of studies, the same intraperitoneal injections as described above were followed. In the second series of studies, each of the intraperitoneal injections contained only one-third of the amount of catalase used for the first series.

The following results were obtained[7,8]: *In vitro* studies show that the K_m for both forms of catalase is 0.55 M. The V of the microencapsulated catalase is 0.5 mM/min, and that of catalase in free solution is 2.5 mM/min. *In vitro* at 4°, the $T_{1/2}$ of soluble catalase was about 15 days; that of catalase microencapsulated with 10 gm/100 ml hemoglobin remained at 90% after 45 days. At 37°, the $T_{1/2}$ of soluble catalase was 1 day; microencapsulated catalase with hemoglobin treated with glutaraldehyde was 6 days. The *in vivo* enzyme kinetics were studied using Feinstein's acatalasemic mice, which have a hereditary deficiency in catalase. These *in vivo* studies show that microencapsulated catalase is as effective as the same assayed amount of catalase in free solution in reducing total body substrates in the form of exogenous perborates. Both types of catalase were effective in protecting the enzyme-deficient animals from the adverse effects of perborate. After injection, microencapsulated catalase with an *in vivo* half-life of 4.4 days is more stable than catalase in free solution, which has an *in vivo* half-life of 2.0 days. *In vitro* studies show that the enclosing membranes of semipermeable microcapsules are impermeable to beef catalase antibodies. Repeated injection of beef catalase solution into acatalasemic mice sensitized the mice to a large dose of the enzyme in solution, but not to the same dose of enzyme immobilized within semipermeable microcapsules. Injection

[36] O. Ouchterlony, *in* "Handbook of Experimental Immunology" (D. M. Weir, ed.), p. 655. Blackwell, Oxford.

of catalase in free solution is removed very rapidly from the circulation of acatalasemic mice immunized to catalase. Catalase in free solution was not effective in reducing total body sodium perborate levels in immunized mice, while microencapsulated catalase acted efficiently when injected into the immunized mice.

Thus, in this model enzyme system, it was demonstrated that immobilized enzyme systems could be used effectively for the replacement of enzyme deficiency conditions. Unlike enzyme in free solution, immobilized enzyme did not cause any immunological and hypersensitivity reactions. In practice, however, one would have to look at each individual situation, since in some situations one may not want to inject immobilized enzymes into the body and in other cases immobilized enzymes may not merely have to act on body fluid, but to be located at specific sites. The following cases are typical examples for solving these specific requirements for practical applications.

Example of Method of Replacement by Extracorporeal Perfusion

The first example involves study carried out to extend the enzyme replacement using an extracorporeal shunt containing microencapsulated catalase.[7] In these studies, microencapsulated catalase was retained in a small extracorporeal shunt chamber. A small volume of fluid was injected intraperitoneally, and the fluid was allowed to recirculate through the shunt. In this way, the extracorporeal shunt chamber containing the microencapsulated catalase was effective in lowering the perborate injected into the body.

Example of Method of Replacement by Local Applications

The second example demonstrates another specific way in which immobilized enzymes can be applied to specific sites by local application. In human acatalasemia, the major problem is related to oral lesions resulting from hydrogen peroxide produced by bacteria, which causes local anoxia and gangrene formation. Repeated direct application of a catalase solution on an oral lesion may result in absorption and allergic immunological reactions. Furthermore, the enzymes may not be sufficiently stable to last for a sufficient length of time. Catalase immobilized by microencapsulation has been tested experimentally in acatalasemic and normal mice.[32] The blood catalase levels of the acatalasemic mice were measured, and only those with a level of 3–4 perborate units per milliliter were used. The mice were allotted to three groups: normal mice to group 1, and acatalasemic mice groups 2 and 3.

Artificial cells containing microencapsulated catalase were prepared as previously described. A paste was made by removing all the supernatant from the final preparation.

The mice were anesthetized with ether. A small lesion was made with a sharp instrument in the gingiva of each mouse, and 5 µl of different concentrations of H_2O_2 (0.1, 0.2, 0.4, 0.8, 1.6, 3.2, and 6.4%) were applied locally to the lesions. The efficiency of the oxidation of H_2O_2 was judged in terms of the concentration at which sufficient oxygen gas was produced to result in observable foaming. In group 3, just before the application of H_2O_2, 2µl of the microcapsule paste was applied to each of the lesions.

In the normal mice (group 1), sufficient O_2 was produced to result in observable foaming at a H_2O_2 concentration of 0.8%. In the untreated acatalasemic group (group 2), no foaming was observed at a H_2O_2 concentration of up to 6%; in addition, brownish discoloration was observed. However, in acatalasemic mice treated locally with artificial cells containing catalase (group 3), sufficient O_2 was produced to result in observable foaming at a concentration of 1.6%, and no local brownish discoloration was observed.

In a preliminary study in man, the paste was applied locally to different areas of the gingiva.[32] There was no local irritation or other observable adverse effect. It continued to act efficiently on H_2O_2 applied to the gingiva.

Example of Method of Intravenous Injection

In a number of situations, the immobilized enzymes may have to be specifically located at certain intracellular sites. Studies here have demonstrated that nylon microcapsules less than 5 µm in diameter are trapped mostly by the lung capillaries. However, if the surface properties are changed by sulfonating the nylon microcapsules, then most microcapsules are trapped by the liver, and most likely phagocytosed by the liver cells.[3,5,15]

Another system requires the use of liposomes containing enzymes. The fate and site of final location of this system was examined using liposome-entrapped ^{131}I-labeled albumin and ^{3}H-labeled cholesterol. In this study, it was found that liposomes with the enclosed protein were removed rapidly from the circulation and were retained mainly in the liver and, to a smaller extent, in the spleen. More detailed analysis[17] showed that the liposomes containing the enzymes are taken up by the hepatic cells, presumably in the form of pinocytic vacuoles, and that they thus finally enters the lysosomes. Once inside, disruption of liposomes releases the entrapped enzyme. In this study, therefore, the liposomes were found to locate very specifically in the lysosomes, and this, in fact, would allow and facilitate the ability to send immobilized enzymes to the subcellular target of lysosomes, where many of the storage diseases are located. As with semipermeable microcapsules, the effect of changing the surface property of the liposomes on their eventual location in specific organs is also being explored.[17]

Experimental Treatment of Substrate-Dependent Tumors

L-Asparaginase suppresses the growth of certain types of asparagine-dependent lymphomas,[37] presumably by depleting the extracellular supply of asparagine. Asparaginase obtained from bacterial sources is removed rapidly after injection. In addition, with repeated injections, it causes hypersensitivity and immunological reactions. The immunological reaction would also render the injected enzyme ineffective. In the last few years, a number of laboratories have been using different forms of immobilized asparaginase in feasibility studies for therapeutic applications. A detailed example of the first animal experimentation carried out is given below, to be followed by a brief summary of other methods.

Example of Method of Implantation by Injection

The first *in vivo* experiment was tested here in which microencapsulated asparaginase injected intraperitoneally into mice was shown to act effectively to remove asparagine.[19] After this, experiments were carried out here to study the use of immobilized asparaginase in experimental suppression of substrate-dependent tumors in animals.[9,10] In these studies, L-asparaginase was microencapsulated within semipermeable microcapsules.

L-Asparaginase obtained from Sigma Company and Nutritional Biochemical were microencapsulated with collodion or nylon[9,10] microcapsules, as described in the chapter on microencapsulation [16].

C3H/HeJ mice were obtained from Jackson Laboratories, Bar Harbor, Maine. In the *in vivo* studies, four groups of mice were used. As control, each mouse in one group received an intraperitoneal injection of 1 ml of saline. As a further control, each mouse in the second group received an intraperitoneal injection of 1 ml of a suspension of control microcapsules. Each mouse in the third group received an intraperitoneal injection of 1 ml of saline containing 5–10 IU of asparaginase. Each mouse in the fourth group received an intraperitoneal injection of micro-

[37] J. D. Broome, *Nature (London)* **191**, 114 (1961).

encapsulated asparaginase with an assayed activity of 5–10 IU. Body asparaginase, blood asparagine, and asparaginase were then followed.

Blood L-*Asparaginase.* Blood was drawn into fire-polished, heparinized Pasteur pipettes from the subclavian artery of ether-anesthetized mice. The blood was then transferred into small containers, which were immediately stored at −20° until ready for assay in the next few hours.

Body L-*Asparaginase.* The mice were exsanguinated, then the skins were removed and the livers were dissected. The remaining portions of the animals were homogenized 1:4 (w/v) in ice-cold saline using a Waring Blender. The homogenates were filtered through triple-layered cotton gauze, and the filtrate was immediately kept at −20° until ready for assay. Any subsequent mention of the amount of L-asparaginase activity remaining in the "body" after enzyme injection refers to the enzyme activity assayed in the homogenates prepared as described here.

Assays of L-asparaginase activity were carried out using a slight modification of the automated continuous-flow method developed by Schwartz *et al.*[38]

Preparation of plasma for L-asparagine assay is carried out as follows: Freshly drawn blood was centrifuged at 1000 g for 10 min. The plasma was removed and added to a tube containing 100 g of trichloroacetic acid (TCA) per liter plasma to 1 volume of TCA. The contents of the tube were mixed well with a vortex mixer and then centrifuged at 1000 g for 20 min. The supernatant was extracted 5 times with ether. After the final extraction and removal of ether, any residual ether in the supernatant was removed by vacuum.

L-Asparagine assay was performed according to the fluorimetric technique of Cooney *et al.*,[39] which is based on the enzymically coupled oxidation of reduced pyridine nucleotide. In our case, four solutions were sequentially prepared as follows: (a) 2.5 mg of NADH (Calbiochem, Los Angeles, California) was dissolved in 100 ml of Tris buffer (5 mM pH 8); (b) 10 ml of the solution prepared in (a) was diluted with 290 ml of the same Tris buffer; (c) 4 mg of α-ketoglutaric acid (Sigma, St. Louis, Missouri) was added to 250 ml of the solution prepared in (b); (d) 52 IU of glutamic-oxaloacetic transaminase (GOT; porcine heart; Sigma) and 126 IU of malic dehydrogenase (MDH; porcine heart; Calbiochem) was added to the solution prepared in (c). Of the reaction mixture prepared in (d), 4 ml were pipetted into glass fluorometer tubes (12 × 75 mm) (G. K. Turner Associates, Palo Alto, California);

[38] M. K. Schwartz, E. D. Lash, H. F. Oettgen, and F. A. Tomao, *Cancer* (Philadelphia) **25**, 244 (1970).
[39] D. A. Cooney, R. L. Capizzi, and R. E. Handschumacher, *Cancer Res.* **30**, 929 (1970).

0.05 ml of the sample was added to the reaction mixture. After incubation at 22° for 30 min, a reading was taken on the fluorometer (Turner, Model III). After this, 0.1 IU of L-asparaginase was added to the reaction mixture, which was incubated at 22° for 1 hr and then read on the fluorometer.

Experimental treatment of implanted 6C3HED lymphosarcoma in C3H/HEJ mice (Jackson Laboratory) was studied. Into each groin, 500,000 lymphosarcoma cells were implanted. Each mouse was then given one of the following intraperitoneal injections: 1 ml/25 g body weight of saline as a control; 1 ml/25 g body weight of a 50% suspension of controlled microcapsules containing no asparaginase; 1 ml/25 g body weight of microcapsules containing 5 IU of L-asparaginase; and 1 ml/25 g body weight of L-asparaginase solution with the same assayed activity as that of the microencapsulated asparaginase. The animals were observed from the time the tumor first appeared. The results show that L-asparaginase solution given in the amounts stated significantly delayed the appearance of lymphosarcoma. Microencapsulated L-asparaginase suppressed the growth of implanted lymphosarcoma more effectively than L-asparaginase solution, even though the assayed activity of the enzyme in solution and in the microencapsulated form are the same.[9,10]

Asparaginase microencapsulated with hemoglobin was much more stable than the asparaginase in free solution at 4° and at body temperature of 37°. The stability at 37° could be further enhanced by treating the microencapsulated asparaginase with glutaraldehyde.[11] Injection of the microencapsulated form maintained a zero asparagine level three times longer than the same amount of enzyme in solution.[11] After injection of L-asparaginase solution, enzyme activity appeared in the blood very rapidly, the highest concentration occurring after 4 hr. The enzyme was then cleared from the circulation with a half-time of 4.4 hr.[12] In marked contrast, when microencapsulated L-asparaginase was injected intraperitoneally, no significant L-asparaginase activity appeared in the blood for the entire duration of the study,[12] showing that microencapsulated asparaginase did not leak out to any detectable level. The following results were obtained with the total asparaginase activity present in the body. After injection of L-asparaginase solution, the enzyme activity decreased rapidly in the body with a $T_{1/2}$ of 2 hr. On the other hand, after injection of microencapsulated L-asparaginase, the asparaginase level in the body decreased much more slowly, with a $T_{1/2}$ of 60–72 hr. The microencapsulated L-asparaginase in the body still retained about 20% of its original activity 16 days after injection. Plasma L-asparagine was maintained at zero concentration 4 days after injection of the enzyme solution, compared to 8 days after injection of the microencapsulated enzyme. Microencapsulated asparaginase in these studies was capable of causing regression of the tumor in the advanced, well-established stage.[12] The same authors found that, unlike microencapsulated catalase, microencapsulated asparaginase appeared to have antigenetic sites exposed on the microcapsule surface (unpublished data of E. Siu Chong and T. M. S. Chang). These antigenetic sites can be masked by cross-linking appropriate macromolecules on the surface (unpublished data).

Asparaginase microencapsulated by interfacial polymerization has also been carried out by another group.[40,41] They found that the membranes of the microcapsules were resistant to both mechanical shock and chymotrypsin. No leakage of microencapsulated asparaginase was detected by the Ouchterlony technique. It was found that, although there was reaction from the antibodies of asparaginase and asparaginase solution, there was no precipitant line detected between the antibodies of asparaginase and microencapsulated asparaginase. Studies using trypsin, chymotrypsin, and Pronase p show that these did not affect microencapsulated asparaginase; however, they lowered the asparaginase in free solution to 28%, 4%, and 12%, respectively.

Summary of Other Methods

Another study makes use of entrapment of asparaginase in polyacrylamide gel.[13] The gel-entrapped enzymes are fragmented to fine particles for peritoneal or intraveneous injection. Intraperitoneal injections lowered the serum asparagine in rats from 76 to 1 nM/ml of 24 hours. This level is maintained for 2 days.[13]

Another group of investigators[14] studied gel entrapment of L-asparaginase by entrapment in poly-2-hydroxyethylmethacrylate with an activity of 730 IU per gram of dry gel. L-Asparaginase solution or gel-entrapped L-asparaginase, 400 IU, was injected intraperitoneally into C3H mice. After intraperitoneal injection of 5 IU into C3H mice, plasma L-asparagine fell to an undetectable level for 4 days and increased to reach a normal level in 8 days for both the native and immobilized enzymes. A single intraperitoneal injection of asparaginase solution of 2, 4, and 8 IU inhibited subcutaneously transplanted 6C3HED lymphoma on day 14 by 36, 53, and 86%, respectively, as compared to 35, 78, and 100% after the injection of L-asparaginase solution.[14]

[40] T. Mori, T. Sato, Y. Matuo, T. Tosa, and I. Chibata, *Biotechnol. Bioeng.* **14**, 663 (1972).
[41] T. Mori, T. Sato, Y. Matuo, T. Tosa, and I. Chibata, *Biochem Biophys. Acta* **321**, 653 (1973).

Another example is the use of the extracorporeal approach using asparaginase insolubilized on glass plates.[20] Studies were carried out in rabbits, dogs, and in one patient.

Asparaginase has been coupled to polymethylmethacrylate by covalent linkage for use in extracorporeal hemoperfusion.[22] The plates are formed with an average asparaginase activity of 2.5 units/plate. Eight of the plates are each separated by small 1-mm buttons to allow blood to flow between them. Blood passing through the plates comes in direct contact with the enzyme covalently linked to the surface of each of the 8 plates. In addition to lowering the L-asparagine level for possible use in the suppression of lymphomas, there was some evidence that hemoperfusion across this device had some immunological suppressing effect.[22] Further analysis of the asparaginase-linked polymethylmethacrylate was studied.[23] It was found that this type of binding increased the storage stability of the enzyme. It was also found that hemoperfusion lowered the asparagine level rapidly within 1 hr of perfusion. Discontinuation of perfusion resulted in a prompt return to initial value of both asparagine and aspartic acid within 3 hr.

Other studies include the use of L-asparaginase covalently bound to nylon tubing.[24] In this case, the enzyme is fully exposed to the perfusing blood. Asparaginase can be covalently bound to a polycarboxylic gel layer attached to the inner wall of a 1–2 mm internal diameter nylon tube.[25] In this case, the L-asparagine must diffuse into the enzyme gel layer and is not in direct contact with the perfusing blood.[25]

In further studies,[42] L-asparaginase was bound covalently to Dacron vascular prosthesis bearing 10.3 IU of L-asparaginase and installed in the inferior vena cava. This resulted in the fall of L-asparagine and a corresponding increase in aspartic acid. The L-asparagine level returned to normal within 5 days of implantation. Circles of Dacron grafts bearing 1 IU of enzyme were implanted in the peritoneal cavity of BDF1 mice, with controls receiving grafts without enzymes attached.[42] Six hours later, both groups received intraperitoneal injections of 10^5 cells of leukemia 5178Y sensitive to L-asparaginase. No difference in survival time was observed between the two groups.

Another material investigated is collagen–asparaginase complex, which was wound to form a spiral multipore reactor.[26] It was found to have good stability, and further studies should be carried out in *in vivo* tests.

Another approach is the use of fiber-entrapped asparaginase.[28] L-Asparaginase was microencapsulated within cellulose triacetate in fiber form.

[42] D. A. Cooney, H. H. Weetall, and E. Long, *Biochem. Pharmacol.* 24, 503 (1975).

Another approach is the use of dialysis against asparaginase.[27] Here, the asparaginase is placed outside in the dialyzate compartment, and blood was allowed to pass through the blood compartment of the hemodialysis system.

Since the enzyme in solution outside the dialyzate is rather unstable, a new proposal was to combine artificial cells in the form of microencapsulated asparaginase with the artificial capillary system hollow fiber.[29] In these studies, blood passes through the capillary system and microencapsulated asparaginase in the dialyzate compartment was found to act effectively on L-asparagine. The stability of the microencapsulated L-asparaginase was maintained for long periods of time. The combination of the heparinized capillary systems and the long stability might allow for long-term implantation for blood hemoperfusion.

Summary

In summary, a very large amount of work is being carried out on the use of immobilized asparaginase. Although the feasibility of using immobilized asparaginase for tumor suppression has been demonstrated,[9,10] much work is still required to develop and make this system practicable for clinical uses. In the intermittent extracorporeal approach using asparaginase, although the serum asparagine is lowered quickly to near zero during hemoperfusion, with termination of hemoperfusion, the serum asparagine level rapidly increases to normal. With this particular condition, where it is necessary to suppress the L-asparagine for a long period of time, the intermittent use of the extracorporeal approach may not be suitable. It will require a system that can be implanted and can act for long periods of time, as in the case of injected immobilized asparaginase[9,10] or long-acting extracorporeal devices.

Organ Failure and Metabolic Disorders

The initial *in vivo* investigation for the feasibility of using immobilized enzymes for the support of organ failure was based on the following theoretical analysis.[18] Ten milliliters of microcapsules 20 μm in diameter, or 33 ml of microcapsules 100 μm in diameter, have a total surface area of about 2 m^2, which is double that available in a whole standard artificial kidney machine. In addition, the microcapsules have a membrane thickness of less than 500 Å, at least 100 times less than that of the standard hemodialysis machine. On the basis of these parameters, the rate of transfer of a given permeant solute across 33 ml of microcapsules 100 μm in diameter would be at least 100 times faster than in the standard artificial kidney machine. All that would be required would be to place adsorbents or enzymes inside the microcapsules to retain or convert metabolites entering the microcapsules.

Example of Method of Extracorporeal Application

A feasibility test involved the use of 10 ml of microencapsulated urease, 90 μm mean diameter, retained in a shunt chamber, 15 cm in height and 2 cm internal diameter.[18] The microcapsules are retained by a nylon screen on either side. Each shunt, when attached to dogs of 10 kg, was effective in removing blood urea from the systemic circulation so that, after 90 min of hemoperfusion, the systemic blood urea level fell to 50% of the initial level. Blood urea is enzymatically converted to blood ammonia, which can be removed by a combined system of microencapsulated urease and ammonia adsorbent.[18] This demonstration for the feasibility of using immobilized enzymes to remove uremic metabolites has resulted in further analytical and experimental study by a number of other workers. The feasibility of using a column to convert blood urea in the human body was tested by theoretical analysis.[43] The theoretical analysis supported our experimental results that for the removal of urea it is sufficient to use microencapsulated urease in an extracorporeal shunt, 2 cm in diameter and 10 cm in length.

An extension of this approach is the suggestion[18] that, with the enzymes and adsorbents placed in the external dialyzing fluid of an artificial kidney, the enzymes and the ammonia adsorbent could also act on urea diffusing across the standard hemodialysis membrane. However, the main advantage of the microencapsulated system for direct blood hemoperfusion is the much larger surface area and ultrathin membrane that allows for a compact artificial kidney. This suggestion of using enzymes in the dialyzate compartment of the artificial kidney has now been incorporated into the eddy recirculating system. In this system, immobilized urease in combination with ammonia adsorbent is used to remove urea diffusing across the standard hemodialysis machine, thereby cutting down the need for using large amounts of dialyzate.[44]

Uricase has been immobilized by microencapsulation to retain its activity on uric acid with the view of removing uric acid from patients with hyperuricemia.[3,5] Actual *in vivo* tests have now been successfully carried out by another group using glass beads-immobilized uricase.[21] In this study, glass beads-immobilized uricase was tested in an extracor-

[43] S. N. Levine and W. C. LaCourse, *J. Biomed. Mater. Res.* **1**, 275 (1967).
[44] A. Gordon, M. A. Greenbaum, L. B. Marantz, M. J. McArthur, and M. H. Maxwell, *Trans. Am. Soc. Artif. Intern. Organs* **15**, 347 (1969).

poreal shunt system. During a 2-hr control period of extracorporeal circulation using Dalmatian dogs, there was no significant changes in the blood uric acid, urinary allantoin, or urinary uric acid levels. The addition of 6 g of uricase–glass beads to the extracorporeal shunt resulted in a rapid fall of blood uric acid level from about 2.5 mg to 0.5 mg per 100 ml in 4 hr.[21] There was a corresponding increase in urinary allantoin level and decrease in urinary uric acid level. There was no leakage of enzyme. Rabbit antiuricase serum markedly inhibited the enzyme activity of uricase solution, but not that of the glass-immobilized uricase activity. This is another excellent demonstration of the feasibility of immobilized enzymes for the selective conversion of accumulated waste metabolites.

Example of Methods for Administration into the GI Tract

Another route for the use of immobilized enzymes for replacing organ deficiency was tested. Here, microencapsulated urease was administered into the G.I. tract.[5,7,30] It was found in these studies that the immunobilized enzyme, urease, was effective in converting urea into ammonia, and if the ammonia adsorbents are simultaneously given, this will result in the lowering of the blood urea level.[30] Recently, more detailed studies have been carried out along this line. Thus, one group[31] used microencapsulated urease and ammonia adsorbent, feeding this to dogs by mixing with food. They were able to demonstrate that this resulted in a significant decrease of the blood urea levels in the animals. Further extensions along this line were carried out by another group.[45] They used a liquid membrane system in which urease was microencapsulated within a liquid membrane, which also contained buffer systems to allow for a very high internal pH. The urea thus diffuses across the liquid membrane and is converted by the enclosed urease to ammonia. Owing to the high pH of the liquid membrane microencapsulated system, most of the ammonia is present in an ionized form that cannot diffuse out of the liquid membrane. One advantage of the liquid membrane approach is that ammonia adsorbent is not required because the ammonia so formed is entrapped by the liquid membrane.

Treatment of Patients

Other types of microencapsulated systems in extracorporeal systems have been used—for instance, microencapsulated adsorbents in the form of activated charcoal or other systems. Here, albumin has been immobi-

[45] S. W. May and N. N. Li, *Biochem. Biophys. Res. Commun.* **47**, 1179 (1972).

lized to the surface to make the system blood compatible. This approach has already been demonstrated in human studies to be effective for the treatment of patients wih chronic renal failure,[5,46] acute intoxication,[5,46,47] or liver failure.[46]

Other Areas

Immobilized enzymes have also been tested experimentally in other areas related to support of organ failure. Silastic membranes containing immobilized carbonic anhydrase can transport CO_2 1.5 times faster than Silastic membranes without the enzyme.[48] Catalase immobilized in dialysis membrane was tested for converting hydrogen peroxide to oxygen.[49]

In preliminary animal testing, implantable immobilized glucose oxidase electrodes have been assessed for use as the glucose sensor component of a synthetic islet cell for insulin secretion.[50]

Immobilized proteolytic or fibrinolytic enzymes have also been used to prevent thrombus formation on biomaterials in contact with blood.[51,52] As discussed earlier, immobilized albumin has made possible the construction of blood-compatible adsorbent systems for use in humans for the treatment of renal failure, liver failure, and acute intoxication.[5,46]

The use of microencapsulated red blood cell content has been investigated since 1957 as possible red blood cell substitutes.[3,5,15,18,53] However, the main problem is their ability to remain circulating in the blood stream.

Perspectives

Multistep Enzyme Systems and Cofactor Regeneration

For more extensive applications of immobilized enzymes in medicine, one would require multistep enzyme systems [31] and cofactors [61].

[46] T. M. S. Chang, *Kidney Int.* **7**, 5387 (1975).
[47] T. M. S. Chang, J. F. Coffey, P. Barre, A. Gonda, J. H. Dirks, M. Levy, and C. Lister, *Can. Med. Assoc. J.* **108**, 429 (1973).
[48] G. Broun, E. Selegny, C. T. Minh, and D. Thomas, *FEBS Lett.* **7**, 223 (1970).
[49] S. J. Updike, M. C. N. Shults, and J. Magnuson, *J. Appl. Physiol.* **34**, 271 (1973).
[50] S. P. Bessman and R. D. Schultz, *Trans. Am. Soc. Artif. Intern. Organs* **19**, 361 (1973).
[51] B. Kusserow, R. Larrow, and J. Nichols, *Trans. Am. Soc. Artif. Intern. Organs* **17**, 1 (1971).
[52] B. Kusserow, R. Larrow, and J. Nichols, *Trans. Am. Soc. Artif. Intern. Organs* **19**, 8 (1973).
[53] W. Sekiguchi and A. Kondo, *J. Jpn. Soc. Blood Transfusion* **13**, 153 (1966).

Immobilization of enzymes by microencapsulation is especially good for studying multistep enzyme reactions, since any combination, any concentration, and any number of enzymes could easily be microencapsulated. Recent studies[54] show that hexokinase and pyruvate kinase co-immobilized within semipermeable microcapsules recycle both ATP and ADP. The recycling of free or immobilized NAD has also been demonstrated in a microencapsulated multienzyme system.[55] These findings allow one to investigate microencapsulated complex multienzyme systems required for enzyme replacement therapy or organ replacement. Thus, a microencapsulated multienzyme system containing galactokinase and ATP regeneration systems of hexokinase and pyruvate kinase is being used here for the conversion of galactose to galactose-1-PO_4 as a possible replacement of galactokinase deficiency condition.[56]

Biodegradable Immobilized Enzyme Systems

After injection, it will be necessary for the implanted material to be removed after the immobilized enzymes complete their function in the body. We have recently microencapsulated enzymes within biodegradable membranes. For example, polylactic acid was used to microencapsulate asparaginase which retained its ability to act on external asparagine diffusing into the microcapsules.[56] The polylactic acid can be degraded by the body into carbon dioxide and water.

Acknowledgment

This work is supported by MRC-MT-2100 and MRC-SP-4 of the Medical Research Council of Canada.

[54] J. Campbell and T. M. S. Chang, *Biochim. Biophys. Acta* **397**, 101 (1975).
[55] J. Campbell and T. M. S. Chang, *Biochem. Biophys. Res. Commun.* (in press).
[56] T. M. S. Chang (unpublished).

[47] Medical Applications of Liposome-Entrapped Enzymes

By GREGORY GREGORIADIS

Introduction

The use of enzymes in the treatment of inherited metabolic disorders or other diseases in which enzymes can be beneficial is hampered by a variety of problems. For instance, repeated administration of a foreign enzyme will almost certainly elicit antibody response and dangerous anaphylactic reactions. Enzymes can interact with their substrate in sites other than those in need of the enzyme activity, and this is especially true for sugar hydrolases potentially useful in the treatment of some of the lysosomal storage diseases; interaction of such enzymes en route to their destination (lysosomes), with relevant substrates terminally located on blood or cell surface glycoproteins, could lead to a number of untoward effects,[1] and therapeutic hydrolases themselves could, upon their arrival within the lysosomes, become subject to attack by lysosomal cathepsins. However, inability of injected enzymes to reach the diseased tissues is the most common cause of our failure in enzyme therapy.

It is apparent that a proper enzyme-carrier system could circumvent most of the present difficulties of enzyme use. Such a carrier should be nontoxic, nonantigenic, and capable of containing therapeutic agents in such a way as to prevent contact of its contents with the biological environment. The carrier should also be amenable to manipulation so as to "home" to target sites in the body, deliver its contents, and be eventually subjected to degradation.

Liposomes as a Carrier Candidate

It has been shown recently that most of the criteria for an ideal carrier can be met by liposomes.[2] These are lipid vesicles consisting of one or more concentric phospholipid bilayers alternating with aqueous compartments within which enzymes and other water-soluble materials can be entrapped[3] (see this volume [17]). Enzymes are rendered largely latent upon entrapment and measurement of their activity requires previous rupture of the lipid bilayers with a detergent. After injection into the circulation, liposome-entrapped enzymes retain their latency and liposomes and their contents are eventually taken up by tissues, mostly by the liver and spleen.[2] Both major types of liver cells (parenchymal and Kupffer cells) participate in the uptake although at varying rates.[2,4] It is possible to modify the rate of elimination of liposomes from the blood by adjusting their surface charge and their size. Thus, positively charged liposomes are cleared from the blood at a slower rate than that

[1] G. Gregoriadis, D. Putman, L. Louis, and E. D. Neerunjun, *Biochem. J.* **140**, 323 (1974).
[2] G. Gregoriadis, *in* "Enzyme Therapy in Lysosomal Storage Disease" (J. M. Tager, G. J. M. Hooghwinkel, and W. T. Daems, eds.), p. 131. North-Holland Publ., Amsterdam, 1974.
[3] A. D. Bangham, M. W. Hill, and N. G. A. Miller, *in* "Methods in Membrane Biology" (E. D. Korn, ed.), Vol. 1, p. 1. Plenum, New York, 1974.
[4] G. Gregoriadis and E. D. Neerunjun, *Eur. J. Biochem.* **47**, 179 (1974).

observed for neutral and negatively charged liposomes.[4] Likewise, small liposomes are removed at a rate slower than that of large liposomes.[5]

Cellular uptake of liposomes is carried out by endocytosis,[6] which leads to lysosomal localization[2,6] of both liposomes and entrapped materials. Once within lysosomes, liposomes are disrupted[6] (presumably by phospholipases) and liberate their contents, which can then act either locally (e.g., enzymes[7]) or, after their escape from lysosomes, in other cellular compartments. Recent work suggests that, in addition to endocytosis, other mechanisms as well might be involved in the cellular uptake of liposomes. It has been claimed, for instance, that monolamellar liposomes can, under certain conditions, fuse with cells *in vitro*.[8] Presumably liposomal contents are then delivered into the cell's cytoplasm.

Although it may be possible to deliver a liposome-entrapped enzyme to an intracellular site without the enzyme coming into contact with the extracellular environment or with plasma membranes, immune response to a foreign protein cannot be avoided. Indeed, as it happens, antibody production appears to be enhanced when liposome-entrapped proteins are used, and in fact liposomes have been suggested as immunological adjuvants.[9] Fortunately, however, liposome-entrapped proteins do not interact with their antibodies in preimmunized animals to give Arthus or serum sickness reactions, and it therefore seems that administration of enzymes via liposomes might be immunologically safe.[10]

Possible Therapeutic Uses of Liposome-Entrapped Enzymes

There are two ways by which a liposome-entrapped enzyme can act on its substrate. In the first, the enzyme must be initially liberated from liposomes; and in the second, substrates (obviously, of small molecular weight) may penetrate the phospholipid bilayers and meet the enzyme in the aqueous spaces. Both ways can find application in therapy.

Lysosomal Storage Diseases

In a number of storage diseases there is an inherited enzyme deficiency that leads to the accumulation of the relevant substrate.[11] In many of

[5] R. L. Juliano and D. Stamp, *Biochem. Biophys. Res. Comm.* **63**, 651 (1975).
[6] A. W. Segal, E. J. Wills, J. E. Richmond, G. Slavin, C. D. V. Black, and G. Gregoriadis, *Br. J. Exp. Pathol.* **55**, 320 (1974).
[7] G. Gregoriadis and R. A. Buckland, *Nature (London)* **244**, 170 (1973).
[8] D. Papahadjopoulos, E. Mayhew, and G. Poste, *Nature (London)* **252**, 163 (1974).
[9] A. C. Allison and G. Gregoriadis, *Nature (London)* **252**, 252 (1974).
[10] G. Gregoriadis and A. C. Allison, *FEBS Lett.* **45**, 71 (1974).
[11] H. G. Hers, *in* "Lysosomes and Storage Diseases" (H. G. Hers and F. Van Hoof, eds.), p. 148. Academic Press, New York, 1973.

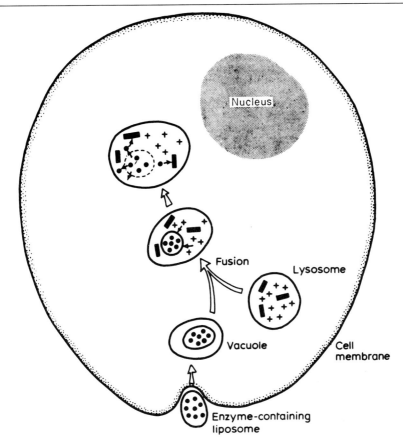

FIG. 1. A liposome containing enzyme molecules (dots) is taken up into a cell by endocytosis. The endocytic vacuole fuses with a lysosome that contains substrate molecules (bars). The lysosomal enzymes (crosses) disrupt the lipid bilayers of the liposome, releasing the entrapped enzyme to act on the stored substrate. From G. Gregoriadis, *New Sci.* **60**, 890 (1973).

these disorders the enzyme, of which the activity is low or altogether missing, is a lysosomal sugar hydrolase and consequently substrates are stored in the lysosomes of a variety of tissues which usually includes the liver (parenchymal and Kupffer cells) and the spleen. Since liposomes can deliver enzymes to the lysosomes of these tissues, treatment of storage diseases with liposomes containing the appropriate enzyme seems an attractive possibility (Fig. 1), and work[7,12] with model lysosomal storage conditions supports this view. It has been shown, for instance, that expo-

[12] C. M. Colley and B. E. Ryman, *Biochem. Soc. Trans.* **2**, 871 (1974).

sure of macrophages or fibroblasts, previously loaded with sucrose in their lysosomes, to liposome-entrapped invertase leads to the uptake of liposomes into the lysosomes of these cells, release of entrapped invertase, and subsequent hydrolysis of sucrose to fructose and glucose, which diffuse out through the lysosomal membranes.[7]

Treatment of storage disease patients with liposome-entrapped enzymes could be hampered by two major problems. First, it is unlikely that enzymes will have any beneficial effect in those patients with involvement of neural tissues (e.g., Tay-Sachs disease), since it is difficult to imagine an enzyme or an enzyme-carrier crossing the blood-brain barrier. Further, in some of these disorders, afflicted tissues include the heart and skeletal muscle (e.g., glycogen storage disease type II),[11] both of which exhibit little endocytic activity. One may hope, however, that depletion of other tissues accessible to therapeutic agents could lead to a shift of some of the material accumulated in the brain and muscle to the rest of the body. The second major obstacle is the unavailability of therapeutic enzymes. For instance, in Gaucher's disease the missing enzyme is glucocerebrosidase, which catalyzes the hydrolysis of glucose from ceramide glucoside. Since bacterial β-glucosidase cannot act on this substrate, it is necessary to extract the enzyme from mammalian tissues, preferably human. Purification[13] of glucocerebrosidase from human placenta is most cumbersome, but it is predicted that the use of affinity chromatography will facilitate preparation and provide realistic quantities of the enzyme.

In regard to the possible toxicity of liposomes, experience in my laboratory suggests that liposomes can be tolerated well by animals subjected to large doses of liposomal lipid and histological examination of a wide selection of tissues has shown no lesions. Further, an ever increasing number of cancer patients receiving liposomes show no obvious adverse effects.[14]

Other Inherited Metabolic Disorders

Attack of substrates by liposome-entrapped enzymes should not necessarily be confined within the lysosomal milieu, where disruption of liposomes occurs. In the case of substrates of small molecular weight, these could penetrate intact lipid bilayers and reach their enzyme. In phenylketonuria and galactosemia, for instance, phenylalanine and

[13] R. O. Brady, P. G. Pentchev, A. E. Gal, S. R. Hibbert, and A. S. Dekaban, *N. Engl. J. Med.* **291**, 989 (1974).
[14] G. Gregoriadis, P. C. Swain, E. J. Wills, and A. S. Tavill, *Lancet* **1**, 1313 (1974).

galactose 1-phosphate cannot be transformed to tyrosine and glucose 1-phosphate, respectively, because of the absence of the two relevant enzymes, i.e., phenylalanine hydroxylase and galactose-1-phosphate uridyltransferase. The use of liposomes as vehicles for the administration of enzymes capable of degrading or transforming phenylalanine and galactose 1-phosphate to substances metabolically acceptable by the organism could be beneficial providing that diffusion of the substrate in the extraliposomal environment (e.g., blood) into the enzyme-containing intraliposomal aqueous spaces occurs rapidly. The use of positively charged liposomes whose circulation in blood is prolonged[4] could contribute to a more efficient elimination of undesirable substances.

Neoplasms

Survival of certain malignant cells is dependent upon continuous exogenous supply of L-asparagine which, in contrast to normal cells, these cells cannot synthesize. For instance, administration of L-asparaginase to mice with EARAD1 leukemia is followed by complete cure.[15] L-Asparaginase has also been used with moderate success in the treatment of patients with acute lymphocytic leukemia, the main problems being the immunogenicity of the enzyme, which is usually of bacterial origin, and toxic reactions, such as fatty metamorphosis of the liver, pancreatitis, and central nervous system depression. In patients treated with asparaginase repeatedly, antiasparaginase antibodies not only inactivate the enzyme, they also promote anaphylactic reactions, and such events could be avoided with liposome-entrapped asparaginase provided that the lipid bilayers surrounding the enzyme are capable of preventing contact of the enzyme with circulating immunoglobulins. Naturally the lipid bilayers should allow rapid penetration and subsequent contact of asparagine molecules with the entrapped enzyme. Isolation of the enzyme within the liposomal walls in the circulation could in addition contribute to a lessening of some of the other toxic effects mentioned above.

The eventual uptake of asparaginase-containing liposomes by the liver, which is the major site of asparagine synthesis, could be an added advantage, assuming that hepatic asparagine is capable of reaching asparaginase carried by liposomes into the cell's lysosomes. Experiments[16] in this laboratory with mice bearing 6C3HED lymphosarcoma have

---

[15] E. A. Boyse, L. J. Old, H. A. Campbell, and L. T. Mashburn, *J. Exp. Med.* **125**, 17 (1967).
[16] E. D. Neerunjun and G. Gregoriadis, *Biochem. Soc. Trans.* **4**, 133 (1976).

shown that, as with the free enzyme, asparaginase entrapped in negatively charged liposomes can also cure diseased mice, although at a concentration higher than that needed with the free enzyme. It is not known as yet to what extent liposomal asparaginase attacks its substrate in the circulation, the liver, and the tumor site.

Intravenous administration of nonentrapped asparaginase to mice already pretreated with the enzyme led within minutes to anaphylactic shock and death. In contrast, similar mice injected with liposome-entrapped asparaginase did not show any signs of illness for at least 48 hours. Pre-immunized tumour-bearing mice injected with a therapeutic dose of nonentrapped asparaginase also died of serum sickness but again, administration of entrapped asparaginase into such animals did not produce anaphylactic shock.[16] This avoidance of dangerous allergic reactions could render the use of asparaginase in cancer treatment immunologically safe.

Liposome-Entrapped Enzymes in the Treatment of Local Disease

In studies on the fate of locally injected liposome-entrapped proteins, it was shown[17] that in contrast to liposomes of small size, which are rapidly cleared from the site of injection (rat testicle) to enter the circulation or the lymphatics, large liposomes persist in the injected area for extended periods of time prior to their disintegration and the liberation of entrapped materials. It was suggested that locally injected liposomes could prove useful in the treatment of local disease, and an example of a disorder that affects localized areas of the body is gout. In this condition uric acid, which is the major end product of purine metabolism in man, accumulates in the plasma and other pools of the body, and inflammatory episodes of acute gout appear to follow interaction of leukocytes with crystals of monosodium urate monohydrate. Previous therapeutic attempts with *Aspergillus flavus* uricase injected intravenously or intramuscularly into gout patients decreased uricemia, and urinary uric acid and typical acute gout attacks were interrupted.[18] It would be of interest to see whether local application of uricase entrapped in large liposomes would have a beneficial effect. Since liver is the main site of uric acid synthesis, it should also be worthwhile to investigate the effect of a mixed population of small uricase-containing liposomes, which are expected to diffuse rapidly and reach the liver,[17] and of a large-sized population leading to a slow release of uricase for local action.

[17] A. W. Segal, G. Gregoriadis, and C. D. V. Black, *Clin. Sci. Mol. Med.* **49**, 99 (1975).
[18] P. Kissel, M. Lamarche, and R. Royer, *Nature (London)* **217**, 72 (1968).

Liposome-Entrapped Enzymes as Immunosuppressants

Treatment of animals with *Escherechia coli* L-asparaginase, glutaminase, or L-arginine iminohydrolase results in a marked suppression of the immune response, especially when the enzyme is given simultaneously or shortly after antigen administration.[19] The effect of enzyme treatment on antibody-forming cells could, as in the case of leukemic cells discussed above, result from substrate (e.g., asparagine) starvation of rapidly dividing cells involved in the immune response. The use of an immunogenic therapeutic agent (drug or enzyme) entrapped in liposomes together with an immunosuppressive enzyme could not only be more effective, but at the same time diminish the formation of antibodies against the administered agent. Indeed, the presence of both antigen and, say asparaginase, in the same liposome could ensure an even more effective immunosuppression.

Liposomes as Immunological Adjuvants

Enhancement of the antibody titers in experimental animals vaccinated with a variety of proteins, many of enzymic nature, can be effected by the use of adjuvants such as Freund's incomplete and complete adjuvants. These, because of their composition (nonbiodegradable oil) and the granulomas they form, cannot be used in man. Recent work[9] suggests that the need for a safe and effective adjuvant for use in human immunization programs could be satisfied by liposomes. For instance, serum antibody response of mice to liposome-entrapped diphtheria toxoid is severalfold greater than when the antigen is given in the free form (see the table). In other experiments an enhanced immune response in guinea pigs was obtained with liposome-entrapped cold-virus hemagglutinin and neuraminidase vis-à-vis the free proteins.[20]

Successful application of liposomes as immunological adjuvants not only would reduce the amount of antigen required for immunization, with corresponding economies especially relevant to the developing countries, but would also eliminate dangerous allergic reactions to the vaccines they contain.[10]

Prospects for the Future

The ultimate goal in creating a carrier for the delivery of enzymes or other agents to diseased sites in the body is specificity. According to

[19] E. M. Hersh, *Transplantation* **12**, 368 (1971).
[20] A. C. Allison and G. Gregoriadis, unpublished observations, 1974.

SERUM ANTIBODY RESPONSES OF MICE TO DIPHTHERIA TOXOID ADMINISTERED
FREE OR IN LIPOSOMES OF DIFFERENT COMPOSITIONS

Experimental group	No. of mice[a]	Mode of administration[b]	Route of administration	Primary Ab response[c]	Secondary Ab response[c]	Probability
a	15	Free	Intravenous	1.8	—	a vs b, $P < 0.01$
b	15	−Liposomes (PA)	Intravenous	10.0	—	
c	6	Free	Subcutaneous	2.5	11.6	c vs d, $P < 0.01$
d	6	−Liposomes (PA)	Subcutaneous	6.7	13.3	d vs e, $P < 0.01$
e	6	+Liposomes	Subcutaneous	0	11.3	e vs c, $P < 0.10$
f	6	Free	Intramuscular	1.7	7.5	f vs g, $P < 0.10$
g	6	−Liposomes (PA)	Intramuscular	3.7	11.0	g vs h, $P < 0.05$
h	6	−Liposomes (DP)	Intramuscular	6.6	12.0	f vs h, $P < 0.01$

[a] Mice were of the TO strain, except in groups c–e, which were of the CBA strain.
[b] Positive (+) liposomes (0.5 mg of lipid) were composed of egg lecithin, cholesterol, and stearylamine in molar ratios 7:2:1; in negative (−) liposomes, phosphatidic acid (PA) or dicetyl phosphate (DP) replaced stearylamine.
[c] Primary antibody responses were measured after 13 or 14 days, booster injections with 20 μg of antigen in the same form were given, and secondary responses were measured after a further 10 days (groups c–e) or 14 days (groups f–h). Antibody responses were measured by indirect hemagglutination and expressed as the $\log_2$ IH titer.

the definition given earlier in this chapter, an ideal carrier should be amenable to modifications so as to "home" to target sites. At present, injected liposomes can carry their contents to a limited variety of cells (e.g., liver cells and spleen) and indeed uptake of liposomes by these cells is so rapid that other cells, also capable of endocytosing liposomes although less actively, fail to take up their share. It would thus seem that inherent to the direction of liposomes to target cells are two problems. First, one must educate liposomes so that they can recognize specific cells, attach onto them, and eventually be endocytized by them; and second, one must prevent the liver and spleen from prematurely removing such liposomes from the circulation.

Delay in the liposome clearance from the blood has been found to increase considerably when positively charged or small liposomes are used and when the capacity of the liver and spleen in liposomal uptake is saturated by the concurrent administration of a large quantity of a different population of liposomes.[4] Clearly, a more sophisticated approach is needed, and in this laboratory the possibility of augmenting the half-life of liposomes in the circulation by the appropriate manipulation of the composition of liposomes at their outermost layer is currently under investigation. It is hoped that specially tailored liposomes will conceal much of their foreignness and thus escape recognition by the reticuloendothelial system. Homing of liposomes to target cells also appears feasible. The approach[21] we have adopted involves the use of macromolecules (probes), such as anti-cell IgG immunoglobulins, tropic hormones, and desialylated glycoproteins, all of which exhibit a specific affinity for the surface of appropriate target cells. The probe is incorporated into the liposome in such a way as to retain its property of interacting with the recognition site on the target cell's surface when liposome and cell are in close proximity. It has now been established that a probe is capable of mediating uptake of the associated liposome and its contents (e.g., therapeutic enzyme) by the target cell *in vitro* and *in vivo*. Indeed, it is anticipated that homing of enzymes via liposomes to target cells will find wide application in both therapeutic and preventive medicine.

Methodology

Enzyme-containing liposomes (see this volume [17]) suspended in 1% NaCl, buffered if necessary with 5 mM sodium phosphate, pH 7.0, are used as such or after being filtered through a Millipore filter (0.22 μm) in the case of sonicated liposomes of appropriate size. For use in man, liposomes are prepared under sterile conditions and subsequently subjected to microbiological examination. When such liposomes are to be stored for more than 24 hr, it is advisable to add benzyl penicillin (about 10000 units per milliliter of suspension). In addition to preparing liposomes under sterile conditions, which simply means use of sterile glassware and solutions, it is also possible to irradiate liposomes after their preparation. Previous experience with liposomes containing *Aspergillus niger* amyloglucosidase has shown that although irradiation with 1.25 Mrad sterilized liposomes, it had little effect on the enzyme activity.

In Vitro Experiments

Prolonged exposure of cells to liposomes requires that liposomal preparations be sterile. It is also important that the amount of the liposomal lipid present should be below toxic levels, which in turn are expected to vary according to the cell type examined. Work[7] with mouse peritoneal macrophages and human embryo lung fibroblasts showed that growth of cells (about 10^6 cells) for up to 3 days in the presence

[21] G. Gregoriadis and E. D. Neerunjun, *Biochem. Biophys. Res. Comm.* **65**, 537 (1975).

of 0.75 mg of liposomal lipid (egg lecithin, cholesterol, and phosphatidic acid, molar ratio 7:2:1) per milliliter of medium was normal. However, the presence of 2.0 mg lipid per milliliter of medium led to cell detachment and death. Conversely, the presence of cells and of the media constituents did not affect the stability of the lipid bilayers, and liposomal enzyme latency was largely preserved for up to 24 hr.

It is often necessary to measure cellular uptake of liposome-entrapped enzymes, and this is carried out after treatment with Triton X-100 (0.1% final concentration), which ruptures not only cells, but also intact liposomes in the interior of the cells.

In Vivo Experiments

Experiments[22] in rats showed that large quantities of liposomes given over a period of several months (1.0 g of lipid per rat, composition as above) did not affect the well-being of the animals, and no toxic reactions, immunological or otherwise, were apparent. Upon histological examination, a large number of tissues appeared to be normal. Further, neutral and negatively or positively charged liposomes have been administered into animals intravenously, intraperitoneally, subcutaneously, intramuscularly, and orally, and no adverse reaction has been observed. However, with nonsonicated liposomes given intravenously, there is a risk of animal death owing to the blocking of lung capillaries with large liposomes. The source and purity of lipids do not appear to be of great importance, and preparations composed of relatively impure egg lecithin, undoubtedly containing traces of proteinaceous material, do not provoke allergic reactions upon repeated administration.

The choice of tissues in the assay of liposomal enzyme activity in animals treated with the liposome-entrapped enzyme will depend upon the route of administration. For instance, in addition to blood, liver and spleen should be assayed after intravenous or intraperitoneal injection. After intramuscular or subcutaneous administration, however, sonicated liposomes localize in the lymph nodes (draining the injected tissue) as well.[17] Unsonicated liposomes will tend to remain in the site of injection for longer periods of time and often disintegrate locally. [17]

Immunogenicity of Liposome-Entrapped Enzymes

It has been established[9] that liposome-entrapped proteins injected into mice elicit antibody response more pronounced than that observed with the nonentrapped proteins. However, such response is related to the

[22] C. D. V. Black and G. Gregoriadis, unpublished observations, 1974.

surface charge and composition of liposomes,[9] and it might depend also on the animal species examined. To ascertain whether animals treated with liposome-entrapped enzyme develop antibodies to the enzyme, 10–50 μl of the preparation containing 50–100 μg of protein is injected intramuscularly into 5–10 mice (preferably inbred). Fourteen days later the animals are bled from the tail vein, immunized again with approximately 20 μg of liposome-entrapped protein (rechromatographed to eliminate traces of free protein possibly released from liposomes), and bled again in 10 days. The serum is then assayed for antibody titer by passive hemagglutination.[23]

The Arthus-Reaction Test

As has already been discussed, formation of antibodies to the liposomal therapeutic enzyme does not necessarily warrant the occurrence of allergic reactions upon repeated administration of the enzyme entrapped in liposomes (see above). To ascertain whether such reactions can occur, the Arthus reaction test is carried out[10]: 10 μl containing 10 μg of the entrapped enzyme are injected into the right foot pad of the hind leg of 5–10 mice. The left foot pad, which serves as control, is injected with 10 μl of the buffer solution used for the suspension of liposomes. Four hours later the thickness of the feet is measured with a micrometer gauge and the (mean) thickness of the right foot pad is divided by that of the left. Ratios higher than unity imply interaction of the enzyme with its antibodies (i.e., positive Arthus reaction) and swelling at the sites of injection 24 or 48 hr later suggests delayed hypersensitivity.

Acknowledgment

I thank Mrs. E. Diane Neerunjun for excellent technical assistance.

[23] W. Page Faulk and V. Houba, *Immunol. Methods* 3, 87 (1973).

[48] Immunoenzymic Techniques for Biomedical Analysis

By STRATIS AVRAMEAS

Since the first publications on the preparation and uses of enzyme-labeled proteins appeared,[1,2] there has been continuous development in

[1] S. Avrameas and J. Uriel, *C. R. Acad. Sci. Ser.* D 262, 2543 (1966).
[2] P. K. Nakane and G. B. Pierce, *J. Histochem. Cytochem.* 14, 929 (1966).

this area. An appreciable number of immunoenzymic techniques, which make use of enzymes coupled to antigens or antibodies, are now available for biomedical analysis. Of these, the following three are the most often employed:

1. As with fluorescent-labeled antibodies, enzyme-labeled antibodies are employed for the localization of cellular constituents.[2] Since specific cytochemical techniques for staining enzymes are also available for electron microscopy, enzyme-labeled proteins are also employed for the ultrastructural detection of antigens and antibodies.[3,4]

2. As with radiolabeled antigen or antibody, enzyme-labeled proteins are employed for the quantitation of humoral constituents.[5-7] In this case the basic principles of the procedures are those developed for quantitative radioimmunoassay, enzyme activity measurement substituting for radioactivity counting.

3. An enzyme-labeled antigen or antibody is used to reveal an immune precipitate, which otherwise could not be visualized.[1] This allows the use of enzyme-labeled antibodies in quantitative immunodiffusion techniques for the measurement of low concentrations of antigens.

In what follows, only the procedures that have been developed and used in the author's laboratory are described. For further information concerning immunoenzymic techniques, recent reviews may be consulted.[8-10]

Coupling of Enzymes to Proteins

Covalent binding of an enzyme to a protein is carried out with glutaraldehyde by a one-step or a two-step procedure.[11,12] The one-step method is applicable to a variety of enzyme–protein systems, whereas the two-step method is at present employed only for the preparation of peroxidase-labeled proteins. A heterogeneous population of enzyme–

[3] P. K. Nakane and G. B. Pierce, *J. Cell Biol.* **33**, 307 (1967).
[4] M. Bouteille and S. Avrameas, *C. R. Acad. Sci. Ser.* D **265**, 2097 (1967).
[5] E. Engvall and P. Perlmann, *Immunochemistry* **8**, 871 (1971).
[6] B. K. Van Weemen and A. H. W. M. Schuurs, *FEBS Lett.* **15**, 232 (1971).
[7] S. Avrameas and B. Guilbert, *C. R. Acad. Sci. Ser.* D **273**, 2705 (1971).
[8] S. Avrameas, *Histochem. J.* **4**, 321 (1972).
[9] G. A. Andres, K. G. Hsu, and B. C. Seegal, in "Handbook of Experimental Immunology" (D. M. Weir, ed.), Vol. 2, p. 34. Alden Press, Oxford, 1973.
[10] L. D. Sternberger (ed.), "Immunocytochemistry." Prentice-Hall, Englewood Cliffs, New Jersey, 1974.
[11] S. Avrameas, *Immunochemistry* **6**, 43 (1969).
[12] S. Avrameas and T. Ternynck, *Immunochemistry* **8**, 1175 (1971).

protein conjugates is obtained with the former procedure, and a homogeneous one with a molar ratio of protein to peroxidase of 1 is obtained with the second.

Conjugation of Enzymes to Antibodies by a One-Step Procedure. Dissolve 10 mg of horseradish peroxidase (RZ = 3, Grade I, ref. 15629, Boehringer, Mannheim), *Aspergillus niger* glucose oxidase (Grade I, ref. 15422, Boehringer, Mannheim), or *Escherichia coli* alkaline phosphatase (Code BAPSF, Worthington, New-Jersey) and 5 mg of the antibody preparation in the volume of 0.1 M phosphate buffer, pH 6.8, given in the table and dialyze overnight at 4°. To the gently stirred solution,

QUANTITIES USED FOR STOCK SOLUTION OF VARIOUS
ENZYMES IN CONJUGATION PROCEDURE

Enzyme	Final volume (ml)	Glutaraldehyde, 1% solution (ml)
Peroxidase	1	0.05
Alkaline phosphatase	2–2.3	0.05
Glucose oxidase	1	0.15

add at room temperature the quantity of glutaraldehyde indicated in the table. Keep the preparation for 3 hr at room temperature, and then dialyze overnight against phosphate-buffered saline (PBS).* Centrifuge at 4° for 20 min at 20,000 rpm (rotor 40 of Spinco ultracentrifuge), and keep this stock solution of enzyme-labeled antibody at 4° until use.

Conjugation of Peroxidase to Antibody or Fab† Fragment of Antibody by a Two-Step Procedure. Dissolve 10 mg of peroxidase in 0.2 ml of a 1% glutaraldehyde solution in 0.1 M phosphate buffer, pH 6.8, and leave the preparation at room temperature for 18 hr. Filter the solution through a Sephadex G-25 fine column (0.9 × 60 cm) equilibrated with 0.15 M NaCl. The brown fractions containing the peroxidase are pooled and, if needed, concentrated to 1 ml. To this solution add 1 ml of 0.15 M NaCl containing 5 mg of the antibody preparation (or 2.5 mg of Fab antibody fragment) and 0.2 ml of 0.5 M carbonate–bicarbonate buffer, pH 9.5. Leave the solution at 4° for 24 hr and then dialyze overnight at 4° against PBS. Keep the preparation at 4° for a week before using.

* 0.15 M NaCl containing 0.01 M phosphate buffer pH 7.5.
† Monovalent Fab antibody fragment (M.W. 50,000) is prepared from whole divalent antibody (M.W. 160,000) by papain digestion according to the method of R. R. Porter, *Biochem. J.* 73, 119 (1959).

Separation of Enzyme-Antibody Conjugates from Unreacted Reagents. After the coupling of the enzyme to the antibody has been carried out, it is advisable to separate enzyme–antibody conjugates from all the other reagents. Isolated conjugates give less background and have greater specific activity than unpurified preparations.

Purification of the conjugates is accomplished by the use of gel filtration through a Sephadex G-200 column. Conjugates (1–2 ml) prepared as described above by the one-step procedure are filtered at 4° on a Sephadex G-200 column (70 × 1.5 cm) equilibrated with PBS. The conjugates that emerge with the void volume of the column are pooled, concentrated if necessary, and read at 280 nm to determine the protein content.

Conjugates (4 ml) prepared by the two-step procedure are applied at 4° to a Sephadex G-200 column (100 × 2.6 cm) equilibrated with PBS. The protein content of the fractions is read at 280 nm and the peroxidase content at 403 nm. The first peak eluting from the column contains antibody labeled with peroxidase (or Fab labeled with peroxidase if Fab was employed instead of whole antibody). The peak fractions are pooled and read at 280 nm and 403 nm to determine the peroxidase-labeled antibody concentration.

Detection of Cellular Constituents with Enzyme-Labeled Antibody

Enzyme-labeled antibody can be employed for light and electron microscopic localization of cellular constituents. In both cases, the best results are obtained with conjugates prepared with antibodies specifically isolated on immunoadsorbents.[8] Peroxidase, alkaline phosphatase, and glucose oxidase-labeled antibodies are used for light microscopy, while peroxidase is employed almost exclusively for electron microscopy. The detection of surface antigens may be carried out on living unfixed cells, but the detection of an intracellular antigen requires the previous fixation of cells. For electron microscopy, the use of Fab antibody fragments (MW 50,000) instead of whole antibody (MW 160,000) allows better penetration of the conjugates into the interior of fixed cells, and consequently more effective ultrastructural staining. In the following sections, the procedures employed for the localization of intracellular antigens at the light and electron microscopic levels are described.

Light Microscopy. Frozen sections of tissues or cell suspensions are plated on microscope slides and fixed for 20 min at −20° in absolute acetone or for 30 min at room temperature in 4% formaldehyde in 0.2 M cacodylate buffer, pH 7.4. The preparations are washed twice for 5 min with PBS and then covered with a solution of PBS containing 50–

100 µg of labeled antibody per milliliter. After 3 hr of incubation in a moist chamber, the slides are washed twice for 5 min in PBS, and enzymic staining is carried out as described below. After staining, the slides are rinsed briefly with distilled water and examined in the light microscope without any special mounting except water.

Peroxidase Staining.[13] Dissolve 5 mg of 3,3'-diaminobenzidine tetrahydrochloride (ref. D8126, Sigma) in 10 ml of 0.1 M Tris-HCl buffer, pH 7.6; check the pH and readjust to 7.6, if necessary, with 1 N NaOH. To this solution, add 0.1 ml of a 3% aqueous solution of H_2O_2. Incubate the preparation for 5–10 min at room temperature, and then rinse with distilled water.

Alkaline Phosphatase Staining.[14] Dissolve 2 mg of Naphthol AS-Mx phosphoric acid disodium salt (ref. N-500, Sigma) in 5 ml of 0.2 M Tris-HCl buffer, pH 8. To this solution add 5 ml of distilled water containing 30 mg of Fast Red TR salt (ref. F-1500, Sigma). The resulting turbid solution is clarified by filtration with filter paper. Incubate the preparation at room temperature for 20–30 min and then rinse with distilled water.

Glucose Oxidase Staining.[14] Dissolve 5 mg of MTT-tetrazolium salt (ref. M-2128, Sigma) and 150 mg of D-glucose in 10 ml of 0.1 M phosphate buffer, pH 7.2. To this solution add 2.5 mg of phenazine methosulfate (ref. P-9625, Sigma) and filter immediately onto the preparation. Incubate in the absence of light for 20–30 min, and rinse with distilled water.

Electron Microscopy.[15] Tissues are sliced with razor blades into blocks and fixed at 4° for 24 hr with 4% formaldehyde or for 1 hr with a solution containing 1% formaldehyde and 1% glutaraldehyde. Fixatives are prepared in 0.2 M cacodylate buffer, pH 7.2. After fixation, free aldehyde groups are blocked by incubating the blocks for 30 min in 0.1 M lysine buffered at pH 7.5. Tissue blocks are reduced to small fragments with a razor blade and from these fragments 10–40 µm frozen sections are cut. Sections are incubated for 2 hr at room temperature with cacodylate buffer containing per milliliter 500 µg of peroxidase-labeled Fab antibody fragment prepared as described above. After incubation, sections are rinsed briefly in buffer followed by three washings of 10–15 min each. Peroxidase activity is revealed by incubating the sections for 20 min at room temperature in the medium described for peroxidase staining. After washing in cacodylate buffer, sections are postfixed for

[13] R. C. Graham and M. J. Karnovsky, *J. Histochem. Cytochem.* **14**, 291 (1966).
[14] S. Avrameas, B. Taudou, and T. Ternynck, *Int. Arch. Allergy* **40**, 161 (1971).
[15] W. D. Kuhlmann, S. Avrameas, and T. Ternynck, *J. Immunol. Methods* **5**, 33 (1974).

1 hr in the same buffer containing 2% OsO_4. The sections are then dehydrated in ascending concentrations of alcohol and finally embedded in Epon.

Quantitation of Humoral Constituents by Enzyme Immunoassay

Because the quantity of an enzyme can be accurately determined by appropriate enzymic techniques, enzyme-labeled antigens and antibodies can be used for the quantitation of humoral constituents.

For the measurement of an antigen, the method involves the incubation of a determined quantity of enzyme-labeled antigen with increasing concentrations of unlabeled antigen, in the presence of a given amount of antibody directed against the antigen. The extent of binding of the enzyme-labeled antigen in the presence of known amounts of unlabeled antigen allows the establishment of a reference curve from which unknown concentrations of the antigen in samples can be determined.

For quantitation of antibodies present in an antiserum, increasing dilutions of the antiserum are incubated with a constant amount of the insolubilized homologous antigen. After incubation, the insoluble antigen, to which the homologous antibody is now bound, is washed by centrifugation and incubated with an enzyme-labeled antibody directed against the immunoglobulins present in the antiserum. After incubation and washing, the enzyme activity is measured in the suspension. The use of an antiserum containing known quantities of antibody allows the establishment of a reference curve from which unknown concentrations of antibody can be determined.

In the following the use of rat IgG-peroxidase conjugates for the measurement of IgG in rat serum is given as an example of the procedures employed. In this procedure peroxidase-labeled rat IgG and unlabeled IgG are first allowed to react with soluble rabbit antiserum directed against rat IgG. The antigen–antibody complexes formed are then precipitated with an insoluble immunoadsorbent of sheep antibody directed against the immunoglobulins of the antiserum.[16]

Procedure. A sufficiently sensitive assay can be obtained only when the soluble rabbit antiserum is used at dilutions higher than 1:1000. Consequently, in the first step of the procedure, one has to establish the amount of insoluble sheep antibody required to precipitate all the peroxidase-labeled IgG fixed by the rabbit antibody present in the diluted antiserum. This is done by adding 0.1 ml of peroxidase-labeled IgG (peroxidase activity corresponding to an optical density of 4.0–6.0) and 0.5 ml of rabbit antiserum antirat IgG diluted 1000-fold to a series of

[16] S. Avrameas and B. Guilbert, *Biochimie* **54**, 837 (1972).

duplicate tubes and incubating them at room temperature for 30 min. Increasing amounts of sheep antirabbit IgG immunoadsorbent are added (final concentration of insoluble antibody from 0.050 to 0.5 mg/ml), and the final volume is brought to 1 ml. The tubes are then incubated, under rotating agitation, for 2 hr at room temperature, centrifuged, and finally the peroxidase activity is measured in 0.1 ml of the supernatant. From the values obtained, the amount of insoluble antibody that precipitates all the labeled IgG fixed by the antibody is determined and used in the steps that follow.

In the second step of the procedure, the optimal dilution of the rabbit antirat IgG antiserum has to be established. For this, a series of duplicate tubes containing 0.1 ml of peroxidase-labeled IgG and 0.5 ml of rabbit antiserum antirat IgG diluted from 1000 to 50,000 times are incubated at room temperature for 30 min; then the optimal amount of insoluble anti-rabbit IgG antibody previously determined is added, and the final volume is brought to 1 ml. Incubation of this suspension, centrifugation, and determination of enzymic activity are then carried out as described above. A curve is plotted using the different dilutions of the antiserum and, from this curve, a dilution of antiserum sufficient to bind 30–50% of the peroxidase-labeled IgG is chosen.

To measure IgG, 1 to 10^4 ng of unlabeled IgG in 0.1 ml of the diluting medium are added to a series of duplicate tubes containing 0.5 ml of the previously determined dilution of the antiserum. The tubes are incubated at room temperature for 30 min, then 0.1 ml of the labeled IgG is added and the solutions are incubated at room temperature for an additional 30 min. After this time, the previously determined quantity of insoluble antibody is added, and the suspensions are then incubated with rotation for 2 hr at room temperature. Subsequent centrifugation and determination of enzymic activity are performed as before. By this procedure, amounts ranging from 5 to 200 ng of IgG per milliliter can be quantitated.

Normal Rat IgG. Rat IgG is isolated on DEAE-cellulose according to Levy and Sober.[17]

Rabbit Antiserum Antirat IgG. The antiserum is obtained by hyperimmunizing rabbits with IgG according to described procedures.[14]

IgG Labeled with Peroxidase. The coupling is carried out by employing the two-step procedure. The IgG-peroxidase conjugate is isolated from the other reactants by passage through Sephadex G-200 as described above.

Insoluble Sheep Antibody Antirabbit Immunoglobulin. This immunoadsorbent is prepared by immobilizing isolated sheep antibody on Bio-

[17] H. B. Levy and H. A. Sober, *Proc. Soc. Exp. Biol. Med.* **103**, 250 (1960).

Gel P-300 minus 400-mesh beads according to procedures described elsewhere.[18]

Dilution Medium. All dilutions are made with 0.1 M phosphate buffer, pH 6.8, containing 1% bovine serum albumin.

Assay for Peroxidase Activity. Peroxidase activity is determined by using H_2O_2 as the substrate and *o*-dianisidine as the hydrogen donor.[16] The reagents are added to 0.6 ml of 1 M phosphate buffer, pH 6, in the following order: 0.6 ml of a 0.3% aqueous solution of H_2O_2, 58.8 ml of distilled water, 0.5 ml of a 1% solution of *o*-dianisidine in methanol (continuous stirring). For the measurement, 0.1 ml of the enzyme solution containing 1–5 ng of peroxidase is added to 2.9 ml of the substrate solution and the reaction mixture is allowed to incubate at room temperature (25°). After 1 hr, the reaction is stopped with 1 drop of 5 N HCl with thorough mixing on a Vortex vibrator, and the color development is read at 400 nm.

The Use of Enzyme-Labeled Antibodies in Quantitative Immunodiffusion and Electroimmunodiffusion Techniques

Like radiolabeled antibodies,[19] enzyme-labeled antibodies can be employed to enhance the visualization of immune precipitates in the quantitative single radial diffusion or electroimmunodiffusion techniques. This increases the sensitivity of the techniques 10- to 20-fold. By these procedures, concentrations ranging from 20 to 1000 ng of proteins per milliliter can be measured. In all cases, the best results are obtained when glucose oxidase-labeled antibodies are used. In the following section, the use of glucose oxidase labeled sheep antibody to rabbit immunoglobulins is used for the measurement of mouse IgG.[20]

Procedure. Agarose plates (8.5 × 10 × 0.15 cm) containing adequately diluted rabbit antiserum to mouse IgG, 0.5% bovine serum albumin (BSA), and 1:5000 Merthiolate (used as a preservative) are prepared. The wells are filled with 5 µl of appropriate dilutions of mouse IgG in PBS containing 0.5% BSA, and the diffusion is allowed to proceed for 48 hr at room temperature in a humid chamber. After diffusion, the slides are washed for 3 hr at room temperature in PBS and then incubated face down in PBS containing 10 µg/ml of glucose oxidase labeled sheep antibody directed against rabbit immunoglobulins. Two hours later, the incubation medium is removed and the plates are postincu-

---

[18] T. Ternynck and S. Avrameas, *FEBS Lett.* **23**, 24 (1972).
[19] D. S. Rowe, *Bull. W.H.O.* **40**, 613 (1969).
[20] J. L. Guesdon and S. Avrameas, *Immunochemistry* **11**, 595 (1974).

bated overnight at room temperature in a moist chamber. The plates are then washed for 20 hr at 4° with several changes of PBS and dried under filter paper. Enzyme activity is revealed by incubating the plates in the dark for 20 min at 37° in the glucose oxidase staining solution described above. The plates are then rinsed in water and the diameter of the stained precipitates is measured.

[49] Immobilized-Enzyme Reactors

By M. D. LILLY and P. DUNNILL

In previous chapters a wide range of methods for the immobilization of enzymes and their subsequent characterization is described. The development of these techniques has produced novel forms of biochemical catalyst, which can be used in reactors to carry out conversions of industrial interest. Already at least three enzymes are being used in an immobilized form on a large scale. These systems, using penicillin amidase, aminoacylase, and glucose isomerase, are described elsewhere in this volume. Their successful commercial operation is encouraging, and there are many other reactions of interest to industry where immobilized enzymes could be used as the catalyst. Nevertheless, in considering any reaction it is essential to decide whether or not the choice of an immobilized enzyme as catalyst is reasonable before undertaking the design of a suitable immobilized-enzyme reactor. Therefore when evaluating a particular biochemical conversion it is important, first, to examine what types of catalyst could be used. Possible biochemical catalysts are listed in Table I. The factors affecting the choice between these different types of catalyst have been discussed elsewhere[1] and will be summarized here (Table I). Most biochemical catalysts are used for conversions of biological materials, but there is a growing awareness of their potential applications in chemical conversions.[2] Conversely, biochemical conversions may be done by chemical catalysts. The development of enzyme analog catalysts soon will link the methodologies of biochemical and chemical catalysis.

The dominant catalysts in the biological industry have for many years been metabolizing microorganisms. The great versatility of microorganisms has led to the development of the present sophisticated fer-

[1] M. D. Lilly and P. Dunnill, *Process Biochem.* 6(8), 29 (1971).
[2] C. J. Suckling and K. E. Suckling, *Chem. Soc. Rev.* 3, 387 (1974). See also Section IX, D of this volume.

TABLE I
Types of Catalyst, Immobilization Method, Support and Reactor and the Factors Affecting the Choice for a Particular Process

	Catalyst	Immobilization method	Support	Reactor

Types	Cells
	  Metabolizing
	  Nonmetabolizing
	  Modified
	  Immobilized
	Enzymes
	  Soluble
	  Stabilized
	  Soluble immobilized
	  Insoluble immobilized
		Entrapment
		  Matrix
		  Capsule
		Binding
		  Absorption
		  Covalent
			Soluble
			  Low molecular weight
			  High molecular weight
			Insoluble
			  Porous/Nonporous
			  Particles
			  Sheets
			  Tubes
			  Fibers
				Batch
				  Single use
				  Reuse
				Continuous
				  Open
				    Tubular
				    Stirred
				  Closed
				    Tubular
				    Stirred

Factors influencing choice	Nature of conversion Product yield and purity Stability and reuse Modification of kinetics	Safety Cost Retention of activity Stability	Method of immobilization Nature of substrate Type of reactor Mechanical properties	Operational requirements Reactor costs and utilization Catalyst replacement, regeneration Reaction kinetics

mentation industry. During the fermentation process, large numbers of microorganisms are formed from nutrients present in the fermentation liquor. These microorganisms, either during or after the growth phase, act as catalysts for the synthesis of products, which may be small molecules, such as ethanol, or extremely complex polymers, such as enzymes or polysaccharides. For product formation not only must the enzymes directly involved in the synthesis be present, but the microbial cells must still be metabolizing nutrients to form the energy required for product synthesis. In some cases, however, only a single enzyme in the microorganism is required for the desired conversion, e.g., for glucose isomerization,[3] and nonmetabolizing cells that may no longer be viable are adequate. Many such transformations can be done with fungal spores.[4] If the conversion process is separated from the production (fermentation) step, it is possible to use cells under conditions that could not otherwise be considered. For instance, nongrowing suspensions of *Agrobacterium tumefaciens* convert sucrose to 3-ketosucrose more rapidly in the presence of ferricyanide.[5] It is also possible to use solvents other than water in the reactor. We have used thick pastes of *Nocardia* containing a high level of cholesterol oxidase suspended in water-immiscible solvents, such as carbon tetrachloride, for the rapid conversion of high concentrations of cholesterol to cholestenone.[6]

In most modern fermentation processes the microorganisms are suspended in the fermentation liquor. There are other processes where the microorganisms are grown on the surface of solid substrates[7] or as films on the surface of inert solids as in trickle filter beds used for sewage treatment. Unfortunately "natural" immobilization of this kind is not suitable for some types of microorganism and industrial situations. Other ways have been devised of immobilizing whole cells, especially by fixing them in gels, such as polyacrylamide,[8] collagen[3] or cellulosic fibers.[9] The immobilized cells may then be used until their activity has decreased to a level where further use is uneconomic. The conversion of pantothenic acid to CoA, involving five enzymic steps, by *Brevibacterium*

---

[3] Vieth and Venkatasubramanian [53] and Poulsen and Zittan [56] in this volume.
[4] C. Vezina, S. N. Sehgal, and K. Singh, *Adv. Appl. Microbiol.* **10**, 221 (1968).
[5] A. H. Fensom, W. H. Kurowski, and S. J. Pirt, *J. Appl. Chem. Biotechnol.* **24**, 457 (1974).
[6] B. C. Buckland, P. Dunnill, and M. D. Lilly, *Biotechnol. Bioeng.* **17**, 815 (1975).
[7] C. W. Hesseltine, *Biotechnol. Bioeng.* **14**, 517 (1972).
[8] K. Yamamoto, T. Sato, T. Tosa, and I. Chibata, *Biotechnol. Bioeng.* **16**, 1589 and 1601 (1974). See also Chibata *et al.* [50] and Larsson and Mosbach [13] in this volume.
[9] Dinelli *et al.*, this volume [18].

ammoniagenes entrapped in polyacrylamide has been reported.[10] One of the disadvantages of whole cells is the likelihood of interfering side reactions resulting in lower yields of product and reduced purity of product. This can be prevented in some cases by modification of the substrate[11] or by deliberate inactivation of an interfering enzyme.[8]

Many enzymes produced extracellularly by microorganisms are available in large quantities and are used in various industrial processes. The techniques for the large-scale extraction and isolation of enzymes present inside microorganisms and other living cells have now improved, and some at least of these enzymes are produced in large amounts.[12] At the present time it is unrealistic to think of using isolated enzymes in biochemical conversions involving the participation of a large number of enzymes, but the situation may change rapidly. The additional costs of enzyme isolation and immobilization must be balanced against the potential advantages of isolated enzymes. Compared to microorganisms, they may give higher yields of the desired product with less contamination by side products and lysis of cells. Also there is the possibility of modifying the kinetics of an enzyme when it is immobilized. In current industrial practice, immobilized enzymes are reused at least 10–15 times in batch processes (or the equivalent time in continuous processes) and in some cases for much longer (over 50 times or the equivalent). So far, immobilized cells normally have been used for fewer batches or shorter periods before being discarded.

The choice between use of the soluble free enzyme and the immobilized enzyme depends on the cost of the enzyme, the nature of the conversion process, and the relative operational stabilities of the two forms. By their nature, some food processes, such as meat-tenderization and baking, involve the addition of the enzymes at the final processing stage, making reuse impossible. In some instances the ability to remove the immobilized enzyme from the product stream, ensuring minimal contamination by protein and to modify the reaction kinetics, may influence the choice, but the main factor is likely to be the operational stability of the enzyme. If the enzyme can be stabilized by modification[13] or immobilization then reuse of the enzyme may be worthwhile. The direct sta-

[10] S. Shimizu, H. Morioka, Y. Tani, and K. Ogata, *J. Fermentn. Technol.* **53**, 77 (1975).
[11] J. de Flines, in "Fermentation Advances" (D. Perlman, ed.), p. 385. Academic Press, New York, 1969.
[12] C. L. Cooney, A. L. Demain, P. Dunnill, A. E. Humphrey, M. D. Lilly, and D. I. C. Wang, "Fermentation and Enzyme Technology." Wiley, New York, in press.
[13] O. R. Zaborsky, in "Enzyme Engineering" (E. K. Pye and L. B. Wingard, Jr., eds.), Vol. 2, p. 115. Plenum, New York, 1973.

bilization of enzymes as distinct from stabilization resulting from immobilization is an approach that will be of increasing interest in industrial enzyme technology.

Types of Immobilized Enzyme

From previous chapters it is obvious that a wide range of immobilization techniques and supports is now available. For convenience immobilization methods have been classified into four main groups[14] (Table I), but in many procedures more than one kind of immobilization can occur. Thus during covalent binding of an enzyme to a support, some of the enzyme molecules may be adsorbed but not covalently bound. Usually these may be removed by washing the preparation in high ionic-strength solution. Enzyme trapped inside a matrix may also be ionically bound to the matrix. The support may be particulate or be a flat sheet or block, an annular cylinder or coil, a tube or rod (Table I). In each case the support may be porous, semipermeable, or nonporous. With nonparticulate support materials these are normally an integral part of the reactor and must have sufficient mechanical strength to withstand any pressure gradients that exist in the reactor. Particulate support materials vary in shape, size, and density, and these parameters influence the behavior of the particles in reactors. Some materials, such as porous glass and Sephadex derivatives, are available as spherical beads whereas others may be more rod-shaped or fibrous. Most particulate materials, when first produced, consist of a range of sizes. It may be necessary to use a certain size fraction, but this inevitably raises the cost. The removal of "fines" is often important since their presence in the reactor may markedly influence its operational performance.

Most support materials that have been used for immobilization of enzymes have apparent densities not much greater than water itself when suspended in an aqueous medium. Glass or metallic materials have densities in the region of 2500 and 8000 kg/m^3, respectively, but when made into porous supports these have much lower apparent densities. Some metallic materials have ferromagnetic properties so that the support may be magnetized.[15] Apart from their greater density, these particles containing ferromagnetic material have some interesting characteristics. When stirred in a reactor the particles separate, but when the stirrer is stopped the particles aggregate and therefore settle at a much faster rate than do individual particles.

[14] Recommendations for standardization of nomenclature in enzyme technology, in "Enzyme Engineering" (E. K. Pye and L. B. Wingard, Jr., eds.), Vol. 2, Appendix I. Plenum, New York, 1973.
[15] P. J. Robinson, P. Dunnill, and M. D. Lilly, *Biotechnol. Bioeng.* **15**, 603 (1973).

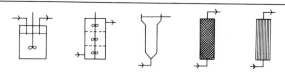

Stirred tank   Multistage   Fluidized bed   Packed bed   Tubular

FIG. 1. Examples of continuous-flow immobilized enzyme reactor systems.

In addition to water-insoluble supports, polymers of high molecular weight have been used for enzyme immobilization.[16] The soluble immobilized enzymes possess many of the properties of the insoluble form, including modification of the kinetics by electrostatic effects and alteration of the thermal stability of the enzyme.[16] In addition, soluble immobilized enzymes are more effective than insoluble forms in acting upon large substrates, such as proteins. Recently the immobilization of lysozyme to alginic acid, giving a preparation that was soluble above pH 4 and essentially insoluble below pH 3, has been reported.[17] After use, the immobilized enzyme could be recovered by lowering the pH to 2.5, although it must be remembered that many enzymes are unstable at these low pH values. The attachment to enzymes of low-molecular-weight compounds, such as fatty acid derivatives, may stabilize some enzymes[18] (Table I).

Types of Reactor

Many different types of reactor have been employed with immobilized enzymes. These may be classified according to mode of operation and the flow pattern in the reactor. The most common system is the stirred tank normally operated as a batch system, but also semicontinuously by repeatedly drawing off part of the reaction liquor at intervals and refilling with fresh substrate solution. Some examples of continuous-flow systems are shown in Fig. 1. The flow patterns range from the "well-mixed" continuous-flow stirred tank reactor (CSTR) to the "ideal" plug-flow or tubular reactor. The basic characteristics of these "ideal" reactor systems are summarized in Table II. Several types of reactor where the flow pattern approximates to plug-flow have been used. One variation is a reactor consisting of semipermeable hollow fibers inside which the substrate solution passes through the reactor with the enzyme solution in

[16] J. R. Wykes, P. Dunnill, and M. D. Lilly, *Biochim. Biophys. Acta* **250**, 522 (1971).
[17] M. Charles, R. W. Coughlin, and F. X. Hasselburger, *Biotechnol. Bioeng.* **16**, 1553 (1974).
[18] I. Urabe and H. Okada, in "Fermentation Technology Today" (G. Terui, ed.), p. 367. Society of Fermentation Technology, Osaka, 1972.

TABLE II
THE CHARACTERISTICS OF "IDEAL" ENZYME REACTOR SYSTEMS

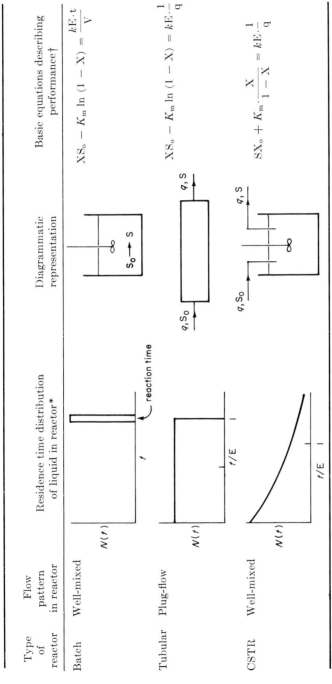

Type of reactor	Flow pattern in reactor	Residence time distribution of liquid in reactor*	Diagrammatic representation	Basic equations describing performance†
Batch	Well-mixed			$XS_o - K_m \ln(1-X) = \dfrac{kE \cdot t}{V}$
Tubular	Plug-flow			$XS_o - K_m \ln(1-X) = kE \cdot \dfrac{1}{q}$
CSTR	Well-mixed			$SX_o + K_m \cdot \dfrac{X}{1-X} = kE \cdot \dfrac{1}{q}$

* $N(t)$ is the number of elements of liquid in the reactor with a residence time, t, and $\bar{t}$ is the mean residence time of liquid leaving the reactor ($= V/q$).
† Derived for an enzyme acting on a single substrate where the reaction rate is described by the Michaelis-Menten equation, kE is the total enzyme activity in the reactor (volume, V), and X is the proportion of the initial or ingoing substrate converted to product ($= S_o - S/S_o$).

the space surrounding the fibers.[19] Alternatively, a tube may be packed with a membrane sheet containing immobilized enzyme wound in a spiral about spacers placed centrally down the tube.[20] In addition to the widely used packed beds of immobilized enzyme particles or fibers, other types of porous bed reactor constructed from porous sheets or blocks have been tried.[21] When the substrate solution is pumped upward through a particulate bed, it is possible to operate with the bed in an expanded state.[22] If the flow rate is increased, the system becomes fluidized and the flow pattern changes to one intermediate between the two ideal flow patterns. Experience in the water demineralization field with ion-exchangers suggests that hindered settling, as distinct from fluidization, is more readily attained in columns of diameter above a meter than in smaller columns where wall effects are significant. With reactions where mixing is required it is still possible to achieve a flow pattern close to plug-flow by using a multistage tank system.

Since one of the main objects of immobilizing an enzyme is to permit its reuse, the reactor system must be a closed one[1]; that is, the catalyst must be retained in the reactor or recovered from the product stream. With immobilized enzyme supports in the form of porous sheets or blocks, membranes, tubes, or fibers that are an integral part of the reactor, no retention system is required. Ultrafiltration membranes have been used to retain free[23] and both water-soluble[24] and water-insoluble immobilized enzymes. For particulate immobilized enzymes, packed beds are a convenient way of retaining the catalyst in the reactor. For other systems where the particles are in suspension, an outlet filter may be fitted. If the immobilized enzyme particles sediment sufficiently fast, then they can be retrieved in a settling tank. When the reactor is operated with upward flow as in expanded or fluidized beds, no separate sedimentation stage is necessary if the correct flow is chosen. The rate of settling can be increased by the incorporation of a metallic component into the particle.[25] In addition, if the particles can be magnetized they will aggregate under conditions of low shear.[15] Magnetized immobilized enzyme particles may be retained or recovered using a magnetic field.[15] Particulate immobi-

[19] J. C. Davis, *Biotechnol. Bioeng.* **16**, 1113 (1974). See also Chambers this volume [23].
[20] W. R. Vieth and K. Venkatasubramanian, *CHEMTECH* **4**(7), 434 (1974).
[21] M. D. Lilly and P. Dunnill, *Biotechnol. Bioeng. Symp.* **3**, 221 (1972).
[22] S. P. O'Neill, P. Dunnill, and M. D. Lilly, *Biotechnol. Bioeng.* **13**, 337 (1971).
[23] T. A. Butterworth, D. I. C. Wang, and A. J. Sinskey, *Biotechnol. Bioeng.* **12**, 615 (1970).
[24] S. P. O'Neill, J. R. Wykes, P. Dunnill, and M. D. Lilly, *Biotechnol. Bioeng.* **13**, 319 (1971).
[25] M. Charles, R. W. Coughlin, R. Tedman, and K. W. Beard, *Biotechnol. Bioeng.* **16**, 1549 (1974).

lized enzymes may also be recovered from the product stream by filtration or centrifugal sedimentation. Ion-exchange particles (20 μm in diameter) can be recovered without damage in basket centrifuges using skimmer discharge.

Choice of Immobilization Method and Support

The main factors affecting the choice of immobilization method are listed in Table I. As in all other types of operation, safety must be a prime consideration, both during the preparation of immobilized enzymes and their subsequent use. Apart from adsorption and certain types of entrapment, all other immobilization methods involve chemical reactions. Reagents used for covalent binding, especially bifunctional agents, are intrinsically protein-binding agents and must be treated with care. Some reagents produce poisonous side products during reaction. Since some applications of immobilized enzymes are, or will be, in the food and pharmaceutical areas, compliance with any food and drug regulations is essential. Decision on the purity of enzyme used and method of immobilization may be influenced by safety criteria. It is important to point out that use of an immobilized enzyme will nevertheless usually be safer than employing other biochemical or chemical catalysts.

Many immobilization techniques done in the laboratory are unsuitable for production of large quantities of immobilized enzyme, either because of the difficulty of scale-up or the very high cost. The choice will be determined to a great extent by the cost and stability of the immobilized enzyme. An expensive method of immobilization is likely to be justified for a cheap enzyme only if the operational life of the enzyme is very long or there are important process advantages. Similarly, an expensive support may be justified only by prolonged immobilized enzyme operating life or if the support is reusable. However, the cost of production of the immobilized enzyme may represent a very small proportion of the total product cost, especially where marketing costs are high. Some methods of immobilization are covered by patents, but licensing is not worthwhile unless a reasonable amount of technological "know-how" is associated with the license, since most enzymes can be immobilized satisfactorily by more than one method. It is interesting that all the main methods of immobilization are being used commercially at the present time. Despite the very large number of methods reported in the literature, in many cases information is limited concerning the influence of the experimental conditions on the amount of enzyme immobilized and its retention of activity. There are very few data on the effect of the conditions chosen on the stability of the immobilized enzyme. This is particularly true for covalent binding, where the number of bonds between enzyme and support may affect both retention of activity and stability.

One of the most notable features of recent studies has been the diversity of surfaces to which enzymes may be bound. Most classical studies were restricted to hydrophilic materials of poor mechanical quality. The extended range of supports, which encompasses relatively hydrophobic polymers and inorganic materials, favors further technological developments. Materials already used in substantial amounts in other industrial processes will have an economic advantage, as will waste materials. Ion-exchange particles, Keiselguhr and conventional catalyst supports, such as metal oxides, illustrate the first category, and stainless steel filings the second. Apart from cost, the choice of support material depends on several factors (Table I) and is inseparable from the method of immobilization. Clearly, for immobilization by ionic or covalent binding the necessary chemical groups must be present.

There has been great interest in the effects of electrostatic interactions between charged groups on the support and charged molecules in solution on the kinetics of immobilized enzymes.[26] In some cases where enzymes have been immobilized to charged supports, shifts of 1-2 units in the optimum pH for activity and 10-fold changes in K_m at low ionic strengths have been observed. By choice of the right support material, these modifications to the kinetic behavior of the enzyme may be used to advantage in a reactor. However, it is important to realize that in industrial systems the highest possible substrate concentrations, which may be much greater than K_m, are likely to be used, so that changes in K_m have little effect. Also the concentrations of the substrate and products are likely to be sufficiently high that if they are charged, as in the deacylation of benzylpenicillin, the resulting ionic strength will override any electrostatic effects.

The nature of the substrate may limit the support materials that can be considered. The activities of enzymes toward high-molecular-weight substrates when bound to water-insoluble supports are often low owing to steric hindrance. Despite some reports to the contrary, an entrapped enzyme if correctly made should not be able to act on high-molecular-weight substrates of similar size to the enzyme unless the substrate is in an unfolded configuration. Thus it is preferable with large substrates to use either the free enzyme or an enzyme immobilized to a water-soluble polymer.[12,16] Another situation that may arise is the presence of undissolved solids in the substrate liquor. To separate the immobilized enzyme from these solids at the end of the reaction, it is

[26] L. Goldstein, this volume [29].

necessary to have a support material with a higher density than the other solids or containing magnetizable material.

In the design of a reactor the amount of catalytic activity per unit volume of reactor is often an important factor, since this determines the size of the reactor. The reaction rate per unit volume will be a function of the weight of immobilized enzyme preparation per unit volume of reactor, the protein content of the preparation, the specific activity of the immobilized enzyme protein and the efficiency of utilization of that activity. The amount of immobilized enzyme preparation per unit volume will vary with the type of reactor. For instance, with particulate immobilized enzymes, the maximum value will be achieved in a packed bed where the liquid void volume may be only 30% of the total reactor volume, whereas in a stirred tank it may only be possible to have a suspension occupying 10% or less of the reactor volume. Thus in stirred tank systems there will be a considerable premium on high activity per support volume. Not only will a large total support volume reduce the product flux per reactor volume, but the separation of product liquor and immobilized enzyme may be more difficult and will certainly be slower. The down time between batches can become a significant proportion of the operating time. With nonparticulate support materials, the proportion of the total volume occupied by the catalyst will depend greatly on the design of the reactor.

For nonporous supports the enzyme activity is a function of the external surface area, which for a spherical particle is inversely related to the particle diameter. If it is assumed that a spherical protein is closely packed as a monolayer on the surface of a support, then it is possible to estimate the amount of enzyme per unit surface area. For chymotrypsin (MW 24,000) with a diameter of about 4 nm, the maximum loading on the surface is about 3 mg/m^2. For a spherical particle with a density of 1000 kg/m^3 this loading corresponds to about 4 and 0.04 mg/g support for particles of 10^{-6} and 10^{-4} m diameter (1–100 µm), respectively. It is possible to get higher loadings by forming more than one layer on the surface. Much higher protein contents can be achieved with porous supports, but if the activity per unit volume of support is too high, the efficiency of utilization of this activity will be less because of diffusional limitations. This is illustrated in Fig. 2, which shows the theoretical relationship between the effectiveness factor and the radius of porous spherical particles for different enzymic activity contents.[27] These curves, which cover the normal range of activities of immobilized enzymes, show that even with a high external substrate concentration, particle diameters below 10^{-4} m (100 µm) are needed to achieve an effectiveness factor

[27] D. L. Regan, M. D. Lilly, and P. Dunnill, *Biotechnol. Bioeng.* **16**, 1081 (1974).

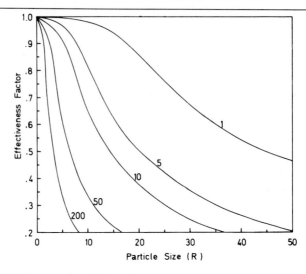

FIG. 2. The influence of particle radius (μm) on the effectiveness factor of a spherical porous particle with various enzyme contents (mg·cm^{-3}, values shown beside curves). Enzyme specific activity was 100 μmoles min^{-1}·mg^{-1} enzyme; effective substrate diffusion coefficient = 4×10^{-4} cm^2 min^{-1}; external surface substrate concentration = $10 \times K_m$. From D. L. Regan, M. D. Lilly, and P. Dunnill, *Biotechnol. Bioeng.* **16**, 1081 (1974).

greater than 0.5. Clearly the smaller the particle size the better, but the lower limit is determined by the ability to retain these small particles in the reactor. For a given particle size, the optimal enzyme content of the particle will be an economic compromise between (1) a high content, which will give a high activity per unit volume of support, but a low effectiveness factor, and (2) a low content, which will give a lower activity, but a more effective use of an expensive enzyme.

The data given in Fig. 2 assume a uniform distribution of enzyme throughout the porous support. For entrapped enzymes this will be a reasonable assumption. For bound enzymes there is also evidence for this where the enzyme content is high.[27,28] However, with a low enzyme content, it is likely that the enzyme concentration will be higher near the surface, in which case the effectiveness factor will be higher than that predicted from the figure. This will also be true for porous support materials with a nonporous core.

In practice it is difficult to get support materials of a constant particle size. This may lead to an uneven distribution of activity among the particles. For instance, when β-galactosidase was immobilized to AE-cellulose, a greater proportion of the measured activity was found in the small particles because of greater diffusional limitation of the large particles even though all the particles contained about the same amount of enzyme.[28]

Spherical particles are advantageous both for packed beds and stirred-tank reactors. In both systems damage by compression and attrition will be less than with irregular particles. For example, in large-scale demineralization of water by ion-exchange, once attrition of spherical particles commences further damage rapidly ensues. However, production of spherical particles is often more expensive and the extra cost must be justified by better operational performance. While small particles have many advantages, they are more susceptible to mechanical stress, and in general the breaking load decreases with particle diameter.[29]

Choice of Reactor Type

The nature of the reaction will sometimes impose particular requirements that influence the choice of reactor. For most biochemical reactions it is essential to control both the operating temperature and pH. There are many instances where hydrogen ions are either removed or produced as a result of the reaction. Although in the laboratory it is often possible to use a strong buffer to counteract this effect, on a large scale this is usually ruled out because of the cost or because the buffer interferes with subsequent processing of the product. It is therefore necessary to control the pH by addition of acid or alkali to a well-mixed reactor system. Sometimes substrate must be fed serially or intermittently to the reactor as the reaction proceeds because the substrate is unstable or the enzyme is inhibited by high substrate concentrations. This can be done without difficulty in batch or tubular reactors, but for CST systems the use of several reactors in series is required. If oxygen is consumed during the reaction a well-mixed gassed system is essential to provide this reactant at a sufficient rate. If the feed to the reactor contains undissolved solids then a packed bed cannot be used. This will also be true if microorganisms are present and cannot be flushed through. If any undissolved solids are not solubilized in the reactor, then a closed system with a physical barrier such as a filter as the means of retention is not satisfactory. In these circumstances, high-density or magnetized supports may be essential for separation from the product stream.

The choice of reactor system will also depend on other factors, such as utilization and cost. The required output from the reactor will influence the choice of mode of operation. Where output is small, continuous operation may be uneconomic. Batch operation, especially of

[28] D. L. Regan, P. Dunnill, and M. D. Lilly, *Biotechnol. Bioeng.* **16**, 333 (1974).
[29] L. S. Golden and J. Irving, *Chem. Ind.*, Nov. 4, p. 837 (1972).

stirred tank reactors, gives a relatively cheap and flexible system that can be used for a wide range of processes. For quality control reasons, separation of the product into batches may necessitate this mode of operation. Continuous operation usually requires the design of a reactor specifically for a particular process and will involve greater capital expenditure. On the other hand, continuous operation has several advantages: diminished labor costs; ease of automatic control; and greater constancy in reaction conditions and hence a more reliable product.

It is also necessary to consider the problem of catalyst regeneration or replacement. This is easy to do with a batch-operated system but may be more difficult with continuous-flow systems. With a hollow-fiber reactor the enzyme can be replaced during operation, but for most other systems the process must be stopped to allow for replacement with new immobilized enzyme or adsorption of fresh enzyme.

All the criteria outlined above require knowledge of the specific process. However, the effect of the reaction kinetics on the choice of reactor can be described in general terms. It is useful therefore to compare the three basic ideal reactor systems: the batch stirred-tank reactor, the continuous tubular reactor, and the well-mixed continuous stirred-tank reactor. The basic equations describing the performance of these reactors, assuming the enzyme obeys the Michaelis–Menten equation, are given in Table II. The kinetics of the batch and tubular reactors are identical in form so that the sole difference in their productivities will be the "down times" of each reactor, which for the tubular reactor should ideally be nil. It is convenient therefore to compare directly the productivities of the tubular and CST reactors. Since the reaction process is independent of reactor size (Table II), the reactors must be compared on the basis of the quantities of enzyme needed in each reactor to carry out a particular reaction. This is illustrated in Fig. 3, in which the ratio of the amounts of enzyme needed in the two reactors (E_{CST}/E_{PF}) is plotted against the conversion (X) for various values of S_o/K_m,[30] where S_o is the feed substrate concentration. When $S_o \gg K_m$ the two equations become identical, but when $S_o \ll K_m$ the relative performances of the two reactors are very different. For instance, when $X = 0.99$ and $S_o = 0.1\ K_m$ then, at a given flow rate through the reactors, 21 times more enzyme is required in the CSTR than in the tubular reactor. The extreme values of the ratio (E_{CST}/E_{PF}) are identical to the ratios of the sizes of CST and tubular reactors required for zero and first-order isothermal chemical reactions where there are no changes in the volume during the reactions.[31]

[30] M. D. Lilly and A. K. Sharp, *Chem. Eng.* (*London*) **215**, CE12 (1968).
[31] O. Levenspiel, "Chemical Reaction Engineering," 2nd ed. Wiley, New York, 1972.

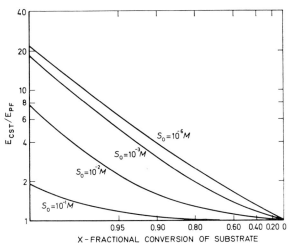

FIG. 3. Variation in the ratio (E_{CST}/E_{PF}) with conversion (X) for various values of feed substrate concentration (S_o), at the same K_m ($= 10^{-3}$ M) and constant flow rate. From M. D. Lilly and A. K. Sharp, *Chem. Eng.* (*London*) **215**, CE12 (1968).

As in other reaction systems, productivity and yield from an immobilized enzyme process are interrelated. In practical processes the usual range of conversion values (X) is 0.8–0.99. The increase in the amount of enzyme needed to give a small increase in X is very dependent on S_o/K_m. For an enzyme obeying the Michaelis–Menten equation, the amount of enzyme required in any of the reactors is proportional to X when S_o/K_m is very large. When S_o/K_m is low, e.g., 0.1, an increase in X from 0.95 to 0.98 requires an extra 30% more enzyme activity in a batch or plug-flow reactor and 250% more in a CSTR. Under these conditions a small increase in yield results in a big decrease in productivity. Thus the choice of X must be made with care when designing a reactor system. For conversions such as steroid transformations and benzylpenicillin deacylation, where the substrates are expensive and the separation of product is difficult, the emphasis will be on yield rather than on productivity, especially if the immobilized enzyme has a high operational stability.

In the preceding comparison it was assumed that the enzyme was not inhibited by either substrate or products of the reaction. Since many enzymes are subject to inhibition of some kind, it is necessary to consider the effect of inhibition on the kinetics of reactors. When the enzyme is subject to inhibition by excess substrate, this is most serious in a batch or plug-flow system but may be reduced by continuous or intermittent addition of substrate to a batch reactor or feeding of substrate at several points along a tubular reactor. The effect of substrate inhibition is far less in a CSTR because all the enzyme is working at a concentration of substrate equal to that in the exit product stream. However, it has been shown theoretically that when substrate inhibition is severe, more than one steady state is possible in a CSTR under certain operating conditions, and some values of conversion (X) may be unobtainable.[32] The effect of product inhibition is much more serious, especially when high conversions are desired. It is more pronounced in a CSTR than a plug-flow reactor since the enzyme is exposed always to the final product concentration, but even in a plug-flow reactor, product inhibition may drastically reduce the productivity. When for operational reasons a well-mixed system must be used, the effect of product inhibition on performance can be reduced by using a multistage stirred-tank system.

So far it has been assumed that the immobilized enzyme obeys the normal enzyme rate equations. However, when substrate diffusion is rate-limiting, the kinetic behavior of the reactor system is different from that which prevails when the enzyme reaction is rate-determining. Diffusional limitation of immobilized enzymes is considered in detail elsewhere in this volume.[26] Essentially limitation of the reaction rate by the rate of diffusion of the substrate is reflected in an increased dependence of the reaction rate on the bulk substrate concentration and the observed reaction becomes closer to first order with respect to that substrate. In those cases where the kinetics of the immobilized enzyme are still described approximately by the Michaelis–Menten equation,[33,34] there will be an apparent increase in the K_m. This will mean that CST reactors will become even more unfavorable than batch or plug-flow reactors (Fig. 3) except for substrate-inhibited systems, where the reverse will be true. Nearly all mathematical treatments of diffusional limitation of immobilized enzymes are done assuming that the external substrate concentration remains constant with time, which is the situation in a CSTR. In a batch or plug-flow reactor, the substrate concentration—and therefore the effect of diffusional limitation—varies with time or distance, respectively. Lee and Tsao[35] concluded that the effectiveness factor calculated using the log mean of the initial and final substrate concentrations gave a reasonable estimate of the average effectiveness factor.

Limitation of the reaction rate by the rate of diffusion of substrate from the bulk of the solution to the surface of the immobilized enzyme

[32] S. P. O'Neill, M. D. Lilly, and P. N. Rowe, *Chem. Eng. Sci.* **26**, 173 (1971).
[33] W. E. Hornby, M. D. Lilly, and E. M. Crook, *Biochem. J.* **107**, 669 (1968).
[34] T. Kobayashi and M. Moo-Young, *Biotechnol. Bioeng.* **13**, 893 (1971).
[35] Y. Y. Lee and G. T. Tsao, *J. Food Sci.* **39**, 667 (1974).

has been observed in stirred tanks[30] and in packed beds,[36–38] where this effect was significant at linear flow rates less than 1–2 cm/min. At higher flow rates the liquid film resistance will be small, but the increased pressure gradient through the column may become a problem if the support material is compressible. In one process using glucose isomerase adsorbed to DEAE-cellulose,[39] compression problems have been overcome by employing stacked shallow columns as in chromatographic separations.[40] Deeper beds of DEAE-Sephadex A-25, which was shown to be noncompressible over the flow rate range examined,[37] are used in the hydrolysis of acetylated amino acids. Diffusional limitation of the reaction rate inside a porous immobilized enzyme support can be reduced only by (1) decreasing the activity/volume of support, (2) increasing the substrate concentration, or (3) decreasing the thickness or diameter of the immobilized enzyme support.

In our comparison so far of reactor types, we have assumed that the fluid flow patterns in these reactors were ideally well-mixed or plug-flow. Although real systems never satisfy completely these criteria, some reactors have flow patterns sufficiently close to the ideal that they can be treated as ideal systems without serious error. However, in systems where channeling of fluid is occurring or stagnant regions of fluid exist, the deviations from the ideal flow pattern will be large and the performance of the reactor will be lower than predicted. Also the presence of back-mixing in a tubular reactor may reduce significantly the output. It is essential that the flow through a packed bed is shown to be close to plug-flow before the integrated rate equation (Table II) is applied to the experimental data.[36] The effect of back-mixing on the degree of substrate conversion in an immobilized enzyme tubular reactor has been considered theoretically by Kobayashi and Moo-Young.[34] In the simplest case, where mass transfer resistance within the liquid film may be neglected and there is no partitioning of the substrate between the liquid and support phases due to electrostatic effects, then, when S_o/K_m is large and the reaction is almost zero order, back-mixing has little effect on the output of the reactor as expected from Fig. 3. For lower substrate concentrations an equation was derived that can be solved analytically only when S_o/K_m is very small. Numerical solutions for $S_o/K_m = 2$ are shown in Fig. 4, which allows comparison of the reactor sizes (i.e., en-

[36] M. D. Lilly, W. E. Hornby, and E. M. Crook, *Biochem. J.* **100**, 718 (1966).
[37] T. Tosa, T. Mori, and I. Chibata, *J. Fermentn. Technol.* **49**, 552 (1971).
[38] T. Kobayashi and M. Moo-Young, *Biotechnol. Bioeng.* **15**, 47 (1973).
[39] B. J. Schnyder, *Die Stärke*. **26**, 409 (1974).
[40] J.-C. Janson and P. Dunnill, *in* "Industrial Aspects of Biochemistry" (B. Spencer, ed.), p. 81. FEBS, Amsterdam, 1974.

zyme activity) with and without back-mixing in a manner previously developed by Levenspiel.[31]

An example of nonideal flow on reactor performance was reported by O'Neill et al.,[22] who found with a packed bed of immobilized amyloglucosidase fed with maltose solution that, in contrast to the theoretical predictions, the output of product was less than for a CSTR using the same immobilized enzyme preparation (Fig. 5). This reduction in output was reflected in a smaller value of the measured maximum reaction rate (kE) at low flow rates. Since the flow pattern through the packed bed was very close to plug-flow, back-mixing was not the cause. The value of kE at low flow rates increased when the bed was operated in an expanded form, indicating that poor mass transfer to some regions in the packed bed was the reason.

Poor mixing in stirred tanks may also cause problems. For instance, a badly designed method of acid or alkali addition to maintain constant the pH may lead to localized regions of low or high pH, especially in large vessels where the mixing times are usually greater. This may cause enzyme inactivation or hydrolysis of the substrates and products if these are labile, as is the case with benzylpenicillin and 6-aminopenicillanic acid. More efficient mixing will minimize these effects but may lead to attrition of immobilized enzyme particles.

From the above considerations it is clear that the consequences of the practical mass transfer properties of both supports and reactors are of great importance to those concerned with the methodology of immobilization and with support materials.

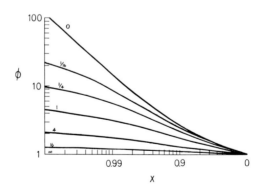

Fig. 4. The proportional increase in reactor size (ϕ) required to compensate for deviation from plug-flow at various conversions (X) and $S_o/K_m = 2$. The extent of back-mixing is given by the Bodenstein numbers shown beside each curve [redrawn from T. Kobayashi and M. Moo-Young, *Biotechnol. Bioeng.* **13**, 893 (1971)], ranging from zero (well-mixed) to infinity (no back-mixing, i.e., plug-flow).

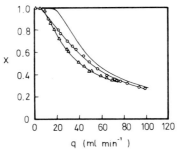

FIG. 5. Substrate conversion (X) as a function of flow rate (q) in continuous-flow stirred-tank reactors (○) and packed-bed reactors (△) containing the same amount of immobilized amyloglucosidase and operated under identical conditions. The top solid line represents the predicted conversions for the packed bed using values of K_m and kE obtained from measurements in the CSTR. From S. P. O'Neill, P. Dunnill, and M. D. Lilly, *Biotechnol. Bioeng.* **13**, 337 (1971).

Reactor Operation

In the preceding sections we have described many of the factors that have to be taken into account to design an immobilized enzyme reactor. In most cases the initial performance of the reactor should be close to that predicted unless it has not been possible to estimate the effect of some of these factors. What is much more difficult to predict is the long-term operational behavior of the system. The productivity of the reactor may change for many reasons, some of which are difficult to estimate without carrying out long-term trials and are listed in Table III.

The immobilized enzyme may be inactivated with time either by denaturation or by poisoning. Denaturation may result from oxidation, adverse pH, or thermal effects. The choice of operating temperature is

TABLE III
REASONS FOR LOSS OF PRODUCTIVITY

A. Loss of enzyme activity
  1. Denaturation
  2. Poisoning
B. Reduced enzyme—substrate contact
  1. Change in flow pattern/enzyme distribution
  2. Fouling
C. Loss of enzyme
  1. Solubilization of enzyme or support
  2. Attrition of support
D. Microbial contamination

critical. High temperatures will increase the initial output from the reactor and reduce the chance of microbial contamination. However, it should be noted that the activation energies for thermal denaturation of enzymes are much higher than for the enzymic reactions and are therefore more temperature-dependent. For instance, typical values for the activation energies of enzyme reaction and denaturation are 11 and 70 kcal/g.mole, respectively, which correspond to increases of 1.8 $\times$ and 60 $\times$ in the rates of catalysis and denaturation when the temperature is raised from 30° to 40°. At higher temperatures the proportional increases in each rate are lower. Small variations in temperature with time or temperature gradients in the reactor can lead to significant changes in the operating life of the immobilized enzyme. The enzyme may be poisoned by inhibitors such as heavy metals in the feed liquor, in which case it may be possible to add a metal-chelating agent such as EDTA.[41] The relative contributions of thermal denaturation and poisoning on loss of activity will depend on the operating conditions. For instance, with β-galactosidase covalently bound to a porous sheet, the main cause at 25° was poisoning, whereas at 50° thermal denaturation was the dominant factor.[41] There is widespread evidence of stabilization of enzymes by their substrates, e.g., the enhanced thermal stability of glucoamylase in the presence of high concentrations of glucose syrups.[12] Similarly, the activities of cell-bound enzymes may decline more rapidly when substrate is not present. For instance, the rate of loss of cholesterol oxidase activity of *Nocardia* was greater between batch reactions than during the reaction.[5] Thus with batch operations, the down-time between batches may be critical.

The output of the reactor may also decrease if the flow pattern or distribution of enzyme in the reactor changes. In a packed bed, channeling may occur or partial blocking alter the flow pattern. This can also result from gas-bubble formation in the reactor if the feed solution has not been preheated to the operating temperature of the reactor. In reactors with a physical retention barrier, immobilized enzyme particles may accumulate at the outlet filter, giving an unequal distribution of enzyme throughout the reactor. One way of overcoming this problem is to fit two filters to the vessel, passing the feed solution in through one and out of the reactor through the other. At regular intervals the flows are switched, the outlet filter now being back-flushed with the ingoing feed solution and the inlet filter now becoming the outlet filter. Reduced contact between the substrate and enzyme can result from coating of the immobilized enzyme by fats or gums present in the reactor feed liquor. It is not

---

[41] A. K. Sharp, G. Kay, and M. D. Lilly, *Biotechnol. Bioeng.* **11**, 363 (1969).

clear whether fouling of this kind will be more serious with porous supports than with nonporous supports.

There are several ways in which the enzyme can be lost from the reactor. When enzymes are immobilized by adsorption, this is usually done under conditions where the equilibrium favors binding of the enzyme. In the reactor the immobilized enzyme is continually exposed to large amounts of reaction liquor which will slowly reduce the amount of enzyme bound. The rate of desorption will increase in the presence of high concentrations of substrate or salts. Tosa et al.[42] found that 0.2 M acetyl-DL-methionine was the highest concentration that could be used with aminoacylase adsorbed to DEAE-Sephadex. It seems likely that the major part of the operational loss of activity (about 30% after 3 weeks) is due to enzyme desorption. It should also be noted that not all covalent bonds between enzymes and support materials are stable over long periods. Thus, more knowledge of the number of covalent bonds between enzyme and support could be important to reactor design. If enzyme leaks at a rate of 0.1% per day when coupled by a single bond and cleavage of one bond does not influence the rate of cleavage of a second bond, then an enzyme linked by two bonds should leak at only 0.0001% per day. Some support materials also solubilize with time. The initial problems with porous glass in alkaline solution were overcome by forming a metallic oxide coating on the surfaces of the porous glass. Some early preparations of CM-cellulose had localized regions of high substitution which went into solution after prolonged use. When particulate immobilized enzymes are used in reactors, disintegration or attrition of the support material will lead to fine particles,[28] which may then be lost from the reactor. This size reduction is commonly ignored when assessing experimentally the stability of diffusion-limited immobilized enzymes. Regan et al.[28] have shown that the measured activity of immobilized β-galactosidase, which was diffusion-limited, increased by 15–24% after exposure to shear stress for 5 hr. The increased activity of immobilized enzymes after attrition may also explain some results with packed beds. For example, the output from a packed bed of immobilized amyloglucosidase acting on maltose increased during the first 200 hr of operation and then declined.[22] It is possible that the DEAE-cellulose support was breaking down into smaller particles with a corresponding increase in activity. Even without attrition, Ollis[43] has pointed out that for a strongly diffusion-limited immobilized enzyme catalyzing a first-order reaction the measured decay rate will be only half that of the real rate.

[42] T. Tosa, T. Mori, N. Fuse, and I. Chibata, *Biotechnol. Bioeng.* **9**, 603 (1967).
[43] D. F. Ollis, *Biotechnol. Bioeng.* **14**, 861 (1972).

All the above reasons for loss of output are concerned with the behavior of the immobilized enzyme itself. It is also possible to lose output as a result of microbial contamination. This may be due to consumption of the substrate or product by the microorganisms or by digestion of the support or enzyme by enzymes produced by the microorganisms. Contamination of the reactor by microorganisms may be minimized or eliminated by operation at temperatures above 45° and pH values away from neutrality. Where this is not possible or is not adequate, the feed may be prefiltered[41] or treated with toluene or formaldehyde. Harper et al.[44] investigated the problem of microbial growth in columns containing β-galactosidase immobilized to porous glass, fed continuously with whey. At pH 6.6 growth was rapid but columns could be operated continuously at 50°–60° and pH 3.5 without loss of enzyme activity with a cleaning and sanitizing cycle every 48 hr. Dinelli and Morisi[45] reported that microbial contamination of enzymes entrapped in fibers can be controlled by intermittent washing of the fibers with 50% aqueous glycerol.

Finally, the susceptibility of immobilized enzyme reactor systems to loss of activity for the above reasons makes rigorous control of the process essential. This may involve supervision by personnel of a higher level of competence or the installation of more instrumentation than when free enzymes are used.

Concluding Remarks

From the above comments, it will be realized that the development of an immobilized enzyme process is dependent on an overall understanding of enzyme immobilization, reactor design, and operation. In particular, it is the solution of some of the operational problems and a more integrated approach to the various aspects which will define the rate of growth of this exciting area.

In this chapter we have concentrated on the problems that may arise in designing and operating an immobilized enzyme reactor, whether it be small- or large-scale. Although these problems may seem daunting, it should be emphasized that many immobilized enzyme reactors are being operated successfully both in the laboratory and in industry. This type of catalyst system can be expected to take its place among the important types available for practical application.

[44] W. J. Harper, E. Okos, and J. L. Blaisdell, in "Enzyme Engineering" (E. K. Pye and L. B. Wingard, Jr., eds.), Vol. 2, p. 287. Plenum, New York, 1973.
[45] D. Dinelli and F. Morisi, in "Enzyme Engineering" (E. K. Pye and L. B. Wingard, Jr., eds.), Vol. 2, p. 293. Plenum, New York, 1973.

[50] Production of L-Aspartic Acid by Microbial Cells Entrapped in Polyacrylamide Gels

By Ichiro Chibata, Tetsuya Tosa, and Tadashi Sato

Aspartic acid is widely used for medicines and food additives. It has been industrially produced from fumaric acid and ammonia by the fermentative or enzymic method employing aspartase as catalyst.

$$HOOCCH=CHCOOH + NH_3 \xrightarrow{aspartase} HOOCCH_2CHCOOH$$
$$| \atop NH_2$$

However, these procedures have some disadvantages in industrial use because the batch process involves incubation of a mixture containing substrate and soluble enzyme. To overcome this difficulty we studied the continuous production of L-aspartic acid using immobilized aspartase.[1] The immobilized aspartase was obtained by entrapping it in a polyacrylamide gel lattice, but its activity yield and operational stability were not satisfactory for industrial application. Further, it was necessary to extract the enzyme from *Escherichia coli* for immobilization, because it is an intracellular enzyme.

If whole microbial cells having higher aspartase activity could be immobilized, these disadvantages might be overcome. Therefore we studied the immobilization of whole microbial cells[2-4] and succeeded in refining an industrial technique in 1973 (see also [13], [53]).

In this article, methods are described for the immobilization of *Escherichia coli* having aspartase activity and for the continuous production of L-aspartic acid from ammonium fumarate by using a column packed with the immobilized *E. coli* cells. These methods are essentially same as those reported by Chibata *et al.*[2,4] and Tosa *et al.*[3]

Assay Methods

Estimation of L-Aspartic Acid. L-Aspartic acid is measured by bioassay using *Leuconostoc mesenteroids* P-60 according to the method of Henderson and Snell.[5]

---

[1] T. Tosa, T. Sato, T. Mori, Y. Matuo, and I. Chibata, *Biotechnol. Bioeng.* **15**, 69 (1973).
[2] I. Chibata, T. Tosa, and T. Sato, *Appl. Microbiol.* **27**, 878 (1974).
[3] T. Tosa, T. Sato, T. Mori, and I. Chibata, *Appl. Microbiol.* **27**, 886 (1974).
[4] I. Chibata, T. Tosa, T. Sato, T. Mori, and K. Yamamoto, *in* "Enzyme Engineering" (E. K. Pye and L. B. Wingard, Jr., eds.), Vol. 2, p. 303. Plenum, New York, 1974.
[5] L. M. Henderson and E. E. Snell, *J. Biol. Chem.* **172**, 15 (1948).

Assay of Aspartase Activity. (i) INTACT CELLS. A mixture of 1 g (wet weight) of intact cells and 30 ml of 1 M ammonium fumarate (adjusted to pH 8.5 with ammonia; containing 1 mM Mg^{2+}) is incubated at 37° for 1 hr with shaking. After the reaction is stopped by immersion in a boiling-water bath for 5 min, the cells are removed by centrifugation. L-Aspartic acid in the supernatant is measured by the above-mentioned method. Its activity is expressed as micromoles of L-aspartic acid produced per hour.

(ii) IMMOBILIZED CELLS. A reaction mixture of immobilized cells (amounts corresponding to 1 g of intact cells) and 30 ml of 1 M ammonium fumarate (adjusted to pH 8.5 with ammonia; containing 1 mM Mg^{2+}) is incubated at 37° for 1 hr with shaking. After the reaction, the immobilized cells in the reaction mixture are filtered off. L-Aspartic acid in the filtrate is measured by the above-mentioned method. Its activity is expressed as micromoles of L-aspartic acid produced per hour.

In some cases the enzyme activities of both intact and immobilized cells can be determined by measuring spectrophotometrically the remaining fumaric acid in the enzyme reaction mixtures.[6]

Culture of *Escherichia coli*

Escherichia coli ATCC 11303 is cultured under aerobic conditions at 37° for 16–24 hr in 100 liters of medium (pH 7.0) containing ammonium fumarate (3 kg), K_2HPO_4 (0.2 kg), $MgSO_4 \cdot 7H_2O$ (50 g), $CaCO_3$ (50 g), and corn steep liquor (4 kg). The cells are collected by centrifugation. About 1 kg (wet weight) of *E. coli* cells is obtained from 100 liters of broth, and its aspartase activity is approximately 1500–2000 μmoles per hour per gram of wet cells in fresh state.

Preparation of Immobilized *E. coli* Cells

Among several methods tested,[2] the most active immobilized *E. coli* cells were obtained by entrapping the cells into a polyacrylamide gel lattice. To prepare the most efficient immobilized cell preparation, the type and concentration of bifunctional reagents, the concentration of acrylamide, and the amount of cells to be entrapped are important factors. Optimum conditions for immobilization of *E. coli* cells are now described.

E. coli Cell Suspension. The *E. coli* cells (1 kg wet weight) are suspended in 2 liters of physiological saline or resuspended in the broth after removing and weighing the cells, and the cell suspension is cooled to 8°.

[6] E. Racker, *Biochim. Biophys. Acta* **4**, 211 (1950).

Monomer Solution. Acrylamide monomer (750 g) and N,N'-methylenebisacrylamide (40 g) are dissolved in 2.4 liters of water, and the monomer solution is cooled to 8°.

Immobilization of E. coli Cells. The *E. coli* cell suspension and monomer solution prepared as described above are mixed at 8°. To the mixture 100 ml of 25% (v/v) β-dimethylaminopropionitrile (as an accelerator of polymerization) and 500 ml of 1% potassium persulfate (as an initiator of polymerization) are added, and the reaction mixture is allowed to stand at 20°–25°. The polymerization reaction starts after about 5 min, and the temperature of the reaction mixture increases as polymerization proceeds. When the temperature has increased to 30°, the resulting stiff gel is rapidly cooled with ice-cold water, and is allowed to stand for 15–20 min to complete the polymerization. In this process, it is very important to keep the temperature of the gel below 50°.

The gel is made into granules about 3–4 mm in diameter for industrial use and washed with water. Its aspartase activity is generally 1300–1800 μmoles/hr for 10 ml of gel (corresponding to 1 g of wet cells) in this state.

Activation of Immobilized E. coli Cells. When the immobilized *E. coli* cells are suspended in 1 M ammonium fumarate (pH 8.5) containing 1 mM Mg^{2+} at 37° for 24–48 hr, its activity increases 9–10 times higher and is 12,000–17,000 μmoles/hr for 10 ml of gel. This activation phenomenon is due to an increase in membrane permeability for the substrate and/or product owing to the autolysis of *E. coli* cells in the gel lattice.[2,4]

Enzymic Properties of Immobilized *E. coli* Cells

The pH-activity profile of the aspartase reaction is different for intact and immobilized cells, as shown in Fig. 1. The immobilized cells show an optimal activity at pH 8.5 (the same as native aspartase), whereas the optimal pH of the intact cells is 10.5.

The optimal temperature for the formation of L-aspartic acid is 50° in intact and immobilized cells,[2,4] and the activation energy of the immobilized *E. coli* cells is calculated to be 12.36 kcal/mole.[7]

The aspartase activities of intact and immobilized cells are not activated by Mn^{2+}, which activates the native and immobilized aspartases. The heat stability of the immobilized cells is somewhat higher than that of the intact cells. Bivalent metal ions such as Mn^{2+}, Mg^{2+}, and Ca^{2+} protect against thermal inactivation of the aspartase activity of the intact and immobilized cells.[2]

[7] T. Sato, T. Mori, T. Tosa, I. Chibata, M. Furui, K. Yamashita, and A. Sumi, *Biotechnol. Bioeng.* **17,** 1797 (1975).

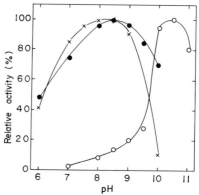

FIG. 1. Effect of pH on the aspartase reaction. The enzyme assay is carried out under standard conditions except for the substrate employed. The substrate solution contains 0.5 M fumaric acid and 0.7 M NH$_4$Cl solution (adjusted to the indicated pH with 5 N NaOH) containing 1 mM MgCl$_2$. ×——×, Aspartase extracted from cells; ○——○; intact cells, ●——●, immobilized cells.

Design of the Column

In order to design the most efficient column for continuous enzyme reactions, it is necessary to satisfy a number of factors. The following are important factors when preparing immobilized microbial cells.

Pressure Drop of the Column. The intrinsic pressure drop of the im-

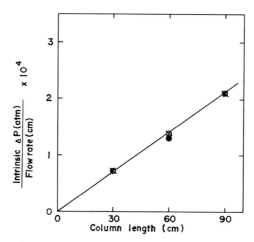

FIG. 2. Pressure drop of immobilized cell column. A substrate solution is passed through various lengths of the immobilized cell columns at the specified flow rate at 37°, and the pressure drop of columns is measured by a mercury manometer. Flow rate: ●——●, 2 cm/min; ○——○, 4 cm/min; ×——×, 8 cm/min.

mobilized *E. coli* cell column is shown in Fig. 2. It is very low, that is, about 7 times lower than that of immobilized aminocylase (see [51] on aminoacylase) prepared by ionic binding the enzyme to DEAE-Sephadex A-25.

Effect of Column Dimension on the Reaction Rate. The reaction rate for the formation of L-aspartic acid is independent of the column dimension when the concentration of substrate is about 1 M.[7]

Exothermic Reactions. Since the aspartase reaction is exothermic, an increase of 6° is observed during the reaction for a 1 M concentration of substrate. Therefore, it is desirable to remove the accumulated heat from the column. Therefore the aspartase reactor system using immobilized *E. coli* cells is designed essentially in the same way as that for the immobilized aminoacylase system (see Fig. 4 in [51]).

Continuous Enzyme Reaction

The basic conditions for continuous production of L-aspartic acid from ammonium fumarate using a column packed with the immobilized *E. coli* cells are as follows.

Preparation of Substrate Solution. Fumaric acid (11.6 kg) and $MgCl_2 \cdot 6H_2O$ (20.3 g) are dissolved in 25% ammonia–water solution (15 liters) and water (60 liters). Then the solution is adjusted to pH 8.5 with 25% ammonia–water, and made up to 100 liters with water.

Effect of Flow Rate of Substrate Solution on Formation of L-*Aspartic Acid.* The relationship between flow rate of substrate and formation

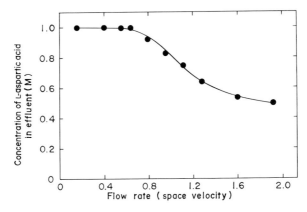

FIG. 3. Relationship of flow rate of substrate to the production of L-aspartic acid from ammonium fumarate. A substrate solution is applied to the immobilized cell column (10 × 100 cm) at 37° at the indicated flow rates. The L-aspartic acid produced in the effluent is measured.

of L-aspartic acid from ammonium fumarate is shown in Fig. 3. The data show that the flow rate of space velocity[8] = 0.8 is maximal for the complete conversion of ammonium fumarate to L-aspartic acid.

Stability of Immobilized E. coli Cell Column. Data on the stability of immobilized *E. coli* cell column are shown in Fig. 4. These data are obtained by passing a substrate solution through the column (10 × 100 cm) for long periods at the maximal flow rate possible for the complete conversion of ammonium fumarate to L-aspartic acid at 37°, 42°, and 45°. The results show that decay of the activity depends on temperature. The column is very stable, i.e., its half-life is 120 days at 37°.

On the other hand, the stability of the immobilized aspartase prepared by the polyacrylamide gel method is not very good, i.e., its half-life is 27 days at 37°.[1]

Industrial Production of L-Aspartic Acid. A substrate solution is passed through a column packed with immobilized *E. coli* cells at a flow rate of space velocity = 0.6 at 37°. The effluent of appropriate volume is adjusted to pH 2.8 with 60% H_2SO_4 at 90° and then cooled at 15°. L-Aspartic acid which crystallized out is collected by centrifugation or by filtration and washed with water. The yield is 90–95% (theoretical). $[\alpha]_D^{20} = +25.5°$ ($c = 8$ in 6 N HCl). If we use a 1000-liter column, about 1700 kg of pure L-aspartic acid per day can be obtained from the column effluent.

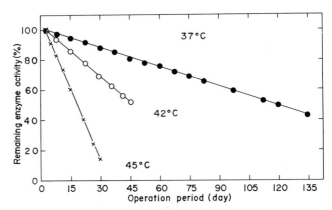

FIG. 4. Stability of the immobilized *Escherichia coli* cell column. A substrate solution is applied to the immobilized cell column (10 × 100 cm) for long periods at indicated temperature and flow rate; i.e., space velocity = 0.8 for 37°, space velocity = 1.0 for 42°, and space velocity = 1.1 for 45°. The L-aspartic acid produced in the effluent is measured.

[8] Space velocity is defined as the volume of liquid passing through a given volume of immobilized cells or immobilized enzyme in 1 hr divided by the latter volume.

Economy

Since the autumn of 1973 we have been operating this new system industrially. The overall production cost of this immobilized cell system is reduced by about 60% compared to the conventional batch process using intact cells or the fermentation process.

Other Applications of Immobilized Microbial Cells

Besides the continuous production of L-aspartic acid, we have studied the continuous production of useful organic compounds, such as L-citrulline,[9] urocanic acid,[10] and 6-aminopenicillanic acid (6-APA) (see also [52]),[11] using immobilized microbial cells and have obtained satisfactory results.

L-Citrulline is used for medicines. It has been produced from L-arginine by the action of microbial L-arginine deiminase. The immobilization of *Pseudomonas putida* ATCC 4539, having a high enzyme activity, was carried out by the polyacrylamide gel method. The stability of a column packed with the immobilized *P. putida* cells is very high, i.e., its half-life is 140 days at 37°, and pure L-citrulline is obtained in a high yield from the column effluent.[9]

Urocanic acid is used as a sun-screening agent and is produced from L-histidine by the action of microbial L-histidine ammonia-lyase. We tested the immobilization of several microorganisms having this enzyme activity using the polyacrylamide gel method and chose *Achromobacter liquidum* IAM 1667, which showed the highest activity after immobilization. From a column packed with the immobilized cells, pure urocanic acid is continuously obtained in a high yield. The enzyme activity of the column is very stable in the presence of Mg^{2+}, i.e., its half-life is 180 days at 37°.[10]

This technique is also applied for the production of 6-APA from penicillins by using immobilized *E. coli* ATCC 9637 cells having high penicillin amidase activity. The immobilized cell column is fairly stable; its half-life is 17 days at 40° and 42 days at 30°.[11]

Conclusion

As described above, several kinds of microbial cells having high enzyme activity can be easily immobilized and stabilized by the entrap-

[9] K. Yamamoto, T. Sato, T. Tosa, and I. Chibata, *Biotechnol. Bioeng.* **16**, 1589 (1974).
[10] K. Yamamoto, T. Sato, T. Tosa, and I. Chibata, *Biotechnol. Bioeng.* **16**, 1601 (1974).
[11] T. Sato, T. Tosa, and I. Chibata, *Europ. J. Appl. Microbiol.*, in press (1976).

ping method using polyacrylamide gel. The continuous enzyme reaction using the immobilized microbial cells can be employed advantageously in the following cases: (1) when enzymes are intracellular, (2) when enzymes extracted from microbial cells are unstable, (3) when enzymes are unstable during and after immobilization, (4) when the microorganism contains no other enzymes that catalyze interfering side reaction, or those interfering enzyme are readily inactivated or removed, and (5) when substrates and products are not high-molecular-weight compounds and can easily pass through the gel lattice. Further, the volume of fermentation broth for the unit production of desired compound is much smaller for the continuous method using immobilized cells than for the conventional fermentative method. The continuous method is very advantageous from the point of view of industrial water pollution.

[51] Production of L-Amino Acids by Aminoacylase Adsorbed on DEAE-Sephadex

By ICHIRO CHIBATA, TETSUYA TOSA, TADASHI SATO, and TAKAO MORI

Recently, fermentative and organic synthetic methods have been employed for the industrial production of L-amino acids instead of the conventional isolation method from protein hydrolyzate. Amino acids produced by the fermentative method are L-form, whereas amino acids produced by chemical synthetic method are optically inactive racemic mixtures of L- and D-isomers. To obtain the L-amino acid from the chemically synthesized DL-form, optical resolution is necessary. The enzymic method for optical resolution using mold aminoacylase (EC 3.5.1.14) is one of the most advantageous procedures for industrial production of optically pure L-amino acids.[1] The reaction is catalyzed by the aminoacylase.

$$\text{DL-R-CHCOOH} \atop \text{NHCOR'} + H_2O \xrightarrow{\text{aminoacylase}} \text{L-R-CHCOOH} \atop \text{NH}_2 + \text{D-R-CHCOOH} \atop \text{NHCOR'}$$

racemization

+ R'COOH

This mold aminoacylase has versatile substrate specificity and can

[1] I. Chibata, T. Ishikawa, and S. Yamada, *Bull. Agric. Chem. Soc. Jpn.* **21**, 300 (1957).

be used for the production of many kinds of L-amino acids. From 1954 to 1969, this enzymic resolution method had been employed in Tanabe Seiyaku Co., Osaka, Japan, for the industrial production of several L-amino acids.

However, the enzyme reaction had been carried out in a batch process by incubating a mixture containing substrate and soluble enzyme, and the procedure had some disadvantages for industrial use. To overcome these disadvantages, the immobilization of aminoacylase and the continuous optical resolution of DL-amino acids using a column packed with the immobilized enzyme have been studied.[2-12] As a result, the efficient and automatically controlled enzyme reactor system was devised, and since 1969 we have been operating several series of these reactors in our industrial plants.

In this article, the preparation, the enzymic properties, and the industrial application of immobilized aminoacylases are presented.

Enzyme Assay

Native Aminoacylase. The standard enzyme assay of native aminoacylase is carried out as follows. A reaction mixture of 1 ml of 0.1 M phosphate buffer (pH 7.5), 1 ml of 0.1 M acetyl-DL-methionine (pH 7.5) containing 1.5 mM $CoCl_2$, and 1 ml of aminoacylase solution is incubated for 30 min at 37°. The liberated L-methionine is measured by the ninhydrin colorimetric method using the Technicon Auto-Analyzer system.[13] The enzyme activity of aminoacylase is expressed in micromoles of L-methionine liberated per hour.

Immobilized Aminoacylase. The standard enzyme assay of immobilized aminoacylase is carried out as follows. A reaction mixture of

---

[2] T. Tosa, T. Mori, N. Fuse, and I. Chibata, *Enzymologia* **31**, 214 (1966).
[3] T. Tosa, T. Mori, N. Fuse, and I. Chibata, *Enzymologia* **31**, 225 (1966).
[4] T. Tosa, T. Mori, N. Fuse, and I. Chibata, *Enzymologia* **32**, 153 (1967).
[5] T. Tosa, T. Mori, N. Fuse, and I. Chibata, *Biotechnol. Bioeng.* **9**, 603 (1967).
[6] T. Tosa, T. Mori, N. Fuse, and I. Chibata, *Agric. Biol. Chem.* **33**, 1047 (1969).
[7] T. Tosa, T. Mori, and I. Chibata, *Agric. Biol. Chem.* **33**, 1053 (1969).
[8] T. Tosa, T. Mori, and I. Chibata, *Enzymologia* **40**, 49 (1971).
[9] T. Tosa, T. Mori, and I. Chibata, *J. Fermentn. Technol.* **49**, 522 (1971).
[10] T. Sato, T. Mori, T. Tosa, and I. Chibata, *Arch. Biochem. Biophys.* **147**, 788 (1971).
[11] T. Mori, T. Sato, T. Tosa, and I. Chibata, *Enzymologia* **43**, 213 (1972).
[12] I. Chibata, T. Tosa, T. Sato, T. Mori, and Y. Matuo, in "Fermentation Technology Today" (*Proc. Int. Fermentn. Symp 4th*), p. 383. Society of Fermentation Technology, Osaka, Japan, 1972.
[13] N. G. Cadavid and A. C. Paladini, *Anal. Biochem.* **9**, 170 (1964).

10 ml of 0.1 M phosphate buffer (pH 7.0), 10 ml of 0.1 M acetyl-DL-methionine (pH 7.0) containing 1.5 mM $CoCl_2$, 10 ml of water, and 100 mg of the immobilized aminoacylase is incubated for 30 min at 37° with shaking. After the reaction, the immobilized aminoacylase in the reaction mixture is filtered off, and the liberated L-methionine in the filtrate is measured by the method described above.[13] The enzyme activity is expressed in micromoles of L-methionine liberated per hour.

Preparation of Native Aminoacylase[2,10]

A culture medium made up of 1000 g of wheat bran, 500 g of rice hulls, and 750 ml of tap water is inoculated with *Aspergillus oryzae* No. 9, which produces higher aminoacylase activity.[1] The culture is incubated for 5 days at 30°. Then it is extracted by stirring with 15 liters of water for 2 hr and is filtered with the aid of Celite. The filtrate is adjusted to pH 7.0 with 1 N NaOH and fractionated using ammonium sulfate. The protein fraction precipitating between 30% and 50% saturation is collected, dissolved in water, and dialyzed against cold water for 16 hr. The resulting solution is lyophilized. The activity of the preparation is 20 µmoles/hr per milligram of the preparation under standard assay conditions.

Other enzyme preparations produced from *Aspergillus oryzae* may also be employed.

Immobilization of Aminoacylase

Various immobilization methods were investigated,[12] and relatively active and stable immobilized aminoacylases were obtained by ionic binding to DEAE-Sephadex,[5] covalent binding to iodoacetylcellulose,[10] and entrapping into polyacrylamide gel lattices.[11] Typical immobilization methods for mold aminoacylase are as follows.

Immobilization by Ionic Binding to DEAE-Sephadex.[5] At room temperature (20°–25°), 100 g of DEAE-Sephadex A-25 (3.5 ± 0.5 meq/g, bead type) suspended in 1 liter of distilled water are added to 5 liters of 0.1 N NaOH, and the suspension is stirred for 3 hr. After filtration, the precipitate is washed with water until the washing solution is neutralized, stirred with 5 liters of 0.1 M phosphate buffer (pH 7.0) for 3 hr and kept standing for 16 hr. The suspension is filtered, and the resulting precipitate is stirred with 1.67 liters of aminoacylase solution (total enzyme activity: 167,000 µmoles/hr) for 3 hr at room temperature. After filtration, 5 liters of distilled water are added to the precipitate; the preparation is stirred for 1 hr, then filtered. For further washing, 5 liters of 0.2 M

sodium acetate are added to the precipitate; the preparation is stirred for 1 hr, then filtered. The precipitate is washed with distilled water until no protein is detected in the washing solution. These washing procedures are carried out at room temperature. The resulting precipitate can be used for the enzyme reaction.

In order to store the immobilized preparation for a long period or to measure its enzyme activity, it is suspended in 2.5 liters of distilled water and lyophilized. Using this procedure, 106 g of DEAE-Sephadex–aminoacylase is obtained. The activity of the preparation is about 700 μmoles/hr per gram of preparation under standard assay conditions.

Immobilization by Covalent Binding to Iodoacetylcellulose.[10] (i) PREPARATION OF IODOACETYLCELLULOSE. Iodoacetylcellulose is prepared by the exchange reaction of bromoacetylcellulose with iodine according to the method of Patchornik.[14] Bromoacetylcellulose is prepared as follows. Ten grams of cellulose powder (100 ∼ 200 mesh) are dispersed in a solution of bromoacetic acid (100 g) dissolved in 30 ml of dioxane, and the mixture is stirred for 16 hr at 30°. To the mixture 75 ml of bromoacetylbromide are added, and the stirring is continued at 30° for 7 hr. The mixture is poured into 2 liters of ice-water; the resulting white solid bromoacetylcellulose is collected by filtration and exhaustively washed with 0.1 M sodium bicarbonate and water.

To 6 g of the bromoacetylcellulose, 36 g of sodium iodide and 300 ml of 95% alcohol are added, and the mixture is stirred for 20 hr at 30°. After filtration, the resulted iodoacetylcellulose is exhaustively washed with alcohol, 0.1 M sodium bicarbonate, water, and alcohol, and dried. The cellulose derivative contains about 30% iodine by weight. This indicates that the derivative contains approximately 1 iodine for every 2–3 glucose residues, or 2.3–2.5 mmoles of iodine per gram of the preparation.

(ii) BINDING OF AMINOACYLASE TO IODOACETYLCELLULOSE. To a solution of aminoacylase (total enzyme activity: 5000 μmoles/hr) dissolved in 50 ml of 0.2 M phosphate buffer (pH 8.5), 1 g of iodoacetylcellulose and 10 g of ammonium sulfate[15] are added. The mixture is stirred for 24 hr at 7° and then centrifuged. The resulting iodoacetylcellulose–aminoacylase complex is washed three times with 200 ml of physiological saline and twice with 200 ml of 0.2 M phosphate buffer (pH 8.5), and can be used for enzyme reaction.

In order to store the immobilized preparation for a long period or to measure its enzyme activity, it is lyophilized. The activity of the

---

[14] A. Patchornik, U.S. Patent No. 3,278,392 (1966).
[15] Addition of ammonium sulfate in the reaction mixture is essential for obtaining the active immobilized aminoacylase.

preparation is about 2100 μmoles/hr per gram of the preparation under standard assay conditions.

Immobilization by Entrapping into a Polyacrylamide Gel Lattice.[11] To 15 ml of aqueous solution containing 3.75 g of acrylamide monomer and 0.2 g of N,N'-methylenebisacrylamide, 5 ml of a solution of aminoacylase (5000 μmoles/hr) dissolved in 0.1 M phosphate buffer (pH 7.0) are added, and to the mixture 2.5 ml of 5% β-dimethylaminopropionitrile and 1% potassium persulfate solution are added. The reaction mixture is incubated at 23° for 10 min. The polymerization reaction occurs, and a stiff gel similar to agar is obtained. In some cases, the temperature in the gel increases to nearly 50° during polymerization. As the aminoacylase is inactivated by exposure to heat, polymerization must be carried out below 50°. The resulting gel is immediately immersed in ice-water and cut into small blocks to accelerate cooling. Then the gel particles are thoroughly washed with 0.2 M phosphate buffer (pH 7.0) and water. The resulting polyacrylamide–aminoacylase is made into granules 2–4 mm in diameter and can be used for enzyme reaction.

In order to store the gel particles for a long period, to measure their enzyme activity, or to make uniform gel particles, they are lyophilized. The lyophilized gel particles are passed through a 42–100 mesh sieve. By this procedure, 5 g of immobilized aminoacylase are obtained. The activity of the preparation is about 300 μmoles/hr per gram of the preparation under standard assay conditions.

Enzymic Properties of Immobilized Aminoacylase[7,10,11]

The enzymic properties of the three immobilized aminoacylases prepared by the above methods are summarized in Table I with those of the native enzyme.

The optimum pH of the DEAE-Sephadex–aminoacylase shifts about 0.5–1.0 pH unit more to the acid side than that of the native enzyme. This shift may be explained by the redistribution of hydrogen ions between the positively charged DEAE-Sephadex and the surrounding aqueous medium. This shift is also found for polyacrylamide–aminoacylase, but the reason is not clear in this case.

A significant difference is observed between the optimum temperature and the activation energy of the immobilized enzymes and the native one. The DEAE-Sephadex–aminoacylase complex shows the highest optimum temperature.

There is no marked difference on the effect of metal ions, inhibitors, substrate specificity, optical specificity, and kinetic constants between the immobilized enzymes and the native one.

TABLE I
COMPARISON OF ENZYMIC PROPERTIES OF NATIVE AND
IMMOBILIZED AMINOACYLASES[a]

Properties	Native aminoacylase	Immobilized aminoacylases		
		Ionic binding to DEAE-Sephadex	Covalent binding to iodoacetyl-cellulose	Entrapping by poly-acrylamide
Optimum pH on the reaction	7.5 ~ 8.0	7.0	7.5 ~ 8.5	7.0
Optimum temperature on the reaction (°C)	60	72	55	65
Activation energy (kcal/mole)	6.9	7.0	3.9	5.3
Optimum concentration of Co^{2+} (mM)	0.5	0.5	0.5	0.5
K_m (mM)	5.7	8.7	6.7	5.0
V_{max} (μmoles/hr)	1.52	3.33	4.65	2.33
Operation stability, half-life[b] (days)	—	65(50°)	—	48(37°)

[a] Data for acetyl-DL-methionine.
[b] The time required for 50% of the enzyme activity to be lost.

Heat stability of the enzyme is an important factor in the industrial application of an immobilized enzyme. The effect of temperature on the stability of the immobilized enzymes and the native ones is shown in Fig. 1. Among the immobilized aminoacylases, DEAE-Sephadex–aminoacylase complex shows the highest stability. It also shows a strong resistance toward proteases, organic solvents, and protein denaturing agents.[7,8]

Selection of Immobilized Aminoacylase Suitable for Industrial Application

For the industrial application of immobilized enzymes, it is necessary to satisfy many conditions. Two factors are most important in the case of immobilized aminoacylase, because aminoacylase and the carriers for immobilization are relatively expensive: (1) operation stability of immobilized enzymes, and (2) regenerability of decayed immobilized aminoacylase columns after long periods of operation.

In the case of covalent binding to iodoacetylcellulose, the immobilization technique is difficult, its cost is high, and regeneration of decayed

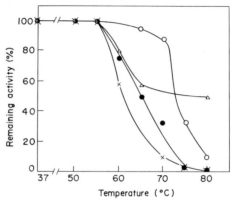

Fig. 1. Comparison of heat stability of native aminoacylase and immobilized ones. In the case of native aminoacylase (×——×), a mixture of 1 ml of 0.1 M Veronal buffer (pH 7.0) and 1 ml of native aminoacylase is incubated for 10 min at the specified temperature, then rapidly cooled. The remaining enzyme activity in the mixture is determined under standard conditions. In the case of immobilized aminoacylases, a mixture of 10 ml of $M/15$ Veronal buffer (pH 7.0) and 40 mg of immobilized aminoacylases is incubated with shaking for 10 min at the specified temperature, and rapidly cooled. The remaining enzyme activity in the mixture is also determined under standard conditions. The values obtained at the preincubation temperature of 37° are taken as 100%. ○——○, DEAE-Sephadex-aminoacylase; △——△, iodoacetylcellulose–aminoacylase; ●——●, aminoacylase entrapped in polyacrylamide. From I. Chibata, T. Tosa, T. Sato, T. Mori, and Y. Matuo, in "Fermentation Technology Today" (*Proc. Int. Fermentn. Symp. 4th*), p. 383. Society of Fermentation Technology, Osaka, Japan, 1972.

immobilized enzyme column is impossible. In the case of entrapping by polyacrylamide, immobilization costs are not so high, but operation stability is lower than in the case of ionic binding to DEAE-Sephadex as shown in Table I.

On the other hand, in the case of ionic binding to DEAE-Sephadex, the preparation is easy, the operation stability is highest, and the regeneration of decayed immobilized preparation is possible. Thus, we chose immobilized DEAE-Sephadex–aminoacylase as the most advantageous enzyme preparation for the industrial production of L-amino acids.

Enzyme Reactor for Continuous Aminoacylase Reaction

For design of the most efficient enzyme column, the following are important factors: (1) effect of flow rate of substrate on the reaction rate, (2) flow system of the substrate solution, (3) effect of column dimension on the reaction rate, and (4) pressure drop of the column.

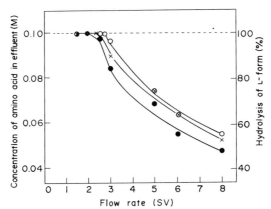

FIG. 2. Extent of hydrolysis of acetyl-DL-amino acids by immobilized DEAE-Sephadex-aminoacylase. A solution of 0.2 M acetyl-DL-amino acids (pH 7.0, containing 5×10^{-4} M CoCl$_2$) is passed through the column at flow rates of space velocity = 1.5–8.0 at 50°. ○——○, acetyl-DL-methionine; ●——●, acetyl-DL-phenylalanine; ×——×, acetyl-DL-valine.

Flow Rate of the Substrate Solution. For the optical resolution of acyl-DL-amino acids, it is desirable to complete the asymmetric hydrolysis. Data on the relationship between the flow rate of the substrate solution and the extent of the reaction are shown in Fig. 2. At the flow rates of space velocity (SV) = 2.8 for acetyl-DL-methionine, SV = 2.0 for acetyl-DL-phenylalanine, and SV = 2.4 for acetyl-DL-valine, the reaction ceases at the cleavage of 0.1 M substrate, that is, 50% hydrolysis of racemic substrates or 100% hydrolysis of the L-form of substrates. This indicates that the immobilized DEAE-Sephadex–aminoacylase column is optically specific.

Table II shows the space velocity of a substrate solution for complete hydrolysis in the column. These data are for freshly immobilized aminoacylase; in the case of decayed immobilized enzyme it is necessary to

TABLE II
MAXIMAL FLOW RATE OF SUBSTRATES FOR
COMPLETE ASYMMETRIC HYDROLYSIS

Substrates	Flow rate (space velocity)
Acetyl-DL-alanine	1.4
Acetyl-DL-methionine	2.8
Acetyl-DL-phenylalanine	2.0
Acetyl-DL-tryptophan	1.2
Acetyl-DL-valine	2.4

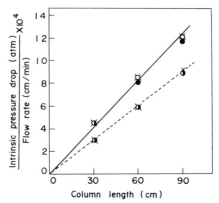

Fig. 3. Pressure drop of immobilized DEAE-Sephadex–aminoacylase column. A solution of 0.2 M acetyl-DL-methionine (pH 7.0, containing 5×10^{-4} M $CoCl_2$) is passed through various lengths of the aminoacylase columns at the specified flow rate at 37° and 50°. The pressure drop of columns is measured by a mercury manometer. Flow rate: ●——●, 2 cm/min; ○——○, 4 cm/min; ×——× 8 cm/min. Temperature: ——, 37°; ----, 50°

slow down the space velocity of a substrate solution according to the activity of the column.

Flow System of Substrate Solution. The activities of aminoacylase columns are equal in both upward and downward flows.[7] In practice, downward flow is employed to prevent channeling of the column because air bubbles evolved from the warmed substrate solution can be easily separated in the top part of the column.

Effect of Column Dimension. The difference in reaction rates due to the variation in column dimensions is scarcely observed in the DEAE-Sephadex–aminoacylase column.[7] Therefore, if the immobilized aminoacylase is uniformly packed into the column and the substrate solution is smoothly passed through the aminoacylase column, the enzyme reaction proceeds at the same reaction rate regardless of the column dimensions.

Pressure Drop. To measure the pressure drop of the DEAE-Sephadex–aminoacylase column, the Kozeny and Carman equation can be applied. The pressure drop is proportional to the flow rate and column length at a specified temperature (data are shown in Fig. 3). From Fig. 3 the pressure drop of the enzyme column can be calculated.[9]

Design of the Reactor System. Taking into consideration the above factors, the aminoacylase reactor system was designed for continuous production of L-amino acid by immobilized aminoacylase.[12] The flow diagram is shown in Fig. 4. This system is automatically controlled, and the enzyme reaction proceeds continuously.

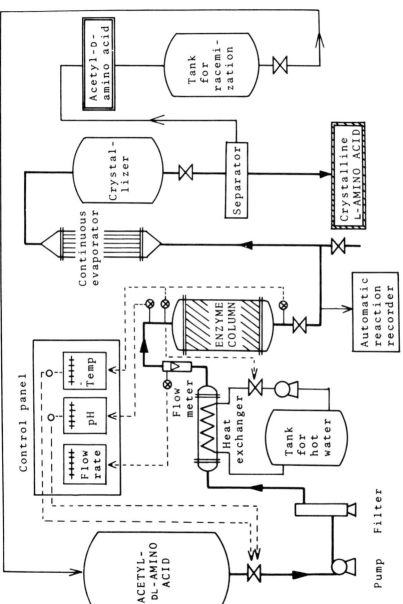

Fig. 4. Flow diagram for continuous production of L-amino acid by immobilized aminoacylase.

TABLE III
PRODUCTION OF L-AMINO ACIDS BY DEAE-SEPHADEX
AMINOACYLASE COLUMN OF 1000 LITER-VOLUME

		Yield (theory) of L-amino acids per	
L-Amino acids	Space velocity[a]	24 hr (kg)	20 days (kg)
L-Alanine	1.0	214	6,420
L-Methionine	2.0	715	21,450
L-Phenylalanine	1.5	594	17,820
L-Tryptophan	0.9	441	13,230
L-Valine	1.8	505	15,150

[a] Space velocity is the volume of liquid passing through a given volume of immobilized cells or immobilized enzyme in 1 hr divided by the latter volume. As the values are for continuous operation for 30 days without change of flow rate, they are somewhat lower than those of Table II.

Preparation of Optically Active Amino Acids

Optically active amino acids can be produced on a laboratory scale by employing a column packed with the immobilized aminoacylase. DEAE-Sephadex–aminoacylase prepared by the method given in the preceding section is suspended in water, and the suspension is poured into a glass column of appropriate volume having an outer jacket for circulating water to maintain the temperature at 50°. At 50°, a solution of 0.2 M[16] acetyl-DL-amino acids (containing 5×10^{-4} M $CoCl_2$; pH adjusted to 7.0 with NaOH) is passed through the column at a specified flow rate suitable for complete hydrolysis of acetyl-L-amino acid as shown in Table II.

The concentration of liberated L-amino acids in the column effluent is determined by the ninhydrin colorimetric method. The isolation procedures for L-amino acids and acetyl-D-amino acids from the column effluent are given in Table III.

Alanine. One liter of the column effluent is evaporated *in vacuo* to syrup. To the condensed residue 100 ml of ethanol are added, and the residue is cooled to 10°. The separated crude L-alanine is collected by filtration and recrystallized from water. The yield is 6.2 g (69.7% of the theoretical), $[\alpha]_D^{25} = +12.9$ ($c = 2$ in $5 N$ HCl).

After filtration of the crude L-alanine, the mother liquor is evapo-

[16] This concentration is very important. If a concentration of a substrate solution higher than 0.2 M is charged into the column, the aminoacylase leaks from the column and the enzyme activity of the column rapidly decreases.

rated to remove ethanol, and made up to 100 ml with water. The solution is passed through a column packed with Amberlite IR-120 (H^+ form). The column is washed with water and the combined eluent is concentrated *in vacuo*. The separated crude acetyl-D-alanine is collected by filtration and is recrystallized from water. The yield is 9.2 g (70.0% of the theoretical), m.p. 125°, $[\alpha]_D^{16} = +67.0°$ ($c = 5$ in water).

Methionine. One liter of the column effluent is evaporated *in vacuo* to about 50 ml, and the condensed residue is cooled to 10°. The separated crude L-methionine is collected by filtration and recrystallized from water. The yield is 12.0 g (80.5% of the theoretical), $[\alpha]_D^{25} = +23.4°$ ($c = 3$ in 1 N HCl).

After filtration of crude L-methionine, the mother liquor is made up to 100 ml with water, and passed through a column packed with Amberlite IR-120 (H^+ form). The procedures that follow are the same as those for acetyl-D-alanine. The yield of recrystallized acetyl-D-methionine is 16.8 g (87.9% of the theoretical), m.p. 104°, $[\alpha]_D^{25} = +20.1°$ ($c = 3$ in water).

Phenylalanine. One liter of the column effluent is treated using same procedures as for L-methionine, and recrystallized L-phenylalanine is obtained. The yield is 14.1 g (85.5% of the theoretical), $[\alpha]_D^{25} = -34.5°$ ($c = 1$ in water).

After filtration of crude L-phenylalanine, the mother liquor is adjusted to pH 1.8 with concentrated HCl, then cooled to 10°.

The separated crude acetyl-D-phenylalanine is collected by filtration, then recrystallized from water. The yield is 17.9 g (86.6% of theoretical), m.p. 170°, $[\alpha]_D^{27} = -50.5°$ ($c = 1$ in absolute ethanol).

Tryptophan. One liter of the column effluent is evaporated *in vacuo* to about 20 ml, and the same procedures are followed as in the case of L-methionine. The yield of recrystallized L-tryptophan is 14.7 g (72.0% of the theoretical), $[\alpha]_D^{25} = -33.7°$ ($c = 1$ in water).

The recrystallized acetyl-D-tryptophan is obtained by the same procedures used for acetyl-D-phenylalanine. The yield is 23.1 g (93.9% of the theoretical), m.p. 189°, $[\alpha]_D^{25} = -26.1°$ ($c = 1$ in methanol).

Valine. One liter of the column effluent is evaporated *in vacuo* to about 40 ml, and the same procedures are followed as for L-methionine. The yield of recrystallized L-valine is 9.5 g (81.2% of the theoretical), $[\alpha]_D^{25} = +28.3°$ ($c = 1$ in 5 N HCl).

After filtration of crude L-valine, the mother liquor is made up to 200 ml with water, and the same procedures are followed as for acetyl-D-methionine. The yield of recrystallized acetyl-D-valine is 15.6 g (98.1% of the theoretical), m.p. 168°, $[\alpha]_D^{25} = +20.1°$ ($c = 4$ in water).

To obtain optically pure acetyl-D-amino acid from the aminoacylase column effluent, it is important that the enzyme reaction be complete, i.e., that acetyl-L-amino acid does not remain in the effluent. Further, to obtain D-amino acid from acetyl-D-amino acid, chemical hydrolysis should be carried out without racemization, employing hydrochloric acid.

Stability and Regeneration of the Aminoacylase Column

The stability of the DEAE-Sephadex–aminoacylase column in an industrial operation for a long period at 50° is shown in Fig. 5. The column maintains more than 60% of the initial activity after over 30 days of operation, and the half-life (the time required for 50% of the enzyme activity to be lost) of the column is estimated to be about 65 days.

As described in a previous section, regeneration of the decayed column is very important for industrial use. In this case, the aminoacylase column decayed by a long period of operation can be completely reactivated by the addition of an amount of aminoacylase corresponding to the decayed activity as shown in Fig. 5.[6]

Since DEAE-Sephadex is much more stable than we had expected, it has been used for over 5 years without significant loss of binding activity or physical decomposition.

Economy

The economic aspects of production of L-amino acids by the immobilized aminoacylase are many. By using immobilized enzyme, purifi-

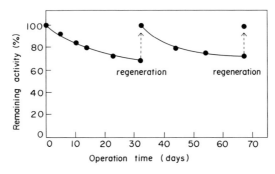

FIG. 5. Stability and regeneration of immobilized DEAE-Sephadex–aminoacylase column. A solution of 0.2 M acetyl-DL-methionine (pH 7.0, containing 5×10^{-4} M $CoCl_2$) is continuously applied to the column at 50° at a flow rate of space velocity = 2 for a long period. The activity is determined by applying the substrate solution to the column at 50° at a flow rate of space velocity = 10 and measuring the concentration of L-methionine in the effluent. Regeneration of the decayed column is carried out as described in the text.

cation procedure of reaction product becomes simple and the yield is higher than in the case of native enzyme. Therefore, less substrate is required for the production of unit amount of L-amino acid. As shown in Fig. 5, the immobilized aminoacylase is very stable. Thus, the cost of the enzyme is markedly reduced from that of native enzyme. In the case of the immobilized enzyme, the process is automatically controlled. Therefore, the labor cost is also dramatically reduced. The overall operating cost of the immobilized enzyme process is about 60% that of the conventional batch process using the native enzyme.[12]

Other Immobilized Aminoacylase Preparations

Besides the mold aminoacylase previously described, bacterial and animal preparations have been also successfully immobilized.

A bacterial aminoacylase has been covalently bound to diazotized polyaminostyrene,[17] N-carboxyanhydride of amino acid,[18] and nitrated copolymer of methacrylic acid and methacrylic acid-3-fluoroanilide.[19] An aminoacylase from pig kidney has been immobilized by ionic binding to DEAE-cellulose,[20] by covalent binding to diazotized Enzacryl AA (produced from Koch-Light Laboratories Ltd.),[21] and by entrapping into cellulose nitrate.[22]

These preparations can be employed on the laboratory scale for continuous preparation of L-amino acids from acyl-DL-amino acids.

[17] S. Kudo and H. Kushiro, Jpn. Patent No. 531,343 (1968).
[18] J. Kirimura and T. Yoshida, Jpn. Patent No. 451,517 (1965).
[19] M. Tanaka, N. Nakamura, and K. Mineura, Jpn. Patent No. 544,960 (1969).
[20] T. Barth and H. Mašková, *Collect. Czech. Chem. Commun.* **36**, 2398 (1971).
[21] H. Mašková, T. Barth, B. Jírovský, and I. Rychlík, *Collect. Czech. Chem. Commun.* **38**, 943 (1973).
[22] F. Leuschner, Br. Patent No. 953,414 (1964).

[52] Production of 6-Aminopenicillanic Acid with Immobilized *Escherichia coli* Acylase

By E. Lagerlöf, L. Nathorst-Westfelt, B. Ekström, and B. Sjöberg

6-Aminopenicillanic acid (6-APA), the key compound in the preparation of semisynthetic penicillins, is produced on a large scale from benzylpenicillin. Two processes, one enzymic and one chemical, are available.

The former originated in the Bayer laboratories in 1959.[1] It utilizes an enzyme produced by *Escherichia coli*, which hydrolyzes off the side chain of benzylpenicillin to give phenylacetic acid and 6-aminopenicillanic acid (6-APA). The reaction is performed at about pH 7.8 in the presence of an added base to neutralize the acid formed.

$$C_6H_5CH_2-CONH-\underset{\text{Benzylpenicillin}}{\begin{array}{c}H\\C\\|\\OC\end{array}\begin{array}{c}H\\C\\|\\N\end{array}\begin{array}{c}S\\\\ \\ \end{array}\begin{array}{c}CH_3\\C\\|\\C\end{array}\begin{array}{c}CH_3\\\\\\COOH\end{array}} \xrightarrow[\text{pH 7.8}]{E.\ coli\ \text{acylase}} \underset{\text{Phenylacetic acid}}{C_6H_5CH_2COOH} + \underset{\substack{\text{6-Aminopenicillanic acid}\\\text{(6-APA)}}}{H_2N-\begin{array}{c}H\\C\\|\\OC\end{array}\begin{array}{c}H\\C\\|\\N\end{array}\begin{array}{c}S\\\\\\ \end{array}\begin{array}{c}CH_3\\C\\|\\C\end{array}\begin{array}{c}CH_3\\\\\\COOH\end{array}}$$

The assay of the penicillin acylase used for enzymic cleavage of benzylpenicillin, its preparation, purification, and properties have been described in detail in the preceding volume of this series.[2] In addition, a method for immobilizing the enzyme by covalent coupling to cyanogen bromide-activated Sepharose 4B was described.[3]

An alternative procedure for purification of the enzyme, adapted for work on a production scale, was developed in our laboratories.[4] This purification method gives higher yields, as well as a method for producing the enzyme covalently bound (CVB) to Sephadex G-200 and a basic production method for 6-APA using this Sephadex G-200 CVB-*E. coli* acylase; it is described below.

Assay Methods

Principle. The penicillin acylase catalyses the conversion of benzylpenicillin to 6-aminopenicillanic acid and phenylacetic acid. Assay methods have been developed based on either of these products.[2] Photometric methods using hydroxylamine[5] or *p*-dimethylaminobenzaldehyde[6] as reagents for the determination of the amount of 6-APA formed appear to

---

[1] W. Kaufmann and K. Bauer, German Patent No. 1 111778 (1961).
[2] M. Cole, T. Savidge, and H. Vanderhaeghe, this series Vol. 43, p. 698.
[3] T. Savidge and M. Cole, this series Vol. 43, p. 705.
[4] S. Delin, B. Ekström, L. Nathorst-Westfelt, B. Sjöberg, and H. Thelin, U.S. Patent No. 3 736230 (1973).
[5] F. R. Batchelor, E. B. Chain, T. L. Hardy, K. R. Mansford, and G. N. Rolinson, *Proc. R. Soc. London, Ser. B* **154**, 498 (1961).
[6] J. Bomstein and W. G. Evans, *Anal. Chem.* **37**, 576 (1965).

be best suited for analysis of large series of samples, using for example, automatic equipment (for further readings on penicillin analysis in general see also this volume [41] and [45]).

The method preferred by us is to determine the phenylacetic acid by titration at constant pH with a pH stat. It is well suited for single analyses, is rapid and easy to carry out, and gives high accuracy as well as reproducibility.

Definitions of Activity. The activity of an enzyme sample is expressed in units (U). One penicillin acylase unit is defined as the activity capable of producing 1 μmole of 6-APA per minute at 37° and pH 7.8 from a 2% aqueous solution of potassium benzylpenicillin.

The specific activity of solutions of the acylase are expressed in A/E values where A is the activity in units/ml and E is the optical density of the solution at 280 nm. The optical density at this wavelength is taken as a measure of the protein content of the solution.

Reagents and Apparatus

Potassium benzylpenicillin (Fermenta AB, Strängnäs, Sweden)
Sodium hydroxide 0.1 M
Enzyme sample, e.g., bacterial cells, bacterial cell extracts, CVB-acylase
An apparatus for the analyses; a suitable one is a Combititrator (Metrohm AG 3D E512/E425/E473)

Procedure. The enzyme sample, e.g., *E. coli*-CVB-acylase ($c = 0.1$ g), is suspended in tap water (40 ml) in the reaction vessel and thermostated at 37.0°. After addition of potassium benzylpenicillin (1 g, 2.7 mM) in water (10 ml) the pH is adjusted to 7.8, and kept there by addition of 0.1 M NaOH by means of the automatic titration equipment with vigorous stirring.

The consumption of NaOH is plotted automatically, and the activity is calculated as the consumption (a) per minute during the linear phase. The activity (A) is calculated as $A = 100 \times (a)$ units.

A suitable sample activity is 20–40 U, and a normal time for the analysis should be 5–10 min.

Purification of Penicillin Acylase

The concentrated slurry of *E. coli* cells, separated from the fermentation broth and partly disrupted owing to the release of pressure by the periodical expulsions of the slurry through the narrow slits in the bowl of a solids-ejecting centrifugal separator (De Laval type BRPX 213) on separation from the fermentation broth, is the starting material for the purification of the penicillin acylase.

In a typical example, a bacterial slurry (800 kg, 64 U/g) is treated in a homogenizer (Manton-Gaulin Model 800 KF 12X-5PS) in order to complete the disintegration of the cells. The slurry is suspended in water (4000 liters) and treated with a flocculent in order to separate the bacterial cells in the solids-ejecting centrifugal separator. The aqueous extract is treated with a filtering aid (Celite 505 and Hyflo Super Cel) and filtered on a filter press (Seitz, 40 × 40 cm). The clear solution is concentrated *in vacuo* to a volume of about 250 liters in a thin-layer evaporator (LUWA) and filtered again in order to remove some precipitated inactive material.

The clear concentrated cell extract (290 kg, 150 U/g) is adjusted to pH 8.6 by means of dilute ammonia to precipitate further amounts of inactive substances, which are removed by filtration. The filtrate is adjusted to pH 5.7 with dilute acetic acid, and a 20% aqueous tannin solution is added with stirring to precipitate the acylase until only a small amount of the activity (1–2%) remains in the mother liquor. The tannin–acylase complex is filtered and washed with deionized water (120 liters) to give 50–60 kg of a moist paste.

The tannin-acylase precipitate in portions of 10–12 kg is treated in a homogenizer (Mannesmann type F3N), an equal amount of cold acetone ($-8°$ to $-10°$) being added with vigorous agitation to dissolve the main part of the precipitate. On addition of further amounts of cold acetone to a total volume of 45–50 liters, the enzyme is precipitated, leaving the tannin in solution. The precipitated acylase is collected by filtration, washed with dry acetone, and suspended in deionized water (75 liters). The mixture is adjusted to pH 5.5–5.7 by means of acetic acid and stirred for an additional period of 3 hr. A filtering aid is added, and the insoluble residue is filtered and washed with deionized water (2 × 40 liters).

In all, a total of 170 kg of a clear solution of crude acylase (185 U/g) is obtained. The total yield is 31.4×10^6 U, corresponding to a recovery of 61% of the activity of the bacterial slurry.

The further purification of the crude penicillin acylase is done by means of a carboxymethyl cellulose cation exchanger (Whatman CM-32) as follows.

Precycling. The ion exchanger (CM-32, 4 kg) is precycled by treatment with 2% sodium hydroxide solution (75 liters). After stirring for 0.5–2 hr, the exchanger is filtered and washed with water (300–400 liters) to neutrality. The same procedure is then carried out twice with 2% hydrochloric acid (50 liters) with stirring for 30 min.

TABLE I
Purification of Penicillin Acylase from *Escherichia coli*

Material	Volume (liters)	Total enzyme activity units[a]	Specific activity (A/E)[b]	Yield (%)
Disrupted cell suspension	4000	51.2×10^6	—	100
Concentrated cell extract	290	43.5×10^6	0.77	85
Crude acylase solution	170	31.4×10^6	2.57	61
Purified acylase solution	86	26.7×10^6	10.80	52

[a] A unit of enzyme activity is defined as the amount of enzyme required to produce 1 µmole of 6-aminopenicillanic acid per minute from a 2% solution of potassium benzylpenicillin at 37°, pH 7.8.

[b] Specific activity is expressed as A/E, where A stands for the activity in U/ml and E for the optical density at 280 nm.

Equilibration. The ion exchanger is equilibrated by stirring it in 0.01 M sodium acetate (60 liters) for 30 min. The exchanger is filtered and resuspended in new 60-liter portions of 0.01 M sodium acetate until the pH of the filtrate has reached 5.45 or higher. The same procedure is then repeated with 0.01 M sodium acetate buffer, pH 5.50, until the pH and conductivity of the filtrate after two consecutive treatments are the same as in the original buffer solution.

Adsorption. The equilibrated CM-32 ion exchanger is added to the clear solution of crude acylase, previously adjusted to pH 5.3 ± 0.1 and to the same conductivity value as in the buffer solution. The mixture is stirred gently for 30 min and samples are taken for control of the adsorption. If more than 5% of the original activity is left in solution, further amounts of ion exchanger are added in small portions until the remaining activity is 5% or lower. The mixture is filtered and the exchanger is washed with 0.01 M acetate buffer, pH 5.50 (3×40 liters).

Elution. The ion exchanger is suspended in 0.01 M sodium phosphate buffer pH 8.0 (20 liters), gently stirred for 20 min and filtered. The procedure is repeated twice with 0.01 M buffer and then four times with 0.03 M sodium phosphate buffer, pH 8.0.

All eluates having a pH lower than 5.65 are discarded. The remaining eluates are pooled to give 86 liters of an acylase solution containing in total 26.7×10^6 U, corresponding to a recovery of 52% from the bacterial slurry.

The results of the purification are summarized in Table I.

Immobilized Enzyme

Various methods to immobilize the enzyme have been tried. In the attempt to bind the enzyme covalently, the main problem is to get a

TABLE II
COUPLING OF *Escherichia coli* ACYLASE TO VARIOUS SUPPORTS

Support	Reactive group/ activating agent	Acylase activity used for coupling (U)	Specific activity of wet product (U/g)	Coupling yield (%)
Carboxymethyl cellulose (Enzite)	$-CON_3$	330	3.96	1.2
Polyacrylamide (Enzacryl AA)	$-N_2^+Cl^-$	3150	3.06	0.1
Polyacrylamide (Enzacryl AH)	$-CON_3$	1276	6.98	1.4
Cellulose (Avicel)	$CH_2BrCOBr$	100	0.9	7.9
Sephadex G-200	$CH_2BrCOBr$	100	0.9	11.2
Sephadex G-200	CNBr	960	117.2	12.2
Anhydrid-Acrylharz-perlen (2828 A)[a]	$-COOOC-$	2820	29.1	7.3
Oxiran-Acrylharzperlen (2878 A)[a]	$-CH\overset{O}{\underset{}{\diagdown\diagup}}CH_2$	637	6.9	5.4

[a] Röhm Pharma AG, Darmstadt, West Germany.

sufficiently high specific activity, and of the methods tried (Table II) the best results were obtained with the CNBr method, first described by Axén, Porath, and Ernback[7] (see also this volume [3]). This method was investigated further with a number of different polysaccharide derivatives. In addition to the specific activity of the product and the yield in the coupling reaction, other parameters like filtration properties, stability of the product and cost of the support were considered. Cellulose, Sepharose 4B, and Sephadex G-200 were chosen as the most promising products for further experiments (Table III). When, finally, these three supports were compared, it was found that the specific activity obtainable with cellulose was too low, and the stability of Sepharose 4B when used for repeated conversions of benzylpenicillin in a batch process was clearly inferior to the stability of Sephadex G-200. (See also chapter [14] for a few additional data on immobilized acylase.)

In the preferred method the following procedure is carried out: Sephadex G-200 (500 g) is swelled overnight in deionized water (20 liters) at $+4°$. NaOH (2.4 kg) in deionized water (10 liters, $+4°$) is added, and the viscous suspension is kept at $+4°$. CNBr (2 kg) dissolved in deionized water (30 liters) is added with vigorous stirring and cooling. The

[7] R. Axén, J. Porath, and S. Ernback, *Nature* (London) **214**, 1302 (1967).

TABLE III
Coupling of *Escherichia coli* Acylase to Various Polysaccharide Derivatives with CNBr as Activating Agent

Support	Specific activity of product, dry weight (U/g)	Coupling yield (%)	Filtration properties (rating 0–5)[a]
Amylose	22.4	1.5	5
α-Glucan	736	47.8	2
Dextran	357	23.0	2
Carboxymethyl cellulose	17.2	11.1	5
Sephadex G-200	468	30.3	4
Carboxymethyl-Sephadex	103	6.6	0
Sulfoethyl-Sephadex	249	16.1	2
DEAE-Sephadex	0	0	1
Sepharose 4B	694	45.1	5
Cellulose	70	3.9	3

[a] Filtration properties: The comparison has been made with a polymer wet weight corresponding to 0.5 g of dry polymer. A sintered-glass filter, type G 3, with a diameter of 3.5 cm was used for the experiments. Rating: 0, it was quite impossible to filter the product; 1, one or two drops per minute could be sucked through; 5, about 25 ml of filtrate could be obtained in 1 min.

temperature rises to 20–25° within about 1 min and then gradually decreases to 4–5° in about 10 min. After an additional period of 20 min, Hyflo Super Cel (1.9 kg) is added and the activated polymer is filtered and washed with deionized water (200 liters, +4°) and a saturated solution of borax (1125 g) in deionized water (90 liters). The filtered product is mixed with a purified acylase solution (600 U/ml, 6.17×10^6 U) and borax (375 g), and the mixture is agitated overnight at +4°. The Sephadex-CVB-acylase is then filtered and washed with deionized water (75 liters).

The yield is about 13 kg of wet polymer with an activity of about 225 U/g, a total of 2.93×10^6 U.

Production of 6-Aminopenicillanic Acid

Principle. 6-APA can be produced on a production scale in a way similar to that described above under assay methods, that is, by treating a solution of potassium benzylpenicillin with the enzyme in a thermostated vessel under controlled pH conditions. The reaction can be carried out either by suspending the CVB-enzyme in the penicillin solution, a batch process, or by circulating penicillin solution through a bed of the immobilized acylase, a recirculation process. In the batch process, the enzyme is removed by filtration after the reaction and the 6-APA solution can be directly acylated to different semisynthetic penicillins.

Batch Process. The conversion can be performed in a practical way as follows (Fig. 1). Sodium dihydrogen phosphate (20.5 kg) is dissolved in tap water (2900 liters) in a vessel (a) thermostated to 35°. The pH is adjusted to 7.8 with sodium hydroxide and a suspension in water (30 liters) of the CVB-acylase (3.7×10^6 U, wet weight 16.5 kg) is added. When the temperature and pH have been stabilized, potassium benzylpenicillin (100 kg) is added to the reaction vessel in solid form. The mixture is vigorously stirred and the pH is kept constant at 7.8 by the addition of sodium hydroxide solution from a funnel (b) connected with a magnetic valve (m), which is regulated by a pH stat (c). The conversion is followed by a recorder (r). Filtration is performed in a filter press (d) (Seitz, 40×40 cm) equipped with a washable type of filter plate. The solution (h) is transferred to another vessel; the CVB-acylase is collected in the trough (e) and brought back to (a) by means of a pump (f).

Provided that the operations are carefully performed, the CVB-acylase can be used for more than a hundred times without addition of fresh enzyme.

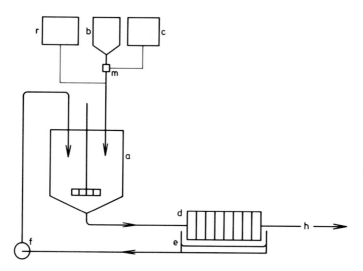

Fig. 1. Batch process for conversion of benzylpenicillin. a, Thermostated vessel; b, funnel; c, pH stat; d, filter press; e, trough; f, pump; h, 6-aminopenicillanic acid solution; m, magnetic valve; r, recorder.

Recirculation Process. Enzymic conversion is carried out with equipment diagrammed in Fig. 2, and can be illustrated by the following run on a pilot scale.

A suspension in water (3 liters) of the immobilized acylase (0.12×10^6 U, 500 g) is poured in one portion into a funnel (E) equipped with a filter plate placed on a perforated disk. When the water has penetrated the plate, another plate is placed above the settled CVB-acylase layer (G; 7–8 mm thick) and the system is closed.

To a vessel (A) equipped with an efficient stirrer and jacket, water (90 liters, 37°) and monosodium phosphate (600 g) are added. The pH is adjusted to 7.8 with a 10% aqueous solution of sodium hydroxide from a funnel (B) connected with a magnetic valve (M) regulated by a pH-stat (C). Potassium benzylpenicillin (3000 g) is added, and the pump (D) is started. The air in funnel E is replaced by the penicillin solution, which is circulated at 37° and pH 7.8 until the regulated consumption of sodium hydroxide ceases.

The aqueous solution obtained (H) is transferred to an extraction vessel adjusted to pH 2.5 and treated with about one-half the volume of methyl isobutyl ketone, to remove phenylacetic acid. The aqueous phase is separated, adjusted to pH 7.8, and concentrated *in vacuo* to a volume of about 10 liters. The 6-APA is precipitated by adjusting the pH to 4.5 with hydrochloric acid. The substance is filtered, washed with

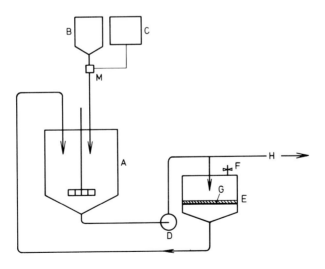

FIG. 2. Recirculation process for conversion of benzylpenicillin. A, Thermostated vessel; B, funnel; C, pH stat; D, pump; E, funnel; F, air vent; G, covalently-bound-acylase layer; H, 6-aminopenicillanic acid solution; M, magnetic valve.

cold water and dried *in vacuo* to give 6-APA (1600 g, purity 98%), corresponding to a total yield of 90%.

Since early 1973 the Sephadex-CVB-acylase has been used for regular production of 6-APA in the batch process. Compared with the older method using whole bacterial cells for the cleavage of benzylpenicillin, this new procedure has resulted in higher yields of purer products, easier handling of the enzyme, and better economy.

The recovery and recharging of the immobilized enzyme are steps that must be carried out with some care to avoid losses of enzyme or inactivation of it owing to contamination with microorganisms having proteolytic activity. These steps are omitted in the recirculation process now being introduced in the factory. According to the studies in the pilot plant, the latter should also have further advantages, such as higher production capacity and lower labor costs, compared to the batch process.

[53] Process Engineering of Glucose Isomerization by Collagen-Immobilized Whole Microbial Cells[1]

By W. R. VIETH and K. VENKATASUBRAMANIAN

The enzymic isomerization of D-glucose to D-fructose is rapidly becoming of major industrial significance (see also this volume [56]). The impetus for this arises mainly from the surging sugar (sucrose) prices in the world market. As an alternative to sucrose, high-fructose syrup (HFS) is gaining in importance as a commerically marketed product. It contains about 50% glucose and 42% fructose and the rest comprises other saccharides. The presence of fructose imparts increased sweetness to the syrup. An immobilized enzyme process is in commerical operation in the United States.[2] This process uses glucose isomerase enzyme, immobilized on DEAE-cellulose by adsorption, in cylindrical reactors containing multiple, shallow-beds of immobilized enzyme. In this chapter, we describe the process development, process engineering, and economic analysis of a glucose isomerization process based on collagen-immobilized whole microbial cells.

We have concentrated our efforts on the use of immobilized whole cells rather than the enzyme (glucose isomerase) itself, prompted by

---

[1] The authors thank Professor A. Constantinides and Dr. R. Saini for their assistance.
[2] B. J. Schnyder and R. M. Logan, Paper No. 9C, A.I.Ch.E. 77th National Meeting, Philadelphia, 1974.

several important considerations (see also this volume [50] and [13]). Isolation and purification of intracellular enzymes can be tedious and difficult, resulting in low overall yields and increased cost of the enzyme. The use of immobilized whole cells will obviate such cost and, more important, can improve the enzyme's stability. By now, the evidence is mounting that conjugated whole cells have superior stabilities relative to conjugated enzymes.[3,4]

In general, whole cell immobilization is also preferred when a number of enzymes and cofactors are required to mediate a given process. One of the conceivable shortcomings of this approach may be the additional transport resistance of the cell wall and cell membrane. However, calculations show that these effects are relatively weak. Moreover, glucose isomerase is a relatively "sluggish" enzyme catalyzing a reversible reaction (equilibrium constant at 70° $\sim$ 1.3). Therefore, it is unlikely that diffusion would be the controlling step in the overall reaction. This statement is indeed valid, as shown later. A further consideration in favor of an immobilized-whole-cells process for glucose isomerization is the observation that certain types of cells can be easily fixed in a stationary phase by a simple thermal treatment.[5]

Process Design Parameters

For the efficient design of a commerical reaction system, it is necessary to establish the relationships between several key process variables. The most important process design parameters are the enzyme loading factor, catalyst packing density, operational stability of the catalyst, external and internal mass transport efficiencies, reactor contact efficiency, and residence time distribution. The interaction of these variables is manifested in the expressed catalytic activity of the reactor, which would in turn govern the extent of substrate conversion, the reaction throughput, and productivity. Presented below is a brief consideration of these parameters as they pertain to the glucose isomerization process by collagen-whole cell membrane reactors.

Enzyme Loading Factor

This parameter is a measure of the amount of whole cells (and therefore the amount of catalytic activity) present in the immobilized prepa-

---

[3] I. Chibata, T. Tosa, T. Sato, T. Mori, and K. Yamamoto, *in* "Enzyme Engineering" (K. E. Pye and L. B. Wingard, Jr., eds.), Vol. 2, p. 303. Plenum, New York, 1974.

[4] K. Shoda and M. Shoda, *Biotechnol. Bioeng.* **17**, 481 (1975).

[5] Y. Takasaki, Y. Kosugi, and A. Kanbayashi, *in* "Fermentation Advances" (D. Perlman, ed.), p. 561. Academic Press, New York, 1969.

ration. It is expressed in units per gram of complex, where 1 unit of glucose isomerase activity is defined as that which produces 1 mg of fructose per hour at 70° and pH 7. Obviously, the loading factor would be largely determined by the immobilization process itself, which is described in detail in this volume [19]. The most important variables in the immobilization process are the relative amounts of cells and collagen, the extent of enzyme inactivation, and the "fixing" of cells. We have found[6] that (1) a cells-to-collagen ratio of 1:1 (dry weight basis) yields a membrane of good activity and strength; (2) when cells are immobilized at pH 11, this adverse pH condition causes a loss of about 64% of the initial activity. The tanning step accounts for the loss of another 11% of initial activity, thus the final activity of the preparation corresponds to about 25% of the original activity of the unbound cells. However, this activity loss can be substantially reduced by a recent modification of the complexation procedure, utilizing lower pH levels and direct addition of a small amount of glutaraldehyde (as outlined in chapter [19]), which results in membranes that recover up to 50–55% of the activity during immobilization. It should also be pointed out that in both cases, the membranes have prolonged operational stabilities. Other aspects of the immobilization process have been discussed in detail elsewhere.[6]

The specific activity of the starting microorganism also influences the enzyme loading factor. When *Streptomyces venezuelae* cells are used, the final loading factor is about 500 units per gram of catalyst. This value can be increased to about 1000 units/g by using a *Bacillus* strain (marketed by Novo Enzyme Corporation, New York; Product No. SP-103).

Catalyst Packing Density

The amount of collagen–cells complex that can be packed in a unit reactor volume is given by the catalyst packing density, expressed as grams of complex per liter of reactor. This parameter would primarily be a function of the reactor configuration, membrane thickness, and allowable pressure drop across the catalyst bed. Collagen-whole cell membranes could be used in a spiral-wound biocatalytic reactor module (see [19]) or in a packed bed in the form of chips. For this particular application, these two reactor types perform about equally well. Membrane thickness usually varies between 50 μm and 250 μm, or 2–10 mils, respectively. Only a small pressure drop is encountered in these reactor systems. Our experience indicates that we are able to pack about 25–30% of the gross reactor volume with solid catalyst, achieving packing levels of approximately 10 lb/ft^3 or about 200 g per liter of gross reactor volume.

[6] R. Saini and W. R. Vieth, *J. Appl. Chem. Biotechnol.* **25**, 115 (1975).

Operational Stability

This is perhaps the most important factor governing reactor design. By modeling the catalyst activity decay as a first-order process, its operational half-life—the time required for 50% of the enzyme activity to be lost—can be easily calculated. The observed half-lives for collagen–whole cell complexes are 50 days and 75 days, with *Streptomyces* and *Bacillus* strains, respectively.

External Mass Transport

The transport of substrate from the bulk fluid to the membrane surface could become a rate-limiting step in a catalytic process, if it is considerably slower than the intrinsic reaction rate. Whether or not such a limitation exists for a collagen–cell reactor system was investigated experimentally. It is well known that the boundary layer resistance to mass transfer is markedly reduced as the fluid velocity through the packed bed is increased. However, variation of substrate mass velocity over a wide range (15–750 g/cm^2 per hour) had no appreciable effect on the observed initial reaction rate. Furthermore, using well-known mass-transfer correlations, the concentration drop between the bulk fluid and the membrane surface was estimated to be only 0.005%. Therefore, external diffusional limitations can be ruled out.

Internal Transport Resistance

The relative rates of intramembrane transport and the biochemical reaction were also examined to ascertain whether internal diffusion could be rate-limiting. Using zero-order and first-order kinetic models, the effectiveness factors were found to be 0.95 and 1.0.[6] It was also experimentally observed that the reaction rate remained constant even with a 5-fold increase in membrane thickness. This implies that the effectiveness factor approaches unity, which is consistent with the value predicted by the model. Thus, intramembrane transport resistance is insignificant. These results are not unexpected considering that glucose isomerization is a relatively slow, reversible reaction and that the highly swollen, open structure of the matrix promotes efficient enzyme–substrate contact.

Reactor Contact Efficiency

It was found that collagen–whole cell complex membrane can be packed in a properly designed reactor in such a way that the factors contributing to inefficient reactor operation, viz., channeling and longitudinal dispersion, are minimized. In addition, the mesh spacer used in the spiral-wound reactor modules promotes good radial mixing (see [19]). Under these conditions, the fluid flow through the reactor approaches the idealized plug-flow condition.

Reactor Design Equation

A reactor design equation has been developed,[6] and the experimental data given in Fig. 1, for a particular combination of space time (based on reactor fluid volume) and inlet substrate concentration, agree well with the predicted values. Thus the equation is quite satisfactory for scale-up purposes.

Reactor Scale-Up and Performance

Reactor studies were performed initially in small laboratory-scale reactors (reactor volume ~ 50 ml) with chemically pure dextrose solution at 1 M concentration as substrate. A reactor design equation was developed based on data obtained from such reactors.[6] The next phase of testing was carried out at the Research and Development Laboratories of a commercial corn wet milling operation active in the isomerization

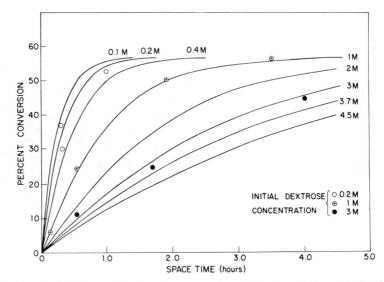

Fig. 1. Simulation profiles, based on experimental data and theoretical calculations for glucose isomerization process.

field. Several scaled-up reactors with sizes ranging from 100 ml to 3 liters were tested. In all cases, actual starch plant dextrose streams, containing 50% dissolved solids (93% dextrose), 0.001 M cobalt chloride, and 0.01 M magnesium sulfate, were used as substrates. The reactors were operated continuously for periods ranging from 20 days to 60 days. In each case, the catalytic activity of the column and its operational stability were determined. Operational stability of a typical reactor is shown in Fig. 2.

Experimental data obtained from these reactors were in reasonably good agreement with the performance predicted by the design equation

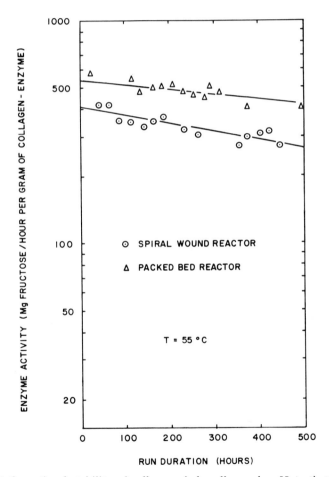

FIG. 2. Operational stability of collagen–whole cell complex. Note that the reactor was operated at 55°; the observed catalytic activities are, therefore, somewhat lower than what would be estimated at higher temperatures.

mentioned above, thus validating its applicability in reactor scale-up. In addition to evaluating the catalytic activity and the operational stability of the system, other important parameters were also monitored. For instance, pressure drop across the catalyst bed was found to be negligible. Product quality in terms of color, taste, and odor was found to be satisfactory. The best half-life values obtained were 500 hr and 1800 hr, respectively, for *Streptomyces* and *Bacillus* organisms.

Based on an initial activity of 1000 units per gram of catalyst and a half-life of 1800 hr, we have carried out a detailed process design for a commercial process to produce 300 million pounds of high-fructose syrup (HFS) per year. In this process design, the entire route from starch to fructose, i.e., starch hydrolysis by α-amylase and glucoamylase to dextrose, followed by isomerization to fructose, has been considered. The salient design features pertaining to the isomerization step are summarized in Table I.

In this design, the cell–collagen complex, in the form of chips, is packed into 6 ft internal diameter × 10 ft high reactors in layers 5 inches thick supported by fine wire mesh. Each column has 15 such

TABLE I
Design Parameters for a Commercial-Scale Glucose Isomerization Process

Capacity: 300,000,000 pounds of syrup per year
Product composition: 71% dissolved solids (50% dextrose, 42% fructose)
Substrate: 93% dextrose solution (dry solids basis)[a]
Catalyst: Collagen–whole cell complex
Cells: *Bacillus* species (Novo Enzyme Corporation, SP-103)
Collagen:cell ratio: 1:1 (dry weight basis)
Reactor: Packed columns with chips of collagen–cell complex
Initial activity: 1000 units[b] per gram of catalyst at 60°
Catalyst half-life: 1800 hr
Catalyst replacement: Replaced after 3 half-lives (5400 hr)
Average activity: 421 units[b] per gram of catalyst
Number of reactors required: 21
Reactor size: 283 ft³
Reactor dimensions: 6 ft diameter × 10 ft high
Void volume: 101 ft³
Cumulative reactor space time: 3.9 hr
Amount of catalyst in each reactor: 1890 pounds
Yearly catalyst requirement: 60,000 pounds/year
Reactor productivity: 265 pounds (solids)/hr/ft³ reactor
Operating temperature: 60°

[a] Also containing 5 mM MgSO$_4$ and 1 mM COCl$_2$.
[b] One unit produces 1 mg of fructose per hour at 70° and pH 7.0.

layers of catalyst with empty spaces of 2 inches between each catalyst layer. Twenty-one such reactors are required, operating in a sequential mode. The catalyst is to be replaced after a period of three half-lives (5400 hr). It is assumed that the recharging of the reactors is staggered in such a way that the average activity of the entire catalyst mass at any time is equivalent to 421 units/g. The cumulative reaction space time is 3.9 hr. The total amount of cell–collagen complex present in the 21 reactors is 39,700 pounds, and the annual catalyst requirement is 60,000 pounds.

Economic Analysis

An economic analysis of the above-mentioned 300 million pounds per year high-fructose syrup production process is presented in Table II. The total capital investment is 40.2 million dollars. Based on a selling price of 15¢ per pound of syrup, the post-tax return on investment is estimated to be 16.2%. (This corresponds to about 31% return on capital prior to taxes.) The transfer price of the high-fructose syrup product is directly proportional to the catalyst cost, as illustrated in Table III. It can be noted that even a 10-fold increase in catalyst cost would not increase the transfer price by more than 12%. Based on these calculations, it appears that the glucose isomerization process by collagen–cells complex is an economically attractive commercial proposition.

Epilog

In this chapter, we have discussed the process engineering aspects of an immobilized whole cells system for glucose isomerization. The

TABLE II
Economic Analysis of Commercial-Scale Glucose Isomerization Process with Cells-Collagen Complex[a]

Capacity: 300 million pounds of high-fructose syrup per year
Total fixed capital: 29.4 million dollars
Working capital: 10.8 million dollars
Total capital: 40.2 million dollars
Gross income (@ 15¢/pound of syrup): 45.0 million dollars
Cost of production: 32.5 million dollars
Net income: 12.5 million dollars
Taxes (@ 48%): 6.0 million dollars
Profit: 6.5 million dollars
Net return on investment: 16.2%

[a] U.S. Gulf Coast location, first quarter, 1975.

TABLE III
EFFECT OF CATALYST COST ON TRANSFER PRICE OF HIGH-FRUCTOSE SYRUP

Catalyst cost		Transfer price
$/Pound catalyst	$/100 pounds syrup	(¢/pound syrup)
10	0.20	14.64
20	0.40	14.84
30	0.60	15.04
40	0.80	15.24
50	1.00	15.44
60	1.20	15.64
100	2.00	16.44

development of the process from inception through commercial-scale design has been outlined. Based on the tests conducted under actual plant operating conditions, the efficiency of the collagen–cell complex reactor system has been established. We believe that the technology has been developed to the point where now the system is ready for a large-scale demonstration test.

Preliminary economic analysis indicates the commercial viability of the isomerization process. Improvement of the loading factor and the expressed activity of the catalyst will further strengthen this point.

[54] Scale-Up Studies on Immobilized, Purified Glucoamylase, Covalently Coupled to Porous Ceramic Support

By H. H. WEETALL, W. P. VANN, W. H. PITCHER, JR., D. D. LEE, Y. Y. LEE, and G. T. TSAO

Choice of Carrier

The economic production of dextrose from corn syrup, utilizing glucoamylase (GA) immobilized on controlled-pore glass, has at least two serious disadvantages: (1) high cost of the carrier and (2) dissolution of glass with time.[1] The development of oxide-coated glass has increased durability and half-life, which are important economic advantages. However, the cost of the carrier still exceeds the value of the product.

[1] H. H. Weetall and N. B. Havewala, *Biotechnol. Bioeng. Symp.* **3**, 241 (1972).

TABLE I
Physical and Chemical Parameters of Support Materials

Designation and composition	Mesh	Pore diameter range (Å)	Pore diameter average (Å)	Pore volume (ml/g)
A (SiO_2 75%, TiO_2 25%)	30/45	875–205	465	0.76
B (SiO_2 90%, ZrO_2 10%)	30/60	700–185	435	0.76
C (SiO_2 100%)	30/60	700–185	435	0.76
D (SiO_2 89.3%, ZrO_2 15.7%)	30/60	575–110	235	1.30
E (SiO_2 75%, Al_2O_3 25%)	30/60	575–205	435	0.89
F (TiO_2 98%, MgO 2%)	30/45	500–205	410	0.53

This report describes half-life studies of six porous ceramic carriers covalently bound[2,3] to purified GA. All the carriers are less expensive to produce than porous glass and could make a presently uneconomic process more attractive. One of the carriers was utilized for a scale-up to a 1 ft³ column. Processing cost estimates based on the performance of this scaled-up column were made.

Materials and Methods

Support Materials

Physical and chemical data for the six support materials (A–F) used in this study are summarized in Table I.

Alkylamine Derivative Preparation

Fifty grams of support material was added to enough 10% γ-aminopropyltriethoxysilane (Union Carbide A1100 Silane) in distilled water so that the carrier was covered. The mixture was then adjusted to pH 3.45 with a 25% solution of NaOH. The mixture was allowed to react in a 75° water bath for 3 hr, then washed with 4 volumes of distilled water and oven-dried at 120° for 15 hr.

Glucoamylase

Crude GA was obtained from NOVO Industries A/S, Copenhagen, Denmark and purified using isopropanol precipitation techniques. Activ-

---

[2] R. A. Messing, *Res. Dev.* **25**, 32 (1974).
[3] R. A. Messing, in "Immobilized Enzymes in Food and Microbial Processes" (A. C. Olson and C. L. Cooney, eds.), pp. 129–156. Plenum, New York, 1974.

ity of the purified enzyme was 65,280 IU/g. One unit represents the production of 1 μmole of dextrose/minute at 60°.

Preparation of Immobilized Glucoamylase

GA was immobilized onto carrier using Schiff base coupling.[4] The alkyamine derivative (available from Pierce Chemical Co., Rockford, Illinois) (usually 30–50 g) was added to a sufficient volume of 2.5% glutaraldehyde in 0.1 M phosphate buffer at pH 7 so that the derivatized support material was covered (see also this volume [10]). The reaction was performed at room temperature for 1 hr, which included 30–45 min in a vacuum desiccator with occasional stirring. The carrier was then washed with 3–4 liters of distilled water. For immobilization, 100 mg of purified enzyme was offered per gram of derivatized support material. The enzyme was dissolved in cold 0.1 M phosphate buffer at pH 7 and added to the support material in an ice bath with occasional stirring. The solution pH was maintained at 6.8–7.0, and the reaction was continued for 2 hr. The immobilized enzyme (IME) was washed with 3–4 liters of distilled water and stored at 4° as a wet cake.

Determination of Immobilized GA Activity

Initial immobilized GA activity was assayed at 60° for 1 hr at pH 4.5. Substrate was 50 ml of 30% corn syrup solids, prepared from 22 Dextrose Equivalent (DE) spray-dried acid-enzyme thinned cornstarch (A. E. Staley Company, Decatur, Illinois), to which a known quantity of immobilized enzyme was added. Total glucose was determined with a Glucostat Special Kit (Worthington Biochemical Corp., Freehold, New Jersey).

Substrate for Columns

Substrate used for this study was the same as for the assay. The substrate feed pH was 4.5. Determinations of DE were performed on the substrate whenever a new bag of material was employed.

Immobilized Glucoamylase Columns

With the exception of support material B (Table I), 10 g wet weight (4.4–5.9 g dry weight, depending on support material) of each immobilized enzyme was put into 3 water-jacketed columns at 50°, 60°, and

[4] H. H. Weetall, N. B. Havewala, W. H. Pitcher, Jr., C. C. Detar, W. P. Vann, and S. Yaverbaum, *Biotechnol. Bioeng.* **16**, 295 (1974).

65°. The 65° column for support B was omitted because of a limited supply of carrier. Each column was assayed with the Glucostat Special Kit once each day for glucose produced. Column flow rates were adjusted where possible to yield initially about 70% or less glucose production. Half-life determinations and regression analysis plots were performed on column data with a Model 10 Hewlett-Packard calculator program.

When half-life determinations were completed, column flow rate was set as low as possible, and dextrose equivalent (DE) determinations were performed on the products.

Results and Conclusions

Initial Activity of Immobilized Supports

Tables I and II summarize physical and chemical data and column half-life results of the various support materials. Activities ranged from 423 to 2400 IU/g IME on a dry weight basis. Supports B and D had the lowest activity. Pore exclusion of enzyme would account for the low activity of support D (pore diameter 235 Å) but not support B (pore diameter 435 Å). Perchloric acid titration of silanized, zirconia-coated supports has shown less-reactive amine groups present as compared to non-zirconia-coated supports. Also, reaction of β-naphthol with diazotized zirconia-coated arylamine support material indicated less color development or fewer diazonium sites available for coupling. Both B and D were zirconia coated, which probably contributed to the low activity of both supports.

Dextrose Equivalent Determinations at 50°

Table II summarizes maximum DE observed for columns operated at 50°. Substrate DE was actually 24.8. Supports A, B, C, E, and F had product DE's of 87.1–92.0. Product DE for support D was 79.9. This value was probably attributable to the low initial activity and/or less than optimum flow rate for this material. It should be stressed that DE values may or may not represent maximum values. Pumps were set to provide a minimum flow rate, which varies considerably from pump to pump. No attempt was made to optimize flow rate vs DE.

Half-Lives of Immobilized Preparations

Half-life data are summarized in Table II. In Fig. 1 one example is given of computer-calculated regression analysis plots of 50°, 60°, and

TABLE II
Column Half-Life Results

Support material	Initial activity[a] (IU/g IME)	Temp. (°C)	Days operated	Half-life (days) $t_{1/2}$	LCL[b]	UCL[b]	Maximum DE[c] observed at 50°	Deactivation energy (kcal/g mole)
A	1750	50	40	46.6	31.8	86.9	89.7	32.4
		60	19	8.8	7.6	10.5		
		65	14	5.6	4.6	7.1		
B	658	50	31	53.3	38.3	87.4	92.0	26.7[d]
		60	14	15.9	10.9	29.0		
		65	—	—	—	—		
C	2400	50	59	51.2	40.0	71.4	88.9	41.4
		60	16	13.8	10.1	22.0		
		65	9	3.4	2.7	4.7		
D	423	50	31	77.1	46.9	215.9	79.9	35.6
		60	12	11.2	9.7	13.3		
		65	12	7.5	5.9	10.4		
E	1729	50	59	113.3	76.5	217.6	89.6	53.9
		60	16	16.2	12.7	22.4		
		65	9	4.2	3.4	5.4		
F	1607	50	40	65.3	42.2	144.4	87.1	37.3
		60	19	10.1	8.3	12.8		
		65	14	5.7	5.0	6.5		

[a] One unit of activity equals the production of 1.0 μmole of dextrose per minute at 60° IME, immobilization enzyme.
[b] LCL, lower confidence limit; UCL, upper confidence limit.
[c] DE, dextrose equivalent.
[d] Calculated between 50° and 60°; all others 50°–65°.

65° column half-life data for support material C. Columns operated at 50°C had appreciably longer half-lives than columns operated at 60°C and 65°C. Longest half-life obtained was 113 days with support E. Half-lives of the remaining columns operated at 50° were ≤68% (46.6–77.1 days) of the longest half-life obtained.

The maximum DE achieved did not exceed 92. The use of an acid-enzyme spray-dried substrate may be the causative factor in the inability to achieve DE's of greater than 95. This is borne out by the fact that GA studies on porous-glass columns with freshly prepared enzyme-thinned cornstarch have given DE values in excess of 95 in the past. Studies are in preparation to examine the substrate problem in greater detail.

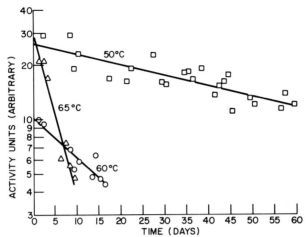

Fig. 1. Denaturation of glucoamylase immobilized on support C (SiO_2, 100%).

A large batch of ceramic support material designation C was chosen for the scale-up at Iowa State University.

This section of the report summarizes data collected at the Iowa State University pilot plant for the scale-up of immobilized glucoamylase. The design of the plant is illustrated on Fig. 2. For these studies only a single reactor was used.

The scale-up began when approximately 32 pounds of SiO_2 carrier, previously silanized at Corning Glass Works, was placed in the column and coupled *in situ*. The coupling data are presented in Table III. The entire process was completed in less than 10 hr.

The calculated process details and residence times vs flow rates are given in Table IV.

TABLE III
SCALE-UP IMMOBILIZATION DETAILS

Enzyme support: SiO_2 30-45 mesh, 400 Å pore
Enzyme: Glucoamylase, Novo, 60,000 units/g (1 unit = 1 μmole of glucose per minute)
Enzyme offered: 4-31.8 lb of carrier (1 ft³ bulk volume)
  Bulk enzyme activity before immobilization = 12,700 units/ml
  Bulk enzyme activity after immobilization = 1400 units/ml
Enzyme attached: 3.56 lb/31.8 lb of carrier or 112 mg/g carrier
Immobilized enzyme activity: 3000 units/g carrier
Enzyme bonding method: Silanization of SiO_2 followed by glutaraldehyde coupling of enzyme[a]

[a] SiO_2 was silanized in a batch, and glutaraldehyde and enzyme solutions were recirculated through the reactor, and coupled *in situ*.

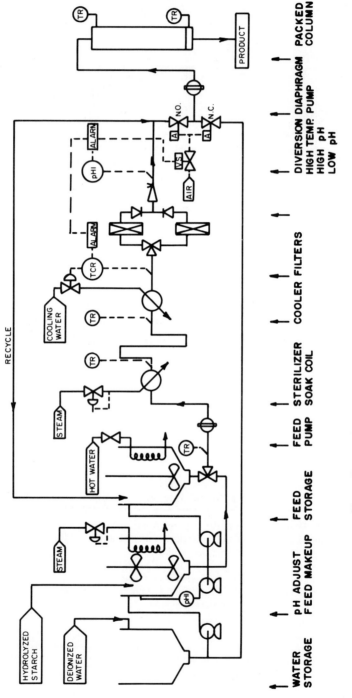

Fig. 2. Process flow diagram of Iowa State glucoamylase pilot plant.

TABLE IV
Process Details

Reactor: Dimensions = 6 inches i.d. × 5 ft high
Volume = 0.98 ft^3
Estimated reactor void volume with packing = 0.3 ft^3 (pore volume excluded)

Reactor Residence Time

Production rate (lb glucose/day)	Flow rate (gpm)[a]	Nominal residence time[b] (min)	Real residence time[c] (min)
1200	0.30	24.4	7.48
1000	0.25	29.3	8.99
800	0.20	36.6	11.22
500	0.13	56.4	17.30

[a] WW glucose per minute.
[b] Reactor volume divided by flow rate.
[c] Reactor void volume divided by flow rate.

The original intention was to operate at 50°–55°. However, because of the high activity of the column and the difficulty of handling substrate, we were forced to operate at temperatures ranging from 38° to 42°. Before starting, a saturated aqueous solution of chloroform was pumped into the entire system. The chloroform solution was replaced by the sterile substrate as it passed into the column.

Pilot Plant Operation

The initial operation of the pilot plant was conducted with a starch hydrolyzate of 24 DE called STAR-DRI 24-R, which is a partially acid-hydrolyzed cornstarch. This was purchased from A. E. Staley. Feed solutions were made up to 30% by weight and the pH was adjusted to approximately 4.5 by using hydrochloric acid. The initial production rate was about 1200 lb dry weight per day glucose at a volumetric flow rate of 1100 ml/min through the enzyme column. Flow was from top to bottom, and the temperature was maintained at 37° at the inlet with an outlet temperature of 38°–39°.

The analysis of the product showed a glucose production of 89.6–90.0% by weight of the solids produced and a dextrose equivalent DE of 92.4–93.0. The enzyme activity during the 70 days of operation showed no detectable change when compared to the original activity.

The values obtained for the glucose yield are lower than desirable from an industrial standpoint. (Discussions with corn processors indicate that desired glucose levels would be 95%. However, most would settle for 94% glucose.) Equipment is being purchased and installed to enable the production of glucose using pearl starch hydrolyzed in a jet-cooker with α-amylase. It is hoped that this will significantly decrease the retrogradation and formation of nonhydrolyzable products that now reduce the glucose yield. Batch data with both free enzyme and immobilized enzyme showed that a maximum of 90–91% glucose could be obtained using the Staley 24 DE substrate.

The biological contamination of the enzyme column and the product stream was measured by taking plate counts of the product stream for bacterial and mold contamination. After 30 days of operation, the product stream showed a bacterial count of 36/ml, recycle stream 93/ml, and feed 220/ml with no mold. At 45 days the count was 6/ml in the product, 25/ml in the recycle, and 25/ml in the feed. At 70 days the bacterial count was 1600/ml in the product, 4400/ml in the recycle, and 20/ml in the feed. The increase followed the introduction of unsterilized air into the column when the feed pump was blocked. Five days later, the count was 41,000/ml in the product (Table V).

The column was washed with water by using bottom-to-top flow for 1 day, and this was followed by a wash with a saturated chloroform solution through the entire system, including the column, for 3 hr. The chloroform solution was pumped out with sterile feed. After 4 days, the count was 300/ml in the product. The column was washed with a chloroform solution and placed in the cold room at 4° for 3 weeks. The plant

TABLE V
BACTERIAL CONTAMINATION

Operation time (days)	Bacterial count/ml fluid		
	Feed	Recycle	Product
30	220	93	36
45	25	25	6
70[a]	20	4400	1,600
75	—	—	41,000
After flush	—	—	300

[a] After column was contaminated with air. No molds were detected at any time. After an additional 3 weeks of storage, the column was operated for an additional 12 hr with no increase in count.

was started up at that time and run for 12 hr; the bacterial count at 4 hr was 330/ml of product, and it was 910/ml of product at 10 hr. The column was then stored again after washing with the chloroform solution. No deactivation was noticed in samples of product analyzed during the 12-hr run compared to the samples taken at the start of the test.

Problems

The most serious problem in operating the pilot plant was with the inlet strainer on the main feed pump. This caused a loss of flow to the column pump resulting in the pumping of air drawn into the recycle line through the column. A check-valve was installed in the line to prevent its recurrence.

Air introduced into the column in this manner was not sterile and contaminated the enzyme column. The air was removed by reversing flow in the column (bottom to top) for the time necessary to remove bubbles present in the product stream. No difference was observed in column performance when flow was reversed. When flow rates were similar, glucose content of the product was the same regardless of flow direction. The only difference noted when flow was returned to the original downflow was a reduction in the pressure drop across the column to about 50–75% of the flow prior to the upset. The pressure drop increased over a period of days to the original operating value of 20–30 psi.

During one period of operation several weeks into the run, it was found that the column temperature could not be controlled because the cooling water to the process cooler had eroded the valve. A pressure regulator and a new valve were installed and remedied the problem.

During the 70 days of operation, the pilot plant was run on Staley STAR-DRI 24-R most of the time. On two occasions, however, this was not available, so Staley STAR-DRI 35-R and Staley STAR-DRI 42-R were substituted to keep the plant operating. The maximum conversion with 35-R was 87.2% glucose, and for 42-R was 86.7% glucose. The lower conversion is attributable to retrogradation of the hydrolyzed starch and, because these are acid hydrolyzed, the formation products that cannot be broken down by the enzyme. The chromatograms of products of both starches show about 5% material in peaks other than glucose, maltose–isomaltose, maltotriose, and starch, which with 5% maltose–isomaltose and 2% starch limits the conversion to glucose.

Tests were run on both these starches (35-R, 42-R) in a batch immobilized-enzyme reactor and in a batch free-enzyme reactor, the results being about the same in both cases, and about equal to the values obtained for the pilot plant starches. The operation of the pilot plant could be monitored by comparing an analysis of the product with the values obtained in the batch runs. Plotting this on an extent of reaction curve shows how the plant is operating.

The analysis of low-flow-rate product showed a large percentage of maltoses, while the amount of starch decreased slightly and the amount of glucose stayed about the same as the values obtained under high-flow-rate conditions. Practically, the difference in the product analysis between the low and high flow rates was in DE, and was primarily affected only by changing maltose concentrations, the glucose concentration remaining relatively unchanged.

Because the conversion in the pilot plant was lower than was desired, a series of tests was conducted on several different hydrolyzed starches to determine whether, indeed, the beginning substrate limited the conversion to glucose (Figs. 3–5). Batch runs with free enzymes and with immobilized enzymes were conducted using acid-thinned Staley 24, 35, and 42 DE hydrolyzed starches and GPC's Maltrins 10, 15, 20, 25, and 42 DE syrup—the Maltrins being enzyme thinned, and the syrup acid thinned. In all these trials, the lower DE starting material results in slightly higher production of glucose in the free-enzyme reaction, producing between 1 and 2% higher glucose yields than the immobilized enzyme reaction. The enzyme-thinned substrates (GPC) produced from 2% to 5% higher yields of glucose than the acid-enzyme-thinned starches obtained from Staley (Table VI).

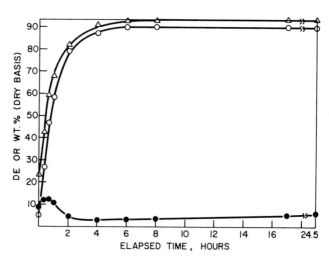

FIG. 3. Formation of glucose from Staley STAR-DRI 24-R in a recirculated batch reactor with 5.2 ml of immobilized glucoamylase in 115 ml of 30 wt.% Staley STAR-DRI 24-R at 45°, pH 4.4. ○, Glucose; △, dextrose equivalent (DE); ●, disaccharides.

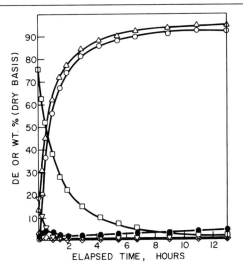

Fig. 4. Formation of glucose from GPC Maltrin-10 in a recirculated batch reactor with 5.6 ml of immobilized glucoamylase in 115 ml, 30 wt. % GPC Maltrin-10 at 45°, pH 4.4. ○, Glucose; △, dextrose equivalent (DE); ●, disaccharides; □, G_{9+}; ▽, G_3; ◇, G_4–G_9.

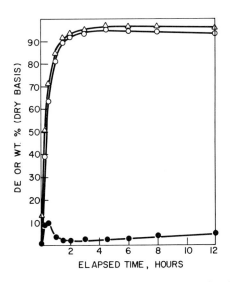

Fig. 5. Formation of glucose from GPC Maltrin-10 in a batch reactor with 0.2 g of free glucoamylase in 115 ml of 30 wt. % GPC Maltrin-10 at 45°, pH 4.4. ○, Glucose; △, dextrose equivalent (DE); ●, disaccharides.

TABLE VI
Maximum Glucose Yields[a]

Substrate DE	Maximum glucose (wt. %)		Maximum DE (calculated)	
	IME[b]	Soluble	IME[b]	Soluble
10	92.8	95.1	95.2	97.1
15	92.0	93.0	94.9	95.8
20	92.2	93.4	94.9	95.8
25	92.8	93.0	95.4	95.4
24	90.5	90.7	93.9	93.8

[a] All starches except the 24 dextrose equivalent (DE) material were prepared by enzyme hydrolysis and spray dried.
[b] Immobilized enzyme.

Tests were conducted with Penick & Ford pearl starch using α-amylase for liquefication and then adding glucoamylase in a free-enzyme batch reaction. This resulted in a 94% yield of glucose. The same procedure for liquefying was then used, but the solution was filtered and allowed to react in the immobilized enzyme batch reactor, resulting in 92.1% glucose. The thinned starch was also run through a continuous column (single-pass) resulting, after product filtration, in 93.5% glucose (a glucose content similar to that of the soluble enzyme). It is believed that by starting with pearl starch, liquefying with α-amylase, then filtering, treating with activated charcoal, and running through the immobilized enzyme column in a continuous manner, 93–94% glucose in the product can be obtained with little difficulty.

The general observation that dextrose levels for both immobilized and soluble systems are similar is very encouraging. It is our belief that by using freshly prepared substrate under optimized conditions a 95+ dextrose is possible. In fact, this would be required to make the immobilized GA a commercially viable approach.

Processing Cost Estimate

In order to evaluate the commercial potential of immobilized glucoamylase, an estimate was made of the cost involved in the saccharification of liquefied starch or low DE corn syrup. A proposed reactor system for just this hydrolysis step, as shown in Fig. 6, was designed to allow maintenance of conversion at a constant level by decreasing flow rate as activity decreases. By utilizing a multiple column system and staggering the column start-up or reloading times, the resulting variation

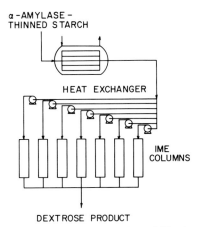

FIG. 6. Reactor system. IME, immobilized enzyme.

in production rate can be kept within any desired limits. The number of columns necessary to assure a production rate within given limits is a function of these limits and the number of half-lives the IME is used.[5] For this system, production rate variation limits of ±10% were set resulting in at least seven columns for two half-life IME use. It would also be desirable to have an additional column available to provide uninterrupted operation during column loading.

Another possible method of operation would be to maintain conversion at a constant level by raising reactor temperature to increase the reaction rate to compensate for activity loss.

Cost estimates cover only the saccharification step (including pH control) and the liquefaction or any subsequent purification steps. Processing costs are given on the basis of 100-pound (dry weight) quantities of solids processed.

Column sizes and IME requirements were based on experimentally observed performance. The flow rate required in the 1 ft³ column for maximum conversion at 40° can be estimated at 500 ml/min, which translates into 16.8 lb of solids per pound of IME per day. IME life was estimated from the results for support material C as shown in Table II (13.8 days at 60°, 51.2 days at 50°, 402 days projected for 40°). The relative reaction rates at 40°, 50°, and 60° for glucoamylase IME were reported by Havewala and Pitcher[6] to be 1.00, 1.87, and 2.78, respectively.

[5] N. B. Havewala and W. H. Pitcher, Jr., in "Enzyme Engineering" (E. K. Pye and L. B. Wingard, Jr., eds.), Vol. 2, p. 315. Plenum, New York, 1974.

[6] N. B. Havewala and W. H. Pitcher, Jr., "Glucose Production from Cornstarch: Reactor Parameter Study," to be published.

TABLE VII
Plant Cost Estimates[a]

Operating temperatures (°C)	Number half-lives enzyme utilization	Column capacity (ft³)	Plant cost ($)
40	3	1247	798,000
40	2	970	634,000
40	1	728	467,000
50	3	668	658,000
50	2	519	512,000
50	1	390	368,000
60	3	448	576,000
60	2	349	442,000
60	1	262	312,000

[a] Basis: 10^8 lb/yr plant, 3rd quarter-1974 prices.

Equipment costs were estimated from standard literature sources[7-10] and updated to third-quarter 1974 using the Marshall and Stevens Index. It should be noted that these are preliminary estimates only. However, a contingency of 10% is included in the final plant cost. Estimated column capacities and plant costs for plants producing 100 million pounds per year under various operating conditions are shown in Table VII.

Labor costs were estimated to be about 11¢/cwt for a plant of this size. Processing costs including labor, IME, and capital costs are shown in Table VIII. Capital costs are reflected in processing costs by a percentage, 20%, allotted annually for depreciation (10%), maintenance (3–5%), insurance (1%), and interest and taxes (4–6%).

Since the estimated cost of purified glucoamylase enzyme is less than 2 dollars per pound of IME, most of the IME cost will be attributable to carrier and immobilization costs. IME costs could reasonably be expected to be in the 5–20 dollars per pound range.

The variation of cost with respect to operating temperature, IME cost, number of half-lives of enzyme utilization (before columns are repacked with new IME), and plant capacity was determined. Increased plant size or capacity above 100 million pounds per year would only slightly decrease costs, primarily through labor cost savings. The effect

[7] H. Popper, "Modern Lost-Engineering Techniques." McGraw-Hill, New York, 1970.
[8] J. W. Drew and A. F. Ginder, *Chem. Eng.* **1970**, Feb. 9, 100 (1970).
[9] B. G. Liptak, *Chem. Eng.*, **1970**, Sept. 7, 60 (1970).
[10] S. P. Marshall and J. L. Brandt, *Chem. Eng.* **1971**, August 23, 68 (1971).

TABLE VIII
Cost of Saccharification by Immobilized Glucoamylase[a]

Operating temperature (°C)	Number half-lives enzyme utilization	Processing cost (¢/cwt dry solids)			
		IME Cost ($/lb): 5	10	15	20
40	3	32.8	38.7	44.6	50.4
40	2	30.5	37.4	44.2	51.0
40	1	30.6	40.9	51.1	61.4
50	3	37.4	50.6	63.8	77.0
50	2	36.6	52.0	67.5	82.9
50	1	41.5	64.6	87.7	110.8
60	3	63.7	105.0	146.2	187.5
60	2	68.0	116.1	164.3	212.3
60	1	89.5	161.7	233.9	306.2

[a] Basis: 10^8 lb/yr plant (20%/yr plant cost for maintenance, depreciation, interest, taxes, etc.).

of the other three variables is shown in Table VIII. Figure 7 is a contour plot of processing cost as a function of operating temperature and IME cost. From the standpoint of processing cost operation at lower temperatures with 2 or 3 half-lives, enzyme utilization would be necessary to compete with the 49¢/cwt (12¢/cwt enzyme cost, 11¢/cwt labor cost and 26¢/cwt capital cost) estimated cost of the soluble enzyme process if installed in a new plant.

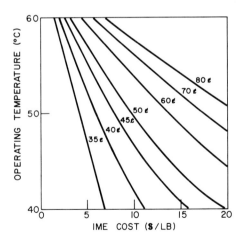

FIG. 7. Processing cost as a function of operating temperature and immobilized enzyme (IME) cost. Plant capacity: 10^8 lb/yr. IME utilization: 3 half-lives. Curves show processing costs as cents per 100 lb.

From Fig. 7 it can be seen that at the lower temperatures IME costs as high as 20 dollars per pound could be competitive with the soluble batch process, while at 60°C an unrealistically low IME cost of 3 dollars per pound would be necessary. The low-temperature operation demonstrated in this scale-up is necessary for the economic feasibility of this system.

Acknowledgments

We are grateful to Mr. D. L. Eaton of Corning Glass Works for supplying the porous ceramic carriers and Dr. D. J. Lartigue of Corning Glass Works for his technical assistance in preparing the purified enzyme. We thank Novo Enzyme Corp. for the donation of 20 pounds of purified glucoamylase. This enzyme was used for scale-up studies. We acknowledge the excellent technical support of Mr. R. E. Lindner of Corning Glass Works, who performed all DE determinations and supplied the 60° column data for support B. The authors also gratefully acknowledge support from National Science Foundation Grant GI-38101 (ATA73-07783) for the scale-up studies at Iowa State University.

[55] The Preparation, Characterization, and Scale-Up of a Lactase System Immobilized to Inorganic Supports for the Hydrolysis of Acid Whey

By WAYNE H. PITCHER, JR., JAMES R. FORD, and HOWARD H. WEETALL

Whey, the major by-product of cheese manufacturing, is comprised chiefly of water and lactose, the latter a rather bland-tasting disaccharide, composed of two monosaccharides, glucose and galactose.

With the increasing cost of sweeteners and the emphasis on whey utilization, a number of efforts have been made to hydrolyze the lactose in whey to obtain sweeter, more soluble sugars for use in dairy products.[1-6]

[1] H. H. Weetall, N. B. Havewala, W. H. Pitcher, Jr., C. C. Detar, W. P. Vann, and S. Yaverbaum, *Biotechnol. Bioeng.* **16**, 295 (1974).
[2] H. H. Weetall, N. B. Havewala, W. H. Pitcher, Jr., C. C. Detar, W. P. Vann, and S. Yaverbaum, *Biotechnol. Bioeng.* **16**, 689 (1974).
[3] A. C. Olson and W. L. Stanley, *J. Agric. Food. Chem.* **21**, 440 (1973).
[4] W. L. Stanley and R. Palter, *Biotechnol Bioeng.* **15**, 597 (1973).
[5] M. Charles, R. W. Coughlin, B. R. Allen, E. K. Paruchuri, and F. X. Hasselberger, "Increasing Economic Value of Whey Wastewaters Using Immobilized Lactase," paper 17b, presented at AIChE 66th Annual Meeting, Philadelphia, 1973.
[6] W. H. Pitcher, Jr., *Dairy Scope.* p. 18 (March, 1974).

Acid whey, a by-product of cottage cheese manufacture, has relatively little value and frequently poses a disposal problem. This report reviews the development of an immobilized enzyme system for the hydrolysis of the lactose in whey (see also this volume [57] for hydrolysis of lactose).

Methods and Materials

Preparation of Immobilized Lactase

Lactase enzymes were immobilized on porous inorganic supports by the aqueous silane–glutaraldehyde method (see also this volume [10]).[7] Porous silica bodies, of 30/45 mesh size and 370 Å average pore diameter, were made by D. L. Eaton, Corning Glass Works, using a procedure developed by R. A. Messing, also of Corning Glass Works. Titania porous bodies were prepared in a similar fashion (400 Å average pore diameter, 1 mm particle diameter). Zirconia-coated controlled-pore glass (CPG) particles, 550 Å average pore diameter and 40/80 mesh particle size, were also used as supports.

Lactase enzymes used were a fungal lactase from Miles Laboratories (called here Lactase-M), a yeast lactase (lactase-Y) supplied by Kyowa Hakko Kogyo Company, Tokyo, Japan, and Wallerstein Company's Lactase-LP, a β-galactosidase from *Aspergillus niger*. Lactase-Y and lactase-M were immobilized only on the zirconia-coated CPG.

Assay Procedure

The soluble lactase-Y was assayed at 60° in a 20% lactose solution containing 0.025 M sodium phosphate buffer, 0.5 mM manganese chloride, and 0.5 mM cobalt chloride at pH 6.5. Lactases M and W were assayed in the same solution without the metal salts. Similar procedures were used at optimal pH to assay immobilized enzyme (IME) samples. Later assays of lactase-W and IME were carried out at 40° or 50°. One activity unit is defined as the production of 1 μmole of glucose per minute at optimum pH at the specified temperature. Glucose concentration was measured using Worthington Glucostat.

Determination of pH Profile

The optimum pH values for soluble and immobilized lactases were determined by using the assay procedure in the preceding section. The substrate was adjusted to the desired pH level before each assay by the addition of HCl or NaOH.

[7] H. H. Weetall and N. B. Havewala, *Biotechnol. Bioeng. Symp.* 3 (1972).

Determination of Temperature Effects

The apparent activities of IME and soluble enzymes at various temperatures were measured by assays or column operation at the appropriate temperatures.

Determination of Kinetic Constants

Michaelis constants (K_m) and the inhibition constants (K_i) were calculated for both soluble and immobilized lactase. K_i values were determined from assays in the presence of known amounts of the inhibitor, galactose.

Reactor Studies

Lactose and whey feeds, from 3.3 to 10% lactose by weight, were passed continuously through columns 1.5 or 2.5 cm in diameter packed with 5–10 g of IME. Columns were water-jacketed to maintain the desired temperature. Larger columns, including one 4 inches in diameter, containing about 6 pounds of IME, were also operated. Conditions of operation for specific studies are discussed in subsequent sections.

Feed or substrate materials included acid whey and deproteinized acid whey (ultrafiltrate permeate), supplied by Crowley Foods, Inc. and Kraft edible lactose. Benzalkonium chloride was used at levels of 150 ppm to prevent microbial growth in some of the edible lactose feed solutions.

The lactose content of the whey feeds was determined by the Lane and Eynon[8] procedure for reducing sugars. Glucose concentration was measured using Worthington Glucostat.

Results

Soluble and Immobilized Enzyme Activity

The soluble enzymes when assayed gave the following activities per gram of solids:

Lactase-M	1,180 units	(60°C)
Lactase-Y	9,630 units	(60°C)
Lactase-W	10,550 units	(60°C)
	10,000 units	(50°C)
	6,000 units	(40°C)

[8] "Standard Analytical Methods of the Member Companies of the Corning Refiners Association," first revision, 5-27-58, E-26.

TABLE I
COUPLING EFFICIENCIES

Enzyme	Carrier	Activity offered	Units (60°C)/g IME	Coupling efficiency (%)
Lactase-M	ZrO_2-CPG	118	75	64
Lactase-Y	ZrO_2-CPG	963	850	88
Lactase-W	ZrO_2-CPG	1055	800	76
Lactase-W	TiO_2	1055	500	47
Lactase-W	SiO_2	1055	800	76
Lactase-W	SiO_2	1055	650	62

Table I shows coupling efficiencies for various immobilized lactases. Coupling efficiency is defined as

$$E_{IME}/(E_{initial} - E_{recovered})$$

where E is enzyme activity, the percentage of enzyme not recovered from the enzyme solution after attachment, that is observed as active IME. For lactase $E_{recovered}$ has normally been negligible. Table II shows the effect of the amount of enzyme offered on the coupling efficiency. Higher loadings typically lead to lower coupling efficiencies.

pH Profile

The pH profiles of lactase-M and lactase-W (Figs. 1 and 2), immobilized as well as soluble, are similar, with a downward shift in pH for the immobilized forms. Soluble lactase-Y has a much higher pH opti-

TABLE II
COUPLING EFFICIENCY AS A FUNCTION OF ENZYME OFFERED

Enzyme	Carrier	Activity offered	Units (40°C)/g IME	Coupling efficiency (%)
Lactase-W	ZrO_2-CPG	621	360	58
Lactase-W	SiO_2	620	400	65
Lactase-W	SiO_2	615	294	48
Lactase-W	SiO_2	900	359	40
Lactase-W	SiO_2	1230	473	38
Lactase-W	SiO_2	1860	579	31

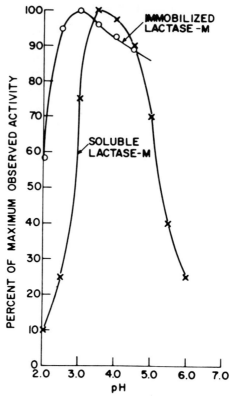

FIG. 1. Lactase-M pH profile at 60°. From H. H. Weetall, N. B. Havewala, W. H. Pitcher, Jr., C. C. Detar, W. P. Vann, and S. Yaverbaum, *Biotechnol. Bioeng.* **16**, 295 (1974).

mum, about 6.2, but this optimum is also shifted much lower, to about 3.0, when the enzyme is immobilized (Fig. 3).

Activation Energy

The activation energies calculated from plots obtained on the variation of reaction rate with temperature are summarized in Table III. The low activation energy for lactase-W coupled to titania is probably the result of diffusion limitations.

Kinetic Constants

Kinetic data such as the Michaelis constant K_m and the inhibition constant K_i for various immobilized and soluble lactases are reported in Table IV.

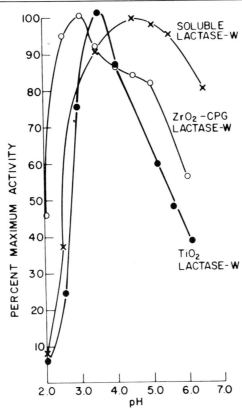

Fig. 2. Lactase-W pH profile at 60°C. From H. H. Weetall, N. B. Havewala, N. B. Pitcher, Jr., C. C. Detar, W. P. Vann, and S. Yaverbaum, *Biotechnol. Bioeng.* **16**, 689 (1974).

TABLE III
ACTIVATION ENERGIES

Enzyme	Activation energy (kcal/g-mole)
Lactase-Y (soluble)	10.5
Lactase-Y (IME)	11.3
Lactase-M (IME)	6.5
Lactase-W (IME)[a]	7.8
Lactase-W (IME)[b]	12.0

[a] Immobilized on 1 mm TiO$_2$ particles.
[b] Immobilized on 30/45 mesh SiO$_2$ particles.

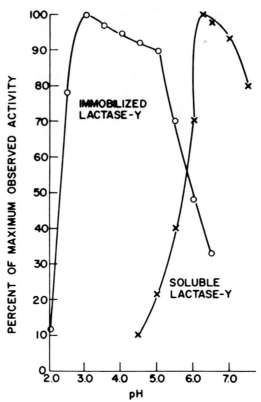

Fig. 3. Lactase-Y pH profile at 60°C. From H. H. Weetall, N. B. Havewala, W. H. Pitcher, Jr., C. C. Detar, W. P. Vann, and S. Yaverbaum, *Biotechnol. Bioeng.* **16**, 295 (1974).

Rate Equation

Weetall et al.[1] discussed the kinetics of lactose hydrolysis, and gave the rate equation as

$$v = \frac{kES}{S + K_m[1 + (P/K_i)]} \qquad (1)$$

where v = reaction rate, k = turnover number, E = amount of enzyme, $kE = V_{max}$ (the maximum rate achieved when $S \gg K_m$), S = substrate concentration, and P product concentration.

The integrated form of Eq. (1) for a batch or plug-flow reactor, assuming no product initially present, is

$$Et/V \text{ or } E/F = 1/k[(K_1 - K_m)/K_i](S_0 - S) \\ + [(K_m S_0/K_i) + K_m] \ln (S_0/S) \qquad (2)$$

TABLE IV
Kinetic Data

	Kinetic constants, at 60°	
Enzyme	Apparent K_m (M)	Apparent K_i (mM)
Lactase-Y (soluble)	0.112	0.67
Lactase-Y (IME)[a]	0.071	3.27
Lactase-M (soluble)	0.04	1.25
Lactase-M (IME)	0.05	3.22
Lactase-W (soluble)	0.05	3.9
Lactase-W (IME)[b]	0.07	3.8
Lactase-W (IME)[c]	0.05	5.0
Lactase-W (IME)[d]	0.053	5.4[e]

[a] IME, immobilized enzyme.
[b] ZrO_2-CPG carrier.
[c] SiO_2 carrier.
[d] SiO_2 carrier.
[e] 40°C.

where F = volumetric flow rate of substrate, V = volume of substrate concentration, and S = substrate concentration at time t or in the reactor effluent.

A theoretical curve relating lactose conversion to normalized residence time (IME activity/flow rate or E/F) was calculated from Eq. (2) using K_m and K_i values (the 40° data in Table IV) as reported by Ford and Pitcher.[9] Data from column operation agreed closely with the theoretical curve as shown in Fig. 4. Thus Eq. (2) could be used to calculate enzyme activity from conversion and flow rate data, allowing convenient monitoring of column performance.

Half-Life Studies

A number of IME samples were evaluated for operational durability by continuous use in columns. Activities were calculated from Eq. (2) or from plots similar to Fig. 4. Examples of activity plotted as a function of time are shown in Figs. 5 and 6. The fact the logarithm of activity versus time yields a straight line is consistent with the normally observed exponential decay of enzyme activity. Half-lives were calculated from linear regressions (least squares) of log E versus time and are listed in Table V.

[9] J. R. Ford and W. H. Pitcher, Jr., "Immobilized Lactase Systems," Whey Products Conference, Chicago, September, 1974.

TABLE V
Half-Life of Immobilized Lactases with 10% Lactose Substrate[a]

Enzyme	Temp. (°C)	Days of operation	Mean $t_{1/2}$ (days)	UCL[b] (days)	LCL[b] (days)
Lactase-M	30	55	198	465	126
Lactase-M	40	35	38.3	48.2	31.8
Lactase-W (ZrO$_2$-CPG)	30	36	43.9	63.0	33.6
Lactase-W (ZrO$_2$-CPG)	40	24	34.7	47.0	27.4
Lactase-W (ZrO$_2$-CPG)	50	4	2.7	4.9	1.7
Lactase-W (NaB-Treated)	40	23	35.8	55.9	26.4
Lactase-W (porous TiO$_2$)					
34/45 mesh	40	33	39.8	57.7	30.1
1 mm particles	40	37	77.9	93.6	69.3
1 mm particles	40	37	66.6	92.4	53.3
1 mm particles[c]	40	45	58.2	96.3	41.3
Lactase-M	70	2	2.7	12.0	—
Lactase-M	60	15	2.6	3.8	2.0
Lactase-M	50	15	14.6	27.4	10.0
Lactase-M	50	15	14.5	27.6	9.8
Lactase-Y	70	2	0.40	0.21	3.18
Lactase-Y	60	15	0.55	0.43	0.75
Lactase-Y	50	25	4.8	4.2	5.7
Lactase-Y	40	25	9.6	7.5	13.2

[a] All studies are from columns maintained at constant flow rate.
[b] The correlation of these data with exponential decay in all cases exceeded 0.85. UCL, upper 95% confidence limits; LCL, lower 95% confidence limits.
[c] These data were obtained from a column operated at pH 6.5.

Table VI shows the effect of feed composition on IME half-life, with the higher purity feeds yielding longer half-lives.

From Table V it can be seen that increased temperatures result in decreased half-lives. Figure 7 shows the effect of temperature on half-life in an Arrhenius-type plot.

Mass Transfer Studies

Internal mass transfer or pore diffusion limitations appear to be minimal, at least up to 60° with 400 units (40°) per gram IME (30/45 mesh silica carrier) for the following reason. Assume, as the extreme case, that the 7.8 kcal/g-mole activation energy for TiO$_2$-lactase-W from Table III was the result of the reaction rate being entirely diffusion con-

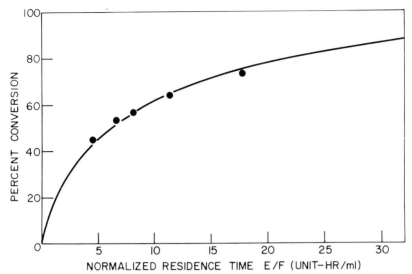

Fig. 4. Conversion versus normalized residence time. Theoretical curve: $K_m = 0.0528$ M, $K_i = 0.0054$ M. Lactase immobilized enzyme at 40°; 5% lactose. From J. R. Ford and W. H. Pitcher, Jr., "Immobilized Enzyme Systems," Whey Products Conference, Chicago, September, 1974.

TABLE VI
Effect of Feed Composition on Half-Life

Feed	Half-life at 50° (days)		
	Mean	95%	Confidence limits
Whole acid whey	7	5	8
Deproteinized acid whey	10	7	17
5% Lactose with 0.5% NaCl	13	10	17
Deionized, deproteinized acid whey	62	43	108
5% Lactose	89	73	116

trolled. In that case the apparent activation energy should have been equal to half the intrinsic activation energy (see Levenspiel[10]) plus 1 or 2 kcal/g-mole for diffusivity effects. An intrinsic activation energy of 13 or 14 kcal/g-mole would be estimated. This value would be less if the TiO_2-lactase-W IME reaction rate had been only partially diffusion

[10] O. Levenspiel, "Chemical Reaction Engineering." Wiley, New York, 1962.

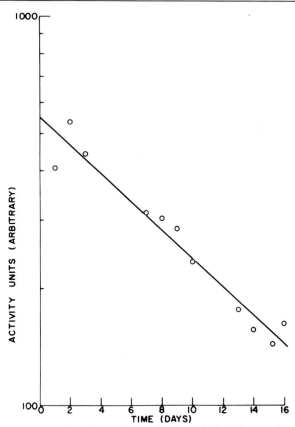

FIG. 5. Activity as a function of time for immobilized lactase-M at 50°. From H. H. Weetall, N. B. Havewala, W. H. Pitcher, Jr., C. C. Detar, W. P. Vann, and S. Yaverbaum, *Biotechnol. Bioeng.* **16**, 295 (1974).

limited. The straight-line plot obtained on varying reaction rate with temperature, coupled with the fact that the intrinsic activation energy was 14 kcal/g-mole or less, indicates the absence of pore diffusion effects.

External mass transfer or film diffusion effects were evaluated by operating columns containing IME beds ranging from 6 cm to 71 cm in height. These reactors were operated with the same normalized residence times and resulted in identical conversion levels. The 10-fold increase in linear velocity evidently did not affect the conversion-rate, implying that external mass transfer limitations were negligible.

Potential effects of back-mixing can be estimated from a dispersion number calculation described by Levenspiel.[10] The dispersion number is defined as D/uL, where D is the dispersion coefficient, u is the fluid

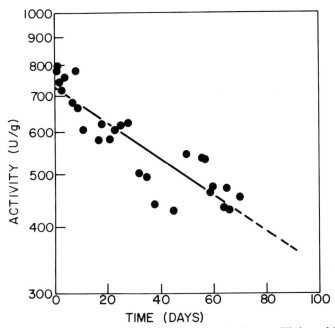

Fig. 6. Activity at 50° as a function of time for lactase-W immobilized on silica. Half-life = 90 days. From J. R. Ford and W. H. Pitcher, Jr., "Immobilized Enzyme Systems," Whey Products Conference, Chicago, September, 1974.

velocity, and L is the bed height. This dispersion number can be related to the Peclet number,

$$N_{Pe} = u d_p / D \epsilon$$

where d_p = particle diameter and ϵ = bed void fraction, by

$$D/uL = (N_{Pe})(-1 d_p / \epsilon L) \qquad (3)$$

At Reynolds numbers less than 10 in a packed bed, the Peclet number is approximately equal to 0.5. Even for a small packed bed, 6 cm in height, containing 30/45-mesh particles (0.46 mm in diameter), with $\epsilon = 0.35$,

$$D/uL = (1/0.5)[0.046/(35)(6)] = 0.044$$

From a graph given by Levenspiel for first-order kinetics, used as an approximation (actual kinetics should be less sensitive to back-mixing), the difference between real and ideal reactor volumes at 80% conversion is less than 8%. At lower conversion or greater bed heights, the difference becomes even smaller.

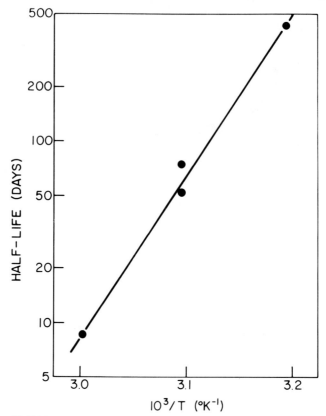

Fig. 7. Half-life as a function of temperature for lactase-W immobilized on silica. From J. R. Ford and W. H. Pitcher, Jr., "Immobilized Enzyme Systems," Whey Products Conference, Chicago, September, 1974.

Column Back-Flushing

In order to prevent microbial growth in the IME beds when whey feeds were used, the columns were back-flushed (fluidized) with distilled water brought to pH 4.0 with acetic acid. Columns were back-flushed once daily for about 0.5 hour. No visible growths have appeared in columns operated in this manner for over a month.

Reactor Scale-Up

A column reactor 4 inches in diameter containing a 28-inch-deep bed of silica–lactase-W IME was operated at 38°. Activity was calcu-

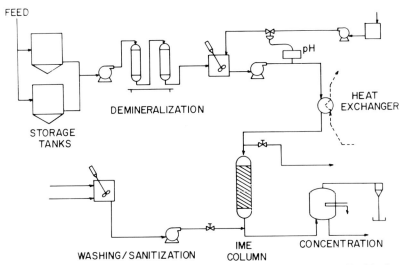

Fig. 8. Process flow sheet. IME, immobilized enzyme. From J. R. Ford and W. H. Pitcher, Jr., "Immobilized Enzyme Systems," Whey Products Conference, Chicago, September, 1974.

lated at 380 units/g (at 38°) equivalent to 400 units/g at 40°, identical to the activity normally observed for laboratory-scale columns.

Pressure drop, one remaining concern for additional scale-up, can be estimated from Leva's correlation as given by Perry.[11]

$$\Delta P = [2f_m G^2 L(1-\epsilon)^{3-n}]/d_p g_c \rho \phi_s^{3-n} \epsilon^3 \qquad (4)$$

where ΔP = pressure (lb-force/ft^2); L = bed depth (ft); G = fluid superficial mass velocity (lb/sec-ft^2); $f_m = 100/N_{Re} = 100 \ \mu/d_p G$ (for $N_{Re} < 10$); ϵ = void fraction; $n = 1$ (for $N_{Re} < 10$); d_p = particle diameter (ft); $g_c = 32.17$ (lb-ft/lb force-sec^2); ϕ_s = shape factor (0.95 for spherical sand); μ = viscosity; ρ = fluid density (lb/ft^3); and N_{Re} = Reynolds number.

For the case of a 10-foot-deep IME bed operating at 35° and 50% hydrolysis, a pressure drop of about 75 psi was calculated. For a 6-foot-high column under identical conditions, a pressure drop of about 27 psi is predicted.

Preliminary System Design

In order to make preliminary cost estimate for the hydrolysis of deproteinized acid whey, a plant design as shown in Fig. 8 was developed.

[11] J. H. Perry, "Chemical Engineers' Handbook," 4th ed. McGraw-Hill, New York, 1963.

TABLE VII
Equipment Cost Estimate

Equipment (10,000 lb/day lactose, 50% hydrolysis)	No.	Cost ($)	Cost ratio	Plant cost ($)
Column	1	4,300	4.0	17,200
Storage tanks (12,500 gal)	2	18,000	2.0	36,000
Process tanks				
100 gal	1	400	2.0	800
300 gal (w/agitator)	1	1,300	4.1	5,330
Pumps				
Centrifugal (20 gpm)	3	2,400	8.0	19,200
Metering	1	600	7.0	4,200
Heat exchanger	1	1,500	4.8	7,200
Instruments		4,000	4.0	16,000
				105,930
Contingency (10%)				10,590
			Total:	116,520

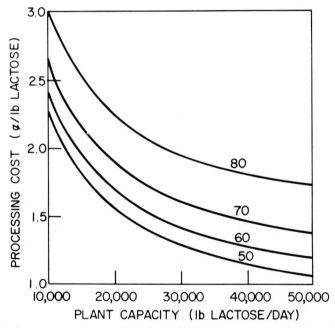

Fig. 9. Processing cost as a function of plant capacity for demineralized, deproteinized acid whey. Immobilized enzyme cost = 75 dollars per pound. Curves: 50, 60, 70, 80% hydrolysis. From J. R. Ford and W. H. Pitcher, Jr., "Immobilized Enzyme Systems," Whey Products Conference, Chicago, September, 1974.

TABLE VIII
OPERATING COST ESTIMATE

	(man-hr/day)	
Labor		
Backflushing	3	
Monitoring	3	
Laboratory	2	
	8 @ $4.50/hr =	$ 36.00
Supervisor	3 @ $6.00/hr =	18.00
		$ 54.00
Overhead and fringes		54.00
Supplies (acid, etc.)		20.00
Cooling costs		3.00
		$131.00
Cost per pound of lactose (10,000 lb lactose/day)		1.3¢

This flow sheet includes ion-exchange and concentration equipment, which will be discussed independently. The sizes for the hydrolysis system were based on use of lactase IME with an apparent activity of 300 units/g at 35°. An example of equipment cost estimates is given in Table VII.

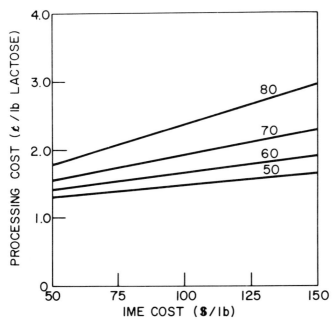

FIG. 10. Processing cost as a function of immobilized enzyme (IME) cost for deproteinized, demineralized acid whey. Plant size: 25,000 lb of lactose per day. Curves: 50, 60, 70, 80% hydrolysis. From J. R. Ford and W. H. Pitcher, Jr., "Immobilized Enzyme Systems," Whey Products Conference, Chicago, September, 1974.

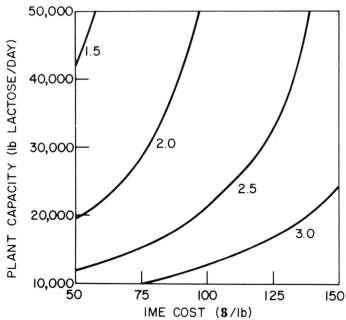

FIG. 11. Processing cost as a function of immobilized enzyme (IME) cost and plant capacity for deproteinized, demineralized acid whey. Curves: 80% hydrolysis processing cost: 1.5, 2.0, 2.5, 3.0 cents per pound of lactose. From J. R. Ford and W. H. Pitcher, Jr., "Immobilized Enzyme Systems," Whey Products Conference, Chicago, September, 1974.

The projected operating conditions were set somewhat arbitrarily, the reactor temperature, initially at 35°, being raised as necessary to maintain the initial conversion level until 50° was reached. From a half-life of 62 days at 50° determined experimentally for deproteinized, deashed acid whey and deactivation energy of 40.6 kcal/g-mole, the time required for the 35° to 50° cycle was calculated at 559 days. The number of pounds of lactose processed per pound of IME was then calculated from the cycle time and reactor size.

Processing costs included labor and supplies cost as shown in Table VIII, capital or equipment costs taken at 20% annually (included depreciation, maintenance, taxes) and IME cost. Total costs are reflected in Figs. 9–11. This total cost does not include the cost of deashing or concentration and assumes the cost of deproteinized whey to be zero.

Processing costs appear to be 1 to 4 cents per pound of lactose, depending on plant size, IME cost, and percentage of hydrolysis. Figure 9 shows the dependence of processing cost on plant size at various hydrolysis levels. The effect of IME cost on processing cost is shown in

Fig. 10. Higher hydrolysis levels obviously are more sensitive to IME costs. Figure 11 shows processing cost as a function of plant size and IME cost. This type of plot can be useful for determining what IME cost must be achieved to meet given capacity and cost objectives.

Reducing the salt level by ion exchange or electrodialysis results in projected costs of 2–5 cents per pound, including capital costs comparable to those for the hydrolysis system. Electrodialysis appears to be economically slightly more favorable at this point, perhaps costing in the vicinity of 3 cents per pound of lactose for 90% salt removal. Similar additional costs will be encountered for concentrating the product from 5% solids to the 50% or higher solids level necessary for sweetener substitution.

Overall costs of 8–10 cents per pound, including royalties, begin to look attractive at the March 1975 17-cent-per-pound (dry basis) price for corn syrups.[12]

[12] *Chemical Marketing Reporter*, March 17 (1975).

[56] Continuous Production of High-Fructose Syrup by Cross-Linked Cell Homogenates Containing Glucose Isomerase[1]

By POUL BØRGE POULSEN and LENA ZITTAN

For many years it was the practice to hydrolyze corn and potato starch to glucose with acid. The by-product formation was relatively high, but this was largely overcome by the introduction of enzymes 10–20 years ago. Some food applications, however, require more sweetness than can be provided by glucose. This need has until now been supplied by sucrose, which is sweeter than glucose.

A breakthrough in the development of sweeter corn syrups occurred when specific organisms were isolated that produced enzymes capable of isomerizing glucose to fructose. These glucose isomerases made it possible to manufacture corn syrups containing sufficient fructose to make them comparable with sucrose in sweetness.

Since 1957, when glucose isomerase was reported for the first time,[2] much research and development effort has been concentrated on making

[1] This work was carried out at Novo Industri A/S in Copenhagen; Research Director, M. Hilmer Nielsen.
[2] R. O. Marshall and E. R. Kooi, *Science* **125**, 648 (1957).

the process economical for large-scale commercial production of high-fructose corn syrup. But even today only a few glucose isomerases are available on the market in industrial quantities. This indicates the problems of transferring the laboratory results into industrial production and application.

Glucose isomerase is an intracellular enzyme, which is produced in large quantities only under carefully controlled conditions. In order to achieve good production economy, precise regulation of very large fermentors must therefore be possible. It is also necessary that the immobilization technique is inexpensive and that the specific volume activity is as high as possible. Furthermore, the engineering aspects of the isomerization process, for example, pressure drop and heat transfer in large reactors, need special attention before the industrial production of high fructose syrups is feasible.

The process to be described here is presently in operation in pilot plant scale and the results are so promising that industrial reactors of more than 10 m³ capacity are expected to be on stream within a year. In the following, the immobilization technique and the chemistry of the enzyme catalyzed isomerization are described. Finally an outline of the process layout will be given.

Kinetics and Definitions

The reaction mechanism for the isomerization of glucose to fructose has been described[3,4] as shown in the following sequence.

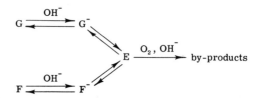

G: D-glucose
F: D-fructose
E: Enolate-ion

As the step between G⁻ and E is known to be rate-determining for the isomerization, we assume that this is the step that is catalyzed by the enzyme. It has been reported several times that glucose isomerase obeys apparent first-order kinetics,[5] so our models are based on this assumption. The following formula can be used to determine the activity, when the

[3] H. G. J. De Wilt and B. F. M. Kuster, *Carbohydr. Res.* **19**, 5 (1971).
[4] H. G. J. De Wilt and B. F. M. Kuster, *Carbohydr. Res.* **23**, 333 (1972).
[5] N. B. Havewala and W. H. Pitcher, Jr., in "Enzyme Engineering" (E. K. Pye and L. B. Wingard, Jr., eds.), pp. 315–328. Plenum, New York, 1974.

degree of conversion is greater that zero; the validity of the formula is limited to plug-flow isomerization and to a substrate concentration in the neighborhood of 40% w/w.

$$\text{Activity} = F[RfX - K \ln(1 - X/X_e)]/kwa(X)$$

where F = flow of substrate; Rf = glucose in the substrate determined by refractometry, % w/w; X = degree of conversion (fructose)/[(fructose) + (glucose)]; X_e = degree of conversion at equilibrium; K = constant (experimentally found value: 58% w/w, Rf = 40% w/w, 65°); w = weight of immobilized enzyme preparation; $a(X)$ = experimentally found correction function; k = constant; a value of 2.65 is required to fulfill the unit definition (see below).

The heat reaction at constant pressure has been calculated to be 0.91 kcal/g mole.[5] From this value it is concluded that an adiabatic large-column operation in effect will be nearly isothermal.[5] In practice a temperature drop of less than 1° is detected in 60-liter columns.

Activity. One unit of activity (1 IGIC) is defined as the amount of enzyme that initially converts 1 μmole of glucose to fructose per minute under standard assay conditions.

The standard conditions for activity determination in a column are glucose, 40% w/w; pH, 8.5; temperature, 65.0°; Mg, 0.004 M.

Stability. The stability or activity decay is expressed as time to 50% and 25% residual activity.

Productivity. Productivity is the combined effect of activity and stability. It is defined as the amount of glucose (DS), in kilograms, which at a given time has been converted to a mixture of 45% fructose and 55% glucose per kilogram of enzyme.

Enzyme Source and Immobilization Technique

The first glucose isomerase source described was *Pseudomonas hydrophila*.[2] Later the enzyme produced turned out to be a glucose-6-phosphate isomerase, which requires arsenate to act on nonphosphorylated glucose. Apparently a genuine, specific D-glucose isomerase does not exist, but D-xylose isomerase (EC 5.3.1.5) does isomerize glucose without an arsenate requirement.

We have found that organisms of the genus *Bacillus* produce a glucose (D-xylose) isomerizing enzyme with interesting characteristics. The highest yields have so far been obtained from an atypical strain of *Bacillus coagulans*. These organisms can be cultivated in relatively simple media by submerged fermentation under aerobic conditions.[6]

[6] H. Outtrup, Belgian Patent No. 832.852.

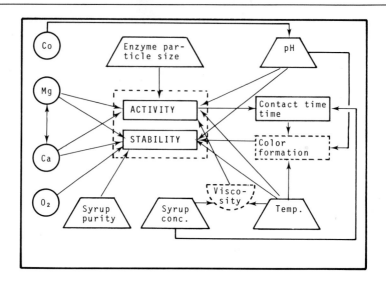

FIG. 1. Influence of process parameters on the activity and stability of immobilized glucose isomerase.

An example of a laboratory procedure for immobilization of glucose isomerase is given below[7] (other immobilization techniques are described by Amotz et al.[8]).

After a pH adjustment to 6.3, the cells are recovered from the fermentation broth by centrifugation. The concentrate, which is estimated to contain approximately 12% dry matter, is homogenized and spray-dried. After standardization the spray-dried powder is wetted with half its weight of water and then extruded in an axial extruder (700 μm). To 1 kg of the fluid bed dried extruded pellets, 1 liter of 20% glutaraldehyde (Merck, technical grade) (pH 7.0) is added with sufficient stirring to ensure thorough mixing. After 60 min at 25° the pellets are washed with 5 liters of deionized water and dried in a rotary drier.

Enzyme recovery varies from batch to batch between 20 and 30%.[7]

Influence of the Individual Process Parameters on Activity/Stability of the Enzyme

Figure 1 gives an overall outline of process parameters and their mutual influence on the activity and stability of the enzyme.

[7] S. Amotz, D.O.S. 2,345,185.
[8] S. Amotz, T. K. Nielsen, and N. O. Thiesen, Belgian Patent No. 809546.

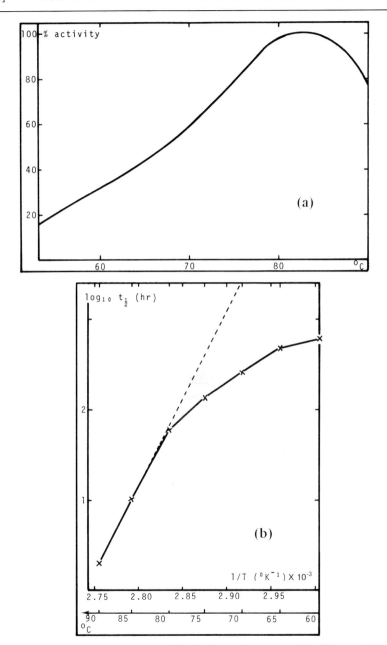

Fig. 2a. Temperature/activity profile of glucose isomerase. Conditions: 40% glucose, w/w; pH 8.5; Mg, 4 mM; column, 150 ml.

Fig. 2b. Temperature/stability profile of glucose isomerase. Conditions: 40% glucose, w/w; pH 8.5; Mg, 4 mM; column, 150 ml.

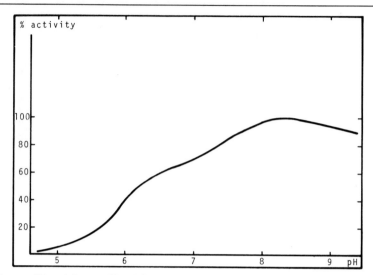

FIG. 3. pH/activity profile of isomerase glucose. Conditions: 40% dextrose syrup, w/w; 65°; Mg, 4 mM; column, 150 ml.

Contact Time. The lower the activity, the longer the contact time needed to achieve a given conversion with a given amount of enzyme. With long contact times, by-product and color formation is increased. This will contribute to increased process costs. The consumption of carbon for decolorizing the syrup grows with increasing color formation. Furthermore by-products may block the active center of the enzyme and thus decrease its productivity.

Temperature. The temperature at which the isomerization is performed has an influence on the activity and the stability of the enzyme as well as on the color formation. Figure 2 shows the relationship between temperature and activity (Fig. 2a) and between temperature and time to 50% residual activity (Fig. 2b).[9] It can be seen that the enzyme is active up to 90° but that the stability at that temperature is too low for practical applications. The highest productivity of the enzyme will be found at a temperature of approximately 65°. A lower limit for isomerization operation temperature is established to minimize the risk of microbial infection and is set at 60°.

pH. Figure 3 shows the relationship between pH and activity. The optimum range is pH 8.0–8.5. Below pH 5 the enzyme is irreversibly denatured. An important pH-related factor is the concentration of cobalt needed to provide optimum activity.

[9] P. B. R. Poulsen and L. E. Zittan, U.S. Patent Appl. SN 558.001 (to be published).

TABLE I
INFLUENCE OF COBALT ON THE STABILITY OF GLUCOSE ISOMERASE[a]

Parameter	pH 7.6		pH 8.5	
	$3.5 \times 10^{-4} M$ Co	No Co	$3.5 \times 10^{-4} M$ Co	No Co
Productivity (kg/kg enzyme)	615	560	590	630
Residual activity (%)	41	28	42	52

[a] Process parameters: syrup, 40% w/w glucose; temperature, 65.0°; Mg, $8 \times 10^{-3} M$; activity, 150 IGIC/g; isomerization time, 450 hr.

Cobalt. Cobalt is a well-known activator of glucose isomerase enzymes. Danno et al.[10] in their basic investigations of glucose isomerases from *Bacillus coagulans* found that the addition of $\geq 10^{-3} M$ cobalt was essential in order to obtain maximum activity from crude enzyme preparations. It is therefore striking that our results with the immobilized enzyme show that addition of Co^{2+} to the feed can be omitted provided the isomerization is performed at a pH above 8. Examples are given in Table I.

It can be seen that at low pH values cobalt addition is essential for the total productivity, whereas no improvement is seen at high pH.

Magnesium. Mg^{2+} is another activator of glucose isomerases; the optimum level is normally reported to be $10^{-2} M$.[10] Again it is interesting that the IGI produced according to our method has a much lower demand for Mg^{2+} (Table II).

TABLE II
INFLUENCE OF THE ADDITION OF MAGNESIUM TO THE SYRUP ON THE STABILITY OF GLUCOSE ISOMERASE[a]

Parameter	No Mg	4×10^{-4} M Mg	8×10^{-4} M Mg	4×10^{-3} M Mg	8×10^{-3} M Mg
Productivity (kg/kg enzyme)	590	630	635	620	630
Residual activity (%)	50	50	50	48	52

[a] Process parameters: syrup, 40% w/w glucose; temperature, 65.0°; pH, 8.5; activity, 150 IGIC/g; isomerization time, 450 hr.

These results show that the addition of Mg^{2+} does not affect the productivity or the residual activity of IGI. But as shown below, this conclusion is valid only if no calcium is present in the syrup.

[10] G. Danno, S. Yoshimura, and M. Natake, *Agric. Biol. Chem.* **31**, 284 (1967).

Calcium. Calcium is an inhibitor of glucose isomerases. Experimental work has shown, however, that Mg^{2+} and Ca^{2+} are competitors and that the inhibitory effect of calcium can be overcome by extra addition of Mg^{2+}, as seen in Fig. 4.

On the basis of experiments of this type, it is concluded that Ca has no effect on productivity provided that the molar ratio of Mg:Ca exceeds 10. As additional Mg increases both isomerization and purification costs, an upper limit of 10^{-2} M Mg corresponding to approximately 35 ppm of Ca^{2+} has been established.

Oxygen. The enzyme is sensitive to oxygen, especially at elevated temperatures. Exposure to oxygen at process temperatures causes inactivation of the enzyme.

Size of the Enzyme Particles. The size of the enzyme particles influences the measured activity of the enzyme but has no influence on enzyme stability. The relationship between particle size and activity is also dependent on particle shape and structure. The activity of the extruded particles, produced by the immobilization technique described above, is influenced by size as shown in Table III.

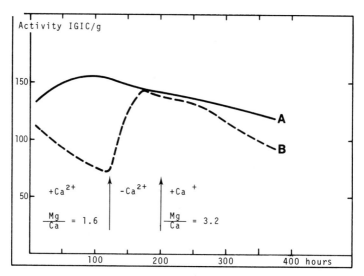

FIG. 4. Influence of magnesium and calcium on the activity of glucose isomerase. Curve A, control column; curve B, Ca-treated column.

Process parameters: syrup, 40% glucose, w/w; temperature, 65°; pH, 8.5; Mg (column A), 4×10^{-3} M.

Column B: 0–120 hr: 4×10^{-3} Mg^{2+}; 2.5×10^{-3} M Ca^{2+}
            120–200 hr: 8×10^{-3} M Mg^{2+}; 0 Ca^{2+}
            200–363 hr: 8×10^{-3} M Mg^{2+}; 2.5×10^{-3} M Ca^{2+}

TABLE III
INFLUENCE OF THE SIZE OF THE ENZYME PARTICLES ON THE ACTIVITY OF
IMMOBILIZED GLUCOSE ISOMERASE (IGI)

Sieve fraction	% Activity
Powdered IGI	100
Fraction 0–150 μm	100
150–297 μm	96
297–500 μm	90
500–1000 μm	75
1000–1410 μm	60
1410–2000 μm	46
2000–2800 μm	36

Purity of the Syrup. In industrial applications of glucose isomerase, the feed syrup will be obtained by degradation of starch. The glucose content of such syrup varies between 90 and 96% based on dry matter. The remaining solids consist mainly of maltose, isomaltose, and other oligosaccharides, but the syrup contains other impurities, such as protein, fat, and metal ions. These impurities have to be removed to obtain maximum stability of the isomerizing enzyme. Therefore the syrup should be purified by filtration, carbon treatment, and ion exchange before it is isomerized.[11]

Concentration of the Syrup. The apparent K_m for IGI—determined by a Hanes plot—is approximately 2.25 M. Thus the glucose concentration should be high in order to favor formation of the enzyme–substrate complex. However, very concentrated solutions have high viscosity so that pore resistance increases and productivity falls. In practice a compromise must be reached, and at 65° the optimal substrate concentration has been found to be 2.6–3.0 M glucose.

Conclusion. Optimal parameter values (contact time, 0.5–3 hr): temperature, 65°; pH, 8.5; magnesium, 4×10^{-4} M Mg, Mg/Ca > 10; oxygen, 0; purity of the syrup—filtered, carbon treated, ion exchanged; concentration of syrup, 2.6–3.0 M.

Process Layout—Continuous Process

Process Design. A continuous operation in which the enzyme is loaded into a column and a glucose syrup is passed through the enzyme bed is characterized by a short contact time between enzyme and syrup. This means that both temperature and pH can be kept in the optimum range without significant formation of by-products.

[11] L. E. Zittan, P. B. R. Poulsen, and S. H. Hemmingsen, *Staerke* **27**(7), 236 (1975).

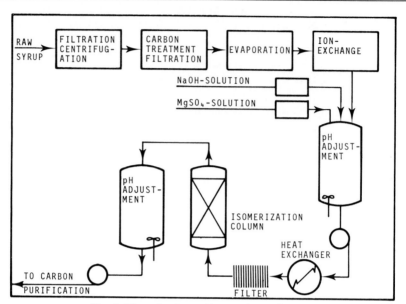

Fig. 5. Flow sheet: continuous isomerization.

Figure 5 gives a flow sheet for the continuous isomerization operation. The glucose syrup produced from starch is purified by filtration and carbon treatment. It is then evaporated to a solids content of 40–45%. After ion exchange, it is taken to a stirred vessel where magnesium is added and the pH is adjusted to 8.5.

With the small pH drop of 0.1–0.5 unit, which is observed over the enzyme bed, the total isomerization process is carried out in the optimum pH range. The main advantage of performing the isomerization at alkaline pH values is that the addition of Co^{2+} can be omitted. This will influence the overall isomerization cost. The cost of additives will be reduced, but it is more important that the cost of postisomerization purification will be lower.

Heating to process temperature (65°) is carried out in a heat exchanger; after a check filtration, the syrup is passed through the enzyme bed in the column. The flow rate of syrup through the column is adjusted to give the desired conversion to fructose. This is controlled by an automatic polarimeter. As the activity of the enzyme slowly decreases with time, the correct conversion is maintained by reducing the flow rate.

From the isomerization column the isosyrup is fed to an agitated vessel, where the pH of the isosyrup is adjusted to 4.5 before refining. In most cases carbon purification to remove the last traces of color will

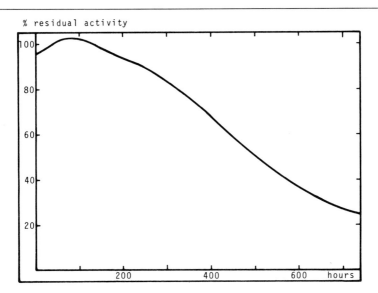

FIG. 6. Stability of immobilized glucose isomerase (IGI)—plug-flow operation. Conditions: 40% glucose, w/w; pH, 8.5; 65°; Mg, 0.4 mM.

suffice. If a specially low ash content of the isosyrup is desired, the syrup should be cation-exchanged, but here the load on the ion-exchanger will be low. It should be mentioned that under the process conditions described above, no psicose or mannose can be detected by thin-layer chromatography.[12]

Figure 6 shows the relationship between time and activity for IGI when used in a continuous process under the above process conditions. After 500 hr of isomerization, the enzyme still has about 50% of its activity left. After 700–750 hr of operation the residual activity will be approximately 25%.

The total productivity obtained with 1 kg of a 150 IGIC/g preparation has been found to be approximately 1000 kg of glucose (DS) converted to a mixture of 45% fructose and 55% glucose within 750 hr (25% residual activity).

Upscaling. Most of the results discussed above were obtained from experiments using 150-ml columns. The validity of these results under large-scale operation has been demonstrated by comparing 150-ml, 1-liter, and 60-liter columns. Such an experiment is described in Table IV. There is no essential deviation between the productivities obtained.

[12] S. A. Hansen, *J. Chromatogr.* **107**, 224 (1975).

TABLE IV
Influence of Column Size on Productivity and Activity of Enzymes[a]

Parameter	Column size		
	150-ml	1-liter	60-liter
Dimensions (height × diameter) cm	32 × 2.5	38 × 5.8	188 × 20
Enzyme (g)	30	200	12000
Productivity (kg/kg enzyme)	920	950	940
Residual activity	34%	36%	40%

[a] Process parameters: syrup, 40% glucose, w/w; temperature, 65.0°; pH, 8.5; Mg^{2+}, 4×10^{-4} M; activity, 150 IGIC/g; isomerization time, 650 hr.

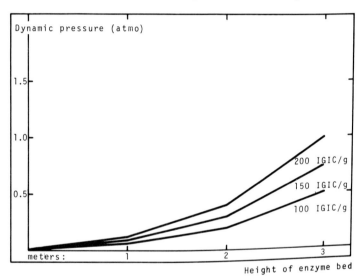

Fig. 7. Pressure drops in relation to height of the enzyme bed.

Pressure Drops. An important factor to be considered when scaling up the continuous fixed-bed isomerization is the pressure drop over the enzyme bed. The linear velocity resulting in the required degree of conversion is dependent on enzyme activity and enzyme bulk density, but for a given enzyme preparation, it is proportional to the height of the enzyme bed. An increase in linear velocity above a certain level causes compression of the enzyme particles and thereby decreases the porosity of the bed. This results in further increase in the pressure drop. In Fig. 7 the initial pressure drops are estimated[13] for linear velocities that will

[13] J. Kozeny, *Ber. Wien. Akad.* **136a**, 271 (1927).

TABLE V
COMPARISON OF BATCH AND CONTINUOUS PROCESSES FOR ISOMERIZATION[a]

Parameter	Batch (GI)	Batch (IGI)	Continuous (IGI)
Reactor volume (m^3)	750	750	30
Enzyme consumption (t)	150–175	17–20	10
MgSo$_4$·7H$_2$O consumption (t)	43	43	2.2
CoSO$_4$·7H$_2$O consumption (t)	2.2	2.2	0
Color formation at OD$_{420\ nm}$	0.20–0.25	0.05–0.10	0–0.02
Psicose formation (%)	<0.1	<0.1	<0.1
Product refining	Filtration, carbon treatment, cation and anion exchange	Carbon treatment, cation and anion exchange	Carbon treatment

[a] Based on monthly production of 10,000 tons DS 42% fructose, DX ≃ 93. Activity 150 IGIC/g, bulk density (IGI) 350 kg/m^3.

give the desired conversion in the stated column for three different enzyme activities.

It can be clearly seen that a more active preparation will give a higher pressure drop in the same column because the total volume activity is increased. To obtain optimal utilization, it is therefore necessary to design the enzyme preparation for the particular columns to be used.

Conclusion

Comparison between Batch and Column Processing. For several years soluble glucose isomerase has been employed in batch isomerization processes. More recently IGI has also been utilized in the same equipment, but the real advantage of IGI is that it is suitable for continuous operation. Table V compares batch isomerization using GI and IGI with the continuous column process using IGI. The advantages obtained by using IGI instead of soluble GI are that (1) IGI can be used in continuous processes; (2) the continuous processes are preferred to batch processes because of lower enzyme dosage, lower costs of additives, lower purification costs, and lower reactor volume.

[57] Lactose Reduction of Milk by Fiber-Entrapped β-Galactosidase. Pilot-Plant Experiments

MAURO PASTORE and FRANCO MORISI

Introduction

During the last few years, interest in enzymatic hydrolysis of milk lactose has increased with the appearance, in the medical literature, of many papers on the occurrence of lactose intolerance in large groups of the population. Lactose intolerance, also called lactose malabsorption or lactase deficiency, is very complex. Excellent reviews have been written on this problem.[1,2] For those suffering from lactose intolerance, milk consumption, in any form, results in gastrointestinal disturbances, the clinical signs of which are abdominal pain, diarrhea, flatulence and bloating among others. There is a high incidence of lactose intolerance among the adult populations of Asia, Africa, Latin America, and the Middle East. Based on these geographic and ethnic differences, genetic and/or adaptation theories have been suggested.[3] Because a majority of the world's population suffers from lactose intolerance, milk consumption has been reduced, although milk is still considered a practically complete food especially useful where protein deficiency is more widespread, for example, in developing countries. Therefore it is of great importance to have good processes for preparing dairy foods in which lactose is partially or completely removed. Many methods for reducing the lactose content of milk have been proposed, but only few of them have industrial application. These methods include chemical, physical and enzymatic procedures.

Milk treated by enzymatic methods retains its original nutritional value because glucose and galactose, the products of lactose hydrolysis, are not removed. Moreover, the enzymatic process is particularly well adapted for immobilized lactase preparations. Because milk is a food, it is of the utmost importance not to transform any of its components, in this case lactose, into harmful substances, nor to add anything to the processed milk, nor to change the properties of milk. Furthermore, because of widespread lactose intolerance a very inexpensive industrial process is needed to increase milk consumption. Otherwise, the low lactose dairy products will be available for only a limited few.

[1] N. Kretchmer, *Sci. Amer.*, **227**(4), 71 (1972).
[2] A. Dahlqvist and B. Lindquist, *Acta Paediat. Scand.*, **60**, 488 (1971).
[3] R. D. McCracken, *Current Anthropology*, **12**, 479 (1971).

At the present time, the only inexpensive process involves using immobilized enzymes. Various immobilized lactase preparations are described in the literature.[4-12]

At the "Centrale del Latte" in Milan, Italy, most of our work concerns the use of fiber-entrapped β-galactosidase. First, the properties of enzymes from two different sources, *E. coli* and yeast, were studied and compared;[13] then the behavior of some fiber-entrapped preparations and the problems relating to industrial application were investigated.[14] Finally pilot-plant experiments were carried out to establish optimal operating conditions.[15]

The purpose of this chapter is to give the reader a more detailed insight into the problems involved in the utilization of fiber-entrapped lactase for the industrial production of milk having a low lactose content.

Materials

Buffer B. For dissolving the enzyme and washing enzyme fibers prepare a buffer containing 0.01 M sodium phosphate buffer, pH 7.2, 1 mM ethylenediamine tetraacetic acid (EDTA), and 2 mM MgSO$_4$.

Milk. Use sterilized skimmed or partially skimmed (lipid content, 1.1%) milk.

Enzyme. β-galactosidase from yeast is employed. In the experiments carried out at the "Centrale del Latte" a purified commercial preparation Maxilact (from Gist-Brocades, Delft, Holland) was used. One unit of enzyme activity is defined as the amount of enzyme required to hydrolyze 1 μmole of *o*-nitrophenyl-β-D-galactopyranoside per min (0.01 M sodium phosphate buffer, pH 7.2; 25°C).

---

[4] J. H. Woychik and M. V. Wondolowski, *Biochim. Biophys. Acta*, **289**, 347 (1972).
[5] J. H. Woychik and M. V. Wondolowski, *J. Milk Technol.*, **36**, 31 (1973).
[6] A. C. Olson and W. L. Stanley, *J. Arg. Food Chem.*, **21**, 440 (1973).
[7] A. Dahlqvist, B. Mattiasson, and K. Mosbach, *Biotechnol. Bioeng.*, **15**, 395 (1973).
[8] W. L. Wendorff, C. H. Amundson, N. F. Olson and J. C. Garver, *J. Milk Food Technol.*, **34**, 294 (1971).
[9] A. K. Sharp, G. Kay, and M. D. Lilly, *Biotechnol. Bioeng.*, **11**, 363 (1969).
[10] H. H. Weetall, N. B. Havewala, W. H. Pitcher, Jr., C. C. Detar, W. P. Vann, and S. Yaverbaum, *Biotech. Bioeng.* **16**, 295 (1974).
[11] I. Hinberg, R. Korus, and K. F. O'Driscoll, *Biotech. Bioeng.*, **16**, 943 (1974).
[12] L. E. Wierzbicki, V. H. Edwards, and F. V. Kosikowski, *Biotech. Bioeng.*, **16**, 397 (1974).
[13] F. Morisi, M. Pastore, and A. Viglia, *J. Dairy Sci.*, **56**, 1123 (1973).
[14] M. Pastore, F. Morisi, and A. Viglia, *J. Dairy Sci.*, **57**, 269 (1974).
[15] M. Pastore, F. Morisi, and D. Zaccardelli, *in* "Insolubilized Enzymes" (M. Salmona, C. Saronio, and S. Garattini, eds.) p. 211. Raven Press, New York, 1974.

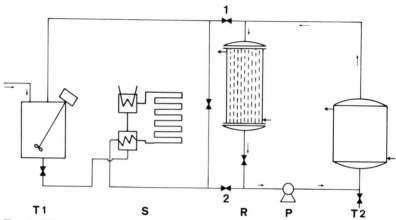

FIG. 1. Diagram of the pilot plant used for the hydrolysis of milk lactose by β-galactosidase fibers: $T1$, stirred vessel for preparing washing solutions or for receiving milk; S, sterilizer; R, enzyme reactor; P, recycling pump; $T2$, reservoir; 1, 2, valves.

Fiber-entrapped enzyme. To prepare the fiber-entrapped β-galactosidase, suspend the enzyme powder in buffer B and allow it to stand for 24 hr at 4°C with gentle stirring. Insoluble material is removed by centrifugation and a 30% by weight solution of clarified extract in glycerol is prepared. The water-glycerol solution of β-galactosidase is entrapped in cellulose triacetate according to the procedure described by Dinelli.[16] (See this volume [18].) The final enzyme density in the fibers is about 1.500 Units/g (dry weight basis).

Equipment

A flow diagram of the pilot plant used for the reduction of milk lactose by fiber-entrapped β-galactosidase is shown in Fig. 1. (1) A stainless steel tank $T1$, large enough to hold 300 liters of milk or buffer, is equipped with a laboratory mixer for preparing the buffer solution needed to wash the catalytic bed. (2) A continuous sterilizer S (CHERRY BURREL, Pilot, Cherry Burrel Co., Cedar Rapids, Iowa, 52404, USA) for sterilizing milk and buffer operates at 142°C, at a rate of 120 liters per hr. The residence time of the fluid at the maximum temperature is about 3 sec. (3) An enzyme reactor R consists of a jacketed glass column (internal diameter, 15 cm; height, 56 cm) packed with enzyme fibers. (4) A centrifugal pump P recirculates milk or the washing

---

[16] D. Dinelli, *Process Biochem.*, **7**, 9 (1972).

solution through the reactor. (5) An aseptic jacketed tank *T2* of about 500-liter capacity serves as a reservoir for the total amount of milk recirculating through the enzyme reactor per batch.

Enzyme reactor setup. β-Galactosidase fibers, 1.5 kg (corresponding to about 0.5 kg of dry material), in the form of skeins having the same length as the glass column, are placed in the column and are fixed at both ends. A very simple method of fixation is schematically shown in Fig. 2.

The fibers skeins are held taut in the column by two stainless steel bars fixed at either end by collars. In this way the fibers are kept parallel to the long axis of the column. The packing density is not very high in order to avoid plugging and to permit high flow rates.

Operations

Fibers washings. Before using the enzyme fibers wash them with sterile buffer *B* to remove traces of organic solvents, enzyme, and other contaminant proteins loosely adsorbed onto the external surface of the fibers and other impurities present in the entrapped enzyme solution. Buffer *B* is prepared in the tank *T1* by dissolving 356 g $Na_2HPO_4 \cdot 2H_2O$,

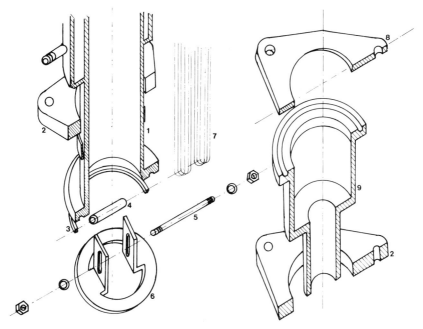

FIG. 2. Diagram showing how to pack a column with enzyme fibers: *1*, jacketed glass column; *2*, flange; *3*, PTFE O-Ring; *4*, metallic bar; *5*, coupling pin; *6*, fibers support; *7*, fibers skeins; *8*, PTFE gasket; *9*, column head.

74.5 g EDTA sodium salt, and 98.6 g. $MgSO_4 \cdot 7H_2O$ in 200 liter deionized water; the pH is adjusted to 7.2 by adding phosphoric acid. The solution is then passed through the sterilizer S, collected in the aseptic tank $T2$, and cooled to 25°C. Valves *1* and *2* are closed and pump P is turned on. The buffer solution is recirculated through the fibers and back to tank $T2$; the recirculation is continued for three hrs, at 25°C.

Then the solution is discarded and the washing repeated twice. After washing the reactor is allowed to stand so that any liquid is completely drained from the fibers.

Lactose hydrolysis. Lactose hydrolysis is carried out using a batchwise procedure. Skimmed milk is fed to tank $T1$, passed through the sterilizer S, collected in tank $T2$, and rapidly cooled to about 5°C. The temperature in the jackets of the reactor R and tank $T2$ is maintained at about 7°C. As soon as 200 liters of milk are collected in tank $T2$, valves *1* and *2* are closed and pump P is turned on to circulate milk downward through the enzyme fibers and back to tank $T2$. The recirculation is continued at a rate of about 7 liters per min for 20 hrs. Then the treated milk is removed and conveyed to the packing equipment. Samples are drawn to measure the degree of hydrolysis and the bacteriological count. At the end of the batch hydrolysis, sterile buffer B is recycled through the reactor to wash the fibers, tank $T2$, and piping, or another run is started.

Decontaminating treatments. To keep bacterial contamination under control, it is necessary to decontaminate enzyme fibers and apparatus periodically. Washing the enzyme fibers with a 0.2% solution of Steramine H (Formenti, Milan, Italy) in buffer B has proved satisfactory because of the bactericidal and cleansing action of the quaternary ammonium compound. Sometimes the use of a combined solution of benzylpenicillin and streptomycin may be necessary. These antibiotics must be carefully removed by washing to prevent their presence in the treated milk. The pilot-plant apparatus must be assembled in such a way that all the piping and vessels (except the enzyme reactor) can be sterilized by steam.

Analytical Determinations

Lactose. Employing the method of Somogyi[17] the concentration of lactose originally present in the milk can be calculated from a standard curve. The polarimetric AOAC method[18] can also be used.

---

[17] M. Somogyi, *J. Biol. Chem.*, **160**, 61 (1945).
[18] AOAC. "Official Methods of Analysis" 11th ed. Assoc. Off. Anal. Chem., Washington, D.C., 1970.

Glucose. The glucose produced by hydrolysis is estimated enzymically using the glucose oxidase-peroxidase system.

Bacterial count. Standard methods[19] set up by the United States Public Health Association are used for the bacteriological analysis of the treated milk.

Performance of the Enzyme Reactor

Source of enzyme. The sources of β-galactosidase include yeasts, molds and bacteria. In our experiments yeast enzymes were used for several reasons. First, the addition of active cells of *Saccharomyces* sp. or partially purified preparations of yeast β-galactosidase to milk is already used in some existing processes for the hydrolysis of lactose. Yeast enzyme is considered GRAS (generally recognized as safe) and no toxicological problems arise causing contamination of milk, as will be shown later. Second, yeast β-galactosidase shows approximately the same activity on pure and milk lactose, whereas the activity of enzymes from other sources, e.g., from *E. coli*, is dramatically reduced in milk probably due to the presence of some inhibitors. Finally, yeast enzyme shows an optimum pH value for activity quite near to the pH of milk. However, yeast enzyme does have some disadvantages, the principal one being the absolute requirement of magnesium ions for activity. When the enzyme fibers come in contact with aqueous solutions, a continuous dialytic removal of low molecular weight compounds from the microcavities of the fibers takes place. If Mg^{++} ions are not present in the external solution β-galactosidase looses the loosely bound magnesium, resulting in an irreversible deactivation of the enzyme. The poor thermal stability of yeast β-galactosidase is obviously another disadvantage, but these effects are negligible as this process is operated at low temperatures.

Enzyme leakage. The fact that the catalyst system (enzyme + support) may release some undesirable substances in the treated milk must be considered. The material, which the fibers are made of, i.e., cellulose triacetate, is quite stable and does not undergo chemical modification under operating conditions. Other supports used for enzyme immobilization have been known to cause toxicological problems. The leakage of enzyme and other substances from the fibers is not a problem from a toxicological point of view, because the enzyme preparation is recognized as GRAS. By analyzing the protein content of β-galactosidase fibers, it has been shown that no detectable leakage of protein materials occurs.[15] This prevents the milk from acquiring the unpleasant taste caused by

[19] Standard Methods of Examination of Dairy Products. Amer. Public Health Ass. Inc. 1960.

free enzymes. Of course, it must be stressed that the enzyme fibers should be washed carefully before using.

Activity of fiber-entrapped β-galactosidase. The activity of the immobilized enzyme is of great importance in this low-temperature process. The technical and economical feasibility of the process depends largely on the activity of the catalyst which influences the ratio of milk to enzyme fibers and therefore, the duration of batch hydrolysis. The technique of enzyme entrapment in fibers enables the preparation of immobilized enzymes having a wide range of concentration. The optimal loading of enzyme in fibers is chosen on the basis of technical and economical considerations. Because of the restricted diffusion of the substrate inside the porous enzyme support, the β-galactosidase fibers display only a fraction of their enzyme activity and this fraction decreases by increasing the enzyme density in the support. For the best utilization of the enzyme, it is desirable to compromise between the activity displayed by the fibers and their enzyme content, i.e., the recovered activity. A typical time course for the hydrolysis of milk lactose, under the conditions employed in the pilot plant, is shown in Fig. 3. Within 20 hours, about 75% of the original lactose is converted to glucose and galactose; such a conversion is considered sufficient for a dairy product to be used by those suffering from lactose intolerance.

The activity of the enzyme reactor in the pilot plant is approximately the same as that obtained by using small laboratory columns. The flow rate and the packing density play an important role in determining the

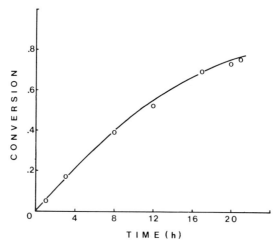

Fig. 3. Typical time course for the hydrolysis of milk lactose obtained in the pilot-plant experiments.

best efficiency of the reactor. At a recirculation flow rate of 7 liters per min and a packing density of 0.15 g catalyst per ml, channeling, film diffusion limitations, and pressure drops are minimized.

Stability of enzyme activity. The stability of fiber-entrapped yeast β-galactosidase depends on operating temperature, the presence of magnesium ions in the external solution, and bacterial contamination. The enzyme is stable in the range of 0°–25°C; it follows that the process temperature, 4°–7°C, does not cause a loss of activity.

Magnesium ions are necessary for activity; the traces of magnesium ions contained in cow milk have proved to be sufficient for preventing enzyme deactivation. Care must be taken to add a small amount of magnesium ions to the washing solutions. β-Galactosidase differs from many other entrapped enzymes in that it looses activity when bacterial contamination occurs, most probably owing to some low molecular weight metabolite which poisons the enzyme.

By operating under conditions designed to avoid enzyme deactivation, fiber-entrapped β-galactosidase can be used for a large number of batches, which is economically beneficial. The table shows some results obtained with the pilot-plant reactor. Although the degree of hydrolysis is not the best parameter to demonstrate the stability of an immobilized enzyme, these data clearly indicate that the rate of enzyme deactivation is undoubtedly low. For a 0.5 kg sample of enzyme fibers which processed about 10,000 liters of milk in 50 batches, the loss in activity was less than 10%.

Microbial contamination. The growth of microorganisms is the most serious problem encountered in the use of immobilized enzymes, because the solution to be treated generally constitutes a good growth medium. This is particularly true for milk, wherein microbial contamination adversely affects both the product and the enzyme.

Operating at low temperatures does not completely solve the problem of microbial contamination, because after a few days psycrophilic microorganisms adapt to growth under these conditions. The enzyme fibers have to be washed carefully and periodically.

During the pilot-plant experiments, at the "Centrale del Latte" of Milan, suitable washing solutions were developed, which have efficient antimicrobial and detergent action. In particular, the use of ammonium quaternary salts, such as Steramine H, proved efficient and they do not negatively affect the activity of the entrapped enzyme.

The table shows the bacterial count per ml of treated milk after incubation of the sample at 32°C or 7°C. Except in a few cases, where different operating temperatures and different decontaminating treatments were tested, the number of colonies was always very low.

CONVERSION AND BACTERIAL COUNT DATA OBTAINED IN
PILOT-PLANT EXPERIMENTS

Batch number	Conversion	Plate count per ml of processed milk	
		32°C	7°C
1	0.77	200	70
3	0.73	40	0
6	0.77	125	0
10	0.79	200	70
12	0.81	260	20
15	0.73	50	40
18	0.79	145	730
20	0.75	15	10
21	0.75	850	360
24	0.71	100	85
27	0.75	300	240
29	0.71	72,000	24,000
31	0.64	20	10
34	0.71	60	40
37	0.77	20	35
40	0.75	2,400	2,600
42	0.73	1,500	2,000
44	0.73	55,000	45,000
48	0.71	10	10
50	0.71	60	500

Quality of the processed milk. The proper choice of operating conditions and the control of bacterial contamination contribute to the quality of the treated milk. During pilot-plant experiments attention has been focused on the organoleptic properties and characteristics of the product. The processed milk retains its original organoleptic properties exhibiting only a slightly sweeter taste.

The presence of glucose and galactose instead of lactose slightly affects the shelf-life characteristics of the product. An industrial plant having a minimum capacity of 8000 liters/day is now in operation.

[58] On the Potential of Soluble and Immobilized Enzymes in Synthetic Organic Chemistry

By J. BRYAN JONES

As will soon be apparent, this contribution deviates from the normal format and approach of the other sections of this volume in that it does not address itself to an area where immobilized enzymes have *yet* become part of the generally accepted methodology. However, the use of enzymes for organic synthesis is increasing rapidly in importance, and it is an area that could be stimulated to a considerable degree by the availability of immobilized enzymes possessing the appropriate properties. Accordingly, this chapter has been written with a view to drawing attention to the types of organic chemical problems in which enzymes can be used with advantage, and to the often demanding (in a biochemical sense) reaction conditions which the organic chemist may be required to use, in the hope that this will help to stimulate the development of immobilized enzymes possessing the desired characteristics. It is felt that the soluble enzymes used in the examples given here can be replaced with immobilized enzyme preparations without much modification.

From the point of view of the synthetic organic chemist, the main attractions of enzymes as catalysts lie in their abilities to discriminate between enantiomers and to distinguish enantiotopic groups and faces of molecules possessing prochiral centers.[1] Reagents and catalysts capable of making distinctions of this kind are receiving considerable attention from the rapidly increasing number of chemists involved in asymmetric synthesis.[2,3]

Strategy

Two basic types of applications of enzymes in synthesis can be recognized.

A. Situations where catalysis of a specific reaction leading to one (or more) narrowly defined product is required. Most industrial applications of immobilized enzymes are in this category. The examples available

---
[1] J. B. Jones, *in* "Applications of Biochemical Systems in Organic Chemistry" (J. B. Jones, C. J. Sih, and D. Perlman, eds.), Chapters 1 and 6. Wiley, New York, 1976. In Press.
[2] J. D. Morrison and H. S. Mosher, "Asymmetric Organic Reactions." Prentice-Hall, Englewood Cliffs, New Jersey, 1971.
[3] J. W. Scott and D. Valentine, *Science* **184**, 943 (1974).

are well documented,[4] and most of them have been described in detail elsewhere in this volume: e.g., amylase [54], acylase [51], lactase [55,57], penicillin amidase [52], fumarase, aspartase [50], asparaginase, tyrosinase, tryptophanase, glucose isomerase [53,56]. In such cases, whether or not an immobilized enzyme process is viable will be determined solely by the overall economic factors involved. The successes already achieved in this area will ensure that this type of application of immobilized enzyme technology will be considered routinely whenever feasible. Thus no further external stimulation of this aspect would appear to be necessary.

B. Applications in research laboratories. Synthetic chemists, especially those engaged in making biologically active moleceules (including natural products), and in biosynthetic and stereochemical studies, often need to effect certain reactions stereospecifically. The topics and compounds on which they work are quite diverse and there is no secure way of predicting their requirements from one month to another. This poses particular problems with respect to the use of enzymes as practical catalysts and it is to this aspect that the rest of this article will be devoted.

For enzymes to become generally accepted as routine catalysts for stereospecific organic transformations, the following criteria need to be met for both soluble and immobilized enzymes.

1. The enzymes must catalyze a reaction of general preparative interest to chemists. Enzymes possessing broad constitutional specificity and narrow stereospecificity are the most desirable.

2. The enzyme should be commercially available. The vast majority of chemists will want to treat both soluble and immobilized enzymes as "shelf reagents." They would almost certainly reconsider their decision to evaluate an enzymic procedure if they discovered that it required them to make their own enzyme preparations.

3. The enzyme preparation should be relatively stable since reaction periods of up to several days may be required if the substrates used are poor ones. Slow reactions must be anticipated since the structures of the compounds encountered by the chemist are likely to be quite unrelated to the natural or biochemically important substrates of the enzyme.

4. Sufficient specificity data should be available to enable reliable predictions to be made for any previously unevaluated potential substrates. Organic chemists have become accustomed to being able to forecast the course and rate of reaction of a given structure with a chemical

[4] J. B. Jones and J. F. Beck, *in* "Applications of Biochemical Systems in Organic Chemistry" (J. B. Jones, C. J. Sih, and D. Perlman, eds.), Chapter 4. Wiley, New York, 1976. In press.

reagent, and also to be provided with information on the limitations of the method. Realistically or not, they will expect to be able to do the same with enzymic catalysts.

5. If the costs of an enzyme or coenzyme are high, then the benefits of being able to operate under mild reaction conditions with high stereo- and/or regiospecificity should be such that the use of an expensive system is justifiable.

6. A convenient experimental procedure must be available. At the present time only a few enzymes meet these criteria. Nevertheless, a large number of examples where enzymes have been applied to problems of organic chemical interest have now been documented,[4] and the organic chemist is becoming increasingly aware of the potential of the approach. The types of problems being tackled, and the experimental procedures involved, will be illustrated by considering representative applications of the use of α-chymotrypsin (as an example of an enzyme catalytically active in its own right) and of horse liver alcohol dehydrogenase (as one requiring an expensive cofactor).

Although, of necessity, the experimental details available are those for the soluble enzymes, the procedures described provide a basis for discussing the benefits, or otherwise, of employing an immobilized enzyme for the same reactions. While such comparisons of the advantages and disadvantages of immobilized catalysts will focus to a considerable extent on the above two enzymes, the general conclusions drawn will be relevant to organic chemical applications of many other immobilized enzymes.

α-Chymotrypsin

Resolution (see also this volume [51])

α-Chymotrypsin has been used to resolve (either completely or partially) over a hundred acids *via* its catalysis of the hydrolysis of their ester derivatives.[4] (For immobilization of this enzyme see, e.g., this volume [3].) The ester substrates vary considerably in their structures. In addition to the biochemically familiar amino acid derivatives, aliphatic and aromatic acyclic, mono-, bi-, tri-, and tetracyclic esters have also proved to be acceptable substrates. A further feature of the enzyme which is attractive to chemists is that the stereospecificity of hydrolysis of chiral substrates, and the enantiotopic specificity with respect to diesters containing a prochiral center, is rationalizable in terms of a simply applied active-site model.[4]

The resolution of methyl (±)-1,2-dihydronaphtho[2,1-b]furan-2-carboxylate (I) is representative.[5]

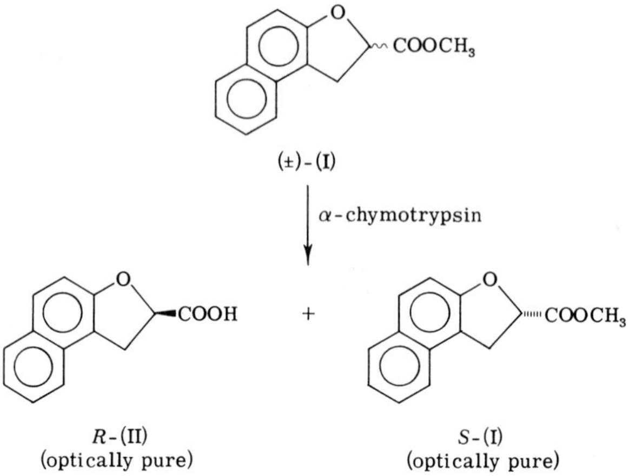

(±)-(I)

↓ α-chymotrypsin

R-(II)
(optically pure)

+

S-(I)
(optically pure)

Reagents

(±)-Ester (I)
α-Chymotrypsin (3 × crystallized)
Aqueous dimethyl sulfoxide, 50%
Aqueous sodium chloride, 1.2 M
Aqueous sodium hydroxide 0.1 N
Work-up chemicals: diethyl ether, anhydrous magnesium sulfate, methanol, concentrated hydrochloric acid, ethanol

Procedure (from Hayashi and Lawson[5]). A 5-liter beaker into which a pH electrode has been inserted is suitable for carrying out this reaction. All operations are performed at room temperature (20°). A solution of 3 g (13.2 mmoles) of the racemic ester (I) in 2.4 liters of 50% aqueous dimethyl sulfoxide is placed in the beaker, together with 60 ml of 1.2 M aqueous sodium chloride, and the mixture is stirred efficiently (without cavitation) using a magnetic stirrer. A freshly prepared solution of 180 mg of α-chymotrypsin in 60 ml of water is then added, and, if necessary, the apparent pH of the reaction mixture is adjusted to 7 by the addition of 0.1 N aqueous sodium hydroxide contained in a burette. The apparent pH is maintained at 7 as the hydrolysis proceeds by periodic additions of the 0.1 N base. After several hours, when 70–75 ml of the base (equivalent to ∼50–55% hydrolysis of the ester substrate) has been added, the reaction practically stops. At this stage, the reaction mixture is diluted with 1 liter of water and extracted four times with 250-ml portions of

[5] Y. Hayashi and W. B. Lawson, *J. Biol. Chem.* **244**, 4158 (1969).

diethyl ether. Both the ether and aqueous solutions are retained. The combined ether extracts are dried with anhydrous magnesium sulfate, the solution is filtered, and the ether is evaporated *in vacuo* with a rotary evaporator. The residual product is the unhydrolyzed ester S-(I) (1.24 g, 41% yield) which is recrystallized from methanol to give optically pure material, m.p. 82°–83°, $[\alpha]_D^{28} + 116.2°$ (c 1.1, ethanol).

In order to obtain the other desired product, viz. the acid R-(II), the aqueous reaction solution retained after the ether extraction stage is acidified to pH 2 with concentrated hydrochloric acid and reextracted with 4×250-ml portions of ether. The combined ether extracts are dried over magnesium sulfate and filtered, and the solvent is rotary-evaporated as before. The acid R-(II) (1.44 g, 51% yield) obtained has m.p. 155°–162°; recrystallization from 70% aqueous ethanol gives the optically pure material, m.p. 165°–168°, $[\alpha]_D^{28} - 75.5°$ (c 1, ethanol).

Selective Ester Hydrolysis in Protecting Group Chemistry

Organic chemists very often require to operate on only one of several chemically similar functional groups in a molecule. Use of an enzyme with appropriate specificity provides an attractive and flexible approach to solving a variety of such problems. For example, the aromatic group specificity of α-chymotrypsin permits the enzyme to be used for this purpose in compounds containing derivatized hydroxyl or amino functions. One illustration of such an application is provided by hydrolyses of tropane diesters such as (III).

In this compound the two ester groups are chemically very similar. However, for steric reasons, hydrolysis of (III) with hydroxide ion occurs selectively at the 6-position to give the 6β-hydroxy compound (IV). On the other hand, as a consequence of the aromatic specificity of α-chymotrypsin, enzymic hydrolysis of (III) exposes only the 3α-hydroxy group, and (V) is produced exclusively. Thus either the 6β- or the 3α-hydroxyl groups can be exposed at will by choosing the appropriate method of hydrolysis.

Reagents

(±)-Diester (III) (as its hydrochloride salt)
Aqueous potassium chloride, 0.1 M
Aqueous sodium hydroxide, 1.0 N
α-Chymotrypsin (3 × crystallized)
Work-up chemicals: potassium chloride, potassium carbonate, anhydrous magnesium sulfate, diethyl ether

Procedure (from Lin and Jones[6]). A radiometer pH stat is used in conjunction with a magnetically stirred 100-ml reaction vessel fitted with a rubber stopper through which a combination electrode, the needle of the pH stat syringe (filled with 0.1 N aqueous sodium hydroxide), and a nitrogen line are introduced. All operations are carried out at room temperature (20°). Hydrochloride salt of the (±)-diester (III) (1.55 g, 4.22 mmoles) is placed in the reaction vessel together with 60 ml of 0.1 M aqueous potassium chloride. The stirred solution is adjusted to pH 7.8 by the addition (via a separate syringe) of ~2.3 ml (~2.3 mmoles) of 1.0 N aqueous sodium hydroxide. The pH stat is now set to maintain pH 7.8. The reaction is initiated by adding 60 mg of α-chymotrypsin, and the hydrolysis proceeds at pH 7.8. After ~4 hr, ~2 ml (~0.2 mmoles) of 0.1 N base have been taken up and the reaction stops. The reaction mixture is saturated with potassium carbonate and is then continuously extracted with ether for 24 hr. The ether extract is dried over magnesium sulfate, filtered, and rotary evaporated. The residual oil (840 mg, quantitative yield) is distilled to give (±)-6β-acetoxy-3α-hydroxytropane (V), b.p. 122°.

Horse Liver Alcohol Dehydrogenase

As a result of the mixture of EE, ES, and SS isozymes present in commercial horse liver alcohol dehydrogenase, the purchased enzyme has a very broad structural specificity, and it catalyzes the oxidoreductions of many substrates of organic chemical interest (for immobilization of this enzyme, see the various procedures in this volume and in Gestrelius

[6] Y. Y. Lin and J. B. Jones, *J. Org. Chem.* **38**, 3575 (1973).

et al.[7]). These compounds include a variety of aldehydes, mono- and bicyclic, and steroidal ketones and their corresponding primary and secondary alcohols. Furthermore, the oxidoreductions are stereospecific to a large degree, with both enantiomeric and enantiotopic distinctions being observed. As with α-chymotrypsin, a very positive feature from the chemical viewpoint is that a reliable active site model has been developed that permits the substrate activity and stereospecificity to be predicted for the majority of the known substrates of the enzyme.[4]

The examples of carbonyl reduction and alcohol oxidation described below are illustrative of the demands that would be made on the enzyme in the organic laboratory. It will be noted that in each case a coenzyme recycling system is used. This is an important aspect of the method. It has both chemical and economic significance and represents a problem that has not been satisfactorily solved for routine applications of horse liver (or any other) alcohol dehydrogenase. It appears possible that applications of soluble immobilized coenzymes will help to overcome some of the remaining difficulties. However, in view of the fact that the overall coenzyme recycling field has recently been reviewed,[4] and also receives some attention in this volume (see chapter [61]), no further coverage will be accorded it in this section.

Stereospecific Reduction of a Ketone

The example chosen is the stereospecific reduction of (±)-*trans*-2-decalone (VI) to give 2S,9R,10R-*trans*-2-decalol (VII), with the unreactive 9S,10S enantiomer of (VI) recovered unchanged. The reduction is also enantiotopically specific for the *re*-face of the carbonyl group of 9R,10R-(VI).

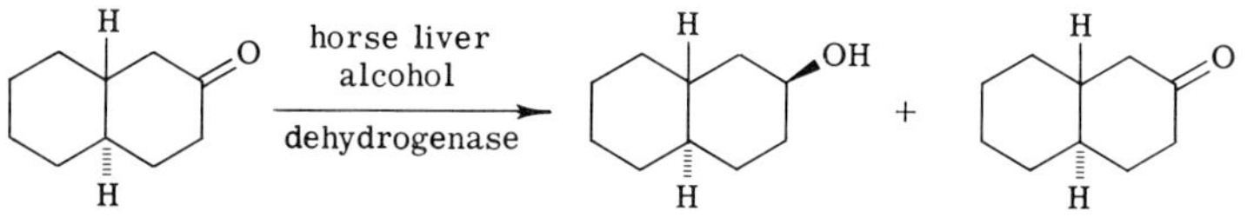

(±)-(VI) → 2S,9R,10R-(VII) (optically pure) + 9S,10S-(VI) (optically pure)

Reagents

(±)-*trans*-2-Decalone (VI)
Cyclohex-2-en-1-ol
NAD⁺

[7] S. Gestrelius, M.-O. Månsson, and K. Mosbach, *Eur. J. Biochem.* **57**, 529 (1975).

Horse liver alcohol dehydrogenase (crystalline suspension containing 10 mg/ml)

Aqueous potassium phosphate buffer, 0.05 M pH 7

Work-up chemicals: diethyl ether, n-pentane, methylene chloride, ethyl acetate, anhydrous sodium sulfate, UV fluorescent silica gel for thin-layer chromatography

Procedure (from Mislin[8]). ($\pm$)-*trans*-2-Decalone (VI) 1.92 g, (12.6 mmoles) and 2.6 g (26.5 mmoles) of cyclohex-2-en-1-ol (for recycling of the cofactor as coupled substrate) are placed in a 500-ml round-bottom flask. The reaction flask is then almost filled with 0.05 M aqueous potassium phosphate buffer of pH 7. Then 100 mg (0.15 mmoles) of NAD$^+$ and 5 ml (50 mg) of the suspension of horse liver alcohol dehydrogenase are added, and the now completely full flask is stoppered (with a polyethylene plug). The resulting emulsion is magnetically stirred at room temperature for $\sim$40 hr, after which time 50% reaction has taken place. The reaction mixture is then extracted with 1 liter of ether in a Vibromixer. The ether extract is dried over sodium sulfate and the filtered ether solution is evaporated using a fractionating column in order to reduce the loss of the relatively volatile 2-decalone. The yield of crude product is 4.26 g. Purification of a methylene chloride solution of 2.09 g of this material by preparative thin-layer chromatography on 20 $\times$ 100-cm plates of silica gel PF$_{254+366}$, using two plate developments with n-pentane–methylene chloride–ethyl acetate (60:35:5), gives three fractions. Molecular distillation of fraction 1 at 120°/10 Torr gives 589 mg (45% yield) of optically pure 9S,10S-2-decalone (VI), $[\alpha]_D + 50°$ (c 3.5, benzene). Fraction 2 (860 mg) and 3 (1.202 g) contain 2S,9R,10R-*trans*-2-decalol (VII) mixed with cyclohex-2-en-1-one and cyclohex-2-en-1-ol, respectively. Recrystallization of fraction 2 from n-pentane at $-20°$ followed by sublimation in high vacuum gives 295 mg (23% yield) of optically pure 2S,9R,10R-(VII) mp 76°, $[\alpha]_D -8.0°$ (c 1.2, ethanol). [The compound (VII) present in fraction 3 was not isolated in this experiment.]

Oxidation of an Alcohol with Enantiotopic Specificity

Specific operation on one of two enantiotopic groups is of great importance in asymmetric synthesis. It would be impossible at present to duplicate by any nonenzymic method the example given below, where the diol (VIII) is converted to the lactone (IX), via virtually complete specificity for the *pro*-S hydroxyethyl group.

[8] R. Mislin, Enzymatische Reduktionen von Dekalonen-(1) und -(2). Ph.D. Dissertation 4169, Eidgenössische Technische Hochschule, Zurich, 1968.

[58] IMMOBILIZED ENZYMES IN ORGANIC CHEMISTRY 839

(96% optically pure)
(IX)

Reagents

3-Methylpentan-1,5-diol (VIII)
Flavin mononucleotide (commercial grade)
NAD⁺
Horse liver alcohol dehydrogenase (1 × crystallized)
Aqueous glycine-sodium hydroxide buffer, 0.05 M, pH 9
Aqueous sodium hydroxide, 5 N
Aqueous hydrochloric acid, 6 N
Work-up chemicals: chloroform, decolorizing charcoal, anhydrous magnesium sulfate

Procedure (from Jones and Beck[4]). A 250-ml round-bottom flask with an easily stoppered standard-taper neck is a suitable reaction vessel. In it is placed a solution of 1.5 g (12.7 mmoles) of 3-methylpentane-1,5-diol (VIII), 840 mg (1.27 mmoles) of NAD⁺, 11.2 g (24 mmoles) of flavin mononucleotide (for reoxidation of NADH formed), and 40 mg of horse liver alcohol dehydrogenase in 200 ml of 0.05 M aqueous glycine-sodium hydroxide buffer, pH 9. The resulting mixture is magnetically stirred at room temperature (20°), and the pH is readjusted to 9 by the dropwise addition of 5 N aqueous sodium hydroxide. The flask is then stoppered, and the reaction solution is stirred until no starting diol remains (~24 hr). The pH is then raised to 12 with the 5 N base, and the mixture is continuously extracted with chloroform for 2 days to remove unreacted diol (VIII). (This chloroform extract is subsequently discarded.) The aqueous solution is now acidified to pH 3 with 6 N hydrochloric acid and again continuously extracted with chloroform for 24 hr. This chloroform extract is decolorized with charcoal, filtered, dried over anhydrous magnesium sulfate, filtered, and then rotary evaporated *in vacuo* at 20°. The residual oil is distilled to give 1.02 g (68% yield) of 96% optically pure 3S-5-hydroxy-3-methyl pentanoic acid lactone (IX), b.p. 93°–94°/0.02 Torr, $[\alpha]_D^{25}$ −24.8° (c 5.6, chloroform).

Discussion

It is clear that replacement of the soluble enzymes by their immobilized counterparts could be beneficial in the above procedures. The parameters 1–6 outlined earlier in the section on strategy provide a convenient framework for analyzing some of the factors that would affect the outcome of experiments in which immobilized enzyme preparations were used.

1. *Preparation Utility Aspects.* Although, as this volume demonstrates, a large number of enzymes have now been immobilized, relatively few catalyze reactions of broad preparative value to chemists. While immobilized proteolytic enzymes, such as α-chymotrypsin, are well documented, satisfactory preparations of immobilized horse liver alcohol dehydrogenase, and other oxidoreductases of general synthetic value, have not yet been developed. In organic synthesis, chemists require to make and break a variety of bonds of which those of the C-X type, where X = C, H, N, O, P, S, halogen, etc., are the most important. Enzymes catalyzing such transformations are to be found within the Enzyme Commission classes 1–6. Unfortunately, the narrow constitutional specificities of most of these enzymes excludes the majority of them from serious practical consideration at this time. Among the immobilized enzymes that would have immediate appeal are oxidoreductases of the general EC group 1, which are of operating of organically interesting structures. The successes achieved with an immobilized steroid 11β-monoxygenase and Δ¹-dehydrogenase[9] and cholesterol oxidase[10] are encouraging in this regard (see also chapter [13]).

2. *Commercial Availability.* Although the number of commercially available immobilized enzymes is increasing, the selection is still very limited and the most useful enzymes such as horse liver alcohol dehydrogenase have not yet been listed in any catalog. (However, the various procedures given in this volume should be applicable to most enzymes.)

3. *Stability of Enzymes and Related Factors.* The long-term stability of the immobilized preparation is of major importance in view of the long reaction periods an organic transformation may require. With proteolytic enzymes such as α-chymotrypsin, the increased stabilization which

---

[9] K. Mosbach and P.-O. Larsson, *Biotechnol. Bioeng.* **12**, 19 (1970).
[10] B. C. Buckland, P. Dunnill, and M. D. Lilly, *Biotechnol. Bioeng.* **17**, 815 (1975).

results from the reduced rate of autolysis of the immobilized preparations is of great practical value. In addition to its contribution in extending the life of the enzyme, the reduced autolysis rate also simplifies the product purification procedure, since the quantities of L-amino acids produced when high concentrations of soluble α-chymotrypsin are in use for long periods can affect the optical rotations of chiral products unless care is taken to exclude them during work-up of the reaction. This can sometimes be troublesome. Any enzyme preparation that contains protease activity, such as horse liver alcohol dehydrogenase, should also be at least partially stabilized by immobilization. Disappointingly, however, the cases where immobilization has been demonstrated to confer significant additional stability on nonhydrolytic enzymes are relatively few.[11-13]

The horse liver alcohol dehydrogenase preparations which have been documented so far are limited in number. The preparations in polyacrylamide gel[14] and on Sepharose (via cyanogen bromide coupling)[15] lose 10–15% of their activity in 1 month at 4°. They are even more rapidly inactivated at room temperature and when the same preparation is recovered and reused in a subsequent experiment.[13] In contrast, immobilized yeast alcohol dehydrogenase and lactate dehydrogenases are stable enough to be available commercially. It is unfortunate that these enzymes do not have the broad specificity desired for routine organic chemical use, since the immobilized forms available possess the types of stability properties that are needed, viz., active for months at 0° with K_m values remaining similar to those of the soluble enzyme[13] and capable of multiple reuses in column operations.[16]

The stabilities of substrates and coenzymes under the reaction conditions must also be considered. For example, the optimum pH for α-chymotrypsin catalyzed hydrolyses is ~7.8. At this pH, many of the problems of acid- and base-catalyzed epimerization, isomerization, racemization, rearrangement, etc., that chemists encounter with sensitive molecules can be avoided. However, most of the immobilized α-chymotrypsin preparations reported operate optimally in an external pH of ~9.[11] (It should be stressed though, that, provided enzyme activity is retained, these reactions need not be carried out under pH-optimum conditions.) Under

[11] O. Zaborsky, "Immobilized Enzymes." Chem. Rubber Publ. Co., Cleveland, Ohio, 1972.
[12] J. R. Wykes, P. Dunnill, and M. D. Lilly, *Biotechnol. Bioeng.* **17**, 51 (1975).
[13] J. E. Dixon, F. E. Stolzenbach, J. A. Berenson, and N. O. Kaplan, *Biochem. Biophys. Res. Commun.* **52**, 905 (1973); J. E. Dixon, F. E. Stolzenbach, C.-I. T. Lee, and N. O. Kaplan, *Isr. J. Chem.* **12**, 529 (1974).
[14] A.-C. Johansson and K. Mosbach, *Biochim. Biophys. Acta* **370**, 348 (1974).
[15] J. B. Jones and H. M. Schwartz, unpublished results.
[16] K. Das, P. Dunnill, and M. D. Lilly, *Biochim. Biophys. Acta* **397**, 277 (1975).

such conditions the chemical advantages of operating under near-neutral conditions are lost. Furthermore, chemical hydrolysis of many of α-chymotrypsin's ester substrates becomes significant at this pH. In the case of catalysis of alcohol oxidation by the horse liver enzyme, use of solutions of pH 9 helps to displace the $C{=}O \rightleftharpoons CHOH$ equilibrium in the desired direction. This pH is not a wholly desirable one, since the rate of pseudobase formation from NAD^+ becomes significant. Furthermore, base-sensitive substrates cannot be used. Such disadvantages could be avoided, or at least minimized, by choosing a support material of the appropriate composition and polarity (see chapters [29] and [51]) to enable the reactions to be carried out in solutions at near-neutral pH while maintaining the active site region pH at the optimum desired. Similar tailoring of soluble supports for nicotinamide coenzymes could reduce the deleterious effects of hydration (at pH <7) and pseudobase formation (at pH >8) on the reduced and oxidized forms, respectively. Alternatively, the binding of the binary enzyme–coenzyme complex to a tailored polymer support may provide a partial solution to some of these problems.[7]

4. *Availability of Specificity Data.* The data bases for α-chymotrypsin and horse liver alcohol dehydrogenase are sufficiently broad to enable the organic chemist to use them predictively with confidence for the soluble enzymes. The aspects which have been documented include specificity with respect to substrates and inhibitors, and effects of organic solvents and of changes in temperature, pH, and electrolyte concentration.[4] Extensive data of this type are not yet available for the immobilized preparations of these, or other, enzymes. Whereas both the soluble and immobilized enzymes can be expected to behave similarly in many respects, more thorough documentation of the above aspects will provide an important additional degree of security to the chemist wishing to predict the course of catalysis for an immobilized catalyst.

5. *Costs.* Bearing in mind that the research chemist would prefer to purchase his catalyst, the high costs of the commercially available immobilized enzymes will act as a severe deterrent in the short term. In the long term the chemist must be convinced that immobilization is usually not a difficult procedure and that it can be relatively easily and economically accomplished provided that the native enzyme is readily obtainable. The detailed procedures provided in this volume will assist greatly in achieving this realization.

When the soluble enzyme is cheap, as is the case with α-chymotrypsin ($\sim$\$10/g), the use of an immobilized preparation for routine purposes is not worthwhile at present. In contrast, horse liver alcohol dehydrogenase is very expensive ($\sim$\$650/g). Even with high payoff factors in terms of regiospecificity and stereospecificity of catalysis on expensive substrates, perhaps also involving stereospecific hydrogen isotope introduction, the expense of soluble enzyme-catalyzed experiments would be considered high by most chemists at present. The economies possible from multiple reuse of a recoverable immobilized preparation would therefore make an important contribution to the general viability of the method. As noted previously, efficient nicotinamide coenzyme recycling is also needed before large-scale applications of alcohol dehydrogenases become economical,[4] although on the research laboratory scale the oxidoreductase is now the most costly single component of most reactions.

6. *Experimental Procedure.* Sophisticated enzyme-handling equipment is seldom available in the organic chemical laboratory and fiber-entrapped, microencapsulated, etc., enzymes are not likely to be easily handled in the apparatus available. On the other hand, enzymes immobilized on, or entrapped in, insoluble supports that can be used in batch or column processes are likely to be adopted with little resistance, particularly if the preparations are mechanically and chemically stable and possess high activities.

Column operations have obvious attractions. For hydrolytic enzymes they permit thermodynamically unfavorable but synthetically valuable reverse reactions, such as catalysis of the formation of optically active esters in the case of α-chymotrypsin,[17] to become practicable. The successes achieved in nucleotide synthesis using this approach[18,19] have already begun to attract the attention of aware organic chemists. Column procedures would not be as suitable for oxidoreductase catalyzed reactions where cofactors are required. Furthermore, many α-chymotrypsin and horse liver alcohol dehydrogenase mediated reactions are carried out on substrates which have very limited solubilities in aqueous solutions.[4] This is not an insurmountable problem with a soluble enzyme since heterogeneous reaction mixtures are often quite satisfactory. However, column operations of immobilized enzymes would be precluded in such situations. On the whole, batch procedures seem best suited for routine chemical use even though clogging of the pores of the support with undissolved substrate could also be serious in such operations.

The general problems of solubility might be overcome by adding organic solvents or surfactants.[4] As yet only very limited data on the effects of organic solvents on immobilized enzymes are available.[20,21] Al-

[17] N. S. Isaacs and C. Niemann, *Biochim. Biophys. Acta.* **44**, 196 (1960).
[18] H. G. Gassen and R. Nolte, *Biochem. Biophys. Res. Commun.* **44**, 1410 (1971).
[19] C. H. Hofmann, E. Harris, S. Chedroff, S. Michelson, J. W. Rothrock, E. Peterson, and W. Reuter, *Biochem. Biophys. Res. Commun.* **41**, 710 (1970).
[20] K. Tanizawa and M. L. Bender, *J. Biol. Chem.* **249**, 2130 (1974).
[21] H. Wan and C. Horvath, *Biochim. Biophys. Acta,* **410**, 135 (1975).

ternatively, when acidic or basic functions are present in a substrate, its solubility in aqueous media can be manipulated by altering the pH. This aspect was exploited to good effect with the basic substrate (III).

The simplified work-up procedures that become possible when immobilized enzymes are used in a batch or column process will also influence the rate of adoption of the technique. For example, on occasion aqueous reaction mixtures containing relatively high concentrations of α-chymotrypsin foam excessively when a separatory funnel product extraction process is used. The ability to remove the catalyst by filtration would obviate this difficulty.

Prognosis

There are evidently a number of problems that need to be solved before immobilized enzymes can be applied routinely in the organic laboratory. Nevertheless, the use of immobilized enzymes as catalysts in organic synthesis is an area for which the long-term outlook is very favorable, with their acceptance likely to be a steady, rather than a spectacular, process.

[59] The Synthesis of Porphobilinogen by Immobilized δ-Aminolevulinic Acid Dehydratase[1]

By Daniela Gurne and David Shemin

δ-Aminolevulinic acid dehydratase catalyzes the synthesis of porphobilinogen from 2 molecules of δ-aminolevulinic acid. The pyrrole is the precursor of porphyrins, chlorophyll, and the corrin ring of vitamin B_{12}. δ-Aminolevulinic acid dehydratase of bovine liver has a molecular weight of 285,000 and is composed of eight subunits, each of which has a molecular weight of 35,000.[2] It is also worth noting that the substrate forms a Schiff base with the enzyme.[3] These data are mentioned for they are particularly pertinent in an evaluation of the characteristics of the immobilized derivative of this enzyme.

---

[1] This investigation was supported by the National Institute of Arthritis and Metabolic Diseases, National Institutes of Health, United States Public Health Service (Grants AM 13091 and AM 16484) and by a grant from the American Cancer Society (BC-18D) and a grant from the National Science Foundation (GB-43916X).

[2] W. H. Wu, D. Shemin, K. E. Richards, and R. C. Willliams, *Proc. Natl. Acad. Sci. U.S.A.* **71**, 1767 (1974).

[3] D. Nandi and D. Shemin, *J. Biol. Chem.* **243**, 1236 (1968).

$$\underset{\text{δ-Aminolevulinic acid}}{\underset{\substack{|\\ NH_2}}{\overset{COOH}{\underset{|}{\overset{|}{C}}}}\underset{|}{\overset{|}{CH_2}}\underset{|}{\overset{|}{CH_2}}\underset{|}{\overset{|}{C=O}}\underset{|}{\overset{|}{CH_2}}} + \underset{\substack{|\\ H_2N}}{\overset{COOH}{\underset{|}{\overset{|}{CH_2}}}\underset{|}{\overset{|}{CH_2}}\underset{|}{\overset{|}{C=O}}\underset{|}{\overset{|}{CH_2}}} \xrightarrow{-2H_2O} \underset{\text{Porphobilinogen}}{\text{pyrrole structure}}$$

Materials

Cyanogen bromide-activated Sepharose 4B was either purchased from Pharmacia Fine Chemicals or formed from Sepharose 4B by published methods.[4-6] The resin AG1-8X (100–200 mesh) in its acetate form was purchased from BioRad, and δ-aminolevulinic acid was purchased from Sigma. A 6 mM solution of δ-aminolevulinic acid[7] contained 0.03 M potassium phosphate, pH 6.9, and 5 mM β-mercaptoethanol.

Assay Method

The enzyme activity is determined by measuring the amount of porphobilinogen formed at 37° in a 1.0 ml reaction mixture containing 100 μmoles of potassium phosphate, pH 6.8; 5 μmoles of dithiothreitol; 10 μmoles of δ-aminolevulinic acid neutralized with KOH; and an appropriate amount of enzyme. The assay mixture without the substrate is first preincubated for 30 min at 37° and then incubated for an additional 5 min after the addition of the substrate. The reaction is stopped by the addition of 0.5 ml of 0.1 M $HgCl_2$ in 20% trichloroacetic acid. The formed porphobilinogen is determined by the addition of 1.5 ml of modified Ehrlich reagent.[8] After centrifugation for 15 min, the absorbance at 555 nm was measured and the amount of porphobilinogen formed was calculated from the molar absorbance of 6.1×10^4.

A unit of enzyme activity is that amount of enzyme which produces 1

---

[4] J. Porath, R. Axén, and S. Ernback, *Nature (London)* **215**, 1491 (1967).
[5] P. Cuatrecasas, M. Wilchek, and C. B. Anfinsen, *Proc. Natl. Acad. Sci. U.S.A.* **61**, 636 (1968).
[6] J. Porath, K. Asperg, H. Drevin, and R. Axén, *J. Chromatogr.* **86**, 53 (1973).
[7] The δ-aminolevulinic acid hydrochloride is first neutralized with KOH before it is added to the phosphate buffer.
[8] D. Shemin, this series Vol. 17, p. 205.

μmole of porphobilinogen in 60 min at 37°. Specific activity is defined as units of enzyme activity per milligram of protein.

On monitoring the column containing the immobilized enzyme, aliquots (usually 10 μl) of the eluate are diluted to 1 ml and then assayed as described above.

Enzyme Preparation

The enzyme of bovine liver purified to homogeneity in our laboratory[9] usually had a specific activity of 15–22 units per milligram of protein. Although our method is as yet unpublished, the published methods[10-12] for the isolation of the enzyme are very adequate for obtaining a preparation of the enzyme to be used as an immobilized derivative. The enzyme preparation used in the work described below had a specific activity of 15 units per milligram of protein.

Preparation of Immobilized Enzyme and Porphobilinogen

Three grams of the dry activated Sepharose 4B was swollen for 15 min with 600 ml of cold 10^{-3} M HCl and filtered on a coarse fritted-glass funnel. The gel was rapidly washed on the filter with 600 ml of ice-cold water. To this gel, transferred to a beaker, 30 mg of enzyme (450 units of enzyme activity) in 4 ml of 0.1 M potassium phosphate, pH 6.8, was added, and the pH adjusted to 9.5 with 1 M Na_2CO_3. As soon as this pH was reached, the pH of the mixture was readjusted to 8 with 1 M KH_2PO_4. The mixture was stirred very gently at 4° overnight or for 2 hr at room temperature. The mixture was then washed on a coarse fritted-glass filter successively with 50 ml of 0.1 M potassium phosphate, pH 7.0; 50 ml of a solution containing 0.5 M KCl and 0.5 M Tris buffer, pH 8.0, over a period of about 1 hr, by gravity filtration; and finally with 100 ml of 0.1 M potassium buffer, pH 7.0, containing 5 mM dithiothreitol. An aliquot of the Sepharose-bound enzyme was assayed for enzymic activity. The total enzyme activity usually bound under these conditions was found to be about 55–65%. In this particular example, 270 units of the 450 units added were found on the Sepharose. Since it appears, from protein determinations, that more than 60% of the protein

---

[9] Thomas Ott, a graduate student, has our sincere thanks for the preparation.
[10] A. M. del C. Batlle, A. M. Ferramola, and M. Grinstein, this series Vol. 17, p. 216.
[11] E. L. Wilson, P. E. Burger, and E. B. Dowdle, *Eur. J. Biochem.* **29**, 563 (1972).
[12] P. E. Gurba, R. E. Sennett, and R. Kobes, *Arch. Biochem. Biophys.* **150**, 130 (1972).

was bound, it would seem that the specific activity of the bound enzyme is somewhat lower than that of the native enzyme.

The gel containing the 270 units of enzyme activity was packed onto a water-jacketed column 1×13.5 cm and maintained at 31°.[13] The δ-aminolevulinic acid solution[14] was passed through the column at a flow rate of 18–80 ml/hr. The flow rate was regulated by a peristaltic pump. The eluate was passed directly through a column (2.5×25 cm) containing 100 ml of the resin AG1-8X in its acetic form. The resin retained the porphobilinogen and very little of the δ-aminolevulinic acid. The column was replaced by a fresh resin column when it became saturated with porphobilinogen. Usually the column retained all the porphobilinogen formed from 3 liters of a 6 mM δ-aminolevulinic acid solution.

The percent conversion of the δ-aminolevulinic acid to porphobilinogen and the amount of porphobilinogen formed per hour depended on the flow rate. This was determined by analysis of samples of the eluate of the enzyme-gel column.

The table summarizes our findings. The choice of flow rate will depend on the particular needs of the investigator.

The Percent Conversion of δ-Aminolevulinic Acid (6 mM) to Porphobilinogen and the Amount of Porphobilinogen Formed per Hour of Immobilized δ-Aminolevulinic Acid Dehydratase at 31°

Flow rate (ml/hr)	Conversion (%)	Yield of porphobilinogen (μmoles/hr)
18	80	43
28	83	70
38	65	74
60	58	105
80	48	116

Isolation of Porphobilinogen

The resin column was washed with distilled water and then treated with 3–4 volumes of 1 M acetic acid and with 0.1 M acetic acid until no more porphobilinogen was eluted[15] The porphobilinogen is obtained as the

[13] Although the rate of synthesis is somewhat less at 31° than at 37°, the enzyme has greater stability at this temperature.
[14] Since δ-aminolevulinic acid is somewhat unstable at neutrality in oxygen, we ordinarily gas the solution with N_2 and set up a simple set of flasks ensuring minimum access of the solution to air.
[15] At times a portion of the porphobilinogen as the hydrate crystallizes out from the 1 M acetic acid solution.

acetate ($C_{10}H_{14}O_4N_2 \cdot CH_3COOH$) by lyophilization of the acetic acid solution. The material is essentially pure without recrystallization. The porphobilinogen acetate was crystallized as the hydrate ($C_{10}H_{14}O_4N_2 \cdot H_2O$) by dissolving the material in 0.1 M ammonium hydroxide and then adding acetic acid for precipitation.

Stability of the Immobilized Enzyme Preparation

The immobilized δ-aminolevulinic acid dehydratase appears to be exceedingly stable. Previously we had synthesized an immobilized preparation of the dehydratase of *Rhodopseudomonas spheroides* and operated it continuously for 27 days at 36° and found only a 30% loss of activity after this time.[16] The preparation described above was operated continuously for 9 days at 31° at a flow rate of 30 ml/hr with little, if any, loss of activity. During this time 16 mg of porphobilinogen was formed per hour, which represented a yield of 80–85%.

Presumably the enzyme is linked to the Sepharose through one subunit, or two. This value is arrived at on consideration of the probability of having both the required position of activated centers on the Sepharose and the amino groups on the enzyme spatially arranged in such a way for linkage to occur without distorting, and thus possibly inactivating, the enzyme. If this is so, it would appear that the binding forces among the subunits are exceedingly strong, for as mentioned above the enzyme preparation was fully active after many days at elevated temperatures. However, one should also consider an alternative hypothesis that the protein containing fewer than eight subunits is enzymically active.

Immobilized Dehydratase from Rhodopseudomonas spheroides

The dehydratase of *Rhodopseudomonas spheroides*[8] can readily be attached to Sepharose 4B by a similar procedure.[16] The advantage of this enzyme is that its specific activity is 6 times greater than that of the bovine liver and that it readily can be linked to the Sepharose even when it exists in a crude extract. The disadvantage of this source is that it is not as readily available as that from the liver.

General Discussion

This method of synthesizing porphobilinogen has several distinct advantages. Since the immobilized enzyme is quite stable, one can accumu-

[16] D. Gurne and D. Shemin, *Science* **180**, 1188 (1973).

late relatively large amounts of the pyrrole by operating the column continuously; on the other hand, one can put the column in a refrigerator after a run and subsequently use it for another preparation of either labeled or unlabeled porphobilinogen. Another advantage of this column is that it can be utilized for the synthesis of porphobilinogen labeled in particular atoms with either a radioactive or a stable isotope. Furthermore the material labeled with isotopes can be converted to porphyrins and related compounds, and the molecules can be labeled with a very high concentration of ^{13}C or ^{15}N, utilized readily in modern spectroscopic studies such as nuclear magnetic resonance or laser Raman spectroscopy.

It may be well to point out that since the enzyme contains eight subunits, and that one subunit, or two, is covalently linked, studies to determine the number of subunits needed for enzyme activity can be carried out with this immobilized preparation.

A relatively small disadvantage of this method is that the porphobilinogen synthesized from δ-aminolevulinic acid labeled in a specific carbon atom will have two atoms with the isotopic carbon atom.

[60] Properties and Applications of an Immobilized Mixed-Function Hepatic Drug Oxidase

By L. L. POULSEN, S. S. SOFER, and D. M. ZIEGLER

A mixed-function oxidase [EC 1.14.13.8, dimethylaniline monooxygenase (N-oxide-forming)] that catalyzes the N- and S-oxidation of a variety of nonnutritive compounds is present in hepatic tissue from all vertebrates examined.[1] The concentration of this oxidase is exceptionally high in human[2] and hog[3] liver tissue, and it has been purified to homogeneity from the latter source.[3] The pig liver enzyme isolated by the method described elsewhere[3] appears to consist predominantly of octomers of identical FAD-containing 65,000 MW subunits. Other than flavin, the preparation is free from metals, cytochromes, and chromophores that absorb in the visible region of the spectrum.

Although studies on the substrate specificity of the purified hog liver oxidase are not complete, it is known (as shown by the reactions in Fig. 1) that the isolated oxidase can catalyze the NADPH- and oxygen-de-

[1] E. Heinze, P. Hlavica, M. Kiese, and G. P. Lipowsky, *Biochem. Pharmacol.* **19**, 641 (1969).
[2] M. S. Gold and D. M. Ziegler, *Xenobiotica* **3**, 179 (1973).
[3] D. M. Ziegler and C. H. Mitchell, *Arch. Biochem. Biophys.* **150**, 116 (1972).

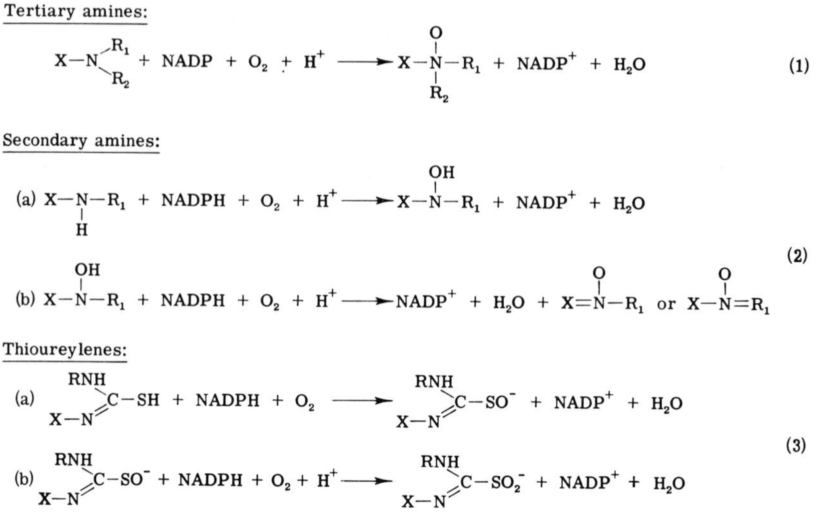

Tertiary amines:

$$X-N\begin{matrix}R_1\\R_2\end{matrix} + NADP + O_2 + H^+ \longrightarrow X-\underset{R_2}{\overset{O}{\underset{|}{N}}}-R_1 + NADP^+ + H_2O \quad (1)$$

Secondary amines:

$$(a)\ X-\underset{H}{\overset{|}{N}}-R_1 + NADPH + O_2 + H^+ \longrightarrow X-\overset{OH}{\underset{|}{N}}-R_1 + NADP^+ + H_2O$$

(2)

$$(b)\ X-\overset{OH}{\underset{|}{N}}-R_1 + NADPH + O_2 + H^+ \longrightarrow NADP^+ + H_2O + X=\overset{O}{\underset{|}{N}}-R_1\ \text{or}\ X-\overset{O}{\underset{|}{N}}=R_1$$

Thioureylenes:

$$(a)\ \begin{matrix}RNH\\X-N\end{matrix}\!\!>\!\!C-SH + NADPH + O_2 \longrightarrow \begin{matrix}RNH\\X-N\end{matrix}\!\!>\!\!C-SO^- + NADP^+ + H_2O$$

(3)

$$(b)\ \begin{matrix}RNH\\X-N\end{matrix}\!\!>\!\!C-SO^- + NADPH + O_2 + H^+ \longrightarrow \begin{matrix}RNH\\X-N\end{matrix}\!\!>\!\!C-SO_2^- + NADP^+ + H_2O$$

FIG. 1. Types of reactions catalyzed by the purified microsomal mixed-function amine oxidase. X = any lipophilic alkyl or aryl group. R_1 and R_2 = usually ethyl or methyl groups, but one or both substituents can be part of a ring system. R = hydrogen or alkyl group; it may also be part of ring with X (e.g., imidazole).

pendent N-oxidation of N-substituted amines[3] and hydrazines[4] and the S-oxidation of thioureylenes, and thiols.[5] Tertiary amines are oxidized to the corresponding N-oxides, whereas secondary amines are oxidized first to the N-hydroxy amines and then at a slower rate to the corresponding nitrones.[6] Sulfur-containing substrates all appear to be oxidized initially to the sulfenates, and the final products obtained depend to a large extent on the properties of the sulfenate derivative of a specific substrate. The sulfenates of all thioureylenes tested are rapidly oxidized to the sulfinates, but other sulfur-containing substrates can yield disulfides, in addition to the sulfinates. In general only lipid soluble sulfur- or nitrogen-containing compounds that do not have a strongly polar group within a two-carbon radius of the atom oxidized are substrates for the pig liver oxidase.

The pig liver oxidase can use either NADH or NADPH as reductant, but the concentration of NADPH required for one-half maximal velocity, $[S]_{1/2}$, is approximately one-tenth of that for NADH, and for this reason NADPH is the preferred reductant. Although various enzymic NADPH-

[4] R. A. Prough, *Arch. Biochem. Biophys.* **158**, 442 (1973).
[5] L. L. Poulsen, R. M. Hyslop, and D. M. Ziegler, *Biochem. Pharmacol.* **23**, 3431 (1974).
[6] L. L. Poulsen, F. F. Kadlubar, and D. M. Ziegler, *Arch. Biochem. Biophys.* **164**, 774 (1974).

generating systems could be used to provide the NADPH required for the oxidations catalyzed by the mixed-function oxidase, the *Leuconostoc mesenteroides* glucose-6-phosphate dehydrogenase has a number of advantages for this purpose. It is available commercially from Worthington Biochemicals, and this dehydrogenase also does not require metal ions for activity. This latter property not only simplifies the reaction medium but also decreases the possibility of metal-catalyzed decomposition of substrates or products and makes it much easier to separate the metabolites from the medium at the end of the reaction.

Immobilization

With glutaraldehyde as a carbonyl intermediate, the mixed-function oxidase has been linked by Schiff's base coupling to nylon (tubing and powder) and zirconia-clad glass beads.[7] The enzyme has also been mounted on Sepharose by the method of Axén and Ernback[8]; however, the lipophilic nature of the drug substrates for the oxidase restricts the choice of the supporting matrix to one that is essentially lipophobic, and most of our studies with the insolubilized oxidase have been carried out with oxidase attached to glass beads.

The oxidase can be bound to glass beads as described earlier,[7] but a nearly 2-fold increase in the amount of enzyme bound per gram of beads can be obtained by the following method. The zirconia-clad, 1350 Å glass beads (Pierce Chemical Co.) are washed thoroughly with water, transferred to 0.25 M glutaraldehyde, and deaerated under vacuum for 30 min. After 60 min, the beads are transferred to a glass column and the excess glutaraldehyde is removed by flushing the column with at least 20 volumes of water over a 1–2 hr period. The column is then equilibrated with 0.025 M phosphate buffer, pH 7.4 at 4°. At the same temperature 2 ml of the purified oxidase (20 mg/ml in 0.025 M phosphate, pH 7.4) is added for each gram of glass beads in the column. The solution of the oxidase is allowed to penetrate into the beads with the flow rate so adjusted that the free oxidase moves through the beads at a rate no greater than 0.5 ml/hr per gram of beads. When the free oxidase reaches the bottom of the column, the flow rate is stopped, the excess enzyme solution at the top of the column is removed, and the column is allowed to stand for an additional 30 min at 4°. The excess oxidase is eluted from the column with 0.025 M phosphate buffer and saved. The beads containing the bound oxidase are washed, in the column, with about five column volumes of

---

[7] S. S. Sofer, D. M. Ziegler, and R. P. Popovich, *Biochem. Biophys. Res. Commun.* **17**, 74 (1974).

[8] R. Axén and S. Ernback, *Eur. J. Biochem.* **18**, 351 (1971).

buffer. The bead-mounted oxidase can be stored at 0°–4° 0.025 M phosphate buffer for several months with little or no loss in activity.

Assays

The amount of oxidase attached to the beads is most conveniently measured by determining the amount of acid-extractable flavin per gram of beads. Approximately 0.5 g of beads is spread on a strip of filter paper to remove virtually all the water and then air-dried at room temperature for about 1 hr. The dried beads are weighed and transferred quantitatively into a glass centrifuge tube. Two milliliters of 0.3 M trichloroacetic acid are added and, after mixing on a test tube mixer, followed by centrifugation in a table top centrifuge, a 0.5-ml aliquot of the clear supernatant fraction is transferred to a 1-ml cuvette containing 0.1 ml of 1 M K_2HPO_4 and 0.02 ml of 6 M KOH. The concentration of flavin in this solution is measured spectrophotometrically,[9] and the concentration of flavin per gram of beads calculated. Since the concentration of flavin in the soluble oxidase is readily measured (the homogeneous oxidase contains 15.0–15.4 nmoles of FAD per milligram of protein) the amount of oxidase covalently linked to the beads can be calculated. Based on the flavin concentration, between 8 and 10 mg of oxidase can be covalently linked per gram of zirconia-clad glass beads.

The activity of the bead-mounted oxidase is measured by following the substrate-dependent increase in oxygen consumption in a sealed, temperature-controlled, 1.9-ml reaction vessel fitted with a Clark-type electrode. The signal from the electrode is recorded on a strip chart recorder. The reaction medium contains 0.2 mM NADP$^+$, 1.5 mM glucose 6-phosphate, two international units of *L. mesenteroides* glucose-6-phosphate dehydrogenase, and buffer, 0.02 M phosphate–0.02 M pyrophosphate, pH 7.6. The medium is stirred with a Teflon-coated stirring bar powered by a variable-speed magnetic stirrer set at a speed to produce turbulence such that further increases in stirrer speed do not change the rate of oxygen consumption. After an initial temperature equilibration of 2–3 min, 2–5 mg of glass beads containing the oxidase are added, and the endogenous rate of oxygen consumption is recorded for 1–2 min.

The reaction is started by injecting 1–2 μmoles of substrate dissolved in 20 μl or less of water, and the rate of oxygen consumption is recorded for another 1–2 min. At the end of the measurement the beads are quantitatively transferred to a test tube, rinsed two or three times with acetone, dried at 60°, and weighed. The rate of the reaction expressed as micromoles of substrate oxidized per minute per gram of beads is calculated

[9] B. de Bernard, *Biochim. Biophys. Acta* **23**, 510 (1957).

from the stoichiometry for a given substrate (Fig. 1). At pH 7.6 the endogenous rate of oxygen uptake is never more than 5% of the rate obtained in the presence of substrate. A high endogenous rate usually indicates substrate contamination of the reaction vessel. Lipophilic substrates adhere to (or are absorbed in) the Teflon stirring bar and the polyethylene membrane covering the oxygen electrode and are difficult to remove by repeated rinsing with water. These compounds are, however, readily removed by rinsing the reaction chamber with two changes of reagent grade methanol, followed by several rinses with distilled water between measurements.

This method for estimating the specific activity of the catalyst takes less than 30 minutes after the equipment is assembled, and measurements are reproducible to within less than ±3% of the average specific activity measured 10 or more times on the same batch of catalyst over a 1-month period. Although specific activity measurements determined by this method are useful in following the efficiency of coupling and stability of the catalyst, the rates are the maximum possible for any specific substrate, but in using the catalyst for the synthesis of even moderate amounts of a metabolite, the rate of the reaction is frequently limited by mass transfer to the catalyst surface or by oxygen concentration. The limits imposed on reaction rate by mass transfer can be minimized by increasing mixing rates in batch reactors or flow rates in column reactors, but the decrease in rate as a function of oxygen concentration is not as readily solved. At pH 7.6 the concentration of O_2 required for half-maximal velocity is 0.075 mM, so that, even in an aqueous medium saturated with air, the O_2 concentration is less than maximal. However, even a 50% higher concentration of oxygen leads to the rapid destruction of the oxidase, and initial oxygen concentrations in excess of 0.25 mM should be avoided. Although the problems imposed by oxygen concentration could limit the use of the insolubilized oxidase in large reactors, it is not serious in most applications.

Properties of the Bead-Mounted Oxidase

The kinetic properties of the glass-bead mounted enzyme have not been studied exhaustively, but the substrate specificity, the activity per milligram of enzyme, and the $[S]_{1/2}$ for O_2, NADPH, and drug substrates are not significantly different from the soluble oxidase. On the other hand, the thermal stability of the oxidase is increased more than 100-fold after covalent coupling to insoluble matrices.[7] The half-life of the glass-bead mounted oxidase at 38° (at pH 7.6) is approximately 5 hr, whereas the half-life of the soluble oxidase under the same conditions is less than 10 min. The optimum operating pH and temperature have been determined for the glass-bead mounted oxidase for the synthesis of the amine oxide of N,N-dimethylaniline and the maximum yield of product per gram of catalyst is obtained by operating between 25° and 28° at pH 7.6.[10] To what extent data collected with this tertiary amine can be extrapolated to the synthesis of metabolites of the more complex drugs needs to be determined.

Applications

While it should be possible to synthesize the metabolites of all (Fig. 1) substrates for the amine oxidase with the bead-mounted oxidase as catalyst, in practice only the amine oxides of tertiary amines can be easily synthesized. One to two milligrams of the N-hydroxy derivatives of a few disubstituted amines have been prepared, but in every case the reaction had to be terminated after only 20% of the parent substrate was oxidized. The nitrones formed upon the further oxidation of N-hydroxyamines hydrolyze almost as rapidly as they are formed. Hydrolysis of the nitrones produces two additional N-hydroxyamines and two aldehydes, which complicate the final separation of the desired product from the reaction medium. However, further oxidation of the first product is negligible as long as the parent secondary amine is present in at least a 5-fold molar excess. When the reaction has proceeded to this point, both the parent substrate and its N-hydroxy derivative were extracted into chloroform, concentrated under vacuum, and separated by thin-layer chromatography.

At the present time the synthesis of either the sulfenate or sulfinate derivative of thioureylenes with the bead-mounted oxidase is not practical. With the exception of these two derivatives of phenylthiourea, the sulfenate and/or sulfinates of all other thiourylenes tested are extremely unstable and very reactive compounds. Catalyst destruction by these products is invariably encountered which limits the use of the bead-mounted oxidase for the synthesis of these compounds.

The greatest potential use of the glass bead-mounted oxidase is as a catalyst for the synthesis of milligram quantities of a specific metabolite of a large number of different tertiary amine drugs. Many of these metabolites, readily synthesized with the use of the bead-mounted oxidase, are difficult or impossible to synthesize by direct chemical methods.

Example. The synthesis of 10–30-mg quantities of a metabolite does not require complex reactors, and in this laboratory these syntheses are carried out in an open beaker at room temperature. The reaction is car-

[10] S. S. Sofer, D. M. Ziegler, and R. P. Popovich, *Biotechnol. Bioeng.* **17**, 107 (1975).

ried out in 0.05 M phosphate buffer, pH 7.6, containing 0.2 mM NADP⁺, glucose-6-phosphate dehydrogenase, the drug substrate (0.5–1.0 μmole/ml), and a 1.5- to 2-fold molar excess of glucose 6-phosphate relative to the drug substrate. The drug substrate can be added as the powder and need not dissolve completely. About 1 g of glass beads containing the oxidase is added for each 50 ml of solution, and the medium is mixed slowly (to prevent excessive breaking of the glass beads) with any suitable stirrer for several hours (usually overnight) to carry the reaction to completion. By this method one can synthesize several milligrams of the N-oxide metabolites of tertiary amine antihistamines and tertiary amine phenothiazine drugs. The $[S]_{1/2}$ of all these tertiary amines is 10^{-4} M or less, and their N-oxide metabolites are quite stable compounds. After the reaction is complete the medium is decanted from the catalyst, adjusted to pH 8–10, and extracted 3–4 times with an equal volume of chloroform. In slightly alkaline solutions virtually all drug N-oxides will partition between water and chloroform. The combined chloroform extracts are taken to dryness under reduced pressure. The residue is resuspended in a few milliliters of dry chloroform, insoluble material is removed, and the N-oxide is precipitated from chloroform with dry diethyl ether. The N-oxide is recrystallized one or two times from anhydrous chloroform or ethanol with ether.

Comments. In most laboratories studying the biotransformations of drugs, only a few milligrams of a specific metabolite are required, primarily as a reference for the identification of that metabolite in biological fluids or tissues, and the batch reactor described above will satisfy most of the anticipated uses of the glass bead-mounted oxidase. For the synthesis of larger quantities (0.5–1 g) of a specific N-oxide metabolite, the reactor configuration shown in Fig. 2 is adequate. The flow rate is adjusted so that the oxygen concentration in the medium leaving the catalyst bed is approximately one-half that of the medium equilibrated with a stream of filtered air. Since the concentration of oxygen in the medium equilibrated with air remains relatively constant, the drop in oxygen concentration, measured at the point indicated, provides a convenient method for following the reaction. An increase in oxygen concentration indicates that one or more of the reactants is limiting (usually the glucose-6-phosphate dehydrogenase or the tertiary amine substrate). The limiting component can be easily identified by removing an aliquot of the reaction mixture and carrying out the appropriate assays. The more water-insoluble drugs, such as the phenothiazine drugs, are most conveniently added in increments during the course of the reaction. The N-oxide metabolites are more polar than the parent tertiary amines; all of those that we have synthesized to date are soluble up to at least 1.5 mM, and

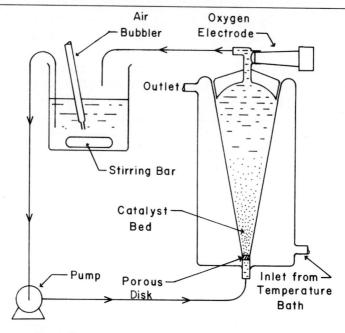

FIG. 2. Reactor configuration for a fluidized bed reactor.

precipitation of product from the reaction medium has not been encountered. The reaction is carried to completion after the last addition of substrate, and the product is isolated as described with the batch reactor.

Section X
Immobilized Coenzymes

[61] Immobilized Coenzymes

By Klaus Mosbach, Per-Olof Larsson, and Christopher Lowe

The immobilization of coenzymes has received increasing attention over the last few years because they have several important applications: (1) as "active immobilized coenzymes," (2) as immobilized general ligands in affinity chromatography (this applies in particular to the various nucleotides such as AMP), and (3) a few reports have recently appeared in which they have been applied to more fundamental enzymological studies. The aspects of affinity chromatography have been covered in a recent volume of this series[1] and are not dealt with here. In this volume the main emphasis is on their use as active coenzymes together with a brief account of their application in basic enzymology. The methodological part will be centered around the various adenine nucleotides, NAD^+, $NADP^+$, ATP, ADP, whereas work on other coenzymes will be treated only in a summary fashion. The various aspects will be treated as outlined below: (1) synthesis of a number of adenine nucleotide coenzymes, (2) coupling to matrices, (3) coenzymic activity, (4) application in enzyme technology and analysis, (5) application in enzymological and protein studies, (6) other immobilized coenzymes, (7) general discussion.

Synthesis of N^6-carboxymethyl- and N^6-[(6-aminohexyl) carbamoylmethyl] Derivatives of NAD^+, $NADP^+$, ATP, and ADP

Principle

The procedures given below follow the same general outline and are summarized in Fig. 1a–c: (a) alkylation with iodoacetic acid to give 1-carboxymethyl nucleotides; (b) rearrangement to the N^6-carboxymethyl nucleotides; (c) condensation with 1,6-diaminohexane to give N^6-[(6-aminohexyl)carbamoylmethyl] nucleotides.

In the alkylation step a 3- to 10-fold excess of iodoacetic acid is employed. The excess is necessary to ensure complete conversion of the parent nucleotide, since in aqueous media a substantial amount of the alkylation agent is hydrolyzed to hydroxyacetic acid. It is also beneficial to use the smallest reaction volume conveniently possible to enhance the rate of the alkylation reaction.

[1] This series Vol. 34 (1974).

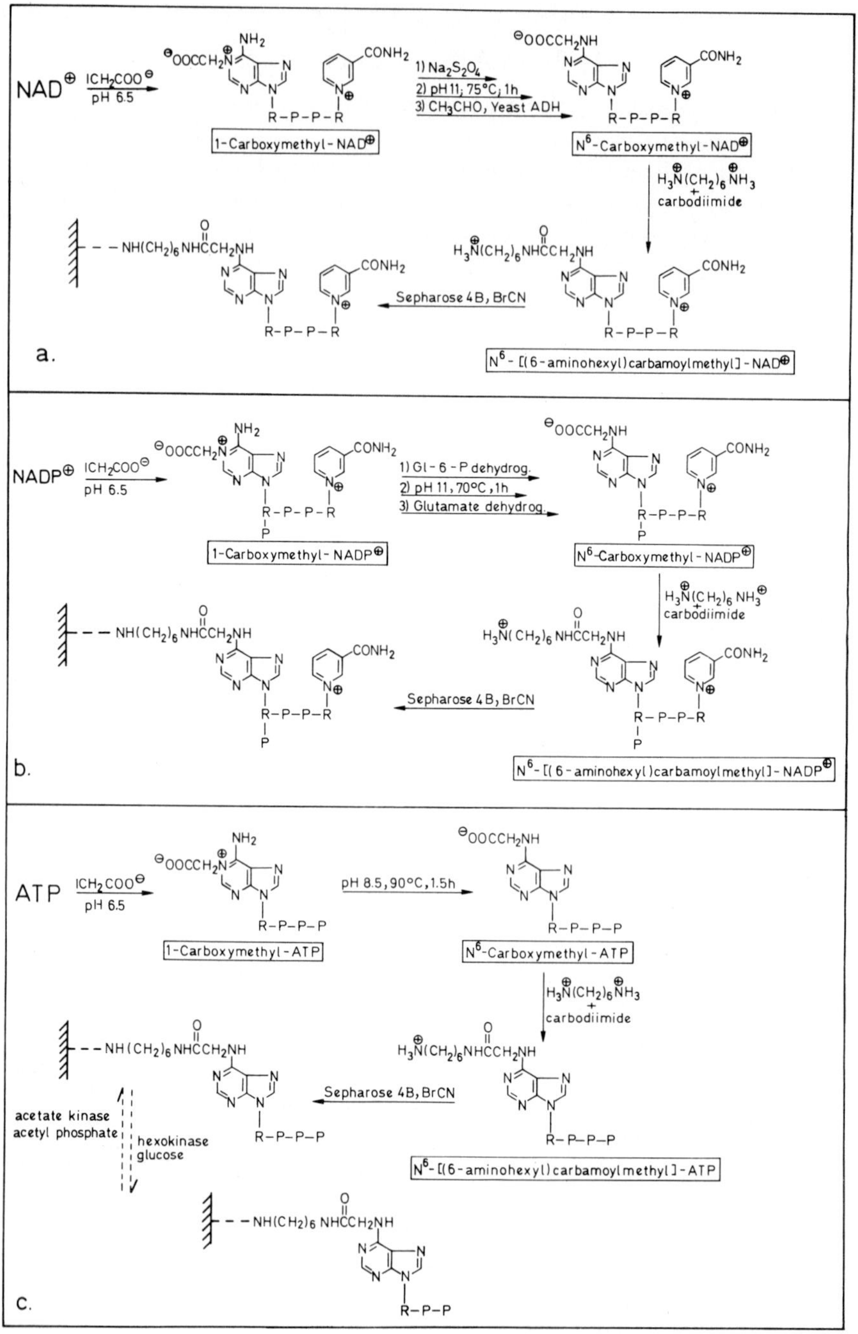

NAD⊕
ICH2COO⊖
pH 6.5
1-Carboxymethyl-NAD⊕
1) Na2S2O4
2) pH 11; 75°C; 1h
3) CH3CHO, Yeast ADH
N6-Carboxymethyl-NAD⊕
H3N(CH2)6NH3 + carbodiimide
Sepharose 4B, BrCN
N6-[(6-aminohexyl)carbamoylmethyl]-NAD⊕
a.
NADP⊕
1-Carboxymethyl-NADP⊕
1) Gl-6-P dehydrog.
2) pH 11, 70°C, 1h
3) Glutamate dehydrog.
N6-Carboxymethyl-NADP⊕
N6-[(6-aminohexyl)carbamoylmethyl]-NADP⊕
b.
ATP
1-Carboxymethyl-ATP
pH 8.5, 90°C, 1.5h
N6-Carboxymethyl-ATP
N6-[(6-aminohexyl)carbamoylmethyl]-ATP
acetate kinase
acetyl phosphate
hexokinase
glucose
c.

By-products are virtually absent until the alkylation has reached a level of about 80%. If higher conversions are attempted, significant amounts of nucleotide by-products start to accumulate. The alkylation reaction should thus be terminated when a conversion of about 90% has been reached in order to obtain a maximum yield of 1-carboxymethyl nucleotide.

No attempt has been made to optimize the alkylation conditions for each of the nucleotides described, although the following may serve as a general rule of thumb for the various nucleotides: 5 days in the dark at room temperature and adjustment of pH once a day with 2 M LiOH. (Recent data obtained with AMP and NAD$^+$ suggest that on raising the temperature to 40°, the time required for alkylation can be reduced to 1 day provided pH adjustments are made more frequently.)

The 1-carboxymethyl nucleotides are intrinsically unstable, especially in alkaline media at elevated temperatures, and rearrange by a Dimroth mechanism to N^6-substituted nucleotides.[2] The 1-carboxymethyl derivatives of ADP and ATP are thus easily converted to N^6 derivatives merely by heating at 90°, pH 8.5, for 1.5 hr. These conditions are, somewhat surprisingly, not accompanied by hydrolytic release of phosphate or pyrophosphate. The 1-carboxymethyl derivatives of NAD$^+$ or NADP$^+$, on the other hand, cannot be satisfactorily rearranged in this direct fashion owing to their marked lability in alkali. However, reduction, either enzymically or with dithionite, yields the alkali-stable reduced nucleotide, which is rearranged subsequently by heating in alkali. The resulting N^6-carboxymethyl-NAD(P)H is then reoxidized enzymically. This reoxidation is necessary to permit purification by ion-exchange chromatography in acid media, which would otherwise destroy the extremely acid-labile reduced analogs. The enzymic reoxidation is preferable to other oxidation methods, since it is highly selective and will oxidize only those nucleotides that are correctly substituted and have the proper configuration. The beta form of N^6-carboxymethyl NAD(P)H is thus oxidized whereas the corresponding alpha form, which may be formed as an undesirable by-product during the rearrangement, will re-

---

FIG. 1a–c. Synthesis of the 1-carboxymethyl and N^6-[(6-aminohexyl)carbamoylmethyl] derivatives of NAD$^+$ (a), NADP$^+$ (b), and ATP (c). Several types of binding are reported for molecules bound to cyanogen bromide-activated gels. The linkage is indicated with a dashed line. R = ribose, P = phosphate, ADH = alcohol dehydrogenase, Gl-6-P dehydrog. = glucose-6-phosphate dehydrogenase.

In Fig. 1c is also given the enzymic interconversion of ATP ↔ ADP obtained on solid phase.

[2] M. H. Wilson and J. A. McCloskey, *J. Org. Chem.* **38**, 2247 (1973).

main reduced. Those nucleotide molecules that remain reduced are rapidly destroyed in the subsequent acid media and are converted into compounds that are easily separated from the desired nucleotide by ion-exchange chromatography.

The N^6-carboxymethyl adenine nucleotides may be bound covalently to various matrices, or, alternatively, condensation of the N^6-carboxymethyl nucleotides with 1,6-diaminohexane generates N^6-[(6-aminohexyl)carbamoylmethyl] nucleotides bearing a terminal amino group suitable for direct attachment to supports. The latter analogs have been applied primarily by the authors. The condensation is efficiently promoted by water-soluble carbodiimides and, provided the diamine is present in 10-fold or higher excess, negligible formation of bis-nucleotides occurs.

A strongly red by-product is formed in small amounts, but may be conveniently removed by ethanol precipitation or absorption on an ion exchanger.

Examples

N^6-Carboxymethyl-NAD^+

The synthesis described below is essentially identical to a previously published procedure[3,4] (Fig. 1a).

NAD^+ (10 g, 13.6 mmoles Sigma grade AA) is added to an aqueous solution containing 30 g of iodoacetic acid (160 mmoles) neutralized with 2 M LiOH, and the pH of the mixture is readjusted to 6.5. The solution (total volume 135 ml) is kept in the dark at room temperature and the pH is adjusted to 6.5 with 2 M LiOH once every day. The alkylation is followed by thin-layer chromatography and, when the conversion of NAD^+ is approximately 90% (after 6–7 days), the reaction is terminated. The pH is lowered to 3.0 with 6 M HCl, two volumes of ethanol are added, and the solution is poured into 1.5 liters of vigorously stirred ethanol (0°). The precipitated crude 1-carboxylmethyl-NAD^+ is filtered off and washed with ethanol and ether and dried in a vacuum. Approximately 12 g of a faintly pink powder is obtained showing a purity with respect to nucleotide of about 90%. The crude 1-carboxymethyl-NAD^+ is dissolved in 300 ml of 2% $NaHCO_3$, the pH is adjusted to 8.5, and the solution is deaerated with 95% N_2–5% CO_2. The reducing agent sodium dithionite (5 g) is added, and, after 2 hr at room temperature in the dark,

---

[3] M. Lindberg, P. O. Larsson, and K. Mosbach, *Eur. J. Biochem.* **40**, 187 (1973).

[4] K. Mosbach, this series Vol. 34, p. 233 (1974).

the reduction is terminated by oxygenation for 10 min followed by a brief treatment with nitrogen gas.

The reduced compound is subsequently rearranged to N^6-carboxymethyl-NADH by adjusting the pH of the solution to 11.5 with NaOH and heating at 75° for 1 hr.

Finally an enzymic reoxidation step is undertaken at room temperature by adding the following: 20 ml of 2 M Tris, 5 ml of redistilled acetaldehyde, 3 M HCl to a final pH of 7.5, and 5 mg of yeast alcohol dehydrogenase (2500 U). The oxidation is followed spectrophotometrically at 340 nm; it is considered complete when no further decrease of absorbance occurs (about 30 min). The solution is acidified with HCl to pH 3.5, and 1 volume of ethanol is added. The solution is then poured into 10 volumes of vigorously stirred ethanol, and the resulting suspension is kept in the refrigerator overnight. The crude N^6-carboxymethyl-NAD$^+$ is filtered off, washed with ethanol and ether, and dried. The product is purified by ion-exchange chromatography on Dowex AG 1X-2 (200–400 mesh, chloride). The column (2.5 × 75 cm) is loaded with a solution of the crude substance (pH 8) and then washed with water (500 ml) and 0.005 M CaCl$_2$, pH 2.7, until the pH of the effluent is about 2.8 (2.5 liters). A linear gradient, 0.005 M CaCl$_2$, pH 2.7, to 0.05 M CaCl$_2$, pH 2.0 (total volume 8 liters), is applied. The effluent between approximately 3.3 liters and 5.2 liters contains the desired compound. The eluate is neutralized with Ca(OH)$_2$ and concentrated on a rotary evaporator. Precipitation with ethanol as above gives pure N^6-carboxylmethyl-NAD$^+$ in a yield of 40–45% (5.6 g).

N^6-[(6-Aminohexyl)carbamoylmethyl]-NAD$^+$

N^6-Carboxymethyl-NAD$^+$ (4.5 g, 5 mmoles) is dissolved in 60 ml of 2 M 1,6-diaminohexane dihydrochloride, and the solution is kept in an ice-bath. 1-Ethyl-3-(3-dimethylaminopropyl)carbodiimide (1.4 g, 7.5 mmoles) dissolved in 5 ml water is added, and the pH is kept at 4.8 by adding 1 M HCl or 1 M LiOH. After 10 min, when the reaction is slowing down, the ice-bath is removed and the condensation is allowed to proceed for a further 60 min at room temperature. The pH is then increased to 6.5, and 1 volume of 0.5 M LiCl in ethanol is added; the resulting red precipitate is filtered off and discarded. The supernatant is slowly poured into a 10-fold volume of vigorously stirred ethanol (0°C), and the precipitated N^6-[(6-aminohexyl)carbamoylmethyl]-NAD$^+$ is collected. The almost pure compound is dissolved in 75 ml of water (pH 8) and the solution is passed through a column containing 5 g of lithium-charged Dowex 50 W-X4. The effluent is adjusted to pH 6 with HCl and passed

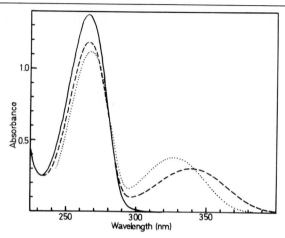

FIG. 2. Ultraviolet spectra of dextran-bound N^6-[(6-aminohexyl)carbamoylmethyl]-NAD$^+$ in 0.10 M Tris-HCl buffer, pH 7.5 (———), of the reduced form obtained after reduction for 15 min with yeast ADH (25 μg/ml) in 0.33 M ethanol and 0.10 M glycine-NaOH buffer, pH 9.5 (---), and of the KCN complex in 1.0 M KCN (····). The reference cells contained the same ingredients as the sample cells except that dextran-NAD(H) was replaced by blank dextran. The spectra of the corresponding free NAD$^+$-derivatives were identical with the exception of the enzymatically reduced NAD$^+$ analog, which in its unbound state showed a higher ratio A_{340nm}/A_{266nm} due to complete reduction.

through a column packed with 5 g of chloride-charged Dowex 1-X4. The final effluent is adjusted to 6.8 and concentrated to about 25 ml. Pure N^6-[(6-aminohexyl)carbamoylmethyl]-NAD$^+$ is obtained by ethanol precipitation carried out as above. The yield is about 80% (3.3 g), the overall yield from NAD$^+$ being approximately 35%. Based on phosphate determinations and UV measurements at 266 nm in Tris-HCl, pH 7.5, the molar absorption coefficient is determined as $\epsilon_{266\,nm}^{pH\,7.5} = 21,700\ M^{-1}\ cm^{-1}$ (Fig. 2).

N^6-Carboxymethyl-NADP$^+$

The synthesis described below is essentially identical to a previously published procedure[5] (Fig. 1b).

Other methods of synthesis based on the condensation of nicotinamide mononucleotide with an appropriate adenosine-2',5'-bisphosphate analog were considered inappropriate since cumbersome separation problems would be anticipated.[6,7] NADP$^+$ (1.0 g, 1.3 mmoles) is added to an aqueous solution of iodoacetic acid (1.0 g, 5.4 mmoles) neutralized with standard 2 M LiOH. The pH is readjusted to 6.5, and the solution (total

[5] C. R. Lowe and K. Mosbach, *Eur. J. Biochem.* **49**, 511 (1974).
[6] D. B. Craven, M. J. Harvey, and P. D. G. Dean, *FEBS Lett.* **38**, 320 (1974).
[7] N. A. Hughes, G. W. Kenner, and A. Todd, *J. Chem. Soc.* 3733 (1957).

volume 10.0 ml) is kept in the dark at 26°–27°. The pH is checked daily and readjusted to 6.5 with 2 M LiOH when necessary. The progress of the reaction may be followed by thin-layer chromatography on poly-(ethyleneimine) cellulose using 0.5 M LiCl as eluent. After 10 days, most of the NADP$^+$ is converted to N^1-carboxymethyl-NADP$^+$ whence the reaction mixture is acidified to pH 3.0 with 3 M HCl, cooled to 0°, and slowly added to a 10-fold excess of ethanol at $-20°$. The flocculent pale pink precipitate is collected, washed extensively with cold ethanol and ether, and finally air-dried.

N^1-Carboxymethyl-NADP$^+$ is intrinsically unstable and rearranges under alkaline conditions and elevated temperatures to the N^6-substituted form. To prevent hydrolysis during these steps the nucleotide is enzymically converted to the alkali-stable reduced form. Glucose 6-phosphate (612.3 mg, 2 mmoles), N^1-carboxymethyl-NADP$^+$ (765.0 mg, 1 mmole), and yeast glucose-6-phosphate dehydrogenase (250 units) are added to 30 ml 0.1 M Tris-HCl buffer pH 7.4 containing 5 mM Mg Cl$_2$ and incubated for 1.5 hr at 25°. The progress of the reduction is followed by the increase in absorbance at 340 nm and is essentially complete after 30 min. The rearrangement is effected by adjusting the pH of the solution to 11.0 with 1 M NaOH and heating for 1 hr at 70°. The solution is cooled to 25°, and the pH is readjusted to 7.4. Ammonium acetate (246.6 mg, 3.2 mmoles), 2-oxoglutarate (116.9 mg, 0.8 mmole), EDTA (3 mg), and beef liver glutamate dehydrogenase (100 units) are added, and the reoxidation allowed to proceed for 1.5 hr at 25°. The reaction is monitored by the decrease in absorbance at 340 nm; on completion, the solution is cooled in ice, poured slowly into cold ethanol ($-20°$), and allowed to stand overnight at $-20°$. An oily precipitate forms and sediments overnight; it is collected, washed with cold ethanol and ether, and finally air-dried. The crude powder (1.3 g) is dissolved in 400 ml of distilled water followed by adjustment of the pH to 7.5 with 0.1 M NaOH. The solution is applied to a Dowex 1-X2 column (Cl$^-$; 1.5 $\times$ 25 cm) that has previously been equilibrated with distilled water. The column is washed with water (30 ml) and then with 1 mM CaCl$_2$, pH 3.0, until the effluent no longer contains UV-absorbing material and the pH has attained a value of 3.0. The nucleotides are eluted with a linear gradient of CaCl$_2$, 1 mM CaCl$_2$, pH 3.0, to 100 mM CaCl$_2$, pH 2.0, 800 ml total volume, at a flow rate of 70–80 ml/hour. Fractions (5 ml) are collected, and those in the two major peaks containing nucleotide are pooled independently. The pooled peak fractions are adjusted to pH 7.5 with 2 M NaOH, diluted 4-fold with distilled water and concentrated on a small Dowex 1-X2 column (Cl; 0.7 $\times$ 3 cm) equilibrated with distilled water. Nucleotide is eluted with 100 mM CaCl$_2$ pH 2.0 and the concentrated pale-straw-colored solution is desalted on a column of Sephadex G-10 (3 $\times$ 41 cm) equilibrated with water. Elution is effected with water at a flow rate of 3.2 ml/minute, and the desalted fractions containing UV-absorbing material are lyophilized. The two major peaks of nucleotide eluted from the first Dowex-1 column are thus obtained as white salt-free lyophilizates homogeneous by thin-layer chromatography in several systems and shown to be N^1-carboxymethyl-NADP+ and N^6-carboxymethyl-NADP+, respectively. The yield of the latter varies from preparation to preparation, but based on phosphate determinations and UV measurements at 265 nm in Tris-HCl buffer pH 7.5, the molar absorption coefficient is 21,800 $M^{-1} \times cm^{-1}$, and the yield is in the range 15–25% (100–200 mg). The purified N^1-carboxymethyl-NADP+ generated at this stage can be recycled through the rearrangement to produce more N^6-carboxymethyl-NADP+.

N^6-[(6-Aminohexyl)carbamoylmethyl]-NADP+

N^6-Carboxymethyl-NADP+ (37.5 mg, 45.6 μmoles) is dissolved in 25 ml of an 1 M aqueous solution of 1,6-diaminohexane dihydrochloride, and the pH is adjusted to 4.7. The water-soluble carbodiimide, 1-ethyl-3-(3-dimethylaminopropyl)carbodiimide hydrochloride (67.9 mg, 350 μmoles) is added slowly in portions, and the pH is adjusted to 4.7 with 0.1 M HCl after each addition. The reaction is allowed to proceed for 20 hr at 21°–22° with periodic readjustments of the pH to 4.7, and then freed from carbodiimide and 1,6-diaminohexane by passage through a Sephadex G-10 column (3 × 41 cm) equilibrated with distilled water. The pooled nucleotide-containing fractions are applied directly to a Dowex 1-X2 column (Cl−; 3.1 × 7.7 cm) and subsequently eluted with a linear gradient of CaCl$_2$, 1 mM CaCl$_2$, pH 3.0, to 100 mM CaCl$_2$, pH 2.0, 20 ml total volume. Fractions (1.5 ml) are collected, and those containing N^6-[(6-aminohexyl)carbamoylmethyl]-NADP+ pooled, desalted by passage through a Sephadex G-10 column as above, and finally lyophilized. The product is essentially salt-free and homogeneous in several thin-layer chromatography systems. Under the above conditions, 80% of the N^6-carboxymethyl-NADP+ is converted to N^6-[(6-aminohexyl)carbamoylmethyl]-NADP+ based on phosphate analysis and UV data ($\epsilon_{265\ nm}^{pH\ 7.5}$ = 21,700 $M^{-1} \times cm^{-1}$). Excess carbodiimide should be avoided because some nucleotide 2′,3′-cyclic phosphate might be formed which could contaminate the 3′-isomer of the NADP+-analog.

N^6-Carboxymethyl-ATP

The synthesis described below is essentially identical to a previously published procedure[8] (Fig. 1c).

[8] M. Lindberg and K. Mosbach, *Eur. J. Biochem.* **53**, 481 (1975).

ATP (5 g, 8.3 mmoles) is added to an aqueous solution of iodoacetic acid (15 g, 80 mmoles) neutralized with 2 M LiOH, and the pH is readjusted to 6.5. The solution (total volume 60 ml) is kept in the dark at 30°, and the pH is kept constant by adjustment with 2 M LiOH. After 5 days most of the ATP is converted to 1-carboxymethyl-ATP as indicated by thin-layer chromatography. The product is precipitated by slowly adding 8 volumes of chilled ethanol (−20°) to the vigorously stirred reaction mixture; the precipitate is filtered off and washed with ethanol. To effect rearrangement to N^6-carboxymethyl-ATP, the precipitate is dissolved in 100 ml of water, the pH is adjusted to 8.5, and the solution is heated at 90° for 1.5 hr. The pH value is checked intermittently and adjusted when necessary with 1 M LiOH.

The solution containing the rearranged compound is cooled, adjusted to pH 2.75 with 1 M HCl, and applied to a Dowex 1-X2 column (200–400 mesh, Cl⁻, 4 × 30 cm). The column is washed with 0.3 M LiCl, pH 2.75, until the effluent contains no ultraviolet-absorbing material. A linear LiCl gradient is then applied, 0.3 M LiCl, pH 2.75, to 0.5 M LiCl, pH 2.0 (total volume, 4 liters). The pooled fractions comprising the main peak are neutralized with 1 M LiOH and concentrated on a rotary evaporator to a final volume of 80 ml. The yield is approximately 50% based on a molar absorption coefficient of $\epsilon_{267\,nm}^{pH\,7.5} = 17{,}300\ M^{-1}\ cm^{-1}$ determined for a similar analog, N^6-(6-aminohexyl)-AMP.[9]

N^6-[(6-Aminohexyl)carbamoylmethyl]-ATP

To produce the N^6-[(6-aminohexyl)carbamoylmethyl]-ATP, 80 ml of a 1 M 1,6-diaminohexane dihydrochloride solution are added to the above solution. 1-Ethyl-3-(3-dimethylaminopropyl)carbodiimide (1.5 g, 8 mmoles) dissolved in 5 ml of water is added dropwise, and the pH is maintained at 4.7 by adding 0.5 M HCl. After 1 hr at room temperature, thin-layer chromatography indicates practically total conversion of the N^6-carboxymethyl-ATP. The product is collected by precipitation with 8 volumes of chilled ethanol–acetone (1:1) and is subsequently applied at pH 5 to a Dowex 1-X2 column (200–400 mesh, Cl⁻, 4 × 30 cm). The column is washed with 100 ml of 0.05 M LiCl, pH 5, and then eluted with a linear LiCl gradient, 0.05 M LiCl, pH 5, to 0.35 M LiCl, pH 2 (total volume, 4 liters). The pooled fractions of the main peak are neutralized and concentrated on a rotary evaporator to a final volume of 50 ml. The product is precipitated with 10 volumes of cold acetone–ethanol mixture (1:1) and is made essentially salt-free by repeated

[9] H. Guilford, P. O. Larsson, and K. Mosbach, *Chem. Scripta* **2**, 165 (1972).

washings with the above mixture. Finally, N^6-[(6-aminohexyl)carbamoylmethyl]-ATP is dissolved in a small amount of water and precipitated by adding 8 volumes of chilled ethanol. The yield is 75% based on the amount of N^6-carboxymethyl-ATP.

N^6-[(6-Aminohexyl)carbamoylmethyl]-ADP

N^6-[(6-Aminohexyl)carbamoylmethyl]-ADP is prepared according to the above procedure using N^6-carboxymethyl-ADP as an intermediate. The overall yield is the same, i.e., about 40%.

N^6-[(6-Aminohexyl)carbamoylmethyl]-AMP

N^6-[(6-Aminohexyl)carbamoylmethyl]-AMP is prepared according to the above procedure using N^6-carboxymethyl-AMP as an intermediate in an overall yield of 45%.

Characterization

The method of synthesis and the properties of the intermediate and products all support the structures assigned to the N^6-carboxymethyl and N^6-[(6-aminohexyl)carbamoylmethyl] derivatives of ADP, ATP, NAD^+, and $NADP^+$. The 1-alkylated nucleotides thus have absorption maximum at 259 nm and a broad shoulder with inflection point at 290 nm. Rearrangement is accompanied by a characteristic shift from 260 nm of the absorption maximum to around 266 nm and disappearance of the shoulder at 290 nm. These spectral properties are in accord with the known properties of analogous hydroxyethyl-substituted[10] and benzyl-substituted[11] nucleotides.

Chemical and enzymic reduction of the NAD^+ and $NADP^+$ analogs gives the expected peak at 340 nm, and incubation with 1 M KCN gives a peak at 325 nm typical of quaternary nicotinamide-cyanide addition compounds (Fig. 2).

Treatment of the NAD^+-analogs with nucleotide pyrophosphatase yields, besides NMN, fragments identical with N^6-carboxymethyl-AMP or N^6-[(6-aminohexyl)carbamoylmethyl]-AMP, both prepared by a different method of synthesis (from 6-chloropurine riboside phosphate).[9]

The various nucleotides show R_f values in good agreement with those expected from their net charges and are summarized in Table I.

The ADP, ATP, and $NADP^+$ derivatives are readily hydrolyzed by

[10] H. G. Windmueller and N. O. Kaplan, *J. Biol. Chem.* **236**, 2716 (1961).
[11] J. W. Jones and R. K. Robins, *J. Am. Chem. Soc.* **85**, 193 (1963).

TABLE I
Thin-Layer Chromatography

	R_f values in chromatography systems[a]						
	Silica	Cellulose		Polyethyleneimine cellulose			
Compound	A	A	B	C	D	E	F
AMP	—	0.46	0.25	—	—	—	0.60
N^6-R″-AMP[b]	—	0.67	0.42	—	—	—	0.89
ADP	—	0.41	0.30	—	—	—	0.30
N^6-R″-ADP	—	0.61	0.47	—	—	—	0.78
ATP	—	0.35	0.37	—	—	—	0.13
N^6-R″-ATP	—	0.56	0.56	—	—	—	0.51
NAD^+	0.44	0.45	—	0.80	0.51	0.74	—
N^6-R′-NAD^+[c]	0.22	0.20	—	0.17	0.17	0.81	—
N^6-R″-NAD^+	0.44	0.55	—	>0.95	>0.95	>0.95	—
$NADP^+$	0.31	0.32	—	0.07	0.02	0.30	—
N^6-R′-$NADP^+$	0.14	0.16	—	0.01	0.01	0.20	—
N^6-R″-$NADP^+$	0.34	0.48	—	0.59	0.69	>0.95	—

[a] Solvent systems: A, isobutyric acid—1 M aqueous ammonia (5:3, v/v), solvent saturated with disodium EDTA; B, 0.1 M potassium phosphate, pH 6.8–ammonium sulfate–1-propanol (100:60:2, v/w/v); C, 3 M acetic acid; D, 0.1 M LiCl; E, 0.5 M LiCl; F, 1 M LiCl.
[b] R″ = (6-aminohexyl)carbamoylmethyl.
[c] R′ = carboxymethyl.

alkaline phosphatase, proving that no alkylation of the phosphate entities by iodoacetic acid has occurred and the carbodiimide-promoted attachment of the 1,6-diaminohexane spacer has not resulted in phosphoramidate formation. This is further evidenced by the fact that the N^6-[(6-aminohexyl)carbamoylmethyl] nucleotides contain just one aliphatic amino group (trinitrobenzene sulfonic acid test) per molecule. Furthermore, the properties of the analogs as active cofactors and as affinity chromatography ligands with several dehydrogenases and kinases are also in agreement with their assigned structures.

Binding to Matrices

For affinity chromatography purposes, cofactor and cofactor analogs have been almost exclusively attached to water-insoluble supports, such as Sepharose and porous glass, to allow use of convenient column techniques. Important requirements of the matrix are sufficient porosity to allow unimpeded passage of macromolecules and lack of nonspecific sites that might adversely interfere with the separation processes. These properties are also necessary for carriers of active cofactors, but the need for fast and efficient interaction between polymer-bound cofactor and enzyme becomes a dominant feature. In this respect the solid supports are far from ideal. The severely restricted mobility imposed on the bound cofactor will slow down its interaction with the enzyme. Experiments with Sepharose-bound N^6-[(6-aminohexyl)carbamoylmethyl]-NAD$^+$ as cofactor for yeast alcohol dehydrogenase and lactate dehydrogenase[12] clearly show that the efficiency of the bound cofactor is very low compared to corresponding free systems. Even when a very mild activation with CNBr of the polymer was undertaken to minimize cross-linking of the gel, only about 30% of the bound cofactor molecules could be enzymically reduced. The remaining 70% of the analogs are obviously bound to the matrix in such a way that proper interaction with the enzyme is prohibited owing to shielding effects imposed by the matrix backbone.

In contrast, with soluble carriers like dextran, about 80% of the cofactor was enzymically reducible indicating a less sterically hindered enzyme–coenzyme interaction (Fig. 2).

The nucleotides described above are provided either with a terminal carboxyl function or an amino function. The well-known coupling procedures involving, for instance, carbodiimides or cyanogen bromide (CNBr) are thus suitable for binding the analogs to soluble or insoluble polymers. The carboxymethyl derivatives may be coupled to carriers functionalized with extension arms bearing terminal amino groups using carbodiimide. This approach suffers from the inherent limitation that the resulting cofactor polymer will contain an excess of unsubstituted extension arms after coupling that might seriously impede proper coenzyme–enzyme interaction. This has been pointed out in affinity chromatography on several occasions, but in our experience it is also relevant when designing immobilized active cofactors. It was thus found that N^6-carboxymethyl-NAD$^+$ when coupled to 1,6-diaminohexane-substituted soluble dextran showed a very erratic behavior as coenzyme for yeast alcohol dehydrogenase. Several experiments, including tests with 1,6-diaminohexane-substituted dextran together with the free NAD$^+$ or NAD$^+$-analog, led to the conclusion that the enzyme gradually becomes quite tightly bound to the polymer and in such a manner that catalytic activity is almost entirely abolished. Furthermore, soluble polyethyleneimine, which also has an abundance of charged groups, showed a similarly disappointing behavior as cofactor carrier in our hands with yeast alcohol dehydrogenase. The liver enzyme, however, was insensitive to the excess of charged groups on the carrier, and it may well be that other enzymes and

[12] P. O. Larsson and K. Mosbach, *FEBS Lett.* **46**, 119 (1974).

supports will also yield useful preparations with the short N^6-carboxymethyl adenine nucleotides described. At present we suggest the use of a preformed coenzyme-spacer entity of the type N^6-[(6-aminohexyl)carbamoylmethyl]-NAD$^+$, and then attach it, for instance, to CNBr-activated polymers (concerning matrix-binding, see further below). Such preparations will not carry unsubstituted spacers.

Some precautions must be observed on coupling analogs to soluble dextran if preparations of high efficiency are desired. Activation with a high concentration of cyanogen bromide results in extensive cross-linking of the dextran, which might lead to insoluble derivatives, especially if the activated dextran is freed from excess cyanogen bromide by precipitation with organic solvents. In contrast, low CNBr concentrations cause little cross-linking, and, since all cyanogen bromide is rapidly consumed, it is not necessary to purify the activated dextran prior to cofactor addition. The coupling yield is in the range of 30–50%, and the resulting products have excellent cofactor properties and are enzymically reducible to the extent of about 80%. If higher coupling yields are attempted by increasing the degree of CNBr activation, a substantial portion of the cofactor cannot interact with the enzymes, probably owing to shielding effects from the cross-linked polymer.

One other problem associated with CNBr activation has recently become apparent. Activation of Sepharose with CNBr leads to attachment of the ligand through predominantly N-substituted isourea bonds. These linkages are not completely stable, and the conjugates exhibit a small but constant "leakage" from the solid matrix. This leakage is particularly acute at high temperatures, at extreme pH values, and if nucleophiles are present in the surrounding media (under normal pH and temperature conditions, however, and, e.g., when applied in phosphate buffer, "leakage problems" can be disregarded). The problem may be circumvented by using alternative methods of coupling of the ligand to the matrix (e.g., to epoxy-gels) and/or other solid support materials. Full details of these alternative procedures are given in this volume.

A few examples of coupling the nucleotides to the carriers dextran and Sepharose are given below with, for the above reasons, greater emphasis placed on soluble matrices. The procedures, however, are likely to be generally applicable to all the N^6-[(6-aminohexyl)carbamoylmethyl] nucleotides.

Preparation of Dextran-NAD$^+$ and -NADP$^+$

A more detailed procedure is given below for dextran-NAD$^+$ essentially following a published procedure.[12] A shorter description on dextran-

NADP$^+$ has also been published previously.[5] It is likely that the somewhat different procedure given for NAD$^+$ is also applicable for NADP$^+$.

Dextran-NAD$^+$. Dextran T 40 (Pharmacia, Uppsala, Sweden) (5.0 g) is dissolved in 50 ml of water (20°); a solution of 0.25 g of CNBr in 5 ml of water is added, and the pH is maintained at 10.8 by continuous addition of 1 M NaOH. After approximately 5 min all CNBr is consumed (no further consumption of NaOH and no cyanide smell), and the activation is judged complete. The pH is lowered to 8.5 with 0.1 M HCl, and 0.90 mmoles of N^6-[(6-aminohexyl)carbamoylmethyl]-NAD$^+$ dissolved in 5 ml of water is added. The coupling is allowed to proceed at pH 8.5 for 12 hr at room temperature, whence any residual active groups are quenched by treatment with 0.2 M ethanolamine-HCl buffer, pH 8.0, for 1 hr at room temperature. The reaction mixture is diluted to 500 ml with 0.1 M LiCl, adjusted to pH 6.8 with HCl, and applied to a Sephadex G-50 column (5 $\times$ 85 cm). The dilution lowers the viscosity enough to prevent anomalous gel filtration behavior. Elution is performed with 0.1 M LiCl to suppress any ion pair formation between polymer and unbound nucleotide and the effluent is collected as 20-ml fractions. Fractions 25–50 contain dextran-NAD$^+$, and fractions 65–100 uncoupled analog. The dextran derivative is concentrated on a rotary evaporator (30°) to approximately 100 ml and then pipetted into 1.5 liters of vigorously stirred ethanol. The precipitate is filtered off, washed with ethanol and ether, and dried in a vacuum. The yield is 5.1 g of a white powder.

Ultraviolet measurements (see also Fig. 2) indicate a nucleotide content of 65 μmoles per gram of dry dextran and a coupling yield of about 40% (assuming a molar absorption coefficient of 21,700 $M^{-1} \cdot cm^{-1}$). The uncoupled nucleotide may be recovered after concentration of fractions 65–100 to approximately 50 ml and precipitation with 500 ml of ethanol. A "blank" dextran is prepared by following the same procedure. The nucleotide in this case is replaced by, for example, 0.90 mmole of butylamine.

Dextran-NADP$^+$. Dextran-bound NADP$^+$ is prepared by dissolving Dextran T 40 (0.5 g) in 10 ml of distilled water and activating at pH 11.0 $\pm$ 0.2 by the stepwise addition of 25 mg of CNBr in 1 ml of water and continuous titration with 0.5 M NaOH at 25°. On completion of the reaction, 1 ml of the activated dextran solution is added to approximately 10 μmoles of N^6-[(6-aminohexyl)carbamoylmethyl]-NADP$^+$ in 50 μl of 0.1 M NaHCO$_3$, pH 8.5, and allowed to stand overnight at 4°. The solution is applied to a column of Sephadex G-50 (3 $\times$ 40 cm) equilibrated with water and eluted at a flow rate of 150 ml/hr. The fractions (2.0 ml) eluted immediately after the void volume and showing absorbance at 265 nm are free of uncoupled NADP$^+$ analog and are used in all subsequent studies. The resulting dextran contains up to 85 µmoles of NADP⁺ analog per gram of dry dextran when based on absorbance measurements at 265 nm.

Preparation of Sepharose-NAD⁺, -NADP⁺, and -ATP

Coupling of the NAD⁺ analog to Sepharose has been described previously[3,4] and is essentially identical to that described below in more detail for NADP⁺. Coupling of the ATP analog is also equivalent and has been described.[8]

Sepharose-NADP⁺. Sepharose 4B is activated by the CNBr method using 30 mg of CNBr per milliliter of gel. The activated gel is thoroughly washed with ice cold 0.1 M NaHCO$_3$, pH 8.5, and 1 g is added to 20 µmoles of N^6-[(6-aminohexyl)carbamoylmethyl]-NADP⁺ in 0.5 ml of the same buffer. Coupling is allowed to proceed for 20 hr at 0°–4°, whence the gel is exhaustively washed with 0.1 M NaHCO$_3$, pH 8.5, 2 M KCl, and distilled water. Under these conditions the amount of bound nucleotide is approximately 1.1 µmole per gram of wet gel or 27.5 µmoles per gram of dry gel according to UV measurements. The immobilized nucleotide is stable as a packed gel at 4° for several months.

Coenzymic Activities of the Various NAD⁺, NADP⁺, and ATP Analogs Described

Of obvious importance for the utilization of immobilized coenzymes is the fact that the "coenzymic activity" of the analogs compared to the unmodified coenzymes is retained or not too severely diminished. Needless to say, this same rigid requirement does not necessarily apply when these cofactors are used as binding ligands in affinity chromatography. Substitution with a spacer group at the exocyclic N^6 group of the adenine moiety has been chosen in all the analogs described here, since this position appeared to be the most suitable as judged from X-ray crystallographic studies on NAD⁺-dependent dehydrogenases and the good coenzymic activities reported from a similarly substituted hydroxyethyl NAD⁺ analog.[10]

In Tables II, III, and IV the coenzymic activities of the NAD⁺-, NADP⁺-, and ATP-analogs with several enzymes are given relative to their parent nucleotides under identical conditions. A comparison of the kind given, combining V_{max} and K_m, is of value, since it will provide information as to coenzymic activity under conditions rather similar to those prevailing in practical use. Alternatively, the k_{cat}, V_{max}, and K_m of the analogs could be given separately.

TABLE II
Coenzymic Activity of NAD+ Analogs

Enzyme	Rate of reduction of[a]		
	N^6-R'-NAD^{+b}	N^6-R''-NAD^{+c}	N^6-R''-NAD$^+$-carrier
Alcohol dehydrogenase (liver)	55	100	—
Lactate dehydrogenase (beef heart)	65	50	—
Malate dehydrogenase	75	75	—
Alcohol dehydrogenase (yeast)[d,e]	—	61	16 (Dextran T 10) 0.7 (Sepharose 4B)

[a] The enzymic reduction of NAD$^+$ and NAD$^+$ analogs was measured by following the increase in absorbance at 340 nm³. All assays were performed at 24° in 0.1 M Tris-HCl buffer, pH 8.5, in a total volume of 1.0 ml. Incubation mixtures contained 100 μmoles of substrate and 0.50 μmole coenzyme or analog. Sufficient enzyme was added to obtain an initial reduction rate of 5–10 % of the coenzyme per minute. Rates are in relation to NAD$^+$ = 100.
[b] R' = carboxymethyl.
[c] R'' = (6-aminohexyl)carbamoylmethyl.
[d] Cofactor concentration = 0.1 mM.
[e] From P. O. Larsson and K. Mosbach, *FEBS Lett.* **46**, 119 (1974).

Because of the great variations of coenzymic activity found with the different NADP$^+$ analogs, as shown in Table III, this aspect is discussed in somewhat more detail below.

The ability of the NADP$^+$ analogs to function as coenzymes was examined by comparing the rates of reduction of the analogs with the unmodified rates for NADP$^+$ in identical molar concentrations. Table IV shows the relative rates of reduction with several NADP$^+$-dependent enzymes and with a typical NAD$^+$-dependent enzyme. Most of the NADP$^+$-dependent enzymes are active with both N^1- and N^6-substituted carboxymethyl analogs of NADP$^+$, with some preference for the N^1-substituted derivatives. Altering the charge and bulk of the substituent on the adenine from -CH$_2$COO$^-$ to -CH$_2$CONH(CH$_2$)$_6$NH$_3^+$ significantly decreased the coenzymic activity of the analogs. These observations contrast with the coenzymic activities of correspondingly substituted derivatives of NAD$^+$, where substantial coenzymic activity was obtained only with the N^6-substituted derivatives and where altering the charge and bulk of the substituent had little effect on the rate of reduction. L-Glutamate dehydrogenase utilizes both NAD$^+$ and NADP$^+$ as coenzyme and exhibits activity with both N^1- and N^6-carboxymethyl derivatives of NADP$^+$ and, to a lesser extent, with the corresponding N^6-[(6-aminohexyl)carbamoylmethyl] analogs. Isocitrate dehydrogenase was totally

TABLE III
COENZYMIC ACTIVITY OF NADP$^+$ ANALOGS[a,b]

Enzymes	NAD$^+$	NADP$^+$	N^1-R'-NADP^{+c}	N^1-R''-NADP^{+d}	N^6-R'-NADP^{+c}	N^6-R''-NADP^{+d}	N^6-R''-NADP$^+$-dextran[d]
Glucose-6-phosphate dehydrogenase	0	100	145	>5	65	<5	35
6-Phosphogluconate dehydrogenase	0	100	60	>5	80	5	10
L-Glutamate dehydrogenase	130	100	50	10	35	10	15
threo-D$_s$-Isocitrate dehydrogenase	0	100	70	0	0	0	0
Yeast alcohol dehydrogenase	100	0	0	0	0	0	0

[a] From C. R. Lowe and K. Mosbach, *Eur. J. Biochem.* **49**, 511 (1974).
[b] The enzymic reduction of NADP$^+$, NAD$^+$, and the NADP$^+$ analogs was measured spectrophotometrically by following the increase in absorbance at 340 nm in 0.1 M Tris-HCl buffer, pH 7.5 (total volume 1 ml) at 25°. Standard assay procedures were used throughout except for beef liver glutamate dehydrogenase, which was assayed in the unfavorable direction using 10 µmoles of L-glutamate as substrate. Sufficient NADP$^+$ was added to give an absorbance of 1.00 at 260 nm, and enzyme was added to give a ΔA_{340}/min of about 0.1. The same molar concentration of NADP$^+$ analog was achieved by suitably adjusting the absorbance in the cuvette. All values are quoted relative to NADP$^+$ (100) and represent the mean of three determinations.
[c] R' = carboxymethyl.
[d] R'' = (6-aminohexyl)carbamoylmethyl.

TABLE IV
CoENZYMIC ACTIVITY OF ATP ANALOGS

Enzyme	Activity on			
	ATP	N^1-R'-ATP^a	N^6-R'-ATP^a	N^6-R''-ATP^b
Hexokinase^{c,d}	100	35	65	95
Glycerokinase^e	100	<5	—	20
Phosphoglycerate kinase^f	100	0	0	0

^{a} R' = carboxymethyl.
^{b} R'' = (6-aminohexyl)carbamoylmethyl.
^{c} Standard assay procedures scaled down to 1 ml were used throughout. In each case where ATP was used, its concentration was 10 K_m for the enzyme in question; the ATP analogs were tested at the same molar concentrations. All values are quoted relative to ATP (100) and represent the means of three determinations.
^{d} Hexokinase activity was determined in a coupled assay with glucose-6-phosphate dehydrogenase; the production of NADPH was measured at 340 nm.
^{e} Glycerokinase activity was determined in a coupled assay with glycerol-3-phosphate dehydrogenase-triosephosphate isomerase; the production of NADH was measured at 340 nm.
^{f} Phosphoglycerate kinase was assayed in a coupled system with glyceraldehyde-3-phosphate dehydrogenase; the formation of NAD$^+$ was measured at 340 nm.

inactive with N^6-substituted derivatives. The NAD$^+$-dependent enzyme, alcohol dehydrogenase, was inactive with NADP$^+$ and all the analogs tested. Table III also lists the rate of reduction of dextran-bound NADP$^+$ relative to free unmodified NADP$^+$ for several enzymes. The higher rate of reduction of the dextran-bound NADP$^+$ compared to the free N^6-[(6-aminohexyl)carbamoylmethyl] analog may reflect the suppression *per se* of the positive charge on the terminal amino group of the spacer arm assembly or, alternatively, the inhibition of its potential intramolecular interaction with the negative charge on the 2'-phosphate. The dextran-bound NADP$^+$ analog is inactive with isocitrate dehydrogenase and alcohol dehydrogenase.

Applications in Enzyme Technology and Analysis

The development of efficient enzyme-catalyzed processes is a prerequisite to the use of enzyme technology on a large scale. Many of these chemical processes will be catalyzed by enzymes that require the participation of readily dissociable coenzymes. Such coenzymes are often expensive, and thus their economical utilization would require methods both to retain them in the reaction mixture and to regenerate them. The immobilization of a coenzyme to a carrier macromolecule and its subsequent retention within a limiting membrane provide one solution to the problem.

Two examples where this approach has been applied are given below and utilize the dextran-bound NAD^+ analog, N^6-[(6-aminohexyl)carbamoylmethyl]-NAD^+.

Enzyme Electrode

Immobilized coenzymes appear to be well suited to applications in enzyme electrodes (see chapter [41] in this volume). Enzymes and coenzymes can be coretained within a limiting membrane around, for instance, an ion-selective electrode and used to determine in a continuous or serial fashion metabolite concentrations under conditions that do not perturb the media to be analyzed. This will permit unrestricted monitoring of metabolite concentrations in physiological fluids *in vivo*, where it would normally not be possible to add a free coenzyme, and in addition the possible reuse of the immobilized coenzyme will make such analyses cheaper.

The coentrapment of coenzymically active dextran-bound NAD^{+13} together with soluble lactate dehydrogenase/glutamate dehydrogenase within an enzyme electrode has been used to determine glutamate concentrations. The presence of glutamate in an assay medium containing pyruvate generated NH_4^+ which was recorded by a NH_4^+-sensitive electrode (Scheme 1).

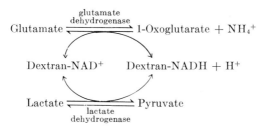

SCHEME 1. The generation of NH_4^+ by glutamate.

Alternatively the lactate dehydrogenase/glutamate dehydrogenase electrode can be used to determine pyruvate concentrations.

Example

Materials and Methods

The dialysis membranes were manufactured by the Union Carbide Corporation (Chicago, Illinois). They were treated before use by boiling for 1 hr in 1 mM EDTA (sodium salt) (pH 7.0) and 1 hr in distilled water. The dextran-bound NAD$^+$, N^6-[(6-(aminohexyl)carbamoylmethyl]-NAD$^+$, contained 30 μmoles of nucleotide per gram dry wt of dextran. The NH$_4^+$-sensitive glass electrode (Beckman 39137 cation-sensitive electrode) used in these assays responds to protons. Therefore all solutions used in determinations with this electrode were adjusted to pH 8.0 with Tris base at 25°. Since the electrode can also detect monovalent ions like Na$^+$ and K$^+$, all reagents used with the electrode were converted to their respective Tris salts. Sodium pyruvate was first converted to the free acid by passage through a column of Dowex 50 and then adjusted with Tris base to pH 8.0. ADP (sodium salt) was directly converted to the Tris salt by passage through Dowex 50 in the Tris form followed by adjustment to pH 8.0 as above.

The enzyme electrode was prepared by enclosing soluble glutamate dehydrogenase and rabbit muscle lactate dehydrogenase, and dextran-bound NAD$^+$, in a piece of dialysis membrane stretched around the bulb of the NH$_4^+$-sensitive glass electrode (see also chapter [41]). Prior to formation of the lactate dehydrogenase/glutamate dehydrogenase electrode the two enzymes, glutamate dehydrogenase (360 units), and lactate dehydrogenase (410 units), were combined and dialyzed against 50 mM Tris-HCl buffer (pH 8.0) containing 10 μM Tris-EDTA and 100 μM Tris-ADP to remove NH$_4^+$. The enzyme solution contained within the dialysis bag was then concentrated against sucrose to give a volume of 150–200 μl (compared to 350 μl before dialysis). Dextran-bound NAD$^+$ (25 mg dry weight) was dissolved in with the enzymes, and the resulting viscous solution was placed on a 5-cm square of dialysis membrane in contact with the tip of the electrode. After formation of the lactate dehydrogenase/glutamate dehydrogenase electrode about three-quarters of the enzyme solution remained entrapped by the membrane.

Immediately after its preparation and each time before use, the lactate dehydrogenase/glutamate dehydrogenase electrode was equilibrated in 50 mM Tris-HCl buffer (pH 8.0) made 10 μM in Tris-EDTA and 100 μM in Tris-ADP. When not in use, the enzyme electrode was kept at 4° in this same equilibration solution.

Glutamate and pyruvate concentrations were routinely determined in 50 ml of equilibration solution at 25°. Glutamate was determined in the presence of 2 mM pyruvate; pyruvate was measured in the presence of 10 mM glutamate. The millivoltage deflections caused by the serial addition of aliquots of 1 M glutamate or 0.2 M pyruvate to the stirred solutions were measured on a Radiometer ion meter (PHM 53) connected to a recorder. A standard fiber-junction saturated calomel electrode was used as a reference electrode.

As a control, the enzymes of the lactate dehydrogenase/glutamate dehydrogenase electrode were inactivated by soaking the electrode in 8 M urea at 25° for 40 min. The electrode was then rinsed in distilled water and equilibrated in 50 mM Tris-HCl buffer (pH 8.0) to remove urea before being retested.

Discussion

As seen in Fig. 3a the millivoltage deflection of the electrode was directly proportional to the logarithm of the glutamate concentration in the range 10^{-4} to 10^{-3} M. Similar results were obtained in the determination of pyruvate (Fig. 3b). This enzyme electrode has potential application in the assay of 1-oxoglutarate and L-lactate by registering the uptake of NH_4^+ by the reverse of the above reactions (Scheme 1). Furthermore, other recycling systems comprising dextran-bound NAD^+ or $NADP^+$ could be envisaged for use as potential enzyme electrodes. Thus a system comprising coentrapped dextran-bound $NADP^+$, glucose-6-phosphate dehydrogenase, and glutamate dehydrogenase[5] might find application in the assay of physiological levels of glucose 6-phosphate.

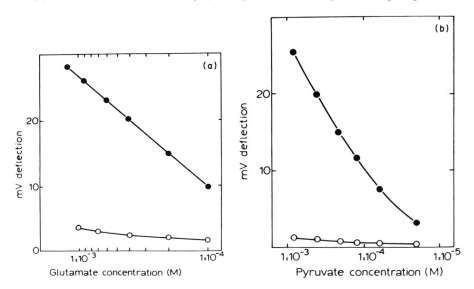

FIG. 3. Response curves for the enzyme/coenzyme electrode. (a) Estimation of glutamate using the lactate dehydrogenase/glutamate dehydrogenase electrode. (b) Estimation of pyruvate with the lactate dehydrogenase/glutamate dehydrogenase electrode.

A representative curve (●) made by plotting millivoltage deflection against (glutamate) and (pyruvate), respectively, is compared to a curve (○) obtained after denaturation of the lactate dehydrogenase/glutamate dehydrogenase electrode in 8 M urea. Reproduced with permission from P. Davies and K. Mosbach, *Biochim. Biophys. Acta* **370**, 329 (1974).

The response time of the lactate dehydrogenase/glutamate dehydrogenase electrode to additions of glutamate or pyruvate was approximately 3–4 min. The response of the electrode decreased by approximately 60% on storage at 4° over a period of 15 days although it was possible to recalibrate the electrode and determine metabolite concentrations as before. This instability of the enzyme electrode may reflect in part the inactivation of the enzymes themselves and in part their binding to the immobilized nucleotide. Clearly, refinements in the proportions of the components of the electrode and in stabilizing the enzymes could improve both the response time and the longevity of the system.

The use of immobilized coenzymes in enzyme electrodes considerably extends the range of potential substrates that can be assayed and, with careful optimization of the parameters discussed above, efficient and stable enzyme electrodes could be developed.

"Enzyme Reactor"

The availability of enzymically regenerable immobilized coenzymes permits an extension in the scope of enzyme reactors (see in particular chapter [49] in this volume) from simple hydrolytic reactions to those such as dehydrogenases which utilize stoichiometric amounts of coenzymes in conversion of their substrates. Thus, addition of immobilized coenzymes obviates the economic burden of recovery of the coenzyme or of purifying the product.

Two model enzyme reactions[13] are described below to generate alanine from pyruvate by the action of alanine dehydrogenase. The two systems (Scheme 2a, b) differ in the enzymic means of generating dextran-bound NADH, with one utilizing galactose dehydrogenase (Scheme 2a) and the other lactate dehydrogenase (Scheme 2b).

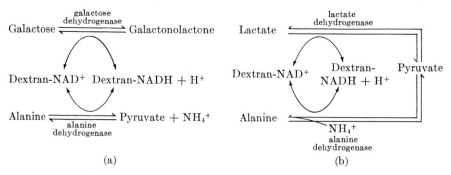

Scheme 2. Model reactor for production of alanine.

[13] P. Davies and K. Mosbach, *Biochim. Biophys. Acta* **370**, 329 (1974).

Example

Materials and Method

The reaction mixture used in the enzyme reactor containing β-galactose dehydrogenase (2 units, Boehringer Corp., Tutzing, W. Germany) and L-alanine dehydrogenase (3 units) consisted of 50 mM sodium pyrophosphate buffer made 10 mM in galactose, 2 mM in sodium pyruvate, 200 mM in NH_4Cl, and 0.1 mM in EDTA. The mixture was adjusted to pH 8.8 with NaOH. The amount of dextran-bound NAD^+ analog was 10 mg dry weight.

The reaction mixture used in the other enzyme reactor containing beef-heart lactate dehydrogenase (85 units) and alanine dehydrogenase (24 units) consisted of 100 mM Tris base made 100 mM in L-lactic acid, 200 mM in NH_4Cl and 0.01 mM in EDTA. The pH of this mixture was adjusted to 8.5. The amount of dextran-bound NAD^+ analog was 12 mg dry weight.

Pyruvate and alanine were determined spectrophotometrically using methods described by Lowry and Passonneau.[14] Rabbit muscle lactate dehydrogenase was used for the assay of pyruvate. Alanine was measured in a coupled assay by using glutamate-pyruvate transaminase in the presence of α-ketoglutarate to convert alanine to pyruvate, which in turn was assayed by using rabbit muscle lactate dehydrogenase. Alanine was determined in the presence of pyruvate by correcting for the contribution made to the reaction by the latter substance.

Enzyme reactor experiments were carried out in a Model 8 MC ultrafiltration apparatus (Amicon Corp., Lexington, Massachusetts) fitted with a PM 10 ultrafiltration membrane. The enzyme pairs together with dextran-bound NAD^+ were present in the small chamber of the apparatus in 2 ml of reaction mixture which contained the buffered substrates. Up to 100 ml more of the same reaction mixture were placed in the reservoir chamber. The ultrafiltration was done at 40–50 psi (which gives a flow rate of about 9 ml/hr) with the apparatus setting on "push liquid" such that the volume in the small chamber remained constant at 2 ml throughout the experiment. The filtrate was collected in 5-ml fractions for analysis.

Discussion

The galactose dehydrogenase/alanine dehydrogenase reactor converted 30% of the added pyruvate (1.5 mM) to alanine at a constant

[14] O. H. Lowry and J. V. Passonneau, "A Flexible System of Enzymatic Analysis." Academic Press, New York, 1972.

rate over a period of 6.5 hr. In contrast, the conversion of pyruvate to alanine in the lactate dehydrogenase/alanine dehydrogenase reactor was initially 60–70% but fell to about 20% after 5.5 hr had lapsed. The immobilized NAD^+ was recycled at rates of 14 times per hour and 33 times per hour, respectively, within these reactors, assuming that all the NAD^+ was sterically available to the enzymes.

The two enzyme reactors described here represent model systems only. Clearly, with suitable adjustment of enzyme and dextran-bound NAD^+ concentrations and flow rate, a conversion rate to product approaching 100% could be realized.

Applications in Enzymology and Protein Studies

To illustrate the wider potential of immobilized cofactors two examples of their application to fundamental studies will be given. In one case unmodified ATP was immobilized to Sepharose[15] using the cyanogen bromide method. The resulting preparation, of which the exact mode of binding of ATP remains to be established, was used to demonstrate the presence of an obligatory phosphoryl enzyme intermediate in the reaction catalyzed by succinyl-CoA synthetase. (The authors of this chapter recommend strongly the use of coenzyme analogs with functional groups suitable for binding like the ATP-analog described in this chapter.)

In another study, ATP bound to Sepharose was used to investigate the reaction between heavy meromyosin and ATP.[16] It was prepared by treating Sepharose with adipic or sebacic acid dihydrazide[16,17] and subsequently binding periodate-oxidized ATP. Both Mg^{2+} and Ca^{2+} activate the cleavage of bound ATP, some 40–50% of the total ATP being cleaved after 3.5 hr at 25°. Similarly, myosin and one-headed myosin[17] were bound to immobilized ATP, but only under conditions where, in the presence of Ca^{2+} or Mg^{2+}, ATP was cleaved. Myosin could thus be resolved from modified myosin by virtue of differences in ATPase activity. In other words, binding is not a reversible process, but intimately associated with the occurrence of a chemical process. Similar studies using N^6-ADP-agarose derivatives have been reported.[18]

Other Immobilized Coenzymes

Apart from the adenine nucleotide analogs described here, other derivatives have been reported where an intact demonstrably coenzy-

[15] E. A. Wider de Xifra, S. Mendiara, and A. M. del C. Batlle, *FEBS Lett.* **27**, 275 (1972).
[16] R. Lamed, Y. Levin, and A. Oplatka, *Biochim. Biophys. Acta* **305**, 163 (1973).
[17] R. Lamed and A. Oplatka, *Biochemistry* **13**, 3137 (1974).
[18] I. P. Trayer, H. R. Trayer, D. A. P. Small, and R. C. Bottomley, *Biochem. J.* **139**, 609 (1974).

mically active entity has been immobilized. One of the first reports describes the preparation of ϵ-aminohexanoyl-NAD-Sepharose obtained by carbodiimide coupling of NAD^+ to Sepharose substituted with ϵ-aminocaproic acid.[19] The coupling of NAD^+ to glass was published[20] independently. Both preparations, however, showed low coenzymic activity and since the latter preparation showed poor stability, another derivative, N^6-aminoethyl, was synthesized by a similar alkylation step to that described in this chapter using aziridine.[21] In its immobilized dextran form, this preparation showed 2–10% of the maximum turnover number (k_{cat}) of unmodified NAD^+. Whereas to our knowledge no applications of this NAD^+ derivative have been reported, another NAD^+ derivative, N^6-succinyl-NAD^+, coupled to a polyethyleneimine carrier has been applied in a two-enzyme recycling system[22] comprising alcohol dehydrogenase and lactate dehydrogenase. Furthermore, this immobilized NAD^+ derivative was shown to be reducible by alcohol dehydrogenase bound to aminoethyl cellulose. Preliminary studies with this analog bound to aminodextran and present in a hollow fiber have also been reported.[23] However, the somewhat dubious stability of this acyl-linked derivative of NAD^+ above pH 7 seems to limit its general applicability. In contrast, the more stable alkyl linkage of the dextran-bound NAD^+ derivative described in detail here seems to make it more suitable for use in prospective enzyme reactors.

Other recently synthesized NAD^+ analogs which have been successfully applied in affinity chromatography should also be coenzymically active. These are a nicotinamide-6-mercaptopurine dinucleotide[24] as well as N^6-(6-aminohexyl)-NAD^{+6}. All the above derivatives except for one[20] are either substituted at the N^6-position or at "S.6" Substitution at position 8 in the adenine moiety, as in the 8-(6-aminohexyl)-amino derivatives of cAMP,[9] AMP,[25,26] and in particular NAD^{+26} and $NADP$,[+27] pro-

[19] P. O. Larsson and K. Mosbach, *Biotechnol. Bioeng.* **13**, 393 (1971).
[20] M. K. Weibel, H. H. Weetall, and H. J. Bright, *Biochem. Biophys. Res. Commun.* **44**, 347 (1971).
[21] M. K. Weibel, C. W. Fuller, J. M. Stadel, A. F. E. P. Buckmann, T. Doyle, and H. J. Bright, *in* "Enzyme Engineering" (E. K. Pye and L. B. Wingard, eds.), Vol. 2, p. 203. Plenum, New York, 1974.
[22] J. R. Wykes, P. Dunnill, and M. D. Lilly, *Biochim. Biophys. Acta* **286**, 260 (1972).
[23] R. P. Chambers, J. R. Ford, J. H. Allender, W. H. Baricos, and W. Cohen, *in* "Enzyme Engineering" (E. K. Pye and L. B. Wingard, eds.), Vol. 2, p. 195. Plenum, New York, 1974.
[24] S. Barry and P. O'Carra, *FEBS Lett.* **37**, 134 (1973).
[25] B. Jergil, H. Guilford, and K. Mosbach, *Biochem. J.* **139**, 441 (1974).
[26] C.-Y. Lee, D. A. Lappi, B. Wermuth, J. Everse, and N. O. Kaplan, *Arch. Biochem. Biophys.* **163**, 561 (1974).
[27] C.-Y. Lee and N. O. Kaplan, *Arch. Biochem. Biophys.* **168**, 665 (1975).

vide alternative choices as do other related 8-substituted NAD⁺ derivatives.[28]

The immobilization of several coenzymes and coenzyme analogs for affinity chromatography has been described in detail in this series[1] and elsewhere,[29,30] and in the following only those examples where an intact, demonstrably coenzymically active entity other than NAD⁺, NADP⁺, or ATP has been immobilized will be considered here.

Adenosine-3',5'-cyclic monophosphate (cAMP) has been attached to Sepharose and shown to be effective for the biospecific adsorption of the cAMP-stimulated protein kinases.[25,31] The cAMP seems to act by promoting dissociation of the enzyme into a regulatory, catalytically inactive, cAMP-binding subunit and an enzymically active catalytic subunit.[32,33] Resolution of the catalytic and regulatory subunits can be accomplished on casein- or histone-Sepharose columns[34,35] or on cAMP columns.[25,31] Thus, protein kinase from trout testis (protamine kinase)[34,35] could be eluted from 8-(6-aminohexyl)amino-cAMP-Sepharose by a pulse of the inhibitor AMP, but was no longer activated by cAMP.[25]

Flavin mononucleotide (FMN) has been coupled to hexanoyl-Sepharose with dicyclohexylcarbodiimide in aqueous pyridine[36] by a procedure essentially analogous to that described for the preparation of hexanoyl-NAD⁺-Sepharose.[19] The resulting immobilized FMN binds bacterial luciferase and when reduced, either catalytically with H_2 or with dithionite, is effective as a substrate in the luminescence reaction. The maximum rate of oxidation of immobilized $FMNH_2$ was only approximately 2% of that of free $FMNH_2$, although, surprisingly, the turnover time for the reaction initiated by the immobilized $FMNH_2$ was identical to the soluble flavin, using both decanal and dodecanal as aldehyde cosubstrates. Substitutions on the flavin molecule and the chain length of the aldehyde generally significantly influence the turnover time. The low rate of enzymic oxidation of $FMNH_2$-Sepharose by bacterial

---

[28] P. Zappeli, A. Rossodivita, G. Prosperi, R. Pappa, and L. Re, *Eur. J. Biochem.* **62**, 211 (1976).

[29] C. R. Lowe and P. D. G. Dean, "Affinity Chromatography," pp. 92–129. Wiley, New York, 1974.

[30] K. Mosbach, *Biochem. Soc. Trans.* **2**, 1294 (1974).

[31] M. Wilchek, Y. Salomon, M. Lowe, and Z. Sellinger, *Biochem. Biophys. Res. Commun.* **45**, 1177 (1971).

[32] G. N. Gill and L. D. Garren, *Biochem. Biophys. Res. Commun.* **39**, 335 (1970).

[33] M. Tao, M. L. Salas, and F. Lipmann, *Proc. Natl. Acad. Sci. U.S.A.* **67**, 408 (1970).

[34] J. D. Corbin, C. O. Brostrom, R. L. Alexander, and E. G. Krebs, *J. Biol. Chem.* **247**, 3736 (1972).

[35] J. D. Corbin, C. O. Brostrom, C. A. King, and E. G. Krebs, *J. Biol. Chem.* **247**, 7790 (1972).

[36] C. A. Waters, J. R. Murphy, and J. W. Hastings, *Biochem. Biophys. Res. Commun.* **57**, 1152 (1974).

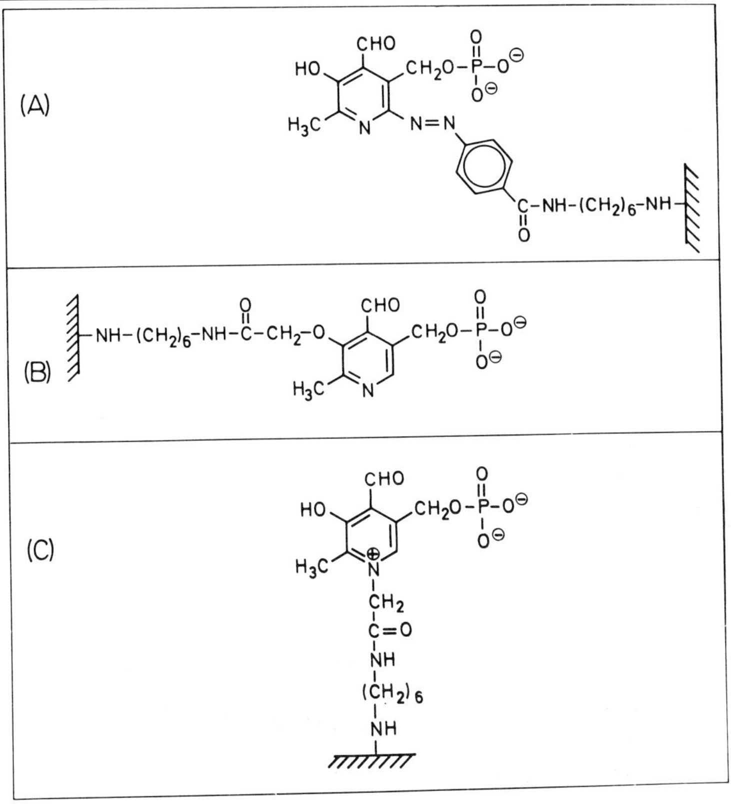

FIG. 4. Three immobilized pyridoxal 5'-phosphate derivatives. (A) 6-Immobilized diazo derivative; (B) 3-O-immobilized; and (C) N-immobilized.

luciferase may possibly be ascribed to the steric restraints imposed by immobilization and to the derivatization of FMN with a hexanoyl group.

Pyridoxal 5'-phosphate has been coupled to diazotized p-aminobenzamidohexyl-Sepharose[37] and catalyzes the cleavage of tryptophan in the presence of Cu^{2+}, as does free pyridoxal 5'-phosphate. This system can be utilized for the immobilization of the enzyme since the immobilized pyridoxal 5'-phosphate retained functional groups for binding to the apoproteins of B_6-dependent enzymes. The resulting pyridoxal 5'-phosphate-Sepharose-apotryptophanase complex was reduced with sodium borohydride to generate a derivative which retained approximately 60% of the catalytic activity of the free tryptophanase used.

This approach was later extended[38] with the introduction of two other Sepharose-bound analogs (Fig. 4). Both N-immobilized and 3-O-

[37] S.-I. Ikeda and S. Fukui, *Biophys. Res. Commun.* **52**, 482 (1973).
[38] S.-I. Ikeda, H. Hara, and S. Fukui, *Biochim. Biophys. Acta* **372**, 400 (1974).

immobilized pyridoxal 5′-phosphate were synthesized by reaction of pyridoxal 5′-phosphate with a bromoacetyl derivative of Sepharose in 50% (v/v) dimethylformamide and in potassium phosphate buffer, pH 6.0, respectively, for approximately 70 hr at room temperature. The catalytic activities of these derivatives were tested in the nonenzymic cleavage of tryptophan. The N-immobilized pyridoxal 5′-phosphate analog displayed catalytic activity, but the 3-O-immobilized derivative did not exhibit appreciable activity. This behavior, however, is not reflected in the ability of the two derivatives to bind apotryptophanase. Alkylation of the pyridine nitrogen of pyridoxal 5′-phosphate leads to a decreased affinity for the apoprotein, but attachment to Sepharose via the 3-hydroxyl group does not impede binding since this analog maintains all the main functional groups necessary for binding to the apoenzyme. Both derivatives bind apotryptophanase in such a way as to retain 50–60% of the specific activity of the starting material. Tyrosine phenol-lyase (β-tyrosinase) from *Escherichia intermedia* has been similarly immobilized.[39] This method of enzyme immobilization was found to be superior to other methods commonly used for preparation of immobilized enzymes.

General Discussion

In summarizing, the following can be said. Two conditions for the practical application of enzyme systems requiring expensive cofactors have to be met, i.e., their retention and regeneration. The regeneration may be accomplished either enzymically as described here, chemically with the participation of artificial electron acceptors or donors,[23] or electrochemically[40] (see also chapter [58] this volume). Depending on the type of application, either one of these may be the best choice. Enzymic regeneration appears to be the procedure of choice in the application of enzyme–coenzyme systems in the potential treatment of enzyme deficiency diseases whereby the catalyst system may be used entrapped in microcapsules or polymer beads and which are either placed in extracorporal shunts or implanted *in vivo*.

With regard to retention, it appears advantageous to have the cofactors immobilized to macromolecular supports although attempts have been made to use unmodified NAD^+ in hollow fibers as well.[23] However,

---

[39] S. Fukui, S.-I. Ikeda, M. Fujimura, H. Yamada, and H. Kumagai, *Eur. J. Biochem.* **51**, 155 (1975).

[40] M. Aizawa, R. W. Coughlin, and M. Charles, *Biochim. Biophys. Acta* **385**, 362 (1975).

the requirement for tight membranes to keep the NAD⁺ entrapped leads to overall poor permeability for substrate/product. Another possibility should be mentioned: for some applications, such as medical and some analytical procedures, which do not require a separation of the catalyst system, including the coenzyme, from the solution in which they are applied, enzymic recycling of the unmodified coenzyme may be the procedure of choice provided the number of cycles is high enough.

Finally an additional possible solution to the problem of coenzyme retention may be found with the following approach. In a previous study glycogen phosphorylase b has been immobilized in its allosterically activated form.[41] This had been accomplished by coimmobilization of the enzyme and its positive allosteric effector, AMP, using the AMP analog N^6-(6-aminohexyl)-AMP, to CNBr-activated Sepharose. In extending these studies an alcohol dehydrogenase–NAD(H)–Sepharose complex was prepared showing no requirement of soluble coenzyme for its activity, at the same time with the coenzyme susceptible to recycling.[42] This was accomplished in a similar fashion by coupling a preformed enzyme–coenzyme binary complex to an activated matrix.

[41] K. Mosbach and S. Gestrelius, *FEBS Lett.* **42**, 200 (1974).
[42] S. Gestrelius, M. O. Månsson, and K. Mosbach, *Eur. J. Biochem.* **57**, 529 (1975).

[62] Covalent Immobilization of Adenylate Kinase and Acetate Kinase in a Polyacrylamide Gel: Enzymes for ATP Regeneration[1]

By GEORGE M. WHITESIDES, ANDRE LAMOTTE,
ORN ADALSTEINSSON, and CLARK. K. COLTON

Adenylate kinase (AMP:ATP phosphotransferase, EC 2.7.4.3) and acetate kinase (ATP:acetate phosphotransferase, EC 2.7.2.1) form the basis for a procedure for the regeneration of ATP from AMP and/or ADP, using the readily available acetyl phosphate (AcP)[2] as the ultimate phosphorylating agent.[3] Both adenylate kinase and acetate kinase

[1] Supported by the National Science Foundation, (RANN) Grant GI 34284.
[2] G. M. Whitesides, M. Siegel, and P. Garrett, *J. Org. Chem.* **40**, 2516 (1975).
[3] C. R. Gardner, C. K. Colton, R. S. Langer, B. K. Hamilton, M. C. Archer, and G. M. Whitesides, *in* "Enzyme Engineering" (E. K. Pye and L. B. Wingard, Jr., eds.), Vol. 2, p. 209. Plenum, New York, 1974; G. M. Whitesides, A. Chmurny, P. Garrett, A. Lamotte, and C. K. Colton, *ibid.*, p. 217.

contain structurally important, reactive, cysteine residues close to their

$$CH_3COCH_3 \xrightarrow{\Delta} CH_2=C=O \xrightarrow[(2)~NH_3]{(1)~H_3PO_4} \underset{AcP}{CH_3\overset{O}{\overset{\|}{C}}O\overset{O}{\overset{\|}{P}}(O^-NH_4^+)_2}$$

$$AMP + ATP \xrightarrow{\text{adenylate kinase}} 2ADP$$

$$2ADP + 2AcP \xrightarrow{\text{acetate kinase}} 2ATP + 2Ac$$

active sites,[4] and their successful immobilization depends primarily on the use of a procedure that permits these residues to emerge from the immobilization unmodified. A solution containing, *inter alia*, enzyme,

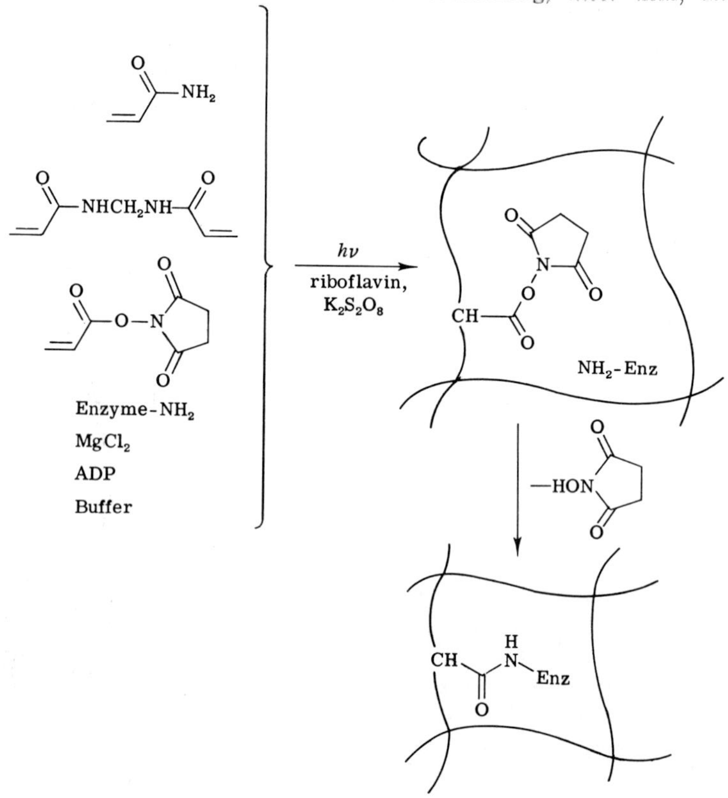

Scheme 1

[4] L. Noda, in "The Enzymes" (P. Boyer, ed.), 3rd ed., Vol. 8, p. 279. Academic Press, New York, 1973; O. H. Callaghan, *Biochem. J.* **67**, 651 (1957); G. E. Schulz, M. Elzinga, F. Marx, and R. H. Schirmer, *Nature (London)* **250**, 120 (1974); I. A. Rose, M. Grunberg-Manago, S. T. Korey, and S. Ochoa, *J. Biol. Chem.* **211**, 734 (1954).

acrylamide monomer, cross-linking agent, N-acryloxysuccinimide, and a photochemical free-radical initiation system is irradiated. Copolymerization of the vinyl monomers results in initial physical entrapment of the enzyme in a polyacrylamide gel containing active ester functionalities. A subsequent, slower step results in covalent coupling of the enzyme to the polymer backbone by reaction of nucleophilic groups on the enzyme (particularly lysine γ-amino groups) with these active esters (Scheme 1).

Immobilization Procedures

Enzymes. Adenylate kinase (porcine muscle) is commonly purchased as a suspension in 3.2 M ammonium sulfate (Sigma). Its specific activity after treatment with dithiothreitol (DTT) is 2050 IU. Activation is carried out by centrifuging 1 ml of the commercial suspension (5 mg of adenylate kinase per milliliter) for 20 min at 27,000 g. The supernatant is discarded and the precipitate is resuspended in degassed HEPES buffer (50 mM, pH 7.50) with final volume adjusted to 1 ml. This suspension is added to 9 ml of degassed HEPES buffer (50 mM) containing 20 mM DTT (pH 7.80); after mixing, the pH drops to 7.50. The enzymic activity is monitored during activation at 25°: It normally increases to a stable plateau in 2 hr. This solution is dialyzed using a hollow-fiber dialysis unit under argon at 4° against two 250-ml charges (1 hr each) of degassed HEPES buffer (0.05 M, pH 7.5) to remove ammonium sulfate. The dialyzed adenylate kinase is transferred under argon to a storage tube and kept at 4°. Typically this solution contains about 2400 U of adenylate kinase per milliliter. The enzyme is stored at 4°C.

Acetate kinase (*Escherichia coli*) is also ordinarily obtained as a suspension in 3.2 M ammonium sulfate (Sigma). Its specific activity after treatment with DTT is 330 IU. Acetate kinase is activated using a procedure analogous to that described for adenylate kinase except that MOPS buffer (0.05 M, pH 6.2) is used throughout and that the mixture of acetate kinase and DTT requires approximately 4 hr at 25° to reach constant activity. Typically the solution resulting from activation with DTT and dialysis contains about 170 U of acetate kinase per milliliter.

Reagents Solutions for Immobilization of Adenylate Kinase. Seven stock solutions (**S**) are required.

S1: HEPES buffer [4-(2-hydroxyethyl)-1-piperazineethanesulfonic acid] 0.2 M, pH 7.0, containing acrylamide (0.475 g/ml) and N,N'-methylenebisacrylamide (0.025 g/ml)

S2: Water, containing riboflavin (2 mg/ml as a fine suspension)

S3: Water, pH 7.6, containing potassium persulfate (50 mg/ml)
S4: N-Acryloyloxysuccinimide (5 M in dimethyl sulfoxide, 845 mg/ml) (preparation follows)
S5: HEPES buffer, 0.05 M, pH 7.5, containing dithiothreitol (10 mM), MgCl$_2$ (30 mM), and ADP (10 mM)
S6: HEPES buffer, 0.05 M, pH 7.5, containing dithiothreitol (10 mM) and ammonium sulfate (0.5 M)
S7: Adenylate kinase (ca 2400 U/ml) in HEPES buffer, 0.05 M, pH 7.5

Immobilization of Adenylate Kinase. **S5** (1.4 ml) and **S2** (100 µl) are placed in a 5-ml beaker containing a small magnetic stirring bar; **S1** (500 µl) and **S3** (50 µl) are transferred into two separate 15-ml centrifuge tubes; **S4** is stored in a 1-ml test tube.[5] Containers are capped with serum stoppers and swept with a stream of argon for 20 min to remove molecular oxygen. **S1, S2,** and **S5** are stored at room temperature; **S3** and **S4** are stored at 0°. **S4** is warmed briefly on a steam bath to room temperature to dissolve precipitated N-acryloyloxysuccinimide, and 10 µl is added to the beaker containing the vigorously stirred mixture of **S2** and **S5**. A portion of the N-acryloyloxysuccinimide initially precipitates as a fine powder on addition to the aqueous solution, but redissolves in about 30 sec.[6] When the ester has dissolved, the solution is cooled to 0° by immersion in an ice-salt bath. **S1** and **S3** are rapidly transferred to the

---

[5] Transfers of less than 10 µl are usually accomplished with a syringe; larger volumes are transferred by forced siphon through a stainless steel cannula under argon. The degassed solutions must be carefully protected from contamination by atmospheric oxygen in order for polymerization behavior, gel times, and gel properties to be reproducible. Even small quantities of oxygen introduced into these solutions by accident can result in unacceptably long gel times and poor gel physical characteristics. Techniques useful in anaerobic transfers using cannulas and serum stoppers are described in useful detail in H. C. Brown, G. W. Kramer, A. B. Levy, and M. M. Midland, *in* "Organic Syntheses via Boranes." Wiley, New York, 1975.

[6] All steps after addition of **S4** to the aqueous solution must be carried out rapidly and in a reproducible way, to minimize destruction of the active ester by hydrolysis. The pseudo first-order rate constant for hydrolysis of N-acryloyloxysuccinimide at pH 7 at 24.5° is 0.002 min^{-1} (half-life for hydrolysis (24.5°) $\simeq$ 220 min); at 0° it can be estimated to be approximately 0.00013 min^{-1} (half-life for hydrolysis (0°) $\simeq$ 5500 min). These hydrolyses are accelerated to a small extent by amine-containing buffers: for a discussion of the related hydrolysis of p-nitrophenyl acetate, see H. J. Goren and M. Fridkin, *Eur. J. Biochem.* **41,** 263 (1974). N-Acryloyloxysuccinimide reacts rapidly with primary and secondary amines. Reagents containing these functional groups, and species containing ammonium ion, will compete with enzymes for N-hydroxysuccinimide active ester moieties, and should be avoided if possible.

beaker. As soon as the mixture in the beaker has reached 0° (about 1 min) polymerization is initiated by irradiation. In this work, irradiaton was accomplished using a high-intensity ultraviolet polymerization lamp (Polysciences Catalog No. 0222) delivering 8.4 mW/cm^2 (measured at a distance of 18 inches) in the active region for initiation (360 nm). An aliquot of the solution of adenylate kinase (22 µl, 54 U) is added to the polymerizing mixture from a syringe 5 sec before the gel point (in our experiments, 32 ± 4 sec).[7] The total time the mixture is irradiated is 60 sec.

The beaker, containing a yellow gel (ca. 2 ml), is removed from the ice bath and permitted to stand at room temperature for 10 ml to allow the enzyme to couple with the polymer. The gel is then transferred to a mortar, which has been precooled to −15°, and is broken up rapidly by grinding with a pestle. Two minutes of vigorous grinding give irregular particles having an average particle size of approximately 20–30 µm. These gel particles are immediately washed into a 50-ml centrifuge tube using a total of about 10 ml of **S6**. The tube is capped and stirred for 15 min at room temperature. The gel particles are separated by centrifugation and resuspended in a second 10-ml aliquot of **S6**. The washing procedure is repeated twice, or until no enzymic activity is detected in the washes. Assay of the gel indicates the presence of 27 U of adenylate kinase in the gel (50%). With experience, higher yields can be obtained.

Reagent Solutions for Immobilization of Acetate Kinase

S′1: MOPS buffer (4-morpholinepropanesulfonic acid), 0.2 M, pH 6.2, containing acrylamide (0.475 g/ml) and N,N'-methylenebisacrylamide (0.025 g/ml)

S′2: Water, containing riboflavin (4 mg/ml as a suspension)

S′3: Water, pH 7.6, containing potassium persulfate (50 mg/ml)

S′4: N-Acryloyloxysuccinimide (5 M in dimethyl sulfoxide) (preparation follows)

S′5: MOPS buffer, 0.05 M, pH 6.2, containing MgCl$_2$ (30 mM), ADP (5 mM), acetyl phosphate (5 mM)

S′6. HEPS buffer, pH 7.5, 0.05 M, containing dithiothreitol (10 mM), and ammonium sulfate (0.5 M)

S′7: Acetate kinase (ca. 170 U/ml) in MOPS buffer, 0.05 M, pH 6.2, containing DTT (2 mM)

---

[7] The gel point is defined as the point at which the polymerization has proceeded to the stage at which the stirring bar stops turning. With care it is reproducible to ±10% (i.e., ca. ±3–4 sec).

Immobilization of Acetate Kinase. The sequence of steps is analogous to that described for the immobilization of adenylate kinase. Each solution is degassed by sweeping for 20 min with a stream of argon, and stored under argon. **S'5** (1.4 ml), **S'2** (50 µl), and **S'1** (500 µl) are added to a capped 5-ml beaker. **S'4** (10 µl) is added, and, as soon as the N-acryloyloxysuccinimide has dissolved, the solution is rapidly transferred to an ice-salt bath and cooled to 0°. **S'3** (50 µl) is added by syringe, and the resulting solution is stirred for 2 min at 0°. Polymerization is initiated by irradiation, and the enzyme-containing solution (**S'7**, 30 µl, 5.40 U) is injected into the solution 5 sec before the gel point. Irradiation is continued for 25 sec. The beaker containing the resulting gel (about 2 ml) is removed from the ice bath and allowed to stand at room temperature for 10 min. The gel is broken up by grinding in a mortar precooled to −15°, transferred to a centrifuge tube with about 10 ml of **S'6**, and washed with 10-ml aliquots of **S'6**. Assay of the gel typically indicates activity of 1.14 U (21%).

Assay Procedures

Adenylate kinase is assayed in homogeneous solution by coupling production of ATP from ADP to the production of NADPH, by first phosphorylating glucose with this ATP using hexokinase, and then oxidizing the resulting glucose 6-phosphate using glucose-6-phosphate dehydrogenase (G-6-PDH) and $NADP^+$. The following stock solutions are required.

Solution I: 0.2 M Tris-HCl buffer, pH 7.5, containing 5 mM glucose, 30 mM $MgCl_2$, hexokinase (2500 U/liter), and G-6-PDH (1250 U/liter). The buffer, glucose, and $MgCl_2$ are mixed, the pH is adjusted to 7.5, and the enzymes are added. The resulting solution is stable at 0° for several months.

Solution II: ADP (disodium salt), 0.5 M in water, pH 6.8. This solution is stable at <5° for several weeks.

Solution III: 62.5 mM $NADP^+$ (sodium salt) in water, no pH adjustment. This solution is also usable after storage at <5° for several weeks.

In a typical assay, 1 ml of solution I is mixed with 20 µl of solution II and 10 µl of solution III. The mixture is equilibrated for 3 min at 25° to destroy ATP present as an impurity in the ADP. An aliquot of the solution to be assayed is then added; the size of this aliquot is adjusted so that the final solution contains less than 0.01 U per milliliter of adenylate kinase. The solution is mixed and poured into a 1-cm quartz cuvette, and the rate of appearance of NADPH is followed spectrophotometrically at 340 nm (30°).

This assay is a compromise between accuracy and economy. The Michaelis constant for binding of ADP to porcine adenylate kinase is $K_{M,ADP}$ = 1.58 mM.[4,8] In these solutions, the concentration of ADP is 10 mM; this concentration gives rates of approximately 0.9 V_{max}. Values closer to V_{max} would be obtainable at higher ADP concentrations, but at greater expense. Enzyme concentrations for adenylate kinase, hexokinase, and G-6-PDH are chosen so that the adenylate kinase-catalyzed reaction is overall rate-limiting. Experimentally, the minimum ratio of hexokinase to adenylate kinase activities for which adenylate kinase activity is rate-determining is 100; in this assay, hexokinase/adenylate kinase = 250, to provide for losses on storage. The optimum ratio of activities for hexokinase and G-6-PDH is 2. For a sample aliquot such that the final activity of adenylate kinase in the assay solution is less than 0.01 U/ml, less than 1% of the total ADP is converted to ATP per minute, and the change in absorbance is linear with time. The assay is reproducible to ±5%, and its accuracy is approximately −10%.

Immobilized adenylate kinase is assayed using the same procedure, with care taken that the enzyme-containing polyacrylamide particles are small (<25 µm average). The washed, enzyme-containing gel particles, suspended in a known volume of HEPES buffer from their preparation, are stirred vigorously. An aliquot of this suspension (10 µl, about 0.005 U) is withdrawn using an Eppendorf pipette, and used for the assay. If the gel particles are small and the stirring vigorous, this aliquot is representative of the total suspension. This aliquot of suspended gel is added to the assay solution, mixed, and poured into the spectrophotometer cuvette, as described for the homogeneous assay. The size and density of the particles are such that they do not settle appreciably during the time of the assay. Diffraction does not interfere with the assay both because the size and concentration of the particles are small, and because they are transparent.

Acetate kinase is assayed using a modification of the adenylate kinase assay. A fourth stock solution is prepared (Solution IV: acetyl phosphate in water, 0.5 M, without pH adjustment, usable for several weeks after storage at 0°).[2] Solutions I, II, and III are mixed and incubated as described above, and 10 µl of Solution IV is added. An aliquot of the acetate kinase-containing solution is added, such that the final activity in the assay solution is less than 0.01 U/ml. This procedure gives

---

[8] P. DeWeer and A. G. Lowe, *J. Biol. Chem.* **248**, 2829 (1973) and references cited therein.

[AcP] $\simeq$ 5 mM in the assay solution, in adequate excess over $K_{M,AcP}$ = 0.34 mM.[9]

N-Acryloyloxysuccinimide.[10] N-Hydroxysuccinimide (11.5 g; 0.10 mole) and sodium bicarbonate (16.8 g; 0.20 mole) are dissolved in 200 ml of water, and the solution is stirred vigorously in an ice bath at 0°. Acryloyl chloride (9.1 g; 0.10 mole) is added in drops over a 10-min interval, and subsequent stirring for 3 min at 0° leaves no odor of acryloyl chloride. Cooling the flask for 5 min in a $-5°$ ice-salt bath completes the precipitation of the acrylate ester. The resulting white precipitate is filtered with suction, washed with 50 ml of ice-cold water, and dried at reduced pressure (ca. 30 min). The product is dissolved in 40 ml of ethyl acetate, dried with a small amount of MgSO$_4$, and filtered into a clean, dry flask. The remaining precipitate is extracted with two 20-ml portions of ethyl acetate and filtered into the flask to give a clear, colorless solution. Hot hexane (150 ml) is added, and the solution is warmed gently until it is clear. Slow cooling to 0° gives a white precipitate, which is filtered and washed with a small amount of hexane, yielding 8.2 g of thin plates, mp 60°–62°. Concentration of the mother liquor affords an additional 3.0 g of product (66% combined yield). A further recrystallization from ethyl acetate:hexane (20:80) gave 8.7 g (52%) of large, thin, colorless plates, m.p. 69–70°, having NMR (CDCl$_3$)δ 2.87 (s, 4H), 6.0–7.0 (mult., 3H).

Discussion

Polyacrylamide has many advantages as a matrix for enzyme immobilization (see also chapters [12]–[15]). It is inexpensive and resistant to biological attack. Its low mechanical strength can be increased, and other physical properties modified, by varying the extent of cross-linking,[11] or by incorporating other, more hydrophobic, vinyl monomers. It can be formed into beads by suspension polymerization[12] (see also chapter [5], [6] and [14] or rendered susceptible to recovery by magnetic

---

[9] C. A. Janson and W. W. Cleland, *J. Biol. Chem.* **249**, 2567 (1974).

[10] This preparation was developed by Mr. Michael Wilson. The properties of the N-acryloyloxysuccinimide prepared in this procedure are similar to those reported by R. L. Schnaar and C. Y. Lee, *Biochemistry* 1975, 1535 (1975).

[11] D. Rodbard and A. Crambach, *Anal. Biochem.* **40**, 95 (1971); A. Crambach and D. Rodbard, *Science* **172**, 440 (1971).

[12] K. E. J. Barrett, *in* "Dispersion Polymerization in Organic Media." Wiley, New York, 1975.

filtration by including colloidal magnetite in the gel.[3,13] The procedure illustrated here is designed to circumvent disadvantages sometimes encountered in other polyacrylamide gel immobilization procedures. Incorporation of active ester groups into the polymer permits covalent coupling of enzyme to the gel and prevents the slow leakage of enzyme from gel characteristic of noncovalent physical entrapment. Since the gel is formed with enzyme initially uniformly distributed throughout, the low immobilization yields sometimes encountered on reaction of a preformed, reactive gel with enzyme can, in some instances, be avoided.

Although this procedure is versatile, it requires careful attention to certain details to be successful. The two major concerns are protection of enzyme from deactivation during polymerization, and choice of the optimum concentration of active ester functionalities in the gel. Studies of the stability of adenylate kinase and acetate kinase in polymerizing polyacrylamide have established that four reactions are responsible for most of the deactivation observed[13]: first, Michael addition (1,4-conjugate addition) of enzyme -SH and -NH$_2$ groups to acrylamide; second, autooxidation of -SH groups by molecular oxygen, in a reaction strongly catalyzed by transition metal ions; third, oxidation of the enzyme by singlet oxygen ($^1O_2$), generated by energy transfer from excited riboflavin to $^3O_2$ during irradiation; fourth, attack on enzyme by SO_4^- and other free radicals. These reactions are controlled by excluding oxygen, by carrying out reactions at 0° and relatively low pH, by minimizing contact of the enzyme with acrylamide and N-acryloyloxysuccinimide, and by protecting the enzyme active site with saturating concentrations of substrate. The only one of these considerations whose experimental implementation is not obvious is the third: it is an important initial part of the procedure for developing immobilization conditions for a new enzyme, or for reproducing an established procedure, to establish a reproducible polymerization procedure and to establish a gel point ($\pm 10\%$) in the *absence* of enzyme. The enzyme can then be immobilized, using this procedure, by injecting enzyme into the polymerizing mixture just before (about 5 sec) this previously established gel point. This procedure effectively eliminates deactivation of the enzyme by reaction with the acrylic monomers. Experimentally, the gel point is not significantly influenced by the presence of the enzyme in solution.

[13] A. Lamotte, O. Adalsteinsson, R. F. Baddour, C. K. Colton, and G. M. Whitesides, unpublished results; A. Lamotte, Ph.D. Thesis, Massachusetts Institute of Technology, Cambridge, Massachusetts, 1975.

The presence of DTT or other mercaptans in the polymerization mixture helps to protect the enzyme against attack by radicals, but increases the gel times and requires longer irradiation. Certain enzymes are immobilized in higher yield in the presence of DTT, certain others in its absence.

For a new enzyme, it is also necessary to determine an optimum N-acryloyloxysuccinimide concentration for a given set of conditions. Concentrations either higher or lower than an optimal concentration lead to reduced immobilization yields for adenylate kinase and acetate kinase (Fig. 1). Certain enzymes (e.g., horseradish peroxidase) are, however, relatively insensitive to high concentrations of active ester, and plots

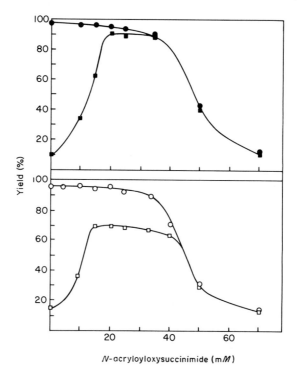

FIG. 1. Immobilized (■) and total recovered yields (●) for adenylate kinase as a function of the starting concentration of N-acryloyloxysuccinimide. The latter yield represents the sum of the enzymic activity immobilized in the gel and that free in the wash solutions. Reaction conditions are those described in the experimental procedure. Analogous data for acetate kinase (□, ○) are also summarized. The maximum yields obtained in these experiments are higher than those described in the experimental procedure, and represent optimized data collected by an experienced individual. The lower yields described in the procedure are those expected with only routine experience in this technique.

analogous to Fig. 1 for these enzymes show either a plateau or a broad maximum instead of the relatively sharp maximum observed for adenylate kinase and acetate kinase.

This procedure is the most efficient one presently available for regeneration of ATP from ADP or AMP.[14] The specific activities of the acetate kinase and adenylate kinase preparations used here (about 300 and 2000 IU, respectively) are such that, with the immobilization yields reported, a reactor utilizing the immobilized enzyme preparations obtained starting from 30 mg of acetate kinase and 2 mg of adenylate kinase should in principle be capable of regenerating about 2 moles of ATP (1 kg) from AMP and acetyl phosphate per day. The actual productivity of such a reactor depends on the success with which its design permits the enzymes to operate close to V_{max}.

[14] An alternative procedure for transformation of ADP to ATP depends on carbamyl phosphate ($H_2NCO_2PO_3^{2-}$) as phosphate donor, and carbamylphosphokinase as catalyst. Although carbamyl phosphate is readily prepared *in situ*, the equilibrium constant for phosphate transfer to ADP is less favorable than that for acetyl phosphate. See D. L. Marshall, *Biotechnol. Bioeng.* **15**, 447 (1973).

Section XI

Miscellaneous

[63] Artificial Enzyme Membranes

By D. THOMAS and G. BROUN

Within the living cell most enzymes are attached to membrane structures or contained in cell organelles. When the enzymes are isolated, they are removed from these structures and are usually studied in homogeneous solution. Under such conditions they may be far from their natural state and are quite often highly unstable. Consequently, the attachment of purified enzymes to synthetic matrices offers an array of exciting possibilities in both fundamental and applied studies.

Various procedures reviewed in this volume can be used to bind enzymes to insoluble matrices without substantially impairing their activities. Most of the work reviewed describes the preparation and properties of *particles* containing immobilized enzyme. In recent years, however, there has been greater interest in *membranes* containing immobilized enzymes, and in this chapter we shall restrict our considerations to the latter. The importance of enzyme membranes will be discussed in relation to two major fields.

1. Kinetic modelization of immobilized enzymes. Fundamental studies of heterogeneous-phase enzyme kinetics have shown that diffusion and accessibility constraints modify enzyme activity. The membrane configuration is particularly convenient for an analysis of such kinetics and is in many ways more useful than particle configuration. It favors a one-dimensional analysis, which facilitates theoretical considerations. At the same time, experimental determinations of parameters such as diffusion coefficients and Thiele moduli are much more readily performed on membranes than on particles.

2. New properties due to the membrane shape. The membrane configuration is especially appropriate for modeling biological membranes. Synthetic enzyme membranes are well adapted for an analysis of the influence of local and global structure on enzyme activity, and they permit an examination of the symmetry properties of the system. Such membranes can be utilized in principle for the preparation of simple models of facilitated or active transport. These systems can be readily characterized experimentally and analyzed theoretically. Comparisons of experimental observation and mathematical simulation are easily made and usually prove to be enlightening.

Methods of Preparation of Enzyme Membranes

Homogeneous Enzyme Distribution

Entrapment in Gel Films

The earliest study of this method was published by Bernfeld and Wan.[1] These authors described methods for the entrapment of the following proteins in polyacrylamide gels: ribonuclease, trypsin, α-chymotrypsin, papain, α-amylase, β-amylase, and aldolase. The enzymes retained about 1% of their original activity.

It is difficult to consider these gels as true membranes. However, most of the subsequent studies employing enzyme entrapment were directed toward using the material as an analytical tool. For such applications the gels were prepared in the form of layers, if not actual membranes, by Updike and Hicks,[2] and Guilbault and Montalvo.[3] Van der Ploeg and Van Duijn[4] used peroxidase entrapped within a polyacrylamide film as a model system, and they concluded that such models could be used to study histochemical reactions quantitatively.

The inclusion of enzymes in a gel such as polyacrylamide or starch is carried out under relatively mild reaction conditions without significant alteration of the protein. Disadvantages of this process, however, are the relative inaccessibility of the enzymes to substrates and products and the possible loss of enzyme from loosely cross-linked gels. In addition, there is no chemical stabilization of the enzymes, and the mechanical properties of this type of film are very poor (for further reading on gel entrapment see Chapters 12–15, 41, and 43).

Entrapment in Collagen Membrane

Karube and Suzuki[5] developed an electrochemical method for preparing a urease–collagen membrane. The membrane is formed by electrolysis on the surface of a cathode, the urease content of the membrane being almost equal to its content in the electrolyte. The relative activity of the membrane was found to be 51% of the native urease. The preparation requires only a few minutes, and the mechanical properties of the film are as good as those of cellophane (see also Chapter 19).

[1] P. Bernfeld and J. Wan, *Science* **142**, 678 (1963).
[2] S. J. Updike and G. P. Hicks, *Nature (London)* **214**, 896 (1967).
[3] G. G. Guilbault and J. Montalvo, *J. Am. Chem. Soc.* **91**, 2164 (1969).
[4] M. Van der Ploeg and P. Van Duijn, *J. R. Microsc. Soc.* **83**, 415 (1964).
[5] I. Karube and S. Suzuki, *Biochem. Biophys. Res. Commun.* **47**, 51 (1972).

Entrapment in Liquid Membrane

May and Li[6] used hydrocarbon-based liquid-surfactant membranes to immobilize urease, and the enzyme was shown to retain its catalytic activity. Furthermore, no significant leakage of enzyme into the external aqueous solution was detected.

Co-cross-linking Enzymes with Inactive Protein

This method has been developed extensively, and its use in the immobilization of several enzymes has been described by Broun et al.[7-9] and Thomas et al.[10-12] (see also Chapter 20). A bifunctional agent, such as glutaraldehyde, and a bulk protein, such as plasma protein, are mixed with one or several enzymes. Upon reacting, total insolubilization of the protein material is achieved and it is possible to prepare water-insoluble active enzymes with a variety of different physical properties, particularly in the form of membranes. All enzyme molecules introduced into the medium are immobilized, the chemical yield being 100%. This procedure yields a homogeneous distribution of active sites within the membrane. The films obtained are transparent and show good mechanical properties. They are characterized by diffusion coefficients of the same magnitude as those of cellophane films. The protein concentration in these films, after drying, is 95–97% by weight.

Variation in the enzymic activity is rather easily achieved by starting with different amounts of enzyme. Enzymes with very different catalytic activities, molecular weights, or isoelectric points may be immobilized with high activity yields. Fragile enzymes, such as lactate dehydrogenase or alcohol dehydrogenase, also yield good activity ratios after immobilization, providing NAD is used as a protector. Multienzyme systems, such as hexokinase–phosphatase or β-glucosidase–glucose oxidase–catalase, can be prepared.

EXAMPLE: 400 units of catalase are added to 5 ml of a 60-mg/ml bovine serum albumin solution in 0.02 M phosphate buffer at pH 6.8. Glutaraldehyde (2.5% solution) is added up to a final concentration of

---

[6] S. W. May and N. N. Li, *Biochem. Biophys. Res. Commun.* **47**, 1179 (1972).

[7] G. Broun, E. Sélegny, and D. Thomas, *Protides Biol. Fluids, Proc. Colloq.* **18**, 425 (1970).

[8] G. Broun, C. Tran Minh, D. Thomas, D. Domurado, and E. Sélegny, *Trans. Am. Soc. Artif. Intern. Organs* **17**, 341 (1971).

[9] G. Broun, D. Thomas, G. Gellf, D. Domurado, A. M. Berjonneau, and C. Guillon, *Biotechnol. Bioeng.* **15**, 359 (1973).

[10] D. Thomas, G. Broun, and E. Sélegny, *Biochimie* **54**, 229 (1972).

[11] D. Thomas, C. Tran Minh, G. Gellf, D. Domurado, B. Paillot, R. Jacobsen, and G. Broun, *Biotechnol. Bioeng. Symp.* **3**, 299 (1972).

[12] D. Thomas and G. Broun, *Biochimie* **55**, 975 (1973).

0.7%. When the viscosity of the mixture begins to increase, the solution is spread on a perfectly plane glass plate, inside the limits of a square drawn with a glass pencil. The presence of air bubbles should be avoided. A yellowish flat membrane is obtained after a couple of hours at 20° (or overnight in a refrigerator), which can be removed easily from the glass plate by soaking the plate in water for 1 or 2 hr, in a 0.1 M lysine solution, and then in distilled water with gentle stirring. The membrane is checked to exclude enzyme leaking (by an enzymic test in the rinse water) and assayed for activity by measuring the decrease in absorption of hydrogen peroxide at 250 nm. The membrane can be stored for months in a petri dish in a refrigerator. Desiccation is to be avoided.

Heterogeneous Enzyme Distribution

Cross-linking inside Collodion and Cellophane Membrane

Enzymically active membranes containing papain and alkaline phosphatase were prepared by Goldman et al.[13-16] by impregnating preformed porous collodion films with enzyme, followed by treatment with the cross-linking agent bisdiazobenzidine-2,2-disulfonic acid. The papain membrane retains much of its enzymic activity when activated with sulfhydryl compounds. There is no evidence of desorption from the membrane, either in the presence of substrate or after 4 months of storage under water at 4°. Enzyme membranes can be prepared with two-layer or three-layer structures. In the latter an inner layer of inactive membrane is sandwiched between two layers containing enzyme. The papain membrane is permeable to sucrose and to substrates such as benzoyl-L-arginine ethyl ester (BAEE), as is the unmodified collodion membrane from which it is derived. At pH 6 it possesses about 5% of the activity of an equivalent amount of crystalline enzyme with respect to BAEE, and about 40% with respect to benzoyl-L-arginine amide (BAA). The behavior of this membrane, which is characterized by a shift in the pH optimum of the enzyme, will be discussed in detail later.

Broun et al.[17] and Thomas et al.[10] prepared cellophane glucose oxidase membranes using glutaraldehyde as the cross-linking agent. Cellophane-supported preparations retain 12% of the specific activity of the enzyme.

---

[13] R. Goldman, H. I. Silman, S. R. Caplan, O. Kedem, and E. Katchalski, *Science* **150**, 758 (1965).

[14] R. Goldman, O. Kedem, H. I. Silman, S. R. Caplan, and E. Katchalski, *Biochemistry* **7**, 486 (1968).

[15] R. Goldman, O. Kedem, and E. Katchalski, *Biochemistry* **7**, 4518 (1968).

[16] R. Goldman, O. Kedem, and E. Katchalski, *Biochemistry* **10**, 165 (1971).

[17] G. Broun, S. Avrameas, E. Sélegny, and D. Thomas, *Biochim. Biophys. Acta* **185**, 260 (1969).

There is no change in the optimum pH. The thermal dependence of the enzyme membrane at its optimum pH is greater than that of glucose oxidase in solution, although the insoluble preparation is thermally more stable than the solution. Both forms exhibit the same Michaelis constant, and their substrate specifities are identical. These authors reported applications of the same technique to trypsin, chymotrypsin, urease, carbonic anhydrase, and urate oxidase.

Cross-linking within a preformed membrane has many applications, but activity yields higher than 15% are obtained only with hydrolases, such as papain, trypsin, or chymotrypsin. It is difficult to determine the distribution and concentration of enzyme within the membrane, and consequently it is impossible to obtain a well-defined variation of enzymic activity (per unit volume of membrane), nor has it proved possible to immobilize enzymes of high molecular weights, such as urease, by this method. The binding of glucose oxidase into a cellophane matrix is given as an example for producing a glucose oxidase-bearing membrane.

EXAMPLE: Cellophane sheets are impregnated with 6 mg of protein solution per milliliter containing 78 units of glucose oxidase per milligram in a 0.02 M phosphate buffer, pH 6.8. Three milliliters of this solution cover 100 cm^2 of a 30 μm-thick cellophane sheet. The sheet is desiccated at 4°. More active preparations can be obtained with two or three further impregnations. After desiccation, the sheets are impregnated with 2.5% glutaraldehyde solution in the same phosphate buffer. Cross-linking proceeds at 4° until complete desiccation is accomplished. Membranes are rinsed in a stream of distilled water until no further absorbance is read at 280 nm, which is a speak for enzyme and glutaraldehyde. The absence of enzymic activity in the rinse water is checked. For glucose oxidase, a mixture of glucose, phenazine methosulfate, and 3(4,5-dimethyl thiazolyl-2)-2,5-diphenyl tetrazolium bromide (MTT), which turns purple in the presence of an enzymic reaction, is added to the control (300 mg of glucose is mixed with 20 mg of MTT and 3 mg of phenazine methosulfate for 20 ml of rinse water). The test must be negative at the end of rinsing (as opposed to the membrane, which rapidly becomes purple when the same test is used). Rinsing is continued for 3–4 hr. Glutaraldehyde functions trapped in the insoluble structure without participating in cross-linking are blocked by an aqueous solution of 10 mg of glycine per milliliter. These membranes can be kept at 4° for several months without notable loss of activity.

Khaiat and Miller[18] have described the adsorption of ribonuclease at the air–water interface and on phospholipid monolayers. At the air–

---

[18] A. Khaiat and I. R. Miller, *Biochim Biophys. Acta* **183**, 309 (1969).

water interface the ribonuclease is adsorbed in two distinctly different (and not necessarily monomolecular) layers; on a condensed phospholipid monolayer, only one layer of native molecules is adsorbed. The adsorption is followed using a tritiated acetic anhydride-labeled protein, but no information is given concerning the enzyme activity after adsorption.

Goldman and Lenhoff[19] described the adsorption of glucose-6-phosphate dehydrogenase in a collodion membrane. The method is especially convenient for oxidoreductases with mobile NAD cofactors because of their high sensitivity to chemical reagents. Vieth et al.[20] described the preparation of a collagen–urease complex membrane by "interdiffusional penetration" (i.e., impregnation). The binding mechanism is related to protein–protein interaction.

Direct Covalent Binding on a Preformed Membrane

Covalent linkage of enzymes to solid matrices is not reversed by changes in pH or ionic strength or by the presence of substrates, products, or cofactors. Enzymes attached by covalent binding are thereby stabilized, and their susceptibility to thermal denaturation is low. However, covalent binding does alter the chemistry of the enzymes and tends to decrease their activity. In extreme cases all activity may be lost. Although most studies on enzyme immobilization have utilized covalent binding, for some reason this is not true of studies on enzyme membranes.

Kay et al.[21] described the preparation of porous sheets bearing proteolytic covalently bound enzymes. Wilson et al.[22] described the preparation of membranes containing lactate dehydrogenase (LDH) covalently attached to anion-exchange cellulose sheets by the use of a dichloro-*sym*-triazinyl dyestuff, Procion brilliant orange M.G.S. (see also Chapter 4). The immobilized LDH is more stable to high temperatures than is native LDH, but the activity yield is very low. Wilson et al.[23] also described the covalent binding of pyruvate kinase to filter paper and measured the properties of the resulting enzymically active porous sheets. In this case also, the yield after binding seems to be very low. A two-enzyme reactor was constructed by combining these membranes with the LDH membranes.

[19] R. Goldman and H. M. Lenhoff, *Biochim. Biophys. Acta* **242**, 514 (1971).
[20] W. R. Vieth, S. S. Wang, and S. G. Gilbert, *Biotechnol. Bioeng. Symp.* **3**, 285 (1972).
[21] G. Kay, M. D. Lilly, A. K. Sharp, and R. J. H. Wilson, *Nature (London)* **217**, 641 (1968).
[22] R. J. H. Wilson, G. Kay, and M. D. Lilly, *Biochem. J.* **108** (1968).
[23] R. J. H. Wilson, G. Kay, and M. D. Lilly, *Anal. Chem.* **109**, 137 (1968).

In order to determine the regulatory effect exerted by the membrane on the oxidation of glutamate by glutamate dehydrogenase (GDH) in the natural membrane-integrated enzymic system, the binding of GDH to a solid collagen matrix was studied by Julliard et al.[24] GDH is coupled to thin films of acid-soluble collagen (to simulate natural membranes) by the azide method (Micheel and Ewers[25]) in the presence of various protectors. The study has yielded a great deal of information on polymeric enzymes covalently bound to solid supports. The fact that the regulatory properties of GDH are preserved after coupling to the collagen suggests that the enzyme continues to possess an oligomeric structure within the matrix (see also Chapter 34). Ultrathin films of collagen have also been used to support covalently bound aspartate aminotransferase, in an effort to mimic biological membranes (Coulet et al.[26]).

Summary

A classified summary of methods used for the preparation of artificial membranes, with literature references,[1-11,13-24,26-33a] is given in the table.

Kinetic Modelization of Immobilized Enzymes: Symmetrical Systems (See also Chapter 25)

Methods for Studying Artificial Enzyme Membranes

Diffusion Cells

In practically all the cases discussed in this chapter, the membranes were studied in classic diffusion cells,[14] in which the membrane separates two compartments containing solutions of substrate, product, and effectors. The solutions are generally stirred and thermostated. Both the fluxes entering and those leaving the membrane can be readily deter-

---

[24] J. H. Julliard, C. Godinot, and D. C. Gautheron, *FEBS Lett.* **14**, 185 (1971).
[25] F. Micheel and J. Ewers, *Makromol. Chem.* **3**, 200 (1949).
[26] P. Coulet, J. H. Julliard, and D. C. Gautheron, *Biotechnol. Bioeng.* **16**, 1005 (1975).
[27] G. G. Guilbault and G. J. Lubrano, *Clin. Chem. Acta* **60**, 254 (1972).
[28] G. G. Guilbault and E. Hrabankova, *Anal. Chem.* **42**, 1779 (1970).
[29] G. G. Guilbault and E. Hrabankova, *Anal. Chim. Acta* **56**, 285 (1971).
[30] J. N. Barbotin, and D. Thomas, *J. Histochem. Cytochem.* **22**, 1408 (1974).
[31] G. Broun, D. Thomas, and E. Sélegny, *J. Membr. Biol.* **8**, 313 (1972).
[32] A. M. Berjonneau, D. Thomas, and G. Broun, *Pathol. Biol.* **22**, 497 (1974).
[33] G. Broun, E. Sélegny, C. Tran Minh, and D. Thomas, *FEBS Lett.* **7**, 223 (1970).
[33a] M. D. Legoy and D. Thomas, *Biochem. J.*, in press.

908   MISCELLANEOUS   [63]

SYNTHETIC ENZYME MEMBRANES

EC number	Enzymes	Methods	References[a]
Oxidoreductases			
1.1.1.1	Alcohol dehydrogenase	Co-cross-linking	Legoy and Thomas[33a]
1.1.1.27	Lactate dehydrogenase	Direct covalent binding (cellulose sheets)	Wilson et al.[22]
		Co-cross-linking	Broun et al.[9]
1.1.1.49	Glucose-6-phosphate dehydrogenase	Adsorption on collodion	Goldman and Lenhoff[19]
1.1.3.4	Glucose oxidase	Entrapment in gel	Updike and Hicks,[2] Guilbault and Lubrano,[27]
		Cross-linking inside a support (cellophane)	Broun et al.[17]
		Co-cross-linking	Thomas et al.[10]
1.2.3.2	Xanthine oxidase	Co-cross-linking	Thomas et al.[11]
1.4.1.3	Glutamate dehydrogenase	Direct covalent binding (collagen film)	Julliard et al.[24]
1.4.3.2	L-Amino acid oxidase	Entrapment in gel	Guilbault and Hrabankova[28]
		Co-cross-linking	Thomas et al.[10]
1.4.3.3	D-Amino acid oxidase	Entrapment in gel	Guilbault and Hrabankova[29]
1.7.3.3	Urate oxidase	Cross-linking inside a support (cellophane)	Broun et al.[17]
1.11.1.6	Catalase	Co-cross-linking	Broun et al.[8]
1.11.1.7	Peroxidase	Entrapment in gel	Van der Ploeg and Van Duijn[4]
		Co-cross-linking	Barbotin and Thomas[30]
Transferases			
2.6.1.1	Aspartate aminotransferase	Direct covalent binding (collagen film)	Coulet et al.[26]
2.7.1.1	Hexokinase	Co-cross-linking	Broun et al.,[31] Thomas et al.[11]
2.7.1.40	Pyruvate kinase	Direct covalent binding (sheets)	Wilson et al.[23]
Hydrolases			
3.1.3.1	Alkaline phosphatase	Absorption in a collodion membrane	Goldman et al.[16]
3.1.4.23	Ribonuclease II	Adsorption on phospholipid monolayer	Khaiat and Miller[18]
		Entrapment in gel	Bernfeld and Wan[1]

EC #	Enzyme	Method	Reference
3.2.1.1	α-Amylase	Entrapment in gel; cross-linking inside a support (silk)	Bernfeld and Wan[1]
3.2.1.23	β-Galactosidase	Co-cross-linking	Broun et al.[9]
3.4.21.1	Chymotrypsin	Direct covalent binding	Kay et al.[21]
		Cross-linking inside cellophane	Thomas et al.[11]
3.4.21.4	Trypsin	Direct covalent binding	Kay et al.[21]
		Entrapment in gel	Bernfeld and Wan[1]
		Cross-linking inside a support cellophane	Broun et al.[17]
		Cross-linking inside a support (collodion)	Goldman et al.[13–15]
3.4.22.2	Papain	Co-cross-linking	Broun et al.[7]
3.5.1.1	Asparaginase	Entrapment in gel	Guilbault and Montalvo[3]
3.5.1.5	Urease	Adsorption in collagen film	Vieth et al.[20]
		Electrochemical inclusion in collagen film	Karube and Suzuki[5]
		Inclusion in liquid membrane	May and Li[6]
		Co-cross-linking	Thomas et al.[10]
Lyases			
4.1.1.25	Tyrosine decarboxylase	Co-cross-linking	Berjonneau et al.[32]
4.1.1.53	Phenylalanine decarboxylase	Co-cross-linking	Berjonneau et al.[32]
4.1.2.13	Aldolase	Entrapment in gel	Bernfeld and Wan[1]
4.2.1.1	Carbonic anhydrase	Cross-linking on a support	Broun et al.[33]
Isomerases			
5.3.1.1	Triosephosphate isomerase	Co-cross-linking	Thomas et al.[10]
5.3.1.9	Glucose-6-phosphate isomerase	Co-cross-linking	Thomas et al.[11]

[a] Superscript numbers refer to text footnotes giving full reference.

mined. The diffusion cells (Schemes 1a and 1b) are made of quartz and Altuglass and have useful areas ranging from 0.1 cm² to 50 cm².

Biochemical Reactors

Artificial enzyme membranes have also been tested in batch reactors. The processed solutions flow continuously through a cuvette, and concentrations are continuously measured in a spectrophotometer.[30] The use of continuously stirred tank reactors offers the possibility of a zero method,[34] which allows kinetic determinations at constant concentrations of substrate, products, and effectors (see Chapter 49).

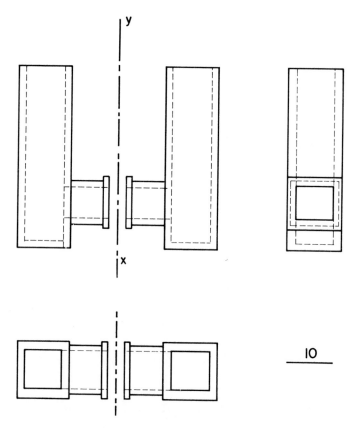

SCHEME 1a. Quartz diffusion cells allowing direct and continuous absorption measurements. The enzyme membrane separates two compartments. The useful area is 0.49 cm². The scale line indicates 10 mm. (The diffusion cell was produced by Ellma.) The device can be used directly in a double-beam spectrophotometer.

[34] J. M. Lemoullec, J. Wetzer, G. Gellf, and D. Thomas, *Anal. Biochem.*, in press.

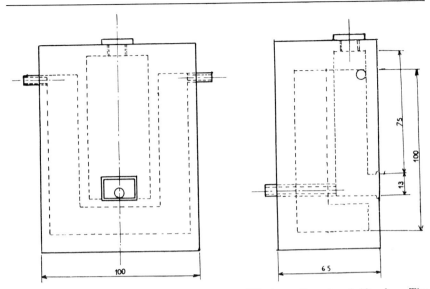

SCHEME 1b. Cross and lateral sections of a diffusion cell made of Altuglass. The membrane is placed between two such parts. The device is thermostated. Dimensions given on the diagrams are in millimeters. With this device it is possible to modulate the ratio between the volume of solution and the interface area.

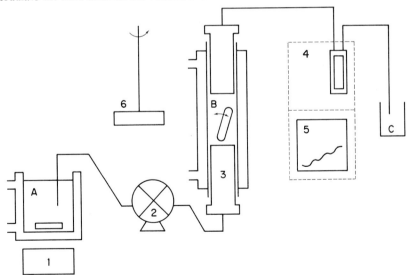

SCHEME 2. Diagram of continuously stirred tank reactor setup. 1, Magnetic stirrer (Amicon); 2, pump (Desaga); 3, chromatographic-type column (Whatman); 4, double-beam spectrophotometer (T. Beckman); 5, recorder (Riken Denshi); 6, magnetic bar; A, substrate reservoir; B, continuously stirred tank reactor; C, output.

Double-Beam Laser Spectrophotometer

Because of the diffusion limitations in immobilized enzyme systems, especially enzyme membranes, kinetic behavior is determined by local concentration and concentration profiles. Quite recently, an instrument for studying local concentrations in artificial enzyme membranes was built by Graves.[35] This is a double-beam laser spectrophotometer that uses a 10 μm^2 laser beam to measure the optical density in the thickness of the membrane itself. A dye is adsorbed within the membrane for the determination. The concentrations are studied both as a function of time at a given point in the membrane during a transient state and as a function of the abscissa inside the membrane under steady-state conditions.

Electron Microscopy

The use of an enzyme system giving an insoluble product, such as the DAB–peroxidase system (described below as an example), allows the visualization of the concentration profiles in artificial enzyme membranes.[30] Electron microscopic and kinetic studies simultaneously performed on artificial enzyme membranes gave much information on the behavior of immobilized enzyme systems.

EXAMPLE: The peroxidase–DAB system utilizes horseradish peroxidase (HRP; type VI), and 3,3′-diaminobenzidine tetra-HCl (DAB), all purchased from Sigma Chemical Company. Enzymes and chemicals are used without further purification.

Membrane Production. The membrane is produced by using the previously described co-cross-linking method.[2] A solution of 0.5–10 mg of HRP per milliliter, 5% serum albumin, and 0.75% glutaraldehyde in 0.02 M phosphate buffer, pH 6.8, is spread on a plane glass surface. After 10 hr at 4° complete insolubilization occurs and a 50-μm-thick protein membrane is produced. The membrane is rinsed with a buffer solution until no glutaraldehyde is noted in the eluted water (which is checked at 280 nm). No activity is found in the rinse water. The membrane is then immersed in a glycine solution (10 mg/ml) in order to block free glutaraldehyde function. The immobilized enzyme remains unaltered for several months at 4°.

HRP Activity Measurement. The enzyme is studied in solutions of DAB ranging from 0.03 to 2.0 mM in a 0.1 M, pH 4–8, citrate-phosphate buffer or in a 0.05 M, pH 7.6, Tris-HCl buffer. The peroxide concentration is 3 mM. Optical density variations of the solution at 309 and 332 nm are continuously recorded on a Beckman DBT spectrophotometer. Nitrogen bubbling ensures efficient stirring and avoids DAB autooxida-

---

[35] D. Graves, *in* "Enzyme Engineering" (L. B. Wingard and K. Pye, eds.), Vol. 2, p. 253. Plenum, New York, 1974.

tion. Kinetic studies are performed with both free and immobilized enzymes.

When free enzymes are used, the insoluble product is trapped on a column (Whatman PC 1520) containing silica (Prolabo) before flowing through the quartz cuvette. With the enzyme membrane system this column is not needed because there is no leakage of the insoluble product from the membrane to the bulk solution. It is absolutely necessary to work in a system free from metal ions of any kind; for instance, traces of Cu^{2+} can produce an insoluble product in the membrane. Kinetic measurements are performed at 25°.

Spectrophotometric Measurements Performed Directly on the Membranes. After 15 min, membranes (bearing an insoluble product) are sandwiched between two glass slides. The optical density is measured at 420–470 nm using a membrane free of insoluble product (without previous DAB incubation) as a reference.

Electron Microscopy. After the incubation period, membranes are postfixed with 1% osmium tetroxide and are dehydrated with alcohol and propylene oxide. Membranes are then embedded in the Epon according to Luft's method. Some thin sections obtained by using an LKB Ultrotome with glass knives are then mounted onto 400-mesh grids and examined without coloration in a Siemens 101 A electron microscope operating at 80 or 100 kV. For scanning microscopy, the membranes, coated with a gold or carbon layer, are observed in a JEOL 100 B microscope at 20 kV with the electron beam incident at about 50°.

Results are described in detail by Barbotin and Thomas.[30]

Enzyme Electrodes as Tools for Studying Enzyme Membranes

Enzyme electrodes not only are efficient analytical tools (see Chapter 41) but also are useful for studying intramembrane concentrations, which are so important for the study of global membrane behavior.[36] Concentrations measured by electrodes are directly linked to the enzyme kinetics.

Artificial Enzyme Membrane Kinetics (See also Chapter 29)

Kinetics of an Irreversible Monoenzyme System

Basic Equations.[10,12,14,37–39] The Michaelis–Menten relation is, by definition, valid only for homogeneous and isotropic media. It cannot be

[36] A. Naparstek, J. L. Romette, J. P. Kernevez, and D. Thomas, *Nature* (*London*) **249**, 490 (1974).
[37] E. Sélegny, S. Avrameas, G. Broun, and D. Thomas, *C. R. Acad. Sci. Ser. D* **266**, 1431 (1968).
[38] W. J. Blaedel, T. R. Kissel, and R. C. Boguslaski, *Anal. Chem* **44**, 2030 (1972).
[39] J. A. De Simone and S. R. Caplan, *J. Theor. Biol.* **39**, 523 (1973).

used to describe overall kinetics of a membrane system because the substrate and product concentrations are not constant within the membrane. However, the Michaelis–Menten relation remains valid for any volume element small enough to be considered homogeneous with regard to substrate concentration. The rate of change of substrate concentration with time in such a local volume element, $\partial S/\partial t$, depends both on substrate diffusion in the membrane matrix and on the enzyme activity:

$$\partial S/\partial t = (\partial S/\partial t)_{\text{diffusion}} + (\partial S/\partial t)_{\text{reaction}} \qquad (1)$$

Considering variations in concentrations to take place only in the x direction, perpendicular to the plane of the membrane, and introducing Fick's second law and the Michaelis–Menten relation, Eq. (1) becomes

$$\partial S/\partial t = D_S(\partial^2 S/\partial x^2) - [V_m S/(K_m + S)] \qquad (2)$$

where the local diffusion coefficient of the substrate, D_s, is assumed to be concentration independent, and the maximum local reaction rate, V_m, depends only on the (uniform) local concentration of enzyme. Similarly, we can write for the product

$$\partial P/\partial t = D_P(\partial^2 P/\partial x^2) + [V_m S/(K_m + S)] \qquad (3)$$

Thus, the system is ruled by nonlinear partial differential equations of the second order. The use of dimensionless parameters—e^2/D as time unit, e (thickness) as space unit, and K_m as concentration unit—gives, from Eq. (2):

$$\partial s/\partial t = (\partial^2 s/\partial x^2) - \sigma[s/(1 + s)] \qquad (4)$$

with

$$\sigma = (e^2 V_m)/(D_s K_m) \qquad (5)$$

From Eq. 5 it appears that the system is ruled by the dimensionless parameter σ similar to the "Thiele modulus."[40] The higher the σ value, the higher the diffusion limitations in the membrane.

For *steady-state conditions* ($\partial s/\partial t = 0$), analytical solutions of Eq. (4) have been obtained for the following cases: zero-order kinetics, i.e., S large in comparison with K_m,[37] and pseudo first-order kinetics, i.e., S small in comparison with K_m,[10,15,39] symmetrical boundary conditions.[15]

Numerical solutions can be made by computer in any case.[41] The solutions to Eqs. (3) or (4) are obtained as expressions for the fluxes and concentration profiles of substrate and product within the membrane.

For *non-steady-state conditions*, the evolution of concentration profiles as a function of time is also obtained by computer simulation (see

[40] E. W. Thiele, *Ind. Eng. Chem.* **31,** 916 (1939).
[41] J. P. Kernevez and D. Thomas, *J. Appl. Math. Optim.* **1,** 222 (1975).

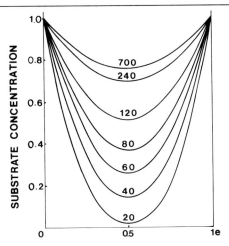

Fig. 1. Substrate concentration as a function of the abscissa of each point inside the membrane. Several concentration profiles are given during the transient state for a membrane empty of substrate immersed in a substrate solution ($\sigma = 1.4$). After J. P. Kernevez and D. Thomas, *J. Appl. Math. Optim.* **1**, 222 (1975).

Kernevez and Thomas[41] (Fig. 1). Such calculations are necessary for studies of regulation and optimization, both in biology and biotechnology.

Experimental Evidence. The global kinetics of diffusion–reaction systems are not Michaelian, and the usual techniques for determining K_m are not applicable to enzymes located within membranes. Although an apparent K_m can be calculated, the substrate concentration corresponding to a half-maximal reaction velocity certainly does not reflect the true value of the Michaelis constant within the membrane. Sélegny et al.[42] developed a relation based on the steady-state integration of Eq. (3) which enables them to determine K_m by the simultaneous measurement of boundary concentrations and fluxes. For numerous enzyme membranes, this value is equal to the K_m in solution and differs from the apparent K_m.

The higher the diffusion limitations, that is to say the higher the σ value, the greater the difference between both "true" and apparent K_m values. The expression σ is formed with a diffusion (e^2/D_s) and a reaction (V_m/K_m) term, so that the *diffusion limitations are linked not only to a slow diffusion process, but also to a high reaction rate.* Thomas and Broun[12] studied the influence of reaction rate on the diffusion-limiting effect using catalase. This enzyme is particularly convenient for such a study, since its specific activity is high and the catalyzed reaction is

[42] E. Sélegny, G. Broun, G. Geffroy, and D. Thomas, *J. Chim. Phys.* **66**, 391 (1969).

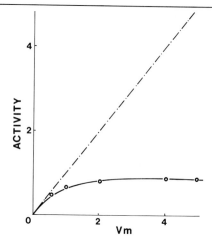

Fig. 2. Overall catalase activity measured as a function of the amount of enzyme activity introduced per unit volume of membrane (○——○). The same relationship in solution is given (·—·—). After D. Thomas and G. Broun, *Biochimie* **55**, 975 (1973).

first order. The authors examined the overall catalase activity as a function of enzyme potential activity present in the membrane (per unit volume) (Fig. 2). In homogeneous solution, the measured activity increases linearly with the enzyme concentration, as would be expected. In the membrane, however, an activity plateau is reached, because the diffusion limitation becomes increasingly significant as the active site density increases.

The same phenomenon was shown in an electron microscope study of a peroxidase membrane using DAB as a substrate.[30]

Effector and Reversibility Effects on the Kinetics of Artificial Enzyme Membranes

Thomas and Broun[12] studied the kinetics of membrane-bound catalase in the presence of an inhibitor (azide). Activity was measured as a function of inhibitor concentration both for the free enzyme and for a series of enzymes membranes with different Thiele moduli. The bound enzyme is less sensitive to the inhibitor, and, as the Thiele modulus decreases, the curves progressively approach those obtained with the free enzyme. The effect of competitive and noncompetitive inhibitors was also studied by Thomas et al.[43] both by simulation and by experimenta-

[43] D. Thomas, C. Bourdillon, G. Broun, and J. P. Kernevez, *Biochemistry* **13**, 2995 (1974).

tion. Competitive inhibitors were found to behave like noncompetitive inhibitors, and noncompetitive inhibitors like anticompetitve inhibitors.

In the same paper the authors were able to show, with a membrane bearing immobilized glucose-6-phosphate isomerase, an increase in the feedback effect of the reversibility on the product accumulation. Owing to the diffusion limitations, the concentration of product is higher in the membrane than outside in the bulk solution and its effect on enzyme kinetics is quicker and greater. The same phenomenon is still more important in multienzyme membranes.

Multienzyme Membranes (See also Chapter 31)

Goldman and Katchalsky[44] gave a theoretical analysis of a bienzyme membrane carrying out two consecutive reactions. Their results show that in such a system diffusion-limited behavior can produce an accelerated reaction rate. The behavior of the membrane is compared to that of an analogous system consisting of the two enzymes in solution. It is assumed that a quasi-stationary state is established within the unstirred layer. For all hypothetical systems analyzed, the concentrations of the products of the two enzymes in the bulk solution increase linearly with time for at least the first 10 minutes of reaction; in the corresponding homogeneous solution, there is an initial lag period in the appearance of the final product. The rate of production of the end product in the first stage of the reaction is markedly higher in the enzyme membrane.

Lecoq *et al.*[45] demonstrated the influence of membrane structure, in a two-enzyme system, on a feedback effect.

EXAMPLE: The membrane is prepared by the co-cross-linking method and incorporates β-glucosidase, structure (I), and glucose oxidase, structure (II). Membranes are produced by a previously described method.[6] A solution [containing a bifunctional agent, such as glutaraldehyde (1%), a bulk protein, such as plasma protein (4%), 2.5 units (0.55 mg/ml) of glucose oxidase bearing an excess of catalase activity, and 4 units (1.5 mg/ml) of β-glucosidase in a 0.02 M, pH 6.8 phosphate buffer] was spread on a plane glass surface. After polymerization (24 hr at 4°), total insolubilization occurs. The film is rinsed in phosphate buffer. The activity yield is 80% for glucose oxidase and 40% for β-glucosidase. Enzyme and albumin molecules are homogeneously distributed inside the matrix in this process.

Using salicin (Sigma) as a substrate for (I), the end product is

---

[44] R. Goldman and E. Katchalski, *J. Theor. Biol.* **32**, 243 (1971).
[45] D. Lecoq, J. F. Hervagault, G. Broun, G. Joly, J. P. Kernevez, and D. Thomas, *J. Biol. Chem.* **250**, 5496 (1975).

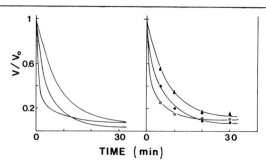

Fig. 3. Simulated and experimental behavior of a β-glucosidase/glucose oxidase system: normalized reaction velocity as a function of time. Simulated and experimental curves (b_1 and b_2, respectively) for enzymes in homogeneous solution (1), (—●—); for enzymes in a single membrane (2), (—○—); and for enzymes in two different membranes (3), (—△—). After D. Lecoq, J. F. Hervagault, G. Broun, G. Joly, J. P. Kernevez, and D. Thomas, *J. Biol. Chem.* **250**, 5496 (1975).

gluconolactone, an inhibitor for (I). Saligenin (Sigma) is neither an inhibitor nor an activator, nor is it a substrate for (II), and consequently it provides a convenient method for following the kinetics of (I) and the efficiency of inhibition. Experiments are performed both with the enzyme membrane and with the enzymes in solution. The feedback inhibition turns out to be more pronounced in the membrane system then in homogeneous solution. The kinetic behavior of the system was examined by computer simulation using its basic parameters, and good agreement is found between the experimental and simulated curves, as shown in Fig. 3.

The inhibition of (I) is more pronounced in the membrane because the inhibitor appears *in situ*. According to the simulation, the inhibitor concentration is still practically zero in the bulk solution after a few seconds, while it is already high inside the membrane. Thus the inhibitory effect in the membrane appears immediately.

Hervagault *et al.*[46] have studied the influence of membrane structure on a bienzyme system showing an apparent kinetic incompatibility.

EXAMPLE: The membrane is prepared by the co-cross-linking method and incorporates xanthine oxidase and urate oxidase. The cross-linking process is performed with 26.5 mg of plasma albumin, 2 mg of glutaraldehyde, 1.25 units of uricase, and 0–2.5 mg of xanthine oxidase per milliliter in 0.02 M P_i buffer.

Xanthine, the first substrate, is an competitive inhibitor for the second enzyme, urate oxidase. Experiments are performed both with the enzyme membrane and with the enzymes in solution. In the membrane

[46] J. F. Hervagault, G. Joly, and D. Thomas, *Eur. J. Biochem.* **51**, 19 (1975).

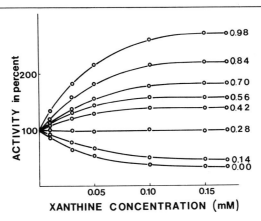

FIG. 4. Uricase activity as a function of xanthine concentration for enzyme membranes bearing different xanthine oxidase activities. These activities are expressed as a fraction of the activity in the absence of xanthine. Uricase activities without xanthine are the same for any membrane. After J. F. Hervagault, G. Joly, and D. Thomas, *Eur. J. Biochem.* **51**, 19 (1975).

at the steady state the intramembrane, xanthine concentration is less than in the bulk solution and the uric acid concentration is higher. The uricase reaction rate is greater than with the free enzymes in solution.

If the uricase activity in the bienzyme membrane is less inhibited than in solution and under other conditions, an activation effect is observed (Fig. 4).

Discussion

Artificial enzyme membranes are tools not only for modeling biological systems (to be described in the next section) but also for modeling other immobilized enzyme systems. In general, the particles bearing immobilized enzymes are not defined in shape, size, or active site distribution. In contrast, membranes are homogeneous in thickness and enzyme distribution.[30] The active sample for kinetic studies is *one* membrane, whereas in other cases the kinetic properties are studied in any number of particles.

The enzyme membranes are convenient from both an experimental and a theoretical point of view. It is possible to study the diffusion properties of cells in the insoluble phase under asymmetrical steady-state conditions. A quantitative determination of the diffusion coefficients or Thiele moduli is possible under these conditions. The spectrophotometer laser double beam of Graves,[35] a unique tool for studying immobilized enzyme systems, is only efficient using the membrane shape. The visualization of the concentration profiles by electron microscopy is feasible only with a membrane.[30] Finally specific electrodes are tools for the measurement of local concentrations in coatings or membranes, but not in spherical particles. From a theoretical point of view, the membrane shape allows a monodimensional calculation and the boundary conditions are quite well defined. Moreover, a comparison between experimental values and numerical simulation is possible. The enzyme membranes are a means of checking quantitatively the validity of a kinetic model for immobilized enzymes. Simulation is used as a guide for experimentation.

New Properties of Immobilized Enzymes Due to the Membrane Shape-Asymmetrical Systems

Heterogeneous Carbonic Anhydrase Membranes: Facilitated Transport of Carbon Dioxide

Broun et al.[8,33] described a system in which a hydrophobic dimethyl silicone membrane separates two compartments containing buffer solutions.

EXAMPLE: The membrane used is permeable to gases such as CO_2 but impermeable to water and electrodes. The CO_2 diffusion rate is measured in the unmodified membrane, in a membrane with carbonic anhydrase attached on the donor side only, and in a membrane with carbonic anhydrase attached on both sides. For every square centimeter of "dimethyl silicone single-backed" membrane (General Electric), 50 ml of freshly prepared solution containing 100 units of carbonic anhydrase (Sigma) and 0.8% glutaraldehyde solution in 0.02 M phosphate buffer (pH 6.8) are deposited on the silicone-treated side. After bonding, the membrane is rinsed with distilled water and 0.133 M glycine.

Twofold and fourfold increases in permeability, respectively, are observed for the enzyme membranes as compared with the unmodified membrane. Typical experimental results are shown in Fig. 5.

The explanation given by the authors is based on the existence of unstirred layers. On the donor side the CO_2 concentration in the unstirred layer is lower than in the bulk solution, whereas on the receptor side it is higher. This concentration polarization decreases the CO_2 gradient within the membrane, and hence the apparent permeability. In the presence of attached enzyme the chemical transformation of bicarbonate ion to CO_2 (and vice versa) is much quicker, and the CO_2 concentration is increased and, with it, the apparent permeability. The bicarbonate ion on the unstirred layers acts as a "carrier" for CO_2. Steady-state concen-

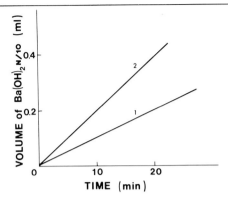

FIG. 5. Facilitated transport of CO_2: the effect of carbonic anhydrase on the diffusion of CO_2 through a silicone rubber membrane. pH-stat titration of CO_2 with 0.1 N $Ba(OH)_2$ added to the receptor compartment. Curve 1, Unmodified membrane; curve 2, enzyme membrane. After G. Broun, C. Tran Minh, D. Thomas, D. Domurado, and E. Sélegny, *Trans. Am. Soc. Artif. Intern. Organs* 17, 341 (1971).

tration profiles are calculated for CO_2 and HCO_3^- both with and without enzyme.

Quite recently Donaldson and Quinn[47] described the use of a membrane bearing carbonic anhydrase produced by the co-cross-linking method between two gas compartments.

Active Transport Effect

It is possible to construct a multienzyme model having an asymmetrical distribution of active sites in the membrane. Such a system has been studied by Broun et al.[31] and Thomas et al.[11] Their work was inspired by Mitchell's conclusion,[48] at a symposium during which different possible modes of active transport were discussed, that "The analysis of the processes of primary or metabolic transport at the molecular level of dimensions will begin to proceed on a sure footing when we learn to produce in vitro transport systems of known chemical and spatial composition comparable to the 'synthetic' multi-enzyme systems of classical biochemistry."

Broun and Thomas and their co-workers described a structured multilayer bienzyme membrane composed of two active protein layers adjacent to one another, sandwiched between two selective films. The enzyme layers carried, respectively, hexokinase and phosphatase co-cross-linked with an inert protein (Scheme 3); both were impregnated

[47] T. Donaldson and J. Quinn, *Proc. Natl. Acad. Sci. U.S.A.* 71, 4995 (1974).
[48] P. Mitchell, *Membr. Transp. Metab. Symp. 1960.* Vol. 1, p. 22 (1961).

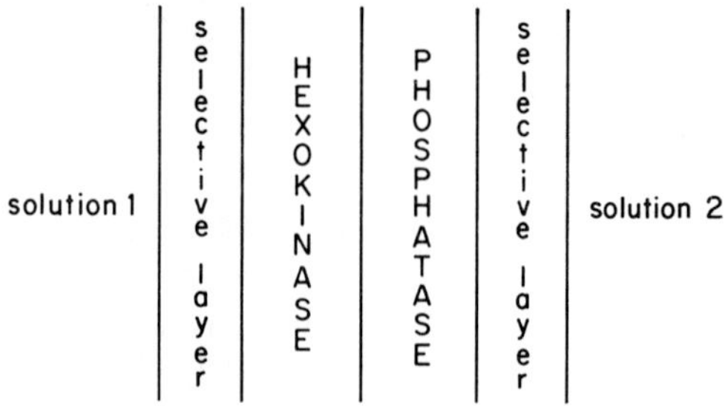

| solution 1 | selective layer | HEXOKINASE | PHOSPHATASE | selective layer | solution 2 |

SCHEME 3. Diagram of a double-layer membrane covered on its external side by two selective films. The system gives an active transport effect for glucose with consumption of ATP.

with ATP and covered on their external surfaces by selective films permeable to glucose but impermeable to glucose 6-phosphate.

The selective films are obtained by the cross-linking of a nonenzyme protein (albumin) in a cellulose matrix according to the previously described method. There, films are permeable to glucose but impermeable to glucose 6-phosphate. This selectivity is due to a "molecular sieve" effect produced by the films and the electrical charges of the cross-linked protein. Both glucose 6-phosphate (with pK_a values 0.94 and 6.11) and albumin are negatively charged at the pH values at which the experiments are carried out.

The "active enzymic films" bear hexokinase and phosphatase reticulated on an inert protein (albumin); both are impregnated with ATP and covered on their external sides by two "selective films." In this asymmetrical membrane, glucose is temporarily phosphorylated:

$$\text{Glucose} + \text{ATP} \xrightarrow{\text{hexokinase}} \text{glucose 6-phosphate} + \text{ADP}$$
$$\text{Glucose 6-phosphate} \xrightarrow{\text{phosphatase}} \text{glucose} + \text{inorganic phosphate}$$

Since the sum of the two reactions is ATP → ADP + inorganic phosphate, the system behaves chemically as a simple ATPase.

The asymmetrical membrane exemplifies a spatially and metabolically sequential enzyme array. When it is placed between glucose solutions, glucose entering the system is transiently phosphorylated and the system behaves chemically as a simple ATPase. In the hexokinase layer the glucose is phosphorylated to glucose 6-phosphate, which diffuses into the phosphatase layer, where it is promptly dephosphorylated.

Examination of the glucose and the glucose 6-phosphate concentration profiles illustrated in Fig. 6 shows that this membrane can act as a glucose pump. The time evolution of the glucose concentration in the receptor compartment is shown in Fig. 7. That in the donor compartment

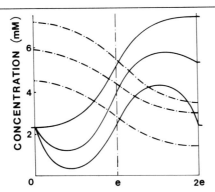

Fig. 6. "Active transport" of glucose (hexokinase/phosphatase system). The membrane separates two compartments; the glucose concentration (S_1) within the first (donor) compartment can be considered constant, the volume being far larger than that of the second (receptor) compartment, in which the concentration (S_2) changes. Glucose concentration profiles (—) and glucose 6-phosphate concentration profiles (·—·—). The lowest profiles represent initial conditions ($S_1 = S_2$), evolving over a period of time; the uppermost profiles represent stationary state conditions (net fluxes, zero). After D. Thomas, C. Tran Minh, G. Gellf, D. Domurado, B. Paillot, R. Jacobsen, and G. Broun, *Biotechnol. Bioeng. Symp.* **3**, 299 (1972).

remains constant. The receptor concentration increases regularly at first; it then attains a plateau, which can be explained in terms of the concentration profiles. However, the concentration falls again once the initial ATP charge of the membrane is depleted. The pumping effect may be reduced by the addition of fructose, which competes with glucose for the hexokinase, as shown in Fig. 7. For an idealized glucose pump made up of

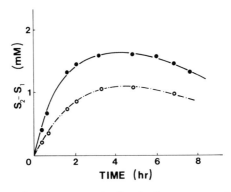

Fig. 7. Evolution of glucose concentration in the presence of fructose (○—·—○) and in the absence of fructose (●——●). Glucose concentration (S_1) = 5.6 mM; fructose concentration = 2.8 mM. After G. Broun, D. Thomas, and E. Sélegny, *J. Membr. Biol.* **8**, 313 (1972).

four monomolecular layers, computer simulation[41] predicts a 130-fold maximum increase in concentration. The possibility of chemodiffusional coupling in this and similar systems is clearly dictated by the global analog of the Curie principle.

Interaction between Enzyme Activity and the Potential Difference at the Boundaries of the Membrane

For the first time, the use of artificial enzyme membranes allows the study of the interaction between enzyme activity and membrane potential in a well-defined context. Before the recent progress in manufacturing artificial membranes bearing immobilized enzyme, Blumenthal *et al.*[49] described a system in which a papain solution was sandwiched between two cation and anion exchange membranes. Under short-circuit conditions the system was able to generate a current. A nonequilibrium thermodynamic analysis was developed by the authors.

David *et al.*[50] have described some studies dealing with a urease membrane. The enzyme transforms a neutral molecule, urea, into two ions, ammonium and carbonate. The reaction gives both a modification of the ion concentration and an increase in pH. The membrane is studied between two compartments containing 1 mM and 10 mM phosphate solution at pH 5.1. The potential is recorded at the boundaries of the membrane after introducing the substrate into both compartments. A jump in the potential difference is observed. The potentials are measured using a UVA-type microvoltmeter (AOIP, Paris) with a high impedance for all scales, between two reference saturated KCl electrodes (Radiometer K 401, Copenhagen). The membrane area in the diffusion cells used for these experiments is only 0.5 cm^2 in order to avoid too rapid variations of ionic concentrations during, and between, measurements. The determination of the effect of urease activity is obtained by using the same membrane both with and without urea and by comparing the resulting potential differences. In these experiments, urea concentration is 0.25 M in both compartments. After introduction of an enzyme inhibitor, the potential difference returns to the initial value.

The phenomenon of potential modification is linked to the enzyme reaction. The steady-state membrane potential is studied both at constant substrate concentration as a function of the ion concentration gradient and at constant ion gradient as a function of the substrate concentration. It is especially important to discuss the latter point. The relationship between the potential and the substrate concentration is a

---

[49] R. Blumenthal, S. R. Caplan, and O. Kedem, *Biophys. J.* **7**, 735 (1967).
[50] A. David, M. Metayer, D. Thomas, and G. Broun, *J. Membr. Biol.* **18**, 113 (1974).

sigmoid curve recalling the behavior of some biological membranes.[51] The coupling between a simple enzyme reaction and a structure explains the "cooperative" effect observed with excitable membranes.

Memory in Artificial Enzyme Membranes

Memory in enzyme membranes was described by Naparstek et al.[36] The problem of information storage in the phase of "short-term memory" has attracted considerable attention. A plausible mechanism for the process is based on an all-or-none process, assuming the existence of metastable states.[51] Such phase transitions are thought to require structural changes in macromolecular components or in the membrane permeability of the information storage unit.[52] The rapid development of methods for manufacturing artificial enzyme membranes provides a new tool for demonstrating the existence of hysteretic phenomena that do not occur as a result of structural changes in a system. The authors demonstrate the existence of hysteretic phenomena through simple kinetic mechanisms that are shared by most enzymes. They refer to the two simple cases: autocatalysis by the product and inhibition by excess substrate. Both are examples of systems that do not satisfy the Le Chatelier–Braun principle, and the phenomenon never takes place along the whole range of concentration values, thus resulting in maximal activity. When coupled with diffusion of the product and substrate, respectively, there is, because of the nonlinearities in the reaction rate, the possibility of three stationary states—two of them stable and one unstable—for a limited range of parameter values.

EXAMPLE: A coating bearing one enzyme (papain) is produced on the surface of a glass pH electrode by the method previously described (cocross-linking).[37] A solution containing, per milliliter, 40 mg of albumin (Sigma Chemical Co., St. Louis, Missouri; fraction V bovine serum albumin), 10 mg of nonactivated papain (Miles-Yeda Ltd, Rehovot, Israel; specific activity 19 units of BAEE) and 5 mg of glutaraldehyde in phosphate buffer (0.02 M, pH 6.8) is poured over the surface of the bulb of a "combined" pH electrode (Radiometer, Copenhagen, No. GK 2311 C). During this procedure the electrode is rotated horizontally about its axis of revolution at 100 rpm. After 15 min, complete insolubilization occurs and the bulb is covered with a thin protein coating. The coating is rinsed in distilled water or phosphate buffer, and the immobilized enzyme is activated by immersion in a solution of EDTA (1 mg/ml) and

---

[51] J. P. Changeux and J. Thiery, in "Regulatory Function of Biological Membranes" (J. Järnefelt, ed.), p. 11. Elsevier, Amsterdam, 1968.
[52] A. Katchalsky and A. Oplatka, Neurosci. Res. Symp. 1, 352 (1966).

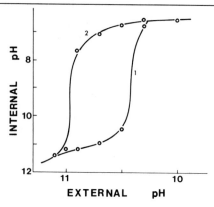

Fig. 8. pH determinations made with a papain membrane-coated electrode. Stationary state description of internal pH as a function of increasing (curve 1) and decreasing (curve 2) external pH values. The pH was decreased at the start of the experiment, and then increased. After A. Naparstek J. L. Romette, J. P. Kernevez, and D. Thomas, *Nature (London)* **249**, 490 (1974).

cysteine (1 mg/ml). The active membrane-coated electrode is then introduced into a solution of KCl (0.1 M) prepared from doubly distilled water, and the pH at the membrane–glass interface is determined.

The papain reaction decreases the pH, and the pH–activity variation gives an autocatalytic effect for pH values greater than the optimum. For zero-order kinetics for the substrate (BAEE), the pH inside the membrane is studied as a function of the pH in the bulk solution in which the electrode is immersed (Fig. 8). A hysteresis effect is observed, and *the enzyme reaction rate depends not only on the metabolite concentrations, but also on the history of the system.*

Oscillatory Behavior and Spontaneous Structuration in Space in Artificial Enzyme Membranes[53]

Oscillations and spontaneous structurations were intensively studied by Prigogine's school and classified as "dissipative structures." Experimentally, sustained oscillations have been observed and established beyond doubt for glycolysis. Most of the experiments have been carried out by Hess[54] and Betz and Chance.[55] Theoretical studies attribute these oscillations to the enzyme phosphofructokinase, which is an allosteric

[53] D. Thomas, in "Membranes, Dissipative Structures and Evolution" (I. Prigogine, ed.), p. 113. Wiley, New York, 1975.
[54] B. Hess, in "Funktionelle und morphologische Organisation der Zelle." Springer-Verlag, Berlin and New York, 1962.
[55] A. Betz and B. Chance, *Arch. Biochem. Biophys.* **109**, 585 (1965).

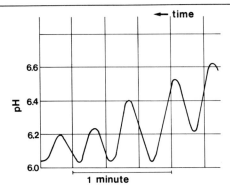

FIG. 9. Time dependence of pH at the enzyme membrane–glass electrode interface. Nominal membrane thickness 10 μm with gentle stirring. Substrate concentration and pH in the bulk solution were initially 4.5 and 9.3 mM, respectively. After A. Naparstek, D. Thomas, and S. R. Caplan, *Biochim. Biophys. Acta* **323**, 643 (1973).

enzyme.[56,57] Naparstek et al.[58] demonstrated the existence of oscillatory behavior in an artificial enzyme membrane linked to the diffusion limitations.

When the papain electrode described in the previous section is introduced in a solution with a high pH (i.e., pH 10) and a lower substrate concentration (under first-order kinetics), a variation in time of the measured pH inside the membrane spontaneously occurs (Fig. 9). This enzyme, which has been extensively studied, does not give oscillations for any conditions of pH and substrate concentration. The period of oscillation is about 0.5 min, and the oscillation is abolished by introducing an enzyme inhibitor. The phenomenon can be explained by the autocatalytic effect and by the feedback action of OH^- diffusing in from the outside solution. There is a qualitative agreement between the computer simulation and the experimental results.

Thomas et al.[59] have presented a system in which the nonlinearity of the enzyme reaction can spontaneously produce an inhomogeneous distribution of one of the products beyond an instability.

An artificial membrane bearing two different enzymes (glucose oxidase and urease), in a spatially homogeneous fashion, is produced by using the previously described co-cross-linking method. The glucose oxidase reaction decreases the pH, and the urease increases the pH. The

[56] J. Higgins, *Proc. Natl. Acad. Sci. U.S.A.* **51**, 989 (1964).
[57] A. Goldbeter and R. Lefever, *Biophys. J.* **12**, 1302 (1972).
[58] A. Naparstek, D. Thomas, and S. R. Caplan, *Biochim. Biophys. Acta* **323**, 643 (1973).
[59] D. Thomas, A. Goldbeter, and R. Lefever, *Biophys. Chem.* in press.

pH activity profiles show an autocatalytic effect for the glucose oxidase in the range of pH values greater than the optimum pH, and for urease, lower than the optimum pH. When both enzymes are mixed together, the global pH variation is zero for a well-defined pH value.

The active membrane separates two compartments. It is possible to get a uniform pH value throughout the system, in the presence of both substrates (glucose and urea), by the transient use of a buffer. Outside pH values are controlled and H^+ fluxes are measured by pH-stat systems. After small asymmetrical perturbations of the pH values at the boundaries (0.05), an inhomogeneous pH distribution arises spontaneously inside the membrane (similar to profile B, Fig 6). The initial perturbations are amplified and the pH values in the compartments tend to evolve in opposite directions. The H^+ fluxes entering and leaving the membrane can be determined by pH-stat measurements. If the boundary pH values are not kept constant by a pH stat, the system evolves to a new steady state characterized by a pH gradient of 2 pH units across the membrane.

This experimental system is similar to the fictitious system of Turing.[60] In his system, the simplest case consists of two cells connected by a permeable wall. In each cell, a chemical reaction takes place in which one substance is converted into an another by an autocatalytic process. In both systems, a positive fluctuation in the concentration in one compartment does not result in a reactional negative value of the derivative, but in a positive value. In such systems the initial homogeneous state is unstable, and the system spontaneously tends toward a non-homogeneous concentration distribution.

Discussion

In considering descriptive membrane models, Stein[61] wrote that such models give rise to new functional hypotheses as well as fruitful discussions. However, purely descriptive models offer little means of analyzing the physicochemical mechanisms underlying enzyme-mediated transport phenomena or structure-dependent regulation—hence the necessity for physicochemical membrane models, which can be experimentally checked as well as mathematically analyzed. Structural models conforming to these requirements have already been studied extensively, especially in the area of lipid bilayers. The important step we now face is to construct a variety of meaningful functional models, which can be expected not

---

[60] A. M. Turing, *Philos. Trans. R. Soc. London, Ser. B* **237**, 37 (1952).
[61] W. D. Stein, *in* "The Movement of Molecules across Cell Membranes." Academic Press, New York, 1967.

only to have heuristic significance, but also to provide new and fundamental insights.

Classical enzymology deals with homogeneous and isotropic media. The introduction of an intermediate stage between classical enzymology and the enzymology of complex biological systems is essential. The study of well-defined preparations of membrane-supported enzymes allowing asymmetrical conditions should provide the intermediate stage needed. With relatively simple systems of this kind, it should be possible to develop a detailed picture of the influence of an anisotropic structure on enzymic function. Experimental evidence is accumulating rapidly that local concentration profiles within enzyme membranes may be far from simple in character. *Such local concentrations of substrates and products not only modulate enzyme kinetic properties* (K_m values, feedback effects, pH-activity curves, etc.), but also introduce totally new phenomena, such as asymmetrical behavior, memory, and oscillations. For example, it is only with an enzyme membrane that it becomes possible to study the influence of reactions on electrical potential difference[49,50] or the influence of electrical potential difference on enzyme kinetics, which may be of crucial importance in many biological systems.

[64] Immobilization of Lactic Dehydrogenase[1]

By FRANCIS E. STOLZENBACH and NATHAN O. KAPLAN

Various methods of immobilizing lactic dehydrogenase have yielded products that are linked to the matrix by different amino acid residues of the protein. Some methods have also yielded material in which only one subunit of the enzyme is attached to the matrix. Levi[2] has attached chicken H_4 lactic dehydrogenase to Sepharose and demonstrated that only one subunit was bound to the matrix. This was achieved by activating Sepharose 4B with cyanogen bromide as described by Axén[3] and allowing the activated Sepharose to react with a large excess of enzyme so that only a single subunit of the tetramer is capable of being attached. The detailed procedure will be given.

A major portion of the work on enzyme immobilization in our laboratory has been done using porous-glass beads that have been

---

[1] This work was supported in part by a grant from the National Science Foundation (NSF GI 36249).
[2] A. Levi, *Arch. Biochem. Biophys.* in press (1975).
[3] R. Axén, J. Porath, and S. Ernback, *Nature (London)* **214**, 1304 (1967).

derivatized to yield alkylamine, arylamine, and succinyl glass. Alkylamine glass was also used with a bifunctional reagent to cross-link the enzyme to the glass. These procedures were carried out under conditions where there was not a great excess of enzyme and yielded products in which the enzyme was linked by all four subunits, and most likely at several residues on each subunit. The physical and catalytic properties of these enzymes exhibit very interesting and different characteristics from the native soluble enzyme.

The binding of the alkylamine and succinyl glass bead derivatives using carbodimide presumably involves carboxyl groups or amino groups, respectively, on the enzyme. Binding to the arylamine glass by diazotization and binding to Sepharose appear to involve only tyrosine residues. Enzyme cross-linked to alkylamine glass by glutaraldehyde is bound by the lysine residues of the enzyme.

Details of binding to Sepharose and porous-glass derivatives and the preparation of the derivatized glass are covered in detail in this paper.

Activation and Coupling to Sepharose

Sepharose 4B (purchased from Pharmacia) is stirred at 0°–5° in a fume hood. To the stirred Sepharose, solid cyanogen bromide (10–15 mg per milliliter of Sepharose) is added slowly with aliquots of 3.5 N sodium hydroxide added to maintain the pH of the mixture between pH 9 and pH 10. During the first cyanogen bromide additions, the pH drops markedly, and care should be taken to control this early pH fluctuation. As the reaction proceeds the pH changes on cyanogen bromide additions, and the pH changes become less marked. After stirring for an additional 10 min to ensure that no further changes in pH occur, the mixture is filtered through a sintered-glass funnel (medium or coarse) and washed exhaustively with cold distilled water until all the cyanogen bromide is removed. Care should be taken at this point to ensure the removal of all residual solid cyanogen bromide. The activated Sepharose can be stored at 0°–5°C and used as needed. Although the Sepharose in this form is reported to be fairly stable our best coupling results have been obtained with freshly activated Sepharose.

To each milliliter of activated Sepharose 4B, 1 ml of 5–10 mg per milliliter of solution of chicken H_4 lactic dehydrogenase is added in 100 mM potassium phosphate at pH 8.5. The coupling is allowed to proceed at 0°–5° for 16 hr. The uncoupled enzyme is removed by gentle centrifugation at 1000 g and decantation from the packed Sepharose. Several washings with 100 mM potassium phosphate, pH 8.5, removes the residual uncoupled enzyme. When all the noncovalent attached

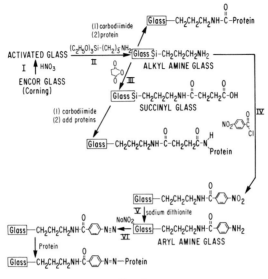

Fig. 1.

lactic dehydrogenase activity is washed off, the coupled enzyme can be stored moist at 0°–5° for at least 6 months with no appreciable loss of activity.

Enzyme immobilized by this method yields an enzyme that is bound by only one subunit. The characteristics of this enzyme will be discussed in the summary.

Derivitization of Porous Glass Beads

The derivitization of the glass beads is carried out essentially by the method described by Weetall[4] with some minor variations (see also this volume [10]). Porous glass beads 40–80 mesh with a pore size of 500 Å (Encorglass purchased from Corning Glass) are boiled for 45 min in a solution of 5% nitric acid. Excess acid is then removed by filtration through a coarse sintered-glass funnel and washed with distilled water until the pH of the wash remains unchanged. This is the activated glass (Fig. 1, reaction I) and may be dried at 200°F for 4 hr and stored as such, or the moist beads may then be used immediately to form the alkylamine glass as described in Fig. 1, reaction II. This is accomplished by suspending the glass beads in a 10% aqueous solution of γ-aminopropyltriethoxysilane (A-1100 solution purchased from Union Carbide Corpora-

---

[4] Information obtainable from H. H. Weetall, Corning Glass Works, Corning, New York.

tion), adjusting the pH to 3.45 with nitric acid, and allowing it to stir gently for 2¾ hr at 75°. The reacted glass is then filtered through a coarse sintered-glass funnel and washed with an equal volume of distilled water. This product is alkylamine glass and may be used for the binding of enzymes or for the preparation of succinyl glass and arylamine glass.

Succinyl glass is prepared from alkylamine glass as depicted in Fig. 1, reaction III. One and a half grams of alkylamine glass are placed in a 100-ml round-bottom flask with 2 g of succinic anhydride, 3 ml of triethylamine, and 50 ml of $HCCl_3$. The mixture is then shaken for 3 hr at room temperature. During the course of the reaction, the solution turns from milky white to a relatively clear solution, which indicates the completion of the reaction. After completion of the reaction the glass beads are filtered through a coarse sintered-glass funnel, washed with several volumes of $HCCl_3$ and finally with several volumes of distilled water. The glass beads can then be air-dried and stored for later use.

Arylamine glass is derived from alkylamine glass as shown in Fig. 1, reactions IV and V. For 100 g of alkylamine glass, 286 mg of p-nitrobenzoylchloride and 10% triethylamine in chloroform are added together and allowed to reflux overnight. The amount of the triethylamine chloroform is not critical as long as there is enough volume to keep the glass covered during the refluxing. This reaction forms a glass derivative with a nitro group (Fig. 1, reaction IV), which can then be reduced with dithionite (Fig. 1, reaction V) to form arylamine glass. The refluxed glass is filtered through a coarse sintered-glass funnel, washed with several volumes of chloroform, and then allowed to air dry. Reduction of the nitro group is achieved by boiling a sample of the glass beads in an aqueous solution of sodium dithionite (10% w/v) for 10 min. After the material has reacted, the glass is again filtered through a coarse sintered-glass funnel and washed with distilled water until the washes show no pH change. The glass beads are then air-dried and can be stored at room temperature for future use.

Used glass may be recycled by placing in a 600°F oven for 12 hr and then resilanizing using the same procedure as described for the unused glass.

Coupling to Alkylamine and Succinyl Glass Beads Using Soluble Carbodiimide

To 0.5 g of glass beads, 25 mg of chicken H_4 lactic dehydrogenase, and 5 ml of an aqueous solution containing 30 mg of 1-cyclohexyl-3-(2-morpholinoethyl) carbodiimide metho-p-toluene sulfanate (purchased from Aldrich Company) are added and the pH is adjusted to pH 4.5 using hydrochloric acid. The mixture is placed under vacuum and allowed to react at 0° overnight. The glass-bound enzyme is then filtered through a coarse sintered-glass funnel and washed with 1 M sodium chloride (to remove noncovalently bound enzyme) and then exhaustively with cold distilled water. The covalently bound enzyme is stored as a moist cake at 0°–5°.

Coupling to Arylamine Glass by the Diazotization Procedure

Arylamine glass was diazotized as shown in Fig. 1, reaction IV, and enzyme could then be covalently bound to the activated glass. Ten milliliters of 2 M hydrochloric acid is combined with 0.5 g of alkylamine glass and 0.25 g of sodium nitrite, mixed thoroughly, and placed under constant vacuum at 0° for 20 min. The reacted glass is then filtered through a coarse sintered-glass funnel and washed with 100 ml of cold 1% sulfamic acid and then with 1000 ml of cold distilled water. Chicken H_4 lactic dehydrogenase (25 mg) in 10 ml of 0.1 M potassium phosphate buffer, pH 7.0, is then added to the activated glass and allowed to react at 5° overnight. The glass-bound enzyme is then filtered through a coarse sintered-glass funnel, washed with 100 ml of 1 M sodium chloride and then 1000 ml of cold distilled water. Noncovalently bound enzyme not removed by the sodium chloride wash may be removed by suspending the glass beads in 1% albumin for 24 hr. When the moist beads are stored at 5°, they will maintain their full activity in the immobilized form for at least 6 months to a year.

Coupling to Alkylamine Glass by Glutaraldehyde Cross-Linking

A mixture of 1 ml of 5% glutaraldehyde in 0.1 M potassium phosphate buffer pH 7.0 and 0.5 g of alkylamine glass is allowed to react at 25° under vacuum for 1 hr. Excess glutaraldehyde is removed by filtration through a coarse sintered-glass funnel and exhaustive washings with cold distilled water. To the reacted glass, 25 mg of chicken H_4 lactic dehydrogenase in 2 ml of 0.1 M potassium phosphate buffer, pH 7.0, are added and allowed to react at 5° overnight. After reacting overnight, the glass beads are filtered through a coarse sintered-glass funnel and washed with 100 ml of cold distilled water to remove any extraneous enzyme.[5] This is followed by a wash of 100 ml of 1 M sodium chloride (to remove non-

---

[5] Only one or two lysines per subunit appear to be present in the attachment of the lactate dehydrogenase to the glutaraldehyde glass.

covalently bound protein) and a wash of 1000 ml of cold distilled water to remove the residual sodium chloride. As in the case of arylamine-bound enzyme, a 24-hr incubation with 1% albumin may be used to completely eliminate the noncovalently attached enzyme. Glass-bound enzyme by this method is stable at 5° for at least 6 months to a year.

Comparison of Some Physicochemical Properties of Immobilized Lactic Dehydrogenase

Chicken H_4 lactic dehydrogenase bound to alkylamine glass with soluble carbodiimide yielded an immobilized enzyme with very low catalytic activity even though analysis indicated that the reaction procedure did not denature any enzyme, and the amount of enzyme bound was comparable to that bound by the other methods of coupling. Chou and Swaisgood[6] have reported that by manipulation one may bind lactic dehydrogenase to succinyl glass by a single subunit; however, the enzyme bound to succinyl glass by the techniques described in the text yielded an enzyme with less than 25% of the activities achieved with enzyme bound to arylamine glass by the diazotization procedure and enzyme cross-linked with glutaraldehyde to alkylamine glass beads. This enzyme appears to be linked by all four subunits. Owing to the lower activities, the carbodiimide glass-bound enzymes were not used in the detailed comparison of the physicochemical properties of the immobilized lactic dehydrogenase. A comparison of the properties of chicken H_4 lactic dehydrogenases bound to Sepharose, arylamine glass beads, and cross-linked to alkylamine glass beads are summarized in the table.

The apparent K_m for both lactate and pyruvate shows no significant change from the native enzyme for any of the immobilized forms of the enzyme. This is shown in lines 1 and 2 of the table. High levels of pyruvate give an activity of 30–40% maximal rate for the native chicken H_4 lactic dehydrogenase. This is caused by the formation of an abortive ternary complex between the NAD formed in the reaction as well as pyruvate with the enzyme. As can be seen in line 3 of the table, the immobilized enzymes show a marked resistance to the formation of this complex. Sepharose-bound enzyme appears most active at high levels of pyruvate with activity of 2–2.5 times that of native enzyme while the glass-bound enzymes are 1.5–2 times as active as the native enzyme.

One of the more interesting features of immobilized enzymes was the shifting, in most of the cases reported, of the pH optimum for their reaction. When this was examined for the immobilized lactic dehydro-

---

[6] I. C. Chou and H. E. Swaisgood, *Biochim. Biophys. Acta* **258**, 675 (1972).

TYPE OF BINDING[a]

	Native	Sepharose bound[b]	Arylamine diazo bound	Alkylamine glutaraldehyde bound
K_m Lactate (M)	8.0×10^{-3}	4×10^{-3}	5.9×10^{-3}	8.0×10^{-3}
K_m Pyruvate (M)	1.25×10^{-4}	8×10^{-5}	3.9×10^{-4}	4.0×10^{-4}
Substrate inhibition 10^{-2} M pyruvate	40% Maximal rate	80% Maximal rate	60% Maximal rate	70% Maximal rate
pH optimum	7.4	7.0	$5.5 \to 9.5$	$5.5 \to 9.5$
pH stability	55% Denatured 15 min at pH 4.1	—	55% Denatured 30 min at pH 4.1	No loss of activity 30 days, pH 3.1
Heat denaturation	50% Denatured 20 min at 75°	—	50% Denatured 65 min at 75°	No loss for 90 min, 40% denatured after 5 hr, 75°
Proteolysis with subtilisin	50% Activity remaining at 50 min	—	80% Activity remaining at 50 min	95% Activity remaining at 50 min

[a] Pyruvate → lactate assay.
[b] Data from Dr. A. Levi.

genase, the following results (see line 4 of the table) were obtained. The pH optimum for Sepharose-bound enzyme showed a shift from 7.4 for the native enzyme to 7.0 for the Sepharose-bound enzyme; however, there was no difference in rate between pH 5.5 and 9.5 for the arylamine- and the glutaraldehyde-bound enzyme. A marked stability to acid inactivation was also noted (see line 5 of the table). At pH 4.1 the native enzyme retains only 45% activity after 15 min whereas the arylamine-bound enzymes requires 30 min to reach this degree of denaturation. While examining the glutaraldehyde-bound enzyme, we found that there was no loss of activity at pH 4.1 throughout the course of the experiment. The pH was lowered to 3.1, and whereas the native enzyme was 70% inactivated in 5 min, the glutaraldehyde-bound enzyme was completely active for 1 month. Chicken M_4 lactic dehydrogenase immobilized by this method exhibits the same characteristics. Although no data on Sepharose-bound chicken H_4 lactic dehydrogenase are available, dogfish M_4 lactic dehydrogenase immobilized on Sepharose, as described in the text, retained 30% of its activity after exposure to pH 4.1 for 2 min whereas the native dogfish enzyme retained only 4%.

Line 6 of the table indicates that 50% inactivation at 75° requires 20 min for the native enzyme, 65 min for the arylamine-bound enzyme, and 4.5 hr for the glutaraldehyde-immobilized enzyme. Unfortunately, no data for the Sepharose chicken H_4 enzyme are available, but the dogfish M_4 lactic dehydrogenase upon 10 min exposure at 60° retains only 8% activity whereas the Sepharose-immobilized enzyme retains 80% of its original activity under these conditions.

The alkylamine-immobilized enzyme and the glutaraldehyde-immobilized enzyme are also more resistant to proteolysis (as shown in line 7 of the table) with 80% and 95%, respectively, of their initial activity remaining after 50 min exposure to subtilisin. Under the same conditions, only 50% of the activity of the native enzyme remains. Treatment with 5.3 M lithium chloride indicates that glutaraldehyde-immobilized enzyme is more resistant to dissociation and is more readily reassociated. Whereas the native and arylamine-bound enzyme reassociate to an activity of between 18 and 33% of their original values, the glutaraldehyde-immobilized enzyme exhibits quantitative reassociation.

It is obvious that immobilization of enzymes is of great value in the stabilizing and engineering of enzymes and enzyme reactors.

Author Index

Numbers in parentheses are footnote reference numbers and indicate that an author's work is referred to although his name is not cited in the text.

A

Adalsteinsson, O., 895
Adams, R. W., 390
Agranat, E. A., 310
Agrawal, K. M. L., 101
Aizawa, M., 886
Åkerlund, A., 585(16), 587, 588(16), 593(16), 597(16), 600(16), 611(16), 616(16), 617(16)
Åkerström, S., 28, 34, 41(28)
Albu-Weissenberg, M., 436, 437(123)
Alexander, R. L., 884
Allen, B. R., 341, 792
Allender, J. H., 302, 306(11), 315, 316(11), 883, 886(23), 887(23)
Allison, A. C., 221, 227, 700, 705(9, 10), 708(9), 709(9, 10)
Allison, J. P., 120, 677, 680, 682, 693(24)
Amotz, S., 812
Amundson, C. H., 823
Anapakos, P. G., 215
Anderson, N. G., 637
Anderson, S. R., 504
Andersson, L.-O., 35, 37
Anderton, B. H., 384
Andres, G. A., 710
Anfalt, T., 588, 601, 608, 609, 613, 616(22)
Anfinsen, C. B., 38, 72, 516, 517, 518(17, 18), 521(17), 522(17), 845
Anson, M. L., 549, 550
Antonini, E., 539, 540, 541, 543, 545
Antonov, V. K., 366, 367(8), 535, 559
Apple, M., 211, 215, 677, 680, 694(27)
Archer, M. C., 887, 895(3)
Aris, R., 400, 408(227), 413, 414(22, 51)
Armiger, W., 444
Arora, A. S., 132
Asher, W. J., 215
Aspberg, K., 28, 493, 845
Atallah, M. T., 480, 487
Atlas, D., 436, 437(126)

Avrameas, S., 55, 60, 86, 264, 265, 267, 268(8), 270, 277, 709, 710(1), 712(8), 713, 714, 715(14), 716, 904, 907(17), 908(17), 909(17), 913, 914(37), 925(37)
Axén, R., 16, 17, 20, 24, 28(3, 6), 29(19), 30(13), 38, 39(6), 40(43), 42, 43, 44(51), 48, 52, 73, 75, 77(16), 127, 351, 353, 362, 373, 383, 385, 386, 387, 388(17), 389, 392, 412, 414(39), 420(39), 421(39), 424(39), 425(39), 427, 428(91), 436, 441(39), 493, 520, 527, 529, 535, 537(3), 764, 845, 851, 929

B

Bach, J. A., 639
Baddour, R. F., 895
Bahl, O. P., 101
Baker, H., 436, 437(129)
Balaji, S., 258
Balmer, G., 115
Badulnur, S., 558
Bangham, A. D., 218, 220(1), 222(1), 699
Barbotin, J. N., 907, 908, 910(30), 912(30), 913, 916(30), 919(30), 920(30)
Bar-Eli, A., 326, 327(18), 383, 436, 437(122), 568
Baricos, W. H., 298, 302, 305, 306(11), 311, 312, 315, 316(11), 317, 883, 886(23), 887(23)
Barker, S. A., 84, 86(7), 87(7), 88, 92(7), 98, 100, 166, 341, 384
Barndt, R. L., 259
Barre, P., 697
Barrett, K. E. J., 894
Barrnett, R. J., 546, 553(4)
Barry, S., 883
Barth, T., 759
Bartling, G. J., 289
Bartoli, F., 228, 241, 242, 638
Bartosek, I., 242, 677, 680, 682, 693(28)
Batchelor, F. R., 760

Batlle, del C., A. M., 846, 882
Bauer, K., 760
Bauliew, E. E., 188
Baum, H., 568
Bauman, E. K., 178, 647
Baumber, M. E., 548, 555(8), 556(8)
Beard, K. W., 724
Beck, J. F., 832, 833(4), 837(4), 839, 842(4), 843(4)
Beck, S. R., 180
Becker, R. R., 326
Begue-Canton, M. L., 567, 572
Bello, J., 516
Bender, M. L., 538, 559, 567, 572, 843
Benesch, K., 546, 553(4)
Benson, R. E., 123
Bentley, R., 478, 479(1)
Benzinger, T. H., 659
Berenson, J. A., 841
Berezin, I. V., 398, 414(7), 420(7), 426(7), 437(7), 439(7), 559, 560, 561(18), 563(18), 564(18), 566(18), 568, 569(30), 571, 572(4), 573(3, 4), 574(3, 4), 575(3, 4), 667
Berger, R. L., 659
Bergman, R., 373, 412, 414(39), 420(39), 421(39), 424(39), 425(39), 441(39)
Bergmeyer, H. U., 637
Bergwall, M., 412, 414(40), 420(40), 421(40), 424(40), 425(40), 528, 529(11), 535(11)
Berjonneau, A. M., 264, 268(6), 277, 903, 907(9), 908(9), 909(9)
Berliner, L. J., 392, 435
Bernath, F. R., 149, 153, 254, 259(13), 677, 680, 682, 693(26)
Bernfeld, P., 98, 171, 191, 443, 902, 907(1), 908, 909
Bessman, S. P., 682, 697
Bessmertnaya, L., 366, 367(8)
Bethge, P. H., 557
Bettonte, M., 241
Betz, A., 926
Bieber, R. E., 443
Bischoff, K. B., 413, 423(45)
Bishop, W. H., 548, 549, 551(12), 552(9), 554(9), 556(12), 557
Bissell, S., 294, 302(1)
Black, C. D. V., 219, 700, 704, 708
Black, M. J., 479

Blackburn, H., 255
Blaedel, W. J., 414, 420(63), 421(63), 604, 913
Blaisdell, J. L., 738
Blake, C. C. F., 546
Blake, R., 282, 284(4), 288(4)
Blatt, W. F., 300
Bleakeley, R. L., 567, 572
Bloch, R., 436, 438(134)
Blum, J. J., 414, 420(58, 59), 421(58)
Blumberg, S., 436, 437(126)
Blumenthal, R., 924, 929(49)
Boguslaski, R. C., 215, 414, 420(63), 421(63), 604, 913
Bohak, Z., 520
Bollen, A., 556
Bomstein, J., 760
Bonet-Maury, P., 685
Bonnichsen, R. K., 479
Bonté, A., 84
Borgerud, A., 667, 668(4), 670(4), 671(4)
Botsch, H., 195
Bottomley, R. C., 882
Boudrant, J., 278, 325, 341
Bouin, J. C., 480, 487
Bourdillon, C., 916
Bourdillon, D., 414, 420(70), 421(70), 427(70), 428(70), 434(70), 435, 505
Bouteille, M., 710
Bovée, K. C., 215
Bowcott, J., 151
Bowes, J. H., 122, 553
Boyer, P. D., 568
Boyse, E. A., 703
Brady, R. O., 702
Brand, K., 497(14), 499(14), 500, 502(14)
Brandt, J., 35, 37
Brandt, J. L., 790
Brandt, R. B., 479
Breslau, B. R., 310, 313, 314
Bright, H. J., 316, 347, 426, 479, 482(4), 636, 883
Brill, A. S., 479
Brocklehurst, K., 40, 41(45), 42, 48, 49(11), 52, 53, 384, 385(12), 389, 400, 403(24), 406(24), 407, 408(24), 436, 437(124), 438(24)
Broome, J. D., 689
Brostrom, C. O., 884
Broughton, P. M. G., 645

Broun, G., 264, 268(6), 270, 272, 277, 398, 412(12), 414(41, 42), 415(42), 420(12, 41, 42, 68, 69, 70), 421(42, 68, 69, 70), 423(42, 68), 424(41, 42), 427(12, 42), 428(12, 42, 70), 429(42), 430(42), 432 (68), 433(102), 434(70), 435, 441(42, 105), 505, 559, 682, 697, 903, 904(10), 907(7, 8, 9, 10, 11), 908(10, 11), 909 (10, 11), 913(10, 12), 914(10, 37), 915, 916, 917, 918, 920, 921(11), 923, 924, 925(37), 929(50)
Brown, E., 84
Brown, F. S., 336
Brown, H. C., 890
Brown, H. D., 6, 178, 289, 471, 659
Brown, J. C., 517, 518(26), 520(26, 27), 522(27), 523(27), 524(38)
Brown, S. P., 85
Brown, W. E., 188
Brubacher, L. J., 567, 572
Brümmer, W., 55
Brunori, M., 539, 540, 541, 545
Brusca, D. R., 328
Bryant, W. M. D., 105
Buchholz, K., 443, 528
Buckland, B. C., 719, 840
Buckland, R. A., 700, 701(7), 702(7), 707 (7)
Buckmann, A. F. E. P., 316, 883
Buennig, K., 171
Bukatina, A. E., 571
Bunn, C. W., 548, 555(8), 556(8)
Bunting, P. S., 120, 401, 413(26), 421(26), 425, 426(81)
Burger, P. E., 846
Burgess, A. W., 527, 533(9), 537(9)
Bustin, M., 519, 520(31), 521(31)
Butler, L., 290
Butterworth, T. A., 724
Buttolph, M. A., 645

C

Cadavid, N. G., 747, 748(13)
Cairns, T. L., 123
Callaghan, O. H., 888, 893(4)
Calvet, E., 659
Camerman, N., 548, 555(8), 556(8)
Campbell, D. H., 38, 48, 49
Campbell, H. A., 703

Campbell, J., 124, 126(20), 215, 216, 322, 639, 698
Canning, L. M., Jr., 667
Capizzi, R. L., 690
Caplan, S. R., 270, 353, 398, 412(8, 11), 414(8, 34, 43), 420(8, 11, 34, 43), 421 (34), 424(43), 425(34), 427(34), 429 (34), 430, 431(34), 432(8, 11), 433(8), 436(95), 438(34), 439(34), 441(34, 43), 904, 907(13, 14), 909(13, 14), 913(14), 914(39), 924, 927, 929(49)
Carlsson, J., 34, 40, 41(28, 45), 42, 351, 384, 355(12), 389, 392, 427, 428
Carr, P. W., 667
Carter, C., 122
Carter, R., Jr., 352
Cassman, M., 507
Cater, C. W., 553
Cecere, F., 241
Cepure, A., 319, 478
Chain, E. B., 760
Chambers, R. P., 253, 298, 302, 305, 306, 311, 312, 314, 315, 316(11), 317, 340, 391, 436, 883, 886(23), 887(23)
Chan, W. W.-C., 336, 337, 338(2), 383, 491, 495, 497(14, 15, 16), 498, 499(11, 12, 14, 15, 16), 500, 502(14), 503, 559
Chance, B., 371, 372, 479, 480, 926
Chance, E. M., 372
Chang, T. M. S., 118, 201, 203, 206(4, 5, 6), 207(3, 4, 5, 6), 208(4, 5, 6), 209 (6), 210(2, 3, 4, 5, 6), 212(2, 3, 4, 5, 6), 214, 215(2, 3, 4, 5, 6), 216(3, 4, 5, 6), 217(6), 676, 677, 679(3), 680(7, 18), 681(5, 6, 7, 15), 682(4), 683(5, 6, 15), 684(3, 4, 5, 7, 8, 18, 19), 685(3, 4, 5, 7), 686(7, 8, 68, 71), 688 (3, 5, 15, 32), 689(9, 10, 19), 691(9, 10, 11, 12), 692(12), 694(9, 10, 18, 29), 695 (3, 5, 18), 696(5, 7, 30), 697(5, 15, 18), 698
Changeux, J. P., 504, 512(2), 925
Chantler, P. D., 390
Chaplain, R. A., 571
Charles, M., 341, 722, 724, 792, 886
Charm, S. E., 184
Chase, T., 255, 389, 390
Chattopadhyay, S. K., 178, 289, 471
Chedroff, S., 843
Chiancone, E., 543

Chibata, I., 164, 184, 202, 215, 384, 443, 682, 692, 719, 733, 737, 739, 740(2), 741(2, 4), 743(7), 744(1), 745, 746, 747, 748(1, 2, 5, 10, 11, 12), 749(10), 750(7, 10, 11), 751(7, 8), 752, 754(7, 9, 12), 758(6), 759(12), 769
Chin, C. C. Q., 134
Chirpich, C., 282, 284(4), 288(4)
Chmurny, A., 887, 895(3)
Cho, I. C., 342, 369, 392, 436, 514, 934
Chong, D. K. K., 336, 337, 338(2), 495, 498, 499(11)
Ciegler, A., 184
Clark, J. L., 517
Clark, L. C., 585, 586, 617(8, 10)
Cleland, W. W., 894
Cleveland, D. S., 297
Coffey, J. F., 697
Cohen, C., 436, 437(128)
Cohen, G., 479
Cohen, W., 253, 298, 302, 305, 306(11), 311, 312, 315, 316(11), 317, 340, 391, 436, 883, 886(23), 887(23)
Cole, M., 760
Colley, C. M., 701
Colman, R. F., 504, 512(8)
Colosimo, A., 539, 540, 545
Colton, C. K., 412, 413(44), 414(44), 420 (44), 421(44), 424(44), 426(44), 436, 438(132), 439(44), 441(44), 887, 895 (3)
Constantinides, A., 249, 253(6), 254(6), 258, 259(6), 260, 261(28), 768
Conte, A., 55, 85
Converse, C. A., 282
Conway-Jacobs, A., 519, 520(31), 521(31)
Cooney, C. L., 667, 720, 726(12), 736(12)
Cooney, D., 677, 680, 682, 690, 693
Corbin, J. D., 884
Cordes, E. H., 552
Corey, E. J., 130
Cornish-Bowden, A., 504, 515
Corno, C., 241
Coughlin, R. W., 341, 722, 724, 792, 886
Coulet, P., 907, 908
Čoupek, J., 60, 66, 68, 70(6), 71(3), 72(8), 76(6), 85
Couwenbergs, C., 373, 376(12)
Crambach, A., 894
Craven, D. B., 864

Crawford, V. L., 557
Cresswell, P., 390, 436, 437
Crochet, K. L., 593, 617(42)
Crook, E. M., 37, 40, 41(45), 42, 47, 48, 49(6, 11), 52, 53, 338, 339, 384, 385 (12), 389, 400, 403(23, 24), 406(24), 407, 408(23, 24), 414, 415(52), 419, 420(75), 436(23), 437(23, 124), 438 (23, 24), 523, 732, 733
Cross, R. A., 297, 298, 310
Csorba, I., 222
Cuatrecasas, P., 26, 28(18), 38(18), 72, 76, 493, 845
Cullin, L. F., 585(39), 593, 614(39), 615 (39), 617(39)

D

Dahlgren Caldwell, K., 43, 44(51)
Dahlqvist, A., 177, 178, 460, 822, 823
Dale, E. C., 336
Danielsson, B., 472, 598, 667, 668(3, 4), 669(5), 670(3, 4), 671(4), 673(5), 674 (5), 675(5)
Danno, G., 815
Das, J., 178
Das, K., 841
Das Gupta, U., 282(16), 283
Datta, R., 444, 449
David, A., 559, 924, 929(50)
Davidson, A., 120
Davidson, B., 260, 261(25, 28)
Davidson, L., 677, 680, 682, 693(24)
Davies, P., 503, 879, 880
Davis, A. W., 647, 652(5)
Davis, J. C., 211, 306, 307, 308, 309, 724
Dayhoff, M., 13
Dean, P. D. G., 864, 884
De Flines, J., 185, 190(17), 720
Degani, Y., 173, 177, 436, 573, 574(7)
De Gier, J., 220
De Jarnette, E., 516
Dekaban, A. S., 702
de Laplace, P. S., 659
Delin, S., 760
Della Penna, G., 241
De Lorenzo, F., 517
Demain, A. L., 720, 726(12), 736(12)
Dembiec, D., 479

AUTHOR INDEX

De Simone, J. A., 412, 414(43), 420(43), 424(43), 441(43), 913, 914(39)
Detar, C. C., 778, 792, 796, 797, 798(1), 802, 823
Determann, H., 171
DeVault, D., 371
DeWeer, P., 893
De Wilt, H. G. J., 810
Diegelman, M. A., 358, 472
Dinelli, D., 211, 228, 738, 824
Dinish, K. N., 180
Dintzis, H. M., 55, 65, 85, 93, 103(21), 445, 446, 447
Dirks, J. H., 697
Dixon, J. E., 344, 345, 352(14), 508, 677, 680, 682, 695(21), 696(21), 841
Dixon, M., 401, 413(25), 470
Doig, A. R., 585(11), 586, 591, 617
Domurado, D., 264, 268(6), 903, 907(8, 9, 11), 908(8, 9, 11), 909(9, 11), 920(8), 921(11), 923
Donaldson, T., 921
Dorfman, R. I., 188
Dowdle, E. B., 846
Downer, N. W., 369
Doyle, T., 316, 883
Drahoslav, L., 85
Drarid, A., 300
Drenan, J. W., 179
Drevin, H., 28, 34, 41(28), 42, 493, 845
Drew, J. W., 790
Dritschilo, W., 636
Dunnill, P., 51, 52, 53(19), 310, 311, 316, 325, 326(13), 352, 384, 470, 559, 717, 719, 720, 721, 722, 724(1, 15), 726(12), 727, 728(27, 28), 729, 733, 734, 735, 736(12, 22), 737(28), 840, 841, 883
Durst. R. A., 582
Duysens, L. N. M., 389

E

Eaton, D. L., 135
Edelman, G. M., 118, 531
Edwards, B. F. P., 557
Edwards, V. H., 201, 677, 823
Eholzer, U., 130
Ehrlich-Rogozinski, S., 131, 384, 386(15)
Eisele, G., 557

Eisenberg, H., 507, 512(20)
Ekström, B., 760
Ellis, M. J., 553
Ellman, G. L., 522
Elmore, D. T., 390
Elving, P. J., 114
Elzinga, M., 888, 893(4)
Emery, A. N., 48, 52, 166
Emmerling, D. C., 677, 681, 682
Engasser, J. M., 179, 373, 378, 398, 400, 408(21), 412(1), 413(32, 33), 414(1, 32, 33), 415(1, 32, 33), 416(33), 417, 418, 419(33), 420(1, 32, 33), 421(32, 33), 422, 423(1, 33), 424(1, 32, 33), 425, 426(1), 427(1), 428(89), 429(89), 435(1), 437(1), 438(1), 439(1, 32, 33), 440, 441(1, 32, 33), 442(32), 466, 509, 510
Englander, S. W., 369
Engvall, E., 710
Entenmann, C., 94
Epstein, C. J., 516, 517, 518(17, 18), 521(17), 522(7)
Epton, J., 88
Epton, R., 84, 86(7), 87(7), 88, 89(23), 90, 91(23), 92(7), 94, 97, 98, 100, 101(30), 102(30), 103(30), 104(29), 341, 384
Erdei, L., 222
Erecinska, M., 372
Erlanger, B. G., 85
Ernback, S., 16(8), 17, 20, 28(3, 6), 29(19), 39(6), 48, 52, 73, 75, 77(16), 383, 385, 436, 493, 520, 529, 764, 851, 929
Eskamani, A., 255, 389
Estabrook, R. W., 347
Esterman, E. F., 399, 400(17), 403(17), 408(17), 438(17)
Evans, D. R., 557
Evans, W. G., 760
Everse, J., 659, 883
Ewers, J., 48, 518, 907

F

Fahimi, H. D., 555
Fahrney, D. E., 567
Fajszi, C., 222
Falb, R. D., 215, 342, 677, 681, 682
Faller, D. V., 667

Farr, A. L., 110, 290
Farrell, P. C., 436, 438(12)
Fasold, H., 384
Fasella, P., 539
Faulk, W. P., 709
Fawcett, J. S., 56
Feder, J., 89, 567, 571
Feeney, R. E., 14, 444
Fein, M. L., 548
Feinstein, R., 479, 685
Feldman, K., 501
Fensom, A. H., 719, 736(5)
Fernandes, P. M., 249, 253(6), 254(6), 259(6), 260
Ferramola, A. M., 846
Fetherolf, K., 517, 519(16)
Fetzer, U., 130
Field, G. F., 583
Fields, R., 533
Fieser, L. F., 555
Fieser, M., 555
Filacrione, E. M., 548
Filippusson, H., 120, 408, 438(29), 638, 639
Filmer, D., 504
Fink, D. J., 297, 306, 307, 316, 373, 414, 420(67), 421(67)
Fisch, H.-U., 29
Fleet, G. W. J., 281, 282(1, 2), 284
Flodin, P., 20, 43(1)
Ford, J. R., 253, 297, 302, 306(11), 314, 315, 316(11), 340, 391, 436, 799, 801, 803, 804, 805, 806, 807, 808, 883, 886(23), 887(23)
Fornstedt, N., 43
Forrester, L. J., 659
Forster, H., 84, 108, 112(3), 113(5)
Frankenfeld, J. W., 215
Frank-Kamenetskii, D. A., 415
Franks, N. E., 184
Freeman, A., 128(27, 28), 129, 132
Freudenberger, J., 389
Fridkin, M., 890
Frieden, C., 504, 507, 508(17), 509(17), 510, 512(8, 17, 25, 26), 513(17, 25)
Fromm, H. J., 515
Frost, A. A., 487
Fruendenberger, J., 255, 258
Fuganti, C., 242

Fujimura, M., 886
Fukui, S., 471, 472, 559, 885
Fuller, C. W., 316, 883
Furui, M., 741, 743(7)
Fuse, N., 165, 737, 747, 748(2, 5), 758(6)

G

Gabel, D., 24, 351, 362, 363, 364, 365, 383, 386, 391, 427, 428(91), 435, 436(108), 524, 527, 530(4), 531(6, 7), 532(8), 533(6, 7, 9), 534(7), 535(4, 5), 536(4), 537(3, 8, 9), 559
Gaertner, F. H., 465
Gal, A. E., 702
Galli, G., 241
Ganelina, L. S., 571
Garattini, S., 242
Gardner, C. R., 412, 413(44), 414(44), 420(44), 421(44), 424(44), 426(44), 439(44), 441(44), 887, 895(3)
Gardner, D. L., 215, 677, 681, 682
Garren, L. D., 884
Garrett, P., 887, 893(2), 895(3)
Garver, J. C., 823
Gassen, H. G., 843
Gautheron, D. C., 507, 907, 908(24, 26)
Gavalas, G. R., 440
Geffroy, G., 915
Geffroy, J., 412, 414(41), 420(41), 424(41)
Gellf, G., 264, 268(6), 278, 325, 341, 903, 907(9, 11), 908(9, 11), 909(9, 11), 910, 921(11), 923
Gelotte, B., 43
Gerring, M., 290
Gestrelius, S., 191, 350(34), 353, 388, 389(19), 432, 433, 434(103), 460, 469, 475, 477, 559, 667, 836, 837, 842(7), 887
Ghizinghelli, D., 242
Giacin, J. R., 259, 261
Giacometti, G. M., 539
Giacomozzi, E., 241
Giangrasso, D., 242
Gibson, K., 594
Gilbert, S. G., 244, 248, 255, 261, 389, 906, 907(20), 908(20)
Gilden, R. V., 48
Gill, G. N., 884
Gilmore, R. B., 260, 261(25)
Ginder, A. F., 790

Giovenco, S., 241
Givol, D., 517, 519
Glassmeyer, C. K., 51, 52, 436, 437(121), 568
Glazer, A. N., 326, 327
Glew, R. H., 212
Gadinot, C., 507, 907, 908(24)
Goheer, M. A., 340, 341
Gokel, G., 127, 130
Gold, A. M., 567
Gold, M. S., 849
Goldberger, R. F., 516, 517
Goldbeter, A., 398, 412(11), 420(11), 432 (11), 927
Golden, L. S., 729
Goldin, B. R., 507, 508(17), 509(17), 512 (17), 513(17)
Goldmacher, V. S., 559, 571, 572(4), 573 (4), 575(4), 576(4)
Goldman, D. E., 568
Goldman, R., 16(7), 17, 18, 97, 270, 353, 362, 398, 399(2, 4), 400(2, 4), 401(2), 403(2), 408(2), 412(2, 4, 10), 414(34, 35, 36), 419(36), 420(2, 4, 10, 34, 35, 36), 421, 423(10, 36), 424, 425(36), 426(2, 36), 427(34, 35, 36), 429, 430, 431, 432, 433(99), 435(2, 4), 436(4, 95), 437(2, 4), 438(2, 34, 35, 36), 439 (4, 34, 35, 36), 441(34, 35, 36), 444, 462, 486, 487(25), 519, 537, 904, 906, 907(13, 14, 15, 16, 19), 908, 909, 913 (14), 914(15), 917
Goldstein, A., 282(15), 283
Goldstein, D. B., 480
Goldstein, L., 16(7), 17, 18(7), 85, 95 (19), 128(27, 28), 129, 132, 346, 352 (17), 353(17), 384, 388(14), 398, 399 (2, 3), 400(2, 3, 16), 401(2, 16), 402, 403(2, 16), 404, 405, 408(2, 16), 409 (19, 20), 410(19), 411(20), 412(2, 20), 420(2, 3), 421(2), 426(2), 435(2, 3), 436(109), 437(2, 3, 107, 109, 125, 126, 128), 438(2, 16, 19, 20), 444, 519, 528, 529(10), 559, 568, 726
Golovina, T. O., 502
Gomori, G., 481
Gonda, A., 697
Goodson, L. H., 178, 637, 647, 652(5)
Goodwin, T. W., 571
Gordon, A., 682, 695

Gorecki, M., 519
Goren, H. J., 890
Gould, B. J., 340, 341
Gowenlock, A. H., 645
Grafflin, M. W., 46
Graham, R. C., 713
Granelli, A., 588, 601(22), 608(22), 609 (22), 613, 616(22)
Grasselli, P., 242
Grassetti, D. R., 385
Gratzer, W. B., 390
Graves, D., 370, 374, 376(17), 912, 919
Gray, D., 586
Green, D. E., 568
Green, N. M., 492, 499
Greenbaum, M. A., 682, 695
Greenwood, C., 368
Gregoriadis, G., 212, 215, 218, 219, 220 (2), 221, 224, 225, 226, 227(10), 677, 679, 682, 688(17), 689(17), 699, 700 (4), 701(7), 702(7), 703(4), 704(16), 705(9, 10), 706(4), 707(17), 708(9), 709(9, 10)
Greville, G. D., 220
Griffith, L., 294, 302(1)
Grinstein, M., 846
Gross, D. R., 527
Grunberg-Manago, M., 888, 893(4)
Guesdon, J. L., 716
Guilbault, G. G., 178, 347, 584, 585(13, 21, 23, 25, 27, 28, 30, 33, 34), 586, 587, 588, 589, 590(28), 591(30, 31), 594, 595(25, 26, 28, 30, 31), 597, 598(30, 33, 34), 599(13), 600(13, 17, 18, 19, 20, 21, 23, 26, 27, 28, 29, 30, 31), 602, 603, 604 (17, 18), 606, 607(13), 608, 609, 610, 611(12, 13, 23, 30, 31, 33, 34), 612, 613, 614(33, 34), 615(13, 15, 17, 18, 26, 28, 30, 31, 32, 33), 616(15, 17, 18, 20, 21, 23, 25), 617(12, 13, 25, 26, 27, 28, 29, 30, 31, 33, 34), 618, 623, 626(50), 627, 628, 630, 631, 632, 633, 636, 647, 902, 907(3), 908, 909
Guilbert, B., 277, 710, 714
Guilford, H., 867, 868(9), 883(9), 884(25)
Guillon, C., 264, 268(6), 903, 907(9), 908 (9), 909(9)
Guire, P., 282, 284(4), 288(4)
Gulinelli, S., 211, 228, 241, 242, 638
Gunter, C. R., 567, 572

Gunzel, G., 84
Günzel, G., 108, 112(2), 113(4, 5)
Gurba, P. E., 846
Gurne, D., 848
Gurvich, A. E., 38
Gutierrez-Hartman, A., 120, 677, 680, 682, 693(24)

H

Haas, D. J., 548, 555(7)
Habeeb, A. F. S. A., 48, 264
Haber, E., 516
Hagen, A., 637
Haley, B. E., 282
Hallén, A., 25
Hamilton, B. K., 412, 413(44), 414(44), 420(44), 421(44), 424(44), 426(44), 439(44), 441(44), 887, 895(3)
Hamilton, R. W., 215
Han, M. H., 529
Handschumacher, R. E., 690
Hansen, E. H., 588, 612
Hansen, S. A., 819
Hanstein, W. G., 282
Hara, H., 885
Hardegen, F. J., 290
Hardy, P. M., 122, 553, 555(19)
Hardy, T. L., 760
Harker, D., 516
Harper, W. J., 738
Harrington, W. F., 516
Harris, D. L., 94
Harris, D. R., 184
Harris, E., 548, 843
Hartsuck, J. A., 557
Harvey, M. J., 864
Hasselberger, F. X., 16, 341, 722, 792
Hastings, J. W., 884
Havashi, K., 322
Havewala, N. B., 134, 776, 778, 789, 792, 793, 796, 797, 798(1), 802, 810, 811(5), 823
Hayano, M., 188
Hayashi, Y., 834
Haynes, R., 436, 437(120), 568
Hedén, C.-G., 324
Heinze, E., 849
Helfferich, F., 463
Helmreich, E., 501

Hemmingsen, S. H., 817, 820
Henderson, L. M., 739
Henderson, P. G., 215
Hennrich, N., 55
Hepp, K. D., 493
Herbert, D., 479
Herbert, Y. A. L., 385
Hers, H. G., 700, 702(11)
Hersh, E. M., 705
Hersh, L. S., 134, 347, 480, 677, 680, 682, 693(22, 23)
Hertler, W. R., 130
Hervagault, J. F., 432 433(102), 917, 918, 919
Hess, B., 926
Hess, G. P., 410, 447
Hesseltine, C. W., 719
Hew, C. L., 282
Hibbert, B. L., 88, 90, 91, 101(30), 102(30), 103(30)
Hibbert, S. R., 702
Hicks, G. P., 171, 181, 585, 600(9), 617(9), 636, 902, 907(2), 908, 912(2)
Higgins, J., 927
Hill, M. W., 218, 220(1), 222(1), 699
Hinberg, I., 823
Hinrichs, E. V., 132
Hiramoto, R., 264
Hjertén, S., 20, 43, 56, 191
Hlavica, P., 849
Hoare, J. P., 104
Hochstadt, H. R., 246
Hoffman, P., 130
Hofmann, C. H., 843
Hofstee, B. H. J., 43
Hofsten, B. V., 436
Holt, P., 151
Holtzapple, P. G., 215
Hopwood, D., 555
Horecker, B. L., 495
Horisberger, M., 325
Hornby, W. E., 52, 53, 119, 120, 121, 124, 126(20), 127(6, 9), 322, 338, 339, 353, 400, 403(23), 408(23), 414, 415(52), 436(23), 437(23), 438(23, 29), 472, 523, 560, 638, 639, 732, 733
Horton, H. R., 436, 506(22), 507, 508(22), 512(22), 517, 518(26), 520(26, 27), 522(27), 523(27), 524(39), 525(39)
Horton, R., 344

Horvath, C., 119, 120, 179, 373, 378, 398, 400, 408(21), 412(1), 413(32, 33), 414 (1, 32, 33), 415(1, 32, 33), 416(33), 417, 418, 419, 420(1, 32, 33), 421(32, 33), 422, 423(1, 33), 424(1, 32, 33), 425, 426(1), 427(1), 428(89), 429(89), 435(1), 437(1), 438(1), 439(1, 32, 33), 440, 441(1, 32, 33), 442(32), 466, 509, 510, 677, 680, 682, 693(25), 843
Houba, V., 709
Hough, J. S., 48, 52, 166
Hrabankova, E., 585(28), 587, 589, 590 (28), 595(26, 28), 600(19, 26, 28), 603, 608, 612, 615(26, 28), 617(26, 28), 907
Hradil, J., 72
Hsu, C. J., 48
Hsu, K. G., 710
Huang, C. Y., 512
Huang, N. F., 633
Huang, R. Y. M., 179
Hubálková, O., 60, 70, 72(8), 85
Huelck, V., 180
Huggett, A. St. G., 479
Hughes, N. A., 864
Hulla, F. W., 384
Hultin, H. O., 480, 487
Hummel, B. C. W., 445(9), 446, 447
Humphrey, A. E., 201, 636, 677, 720, 726 (12), 736(12)
Hunt, L. T., 13
Hunter, M. J., 321, 556
Hussein, W. R., 618
Hustad, G. O., 343
Hydén, H., 677, 680, 693(20)
Hyslop, R. M., 850

I

Ihler, G. M., 212
Ikeda, S.-I., 471, 472, 559, 885
Ikegami, A., 179
Ikenberry, L. D., 176
Imai, K., 517
Imai, N., 179
Inman, D. J., 119, 120, 121, 124, 127(6), 322, 638, 639
Inman, J. K., 55, 65, 85, 93, 103(21), 445, 446, 447, 472
Irving, J., 729
Isaacs, N. S., 843

Isambert, M. F., 85
Isemura, T., 517
Ishikawa, T., 746, 748(1)
Izu, M., 175

J

Jacobs, W. B., 637, 647, 652(5)
Jacobsen, C. F., 111
Jacobsen, R., 903, 907(11), 908(11), 909 (11), 921(11), 923
Jagendorf, A. T., 36, 49
Jagner, D., 588, 601(22), 608(22), 609(22), 613, 616(22)
Jahnke, M., 667
Jakubowski, J., 261
Jameson, G. W., 390
Janik, A. N., 215
Janolino, V. G., 525, 526(43)
Janson, C. A., 894
Janson, E. F., 549, 552(11), 553(11)
Janson, J.-C., 21, 22(9), 23(7), 24, 30(13), 34(9), 39(7), 425, 535, 733
Jaroslow, B. N., 685
Jarvis, N. R., 179
Jaworek, D., 195
Jencks, W. P., 14, 552
Jender, D. J., 414, 420(59)
Jensen, P. K., 480
Jergil, B., 883, 884(25)
Jírovský, B., 759
Johansson, A. C., 55, 57, 58(7), 60(7), 63, 399, 406, 407(18, 28), 409(28), 438 (28), 473, 659, 661, 662, 663(5), 664 (6), 665, 666, 841
Johnson, D. E., 184
Johnson, L. J., 212, 214, 215, 677, 680, 682, 684(19), 689(19)
Johnson, M. J., 290
Joly, G., 432, 433(102), 917, 918, 919
Jones, J. B., 831, 832, 833(4), 836, 837 (4), 839, 841, 842(4), 843(4)
Jones, J. W., 868
Jones, P., 479, 483(12)
Joó, F., 222
Josephs, R., 507, 512(20)
Joustra, M., 28
Joyeau, R., 84
Juliano, R. L., 700
Julliard, J. H., 507, 907, 908(26)

K

Kaback, H. R., 282
Kadlubar, F. F., 850
Kågedahl, L., 28
Kahan, L., 556
Kaiser, C., 498, 499(12)
Kaiser, E. T., 390
Kálal, J., 68, 71(3), 72
Kallen, R., 358, 472
Kanbayashi, A., 769
Kaplan, N. O., 201, 316, 317(27), 344, 345, 352(14), 508, 659, 677, 680, 682, 695(21), 696(21), 841, 868, 873(10), 883
Kapune, A., 528
Karlin, A., 353
Karnovsky, M. J., 713
Karube, I., 248, 258(5), 902, 907(5), 909
Kasche, V., 373, 412, 414(39, 40), 420(39, 40), 421(39, 40), 424(39, 40), 425(39, 40), 441(39), 527, 528, 529(11), 531(6, 7), 533(6, 7), 534(7), 535(11), 559
Kastl, P. R., 317
Katchalski, E., 16(7), 17, 18(7), 84, 95(6), 97, 201, 263, 270, 326, 327(18), 346, 352(17), 353(17), 362, 363(7), 364(7), 365(7), 383, 384, 388(14), 398, 399(2, 3, 4), 400(2, 3, 4, 16), 401(2, 16), 402(16), 403(2, 16), 404, 405, 408(2, 16), 412(2, 4), 414(34, 35, 36), 420(2, 3, 4, 34, 35, 36), 421(2, 4, 34, 35, 36), 423(36), 424(35, 36), 425(34, 35, 36), 426(2, 36), 427(34, 35, 36), 429(34, 35, 36), 431(34), 432, 433(99), 435(2, 3, 4), 436(4, 95, 108), 437(2, 3, 4, 107, 122, 123), 438(2, 16, 34, 35, 36, 133), 439(4, 34, 35, 36), 441(34, 35, 36), 444, 462, 486, 487(25), 519, 524, 527, 530(4), 535(4), 536(4), 537, 568, 677, 904, 907(13, 14, 15, 16), 908(16), 909(13, 14, 15), 913(14), 914(15), 917, 925
Katzenellenbogen, J. A., 282(17), 283
Kaufmann, W., 760
Kauzman, W. J., 558
Kay, C. M., 436, 437(130)
Kay, G., 37, 47, 49(6), 50(7), 52, 53(7), 384, 419, 420(76, 77), 441(76), 559, 736, 738(41), 823, 906, 907(21, 22, 23), 908(22, 23)

Kay, R. E., 94
Kedem, O., 270, 353, 412, 414(34, 35, 36), 419(36), 420(34, 35, 36), 421(34, 35, 36), 423(36), 424(35, 36), 425(34, 35, 36), 426(36), 427(34, 35, 36), 429(34, 35, 36), 430, 431(34), 436(95), 438(34, 35, 36, 133, 134), 439(34, 35, 36), 441(34, 35, 36), 537, 904, 907(13, 14, 15, 16), 908(16), 909(13, 14, 15), 913(14), 914(15), 924, 929(49)
Kédzy, F. J., 390
Kelleher, G., 48
Kempner, E. S., 453
Kendrew, J. C., 546
Kendrick, R. E., 527
Kennedy, J. F., 88
Kenner, G. W., 864
Kern, W., 116
Kernevez, J. P., 373, 414, 420(70), 421(70), 427(70), 428(70), 432, 433(102), 434(70), 435, 505, 913, 914, 915, 916, 917, 918, 924(41), 925(36), 926
Kezdy, F. J., 567, 572
Khaiat, A., 905, 907(18), 908
Kiefer, H., 282(14), 283
Kierstan, M. P. J., 40, 41(45), 42, 384, 385(12)
Kiese, M., 849
Kilcullen, B. M., 313, 314
Killheffer, J. V., 567, 572
Kim, B. C., 215, 677, 681, 682
Kim, B. U., 294, 302(1)
Kimmel, J. R., 110
Kimelberg, H. K., 221
King, C. A., 884
Kinsky, S. C., 226
Kirimura, J., 327, 759
Kissel, P., 704
Kissel, T. R., 414, 420(63), 421(63), 604
Kitajima, M., 210, 214, 215
Kitawa, K., 319
Kitto, G. B., 120, 677, 680, 682, 693(24)
Kittrell, J. R., 419
Kitzinger, C., 659
Kivisilla, K., 191
Klee, W. A., 537
Kleppe, K., 319, 478
Kleyn, D. H., 259, 261
Klibanov, A. M., 398, 414(7), 420(7), 426(7), 437(7), 439(7), 559, 560, 561, 563

(18), 564(18), 566, 568, 569(30), 571, 572(4), 573(3, 4), 574(3, 4), 575(3, 4), 576(4), 667
Klockow, M., 55
Klose, S., 637
Knights, R. J., 390, 391(28), 392(28), 528, 535, 568
Knowles, C. O., 264
Knowles, J. R., 122, 281, 282(1, 2, 3), 284, 552, 553
Knudsen, C. W., 413, 421(46), 423(46)
Knupfer, H., 130
Kobayashi, T., 412, 413(38), 414(38), 415 (38, 54), 419, 420(38, 62, 65, 80), 421 (38, 54), 424(38), 425(80), 427(65), 439(38), 441(38), 442(38), 443(38), 510, 732, 733, 734
Kobes, R., 846
Koelsch, R., 373
Koening, D. F., 546
Koishi, M., 215
Kojima, S., 115
Kolb, H. J., 493
Kolthoff, I. M., 92
Kondo, A., 210, 214, 215, 677, 682, 697
Kondo, F., 185
Kondo, T., 215
Konecny, J., 429, 431(94)
Konigsberg, W. H., 282
Kooi, E. R., 809, 811(2)
Kools, J. W., 639
Korenzecher, R., 108, 114(7)
Korey, S. T., 888, 893(4)
Korosi, A., 585(11), 591, 617
Korus, R., 175, 176, 177(7), 183, 677, 679, 682, 692(14), 823
Koshland, D. E., Jr., 504, 515
Kosikowski, F. V., 823
Koslov, L. V., 535
Kostick, J., 165
Köstner, A., 191
Kosugi, Y., 769
Kovács, J., 116
Kozeny, J., 820, 821(12)
Kozlov, L. V., 366, 367(8)
Kramer, D. N., 178, 647
Kramer, G. W., 890
Krebs, E. G., 884
Kremer, M. L., 479
Kretchmer, N., 822

Krisam, G., 661, 663, 664
Kristiansen, T., 29
Krivakova, M., 60, 66, 68, 70, 71, 72(8), 85
Kuan, J. W., 630, 632, 633
Kuan, S. S., 585(21), 588, 600(21), 616 (21), 632
Kudo, S., 759
Kuhlmann, W. D., 713
Kuhn, W., 115
Kukhareva, L. V., 571
Kumagai, H., 886
Kunitz, M., 89
Kurowski, W. H., 719, 736(5)
Kushiro, H., 759
Kusserow, B., 682, 697
Kuster, B. F. M., 810
Kwong, A., 141, 362, 389
Kyle, W. S. A., 390

L

Låås, T., 21, 22(9), 23(7), 34(9), 39(7), 44
Labouesse, B., 447
Labský, J., 68, 71(3)
LaCourse, W. C., 214, 682, 695
Laidler, K. J., 104, 120, 398, 401, 412(6), 413(26, 37, 38), 414(38), 415(6, 37, 38, 54), 420(6, 37, 38), 421(38, 54), 424(37, 38), 425, 426(81), 435(6), 437 (6), 439(6, 38), 441(38), 442(38), 443 (38), 510
Lamarche, M., 704
Lamaze, C. E., 175, 176
Lambert, A. H., 253, 298, 305, 340
Lamed, R., 882
Lamotte, A., 887, 895(3)
Landau, J., 297, 298
Landfear, S. M., 557
Lang, H., 55
Langer, R. S., 887, 895(3)
Langley, T. J., 507
Lapka, R., 68, 70(6), 76(6)
Lappi, D. A., 883
Larrow, R., 682, 697
Larsson, P. O., 184, 188, 190(14), 316, 389, 840, 862, 867, 868(9), 870, 871(12), 873(3), 874, 883(9), 884(19)
Lasch, J., 366, 367, 373, 414, 420(64), 421 (64)

Lash, E. D., 690
Lau, H., 628, 630, 631
Lavoisier, A. L., 659
Lawford, G. R., 498, 499(12)
Lawrence, R. L., 181
Lawson, W. B., 834
Leathwood, P. D., 212, 215, 224
Lecoq, D., 432, 433(102), 917, 918
Lee, B. K., 188
Lee, C. P., 372
Lee, C. Y., 316, 317(27), 883, 894
Lee, Y. Y., 414, 420(60), 421(60), 425(60), 732
Leebrick, J. R., 289
Leeder, J. G., 259, 261
Lefever, R., 927
Legoy, M. D., 907, 908
Lehmann, K., 55, 85
Lehninger, A. L., 571
Lemoullec, J. M., 910
Lenhoff, H. M., 906, 907(19)
Lennox, E. S., 282(14), 283
Lenscher, E., 38
Léonis, J., 111
Lerman, L. S., 38, 49
Leuschner, F., 759
Levenspiel, O., 730, 734, 801, 802
Levi, A., 929, 935
Levich, V. G., 415
Levin, Y., 201, 346, 352(17), 353(17), 384, 388, 399, 400(16), 401(16), 402(16), 403(16), 405, 408(16), 435, 436, 437(107, 126), 438(16), 568, 677, 882
Levine, S. N., 214, 682, 695
Levinthal, C., 517, 519(16)
Levitzki, A., 502, 504
Levy, A. B., 890
Levy, D., 282, 284
Levy, H. B., 715
Levy, M., 697
Li, N. N., 184, 211, 214, 215, 328, 330, 682, 696, 903, 907(6), 909, 917(6)
Lieberman, E. R., 246
Liberman, R., 447
Liberti, A., 585(37), 592, 600(37), 602, 604, 605(37), 606, 607, 609, 614, 617
Lifter, J., 282
Light, A., 390, 391(28), 392(28), 528, 535, 568
Light, R. T., 426

Lilly, M. D., 47, 50(7), 51, 52, 53(7, 19), 201, 310, 311, 316, 325, 326(13), 338, 339, 352, 353, 384, 400, 403(23), 408(23), 414, 415(52), 419, 420(74, 76, 77), 436(23), 437(23, 74), 438(23), 441(76), 470, 523, 559, 677, 717, 719, 720, 721, 722, 724(1, 15), 726(12, 16), 727, 728(27, 28), 729, 730, 731, 732, 733(30), 734(22), 735, 736(12), 737(22, 28), 738(41), 823, 840, 841, 883, 906, 907(21, 22, 23), 908(21, 22, 23)
Lin, Y. Y., 836
Lindan, O., 215
Lindberg, M., 389, 862, 866, 873(3, 8)
Lindberg, O., 571
Linderstrøm-Lang, K., 111
Lindquist, B., 822
Lindstrom, J., 282(14), 283
Line, W. F., 141, 362, 389
Linschitz, H., 527
Lipmann, F., 884
Lipowsky, G. P., 849
Lipscomb, W. N., 550, 551(14), 555, 557
Liptak, B. G., 790
Lister, C., 697
Litt, M. H., 215
Llenado, R., 585(35), 592, 600(35, 36), 602, 607, 614
Lo, W. K., 294, 302(1, 2), 307
Loa, S. K., 214, 677, 681, 682, 696(30)
Logan, R. M., 768
Long, E., 682, 693
Lopiekes, D. V., 85
Louis, L., 221, 699
Lowe, A. G., 893
Lowe, C. R., 316, 864, 872(5), 875, 879(5), 884
Lowe, M., 884
Lowey, S., 436, 437(129)
Lowry, O. H., 110, 290, 881
Lubrano, G. J., 585(13, 30), 586, 590, 591(30, 31), 595(30), 598(30), 599(13), 600(13, 30, 31), 603, 606, 607(13), 608, 611(12, 13, 30, 31), 614, 615(13, 30, 31), 617(12, 13, 30), 907
Luck, S. M., 436, 437(128)
Lüdke, G., 127
Ludwig, M. L., 321, 550, 551(14), 555, 556, 557(24)
Luescher, E., 49

Lundberg, J., 659, 661(6), 662(6), 664(6), 665, 666(6)
Lundquist, H., 373, 412, 414(39), 420(39), 421(39), 424(39), 425(39), 441(39)
Lyons, C., 585, 617(8)
Lyons, T. P., 48, 52, 166

M

Macdonald, P. D. M., 498, 499(11)
MacIntosh, F. C., 203, 206(5), 207(3, 5), 208(5), 210(3, 5), 212(3, 5), 214, 215 (3, 5), 216(3, 5), 677, 679, 681(6), 682(4), 683(6), 684(4), 685(4)
Maeda, H., 178
Maeda, Y., 517
Maehly, A. C., 479, 480
Maekawa, K., 319
Magnuson, J., 677, 679, 682, 692(13), 697
Maier, J., 195
Mair, G. A., 546
Malave, N., 215, 677, 680, 682, 684(19), 689(19)
Malmqvist, M., 23
Malthouse, Y. P. G., 385
Mandel, M., 165, 191
Manecke, G., 84, 108, 112(1, 2, 3), 113(4, 5, 6), 114(7), 115(8), 116(9), 117(10), 201, 436, 677
Mansford, K. R., 760
Månsson, M. O., 836, 837, 842(7), 887
Marantz, L. B., 682, 695
March, S., 26, 28(18), 38(18)
Marconi, W., 211, 228, 241, 242, 638
Marcus, J., 479
March, S. C., 493
Markowitz, J. M., 114
Marquarding, D., 130
Marr, G., 88, 90, 91, 101(30), 102(30), 103(30)
Marsh, D. R., 414, 420(60), 421(60), 425 (60)
Marshall, D. L., 342, 470, 897
Marshall, R. O., 809, 811(2)
Marshall, S. P., 790
Marshall, T. H., 567, 572
Mårtensson, K., 60, 63(12), 461, 470(8)
Martinek, K., 398, 414(7), 420(7), 426(7), 437(7), 439(7), 559, 560, 561(18), 563 (18), 564(18), 566(18), 568, 569(30), 571, 572(4), 573(3, 4), 574(3, 4), 575 (3, 4), 576(4), 667
Martiny, S. C., 215
Marx, F., 888, 893(4)
Mascini, M., 585(37), 592, 600(37), 602, 604, 605(37), 606, 607, 609, 614, 617
Mashburn, L. T., 703
Mašková, H., 759
Mason, S. G., 203, 206(5), 207(3, 5), 208 (5), 210(3, 5), 212(3, 5), 214, 215(3, 5), 216(3, 5), 677, 682(4), 684(4), 685 (4)
Masuo, E., 185
Matlib, M. A., 468
Matthews, B. W., 547, 548(5)
Mattiasson, B., 177, 178, 181, 191, 342, 344, 350(34), 352(15), 353, 389, 432, 433(97), 434(98, 103), 454, 456, 457, 460, 463, 464, 467, 469, 472, 473, 475, 477, 487, 505, 508, 530, 559, 659, 661 (6), 662(6), 664(6), 665, 666(6), 667, 669(5), 673(5), 674(5), 675(5), 823
Matuo, Y., 164, 215, 682, 692, 739, 744(1), 752
Maurer, H. R., 571
Mawer, H. M., 383, 497(15), 499(15), 500
Maxwell, M. H., 682, 695
May, S. W., 211, 214, 328, 330, 682, 696, 903, 907(6), 909, 917(6)
Mayhew, E., 700
Mazarguil, H., 48
McArthur, M. J., 682, 695
McCapra, F., 583
McCloskey, J. A., 861
McClure, W. O., 531
McCracken, R. D., 822
McDonald, A., 120, 638, 639
McLaren, A. D., 398, 399(13), 400(13, 17), 403(13, 17), 408(13, 17), 435(13), 436(13), 438(13, 17)
McLaren, J. V., 84, 86(7), 87(7), 88, 89 (23), 91(23), 92(7), 97, 100, 104(29), 384
McQueen, R., 595
Means, G., 14, 444
Meerwein, H., 126
Meier, P. M., 215
Meighen, E. A., 505
Mendiara, S., 882

Mendiratta, A. K., 260, 414, 420(66), 421 (66)
Merrill, E. W., 436, 438(132)
Messing, R. A., 134, 136, 147, 149, 150, 152, 153, 155, 156, 157(11), 158, 159, 165, 167, 168, 348, 349, 444, 479, 777, 789(3)
Messinger, S., 310
Metayer, M., 559, 924, 929(50)
Mevkh, A. T., 502
Meyer, J., 441
Michaels, A. S., 296, 300, 313, 426
Micheel, F., 48, 518, 907
Michelson, A. M., 85
Michelson, S., 843
Midland, M. M., 890
Miller, C. G., 567, 572
Miller, I. R., 905, 907(18), 908
Miller, J. H., 453
Miller, N. G. A., 218, 220, 222(1), 699
Miller, S. T., 392, 435
Millette, C. F., 118
Minami, E., 115
Mineura, K., 759
Minh, C. T., 682, 697
Miron, T., 38, 173, 177, 436, 573, 574(7)
Mislin, R., 838
Mitchell, C. H., 849, 850(3)
Mitchell, P., 921
Mitz, M. A., 47, 48, 518
Miura, Y., 210, 214
Miyamoto, K., 210, 214
Miyano, S., 210, 214
Moews, P. C., 548, 555(8), 556(8)
Moffat, J., 130
Mogensen, A. O., 260, 414, 415(55), 419(55), 420(55, 66), 421(66)
Mohan, R. R., 184, 328
Moncrieff, R. W., 228
Monod, J., 504, 512
Montalvo, J., 347, 587, 593, 597, 600(17, 18), 602, 604(17, 18), 609, 610, 611, 612, 615(17, 18), 616(17, 18), 617(40, 42), 902, 907(3), 909
Monzan, P., 48
Moody, G. J., 582
Moore, S., 73, 77(15), 81(15)
Moore, T. A., 368
Moo-Young, M., 414, 419, 420(62, 65, 80), 425(80), 427(65), 732, 733, 734

Mori, T., 164, 184, 215, 384, 443, 682, 692, 733, 737, 739, 741(4), 743(7), 744(1), 747, 748(2, 5, 10, 11, 12), 749(10), 750(7, 10, 11), 751(7, 8), 752, 754(7, 9, 12), 758(6), 759(12), 769
Morioka, H., 720
Morisi, F., 211, 228, 241, 242, 638, 738, 823, 827(15)
Morozov, N. M., 571
Morris, C. J. O. R., 56
Morris, D. L., 124, 126, 322, 639
Morrison, J. D., 831
Mort, J. S., 336, 337, 338, 495, 498, 499 (11)
Mosbach, K., 55, 57, 58(6, 7), 60(7), 63, 85, 171, 177, 178, 181(3), 184, 188, 190(14), 191, 201, 316, 342, 344, 348, 350(34), 352(15), 353, 374, 388, 389(19), 398, 399, 406, 407(18, 28), 409(28), 432, 433(97), 434(98, 103), 438(28), 454, 456, 457, 460, 463, 464, 465, 467, 469, 472, 473, 475, 477, 487, 503, 505, 506(22), 507, 508(22), 512(22), 530, 559, 585(16), 587(16), 588, 593, 597(16), 598, 600(16), 611(16), 614, 616, 617, 659, 661(6), 662(6), 664(6), 665, 666(6), 667, 668(3, 4), 669(5), 670(3, 4), 671(4), 673(5), 674(5), 675(5), 677, 823, 836, 837, 840, 841, 842(7), 862, 864, 866, 867, 868(9), 870, 871(12), 872(5), 873(3, 4, 8), 874, 875, 879(5), 881, 883(9), 884(19, 25), 887
Mosbach, R., 56, 57, 58(6, 7), 60(7), 171, 177, 181(3), 184, 191, 348
Mosher, H. S., 831
Mosolov, V. V., 565, 568(21), 570(21)
Mueller, P., 212
Mukherji, A. K., 585(38), 592, 600(38), 608(38), 614(38), 615(38), 617(38)
Murphy, G. P., 677, 680, 682, 693(22)
Murphy, J. R., 884
Murray, J. F., 385
Mussini, E., 242, 677, 680, 682, 693(28)
Myrin, P. A., 24, 30(13), 425, 535

N

Na, T., 414, 420(67), 421(67)
Nack, H., 215

Naghaski, J., 548
Nagradova, N. K., 502
Nagy, G., 585(21, 27), 586, 587, 588, 589, 600(20, 27), 612, 614, 615(15), 616 (15, 20), 617(27)
Nakagawa, Y., 517, 518(19)
Nakamura, N., 759
Nakane, P. K., 709, 710(2)
Nandi, D., 844
Nanjo, M., 585(33, 34), 591, 598(33, 34), 611(33, 34), 614(33, 34), 615(32, 33), 617(33, 34)
Naparstek, A., 913, 925, 926, 927
Natake, M., 521, 815
Nathorst-Westfelt, L., 760
Neerunjun, E. D., 221, 225, 227(10), 699, 700(4), 703(4), 706(4, 16), 707
Neill, D. W., 645
Nelsen, L., 296, 299, 300
Némethy, G., 504
Nernst, W., 415
Netheler, H. G., 637
Neuberger, A., 445
Newirth, T. L., 358, 472
Newton, M. V., 130
Ngo, T. T., 443
Nicholls, A. C., 122, 553, 555(19)
Nicholls, P., 479
Nichols, J., 682, 697
Niemann, C., 843
Nielsen, M. H., 809
Nielsen, T. K., 812
Nilsson, H., 57, 58(6), 177, 585(16), 587, 588(16), 593(16), 597, 600(16), 611 (16), 614(16), 616(16), 617(16)
Nixon, D. A., 479
Noda, L., 888, 893(4)
Nolte, R., 843
Norris, R., 385
North, A. C. T., 546
Novais, J. M., 48, 52, 166, 341
Noyes, R. M., 574

O

O'Brien, P. J., 468
O'Carra, P., 883
Ochoa, S., 888, 893(4)
O'Driscoll, K. F., 175, 176, 177(7), 183, 677, 679, 682, 692(14), 823

Oettgen, H. F., 690
Ofengand, J., 282
Offermann, K., 130
Ogata, K., 720
Ogle, J. D., 51, 436, 437(121), 568
Ogletree, J., 317
Ohlson, S., 188
Ohnishi, To., 571
Ohnishi, Ts., 571
Ohno, Y., 55, 87
Ohnuma, T., 677, 679, 682, 692(14)
Okada, H., 722
Okamura, S., 322
Okay, V., 181
Okazaki, M., 210, 214
Okos, E., 738
Old, L. J., 703
Öljelund, G., 664
Ollis, D. F., 183, 352, 428, 444, 447, 449, 737
Olson, A. C., 549, 552(11), 553(11), 792, 823
Olson, N. F., 343, 823
Olsson, I., 43
O'Neill, S. P., 310, 311, 414, 415(53), 559, 724, 732, 734, 735, 737(22)
Ong, E. B., 436, 438(131)
Oplatka, A., 882, 925
Oppenheimer, H. L., 447
Orth, H. D., 55
Osborn, M., 474
Ostergaard, J. C. W., 215
Ostwald, M., 373, 376(12)
Otillio, N. F., 43
Ott, E., 46
Ott, T., 846
Ottesen, M., 111
Ouchterlony, O., 686
Outtrup, H., 811

P

Packer, L., 398, 399(13), 400(13), 403 (13), 408(13), 435(13), 436(13), 438 (13)
Påhlman, S., 43
Paillot, B., 272, 903, 907(11), 908(11), 909(11), 921(11), 923
Paladini, A. C., 747, 748(13)
Palter, R., 792

Pansolli, P., 241
Papahadjopoulos, D., 220, 221, 700
Papariello, G. J., 585(38, 39), 592, 593, 600(38), 608, 614, 615(38, 39), 617
Papenmeier, G. J., 130
Papermaster, M., 374
Pappa, R., 884
Parikh, I., 26, 28(18), 38(18), 493
Parizek, R., 165
Park, F., 246
Park, J. T., 290
Paruchuri, E. K., 792
Pascoe, E., 182, 414, 420(57), 425(57), 426(57)
Passonneau, J. V., 881
Pastore, M., 241, 823, 827(15)
Patchornik, A., 36, 49, 131, 201, 384, 386 (15), 677, 749
Patel, A. B., 178, 471
Patel, B. P., 85
Patzelová, M., 68
Paul, I. C., 550, 551(14)
Pawley, G. S., 550, 551(14)
Pazur, J. H., 319, 478
Pearson, R. G., 487
Pecht, M., 346, 352(17), 353(17), 384, 388(14), 435, 436, 437(107, 126), 568
Pecoraro, R., 290
Penasse, L., 188
Pennington, S. N., 178, 659, 667
Pentchev, P. G., 702
Perlmann, G. E., 436, 438(131), 517
Perlmann, P., 710
Perlmutter, D. D., 358
Perry, J. H., 805
Pertoft, H., 25
Peterlin, A., 175
Petersen, E. E., 436, 440(136)
Peterson, E., 843
Peyre, M., 188
Phillips, D. C., 546
Pierce, G. B., 709, 710(2)
Pirt, S. J., 719, 736(5)
Pitcher, W. H., Jr., 778, 789, 792, 796, 797, 798(1), 799, 801, 802, 803, 804, 805, 806, 807, 808, 810, 823
Podrebarac, E., 282, 284(4), 288(4)
Poglazov, B. F., 567
Pohl, R., 108, 116(9)
Pokorny, S., 66, 71, 85

Polakowski, D., 108, 117(10)
Poland, D. C., 535
Polson, A., 20
Pont, A., 215, 677, 680, 682, 684(19), 689 (19)
Popovich, R. P., 851, 853(7), 854
Popper, H., 790
Porath, J., 16(8), 17, 20, 21, 22(9), 23 (7), 24, 26, 28(3, 6, 17), 29, 32(17), 33(27), 34(9, 17), 35, 38, 39(6, 7), 43(1), 44(51), 127, 201, 362, 384, 436, 492, 493, 520, 527, 537(3), 677, 764, 845, 929
Porter, M. C., 296, 299
Porter, R. R., 281, 282(1, 2), 711
Poste, G., 700
Potts, J. T., Jr., 516
Poulsen, L. L., 850
Poulsen, P. B. R., 719, 814, 817, 820
Poznansky, M., 215, 677, 679, 680(7), 681 (7), 682, 684(7, 8), 685(7), 686(7, 8), 687(7), 696(7)
Prat, H., 659
Prater, C. D., 441, 442
Price, S., 85
Prieve, C., 677, 679, 692(13)
Prosperi, G., 884
Prough, R. A., 850
Puchinger, H., 373, 376(12), 379(11)
Pungor, E., 582
Putcha, S., 294, 302(1)
Putnam, D., 221, 699
Puzo, G., 48
Pye, E. K., 201, 357, 472, 677

Q

Quinn, J., 921
Quiocho, F. A., 264, 546, 548(3), 549(3), 550(3), 551(3, 10, 12), 555(3, 10), 556 (10, 12), 557(24)

R

Racker, E., 495, 740
Raçois, A., 84
Ramsden, H. E., 289
Randall, R. J., 110, 290
Ransome, O., 212, 214

Rappuoli, B., 241
Rarachuri, E. K., 341
Rase, H. F., 180
Rattle, H. W. E., 367
Re, L., 243, 884
Rechnitz, G. A., 582, 585(35, 36), 600(35, 36), 602, 607, 614
Reeke, G. N., 557
Regan, D. L., 52, 53, 352, 727, 728(27, 28), 729, 737
Rehak, N., 659
Reiner, R., 368
Reisler, E., 507, 512(20)
Remy, M. H., 272
Renner, R., 493
Reuter, W., 843
Reynolds, J. H., 119, 350, 514
Richards, E. G., 56
Richards, F., 122, 264, 282, 546, 548(3), 549(3), 550(3), 551(3, 12), 552(9), 553, 554(9), 555(3, 10), 556(10, 12), 557
Richards, K. E., 844
Richardson, T., 343
Richmond, J. E., 219, 700
Rieske, J. S., 282(16), 283
Rioual, J., 84
Ristow, S., 517
Roberts, D. V., 390
Roberts, G. W., 413, 414(47), 421(46, 47), 423(46, 47), 441(47), 443(47)
Robertson, C. R., 313, 426
Robins, R. K., 868
Robinson, C., 92
Robinson, P. J., 325, 326, 384, 559, 721, 724(15)
Rochmans, P., 555
Rodbard, D., 894
Rodwell, V. W., 297, 306, 307, 316
Roeske, R. W., 567, 572
Rogers, S., 211
Rolinson, G. N., 760
Romette, J. L., 913, 925(36), 926
Rony, P. R., 211, 215, 294, 302, 307, 414, 420(61), 421(61)
Rorth, M., 480
Rose, I. A., 888, 893(4)
Rosebrough, N. J., 110, 290
Rosengren, J., 43
Rosenthal, J., 114

Rossi Fanelli, A., 539
Rossi Fanelli, M. R., 543
Rossodivita, A., 243, 884
Roston, S., 256
Rothrock, J. W., 843
Rotman, B., 374
Rovito, B. J., 419
Rowe, D. S., 716
Rowe, P. N., 732
Royer, G. P., 386, 392, 435
Royer, R., 704
Rudin, D. O., 212
Rudnick, G., 282
Rupley, J. A., 290
Rusling, J. F., 585(39), 593, 614(39), 615(39), 617(39)
Rüth, W., 443, 528
Rutishauser, U., 118
Ruzicka, J., 588, 612
Rychlík, I., 759
Rydon, H. N., 122, 553, 555(19)
Ryman, B. E., 212, 215, 224, 677, 679, 682, 688(16), 689(16), 701
Ryu, D. Y., 188

S

Sabatini, D. D., 546, 553(4)
Sadar, S., 595
Sadoff, H. L., 639
Saini, R., 184, 244, 248, 252, 258(2), 260, 261, 263(2), 414, 420(66), 421(66), 768, 770, 771(6), 772(6)
Salas, M. L., 884
Salemme, R. M., 215
Salmona, M., 242, 677, 680, 682, 693(28)
Salomon, Y., 884
Salvo, J. M., 498, 499(12)
Samokhin, G. P., 560, 561(18), 563(18), 564(18), 566(18), 568, 569(30)
Sampson, D., 677, 680, 682, 693(22)
Sanderson, A. R., 390, 436, 437
Santopietro Amisano, A., 242
Sardi, A., 677, 680, 682, 693(25)
Saronio, C., 242, 677, 680, 682, 693(28)
Sani, R., 149
Sarma, V. R., 546
Sato, T., 164, 184, 215, 384, 443, 682, 692, 719, 720(8), 739, 740(2), 741(2, 4), 743(7), 745, 747, 748(10, 11, 12), 749

(10), 750(10, 11), 752, 754(12), 759 (12), 769
Satterfield, C. N., 373, 413, 414(47, 48), 421(46, 47, 48), 423(46, 47, 48), 426 (48), 439(48), 441(48), 443(47)
Sauer, F., 441
Savidge, T., 760
Schachman, H. K., 505, 507
Scheraga, H. A., 527, 533(9), 535, 537(9)
Schirmer, R. H., 888
Schleifer, A., 585(39), 593, 614(39), 615 (39), 617(39)
Schmidt, H.-L., 661, 663, 664
Schlünsen, J., 108, 113(6)
Schnaar, R. L., 894
Schnure, F. W., 212
Schnyder, B. J., 47, 733, 768
Schoen, E., 195
Schoener, B., 371
Schonbaum, G. R., 479
Schuck, J. M., 89
Schultz, J. S., 373, 414, 420(67), 421(67)
Schultz, R. D., 682, 697
Schulz, G. E., 888, 893(4)
Schulz, R. C., 116
Schutt, H., 497(74), 499(14), 500, 502(14)
Schuurs, A. H. W. M., 710
Schwartz, H. M., 841
Schwartz, I., 282
Schwartz, M. K., 690
Schwyzer, R., 29
Scott, D., 157
Scott, J. W., 831
Scott, M., 667, 668(4), 670(4), 671(4)
Seegal, B. C., 710
Segal, A. W., 219, 700, 704, 708(17)
Sehgal, S. N., 719
Sekiguchi, W., 677, 682, 697
Sela, M., 36, 49, 516
Sélegny, E., 268, 270, 398, 412(9), 414(41, 42), 415(42), 420(9, 41, 42, 68), 421 (68), 423(42), 424(41, 42), 427(42), 428(42), 429(42), 430(42), 432(9, 68), 434, 435, 441(42, 105), 682, 697, 903, 904(10), 907(7, 8, 10, 17), 908(8, 10, 17, 31, 33), 909(7, 10, 17), 913(10), 914(10, 37), 915, 920(8, 33), 921(31), 923, 925(37)
Sellinger, Z., 884
Sengbusch, G. V., 374, 379(15)

Sennett, R. E., 846
Sernetz, M., 373, 374, 376(12), 378(18), 379(11, 15)
Sessa, G., 212, 215
Ševčík, S., 68, 69(4)
Shaltiel, S., 43
Sharp, A. K., 50, 419, 420(77), 730, 731, 733(30), 736, 738(41), 823, 906, 907 (21), 908(21)
Shaw, E., 390
Shearer, C. M., 585(38), 592, 600(38), 608(38), 614(38), 615(38), 617(38)
Shemin, D., 844, 845, 848(8)
Shendalman, L. H., 426
Shiba, M., 215
Shields, A., 667
Shimizu, S., 720
Shipton, M., 385
Shivaram, K. N., 180
Shnol, S. E., 571
Shoda, F., 769
Shoda, M., 184, 769
Shrago, E., 184
Shu, F., 585(25), 588, 589, 590, 595(25), 600(29), 608, 616(25), 617(25, 29)
Shugar, D., 290
Shuler, M. L., 400, 408(22), 414(22)
Shults, M. C. N., 682, 697
Shyam, R., 253, 260
Siebeneick, H.-U., 368
Siegel, M., 887, 893(2)
Signer, E. R., 517, 519(16)
Siimer, E., 191
Silman, I., 84, 95(6), 97, 263, 270, 353, 362, 398, 399(4), 400(4), 412, 414(34), 420 (4, 34), 421(4, 34), 425(34), 427(34), 429(34), 430, 431(34), 435(4), 436(4, 95), 437(4, 123), 438(34), 439(4, 34), 441(34), 904, 907(13, 14), 909(13, 14), 913(14)
Sinanoglu, O., 558
Singh, K., 719
Singer, S., 108, 112(1), 282(14), 283
Sinskey, A. J., 724
Siu Chong, E., 215, 677, 679, 682, 691(12), 692(12)
Sjöberg, B., 760
Skentelbergy, R. G., 645
Skulachev, V. P., 371
Slanicka, J., 429, 431(94)

Slavin, G., 219, 700
Slayter, H. S., 436, 437(129)
Slowinski, W., 184
Small, D. A. P., 882
Smith, D. M., 105
Smith, E. L., 110, 507
Smith, J. M., 413, 414(49), 421(49), 423 (49), 426(49), 441(49), 443(49)
Smith, K. A., 436, 438(132)
Smith, R. K., 347
Smulders, P., 436, 438(135)
Snamprogetti, S. p. A., 241
Snell, E. E., 739
Sober, H. A., 715
Sofer, S. S., 851, 853(7), 854
Sokolovsky, M., 128(27, 28), 129
Solomon, B. A., 120, 419
Somers, P. J., 84, 86(7), 87(7), 92(7), 98, 100, 384
Somogyi, M., 826
Sovak, M., 179
Spackman, D. H., 73, 77(15), 81(15)
Sparks, R. E., 215
Sperling, L. H., 180
Spruit, C. J. P., 527
Spurlin, H. M., 46
Srere, P. A., 181, 344, 352(15), 398, 432, 434(98), 457, 465, 467, 468, 505, 559, 667
Staab, H., 532
Stadel, J. M., 316, 883
Stahmann, M. A., 55, 87, 326
Štamberg, J., 68, 69
Stamp, D., 700
Stanley, W. L., 792, 823
Stefanini, S., 539, 540
Stefanovic, V., 188
Stegemann, H., 180
Stein, W., 73, 77(15), 81(15), 928
Steinberg, I. Z., 362, 363(7), 364(7), 365 (7), 435, 436(108), 524, 527, 530(4), 535(4), 536(4)
Steiner, D. F., 517
Steinhardt, J., 514
Steitz, T. A., 555, 557(24)
Sternberg, L. D., 710
Steyermark, A., 133
Stinson, H. R., 134, 147, 153
Stockel, A., 110
Stolzenbach, F. E., 841

Stoops, J. K., 567, 572
Stopponi, A., 241
Straub, F. B., 517
Strumeyer, D., 258
Stuchbury, T., 385
Sturtevant, J. M., 664
Suckling, C. J., 717
Suckling, K. E., 717
Suggett, A., 479, 483(12)
Sumi, A., 741, 743(7)
Sumi, Y., 472
Summaria, L. J., 47, 48, 518
Sun, T. T., 556
Sundberg, L., 24, 29, 32, 33(27), 34, 43, 384, 492
Sundaram, P. V., 120, 127(9), 201, 215, 357, 398, 412(6), 413(37), 414(37), 415 (6, 37, 56), 419(56), 420(6, 37, 56), 421, 424(37), 435(6), 437(6), 439(6), 560, 677
Surovtzev, V. I., 535
Suschitzky, H., 385
Suter, H., 685
Suzuki, H., 178
Suzuki, S., 248, 258(5), 902, 907(5), 909
Svantesson, R., 527, 535(5), 536
Švec, F., 72
Swain, P. C., 226, 702
Swaisgood, H. E., 342, 344, 369, 392, 419, 436, 506(22), 507, 508(22), 512(22), 514, 518, 520, 521, 523, 524(39), 525 (39), 526(43), 559, 934
Swan, G. A., 298, 305

T

Takagi, T., 517
Takamatsu, N., 210, 214
Takasaki, Y., 769
Tanaka, M., 759
Tanford, C., 13, 14, 505
Tani, Y., 720
Tanizawa, K., 538, 559, 843
Tannenbaum, S. R., 667
Tanzer, M. L., 262
Tao, M., 884
Tarp, M., 585(23), 588, 600(23), 608, 611 (23), 613, 616(23)
Taudou, B., 713, 715(14)
Tavill, A. S., 226, 702

Taylor, J. B., 419, 559
Tazuke, S., 322
Tchola, O., 495
Tedman, R., 724
Teipel, J. W., 505
Teitelbaum, H., 369
Temple, C. J., 56
Tenenbaum, J., 527
Ternynck, T., 55, 264, 265, 268(8), 710, 713, 715(14), 716
Tesser, G. I., 29
Testa, A. J., 310
Thaer, A., 378(18), 379
Thau, G., 436, 438(134)
Thelin, H., 760
Theorell, H., 479
Thiele, E. W., 413, 414(50), 421(50), 914
Thiery, J., 925
Thiesen, N. O., 812
Thiesen, N. O., 812
Thoma, R. W., 188
Thomas, D., 180, 264, 268(6), 270, 272, 277, 373, 398, 412(8, 12), 414(8, 41, 42), 415(42), 420(8, 12, 41, 42, 68, 69, 70), 421(42, 68, 69, 70), 423(68), 424(41, 42), 427(12, 42, 70), 428(12, 42), 430, 432(8, 68), 433(8, 102), 434, 435, 441 (42, 105), 505, 559, 682, 697, 903, 904, 907(7, 8, 9, 10, 11, 17), 908(8, 9, 17, 31, 32, 33), 909(7, 9, 17), 910(30), 912 (30), 913(10, 12), 914(10, 37), 915, 916, 917, 918, 919(30), 920(8, 30, 33), 921(31), 923, 924(41), 925(36, 37), 926, 927, 929(50)
Thomas, J. D. R., 582
Thomas, T. H., 87, 88, 89(23), 91(23), 94, 97, 104(29)
Thomas, T. T., 384
Thompson, A., 460
Toda, K., 184
Todd, A., 864
Tomao, F. A., 690
Tombs, M. P., 575
Tomimatsu, Y., 549, 552(11), 553(11)
Tomioka, S., 215
Toms, E. J., 492, 499
Tosa, T., 164, 184, 215, 384, 443, 682, 692, 719, 720(8), 733, 737, 739, 740 (2), 741(2, 4), 743(7), 744(1), 745, 747, 748(2, 5, 10, 11, 12), 749(10),
750(7, 10, 11), 751(7, 8), 752, 754(7, 9, 12), 758(6), 759(12), 769
Tran-Minh, C., 277, 903, 907(8, 11), 908 (8, 11, 33), 909(11), 920(8, 33), 921 (11), 923
Traut, R. R., 556
Trayer, H. R., 882
Trayer, I. P., 882
Trosper, T., 282
T'sai, L. T., 548, 555(8), 556(8)
Tsang, Y., 436, 438(131)
Tsao, G. T., 414, 420(60), 421(60), 425 (60), 732
Tsuchiya, H. M., 400, 408(22), 414(22)
Tsu Yen Na, 373
Turing, A. M., 928
Turková, J., 60, 68, 70(6), 71(3), 72(8), 76(6), 85
Tweedale, H., 412, 413(37), 414(37), 415 (37), 420(37), 421(37), 424(37)
Tyson, W. H., 297

U

Udenfriend, S., 379
Ugi, I., 39, 127, 130, 132
Unge, T., 42
Updike, S., 171, 181, 184, 585, 600(9), 617(9), 636, 677, 679, 682, 692(13), 697, 902, 907(2), 908, 912(2)
Urabe, I., 722
Uriel, J., 273, 709, 710(1)
Uy, R., 386, 392, 435

V

Valentine, D., 831
Valentová, O., 68, 70(6), 71(3), 76(6)
van Alteena, C., 310
Van Dedem, G., 414, 420(62)
Vanderhaeghe, H., 760
van der Ploeg, M., 182, 414, 420(57), 425 (57), 426(57), 902, 907(4) 908
van der Waard, W. F., 185, 190(17)
Van Duijn, P., 182, 414, 420(57), 425(57), 426(57), 902, 907(4), 908
Van Leemputten, E., 325
Vann, W. P., 778, 792, 796, 797, 798(1), 802, 823
Van Weemen, B. K., 710

Vaughan, A., 623, 626(50)
Vavrenová, S., 71
Vecchi, A., 242, 677, 680, 682, 693(28)
Venetianer, P., 517
Venkatasubramanian, K., 149, 244, 249, 251(7), 252, 253, 254, 258(2), 259(7, 21), 260(8), 261, 263(2), 414, 420(66), 421(66), 677, 680, 682, 693(26), 719, 724
Venter, B. R., 677, 680, 682, 695(21), 696(21)
Venter, J. C., 677, 680, 682, 695(21), 696(21)
Vesterberg, O., 473
Vezina, C., 719
Vieth, W. R., 149, 153(3), 184, 244, 248, 249, 251(7), 252, 253(6), 254(6), 258(2), 259(6, 7, 13, 21), 260(8), 261(25, 28), 263(2), 414, 415(55), 419(55), 420(55, 66), 421(66), 677, 680, 682, 693(26), 719, 724, 770, 771(6), 772(6), 906, 907(20), 908
Viglia, A., 241, 823
Vinnikov, Y. A., 568
Vitobello, V., 241
Vogt, H.-G., 108, 115(8)
Volkenstein, M. V., 568
Volková, J., 68
von Hofsten, B., 23
von Storp, L. H., 586, 614, 615(15), 616(15)
Vorob'ev, V. I., 571
Voss, H. F., 316, 317(27)
Vretblad, P., 24, 28, 39, 40(43), 127, 362, 386, 388(17), 527, 537(3)

W

Wadsö, I., 659, 661, 664, 667
Wagner, C., 441
Wagner, T., 48
Wall, F. T., 179
Walsh, K. A., 436, 437(120), 569
Walter, J. L., 342
Wan, H., 843
Wan, J., 171, 191, 902, 907(1), 908, 909
Wang, D. I. C., 720, 724, 726(12), 736(12)
Wang, K., 556
Wang, M. F. P., 258

Wang, S. S., 184, 244, 248, 249, 251(7), 254, 258, 259(7), 906, 907(20), 908(20)
Ward, W. J., 302
Warren, S. G., 557
Watanake, Y., 248, 258(5)
Waterland, L. R., 313, 426
Waters, C. A., 884
Watt, G. D., 664
Waxczack, I. M., 677, 679, 682, 692(14)
Weaver, J. C., 667
Webb, E. C., 401, 413(25), 470
Weber, G., 504
Weber, K., 474
Weeds, A. G., 436, 437(129)
Weetall, H. H., 46, 47(1), 48, 134, 141, 201, 347, 362, 389, 435, 444, 480, 636, 677, 682, 693, 776, 778, 792, 793, 796, 797, 798, 802, 823, 883, 931
Weibel, M. K., 316, 347, 426, 479, 482(4), 636, 883
Weil, R., 282
Weinryb, I., 319
Weinstein, L. I., 527, 533(9), 537(9)
Weinstein, Y., 519
Weiss, L., 493
Weissmann, G., 212, 215
Weisz, P. B., 441, 442
Welch, G. R., 465
Weldes, H., 152
Weliky, N., 46, 47(1), 48, 336
Wendorff, W. L., 823
Wermuth, B., 883
Weston, P., 55, 60, 86
Wetlaufer, D. B., 517
Wetzer, J., 910
Wharton, C., 48, 49(11), 53, 400, 403(24), 406(24), 436, 437(124), 438(24)
White, E. W., 583
White, F. H., Jr., 516
White, H. A., 384
White, W. E., 282
Whitesides, G. M., 887, 893(2), 895(3)
Wider de Xifra, E. A., 882
Widmer, F., 344, 345, 352(14), 508
Wieland, O. H., 493
Wieland, T., 171
Wierzbicki, L. E., 823
Wilchek, M., 38, 72, 519, 845, 884
Wiley, D. C., 557
Williams, D. L., 585(11), 586, 591, 617

Williams, R. C., 844
Williams, R. J. P., 479
Wills, E. J., 219, 226, 700, 702
Wilson, B. M., 445
Wilson, E. L., 846
Wilson, J. H., 419, 420(76), 441(76)
Wilson, M., 861, 894
Wilson, R. J. H., 384, 906, 907(21, 22, 23), 908(21)
Windmueller, H. G., 868
Wingard, L. B., 201, 373, 398, 412(15), 414(15), 420(15), 677
Winter, B. A., 282(15), 283
Woermann, D., 441
Wold, F., 134
Wolfrom, M. L., 460
Wolodko, T. W., 436, 437(130)
Wondolowski, M. V., 823
Woods, J. S., 677, 680, 682, 693(25)
Woychik, J. H., 823
Wright, E. M., 436, 438(135)
Wu, W. H., 844
Wykes, J. R., 51, 52, 53(19), 310, 311, 316, 470, 722, 724, 726(16), 841, 883
Wyman, J., 504, 512(2), 539

Y

Yamada, H., 886
Yamada, S., 746, 748(1)
Yamamoto, K., 184, 719, 720(8), 739, 741(4), 745, 769
Yamashita, K., 741, 743(7)
Yamauchi, A., 178
Yaqub, M., 282, 284(4), 288(4)
Yasuda, H., 175, 176
Yaverbaum, S., 792, 796, 797, 798, 802, 823
Yielding, K. L., 282
Yielding, L. W., 282
Yon, R. I., 43
Yoshida, T., 327, 759
Yoshimura, S., 815
Yoshioka, M., 282
Young, D. M., 516
Young, L. J., 181
Yourtree, D. M., 659
Yutani, K., 517

Z

Zaborsky, O. R., 16, 19(6), 215, 263, 317, 319, 322, 323(9), 331, 367, 398, 420(5), 435(5), 437(5), 439(5), 720, 841
Zaccardelli, D., 241, 823, 827(15)
Zaffaroni, P., 241
Zappelli, P., 243, 884
Zeisel, H., 501
Zewe, V., 515
Zidoni, E., 479
Ziegler, D. M., 849, 850(3), 851, 853(7), 854
Zima, J., 68, 70(6), 76(6)
Zimmerman, R., 626, 627
Zittan, L. E., 719, 814, 817, 820
Zuidweg, M. H. J., 185, 190(17)
Zweibel, S., 260, 261(25)

Subject Index

A

Acatalasemia, 684–688
Acetaldehyde, 39–40, 128–131, 133–134, 863
  permeability data, 297, 302
p-Acetaminophenylethoxy methacrylate, 71–72
Acetate, 129, 133
  oxidation, 315–316
Acetate kinase
  activation, 889
  activity assay, 893–894
  cysteine residues, 887–889
  immobilization, 891–892
Acetic acid, 23, 77, 132, 392, 762, 804, 847
  dialytic permeability data, 297
Acetone, 33–34, 41–42, 50, 99, 103, 112–113, 117, 140, 187, 230, 384, 480, 506, 632, 762, 852, 867
Acetonitrile
  assay reagent, 572
  solvent for cyanogen bromide, 28
  trypsin conjugate active site titration, 391
(±)-6β-Acetoxy-3α-hydroxytropane, 836
Acetyl-DL-alanine, maximal reactor space velocity, 753
Acetylcholine
  Reineckate salt, 594
  substrate, of cholinesterase, 593–594, 653
Acetyl Coenzyme A, 458, 474
N-Acetylglucosamine, substrate, of lysozyme, 290, 445
Acetyl L-glutamic acid diamide, substrate, of papain, 425
N-Acetylhomocysteine thiolacetone, 87
N-Acetylimidazole, 532
Acetyl-DL-methionine
  maximal reactor space velocity, 753
  substrate, aminoacylase, 747
N-Acetyl-L-methionine, substrate, of aminoacylase, 199
Acetyl-DL-phenylalanine, maximal reactor space velocity, 753

Acetyl-L-phenylalanine methyl ester, substrate, of chymotrypsin, 409
Acetyl phosphate, 891–894
Acetylthiocholine iodide, substrate, of cholinesterase, 658
Acetyl-L-tyrosinamide, substrate, of chymotrypsin, 410
Acetyltyrosine, 75
N-Acetyl-L-tyrosine ethyl ester substrate, of chymotrypsin, 31, 34, 36, 73–76, 266, 410–411, 530, 536, 561–562
Acetyl-DL-tryptophan, maximal reactor space velocity, 753
N-Acetyl-D-tryptophan, inhibitor, of α-chymotrypsin, 565
Acetyl-DL-valine, maximal reactor space velocity, 753
Achromobacter liquidum, L-histidine ammonia-lyase, 745
Acid chloride coupling, to inorganic supports, 145
Acid hydrolysis, determination, covalent coupling group, 16
Acrylamide, 57, 171–177, 185, 194, 198, 273, 741, 750, 889, 891
  coentrapment procedure, 475–476
  copolymer, with vinylated proteins, 195–201
  entrapment for enzyme electrode, 600
  photochemical polymerization, 571–572
  support
    for trypsin, 662
    for trypsin/chymotrypsin, 571–572
Acrylamide/acrylic acid copolymer, 180
  1-ethyl-3(3-dimethylaminopropyl) carbodiimide activation, 62
  multistep enzyme system, 454–456
  negative charge, advantage of, 63
Acrylamide/1-acryloylamino-2-(4-nitrobenzoylamino)ethane copolymer
  characterization of, 92
  mechanical stability, 92
  reduction of aryl nitro group in, 92
  synthesis, 91–92
  thermolysin immobilization on, 93–94
Acrylamide/N-acryloyl-N'-t-butyloxy-

carbonyl hydrazine copolymer, *see* Enzacryl AH
Acrylamide/*N*-acryloyl-4-carboxymethyl-2,2-dimethylthiazolidine copolymer, 87
Acrylamide/2-hydroxyethylmethacrylate ester copolymer, 85
Acrylamide/methylacrylate copolymer, 87
  hydrophobicity studies with, 63–64
Acrylamide/4-nitroacrylanilide copolymer, *see* Enzacryl AA
Acrylic acid, 57
  anionic copolymers with, 108
Acrylic acid-2,3-epoxypropyl ester, structural formula, 197
Acrylic acid/3-isothiocyanatostyrene copolymer
  polymerization, 113–114
  papain immobilization on, 114
  swellability, 113–114
Acrylic acid-*O*-succinimide ester, structural formula, 197
Acrylic acid-2,3-thioglycidyl ester, structural formula, 197
Acryloylaminoacetaldehyde dimethylacetal, 88
Acryloylaminoacetaldehydedimethylacetal/*N,N'*-methylene*di*acrylamide copolymer, 105
1-Acryloylamino-2-(4-nitrobenzoylamino)ethane, 91
Acryloyl-*N,N*-bis(2,2-dimethyoxyethyl) amine, 88, 90
  preparation, 102
Acryloyl-*N,N*-bis(2,2-dimethoxyethyl) amine/*N,N'*-methylene*di*acrylamide copolymer
  preparation, 102
  β-D-glucosidase immobilization on, 103
N-Acryloyl-*N'*-*t*-butoxycarbonyl hydrazide, 87, 92–93
Acryloyl morpholine, 102
Acryloyl morpholine/acryloyl-*N,N*-bis(2,2-dimethoxyethyl) amine copolymer, 88, 90
  preparation, 102–103
  characterization, 103
  β-D-glucosidase immobilization on, 103
  storage, 103
N-Acryloyloxysuccinimide, 889–891

optimum concentration, 896–897
  synthesis, 894
Acrylyl chloride, 101–102, 196, 198, 894
  structural formula, 197
Active site
  effect on residue reactivity, 15
  involved in adsorption, 153
  protection during coupling, 15
  titration, 390–393, 436
Active transport
  across artificial enzyme membranes, 921–924
  model system, 434–435
Acylation, monomers used for, 197
Acyl azide
  activation mechanism, 19
  thermolysin binding, 95
Acylcholine acyl-hydrolase, 647
Acyl hydrazide group, assay, by titration, 93
Adenosine-2',5'-biphosphate, 864
Adenosine-3',5'-cyclic monophosphate, 884
Adenosine 5'-diphosphate, 507
  allosteric effector, of glutamate dehydrogenase, 507, 512–513
  recycling enzyme system and, 216
  substrate, of adenylate kinase, 892
Adenosine 5'-monophosphate, immobilization, 887
Adenosine 5'-triphosphate
  active transport and, 921–924
  analogs of
    characterization, 868–869
    coenzymic activity, 873–876
    immobilization, 869–873
    synthesis, 866–868
  in hexokinase assay, 342, 455, 457, 476
  immobilization, on Sepharose, 882
  product, adenylate kinase, 892
  regeneration, 887–888
Adenylate kinase
  activation, 889
  activity assay, 892–893
  cysteine residues, 887–889
  immobilization, in polyacrylamide gel, 890–891
Adipic acid dihydrazide, 124, 127, 640
  Sepharose activation, 882
Adsorbents, types of, 43
Adsorption, 148–169

applications, 165–166
bonding mechanisms, 149–153
carrier choice for, 154–160
to DEAE-Sephadex, of aminoacylase, 748–751
definition, 149
effect of enzyme concentration on, 153–154
  of hydrogen ion concentration on, 153
  of ionic strength on, 153
  of temperature on, 45, 154
  of time on, 154
effects of, on enzyme activity, 44–45
enzyme cleavage and, 153
hydrophobic, 43–45
methods, 43–45, 160–163
Aerobacter (Enterobacter) aerogenes, pullulanase from, 62–63
Affi-Gel 703, 311–312
Affinity chromatography, 38
  biospecific, adsorbents for, 32
  separation on aggregated enzyme, 277
AG1-8X, 845
Agar
  amphipathic gels of, 43–45
  biological resistance, 23
  chemical resistance, 23
  derivatization, 23–24
  economic considerations, 24–25
  effect on enzymatic activity, 24
  epichlorohydrin cross-linking, procedure, 23
  matrices, activity on, 24
  mechanical stability, 21–22
  space-fitting properties, 22–23
  support properties of, 20
  uses, 20–21
  vicinal hydroxyl in, 27
Agarose
  activated
    commercial source of, 29
    handling and storage, 29, 33
  activation procedure with bisoxirane, 33–34
  with cyanogen bromide, 28–29
  advantages of, 45
  aminoalkyl, and photochemical immobilization, 284–285
  beads, formation mechanism, 25

chemical resistance, 23
copolymer with polyacrylamide, 273
definition, 21
derivatization, 38–42, 44
  to hexyl agarose, 44
  to mercapto-agarose, 41
  to oxo-agarose, 40
  to p-phenylene diamine agarose, 39–40
immunodiffusion and, 716–717
mechanical stability, 21–22
periodate oxidation of, 40
p-phenylenediamine derivative, preparation, 39–40
uses, 20–21
AGDA, *see* Acetyl L-glutamic acid diamide
Aggregation, 263–280
  applications, 276–279
  literature survey, 263–265
  optimization of, 274–276
  procedure, 265–274
    in frozen state, 269–270, 277–278
    with isolated enzyme, 265–267
    using inert proteic feeder, 268–270
  storage after, 266
AIBN, *see* 2,2′-Azobisisobutyronitrile
Air monitoring, 647–658
  constant current supply, 655
  electrochemical cell, 654
  mechanical components, 656
  platinum electrodes, 654–655
  starch pad, 652–654
Alanine, 13
  production in enzyme reactor, 880–882
  in Sepharose, 497
β-Alanine, derivative, 284
L-Alanine, isolation, 756–757
Alanine dehydrogenase, in enzyme reactor, 880–882
Albumin, *see also* Bovine serum albumin
  immobilization
    in liposomes, 227
    by microencapsulation, 214
Alcohol, enzyme electrode for, 585, 617
Alcohol dehydrogenase
  alcohol oxidation, 838–840
  coenzyme oxidation by, 863
  conjugate
    activity, 271

kinetics, 406–407
stability, 841
hydrophobicity studies with, 63–64
immobilization, 836, 837
  in hollow-fiber membrane device, 307
  on inert protein, 903, 908
  in multistep enzyme system, 315–316
  on nylon tube, 120
  on water-insoluble carriers, 108
  reduction of NAD^+ analogs, 873–874
  of $NADP^+$ analogs, 874–876
  stereospecific reduction of ketone, 837–838
Alcohol oxidase
  in enzyme electrode, 590
  substrates, 615
Aldehyde dehydrogenase, immobilization, in enzyme system, 315–316
Aldehydrol coupling, of urease, 105
Aldehydrol content, assay, 105
Aldolase
  conjugate
    assay, 336–337
    of monomeric derivatives, 495
    tetramer dissociation, 494
  denaturation studies, 498–500
  hollow-fiber retention data, 296
  immobilization
    in polyacrylamide gel, 902, 909
    on Sepharose 4B, 493–501
Alkaline phosphatase
  assay, fluorometric, 626–627
  conjugate
    kinetics, 426
    optimal effective pH, 429
  conjugation, with antibody, 711
  immobilization
    in collodion membrane, 904, 908
    in hollow-fiber membrane, 308–309
    by microencapsulation, 214
    on water-insoluble supports, 108
  reductive denaturation studies, 517
  stain for detection of, 713
Alkylamine coupling, 139–140
  by aqueous silanization, 139
  by organic silanization, 140
Alkylation, monomers used for, 197
Alkyl halide derivative
  limitation, 49
  preparation, 49

Alkyl imidates, cleavable, bifunctional, 555–556
Allantoin, 591
Allosteric regulation, definition, 504
Allyl alcohol, 84, 108
1-Allyloxy-3-(N-aziridine)-2-propanol, 198–199
Allyloxy-2,3-epoxypropane, 196
1-Allyloxy-3-(N-ethyleneimine)-2-propanol
  structural formula, 197
  synthesis, 196
Alumina
  controlled-pore, glucose isomerase immobilization on, 161–164
  hydrogen ion concentration operating range of, 163
Aluminum chlorhydroxide complex, 652–653
Aluminum oxide, 158
  durability, 135
  pore properties, 137
Amberlite IR-120, L-amino acid purification, 757
Amidine, formation, 321
Amine, tertiary
  activation catalyst, 37
  production of amine oxides of, 850, 854–855
Amino acid
  analysis, for bound protein determination, 73, 388–389
  hydrophobic nature, differences in, 13–14
  number of available modification reactions, 14
  production systems, 184
D-Amino acid
  enzyme electrode for, 585, 589–590
  response time, 608
  preparation, 758
L-Amino acid, enzyme electrode for, 585, 589–590
D-Amino acid oxidase
  in enzyme electrode, 589
  immobilization, in gel, 908
  substrates, 615
L-Amino acid oxidase
  conjugate, activity, 271
  in enzyme electrode, 589

immobilization
 in gel, 908
 on inert protein, 908
 substrates, 615
Aminoacylase
 assay, 199–201, 747–748
 entrapment, in polyacrylamide gel, 750
 fiber entrapment, industrial application, 241
 immobilization on DEAE-Sephadex, 748–749
  on iodoacetylcellulose, 749–750
  by protein copolymerization, 198–200
  regeneration of activity during, 164–165
 immobilized, kinetic properties, 750–751
 industrial application, 746–759
 native, preparation, 748
 source of, 196
 supports for, 759
 thermal stability, 751–752
 vinylation, effect of pH on, 199
L-Aminoacylase
 commercial use of, 165
p-Aminobenzyl-cellulose, 46
ε-Aminocaproic acid, 70, 76–77
2-Amino-4,6-dichloro-s-triazine, preparation, 50
Aminoethyl Bio-Gel P, 56
Aminoethyl cellulose, 46, 48
 glutaraldehyde binding, 51
Aminoethyl group, titration, 445
Amino group
 concentration determination, 384, 533
 determination, using [^{36}Cl], 384–385
 introduction of, procedure, 115
 N-terminal, 12
  coupling with activated Sepharose, 17
 primary, reactions, 263
 source of, 115
ε-Amino group, and covalent coupling, 12
N^6-(6-Aminohexyl)-AMP, molar absorption coefficient, 867
N^6-[(6-Aminohexyl)carbamoylmethyl]-ADP, synthesis, 868
N^6-[(6-Aminohexyl)carbamoylmethyl]-AMP, synthesis, 868
N^6-[(6-Aminohexyl)carbamoylmethyl]-ATP, synthesis, 867–868

N^6-[(6-Aminohexyl)carbamoylmethyl]-NAD$^+$
 molar absorption coefficient, 864
 synthesis, 863–864
N^6-[(6-Aminohexyl)carbamoylmethyl]-NADP$^+$
 molar absorption coefficient, 866
 synthesis, 866
2-Amino-2-(hydroxymethyl)-1,3-propanedial, 684
δ-Aminolevulinic acid
 structural formula, 845
 substrate, δ-aminolevulinic acid dehydratase, 844–845
δ-Aminolevulinic acid dehydratase
 immobilization, on Sepharose 4B, 846–847
 properties, 844
2-Amino-2-methylpropanol-HCl buffer, alkaline phosphatase assay, 627
6-Aminopenicillanic acid, 241
 production methods, 759–760, 765–768
  batch process, 766
  recirculation process, 745, 767–768
 structural formula, 760
2-(m-Aminophenyl)-1,3-dioxolane, 115
p-Aminophenylglycidyl ether, 38
N-(3-Aminopropyl)-diethanolamine, 50
α-Aminopropyltriethoxysilane, 480
3-Aminopropyltriethoxysilane, glass bead activation, 325, 506
γ-Aminopropyltriethoxysilane, preparation of alkylamine glass, 139–140, 670, 777, 931
4-Aminosalicyclic acid, 88
5-Aminosalicyclic acid, 88
p-Aminostyrene, glucose oxidase immobilization, on, 320
Ammonia, 50, 80, 92, 104, 117, 231, 762
 blood level, 681–684
Ammonium chloride, 508, 881
Ammonium fumarate, substrate, aspartase, 740–743
Ammonium molybdate, assay reagent, 664
Ammonium peroxydisulfate, 198
Ammonium persulfate, 57–58, 102, 177, 185, 187–188, 193, 196, 455, 458, 476, 662
Ammonium sulfate, 162, 374, 748–749, 889–891

SUBJECT INDEX

interference in glutaraldehyde crosslinking, 551, 641
Ammonium thiocyanate, 132
Amphiphilic gel, 43
Ampicillin, 615
α-Amylase
  assay, spectrophotometric, 98
  conjugate
    activity, 100, 271
    stability, 100–101
    storage, 98
  diazo binding, 98
  immobilization
    by adsorption, effects on activity, 44
    on Enzacryl AA, 98–99
    on Enzacryl AH, 99–100
    in polyacrylamide gel, 902, 909
    on silk, 909
  isothiocyanato coupling, 99
  source, 98
  starch hydrolysis, 784, 788–789
β-Amylase, 63
  assay, spectrophotometric, 98
  conjugate
    activity, 100
    stability, 100–101
    storage, 99
  conjugation with oxo-agarose, procedure, 40
  diazo binding, 98
  entrapment, in N,N'-methylenebisacrylamide, 171
  immobilization on Enzacryl AA, 98–99
    on Enzacryl AH, 99–100
    on hexyl agarose, 44
    in polyacrylamide gel, 902
  immobilized, relative activity, 40
  isothiocyanate coupling, 99
  source of, 98
Amyloglucosidase, see Glucoamylase
Amylo-α-1,4-α-1,6-glucosidase
  conjugate, pH optimum shift, 468–469
  immobilization multistep enzyme system, 460–461
Amylose, support, for penicillin acylase, 765
Anhydrid-Acrylharzperlen, support, for penicillin acylase, 764
Aniline

blocking reagent, of activated support, 494
  inhibitor, β-fructofuranosidase, 408
1-Anilino-8-naphthalene sulfonate, 363
ANP, see 2-Nitro-4-azidophenyl group
ANS, see 1-Anilino-8-naphthalene sulfonate
Antibiotic, see also specific compounds
  in storage procedure, 270
Antibody
  enzyme labeling, 709–717
    procedure, 710–711
    quantitation, 714–716
Antigen
  localization, 712–714
  quantitation, 714–716
Antihistamine, 855
Antiserum, microencapsulation of, 215
6-APA, see 6-Aminopenicillanic acid
APEMA, see p-Acetaminophenylethoxy methacrylate
Arginine
  analysis, for bound protein determination, 389
  chemical properties, 12–18
  enzyme aggregation and, 263
Arginine decarboxylase, conjugate, activity, 271
L-Arginine iminohydrolase, effect on immune response, 705
Argon, 222–223, 889–892
Arsenate, and glucose isomerization, 811
Arthrobacter simplex, immobilization, by polyacrylamide entrapment, 184–190
Artificial cells, 206, 209, 217–218, see also Microencapsulation
Arthus reaction, 709
Aryl amino group, assay, by radioestimation with [^{36}Cl], 92
Aryl nitrene diradical, 282
Ascorbic acid, 256–257
L-Asparaginase
  activity
    assay procedure, native protein, 690–691
    relative, conjugate, 271
    specific, conjugate, 255
  effect on immune response, 705
  encapsulated, exposed antigenic sites, 692

entrapment in polyacrylamide gel, 692
  in poly-2-hydroxyethylmethacrylate, 176, 692
fiber entrapment
  biomedical application, 242
  in cellulose triacetate fiber, 693
  immobilization on collagen, 255, 260, 693
    on Dacron, 693
    on glass plates, 693
    on inert protein, 909
    on nylon tube, 120, 693
    on polymethylmethacrylate, 693
  microencapsulation, 212–214
  storage in solution, 212–213
  therapeutic application, 679, 689–694
L-Asparagine, 13, 242
  enzyme electrode for, 585
L-Aspartase
  pH activity profile, 741–742
  thermal stability, 741
Aspartate aminotransferase
  coimmobilization, 472
  immobilization, on collagen film, 908
Aspartate transcarbamylase, binding studies, 557
L-Aspartic acid
  bioassay, 739
  chemical properties, 12–18
  factors affecting production, 741–744
  industrial production, 744
  uses, 739
Aspergillus niger
  catalase from, 481
  β-galactosidase from, 793
  glucose oxidase from, 196, 268, 481
Aspergillus oryzae, culture, 748
ATEE, *see* N-Acetyl-L-tyrosine ethyl ester
Automation of analysis, *see also* Continuous flow analyzer
  using immobilized enzymes, 633–646
2,2′-Azobisisobutyronitrile, 113
Azo coupling, to inorganic supports, 143–144

B

β, *see* Damköhler number
BAA, *see* Benzoyl-L-arginine amide
Bacitracin
  hollow-fiber dialytic permeability, 297
  hollow-fiber retention data, 296
BAEE, *see* N-α-Benzoyl-L-arginine ethyl ester
Bacillus coagulans, 811, 815
Bacillus subtilis, α-amylase from, 98
Bacillus thermoproteolyticus, 89
Bead polymerization
  advantages of, 64–65, 68
  column procedure, 191–195
  procedure, 57–60, 176–177, 662–663
Bentonite, surface area of, 155
Benzalkonium chloride, preservative, 794
Benzamidine, 212, 529, 532
  inhibitor, of trypsin, 428, 476
Benzene, 131, 141
  effect on Spheron swelling, 69
Benzidine, 120
Benzoquinone, support activation by, procedure, 35
N-α-Benzoyl-L-arginine amide
  substrate, of papain, 110–111, 425
  partitioning effect on, 402–403
N-α-Benzoyl-L-arginine ethyl ester
  substrate of bromelain, 406–408
    of ficin, 338–339
    of papain, 111
    of trypsin, 63, 346, 428, 476, 519, 561, 664
DL-Benzoylarginine *p*-nitroanilide, substrate, of trypsin, 312
N-Benzoyl-L-arginine-*p*-nitroanilide, substrate, of trypsin, 198, 428
Benzoyl-DEAE-cellulose, 46
Benzoylglycine ethyl ester, substrate, of papain, 430
Benzoyl-naphthoyl-DEAE-cellulose, 46
Benzoyl peroxide, 289
Benzoyl-L-tyrosine ethyl ester, substrate, of α-chymotrypsin, 446, 523
Benzylpenicillin, 241
  instrumentation for conversion of, 766–767
  in lactose reduction reactor, 826
  substrate, penicillin acylase, 760–761
  structural formula, 760
Benzylpenicillin hydrolysis, 195
BGEE, *see* Benzolyglycine ethyl ester

Bifunctional reagent, 124–128, see also specific compounds
Binding mechanism, for collagen supports, 261–263
Biofiber 20, 293, 296–298
  hydraulic permeability, 298
Biofiber 50, 293, 296–298, 308, 312
  hydraulic permeability, 298
Biofiber 80, 293, 296–298, 307
  hydraulic permeability, 298
Bio-Gel CM-2, 56
Bio-Gel CM-100, 61–63
Bio-Gel P, 56, 85–86
Bio-Gel P-2, 445–446
Bio-Gel P-300, support, for sheep antibody, 715–716
Bis, see N,N'-Methylenebisacrylamide
Bisacrylamide, 273
Bisdiazobenzidine, 263
Bisdiazobenzidine-2,2-disulfonic acid, 904
N,N-Bis(2,2-dimethoxyethyl)amine, 101–102
Bishydroxyethylaniline, 83
Bisimidate, 125
4,4'-Bis(2-methoxybenzene diazonium) chloride, 147
Bisoxirane, 26
  Sepharose activation by, 32–33
Blood, perfusion, 679–680
Borax, 765
Boron trifluoride diethyl etherate, 126
Boron trifluoride etherate, 117
Bovine serum albumin
  electrophoresis reference, 474
  glutaraldehyde insolubilization, 553
  hollow-fiber dialytic permeability, 297
  hollow-fiber retention data, 296
  in immunoassay, 716
  for inert proteic matrix, 264–265, 268–269, 271, 903–904, 912, 925
  surface-caused enzyme inactivation and, 307
  in ultrafiltration studies, 300
Bridge formation, inorganic, 166–169
Britton–Robinson buffer, 81
Bromelain
  conjugate, kinetics, 406–408, 437
  immobilization
    on carboxymethyl cellulose, 52
    property changes after, 53

Bromoacetyl bromide, 49, 749
Bromoacetyl cellulose, 36
  preparation, 749
p-Bromophenol, 133, 386
Bromophenol blue, 385
3-Bromopropene, structural formula, 197
BTEE, see Benzoyl-L-tyrosine ethyl ester
Buffer
  effect on activity of entrapped enzymes, 187
  heat of protonization, 664
1,4-n-Butanediol diglycidyl ether, 33
n-Butanol, 58, 94, 539–540
  substrate, of alcohol dehydrogenase, 64, 407
n-Butyl acetate, 230
n-Butyl benzoate, 204–206
  effects on membrane permeability, 206
Butyrylthiocholine iodide, substrate, cholinesterase, 637, 648, 655, 658
Butyryl cholinesterase, turnover number, 647

C

Calcium chloride, 89–91, 94, 121, 130, 324, 446, 447, 523, 529, 531, 863, 865, 866
Calcium hydroxide, 284, 863
Calcium ion
  coarctation and, 179
  glucose isomerase inhibitor, 816
Caprolactam, 118, 126
Carbamate, formation, 27
Carbamate pesticide, detection, 648, 657–658
  covalent binding with, 61–62, 87, 118–120, 141–142, 144–145
  mechanism, 17–18
Carbohydrase, immobilization, choice of pore diameter for, 160
Carbohydrates, and whole cell entrapment, 184
Carbon dioxide
  evolved, 348
  inhibitor, 346
  transport studies, 920–921
Carbonic anhydrase
  carbon dioxide transport and, 920–921
  conjugate, activity, 271
  immobilization

on membrane, 905, 909, 920–921
by microencapsulation, 215
therapeutic use, 697
Carbon tetrachloride, 99
N,N'-Carbonyldiimidazole, 288–289
Carbonyl group, concentration determination, 385
Carboxybenzoyl-L-tyrosine-p-nitrophenyl ester, substrate, of chymotrypsin, 523
N-Carboxyl anhydride, 18
Carboxylate, assay, by titration, 92, 93
Carboxyl group
 activation, 17–18, 542
 concentration determination, 131, 384
Carboxylic ester hydrolase, conjugate, microfluorometric assay, 374–376
Carboxylmethyl cysteine, 392
N^6-Carboxymethyl-ATP
 molar absorption coefficient, 867
 synthesis, 866–867
Carboxymethyl-cellulose, 46, 51
 activation mechanism, 18–19
 penicillin acylase purification, 762–763
N^6-Carboxymethyl-NAD$^+$, synthesis, 862–863
N^6-Carboxymethyl-NADP$^+$
 molar absorption coefficient, 866
 synthesis, 864–866
Carboxypeptidase A
 conjugate, thermal stability, 537
 crystal
  immobilization of, 548–551
  substrate binding studies, 557
  X-ray analysis, 555–556
 immobilization, on Sephadex G-200 529, 533
 unit cell dimensions, 550
N-Carboxy-L-tyrosine anhydride, 327
Casein, substrate
 of chymotrypsin, 31, 523
 of thermolysin, 89, 95–97
Caseinolytic activity, unit of, definition, 96
Catalase
 in aggregation procedure, 268
 assay, perborate, 685
 catalytic reaction, 478–479
 commercial source, 481
 conjugate, activity, 156, 200, 271–272, 482–483

 in enzyme thermistor, 674–675
 in glucose oxidase assay, 348–349
 immobilization on collagen, 153, 255
  on controlled-pore ceramics, 152–153, 156–160
  in enzyme system, 315–316
  on glass beads, 671
  with glucose oxidase, 181, 472–473, 478–488
  with inert protein, 903–904, 908
  by microencapsulation, 207, 212–214, 684–685
  in porous matrix, 270–272
 immobilized, effect of inhibitor, 916–917
 kinetics, 483–488, 686
 properties, 479
 soluble, assay, 479, 482
 specific activity, 255
 spin diameter, 160
 storage in solution, 212–213
 therapeutic application, 679–681, 684–688
 unit cell dimensions, 159
Catalyst, pellicular particles, 179
4CC, see Condensation reaction, four-, component
CD, see Circular dichroism
CDAK, 293
Celite 505, 762
Cellophane, commercial source, 270–271
Cellulase, 108, 165–166
Cellulose
 activation with halogens, 36
 aminoethyl, and photochemical immobilization, 284–285
 bridge formation and, 166
 bromoacetyl derivative, preparation, 749
 chemical properties, 46
 composite formation, with iron oxide, 326
 derivatives commercially available, 46
 enzyme attachment sites, 47–49
 fibers of
  elastic support, 560
  preparation, 230–231
 hollow-fiber device, 293
 physical properties, 46
 rehydration procedure, 46–47

substrate, of cellulase, 166
sulfoethyl, 46
support properties of, 20
Cellulose acetate, hollow-fiber device, 293
Cellulose nitrate, microencapsulation with, 204–207, 212
Cellulose triacetate, fibers of, preparation, 230
Ceramic support, see also Glass beads; Glass, controlled pore; Support, inorganic
    comparative half-life data, 779–783
    controlled-pore, surface area, 155–158
    physical parameters, 777
    porous, physical properties, 135–137
    relative cost, 776–777
    types, 777
Charcoal, activated
    glutaraldehyde treatment with, 550–551
    microencapsulation and, 216
Chlorhydrol, see Aluminum chlorhydroxide complex
1-Chloro-2,3-epoxypropane, 126
Chloroform, 58, 141, 145, 177, 207, 222, 480, 662, 839, 855, 932
    sterilizing agent, 783–785
Chloromaleic acid anhydride, 197
p-Chloromercuribenzoate, inhibitor, 615
4-Chlorostyrene, 289
Chlorotriazine, multifunctional agent, 263
Cholinesterase
    assay, fluorometric, 626
    immobilization
        by entrapment, 173, 178
        in starch pad, 652–653
    immobilized
        activity assay, 653
        in monitoring system, 647–658
    pesticide inhibitors, 637
        detection, 647–658
Cholesterol
    assay, fluorometric, 633
    liposome formation and, 222, 227
Cholesterol ester hydrolase, 633
Cholesterol oxidase, in enzyme thermistor, 674
Chromatography, agar and agarose for, 21
Chrome, 248
Chromium sulfate, 248

CHT, see Chymotrypsin
Chymotrypsin
    active site, 410
    activity, 30–31, 34, 36, 200, 266–267, 271, 446–447, 530, 561
    acylation, 447, 562
    conformation studies, 363, 365–366
    conjugate
        active site titration, 391–392
        effect of heat treatment, 535–536
        of support stretching, 562–568
        kinetics, 425
        storage, 75–76
    cyanogen bromide binding, 52, 72–76
    dye binding, 535
    immobilization on activated agarose, 33–36
        on activated Spheron, 72–76, 83
        by adsorption, effects on activity, 44
        by aggregation, 265–267
        on anhydride derivative of hydroxyalkyl methacrylate gel, 83
        in cellophane membrane, 905, 909
        on cellulose, 52
        in hollow-fiber membrane device, 310–311
        on imidoester-containing polyacrylonitrile, 323–324
        on inert protein, 909
        on polyacrylamide beads, 446
        in polyacrylamide gel, 902
        on Sephadex, 30, 529
        using acyl azide intermediate method, 446
        using glutaraldehyde, 265–267
    matrix blocking on Sephadex, 24
    maximum loading of, 727
    modification, 310
        effect on activtiy, 409–412, 448–450
    pH-activity profiles, 404
    resolution of organic acids, 833–835
    selective ester hydrolysis, 835–836
    s-triazine binding, 52
    activation, rate determination, 572
    acquisition of tertiary structure, 523–526
    steric exclusion of trypsin and, 667
α-Chymotrypsinogen, hollow-fiber retention data, 296
Chymotrypsinogen A

conjugate, reductive denaturation studies, 522–526
immobilization on glass beads, 520–522
reductive denaturation studies, 518
Circular dichroism(CD), and conformation studies, 366–367
Citrate synthase
  activity assay, 458–460, 474
  immobilization in multistep enzyme system, 181, 457–460, 473–475
  kinetics, in multistep enzyme system, 434, 466–468
L-Citrulline, continuous production, 745
Clostridium perfringens, neuraminidase from, 223
CM, *see* Carboxymethyl-cellulose
CMC, *see* 1-Cyclohexyl-3-(2-morpholinoethyl) carbodiimide metho-*p*-toluenesulfonate
CPG, *see* Glass, controlled pore
CPK, *see* Creatine phosphokinase
Coagulating solvent
  for cellulose triacetate, 230
  of ethyl cellulose, 231
  for nitrocellulose, 230
Coarctation, 179
Cobalt acetate, 162
Cobalt chloride, 198, 747–748, 774, 793
Cobalt ion, glucose isomerase activation, 815
Coenzyme
  alkylation, 859–862
  alkylated, spectral properties, 868
  analogs, R_f values, 868–869
  derivative synthesis, 859–862
  enzyme electrode and, 603
  [^{19}F]-substituted, 366
  immobilization, 859–887
  support pretreatment with, 155
Collagen
  assay of enzyme bound to, by cysteine analysis, 256–257
  by tryptophan analysis, 255–256
  chemical modification of, 244
  chemical properties, 243
  complexes
    binding mechanism, 261–263
    kinetic behavior of, 259–261
    storage stability, 258
  flexible pore diameter, 155

glutaraldehyde insolubilization, 553
immobilization procedures, 245–250
  electrocodeposition method, 248–250
  macromolecular complexation, 247
  membrane impregnation method, 246–247
  for whole microbial cells, 250
membrane, 902–903
  loading capacity, 254–255
microbial degradation of, 258–259
physical properties, 244
source, 245
support properties, 243–263
  in microcapsules, 684–685
  preparation, 204
support properties, 904, 906
Colorimetric test, for enzyme leakage, 276
Complexation, 149
Concanavalin A, crystal immobilization, 548
Concentration polarization, definition, 299–300
Condensation reaction, four-component, enzyme immobilization on nylon, 127–129, 133–134
Conductivity change, assay using, 348–349
Conformation
  perception and, 567–568, 571
  stretching studies, 560–561
  techniques for studying, 361–370
    fluorescence, 362–366
  transitions of
    in immobilized proteases, 528–538
    kinetics, 528–529
Congo red, 654
Conjugation, *see* Immobilization
Continuous flow analyzer
  bis-acid hydrazide spacers, 645
  carry-over, 644–645
  enzyme stability, 644
  kinetics, 642–643
  method of operation, 638
  sample rate, 644
COOH-Spheron, 76–80
Copoliodal, 84
Copolymer, *see also* specific types
  compositional heterogeneity, 181
Copper sulfate, 387

Cornsteep liquor, 190
Corn syrup solid, substrate, glucoamylase, 778
Cortisol
  assay, by thin-layer chromatograpy, 187
  molar absorption coefficient, 186
  substrate, of 3-ketosteroid-Δ¹-dehydrogenase, 186
Corynebacterium simplex, see *Arthrobacter simplex*
Coupling efficiency, definition, 795
Covalent binding
  adsorption and, 149
  via chelation, 88
  choice of technique, 148
  determination of group involved, 16
  direct, 26–37
  general experimental conditions for, 11–12
  indirect, 37–42
    advantages of, 38
  on inorganic supports, 134–148
  irreversible, 45
  to polyacrylic copolymers, 84–107
  reactive residues for, 12–15
  reversible, 45
Creatine, urinary, assay, fluorometric, 630–631
Creatine phosphokinase
  assay, fluorometric, 628–629
  in creatine assay, 630–631
Creatinine
  hollow-fiber dialytic permeability, 297
  hollow-fiber retention data, 296
Cross-linkage, *see also* Aggregation
  agents for, *see* specific compounds
    bifunctional, 263, 458
    electron beam irradiation, 178–179
    gamma-irradiation, 178
    multifunctional, 263
    optimum concentration, 173
    photochemical, 280–288
    ultraviolet radiation, 248
  effect on swelling capacity, 108
  enzyme bonding heterogeneity and, 358–360
Cuprammonium hydroxide, 326
Curtius azide method, 48
Curvularia lunata, immobilization, by polyacrylamide entrapment, 184–190

Cyanamide gels, 30
Cyanate, lability, 27
Cyanogen bromide activation
  of acrylamide/hydroxyethylmethacrylate polymer, 61–62
  of cellulose, 48, 52
  of Dextran T40, 872
  of hydroxyalkyl methacrylate, 70
  of polysaccharide support, 27–30
  of porous glass, 146–147
  of Sephadex A-50, 540
  of Sephadex DEAE-50, 542
  of Sephadex G-50, 457
  of Sephadex G-100, 540, 542
  of Sephadex G-200, 529, 764–765
  of Sepharose 4B, 16–17, 454, 457–458, 493, 529, 930
  of Spheron gels, 72–73
Cyanuric chloride, cellulose activator, 19, 37, 47, 49–50, 120
Cyclibillin, 615
Cyclohexane, 207
Cyclohexanol, pore distribution in Spheron and, 67
Cyclohex-2-en-1-ol, 837
Cyclohexyl isocyanide, conjugation agent, 40
1-Cyclohexyl-3-(2-morpholinoethyl) carbodiimide metho-*p*-toluene sulfonate, 62, 141, 932–933
Cysteine, 81, 88, 110–111, 476, 926
  assay, for bound enzyme determination, 256–257
  chemical properties, 12–18
  enzyme aggregation and, 263
Cytidine-2,3-cyclic phosphate, substrate, for ribonuclease, 36
Cytochrome *c*
  hollow-fiber retention data, 296
  immobilized, electron transfer by, 545–546
  membrane orientation, 371
  oxidized, conformation studies, 368–369
Cytochrome oxidase, 371–372, 545

D

DAB, *see* 3,3'-Diaminobenzidine tetrahydrochloride
Dacron, support, for L-asparaginase, 693

Damköhler number, 182–183
DBPO, see Dibenzoylperoxide
Deactivation, of enzymes, causes of, 240
DEAE-cellulose, 46
  enzyme modification with, 310
DEAE-Sephadex, support, for penicillin acylase, 765
DEAE-Sephadex A-25, support, for aminoacylase, 748–749
2S, 9R, 10R-trans-2-Decalol, 837
9S, 10S-2-Decalone, 837–838
(±)-trans-2-Decalone, 837
DeeO liquid, 157
$\Delta^{1\text{-}2}$-Dehydrogenase, 471
Denaturation
  conformation studies and, 369–370
  for covalent coupling, 12
  in diffusion controlled systems, 428
  reductive, reversibility studies, 516–526
  resistance to, 383
Denier, definition, 235
Desorption, and ionic strength, 153
Detergent, use in bead (suspension) polymerization, 176–177
Dextran
  effect of solvents on volume, 22
  enzyme modification with, 310
  support for coenzymes, 871–873
    for NAD$^+$, 872
    for NADP$^+$, 872–873
    for penicillin acylase, 765
  properties, 20
Dextran 110, hollow-fiber retention data, 296
Dextran T40, 189
Dextranase
  immobilization, by adsorption, effects on activity, 44
  source of, 30
Dextrose, see Glucose
Dialysis membrane, preparation, 877–878
3,3'-Diaminobenzidine tetrahydrochloride, 713, 912
p,p'-Diaminodiphenylmethane, 129, 132
1,6-Diaminohexane, 76–77, 85, 118, 125, 127, 130, 458, 859–868
1,3-Diaminopropanol derivative, 284
o-Dianisidine, 460, 474, 476, 716
Diaphorase, immobilization, in enzyme system, 315–316

Diastase, 108
Diazinon, detection, 658
Diazobenzene sulfonic acid, 445
Diazo group, concentration determination, 385–386
Diazonium coupling, 18, 38
Diazotization
  capacity for, assay, 133
  with nitrous acid, 88, 91
  with sodium nitrite, 93, 115–117
Dibenzoylperoxide, 110, 112–114
Di(sec-butyl)peroxydicarbonate, 175
Dicetyl phosphate, 222, 227
1,2-Dichloroethane, 231
Dichloromethane, 126, 187, 198
1,3-Dichloro-5-methoxytriazine, 141
2,2-Dichlorovinyl dimethyl phosphate, inhibitor, cholinesterase, 651, 656–657
Dicloxacillin, 615
N,N-Dicyclohexylcarbodiimide, 48, 146, 384
Diethyl adipimidate, 124, 127
N,N-Diethylaminoethyl methacrylate, 71
Diethylene glycol, 117
Diethyl ether, 117, 834–836, 838
Differential conductivity, assay by, 348–349
Diffusion
  activated, 175
  assay and, 351–352
  particle size and, 136, 138
  of substrate
    external resistances to, 413–414, 415–420
    internal resistances to, 414, 420–426
Diffusion cell, and artificial enzyme membranes, 907, 910–911
Diffusion constraint, determination, 269
p,p'-Difluoro-m,m'-dinitrophenylsulfone, 263
N,N'-Diformyl-1,6-diaminohexane synthesis, 130
1,4-Diisocyanatobenzene, 116
1,6-Diisocyanohexane, synthesis, 129–130
Diisopropylfluorophosphate, 15
  titration, 524
Diisopropylphosphofluoridate, see Diisopropylfluorophosphate
p-Dimethylaminobenzaldehyde, 256
  penicillin analysis, 760

β-Dimethylaminopropionitrile, 194, 196, 198, 741, 750
Dimethylaniline monooxygenase conjugate
  activity assay, 852–853
  concentration determination, 852
  properties, 853–854
  immobilization, on glass beads, 851–852
  oxygen sensitivity, 853
  soluble
    properties, 849–851
    substrate specificity, 849–850
N,N-Dimethylformamide, 115–116, 131, 289
  for carboxyl group titration, 384
  trypsin conjugate active site titration, 391
N,N-Dimethyl-1,3-propanediamine, 120–122
Dimethyl suberimidate, 640
Dimethyl sulfate, 124
Dimethyl sulfoxide, 187, 284, 384, 834, 890, 891
  storage of activated gel in, 33
3(4,5-Dimethylthiazolyl-2)-2,5-diphenyl tetrazolium bromide, 713, 905
Dinitrosalicylate reagent, assay, of reducing sugar, 98
Dioxane, 50, 117, 126, 146, 146, 749
Diptheria toxoid, antibody response to, 706
2,2′-Dipyridine disulfide, activation agent, 41
Dispersion number, definition, 802–803
5,5′-Dithiobis(2-nitrobenzoic acid), 458–459, 474, 522
  cysteinyl residue measurement, 16
2,2′-Dithiodipyridine, 385
Dithionite, preparation of arylamine glass, 932
Dithiothreitol, 41, 316, 514, 846, 889–891
  effect on polymerization of polyacrylamide, 896
  reductive denaturing agent, 522
1,4-Divinylbenzene, 84, 110, 112–114
Divinyl ether, 117
Divinyl ketone, 35
Divinyl sulfone
  activation, 26, 34
  cross-linking agent, 22, 492

DMF, see Dimethylformamide
Dodecyl alcohol, and pore distribution in Spheron, 67
Dodecyl sulfate gel electrophoresis, for soluble enzyme aggregate characterization, 474
Double-beam laser spectrophotometer, 912, 919
Dowex 1-X2, 865–867
Dowex 1-X4, 864
Dowex 50, 878
Dowex 50 W-X4, 863
Dursban, detection, 658
DVB, see 1,4-Divinylbenzene

E

ECTEOLA, see Epichlorohydrin triethanolamine-cellulose
EDC, see 1-Ethyl-3(3-dimethylaminopropyl) carbodiimide
Effectiveness factor, 412, 423–426
  definition, 182, 412
  influence of particle radius on, 727–729
  for inhibition, definition, 427
  limiting value, 423
EGDMA, see Ethyleneglycol dimethacrylate
Ehrlich reagent, modified, 845
Electrode
  air-gap, 613
  amino acid, 615, 617
  ammonia gas, 582, 588
  ammonium-ion, 347–348, 582, 587–590, 593
  amygdalin, 585, 592, 617
    interferences, 614
    response time, 604–609
    stability, 602
  bromide ion, 582
  cadmium ion, 582
  calcium ion, 582
  carbon dioxide, 582, 588–589
  chloride ion, 582
  cholinesterase, 593–594
  copper ion, 582
  creatinine, 617
  cyanide ion, 582, 614
  divalent ion, 582

enzyme, see Enzyme electrode
fluoride ion, 582
fluoroborate ion, 582
glucose, 585–587, 602–616
   commercial, 616
   interferences, 614
   response time, 607
   stability, 602
glutamine, response time, 608
hydrogen cyanide gas, 582
hydrogen fluoride gas, 582
hydrogen sulfide gas, 582
hydroxide ion, 582
iodide ion, 582, 586–587
ion-selective, 347–348
   definition, 583
   description, 581–583
   types commercially available, 582
   useful concentration range, 581
lead ion, 582
monovalent ion, 582
nitrate ion, 528
oxygen, 347
   in catalase assay, 482
   commercial source, 582
   in glucose assay, 586
   glucose oxidase assay, 481
   washing, 611
penicillin, see Penicillin electrode
perchlorate ion, 582
platinum, preparation, 654–655
potassium ion, 582
rubidium ion, 582
silver ion, 582
sodium ion, 582
sulfide ion, 582
sulfur dioxide gas, 582
thiocyanate ion, 582
urea
   comparison of types, 616
   interferences, 612–613, 615
   optimum enzyme concentration, 611
   response time, 604, 608–611
urease, 593
Electrode-monitoring, assay by, 346–348
Electrodeposition adsorption method, 160, 248–250
Electron microscopy
   artificial enzyme membranes and, 912–913

localization of cellular constituents and, 713
Electron spin resonance, trypsin active site titration, 392
Electron-transfer, in immobilized cytochrome c, 545–546
EMA, see Ethylene/maleic anhydride copolymer
ENJ-3029, 211, 330
Entrapment, see also Microencapsulation
   aggregation and, 265
   by cross-linking polymers, 177–179
   in gels, 169–183, 902
   of multistep enzyme system, 181–182, 475–478
   parameters effecting, 169–171
   by solution polymerization, 172–176
   by suspension polymerization, 176–177
   of whole cells, 183–190, 740–741, 768–776
   effects of buffering on, 187
Enzacryl, structural features, 86
Enzacryl AA, 56, see also Polyacrylamide gel
   amylase immobilization on, 98–100
   components of, 86, 98
   enzyme coupling with, 87
Enzacryl AH, 56, see also Polyacrylamide gel
   acyl azide derivative, 99
   amylase immobilization on, 99–100
   characterization, 93
   components of, 87
   synthesis, 92–93
   thermolysin immobilization on, 95
Enzacryl polyacetal, 105
   structural features of, 88
Enzacryl polyaldehyde, immobilization reactions with, 89
Enzacryl polyaldehyde A, 105
Enzacryl polyaldehyde B, 105
Enzacryl polythioacetone, 87–88
Enzacryl polythiol, 87
Enzite, see Carboxymethyl cellulose
Enzyme, see also specific types, substances
   action of, mechanistic studies, 560–566
   agar gels for, 22
   aggregation, see Aggregation
   allergic reaction to, 685–687, 689, 692, 699, 704, 708–709

binding mechanism
  to collagen supports, 261–263
  covalent, 25–27, 37–38, 41, 47
commercial source, 584
coupling by γ-radiation, 37
deficiency, treatment, 684–688, 700–703
entrapment in gels, 169–183
fiber-entrapped, 227–242
  assay of, 231, 232, 237
  efficiency parameters, 235–238
  stability, 238–240
immobilization
  on aminochloro-s-triazinyl cellulose, 50–51
  on cellulose-transition metal salt complex, 48
  choice of coupling technique for, 148
  of pore diameter for, 156–160
  coupling time, 144
  effect of concentration on, 143
  on thermal stability, 97–98
  in hollow-fiber membranes, 306–310
  to inert protein, 709–717
  on magnetic particles, 724
immobilized
  active site titration, 390–393
  activity, 3, 4
    parameters effecting, 136–138
  automated analysis and, 633–646
  as biological model systems, 63–64
  bonding heterogeneity, 358–360
  bound protein determination, 386–390
  on cellulose, properties, 46–53
  conformation studies, 361–362
  classification, 676–677
  as commercial products, 584, 596
  definition, 201–202
  effect of organic solvent on, 843
  injection of, 678–679
  kinetic behavior, 397–443
  physical properties, 357–372
  reactivation, 164
  routes of administration, 678–681
  stability, 97–98, 361–362, 643–644
  theoretical enzymology and, 559–560
  therapeutic applications, 676–699
  types of, 202
  use, as analytic reagent, 579–580, see also Enzyme electrode; Organic synthesis; Reactor

immunoassay, 714–716
inhibitors, 615
leakage, test for, 276
loading, definition, 170
microencapsulation, 201–218
  theoretical considerations, 202–203
modification, effect on activity, 448–450
molecular size, effect on immobilization, 27
pad, preparation, 652–654
spin diameter, 156
subunits
  activity, 501–503
  immobilization, 491–503
system, see Multistep enzyme system
thiolation, by 3-mercaptopropioimidate, 42
Enzyme electrode
for analysis of substrate levels, 277–278, 580–593, 877–880
artificial enzyme membranes and, 913
characteristics, 585
construction, 596–601
  for air and water monitoring, 654–655
  of dialysis membrane type, 599–600
  by physical entrapment, 600
effect of interferences, 612–615
  with enzyme, 614–615
  with sensor electrode, 612–614
for glutamate determination, 877–880
for lysine determination, 277–278
for pyruvate determination, 877–880
range of substrate determinable, 611–612
response time, 604–611
  dialysis membrane thickness and, 608–609
  electrode sensor response speed and, 609–610
  enzyme concentration and, 606, 610–611
  gel-layer thickness and, 608–609
  pH and, 607–608
  stirring rate and, 604–605
  substrate concentration and, 605–606
  temperature and, 608
stability, 601
storage, 601–602
theory, 580–581
types, 585

utilizing dextran-bound NAD$^+$, 877–880
washing, 611
Enzyme-substrate complexes, large, agar supports for, 23
Enzyme thermistor, 472, 667–676
  instrumentation for, 668–670
Epichlorohydrin, agar cross-linking, 23, 39, 41
Epichlorohydrin triethanolamine-cellulose, 46
3,4-Epoxybutene, structural formula, 197
Ergosterol, and liposome formation, 227
Erythrocyte, microencapsulation, 206, 210, 212, 214–215
Erythrose-4-phosphate, substrate, transaldolase, 495
Escherichia coli
  culture, 740
  immobilization of, 740–741
  penicillin acylase from, 760
  use in industrial production of L-aspartic acid, 740–745
Escherichia intermedia, 886
ESR, see Electron spin resonance
Ester, selective hydrolysis, 835–836
Esterase, serine residues, reaction with diisopropyl fluorophosphate, 15
Esterolytic activity, assay, 73
Ethane thiol, 88
Ethanol, 34, 44, 92–94, 102, 131, 204, 284, 285, 302 385, 460, 530, 749, 756, 834, 862, 863, 865, 867, 868, 872
  benzoquinone activation, 35
  concentration determination, with enzyme electrode, 590–591
  hollow-fiber dialytic permeability, 297, 302
  oxidation rate, 316
  substrate, alcohol dehydrogenase, 64
Ethanolamine, 73, 266, 273, 375
Ethanolamine hydrochloride, blocking reagent, of activated support, 493
Ether, 102, 130–133, 198, 204, 324, 690, 862, 863, 865
Ethyl acetate, 132, 187, 838, 894
Ethyl cellulose N200, fibers of, preparation, 231
1-Ethyl-3(3-dimethylaminopropyl) carbodiimide, 61–63, 77, 506–507, 521, 524, 860, 863, 866–867

Ethylene, 85
Ethylenediamine, 86
  derivative, 284
Ethylene diaminetetraacetic acid, 41, 42, 106, 111, 197, 313, 326, 337, 347, 494, 495, 508, 823, 865, 878, 881, 925
Ethylene dimethacrylate, 67, 172–175
Ethylene glycol, 91, 93, 320
Ethyleneglycol dimethacrylate, 172
Ethyleneimine, 196
Ethylene oxide, 70
Ethyl formate, 130
Ethylisocyanide, 541–543
N-Ethylmorpholine, 127
N-Ethyl-5-phenylisoxazolium 3'-sulfonate, 85

F

Fab fragment, and antigen localization, 712
FAD, see Flavin adenine dinucleotide
FAGLA, see Furacryloylglycyl-L-leucinamide
Fast Red TR salt, reagent, alkaline phosphatase staining, 713
Feedback inhibition, in multistep enzyme systems, 916–919
Ferric ammonium sulfate, 290
Ferricyanide, in lactic acid determination, 591–592
Fiber
  column packing with, 825
  dry weight determination, 232
  laboratory reactor for, 233–234
  nitrogen content determination, 232
  physical properties, 233–235
  preparation, 228–231
  ultrastructure, 235–236
  use, 232–233, 241–243
Fibrinogen, inert proteic matrix, 268
Ficin
  azide binding, 52
  conjugate, assay, 338–339
  immobilization
    on carboxymethyl cellulose, 52, 338–339
    with collagen, 244
  property changes after, 53
  kinetics, of substrate diffusion, 437

Filter paper, bridge formation and, 166
Filtration aid, see Celite 505; Hyflo supercel
Flavin adenine dinucleotide, 152
  in enzyme electrode storage buffer, 590, 603
  in oxidase determination, 852
  prosthetic group, 389
Flavin mononucleotide, 839
  immobilization, 884–885
Fluorescein, product, in microfluorometry, 374–376
Fluorescence, and conformation studies, 362–366
Fluorimetric assay, 528–538
Fluorocarbon 43, and hollow-fiber membranes, 302–303
Fluorodinitrobenzene, 284
Fluorometer, adaptations for solid surface method, 621–624
Fluorometry
  L-asparaginase activity assay, 690–691
  chymotrypsinogen tertiary structure, 524–526
  of immobilized proteins, 530–531, 533–534
  for microanalysis, 373–379
  procedural advantages, 618
  silicone rubber pad method, 619–625
    advantages, 624–625
    preparation, 619–624
    procedure, 624
  solid surface methods, 618–633
  titration of active sites, 391–392
4-Fluoro-3-nitroaniline, 284
Fluoro-2-nitro-4-azidobenzene
  commercial source, 283–284
  in photochemical immobilization, 281–282
  preparation, 284
Fluorostyrene, 108, 112
3-Fluorostyrene, in copolymer, 84
4-Fluorostyrene, in copolymer, 84
FNAB, see Fluoro-2-nitro-4-azidobenzene
Folin reagent, 110, 387
Formaldehyde, 248
  hollow-fiber membrane sterilization, 308
  tissue fixation, 712–713
Formalin, 110

Formamide, 127, 130
Free radical, 37, 172, 179
β-Fructofuranosidase, see Invertase
Fructose
  from glucose, mechanism, 810–811
  production in reactor, 768, 775–776, 809–821
Fructose 1,6-biphosphate, substrate, aldolase, 495
Fructose 1,6-diphosphate, 337
Fructose-6-phosphate, substrate, transaldolase, 495
Fruit juice, debittering process, 241
Fumarase, 832
Furacryloyglycyl-L-leucinamide, substrate, of thermolysin, 89, 95–97
Furacryloylglycyl-L-leucinamide hydrolase, activity, unit of, definition, 96

G

Galactokinase, 698
Galactose
  product, of α-galactosidase, 350
  substrate, of β-galactose dehydrogenase, 881
β-Galactose dehydrogenase, in enzyme reactor, 880–882
Galactose oxidase, immobilization, by entrapment, 176
α-Galactosidase, conjugate, assay, 350
β-Galactosidase
  conjugate
    activity assay, 269–271, 343, 352
    kinetics, 426
  entrapment in cellulose triacetate fibers, 824
  in polyacrylamide gels, 176–178
  fiber entrapment, industrial application, 241
  glutaraldehyde binding, 52
  immobilization
    on aminoethyl cellulose, 52
    by chemical aggregation, 269–270
    on collagen, 255, 260
    in hollow-fiber membrane device, 313–314
    on inert protein, 909
    in multistep enzyme system, 456–457

SUBJECT INDEX

kinetics, in multistep enzyme system, 465–466
for lactose reduction of milk, 822–830
magnesium requirement, 827
soluble aggregate, activity assay, 474
specific activity, 255
Gamma irradiation, for cross-linking, 37, 178
Gastrointestinal tract, and immobilized enzymes, 680–681, 696
Gel, fragmentation procedure, 174
Gel entrapment, see Entrapment
Gel formation, with polyvalent ions and polyelectrolytes, 179
Glass, controlled pore, 149–169
   bonding mechanisms, 149–153
   comparative properties of types of, 158
   cyanogen bromide coupling to, 146–147
   coupling through bifunctional bis-diazotized reagent, 147
   pH, operating range of, 163
   physical properties, 135
   pore diameter, 155–156
   protein binding by adsorption, 150–152
   relationship between pore size and surface area, 136–137
   relative cost, 776–777
   silanization, 139–140
   surface pH, 155
Glass beads
   activation, 670
   alkylamine
      photochemical immobilization on, 284–285
      preparation, 931–932
   arylamine, preparation, 932
   density, 721
   with gel pellicle, 179
   incorporation of thioester linkage, 524
   succinyl, preparation, 506, 932
Glucoamylase
   commercial source, 777
   conjugate, activity assay, 778
   fiber entrapment, industrial application, 241
   immobilization, 42
      by bridge formation, 167
      on cellulose, 52, 167
      by entrapment, 178, 180
      in liposomes, 224
      on porous ceramics, 136–137, 778
      scale-up data, 781
   industrial production of glucose, 776–792
   kinetics, in enzyme system, 433–434
   support for, choice of, 776–777, 779–783
   titanium chloride binding, 52
α-Glucan, support, for penicillin acylase, 765
Glucocerebrosidase, 702
Gluconic acid, product, of glucose oxidase, 272
Gluconolactone, inhibitor, of β-glucosidase, 433, 918
Glucose
   active transport, 921–924
   assay
      in continuous flow analyzer, 638–641
      by microcalorimetry, 673–675
   automated analysis, comparative discussion, 636–637
   isomerization, 768–776, 809–821
   product, in invertase assay, 232
   production, from corn syrup, 776–792
   substrate, of glucose oxidase, 198, 347, 476
Glucose dehydrogenase
   in continuous flow analyzer, 639, 646
   immobilization, on nylon tube, 640, 646
Glucose isomerase
   activity, 161, 163–164, 200, 237, 811–813
   effects of metal ions on, 815–816
   efficiency
      effect of pH on, 237–238
      of temperature on, 238–239
   immobilization
      by adsorption, 161
      on collagen, 770
      on controlled-pore alumina, 161–164
      in cross-linked cell homogenates, 812
      on DEAE-cellulose, 165
   industrial use of, 47, 165, 241, 809–821
   kinetics, 769
   microbial sources, 770, 811
   oxygen sensitivity, 816
   pH optimum, 164
   productivity, 811
   stability, 811–813
   use in reactor, 769–776
Glucose monohydrate, substrate, of glucose oxidase, 157

Glucose oxidase
  activity assay of immobilized enzyme, 198–200, 347–349, 482–483
    of insoluble aggregate, 268–272
    by microcalorimetry, 663–664
    of native protein, 479–482
    of soluble aggregate, 474
  catalytic reaction, 478
  conjugate
    pH optimum shift, 468–469, 607
    support surface area and activity of, 156
  conjugation, with antibody, 711
  in enzyme electrode, 585, 597
  immobilization by carbohydrate residue activation, 319–320
    in cellophane membrane 904–905, 908
    by co-cross-linking, 268–269, 268–272, 908
    by copolymerization, 198, 200
    on collagen, 260
    on controlled-pore ceramics, 152–153, 156–160
    on glass beads, 347, 671
    by entrapment, 173–176, 908
    with magnetic carrier, 278–279
    by microencapsulation, 214
    in multistep enzyme systems, 181, 460–461, 472–478, 917
    on nylon tube, 124
  kinetics
    of conjugate, 201, 425–426, 483–488
    in multistep enzyme system, 433–434
  loading capacity on collagen membrane, 255
  oxidation, periodate, 318–320
  peroxidase inhibition of, 160
  properties, 478–479
  quantitization, 389
  spontaneous structuration studies and, 927–928
  stain for detection of, 713, 905
  specific activity, 255
  specificity, 583
  substrates, 615
  temperature-activity profile, 259
  use in quantitative immunodiffusion, 716–717
  unit cell dimensions, 159
Glucose 6-phosphate
  active transport and, 921–924
  enzyme electrode for, 879
  substrate, for glucose-6-phosphate dehydrogenase, 340, 455
Glucose 6-phosphate dehydrogenase, 342, 629
  in adenylate kinase activity assay, 892
  conjugate, activity assay, 200, 271, 340–341
  immobilization in collodion membrane, 906, 908
  in multistep enzyme system, 454–457
  kinetics, in enzyme system, 432–433, 463–466
  in NADPH-generating system, 850–851, 865
  reduction of $NADP^+$ analogs, 874–876
Glucose-6-phosphate isomerase, 811
  conjugate
    activity, 271
    kinetics, 917
  immobilization, on inert protein, 909
α-Glucosidase, immobilization, by microencapsulation, 214
β-D-Glucosidase
  acyl azide coupling, 103–104
  in amygdalin electrode, 592, 604–607
  assay, spectrophotometric, 101
  conjugate
    activity, 103–104, 271–272
    storage, 103
    stability, 104
  immobilization in multistep enzyme system, 473–475, 917
    on polyacrylic type copolymers, 103–104
    in porous matrix, 270–272
  kinetics, in enzyme system, 433
  pH optimum, 607
  source, 101
Glutamate, chemical properties, 12–18
L-Glutamate, concentration determination, with enzyme electrode, 877–880
Glutamate dehydrogenase
  allosteric effectors and, 512–513
  conjugate
    assay, 508
    coenzyme binding, 508–510
    denaturation studies, 510–512
  in enzyme electrode, 877

enzymic oxidation of NADPH, 865
immobilization
  by azide method, 907
  on collagen, 907–908
  on porous glass beads, 506
reduction of NADP⁺ analogs, 874–876
structure, 507
Glutamic-oxaloacetic transaminase, in L-asparaginase assay, 690
Glutaminase, effect on immune response, 705
Glutaraldehyde
  for covalent binding of aminoethyl cellulose, 48, 51
  of cellophane and enzyme, 905
  of ceramics, 778, 851
  of enzyme aggregates, 263, 473
  of inert protein and enzymes, 21, 710–712, 903–904, 912, 925
  of inorganic supports, 140–141
  after microencapsulation, 212
  of nylon, 121–127
  enzyme/hair conjugation, 561
  polyacrylamide activation, 60–61, 86
  porous silica alumina pellet activation, 480
  protein crystal insolubilization, 546, 548–555
    mechanism, 551–555
  Sepharose activation, 458
  tanning of collagen membrane, 248 250, 254, 257
  tissue fixation, 713
Glutaryl dichloride, agar gel cross-linker, 22, 37
Glycerokinase, activity on ATP analogs, 876
Glycerol, 157
  control of microbial contamination, 738
α-Glycerophosphate dehydrogenase, 337, 495–496
Glycidyl methacrylate, 72
Glycine
  aggregation and, 272–273
  amounts bound to Spheron gels, 73–74
  blocking agent, 529, 905, 912, 920
  cyanogen bromide binding, 62
  derivative, 284
  1-ethyl-3(3-dimethylaminopropyl) carbodiimide binding, 62

glutaraldehyde binding, 61
imido carbonate deactivation by, 30
vinylketo deactivation by, 36
Glycine ethyl ester, in carboxyl group determination, 384
Glycoenzyme, immobilization, via carbohydrate residue activation, 317–320
Glycogen phosphorylase, 501, 887
Glycol ether, 117
Glycoprotein, immobilization, 320
Glycyl-L-tyrosine, substrate, carboxypeptidase Aα, 557
Glyoxal, 248
20-GM80, 293, 310
  hydraulic permeability, 298
Gout, treatment, 704
GPC Maltrin syrup, 786–787
Grain sizes, effect on binding ability of support, 108
Granulation
  disadvantages of, 176, 191
  procedure
    by chopping, 174
    using homogenizer, 186
    using 30-mesh nylon net, 186
    using metal sieve, 198
    with syringe, 174
Guanidine hydrochloride, aldolase dissociation, 494
Guanidinium chloride, 370
  dissociation, of lactate dehydrogenase, 514
Guanosine triphosphate, allosteric effector, of glutamate dehydrogenase, 507, 512–513

H

Hair, as elastic support, 560–561
  binding procedure, 561
Halogen, activation agent, 36–37
Halohydrins, 32, 36
HDZ, see Hydroxyalkyl methacrylate gel, hydrazide derivative
HEMA, see 2-Hydroxyethyl methacrylate
Hemodialysis, and microcapsules, 694–696
Hemoglobin, 684
  conjugate, ligand equilibria in, 540–543
  dissociation by adsorption, 153
  immobilization

electrostatic, 542–543
  on Sephadex, 542
  inert proteic matrix, 268, 271
  microencapsulation and, 204–208
  substrate, of chymotrypsin, 73–75
    of papain, 82
    of pepsin, 77–79
Hemoprotein conjugate, properties, 538–539
Heparin, 680
L-Heptadecylamine, 151, 227
Hexachloroplatinic acid, 654
Hexamethylene bisiodoacetamide, 263
1,6-Hexamethylenediamine, 70, 77, 207
Hexamethylene diisocyanate, 263
Hexane, 894
Hexanoic acid hydrazide, 126
Hexokinase
  active transport studies and, 921–924
  activity on ATP analogs, 876
  in adenylate kinase activity assay, 892
  in ATP regeneration, 698
  coentrapment, 475–478
  conjugate, activity, 200, 271, 342
  immobilization
    on inert protein, 908
    by microencapsulation, 214
    in multistep enzyme system, 454–457
    kinetics, in multistep enzyme system, 432–435, 463–466
Hexylamine, 126
n-Hexylbromide, 44
Histidine
  chemical properties, 12–18
  enzyme aggregation and, 263
L-Histidine ammonia-lyase, and urocanic acid production, 745
Hollow-fiber devices, 292–295
Hollow-fiber membrane, 291–317
  cleaning procedure, 308–310
  commercial sources, 292–294
  immobilization in, 303–317
    advantages, 291
    enzyme slurry, 310–312
    enzyme system, 314–317
    modified soluble enzyme, 310
    soluble enzyme, 306–310
  leak check, 308
  liquid membrane impregnation of, 303
  permeability changes, 307

permeability measurement, 301–302
permeation selectivity, 302–303
preparation, 302–303
selection factors, 295–300
sterilization, 307–308
Homovanillic acid, 632
Hormone, *see also* specific substances
  microencapsulation of, 215
Hydraulic permeability, definition, 298
Hydrazide Bio-Gel P, 56
Hydrazide group, titration, 446
Hydrazine, 120
Hydrazine hydrate, 446
Hydrogen bonds, 149, 152
Hydrogen exchange, and conformation studies, 369
Hydrogen peroxide
  acatalasemia and, 684–688
  amperometric determination, 586, 590–591, 613–614
  microcalorimetric determination, 674–675
  polarographic monitoring, 617
  reagent, peroxidase stain, 713
  substrate, peroxidase, 716
  in uric acid assay, 631–632, 674
Hydrogen sulfide, 231
Hydrophilicity, 19–20, 84
Hydrophobicity, 84
  of amino acid residues, 13–14
  on glass surface, procedure, 194
Hydroxide ion, cyanogen bromide activation and, 28
Hydroxyalkyl methacrylate gel, 66–83
  activated
    preparation steps, 70
    storage, 70
  amino derivatives, preparation, 76–77
  anhydride derivative, 83
  carboxyl derivatives, preparation, 76–77
  chemical modifications, methods for, 69–70
  chemical properties, 69
  copolymerization with monomers containing functional groups, 71
    containing functional group precursors, 71–72
  hydrazide derivative, 82–83
  mechanical stability, 66
  physical properties, 68–69

polymeranalogous transformation, 70
preparation, 66–68
reaction mechanism, 66–67
surface modification, 70
swelling capacity, 66, 69
2-Hydroxyethyl methacrylate, 57, 67, 172–175
2-Hydroxyethyl methacrylate gel
hydrolytic stability, 69
specific surface area, 68
structural formula of, 67
Hydroxylamine
penicillin analysis, 760
thioester bond cleavage, 524
Hydroxylamine hydrochloride, 105, 385, 532
11-β-Hydroxylase, 184, 471
activity assay, 187
Hydroxyl group, activation, 542
3S-5-Hydroxy-3-methyl pentanoic acid lactone, 839–840
N-Hydroxysuccinimide, 146, 894
Hyflo supercel, 762, 765
Hyperuricemia, 695

I

Imidate salts, of nylon, 122–127
Imido carbonate
formation in support, 27
hydrolysis of, 28
reactive, on Sepharose, 17
Imidoester compound
reactivity, 321
stability, 321–322
synthesis, 322
Imidoester-containing polymer, 118–134, 320–324
Immobilization, see also specific techniques
allosteric regulation studies and, 504–506
anhydrous, 288–290
decrease in activity after, 24
choice of technique, for industrial applications, 725–729
functional groups involved, comparative data, 929–930
methods, classification, 10–11, 330–332
photochemical, 280–288

materials, 283–284
procedure, 284–287
theoretical considerations, 280–283, 287–288
Immunodiffusion, quantitative, and enzyme-labeled antibodies, 716–717
Immunoglobulin G
hollow-fiber retention data, 296
use in antibody quantitation, 714–717
Impregnation, definition, 149–150
In-column loading adsorption method, 160
Indigo White, in uric acid fluorometric assay, 632
Infrared spectroscopy, and conformation studies, 368–369
Inhibition
in diffusionally constrained systems, 427–431
effect of immobilization, 568–570
Insecticide, automated analysis, 637, 647–658
Interpenetrating networks, polyacrylamide, 180
Inulin
hollow-fiber dialytic permeability, 297
hollow-fiber retention data, 296
Invertase
coimmobilization, 472
conjugate, kinetics, 408
efficiency, effect of substrate concentration on, 238
fiber entrapment, 230–231
activity, 237
industrial application, 241
stability, 241
immobilization
by bead polymerization, 194
on collagen, 255, 260
by entrapment, 176, 178
in liposomes, 224, 702
with magnetic carrier, 278–279
by microencapsulation, 214
photochemical, 286–287
specific activity, 255
Iodoacetic acid, 392
alkylation, of coenzymes, 859–867
Iodoacetylcellulose, preparation, 749
4-Iodobutyl methacrylate, 84
5'-Iodocytidine triphosphate, allosteric

inhibitor, aspartate transcarbamylase, 557
2-Iodoethyl methacrylate, 84
Ionic bonds, in adsorption, 149–153
Isobutyral, 128–129, 131–132
threo-D_s-Isocitrate dehydrogenase, reduction of NADP⁺ analogs, 874–876
1-Isocyanato-4-isothiocyanatobenzene, 116
Isocyanide group
  assay, titrimetric, 131–132
  indirect coupling with, 38–40
Isoelectric focusing, for soluble enzyme aggregate characterization, 473
Isoelectric point, adsorption and, 151–152, 154
Isopropanol, 131, 562
Isoxazolium salt, 48
Isothiocyanate, 38, 132, 141
3-Isothiocyanatostyrene, 108, 113
4-Isothiocyanatostyrene, 108
m-Isothiocyanatostyrene, in copolymer, 84

K

μkat, see Microkatal
α-Ketoglutarate, substrate, glutamate dehydrogenase, 508, 690
3-Ketosteroid-Δ¹-dehydrogenase, 184–190
  activity assay, 186
  cofactor system of, 188
Kidney, artificial, 211
Kinetics
  of artificial enzyme membranes, 913–919
  experimental data for parameter determination, 438–443
  of irreversible monoenzyme system, 913–916
  measurement, for immobilized enzymes, 3
  theoretical, 397–443

L

Lactase
  activation energy, 796–797
  activity assay, 793
  coupling efficiencies, 795

  in enzyme thermistor, 674–675
  immodilization
    on collagen, 153
    in hollow-fiber membrane device, 313–314
    by microencapsulation, 212, 214
    on porous ceramics, 793
  inhibition by metal ions, 261
  pH activity profile, 793, 795–798
  rate equation, 798–799
  sources, 793
Lactate, automated analysis, 636
Lactate dehydrogenase
  activity assay, 271, 458–460, 664
  assay, fluorometric, 627–628
  conjugate
    concentration determination, 392–393
    ligand binding, 514–515
    oligomer dissociation, 514
    physicochemical properties, comparative, 934–936
  in enzyme electrode, 591, 877
  immobilization
    by adsorption, effects on activity, 44
    on cellulose sheets, 906, 908
    by diazotization procedure, 933
    on diethylaminoethyl cellulose, 52
    on glass beads, 932–934
    by glutaraldehyde cross-linking, 933–934
    on inert protein, 903, 908
    in multistep enzyme system, 181, 457–460, 471–472
    on nylon tube, 120, 641
    on Sepharose 4B, 930–931
    using soluble carbodiimide, 932–933
  kinetics
    of inhibition, 428
    in multistep enzyme system, 467–468
  reduction of NAD⁺-analogs, 873–874
  s-triazine binding, 52
L-Lactic acid, 246, 249
  enzyme electrode for, 585, 591–592
  substrate, lactate dehydrogenase, 881
β-Lactoglobulin
  crystal immobilization, 548
  solute diffusion studies with, 557–558
γ-Lactoglobulin, 552, 554
Lactose

hydrolysis, kinetics, 798–799
intolerance, 822
substrate, of β-galactosidase, 314, 457
in whey, commercial use, 792–809
Laser microspectrophotometer, 370, see also Double-beam laser spectrophotometer
Lead acetate, 654
Lecithin, 222, 226–227
Leuconostoc mesenteroides
in L-asparatic acid assay, 739
glucose 6-phosphate dehydrogenase from, 851–852
Leukocytes, microencapsulation of, 215
Ligand binding, to hemoglobin conjugate, 540–543
Lipase
immobilization
by microencapsulation, 214
on polyacrylamide, 447
modification, effect on activity, 447–450
Lipid bilayer, charge on surfaces of, 220
Liposome
Arthus reaction and, 709
assay of enzyme activity in, 225–226
clearance, 699–700, 705–707
definition, 69
effect of surface charge, 688–689, 703, 706, 708–709
enzyme entrapment in, 218–227
immunological adjuvant, 227, 700, 705
lipid sources, 708
physical characteristics, 218–220
sterilization, 707
therapeutic application, 688–689, 698–709
toxicity, 708
ultrastructure, 219
Liposome-entrapped enzymes, assay of, 225–226
Liquid-surfactant membrane, definition, 328
Lithium chloride, 550, 863
Lithium hydroxide, 861–867
Lithium lactate, substrate, lactate dehydrogenase, 628
Liver, liposome uptake, 699, 701, 706
Liver cells, microencapsulation of, 215
Luciferase, specificity, 583
Lupersol 225, 175

Lyophilization
of activated agarose gels, 33
effect on immobilized-enzyme activity, 76
Lysine
ε-amino group, 12
coupling with activated Sepharose, 17
assay, with enzyme electrode, 277–278
enzyme aggregation and, 263–273
percent in protein, 13
reaction with anhydrides, 18
with diazonium compounds, 18
with glutaraldehyde, 552, 713, 904
with imidoester compounds, 321
residues, and enzyme immobilization, 930
side chain reactivity, 14–15
Lysine decarboxylase
conjugate, activity, 271
in lysine assay, 277–278
Lysosome, and liposome, interaction, 701–702
Lysozyme
conjugate
activity assay, 290
activity correlations, 444–450
crystal
immobilization, 548
X-ray analysis, 555
diazotization, 445–446
glutaraldehyde insolubilization, 553
immobilization
on collagen, 153, 255, 260
by microencapsulation, 214
on polyacrylamide beads, 444–445
on polystyrene, 288–290
modification, effect on activity, 448–450
reductive denaturation studies, 517
soluble, activity assay, 446
specific activity, 255

M

Macromolecular complexation, 247
Macroreticular gel, 56–57
Magnesium ion
glucose isomerase activation, 815
inhibitor of immobilized lactase, 261
required by β-galactosidase, 827
Magnesium acetate, 162

Magnesium chloride, 313, 340, 342, 454, 456, 476, 549, 653, 793, 865, 890, 892
Magnesium oxide, pore properties, 137
Magnesium sulfate, 132, 163, 327, 823, 834–836, 839, 894
Magnetic iron-manganese oxide particles, commercial source of, 278
Magnetic supports, 324–326
  iron filings in, 216
Magnetite, 325–326
Malate dehydrogenase
  activity assay, 458–460, 474
  in L-asparaginase assay, 690
  in enzyme electrode, 595
  immobilization
    in multistep enzyme system, 181–182, 457–460, 472–475
    on nylon tube, 120
  kinetics, in multistep enzyme system, 434, 466–468
  reduction of NAD^+ analogs, 873–874
Maleate buffer, for lysozyme immobilization, 289–290
Maleic acid anhydride, 85, 108, 114, 179
  structural formula, 197
Maleic acid azide, structural formula, 197
Maleic anhydride/ethylene copolymer, 85
Maleic anhydride/vinyl methylether copolymer, 179
β-Maltose octacetate, 460
Matrix, see Support
McIlvaine buffer, 114
Mechanochemistry, 558–576
  enzyme activity and, 558–571
  protein-protein interaction and, 571–576
Membrane
  collagen, preparation, 246–247
  effect of mechanical deformation, 576
  enzyme, artificial, 901–929
    experimental methods using, 907–913
    kinetics, 913–919
    preparation, 902–907
    models, using insolubilized protein crystals, 558
  for microencapsulation, 203
Membrane impregnation method, 246–247
Membrane potential, 924–925
Memory, in artificial enzyme membranes, 925–926
Menadione, 188

Mercapto-agarose gel
  preparation, 41–42
  storage, 41
β-Mercaptoethanol, 81, 457, 494, 845
  reductive denaturing agent, 519, 522
3-Mercaptopropioimidate, 42
γ-Mercaptopropyltrimethoxysilane, 146
Mercuric chloride, 845
Meromyosin, 882
Merthiolate, preservative, 716
Metal ion, see also specific metal
  inhibitor, of collagen-bound enzyme, 261
  support pretreatment with, 155
Metal-linked enzymes, 166–169
Methacrylanilide, component of copolymers, 71
Methacrylic acid, in copolymer, 57, 84, 107–108, 112–113
Methacrylic acid m-fluoranilide, in copolymer, 84, 107–108, 112
Methacrylic acid/3-fluorostyrene copolymer
  nitration, 113
  papain immobilization on, 113
  polymerization, 113
Methacrylic acid/3-fluorostyrene/divinyl benzene copolymer, 84
Methacrylic acid/4-fluorostyrene/divinyl benzene copolymer, 84
Methacrylic acid/m-isothiocyanatostyrene/divinyl benzene copolymer, 84
Methacrylic acid/methacrylic acid 3-fluoroanilide copolymer
  nitration, 112
  papain immobilized on, 112
  polymerization, 112
Methacrylic acid/methacrylic acid m-fluoranilide/divinyl benzene copolymer, 84
Methanol, 58, 112–113, 116, 121–122, 127, 130, 133, 146, 175, 186, 324, 385, 386, 446, 474, 716, 834
Methanolbenzene, in carboxyl group titration, 384
Methionine, 12–15
  hydrophobicity, 14
  percent in protein, 13
  side chain reactivity, 14
L-Methionine, isolation, 757

SUBJECT INDEX

Methyl acetate, permeability, 302
Methyl alcohol, 231
O-Methylcaprolactim, 126
Methyl Cellosolve, phthalic anhydride solvent, 632
Methyl (±)-1,2-dihydronaphtho[2,1,-b]-furan-2-carboxylate, 834
Methyl-4,4'-dithiobisbutyrimidate, 556
N,N'-Methylenebisacrylamide, 56–57, 86, 91, 171–176, 185, 194–196, 198, 455, 458, 476, 572, 600, 662, 741, 750
  concentration effects, 174–175
Methylene blue, 132
Methylene chloride, 117, 230, 838
N-Methylindoxyl acetate, substrate, cholinesterase, 626
Methyl isobutyl ketone, 767
Methyl methacrylate, 84
Methyl methacrylate/2-iodiethyl methacrylate copolymer, 84
Methyl methacrylate/4-iodobutyl methacrylate copolymer, 84
2-Methyl-1,4-naphthoquinone, 188
3-Methylpentane-1,5-diol, 839
4-Methylumbelliferone, 391
4-Methylumbelliferyl p-(N,N,N-trimethyl ammonium) cinnamate, 391
N-Methylurea, substrate, urease, 595
Mice
  acatalasemic, 684–689
  implanted lymphosarcoma in, 689–694, 703
  leukemia, treatment, 703
Microcalorimetry, 659–667
  glucose oxidase assay, 663–664
  instrumentation, 661
  theory, 659–661
  trypsin assay, 665
Microcapsule, see also Microencapsulation
  biodegradable, 698
  surface area, 694
Micrococcus lysodeikticus, lysis, 445–446
Microencapsulation, 201–218
  advantages of, 218
  in cellulose nitrate, 203–207
  in cellulose polymer material, 210
  chemical method for, 207–209
  in fibers, 211, 227–242
  with heparin-complexed membrane, 212
  with lipid membranes, 212
  with lipid-protein membrane, 212
  in liposomes, 212, 218–227, 698–709
  in liquid membrane emulsions, 211
  magnetic preparations, 216
  membrane surface charge and, 212
  in microspheres, 201–218
  of multistep enzyme systems, 213–217
  in nylon, 207–209
    by drop technique, 207–208
    by emulsification technique, 207–209
  physical method for, 203–207
  with protein membranes, 212
  in red blood cells, 212
  in silicone rubber, 210
  theory, 202–203, 210
  therapeutic application of, 676–709
Microenvironment, kinetic effects, 399–400
Microkatal, definition, 3
Microorganism, see also specific organism
  contamination of glucose reactor, 784–785
  of lactose reactor, 804–826
  immobilization, in poly(acrylamide), 183–190
Milk
  continuous sterilization of, 824
  lactose reduction of, 822–830
Microscopy, light, and localization of cellular constituents, 712–713
Minibeaker, 293
Miniplant, 293
Minitube, 293, 307–308
Mixing-bath loading, see Shaking-bath loading adsorption method
Model enzyme systems, 63–64, 558
Monochloro-s-triazinyl derivative, 50
3-(2-Morpholinoethyl) carbodiimide, 144
MTT-tetrazolium salt, reagent, glucose oxidase staining, 713, 905
Multistep enzyme system
  application, 468–473
  assay, 455–456, 462
  asymmetric for active site distribution, 921–924
  bound protein determination, 462–463
  cofactor regeneration and, 698
  coimmobilization, 454–478

definition, 453
enzyme activity ratio, 461–462
immobilization
  in hollow fiber membranes, 312–314
  on inert protein, 917–919
  technique, choice, 461
kinetics, 432–435, 463–468, 917–919
perturbed pH optima, 468–469
physical characterization, 370
polarimetric assay, 350
soluble aggregates, 473–475
support, choice, 461
Muscle extract, immobilization, by microencapsulation, 214
Myoglobin
  conjugate, conformational transitions, 539–540
  hollow-fiber retention data, 296
  immobilization, on Sephadex, 540
Myosin, 882

N

Naficillin, 615
NAG, see N-Acetylglucosamine
β-Naphthol, color reaction with diazo groups, 83
Naphthol AS-BI phosphate, substrate, alkaline phosphatase, 626–627
Naphthol AS-Mx phosphoric acid disodium salt, reagent, alkaline phosphatase stain, 713
Naringin, 241
Neuraminidase, immobilization, in liposomes, 223
Neutral red, 132, 431
NH_2-Spheron, 76–80
Nicotinamide adenine dinucleotide
  alkylation, 862–863
  analogs of
    characterization, 868–869
    coenzymic activity, 873–876
    immobilization, 869–873, 882–884
    synthesis, 862–866
  binding, to immobilized lactate dehydrogenase, 514–515
  expensive cofactor, 837
  for horse liver alcohol dehydrogenase, 837–840
  fiber entrapment of coenzyme, 243
  immobilization
    on Dextran, 872
    in enzyme system, 216, 243, 315–316
    hollow-fiber dialytic permeability, 297, 302
  pseudobase formation, 842
Nicotinamide adenine dinucleotide, reduced
  binding, to immobilized glutamate dehydrogenase, 508–510
  oxidation, enzymic, 863
Nicotinamide adenine dinucleotide phosphate, oxidized
  enzymic reduction of, 865
  immobilization, on Dextran, 872–873
Nicotinamide adenine dinucleotide phosphate, reduced
  binding, to immobilized glutamate dehydrogenase, 508–510
  oxidation, enzymic, 865
  reductant, for pig liver oxidase, 850–852
Ninhydrin colorimetric assay
  estimation of amino acid components, 94
  of L-methionine, 747
Nitrate reductase, immobilization, by microencapsulation, 214
Nitration, with HNO_3/H_2SO_4, 112
Nitric acid, 112
  glass bead activation, 931
  metal cleaning, 654
4-Nitroacrylanilide, 98
2-Nitro-4-azidophenyl group, in photochemical immobilization, 281–288
p-Nitrobenzoylchloride, 86, 142
  preparation of arylamine glass, 932
Nitrocellulose, fibers of, preparation, 230
Nitrogen
  agarose derivatization, 39
  analysis, for bound protein determination, 387–388
  assay, in fibers, 232
  gaseous, 50, 58, 92, 102, 110–111, 117, 174, 175, 177, 185, 198, 230, 346, 392, 455, 476, 482, 524, 600, 662, 862, 912
  product, in photochemical immobilization, 282, 285, 287
p-Nitrophenol
  concentration determination, by spectroscopy, 101

ester derivatives of hydroxyalkyl methacrylate gels, 80–82
  capacity, 80–81
p-Nitrophenyl acetate, 574
p-Nitrophenyl chloroacetate, 263
O-Nitrophenyl-β-D-galactopyranoside, substrate, of β-galactosidase, 269–270, 313, 823
p-Nitrophenyl β-D-glucopyranoside, substrate, of β-D-glucosidase, 101, 474
p-Nitrophenyl-p'-guanidinobenzoate, 391
p-Nitrophenyl phosphate, substrate, of alkaline phosphatase, 308–309
p-Nitrophenyl trimethylacetate, acylating agent, 562, 567, 572
Nitrous acid, 88, 99, 125
NMR, see Nuclear magnetic resonance
Noradrenochrome, 256
Norepinephrine, 256–257
NPAC gel, see p-Nitrophenol, ester derivative of hydroxyalkyl methacrylate gel
Nuclear magnetic resonance
  conformation studies and, 366
  spectra, glutaraldehyde, 552–553
Nucleophilic groups
  reaction, 263
  rate determination of, 14
Nylon, see also Polyaminoaryl-nylon; Polyisonitrile-nylon
  alkylation of, 122–134
  bridge formation and, 166
  chemical properties of, 118
  in continuous flow analyzer, 639–642
    activation, 640–641
    bore size, 641
    storage, 642
  covalent binding to, 560–561
  elastic support, 560–561
  enzyme conjugates, storage, 121, 133
  enzyme immobilization on, 118–134
  hydrazide-substituted, storage, 124
  imidate salts of, 122–127
  microencapsulation with, 207–209
  partial acid hydrolysis, 120–121, 131
  peptide bond cleavage, 121
  physical properties of, 118
  powder, partially hydrolyzed
    preparation, 131
    storage, 131

Nylon 6, 118, 126
  N-alkylation of, 130–131
  O-alkylation of, 126–127
Nylon 66, 118
Nylon 610, 118

O

ONPG, see o-Nitrophenyl-β-galactoside
Optical rotatory dispersion, and conformation studies, 366–367
Oral lesion, and human acatalasemia, 687
ORD, see Optical rotatory dispersion
Organic acid
  production system, 184, 831–844
  resolution by α-chymotrypsin, 833–835
Organic synthesis, use of enzymes for, 831–844
Organophosphate pesticide, detection, 648, 657–658
Orsellinic acid decarboxylase, immobilized assay, 348
Oscillatory behavior, 926–927
Osmium tetroxide, 714
Ovalbumin
  hollow-fiber retention data, 296
  inert proteic matrix, 268
Oxaloacetate, 181
  substrate, citrate synthase, 458, 474
Oxidase, mixed function, hepatic, see Dimethylaniline monooxygenase
Oxiran-Acrylharzperlen, support, for penicillin acylase, 764
Oxirane, 32–34
  assay, by sodium thiosulfate titration, 34
  stability, 33
  storage, 33
2-Oxoglutarate, 865
Oxy-cellulose, 46
Oxygen
  acrylic polymerization inhibition by, 57
  glucose isomerase inactivation by, 816
Oxyhemoglobin, 544–545

P

P5, 293
  hydraulic permeability, 298
P8, 293

P10, 293, 296
  hydraulic permeability, 298
P30, 297
P100, 293
  hydraulic permeability, 298
Pancreatic trypsin inhibitor, 568–570
Papain
  conjugate
    activity, 110–116, 156
    optimal effective pH, 429–431
    storage, 111
    sulfhydryl group determination, 392
  immobilization
    with collagen, 244, 255
    in collodion membrane, 904, 909
    on controlled-pore glass, 156
    on hydroxyalkyl methacrylate gel derivative, 80–82
    with inert protein, 925
    with magnetic carrier, 278–279
    on nitrated methacrylic acid/3-fluorostyrene, 113
    on nitrated methacrylic acid/methacrylic acid 3-fluoroanilide, 112
    in polyacrylamide gel, 902
    in porous matrix, 270–272
  kinetics, 425, 437
  pH profile, of proteolytic activity, 82
  p-nitrophenol ester binding, 81–82
  oscillatory behavior studies and, 926–927
  short-term memory studies and, 925–926
  specific activity, 255
  sulfhydryl group, protection of, 274
Paper, aminated, 271
Paraffin
  in bead polymerization, 102–103
  for microencapsulation, 328–330
Paraoxon, detection, 658
Particle size, effect on diffusion control, 136, 138
Partition coefficient, 400–402
  definition, 400
Partitioning effect, of charged matrix, 400–408
PBS
  definition, 711
  tissue fixation, 712–713
Peclet number, 803

Penicillin acylase, see Penicillin amidase
Penicillin amidase
  activity assay, 237, 761
  fiber entrapment, industrial application, 241
  immobilization
    by bead polymerization, 194
    on Sephadex G-200, 764–765
    on Sepharose 4B, 760
  purification, 760–763
  supports for, 764
Penicillin analysis, 760–761
Penicillinase
  in enzyme electrode, 592, 597
  immobilization, on glass beads, 670
  substrates, 615
Penicillin electrode, 585, 592–593, 617
  interferences, 615
  response time, 608
Penicillin G, assay, by microcalorimetry, 671–672
n-Pentane, 838
Pepsin
  carbodiimide binding, 77
  concentration dependence of proteolytic activity, 77–79
  conjugation with phenylenediamine-agarose, procedure, 40
  hollow-fiber retention data, 296
  immobilization on activated inorganic support, 141
  on Spheron derivatives, 77–80
  immobilized, relative activity, 40
  pH profile
    for proteolytic activity, 79
    for stability, 79–80
  reductive denaturation studies, 517–518
Pepsinogen
  conjugate, intramolecular activation, 520
  immobilization, on Sepharose, 519–520
Peptone, in reactivation medium, 190
Perborate unit, 685
Perfusion, extracorporeal, 679–680
Periodate oxidation, of glycoenzyme, 317–320
Permeability, measurement, of hollow-fiber membrane, 301–302
Permeation chromatography, 85–86
Peroxidase

activity, 200–201, 271, 336, 716, 912–913
  assay reagent, 460, 474, 476, 632
  conjugation, with antibody, 711
  in enzyme electrode, 589
  immobilization
    on inert protein, 908, 912
    in liposomes, 224
    in polyacrylamide gel, 902, 908
    in immunoassay, 714–716
    stain for detection of, 713
Peroxido sulfate, see Ammonium persulfate
Petroleum ether, 103, 113, 230–231
pH
  activity profile
    changes in, 53, 346, 352–353
    of collagen complexes, 259
    of esterolytic activity of chymotrypsin and chymotrypsin-Spheron 300, 75–76
    of glucoamylase, 180
    of immobilized papain, 111
    partitioning effect and, 401–404
    of proteolytic activity
      of chymotrypsin and chymotrypsin-Spheron 300, 75
      of papain, 82
  carrier choice and, 154–155
  choice of coupling technique and, 148
  partitioning effect and, 401
Phenazine methosulfate
  colorimetric test for tetrazolium derivatives, 276, 905
  reagent, glucose oxidase staining, 713, 905
Phenol
  diazo groups destroyed by, 93
  test, 276
Phenolase, immobilization, by microencapsulation, 214
Phenol-2,4-disulfonyl chloride, 263
Phenolphthalein, 111
Phenothiazine, 855
Phenylacetic acid
  assay, 760–761
  removal, 767
DL-Phenylalanine, 81–82
L-Phenylalanine
  enzyme electrode for, 589, 617–618
  isolation, 757

Phenylalanine decarboxylase
  conjugate, activity, 271
  immobilization, on inert protein, 909
N,N'-(1,2-Phenylene)bismaleimide, 263
p-Phenylenediamine, agarose derivatization with, 39–40
Phenylhydrazine, 117, 144
Phenyl mercury(II) acetate, inhibitor, 615
Phenylmethyl sulfonylfluoride, acylating agent, 562, 567
3-Phenylpropionic acid, substrate analog, of chymotrypsin, 365
Phosgene, 120
Phosphatase
  active transport studies and, 921–924
  kinetics, in enzyme system, 435
Phosphate analysis, for bound protein determination, 388
Phosphatidic acid, 227
Phosphatidylcholine, 218
Phosphatidylethanolamine, 218
Phosphatidylserine, 218
Phospho-cellulose, 46
Phosphofructokinase, conjugate, activity, 271
6-Phosphogluconate dehydrogenase, reduction of $NADP^+$ analogs, 874–876
Phosphoglycerate kinase, activity on ATP analogs, 876
Phosphokinase, assay reagent, 630
Phospholipid
  liposome formation and, 218–227
  monolayer, support, for ribonuclease, 905–906
Phosphorus pentoxide, 131
Photoimmobilization, see Immobilization, photochemical
Phthalic anhydride, in uric acid assay, 632
Phytochrome, conformational transitions, 527
Plasma protein, support, for multistep enzyme system, 917–918
Plug-flow loading, 160–161
20-PM10, 293
  hydraulic permeability, 298
PMR, see Proton magnetic resonance
Polarimetric assay, 350
Poliodal-2, 84
Poliodal-4, 84

Poly 610, 207, see also Nylon
Polyacetal carrier, see Acryloyl-N,N-bis(2,2-dimethoxyethyl)amine
Polyacrolein, 110
  reduction with sodium borohydride, 116
Polyacrylamide-agarose gel, inclusion of aggregated enzyme in, 272–273
Polyacrylic acid, activation sequence, 85
Polyacrylic acid amide, cross-linked, 20
Polyacrylic copolymer, 84–107
Polyacrylic polymer support
  advantages of, 894–895
  artificial membranes and, 902
  as beads, conjugation with enzyme, 445–446
  chemical properties, 53, 85
  effects of solvents on volume, 22
  hollow-fiber devices and, 293, 311–312
  hydrophobicity studies with, 63–64
  multienzyme immobilization in, 181–182
  physical properties, 53, 85
  polymerization
    as beads, 57–60
    as granules by bulk-polymerization, 57
    inhibitor of, 57
    second, 180
    whole cell entrapment and, 185, 740–741, 745
  porosity, factors determining, 56–57
  protein-protein interaction
    effect of gel concentration, 573–574
    of gel deformation, 574–576
  steric exclusion, 666–667
  suitability for entrapment, 171–172
  water content, 172, 174
Polyacrylonitrile, methyl imidoester-containing, 324
Poly[acryloyl-N,N-bis(2,2-dimethoxyethyl)amine], synthesis, 101–103
Polyallyl alcohol, 108, 116
Polyaminoaryl-nylon
  diazotization capacity, 133
  immobilization on, 134
  synthesis, 129, 132–133
Poly(dimethylsiloxane), 178
Polyethylene glycol, 328
  spectrophotometric media, 389

Polyethyleneimine, 124
Poly(hydroxyalkylmethacrylate), 56, 85, see also Spheron
Poly(2-hydroxyethyl methacrylate), 172–175
  entrapment using, 692
  water content, 174–176
Polyisonitrile-nylon
  immobilization on, 133–134
  storage, 131
  synthesis, 128, 130–132
Polylactic acid, 217
  biodegradable microcapsule membrane, 698
Polymerization, see also Bead polymerization
  common ingredients in, 172
  entrapment by, 171–177, 182–183
  general procedure, 111
  initiation of
    agents for, choice of, 173
    photochemical, 173–174, 194
    by redox method, 173, 193–194
  photoactivated, instrumentation for, 174, 194, 891
  second, and interpenetrating networks, 180
  solution, 172–176
  suspension, 176–177
    monomer concentration effects on, 178
  vinyl, characteristics of, 172
Polymethacrylic acid anhydride, 85
Polymethacrylic copolymer, 84–107
Poly-γ-methyl-L-glutamate, fibers of, preparation, 231
Polymethyl methacrylate, 194, 693
Polyolefin, hollow-fiber device, 293
Polyoxyethylene (20) sorbitan trioleate, 102
Polystyrene
  in continuous flow analyzer, 639
  lysozyme immobilization, 288–290
Polysulfone, hollow-fiber membrane, 293
Polyvinyl alcohol, 108, 110, 115–116, 328
  effects of solvents on volume, 22
  entrapment in, 178
Polyvinyl ether, 117
Polyvinylpyrrolidone, 175
  suspension stabilizer in Spheron, 67

Pore diameter
  optimum, 156
  support choice and, 155-158
Porphobilinogen
  isolation, 847-848
  product, δ-aminolevulinic acid dehydratase, 844-847
  structural formula, 845
Potassium carbonate, 836
Potassium cyanide, 290, 347
Potassium ferricyanide, 87, 132, 256-257, 290
Potassium hydroxide, grease removal, 620
Potassium iodide, assay reagent, 664
Potassium permanganate, 685
Potassium persulfate, 102, 173-174, 600, 741, 750, 890, 891
Prednisolone, molar absorption coefficient, 186
Preservative, for proteinaceous membrane, 268
Pressure-drop, see also Reactor
  in collagen supports, 244
  particle size and, 136, 138
Procion brilliant orange M.G.S., 906
Prolabo, 913
Pronase, 390
  kinetics, of substrate diffusion, 437
Protease
  and collagen supports, 244
  immobilization
    choice of pore diameter for, 160
    coupling time and autolysis, 144
  immobilized, conformational transitions, 528-538
Protein
  bound
    colorimetric assay, 387
    determination methods, 94, 386-390
  concentration and activity, 383
  enzyme-binding to applications, 709-710
  reaction with glutaraldehyde, 551-555
  reactive residues, percent composition, 13
  resolution in polyacrylic gels, 56
  support, for enzyme, 903-904
Protein copolymerization, 195-201
  definition, 195
Protein crystal

insolubilization, 548-556
insolubilized, X-ray analysis, 555
properties, 546-548
Protein-protein interaction, and mechanochemistry, 571-576
Proteolytic enzyme
  catalytic constant, 528, 529
  immobilization with diazonium compounds, 18
Proton magnetic resonance, 366-367
Pseudomonas hydrophila, 811
Pseudomonas putida, 745
Pullulanase, immobilized, 470
PVP, see Polyvinylpyrrolidone
PVA, see Polyvinyl alcohol
Pyridine, in carbonyl group determination, 385
2-Pyridine disulfide agarose gel, 41-42
  use of, 42
Pyridoxal 5'-phosphate, immobilization, 885-886
Pyruvate
  concentration determination, with enzyme electrode, 877-880
  substrate, of lactate dehydrogenase, 459
Pyruvate kinase
  in ATP regeneration, 698
  coimmobilization, 472
  conjugate, activity, 271
  immobilization on cellulose sheets, 908
    on filter paper, 906
    by microencapsulation, 214

Q

Quartz diffusion cell, 910
Quinone
  activation agent, 35
  in glucose determination, 586, 617

R

Raffinose
  hollow-fiber dialytic permeability, 297
  hollow-fiber retention data, 296
  optical rotation, 350
  substrate, for α-galactosidase, 350
Reactor
  for alanine production, 880-882
  for L-amino acid production, 752-758

artificial enzyme membranes for, 910–911
batch, 251, 718, 722–723
catalyst packing density and, 770
collagen-invertase, constructional details, 252
column dimension, 743, 754
column pressure drop, 742–743, 754
contact efficiency, 771–772
for continuous aminoacylase production, 752–755
design equation, 772
effect of temperature, 735–736, 738
enzyme loading factor and, 769–770
enzymes for, 721–722
external mass transport, 771
flow patterns, 723, 733–736, 754–755
flow rate, 734, 744, 753–754, 756
fluidized bed, 341, 722, 854–855
for fructose production
  design, 817–821
  effect of metal ions, 815–816
  pressure drop, 820–821
  process parameters, 812–817
  scale-up, 820
for glucose isomerization, 768–776
for glucose production, 776–792
  bacterial contamination of, 784–785
  cost analysis, 788–792
  flow pattern, 782, 789
  operation, 783–792
  process detail, 783
  supports, comparative data, 779–783
immobilization methods for, 725–729
industrial, design parameters, 260–261, 742–743, 769–772, 774
internal transport resistance, 771
for lactose production, 792–809
  column back-flushing, 804
  cost analysis, 805–809
  enzymes, comparative data, 794–805
  half-life studies, 799–801
  mass transfer studies, 800–804
  pressure drop, 805
  scale-up, 804–805
  supports, comparative data, 794–805
  temperature effects, 794
for lactose reduction of milk, 822–830
  enzyme leakage, 827–828
  enzyme source, 827

flow pattern, 824
microbial contamination, 826, 829–830
operation, 825–826, 828–830
operation, general characteristics, 735–738, 756
operational stability, 735–738, 744, 758, 771, 773
packed-bed, 252, 338–340, 722–724
plug-flow, 722–724
for porphobilinogen production, 844–849
for production of amine oxides of tertiary amines, 854–855
recycle, 253–254
regeneration, 758
spiral-wound, 251–253, 724
stirred batch, 722–724
  assay using, 336–338, 342–345
substrate diffusion, 732–733
supports for, choice of, 725–729, 751–752
types of, 722–725, 729–735
Red No. 2, 654
Regeneration, of enzyme and carrier, 164–165
Reichstein's compound S, 187–190, 471
Reinickate salt, 594
Rennin
  crystal
    immobilization, 548
    x-ray analysis, 555–556
  immobilization, with collagen, 244
Rhodopseudomonas spheroides, 848
Riboflavin, initiator photochemical polymerization, 174, 194, 571–572, 600, 889, 891
Ribonuclease
  conformation studies using circular dichroism, 367
    using electron spin resonance, 368
  conjugate, activity, 36, 271
  entrapment in N,N'-methylenebisacrylamide, 171
  hollow-fiber retention data, 296
  immobilization, by adsorption at air-water interface, 905
Ribonuclease A
  conjugate, reductive denaturation studies, 518–519
  digestion, with carboxypeptidase A-Sephadex, 533, 537

immobilization on acrylamide-acrylic acid copolymer, 61–62
   on acrylamide-2-hydroxyethylmethacrylate copolymer, 61–62
   on polyacrylamide beads, 60–61
   reductive denaturation studies, 516, 526
Ribonuclease II, immobilization
   by adsorption on phospholipid monolayer, 905, 908
   in polyacrylamide gel, 902, 908
Ribonucleic acid, degradation, agar supports for, 23

S

$S_{0.5}$, definition, 3
S100N, 328, 330
Safranin O, 654
Salicin, substrate, of β-glucosidase, 272, 433, 917
Saligenin, 918
   product, of β-glucosidase, 272
Salt bridge, 149
Sanger's reagent, 284
Sarin, inhibitor, cholinesterase, 651
Schöniger combustion method, 133
Scott unit, 157
Sebacic acid, 118
Sebacic acid dihydrazide, Sepharose activation, 882
Sebacyl chloride, 207–208
Semicarbazide, 552
Sephadex, see also Dextran
   blocking of chymotrypsin by, 24
   lack of fluorescent impurities, 363–364
   multienzyme system on, 181, 456
Sephadex CM-C50, carboxyl group activation, 542
Sephadex G-10, for desalting, 866
Sephadex G-25, 532, 711
Sephadex G-50, 456, 872
   cyanogen bromide activation, 457
Sephadex G-100, 532
Sephadex G-200
   benzoquinone activation, 35
   cyanogen bromide activation, 30–31, 529, 764–765
   for gel filtration, 712
   as support, 529, 764–765
Sepharose
   amino acid content, 497

   bisoxirane activation, 32–33
   cyanogen bromide activation, 16–17
   as support, 340
Sepharose 4B, see also Agarose
   benzoquinone activation, 35
   cyanogen bromide activation, 28–30, 454, 529, 930
   as support, 336–337, 454, 457–458, 491–494, 529–530, 765, 846–847, 930–931
Sepharose 6B
   hexylation, 44
   for separation of liposomes, 224
Serine
   chemical properties, 12–18
   hydrophobicity, 14
Serum albumin, immobilized, on hydroxyalkyl methacrylate gel derivative, 81
SE-Sephadex, 532, 765
Sevin, detection, 658
Shaking-bath loading adsorption method, 160
Silane, see also specific compound
   commercial source, 139
   coupling techniques, 139–140
Silanization
   aqueous, 139
   organic, 140
Silanol group, ionic bonding to, 151
Silastic, 210
Silica, see Glass
Silica alumina pellet, porous support, for glucose oxidase/catalase, 480
Silicic acid, 151
Silicon dioxide, pore properties, 137
Silicone membrane, 178
Silicone rubber
   microencapsulation and, 210
   pad, preparation, 619–621
Silicon sheet, 271
Silk, 271
Sizing of beads, 195
SLOON, 211
Smectic mesophases, see Liposomes
Soda glass beads, bridge formation and, 166
Sodium acetate, 93, 749, 763
Sodium ampicillin, 592–593
Sodium azide, preservative, 269
Sodium bisulfite, 256–257, 473

Sodium borate, substrate, catalase, 685
Sodium borohydride, 23, 44, 212
Sodium chloride
 benzoquinone activation, 35
 hollow-fiber dialytic permeability, 297
 storage of activated gels in, 33
Sodium citrate, 92, 387, 476
Sodium dithionite, 142, 862
Sodium dodecyl sulfate, 370
Sodium glutamate, 189
Sodium hydrogen carbonate, 99
Sodium hydrosulfite, 210
Sodium hypochlorite, in cleaning solution, 310
Sodium iodide, 749
Sodium lauryl sulfate, 290
Sodium metaborate, 212
Sodium metaperiodate, 40
Sodium methoxide, carboxyl group determination, 131, 384
Sodium nitrite, 93, 95, 98, 103, 115–117, 133–134, 143, 256, 311, 445–446, 476, 933
 preservative, 40
Sodium perborate
 assay by potassium permanganate titration, 685
  by titanium sulfate colorimetric titration, 685
Sodium periodate, viscose fiber activation, 561
Sodium persulfate, 273
Sodium sulfate, anhydrous, 130, 132, 838
Sodium sulfide, 163, 533
Sodium taurocholate, 447
Sodium thiosulfate, 41, 561
 titration of oxirane groups with, 34
Sonication, multilamellar liposome formation and, 223
Sorbitan monooleate, 211, 328, 330
Sorbitan sesquioleate, 58, 177, 476, 662
Sorbitan trioleate, 102, 204, 207
Sorption isotherms, 153
Soybean trypsin inhibitor, 568–570, 662
 conjugate, acetylation, 532
 immobilization, on Sepharose 4B, 529
 steric exclusion, 666–667
Spacer
 role in retaining enzyme activity, 38
 types of, 70

Space velocity
 in L-amino acid reactors, 753
 definition, 756
Span 80, see Sorbitan monooleate
Span 85, see Sorbitan trioleate
Spectrophotometer, double-beam lasar, 912, 919
Spectrophotometric assay, 336–345
Spectrophotometric titration, of active sites, 391
Spheron, 56, 69, 74–82, 85
 monomer components of, 67
 pore distribution in, 67
Spheron 10, 74
Spheron 100, 74
Spheron 200, 74
Spheron 500, 74
Spheron 700, 74
Spheron 1000 BTD, 77–80
Spin diameter, definition, 156
Spinneret, 228–229
Stability
 diffusion effects on, 183
 immobilization effects on, 3
Staley STAR-DRI, 783, 785–786
Stannous chloride, 167–169
Stannous octoate, 178, 210
Starch
 degradation, 23
 immobilization in, 178
 pad, preparation, 652–654
 support properties of, 20
 substrate of α-amylase, 98
  of β-amylase, 40, 44
  of glucoamylase, 778, 783–786
 Zulkowsky, 44
Static procedure adsorption method, 160, 162
Stearylamine, 227
Steramine H, antiseptic, 826, 829
Steroid transformation, 184
STI, see Soybean trypsin inhibitor
Storage disease, lysosomal, therapy, 700–702
Streptomyces albus, 60
Streptomyces venezuelae, glucose isomerase from, 161, 250, 770
Streptomycin, in lactose reduction reactor, 826
Strip chart recorder, 656

Structure-function relationship, *see* Mechanochemistry
Styrene, 289
Styrene/4-vinylbenzoic acid copolymer, synthesis, 289
SU, *see* Sumner units
Substitution, degree of, definition, 46
Substrate
  concentration, and membrane kinetics, 914–915
  concentration determination
    fluorometric silicone rubber pad method, 642
    rate method, 624
  diffusion and, 176, 179, 182–183, 415–426
  size
    effect on coupling, 27
    restrictions on, 580
Subtilisin, 936
Subtilopeptidase A, kinetics, of substrate diffusion, 437
Subunit exchange chromatography, kinetics, 543–545
Succinic anhydride, 144, 506, 932
Succinyl-CoA synthetase, 882
Sucrose
  hollow-fiber retention data, 296
  product, of α-galactosidase, 350
  in reactivation medium, 190
  substrate, in invertase assay, 232, 238–239
  in whole cell storage, 189
Sucrose hydrolase, immobilization, by bead polymerization, 195
Sulfanilic acid, 445
Sulfhydryl group
  concentration determination, 385, 392, 522
  reaction with carbodiimides, 18
Sulfhydryl oxidase, 525
Sulfur, analysis, for bound protein determination, 388
Sumner units, 167
Support, *see also* specific types
  activation
    by direct coupling, 26–37, *see also* Covalent binding
    photochemical, 280–281
    polymeranalogous transformation, 70
  anionic, 107–108
  binding ability, parameters influencing, 108
  choice parameters, 19–20, 134–138
  for coenzymes, 155, 869–873
  comparative data
    surface area, 727–729
    surface charge, 726
    using aminoacylase, 751
    using pencillin acylase, 764–765
  concentration of reactive groups, 383–386
  density of particles, 721
  effect of charged groups, 400–408, 726
  hydrophilic, 270
    effect of solvents on, 22
    synthetic, 20
  hydrophobic, 270
  inorganic, 134–148
    activation with γ-mercaptopropyltrimethoxysilane, 146
    alkylamine derivatives, 139–142
    alkylhalide silane derivatives, 146
    arylamine derivatives, 142–144
    hydration and cleaning, 139
    hydrogen ion concentration operating range of, 163
  lysine-rich inert protein, 264–265
  neutral, 108
  polysaccharide, types of, 20
  porous
    enzyme aggregation and, 265, 270–272
    selection factors, 272
  properties of, 19–20, 107
  regeneration of, 165
  types, 18–20
Surface charge, adsorption rate and, 151
Suspension polymerization, *see* Bead polymerization
Sutan, detection, 658
Sweet potato, β-amylase from, 98
Swelling capacity
  effect of cross-linking on, 108
  pH dependence, 108
Systox, detection, 658

T

Taka-amylase A, reductive denaturation studies, 517
Tannase, specific activity, 255

Tannin, precipitation of penicillin acylase, 762
Tanning, of collagen membrane, 247–248
Tartaric acid dihydrazide, 88, 91, 103
Tartrate, inhibitor, of lactate dehydrogenase, 428
TDH, see Tartaric acid dihydrazide
TEAE, see Triethylaminoethyl-cellulose
Technicon Auto-Analyzer, L-methionine assay, 747
Teflon tubing, in hollow-fiber device, 295
Temik, detection, 658
TEMED, see N,N,N',N'-Tetramethylethylenediamine
Terephthalaldehyde, 115
N,N,N',N'-Tetramethylethylenediamine, 57–58, 177, 185, 187–188, 193–194, 273, 455, 458, 476, 662
N-4-(2,2,6,6-Tetramethyl-1-oxylpiperidinyl) bromacetamide, 368
Tetrazolium test, 276
Thiele modulus, 421
Thioester group, 12–14
  stability, 14
Thioglycolic acid, 524
Thiolnitrobenzoate anion, extinction coefficient, 522
Thionylchloride, 145
Thiophosgene, 86, 99, 141
2-Thiopyridone, 385
Thermistor, 598, 667–676
Thermoanalysis, assay by, 659–676
Thermolysin
  acyl azide binding, 95
  assay, by spectroscopy, 89–91
  conjugates
    activity assay, 94
    bound protein determination, 94
    kinetic behavior, 95
    thermal stability, 97–98
  diazo binding, 93
  immobilization on aminoaryl derivatized copolyacrylamide, 93–94
    on Enzacryl AH, 95
  source of, 89
Threonine, 12–15
  hydrophobicity, 14
Thymol blue, 131
Tissue, preservation of ultrastructure, 553
Titania, see Titanium oxide

Titanic chloride, 48, 166–167
Titanium oxide, 153
  available preparations of, 158
  controlled-pore, urease immobilization on, 167–169
  durability, 135–136
  hydrogen ion concentration operating range of, 163
  pore properties, 137
Titanous chloride, 91–92, 166
Titrator, automatic, 111
Titrimetric assay, 346
TNBS, see Trinitrobenzene sulfonic acid
TNS, see 2-p-Toluidinylnaphthalene-6-sulfonic acid
Toluene, 50, 58, 140, 146, 177, 230, 231, 480, 662
Toluenechloroform, 476
p-Toluene sulfonyl chloride, 130
2-p-Toluidinylnaphthalene-6-sulfonic acid, 535
  conformation studies, of chymotrypsin, 365
Tosyl-L-arginine methyl ester, substrate, trypsin, 530–531, 561–562, 573
Tosylation, 32
Transaldolase
  denaturation studies, 498–500
  immobilization, on Sepharose 4B, 494
Triazine coupling, 47
  to inorganic supports, 141–142
Tributyrin, substrate, of lipase, 447
Trichloroacetic acid, 90, 533, 632, 685, 690, 845, 852
Trichoderma viride, 166
Triethylamine, 132, 141, 384
  in derivatization of glass, 932
Triethylaminoethyl-cellulose, 46
Triethyloxonium tetrafluoroborate, alkylating agent, 123–124, 126, 640
Trinitrobenzene sulfonic acid reagent, 103
Triose phosphate isomerase, 337
  assay reagent, 495–496
  conjugate, activity, 271
  immobilization, on inert protein, 909
1-Tris(hydroxymethyl)aminomethane
  aggregation inhibitor, 266
  use in conversion to Tris salt, 878
Triton X-100, 225–226
  assay reagent, 460, 474

function, 476
in liposome rupture, 708
Tropane diester, hydrolysis, 835–836
Trypsin
  autolysis, rate determination, 572–573
  coentrapment, 475–478
  conformation studies, 363–364, 533–535
  conjugate
    active site titration, 391–392
    activity assay, 198, 200, 267, 271, 346, 530, 561
    change in digestion products, 437–438
    effect of support stretching 562–563, 565
    with polycarboxylic acid, 179
    reductive denaturation studies, 518–519
    stability, 537
  immobilization
    on aminoethyl cellulose, 48, 52
    in cellophane membrane, 905, 909
    by direct covalent binding, 909
    on hydrazide derivative of hydroxyalkyl methacrylate gel, 82–83
    on imidoester-containing polyacrylonitrile, 323–324
    by microencapsulation, 214
    on nylon tube, 120
    in polyacrylamide beads, 60–62, 171, 175–176, 662–663
    in polyacrylamide gel, 902, 909
    on porous glass, 153
    by protein copolymerization, 197–198, 200
    on Sephadex G-200, 529
    using N-carboxy-L-tyrosine, 327
  inhibitors, isolation of, 277
  kinetics
    of inhibition, 428
    perturbed by immobilization, 403–405
    of substrate diffusion, 436–438
    microcalorimetric assay, 664–665
  modification, 311–312
  pH-activity profiles, 403
  reductive denaturation studies, 517–519
  source of, 196
  suitable carrier for, 63
β-Trypsin

  acetylation, 532
  autolysis, 531–532
Tryptophan
  assay, for bound enzyme determination, 255–256, 389
  chemical properties, 12–18
  fluorescence, 364
L-Tryptophan, isolation, 757
Tryptophanase, coimmobilization, 471
Tryptophan synthetase, fiber entrapment
  activity, 237
  in cellulose triacetate, 235
  industrial application, 241
Tumor, substrate-dependent, treatment, 689–694, 703–704
Tween-20, 204, 208, 210
Tween 80, 185, 187, 190
Tween 85, see Polyoxyethylene(20) sorbitan trioleate
β-Tyrosinase, 886
Tyrosine, 49
  chemical properties, 12–18
  enzyme aggregation and, 263
  fluorescence, 364
  residues, and enzyme immobilization, 930
L-Tyrosine, enzyme electrode for, 585, 589
Tyrosine decarboxylase
  conjugate, activity, 271
  in enzyme electrode, 589
  immobilization, on inert protein, 909
Tyrosine phenol-lyase, 886

U

Ultrafiltration
  factors producing, 298
  in hollow-fiber membrane devices, 298–300
Ultraviolet radiation
  collagen membrane immobilization, 248
  polyacrylamide gel polymerization, 92–93
UM-10, 300
Union Carbide A1100, see γ-Aminopropyltriethoxysilane
Urate oxidase, see Uricase
Urea
  assay
    in continuous flow analyzer, 638

fluorometric, 630
  by microcalorimetry, 671–673
denaturation studies with chymotrypsinogen A, 522, 524
  with glutamate dehydrogenase, 510–512
  with monomers and oligomers, 498–500
  with trypsin derivatives, 518–519
effect on carboxypeptidase A crystals, 549
electrode, see Electrode, urea
enzyme deactivation with, 364, 879
enzyme desorption with, 140–141, 147, 152
enzyme electrode for, 585, 587–589
hollow-fiber dialytic permeability, 297
hollow-fiber retention data, 296
substrate, of urease, 104, 348
in washing solution, 77
Urease
  aldehydrol coupling, 105
  assay, spectrophotometric, 104
  attachment to mercapto-agarose gel, 42
  conjugate
    activity assay, 105, 271, 347–348
    kinetic behavior, 105–106
    optimal effective pH, 430–431
    storage stability, 106, 168–169
  in enzyme electrode, 587–589, 597, 600
  fiber entrapment, biomedical application, 242
  in fluorometric urea assay, 630
  immobilization
    by bridge formation, 167–169
    in cellophane membrane, 905
    on collagen, 255, 260, 902, 906, 909
    on controlled-pore ceramic, 167–169
    by entrapment, 176, 903, 909
    in gel, 909
    on glass beads, 670–671
    on inert protein, 909
    by microencapsulation, 214, 330
    on nylon tube, 120
    in polyacrylamide-agarose gel, 272–273
  membrane potential studies and, 924–925
  purification, 389
  source, 104

specific activity, 255
spontaneous structuration studies and, 927–928
therapeutic application, 679–684, 695–696
  by extracorporeal perfusion, 683–684
  by intraperitoneal injection, 681–683
Uric acid
  assay
    in continuous flow analyzer, 638
    fluorometric, 631–633
  enzyme electrode for, 585, 591
  gout and, 704
  hollow-fiber dialytic permeability, 297
Uricase
  activity, effect of xanthine concentration on, 918–919
  conjugate, activity, 200, 271
  in enzyme electrode, 591
  in enzyme thermistor, 674
  immobilization
    in cellophane membrane, 905, 908
    by entrapment, 176
    by microencapsulation, 214
    in multistep enzyme system, 918
    on nylon tube, 120
  therapeutic use, 695–696, 704
Uridine 2′,3′-cyclic monophosphate substrate, ribonuclease A, 519
Urocanic acid, continuous production, 745

V

V_{max}, apparent, definition, 3
Vaccine, microencapsulation of, 215
L-Valine, isolation, 757
Vanacryl P, 84
Vanillin methacrylate, 84
Vanillin methacrylate/allylic alcohol copolymer, 84
Vexar, 251–252
Vicinal hydroxyl
  in agar polysaccharides, 27
  oxirane groups and, 32
4-Vinylbenzoic acid, 289
Vinyl ether, 108
Vinylethyl ether, 117
Vinylmethyl ether, 179
Vinyl monomers, relative reactivities, 181
2-Vinyloxyethyl-4-nitrobenzoate, 117

SUBJECT INDEX

N-Vinylpyrrolidone, 108, 114
N-Vinylpyrrolidone/2-hydroxyethylmethacrylate copolymer, polymerization procedure, 175
N-Vinylpyrrolidone/maleic acid anhydride copolymer
  papain immobilization on, 115
  polymerization, 114–115
Vinyl sulfonylethylene ether, order of reactivity to nucleophilic groups, 34
Viscose fiber
  elastic support, 560
  enzyme binding procedure, 561
Vitamin B_{12}
  hollow-fiber dialytic permeability, 297
  hollow-fiber retention data, 296

W

Warburg apparatus assay, 348
Water monitoring, 647–658, *see also* Air monitoring
Whatman 3 paper, 271
Whatman PC 1520, 913
Whey
  commercial use, 792–809
  source, 792
Whole cells
  immobilization with collagen, 244
    with polyacrylamide gel, 185, 740–741, 745
  in reactor, for glucose isomerization, 768–776
Woodward Reagent K, *see* N-Ethyl-5-phenylisooxazolium 3′-sulfonate

X

X50, 293–298
  hydraulic permeability, 298
  permeability changes, 307
Xanthine, competitive inhibitor, urate oxidase, 918
Xanthine oxidase
  conjugate, activity, 271
  immobilization
    on inert protein, 908
    in multistep enzyme system, 918
20-XM50, 293, 313, 320
45-XM50, 293, 314
D-Xylose isomerase, 811

Z

Zinc chloride, 529, 533
Zirconium oxide
  durability, 135
  pore properties, 137
Zirconium phosphate, 216
Zymase complex (yeast), immobilization, by microencapsulation, 214

156535
v. 44 -